FORMULAS/EQUATIONS

Distance Formula

If $P_1 = (x_1, y_1)$ and $P_2 = (x_2, y_2)$, the distance from P_1 to P_2 is

$$d(P_1, P_2) = \sqrt{(x_2 - x_1)^2 + (y_2 - y_1)^2}$$

Standard Equation of a Circle

The standard equation of a circle of radius r with center at (h, k) is

$$(x - h)^2 + (y - k)^2 = r^2$$

Slope Formula

The slope m of the line containing the points $P_1 = (x_1, y_1)$ and $P_2 = (x_2, y_2)$ is

$$m = \frac{y_2 - y_1}{x_2 - x_1} \qquad \text{if } x_1 \neq x_2$$

$$m \text{ is undefined} \qquad \text{if } x_1 = x_2$$

Point–Slope Equation of a Line

The equation of a line with slope m containing the point (x_1, y_1) is

$$y - y_1 = m(x - x_1)$$

Slope–Intercept Equation of a Line

The equation of a line with slope m and y-intercept b is

$$y = mx + b$$

Quadratic Formula

The solutions of the equation $ax^2 + bx + c = 0, a \neq 0$, are

$$x = \frac{-b \pm \sqrt{b^2 - 4ac}}{2a}$$

If $b^2 - 4ac > 0$, there are two real unequal solutions.
If $b^2 - 4ac = 0$, there is a repeated real solution.
If $b^2 - 4ac < 0$, there are two complex solutions that are not real.

GEOMETRY FORMULAS

Circle

$r = $ Radius, $A = $ Area, $C = $ Circumference
$A = \pi r^2 \qquad C = 2\pi r$

Triangle

$b = $ Base, $h = $ Altitude (Height), $A = $ area
$A = \frac{1}{2}bh$

Rectangle

$l = $ Length, $w = $ Width, $A = $ area, $P = $ perimeter
$A = lw \qquad P = 2l + 2w$

Rectangular Box

$l = $ Length, $w = $ Width, $h = $ Height, $V = $ Volume
$V = lwh$

Sphere

$r = $ Radius, $V = $ Volume, $S = $ Surface area
$V = \frac{4}{3}\pi r^3 \qquad S = 4\pi r^2$

Right Circular Cylinder

$r = $ Radius, $h = $ Height, $V = $ Volume
$V = \pi r^2 h$

College Algebra

SIXTH EDITION

Michael Sullivan
Chicago State University

Prentice Hall
Upper Saddle River, New Jersey 07458

Library of Congress Cataloging-in-Publication Data

Sullivan, Michael
   College algebra / Michael Sullivan.—6th ed.
      p. cm.
   Includes index.
   ISBN 0-13-091453-3
   1. Algebra.   I. Title.
   QA154.2.S84 2002
   512.9—dc21   00-140090

Editor-in-Chief/Acquisitions Editor: Sally Yagan
Associate Editor: Dawn Murrin
Vice President/Director of Production and Manufacturing: David W. Riccardi
Executive Managing Editor: Kathleen Schiaparelli
Senior Managing Editor: Linda Mihatov Behrens
Production Editor: Bob Walters
Manufacturing Buyer: Alan Fischer
Manufacturing Manager: Trudy Pisciotti
Marketing Manager: Patrice Lumumba Jones
Marketing Assistant: Vince Jansen
Director of Marketing: John Tweeddale
Associate Editor, Mathematics/Statistics Media: Audra J. Walsh
Assistant Managing Editor, Math Media Production: John Matthews
Editorial Assistant: Mary Passafiume
Art Director: Joseph Sengotta
Interior Designer: Judith A. Matz-Coniglio
Cover Designer: Maureen Eide
Creative Director: Carole Anson
Managing Editor Audio/Video Assets: Grace Hazeldine
Director of Creative Services: Paul Belfanti
Photo Editor: Beth Boyd
Cover Photo: Black Sheep Stock Photography/PhotoDisc, Inc./PhotoDisc, Inc.
Art Studio: Artworks

© 2002, 1999, 1996, 1993, 1990, 1987 by Prentice-Hall, Inc.
Upper Saddle River, New Jersey 07458

Printed in the United States of America
10 9 8 7 6 5 4 3 2 1

ISBN: 0-13-091453-3

Prentice-Hall International (UK) Limited, *London*
Prentice-Hall of Australia Pty. Limited, *Sydney*
Prentice-Hall of Canada Inc., *Toronto*
Prentice-Hall Hispanoamericana, S.A., *Mexico*
Prentice-Hall of India Private Limited, *New Delhi*
Prentice-Hall of Japan, Inc., *Tokyo*
Pearson Education Asia Pte. Ltd.
Editora Prentice-Hall do Brasil, Ltda., *Rio de Janeiro*

In Memory of Mary

Contents

Preface to the Instructor

As a professor at an urban public university for over 30 years, I am aware of the varied needs of college algebra students who range from having little mathematical background and a fear of mathematics courses to those who have had a strong mathematical education and are highly motivated. For some of your students, this will be their last course in mathematics, while others may decide to further their mathematical education. I have written this text for both groups. As the author of precalculus, engineering calculus, finite math and business calculus texts, and, as a teacher, I understand what students must know if they are to be focused and successful in upper level mathematics courses. However, as a father of four college graduates, I also understand the realities of college life. I have taken great pains to insure that the text contains solid, student-friendly examples and problems, as well as a clear, seamless, writing style. I encourage you to share with me your experiences teaching from this text.

THE SIXTH EDITION

The Sixth Edition builds upon a solid foundation by integrating new features and techniques that further enhance student interest and involvement. The elements of previous editions that have proved successful remain, while many changes, some obvious, others subtle, have been made. A huge benefit of authoring a successful series is the broad-based feedback upon which improvements and additions are ultimately based. Virtually every change to this edition is the result of thoughtful comments and suggestions made from colleagues and students who have used previous editions. I am sincerely grateful for this feedback and have tried to make changes that improve the flow and usability of the text.

NEW TO THE SIXTH EDITION
Real Mathematics at Motorola

Each chapter begins with 📱 Field Trip to Motorola, a brief description of a current situation at Motorola, followed by 📱 Interview at Motorola, a biographical sketch of a Motorola employee. At the end of each chapter is 📱 Project at Motorola, written by the Motorola employee, that contains a description, with exercises, of a problem at Motorola that relates to the mathematics found in the chapter. It doesn't get more REAL than this.

Preparing for This Section
Most sections now open with a referenced list (by section and page number) of key items to review in preparation for the section ahead. This provides a just-in-time review for students.

Chapter R Review
This chapter, a revision of the old Chapter 1, has been renamed to more accurately reflect its content. It may be used as the first part of the course or as a just-in-time review when the content is required in a later chapter. Specific references to this chapter occur throughout the book to assist in the review process.

Content

The Appendix, Graphing Utilities, has been updated and expanded to include the latest features of the graphing calculator. While the graphing calculator remains an option, identified by a graphing icon 📱, references to the Appendix occur at appropriate places in the text for those inclined to use the graphing calculator features of the text.

Organization

- The discussion on Rational Functions now appears in two sections, Rational Functions I and Rational Functions II: Analyzing Graphs. This division should allow the sections to be covered in one teaching period each.
- The discussion of Polynomial and Rational Inequalities now appears after Polynomial and Rational Functions. This allows us to use information obtained about the graphs to solve the inequalities. Students and instructors will appreciate how easy this usually tough concept is now handled.
- Zeros of a Polynomial Function now appears in a separate chapter following Polynomial and Rational Functions to provide more flexibility in teaching and testing.
- Separate chapters on Sequences; Induction; the Binomial Theorem and Counting and Probability also provide more flexibility in coverage.

FEATURES IN THE 6TH EDITION

- Section ***OBJECTIVES*** appear in a numbered list to begin each section.
- ✏ ━━━━ NOW WORK PROBLEM XX appears after a concept has been introduced. This directs the student to a problem in the exercises that tests the concept, insuring that the concept has been mastered before moving on. The Now Work problems are identified in the exercises using yellow numbers and a ✏ pencil icon.
- Optional Comments, Explorations, Seeing the Concept, Examples, and Exercises that utilize the graphing calculator are clearly marked with a calculator icon. Calculator exercises are also identified by the 📱 icon and green numbers.
- References to Calculus are identified by a ⚠ calculus icon.
- Historical Perspectives, sometimes with exercises, are presented in context and provide interesting anecdotal information.
- Varied applications are abundant both in Examples and in Exercises. Many contain sourced data.
- Discussion, Writing, and Research problems appear in each exercise set, identified by an 🖌 icon and red numbers. These provide the basis for class discussion, writing projects, and collaborative learning experiences.
- An extensive Chapter Review provides a list of important formulas, definitions, theorems, and objectives, as well as a complete set of Review Exercises, with sample test questions identified by blue numbers.

USING THE 6TH EDITION EFFECTIVELY AND EFFICIENTLY WITH YOUR SYLLABUS

To meet the varied needs of diverse syllabi, this book contains more content than expected in a college algebra course. The illustration shows the dependencies of chapters on each other.

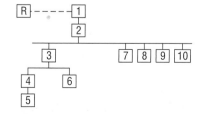

As the chart indicates, this book has been organized with flexibility of use in mind. Even within a given chapter, certain sections can be skipped without fear of future problems.

Chapter R   Review

This chapter, a revision of the old Chapter 1, has been renamed to more accurately reflect its content. It may be used as the first part of the course or as a just-in-time review when the content is required in a later chapter. Specific references to this chapter occur throughout the book to assist in the review process.

Chapter 1   Equations and Inequalities

Primarily a review of Intermediate Algebra topics, this material is prerequisite for later topics. For those who prefer to treat complex numbers and negative discriminants early, Section 5.3 can be covered at any time after Section 1.3.

Chapter 2   Graphs

This chapter lays the foundation. Sections 2.5 and 2.6 may be skipped without adverse effects.

Chapter 3   Functions and Their Graphs

Perhaps the most important chapter. Section 3.6 can be skipped without adverse effects.

Chapter 4   Polynomial and Rational Functions

Topic selection is dependent on your syllabus.

Chapter 5   The Zeros of a Polynomial Function

Topic selection is dependent on your syllabus. Section 5.1 is not absolutely necessary, but its coverage makes some computations easier.

Chapter 6   Exponential and Logarithmic Functions

Sections 6.1–6.5 follow in sequence; Sections 6.6, 6.7, and 6.8 each require Section 6.3.

Chapter 7   Conics

Sections 7.1–7.4 follow in sequence.

Chapter 8   Systems of Equations and Inequalities

Sections 8.1–8.2 follow in sequence; Sections 8.3–8.8 require Sections 8.1 and 8.2, and may be covered in any order. Section 8.9 depends on Section 8.8.

Chapter 9   Sequences; Induction; the Binomial Theorem

There are three independent parts: Sections 9.1–9.3, 9.4, and 9.5.

Chapter 10   Counting and Probability

Sections 10.1–10.3 follow in order.

ACKNOWLEDGMENTS

Textbooks are written by authors, but evolve from an idea into final form through the efforts of many people. Special thanks to Don Dellen, who first suggested this book and the other books in this series. Don's extensive contributions to publishing and mathematics are well known; we all miss him dearly.

I would like to thank Motorola and its people who helped make the projects in this new edition possible. Special thanks to Iwona Turlik, Vice President and Director of the Motorola Advanced Technology Center (MATC), for providing the opportunity to share with students examples of their experience in applying mathematics to engineering tasks.

I would also like to thank the authors of these projects:

- Tomasz Klosowiak, the Automotive and Industrial Electronics Group of Integrated Electronic Systems Segment

- Nick Buris, Brian Classon, Terri Fry, and Margot Karam, Motorola Laboratories, Communications Systems and Technologies Labs (CSTL)

- Bill Oslon, Andrew Skipor, John St. Peter, Tom Tirpak, and George Valliath, Motorola Laboratories, Motorola Advanced Technology Center (MATC)

- Jocelyn Carter-Miller, Corporate Vice President and Chief Marketing Officer

- Sheila MB. Griffen, Vice President and Director, Corporate Strategic Marketing, Chief Marketing Office

- Sue Eddins, Curriculum and Assessment Leader for Mathematics and Chuck Hamberg, Mathematics Faculty of the Illinois Mathematics and Science Academy for their generous help and contributions to Chapters 1 and 2.

Special thanks also go to the following:

- Douglas Fekete, Intellectual Property Department

- Jim Coffiing, Director Communications Future Business and Technology

- Anne Stuessy, Director Communications Future Business and Technology

- Vesna Arsic, Director Corporate Marketing Strategy, Chief Marketing Office

- Rosemarie Broda, Chief Marketing Office

- Rita Browne, CSTL

- David Broth, Vice President and Director of the Communications Systems and Technologies Labs

- Joseph Nowack and Bruce Eastmond, CSTL managers

- Tom Babin, Kevin Jelley and Bill Olson, MATC managers

Last, but not least, for his dedication to this project and the daunting task of managing it, I thank Andrew Skipor.

There are many colleagues I would like to thank for their input, encouragement, patience, and support. They have my deepest thanks and appreciation. I apologize for any omissions …

James Africh, *College of DuPage*
Steve Agronsky, *Cal Poly State University*
Grant Alexander, *Joliet Junior College*
Dave Anderson, *South Suburban College*
Joby Milo Anthony, *University of Central Florida*

James E. Arnold, *University of Wisconsin-Milwaukee*
Carolyn Autray, *University of West Georgia*
Agnes Azzolino, *Middlesex County College*
Wilson P Banks, *Illinois State University*
Sudeshna Basu, *Howard University*

Dale R. Bedgood, *East Texas State University*
Beth Beno, *South Suburban College*
Carolyn Bernath, *Tallahassee Community College*
William H, Beyer, *University of Akron*

Annette Blackwelder, *Florida State University*
Richelle Blair, *Lakeland Community College*
Trudy Bratten, *Grossmont College*
Joanne Brunner, *Joliet Junior College*
Warren Burch, *Brevard Community College*
Mary Butler, *Lincoln Public Schools*
William J. Cable, *University of Wisconsin-Stevens Point*
Lois Calamia, *Brookdale Community College*
Jim Campbell, *Lincoln Public Schools*
Roger Carlsen, *Moraine Valley Community College*
Elena Catoiu, *Joliet Junior College*
John Collado, *South Suburban College*
Nelson Collins, *Joliet Junior College*
Jim Cooper, *Joliet Junior College*
Denise Corbett, *East Carolina University*
Theodore C. Coskey, *South Seattle Community College*
John Davenport, *East Texas State University*
Faye Dang, *Joliet Junior College*
Antonio David, *Del Mar College*
Duane E. Deal, *Ball State University*
Timothy Deis, *University of Wisconsin-Platteville*
Vivian Dennis, *Eastfield College*
Guesna Dohrman, *Tallahassee Community College*
Karen R. Dougan, *University of Florida*
Louise Dyson, *Clark College*
Paul D. East, *Lexington Community College*
Don Edmondson, *University of Texas-Austin*
Erica Egizio, *Joliet Junior College*
Christopher Ennis, *University of Minnesota*
Ralph Esparza, Jr., *Richland College*
Garret J. Etgen, *University of Houston*
W.A. Ferguson, *University of Illinois-Urbana/Champaign*
Iris B. Fetta, *Clemson University*
Mason Flake, *student at Edison Community College*
Timothy W. Flood, *Pittsburg State University*
Merle Friel, *Humboldt State University*
Richard A. Fritz, *Moraine Valley Community College*
Carolyn Funk, *South Suburban College*
Dewey Furness, *Ricke College*
Dawit Getachew, *Chicago State University*
Wayne Gibson, *Rancho Santiago College*
Robert Gill, *University of Minnesota Duluth*
Sudhir Kumar Goel, *Valdosta State University*
Joan Goliday, *Sante Fe Community College*
Frederic Gooding, *Goucher College*
Sue Graupner, *Lincoln Public Schools*
Jennifer L. Grimsley, *University of Charleston*
Ken Gurganus, *University of North Carolina*
James E. Hall, *University of Wisconsin-Madison*
Judy Hall, *West Virginia University*

Edward R. Hancock, *DeVry Institute of Technology*
Julia Hassett, *DeVry Institute-Dupage*
Michah Heibel, *Lincoln Public Schools*
LaRae Helliwell, *San Jose City College*
Brother Herron, *Brother Rice High School*
Robert Hoburg, *Western Connecticut State University*
Lee Hruby, *Naperville North High School*
Kim Hughes, *California State College-San Bernardino*
Ron Jamison, *Brigham Young University*
Richard A. Jensen, *Manatee Community College*
Sandra G. Johnson, *St. Cloud State University*
Tuesday Johnson, *New Mexico State University*
Moana H. Karsteter, *Tallahassee Community College*
Arthur Kaufman, *College of Staten Island*
Thomas Kearns, *North Kentucky University*
Shelia Kellenbarger, *Lincoln Public Schools*
Keith Kuchar, *Manatee Community College*
Tor Kwembe, *Chicago State University*
Linda J. Kyle, *Tarrant Country Jr. College*
H.E. Lacey, *Texas A & M University*
Harriet Lamm, *Coastal Bend College*
Matt Larson, *Lincoln Public Schools*
Christopher Lattin, *Oakton Community College*
Adele LeGere, *Oakton Community College*
Kevin Leith, *University of Houston*
Jeff Lewis, *Johnson County Community College*
Stanley Lukawecki, *Clemson University*
Janice C. Lyon, *Tallahassee Community College*
Virginia McCarthy, *Iowa State University*
Jean McArthur, *Joliet Junior College*
Tom McCollow, *DeVry Institute of Technology*
Laurence Maher, *North Texas State University*
Jay A. Malmstrom, *Oklahoma City Community College*
Sherry Martina, *Naperville North High School*
Alec Matheson, *Lamar University*
James Maxwell, *Oklahoma State University-Stillwater*
Judy Meckley, *Joliet Junior College*
David Meel, *Bowling Green State University*
Carolyn Meitler, *Concordia University*
Sarnia Metwali, *Erie Community College*
Rich Meyers, *Joliet Junior College*
Eldon Miller, *University of Mississippi*
James Miller, *West Virginia University*
Michael Miller, *Iowa State University*
Kathleen Miranda, *SUNY at Old Westbury*
Thomas Monaghan, *Naperville North High School*
Craig Morse, *Naperville North High School*
Samad Mortabit, *Metropolitan State University*
A. Muhundan, *Manatee Community College*

Jane Murphy, *Middlesex Community College*
Richard Nadel, *Florida International University*
Gabriel Nagy, *Kansas State University*
Bill Naegele, *South Suburban College*
Lawrence E. Newman, *Holyoke Community College*
James Nymann, *University of Texas-El Paso*
Sharon O'Donnell, *Chicago State University*
Seth F. Oppenheimer, *Mississippi State University*
Linda Padilla, *Joliet Junior College*
E. James Peake, *Iowa State University*
Thomas Radin, *San Joaquin Delta College*
Ken A. Rager, *Metropolitan State College*
Kenneth D. Reeves, *San Antonio College*
Elsi Reinhardt, *Truckee Meadows Community College*
Jane Ringwald, *Iowa State University*
Stephen Rodi, *Austin Community College*
Bill Rogge, *Lincoln Public Schools*
Howard L. Rolf, *Baylor University*
Phoebe Rouse, *Lousiana State University*
Edward Rozema, *University of Tennessee at Chattanooga*
Dennis C. Runde, *Manatee Community College*
John Sanders, *Chicago State University*
Susan Sandmeyer, *Jamestown Community College*
A.K. Shamma, *University of West Florida*
Martin Sherry, *Lower Columbia College*
Tatrana Shubin, *San Jose State University*
Anita Sikes, *Delgado Community College*
Timothy Sipka, *Alma College*
Lori Smellegar, *Manatee Community College*
John Spellman, *Southwest Texas State University*
Becky Stamper, *Western Kentucky University*
Judy Staver, *Florida Community College-South*
Neil Stephens, *Hinsdale South High School*
Christopher Terry, *Augusta State University*
Diane Tesar, *South Suburban College*
Tommy Thompson, *Brookhaven College*
Richard J. Tondra, *Iowa State University*
Marvel Townsend, *University of Florida*
Jim Trudnowski, *Carroll College*
Robert Tuskey, *Joliet Junior College*
Richard G. Vinson, *University of South Alabama*
Mary Voxman, *University of Idaho*
Jennifer Walsh, *Daytona Beach Community College*
Donna Wandke, *Naperville North High School*
Darlene Whitkenack, *Northern Illinois University*
Christine Wilson, *West Virginia University*
Brad Wind, *Florida International University*
Canton Woods, *Auburn University*
George Zazi, *Chicago State University*

Recognition and thanks are due particularly to the following individuals for their valuable assistance in the preparation of this edition: Sally Yagan, for her continued support and genuine interest; Patrice Jones, for his innovative marketing efforts; Bob Walters, for his organizational skills as production supervisor; Phoebe Rouse, for her specific suggestions for this edition and careful proofreading of page proof; Brad Davis and Teri Lovelace of Laurel Technical Services for their proofreading skill and checking of my answers; and to the entire Prentice-Hall sales staff for their continuing confidence in this book.

Michael Sullivan

Preface to the Student

As you begin your study of College Algebra, you may feel overwhelmed by the number of theorems, definitions, procedures, and equations that confront you. You may even wonder whether or not you can learn all of this material in the time allotted. These concerns are normal. Keep in mind that many elements of College Algebra are all around us as we go through our daily routines. Many of the concepts you will learn to express mathematically, you already know intuitively. For many of you, this may be your last math course, while for others, just the first in a series of many. Either way, this text was written with you in mind. I have taught college algebra courses for over thirty years. I am also the father of four college students who called home from time to time, frustrated and with questions. I know what you're going through. So I have written a text that doesn't overwhelm, or unnecessarily complicate College Algebra, but at the same time it gives you the skills and practice you need to be successful.

This text is designed to help you, the student, master the terminology and basic concepts of College Algebra. These aims have helped to shape every aspect of the book. Many learning aids are built into the format of the text to make your study of the material easier and more rewarding. This book is meant to be a "machine for learning," one that can help you focus your efforts and get the most from the time and energy you invest.

HOW TO USE THIS BOOK EFFECTIVELY AND EFFICIENTLY

First, and most important, this book is meant to be read—so please, begin by reading the material assigned. You will find that the text has additional explanation and examples that will help you. Also, it is best to read the section before the lecture, so you can ask questions right away about anything you didn't understand.

Many sections begin with "Preparing for This Section," a list of concepts that will be used in the section. Take the short amount of time required to refresh your memory. This will make the section easier to understand and will actually save you time and effort.

A list of *OBJECTIVES* is provided at the beginning of each section. Read them. They will help you recognize the important ideas and skills developed in the section.

After a concept has been introduced and an example given, you will see ✏ NOW WORK PROBLEM XX. Go to the exercises at the end of the section, work the problem cited, and check your answer in the back of the book. If you get it right, you can be confident in continuing on in the section. If you don't get it right, go back over the explanations and examples to see what you might have missed. Then rework the problem. Ask for help if you miss it again.

If you follow these practices throughout the section, you will find that you have probably done many of your homework problems. In the exercises, every "Now Work Problem" number is in yellow with a pencil icon ✏. All the odd-numbered problems have answers in the back of the book and worked-out solutions in the Student Solutions Manual supplement. Be sure you have made an honest effort before looking at a worked-out solution.

At the end of each chapter is a Chapter Review. Use it to be sure you are completely familiar with the equations and formulas listed under "Things to Know." If you are unsure of an item here, use the page reference to go back and review it. Go through the Objectives and be sure you can answer "Yes" to the question "I should be able to. . . ." If you are uncertain, a page reference to the objective is provided.

Spend the few minutes necessary to answer the "Fill-in-the-Blank" items and the "True/False" items. These are quick and valuable questions to answer.

Lastly, do the problems identified with blue numbers in the Review Exercises. These are my suggestions for a Practice Test. Do some of the other problems in the review for more practice to prepare for your exam.

Please do not hesitate to contact me, through Prentice Hall, with any suggestions or comments that would improve this text. I look forward to hearing from you.

Best Wishes!

Michael Sullivan

MOTOROLA PROJECTS

Everyone seems to have a cell phone or pager... Focusing on this type of product, we visit the Motorola Corporation. **"Field Trip to Motorola"** highlights an individual's use of mathematics on the job at Motorola. **"Interview at Motorola"** is a short biography chronicling that individual's educational and career path. The **"Project at Motorola"** concludes the chapter, leading you through an assignment like the one described at the beginning of the chapter.

Functions and Their Graphs

Outline

Field Trip to Motorola

During the past decade the availability and usage of wireless Internet services has increased manyfold. The industry has developed a number of pricing proposals for such services. Marketing data have indi- cated that subscribers of wireless Internet services have tended to desire flat rate fee structures as com- pared with rates based totally on usage.

203

Page 203

Interview at Motorola

Jocelyn Carter-Miller is Corporate Vice President and Chief Marketing Officer (CMO) for Motorola, Inc., an over $30 bil- lion global provider of integrated communi- cations and embedded electronics solutions. As CMO she has helped build the Motoro- la brand and image and has developed high- performance marketing organizations and processes. Jocelyn also heads motorola.com, the Motorola electronic commerce and information Web site. In this new role, she and her team have de- veloped a strategy for serving Motorola's broad and diverse constituencies offering a full range of elec- tronic services.

In her previous roles as Vice President–Latin American and Caribbean Operations and Director of European, Middle East and African Operations, Jocelyn headed international wireless data commu- nications operations for Motorola, creating profitable opportunities through strategic alliances, value-added applications, and new product and service launches. She also developed skills in managing complex, high- risk ventures in countries like Brazil and Russia, set- ting standards for her company's practices in emerging markets.

Prior to her career at Motorola, Jocelyn served as Vice President, Marketing and Product Develop- ment for Mattel, where she broke new ground, driv- ing record sales of Barbie and other toys using

integrated product, entertainment, promo- tional, and licensing programs.

Jocelyn builds strong relationships and new opportunities through her involvement on outside boards and community organiza- tions. She serves on the board of the Princi- pal Financial Group and on the nonprofit boards of the Association of National Ad- vertisers, the University of Chicago Women's Business Group Advisory Board, and the Smart School Charter Middle School.

Jocelyn holds a Master of Business Administra- tion degree in marketing and finance from the Uni- versity of Chicago and a Bachelor of Science degree in accounting from the University of Illinois at Ur- bana–Champaign and is a Certified Public Accoun- tant. She is married to Edward Miller, President of Edventures, an educational reform development firm, and has two daughters, Alexis and Kennedy.

Jocelyn has won numerous awards, is featured in national publications, and regularly addresses busi- ness and community groups. She has coauthored with Melissa Giavagnoli the book *Networking: The Art of Relationships and Opportunities for Professionals* that was published in June 2000 by Jossey-Bass, Inc. Through their Web site Networlding.com, Jocelyn and Melissa facilitate meaningful connections and mutu- ally beneficial opportunities for new and experienced Networlders alike.

Page 204

Project at Motorola

During the past decade the availability and usage of wireless Internet services have increased. The in- dustry has developed a number of pricing proposals for such services. Marketing data have indicated that subscribers of wireless Internet services have tended to desire flat fee rate structures as compared with rates based totally on usage. The Computer Resource Department of Indigo Media (hypothetical) has en- tered into a contractual agreement for wireless Internet services. As a part of the contractual agree- ment, employees are able to sign up for their own wireless services. Three pricing options are available:

Silver Plan: $20/month for up to 200 K-bytes of service plus $0.16 for each addi- tional K-byte of service

Gold Plan: $50/month for up to 1000 K-bytes of service plus $0.08 for each addi- tional K-byte of service

Platinum Plan: $100/month for up to 3000 K-bytes of service plus $0.04 for each addi- tional K-byte of service

You have been requested to write a report that answers the following questions in order to aid em- ployees in choosing the appropriate pricing plan.

(a) If C is the monthly charge for x K-bytes of serv- ice, express C as a function of x for each of the three plans.

(b) Graph each of the three functions found in part (a).

(c) For how many K-bytes of service is the Silver Plan the best pricing option? When is the Gold Plan best? When is the Platinum Plan best? Explain your reasoning.

(d) Write a report that summarizes your findings.

Page 279

CLEAR WRITING STYLE

Sullivan's **accessible writing style** is apparent throughout, often utilizing various approaches to the same concept. An author who writes clearly makes potentially difficult concepts intuitive, making class time more productive.

Sometimes it is helpful to think of a function f as a machine that receives as input a number from the domain, manipulates it, and outputs the value. See Figure 6.

Figure 6

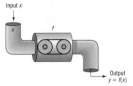

The restrictions on this input/output machine are as follows:

1. It only accepts numbers from the domain of the function.
2. For each input, there is exactly one output (which may be repeated for different inputs).

For a function $y = f(x)$, the variable x is called the **independent variable**, because it can be assigned any of the permissible numbers from the domain. The variable y is called the **dependent variable**, because its value depends on x.

Pages 208-209

Page 204

PREPARING FOR THIS SECTION

Before getting started, review the following:

✓ Intervals (Section 1.5, pp. 125–126)

✓ Evaluating Algebraic Expressions, Domain of a Variable (Review, Section 2, pp. 19–21)

✓ Intercepts (Section 2.2, pp. 157–158)

✓ Scatter Diagrams; Linear Curve Fitting (Section 2.5, pp. 185–189)

3.1 FUNCTIONS

OBJECTIVES
1. Determine Whether a Relation Represents a Function
2. Find the Value of a Function
3. Find the Domain of a Function
4. Identify the Graph of a Function
5. Obtain Information from or about the Graph of a Function

PREPARING FOR THIS SECTION

The **"Preparing for this Section"** feature provides you and your instructor with a list of skills and concepts needed to approach the section, along with page references. You can use the feature to determine what you should review before tackling each section.

STEP-BY-STEP EXAMPLES

Step-by-step examples ensure that you follow the entire solution process and give you an opportunity to check your understanding of each step.

EXAMPLE 5   Analyzing the Graph of a Rational Function with a Hole

Analyze the graph of the rational function: $R(x) = \dfrac{2x^2 - 5x + 2}{x^2 - 4}$

Solution   We factor R and obtain

$$R(x) = \frac{(2x - 1)(x - 2)}{(x + 2)(x - 2)}$$

In lowest terms,

$$R(x) = \frac{2x - 1}{x + 2}, \qquad x \neq -2$$

STEP 1: The domain of R is $\{x \mid x \neq -2, x \neq 2\}$.

STEP 2: The graph has one x-intercept: $1/2$. The y-intercept is $R(0) = -1/2$.

STEP 3: Because

$$R(-x) = \frac{2x^2 + 5x + 2}{x^2 - 4}$$

we conclude that R is neither even nor odd. Thus, there is no symmetry with respect to the y-axis or the origin.

STEP 4: The graph has one vertical asymptote, $x = -2$, since $x + 2$ is the only factor of the denominator of $R(x)$ *in lowest terms*. However, the rational function is undefined at both $x = 2$ and $x = -2$.

STEP 5: Since the degree of the numerator equals the degree of the denominator, the graph has a horizontal asymptote. To find it, we either use long division or form the quotient of the leading coefficient of the numerator, 2, and the leading coefficient of the denominator, 1. Thus, the graph of R has the horizontal asymptote $y = 2$. To find out whether the graph of R intersects the asymptote, we solve the equation $R(x) = 2$.

$$R(x) = \frac{2x - 1}{x + 2} = 2$$

$$2x - 1 = 2(x + 2)$$

$$2x - 1 = 2x + 4$$

$$-1 = 4 \qquad \text{Impossible}$$

The graph does not intersect the line $y = 2$.

Pages 335-336

REAL-WORLD DATA

Real-world data is incorporated into examples and exercise sets to emphasize that mathematics is a tool used to understand the world around us. As you use these problems and examples, you will see the relevance and utility of the skills being covered.

TABLE 1	
Date	Closing Price ($)
8/31/99	41.09
9/30/99	37.16
10/31/99	38.72
11/30/99	38.34
12/31/99	41.16
1/31/00	49.47
2/29/00	56.50
3/31/00	65.97
4/30/00	63.41
5/31/00	62.34
6/30/00	66.84
7/31/00	66.75
8/31/00	74.88
Courtesy of A.G. Edwards & Sons, Inc.	

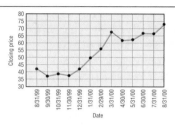

Figure 8
Monthly closing prices of Intel stock 8/31/99 through 8/31/00

We can see from the graph that the price of the stock was rising rapidly from 11/30/99 through 3/31/00 and was falling slightly from 3/31/00 through 5/31/00. The graph also shows that the lowest price occurred at the end of September, 1999, whereas the highest occurred at the end of August, 2000. Equations and tables, on the other hand, usually require some calculations and interpretation before this kind of information can be "seen."

Look again at Figure 8. The graph shows that for each date on the horizontal axis there is only one price on the vertical axis. Thus, the graph represents a function, although the exact rule for getting from date to price is not given.

Page 212

EXAMPLE 10 The Golden Gate Bridge

The Golden Gate Bridge, a suspension bridge, spans the entrance to San Francisco Bay. Its 746-foot-tall towers are 4200 feet apart. The bridge is suspended from two huge cables more than 3 feet in diameter; the 90-foot-wide roadway is 220 feet above the water. The cables are parabolic in shape and touch the road surface at the center of the bridge. Find the height of the cable at a distance of 1000 feet from the center.

Solution We begin by choosing the placement of the coordinate axes so that the x-axis coincides with the road surface and the origin coincides with the center of the bridge. As a result, the twin towers will be vertical (height $746 - 220 = 526$ feet above the road) and located 2100 feet from the center. Also, the cable, which has the shape of a parabola, will extend from the towers, open up, and have its vertex at $(0, 0)$. As illustrated in Figure 14, the choice of placement of the axes enables us to identify the equation of the parabola as $y = ax^2$, $a > 0$. We can also see that the points $(-2100, 526)$ and $(2100, 526)$ are on the graph.

Figure 14

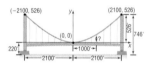

Based on these facts, we can find the value of a in $y = ax^2$.

$$y = ax^2$$
$$526 = a(2100)^2 \quad y = 526; x = 2100$$
$$a = \frac{526}{(2100)^2}$$

The equation of the parabola is therefore

$$y = \frac{526}{(2100)^2} x^2$$

The height of the cable when $x = 1000$ is

$$y = \frac{526}{(2100)^2} (1000)^2 \approx 119.3 \text{ feet}$$

The cable is 119.3 feet high at a distance of 1000 feet from the center of the bridge. ∎

NOW WORK PROBLEM 63.

FITTING A QUADRATIC FUNCTION TO DATA

⑤ In Section 2.5 we found the line of best fit for data that appeared to be linearly related. It was noted that data may also follow a nonlinear relation. Figures 15(a) and (b) show scatter diagrams of data that follow a quadratic relation.

"NOW WORK" PROBLEMS

Many examples end with the phrase **"Now Work Problem —."** Sending you to the exercise set to work a similar problem provides the opportunity to immediately check your understanding. The corresponding "Now Work" problem is easily identified in the exercise sets by the pencil icon and yellow exercise number.

63. **Suspension Bridge** A suspension bridge with weight uniformly distributed along its length has twin towers that extend 75 meters above the road surface and are 400 meters apart. The cables are parabolic in shape and are suspended from the tops of the towers. The cables touch the road surface at the center of the bridge. Find the height of the cables at a point 100 meters from the center. (Assume that the road is level.)

Page 297

SOLUTIONS

Solutions, both algebraic and graphical, are clearly expressed throughout the text.

EXAMPLE 1

Graphing a Quadratic Function Using Trasformations

Graph the function $f(x) = 2x^2 + 8x + 5$. Find the vertex and axis of symmetry.

Solution   We begin by completing the square on the right side.

$$f(x) = 2x^2 + 8x + 5$$

$$= 2(x^2 + 4x) + 5 \qquad \text{Factor out the 2 from } 2x^2 + 8x.$$

$$= 2(x^2 + 4x + 4) + 5 - 8 \qquad \begin{array}{l}\text{Complete the square of } 2(x^2 + 4x).\\ \text{Notice that the factor of 2 requires}\end{array}$$

$$= 2(x + 2)^2 - 3 \qquad \text{that 8 be added and subtracted.} \qquad \textbf{(2)}$$

The graph of f can be obtained in three stages, as shown in Figure 6. Now compare this graph to the graph in Figure 5(a). The graph of $f(x) = 2x^2 + 8x + 5$ is a parabola that opens up and has its vertex (lowest point) at $(-2, -3)$. Its axis of symmetry is the line $x = -2$.

Figure 6

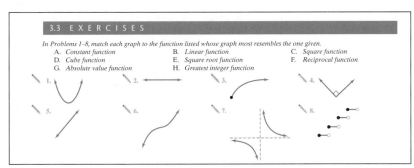

(a) $y = x^2$ Multiply by 2; Vertical stretch

(b) $y = 2x^2$ Replace x by $x + 2$; Shift left 2 units

(c) $y = 2(x + 2)^2$ Subtract 3; Shift down 3 units

(d) $y = 2(x + 2)^2 - 3$ Vertex $(-2, -3)$

Check: Graph $f(x) = 2x^2 + 8x + 5$ and use the MINIMUM command to locate its vertex.

NOW WORK PROBLEM **17**.

The method used in Example 1 can be used to graph any quadratic function $f(x) = ax^2 + bx + c, a \neq 0$, as follows:

$$f(x) = ax^2 + bx + c$$

$$= a\left(x^2 + \frac{b}{a}x\right) + c \qquad \text{Factor out } a \text{ from } ax^2 + bx.$$

$$= a\left(x^2 + \frac{b}{a}x + \frac{b^2}{4a^2}\right) + c - a\left(\frac{b^2}{4a^2}\right) \qquad \begin{array}{l}\text{Complete the square by adding}\\ \text{and subtracting } a(b^2/4a^2).\\ \text{Look closely at this step!}\end{array}$$

$$= a\left(x + \frac{b}{2a}\right)^2 + c - \frac{b^2}{4a}$$

$$= a\left(x + \frac{b}{2a}\right)^2 + \frac{4ac - b^2}{4a} \qquad c - \frac{b^2}{4a} = c \cdot \frac{4a}{4a} - \frac{b^2}{4a} = \frac{4ac - b^2}{4a}$$

Based on these results, we conclude the following:

GRAPHING UTILITIES

Graphing utilities are optional in this text and their use is clearly identified by the use of the graphing utility icon.

3.3 EXERCISES

In Problems 1–8, match each graph to the function listed whose graph most resembles the one given.

A. *Constant function*   B. *Linear function*   C. *Square function*
D. *Cube function*   E. *Square root function*   F. *Reciprocal function*
G. *Absolute value function*   H. *Greatest integer function*

1.   2.   3.   4.

5.   6.   7.   8.

END-OF-SECTION EXERCISES

Sullivan's exercises are unparalleled in terms of thorough coverage and accuracy. Each **end-of-section exercise** set begins with visual- and concept-based problems, starting you out with the basics of the section. Well-thought-out exercises better prepare you for exams.

MODELING

Many examples and exercises connect real-world situations to mathematical concepts. Learning to work with **models** is a skill that transfers to many disciplines.

with his or her age. The following table shows the median income I of individuals of different age groups within the United States for 1996. For each age group, the class midpoint represents the independent variable, x. For the age group "65 years and older," we will assume that the class midpoint is 69.5.

Age	Class Midpoint, x	Median Income, I
15–24 years	19.5	$21,438
25–34 years	29.5	$35,888
35–44 years	39.5	$44,420
45–54 years	49.5	$50,472
55–64 years	59.5	$39,815
65 years and older	69.5	$19,448

Source: U.S. Census Bureau

(a) Draw a scatter diagram of the data. Comment on the type of relation that may exist between the two variables.
(b) The quadratic function of best fit to these data is

$$I(x) = -44.8x^2 + 4009x - 41392$$

Page 335

ports of crude oil (1000 barrels per day) for the years 1980–1997.

Year, x	Imports, I	Year, x	Imports, I
1980	5263	1989	5843
1981	4396	1990	5894
1982	3488	1991	5782
1983	3329	1992	6083
1984	3426	1993	6787
1985	3201	1994	7063
1986	4178	1995	7230
1987	4674	1996	7508
1988	5107	1997	7996

Source: U.S. Energy Information Administration

(a) Draw a scatter diagram of the data. Comment on the type of relation that may exist between the two variables.
(b) The quadratic function of best fit to these data is

$$I(x) = 18.04x^2 - 71,495.6x + 70,831,298$$

Use this function to determine the year in which imports of crude oil were lowest.
(c) Use the function found in part (b) to predict the number of barrels of imported crude oil in 1998.
(d) Use a graphing utility to verify that the function given in part (b) is the quadratic function of best fit.
(e) With a graphing utility, draw a scatter diagram of the data and then graph the quadratic function of best fit on the scatter diagram.

Pages 298-299

The new graphs reflect the behavior produced by the analysis. Furthermore, we observe two turning points, one between 0 and 1 and the other to the right of 4. Rounded to two decimal places, these turning points are (0.52, 0.07) and (11.48, 2.75).

determine the age at which an to earn the most income.
redict the peak income earned.
to verify that the function given dratic function of best fit.
ty, draw a scatter diagram of the he quadratic function of best fit m.

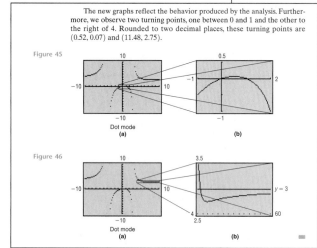

Figure 45

Dot mode
(a)

(b)

Figure 46

Dot mode
(a)

$y = 3$

(b)

GRAPHING UTILITIES AND TECHNIQUES

Increase your understanding, visualize, discover, explore, and solve problems using a **graphing utility.** Sullivan uses the graphing utility to further your understanding of concepts, not to circumvent essential math skills.

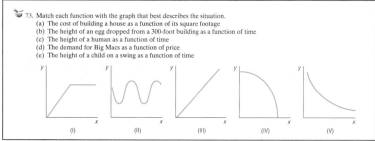

73. Match each function with the graph that best describes the situation.
(a) The cost of building a house as a function of its square footage
(b) The height of an egg dropped from a 300-foot building as a function of time
(c) The height of a human as a function of time
(d) The demand for Big Macs as a function of price
(e) The height of a child on a swing as a function of time

(I) (II) (III) (IV) (V)

Page 221

DISCUSSION WRITING AND READING EXERCISES

These exercises, clearly identified by the notebook icon and/or red numeration, are designed to get you to "think outside the box." These exercises seek to foster an intuitive understanding of key mathematical concepts. It is easy to find these exercises as they are highlighted by the book icon and red exercise number.

LINKS TO CALCULUS

This icon draws attention to the underpinnings of calculus.

If we know the function $C(t)$ that relates the year t to the cost C of tuition and fees, then the average rate of change from 1991 to 1992 may be expressed as

$$\text{Average rate of change} = \frac{C(1992) - C(1991)}{1992 - 1991} = \frac{251}{1} = \$251/\text{year}$$

Expressions like this occur frequently in calculus.

Page 226

CHAPTER REVIEW

The Chapter Review helps check your understanding of the chapter materials in several ways. **"Things to Know"** gives a general overview of review topics. The **"How To"** section provides a concept-by-concept listing of operations you are expected to perform. The **"Review Exercises"** then serve as a chance to practice the concepts presented within the chapter. Several of the Review Exercises are numbered in blue. These exercises can be combined to create the Chapter Test. Since these problems are odd numbered, you can check your answers in the back of the book. The review materials are designed to make you, the student, confident in knowing the chapter material.

CHAPTER REVIEW

Library of Functions

Linear function (p. 235)

$f(x) = mx + b$
Graph is a line with slope m and y-intercept b.

Constant function (p. 235)

$f(x) = b$
Graph is a horizontal line with y-intercept b.
See Figure 22.

Identity function (p. 235)

$f(x) = x$
Graph is a line with slope 1 and y-intercept 0.
See Figure 23.

Cube function (p. 236)

$f(x) = x^3$
See Figure 25.

Reciprocal function (p. 236)

$f(x) = 1/x$
See Figure 27.

Square function (p. 235)

$f(x) = x^2$
Graph is a parabola with intercept at $(0, 0)$.
See Figure 24.

Square root function (p. 236)

$f(x) = \sqrt{x}$
See Figure 26.

Absolute value function (p. 236)

$f(x) = |x|$
See Figure 28.

Greatest integer function (p. 237)

$f(x) = \text{int}(x)$
See Figure 29.

Things To Know

Function (p. 206)

A relation between two sets of real numbers so that each number x in the first set, the domain, has corresponding to it exactly one number y in the second set. The range is the set of y values of the function for the x values in the domain.

x is the independent variable; y is the dependent variable.

A function f may be defined implicitly by an equation involving x and y or explicitly by writing $y = f(x)$.

Pages 273-274

Objectives

You should be able to:

Determine whether a relation represents a function (p. 205)

Find the value of a function (p. 209)

Find the domain of a function (p. 211)

Identify the graph of a function (p. 212)

Obtain information from or about the graph of a function (p. 213)

Find the average rate of a change of a function (p. 224)

Use a graph to determine where a function is increasing, is decreasing, or is constant (p. 227)

Use a graph to locate local maxima and minima (p. 228)

Determine even or odd functions from a graph (p. 229)

Identify even or odd functions from the equation (p. 230)

Graph the functions listed in the library of functions (p. 235)

Graph piecewise-defined functions (p. 238)

Graph functions using horizontal and vertical shifts (p. 242)

Graph functions using compressions and stretches (p. 246)

Graph functions using reflections

Fill-in-the-Blank Items

1. If f is a function defined by the equation $y = f(x)$, then x is called _____ variable.

2. A set of points in the xy-plane is the graph of a function if and only graph in at most one point.

3. The average rate of change of a function equals the _____ graph.

4. A(n) _____ function f is one for which $f(-x) = f(x)$ for ev function f is one for which $f(-x) = -f(x)$ for every x in the domain

5. Suppose that the graph of a function f is known. Then the graph of $y =$

True/False Items

T F   **1.** Every relation is a function.

T F   **2.** Vertical lines intersect the graph of a function in no more than one point.

T F   **3.** The y-intercept of the graph of the function $y = f(x)$, whose domain is all real numbers, is $f(0)$.

T F   **4.** A function f is decreasing on an open interval I if, for any choice of x_1 and x_2 in I, with $x_1 < x_2$, we have $f(x_1) < f(x_2)$.

T F   **5.** Even functions have graphs that are symmetric with respect to the origin.

T F   **6.** The graph of $y = f(-x)$ is the reflection about the y-axis of the graph of $y = f(x)$.

T F   **7.** $f(g(x)) = f(x) \cdot g(x)$.

T F   **8.** The domain of the composite function $(f \circ g)(x)$ is the same as that of $g(x)$.

Review Exercises

Blue problem numbers indicate the author's suggestions for use in a Practice Test.

1. Given that f is a linear function, $f(4) = -5$ and $f(0) = 3$, write the equation that defines f.

2. Given that g is a linear function with slope $= -4$ and $g(-2) = 2$, write the equation that defines g.

3. A function f is defined by $f(x) = \dfrac{Ax + 5}{6x - 2}$. If $f(1) = 4$, find A.

4. A function g is defined by $g(x) = \dfrac{A}{x} + \dfrac{8}{x^2}$. If $g(-1) = 0$, find A.

5. Tell which of the following graphs are graphs of functions.

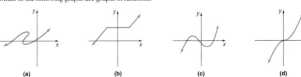

(a)   (b)   (c)   (d)

Sullivan M@thPak
An Integrated Learning Environment

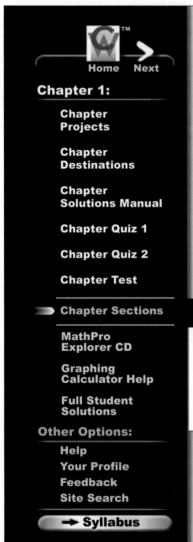

GET IT TOGETHER!

M@THP@K integrates and organizes all major student supplements into an easy-to-use format at a price that can't be beat!

Here's just a sample of what you'll find in MathPak:

- **MultiMedia MathPro 4.0**
  This interactive tutorial program offers unlimited practice on College Algebra content. Watch the author work the problems via videos, view other examples, and see a fully worked out solution to the problem you are working on.

- **Graphing Calculator Manuals**
  Includes step by step procedures and screen shots for working with your TI- 82, TI-83, TI-85, TI-86, TI-89, TI-92, HP48G, CFX-9850 GaPlus, and SharpE 9600c.

- **Full Student Solutions Manual**
  Includes step by step solutions for all the odd numbered exercises in the text.

System Requirements:
- 32MB of random access memory (RAM); 64MB or more recommended
- 200MB free hard disk space
- CD-ROM drive
- QuickTime™ 4.0 or better
- Internet Browser 4.5 or higher
- Internet Access 28.8k or better

Sullivan M@thPak
Helping Students
Get it Together

Additional Media

Sullivan Companion Website

www.prenhall.com/sullivan
This text-specific website beautifully complements
the text. Here students can find chapter tests,
section-specific links, and PowerPoint downloads
in addition to other helpful features.

MathPro 4.0 (network version)

This networkable version of MathPro is *free* to adopters.
Contact your Prentice Hall Sales Representative.

Test Gen-EQ

CD-Rom (Windows/Macintosh)

- Algorithmically driven, text-specific testing program
- Networkable for administering tests and
  capturing grades
- Edit existing test items or add your own questions
  to create a nearly unlimited number of tests and
  drill worksheets

ISBN: 0-13-091457-6

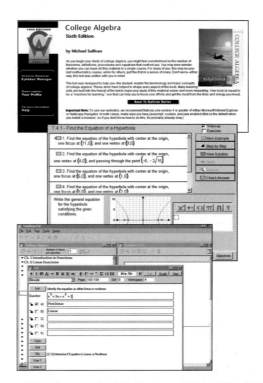

OTHER SUPPLEMENTS

Student Supplements

Student Solutions Manual

Worked solutions to all odd-numbered exercises from the
text and complete solutions for chapter review problems
and chapter tests. ISBN: 0-13-092452-0

Lecture Videos

The instructional tapes, in a lecture format, feature
worked-out examples and exercises taken from each
section of the text. ISBN: 0-13-091458-4

New York Times
Themes of the Times

A *free* newspaper from Prentice Hall and *The New York
Times*. Interesting and current articles on mathematics
which invite discussion and writing about mathematics.

Mathematics on the Internet

Free guide providing a brief history of the Internet,
discussing the use of the World Wide Web, and
describing how to find your way within the Internet
and how to find others on it.

Instructor Supplements

Instructor's Edition with Resource CD

For your convenience, the Instructor Resource CD
contains the full version of TestGen-EQ, additional
Chapter Projects, links to a passcode protected Web site
which has the complete Instructor'sSolutions Manual and
Test Item File. This allows flexibility in preparing for class.

Instructor's Resource Manual

Contains complete step-by-step worked-out solutions
to all even-numbered exercises in the textbook.
ISBN: 0-13-091450-9

Test Item File

Hard copy of the algorithmic computerized testing
materials. ISBN: 0-13-091462-2

List of Applications

Photo and Illustration Credits

Review

PREPARING FOR THIS BOOK

Before getting started, read the Preface to the Student

As the title states, this chapter is a review. The purpose of the chapter is to help you recall facts that you learned in an earlier course. The topics chosen for inclusion here represent information that will be useful for this course.

The word *algebra* is derived from the Arabic word *al-jabr*. This word is a part of the title of a ninth century work, "Hisâb al-jabr w'al-muqâbalah," writ-ten by Mohammed ibn Mûsâ al-Khowârizmî. The word *al-jabr* means "a restoration," a reference to the fact that, if a number is added to one side of an equa-tion, then it must also be added to the other side in order to "restore" the equality. The title of the work, freely translated, means "The science of reduction and cancellation." Of course, today, algebra has come to mean a great deal more.

1 │ REAL NUMBERS

OBJECTIVES ① Classify Numbers
② Evaluate Numerical Expressions
③ Work with Properties of Real Numbers

SETS

When we want to treat a collection of similar but distinct objects as a whole, we use the idea of a **set.** For example, the set of *digits* consists of the collection of numbers 0, 1, 2, 3, 4, 5, 6, 7, 8, and 9. If we use the symbol D to denote the set of digits, then we can write

$$D = \{0, 1, 2, 3, 4, 5, 6, 7, 8, 9\}$$

In this notation, the braces { } are used to enclose the objects, or **elements,** in the set. This method of denoting a set is called the **roster method.** A second way to denote a set is to use **set-builder notation,** where the set D of digits is written as

$$D = \{\quad x \quad | \quad x \text{ is a digit}\}$$

Read as "D is the set of all x such that x is a digit."

EXAMPLE 1 ── Using Set-builder Notation and the Roster Method

(a) $E = \{x \mid x \text{ is an even digit}\} = \{0, 2, 4, 6, 8\}$

(b) $O = \{x \mid x \text{ is an odd digit}\} = \{1, 3, 5, 7, 9\}$ ∎

In listing the elements of a set, we do not list an element more than once because the elements of a set are distinct. Also, the order in which the elements are listed is not relevant. For example, $\{2, 3\}$ and $\{3, 2\}$ both represent the same set.

If every element of a set A is also an element of a set B, then we say that A is a **subset** of B. If two sets A and B have the same elements, then we say that A **equals** B. For example, $\{1, 2, 3\}$ is a subset of $\{1, 2, 3, 4, 5\}$, and $\{1, 2, 3\}$ equals $\{2, 3, 1\}$.

Finally, if a set has no elements, it is called the **empty set,** or the **null set,** and is denoted by the symbol $\varnothing$.

CLASSIFICATION OF NUMBERS

① It is helpful to classify the various kinds of numbers that we deal with as sets. The **counting numbers,** or **natural numbers,** are the numbers in the set $\{1, 2, 3, 4, \ldots\}$. (The three dots, called an **ellipsis,** indicate that the pattern continues indefinitely.) As their name implies, these numbers are often used to count things. For example, there are 26 letters in our alphabet; there are 100 cents in a dollar. The **whole numbers** are the numbers in the set $\{0, 1, 2, 3, \ldots\}$, that is, the counting numbers together with 0.

The **integers** are the numbers in the set $\{\ldots, -3, -2, -1, 0, 1, 2, 3, \ldots\}$.

These numbers are useful in many situations. For example, if your checking account has $10 in it and you write a check for $15, you can represent the current balance as −$5.

Notice that the set of counting numbers is a subset of the set of whole numbers. Each time we expand a number system, such as from the whole numbers to the integers, we do so in order to be able to handle new, and usually more complicated, problems. The integers allow us to solve problems requiring both positive and negative counting numbers, such as profit/loss, height above/below sea level, temperature above/below 0°F, and so on.

But integers alone are not sufficient for *all* problems. For example, they do not answer the question "What part of a dollar is 38 cents?" To answer such a question, we enlarge our number system to include *rational numbers*. For example, $\frac{38}{100}$ answers the question "What part of a dollar is 38 cents?"

> A **rational number** is a number that can be expressed as a quotient a/b of two integers. The integer a is called the **numerator,** and the integer b, which cannot be 0, is called the **denominator.** The rational numbers are the numbers in the set $\{x \mid x = \frac{a}{b}, \text{ where } a, b \neq 0 \text{ are integers}\}$.

Examples of rational numbers are $\frac{3}{4}, \frac{5}{2}, \frac{0}{4}, -\frac{2}{3}$, and $\frac{100}{3}$. Since $a/1 = a$ for any integer a, it follows that the set of integers is a subset of the set of rational numbers.

Rational numbers may be represented as **decimals.** For example, the rational numbers $\frac{3}{4}, \frac{5}{2}, -\frac{2}{3}$, and $\frac{7}{66}$ may be represented as decimals by merely carrying out the indicated division:

$$\frac{3}{4} = 0.75 \qquad \frac{5}{2} = 2.5 \qquad -\frac{2}{3} = -0.666\ldots \qquad \frac{7}{66} = 0.1060606\ldots$$

Notice that the decimal representations of $\frac{3}{4}$ and $\frac{5}{2}$ terminate, or end. The decimal representations of $-\frac{2}{3}$ and $\frac{7}{66}$ do not terminate, but they do exhibit a pattern of repetition. For $-\frac{2}{3}$, the 6 repeats indefinitely; for $\frac{7}{66}$, the block 06 repeats indefinitely. It can be shown that every rational number may be represented by a decimal that either terminates or is nonterminating with a repeating block of digits, and vice versa.

On the other hand, there are decimals that do not fit into either of these categories. Such decimals represent **irrational numbers.** Every irrational number may be represented by a decimal that neither repeats nor terminates. In other words, irrational numbers cannot be written in the form a/b, where a, $b \neq 0$ are integers.

Irrational numbers occur naturally. For example, consider the isosceles right triangle whose legs are each of length 1. See Figure 1. The length of the hypotenuse is $\sqrt{2}$, an irrational number.

Also, the number that equals the ratio of the circumference C to the diameter d of any circle, denoted by the symbol π (the Greek letter pi), is an irrational number. See Figure 2.

> Together, the rational numbers and irrational numbers form the set of **real numbers.**

Figure 1

Figure 2

$\pi = \dfrac{C}{d}$

Figure 3 shows the relationship of various types of numbers.*

Figure 3

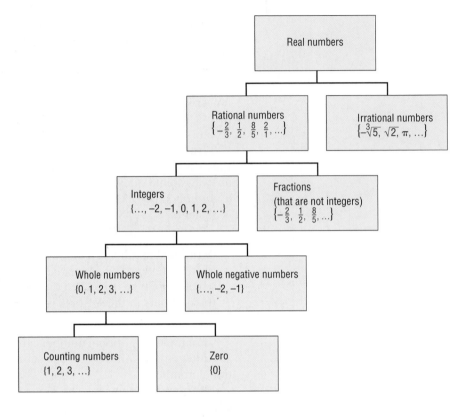

EXAMPLE 2

Classifying the Numbers in a Set

List the numbers in the set

$$\left\{-3, \tfrac{4}{3}, 0.12, \sqrt{2}, \pi, 2.151515 \ldots \text{ (where the block 15 repeats), } 10\right\}$$

that are

(a) Natural numbers   (b) Integers   (c) Rational numbers

(d) Irrational numbers   (e) Real numbers

Solution

(a) 10 is the only natural number.

(b) −3 and 10 are integers.

(c) $-3, \tfrac{4}{3}, 0.12, 2.151515 \ldots$, and 10 are rational numbers.

(d) $\sqrt{2}$ and π are irrational numbers.

(e) All the numbers listed are real numbers.

NOW WORK PROBLEM 3.

APPROXIMATIONS

Every decimal may be represented by a real number (either rational or irrational), and every real number may be represented by a decimal.

---

*The set of real numbers is a subset of the set of complex numbers. We study complex numbers in Section 5.3.

The irrational numbers $\sqrt{2}$ and π have decimal representations that begin as follows:

$$\sqrt{2} = 1.414213\ldots \qquad \pi = 3.14159\ldots$$

In practice, decimals are generally represented by approximations. For example, using the symbol $\approx$ (read as "approximately equal to"), we can write

$$\sqrt{2} \approx 1.4142 \qquad \pi \approx 3.1416$$

In approximating decimals, we either *round off* or *truncate* to a given number of decimal places. The number of places establishes the location of the *final digit* in the decimal approximation.

> **Truncation:**   Drop all the digits that follow the specified final digit in the decimal.
>
> **Rounding:**   Identify the specified final digit in the decimal. If the next digit is 5 or more, add 1 to the final digit; if the next digit is 4 or less, leave the final digit as it is. Now truncate following the final digit.

EXAMPLE 3

Approximating a Decimal to Two Places

Approximate 20.98752 to two decimal places by

(a) Truncating   (b) Rounding

Solution   For 20.98752, the final digit is 8, since it is two decimal places from the decimal point.

(a) To truncate, we remove all digits following the final digit 8. Thus, the truncation of 20.98752 to two decimal places is 20.98.

(b) The digit following the final digit 8 is the digit 7. Since 7 is 5 or more, we add 1 to the final digit 8 and truncate. The rounded form of 20.98752 to two decimal places is 20.99. ■

EXAMPLE 4

Approximating a Decimal to Two and Four Places

Number	Rounded to Two Decimal Places	Rounded to Four Decimal Places	Truncated to Two Decimal Places	Truncated to Four Decimal Places
(a) 3.14159	3.14	3.1416	3.14	3.1415
(b) 0.056128	0.06	0.0561	0.05	0.0561
(c) 893.46125	893.46	893.4613	893.46	893.4612 ■

NOW WORK PROBLEM 7.

CALCULATORS

Calculators are finite machines. As a result, they are incapable of displaying decimals that contain a large number of digits. For example, some calculators are capable of displaying only eight digits. When a number requires more than eight digits, the calculator either truncates or rounds. To see how your calculator handles decimals, divide 2 by 3. How many digits do you see? Is the last digit a 6 or a 7? If it is a 6, your calculator truncates; if it is a 7, your calculator rounds.

There are different kinds of calculators. An **arithmetic** calculator can only add, subtract, multiply, and divide numbers; therefore, this type is not adequate for this course. **Scientific** calculators have all the capabilities of arithmetic calculators and also contain **function keys** labeled ln, log, sin, cos, tan, x^y, inv, and so on. As you proceed through this text, you will discover how to use many of the function keys. **Graphing** calculators have all the capabilities of scientific calculators and contain a screen on which graphs can be displayed.

For those who have access to a graphing calculator, we have included comments, examples, and exercises marked with a 📟, indicating that a graphing calculator is required. We have also included an appendix that explains some of the capabilities of a graphing calculator. The 📟 comments, examples, and exercises may be omitted without loss of continuity, if so desired.

OPERATIONS

In algebra, we use letters such as x, y, a, b, and c to represent numbers. The symbols used in algebra for the operations of addition, subtraction, multiplication, and division are $+$, $-$, $\cdot$, and $/$. The words used to describe the results of these operations are **sum**, **difference**, **product**, and **quotient**. Table 1 summarizes these ideas.

TABLE 1		
Operation	Symbol	Words
Addition	$a + b$	Sum: a plus b
Subtraction	$a - b$	Difference: a less b
Multiplication	$a \cdot b, (a) \cdot b, a \cdot (b), (a) \cdot (b),$ $ab, (a)b, a(b), (a)(b)$	Product: a times b
Division	a/b or $\dfrac{a}{b}$	Quotient: a divided by b

In algebra, we generally avoid using the multiplication sign $\times$ and the division sign $\div$ so familiar in arithmetic. Notice also that when two expressions are placed next to each other without an operation symbol, as in ab, or in parentheses, as in $(a)(b)$, it is understood that the expressions, called **factors,** are to be multiplied.

We also prefer not to use mixed numbers in algebra. When mixed numbers are used, addition is understood; for example, $2\frac{3}{4}$ means $2 + \frac{3}{4}$. In algebra, use of a mixed number may be confusing because the absence of an operation symbol between two terms is generally taken to mean multiplication. The expression $2\frac{3}{4}$ is therefore written instead as 2.75 or as $\frac{11}{4}$.

The symbol $=$, called an **equal sign** and read as "equals" or "is," is used to express the idea that the number or expression on the left of the equal sign is equivalent to the number or expression on the right.

EXAMPLE 5

Writing Statements Using Symbols

(a) The sum of 2 and 7 equals 9. In symbols, this statement is written as
$2 + 7 = 9$.

(b) The product of 3 and 5 is 15. In symbols, this statement is written as
$3 \cdot 5 = 15$. ∎

NOW WORK PROBLEM **19.**

ORDER OF OPERATIONS

(2) Consider the expression $2 + 3 \cdot 6$. It is not clear whether we should add 2 and 3 to get 5, and then multiply by 6 to get 30; or first multiply 3 and 6 to get 18, and then add 2 to get 20. To avoid this ambiguity, we have the following agreement.

> We agree that whenever the two operations of addition and multiplication separate three numbers the multiplication operation always will be performed first, followed by the addition operation.

Thus, for $2 + 3 \cdot 6$, we have

$$2 + 3 \cdot 6 = 2 + 18 = 20$$

EXAMPLE 6 **Finding the Value of an Expression**

Evaluate each expression.

(a) $3 + 4 \cdot 5$ (b) $8 \cdot 2 + 1$ (c) $2 + 2 \cdot 2$

Solution   (a) $3 + 4 \cdot 5 = 3 + 20 = 23$ (b) $8 \cdot 2 + 1 = 16 + 1 = 17$
↑ Multiply first   ↑ Multiply first

(c) $2 + 2 \cdot 2 = 2 + 4 = 6$ ∎

━━ **NOW WORK PROBLEM 31.**

To first add 3 and 4 and then multiply the result by 5, we use parentheses and write $(3 + 4) \cdot 5$. Whenever parentheses appear in an expression, it means "perform the operations within the parentheses first!"

EXAMPLE 7 **Finding the Value of an Expression**

(a) $(5 + 3) \cdot 4 = 8 \cdot 4 = 32$
(b) $(4 + 5) \cdot (8 - 2) = 9 \cdot 6 = 54$ ∎

When we divide two expressions, as in

$$\frac{2 + 3}{4 + 8}$$

it is understood that the division bar acts like parentheses; that is,

$$\frac{2 + 3}{4 + 8} = \frac{(2 + 3)}{(4 + 8)}$$

The following list gives the rules for the order of operations.

RULES FOR THE ORDER OF OPERATIONS

1. Begin with the innermost parentheses and work outward. Remember that in dividing two expressions the numerator and denominator are treated as if they were enclosed in parentheses.
2. Perform multiplications and divisions, working from left to right.
3. Perform additions and subtractions, working from left to right.

EXAMPLE 8

Finding the Value of an Expression

Evaluate each expression.

(a) $8 \cdot 2 + 3$ (b) $5 \cdot (3 + 4) + 2$

(c) $\dfrac{2 + 5}{2 + 4 \cdot 7}$ (d) $2 + [4 + 2 \cdot (10 + 6)]$

Solution

(a) $8 \cdot 2 + 3 = 16 + 3 = 19$

          ↑
       Multiply first

(b) $5 \cdot (3 + 4) + 2 = 5 \cdot 7 + 2 = 35 + 2 = 37$

            ↑   ↑
    Parentheses first   Multiply before adding

(c) $\dfrac{2 + 5}{2 + 4 \cdot 7} = \dfrac{2 + 5}{2 + 28} = \dfrac{7}{30}$

(d) $2 + [4 + 2 \cdot (10 + 6)] = 2 + [4 + 2 \cdot (16)]$

$$= 2 + [4 + 32] = 2 + [36] = 38 \quad \blacksquare$$

▬▬▬▬ NOW WORK PROBLEMS **37** AND **45**.

PROPERTIES OF REAL NUMBERS

③ We have used the equal sign to mean that one expression is equivalent to another. Four important properties of equality are listed next. In this list, a, b, and c represent numbers.

1. The **reflexive property** states that a number always equals itself; that is, $a = a$.
2. The **symmetric property** states that if $a = b$ then $b = a$.
3. The **transitive property** states that if $a = b$ and $b = c$ then $a = c$.
4. The **principle of substitution** states that if $a = b$ then we may substitute b for a in any expression containing a.

Now, let's consider some other properties of real numbers. We begin with an example.

EXAMPLE 9

Commutative Properties

(a) $3 + 5 = 8$ (b) $2 \cdot 3 = 6$
        $5 + 3 = 8$ $3 \cdot 2 = 6$
        $3 + 5 = 5 + 3$ $2 \cdot 3 = 3 \cdot 2$ ■

This example illustrates the **commutative property** of real numbers, which states that the order in which addition or multiplication takes place will not affect the final result.

Commutative Properties

$$a + b = b + a \tag{1a}$$

$$a \cdot b = b \cdot a \tag{1b}$$

Here, and in the properties listed next and on pages 10–12, a, b, and c represent real numbers.

EXAMPLE 10

Associative Properties

(a) $2 + (3 + 4) = 2 + 7 = 9$
$\quad (2 + 3) + 4 = 5 + 4 = 9$
$\quad 2 + (3 + 4) = (2 + 3) + 4$

(b) $2 \cdot (3 \cdot 4) = 2 \cdot 12 = 24$
$\quad (2 \cdot 3) \cdot 4 = 6 \cdot 4 = 24$
$\quad 2 \cdot (3 \cdot 4) = (2 \cdot 3) \cdot 4$ ■

The way we add or multiply three real numbers will not affect the final result. Thus, expressions such as $2 + 3 + 4$ and $3 \cdot 4 \cdot 5$ present no ambiguity, even though addition and multiplication are performed on one pair of numbers at a time. This property is called the **associative property.**

Associative Properties

$$a + (b + c) = (a + b) + c = a + b + c \qquad \textbf{(2a)}$$
$$a \cdot (b \cdot c) = (a \cdot b) \cdot c = a \cdot b \cdot c \qquad \textbf{(2b)}$$

The next property is perhaps the most important.

Distributive Property

$$a \cdot (b + c) = a \cdot b + a \cdot c \qquad \textbf{(3a)}$$
$$(a + b) \cdot c = a \cdot c + b \cdot c \qquad \textbf{(3b)}$$

The **distributive property** may be used in two different ways.

EXAMPLE 11

Distributive Property

(a) $2 \cdot (x + 3) = 2 \cdot x + 2 \cdot 3 = 2x + 6$ Use to remove parentheses.
(b) $3x + 5x = (3 + 5)x = 8x$ Use to combine two expressions. ■

NOW WORK PROBLEM **63**.

The real numbers 0 and 1 have unique properties.

EXAMPLE 12

Identity Properties

(a) $4 + 0 = 0 + 4 = 4$ (b) $3 \cdot 1 = 1 \cdot 3 = 3$ ■

The properties of 0 and 1 illustrated in Example 12 are called the **identity properties.**

Identity Properties

$$0 + a = a + 0 = a \qquad \textbf{(4a)}$$
$$a \cdot 1 = 1 \cdot a = a \qquad \textbf{(4b)}$$

We call 0 the **additive identity** and 1 the **multiplicative identity.**

For each real number a, there is a real number $-a$, called the **additive inverse** of a, having the following property:

Additive Inverse Property

$$a + (-a) = -a + a = 0 \qquad \text{(5a)}$$

EXAMPLE 13 Finding an Additive Inverse

(a) The additive inverse of 6 is -6, because $6 + (-6) = 0$.

(b) The additive inverse of -8 is $-(-8) = 8$, because $-8 + 8 = 0$. ■

The additive inverse of a, that is, $-a$, is often called the *negative* of a or the *opposite* of a. The use of such terms can be dangerous, because they suggest that the additive inverse is a negative number, which it may not be. For example, the additive inverse of -3, or $-(-3)$, equals 3, a positive number.

For each *nonzero* real number a, there is a real number $1/a$, called the **multiplicative inverse** of a, having the following property:

Multiplicative Inverse Property

$$a \cdot \frac{1}{a} = \frac{1}{a} \cdot a = 1 \qquad \text{if } a \neq 0 \qquad \text{(5b)}$$

The multiplicative inverse $1/a$ of a nonzero real number a is also referred to as the **reciprocal** of a.

EXAMPLE 14 Finding a Reciprocal

(a) The reciprocal of 6 is $\frac{1}{6}$, because $6 \cdot \frac{1}{6} = 1$.

(b) The reciprocal of -3 is $\frac{1}{-3}$, because $-3 \cdot \frac{1}{-3} = 1$.

(c) The reciprocal of $\frac{2}{3}$ is $\frac{3}{2}$, because $\frac{2}{3} \cdot \frac{3}{2} = 1$. ■

With these properties for adding and multiplying real numbers, we can now define the operations of subtraction and division as follows:

The **difference** $a - b$, also read "a less b" or "a minus b," is defined as

$$a - b = a + (-b) \qquad \text{(6)}$$

Thus, to subtract b from a, add the opposite of b to a.

If b is a nonzero real number, the **quotient** a/b, also read as "a divided by b" or "the ratio of a to b," is defined as

$$\frac{a}{b} = a \cdot \frac{1}{b} \qquad \text{if } b \neq 0 \qquad\qquad \textbf{(7)}$$

EXAMPLE 15

Working with Differences and Quotients

(a) $8 - 5 = 8 + (-5) = 3$ (b) $4 - 9 = 4 + (-9) = -5$

(c) $\dfrac{5}{8} = 5 \cdot \dfrac{1}{8}$ ∎

For any number a, the product of a times 0 is always 0; that is,

Multiplication by Zero

$$a \cdot 0 = 0 \qquad\qquad \textbf{(8)}$$

For a nonzero number a,

Division Properties

$$\frac{0}{a} = 0 \qquad \frac{a}{a} = 1 \qquad \text{if } a \neq 0 \qquad\qquad \textbf{(9)}$$

NOTE: Division by 0 is *not defined*. One reason is to avoid the following difficulty: $\frac{2}{0} = x$ means to find x such that $0 \cdot x = 2$. But $0 \cdot x$ equals 0 for all x, so there is *no* number x such that $\frac{2}{0} = x$.

Rules of Signs

$$a(-b) = -(ab) \qquad (-a)b = -(ab) \qquad (-a)(-b) = ab$$

$$-(-a) = a \qquad \frac{a}{-b} = \frac{-a}{b} = -\frac{a}{b} \qquad \frac{-a}{-b} = \frac{a}{b} \qquad \textbf{(10)}$$

EXAMPLE 16

Applying the Rules of Signs

(a) $2(-3) = -(2 \cdot 3) = -6$ (b) $(-3)(-5) = 3 \cdot 5 = 15$

(c) $\dfrac{3}{-2} = \dfrac{-3}{2} = -\dfrac{3}{2}$ (d) $\dfrac{-4}{-9} = \dfrac{4}{9}$ (e) $\dfrac{x}{-2} = \dfrac{1}{-2} \cdot x = -\dfrac{1}{2}x$ ∎

If c is a nonzero number, then

Cancellation Properties

$$ac = bc \quad \text{implies} \quad a = b \qquad \text{if } c \neq 0$$

$$\frac{ac}{bc} = \frac{a}{b} \qquad\qquad\qquad \text{if } b \neq 0, c \neq 0 \qquad \textbf{(11)}$$

| EXAMPLE 17 | Using the Cancellation Properties |

(a) If $2x = 6$, then

$$2x = 6$$
$$2x = 2 \cdot 3 \qquad \text{Factor 6.}$$
$$x = 3 \qquad \text{Cancel the 2's.}$$

(b) $\dfrac{18}{12} = \dfrac{3 \cdot \cancel{6}}{2 \cdot \cancel{6}} = \dfrac{3}{2}$

$\qquad\qquad\qquad \uparrow$
$\qquad\qquad$ Cancel the 6's.

NOTE: We follow the common practice of using slash marks to indicate cancellations.

Zero-Product Property

$$\boxed{\text{If } ab = 0, \text{ then } a = 0 \text{ or } b = 0, \text{ or both.} \qquad \textbf{(12)}}$$

| EXAMPLE 18 | Using the Zero-Product Property |

If $2x = 0$, then either $2 = 0$ or $x = 0$. Since $2 \neq 0$, it follows that $x = 0$. ∎

Arithmetic of Quotients

$$\frac{a}{b} + \frac{c}{d} = \frac{ad}{bd} + \frac{bc}{bd} = \frac{ad + bc}{bd} \qquad \text{if } b \neq 0, d \neq 0 \qquad \textbf{(13)}$$

$$\frac{a}{b} \cdot \frac{c}{d} = \frac{ac}{bd} \qquad \text{if } b \neq 0, d \neq 0 \qquad \textbf{(14)}$$

$$\frac{\dfrac{a}{b}}{\dfrac{c}{d}} = \frac{a}{b} \cdot \frac{d}{c} = \frac{ad}{bc} \qquad \text{if } b \neq 0, c \neq 0, d \neq 0 \qquad \textbf{(15)}$$

| EXAMPLE 19 | Adding, Subtracting, Multiplying, and Dividing Quotients |

(a) $\dfrac{2}{3} + \dfrac{5}{2} = \dfrac{2 \cdot 2}{3 \cdot 2} + \dfrac{3 \cdot 5}{3 \cdot 2} = \dfrac{2 \cdot 2 + 3 \cdot 5}{3 \cdot 2} = \dfrac{4 + 15}{6} = \dfrac{19}{6}$

$\qquad\qquad\uparrow$
$\qquad$ By equation (13)

(b) $\dfrac{3}{5} - \dfrac{2}{3} = \dfrac{3}{5} + \left(-\dfrac{2}{3}\right) = \dfrac{3}{5} + \dfrac{-2}{3}$

$\qquad\qquad\quad\uparrow \qquad\qquad\quad\uparrow$
$\qquad\quad$ By equation (6) $\quad$ By equation (10)

$$= \dfrac{3 \cdot 3 + 5 \cdot (-2)}{5 \cdot 3} = \dfrac{9 + (-10)}{15} = \dfrac{-1}{15} = -\dfrac{1}{15}$$

(c) $\dfrac{8}{3} \cdot \dfrac{15}{4} = \dfrac{8 \cdot 15}{3 \cdot 4} = \dfrac{2 \cdot \cancel{4} \cdot \cancel{3} \cdot 5}{\cancel{3} \cdot \cancel{4} \cdot 1} = \dfrac{2 \cdot 5}{1} = 10$

By equation (14)   By equation (11)

NOTE: Slanting the cancellation marks in different directions for different factors, as shown here, is a good practice to follow, since it will help in checking for errors.

(d) $\dfrac{\frac{3}{5}}{\frac{7}{9}} = \dfrac{3}{5} \cdot \dfrac{9}{7} = \dfrac{3 \cdot 9}{5 \cdot 7} = \dfrac{27}{35}$

By equation (15)

By equation (14)   ▬

NOTE: In writing quotients, we shall follow the usual convention and write the quotient in lowest terms; that is, we write it so that any common factors of the numerator and the denominator have been removed using the cancellation properties, equation (11).

$$\frac{90}{24} = \frac{15 \cdot \cancel{6}}{4 \cdot \cancel{6}} = \frac{15}{4}$$

$$\frac{24x^2}{18x} = \frac{4 \cdot \cancel{6} \cdot x \cdot \cancel{x}}{3 \cdot \cancel{6} \cdot \cancel{x}} = \frac{4x}{3} \qquad x \neq 0$$

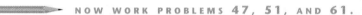 NOW WORK PROBLEMS **47, 51,** AND **61.**

Sometimes it is easier to add two fractions using *least common multiples* (LCM). The LCM of two numbers is the smallest number that each has as a common multiple.

EXAMPLE 20   **Finding the Least Common Multiple of Two Numbers**

Find the least common multiple of 15 and 12.

Solution   To find the LCM of 15 and 12, we look at multiples of 15 and 12.

15,   30,   45,   60,   75,   90,   105,   120,…

12,   24,   36,   48,   60,   72,   84,   96,   108,   120,…

The *common* multiples are in blue. The *least* common multiple is 60.   ▬

EXAMPLE 21   **Using the Least Common Multiple to Add Two Fractions**

Find: $\dfrac{8}{15} + \dfrac{5}{12}$

Solution   We use the LCM of the denominators of the fractions and rewrite each fraction using the LCM as a common denominator. The LCM of the denominators (12 and 15) is 60. Rewrite each fraction using 60 as the denominator.

$$\frac{8}{15} + \frac{5}{12} = \frac{8}{15} \cdot \frac{4}{4} + \frac{5}{12} \cdot \frac{5}{5} = \frac{32}{60} + \frac{25}{60} = \frac{32 + 25}{60} = \frac{57}{60}$$ ▬

NOW WORK PROBLEM **55.**

HISTORICAL FEATURE

The real number system has a history that stretches back at least to the ancient Babylonians (1800 BC). It is remarkable how much the ancient Babylonian attitudes resemble our own. As we stated in the text, the fundamental difficulty with irrational numbers is that they cannot be written as quotients of integers or, equivalently, as repeating or terminating decimals. The Babylonians wrote their numbers in a system based on 60 in the same way that we write ours based on 10. They would carry as many places for π as the accuracy of the problem demanded, just as we now use

$$\pi \approx 3\frac{1}{7} \quad \text{or} \quad \pi \approx 3.1416 \quad \text{or} \quad \pi \approx 3.14159$$

$$\text{or} \quad \pi \approx 3.141592653589$$

depending on how accurate we need to be.

Things were very different for the Greeks, whose number system allowed only rational numbers. When it was discovered that $\sqrt{2}$ was not a rational number, this was regarded as a fundamental flaw in the number concept. So serious was the matter that the Pythagorean Brotherhood (an early mathematical society) is said to have drowned one of its members for revealing this terrible secret. Greek mathematicians then turned away

from the number concept, expressing facts about whole numbers in terms of line segments.

In astronomy, however, Babylonian methods, including the Babylonian number system, continued to be used. Simon Stevin (1548–1620), probably using the Babylonian system as a model, invented the decimal system, complete with rules of calculation, in 1585. [Others, for example, al-Kashi of Samarkand (d. 1424), had made some progress in the same direction.] The decimal system so effectively conceals the difficulties that the need for more logical precision began to be felt only in the early 1800s. Around 1880, Georg Cantor (1845–1918) and Richard Dedekind (1831–1916) gave precise definitions of real numbers. Cantor's definition, although more abstract and precise, has its roots in the decimal (and hence Babylonian) numerical system.

Sets and set theory were a spin-off of the research that went into clarifying the foundations of the real number system. Set theory has developed into a large discipline of its own, and many mathematicians regard it as the foundation upon which modern mathematics is built. Cantor's discoveries that infinite sets can also be counted and that there are different sizes of infinite sets are among the most astounding results of modern mathematics.

HISTORICAL PROBLEMS

The Babylonian number system was based on 60. Thus 2,30 means $2 + \frac{30}{60} = 2.5$, and 4,25,14 means

$$4 + \frac{25}{60} + \frac{14}{60^2} = 4 + \frac{1514}{3600} = 4.42055555\ldots$$

1. What are the following numbers in Babylonian notation?

(a) $1\frac{1}{3}$ (b) $2\frac{5}{6}$

2. What are the following Babylonian numbers when written as fractions and as decimals?

(a) 2,20   (b) 4,52,30   (c) 3,8,29,44

In Problems 1–6, list the numbers in each set that are (a) Natural numbers, (b) Integers, (c) Rational numbers, (d) Irrational numbers, (e) Real numbers.

1. $A = \left\{-6, \frac{1}{2}, -1.333\ldots \text{ (the 3's repeat)}, \pi, 2, 5\right\}$

2. $B = \left\{-\frac{5}{3}, 2.060606\ldots \text{ (the block 06 repeats)}, 1.25, 0, 1, \sqrt{5}\right\}$

3. $C = \left\{0, 1, \frac{1}{2}, \frac{1}{3}, \frac{1}{4}\right\}$

4. $D = \{-1, -1.1, -1.2, -1.3\}$

5. $E = \left\{\sqrt{2}, \pi, \sqrt{2} + 1, \pi + \frac{1}{2}\right\}$

6. $F = \left\{-\sqrt{2}, \pi + \sqrt{2}, \frac{1}{2} + 10.3\right\}$

In Problems 7–18, approximate each number (a) rounded and (b) truncated to three decimal places.

7. 18.9526

8. 25.86134

9. 28.65319

10. 99.05249

11. 0.06291

12. 0.05388

13. 9.9985

14. 1.0006

15. $\dfrac{3}{7}$

16. $\dfrac{5}{9}$

17. $\dfrac{521}{15}$

18. $\dfrac{81}{5}$

In Problems 19–28, write each statement using symbols.

19. The sum of 3 and 2 equals 5.

20. The product of 5 and 2 equals 10.

21. The sum of x and 2 is the product of 3 and 4.

22. The sum of 3 and y is the sum of 2 and 2.

23. The product of 3 and y is the sum of 1 and 2.

24. The product of 2 and x is the product of 4 and 6.

25. The difference x less 2 equals 6.

26. The difference 2 less y equals 6.

27. The quotient x divided by 2 is 6.

28. The quotient 2 divided by x is 6.

In Problems 29–62, evaluate each expression.

29. $9 - 4 + 2$

30. $6 - 4 + 3$

31. $-6 + 4 \cdot 3$

32. $8 - 4 \cdot 2$

33. $4 + 5 - 8$

34. $8 - 3 - 4$

35. $4 + \dfrac{1}{3}$

36. $2 - \dfrac{1}{2}$

37. $6 - [3 \cdot 5 + 2 \cdot (3 - 2)]$

38. $2 \cdot [8 - 3(4 + 2)] - 3$

39. $2 \cdot (3 - 5) + 8 \cdot 2 - 1$

40. $1 - (4 \cdot 3 - 2 + 2)$

41. $10 - [6 - 2 \cdot 2 + (8 - 3)] \cdot 2$

42. $2 - 5 \cdot 4 - [6 \cdot (3 - 4)]$

43. $(5 - 3)\dfrac{1}{2}$

44. $(5 + 4)\dfrac{1}{3}$

45. $\dfrac{4 + 8}{5 - 3}$

46. $\dfrac{2 - 4}{5 - 3}$

47. $\dfrac{3}{5} \cdot \dfrac{10}{21}$

48. $\dfrac{5}{9} \cdot \dfrac{3}{10}$

49. $\dfrac{6}{25} \cdot \dfrac{10}{27}$

50. $\dfrac{21}{25} \cdot \dfrac{100}{3}$

51. $\dfrac{3}{4} + \dfrac{2}{5}$

52. $\dfrac{4}{3} + \dfrac{1}{2}$

53. $\dfrac{5}{6} + \dfrac{9}{5}$

54. $\dfrac{8}{9} + \dfrac{15}{2}$

55. $\dfrac{5}{18} + \dfrac{1}{12}$

56. $\dfrac{2}{15} + \dfrac{8}{9}$

57. $\dfrac{1}{30} - \dfrac{7}{18}$

58. $\dfrac{3}{14} - \dfrac{2}{21}$

59. $\dfrac{3}{20} - \dfrac{2}{15}$

60. $\dfrac{6}{35} - \dfrac{3}{14}$

61. $\dfrac{\dfrac{5}{18}}{\dfrac{11}{27}}$

62. $\dfrac{\dfrac{5}{21}}{\dfrac{2}{35}}$

In Problems 63–74, use the distributive property to remove the parentheses.

63. $6(x + 4)$

64. $4(2x - 1)$

65. $x(x - 4)$

66. $4x(x + 3)$

67. $(x + 2)(x + 4)$

68. $(x + 5)(x + 1)$

69. $(x - 2)(x + 1)$

70. $(x - 4)(x + 1)$

71. $(x - 8)(x - 2)$

72. $(x - 4)(x - 2)$

73. $(x + 2)(x - 2)$

74. $(x - 3)(x + 3)$

---

75. Explain to a friend how the distributive property is used to justify the fact that $2x + 3x = 5x$.

76. Explain to a friend why $2 + 3 \cdot 4 = 14$, whereas $(2 + 3) \cdot 4 = 20$.

77. Explain why $2(3 \cdot 4)$ is not equal to $(2 \cdot 3) \cdot (2 \cdot 4)$.

78. Explain why $\dfrac{4 + 3}{2 + 5}$ is not equal to $\dfrac{4}{2} + \dfrac{3}{5}$.

79. Is subtraction commutative? Support your conclusion with an example.

80. Is subtraction associative? Support your conclusion with an example.

81. Is division commutative? Support your conclusion with an example.

82. Is division associative? Support your conclusion with an example.

83. If $2 = x$, why does $x = 2$?

84. If $x = 5$, why does $x^2 + x = 30$?

85. Are there any real numbers that are both rational and irrational? Are there any real numbers that are neither? Explain your reasoning.

86. Explain why the sum of a rational number and an irrational number must be irrational.

87. What rational number does the repeating decimal $0.9999\ldots$ equal?

2 ALGEBRA REVIEW

OBJECTIVES ① Graph Inequalities

② Find Distance on the Real Number Line

③ Evaluate Algebraic Expressions

④ Determine the Domain of a Variable

THE REAL NUMBER LINE

Figure 4
Real number line

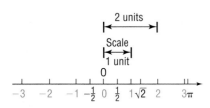

The real numbers can be represented by points on a line called the **real number line.** There is a one-to-one correspondence between real numbers and points on a line. That is, every real number corresponds to a point on the line, and each point on the line has a unique real number associated with it.

Pick a point on the line somewhere in the center, and label it O. This point, called the **origin,** corresponds to the real number 0. See Figure 4. The point 1 unit to the right of O corresponds to the number 1. The distance between 0 and 1 determines the **scale** of the number line. For example, the point associated with the number 2 is twice as far from O as 1 is. Notice that an arrowhead on the right end of the line indicates the direction in which the numbers increase. Figure 4 also shows the points associated with the irrational numbers $\sqrt{2}$ and π. Points to the left of the origin correspond to the real numbers -1, -2, and so on.

> The real number associated with a point P is called the **coordinate** of P, and the line whose points have been assigned coordinates is called the **real number line.**

Figure 5

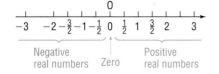

✏️ N O W   W O R K   P R O B L E M   **1.**

The real number line consists of three classes of real numbers, as shown in Figure 5.

> 1. The **negative real numbers** are the coordinates of points to the left of the origin O.
> 2. The real number **zero** is the coordinate of the origin O.
> 3. The **positive real numbers** are the coordinates of points to the right of the origin O.

Negative and positive numbers have the following multiplication properties:

Multiplication Properties of Positive and Negative Numbers

> 1. The product of two positive numbers is a positive number.
> 2. The product of two negative numbers is a positive number.
> 3. The product of a positive number and a negative number is a negative number.

Figure 6

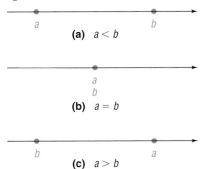

(a) $a < b$

(b) $a = b$

(c) $a > b$

INEQUALITIES

An important property of the real number line follows from the fact that, given two numbers (points) a and b, either a is to the left of b, a is at the same location as b, or a is to the right of b. See Figure 6.

If a is to the left of b, we say that "a is less than b" and write $a < b$. If a is to the right of b, we say that "a is greater than b" and write $a > b$. If a is at the same location as b, then $a = b$. If a is either less than or equal to b, we write $a \le b$. Similarly, $a \ge b$ means that a is either greater than or equal to b. Collectively, the symbols $<, >, \le,$ and $\ge$ are called **inequality symbols.**

Note that $a < b$ and $b > a$ mean the same thing. It does not matter whether we write $2 < 3$ or $3 > 2$.

Furthermore, if $a < b$ or if $b > a$, then the difference $b - a$ is positive. Do you see why?

| EXAMPLE 1 | Using Inequality Symbols |

(a) $3 < 7$ (b) $-8 > -16$ (c) $-6 < 0$

(d) $-8 < -4$ (e) $4 > -1$ (f) $8 > 0$ ∎

In Example 1(a), we conclude that $3 < 7$ either because 3 is to the left of 7 on the real number line or because the difference $7 - 3 = 4$, is a positive real number.

Similarly, we conclude in Example 1(b) that $-8 > -16$ either because -8 lies to the right of -16 on the real number line or because the difference $-8 - (-16) = -8 + 16 = 8$, is a positive real number.

Look again at Example 1. Note that the inequality symbol always points in the direction of the smaller number.

Statements of the form $a < b$ or $b > a$ are called **strict inequalities,** whereas statements of the form $a \le b$ or $b \ge a$ are called **nonstrict inequalities.** An **inequality** is a statement in which two expressions are related by an inequality symbol. The expressions are referred to as the **sides** of the inequality.

Based on the discussion thus far, we conclude that

$a > 0$	is equivalent to a is positive
$a < 0$	is equivalent to a is negative

We sometimes read $a > 0$ by saying that "a is positive." If $a \ge 0$, then either $a > 0$ or $a = 0$, and we may read this as "a is nonnegative."

✎ NOW WORK PROBLEMS **5** AND **15.**

① We shall find it useful in later work to graph inequalities on the real number line.

| EXAMPLE 2 | Graphing Inequalities |

(a) On the real number line, graph all numbers x for which $x > 4$.

(b) On the real number line, graph all numbers x for which $x \le 5$.

Figure 7
x > 4

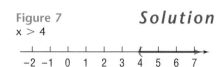

Figure 8
x ≤ 5

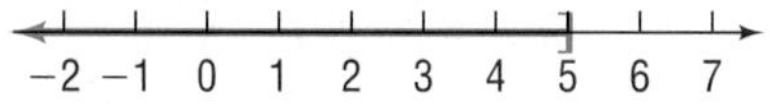

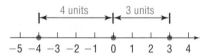

Figure 9

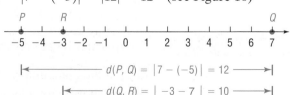

Solution (a) See Figure 7. Notice that we use a left parenthesis to indicate that the number 4 is *not* part of the graph.

(b) See Figure 8. Notice that we use a right bracket to indicate that the number 5 *is* part of the graph. ∎

NOW WORK PROBLEM **21.**

ABSOLUTE VALUE

The *absolute value* of a number a is the distance from 0 to a on the number line. For example, −4 is 4 units from 0; and 3 is 3 units from 0. See Figure 9. Thus, the absolute value of −4 is 4, and the absolute value of 3 is 3.

A more formal definition of absolute value is given next.

> The **absolute value** of a real number a, denoted by the symbol $|a|$, is defined by the rules
>
> $$|a| = a \quad \text{if } a \geq 0 \qquad \text{and} \qquad |a| = -a \quad \text{if } a < 0$$

For example, since $-4 < 0$, the second rule must be used to get $|-4| = -(-4) = 4$.

EXAMPLE 3 **Computing Absolute Value**

(a) $|8| = 8$ (b) $|0| = 0$ (c) $|-15| = -(-15) = 15$ ∎

2 Look again at Figure 9. The distance from −4 to 3 is 7 units. This distance is the difference $3 - (-4)$, obtained by subtracting the smaller coordinate from the larger. However, since $|3 - (-4)| = |7| = 7$ and $|-4 - 3| = |-7| = 7$, we can use absolute value to calculate the distance between two points without being concerned about which is smaller.

> If P and Q are two points on a real number line with coordinates a and b, respectively, the **distance between P and Q,** denoted by $d(P, Q)$, is
>
> $$d(P, Q) = |b - a|$$

Since $|b - a| = |a - b|$, it follows that $d(P, Q) = d(Q, P)$.

EXAMPLE 4 **Finding Distance on a Number Line**

Let $P, Q,$ and R be points on a real number line with coordinates $-5, 7,$ and -3, respectively. Find the distance

(a) between P and Q (b) between Q and R

Solution (a) $d(P, Q) = |7 - (-5)| = |12| = 12$ (see Figure 10)

Figure 10

$$d(P, Q) = |7 - (-5)| = 12$$
$$d(Q, R) = |-3 - 7| = 10$$

(b) $d(Q, R) = |-3 - 7| = |-10| = 10$ ■

NOW WORK PROBLEM **27**.

CONSTANTS AND VARIABLES

As we said earlier, in algebra we use letters such as x, y, a, b, and c to represent numbers. If the letter used is to represent *any* number from a given set of numbers, it is called a **variable.** A **constant** is either a fixed number, such as 5 or $\sqrt{3}$, or a letter that represents a fixed (possibly unspecified) number.

Constants and variables are combined using the operations of addition, subtraction, multiplication, and division to form *algebraic expressions.* Examples of algebraic expressions include

$$x + 3 \qquad \frac{3}{1 - t} \qquad 7x - 2y$$

③ To evaluate an algebraic expression, substitute for each variable its numerical value.

EXAMPLE 5 **Evaluating an Algebraic Expression**

Evaluate each expression if $x = 3$ and $y = -1$.

(a) $x + 3y$ (b) $5xy$ (c) $\dfrac{3y}{2 - 2x}$ (d) $|-4x + y|$

Solution (a) Substitute 3 for x and -1 for y in the expression $x + 3y$.

$$x + 3y = 3 + 3(-1) = 3 + (-3) = 0$$
$$\uparrow$$
$$x = 3, y = -1$$

(b) If $x = 3$ and $y = -1$, then

$$5xy = 5(3)(-1) = -15$$

(c) If $x = 3$ and $y = -1$, then

$$\frac{3y}{2 - 2x} = \frac{3(-1)}{2 - 2(3)} = \frac{-3}{2 - 6} = \frac{-3}{-4} = \frac{3}{4}$$

(d) If $x = 3$ and $y = -1$, then

$$|-4x + y| = |-4(3) + (-1)| = |-12 + (-1)| = |-13| = 13$$ ■

NOW WORK PROBLEMS **29** AND **37**.

④ In working with expressions or formulas involving variables, the variables may be allowed to take on values from only a certain set of numbers. For example, in the formula for the area A of a circle of radius r, $A = \pi r^2$, the variable r is necessarily restricted to the positive real numbers. In the expression $1/x$, the variable x cannot take on the value 0, since division by 0 is not defined.

The set of values that a variable may assume is called the **domain of the variable.**

| EXAMPLE 6 | **Finding the Domain of a Variable** |

The domain of the variable x in the expression

$$\frac{5}{x-2}$$

is $\{x \mid x \neq 2\}$, since, if $x = 2$, the denominator becomes 0, which is not defined. ▪

| EXAMPLE 7 | **Circumference of a Circle** |

In the formula for the circumference C of a circle of radius r,

$$C = 2\pi r$$

the domain of the variable r, representing the radius of the circle, is the set of positive real numbers. The domain of the variable C, representing the circumference of the circle, is also the set of positive real numbers. ▪

In describing the domain of a variable, we may use either set notation or words, whichever is more convenient.

NOW WORK PROBLEM 47.

---

2 EXERCISES

1. On the real number line, label the points with coordinates $0, 1, -1, \frac{5}{2}, -2.5, \frac{3}{4}$, and 0.25.
2. Repeat Problem 1 for the coordinates $0, -2, 2, -1.5, \frac{3}{2}, \frac{1}{3}$, and $\frac{2}{3}$.

In Problems 3–12, replace the question mark by $<, >$, or $=$, whichever is correct.

3. $\frac{1}{2}$? 0 4. 5 ? 6 5. -1 ? -2 6. -3 ? $-\frac{5}{2}$ 7. π ? 3.14
8. $\sqrt{2}$? 1.41 9. $\frac{1}{2}$? 0.5 10. $\frac{1}{3}$? 0.33 11. $\frac{2}{3}$? 0.67 12. $\frac{1}{4}$? 0.25

In Problems 13–18, write each statement as an inequality.

13. x is positive   14. z is negative   15. x is less than 2   16. y is greater than -5
17. x is less than or equal to 1   18. x is greater than or equal to 2

In Problems 19–22, graph the numbers x on the real number line.

19. $x \geq -2$ 20. $x < 4$ 21. $x > -1$ 22. $x \leq 7$

In Problems 23–28, use the real number line below to compute each distance.

23. $d(C, D)$ 24. $d(C, A)$ 25. $d(D, E)$ 26. $d(C, E)$ 27. $d(A, E)$ 28. $d(D, B)$

In Problems 29–36, evaluate each expression if x = −2 and y = 3.

29. $x + 2y$

30. $3x + y$

31. $5xy + 2$

32. $-2x + xy$

33. $\dfrac{2x}{x - y}$

34. $\dfrac{x + y}{x - y}$

35. $\dfrac{3x + 2y}{2 + y}$

36. $\dfrac{2x - 3}{y}$

In Problems 37–46, find the value of each expression if x = 3 and y = −2.

37. $|x + y|$

38. $|x - y|$

39. $|x| + |y|$

40. $|x| - |y|$

41. $\dfrac{|x|}{x}$

42. $\dfrac{|y|}{y}$

43. $|4x - 5y|$

44. $|3x + 2y|$

45. $\|4x| - |5y\|$

46. $3|x| + 2|y|$

In Problems 47–54, determine which of the value(s) given below, if any, must be excluded from the domain of the variable in each rational expression:

(a) $x = 3$ (b) $x = 1$ (c) $x = 0$ (d) $x = -1$

47. $\dfrac{x^2 - 1}{x}$

48. $\dfrac{x^2 + 1}{x}$

49. $\dfrac{x}{x^2 - 9}$

50. $\dfrac{x}{x^2 + 9}$

51. $\dfrac{x^2}{x^2 + 1}$

52. $\dfrac{x^3}{x^2 - 1}$

53. $\dfrac{x^2 + 5x - 10}{x^3 - x}$

54. $\dfrac{-9x^2 - x + 1}{x^3 + x}$

In Problems 55–58, determine the domain of the variable x in each expression.

55. $\dfrac{4}{x - 5}$

56. $\dfrac{-6}{x + 4}$

57. $\dfrac{x}{x + 4}$

58. $\dfrac{x - 2}{x - 6}$

In Problems 59–62, use the formula $C = \frac{5}{9}(F - 32)$ for converting degrees Fahrenheit into degrees Celsius to find the Celsius measure of each Fahrenheit temperature.

59. $F = 32°$

60. $F = 212°$

61. $F = 77°$

62. $F = -4°$

In Problems 63–72, express each statement as an equation involving the indicated variables.

63. Area of a Rectangle   The area A of a rectangle is the product of its length l times its width w.

64. Perimeter of a Rectangle   The perimeter P of a rectangle is twice the sum of its length l and its width w.

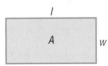

65. Circumference of a Circle   The circumference C of a circle is the product of π times its diameter d.

66. Area of a Triangle   The area A of a triangle is one-half the product of its base b times its height h.

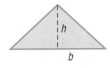

67. Area of an Equilateral Triangle   The area A of an equilateral triangle is $\sqrt{3}/4$ times the square of the length x of one side.

68. Perimeter of an Equilateral Triangle   The perimeter P of an equilateral triangle is 3 times the length x of one side.

69. Volume of a Sphere   The volume V of a sphere is the product of $\frac{4}{3}$ times π times the cube of the radius r.

70. Surface Area of a Sphere   The surface area S of a sphere is the product of 4 times π times the square of the radius r.

71. Volume of a Cube   The volume V of a cube is the cube of the length x of a side.

72. Surface Area of a Cube   The surface area S of a cube is 6 times the square of the length x of a side.

73. Manufacturing Cost   The weekly production cost C of manufacturing x watches is given by the formula $C = 4000 + 2x$, where the variable C is in dollars.
(a) What is the cost of producing 1000 watches?
(b) What is the cost of producing 2000 watches?

74. Balancing a Checkbook   At the beginning of the month, Mike had a balance of $210 in his checking account. During the next month, he deposited $80, wrote a check for $120, made another deposit of $25, wrote two checks for $60 and $32, and was assessed a monthly service charge of $5. What was his balance at the end of the month?

75. U.S. Voltage   In the United States, normal household voltage is 115 volts. It is acceptable for the actual voltage x to differ from normal by at most 5 volts. A formula that describes this is

$$|x - 115| \leq 5$$

(a) Show that a voltage of 113 volts is acceptable.
(b) Show that a voltage of 109 volts is not acceptable.

76. Foreign Voltage   In other countries, normal household voltage is 220 volts. It is acceptable for the actual voltage x to differ from normal by at most 8 volts. A formula that describes this is

$$|x - 220| \leq 8$$

(a) Show that a voltage of 214 volts is acceptable.
(b) Show that a voltage of 209 volts is not acceptable.

77. Making Precision Ball Bearings   The FireBall Company manufactures ball bearings for precision equipment. One of their products is a ball bearing with a stated radius of 3 centimeters (cm). Only ball bearings with a radius within 0.01 cm of this stated radius are acceptable. If x is the radius of a ball bearing, a formula describing this situation is

$$|x - 3| \leq 0.01$$

(a) Is a ball bearing of radius $x = 2.999$ acceptable?
(b) Is a ball bearing of radius $x = 2.89$ acceptable?

78. Body Temperature   Normal human body temperature is 98.6°F. A temperature x that differs from normal by at least 1.5°F is considered unhealthy. A formula that describes this is

$$|x - 98.6| \geq 1.5$$

(a) Show that a temperature of 97°F is unhealthy.
(b) Show that a temperature of 100°F is not unhealthy.

79. Does $\frac{1}{3}$ equal 0.333? If not, which is larger? By how much?

80. Does $\frac{2}{3}$ equal 0.666? If not, which is larger? By how much?

81. Is there a positive real number "closest" to 0?

82. I'm thinking of a number! It lies between 1 and 10; its square is rational and lies between 1 and 10. The number is larger than π. Correct to two decimal places, name the number. Now think of your own number, describe it, and challenge a fellow student to name it.

83. Write a brief paragraph that illustrates the similarities and differences between "less than" ($<$) and "less than or equal to" ($\leq$).

---

3 | GEOMETRY REVIEW

OBJECTIVES   1   Use the Pythagorean Theorem and Its Converse
2   Know Geometry Formulas

In this section we review some topics studied in geometry that we shall need for our study of algebra.

PYTHAGOREAN THEOREM

1   The *Pythagorean Theorem* is a statement about *right triangles*. A **right triangle** is one that contains a **right angle,** that is, an angle of 90°. The side of the triangle opposite the 90° angle is called the **hypotenuse;** the remaining two sides are called **legs.** In Figure 11 we have used c to represent the length of

Figure 11

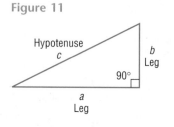

the hypotenuse and a and b to represent the lengths of the legs. Notice the use of the symbol $\ulcorner$ to show the 90° angle. We now state the Pythagorean Theorem.

Pythagorean Theorem   In a right triangle, the square of the length of the hypotenuse is equal to the sum of the squares of the lengths of the legs. That is, in the right triangle shown in Figure 11,

$$c^2 = a^2 + b^2 \qquad\qquad \textbf{(1)}$$

We prove this result at the end of the section.

| EXAMPLE 1 | Finding the Hypotenuse of a Right Triangle |

In a right triangle, one leg is of length 4 and the other is of length 3. What is the length of the hypotenuse?

Solution   Since the triangle is a right triangle, we use the Pythagorean Theorem with $a = 4$ and $b = 3$ to find the length c of the hypotenuse. From equation (1), we have

$$c^2 = a^2 + b^2$$
$$c^2 = 4^2 + 3^2 = 16 + 9 = 25$$
$$c = \sqrt{25} = 5$$

NOW WORK PROBLEM 3.

The converse of the Pythagorean Theorem is also true.

Converse of the Pythagorean Theorem   In a triangle, if the square of the length of one side equals the sum of the squares of the lengths of the other two sides, then the triangle is a right triangle. The 90° angle is opposite the longest side.

We prove this result at the end of the section.

| EXAMPLE 2 | Verifying That a Triangle Is a Right Triangle |

Show that a triangle whose sides are of lengths 5, 12, and 13 is a right triangle. Identify the hypotenuse.

Solution   We square the lengths of the sides.

$$5^2 = 25, \quad 12^2 = 144, \quad 13^2 = 169$$

Figure 12

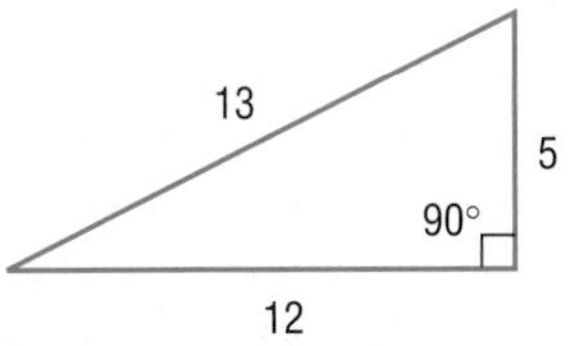

Notice that the sum of the first two squares (25 and 144) equals the third square (169). Hence, the triangle is a right triangle. The longest side, 13, is the hypotenuse. See Figure 12.

NOW WORK PROBLEM 11.

| EXAMPLE 3 | **Applying the Pythagorean Theorem** |

The tallest inhabited building in the world is the Sears Tower in Chicago.* If the observation tower is 1450 feet above ground level, how far can a person standing in the observation tower see (with the aid of a telescope)? Use 3960 miles for the radius of Earth. See Figure 13.

Figure 13

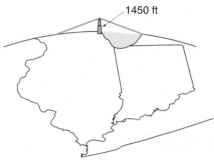

1450 ft

[**NOTE:**   1 mile = 5280 feet]

Solution   From the center of Earth, draw two radii: one through the Sears Tower and the other to the farthest point a person can see from the tower. See Figure 14. Apply the Pythagorean Theorem to the right triangle.

Figure 14

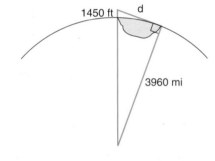

1450 ft   d

3960 mi

Since 1450 feet = 1450/5280 miles, we have

$$d^2 + (3960)^2 = \left(3960 + \frac{1450}{5280} \right)^2$$

$$d^2 = \left(3960 + \frac{1450}{5280} \right)^2 - (3960)^2 \approx 2175.08$$

$$d \approx 46.64$$

A person can see about 47 miles from the observation tower.   ■

   N O W   W O R K   P R O B L E M   **37.**

GEOMETRY FORMULAS

2   Certain formulas from geometry are useful in solving algebra problems. We list some of these formulas next.

For a rectangle of length *l* and width *w*,

| Area = *lw*   Perimeter = 2*l* + 2*w* |

For a triangle with base *b* and altitude *h*,

| Area = $\dfrac{1}{2} bh$ |

For a circle of radius *r* (diameter *d* = 2*r*),

| Area = πr^2 Circumference = $2\pi r = \pi d$ |

Source: Council on Tall Buildings and Urban Habitat (1997): Sears Tower No. 1 for tallest roof (1450 ft) and tallest occupied floor (1431 ft).

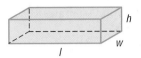

For a rectangular box of length l, width w, and height h,

$$\text{Volume} = lwh$$

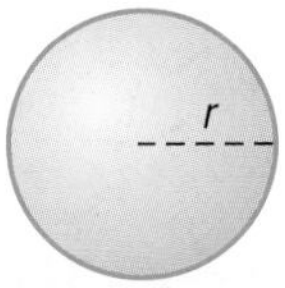

For a sphere of radius r,

$$\text{Volume} = \tfrac{4}{3}\pi r^3 \qquad \text{Surface area} = 4\pi r^2$$

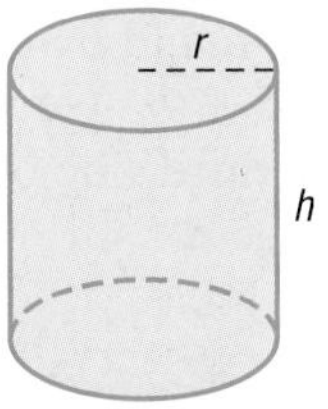

For a right circular cylinder of height h and radius r,

$$\text{Volume} = \pi r^2 h$$

NOW WORK PROBLEM **19.**

EXAMPLE 4   ## Using Geometry Formulas

A Christmas tree ornament is in the shape of a semicircle on top of a triangle. How many square centimeters (cm) of copper is required to make the ornament if the height of the triangle is 6 cm and the base is 4 cm?

Solution   See Figure 15. The amount of copper required equals the shaded area. This area is the sum of the area of the triangle and the semicircle. The triangle has height $h = 6$ and base $b = 4$. The semicircle has diameter $d = 4$, so its radius is $r = 2$.

Figure 15

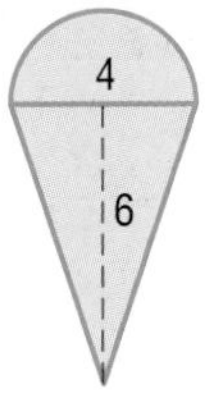

$$\text{Area} = \text{Area of triangle} + \text{Area of semicircle}$$
$$= \tfrac{1}{2}bh + \tfrac{1}{2}\pi r^2 = \tfrac{1}{2}(4)(6) + \tfrac{1}{2}\pi 2^2 \qquad b = 4;\ h = 6;\ r = 2.$$
$$= 12 + 2\pi \approx 18.28 \text{ cm}^2$$

About 18.28 cm² of copper is required.   ■

NOW WORK PROBLEM **33.**

Proof of the Pythagorean Theorem   We begin with a square, each side of length $a + b$. In this square, we can form four right triangles, each having legs equal in length to a and b. See Figure 16. All of these triangles are **congruent** (two sides and their included angle are equal). As a result, the hypotenuse of each is the same, say c, and the color shading in Figure 16 indicates a square with an area equal to c^2. The area of the original square with side

Figure 16

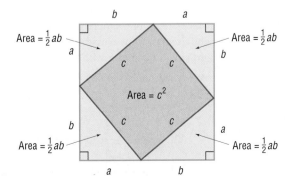

$a + b$ equals the sum of the areas of the four triangles (each of area $\frac{1}{2}ab$) plus the area of the square with side c. Thus,

$$(a + b)^2 = \tfrac{1}{2}ab + \tfrac{1}{2}ab + \tfrac{1}{2}ab + \tfrac{1}{2}ab + c^2$$
$$a^2 + 2ab + b^2 = 2ab + c^2$$
$$a^2 + b^2 = c^2$$

Figure 17

The proof is complete.   ■

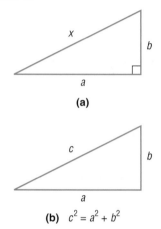

(a)

(b) $c^2 = a^2 + b^2$

Proof of the Converse of the Pythagorean Theorem   We begin with two triangles: one a right triangle with legs a and b and the other a triangle with sides a, b, and c for which $c^2 = a^2 + b^2$ (see Figure 17). By the Pythagorean Theorem, the length x of the third side of the first triangle is

$$x^2 = a^2 + b^2$$

But $c^2 = a^2 + b^2$. Hence,

$$x^2 = c^2$$
$$x = c$$

The two triangles have the same sides and are therefore congruent; hence, corresponding angles are equal. Thus, the angle opposite side c of the second triangle equals 90°.
The proof is complete.   ■

3 EXERCISES

In Problems 1–6, the lengths of the legs of a right triangle are given. Find the hypotenuse.

1. $a = 5$, $b = 12$ **2.** $a = 6$, $b = 8$ **3.** $a = 10$, $b = 24$
4. $a = 4$, $b = 3$ **5.** $a = 7$, $b = 24$ **6.** $a = 14$, $b = 48$

In Problems 7–14, the lengths of the sides of a triangle are given. Determine which are right triangles. For those that are, identify the hypotenuse.

7. $3, 4, 5$ **8.** $6, 8, 10$ **9.** $4, 5, 6$ **10.** $2, 2, 3$
11. $7, 24, 25$ **12.** $10, 24, 26$ **13.** $6, 4, 3$ **14.** $5, 4, 7$

15. Find the area A of a rectangle with length 4 inches and width 2 inches.
16. Find the area A of a rectangle with length 9 centimeters and width 4 centimeters.
17. Find the area A of a triangle with height 4 inches and base 2 inches.
18. Find the area A of a triangle with height 9 centimeters and base 4 centimeters.
19. Find the area A and circumference C of a circle of radius 5 meters.
20. Find the area A and circumference C of a circle of radius 2 feet.
21. Find the volume V of a rectangular box with length 8 feet, width 4 feet, and height 7 feet.
22. Find the volume V of a rectangular box with length 9 inches, width 4 inches, and height 8 inches.
23. Find the volume V and surface area S of a sphere of radius 4 centimeters.
24. Find the volume V and surface area S of a sphere of radius 3 feet.
25. Find the volume V of a right circular cylinder with radius 9 inches and height 8 inches.
26. Find the volume V of a right circular cylinder with radius 8 inches and height 9 inches.

In Problems 27–30, find the area of the shaded region.

27.

28.

29.

30.

31. How many feet does a wheel with a diameter of 16 inches travel after four revolutions?

32. How many revolutions will a circular disk with a diameter of 4 feet have completed after it has rolled 20 feet?

33. In the figure shown, *ABCD* is a square, with each side of length 6 feet. The width of the border (shaded portion) between the outer square *EFGH* and *ABCD* is 2 feet. Find the area of the border.

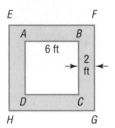

34. Refer to the figure. Square *ABCD* has an area of 100 square feet; square *BEFG* has an area of 16 square feet. What is the area of the triangle *CGF*?

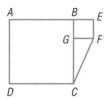

35. Architecture   A Norman window consists of a rectangle surmounted by a semicircle. Find the area of the Norman window shown in the illustration. How much wood frame is needed to enclose the window?

36. Construction   A circular swimming pool, 20 feet in diameter, is enclosed by a wooden deck that is 3 feet wide. What is the area of the deck? How much fence is required to enclose the deck?

In Problems 37–39, use the facts that the radius of Earth is 3960 miles and 1 mile = 5280 feet.

37. How Far Can You See?   The conning tower of the U.S.S. *Silversides*, a World War II submarine now permanently stationed in Muskegon, Michigan, is approximately 20 feet above sea level. How far can you see from the conning tower?

38. How Far Can You See?   A person who is 6 feet tall is standing on the beach in Fort Lauderdale, Florida, and looks out onto the Atlantic Ocean. Suddenly, a ship appears on the horizon. How far is the ship from shore?

39. How Far Can You See?   The deck of a destroyer is 100 feet above sea level. How far can a person see from the deck? How far can a person see from the bridge, which is 150 feet above sea level?

40. Suppose that *m* and *n* are positive integers with $m > n$. If $a = m^2 - n^2$, $b = 2mn$, and $c = m^2 + n^2$, show that a, b, and c are the lengths of the sides of a right triangle. (This formula can be used to find the sides of a right triangle that are integers, such as 3, 4, 5; 5, 12, 13; and so on. Such triplets of integers are called **Pythagorean triples.**)

41. You have 1000 feet of flexible pool siding and wish to construct a swimming pool. Experiment with rectangular-shaped pools with perimeters of 1000 feet. How do their areas vary? What is the shape of the rectangle with the largest area? Now compute the area enclosed by a circular pool with a perimeter (circumference) of 1000 feet. What would be your choice of shape for the pool? If rectangular, what is your preference for dimensions? Justify your choice. If your only consideration is to have a pool that encloses the most area, what shape should you use?

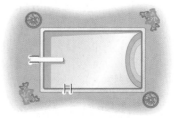

42. The Gibb's Hill Lighthouse, Southampton, Bermuda, in operation since 1846, stands 117 feet high on a hill 245 feet high, so its beam of light is 362 feet above sea level. A brochure states that the light itself can be seen on the horizon about 26 miles distant. Verify the correctness of this information. The brochure further states that ships 40 miles away can see the light and planes flying at 10,000 feet can see it 120 miles away. Verify the accuracy of these statements. What assumption did the brochure make about the height of the ship?

---

4 | INTEGER EXPONENTS

OBJECTIVES
1. Evaluate Expressions Containing Exponents
2. Work with the Laws of Exponents
3. Use a Calculator to Evaluate Exponents
4. Use Scientific Notation

Integer exponents provide a shorthand device for representing repeated multiplications of a real number. For example,

$$3^4 = 3 \cdot 3 \cdot 3 \cdot 3 = 81$$

Additionally, many formulas have exponents. For example,

- The formula for the horsepower rating H of an engine is

$$H = \frac{D^2 N}{2.5}$$

where D is the diameter of a cylinder and N is the number of cylinders.

- A formula for the resistance R of blood flowing in a blood vessel is

$$R = C \frac{L}{r^4}$$

where L is the length of the blood vessel, r is the radius, and C is a positive constant.

If a is a real number and n is a positive integer, then the symbol a^n represents the product of n factors of a. That is,

$$a^n = \underbrace{a \cdot a \cdot \ldots \cdot a}_{n \text{ factors}} \qquad (1)$$

Here it is understood that $a^1 = a$.

In particular, we have

$$a^1 = a$$

$$a^2 = a \cdot a$$

$$a^3 = a \cdot a \cdot a$$

and so on.

①   In the expression a^n, a is called the **base** and n is called the **exponent,** or **power.** We read a^n as "a raised to the power n" or as "a to the nth power." We usually read a^2 as "a squared" and a^3 as "a cubed."

| EXAMPLE 1 | **Evaluating Expressions Containing Exponents** |

(a) $2^3 = 2 \cdot 2 \cdot 2 = 8$ (b) $5^2 = 5 \cdot 5 = 25$

(c) $10^1 = 10$ (d) $(-2)^4 = (-2)(-2)(-2)(-2) = 16$

(e) $-2^4 = -(2 \cdot 2 \cdot 2 \cdot 2) = -16$ ∎

Notice the difference between Examples 1(d) and 1(e): The exponent applies only to the symbol or parenthetical expression immediately preceding it.

We define a raised to a negative power as follows:

> If n is a positive integer and if a is a nonzero real number, then we define
>
> $$a^{-n} = \frac{1}{a^n} \qquad \text{if } a \neq 0 \qquad\qquad \textbf{(2)}$$

Whenever you encounter a negative exponent, think "reciprocal."

| EXAMPLE 2 | **Evaluating Expressions Containing Negative Exponents** |

(a) $2^{-3} = \dfrac{1}{2^3} = \dfrac{1}{8}$ (b) $x^{-4} = \dfrac{1}{x^4}$

(c) $\left(\dfrac{1}{5}\right)^{-2} = \dfrac{1}{\left(\dfrac{1}{5}\right)^2} = \dfrac{1}{\dfrac{1}{25}} = 25$ ∎

✐   **NOW WORK PROBLEM 3.**

> If a is a nonzero number, we define
>
> $$a^0 = 1 \qquad \text{if } a \neq 0 \qquad\qquad \textbf{(3)}$$

Notice that we do not allow the base a to be 0 in a^{-n} or in a^0.

LAWS OF EXPONENTS

2   Several general rules can be used when dealing with exponents. The first rule is used when multiplying two expressions that have the same base.

$$\underset{\substack{\uparrow \quad \uparrow \\ \text{Same base} \quad 2 \text{ factors} \quad 4 \text{ factors}}}{\overset{\text{Multiplying}}{3^2 \cdot 3^4}} = \underbrace{(3 \cdot 3)}_{} \underbrace{(3 \cdot 3 \cdot 3 \cdot 3)}_{} = \underbrace{3 \cdot 3 \cdot 3 \cdot 3 \cdot 3 \cdot 3}_{6 \text{ factors}} = \overset{\text{Sum of powers}}{\underset{\text{Same base}}{3^6}}$$

To multiply two expressions having the same base, retain the base and add the exponents.

If a is a real number and m, n are integers, then

$$a^m a^n = a^{m+n} \qquad \text{(4)}$$

Equation (4) is true whether the integers m and n are positive, negative, or 0. If m, n, or $m + n$ is 0 or negative, then a cannot be 0, as stated earlier.

EXAMPLE 3

Using Equation (4)

(a) $2 \cdot 2^4 = 2^1 \cdot 2^4 = 2^{1+4} = 2^5 = 32$

(b) $(-3)^2(-3)^{-3} = (-3)^{2+(-3)} = (-3)^{-1} = \dfrac{1}{-3} = -\dfrac{1}{3}$

(c) $x^{-3} \cdot x^5 = x^{-3+5} = x^2$

(d) $(-y)^{-1}(-y)^{-3} = (-y)^{-1+(-3)} = (-y)^{-4} = \dfrac{1}{(-y)^4} = \dfrac{1}{y^4}$

∎

NOW WORK PROBLEM 11.

Another law of exponents applies when an expression containing a power is itself raised to a power.

$$(2^3)^4 = \underbrace{2^3 \cdot 2^3 \cdot 2^3 \cdot 2^3}_{4 \text{ factors}} = \underbrace{(2 \cdot 2 \cdot 2)}_{3 \text{ factors}}\underbrace{(2 \cdot 2 \cdot 2)}_{3 \text{ factors}}\underbrace{(2 \cdot 2 \cdot 2)}_{3 \text{ factors}}\underbrace{(2 \cdot 2 \cdot 2)}_{3 \text{ factors}} = 2^{12}$$
$$3 \cdot 4 = 12 \text{ factors}$$

If an expression containing a power is raised to a power, retain the base and multiply the powers.

If a is a real number and m, n are integers, then

$$\left(a^m\right)^n = a^{mn} \qquad \text{(5)}$$

Equation (5) is true whether the integers m and n are positive, negative, or 0. Again, if m or n is 0 or negative, then a must not be 0.

| EXAMPLE 4 | **Using Equation (5)** |

(a) $\left[(-2)^3\right]^2 = (-2)^{3\cdot2} = (-2)^6 = 64$ (b) $\left(3^{-4}\right)^0 = 3^{(-4)(0)} = 3^0 = 1$

(c) $\left(x^{-3}\right)^2 = x^{-3\cdot2} = x^{-6} = \dfrac{1}{x^6}$ (d) $\left(y^{-1}\right)^{-2} = y^{(-1)(-2)} = y^2$ ■

The next law of exponents involves raising a product to a power.
$$(2\cdot5)^3 = (2\cdot5)(2\cdot5)(2\cdot5) = (2\cdot2\cdot2)(5\cdot5\cdot5) = 2^3\cdot5^3$$
If a product is raised to a power, the result equals the product of each factor raised to that power.

If a, b are real numbers and n is an integer, then
$$(a\cdot b)^n = a^n\cdot b^n \qquad\qquad \textbf{(6)}$$

Equation (6) is true whether the integer n is positive, negative, or 0. If n is 0 or negative, neither a nor b can be 0.

| EXAMPLE 5 | **Using Equation (6)** |

(a) $(2x)^3 = 2^3\cdot x^3 = 8x^3$

(b) $(-2x)^0 = (-2)^0\cdot x^0 = 1\cdot1 = 1$ $x \ne 0$

(c) $(ax)^{-2} = a^{-2}x^{-2} = \dfrac{1}{a^2}\cdot\dfrac{1}{x^2} = \dfrac{1}{a^2x^2}$ $x \ne 0, a \ne 0$

(d) $(ax)^{-2} = \dfrac{1}{(ax)^2} = \dfrac{1}{a^2x^2}$ $x \ne 0, a \ne 0$ ■

✎ NOW WORK PROBLEM **29**.

Equations (4) and (6) both involve products. Two similar laws involve quotients.

If a and b are real numbers and if m and n are integers, then
$$\frac{a^m}{a^n} = a^{m-n} = \frac{1}{a^{n-m}} \qquad \text{if } a \ne 0 \qquad \textbf{(7a)}$$
$$\left(\frac{a}{b}\right)^n = \frac{a^n}{b^n} \qquad\qquad \text{if } b \ne 0 \qquad \textbf{(7b)}$$

Equation (7a) is true whether the integers m and n are positive, negative, or 0. Equation (7b) is true whether the integer n is positive, negative, or zero. In (7b), if n is negative or 0, then a cannot be zero.

| EXAMPLE 6 | **Using Equations (7a) and (7b)** |

(a) $\dfrac{2^6}{2^4} = 2^{6-4} = 2^2 = 4$ (b) $\left(\dfrac{2}{3}\right)^4 = \dfrac{2^4}{3^4} = \dfrac{16}{81}$

(c) $\dfrac{x^{-2}}{x^{-5}} = x^{-2-(-5)} = x^3$ (d) $\left(\dfrac{5}{2}\right)^{-2} = \dfrac{1}{\left(\dfrac{5}{2}\right)^2} = \dfrac{1}{\dfrac{5^2}{2^2}} = \dfrac{1}{\dfrac{25}{4}} = \dfrac{4}{25}$ ■

You may have observed the following shortcut for problems like Example 6(d):

$$\left(\frac{a}{b}\right)^{-n} = \left(\frac{b}{a}\right)^{n} \qquad \text{if } a \neq 0, b \neq 0 \qquad \textbf{(8)}$$

EXAMPLE 7 — Using Equation (8)

(a) $\left(\dfrac{2}{3}\right)^{-3} = \left(\dfrac{3}{2}\right)^{3} = \dfrac{27}{8}$

(b) $\left(\dfrac{3}{4}\right)^{-2} = \left(\dfrac{4}{3}\right)^{2} = \dfrac{16}{9}$ ∎

NOW WORK PROBLEM **19.**

EXAMPLE 8 — Using the Laws of Exponents

Simplify each expression. Express the answer so that all exponents are positive.

(a) $\dfrac{x^5 y^{-2}}{(x^3 y)^2}, \qquad x \neq 0, y \neq 0$

(b) $\left(\dfrac{x^{-3}}{3y^{-1}}\right)^{-2}, \qquad x \neq 0, y \neq 0$

Solution   (a) $\dfrac{x^5 y^{-2}}{(x^3 y)^2} = \dfrac{x^5 y^{-2}}{(x^3)^2 y^2} = \dfrac{x^5 y^{-2}}{x^6 y^2} = \dfrac{x^5}{x^6} \cdot \dfrac{y^{-2}}{y^2}$

$$= x^{5-6} y^{-2-2} = x^{-1} y^{-4} = \dfrac{1}{x} \cdot \dfrac{1}{y^4} = \dfrac{1}{xy^4}$$

(b) $\left(\dfrac{x^{-3}}{3y^{-1}}\right)^{-2} = \dfrac{(x^{-3})^{-2}}{(3y^{-1})^{-2}} = \dfrac{x^6}{3^{-2}(y^{-1})^{-2}} = \dfrac{x^6}{\dfrac{1}{9} y^2} = \dfrac{9x^6}{y^2}$ ∎

NOW WORK PROBLEM **47.**

CALCULATOR USE

③ Your calculator should have the key $\boxed{x^y}$ or $\boxed{y^x}$ or $\boxed{\wedge}$, which is used for computations involving exponents. The next example shows how this key is used.

EXAMPLE 9 — Exponents on a Calculator

Evaluate: $(2.3)^5$

Solution   Keystrokes: $\boxed{2.3}$ $\boxed{x^y}$ $\boxed{5}$ $\boxed{=}$

Display: $\boxed{2.3}$ $\boxed{5}$ $\boxed{64.36343}$ ∎

NOW WORK PROBLEM **63.**

SCIENTIFIC NOTATION

4   Measurements of physical quantities can range from very small to very large. For example, the mass of a proton is approximately 0.0000000000000000000000000167 kilogram and the mass of Earth is about 5,980,000,000,000,000,000,000,000 kilograms. These numbers obviously are tedious to write down and difficult to read, so we use exponents to rewrite each.

> When a number has been written as the product of a number x, where $1 \leq x < 10$, times a power of 10, it is said to be written in **scientific notation.**

In scientific notation,

$$\text{Mass of a proton} = 1.67 \times 10^{-27} \text{ kilogram}$$

$$\text{Mass of Earth} = 5.98 \times 10^{24} \text{ kilograms}$$

> ### CONVERTING A DECIMAL TO SCIENTIFIC NOTATION
>
> To change a positive number into scientific notation:
>
> 1. Count the number N of places that the decimal point must be moved in order to arrive at a number x, where $1 \leq x < 10$.
>
> 2. If the original number is greater than or equal to 1, the scientific notation is $x \times 10^{N}$. If the original number is between 0 and 1, the scientific notation is $x \times 10^{-N}$.

| EXAMPLE 10 | Using Scientific Notation |

Write each number in scientific notation.

(a) 9582   (b) 1.245   (c) 0.285   (d) 0.000561

Solution   (a) The decimal point in 9582 follows the 2. Thus, we count

$$9\ \ 5\ \ 8\ \ 2\ \ .$$
$$\ \uparrow\ \ \uparrow\ \ \uparrow\ \ \ \ \ \ \leftarrow$$
$$3\ \ 2\ \ 1$$

stopping after three moves, because 9.582 is a number between 1 and 10. Since 9582 is greater than 1, we write

$$9582 = 9.582 \times 10^3$$

(b) The decimal point in 1.245 is between the 1 and 2. Since the number is already between 1 and 10, the scientific notation for it is $1.245 \times 10^0 = 1.245$.

(c) The decimal point in 0.285 is between the 0 and the 2. Thus, we count

$$0\ \ .\ \ 2\ \ 8\ \ 5$$
$$\llcorner\!\!\rightarrow\ \uparrow$$
$$1$$

stopping after one move, because 2.85 is a number between 1 and 10. Since 0.285 is between 0 and 1, we write

$$0.285 = 2.85 \times 10^{-1}$$

(d) The decimal point in 0.000561 is moved as follows:

$$0 \ . \ 0 \ 0 \ 0 \ 5 \ 6 \ 1$$
$$\quad\quad 1 \ \ 2 \ \ 3 \ \ 4$$

Thus,

$$0.000561 = 5.61 \times 10^{-4}$$

➤ NOW WORK PROBLEM **69**.

| EXAMPLE 11 | **Changing from Scientific Notation to Decimals** |

Write each number as a decimal.

(a) 2.1×10^4 (b) 3.26×10^{-5} (c) 1×10^{-2}

Solution

(a) $2.1 \times 10^4 = 2 \ . \ 1 \ 0 \ 0 \ 0 \ \times 10^4 = 21,000$
$$\quad\quad\quad\quad\quad I \ \ 2 \ \ 3 \ \ 4$$

(b) $3.26 \times 10^{-5} = 0 \ \ 0 \ 0 \ 0 \ 0 \ 3 \ . \ 2 \ \ 6 \times 10^{-5} = 0.0000326$
$$\quad\quad\quad\quad\quad\quad\quad 5 \ \ 4 \ \ 3 \ \ 2 \ \ I$$

(c) $1 \times 10^{-2} = 0 \ \ 0 \ 1 \ . \ \times 10^{-2} = 0.01$
$$\quad\quad\quad\quad\quad 2 \ \ I$$

On a calculator, a number such as 3.615×10^{12} is usually displayed as
3.615E12.

➤ NOW WORK PROBLEM **77**.

| EXAMPLE 12 | **Using Scientific Notation** |

(a) The diameter of the smallest living cell is only about 0.00001 centimeter (cm).* Express this number in scientific notation.

(b) The surface area of Earth is about 1.97×10^8 square miles.[†] Express the surface area as a whole number.

Solution

(a) 0.00001 cm $= 1 \times 10^{-5}$ cm because the decimal point is moved five places and the number is less than 1.

(b) 1.97×10^8 square miles $= 197,000,000$ square miles.

➤ NOW WORK PROBLEM **85**.

---

* *Powers of Ten*, Philip and Phylis Morrison.
[†] *1998 Information Please Almanac.*

SUMMARY

We close this section by summarizing the Laws of Exponents. In the list that follows, a and b are real numbers and m and n are integers. Also, we assume that no denominator is 0 and that all expressions are defined.

Laws of Exponents

$$a^{-n} = \frac{1}{a^n} \qquad a^0 = 1, \qquad a \neq 0$$

$$a^m a^n = a^{m+n} \qquad \left(a^m\right)^n = a^{mn} \qquad (a \cdot b)^n = a^n \cdot b^n$$

$$\frac{a^m}{a^n} = a^{m-n} = \frac{1}{a^{n-m}}, \qquad a \neq 0 \qquad \left(\frac{a}{b}\right)^n = \frac{a^n}{b^n}, \qquad b \neq 0$$

HISTORICAL FEATURE

René Descartes
1596–1650

Our method of writing exponents originated with René Descartes (1596–1650), although the concept goes back in various forms to the ancient Babylonians. Even after its introduction by Descartes in 1637, the method took a remarkable amount of time to become completely standardized, and expressions like $aaaaa + 3aaaa + 2aaa - 4aa + 2a + 1$ remained common until 1750. The concept of a rational exponent (see Section 9) was known by 1400, although inconvenient notation prevented the development of any extensive theory. John Wallis (1616–1703), in 1655, was the first to give a fairly complete explanation of negative and rational exponents, and Sir Isaac Newton's (1642–1727) use of them made exponents standard in their current form.

4 EXERCISES

In Problems 1–24, simplify each expression.

1. 4^2

2. -4^2

3. 4^{-2}

4. $(-4)^2$

5. -4^{-2}

6. $(-4)^{-2}$

7. $4^0 \cdot 2^{-3}$

8. $(-2)^{-3} \cdot 3^0$

9. $2^{-3} + \left(\frac{1}{2}\right)^3$

10. $3^{-2} + \left(\frac{1}{3}\right)^2$

11. $3^{-6} \cdot 3^4$

12. $4^{-2} \cdot 4^3$

13. $\frac{\left(3^2\right)^2}{\left(2^3\right)^2}$

14. $\frac{\left(2^3\right)^3}{\left(2^2\right)^3}$

15. $\left(\frac{2}{3}\right)^{-3}$

16. $\left(\frac{3}{2}\right)^{-2}$

17. $\frac{2^3 \cdot 3^2}{2^4 \cdot 3^{-2}}$

18. $\frac{3^{-2} \cdot 5^3}{3^2 \cdot 5}$

19. $\left(\frac{9}{2}\right)^{-2}$

20. $\left(\frac{6}{5}\right)^{-3}$

21. $\frac{2^{-2}}{3}$

22. $\frac{3^{-2}}{2}$

23. $\frac{-3^{-1}}{2^{-1}}$

24. $\frac{-2^{-3}}{-1}$

In Problems 25–56, simplify each expression. Express the answer so that all exponents are positive. Whenever an exponent is negative or 0, we assume that the base does not equal 0.

25. $x^0 y^2$

26. $x^{-1} y$

27. xy^{-2}

28. $x^0 y^4$

29. $\left(8x^3\right)^{-2}$

30. $\left(-8x^3\right)^{-2}$

31. $-4x^{-1}$

32. $(-4x)^{-1}$

33. $3x^0$

34. $(3x)^0$

35. $\dfrac{x^{-2}y^3}{xy^4}$

36. $\dfrac{x^{-2}y}{xy^2}$

37. $x^{-1}y^{-1}$

38. $\dfrac{x^{-2}y^{-3}}{x}$

39. $\dfrac{x^{-1}}{y^{-1}}$

40. $\left(\dfrac{2x}{3}\right)^{-1}$

41. $\left(\dfrac{4y}{5x}\right)^{-2}$

42. $(x^2y)^{-2}$

43. $x^{-2}y^{-2}$

44. $x^{-1}y^{-1}$

45. $\dfrac{x^{-1}y^{-2}z^3}{x^2yz^3}$

46. $\dfrac{3x^{-2}yz^2}{x^4y^{-3}z^2}$

47. $\dfrac{(-2)^3x^4(yz)^2}{3^2xy^3z^4}$

48. $\dfrac{4x^{-2}(yz)^{-1}}{(-5)^2x^4y^2z^{-2}}$

49. $\dfrac{\left(\dfrac{x}{y}\right)^{-2}\cdot\left(\dfrac{y}{x}\right)^4}{x^2y^3}$

50. $\dfrac{\left(\dfrac{y}{x}\right)^2}{x^{-2}y}$

51. $\left(\dfrac{3x^{-1}}{4y^{-1}}\right)^{-2}$

52. $\left(\dfrac{5x^{-2}}{6y^{-2}}\right)^{-3}$

53. $\dfrac{(xy^{-1})^{-2}}{xy^3}$

54. $\dfrac{(3xy^{-1})^2}{(2x^{-1}y)^3}$

55. $\left(\dfrac{x}{y^2}\right)^{-2}\cdot(y^2)^{-1}$

56. $\dfrac{(x^2)^{-3}y^3}{(x^3y)^{-2}}$

57. Find the value of the expression $2x^3 - 3x^2 + 5x - 4$ if $x = 2$. If $x = 1$.

58. Find the value of the expression $4x^3 + 3x^2 - x + 2$ if $x = 1$. If $x = 2$.

59. What is the value of $\dfrac{(666)^4}{(222)^4}$?

60. What is the value of $(0.1)^3(20)^3$?

In Problems 61–68, use a calculator to evaluate each expression. Round your answers to three decimal places.

61. $(8.2)^6$

62. $(3.7)^5$

63. $(6.1)^{-3}$

64. $(2.2)^{-5}$

65. $(-2.8)^6$

66. $-(2.8)^6$

67. $(-8.11)^{-4}$

68. $-(8.11)^{-4}$

In Problems 69–76, write each number in scientific notation.

69. 454.2

70. 32.14

71. 0.013

72. 0.00421

73. 32,155

74. 21,210

75. 0.000423

76. 0.0514

In Problems 77–84, write each number as a decimal.

77. 6.15×10^4

78. 9.7×10^3

79. 1.214×10^{-3}

80. 9.88×10^{-4}

81. 1.1×10^8

82. 4.112×10^2

83. 8.1×10^{-2}

84. 6.453×10^{-1}

---

85. Distance from Earth to Its Moon The distance from Earth to the Moon is about 4×10^8 meters.* Express this distance as a whole number.

86. Height of Mt. Everest The height of Mt. Everest is 8872 meters.* Express this height in scientific notation.

87. Wavelength of Visible Light The wavelength of visible light is about 5×10^{-7} meter.* Express this wavelength as a decimal.

88. Diameter of an Atom The diameter of an atom is about 1×10^{-10} meter.* Express this diameter as a decimal.

89. Diameter of Copper Wire The smallest commercial copper wire is about 0.0005 inch in diameter.† Express this diameter using scientific notation.

90. Smallest Motor The smallest motor ever made is less than 0.05 centimeter wide.† Express this width using scientific notation.

91. World Oil Production In 1996, world oil production averaged 64,000,000 barrels per day.† How many barrels on average were produced in the month of April 1996? Express your answer in scientific notation.

92. U.S. Consumption of Oil In 1995, the United States consumed 17,640,000 barrels of petroleum products per day.† If there are 42 gallons in one barrel, how many gallons did the United States consume per day? Express your answer in scientific notation.

93. Astronomy One light-year is defined by astronomers to be the distance that a beam of light will travel in 1 year (365 days). If the speed of light is 186,000 miles per second, how many miles are in a light-year? Express your answer in scientific notation.

94. Astronomy How long does it take a beam of light to reach Earth from the Sun, when the Sun is 93,000,000 miles from Earth? Express your answer in seconds, using scientific notation.

95. Look at the summary box where the Laws of Exponents are given. List them in the order of most importance to you. Write a brief position paper defending your ordering.

96. Write a paragraph to justify the definition given in the text that $a^0 = 1$, $a \neq 0$.

* *Powers of Ten*, Philip and Phylis Morrison.
† *1998 Information Please Almanac.*

5 | POLYNOMIALS

OBJECTIVES
1. Recognize Monomials
2. Recognize Polynomials
3. Add and Subtract Polynomials
4. Multiply Polynomials
5. Know Formulas for Special Products
6. Divide Polynomials

We have described algebra as a generalization of arithmetic in which letters are used to represent real numbers. From now on, we shall use the letters at the end of the alphabet, such as x, y, and z, to represent variables and the letters at the beginning of the alphabet, such as a, b, and c, to represent constants. In the expressions $3x + 5$ and $ax + b$, it is understood that x is a variable and that a and b are constants, even though the constants a and b are unspecified. As you will find out, the context usually makes the intended meaning clear.

Now we introduce some basic vocabulary.

A **monomial** in one variable is the product of a constant times a variable raised to a nonnegative integer power. A monomial is of the form

$$ax^k$$

where a is a constant, x is a variable, and $k \geq 0$ is an integer. The constant a is called the **coefficient** of the monomial. If $a \neq 0$, then k is called the **degree** of the monomial.

EXAMPLE 1

Examples of Monomials

1.

Monomial	Coefficient	Degree	
(a) $6x^2$	6	2	
(b) $-\sqrt{2}x^3$	$-\sqrt{2}$	3	
(c) 3	3	0	Since $3 = 3 \cdot 1 = 3x^0$, $x \neq 0$
(d) $-5x$	-5	1	Since $-5x = -5x^1$
(e) x^4	1	4	Since $x^4 = 1 \cdot x^4$

Now let's look at some expressions that are not monomials.

EXAMPLE 2

Examples of Nonmonomial Expressions

(a) $3x^{1/2}$ is not a monomial, since the exponent of the variable x is $\frac{1}{2}$ and $\frac{1}{2}$ is not a nonnegative integer.

(b) $4x^{-3}$ is not a monomial, since the exponent of the variable x is -3 and -3 is not a nonnegative integer.

NOW WORK PROBLEM 1.

Two monomials with the same variable raised to the same power are called **like terms.** For example, $2x^4$ and $-5x^4$ are like terms. In contrast, the monomials $2x^3$ and $2x^5$ are not like terms.

We can add or subtract like terms using the distributive property. For example,

$$2x^2 + 5x^2 = (2 + 5)x^2 = 7x^2 \quad \text{and} \quad 8x^3 - 5x^3 = (8 - 5)x^3 = 3x^3$$

The sum or difference of two monomials having different degrees is called a **binomial.** The sum or difference of three monomials with three different degrees is called a **trinomial.** For example,

$x^2 - 2$ is a binomial.

$x^3 - 3x + 5$ is a trinomial.

$2x^2 + 5x^2 + 2 = 7x^2 + 2$ is a binomial.

A **polynomial** in one variable is an algebraic expression of the form

$$a_n x^n + a_{n-1} x^{n-1} + \cdots + a_1 x + a_0 \tag{1}$$

where $a_n, a_{n-1}, \ldots, a_1, a_0$ are constants* called the **coefficients** of the polynomial, $n \geq 0$ is an integer, and x is a variable. If $a_n \neq 0$, it is called the **leading coefficient,** and n is called the **degree** of the polynomial.

The monomials that make up a polynomial are called its **terms.** If all the coefficients are 0, the polynomial is called the **zero polynomial,** which has no degree.

Polynomials are usually written in **standard form,** beginning with the nonzero term of highest degree and continuing with terms in descending order according to degree.

| EXAMPLE 3 | **Examples of Polynomials** |

Polynomial	**Coefficients**	**Degree**
$3x^2 - 5 = 3x^2 + 0 \cdot x + (-5)$	$3, 0, -5$	2
$8 - 2x + x^2 = 1 \cdot x^2 + (-2)x + 8$	$1, -2, 8$	2
$5x + \sqrt{2} = 5x^1 + \sqrt{2}$	$5, \sqrt{2}$	1
$3 = 3 \cdot 1 = 3 \cdot x^0$	3	0
0	0	No degree ∎

Although we have been using x to represent the variable, letters such as y or z are also commonly used.

$3x^4 - x^2 + 2$ is a polynomial (in x) of degree 4.

$9y^3 - 2y^2 + y - 3$ is a polynomial (in y) of degree 3.

$z^5 + \pi$ is a polynomial (in z) of degree 5.

---

* The notation a_n is read as "a sub n." The number n is called a **subscript** and should not be confused with an exponent. We use subscripts in order to distinguish one constant from another when a large or undetermined number of constants is required.

Algebraic expressions such as

$$\frac{1}{x} \quad \text{and} \quad \frac{x^2 + 1}{x + 5}$$

are not polynomials. The first is not a polynomial because $1/x = x^{-1}$ has an exponent that is not a nonnegative integer. Although the second expression is the quotient of two polynomials, the polynomial in the denominator has degree greater than 0, so the expression cannot be a polynomial.

 N O W   W O R K   P R O B L E M   **11**.

ADDING AND SUBTRACTING POLYNOMIALS

③ Polynomials are added and subtracted by combining like terms.

EXAMPLE 4   ### Adding Polynomials

Find the sum of the polynomials:

$$8x^3 - 2x^2 + 6x - 2 \quad \text{and} \quad 3x^4 - 2x^3 + x^2 + x$$

Solution   We shall find the sum in two ways.

Horizontal Addition: The idea here is to group the like terms and then combine them.

$$\left(8x^3 - 2x^2 + 6x - 2\right) + \left(3x^4 - 2x^3 + x^2 + x\right)$$
$$= 3x^4 + \left(8x^3 - 2x^3\right) + \left(-2x^2 + x^2\right) + \left(6x + x\right) - 2$$
$$= 3x^4 + 6x^3 - x^2 + 7x - 2$$

Vertical Addition: The idea here is to vertically line up the like terms in each polynomial and then add the coefficients.

$$
\begin{array}{cccccc}
 & x^4 & x^3 & x^2 & x^1 & x^0 \\
 & & 8x^3 & - 2x^2 & + 6x & - 2 \\
(+) & 3x^4 & - 2x^3 & + x^2 & + x & \\
\hline
 & 3x^4 & + 6x^3 & - x^2 & + 7x & - 2
\end{array}
$$

We can subtract two polynomials in either of the previous ways.

EXAMPLE 5   ### Subtracting Polynomials

Find the difference: $\left(3x^4 - 4x^3 + 6x^2 - 1\right) - \left(2x^4 + 8x^2 + 6x + 5\right)$

Solution   *Horizontal Subtraction*

$$\left(3x^4 - 4x^3 + 6x^2 - 1\right) - \left(2x^4 - 8x^2 - 6x + 5\right)$$
$$= 3x^4 - 4x^3 + 6x^2 - 1 + \underbrace{\left(-2x^4 + 8x^2 + 6x - 5\right)}$$

Be sure to change the sign of each
term in the second polynomial.

$$= \underset{\uparrow}{\left(3x^4 - 2x^4\right)} + \left(-4x^3\right) + \left(6x^2 + 8x^2\right) + 6x + \left(-1 - 5\right)$$

Group like terms.

$$= x^4 - 4x^3 + 14x^2 + 6x - 6$$

Vertical Subtraction: We line up like terms, change the sign of each coefficient of the second polynomial, and add.

$$
\begin{array}{r}
3x^4 - 4x^3 + 6x^2 - 1 = 3x^4 - 4x^3 + 6x^2 - 1 \\
(-)[2x^4 - 8x^2 - 6x + 5] = (+)-2x^4 + 8x^2 + 6x - 5 \\
\hline
x^4 - 4x^3 + 14x^2 + 6x - 6 \blacksquare
\end{array}
$$

with column labels $x^4 \quad x^3 \quad x^2 \quad x^1 \quad x^0$ above each polynomial.

The choice of which of these methods to use for adding and subtracting polynomials is left to you. To save space, we shall most often use the horizontal format.

➤ NOW WORK PROBLEM **23**.

MULTIPLYING POLYNOMIALS

④ Two monomials may be multiplied using the Laws of Exponents and the Commutative and Associative Properties. For example,

$$
\left(2x^3\right) \cdot \left(5x^4\right) = (2 \cdot 5) \cdot \left(x^3 \cdot x^4\right) = 10x^{3+4} = 10x^7
$$

Products of polynomials are found by repeated use of the Distributive Property and the Laws of Exponents. Again, you have a choice of horizontal or vertical format.

EXAMPLE 6   Multiplying Polynomials

Find the product: $(2x + 5)(x^2 - x + 2)$

Solution   *Horizontal Multiplication*

$$(2x + 5)(x^2 - x + 2) = 2x(x^2 - x + 2) + 5(x^2 - x + 2)$$
↑
Distributive property

$$= \left(2x \cdot x^2 - 2x \cdot x + 2x \cdot 2\right) + \left(5 \cdot x^2 - 5 \cdot x + 5 \cdot 2\right)$$
↑
Distributive property

$$= \left(2x^3 - 2x^2 + 4x\right) + \left(5x^2 - 5x + 10\right)$$
↑
Law of exponents

$$= 2x^3 + 3x^2 - x + 10$$
↑
Combine like terms

Vertical Multiplication: The idea here is very much like multiplying a two-digit number by a three-digit number.

$$
\begin{array}{r}
x^2 - x + 2 \\
2x + 5 \\
\hline
2x^3 - 2x^2 + 4x \\
(+) 5x^2 - 5x + 10 \\
\hline
2x^3 + 3x^2 - x + 10
\end{array}
$$

This line is $2x(x^2 - x + 2)$.
This line is $5(x^2 - x + 2)$.
Sum of the above two lines.   ∎

➤ NOW WORK PROBLEM **35**.

SPECIAL PRODUCTS

⑤ Certain products, which we call **special products,** occur frequently in algebra. We can calculate them easily using the **FOIL** (**F**irst, **O**uter, **I**nner, **L**ast) method of multiplying two binomials.

$$(ax + b)(cx + d) = ax(cx + d) + b(cx + d)$$

$$= \overbrace{ax \cdot cx}^{\text{First}} + \overbrace{ax \cdot d}^{\text{Outer}} + \overbrace{b \cdot cx}^{\text{Inner}} + \overbrace{b \cdot d}^{\text{Last}}$$
$$= acx^2 + adx + bcx + bd$$
$$= acx^2 + (ad + bc)x + bd$$

EXAMPLE 7 **Using FOIL to Find Products of the Form** $(x - a)(x + a)$

$$(x - 3)(x + 3) = x^2 + 3x - 3x - 9 = x^2 - 9$$
$$\quad\quad\quad\quad\quad\quad\quad\text{F}\quad\text{O}\quad\text{I}\quad\text{L}$$ ∎

EXAMPLE 8 **Using FOIL to Find the Square of a Binomial**

(a) $(x + 2)^2 = (x + 2)(x + 2) = x^2 + 2x + 2x + 4 = x^2 + 4x + 4$

(b) $(x - 3)^2 = (x - 3)(x - 3) = x^2 - 3x - 3x + 9 = x^2 - 6x + 9$ ∎

EXAMPLE 9 **Using FOIL to Find the Product of Two Binomials**

(a) $(x + 3)(x + 1) = x^2 + x + 3x + 3 = x^2 + 4x + 3$

(b) $(2x + 1)(3x + 4) = 6x^2 + 8x + 3x + 4 = 6x^2 + 11x + 4$ ∎

✏ NOW WORK PROBLEMS **41** AND **49**.

Some products have been given special names because of their form. The following special products are based on Examples 7 and 8.

Difference of Two Squares

$$(x - a)(x + a) = x^2 - a^2 \tag{2}$$

Squares of Binomials, or Perfect Squares

$$(x + a)^2 = x^2 + 2ax + a^2 \tag{3a}$$

$$(x - a)^2 = x^2 - 2ax + a^2 \tag{3b}$$

EXAMPLE 10 **Using Special Product Formulas**

(a) $(x - 5)(x + 5) = x^2 - 5^2 = x^2 - 25$ Difference of two squares

(b) $(x + 7)^2 = x^2 + 2 \cdot 7 \cdot x + 7^2 = x^2 + 14x + 49$ Square of a binomial ∎

EXAMPLE 11

Using Special Product Formulas

(a) $(2x + 1)^2 = (2x)^2 + 2 \cdot 1 \cdot 2x + 1^2 = 4x^2 + 4x + 1$ Notice that we used $2x$ in place of x in formula (3a).

(b) $(3x - 4)^2 = (3x)^2 - 2 \cdot 4 \cdot 3x + 4^2 = 9x^2 - 24x + 16$ Replace x by $3x$ in formula (3b).

▬▬▬➤   **NOW WORK PROBLEM 61.**

Let's look at some more examples that lead to general formulas.

EXAMPLE 12

Cubing a Binomial

(a) $(x + 2)^3 = (x + 2)(x + 2)^2 = (x + 2)(x^2 + 4x + 4)$ Formula (3a)

$\qquad = (x^3 + 4x^2 + 4x) + (2x^2 + 8x + 8)$

$\qquad = x^3 + 6x^2 + 12x + 8$

(b) $(x - 1)^3 = (x - 1)(x - 1)^2 = (x - 1)(x^2 - 2x + 1)$ Formula (3b)

$\qquad = (x^3 - 2x^2 + x) - (x^2 - 2x + 1)$

$\qquad = x^3 - 3x^2 + 3x - 1$

Cubes of Binomials, or Perfect Cubes

$$(x + a)^3 = x^3 + 3ax^2 + 3a^2x + a^3 \qquad \textbf{(4a)}$$
$$(x - a)^3 = x^3 - 3ax^2 + 3a^2x - a^3 \qquad \textbf{(4b)}$$

▬▬▬➤   **NOW WORK PROBLEM 79.**

EXAMPLE 13

Forming the Difference of Two Cubes

$(x - 1)(x^2 + x + 1) = x(x^2 + x + 1) - 1(x^2 + x + 1)$

$\qquad = x^3 + x^2 + x - x^2 - x - 1$

$\qquad = x^3 - 1$

EXAMPLE 14

Forming the Sum of Two Cubes

$(x + 2)(x^2 - 2x + 4) = x(x^2 - 2x + 4) + 2(x^2 - 2x + 4)$

$\qquad = x^3 - 2x^2 + 4x + 2x^2 - 4x + 8$

$\qquad = x^3 + 8$

Examples 13 and 14 lead to two more special products.

Difference of Two Cubes

$$(x - a)(x^2 + ax + a^2) = x^3 - a^3 \qquad \textbf{(5)}$$

Sum of Two Cubes

$$(x + a)(x^2 - ax + a^2) = x^3 + a^3 \qquad \textbf{(6)}$$

DIVIDING POLYNOMIALS

⑥ The procedure for dividing two polynomials is similar to the procedure for dividing two integers. This procedure should be familiar to you, but we review it briefly next.

EXAMPLE 15 | **Dividing Two Integers**

Divide 842 by 15.

Solution

$$
\begin{array}{r}
56 \quad \leftarrow \text{Quotient} \\
\text{Divisor} \rightarrow \quad 15\overline{)842} \quad \leftarrow \text{Dividend} \\
75 \quad \leftarrow 5 \cdot 15 \ (\text{Subtract}) \\
\hline
92 \\
90 \quad \leftarrow 6 \cdot 15 \ (\text{Subtract}) \\
\hline
2 \quad \leftarrow \text{Remainder}
\end{array}
$$

Thus, $\dfrac{842}{15} = 56 + \dfrac{2}{15}$. ∎

In the long division process detailed in Example 15, the number 15 is called the **divisor,** the number 842 is called the **dividend,** the number 56 is called the **quotient,** and the number 2 is called the **remainder.**

To check the answer obtained in a division problem, multiply the quotient by the divisor and add the remainder. The answer should be the dividend.

$$(\text{Quotient})(\text{Divisor}) + \text{Remainder} = \text{Dividend}$$

For example, we can check the results obtained in Example 15 as follows:

$$(56)(15) + 2 = 840 + 2 = 842$$

To divide two polynomials, we first must write each polynomial in standard form. The process then follows a pattern similar to that of Example 15. The next example illustrates the procedure.

EXAMPLE 16 | **Dividing Two Polynomials**

Find the quotient and the remainder when

$$3x^3 + 4x^2 + x + 7 \quad \text{is divided by} \quad x^2 + 1$$

Solution Each polynomial is in standard form. The dividend is $3x^3 + 4x^2 + x + 7$, and the divisor is $x^2 + 1$.

STEP 1: Divide the leading term of the dividend, $3x^3$, by the leading term of the divisor, x^2. Enter the result, $3x$, over the term $3x^3$, as follows:

$$
\begin{array}{r}
3x \\
x^2 + 1\overline{)3x^3 + 4x^2 + x + 7}
\end{array}
$$

STEP 2: Multiply $3x$ by $x^2 + 1$ and enter the result below the dividend.

$$
\begin{array}{r}
3x \\
x^2 + 1 \overline{\smash{)}3x^3 + 4x^2 + x + 7} \\
3x^3 + 3x
\end{array}
$$

$\leftarrow 3x \cdot (x^2 + 1) = 3x^3 + 3x$

Notice that we align the $3x$ term under the x to make the next step easier.

STEP 3: Subtract and bring down the remaining terms.

$$
\begin{array}{r}
3x \\
x^2 + 1 \overline{\smash{)}3x^3 + 4x^2 + x + 7} \\
3x^3 + 3x \\
\hline
4x^2 - 2x + 7
\end{array}
$$

$\leftarrow$ Subtract (change the signs and add).

$\leftarrow$ Bring down the $4x^2$ and the 7.

STEP 4: Repeat Steps 1–3 using $4x^2 - 2x + 7$ as the dividend.

$$
\begin{array}{r}
3x + 4 \\
x^2 + 1 \overline{\smash{)}3x^3 + 4x^2 + x + 7} \\
3x^3 + 3x \\
\hline
4x^2 - 2x + 7 \\
4x^2 + 4 \\
\hline
-2x + 3
\end{array}
$$

$\leftarrow$ Divide $4x^2$ by x^2 to get 4.

$\leftarrow$ Multiply $x^2 + 1$ by 4; subtract.

Since x^2 does not divide $-2x$ evenly (that is, the result is not a monomial), the process ends. The quotient is $3x + 4$, and the remainder is $-2x + 3$.

Check: (Quotient)(Divisor) + Remainder

$$
\begin{aligned}
&= (3x + 4)(x^2 + 1) + (-2x + 3) \\
&= 3x^3 + 4x^2 + 3x + 4 + (-2x + 3) \\
&= 3x^3 + 4x^2 + x + 7 = \text{Dividend}
\end{aligned}
$$

Thus,

$$
\frac{3x^3 + 4x^2 + x + 7}{x^2 + 1} = 3x + 4 + \frac{-2x + 3}{x^2 + 1}
$$

The next example combines the steps involved in long division.

EXAMPLE 17   **Dividing Two Polynomials**

Find the quotient and the remainder when

$$x^4 - 3x^3 + 2x - 5 \quad \text{is divided by} \quad x^2 - x + 1$$

Solution   In setting up this division problem, it is necessary to leave a space for the missing x^2 term in the dividend.

$$
\begin{array}{r}
x^2 - 2x - 3 \leftarrow \text{Quotient} \\
\text{Divisor} \rightarrow x^2 - x + 1 \overline{\smash{)}x^4 - 3x^3 + 2x - 5} \leftarrow \text{Dividend} \\
\text{Subtract} \rightarrow x^4 - x^3 + x^2 \\
\hline
-2x^3 - x^2 + 2x - 5 \\
\text{Subtract} \rightarrow -2x^3 + 2x^2 - 2x \\
\hline
-3x^2 + 4x - 5 \\
\text{Subtract} \rightarrow -3x^2 + 3x - 3 \\
\hline
x - 2 \leftarrow \text{Remainder}
\end{array}
$$

Check: (Quotient)(Divisor) + Remainder

$$= (x^2 - 2x - 3)(x^2 - x + 1) + x - 2$$

$$= x^4 - x^3 + x^2 - 2x^3 + 2x^2 - 2x - 3x^2 + 3x - 3 + x - 2$$

$$= x^4 - 3x^3 + 2x - 5 = \text{Dividend}$$

As a result,

$$\frac{x^4 - 3x^3 + 2x - 5}{x^2 - x + 1} = x^2 - 2x - 3 + \frac{x - 2}{x^2 - x + 1} \quad \blacksquare \quad \blacksquare$$

The process of dividing two polynomials leads to the following result:

Theorem   Let Q be a polynomial of positive degree and let P be a polynomial whose degree is greater than the degree of Q. The remainder after dividing P by Q is either the zero polynomial or polynomial of degree less than the degree of the divisor Q.

$\blacksquare$

NOW WORK PROBLEM **89.**

POLYNOMIALS IN TWO VARIABLES

A **monomial in two variables** x and y has the form $ax^n y^m$, where a is a constant, x and y are variables, and n and m are nonnegative integers. The **degree** of a monomial is the sum of the powers of the variables.

For example,

$$2xy^3, \quad a^2 b^2, \quad \text{and} \quad x^3 y$$

are monomials, each of which has degree 4.

A **polynomial in two variables** x and y is the sum of one or more monomials in two variables. The **degree of a polynomial** in two variables is the highest degree of all the monomials with nonzero coefficients.

EXAMPLE 18   ### Examples of Polynomials in Two Variables

$$3x^2 + 2x^3 y + 5 \qquad \pi x^3 - y^2 \qquad x^4 + 4x^3 y - xy^3 + y^4$$

Two variables,   Two variables,   Two variables,
degree is 4.   degree is 3.   degree is 4. $\blacksquare$

Multiplying polynomials in two variables is handled in the same way as polynomials in one variable.

EXAMPLE 19   ### Using a Special Product Formula

To multiply $(2x - y)^2$, use the Squares of Binomials (3b) formula with $2x$ instead of x and y instead of a.

$$(2x - y)^2 = (2x)^2 - 2 \cdot y \cdot 2x + y^2 = 4x^2 - 4xy + y^2 \qquad \blacksquare$$

NOW WORK PROBLEM **73.**

5 EXERCISES

In Problems 1–10, tell whether the expression is a monomial. If it is, name the variable(s) and the coefficient and give the degree of the monomial.

1. $2x^3$

2. $-4x^2$

3. $8/x$

4. $-2x^{-3}$

5. $-2xy^2$

6. $5x^2y^3$

7. $8x/y$

8. $-2x^2/y^3$

9. $x^2 + y^2$

10. $3x^2 + 4$

In Problems 11–20, tell whether the expression is a polynomial. If it is, give its degree.

11. $3x^2 - 5$

12. $1 - 4x$

13. 5

14. $-\pi$

15. $3x^2 - \dfrac{5}{x}$

16. $\dfrac{3}{x} + 2$

17. $2y^3 - \sqrt{2}$

18. $10z^2 + z$

19. $\dfrac{x^2 + 5}{x^3 - 1}$

20. $\dfrac{3x^3 + 2x - 1}{x^2 + x + 1}$

In Problems 21–40, add, subtract, or multiply, as indicated. Express your answer as a single polynomial in standard form.

21. $(x^2 + 4x + 5) + (3x - 3)$

22. $(x^3 + 3x^2 + 2) + (x^2 - 4x + 4)$

23. $(x^3 - 2x^2 + 5x + 10) - (2x^2 - 4x + 3)$

24. $(x^2 - 3x - 4) - (x^3 - 3x^2 + x + 5)$

25. $(6x^5 + x^3 + x) + (5x^4 - x^3 + 3x^2)$

26. $(10x^5 - 8x^2) + (3x^3 - 2x^2 + 6)$

27. $(x^2 - 3x + 1) + 2(3x^2 + x - 4)$

28. $-2(x^2 + x + 1) + (-5x^2 - x + 2)$

29. $6(x^3 + x^2 - 3) - 4(2x^3 - 3x^2)$

30. $8(4x^3 - 3x^2 - 1) - 6(4x^3 + 8x - 2)$

31. $(x^2 - x + 2) + (2x^2 - 3x + 5) - (x^2 + 1)$

32. $(x^2 + 1) - (4x^2 + 5) + (x^2 + x - 2)$

33. $9(y^2 - 3y + 4) - 6(1 - y^2)$

34. $8(1 - y^3) + 4(1 + y + y^2 + y^3)$

35. $x(x^2 + x - 4)$

36. $4x^2(x^3 - x + 2)$

37. $-2x^2(4x^3 + 5)$

38. $5x^3(3x - 4)$

39. $(x + 1)(x^2 + 2x - 4)$

40. $(2x - 3)(x^2 + x + 1)$

In Problems 41–58, multiply the polynomials using the FOIL method. Express your answer as a single polynomial in standard form.

41. $(x + 2)(x + 4)$

42. $(x + 3)(x + 5)$

43. $(2x + 5)(x + 2)$

44. $(3x + 1)(2x + 1)$

45. $(x - 4)(x + 2)$

46. $(x + 4)(x - 2)$

47. $(x - 3)(x - 2)$

48. $(x - 5)(x - 1)$

49. $(2x + 3)(x - 2)$

50. $(2x - 4)(3x + 1)$

51. $(-2x + 3)(x - 4)$

52. $(-3x - 1)(x + 1)$

53. $(-x - 2)(-2x - 4)$

54. $(-2x - 3)(3 - x)$

55. $(x - 2y)(x + y)$

56. $(2x + 3y)(x - y)$

57. $(-2x - 3y)(3x + 2y)$

58. $(x - 3y)(-2x + y)$

In Problems 59–82, multiply the polynomials using the special product formulas. Express your answer as a single polynomial in standard form.

59. $(x - 7)(x + 7)$

60. $(x - 1)(x + 1)$

61. $(2x + 3)(2x - 3)$

62. $(3x + 2)(3x - 2)$

63. $(x + 4)^2$

64. $(x + 5)^2$

65. $(x - 4)^2$

66. $(x - 5)^2$

67. $(3x + 4)(3x - 4)$

68. $(5x - 3)(5x + 3)$

69. $(2x - 3)^2$

70. $(3x - 4)^2$

71. $(x + y)(x - y)$

72. $(x + 3y)(x - 3y)$

73. $(3x + y)(3x - y)$

74. $(3x + 4y)(3x - 4y)$

75. $(x + y)^2$

76. $(x - y)^2$

77. $(x - 2y)^2$

78. $(2x + 3y)^2$

79. $(x - 2)^3$

80. $(x + 1)^3$

81. $(2x + 1)^3$

82. $(3x - 2)^3$

In Problems 83–100, find the quotient and the remainder. Check your work by verifying that

$$(Quotient)(Divisor) + Remainder = Dividend$$

83. $4x^3 - 3x^2 + x + 1$ divided by x

84. $3x^3 - x^2 + x - 2$ divided by x

85. $4x^3 - 3x^2 + x + 1$ divided by $x + 2$

86. $3x^3 - x^2 + x - 2$ divided by $x + 2$

87. $4x^3 - 3x^2 + x + 1$ divided by x^2

88. $3x^3 - x^2 + x - 2$ divided by x^2

89. $4x^3 - 3x^2 + x + 1$ divided by $x^2 + 2$

90. $3x^3 - x^2 + x - 2$ divided by $x^2 + 2$

91. $4x^3 - 3x^2 + x + 1$ divided by $2x^3 - 1$

92. $3x^3 - x^2 + x - 2$ divided by $3x^3 - 1$

93. $4x^3 - 3x^2 + x + 1$ divided by $2x^2 + x + 1$

94. $3x^3 - x^2 + x - 2$ divided by $3x^2 + x + 1$

95. $-4x^3 + x^2 - 4$ divided by $x - 1$

96. $-3x^4 - 2x - 1$ divided by $x - 1$

97. $1 - x^2 + x^4$ divided by $x^2 + x + 1$

98. $1 - x^2 + x^4$ divided by $x^2 - x + 1$

99. $x^3 - a^3$ divided by $x - a$

100. $x^5 - a^5$ divided by $x - a$

101. Find the sum of $a, b, c,$ and d if

$$\frac{x^3 - 2x^2 + 3x + 5}{x + 2} = ax^2 + bx + c + \frac{d}{x + 2}$$

---

102. Explain why the degree of the product of two polynomials equals the sum of their degrees.

103. Explain why the degree of the sum of two polynomials of different degrees equals the larger of their degrees.

104. Give a careful statement about the degree of the sum of two polynomials of the same degree.

105. Do you prefer adding two polynomials using the horizontal method or the vertical method? Write a brief position paper defending your choice.

106. Do you prefer to memorize the rule for the square of a binomial $(x + a)^2$ or to use FOIL to obtain the product? Write a brief position paper defending your choice.

6 | FACTORING POLYNOMIALS

OBJECTIVES

1. Factor the Difference of Two Squares and the Sum and the Difference of Two Cubes
2. Factor Perfect Squares
3. Factor a Second-degree Polynomial: $x^2 + Bx + C$
4. Factor by Grouping
5. Factor a Second-degree Polynomial: $Ax^2 + Bx + C$

Consider the following product:

$$(2x + 3)(x - 4) = 2x^2 - 5x - 12$$

The two polynomials on the left side are called **factors** of the polynomial on the right side. Expressing a given polynomial as a product of other polynomials, that is, finding the factors of a polynomial, is called **factoring.**

We shall restrict our discussion here to factoring polynomials in one variable into products of polynomials in one variable, where all coefficients are integers. We call this **factoring over the integers.**

Any polynomial can be written as the product of 1 times itself or as -1 times its additive inverse. If a polynomial cannot be written as the product of two other polynomials (excluding 1 and -1), then the polynomial is said to be **prime.** When a polynomial has been written as a product consisting only of prime factors, it is said to be **factored completely.** Examples of prime polynomials are

$$2, \quad 3, \quad 5, \quad x, \quad x + 1, \quad x - 1, \quad 3x + 4$$

The first factor to look for in a factoring problem is a common monomial factor present in each term of the polynomial. If one is present, use the distributive property to factor it out.

| | EXAMPLE 1 | | Identifying Common Monomial Factors |

Polynomial	Common Monomial Factor	Remaining Factor	Factored Form
$2x + 4$	2	$x + 2$	$2x + 4 = 2(x + 2)$
$3x - 6$	3	$x - 2$	$3x - 6 = 3(x - 2)$
$2x^2 - 4x + 8$	2	$x^2 - 2x + 4$	$2x^2 - 4x + 8 = 2(x^2 - 2x + 4)$
$8x - 12$	4	$2x - 3$	$8x - 12 = 4(2x - 3)$
$x^2 + x$	x	$x + 1$	$x^2 + x = x(x + 1)$
$x^3 - 3x^2$	x^2	$x - 3$	$x^3 - 3x^2 = x^2(x - 3)$
$6x^2 + 9x$	$3x$	$2x + 3$	$6x^2 + 9x = 3x(2x + 3)$

Notice that, once all common monomial factors have been removed from a polynomial, the remaining factor is either a prime polynomial of degree 1 or a polynomial of degree 2 or higher. (Do you see why?)

✎ NOW WORK PROBLEM **1**.

SPECIAL FORMULAS

① When you factor a polynomial, first check whether you can use one of the special formulas discussed in the previous section.

Difference of Two Squares	$x^2 - a^2 = (x - a)(x + a)$
Perfect Squares	$x^2 + 2ax + a^2 = (x + a)^2$
	$x^2 - 2ax + a^2 = (x - a)^2$
Sum of Two Cubes	$x^3 + a^3 = (x + a)(x^2 - ax + a^2)$
Difference of Two Cubes	$x^3 - a^3 = (x - a)(x^2 + ax + a^2)$

| | EXAMPLE 2 | | Factoring the Difference of Two Squares |

Factor completely: $x^2 - 4$

Solution We notice that $x^2 - 4$ is the difference of two squares, x^2 and 2^2. Thus,

$$x^2 - 4 = (x - 2)(x + 2)$$ ■

| | EXAMPLE 3 | | Factoring the Difference of Two Cubes |

Factor completely: $x^3 - 1$

Solution Because $x^3 - 1$ is the difference of two cubes, x^3 and 1^3, we find that

$$x^3 - 1 = (x - 1)(x^2 + x + 1)$$ ■

| | EXAMPLE 4 | | Factoring the Sum of Two Cubes |

Factor completely: $x^3 + 8$

Solution Because $x^3 + 8$ is the sum of two cubes, x^3 and 2^3, we have

$$x^3 + 8 = (x + 2)(x^2 - 2x + 4)$$ ■

EXAMPLE 5

Factoring the Difference of Two Squares

Factor completely: $x^4 - 16$

Solution   Because $x^4 - 16$ is the difference of two squares, $x^4 = \left(x^2\right)^2$ and $16 = 4^2$, we have

$$x^4 - 16 = \left(x^2 - 4\right)\left(x^2 + 4\right)$$

But $x^2 - 4$ is also the difference of two squares. Thus,

$$x^4 - 16 = \left(x^2 - 4\right)\left(x^2 + 4\right) = (x - 2)(x + 2)\left(x^2 + 4\right) \qquad ▬$$

NOW WORK PROBLEMS **11** AND **29**.

2   When the first term and third term of a trinomial are both positive and are perfect squares, such as x^2, $9x^2$, 1, and 4, check to see whether the trinomial is a perfect square.

EXAMPLE 6

Factoring Perfect Squares

Factor completely: $x^2 + 6x + 9$

Solution   The first term, x^2, and the third term, $9 = 3^2$, are perfect squares. Because the middle term $6x$ is twice the product of x and 3, we have a perfect square.

$$x^2 + 6x + 9 = (x + 3)^2 \qquad ▬$$

EXAMPLE 7

Factoring Perfect Squares

Factor completely: $9x^2 - 6x + 1$

Solution   The first term, $9x^2 = (3x)^2$, and the third term, $1 = 1^2$, are perfect squares. Because the middle term $-6x$ is -2 times the product of $3x$ and 1, we have a perfect square.

$$9x^2 - 6x + 1 = (3x - 1)^2 \qquad ▬$$

EXAMPLE 8

Factoring Perfect Squares

Factor completely: $25x^2 + 30x + 9$

Solution   The first term, $25x^2 = (5x)^2$, and the third term, $9 = 3^2$, are perfect squares. Because the middle term $30x$ is twice the product of $5x$ and 3, we have a perfect square.

$$25x^2 + 30x + 9 = (5x + 3)^2 \qquad ▬$$

NOW WORK PROBLEMS **21** AND **89**.

If a trinomial is not a perfect square, it may be possible to factor it using the technique discussed next.

FACTORING A SECOND-DEGREE POLYNOMIAL: $x^2 + Bx + C$

③ The idea behind factoring a second-degree polynomial like $x^2 + Bx + C$ is to see whether it can be made equal to the product of two, possibly equal, first-degree polynomials.

For example, we know that

$$(x + 3)(x + 4) = x^2 + 7x + 12$$

The factors of $x^2 + 7x + 12$ are $x + 3$ and $x + 4$. Notice the following:

$$x^2 + 7x + 12 = (x + 3)(x + 4)$$

the product of 3 and 4 is 12

7 is the sum of 3 and 4

In general, if $x^2 + Bx + C = (x + a)(x + b)$, then $ab = C$ and $a + b = B$.

> To factor a second-degree polynomial $x^2 + Bx + C$, find integers whose product is C and whose sum is B. That is, if there are numbers a, b, where $ab = C$ and $a + b = B$, then
>
> $$x^2 + Bx + C = (x + a)(x + b)$$

EXAMPLE 9

Factoring Trinomials

Factor completely: $x^2 + 7x + 10$

Solution First, determine all integers whose product is 10 and then compute their sums.

Integers whose product is 10	1, 10	−1, −10	2, 5	−2, −5
Sum	11	−11	7	−7

The integers 2 and 5 have a product of 10 and add up to 7, the coefficient of the middle term. Thus,

$$x^2 + 7x + 10 = (x + 2)(x + 5)$$ ∎

EXAMPLE 10

Factoring Trinomials

Factor completely: $x^2 - 6x + 8$

Solution First, determine all integers whose product is 8 and then compute each sum.

Integers whose product is 8	1, 8	−1, −8	2, 4	−2, −4
Sum	9	−9	6	−6

Since −6 is the coefficient of the middle term,

$$x^2 - 6x + 8 = (x - 2)(x - 4)$$ ■

EXAMPLE 11	**Factoring Trinomials**

Factor completely: $x^2 - x - 12$

Solution   First, determine all integers whose product is −12 and then compute each sum.

Integers whose product is −12	1, −12	−1, 12	2, −6	−2, 6	3, −4	−3, 4
Sum	−11	11	−4	4	−1	1

Since −1 is the coefficient of the middle term,

$$x^2 - x - 12 = (x + 3)(x - 4)$$ ■

EXAMPLE 12	**Factoring Trinomials**

Factor completely: $x^2 + 4x - 12$

Solution   The integers −2 and 6 have a product of −12 and have the sum 4. Thus,

$$x^2 + 4x - 12 = (x - 2)(x + 6)$$ ■

To avoid errors in factoring, always check your answer by multiplying it out to see if the result equals the original expression.

When none of the possibilities works, the polynomial is prime.

EXAMPLE 13	**Identifying Prime Polynomials**

Show that $x^2 + 9$ is prime.

Solution   First, list the integers whose product is 9 and then compute their sums.

Integers whose product is 9	1, 9	−1, −9	3, 3	−3, −3
Sum	10	−10	6	−6

Since the coefficient of the middle term in $x^2 + 9 = x^2 + 0x + 9$ is 0 and none of the sums equals 0, we conclude that $x^2 + 9$ is prime. ■

Example 13 demonstrates a more general result:

Theorem   Any polynomial of the form $x^2 + a^2$, a real, is prime.

■

NOW WORK PROBLEMS **35** AND **73**.

FACTORING BY GROUPING

④ Sometimes a common factor does not occur in every term of the polynomial, but in each of several groups of terms that together make up the polynomial. When this happens, the common factor can be factored out of each group by means of the distributive property. This technique is called **factoring by grouping.**

EXAMPLE 14　Factoring by Grouping

Factor completely by grouping: $(x^2 + 2)x + (x^2 + 2) \cdot 3$

Solution　Notice the common factor $x^2 + 2$. By applying the distributive property, we have

$$(x^2 + 2)x + (x^2 + 2) \cdot 3 = (x^2 + 2)(x + 3)$$

Since $x^2 + 2$ and $x + 3$ are prime, the factorization is complete. ■

EXAMPLE 15　Factoring by Grouping

Factor completely by grouping: $x^3 - 4x^2 + 2x - 8$

Solution　To see if factoring by grouping will work, group the first two terms and the last two terms. Then look for a common factor in each group. In this example, we can factor x^2 from $x^3 - 4x^2$ and 2 from $2x - 8$. The remaining factor in each case is the same, $x - 4$. This means that factoring by grouping will work, as follows:

$$\begin{aligned} x^3 - 4x^2 + 2x - 8 &= (x^3 - 4x^2) + (2x - 8) \\ &= x^2(x - 4) + 2(x - 4) \\ &= (x^2 + 2)(x - 4) \end{aligned}$$

Since $x^2 + 2$ and $x - 4$ are prime, the factorization is complete. ■

EXAMPLE 16　Factoring by Grouping

Factor completely by grouping: $3x^3 + 4x^2 - 6x - 8$

Solution　Here, $3x + 4$ is a common factor of $3x^3 + 4x^2$ and of $-6x - 8$. Hence,

$$\begin{aligned} 3x^3 + 4x^2 - 6x - 8 &= (3x^3 + 4x^2) - (6x + 8) \\ &= x^2(3x + 4) - 2(3x + 4) \\ &= (x^2 - 2)(3x + 4) \end{aligned}$$

Since $x^2 - 2$ and $3x + 4$ are prime (over the integers), the factorization is complete. ■

NOW WORK PROBLEM **47.**

FACTORING A SECOND-DEGREE POLYNOMIAL: $Ax^2 + Bx + C$

⑤ To factor a second-degree polynomial $Ax^2 + Bx + C$, when $A \neq 1$ and A, B, and C have no common factors follow these steps:

> **STEPS FOR FACTORING $Ax^2 + Bx + C$, $A \neq 1$, A, B, AND C HAVE NO COMMON FACTORS**
>
> **STEP 1:** Find the value of AC.
>
> **STEP 2:** Find integers whose product is AC that add up to B. That is, find a and b so that $ab = AC$ and $a + b = B$.
>
> **STEP 3:** Write $Ax^2 + Bx + C = Ax^2 + ax + bx + C$.
>
> **STEP 4:** Factor this last expression by grouping.

EXAMPLE 17   **Factoring Trinomials**

Factor completely: $2x^2 + 5x + 3$

Solution   Comparing $2x^2 + 5x + 3$ to $Ax^2 + Bx + C$, we find that $A = 2$, $B = 5$, and $C = 3$.

STEP 1: The value of AC is $2 \cdot 3 = 6$.

STEP 2: Determine the integers whose product is $AC = 6$ and compute their sums.

Integers whose product is 6	1, 6	−1, −6	2, 3	−2, −3
Sum	7	−7	5	−5

The integers whose product is 6 that add up to $B = 5$ are 2 and 3.

STEP 3: $2x^2 + 5x + 3 = 2x^2 + 2x + 3x + 3$

STEP 4: Factor by grouping.

$$2x^2 + 2x + 3x + 3 = (2x^2 + 2x) + (3x + 3)$$
$$= 2x(x + 1) + 3(x + 1)$$
$$= (2x + 3)(x + 1)$$

Thus,

$$2x^2 + 5x + 3 = (2x + 3)(x + 1)$$ ■

EXAMPLE 18   **Factoring Trinomials**

Factor completely: $2x^2 - x - 6$

Solution   Comparing $2x^2 - x - 6$ to $Ax^2 + Bx + C$, we find that $A = 2$, $B = -1$, and $C = -6$.

STEP 1: The value of AC is $2 \cdot (-6) = -12$.

STEP 2: Determine the integers whose product is $AC = -12$ and compute their sums.

Integers whose product is -12	$1, -12$	$-1, 12$	$2, -6$	$-2, 6$	$3, -4$	$-3, 4$
Sum	-11	11	-4	4	-1	1

The integers whose product is -12 that add up to $B = -1$ are -4 and 3.

STEP 3: $2x^2 - x - 6 = 2x^2 - 4x + 3x - 6$

STEP 4: Factor by grouping.

$$2x^2 - 4x + 3x - 6 = (2x^2 - 4x) + (3x - 6)$$
$$= 2x(x - 2) + 3(x - 2)$$
$$= (2x + 3)(x - 2)$$

Thus,

$$2x^2 - x - 6 = (2x + 3)(x - 2)$$

NOW WORK PROBLEM 53.

SUMMARY

We close this section with a capsule summary.

Type of Polynomial	Method	Example
Any polynomial	Look for common monomial factors. (Always do this first!)	$6x^2 + 9x = 3x(2x + 3)$
Binomials of degree 2 or higher	Check for a special product:	
	Difference of two squares, $x^2 - a^2$	$x^2 - 16 = (x - 4)(x + 4)$
	Difference of two cubes, $x^3 - a^3$	$x^3 - 64 = (x - 4)(x^2 + 4x + 16)$
	Sum of two cubes, $x^3 + a^3$	$x^3 + 27 = (x + 3)(x^2 - 3x + 9)$
Trinomials of degree 2	Check for a perfect square, $(x \pm a)^2$.	$x^2 + 8x + 16 = (x + 4)^2$
	Follow the procedures on page 50 or 53.	$6x^2 + x - 1 = (2x + 1)(3x - 1)$
Three or more terms	Grouping	$2x^3 - 3x^2 + 4x - 6 = (2x - 3)(x^2 + 2)$

6 EXERCISES

In Problems 1–10, factor each polynomial by removing the common monomial factor.

1. $3x + 6$
2. $7x - 14$
3. $ax^2 + a$
4. $ax - a$
5. $x^3 + x^2 + x$
6. $x^3 - x^2 + x$
7. $2x^2 - 2x$
8. $3x^2 - 3x$
9. $3x^2y - 6xy^2 + 12xy$
10. $60x^2y - 48xy^2 + 72x^3y$

In Problems 11–18, factor the difference of two squares.

11. $x^2 - 1$
12. $x^2 - 4$
13. $4x^2 - 1$
14. $9x^2 - 1$
15. $x^2 - 16$
16. $x^2 - 25$
17. $25x^2 - 4$
18. $36x^2 - 9$

In Problems 19–28, factor the perfect squares.

19. $x^2 + 2x + 1$
20. $x^2 - 4x + 4$
21. $x^2 + 4x + 4$
22. $x^2 - 2x + 1$
23. $x^2 - 10x + 25$
24. $x^2 + 10x + 25$
25. $4x^2 + 4x + 1$
26. $9x^2 + 6x + 1$
27. $16x^2 + 8x + 1$
28. $25x^2 + 10x + 1$

In Problems 29–34, factor the sum or difference of two cubes.

29. $x^3 - 27$ **30.** $x^3 + 125$ **31.** $x^3 + 27$ **32.** $27 - 8x^3$

33. $8x^3 + 27$ **34.** $64 - 27x^3$

In Problems 35–46, factor each polynomial.

35. $x^2 + 5x + 6$ **36.** $x^2 + 6x + 8$ **37.** $x^2 + 7x + 6$ **38.** $x^2 + 9x + 8$

39. $x^2 + 7x + 10$ **40.** $x^2 + 11x + 10$ **41.** $x^2 - 10x + 16$ **42.** $x^2 - 17x + 16$

43. $x^2 - 7x - 8$ **44.** $x^2 - 2x - 8$ **45.** $x^2 + 7x - 8$ **46.** $x^2 + 2x - 8$

In Problems 47–52, factor by grouping.

47. $2x^2 + 4x + 3x + 6$ **48.** $3x^2 - 3x + 2x - 2$ **49.** $2x^2 - 4x + x - 2$ **50.** $3x^2 + 6x - x - 2$

51. $6x^2 + 9x + 4x + 6$ **52.** $9x^2 - 6x + 2x - 3$

In Problems 53–64, factor each polynomial.

53. $3x^2 + 4x + 1$ **54.** $2x^2 + 3x + 1$ **55.** $2z^2 + 5z + 3$ **56.** $6z^2 + 5z + 1$

57. $3x^2 + 2x - 8$ **58.** $3x^2 + 10x + 8$ **59.** $3x^2 - 2x - 8$ **60.** $3x^2 - 10x + 8$

61. $3x^2 + 14x + 8$ **62.** $3x^2 - 14x + 8$ **63.** $3x^2 + 10x - 8$ **64.** $3x^2 - 10x - 8$

In Problems 65–112, factor completely each polynomial. If the polynomial cannot be factored, say it is prime.

65. $x^2 - 36$ **66.** $x^2 - 9$ **67.** $2 - 8x^2$ **68.** $3 - 27x^2$

69. $x^2 + 7x + 10$ **70.** $x^2 + 5x + 4$ **71.** $x^2 - 10x + 21$ **72.** $x^2 - 6x + 8$

73. $4x^2 - 8x + 32$ **74.** $3x^2 - 12x + 15$ **75.** $x^2 + 4x + 16$ **76.** $x^2 + 12x + 36$

77. $15 + 2x - x^2$ **78.** $14 + 6x - x^2$ **79.** $3x^2 - 12x - 36$ **80.** $x^3 + 8x^2 - 20x$

81. $y^4 + 11y^3 + 30y^2$ **82.** $3y^3 - 18y^2 - 48y$ **83.** $4x^2 + 12x + 9$ **84.** $9x^2 - 12x + 4$

85. $6x^2 + 8x + 2$ **86.** $8x^2 + 6x - 2$ **87.** $x^4 - 81$ **88.** $x^4 - 1$

89. $x^6 - 2x^3 + 1$ **90.** $x^6 + 2x^3 + 1$ **91.** $x^7 - x^5$ **92.** $x^8 - x^5$

93. $16x^2 + 24x + 9$ **94.** $9x^2 - 24x + 16$ **95.** $5 + 16x - 16x^2$ **96.** $5 + 11x - 16x^2$

97. $4y^2 - 16y + 15$ **98.** $9y^2 + 9y - 4$ **99.** $1 - 8x^2 - 9x^4$ **100.** $4 - 14x^2 - 8x^4$

101. $x(x + 3) - 6(x + 3)$ **102.** $5(3x - 7) + x(3x - 7)$ **103.** $(x + 2)^2 - 5(x + 2)$ **104.** $(x - 1)^2 - 2(x - 1)$

105. $(3x - 2)^3 - 27$ **106.** $(5x + 1)^3 - 1$ **107.** $3(x^2 + 10x + 25) - 4(x + 5)$

108. $7(x^2 - 6x + 9) + 5(x - 3)$ **109.** $x^3 + 2x^2 - x - 2$ **110.** $x^3 - 3x^2 - x + 3$

111. $x^4 - x^3 + x - 1$ **112.** $x^4 + x^3 + x + 1$

113. Show that $x^2 + 4$ is prime. **114.** Show that $x^2 + x + 1$ is prime.

115. Make up a polynomial that factors into a perfect square.

116. Explain to a fellow student what you look for first when presented with a factoring problem. What do you do next?

---

7   RATIONAL EXPRESSIONS

OBJECTIVES **1** Reduce a Rational Expression to Lowest Terms

2 Multiply and Divide Rational Expressions

3 Add and Subtract Rational Expressions

4 Use the Least Common Multiple Method

5 Simplify Mixed Quotients

If we form the quotient of two polynomials, the result is called a **rational expression.** Some examples of rational expressions are

(a) $\dfrac{x^3 + 1}{x}$ (b) $\dfrac{3x^2 + x - 2}{x^2 + 5}$ (c) $\dfrac{x}{x^2 - 1}$ (d) $\dfrac{xy^2}{(x - y)^2}$

Expressions (a), (b), and (c) are rational expressions in one variable, x, whereas (d) is a rational expression in two variables, x and y.

Rational expressions are described in the same manner as rational numbers. In expression (a), the polynomial $x^3 + 1$ is called the **numerator,** and x is called the **denominator.** When the numerator and denominator of a rational expression contain no common factors (except 1 and -1), we say that the rational expression is **reduced to lowest terms,** or **simplified.**

The polynomial in the denominator of a rational expression cannot be equal to 0 because division by 0 is not defined. For example, for the expression $\dfrac{x^3 + 1}{x}$, x cannot take on the value 0. The domain of the variable x is $\{x \mid x \neq 0\}$.

A rational expression is reduced to lowest terms by factoring completely the numerator and the denominator and canceling any common factors by using the cancellation property:

$$\frac{a\cancel{c}}{b\cancel{c}} = \frac{a}{b} \qquad \text{if } b \neq 0, c \neq 0 \tag{1}$$

EXAMPLE 1 | **Reducing a Rational Expression to Lowest Terms**

Reduce to lowest terms: $\dfrac{x^2 + 4x + 4}{x^2 + 3x + 2}$

Solution We begin by factoring the numerator and the denominator.

$$x^2 + 4x + 4 = (x + 2)(x + 2)$$

$$x^2 + 3x + 2 = (x + 2)(x + 1)$$

Since a common factor, $x + 2$, appears, the original expression is not in lowest terms. To reduce it to lowest terms, we use the cancellation property:

$$\frac{x^2 + 4x + 4}{x^2 + 3x + 2} = \frac{\cancel{(x + 2)}(x + 2)}{\cancel{(x + 2)}(x + 1)} = \frac{x + 2}{x + 1} \qquad x \neq -2, -1 \qquad \blacksquare$$

WARNING: Apply the cancellation property only to rational expressions written in factored form. Be sure to cancel only common factors! $\blacksquare$

EXAMPLE 2 | **Reducing Rational Expressions to Lowest Terms**

Reduce each rational expression to lowest terms.

(a) $\dfrac{x^3 - 8}{x^3 - 2x^2}$ (b) $\dfrac{8 - 2x}{x^2 - x - 12}$

Solution (a) $\dfrac{x^3 - 8}{x^3 - 2x^2} = \dfrac{\cancel{(x - 2)}(x^2 + 2x + 4)}{x^2\cancel{(x - 2)}} = \dfrac{x^2 + 2x + 4}{x^2} \qquad x \neq 0, 2$

(b) $\dfrac{8 - 2x}{x^2 - x - 12} = \dfrac{2(4 - x)}{(x - 4)(x + 3)} = \dfrac{2(-1)\cancel{(x - 4)}}{\cancel{(x - 4)}(x + 3)} = \dfrac{-2}{x + 3} \qquad x \neq -3, 4$

$\blacksquare$

NOW WORK PROBLEM **1.**

MULTIPLYING AND DIVIDING RATIONAL EXPRESSIONS

② The rules for multiplying and dividing rational expressions are the same as the rules for multiplying and dividing rational numbers. If $\dfrac{a}{b}$ and $\dfrac{c}{d}, b \neq 0, d \neq 0$, are two rational expressions, then

$$\frac{a}{b} \cdot \frac{c}{d} = \frac{ac}{bd} \qquad \text{if } b \neq 0, d \neq 0 \qquad \textbf{(2)}$$

$$\frac{\dfrac{a}{b}}{\dfrac{c}{d}} = \frac{a}{b} \cdot \frac{d}{c} = \frac{ad}{bc} \qquad \text{if } b \neq 0, c \neq 0, d \neq 0 \qquad \textbf{(3)}$$

In using equations (2) and (3) with rational expressions, be sure first to factor each polynomial completely so that common factors can be canceled. Leave your answer in factored form.

EXAMPLE 3   ## Multiplying and Dividing Rational Expressions

Perform the indicated operation and simplify the result. Leave your answer in factored form.

(a) $\dfrac{x^2 - 2x + 1}{x^3 + x} \cdot \dfrac{4x^2 + 4}{x^2 + x - 2}$ (b) $\dfrac{\dfrac{x + 3}{x^2 - 4}}{\dfrac{x^2 - x - 12}{x^3 - 8}}$

Solution   (a) $\dfrac{x^2 - 2x + 1}{x^3 + x} \cdot \dfrac{4x^2 + 4}{x^2 + x - 2} = \dfrac{(x - 1)^2}{x(x^2 + 1)} \cdot \dfrac{4(x^2 + 1)}{(x + 2)(x - 1)}$

$$= \frac{(x - 1)^2 (4)\cancel{(x^2 + 1)}}{x\cancel{(x^2 + 1)}(x + 2)\cancel{(x - 1)}}$$

$$= \frac{4(x - 1)}{x(x + 2)}, \qquad x \neq -2, 0, 1$$

(b) $\dfrac{\dfrac{x + 3}{x^2 - 4}}{\dfrac{x^2 - x - 12}{x^3 - 8}} = \dfrac{x + 3}{x^2 - 4} \cdot \dfrac{x^3 - 8}{x^2 - x - 12}$

$$= \frac{x + 3}{(x - 2)(x + 2)} \cdot \frac{(x - 2)(x^2 + 2x + 4)}{(x - 4)(x + 3)}$$

$$= \frac{\cancel{(x + 3)}\,\cancel{(x - 2)}(x^2 + 2x + 4)}{\cancel{(x - 2)}(x + 2)(x - 4)\cancel{(x + 3)}}$$

$$= \frac{x^2 + 2x + 4}{(x + 2)(x - 4)} \qquad x \neq -3, -2, 2, 4$$

➤ NOW WORK PROBLEMS **13** AND **21**.

ADDING AND SUBTRACTING RATIONAL EXPRESSIONS

③ The rules for adding and subtracting rational expressions are the same as the rules for adding and subtracting rational numbers. Thus, if the denominators of two rational expressions to be added (or subtracted) are equal, we add (or subtract) the numerators and keep the common denominator.

If $\dfrac{a}{b}$ and $\dfrac{c}{b}$ are two rational expressions, then

$$\frac{a}{b} + \frac{c}{b} = \frac{a+c}{b} \qquad \frac{a}{b} - \frac{c}{b} = \frac{a-c}{b} \qquad \text{if } b \neq 0 \qquad \textbf{(4)}$$

EXAMPLE 4   **Adding and Subtracting Rational Expressions with Equal Denominators**

Perform the indicated operation and simplify the result. Leave your answer in factored form.

(a) $\dfrac{2x^2 - 4}{2x + 5} + \dfrac{x + 3}{2x + 5} \qquad x \neq -\dfrac{5}{2}$ (b) $\dfrac{x}{x - 3} - \dfrac{3x + 2}{x - 3} \qquad x \neq 3$

Solution (a) $\dfrac{2x^2 - 4}{2x + 5} + \dfrac{x + 3}{2x + 5} = \dfrac{(2x^2 - 4) + (x + 3)}{2x + 5}$

$$= \frac{2x^2 + x - 1}{2x + 5} = \frac{(2x - 1)(x + 1)}{2x + 5}$$

(b) $\dfrac{x}{x - 3} - \dfrac{3x + 2}{x - 3} = \dfrac{x - (3x + 2)}{x - 3} = \dfrac{x - 3x - 2}{x - 3}$

$$= \frac{-2x - 2}{x - 3} = \frac{-2(x + 1)}{x - 3}$$ ∎

EXAMPLE 5   **Adding Rational Expressions Whose Denominators Are Additive Inverses of Each Other**

Perform the indicated operation and simplify the result. Leave your answer in factored form.

$$\frac{2x}{x - 3} + \frac{5}{3 - x} \qquad x \neq 3$$

Solution Notice that the denominators of the two rational expressions are different. However, the denominator of the second expression is just the additive inverse of the denominator of the first. That is,

$$3 - x = -x + 3 = -1 \cdot (x - 3) = -(x - 3)$$

Then,

$$\frac{2x}{x - 3} + \frac{5}{3 - x} \underset{\substack{\uparrow \\ 3-x=-(x-3)}}{=} \frac{2x}{x - 3} + \frac{5}{-(x - 3)} \underset{\substack{\uparrow \\ \frac{a}{-b}=\frac{-a}{b}}}{=} \frac{2x}{x - 3} + \frac{-5}{x - 3}$$

$$= \frac{2x + (-5)}{x - 3} = \frac{2x - 5}{x - 3}$$ ∎

✏ NOW WORK PROBLEMS **33** AND **39**.

If the denominators of two rational expressions to be added or subtracted are not equal, we can use the general formulas for adding and subtracting quotients.

$$\frac{a}{b} + \frac{c}{d} = \frac{a \cdot d}{b \cdot d} + \frac{b \cdot c}{b \cdot d} = \frac{ad + bc}{bd} \qquad \text{if } b \neq 0, d \neq 0 \qquad \textbf{(5a)}$$

$$\frac{a}{b} - \frac{c}{d} = \frac{a \cdot d}{b \cdot d} - \frac{b \cdot c}{b \cdot d} = \frac{ad - bc}{bd} \qquad \text{if } b \neq 0, d \neq 0 \qquad \textbf{(5b)}$$

EXAMPLE 6

Adding and Subtracting Rational Expressions with Unequal Denominators

Perform the indicated operation and simplify the result. Leave your answer in factored form.

(a) $\dfrac{x-3}{x+4} + \dfrac{x}{x-2} \qquad x \neq -4, 2$ (b) $\dfrac{x^2}{x^2-4} - \dfrac{1}{x} \qquad x \neq -2, 0, 2$

Solution (a) $\dfrac{x-3}{x+4} + \dfrac{x}{x-2} \underset{\underset{\text{(5a)}}{\uparrow}}{=} \dfrac{x-3}{x+4} \cdot \dfrac{x-2}{x-2} + \dfrac{x+4}{x+4} \cdot \dfrac{x}{x-2}$

$$= \frac{(x-3)(x-2) + (x+4)(x)}{(x+4)(x-2)}$$

$$= \frac{x^2 - 5x + 6 + x^2 + 4x}{(x+4)(x-2)} = \frac{2x^2 - x + 6}{(x+4)(x-2)}$$

(b) $\dfrac{x^2}{x^2-4} - \dfrac{1}{x} \underset{\underset{\text{(5b)}}{\uparrow}}{=} \dfrac{x^2}{x^2-4} \cdot \dfrac{x}{x} - \dfrac{x^2-4}{x^2-4} \cdot \dfrac{1}{x} = \dfrac{x^2(x) - (x^2-4)(1)}{(x^2-4)(x)}$

$$= \frac{x^3 - x^2 + 4}{(x-2)(x+2)(x)}$$ ∎

NOW WORK PROBLEM 43.

LEAST COMMON MULTIPLE (LCM)

④ If the denominators of two rational expressions to be added (or subtracted) have common factors, we usually do not use the general rules given by equations (5a) and (5b). Just as with fractions, we apply the **least common multiple (LCM) method.** The LCM method uses the polynomial of least degree that has each denominator polynomial as a factor.

THE **LCM** METHOD FOR ADDING OR
SUBTRACTING RATIONAL EXPRESSIONS

The Least Common Multiple (LCM) Method requires four steps:

STEP 1: Factor completely the polynomial in the denominator of each rational expression.

STEP 2: The LCM of the denominator is the product of each of these factors raised to a power equal to the greatest number of times that the factor occurs in the polynomials.

STEP 3: Write each rational expression using the LCM as the common denominator.

STEP 4: Add or subtract the rational expressions using equation (4).

Let's work an example that only requires Steps 1 and 2.

EXAMPLE 7   Finding the Least Common Multiple

Find the least common multiple of the following pair of polynomials:

$$x(x - 1)^2(x + 1) \quad \text{and} \quad 4(x - 1)(x + 1)^3$$

Solution   STEP 1: The polynomials are already factored completely as

$$x(x - 1)^2(x + 1) \quad \text{and} \quad 4(x - 1)(x + 1)^3$$

STEP 2: Start by writing the factors of the left-hand polynomial. (Or you could start with the one on the right.)

$$x(x - 1)^2(x + 1)$$

Now look at the right-hand polynomial. Its first factor, 4, does not appear in our list, so we insert it.

$$4x(x - 1)^2(x + 1)$$

The next factor, $x - 1$, is already in our list, so no change is necessary. The final factor is $(x + 1)^3$. Since our list has $x + 1$ to the first power only, we replace $x + 1$ in the list by $(x + 1)^3$. The LCM is

$$4x(x - 1)^2(x + 1)^3 \qquad ■$$

Notice that the LCM is, in fact, the polynomial of least degree that contains $x(x - 1)^2(x + 1)$ and $4(x - 1)(x + 1)^3$ as factors.

NOW WORK PROBLEM **49.**

EXAMPLE 8   Using the Least Common Multiple to Add Rational Expressions

Perform the indicated operation and simplify the result. Leave your answer in factored form.

$$\frac{x}{x^2 + 3x + 2} + \frac{2x - 3}{x^2 - 1} \qquad x \neq -2, -1, 1$$

Solution STEP 1: Factor completely the polynomials in the denominators.

$$x^2 + 3x + 2 = (x + 2)(x + 1)$$

$$x^2 - 1 = (x - 1)(x + 1)$$

STEP 2: The LCM is $(x + 2)(x + 1)(x - 1)$. Do you see why?

STEP 3: Write each rational expression using the LCM as the denominator.

$$\frac{x}{x^2 + 3x + 2} = \frac{x}{(x + 2)(x + 1)} = \frac{x}{(x + 2)(x + 1)} \cdot \frac{x - 1}{x - 1} = \frac{x(x - 1)}{(x + 2)(x + 1)(x - 1)}$$

Multiply numerator and
denominator by $x - 1$ to get the
LCM in the denominator.

$$\frac{2x - 3}{x^2 - 1} = \frac{2x - 3}{(x - 1)(x + 1)} = \frac{2x - 3}{(x - 1)(x + 1)} \cdot \frac{x + 2}{x + 2} = \frac{(2x - 3)(x + 2)}{(x - 1)(x + 1)(x + 2)}$$

Multiply numerator and
denominator by $x + 2$ to get
the LCM in the denominator.

STEP 4: Now we can add by using equation (4).

$$\frac{x}{x^2 + 3x + 2} + \frac{2x - 3}{x^2 - 1} = \frac{x(x - 1)}{(x + 2)(x + 1)(x - 1)} + \frac{(2x - 3)(x + 2)}{(x + 2)(x + 1)(x - 1)}$$

$$= \frac{(x^2 - x) + (2x^2 + x - 6)}{(x + 2)(x + 1)(x - 1)}$$

$$= \frac{3x^2 - 6}{(x + 2)(x + 1)(x - 1)} = \frac{3(x^2 - 2)}{(x + 2)(x + 1)(x - 1)}$$ ∎

EXAMPLE 9   **Using the Least Common Multiple to Subtract Rational Expressions**

Perform the indicated operations and simplify the result. Leave your answer in factored form.

$$\frac{3}{x^2 + x} - \frac{x + 4}{x^2 + 2x + 1} \qquad x \neq -1, 0$$

Solution STEP 1: Factor completely the polynomials in the denominators.

$$x^2 + x = x(x + 1)$$

$$x^2 + 2x + 1 = (x + 1)^2$$

STEP 2: The LCM is $x(x + 1)^2$.

STEP 3: Write each rational expression using the LCM as the denominator.

$$\frac{3}{x^2 + x} = \frac{3}{x(x + 1)} = \frac{3}{x(x + 1)} \cdot \frac{x + 1}{x + 1} = \frac{3(x + 1)}{x(x + 1)^2}$$

$$\frac{x + 4}{x^2 + 2x + 1} = \frac{x + 4}{(x + 1)^2} = \frac{x + 4}{(x + 1)^2} \cdot \frac{x}{x} = \frac{x(x + 4)}{x(x + 1)^2}$$

STEP 4: Subtract, using equation (4).

$$\frac{3}{x^2 + x} - \frac{x + 4}{x^2 + 2x + 1} = \frac{3(x + 1)}{x(x + 1)^2} - \frac{x(x + 4)}{x(x + 1)^2}$$

$$= \frac{3(x + 1) - x(x + 4)}{x(x + 1)^2}$$

$$= \frac{3x + 3 - x^2 - 4x}{x(x + 1)^2}$$

$$= \frac{-x^2 - x + 3}{x(x + 1)^2}$$

NOW WORK PROBLEM 59.

MIXED QUOTIENTS

⑤ When sums and/or differences of rational expressions appear as the numerator and/or denominator of a quotient, the quotient is called a **mixed quotient**.* For example,

$$\frac{1 + \dfrac{1}{x}}{1 - \dfrac{1}{x}} \quad \text{and} \quad \frac{\dfrac{x^2}{x^2 - 4} - 3}{\dfrac{x - 3}{x + 2} - 1}$$

are mixed quotients. To **simplify** a mixed quotient means to write it as a rational expression reduced to lowest terms. This can be accomplished in either of two ways.

SIMPLIFYING A MIXED QUOTIENT

METHOD 1: Treat the numerator and denominator of the mixed quotient separately, performing whatever operations are indicated and simplifying the results. Follow this by simplifying the resulting rational expression.

METHOD 2: Find the LCM of the denominators of all rational expressions that appear in the mixed quotient. Multiply the numerator and denominator of the mixed quotient by the LCM and simplify the result.

We will use both methods in the next example. By carefully studying each method, you can discover situations in which one method may be easier to use than the other.

* Some texts use the term **complex fraction**.

EXAMPLE 10   **Simplifying a Mixed Quotient**

Simplify: $\dfrac{\dfrac{1}{2} + \dfrac{3}{x}}{\dfrac{x+3}{4}}$ $x \neq -3, 0$

Solution   METHOD 1: First, we perform the indicated operation in the numerator, and then we divide.

$$\dfrac{\dfrac{1}{2} + \dfrac{3}{x}}{\dfrac{x+3}{4}} = \dfrac{\dfrac{1 \cdot x + 2 \cdot 3}{2 \cdot x}}{\dfrac{x+3}{4}} = \dfrac{\dfrac{x+6}{2x}}{\dfrac{x+3}{4}} = \dfrac{x+6}{2x} \cdot \dfrac{4}{x+3}$$

Rule for adding quotients   Rule for dividing quotients

$$= \dfrac{(x+6) \cdot 4}{2 \cdot x \cdot (x+3)} = \dfrac{2 \cdot 2 \cdot (x+6)}{2 \cdot x \cdot (x+3)} = \dfrac{2(x+6)}{x(x+3)}$$

Rule for multiplying quotients

METHOD 2: The rational expressions that appear in the mixed quotient are

$$\dfrac{1}{2}, \quad \dfrac{3}{x}, \quad \dfrac{x+3}{4}$$

The LCM of their denominators is $4x$. We multiply the numerator and denominator of the mixed quotient by $4x$ and then simplify.

$$\dfrac{\dfrac{1}{2} + \dfrac{3}{x}}{\dfrac{x+3}{4}} = \dfrac{4x \cdot \left(\dfrac{1}{2} + \dfrac{3}{x}\right)}{4x \cdot \left(\dfrac{x+3}{4}\right)} = \dfrac{4x \cdot \dfrac{1}{2} + 4x \cdot \dfrac{3}{x}}{\dfrac{4x \cdot (x+3)}{4}}$$

Multiply the numerator   Distributive property in numerator
and denominator by $4x$

$$= \dfrac{2 \cdot 2x \cdot \dfrac{1}{2} + 4x \cdot \dfrac{3}{x}}{\dfrac{4x \cdot (x+3)}{4}} = \dfrac{2x + 12}{x(x+3)} = \dfrac{2(x+6)}{x(x+3)}$$

Simplify   Factor ■

EXAMPLE 11   **Simplifying a Mixed Quotient**

Simplify: $\dfrac{\dfrac{x^2}{x-4} + 2}{\dfrac{2x-2}{x} - 1}$

Solution   We will use Method 1.

$$\frac{\dfrac{x^2}{x-4}+2}{\dfrac{2x-2}{x}-1} = \frac{\dfrac{x^2}{x-4}+\dfrac{2(x-4)}{x-4}}{\dfrac{2x-2}{x}-\dfrac{x}{x}} = \frac{\dfrac{x^2+2x-8}{x-4}}{\dfrac{2x-2-x}{x}}$$

$$= \frac{\dfrac{(x+4)(x-2)}{x-4}}{\dfrac{x-2}{x}} = \frac{(x+4)\cancel{(x-2)}}{x-4}\cdot\frac{x}{\cancel{x-2}}$$

$$= \frac{(x+4)\cdot x}{x-4} \qquad\blacksquare$$

NOW WORK PROBLEM **69.**

EXAMPLE 12

Solving an Application in Electricity

Figure 18

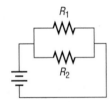

R_1

R_2

An electrical circuit contains two resistors connected in parallel, as shown in Figure 18. If the resistance of each is R_1 and R_2 ohms, respectively, their combined resistance R is given by the formula

$$R = \frac{1}{\dfrac{1}{R_1}+\dfrac{1}{R_2}}$$

Express R as a rational expression; that is, simplify the right-hand side of this formula. Evaluate the rational expression if $R_1 = 6$ ohms and $R_2 = 10$ ohms.

Solution   We will use Method 2. If we consider 1 as the fraction $\dfrac{1}{1}$, then the rational expressions in the mixed quotient are

$$\frac{1}{1}, \quad \frac{1}{R_1}, \quad \frac{1}{R_2}$$

The LCM of the denominators is $R_1 R_2$. We multiply the numerator and denominator of the mixed quotient by $R_1 R_2$ and simplify.

$$\frac{1}{\dfrac{1}{R_1}+\dfrac{1}{R_2}} = \frac{1\cdot R_1 R_2}{\left(\dfrac{1}{R_1}+\dfrac{1}{R_2}\right)\cdot R_1 R_2} = \frac{R_1 R_2}{\dfrac{1}{R_1}\cdot R_1 R_2+\dfrac{1}{R_2}\cdot R_1 R_2} = \frac{R_1 R_2}{R_2+R_1}$$

Thus,

$$R = \frac{R_1 R_2}{R_2+R_1}$$

If $R_1 = 6$ and $R_2 = 10$, then

$$R = \frac{6\cdot 10}{10+6} = \frac{60}{16} = \frac{15}{4}\text{ ohms} \qquad\blacksquare$$

7 EXERCISES

In Problems 1–12, reduce each rational expression to lowest terms.

1. $\dfrac{3x + 9}{x^2 - 9}$

2. $\dfrac{4x^2 + 8x}{12x + 24}$

3. $\dfrac{x^2 - 2x}{3x - 6}$

4. $\dfrac{15x^2 + 24x}{3x^2}$

5. $\dfrac{24x^2}{12x^2 - 6x}$

6. $\dfrac{x^2 + 4x + 4}{x^2 - 16}$

7. $\dfrac{y^2 - 25}{2y^2 - 8y - 10}$

8. $\dfrac{3y^2 - y - 2}{3y^2 + 5y + 2}$

9. $\dfrac{x^2 + 4x - 5}{x^2 - 2x + 1}$

10. $\dfrac{x - x^2}{x^2 + x - 2}$

11. $\dfrac{x^2 + 5x - 14}{2 - x}$

12. $\dfrac{2x^2 + 5x - 3}{1 - 2x}$

In Problems 13–30, perform the indicated operation and simplify the result. Leave your answer in factored form.

13. $\dfrac{3x + 6}{5x^2} \cdot \dfrac{x}{x^2 - 4}$

14. $\dfrac{3}{2x} \cdot \dfrac{x^2}{6x + 10}$

15. $\dfrac{4x^2}{x^2 - 16} \cdot \dfrac{x - 4}{2x}$

16. $\dfrac{12}{x^2 - x} \cdot \dfrac{x^2 - 1}{4x - 2}$

17. $\dfrac{4x - 8}{-3x} \cdot \dfrac{12}{12 - 6x}$

18. $\dfrac{6x - 27}{5x} \cdot \dfrac{2}{4x - 18}$

19. $\dfrac{x^2 - 3x - 10}{x^2 + 2x - 35} \cdot \dfrac{x^2 + 4x - 21}{x^2 + 9x + 14}$

20. $\dfrac{x^2 + x - 6}{x^2 + 4x - 5} \cdot \dfrac{x^2 - 25}{x^2 + 2x - 15}$

21. $\dfrac{\dfrac{6x}{x^2 - 4}}{\dfrac{3x - 9}{2x + 4}}$

22. $\dfrac{\dfrac{12x}{5x + 20}}{\dfrac{4x^2}{x^2 - 16}}$

23. $\dfrac{\dfrac{8x}{x^2 - 1}}{\dfrac{10x}{x + 1}}$

24. $\dfrac{\dfrac{x - 2}{4x}}{\dfrac{x^2 - 4x + 4}{12x}}$

25. $\dfrac{\dfrac{4 - x}{4 + x}}{\dfrac{4x}{x^2 - 16}}$

26. $\dfrac{\dfrac{3 + x}{3 - x}}{\dfrac{x^2 - 9}{9x^3}}$

27. $\dfrac{\dfrac{x^2 + 7x + 12}{x^2 - 7x + 12}}{\dfrac{x^2 + x - 12}{x^2 - x - 12}}$

28. $\dfrac{\dfrac{x^2 + 7x + 6}{x^2 + x - 6}}{\dfrac{x^2 + 5x - 6}{x^2 + 5x + 6}}$

29. $\dfrac{\dfrac{2x^2 - x - 28}{3x^2 - x - 2}}{\dfrac{4x^2 + 16x + 7}{3x^2 + 11x + 6}}$

30. $\dfrac{\dfrac{9x^2 + 3x - 2}{12x^2 + 5x - 2}}{\dfrac{9x^2 - 6x + 1}{8x^2 - 10x - 3}}$

In Problems 31–48, perform the indicated operations and simplify the result. Leave your answer in factored form.

31. $\dfrac{x}{2} + \dfrac{5}{2}$

32. $\dfrac{3}{x} - \dfrac{6}{x}$

33. $\dfrac{x^2}{2x - 3} - \dfrac{4}{2x - 3}$

34. $\dfrac{3x^2}{2x - 1} - \dfrac{9}{2x - 1}$

35. $\dfrac{x + 1}{x - 3} + \dfrac{2x - 3}{x - 3}$

36. $\dfrac{2x - 5}{3x + 2} + \dfrac{x + 4}{3x + 2}$

37. $\dfrac{3x + 5}{2x - 1} - \dfrac{2x - 4}{2x - 1}$

38. $\dfrac{5x - 4}{3x + 4} - \dfrac{x + 1}{3x + 4}$

39. $\dfrac{4}{x - 2} + \dfrac{x}{2 - x}$

40. $\dfrac{6}{x - 1} - \dfrac{x}{1 - x}$

41. $\dfrac{4}{x - 1} - \dfrac{2}{x + 2}$

42. $\dfrac{2}{x + 5} - \dfrac{5}{x - 5}$

43. $\dfrac{x}{x + 1} + \dfrac{2x - 3}{x - 1}$

44. $\dfrac{3x}{x - 4} + \dfrac{2x}{x + 3}$

45. $\dfrac{x - 3}{x + 2} - \dfrac{x + 4}{x - 2}$

46. $\dfrac{2x - 3}{x - 1} - \dfrac{2x + 1}{x + 1}$

47. $\dfrac{x}{x^2 - 4} + \dfrac{1}{x}$

48. $\dfrac{x - 1}{x^3} + \dfrac{x}{x^2 + 1}$

In Problems 49–56, find the LCM of the given polynomials.

49. $x^2 - 4, \quad x^2 - x - 2$

50. $x^2 - x - 12, \quad x^2 - 8x + 16$

51. $x^3 - x, \quad x^2 - x$

52. $3x^2 - 27, \quad 2x^2 - x - 15$

53. $4x^3 - 4x^2 + x, \quad 2x^3 - x^2, \quad x^3$

54. $x - 3, \quad x^2 + 3x, \quad x^3 - 9x$

55. $x^3 - x, \quad x^3 - 2x^2 + x, \quad x^3 - 1$

56. $x^2 + 4x + 4, \quad x^3 + 2x^2, \quad (x + 2)^3$

In Problems 57–68, perform the indicated operations and simplify the result. Leave your answer in factored form.

57. $\dfrac{x}{x^2 - 7x + 6} - \dfrac{x}{x^2 - 2x - 24}$

58. $\dfrac{x}{x - 3} - \dfrac{x + 1}{x^2 + 5x - 24}$

59. $\dfrac{4x}{x^2 - 4} - \dfrac{2}{x^2 + x - 6}$

60. $\dfrac{3x}{x - 1} - \dfrac{x - 4}{x^2 - 2x + 1}$

61. $\dfrac{3}{(x - 1)^2(x + 1)} + \dfrac{2}{(x - 1)(x + 1)^2}$

62. $\dfrac{2}{(x + 2)^2(x - 1)} - \dfrac{6}{(x + 2)(x - 1)^2}$

63. $\dfrac{x + 4}{x^2 - x - 2} - \dfrac{2x + 3}{x^2 + 2x - 8}$

64. $\dfrac{2x - 3}{x^2 + 8x + 7} - \dfrac{x - 2}{(x + 1)^2}$

65. $\dfrac{1}{x} - \dfrac{2}{x^2 + x} + \dfrac{3}{x^3 - x^2}$

66. $\dfrac{x}{(x - 1)^2} + \dfrac{2}{x} - \dfrac{x + 1}{x^3 - x^2}$

67. $\dfrac{1}{h}\left(\dfrac{1}{x + h} - \dfrac{1}{x}\right)$

68. $\dfrac{1}{h}\left[\dfrac{1}{(x + h)^2} - \dfrac{1}{x^2}\right]$

In Problems 69–78, perform the indicated operations and simplify the result. Leave your answer in factored form.

69. $\dfrac{1 + \dfrac{1}{x}}{1 - \dfrac{1}{x}}$

70. $\dfrac{4 + \dfrac{1}{x^2}}{3 - \dfrac{1}{x^2}}$

71. $\dfrac{x - \dfrac{1}{x}}{x + \dfrac{1}{x}}$

72. $\dfrac{1 - \dfrac{x}{x + 1}}{2 - \dfrac{x - 1}{x}}$

73. $\dfrac{\dfrac{x + 4}{x - 2} - \dfrac{x - 3}{x + 1}}{x + 1}$

74. $\dfrac{\dfrac{x - 2}{x + 1} - \dfrac{x}{x - 2}}{x + 3}$

75. $\dfrac{\dfrac{x - 2}{x + 2} + \dfrac{x - 1}{x + 1}}{\dfrac{x}{x + 1} - \dfrac{2x - 3}{x}}$

76. $\dfrac{\dfrac{2x + 5}{x} - \dfrac{x}{x - 3}}{\dfrac{x^2}{x - 3} - \dfrac{(x + 1)^2}{x + 3}}$

77. $1 - \dfrac{1}{1 - \dfrac{1}{x}}$

78. $1 - \dfrac{1}{1 - \dfrac{1}{1 - x}}$

79. The Lensmaker's Equation The focal length f of a lens with index of refraction n is

$$\frac{1}{f} = (n - 1)\left[\frac{1}{R_1} + \frac{1}{R_2}\right]$$

where R_1 and R_2 are the radii of curvature of the front and back surfaces of the lens. Express f as a rational expression. Evaluate the rational expression for $n = 1.5$, $R_1 = 0.1$ meter, and $R_2 = 0.2$ meter.

80. Electrical Circuits An electrical circuit contains three resistors connected in parallel. If the resistance of each is R_1, R_2, and R_3 ohms, respectively, their combined resistance R is given by the formula

$$\frac{1}{R} = \frac{1}{R_1} + \frac{1}{R_2} + \frac{1}{R_3}$$

Express R as a rational expression. Evaluate R for $R_1 = 5$ ohms, $R_2 = 4$ ohms, and $R_3 = 10$ ohms.

81. The following expressions are called **continued fractions:**

$$1 + \frac{1}{x}, \quad 1 + \frac{1}{1 + \dfrac{1}{x}}, \quad 1 + \frac{1}{1 + \dfrac{1}{1 + \dfrac{1}{x}}}, \quad 1 + \frac{1}{1 + \dfrac{1}{1 + \dfrac{1}{1 + \dfrac{1}{x}}}}, \quad \cdots$$

Each simplifies to an expression of the form

$$\frac{ax + b}{bx + c}$$

Trace the successive values of a, b, and c as you "continue" the fraction. Can you discover the patterns that these values follow? Go to the library and research Fibonacci numbers. Write a report on your findings.

82. Explain to a fellow student when you would use the LCM method to add two rational expressions. Give two examples of adding two rational expressions, one in which you use the LCM and the other in which you do not.

83. Which of the two methods given in the text for simplifying mixed quotients do you prefer? Write a brief paragraph stating the reasons for your choice.

8   SQUARE ROOTS; RADICALS

OBJECTIVES   ① Work with Properties of Square Roots
②   Rationalize a Denominator
③   Simplify nth Roots
④   Simplify Radicals

SQUARE ROOTS

A real number is squared when it is raised to the power 2. The inverse of squaring is finding a **square root.** For example, since $6^2 = 36$ and $(-6)^2 = 36$, the numbers 6 and -6 are square roots of 36.

The symbol $\sqrt{}$, called a **radical sign,** is used to denote the **principal,** or nonnegative, square root. Thus, $\sqrt{36} = 6$.

> In general, if a is a nonnegative real number, the nonnegative number b such that $b^2 = a$ is the **principal square root** of a is denoted by $b = \sqrt{a}$.

The following comments are noteworthy:

1. Negative numbers do not have square roots (in the real number system), because the square of any real number is *nonnegative*. For example, $\sqrt{-4}$ is not a real number, because there is no real number whose square is -4.
2. The principal square root of 0 is 0, since $0^2 = 0$. That is, $\sqrt{0} = 0$.
3. The principal square root of a positive number is positive.
4. If $c \geq 0$, then $\left(\sqrt{c}\right)^2 = c$. For example, $\left(\sqrt{2}\right)^2 = 2$ and $\left(\sqrt{3}\right)^2 = 3$.

EXAMPLE 1   **Evaluating Square Roots**

(a) $\sqrt{64} = 8$ (b) $\sqrt{\dfrac{1}{16}} = \dfrac{1}{4}$ (c) $\left(\sqrt{1.4}\right)^2 = 1.4$ ∎

Examples 1(a) and (b) are examples of square roots of perfect squares, since $64 = 8^2$; and $\frac{1}{16} = \left(\frac{1}{4}\right)^2$.
In general, we have the rule

$$\sqrt{a^2} = |a| \qquad \qquad \textbf{(1)}$$

Notice the absolute value in equation (1). We need it since the principal square root is nonnegative.

NOW WORK PROBLEM **1.**

EXAMPLE 2 | **Square Roots of Perfect Squares**

(a) $\sqrt{(2.3)^2} = |2.3| = 2.3$ (b) $\sqrt{(-2.3)^2} = |-2.3| = 2.3$

(c) $\sqrt{x^2} = |x|$ ∎

PROPERTIES OF SQUARE ROOTS

① We begin with the following observation:

$$\sqrt{4 \cdot 25} = \sqrt{100} = 10 \quad \text{and} \quad \sqrt{4}\sqrt{25} = 2 \cdot 5 = 10$$

This suggests the following property of square roots:

Product Property of Square Roots

If a and b are each nonnegative real numbers, then

$$\sqrt{ab} = \sqrt{a}\sqrt{b} \qquad\qquad \textbf{(2)}$$

When used in connection with square roots, the direction "simplify" means to remove from the square root any perfect squares that occur as factors. We can use equation (2) to simplify a square root that contains a perfect square as a factor, as illustrated by the following examples.

EXAMPLE 3 | **Simplifying Square Roots**

(a) $\sqrt{32} = \underset{\substack{\uparrow \\ \text{16 is a} \\ \text{perfect square.}}}{\sqrt{16 \cdot 2}} = \underset{\substack{\uparrow \\ (2)}}{\sqrt{16}\sqrt{2}} = 4\sqrt{2}$

(b) $\sqrt{5}\sqrt{10} = \underset{\substack{\uparrow \\ (2)}}{\sqrt{5 \cdot 10}} = \sqrt{50} = \sqrt{25 \cdot 2} = \sqrt{25}\sqrt{2} = 5\sqrt{2}$

(c) $-3\sqrt{72} = -3\sqrt{36 \cdot 2} = -3\sqrt{36}\sqrt{2} = -3 \cdot 6\sqrt{2} = -18\sqrt{2}$

(d) $\sqrt{75x^2} = \sqrt{(25x^2)(3)} = \sqrt{25x^2}\sqrt{3} = \underset{\substack{\uparrow \\ (1)}}{\sqrt{25}\sqrt{x^2}\sqrt{3}} = 5|x|\sqrt{3}$ ∎

NOW WORK PROBLEMS 13 AND 27.

Sometimes, to multiply two expressions with square roots, we use the distributive property and the product property of square roots. For example,

$$\sqrt{3}(4\sqrt{2} - \sqrt{12}) = \sqrt{3} \cdot 4\sqrt{2} - \sqrt{3} \cdot \sqrt{12} \qquad \text{Distributive Property}$$

$$= 4\sqrt{3 \cdot 2} - \sqrt{3 \cdot 12} \qquad \text{Product Property of Square Roots}$$

$$= 4\sqrt{6} - \sqrt{36} \qquad \text{Simplify.}$$

$$= 4\sqrt{6} - 6$$

EXAMPLE 4	Multiplying Square Roots

$$(\sqrt{5} - 5)(\sqrt{5} + 2) = \sqrt{5}(\sqrt{5} + 2) - 5(\sqrt{5} + 2)$$
$$= \sqrt{5} \cdot \sqrt{5} + \sqrt{5} \cdot 2 - 5 \cdot \sqrt{5} - 5 \cdot 2$$
$$= 5 + 2\sqrt{5} - 5\sqrt{5} - 10$$
$$= -5 - 3\sqrt{5}$$

∎

NOW WORK PROBLEM **51**.

Another property of square roots is suggested by the following examples:

$$\sqrt{\frac{36}{9}} = \sqrt{4} = 2 \quad \text{and} \quad \frac{\sqrt{36}}{\sqrt{9}} = \frac{6}{3} = 2$$

Quotient Property of Square Roots

If a is a nonnegative real number and b is a positive real number, then

$$\sqrt{\frac{a}{b}} = \frac{\sqrt{a}}{\sqrt{b}} \qquad \textbf{(3)}$$

EXAMPLE 5	Simplifying Square Roots

(a) $\sqrt{\dfrac{81}{25}} = \dfrac{\sqrt{81}}{\sqrt{25}} = \dfrac{9}{5}$

(b) $\dfrac{\sqrt{24}}{\sqrt{3}} = \sqrt{\dfrac{24}{3}} = \sqrt{8} = \sqrt{4 \cdot 2} = \sqrt{4}\sqrt{2} = 2\sqrt{2}$

∎

NOW WORK PROBLEM **31**.

RATIONALIZING

② When square roots occur in quotients, it is customary to rewrite the quotient so that the denominator contains no square roots. This process is referred to as **rationalizing the denominator.**

The idea is to multiply by an appropriate expression so that the new denominator contains no square roots. For example:

If Denominator Contains the Factor	Multiply by	To Obtain Denominator Free of Radicals
$\sqrt{3}$	$\sqrt{3}$	$\left(\sqrt{3}\right)^2 = 3$
$\sqrt{3} + 1$	$\sqrt{3} - 1$	$\left(\sqrt{3}\right)^2 - 1^2 = 3 - 1 = 2$
$\sqrt{2} - 3$	$\sqrt{2} + 3$	$\left(\sqrt{2}\right)^2 - 3^2 = 2 - 9 = -7$
$\sqrt{5} - \sqrt{3}$	$\sqrt{5} + \sqrt{3}$	$\left(\sqrt{5}\right)^2 - \left(\sqrt{3}\right)^2 = 5 - 3 = 2$

In rationalizing the denominator of a quotient, be sure to multiply both the numerator and the denominator by the expression.

| EXAMPLE 6 | **Rationalizing Denominators** |

Rationalize the denominator: $\dfrac{1}{\sqrt{3}}$

Solution The denominator contains the factor $\sqrt{3}$, so we multiply the numerator and denominator by $\sqrt{3}$ to obtain

$$\frac{1}{\sqrt{3}} = \frac{1}{\sqrt{3}} \cdot \frac{\sqrt{3}}{\sqrt{3}} = \frac{\sqrt{3}}{\left(\sqrt{3}\right)^2} = \frac{\sqrt{3}}{3}$$ ■

| EXAMPLE 7 | **Rationalizing Denominators** |

Rationalize the denominator: $\dfrac{5}{4\sqrt{2}}$

Solution The denominator contains the factor $\sqrt{2}$, so we multiply the numerator and denominator by $\sqrt{2}$ to obtain

$$\frac{5}{4\sqrt{2}} = \frac{5}{4\sqrt{2}} \cdot \frac{\sqrt{2}}{\sqrt{2}} = \frac{5\sqrt{2}}{4\left(\sqrt{2}\right)^2} = \frac{5\sqrt{2}}{4 \cdot 2} = \frac{5\sqrt{2}}{8}$$ ■

| EXAMPLE 8 | **Rationalizing Denominators** |

Rationalize the denominator: $\dfrac{\sqrt{2}}{\sqrt{3} - \sqrt{2}}$

Solution The denominator contains the factor $\sqrt{3} - \sqrt{2}$, so we multiply the numerator and denominator by $\sqrt{3} + \sqrt{2}$ to obtain

$$\frac{\sqrt{2}}{\sqrt{3} - \sqrt{2}} = \frac{\sqrt{2}}{\sqrt{3} - \sqrt{2}} \cdot \frac{\sqrt{3} + \sqrt{2}}{\sqrt{3} + \sqrt{2}} = \frac{\sqrt{2}\left(\sqrt{3} + \sqrt{2}\right)}{\left(\sqrt{3}\right)^2 - \left(\sqrt{2}\right)^2}$$

$$= \frac{\sqrt{2}\sqrt{3} + \left(\sqrt{2}\right)^2}{3 - 2} = \sqrt{6} + 2$$ ■

NOW WORK PROBLEM **69.**

nth Roots

> The **principal nth root of a real number a, $n \geq 2$** an integer, symbolized by $\sqrt[n]{a}$, is defined as follows:
>
> $$\sqrt[n]{a} = b \quad \text{means} \quad a = b^n$$
>
> where $a \geq 0$ and $b \geq 0$ if $n \geq 2$ is even and a, b are any real numbers if $n \geq 3$ is odd.

Notice that if a is negative and n is even then $\sqrt[n]{a}$ is not defined. When it is defined, the principal nth root of a number is unique.

③ The symbol $\sqrt[n]{a}$ for the principal nth root of a is sometimes called a **radical;** the integer n is called the **index,** and a is called the **radicand.** If the index of a radical is 2, we call $\sqrt[2]{a}$ the **square root** of a and omit the index 2 by simply writing $\sqrt{a}$. If the index is 3, we call $\sqrt[3]{a}$ the **cube root** of a.

EXAMPLE 9

Simplifying Principal nth Roots

(a) $\sqrt[3]{8} = \sqrt[3]{2^3} = 2$ (b) $\sqrt[3]{-64} = \sqrt[3]{(-4)^3} = -4$

(c) $\sqrt[4]{1/16} = \sqrt[4]{(1/2)^4} = 1/2$ (d) $\sqrt[6]{(-2)^6} = |-2| = 2$

These are examples of **perfect roots,** since each one simplifies to a rational number. Notice the absolute value in Example 9(d). If n is even, the principal nth root must be nonnegative.

In general, if $n \geq 2$ is a positive integer and a is a real number, we have

$$\sqrt[n]{a^n} = a \qquad \text{if } n \geq 3 \text{ is odd} \qquad \textbf{(4a)}$$

$$\sqrt[n]{a^n} = |a| \qquad \text{if } n \geq 2 \text{ is even} \qquad \textbf{(4b)}$$

Radicals provide a way of representing many irrational real numbers. For example, there is no rational number whose square is 2. Using radicals, we can say that $\sqrt{2}$ *is the positive number whose square is 2.*

PROPERTIES OF RADICALS

Let $n \geq 2$ and $m \geq 2$ denote positive integers, and let a and b represent real numbers. Assuming that all radicals are defined, we have the following properties:

Properties of Radicals

$$\sqrt[n]{ab} = \sqrt[n]{a}\,\sqrt[n]{b} \qquad \textbf{(5a)}$$

$$\sqrt[n]{\frac{a}{b}} = \frac{\sqrt[n]{a}}{\sqrt[n]{b}} \qquad \textbf{(5b)}$$

$$\sqrt[n]{a^m} = (\sqrt[n]{a})^m \qquad \textbf{(5c)}$$

4   When used in reference to radicals, the direction to "simplify" will mean to remove from the radicals any perfect roots that occur as factors. Let's look at some examples of how the preceding rules are applied to simplify radicals.

EXAMPLE 10

Simplifying Radicals

(a) $\sqrt[3]{16} = \sqrt[3]{8 \cdot 2} = \sqrt[3]{8} \cdot \sqrt[3]{2} = \sqrt[3]{2^3} \cdot \sqrt[3]{2} = 2\sqrt[3]{2}$

    ↑   ↑
Factor out (5a)
perfect cube.

(b) $\sqrt[3]{-16x^4} = \sqrt[3]{-8 \cdot 2 \cdot x^3 \cdot x} = \sqrt[3]{(-8x^3)(2x)}$

    ↑   ↑
Factor perfect   Combine perfect
cubes inside radical.   cubes.

$$= \sqrt[3]{(-2x)^3 \cdot 2x} = \sqrt[3]{(-2x)^3} \cdot \sqrt[3]{2x}$$

$$= -2x\sqrt[3]{2x} \qquad \text{(5a)}$$

| EXAMPLE 11 | Simplifying Radicals |

$$\sqrt[3]{\frac{8x^5}{27}} = \sqrt[3]{\frac{2^3 x^3 x^2}{3^3}} = \sqrt[3]{\left(\frac{2x}{3}\right)^3 \cdot x^2} = \sqrt[3]{\left(\frac{2x}{3}\right)^3} \cdot \sqrt[3]{x^2} = \frac{2x}{3} \sqrt[3]{x^2}$$ ■

NOW WORK PROBLEMS 19 AND 23.

Two or more radicals can be combined, provided that they have the same index and the same radicand. Such radicals are called **like radicals.**

| EXAMPLE 12 | Combining Like Radicals |

$$-8\sqrt{12} + \sqrt{3} = -8\sqrt{4 \cdot 3} + \sqrt{3}$$
$$= -8 \cdot \sqrt{4}\sqrt{3} + \sqrt{3}$$
$$= -16\sqrt{3} + \sqrt{3} = -15\sqrt{3}$$ ■

| EXAMPLE 13 | Combining Like Radicals |

$$\sqrt[3]{8x^4} + \sqrt[3]{-x} + 4\sqrt[3]{27x} = \sqrt[3]{2^3 x^3 x} + \sqrt[3]{-1 \cdot x} + 4\sqrt[3]{3^3 x}$$
$$= \sqrt[3]{(2x)^3} \cdot \sqrt[3]{x} + \sqrt[3]{-1} \cdot \sqrt[3]{x} + 4\sqrt[3]{3^3} \cdot \sqrt[3]{x}$$
$$= 2x\sqrt[3]{x} - 1 \cdot \sqrt[3]{x} + 12\sqrt[3]{x}$$
$$= (2x + 11)\sqrt[3]{x}$$ ■

NOW WORK PROBLEM 43.

HISTORICAL FEATURE

The radical sign, $\sqrt{}$, was first used in print by Coss in 1525. It is thought to be the manuscript form of the letter r (for the Latin word *radix* = *root*), although this is not quite conclusively proved. It took a long time for $\sqrt{}$ to become the standard symbol for a square root and much longer to standardize $\sqrt[3]{}$, $\sqrt[4]{}$, $\sqrt[5]{}$, and so on. The indexes of the root were placed in every conceivable position, with

$$\sqrt[3]{8}, \quad \sqrt{\text{\textcircled{3}}8}, \quad \text{and} \quad \sqrt[]{8}_3$$

all being variants for $\sqrt[3]{8}$. The notation $\sqrt{\sqrt{16}}$ was popular for $\sqrt[4]{16}$. By the 1700s, the index had settled where we now put it.

The bar on top of the present radical symbol, as follows,

$$\sqrt{a^2 + 2ab + b^2}$$

is the last survivor of the **vinculum,** a bar placed atop an expression to indicate what we would now indicate with parentheses. For example,

$$\overline{ab + c} = a(b + c)$$

8 EXERCISES

In Problems 1–12, evaluate each perfect root.

1. $\sqrt{25}$ **2.** $\sqrt{81}$ **3.** $\sqrt[3]{27}$ **4.** $\sqrt[3]{125}$ **5.** $\sqrt[3]{-64}$

6. $\sqrt[3]{-8}$ **7.** $\sqrt[3]{\dfrac{1}{9}}$ **8.** $\sqrt[3]{\dfrac{27}{8}}$ **9.** $\sqrt{25x^4}$ **10.** $\sqrt[3]{64x^6}$

11. $\sqrt[3]{8(1+x)^3}$ **12.** $\sqrt{4(x+4)^2}$

In Problems 13–38, simplify each expression.

13. $\sqrt{8}$ **14.** $\sqrt{27}$ **15.** $\sqrt{50}$ **16.** $\sqrt{72}$ **17.** $\sqrt[3]{16}$ **18.** $\sqrt[3]{24}$

19. $\sqrt[3]{-16}$ **20.** $-\sqrt[3]{16}$ **21.** $\sqrt{\dfrac{25x^3}{9x}}$, $\;x \neq 0$

22. $\sqrt[3]{\dfrac{x}{8x^4}}$, $\;x \neq 0$ **23.** $\sqrt[4]{x^{12}y^8}$, $\;x \geq 0, y \geq 0$ **24.** $\sqrt[5]{x^{10}y^5}$

25. $\sqrt{36x}$, $\;x \geq 0$ **26.** $\sqrt{9x^5}$, $\;x \geq 0$ **27.** $\sqrt{3x^2}\,\sqrt{12x}$, $\;x \geq 0$

28. $\sqrt{5x}\,\sqrt{20x^3}$, $\;x \geq 0$ **29.** $\dfrac{\sqrt{3xy^3}\,\sqrt{2x^2y}}{\sqrt{6x^3y^4}}$, $\;x > 0, y > 0$ **30.** $\dfrac{\sqrt[3]{x^2y}\,\sqrt[3]{125x^3}}{\sqrt[3]{8x^3y^4}}$, $\;x \neq 0, y \neq 0$

31. $\sqrt{\dfrac{16y^4}{9x^2}}$, $\;x > 0, y \geq 0$ **32.** $\sqrt{\dfrac{9x^4}{16y^6}}$, $\;x \geq 0, y > 0$ **33.** $\left(\sqrt{5}\,\sqrt[3]{9}\right)^2$

34. $\left(\sqrt[3]{3}\,\sqrt{10}\right)^4$ **35.** $\sqrt{\dfrac{2x-3}{2x^4+3x^3}}\,\sqrt{\dfrac{x}{4x^2-9}}$, $\;x > \dfrac{3}{2}$

36. $\sqrt[3]{\dfrac{x-1}{x^2+2x+1}}\,\sqrt[3]{\dfrac{(x-1)^2}{x+1}}$, $\;x \neq -1$ **37.** $\sqrt{\dfrac{x-1}{x+1}}\,\sqrt{\dfrac{x^2+2x+1}{x^2-1}}$, $\;x > 1$ **38.** $\sqrt{\dfrac{x^2+4}{x(x^2-4)}}\,\sqrt{\dfrac{4x^2}{x^4-16}}$, $\;x > 2$

In Problems 39–48, simplify each expression.

39. $3\sqrt{2} + 4\sqrt{2}$ **40.** $6\sqrt{5} - 4\sqrt{5}$ **41.** $-\sqrt{18} + 2\sqrt{8}$

42. $2\sqrt{12} - 3\sqrt{27}$ **43.** $5\sqrt[3]{2} - 2\sqrt[3]{54}$ **44.** $9\sqrt[3]{24} - \sqrt[3]{81}$

45. $\sqrt{8x^3} - 3\sqrt{50x}$, $\;x \geq 0$ **46.** $3x\sqrt{9y} + 4\sqrt{25y}$, $\;y \geq 0$

47. $\sqrt[3]{16x^4y} - 3x\sqrt[3]{2xy} + 5\sqrt[3]{-2xy^4}$ **48.** $8xy - \sqrt{25x^2y^2} + \sqrt[3]{8x^3y^3}$, $\;x \geq 0, y \geq 0$

In Problems 49–62, perform the indicated operation and simplify the results.

49. $\left(3\sqrt{6}\right)\left(4\sqrt{3}\right)$ **50.** $\left(5\sqrt{8}\right)\left(-3\sqrt{6}\right)$ **51.** $\sqrt{3}\left(\sqrt{3} - 4\right)$

52. $\sqrt{5}\left(\sqrt{5} + 6\right)$ **53.** $3\sqrt{7}\left(2\sqrt{7} + 3\right)$ **54.** $\left(2\sqrt{6} + 3\right)\left(3\sqrt{6}\right)$

55. $\left(\sqrt{2} - 1\right)^2$ **56.** $\left(\sqrt{3} + \sqrt{5}\right)^2$ **57.** $\left(\sqrt[3]{2} - 1\right)^3$

58. $\left(\sqrt[3]{4} + 2\right)^3$ **59.** $\left(2\sqrt{x} - 3\right)\left(2\sqrt{x} + 5\right)$, $\;x \geq 0$ **60.** $\left(4\sqrt{x} - 3\right)\left(\sqrt{x} + 3\right)$, $\;x \geq 0$

61. $\sqrt{1-x^2} - \dfrac{1}{\sqrt{1-x^2}}$, $\;-1 < x < 1$ **62.** $\sqrt{1-x^2} + \dfrac{x^2}{\sqrt{1-x^2}}$, $\;-1 < x < 1$

In Problems 63–78, rationalize the denominator of each expression.

63. $\dfrac{2}{\sqrt{5}}$ **64.** $\dfrac{\sqrt{3}}{\sqrt{5}}$ **65.** $\dfrac{8}{\sqrt{6}}$ **66.** $\dfrac{5}{\sqrt{10}}$

67. $\dfrac{1}{\sqrt{x}}$, $\;x > 0$ **68.** $\dfrac{x}{\sqrt{x^2+4}}$ **69.** $\dfrac{3}{5+\sqrt{2}}$ **70.** $\dfrac{2}{\sqrt{7}-2}$

71. $\dfrac{3}{4+\sqrt{7}}$ **72.** $\dfrac{10}{4-\sqrt{2}}$ **73.** $\dfrac{\sqrt{5}}{2+3\sqrt{5}}$ **74.** $\dfrac{\sqrt{3}}{2\sqrt{3}+3}$

75. $\dfrac{\sqrt{3}-\sqrt{2}}{\sqrt{3}+\sqrt{2}}$ **76.** $\dfrac{\sqrt{5}+\sqrt{3}}{\sqrt{5}-\sqrt{3}}$ **77.** $\dfrac{1}{\sqrt{x}+2}$, $\;x \geq 0$ **78.** $\dfrac{1}{\sqrt{x}-3}$, $\;x \geq 0, x \neq 9$

In Problems 79–86, use a calculator to approximate each radical. Round your answer to two decimal places.

79. $\sqrt{2}$

80. $\sqrt{7}$

81. $\sqrt[3]{4}$

82. $\sqrt[3]{-5}$

83. $\dfrac{2 + \sqrt{3}}{3 - \sqrt{5}}$

84. $\dfrac{\sqrt{5} - 2}{\sqrt{2} + 4}$

85. $\dfrac{3\sqrt[3]{5} - \sqrt{2}}{\sqrt{3}}$

86. $\dfrac{2\sqrt{3} - \sqrt[3]{4}}{\sqrt{2}}$

---

87. Calculating the Amount of Gasoline in a Tank   An Exxon station stores its gasoline in underground tanks that are right circular cylinders lying on their sides. See the illustration. The volume V of gasoline in the tank (in gallons) is given by the formula

$$V = 40h^2 \sqrt{\dfrac{96}{h} - 0.608}$$

where h is the height of the gasoline (in inches) as measured on a depth stick.
(a) If $h = 12$ inches, how many gallons of gasoline are in the tank?
(b) If $h = 1$ inch, how many gallons of gasoline are in the tank?

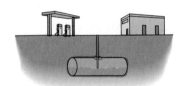

88. Inclined Planes   The final velocity v of an object in feet per second (ft/sec) after it slides down a frictionless inclined plane of height h feet is

$$v = \sqrt{64h + v_0^2}$$

where v_0 is the initial velocity (in ft/sec) of the object.

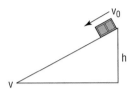

(a) What is the final velocity v of an object that slides down a frictionless inclined plane of height 4 feet? Assume that the initial velocity is 0.
(b) What is the final velocity v of an object that slides down a frictionless inclined plane of height 16 feet? Assume that the initial velocity is 0.
(c) What is the final velocity v of an object that slides down a frictionless inclined plane of height 2 feet with an initial velocity of 4 ft/sec?

In Problems 89–92, use the following information.

Period of a Pendulum   The period T, in seconds, of a pendulum of length l, in feet, may be approximated using the formula

$$T = 2\pi \sqrt{l/32}$$

In the following problems, express your answer both as a square root and as a decimal.

89. Find the period T of a pendulum whose length is 64 feet.
90. Find the period T of a pendulum whose length is 16 feet.
91. Find the period T of a pendulum whose length is 8 inches.
92. Find the period T of a pendulum whose length is 4 inches.
93. Give an example to show that $\sqrt{a^2}$ is not equal to a. Use it to explain why $\sqrt{a^2} = |a|$.

<table>
<tr><td>**9**</td><td>RATIONAL EXPONENTS</td></tr>
</table>

OBJECTIVES   ① Evaluate Expressions with Fractional Exponents
    ② Simplify Radicals Using Rational Exponents

Our purpose in this section is to give a definition for "a raised to the power m/n," where a is a real number and m/n is a rational number. However, we want the definition to obey the laws of exponents stated for integer exponents in Section 4. For example, if the law of exponents $(a^r)^s = a^{rs}$ is to hold, then it must be true that

$$\left(3^{1/2}\right)^2 = 3^{\frac{1}{2} \cdot 2} = 3^1 = 3$$

That is, $a^{1/n}$ is a number that, when raised to the power n, is a. But this was the definition we gave of the principal nth root of a in Section 8. Thus,

$$a^{1/2} = \sqrt{a} \qquad a^{1/3} = \sqrt[3]{a} \qquad a^{1/4} = \sqrt[4]{a}$$

and we can state the following definition:

If a is a real number and $n \geq 2$ is an integer, then

$$a^{1/n} = \sqrt[n]{a} \qquad \textbf{(1)}$$

---

provided that $\sqrt[n]{a}$ exists.

EXAMPLE 1	Writing Expressions Containing Fractional Exponents as Radicals

1
(a) $4^{1/2} = \sqrt{4} = 2$ (b) $8^{1/2} = \sqrt{8} = 2\sqrt{2}$
(c) $(-27)^{1/3} = \sqrt[3]{-27} = -3$ (d) $16^{1/3} = \sqrt[3]{16} = 2\sqrt[3]{2}$ ■

Note that if n is even and $a < 0$ then $\sqrt[n]{a}$ and $a^{1/n}$ do not exist.

We now seek a definition for $a^{m/n}$, where m and n are integers containing no common factors (except 1 and -1) and $n \geq 2$. Again, we want the definition to obey the laws of exponents stated earlier. For example,

$$a^{m/n} = a^{m(1/n)} = \left(a^m\right)^{1/n} \quad \text{and} \quad a^{m/n} = a^{(1/n)m} = \left(a^{1/n}\right)^m$$

If a is a real number and m and n are integers containing no common factors, with $n \geq 2$, then

$$a^{m/n} = \sqrt[n]{a^m} = \left(\sqrt[n]{a}\right)^m \qquad \textbf{(2)}$$

---

provided that $\sqrt[n]{a}$ exists.

We have two comments about equation (2):

1. The exponent m/n must be in lowest terms and n must be positive.
2. In simplifying the rational expression $a^{m/n}$, either $\sqrt[n]{a^m}$ or $\left(\sqrt[n]{a}\right)^m$ may be used, the choice depending on which is easier to simplify. Generally, taking the root first, as in $\left(\sqrt[n]{a}\right)^m$, is easier.

EXAMPLE 2	Using Equation (2)

(a) $4^{3/2} = \left(\sqrt{4}\right)^3 = 2^3 = 8$

(b) $(-8)^{4/3} = \left(\sqrt[3]{-8}\right)^4 = (-2)^4 = 16$

(c) $(32)^{-2/5} = \left(\sqrt[5]{32}\right)^{-2} = 2^{-2} = \dfrac{1}{4}$

(d) $4^{6/4} = 4^{3/2} = \left(\sqrt{4}\right)^3 = 2^3 = 8$ ■

✏— NOW WORK PROBLEM 7.

Based on the definition of $a^{m/n}$, no meaning is given to $a^{m/n}$ if a is a negative real number and n is an even integer.

The definitions in equations (1) and (2) were stated so that the Laws of Exponents would remain true for rational exponents. For convenience, we list again the Law of Exponents.

Laws of Exponents

If a and b are real numbers and r and s are rational numbers, then

$$a^r a^s = a^{r+s} \qquad (a^r)^s = a^{rs} \qquad (ab)^r = a^r \cdot b^r$$

$$a^{-r} = \frac{1}{a^r} \qquad \left(\frac{a}{b}\right)^r = \frac{a^r}{b^r} \qquad \frac{a^r}{a^s} = a^{r-s} = \frac{1}{a^{s-r}}$$

where it is assumed that all expressions used are defined.

2   Rational exponents can sometimes be used to simplify radicals.

EXAMPLE 3   **Simplifying Radicals Using Rational Exponents**

Simplify each expression.

(a) $\left(\sqrt[4]{7}\right)^2$ (b) $\sqrt[9]{x^3}$ (c) $\sqrt[3]{4}\sqrt{2}$

Solution   (a) $\left(\sqrt[4]{7}\right)^2 = \left(7^{1/4}\right)^2 = 7^{\frac{1}{4}\cdot 2} = 7^{\frac{1}{2}} = \sqrt{7}$

(b) $\sqrt[9]{x^3} = x^{3/9} = x^{1/3} = \sqrt[3]{x}$

(c) $\sqrt[3]{4}\sqrt{2} = 4^{1/3}\cdot 2^{1/2} = \left(2^2\right)^{1/3}\cdot 2^{1/2} = 2^{2/3}\cdot 2^{1/2} = 2^{7/6} = 2\cdot 2^{1/6} = 2\sqrt[6]{2}$

➤ **NOW WORK PROBLEM 17.**

The next example illustrates the use of the Laws of Exponents to simplify.

EXAMPLE 4   **Simplifying Expressions Containing Rational Exponents**

Simplify each expression. Express your answer so that only positive exponents occur. Assume that the variables are positive.

(a) $\left(x^{2/3}y\right)\left(x^{-2}y\right)^{1/2}$ (b) $\left(\frac{2x^{1/3}}{y^{2/3}}\right)^{-3}$ (c) $\left(\frac{9x^2y^{1/3}}{x^{1/3}y}\right)^{1/2}$

Solution   (a) $\left(x^{2/3}y\right)\left(x^{-2}y\right)^{1/2} = \left(x^{2/3}y\right)\left[\left(x^{-2}\right)^{1/2}y^{1/2}\right]$

$= x^{2/3}yx^{-1}y^{1/2}$

$= \left(x^{2/3}\cdot x^{-1}\right)\left(y\cdot y^{1/2}\right)$

$= x^{-1/3}y^{3/2}$

$= \dfrac{y^{3/2}}{x^{1/3}}$

(b) $\left(\dfrac{2x^{1/3}}{y^{2/3}}\right)^{-3} = \left(\dfrac{y^{2/3}}{2x^{1/3}}\right)^3 = \dfrac{\left(y^{2/3}\right)^3}{\left(2x^{1/3}\right)^3} = \dfrac{y^2}{2^3\left(x^{1/3}\right)^3} = \dfrac{y^2}{8x}$

↑
Equation (8), page 32

(c) $\left(\dfrac{9x^2y^{1/3}}{x^{1/3}y}\right)^{1/2} = \left(\dfrac{9x^{2-(1/3)}}{y^{1-(1/3)}}\right)^{1/2} = \left(\dfrac{9x^{5/3}}{y^{2/3}}\right)^{1/2} = \dfrac{9^{1/2}\left(x^{5/3}\right)^{1/2}}{\left(y^{2/3}\right)^{1/2}} = \dfrac{3x^{5/6}}{y^{1/3}}$

➤ **NOW WORK PROBLEM 43.**

The next two examples illustrate some algebra that you will need to know for certain calculus problems.

| EXAMPLE 5 | Writing an Expression as a Single Quotient |

Write the following expression as a single quotient in which only positive exponents appear.

$$\left(x^2 + 1\right)^{1/2} + x \cdot \frac{1}{2}\left(x^2 + 1\right)^{-1/2} \cdot 2x$$

Solution $\left(x^2 + 1\right)^{1/2} + x \cdot \dfrac{1}{2}\left(x^2 + 1\right)^{-1/2} \cdot 2x = \left(x^2 + 1\right)^{1/2} + \dfrac{x^2}{\left(x^2 + 1\right)^{1/2}}$

$$= \frac{\left(x^2 + 1\right)^{1/2}\left(x^2 + 1\right)^{1/2} + x^2}{\left(x^2 + 1\right)^{1/2}}$$

$$= \frac{\left(x^2 + 1\right) + x^2}{\left(x^2 + 1\right)^{1/2}}$$

$$= \frac{2x^2 + 1}{\left(x^2 + 1\right)^{1/2}}$$

NOW WORK PROBLEM **51**.

| EXAMPLE 6 | Factoring an Expression Containing Rational Exponents |

Factor: $4x^{1/3}(2x + 1) + 2x^{4/3}$

Solution   We begin by looking for factors that are common to the two terms. Notice that 2 and $x^{1/3}$ are common factors. Thus,

$$4x^{1/3}(2x + 1) + 2x^{4/3} = 2x^{1/3}\left[2(2x + 1) + x\right]$$
$$= 2x^{1/3}(5x + 2)$$

---

9 EXERCISES

In Problems 1–34, simplify each expression.

1. $8^{2/3}$
2. $4^{3/2}$
3. $(-27)^{2/3}$
4. $(-64)^{2/3}$
5. $4^{-3/2}$

6. $(-8)^{-5/3}$
7. $9^{-3/2}$
8. $25^{-5/2}$
9. $\left(\dfrac{9}{4}\right)^{3/2}$
10. $\left(\dfrac{27}{8}\right)^{2/3}$

11. $\left(\dfrac{4}{9}\right)^{-3/2}$
12. $\left(\dfrac{8}{27}\right)^{-2/3}$
13. $4^{1.5}$
14. $16^{-1.5}$
15. $\left(\dfrac{1}{4}\right)^{-1.5}$

16. $\left(\dfrac{1}{9}\right)^{1.5}$
17. $\left(\sqrt{3}\right)^6$
18. $\left(\sqrt[3]{4}\right)^6$
19. $\left(\sqrt{5}\right)^{-2}$
20. $\left(\sqrt[4]{3}\right)^{-8}$

21. $3^{1/2} \cdot 3^{3/2}$
22. $5^{1/3} \cdot 5^{4/3}$
23. $\dfrac{7^{1/3}}{7^{4/3}}$
24. $\dfrac{6^{5/4}}{6^{1/4}}$
25. $2^{1/3} \cdot 4^{1/3}$

26. $9^{1/3} \cdot 3^{1/3}$
27. $\sqrt[3]{3} \cdot \sqrt[3]{27}$
28. $\sqrt[3]{2} \cdot \sqrt[3]{4}$
29. $\left(\sqrt[4]{2}\right)^{-4}$
30. $\left(\sqrt[5]{3}\right)^{-5}$

31. $\left(\sqrt[3]{6}\right)^2$
32. $\left(\sqrt[4]{5}\right)^3$
33. $\sqrt{2} \cdot \sqrt[3]{2}$
34. $\sqrt{5} \cdot \sqrt[3]{5}$

In Problems 35–48, simplify each expression. Express your answer so that only positive exponents occur. Assume that any variables are positive.

35. $\sqrt[8]{x^4}$ **36.** $\sqrt[6]{x^3}$ **37.** $\sqrt{x^3}\sqrt[4]{x}$ **38.** $\sqrt[3]{x^2}\sqrt{x}$

39. $x^{3/2}x^{-1/2}$ **40.** $x^{5/4}x^{-1/4}$ **41.** $(x^3y^6)^{2/3}$ **42.** $(x^4y^8)^{5/4}$

43. $(x^2y)^{1/3}(xy^2)^{2/3}$ **44.** $(xy)^{1/4}(x^2y^2)^{1/2}$ **45.** $(16x^2y^{-1/3})^{3/4}$ **46.** $(4x^{-1}y^{1/3})^{3/2}$

47. $\left(\dfrac{x^{2/5}y^{-1/5}}{x^{-1/3}}\right)^{15}$ **48.** $\left(\dfrac{x^{1/2}}{y^2}\right)^4\left(\dfrac{y^{1/3}}{x^{-2/3}}\right)^3$

In Problems 49–62, write each expression as a single quotient in which only positive exponents and/or radicals appear.

49. $\dfrac{x}{(1+x)^{1/2}} + 2(1+x)^{1/2}, \quad x > -1$

50. $\dfrac{1+x}{2x^{1/2}} + x^{1/2}, \quad x > 0$

51. $2x(x^2+1)^{1/2} + x^2 \cdot \dfrac{1}{2}(x^2+1)^{-1/2} \cdot 2x$

52. $(x+1)^{1/3} + x \cdot \dfrac{1}{3}(x+1)^{-2/3}, \quad x \neq -1$

53. $\sqrt{4x+3} \cdot \dfrac{1}{2\sqrt{x-5}} + \sqrt{x-5} \cdot \dfrac{1}{5\sqrt{4x+3}}, \quad x > 5$

54. $\dfrac{\sqrt[3]{8x+1}}{3\sqrt[3]{(x-2)^2}} + \dfrac{\sqrt[3]{x-2}}{24\sqrt[3]{(8x+1)^2}}, \quad x \neq 2, x \neq -\frac{1}{8}$

55. $\dfrac{\sqrt{1+x} - x \cdot \dfrac{1}{2\sqrt{1+x}}}{1+x}, \quad x > -1$

56. $\dfrac{\sqrt{x^2+1} - x \cdot \dfrac{2x}{2\sqrt{x^2+1}}}{x^2+1}$

57. $\dfrac{(x+4)^{1/2} - 2x(x+4)^{-1/2}}{x+4}, \quad x > -4$

58. $\dfrac{(9-x^2)^{1/2} + x^2(9-x^2)^{-1/2}}{9-x^2}, \quad -3 < x < 3$

59. $\dfrac{\dfrac{x^2}{(x^2-1)^{1/2}} - (x^2-1)^{1/2}}{x^2}, \quad x < -1 \text{ or } x > 1$

60. $\dfrac{(x^2+4)^{1/2} - x^2(x^2+4)^{-1/2}}{x^2+4}$

61. $\dfrac{\dfrac{1+x^2}{2\sqrt{x}} - 2x\sqrt{x}}{(1+x^2)^2}, \quad x > 0$

62. $\dfrac{2x(1-x^2)^{1/3} + \dfrac{2}{3}x^3(1-x^2)^{-2/3}}{(1-x^2)^{2/3}}, \quad x \neq -1, x \neq 1$

In Problems 63–72, factor each expression. Your answer should contain only positive exponents.

63. $(x+1)^{3/2} + x \cdot \dfrac{3}{2}(x+1)^{1/2}, \quad x \geq -1$

64. $(x^2+4)^{4/3} + x \cdot \dfrac{4}{3}(x^2+4)^{1/3} \cdot 2x$

65. $6x^{1/2}(x^2+x) - 8x^{3/2} - 8x^{1/2}, \quad x \geq 0$

66. $6x^{1/2}(2x+3) + x^{3/2} \cdot 8, \quad x \geq 0$

67. $3(x^2+4)^{4/3} + x \cdot 4(x^2+4)^{1/3} \cdot 2x$

68. $2x(3x+4)^{4/3} + x^2 \cdot 4(3x+4)^{1/3}$

69. $4(3x+5)^{1/3}(2x+3)^{3/2} + 3(3x+5)^{4/3}(2x+3)^{1/2}, \quad x \geq -\frac{3}{2}$

70. $6(6x+1)^{1/3}(4x-3)^{3/2} + 6(6x+1)^{4/3}(4x-3)^{1/2}, \quad x \geq \frac{3}{4}$

71. $3x^{-1/2} + \dfrac{3}{2}x^{1/2}, \quad x > 0$

72. $8x^{1/3} - 4x^{-2/3}, \quad x \neq 0$

C H A P T E R R E V I E W

Things To Know

Classification of numbers (p. 2)

Counting numbers	$1, 2, 3 \ldots$
Whole numbers	$0, 1, 2, 3 \ldots$
Integers	$\ldots, -2, -1, 0, 1, 2, \ldots$
Rational numbers	Quotients of two integers (denominator not equal to 0); terminating or repeating decimals
Irrational numbers	Nonrepeating decimals
Real numbers	Rational or irrational numbers

Properties of real numbers (pp. 8–12)

Commutative properties	$a + b = b + a, \qquad a \cdot b = b \cdot a$
Associative properties	$a + (b + c) = (a + b) + c,$
	$a \cdot (b \cdot c) = (a \cdot b) \cdot c$
Distributive property	$a \cdot (b + c) = a \cdot b + a \cdot c$
Identity properties	$a + 0 = a, \qquad a \cdot 1 = a$
Inverse properties	$a + (-a) = 0, \qquad a \cdot \dfrac{1}{a} = 1, \quad \text{where } a \neq 0$
Zero-Product property	If $ab = 0$, then $a = 0$ or $b = 0$ or both.

Absolute value (p. 18)

$$|a| = a \text{ if } a \geq 0, \quad |a| = -a \text{ if } a < 0$$

Pythagorean Theorem (p. 23)

In a right triangle, the square of the length of the hypotenuse is equal to the sum of the squares of the lengths of the legs.

Converse of the Pythagorean Theorem (p. 23)

In a triangle, if the square of the length of one side equals the sum of the squares of the lengths of the other two sides, then the triangle is a right triangle.

Exponents (pp. 28–29)

$$a^n = \underbrace{a \cdot a \cdot \ldots \cdot a}_{n \text{ factors}}, n \text{ a positive integer} \qquad a^0 = 1, a \neq 0 \qquad a^{-n} = \frac{1}{a^n}, a \neq 0, n \text{ a positive integer}$$

Laws of exponents (p. 35)

$$a^m \cdot a^n = a^{m+n} \qquad \left(a^m\right)^n = a^{mn} \qquad (a \cdot b)^n = a^n \cdot b^n \qquad \frac{a^m}{a^n} = a^{m-n} = \frac{1}{a^{n-m}} \qquad \left(\frac{a}{b}\right)^n = \frac{a^n}{b^n}$$

Polynomial (p. 38)

Algebraic expression of the form $a_n x^n + a_{n-1} x^{n-1} + \cdots + a_1 x + a_0$, n a nonnegative integer

Special products/factoring formulas (pp. 41–43)

$$(x - a)(x + a) = x^2 - a^2$$
$$(x + a)^2 = x^2 + 2ax + a^2, \qquad (x - a)^2 = x^2 - 2ax + a^2$$
$$(x + a)(x + b) = x^2 + (a + b)x + ab$$
$$(ax + b)(cx + d) = acx^2 + (ad + bc)x + bd$$
$$(x + a)^3 = x^3 + 3ax^2 + 3a^2x + a^3, \qquad (x - a)^3 = x^3 - 3ax^2 + 3a^2x - a^3$$
$$(x - a)(x^2 + ax + a^2) = x^3 - a^3, \qquad (x + a)(x^2 - ax + a^2) = x^3 + a^3$$

Radicals; Rational exponents (pp. 70 and 75)

$\sqrt[n]{a} = b$ means $a = b^n$, where $a \geq 0$ and $b \geq 0$ if $n \geq 2$ is even and a, b are any real numbers if $n \geq 3$ is odd

$$a^{m/n} = \sqrt[n]{a^m} = \left(\sqrt[n]{a}\right)^m$$

Objectives

You should be able to:

Classify numbers (p. 2)

Evaluate numerical expressions (p. 7)

Work with properties of real numbers (p. 8)

Graph inequalities (p. 18)

Find distance on the real number line (p. 19)

Evaluate algebraic expressions (p. 19)

Determine the domain of a variable (p. 20)

Use the Pythagorean Theorem and its converse (p. 22)

Know geometry formulas (p. 24)

Evaluate expressions containing exponents (p. 29)

Work with the Laws of Exponents (p. 30)

Use a calculator to evaluate exponents (p. 32)

Use scientific notation (p. 33)

Recognize monomials (p. 37)

Recognize polynomials (p. 38)

Add and subtract polynomials (p. 39)

Multiply polynomials (p. 40)

Know formulas for special products (p. 41)

Divide polynomials (p. 43)

Factor the difference of two squares and the sum and difference of two cubes (p. 48)

Factor perfect squares (p. 49)

Factor a second-degree polynomial $x + Bx + C$ (p. 50)

Factor by grouping (p. 52)

Factor a second-degree polynomial $Ax + Bx + C$ (p. 53)

Reduce a rational expression to lowest terms (p. 56)

Multiply and divide rational expressions (p. 57)

Add and subtract rational expressions (p. 58)

Use the least common multiple method (p. 59)

Simplify mixed quotients (p. 62)

Work with properties of square roots (p. 68)

Rationalize a denominator (p. 69)

Simplify nth roots (p. 70)

Simplify radicals (p. 71)

Evaluate expressions with rational exponents (p. 75)

Simplify radicals using rational exponents (p. 76)

Fill-in-the-Blank Items

1. The _____ property of equality states that if $a = b$ then $b = a$.
2. The _____ properties state that $a + b = b + a$ and $ab = ba$.
3. The _____ property states that $a \cdot (b + c) = a \cdot b + a \cdot c$.
4. In the expression 2^4, the number 2 is called the _____, and 4 is called the _____.
5. The polynomial $5x^3 - 3x^2 + 2x - 4$ is of degree _____; the _____ coefficient is 5.
6. In the expression $\sqrt[3]{25}$, the number 3 is called the _____; the number 25 is called the _____.
7. In a right triangle, the side that is longest is called the _____.

True/False Items

T F   **1.** Rational numbers are also real numbers.

T F   **2.** Irrational numbers have decimals that neither repeat nor terminate.

T F   **3.** $(a + b)^2 = a^2 + b^2$

T F   **4.** The polynomial $x^2 + 9$ is prime.

T F   **5.** To rationalize $\dfrac{3}{2 - \sqrt{3}}$ you would multiply the numerator and the denominator by $-2 + \sqrt{3}$.

T F   **6.** The circumference of a circle of diameter d is πd.

Review Exercises

Blue problem numbers indicate the author's suggestions for use in a Practice Test.

In Problems 1–20, perform the indicated operations.

1. $3 - 4 \cdot 5 + 6$

2. $8 + 4 \cdot 2 - 6$

3. $\dfrac{3}{4} - \dfrac{7}{12}$

4. $\dfrac{9}{8} - \dfrac{3}{4}$

5. $\dfrac{\dfrac{15}{2}+\dfrac{1}{4}}{\dfrac{2}{3}}$

6. $\dfrac{\dfrac{9}{2}+\dfrac{3}{4}}{\dfrac{4}{3}}$

7. $5^2 - 3^3 \cdot 2$

8. $2^3 + 3^2 \cdot 4$

9. $\dfrac{2^{-3} \cdot 5^0}{4^2}$

10. $\dfrac{3^{-2} - 4}{2^{-2}}$

11. $(2\sqrt{5} - 2)(2\sqrt{5} + 2)$

12. $(2\sqrt{3} + \sqrt{2})(2\sqrt{3} - \sqrt{2})$

13. $\left(\dfrac{8}{27}\right)^{-2/3}$

14. $\left(\dfrac{4}{9}\right)^{-3/2}$

15. $(\sqrt[3]{2})^{-3}$

16. $(4\sqrt{32})^{-1/2}$

17. $|6 - 8^{1/3}|$

18. $|25^{1/2} - 27^{2/3}|$

19. $\sqrt{|3^2 - 5^2|}$

20. $\sqrt[3]{-|4^2 - 2^3|}$

In Problems 21–32, write each expression so that all exponents are positive. Assume that $x > 0$ and $y > 0$.

21. $\dfrac{x^{-2}}{y^{-2}}$

22. $\left(\dfrac{x^{-1}}{y^{-3}}\right)^2$

23. $\dfrac{(x^2 y)^{-4}}{(xy)^{-3}}$

24. $\dfrac{\left(\dfrac{x}{y}\right)^2}{\left(\dfrac{x}{y}\right)^{-1}}$

25. $\dfrac{\left(\dfrac{x^2}{y}\right)^2}{\left(\dfrac{x}{y^2}\right)^3}$

26. $\dfrac{\left(\dfrac{2x}{3y^2}\right)^{-1}}{\dfrac{4}{y^3}}$

27. $\dfrac{x^{-2}}{x^{-2} + y^{-2}}$

28. $\dfrac{x^{-1} + y^{-1}}{x^{-1} - y^{-1}}$

29. $(25x^{-4/3}y^{-2/3})^{3/2}$

30. $(16x^{-2/3}y^{4/3})^{-3/2}$

31. $\left(\dfrac{2x^{-1/2}}{y^{-3/4}}\right)^{-4}$

32. $\left(\dfrac{8x^{-3/2}}{y^{-3}}\right)^{-2/3}$

In Problems 33–40, perform the indicated operations. Express your answer as a polynomial in standard form.

33. $(2x - 3)(-4x + 2)$

34. $(3x + 4)(-8x - 2)$

35. $4(3x^3 - 2x^2 + 1) - 3(x^3 + 4x^2 - 2x - 3)$

36. $8(1 - x^2 + x^3) - 4(1 + 2x^2 - 4x^4)$

37. $(2x - 5)(3x^2 + 2)$

38. $(1 - 2x^3)(1 - 4x)$

39. $(x + 1)(x + 2)(x - 3)$

40. $(x + 1)(x + 3)(x - 5)$

In Problems 41–50, find the quotient and remainder. Check your work by verifying that

$$(Quotient)(Divisor) + Remainder = Dividend$$

41. $3x^3 - x^2 + x + 4$ divided by $x - 3$

42. $2x^3 - 3x^2 + x + 1$ divided by $x - 2$

43. $-3x^4 + x^2 + 2$ divided by $x^2 + 1$

44. $-4x^3 + x^2 - 2$ divided by $x^2 - 1$

45. $8x^4 - 2x^2 + 5x + 1$ divided by $x^2 - 3x + 1$

46. $3x^4 - x^3 - 8x + 4$ divided by $x^2 + 3x - 2$

47. $x^5 + 1$ divided by $x + 1$

48. $x^5 - 1$ divided by $x - 1$

49. $6x^5 + 3x^4 - 4x^3 - 2x^2 + 2x + 1$ divided by $2x + 1$

50. $6x^5 - 3x^4 - 4x^3 + 2x^2 + 2x - 1$ divided by $2x - 1$

In Problems 51–64, factor each polynomial completely (over the integers). If the polynomial cannot be factored, say it is prime.

51. $x^2 + 5x - 14$

52. $x^2 - 9x + 14$

53. $6x^2 - 5x - 6$

54. $6x^2 + x - 2$

55. $3x^2 - 15x - 42$

56. $2x^3 + 18x^2 + 28x$

57. $8x^3 + 1$

58. $27x^3 - 8$

59. $2x^3 + 3x^2 - 2x - 3$

60. $2x^3 + 3x^2 + 2x + 3$

61. $25x^2 - 4$

62. $16x^2 - 1$

63. $9x^2 + 1$

64. $x^2 - x + 1$

In Problems 65–72, perform the indicated operation and simplify.

65. $\dfrac{2x^2 + 11x + 14}{x^2 - 4}$

66. $\dfrac{x^2 - 5x - 14}{4 - x^2}$

67. $\dfrac{9x^2 - 1}{x^2 - 9} \cdot \dfrac{3x - 9}{9x^2 + 6x + 1}$

68. $\dfrac{x^2 - 25}{x^3 - 4x^2 - 5x} \cdot \dfrac{x^2 + x}{1 - x^2}$

69. $\dfrac{x + 1}{x - 1} - \dfrac{x - 1}{x + 1}$

70. $\dfrac{x}{x + 1} - \dfrac{2x}{x + 2}$

71. $\dfrac{3x + 4}{x^2 - 4} - \dfrac{2x - 3}{x^2 + 4x + 4}$

72. $\dfrac{x^2}{2x^2 + 5x - 3} + \dfrac{x^2}{2x^2 - 5x + 2}$

In Problems 73–78, rationalize the denominator of each expression.

73. $\dfrac{4}{\sqrt{5}}$

74. $\dfrac{-2}{\sqrt{3}}$

75. $\dfrac{2}{1 - \sqrt{2}}$

76. $\dfrac{-4}{1 + \sqrt{3}}$

77. $\dfrac{1 + \sqrt{5}}{1 - \sqrt{5}}$

78. $\dfrac{4\sqrt{3} + 2}{2\sqrt{3} + 1}$

In Problems 79–82, write each expression as a single quotient in which only positive exponents and/or radicals appear.

79. $(2 + x^2)^{1/2} + x \cdot \frac{1}{2}(2 + x^2)^{-1/2} \cdot 2x$

80. $(x^2 + 4)^{2/3} + x \cdot \frac{2}{3}(x^2 + 4)^{-1/3} \cdot 2x$

81. $\dfrac{(x + 4)^{1/2} \cdot 2x - x^2 \cdot \frac{1}{2}(x + 4)^{-1/2}}{x + 4}$, $\quad x > -4$

82. $\dfrac{(x^2 + 4)^{1/2} \cdot 2x - x^2 \cdot \frac{1}{2}(x^2 + 4)^{-1/2} \cdot 2x}{x^2 + 4}$

83. Manufacturing Cost   The weekly production cost C of manufacturing x hand calculators is given by the formula

$$C = 3000 + 6x - \frac{x^2}{1000}$$

   (a) What is the cost of producing 1000 hand calculators?
   (b) What is the cost of producing 3000 hand calculators?

84. Quarterly Corporate Earnings   In the first quarter of its fiscal year, a company posted earnings of $1.20 per share. During the second and third quarters, it posted losses of $0.75 per share and $0.30 per share, respectively. In the fourth quarter, it earned a modest $0.20 per share. What were the annual earnings per share of this company?

85. Design   A window consists of a rectangle surmounted by a triangle. Find the area of the window shown in the illustration. How much wood frame is needed to enclose the window?

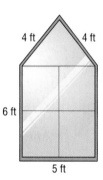

4 ft   4 ft

6 ft

5 ft

86. Construction   A rectangular swimming pool, 20 feet long and 10 feet wide, is enclosed by a wooden deck that is 3 feet wide. What is the area of the deck? How much fence is required to enclose the deck?

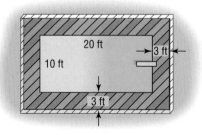

20 ft
3 ft
10 ft
3 ft

87. Construction   A statue with a circular base of radius 3 feet is enclosed by a circular pond as shown in the illustration. What is the surface area of the pond? How much fence is required to enclose the pond?

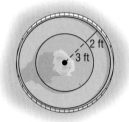

2 ft
3 ft

88. How far can a pilot see?   On a recent flight to San Francisco, the pilot announced we were 139 miles from the city, flying at an altitude of 35,000 feet. The pilot claimed he could see the Golden Gate bridge and beyond. Was he telling the truth? How far could he see?

San Francisco

Alcatraz

89. Use the material in this chapter to create a problem that uses each of the following words:
   (a) Simplify   (b) Factor   (c) Reduce

90. A rational number is defined as the quotient of two integers. When written as a decimal, the decimal will either repeat or terminate. By looking at the denominator of the rational number, there is a way to tell in advance whether its decimal representation will repeat or terminate. Make a list of rational numbers and their decimals. See if you can discover the pattern. Confirm your conclusion by consulting books on number theory at the library. Write a brief essay on your findings.

91. The current time is 12 noon CST. What time (CST) will it be 12,997 hours from now?

92. Both $a/0$ ($a \neq 0$) and $0/0$ are undefined, but for different reasons. Write a paragraph or two explaining the different reasons.

Equations and Inequalities

1

Field Trip To Motorola

First-, second-, and third-degree polynomial equations are widely used at Motorola to model manufacturing processes and product design features, based on measurements made in factories and testing labs. Equipment cycle time characterization is one example. By performing a set of designed experiments, engineers can develop empirical models to predict the assembly time for a given product on a given type of machine. Trade-off analyses, for example, to optimize production costs, are frequently performed using ratios that express the costs and benefits of a particular course of action.

PREPARING FOR THIS SECTION

Before getting started, review the following:

✓ Real Numbers (Review, Section 1, pp. 2–13)

✓ Algebra Review (Review, Section 2, pp. 16–21)

✓ Factoring Polynomials (Review, Section 6, pp. 47–54)

1.1 EQUATIONS

OBJECTIVES
1. Solve an Equation in One Variable
2. Solve a Linear Equation
3. Solve Equations That Lead to Linear Equations

1. An **equation in one variable** is a statement in which two expressions, at least one containing the variable, are equal. The expressions are called the **sides** of the equation. Since an equation is a statement, it may be true or false, depending on the value of the variable. Unless otherwise restricted, the admissible values of the variable are those in the domain of the variable. Those admissible values of the variable, if any, that result in a true statment are called **solutions,** or **roots,** of the equation. To **solve an equation** means to find all the solutions of the equation.

For example, the following are all equations in one variable, x:

$$x + 5 = 9 \qquad x^2 + 5x = 2x - 2 \qquad \frac{x^2 - 4}{x + 1} = 0 \qquad x^2 + 9 = 5$$

The first of these statements, $x + 5 = 9$, is true when $x = 4$ and false for any other choice of x. Thus, 4 is a solution of the equation $x + 5 = 9$. We also say that 4 **satisfies** the equation $x + 5 = 9$, because, when we substitute 4 for x, a true statement results.

Sometimes an equation will have more than one solution. For example, the equation

$$\frac{x^2 - 4}{x + 1} = 0$$

has $x = -2$ and $x = 2$ as solutions.

Usually, we will write the solution of an equation in set notation. This set is called the **solution set** of the equation. For example, the solution set of the equation $x^2 - 9 = 0$ is $\{-3, 3\}$.

Unless indicated otherwise, we will limit ourselves to real solutions, that is, solutions that are real numbers. Some equations have no real solution. For example, $x^2 + 9 = 5$ has no real solution, because there is no real number whose square when added to 9 equals 5.

An equation that is satisfied for every choice of the variable for which both sides are defined is called an **identity.** For example, the equation

$$3x + 5 = x + 3 + 2x + 2$$

is an identity, because this statement is true for any real number x.

Two or more equations that have precisely the same solutions are called **equivalent equations.**

For example, all the following equations are equivalent, because each has only the solution $x = 5$:

$$2x + 3 = 13$$
$$2x = 10$$
$$x = 5$$

These three equations illustrate one method for solving many types of equations: Replace the original equation by an equivalent equation, and continue until an equation with an obvious solution, such as $x = 5$, is reached. The question, though, is "How do I obtain an equivalent equation?" In general, there are five ways to do so.

PROCEDURES THAT RESULT IN EQUIVALENT EQUATIONS

1. Interchange the two sides of the equation:

   Replace $3 = x$ by $x = 3$

2. Simplify the sides of the equation by combining like terms, eliminating parentheses, and so on:

   Replace $(x + 2) + 6 = 2x + (x + 1)$

   by $x + 8 = 3x + 1$

3. Add or subtract the same expression on both sides of the equation:

   Replace $3x - 5 = 4$

   by $(3x - 5) + 5 = 4 + 5$

4. Multiply or divide both sides of the equation by the same nonzero expression:

$$\text{Replace} \qquad \frac{3x}{x-1} = \frac{6}{x-1} \qquad x \neq 1$$

$$\text{by} \qquad \frac{3x}{x-1} \cdot (x-1) = \frac{6}{x-1} \cdot (x-1)$$

5. If one side of the equation is 0 and the other side can be factored, then we may use the Zero-Product Property* and set each factor equal to 0:

$$\text{Replace} \qquad x(x-3) = 0$$

$$\text{by} \qquad x = 0 \quad \text{or} \quad x - 3 = 0$$

WARNING: Squaring both sides of an equation does not necessarily lead to an equivalent equation. ∎

Whenever it is possible to solve an equation in your head, do so. For example:

The solution of $2x = 8$ is $x = 4$.

The solution of $3x - 15 = 0$ is $x = 5$.

Often, though, some rearrangement is necessary.

EXAMPLE 1 **Solving an Equation**

Solve the equation: $3x - 5 = 4$

Solution We replace the original equation by a succession of equivalent equations.

$$3x - 5 = 4$$
$$(3x - 5) + 5 = 4 + 5 \qquad \text{Add 5 to both sides.}$$
$$3x = 9 \qquad \text{Simplify.}$$
$$\frac{3x}{3} = \frac{9}{3} \qquad \text{Divide both sides by 3.}$$
$$x = 3 \qquad \text{Simplify.}$$

The last equation, $x = 3$, has the single solution 3. All these equations are equivalent, so 3 is the only solution of the original equation, $3x - 5 = 4$.

Check: It is a good practice to check the solution by substituting 3 for x in the original equation.

$$3x - 5 = 4$$
$$3(3) - 5 \stackrel{?}{=} 4$$
$$9 - 5 \stackrel{?}{=} 4$$
$$4 = 4$$

The solution checks. ◾ ■

*The Zero-Product Property says that if $ab = 0$ then $a = 0$ or $b = 0$ or both equal 0.

NOW WORK PROBLEM **15.**

In the next examples, we use the Zero-Product Property (procedure 5, listed in the box on p. 86).

| EXAMPLE 2 | **Solving Equations by Factoring** |

Solve the equations: (a) $x^2 = 4x$ (b) $x^3 - x^2 - 4x + 4 = 0$

Solution   (a) We begin by collecting all terms on one side. This results in 0 on one side and an expression to be factored on the other.

$$x^2 = 4x$$

$$x^2 - 4x = 0$$

$$x(x - 4) = 0 \qquad \text{Factor.}$$

$$x = 0 \quad \text{or} \quad x - 4 = 0 \qquad \text{Apply the Zero-Product Property.}$$

$$x = 4$$

The solution set is $\{0, 4\}$.

Check: $x = 0$: $0^2 = 4 \cdot 0$ So 0 is a solution.

$x = 4$: $4^2 = 4 \cdot 4$ So 4 is a solution.   ■

(b) Do you recall the method of factoring by grouping? (If not, review p. 52.) We group the terms of $x^3 - x^2 - 4x + 4 = 0$ as follows:

$$\left(x^3 - x^2\right) - (4x - 4) = 0$$

Factor out x^2 from the first grouping and 4 from the second.

$$x^2(x - 1) - 4(x - 1) = 0$$

This reveals the common factor $(x - 1)$, so we have

$$\left(x^2 - 4\right)(x - 1) = 0$$

$$(x - 2)(x + 2)(x - 1) = 0 \qquad \text{Factor again.}$$

$$x - 2 = 0 \quad \text{or} \quad x + 2 = 0 \quad \text{or} \quad x - 1 = 0 \qquad \text{Set each factor equal to 0.}$$

$$x = 2 \qquad\qquad x = -2 \qquad\qquad x = 1 \qquad \text{Solve.}$$

The solution set is $\{-2, 1, 2\}$.

Check: $x = -2$: $(-2)^3 - (-2)^2 - 4(-2) + 4 = -8 - 4 + 8 + 4 = 0$ −2 is a solution.

$x = 1$: $1^3 - 1^2 - 4(1) + 4 = 1 - 1 - 4 + 4 = 0$ 1 is a solution.

$x = 2$: $2^3 - 2^2 - 4(2) + 4 = 8 - 4 - 8 + 4 = 0$ 2 is a solution.   ■ ■

STEPS FOR SOLVING EQUATIONS

STEP 1: List any restrictions on the domain of the variable.

STEP 2: Simplify the equation by replacing the original equation by a succession of equivalent equations following the procedures listed earlier.

STEP 3: If the result of Step 2 is a product of factors equal to 0, use the Zero-Product Property and set each factor equal to 0 (procedure 5).

STEP 4: Check your solution(s).

NOW WORK PROBLEMS 43 AND 73.

LINEAR EQUATIONS

② *Linear equations* are equations such as

$$3x + 12 = 0 \qquad -2x + 5 = 0 \qquad 4x - 3 = 0$$

A general definition is given next.

A **linear equation in one variable** is equivalent to an equation of the form

$$ax + b = 0$$

where a and b are real numbers and $a \neq 0$.

Sometimes, a linear equation is called a **first-degree equation,** because the left side is a polynomial in x of degree 1.

It is relatively easy to solve a linear equation. The idea is to *isolate* the variable:

$$ax + b = 0$$

$$ax = -b \qquad \text{Subtract } b \text{ from both sides.}$$

$$x = \frac{-b}{a} \qquad \text{Divide both sides by } a, a \neq 0.$$

The linear equation $ax + b = 0$ has the single solution given by the formula $x = -b/a$.

EXAMPLE 3   **Solving a Linear Equation**

Solve the equation: $\frac{1}{2}(x + 5) - 4 = \frac{1}{3}(2x - 1)$

Solution   To clear the equation of fractions, we multiply both sides by 6, the least common multiple of the denominators of the fractions $\frac{1}{2}$ and $\frac{1}{3}$.

$$\frac{1}{2}(x + 5) - 4 = \frac{1}{3}(2x - 1)$$

$$6\left[\frac{1}{2}(x + 5) - 4\right] = 6\left[\frac{1}{3}(2x - 1)\right] \qquad \text{Multiply both sides by 6, the LCM of 2 and 3.}$$

$$3(x + 5) - 24 = 2(2x - 1) \qquad \text{Use the Distributive Property on the left and the Associative Property on the right.}$$

$$3x + 15 - 24 = 4x - 2 \qquad \text{Use the Distributive Property.}$$

$$3x - 9 = 4x - 2 \qquad \text{Combine like terms.}$$

$$3x - 9 + 9 = 4x - 2 + 9 \qquad \text{Add 9 to each side.}$$

$$3x = 4x + 7 \qquad \text{Simplify.}$$

$$3x - 4x = 4x + 7 - 4x \qquad \text{Subtract } 4x \text{ from each side.}$$

$$-x = 7 \qquad \text{Simplify.}$$

$$x = -7 \qquad \text{Multiply both sides by } -1.$$

Check: $\dfrac{1}{2}(x + 5) - 4 = \dfrac{1}{2}(-7 + 5) - 4 = \dfrac{1}{2}(-2) - 4 = -1 - 4 = -5$

$\dfrac{1}{3}(2x - 1) = \dfrac{1}{3}[2(-7) - 1] = \dfrac{1}{3}(-14 - 1) = \dfrac{1}{3}(-15) = -5$ ▪

Since the two expressions are equal, the solution $x = -7$ checks and the solution set is $\{-7\}$. ▪

EXAMPLE 4

Solving a Linear Equation Using a Calculator

Solve the equation: $2.78x + \dfrac{2}{17.931} = 54.06$

Round the answer to two decimal places.

Solution   To avoid rounding errors, we solve for x before using the calculator.

$$2.78x + \frac{2}{17.931} = 54.06$$

$$2.78x = 54.06 - \frac{2}{17.931} \qquad \text{Subtract 2/17.931 from each side.}$$

$$x = \frac{54.06 - (2/17.931)}{2.78} \qquad \text{Divide each side by 2.78.}$$

Now use your calculator. The solution, rounded to two decimal places, is 19.41.

Check: We store the unrounded solution in memory and proceed to evaluate $2.78x + \dfrac{2}{17.931}$.

$$(2.78)(19.405921) + (2/17.931) = 54.06 \qquad \blacksquare \; \blacksquare$$

 NOW WORK PROBLEMS **25** AND **61**.

EQUATIONS THAT LEAD TO LINEAR EQUATIONS

③   The next three examples illustrate equations that do not appear to be linear, but lead to linear equations upon simplification.

| EXAMPLE 5 | **Solving Equations** |

Solve the equation: $(2y + 1)(y - 1) = (y + 5)(2y - 5)$

Solution
$$(2y + 1)(y - 1) = (y + 5)(2y - 5)$$

$$2y^2 - y - 1 = 2y^2 + 5y - 25 \qquad \text{Multiply and combine like terms.}$$

$$-y - 1 = 5y - 25 \qquad \text{Subtract } 2y^2 \text{ from each side.}$$

$$-y = 5y - 24 \qquad \text{Add 1 to each side.}$$

$$-6y = -24 \qquad \text{Subtract } 5y \text{ from each side.}$$

$$y = 4 \qquad \text{Divide both sides by } -6.$$

Check: $(2y + 1)(y - 1) = [2(4) + 1](4 - 1) = (8 + 1)(3) = (9)(3) = 27$
$(y + 5)(2y - 5) = (4 + 5)[2(4) - 5] = (9)(8 - 5) = (9)(3) = 27$

Since the two expressions are equal, the solution $y = 4$ checks.

The solution set is $\{1\}$. ◼ ◼

| EXAMPLE 6 | **Solving Equations** |

Solve the equation: $\dfrac{3}{x - 2} = \dfrac{1}{x - 1} + \dfrac{7}{(x - 1)(x - 2)}$

Solution First, we note that the domain of the variable is $\{x \mid x \neq 1, x \neq 2\}$. We clear the equation of fractions by multiplying both sides by the least common multiple of the denominators of the three fractions, $(x - 1)(x - 2)$.

$$\frac{3}{x - 2} = \frac{1}{x - 1} + \frac{7}{(x - 1)(x - 2)}$$

$$(x - 1)(x - 2) \frac{3}{x - 2} = (x - 1)(x - 2)\left[\frac{1}{x - 1} + \frac{7}{(x - 1)(x - 2)}\right] \qquad \begin{array}{l}\text{Multiply both sides by}\\ (x - 1)(x - 2). \text{ Cancel}\\ \text{on the left.}\end{array}$$

$$3x - 3 = (x - 1)(x - 2) \frac{1}{x - 1} + (x - 1)(x - 2) \frac{7}{(x - 1)(x - 2)} \qquad \begin{array}{l}\text{Use the Distributive}\\ \text{Property on each side;}\\ \text{cancel on the right.}\end{array}$$

$$3x - 3 = (x - 2) + 7$$

$$3x - 3 = x + 5 \qquad \text{Combine like terms.}$$

$$\qquad \qquad \begin{array}{l}\text{Add 3 to each side.}\\ \text{Subtract } x \text{ from each}\\ \text{side.}\end{array}$$

$$2x = 8$$

$$x = 4 \qquad \text{Divide by 2.}$$

Check: $\dfrac{3}{x - 2} = \dfrac{3}{4 - 2} = \dfrac{3}{2}$

$\dfrac{1}{x - 1} + \dfrac{7}{(x - 1)(x - 2)} = \dfrac{1}{4 - 1} + \dfrac{7}{(4 - 1)(4 - 2)} = \dfrac{1}{3} + \dfrac{7}{3 \cdot 2} = \dfrac{2}{6} + \dfrac{7}{6} = \dfrac{9}{6} = \dfrac{3}{2}$

Since the two expressions are equal, the solution $x = 4$ checks.

The solution set is $\{4\}$. ◼ ◼

✏ NOW WORK PROBLEM 55.

The next example is of an equation that has no solution.

EXAMPLE 7	An Equation with No Solution

Solve the equation: $\dfrac{3x}{x-1} + 2 = \dfrac{3}{x-1}$

Solution   First, we note that the domain of the variable is $\{x \mid x \neq 1\}$. Since the two quotients in the equation have the same denominator, $x - 1$, we can simplify by multiplying both sides by $x - 1$. The resulting equation is equivalent to the original equation, since we are multiplying by $x - 1$, which is not 0 (remember, $x \neq 1$).

$$\frac{3x}{x-1} + 2 = \frac{3}{x-1}$$

$$\left(\frac{3x}{x-1} + 2\right) \cdot (x-1) = \frac{3}{x-1} \cdot (x-1) \qquad \text{Multiply both sides by } x - 1; \text{ cancel on the right.}$$

$$\frac{3x}{x-1} \cdot (x-1) + 2 \cdot (x-1) = 3 \qquad \text{Use the Distributive Property on the left side; cancel on the left.}$$

$$3x + 2x - 2 = 3 \qquad \text{Simplify.}$$

$$5x - 2 = 3$$

$$5x = 5 \qquad \text{Add 2 to each side.}$$

$$x = 1 \qquad \text{Divide both sides by 5.}$$

The solution appears to be 1. But recall that $x = 1$ is not in the domain of the variable. The equation has no solution.   ■

NOW WORK PROBLEM **41**.

EXAMPLE 8	Converting to Fahrenheit from Celsius

In the United States we measure temperature in both degrees Fahrenheit (°F) and degrees Celsius (°C), which are related by the formula $C = \frac{5}{9}(F - 32)$. What are the Fahrenheit temperatures corresponding to Celsius temperatures of 0°, 10°, 20°, and 30°C?

Solution   We could solve four equations for F by replacing C each time by 0, 10, 20, and 30. Instead, it is much easier and faster first to solve the equation $C = \frac{5}{9}(F - 32)$ for F and then substitute in the values of C.

$$C = \frac{5}{9}(F - 32)$$

$$9C = 5(F - 32) \qquad \text{Multiply by 9.}$$

$$9C = 5F - 160 \qquad \text{Use the Distributive Property.}$$

$$5F - 160 = 9C \qquad \text{Interchange sides.}$$

$$5F = 9C + 160 \qquad \text{Add 160 to each side.}$$

$$F = \frac{9}{5}C + 32 \qquad \text{Divide both sides by 5.}$$

We can now do the required arithmetic.

$$0°C: \quad F = \frac{9}{5}(0) + 32 = 32°F$$

$$10°C: \quad F = \frac{9}{5}(10) + 32 = 50°F$$

$$20°C: \quad F = \frac{9}{5}(20) + 32 = 68°F$$

$$30°C: \quad F = \frac{9}{5}(30) + 32 = 86°F$$

SUMMARY

Steps for Solving a Linear Equation

To solve a linear equation, follow these steps:

STEP 1: If necessary, clear the equation of fractions by multiplying both sides by the least common multiple (LCM) of the denominators of all the fractions.

STEP 2: Remove all parentheses and simplify.

STEP 3: Collect all terms containing the variable on one side and all remaining terms on the other side.

STEP 4: Check your solution(s).

HISTORICAL FEATURE

The solution of equations is among the oldest of mathematical activities, and efforts to systematize this activity determined much of the shape of modern mathematics.

Consider the following problem and its solution using only words: Solve the problem of how many apples Jim has, given that

"Bob's five apples and Jim's apples together make twelve apples" by thinking,

"Jim's apples are all twelve apples less Bob's five apples" and then concluding,

"Jim has seven apples."

The mental steps translated into algebra are

$$5 + x = 12$$

$$x = 12 - 5$$

$$x = 7$$

The solution of this problem using only words is the earliest form of algebra. Such problems were solved exactly this way in Babylonia in 1800 BC. We know almost nothing of mathematical work before this date, although most authorities believe the sophistication of the earliest known texts indicates that a long period of previous development must have occurred. The method of writing out equations in words persisted for thousands of years, and although it now seems extremely cumbersome, it was used very effectively by many generations of mathematicians. The Arabs developed a good deal of the theory of cubic equations while writing out all the equations in words. About AD 1500, the tendency to abbreviate words in the written equations began to lead in the direction of modern notation; for example, the Latin word *et* (meaning *and*) developed into the plus sign, +. Although the occasional use of letters to represent variables dates back to AD 1200, the practice did not become common until about AD 1600. Development thereafter was rapid, and by 1635 algebraic notation did not differ essentially from what we use now.

1.1   EXERCISES

In Problems 1–8, mentally solve each equation.

1. $7x = 21$
2. $6x = -24$
3. $3x + 15 = 0$
4. $6x + 18 = 0$

5. $2x - 3 = 0$
6. $3x + 4 = 0$
7. $\frac{1}{3}x = \frac{5}{12}$
8. $\frac{2}{3}x = \frac{9}{2}$

In Problems 9–60, solve each equation.

9. $3x + 4 = x$
10. $2x + 9 = 5x$
11. $2t - 6 = 3 - t$

12. $5y + 6 = -18 - y$
13. $6 - x = 2x + 9$
14. $3 - 2x = 2 - x$

15. $3 + 2n = 4n + 7$
16. $6 - 2m = 3m + 1$
17. $2(3 + 2x) = 3(x - 4)$

18. $3(2 - x) = 2x - 1$
19. $8x - (3x + 2) = 3x - 10$
20. $7 - (2x - 1) = 10$

21. $\frac{3}{2}x + 2 = \frac{1}{2} - \frac{1}{2}x$
22. $\frac{1}{3}x = 2 - \frac{2}{3}x$
23. $\frac{1}{2}x - 5 = \frac{3}{4}x$

24. $1 - \frac{1}{2}x = 6$
25. $\frac{2}{3}p = \frac{1}{2}p + \frac{1}{3}$
26. $\frac{1}{2} - \frac{1}{3}p = \frac{4}{3}$

27. $0.9t = 0.4 + 0.1t$
28. $0.9t = 1 + t$
29. $\frac{x + 1}{3} + \frac{x + 2}{7} = 2$

30. $\frac{2x + 1}{3} + 16 = 3x$
31. $\frac{2}{y} + \frac{4}{y} = 3$
32. $\frac{4}{y} - 5 = \frac{5}{2y}$

33. $\frac{1}{2} + \frac{2}{x} = \frac{3}{4}$
34. $\frac{3}{x} - \frac{1}{3} = \frac{1}{6}$
35. $(x + 7)(x - 1) = (x + 1)^2$

36. $(x + 2)(x - 3) = (x + 3)^2$
37. $x(2x - 3) = (2x + 1)(x - 4)$
38. $x(1 + 2x) = (2x - 1)(x - 2)$

39. $z(z^2 + 1) = 3 + z^3$
40. $w(4 - w^2) = 8 - w^3$
41. $\frac{x}{x - 2} + 3 = \frac{2}{x - 2}$

42. $\frac{2x}{x + 3} = \frac{-6}{x + 3} - 2$
43. $x^2 = 9x$
44. $4x^3 = x^2$

45. $t^3 - 9t^2 = 0$
46. $4z^3 - 8z^2 = 0$
47. $\frac{2x}{x^2 - 4} = \frac{4}{x^2 - 4} - \frac{3}{x + 2}$

48. $\frac{x}{x^2 - 9} + \frac{4}{x + 3} = \frac{3}{x^2 - 9}$
49. $\frac{x}{x + 2} = \frac{3}{2}$
50. $\frac{3x}{x - 1} = 2$

51. $\frac{5}{2x - 3} = \frac{3}{x + 5}$
52. $\frac{-4}{x + 4} = \frac{-3}{x + 6}$
53. $\frac{6t + 7}{4t - 1} = \frac{3t + 8}{2t - 4}$

54. $\frac{8w + 5}{10w - 7} = \frac{4w - 3}{5w + 7}$
55. $\frac{4}{x - 2} = \frac{-3}{x + 5} + \frac{7}{(x + 5)(x - 2)}$

56. $\frac{-4}{2x + 3} + \frac{1}{x - 1} = \frac{1}{(2x + 3)(x - 1)}$
57. $\frac{2}{y + 3} + \frac{3}{y - 4} = \frac{5}{y + 6}$

58. $\frac{5}{5z - 11} + \frac{4}{2z - 3} = \frac{-3}{5 - z}$
59. $\frac{x}{x^2 - 1} - \frac{x + 3}{x^2 - x} = \frac{-3}{x^2 + x}$
60. $\frac{x + 1}{x^2 + 2x} - \frac{x + 4}{x^2 + x} = \frac{-3}{x^2 + 3x + 2}$

In Problems 61–64, use a calculator to solve each equation. Round the solution to two decimal places.

61. $3.2x + \frac{21.3}{65.871} = 19.23$
62. $6.2x - \frac{19.1}{83.72} = 0.195$

63. $14.72 - 21.58x = \frac{18}{2.11}x + 2.4$
64. $18.63x - \frac{21.2}{2.6} = \frac{14}{2.32}x - 20$

In Problems 65–76, find the real solutions of each equation by factoring.

65. $x^2 - 7x + 12 = 0$
66. $x^2 - x - 6 = 0$
67. $2x^2 + 5x - 3 = 0$

68. $3x^2 + 5x - 2 = 0$
69. $x^3 = 9x$
70. $x^4 = x^2$

71. $x^3 + x^2 - 20x = 0$
72. $x^3 + 6x^2 - 7x = 0$
73. $x^3 + x^2 - x - 1 = 0$

74. $x^3 + 4x^2 - x - 4 = 0$
75. $x^3 - 3x^2 - 4x + 12 = 0$
76. $x^3 - 3x^2 - x + 3 = 0$

In Problems 77–82, solve each equation. The letters a, b, and c are constants.

77. $ax - b = c, \quad a \neq 0$

78. $1 - ax = b, \quad a \neq 0$

79. $\dfrac{x}{a} + \dfrac{x}{b} = c, \quad a \neq 0, b \neq 0, a \neq -b$

80. $\dfrac{a}{x} + \dfrac{b}{x} = c, \quad c \neq 0$

81. $\dfrac{1}{x - a} + \dfrac{1}{x + a} = \dfrac{2}{x - 1}$

82. $\dfrac{b + c}{x + a} = \dfrac{b - c}{x - a}, \quad c \neq 0, a \neq 0$

83. Find the number a for which $x = 4$ is a solution of the equation

$$x + 2a = 16 + ax - 6a$$

84. Find the number b for which $x = 2$ is a solution of the equation

$$x + 2b = x - 4 + 2bx$$

Problems 85–90 list some formulas that occur in applications. Solve each formula for the indicated variable.

85. Electricity $\dfrac{1}{R} = \dfrac{1}{R_1} + \dfrac{1}{R_2}$ for R

86. Finance $A = P(1 + rt)$ for r

87. Mechanics $F = \dfrac{mv^2}{R}$ for R

88. Chemistry $PV = nRT$ for T

89. Mathematics $S = \dfrac{a}{1 - r}$ for r

90. Mechanics $v = -gt + v_0$ for t

91. One of the steps in the following list contains an error. Identify it and explain what is wrong.

$$x = 2 \tag{1}$$
$$3x - 2x = 2 \tag{2}$$
$$3x = 2x + 2 \tag{3}$$
$$x^2 + 3x = x^2 + 2x + 2 \tag{4}$$
$$x^2 + 3x - 10 = x^2 + 2x - 8 \tag{5}$$
$$(x - 2)(x + 5) = (x - 2)(x + 4) \tag{6}$$
$$x + 5 = x + 4 \tag{7}$$
$$1 = 0 \tag{8}$$

 92. Which of the following pairs of equations are equivalent? Explain.

(a) $x^2 = 9; \quad x = 3$

(b) $x = \sqrt{9}; \quad x = 3$

(c) $(x - 1)(x - 2) = (x - 1)^2; \quad x - 2 = x - 1$

93. The equation

$$\frac{5}{x + 3} + 3 = \frac{8 + x}{x + 3}$$

has no solution, yet when we go through the process of solving it we obtain $x = -3$. Write a brief paragraph to explain what causes this to happen.

94. Make up an equation that has no solution and give it to a fellow student to solve. Ask the fellow student to write a critique of your equation.

1.2 SETTING UP EQUATIONS: APPLICATIONS

OBJECTIVES
1. Translate Verbal Descriptions into Mathematical Expressions
2. Set Up Applied Problems
3. Solve Interest Problems
4. Solve Mixture Problems
5. Solve Uniform Motion Problems
6. Solve Constant Rate Job Problems

Applied (word) problems do not come in the form "Solve the equation ...." Instead, they supply information using words, a verbal description of the real problem. So, to solve applied problems, we must be able to translate the verbal description into the language of mathematics. We do this by using variables to represent unknown quantities and then finding relationships (such as equations) that involve these variables. The process of doing all this is called **mathematical modeling.**

Any solution to the mathematical problem must be checked against the mathematical problem, the verbal description, and the real problem. See Figure 1 for an illustration of the **modeling process.**

Figure 1

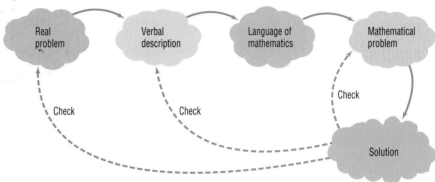

①   Let's look at a few examples that will help you to translate certain words into mathematical symbols.

| **EXAMPLE 1** | **Translating Verbal Descriptions into Mathematical Expressions** |

(a) The area of a rectangle is the product of its length times its width.

Translation: If A is used to represent the area, l the length, and w the width, then $A = lw$.

(b) For uniform motion, the velocity of an object equals the distance traveled divided by the time required.

Translation: If v is the velocity, s the distance, and t the time, then $v = s/t$.

(c) A total of $5000 is invested, some in stocks and some in bonds. If the amount invested in stocks is x, express the amount invested in bonds in terms of x.

Translation: If x is the amount invested in stocks, then the amount invested in bonds is $5000 - x$, since their sum is $x + (5000 - x) = 5000$.

(d) Let x denote a number.

The number 5 times as large as x is $5x$.

The number 3 less than x is $x - 3$.

The number that exceeds x by 4 is $x + 4$.

The number that, when added to x, gives 5 is $5 - x$.   ■

     NOW WORK PROBLEM **1.**

②   Always check the units used to measure the variables of an applied problem. In Example 1(a), if l is measured in feet, then w also must be expressed in feet, and A will be expressed in square feet. In Example 1(b), if v is measured in miles per hour, then the distance s must be expressed in miles and the time t must be expressed in hours. It is a good practice to check units to be sure that they are consistent and make sense.

Although each situation has unique features, we can provide an outline of the steps to follow in setting up applied problems.

STEPS FOR SETTING UP APPLIED PROBLEMS

STEP 1: Read the problem carefully, perhaps two or three times. Pay particular attention to the question being asked in order to identify what you are looking for. If you can, determine realistic possibilities for the answer.

STEP 2: Assign a letter (variable) to represent what you are looking for, and, if necessary, express any remaining unknown quantities in terms of this variable.

STEP 3: Make a list of all the known facts, and translate them into mathematical expressions. These may take the form of an equation (or, later, an inequality) involving the variable. If possible, draw an appropriately labeled diagram to assist you. Sometimes a table or chart helps.

STEP 4: Solve the equation for the variable, and then answer the question, usually using a complete sentence.

STEP 5: Check the answer with the facts in the problem. If it agrees, congratulations! If it does not agree, try again.

Let's look at an example.

EXAMPLE 2   Investments

A total of $18,000 is invested, some in stocks and some in bonds. If the amount invested in bonds is half that invested in stocks, how much is invested in each category?

Solution   STEP 1: We are being asked to find the amount of two investments. These amounts must total $18,000. (Do you see why?)

STEP 2: If we let x equal the amount invested in stocks, then $18,000 - x$ is the amount invested in bonds. [Look back at Example 1(c) to see why.]

STEP 3: We set up a table:

Amount in Stocks	Amount in Bonds	Reason
x	$18,000 - x$	Total invested is $18,000

We also know that

Total amount invested in bonds   is   one-half that in stocks

$$18,000 - x \qquad = \qquad \frac{1}{2}(x)$$

STEP 4: $18,000 - x = \frac{1}{2}x$

$$18,000 = x + \frac{1}{2}x \quad \text{Add } x \text{ to both sides}$$

$$18,000 = \frac{3}{2}x \quad \text{Simplify}$$

$$\left(\tfrac{2}{3}\right)18,000 = \left(\tfrac{2}{3}\right)\left(\tfrac{3}{2}x\right) \quad \text{Multiply both sides by } \tfrac{2}{3}$$

$$12,000 = x$$

Thus, $12,000 is invested in stocks and $18,000 - $12,000 = $6000 is invested in bonds.

STEP 5: The total invested is $12,000 + $6000 = $18,000, and the amount in bonds, $6000, is half that in stocks, $12,000. ■

NOW WORK PROBLEM **11.**

| EXAMPLE 3 | **Determining an Hourly Wage** |

Shannon grossed $435 one week by working 52 hours. Her employer pays time-and-a-half for all hours worked in excess of 40 hours. With this information, can you determine Shannon's regular hourly wage?

Solution   STEP 1: We are looking for an hourly wage. Our answer will be in dollars per hour.

STEP 2: Let x represent the regular hourly wage; x is measured in dollars per hour.

STEP 3: We set up a table:

	Hours Worked	Hourly Wage	Salary
Regular	40	x	$40x$
Overtime	12	$1.5x$	$12(1.5x) = 18x$

The sum of regular salary plus overtime salary will equal $435. From the table, $40x + 18x = 435$.

STEP 4: $40x + 18x = 435$

$$58x = 435$$

$$x = 7.50$$

Shannon's regular hourly wage is $7.50 per hour.

STEP 5: Forty hours yields a salary of $40(7.50) = $300, and 12 hours of overtime yields a salary of $12(1.5)(7.50) = $135, for a total of $435. ■

 NOW WORK PROBLEM 1 **5.**

INTEREST

③ The next example involves **interest.** Interest is money paid for the use of money. The total amount borrowed (whether by an individual from a bank in the form of a loan or by a bank from an individual in the form of a savings account) is called the **principal.** The **rate of interest,** expressed as a percent, is the amount charged for the use of the principal for a given period of time, usually on a yearly (that is, per annum) basis.

> **Simple Interest Formula**
>
> If a principal of P dollars is borrowed for a period of t years at a per annum interest rate r, expressed as a decimal, the interest I charged is
>
> $$I = Prt \qquad\qquad \textbf{(1)}$$

Interest charged according to formula (1) is called **simple interest.**

EXAMPLE 4   **Finance: Computing Interest on a Loan**

Suppose that Juanita borrows $500 for 6 months at the simple interest rate of 9% per annum. What is the interest that Juanita will be charged on this loan? How much does Juanita owe after 6 months?

Solution   The rate of interest is given per annum, so the actual time that the money is borrowed must be expressed in years. The interest charged would be the principal, $500, times the rate of interest ($9\% = 0.09$) times the time in years, $\frac{1}{2}$:

$$\text{Interest charged} = I = Prt = (500)(0.09)\left(\tfrac{1}{2}\right) = \$22.50$$

After 6 months, Juanita will owe what she borrowed plus the interest: $500 + $22.50 = $522.50. ■

EXAMPLE 5   **Financial Planning**

Candy has $70,000 to invest and requires an overall rate of return of 9%. She can invest in a safe, government-insured certificate of deposit, but it only pays 8%. To obtain 9%, she agrees to invest some of her money in noninsured corporate bonds paying 12%. How much should be placed in each investment to achieve her goal?

Solution   STEP 1: The question is asking for two dollar amounts: the principal to invest in the corporate bonds and the principal to invest in the certificate of deposit.

STEP 2: We let x represent the amount (in dollars) to be invested in the bonds. Then $70,000 - x$ is the amount that will be invested in the certificate. (Do you see why?)

STEP 3: We set up a table:

	Principal ($)	Rate	Time (yr)	Interest ($)
Bonds	x	$12\% = 0.12$	1	$0.12x$
Certificate	$70{,}000 - x$	$8\% = 0.08$	1	$0.08(70{,}000 - x)$
Total	$70{,}000$	$9\% = 0.09$	1	$0.09(70{,}000) = 6300$

Since the total interest from the investments is equal to $0.09(70,000) = 6300$, we must have the equation

$$0.12x + 0.08(70,000 - x) = 6300$$

(Note that the units are consistent: the unit is dollars on each side.)

STEP 4: $0.12x + 5600 - 0.08x = 6300$

$$0.04x = 700$$

$$x = 17,500$$

Candy should place \$17,500 in the bonds and $70,000 - \$17,500 = \$52,500$ in the certificate.

STEP 5:   The interest on the bonds after 1 year is $0.12(\$17,500) = \$2100$; the interest on the certificate after 1 year is $0.08(\$52,500) = \$4200$. The total annual interest is \$6300, the required amount.   ■

   NOW WORK PROBLEM 27.

MIXTURE PROBLEMS

④   Oil refineries sometimes produce gasoline that is a blend of two or more types of fuel; bakeries occasionally blend two or more types of flour for their bread. These problems are referred to as **mixture problems** because they combine two or more quantities to form a mixture.

| EXAMPLE 6 | Blending Coffees |

The manager of a Starbucks store decides to experiment with a new blend of coffee. She will mix some B grade Colombian coffee that sells for \$5 per pound with some A grade Arabica coffee that sells for \$10 per pound to get 100 pounds of the new blend. The selling price of the new blend is to be \$7 per pound, and there is to be no difference in revenue from selling the new blend versus selling the other types. How many pounds of the B grade Colombian and A grade Arabica coffees are required?

Solution   Let x represent the number of pounds of the B grade Colombian coffee. Then $100 - x$ equals the number of pounds of the A grade Arabica coffee. See Figure 2.

Figure 2

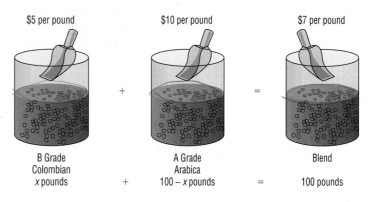

Since there is to be no difference in revenue between selling the A and B grades separately versus the blend, we have

$$\left\{\begin{array}{c}\text{Price per pound}\\ \text{of B grade}\end{array}\right\}\left\{\begin{array}{c}\text{\# Pounds}\\ \text{B grade}\end{array}\right\} + \left\{\begin{array}{c}\text{Price per pound}\\ \text{of A grade}\end{array}\right\}\left\{\begin{array}{c}\text{\# Pounds}\\ \text{A grade}\end{array}\right\} = \left\{\begin{array}{c}\text{Price per pound}\\ \text{of blend}\end{array}\right\}\left\{\begin{array}{c}\text{\# Pounds}\\ \text{blend}\end{array}\right\}$$

$$\$5 \quad \cdot \quad x \quad + \quad \$10 \quad \cdot \ (100 - x) = \quad \$7 \quad \cdot \quad 100$$

We have the equation

$$5x + 10(100 - x) = 700$$

$$5x + 1000 - 10x = 700$$

$$-5x = -300$$

$$x = 60$$

The manager should blend 60 pounds of B grade Colombian coffee with $100 - 60 = 40$ pounds of A grade Arabica coffee to get the desired blend.

Check: The 60 pounds of B grade coffee would sell for $(\$5)(60) = \$300$, and the 40 pounds of A grade coffee would sell for $(\$10)(40) = \$400$; the total revenue, $700, equals the revenue obtained from selling the blend, as desired. ■ ■

➤ NOW WORK PROBLEM 31.

UNIFORM MOTION

⑤ Objects that move at a constant velocity are said to be in **uniform motion.** When the average velocity of an object is known, it can be interpreted as its constant velocity. For example, a bicyclist traveling at an average velocity of 25 miles per hour is in uniform motion.

> **Uniform Motion Formula**
>
> If an object moves at an average velocity v, the distance s covered in time t is given by the formula
>
> $$s = vt \tag{2}$$

That is, Distance = Velocity · Time.

EXAMPLE 7   **Physics: Uniform Motion**

Tanya, who is a long-distance runner, runs at an average velocity of 8 miles per hour (mi/hr). Two hours after Tanya leaves your house, you leave in your Honda and follow the same route. If your average velocity is 40 mi/hr, how long will it be before you catch up to Tanya? How far will each of you be from your home?

Solution   Refer to Figure 3. We use t to represent the time (in hours) that it takes the Honda to catch up to Tanya. When this occurs, the total time elapsed for Tanya is $t + 2$ hours.

Figure 3

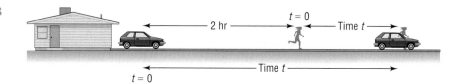

Set up the following table:

	Velocity mi/hr	Time hr	Distance mi
Tanya	8	$t + 2$	$8(t + 2)$
Honda	40	t	$40t$

Since the distance traveled is the same, we are led to the following equation:

$$8(t + 2) = 40t$$
$$8t + 16 = 40t$$
$$32t = 16$$
$$t = \frac{1}{2}\text{ hour}$$

It will take the Honda $\frac{1}{2}$ hour to catch up to Tanya. Each of you will have gone 20 miles.

Check: In 2.5 hours, Tanya travels a distance of $(2.5)(8) = 20$ miles. In $\frac{1}{2}$ hour, the Honda travels a distance of $\left(\frac{1}{2}\right)(40) = 20$ miles. ■ ■

 NOW WORK PROBLEM 35.

CONSTANT RATE JOBS

⑥ This section involves jobs that are performed at a **constant rate.** Our assumption is that, if a job can be done in t units of time, $1/t$ of the job is done in 1 unit of time. Let's look at an example.

EXAMPLE 8

Working Together to Do a Job

At 10 AM Danny is asked by his father to weed the garden. From past experience, Danny knows that this will take him 4 hours, working alone. His older brother, Mike, when it is his turn to do this job, requires 6 hours. Since Mike wants to go golfing with Danny and has a reservation for 1 PM, he agrees to help Danny. Assuming no gain or loss of efficiency, when will they finish if they work together? Can they make the golf date?

Solution

We set up Table 1. In 1 hour, Danny does $\frac{1}{4}$ of the job, and in 1 hour, Mike does $\frac{1}{6}$ of the job. Let t be the time (in hours) that it takes them to do the job together. In 1 hour, then, $1/t$ of the job is completed. We reason as follows:

$$\left(\begin{array}{c}\text{Part done by Danny}\\ \text{in 1 hour}\end{array}\right) + \left(\begin{array}{c}\text{Part done by Mike}\\ \text{in 1 hour}\end{array}\right) = \left(\begin{array}{c}\text{Part done together}\\ \text{in 1 hour}\end{array}\right)$$

TABLE 1

	Hours to Do Job	Part of Job Done in 1 Hour
Danny	4	$\frac{1}{4}$
Mike	6	$\frac{1}{6}$
Together	t	$\frac{1}{t}$

From Table 1,

$$\frac{1}{4} + \frac{1}{6} = \frac{1}{t}$$

$$\frac{3 + 2}{12} = \frac{1}{t}$$

$$\frac{5}{12} = \frac{1}{t}$$

$$5t = 12$$

$$t = \frac{12}{5}$$

Working together, the job can be done in $\frac{12}{5}$ hours, or 2 hours, 24 minutes. They should make the golf date, since they will finish at 12:24 PM. ▬

NOW WORK PROBLEM 39.

1.2 EXERCISES

In Problems 1–10, translate each sentence into a mathematical equation. Be sure to identify the meaning of all symbols.

1. **Geometry**   The area of a circle is the product of the number π times the square of the radius.

2. **Geometry**   The circumference of a circle is the product of the number π times twice the radius.

3. **Geometry**   The area of a square is the square of the length of a side.

4. **Geometry**   The perimeter of a square is four times the length of a side.

5. **Physics**   Force equals the product of mass times acceleration.

6. **Physics**   Pressure is force per unit area.

7. **Physics**   Work equals force times distance.

8. **Physics**   Kinetic energy is one-half the product of the mass times the square of the velocity.

9. **Business**   The total variable cost of manufacturing x dishwashers is $150 per dishwasher times the number of dishwashers manufactured.

10. **Business**   The total revenue derived from selling x dishwashers is $250 per dishwasher times the number of dishwashers sold.

11. **Finance**   A total of $20,000 is to be invested, some in bonds and some in certificates of deposit (CDs). If the amount invested in bonds is to exceed that in CDs by $3000, how much will be invested in each type of investment?

12. **Finance**   A total of $10,000 is to be divided between Sean and George, with George to receive $3000 less than Sean. How much will each receive?

13. **Finance**   An inheritance of $900,000 is to be divided among Scott, Alice, and Tricia in the following manner:

Alice is to receive $\frac{3}{4}$ of what Scott gets, while Tricia gets $\frac{1}{2}$ of what Scott gets. How much does each receive?

14. **Sharing the Cost of a Pizza**   Canter and Carole agree to share the cost of an $18 pizza based on how much each ate. If Carole ate $\frac{2}{3}$ the amount that Canter ate, how much should each pay?
   [**Hint:** Some pizza may be left.]

Carole's portion

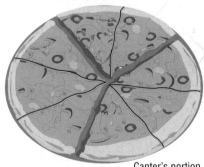

Canter's portion

15. **Computing Hourly Wages**   Sandra, who is paid time-and-a-half for hours worked in excess of 40 hours, had gross weekly wages of $442 for 48 hours worked. What is her regular hourly rate?

16. **Computing Hourly Wages**   Leigh is paid time-and-a-half for hours worked in excess of 40 hours and double-time for hours worked on Sunday. If Leigh had gross weekly wages of $342 for working 50 hours, 4 of which were on Sunday, what is her regular hourly rate?

17. **Computing Grades**   Going into the final exam, which will count as two tests, Brooke has test scores of 80, 83, 71,

61, and 95. What score does Brooke need on the final in order to have an average score of 80?

18. Computing Grades   Going into the final exam, which will count as two-thirds of the final grade, Mike has test scores of 86, 80, 84, and 90. What score does Mike need on the final in order to earn a B, which requires an average score of 80? What does he need to earn an A, which requires an average of 90?

19. Business: Discount Pricing   A builder of tract homes reduced the price of a model by 15%. If the new price is $125,000, what was its original price? How much can be saved by purchasing the model?

20. Business: Discount Pricing   A car dealer, at a year-end clearance, reduces the list price of last year's models by 15%. If a certain four-door model has a discounted price of $8000, what was its list price? How much can be saved by purchasing last year's model?

21. Business: Marking up the Price of Books   A college book store marks up the price that it pays the publisher for a book by 35%. If the selling price of a book is $56.00, how much did the book store pay for this book?

22. Personal Finance: Cost of a Car   The suggested list price of a new car is $12,000. The dealer's cost is 85% of list. How much will you pay if the dealer is willing to accept $100 over cost for the car?

23. Business: Theater Attendance   The manager of the Coral Theater wants to know whether the majority of its patrons are adults or children. During a week in July, 5200 tickets were sold and the receipts totaled $20,335. The adult admission is $4.75, and the children's admission is $2.50. How many adult patrons were there?

24. Business: Discount Pricing   A wool suit, discounted by 30% for a clearance sale, has a price tag of $399. What was the suit's original price?

25. Geometry   The perimeter of a rectangle is 60 feet. Find its length and width if the length is 8 feet longer than the width.

26. Geometry   The perimeter of a rectangle is 42 meters. Find its length and width if the length is twice the width.

27. Financial Planning   Betsy, a recent retiree, requires $6000 per year in extra income. She has $50,000 to invest and can invest in B-rated bonds paying 15% per year or in a certificate of deposit (CD) paying 7% per year. How much money should be invested in each to realize exactly $6000 in interest per year?

28. Financial Planning   After 2 years, Betsy (see Problem 27) finds that she will now require $7000 per year. Assuming that the remaining information is the same, how should the money be reinvested?

29. Banking   A bank loaned out $12,000, part of it at the rate of 8% per year and the rest at the rate of 18% per year. If the interest received totaled $1000, how much was loaned at 8%?

30. Banking   Wendy, a loan officer at a bank, has $1,000,000 to lend and is required to obtain an average return of 18% per year. If she can lend at the rate of 19% or at the rate of 16%, how much can she lend at the 16% rate and still meet her requirement?

31. Blending Teas   The manager of a store that specializes in selling tea decides to experiment with a new blend. She will mix some Earl Grey tea that sells for $5 per pound with some Orange Pekoe tea that sells for $3 per pound to get 100 pounds of the new blend. The selling price of the new blend is to be $4.50 per pound, and there is to be no difference in revenue from selling the new blend versus selling the other types. How many pounds of the Earl Grey tea and Orange Pekoe tea are required?

32. Business: Blending Coffee   A coffee manufacturer wants to market a new blend of coffee that sells for $3.90 per pound by mixing two coffees that sell for $2.75 and $5 per pound, respectively. What amounts of each coffee should be blended to obtain the desired mixture?
[**Hint:** Assume that the total weight of the desired blend is 100 pounds.]

33. Business: Mixing Nuts   A nut store normally sells cashews for $4.00 per pound and peanuts for $1.50 per pound. But at the end of the month the peanuts had not sold well, so, in order to sell 60 pounds of peanuts, the manager decided to mix the 60 pounds of peanuts with some cashews and sell the mixture for $2.50 per pound. How many pounds of cashews should be mixed with the peanuts to ensure no change in the profit?

34. Business: Mixing Candy   A candy store sells boxes of candy containing caramels and cremes. Each box sells for $12.50 and holds 30 pieces of candy (all pieces are the same size). If the caramels cost $0.25 to produce and the cremes cost $0.45 to produce, how many of each should be in a box to make a profit of $3?

35. Physics: Uniform Motion   A motorboat can maintain a constant speed of 16 miles per hour relative to the water. The boat makes a trip upstream to a certain point in 20 minutes; the return trip takes 15 minutes. What is the speed of the current? (See the figure.)

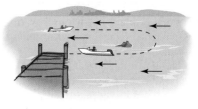

36. Physics: Uniform Motion   A motorboat heads upstream on a river that has a current of 3 miles per hour. The trip upstream takes 5 hours, and the return trip takes 2.5 hours. What is the speed of the motorboat? (Assume that the motorboat maintains a constant speed relative to the water.)

37. Physics: Uniform Motion   A Metra commuter train leaves Union Station in Chicago at 12 noon. Two hours later, an Amtrak train leaves on the same track, traveling at an average speed that is 50 miles per hour faster than the Metra train. At 3 PM the Amtrak train is 10 miles behind the commuter train. How fast is each going?

38. Physics: Uniform Motion Two cars enter the Florida Turnpike at Commercial Boulevard at 8:00 AM, each heading for Wildwood. One car's average speed is 10 miles per hour more than the other's. The faster car arrives at Wildwood at 11:00 AM, $\frac{1}{2}$ hour before the other car. What is the average speed of each car? How far did each travel?

39. Working Together on a Job Trent can deliver his newspapers in 30 minutes. It takes Lois 20 minutes to do the same route. How long would it take them to deliver the newspapers if they work together?

40. Working Together on a Job Patrice, by himself, can paint four rooms in 10 hours. If he hires April to help, they can do the same job together in 6 hours. If he lets April work alone, how long will it take her to paint four rooms?

41. Enclosing a Garden A gardener has 46 feet of fencing to be used to enclose a rectangular garden that has a border 2 feet wide surrounding it (see the figure).
(a) If the length of the garden is to be twice its width, what will be the dimensions of the garden?
(b) What is the area of the garden?
(c) If the length and width of the garden are to be the same, what would be the dimensions of the garden?
(d) What would be the area of the square garden?

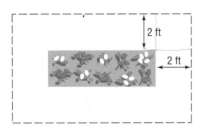

42. Construction A pond is enclosed by a wooden deck that is 3 feet wide. The fence surrounding the deck is 100 feet long.
(a) If the pond is square, what are its dimensions?
(b) If the pond is rectangular and the length of the pond is to be three times its width, what are its dimensions?
(c) If the pond is circular, what is its diameter?
(d) Which pond has the most area?

43. Football A tight end can run the 100 yard dash in 12 seconds. A defensive back can do it in 10 seconds. The tight end catches a pass at his own 20 yard line with the defensive back at the 15 yard line. (See the figure.) If no other players are nearby, at what yard line will the defensive back catch up to the tight end?
[**Hint:** At time $t = 0$, the defensive back is 5 yards behind the tight end.]

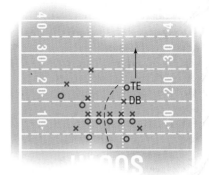

44. Computing Business Expense Therese, an outside salesperson, uses her car for both business and pleasure. Last year, she traveled 30,000 miles, using 900 gallons of gasoline. Her car gets 40 miles per gallon on the highway and 25 in the city. She can deduct all highway travel, but no city travel, on her taxes. How many miles should Therese be allowed as a business expense?

45. Mixing Water and Antifreeze How much water should be added to 1 gallon of pure antifreeze to obtain a solution that is 60% antifreeze?

46. Mixing Water and Antifreeze The cooling system of a certain foreign-made car has a capacity of 15 liters. If the system is filled with a mixture that is 40% antifreeze, how much of this mixture should be drained and replaced by pure antifreeze so that the system is filled with a solution that is 60% antifreeze?

47. Chemistry: Salt Solutions How much water must be evaporated from 32 ounces of a 4% salt solution to make a 6% salt solution?

48. Chemistry: Salt Solutions How much water must be evaporated from 240 gallons of a 3% salt solution to produce a 5% salt solution?

49. Purity of Gold The purity of gold is measured in karats, with pure gold being 24 karats. Other purities of gold are expressed as proportional parts of pure gold. Thus, 18 karat gold is $\frac{18}{24}$, or 75% pure gold; 12 karat gold is $\frac{12}{24}$, or 50% pure gold; and so on. How much 12 karat gold should be mixed with pure gold to obtain 60 grams of 16 karat gold?

50. Chemistry: Sugar Molecules A sugar molecule has twice as many atoms of hydrogen as it does oxygen and one more atom of carbon than oxygen. If a sugar molecule has a total of 45 atoms, how many are oxygen? How many are hydrogen?

51. Running a Race Mike can run the mile in 6 minutes, and Dan can run the mile in 9 minutes. If Mike gives Dan a head start of 1 minute, how far from the start will Mike pass Dan? (See the figure.) How long does it take?

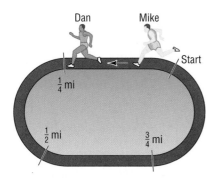

52. Range of an Airplane   An air rescue plane averages 300 miles per hour in still air. It carries enough fuel for 5 hours of flying time. If, upon takeoff, it encounters a head wind of 30 mi/hr, how far can it fly and return safely? (Assume that the wind remains constant.)

53. Emptying Oil Tankers   An oil tanker can be emptied by the main pump in 4 hours. An auxiliary pump can empty the tanker in 9 hours. If the main pump is started at 9 AM, when should the auxiliary pump be started so that the tanker is emptied by noon?

54. Cement Mix   A 20-pound bag of Economy brand cement mix contains 25% cement and 75% sand. How much pure cement must be added to produce a cement mix that is 40% cement?

55. Emptying a Tub   A bathroom tub will fill in 15 minutes with both faucets open and the stopper in place. With both faucets closed and the stopper removed, the tub will empty in 20 minutes. How long will it take for the tub to fill if both faucets are open and the stopper is removed?

56. Using Two Pumps   A 5 horsepower (hp) pump can empty a pool in 5 hours. A smaller, 2 hp pump empties the same pool in 8 hours. The pumps are used together to begin emptying this pool. After two hours, the 2 hp pump breaks down. How long will it take the larger pump to empty the pool?

57. Comparing Olympic Heroes   In the 1984 Olympics, Carl Lewis of the United States won the gold medal in the 100 meter race with a time of 9.99 seconds. In the 1896 Olympics, Thomas Burke, also of the United States, won the gold medal in the 100 meter race in 12.0 seconds. If they ran in the same race repeating their respective times, by how many meters would Lewis beat Burke?

58. Critical Thinking   You are the manager of a clothing store and have just purchased 100 dress shirts for $20.00 each. After 1 month of selling the shirts at the regular price, you plan to have a sale giving 40% off the original selling price. However, you still want to make a profit of $4 on each shirt at the sale price. What should you price the shirts at initially to ensure this? If, instead of 40% off at the sale, you give 50% off, by how much is your profit reduced?

59. Critical Thinking   Make up a word problem that requires solving a linear equation as part of its solution. Exchange problems with a friend. Write a critique of your friend's problem.

60. Critical Thinking   Without solving, explain what is wrong with the following mixture problem: How many liters of 25% ethanol should be added to 20 liters of 48% ethanol to obtain a solution of 58% ethanol? Now go through an algebraic solution. What happens?

PREPARING FOR THIS SECTION

Before getting started, review the following:

✓ Factoring Polynomials (Review, Section 6, pp. 47–54)   ✓ Square Roots (Review, Section 8, pp. 67–70)
✓ Zero-Product Property (p. 12)

1.3   QUADRATIC EQUATIONS

OBJECTIVES   1 Solve a Quadratic Equation by Factoring
2 Know How to Complete the Square
3 Solve a Quadratic Equation by Completing the Square
4 Solve a Quadratic Equation Using the Quadratic Formula
5 Solve Applied Problems

Quadratic equations are equations such as

$$2x^2 + x + 8 = 0$$
$$3x^2 - 5x + 6 = 0$$
$$x^2 - 9 = 0$$

A general definition is given next.

A **quadratic equation** is an equation equivalent to one of the form

$$ax^2 + bx + c = 0 \qquad \text{(1)}$$

where a, b, and c are real numbers and $a \neq 0$.

A quadratic equation written in the form $ax^2 + bx + c = 0$ is said to be in **standard form.**

Sometimes, a quadratic equation is called a **second-degree equation,** because the left side is a polynomial of degree 2. We shall discuss three ways of solving quadratic equations: by factoring, by completing the square, and by using the quadratic formula.

FACTORING

We have already used factoring to solve equations. In particular, when a quadratic equation is written in standard form $ax^2 + bx + c = 0$, it may be possible to factor the expression on the left side as the product of two first-degree polynomials. Then, by using the Zero-Product Property and setting each factor equal to 0, we can solve the resulting linear equations and obtain the solutions of the quadratic equation.

Let's look at an example.

EXAMPLE 1   **Solving a Quadratic Equation by Factoring**

Solve the equation: $x^2 - 5x + 6 = 0$

Solution   The equation is in the standard form specified in equation (1). The left side may be factored as

$$x^2 - 5x + 6 = 0$$
$$(x - 2)(x - 3) = 0$$

We set each factor equal to 0 and solve the resulting first-degree equations.

$$x - 2 = 0 \quad \text{or} \quad x - 3 = 0$$
$$x = 2 \quad \text{or} \quad x = 3$$

The solution set is $\{2, 3\}$.   ∎

EXAMPLE 2   **Solving a Quadratic Equation by Factoring**

Solve the equation: $2x^2 = x + 3$

Solution   We put the equation in standard form by adding $-x - 3$ to both sides.

$$2x^2 = x + 3 \qquad \text{Add } -x - 3 \text{ to both sides.}$$
$$2x^2 - x - 3 = 0$$

The left side may now be factored as

$$(2x - 3)(x + 1) = 0$$

so that

$$2x - 3 = 0 \quad \text{or} \quad x + 1 = 0$$
$$x = \tfrac{3}{2} \qquad\qquad x = -1$$

The solution set is $\left\{-1, \tfrac{3}{2}\right\}$.   ∎

When the left side factors into two linear equations with the same solution, the quadratic equation is said to have a **repeated solution.** We also call this solution a **root of multiplicity 2,** or a **double root.**

EXAMPLE 3	Solving a Quadratic Equation by Factoring

Solve the equation: $9x^2 - 6x + 1 = 0$

Solution   This equation is already in standard form, and the left side can be factored.

$$9x^2 - 6x + 1 = 0$$

$$(3x - 1)(3x - 1) = 0$$

so

$$x = \tfrac{1}{3} \quad \text{or} \quad x = \tfrac{1}{3}$$

This equation has only the repeated solution $\tfrac{1}{3}$.   ▬

NOW WORK PROBLEMS 3 AND 13.

THE SQUARE ROOT METHOD

Suppose that we wish to solve the quadratic equation

$$x^2 = p \tag{2}$$

where $p \geq 0$ is a nonnegative number. We proceed as in the earlier examples.

$$x^2 - p = 0 \qquad \text{Put in standard form.}$$

$$\left(x - \sqrt{p}\right)\left(x + \sqrt{p}\right) = 0 \qquad \text{Factor (over the real numbers).}$$

$$x = \sqrt{p} \quad \text{or} \quad x = -\sqrt{p} \quad \text{Solve.}$$

We have the following result:

If $x^2 = p$ and $p \geq 0$, then $x = \sqrt{p}$ or $x = -\sqrt{p}$.   **(3)**

When statement (3) is used, it is called the **Square Root Method.** In statement (3), note that if $p > 0$ the equation $x^2 = p$ has two solutions, $x = \sqrt{p}$ and $x = -\sqrt{p}$. We usually abbreviate these solutions as $x = \pm\sqrt{p}$, read as "x equals plus or minus the square root of p."

For example, the two solutions of the equation

$$x^2 = 4$$

are

$$x = \pm\sqrt{4} \quad \text{Use the Square Root Method}$$

and, since $\sqrt{4} = 2$, we have

$$x = \pm 2$$

The solution set is $\{-2, 2\}$.

| EXAMPLE 4 | **Solving a Quadratic Equation Using the Square Root Method** |

Solve each equation.

(a) $x^2 = 5$ (b) $(x - 2)^2 = 16$

Solution   (a) We use the Square Root Method to get

$$x^2 = 5$$

$$x = \pm\sqrt{5} \qquad\qquad \text{Use the Square Root Method.}$$

$$x = \sqrt{5} \quad \text{or} \quad x = -\sqrt{5}$$

The solution set is $\{-\sqrt{5}, \sqrt{5}\}$.

(b) We use the Square Root Method to get

$$(x - 2)^2 = 16$$

$$x - 2 = \pm\sqrt{16} \qquad\qquad \text{Use the Square Root Method.}$$

$$x - 2 = \pm 4$$

$$x - 2 = 4 \quad \text{or} \quad x - 2 = -4$$

$$x = 6 \quad \text{or} \qquad x = -2$$

The solution set is $\{-2, 6\}$.   ∎

NOW WORK PROBLEM **23**.

COMPLETING THE SQUARE

② We now introduce the method of **completing the square.** The idea behind this method is to *adjust* the left side of a quadratic equation, $ax^2 + bx + c = 0$, so that it becomes a perfect square, that is, the square of a first-degree polynomial. For example, $x^2 + 6x + 9$ and $x^2 - 4x + 4$ are perfect squares because

$$x^2 + 6x + 9 = (x + 3)^2 \quad \text{and} \quad x^2 - 4x + 4 = (x - 2)^2$$

How do we adjust the left side? We do it by adding the appropriate number to the left side to create a perfect square. For example, to make $x^2 + 6x$ a perfect square, we add 9.

Let's look at several examples of completing the square when the coefficient of x^2 is 1:

Start	Add	Result
$x^2 + 4x$	4	$x^2 + 4x + 4 = (x + 2)^2$
$x^2 + 12x$	36	$x^2 + 12x + 36 = (x + 6)^2$
$x^2 - 6x$	9	$x^2 - 6x + 9 = (x - 3)^2$
$x^2 + x$	$\frac{1}{4}$	$x^2 + x + \frac{1}{4} = \left(x + \frac{1}{2}\right)^2$

Do you see the pattern? Provided that the coefficient of x^2 is 1, we complete the square by adding the square of $\frac{1}{2}$ of the coefficient of x.

Start	Add	Result
$x^2 + mx$	$\left(\dfrac{m}{2}\right)^2$	$x^2 + mx + \left(\dfrac{m}{2}\right)^2 = \left(x + \dfrac{m}{2}\right)^2$

NOW WORK PROBLEM 27.

③   The next example illustrates how the procedure of completing the square can be used to solve a quadratic equation.

EXAMPLE 5

Solving a Quadratic Equation by Completing the Square

Solve by completing the square: $x^2 + 5x + 4 = 0$

Solution   We always begin this procedure by rearranging the equation so that the constant is on the right side.

$$x^2 + 5x + 4 = 0$$
$$x^2 + 5x = -4$$

Since the coefficient of x^2 is 1, we can complete the square on the left side by adding $\left(\frac{1}{2} \cdot 5\right)^2 = \frac{25}{4}$. Of course, in an equation, whatever we add to the left side also must be added to the right side. Thus, we add $\frac{25}{4}$ to *both* sides.

$$x^2 + 5x + \frac{25}{4} = -4 + \frac{25}{4} \qquad \text{Add } \tfrac{25}{4} \text{ to both sides.}$$

$$\left(x + \frac{5}{2}\right)^2 = \frac{9}{4} \qquad \text{Complete the square.}$$

$$x + \frac{5}{2} = \pm\sqrt{\frac{9}{4}} \qquad \text{Use the Square Root Method.}$$

$$x + \frac{5}{2} = \pm\frac{3}{2}$$

$$x = -\frac{5}{2} \pm \frac{3}{2}$$

$$x = -\frac{5}{2} + \frac{3}{2} = -1 \quad \text{or} \quad x = -\frac{5}{2} - \frac{3}{2} = -4$$

The solution set is $\{-4, -1\}$.   ∎

THE SOLUTION OF THE EQUATION IN EXAMPLE 5 ALSO CAN BE OBTAINED BY FACTORING. REWORK EXAMPLE 5 USING THIS TECHNIQUE.

The next example illustrates an equation that cannot be solved by factoring.

EXAMPLE 6

Solving a Quadratic Equation by Completing the Square

Solve by completing the square: $2x^2 - 8x - 5 = 0$

Solution   First, we rewrite the equation.

$$2x^2 - 8x - 5 = 0$$
$$2x^2 - 8x = 5$$

Next, we divide both sides by 2 so that the coefficient of x^2 is 1. (This enables us to complete the square at the next step.)

$$x^2 - 4x = \frac{5}{2}$$

Finally, we complete the square by adding 4 to both sides.

$$x^2 - 4x + 4 = \frac{5}{2} + 4$$

$$(x - 2)^2 = \frac{13}{2}$$

$$x - 2 = \pm\sqrt{\frac{13}{2}} \qquad \text{Use the Square Root Method.}$$

$$x - 2 = \pm\frac{\sqrt{26}}{2} \qquad \sqrt{\frac{13}{2}} = \frac{\sqrt{13}}{\sqrt{2}} = \frac{\sqrt{13}}{\sqrt{2}}\frac{\sqrt{2}}{\sqrt{2}} = \frac{\sqrt{26}}{2}$$

$$x = 2 \pm \frac{\sqrt{26}}{2}$$

The solution set is $\left\{ 2 - \dfrac{\sqrt{26}}{2}, 2 + \dfrac{\sqrt{26}}{2} \right\}$.

NOTE: If we wanted an approximation, say rounded to two decimal places, of these solutions, we would use a calculator to get $\{-0.55, 4.55\}$.

NOW WORK PROBLEM **33.**

THE QUADRATIC FORMULA

④   We can use the method of completing the square to obtain a general formula for solving the quadratic equation.

$$ax^2 + bx + c = 0 \qquad a \neq 0$$

NOTE: There is no loss in generality to assume that $a > 0$, since if $a < 0$ we can multiply by -1 to obtain an equivalent equation with a positive leading coefficient.

As in Examples 5 and 6, we rearrange the terms as

$$ax^2 + bx = -c \quad a > 0$$

Since $a > 0$, we can divide both sides by a to get

$$x^2 + \frac{b}{a}x = -\frac{c}{a}$$

Now the coefficient of x^2 is 1. To complete the square on the left side, add the square of $\frac{1}{2}$ of the coefficient of x; that is, add

$$\left(\frac{1}{2} \cdot \frac{b}{a}\right)^2 = \frac{b^2}{4a^2}$$

to both sides. Then

$$x^2 + \frac{b}{a}x + \frac{b^2}{4a^2} = \frac{b^2}{4a^2} - \frac{c}{a}$$

$$\left(x + \frac{b}{2a}\right)^2 = \frac{b^2 - 4ac}{4a^2} \qquad \frac{b^2}{4a^2} - \frac{c}{a} = \frac{b^2}{4a^2} - \frac{4ac}{4a^2} = \frac{b^2 - 4ac}{4a^2} \qquad \textbf{(4)}$$

Provided that $b^2 - 4ac \geq 0$, we now can use the Square Root Method to get

$$x + \frac{b}{2a} = \pm\sqrt{\frac{b^2 - 4ac}{4a^2}}$$

$$x + \frac{b}{2a} = \frac{\pm\sqrt{b^2 - 4ac}}{2a}$$
The square root of a quotient equals the quotient of the square roots. Also, $\sqrt{4a^2} = 2a$ since $a > 0$.

$$x = -\frac{b}{2a} \pm \frac{\sqrt{b^2 - 4ac}}{2a}$$
Add $-\dfrac{b}{2a}$ to both sides.

$$x = \frac{-b \pm \sqrt{b^2 - 4ac}}{2a}$$
Combine the quotients on the right.

What if $b^2 - 4ac$ is negative? Then equation (4) states that the left expression (a real number squared) equals the right expression (a negative number). Since this occurrence is impossible for real numbers, we conclude that if $b^2 - 4ac < 0$ the quadratic equation has no *real* solution. (We discuss quadratic equations for which the quantity $b^2 - 4ac < 0$ in detail in Section 5.3.)*

We now state the *quadratic formula*.

Theorem

Consider the quadratic equation

$$ax^2 + bx + c = 0 \qquad a \neq 0$$

If $b^2 - 4ac < 0$, this equation has no real solution.

If $b^2 - 4ac \geq 0$, the real solution(s) of this equation is (are) given by the **quadratic formula.**

Quadratic Formula

$$x = \frac{-b \pm \sqrt{b^2 - 4ac}}{2a} \tag{5}$$

The quantity $b^2 - 4ac$ is called the **discriminant** of the quadratic equation, because its value tells us whether the equation has real solutions. In fact, it also tells us how many solutions to expect.

DISCRIMINANT OF A QUADRATIC EQUATION

For a quadratic equation $ax^2 + bx + c = 0$:

1. If $b^2 - 4ac > 0$, there are two unequal real solutions.
2. If $b^2 - 4ac = 0$, there is a repeated solution, a root of multiplicity 2.
3. If $b^2 - 4ac < 0$, there is no real solution.

---

*Section 5.3 may be covered anytime after completing this section without any loss of continuity.

When asked to find the real solutions, if any, of a quadratic equation, always evaluate the discriminant first to see how many real solutions there are.

EXAMPLE 7 | **Solving a Quadratic Equation Using the Quadratic Formula**

Use the quadratic formula to find the real solutions, if any, of the equation

$$3x^2 - 5x + 1 = 0$$

Solution The equation is in standard form, so we compare it to $ax^2 + bx + c = 0$ to find a, b, and c.

$$3x^2 - 5x + 1 = 0$$

$$ax^2 + bx + c = 0 \quad a = 3, b = -5, c = 1$$

With $a = 3$, $b = -5$, and $c = 1$, we evaluate the discriminant $b^2 - 4ac$.

$$b^2 - 4ac = (-5)^2 - 4(3)(1) = 25 - 12 = 13$$

Since $b^2 - 4ac > 0$, there are two real solutions, which can be found using the quadratic formula.

$$x = \frac{-b \pm \sqrt{b^2 - 4ac}}{2a} = \frac{-(-5) \pm \sqrt{13}}{2(3)} = \frac{5 \pm \sqrt{13}}{6}$$

The solution set is $\left\{ \dfrac{5 - \sqrt{13}}{6}, \dfrac{5 + \sqrt{13}}{6} \right\}$. ∎

EXAMPLE 8 | **Solving a Quadratic Equation Using the Quadratic Formula**

Use the quadratic formula to find the real solutions, if any, of the equation

$$\tfrac{25}{2}x^2 - 30x + 18 = 0$$

Solution The equation is given in standard form. However, to simplify the arithmetic, we clear the fractions.

$$\tfrac{25}{2}x^2 - 30x + 18 = 0$$

$$25x^2 - 60x + 36 = 0 \quad \text{Clear fractions; multiply by 2.}$$

$$ax^2 + bx + c = 0 \quad \text{Compare to standard form.}$$

With $a = 25$, $b = -60$, and $c = 36$, we evaluate the discriminant.

$$b^2 - 4ac = (-60)^2 - 4(25)(36) = 3600 - 3600 = 0$$

The equation has a repeated solution, which we find by using the quadratic formula.

$$x = \frac{-b \pm \sqrt{b^2 - 4ac}}{2a} = \frac{60 \pm \sqrt{0}}{50} = \frac{60}{50} = \frac{6}{5}$$

The solution set is $\left\{ \tfrac{6}{5} \right\}$. ∎

EXAMPLE 9

Solving a Quadratic Equation Using the Quadratic Formula

Use the quadratic formula to find the real solutions, if any, of the equation

$$3x^2 + 2 = 4x$$

Solution   The equation, as given, is not in standard form.

$$3x^2 + 2 = 4x$$

$$3x^2 - 4x + 2 = 0 \qquad \text{Put in standard form.}$$

$$ax^2 + bx + c = 0 \qquad \text{Compare to standard form.}$$

With $a = 3$, $b = -4$, and $c = 2$, we find

$$b^2 - 4ac = (-4)^2 - 4(3)(2) = 16 - 24 = -8$$

Since $b^2 - 4ac < 0$, the equation has no real solution.   ■

⬛▬▬▶ **NOW WORK PROBLEMS 43 AND 53.**

Sometimes a given equation can be transformed into a quadratic equation so that it can be solved using the quadratic formula.

EXAMPLE 10

Solving a Quadratic Equation Using the Quadratic Formula

Find the real solutions, if any, of the equation: $9 + \dfrac{3}{x} - \dfrac{2}{x^2} = 0$, $x \neq 0$

Solution   In its present form, the equation

$$9 + \frac{3}{x} - \frac{2}{x^2} = 0$$

is not a quadratic equation. However, it can be transformed into one by multiplying each side by x^2. The result is

$$9x^2 + 3x - 2 = 0$$

Although we multiplied each side by x^2, we know that $x^2 \neq 0$ (do you see why?), so this quadratic equation is equivalent to the original equation.
Using $a = 9$, $b = 3$, and $c = -2$, the discriminant is

$$b^2 - 4ac = 3^2 - 4(9)(-2) = 9 + 72 = 81$$

Since $b^2 - 4ac > 0$, the new equation has two real solutions.

$$x = \frac{-b \pm \sqrt{b^2 - 4ac}}{2a} = \frac{-3 \pm \sqrt{81}}{2(9)} = \frac{-3 \pm 9}{18}$$

$$x = \frac{-3 + 9}{18} = \frac{6}{18} = \frac{1}{3} \quad \text{or} \quad x = \frac{-3 - 9}{18} = \frac{-12}{18} = -\frac{2}{3}$$

The solution set is $\left\{ -\frac{2}{3}, \frac{1}{3} \right\}$.   ■

SUMMARY

Procedure for Solving a Quadratic Equation

To solve a quadratic equation, first put it in standard form:

$$ax^2 + bx + c = 0$$

Then:

STEP 1: Identify a, b, and c.

STEP 2: Evaluate the discriminant, $b^2 - 4ac$.

STEP 3: (a) If the discriminant is negative, the equation has no real solution.
(b) If the discriminant is zero, the equation has one real solution, a repeated root.
(c) If the discriminant is positive, the equation has two distinct real solutions. If you can easily spot factors, use the factoring method to solve the equation. Otherwise, use the quadratic formula or the method of completing the square.

APPLICATIONS

⑤ Many applied problems require the solution of a quadratic equation. Let's look at one that you will probably see again in a slightly different form if you study calculus.

EXAMPLE 11 **Constructing a Box**

From each corner of a square piece of sheet metal, remove a square of side 9 centimeters. Turn up the edges to form an open box. If the box is to hold 144 cubic centimeters (cm^3), what should be the dimensions of the piece of sheet metal?

Solution We use Figure 4 as a guide. We have labeled by x the length of a side of the square piece of sheet metal. The box will be of height 9 centimeters, and its square base will have $x - 18$ as the length of a side. The volume (Length $\times$ Width $\times$ Height) of the box is therefore

$$9(x - 18)(x - 18) = 9(x - 18)^2$$

Figure 4

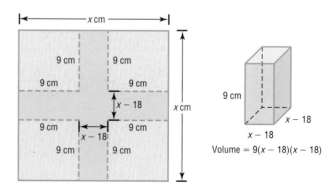

Since the volume of the box is to be 144 cm³, we have

$$9(x - 18)^2 = 144$$

$$(x - 18)^2 = 16 \qquad \text{Divide each side by 9}$$

$$x - 18 = \pm 4 \qquad \text{Use the Square Root Method.}$$

$$x = 18 \pm 4$$

$$x = 22 \quad \text{or} \quad x = 14$$

We discard the solution $x = 14$ (do you see why?) and conclude that the sheet metal should be 22 centimeters by 22 centimeters.

Check: If we begin with a piece of sheet metal 22 centimeters by 22 centimeters, cut out a 9 centimeter square from each corner, and fold up the edges, we get a box whose dimensions are 9 by 4 by 4, with volume $9 \times 4 \times 4 = 144$ cm³, as required.   ■ ■

NOW WORK PROBLEM **89**.

| EXAMPLE 12 | **Physics: Uniform Motion** |

A motorboat heads upstream a distance of 24 miles on a river whose current is running at 3 miles per hour (mi/hr). The trip up and back takes 6 hours. Assuming that the motorboat maintained a constant speed relative to the water, what was its speed?

Solution See Figure 5. We use v to represent the constant speed of the motorboat relative to the water. Then the true speed going upstream is $v - 3$ mi/hr, and the true speed going downstream is $v + 3$ mi/hr. Since Distance = Velocity × Time, then Time = Distance/Velocity. We set up a table.

Figure 5

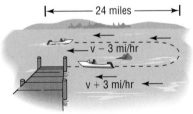

	Velocity mi/hr	Distance mi	Time = Distance/Velocity hr
Upstream	$v - 3$	24	$\dfrac{24}{v - 3}$
Downstream	$v + 3$	24	$\dfrac{24}{v + 3}$

Since the total time up and back is 6 hours, we have

$$\frac{24}{v - 3} + \frac{24}{v + 3} = 6$$

$$\frac{24(v + 3) + 24(v - 3)}{(v - 3)(v + 3)} = 6 \qquad \text{Add the quotients on the left.}$$

$$\frac{48v}{v^2 - 9} = 6 \qquad \text{Simplify.}$$

$$48v = 6(v^2 - 9)$$

$$6v^2 - 48v - 54 = 0 \qquad \text{Place in standard form.}$$

$$v^2 - 8v - 9 = 0 \qquad \text{Divide by 6.}$$

$$(v - 9)(v + 1) = 0 \qquad \text{Factor.}$$

$$v = 9 \quad \text{or} \quad v = -1 \qquad \text{Apply the Zero-Product Property and solve.}$$

We discard the solution $v = -1$ mi/hr, so the speed of the motorboat relative to the water is 9 mi/hr. ∎

HISTORICAL FEATURE

Problems using quadratic equations are found in the oldest known mathematical literature. Babylonians and Egyptians were solving such problems before 1800 BC. Euclid solved quadratic equations geometrically in his *Data* (300 BC), and the Hindus and Arabs gave rules for solving any quadratic equation with real roots. Because negative numbers were not freely used before AD 1500, there were several different types of quadratic equations, each with its own rule. Thomas Har-

riot (1560–1621) introduced the method of factoring to obtain solutions, and François Viète (1540–1603) introduced a method that is essentially completing the square.

Until modern times it was usual to neglect the negative roots (if there were any), and equations involving square roots of negative quantities were regarded as unsolvable until the 1500s.

HISTORICAL PROBLEMS

1. *One of al-Khowârizmî's solutions* We solve $x^2 + 12x = 85$ by drawing the square shown. The area of the four white rectangles and the yellow square is $x^2 + 12x$. We then set this expression equal to 85 to get the equation $x^2 + 12x = 85$. If we add the four blue squares, we will have a larger square of known area. Complete the solution.

2. *Viète's method* We solve $x^2 + 12x - 85 = 0$ by letting $x = u + z$. Then

$$(u + z)^2 + 12(u + z) - 85 = 0$$

$$u^2 + (2z + 12)u + (z^2 + 12z - 85) = 0$$

Now select z so that $2z + 12 = 0$ and finish the solution.

3. *Another method to get the quadratic formula* Look at equation (4) on page 110. Rewrite the right side as $\left(\sqrt{b^2 - 4ac}/2a\right)^2$ and then subtract it from each

side. The right side is now 0 and the left side is a difference of two squares. If you factor this difference of two squares, you will easily be able to get the quadratic formula, and, moreover, the quadratic expression is factored, which is sometimes useful.

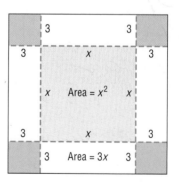

1.3 EXERCISES

In Problems 1–20, solve each equation by factoring.

1. $x^2 - 9x = 0$

2. $x^2 + 4x = 0$

3. $x^2 - 25 = 0$

4. $x^2 - 9 = 0$

5. $z^2 + z - 6 = 0$

6. $v^2 + 7v + 6 = 0$

7. $2x^2 - 5x - 3 = 0$

8. $3x^2 + 5x + 2 = 0$

9. $3t^2 - 48 = 0$

10. $2y^2 - 50 = 0$

11. $x(x - 8) + 12 = 0$

12. $x(x + 4) = 12$

13. $4x^2 + 9 = 12x$

14. $25x^2 + 16 = 40x$

15. $6(p^2 - 1) = 5p$

16. $2(2u^2 - 4u) + 3 = 0$

17. $6x - 5 = \dfrac{6}{x}$

18. $x + \dfrac{12}{x} = 7$

19. $\dfrac{4(x-2)}{x-3} + \dfrac{3}{x} = \dfrac{-3}{x(x-3)}$

20. $\dfrac{5}{x+4} = 4 + \dfrac{3}{x-2}$

In Problems 21–26, solve each equation by the Square Root Method.

21. $x^2 = 25$

22. $x^2 = 36$

23. $(x-1)^2 = 4$

24. $(x+2)^2 = 1$

25. $(2x+3)^2 = 9$

26. $(3x-2)^2 = 4$

In Problems 27–32, what number should be added to complete the square of each expression?

27. $x^2 + 8x$

28. $x^2 - 4x$

29. $x^2 + \frac{1}{2}x$

30. $x^2 - \frac{1}{3}x$

31. $x^2 - \frac{2}{3}x$

32. $x^2 - \frac{2}{5}x$

In Problems 33–38, solve each equation by completing the square.

33. $x^2 + 4x = 21$

34. $x^2 - 6x = 13$

35. $x^2 - \frac{1}{2}x - \frac{3}{16} = 0$

36. $x^2 + \frac{2}{3}x - \frac{1}{3} = 0$

37. $3x^2 + x - \frac{1}{2} = 0$

38. $2x^2 - 3x - 1 = 0$

In Problems 39–58, find the real solutions, if any, of each equation. Use the quadratic formula.

39. $x^2 - 4x + 2 = 0$

40. $x^2 + 4x + 2 = 0$

41. $x^2 - 4x - 1 = 0$

42. $x^2 + 6x + 1 = 0$

43. $2x^2 - 5x + 3 = 0$

44. $2x^2 + 5x + 3 = 0$

45. $4y^2 - y + 2 = 0$

46. $4t^2 + t + 1 = 0$

47. $4x^2 = 1 - 2x$

48. $2x^2 = 1 - 2x$

49. $4x^2 = 9x$

50. $5x = 4x^2$

51. $9t^2 - 6t + 1 = 0$

52. $4u^2 - 6u + 9 = 0$

53. $\dfrac{3}{4}x^2 - \dfrac{1}{4}x - \dfrac{1}{2} = 0$

54. $\dfrac{2}{3}x^2 - x - 3 = 0$

55. $4 - \dfrac{1}{x} - \dfrac{2}{x^2} = 0$

56. $4 + \dfrac{1}{x} - \dfrac{1}{x^2} = 0$

57. $3x = 1 - \dfrac{1}{x}$

58. $x = 1 - \dfrac{4}{x}$

In Problems 59–66, find the real solutions, if any, of each equation. Use the quadratic formula and a calculator. Express any solutions rounded to two decimal places.

59. $x^2 - 4.1x + 2.2 = 0$

60. $x^2 + 3.9x + 1.8 = 0$

61. $x^2 + \sqrt{3}x - 3 = 0$

62. $x^2 + \sqrt{2}x - 2 = 0$

63. $\pi x^2 - x - \pi = 0$

64. $\pi x^2 + \pi x - 2 = 0$

65. $3x^2 + 8\pi x + \sqrt{29} = 0$

66. $\pi x^2 - 15\sqrt{2}x + 20 = 0$

In Problems 67–78, find the real solutions, if any, of each quadratic equation. Use any method.

67. $x^2 - 5 = 0$

68. $x^2 - 6 = 0$

69. $16x^2 - 8x + 1 = 0$

70. $9x^2 - 6x + 1 = 0$

71. $10x^2 - 19x - 15 = 0$

72. $6x^2 + 7x - 20 = 0$

73. $2 + z = 6z^2$

74. $2 = y + 6y^2$

75. $x^2 + \sqrt{2}x = \frac{1}{2}$

76. $\frac{1}{2}x^2 = \sqrt{2}x + 1$

77. $x^2 + x = 4$

78. $x^2 + x = 1$

In Problems 79–84, use the discriminant to determine whether each quadratic equation has two unequal real solutions, a repeated real solution, or no real solution, without solving the equation.

79. $2x^2 - 6x + 7 = 0$

80. $x^2 + 4x + 7 = 0$

81. $9x^2 - 30x + 25 = 0$

82. $25x^2 - 20x + 4 = 0$

83. $3x^2 + 5x - 8 = 0$

84. $2x^2 - 3x - 7 = 0$

---

85. **Dimensions of a Window**   The area of the opening of a rectangular window is to be 143 square feet. If the length is to be 2 feet more than the width, what are the dimensions?

86. **Dimensions of a Window**   The area of a rectangular window is to be 306 square centimeters. If the length exceeds the width by 1 centimeter, what are the dimensions?

87. **Geometry**   Find the dimensions of a rectangle whose perimeter is 26 meters and whose area is 40 square meters.

88. **Watering a Field**   An adjustable water sprinkler that sprays water in a circular pattern is placed at the center of a square field whose area is 1250 square feet (see the figure). What is the shortest radius setting that can be

used if the field is to be completely enclosed within the circle?

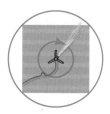

89. Constructing a Box   An open box is to be constructed from a square piece of sheet metal by removing a square of side 1 foot from each corner and turning up the edges. If the box is to hold 4 cubic feet, what should be the dimensions of the sheet metal?

90. Constructing a Box   Rework Problem 89 if the piece of sheet metal is a rectangle whose length is twice its width.

91. Physics   A ball is thrown vertically upward from the top of a building 96 feet tall with an initial velocity of 80 feet per second. The distance s (in feet) of the ball from the ground after t seconds is $s = 96 + 80t - 16t^2$.
(a) After how many seconds does the ball strike the ground?
(b) After how many seconds will the ball pass the top of the building on its way down?

92. Constructing a Coffee Can   A 39 ounce can of Hills Bros.® coffee requires 188.5 square inches of aluminum. If its height is 7 inches, what is its radius? (The surface area A of a right cylinder is $A = 2\pi r^2 + 2\pi rh$, where r is the radius and h is the height.)

93. Business: Large Order Discounts   A company charges $200 for each box of tools on orders of 150 or fewer boxes. If a customer orders x boxes in excess of 150, the cost for each box ordered is reduced by x dollars. If a customer's bill came to $30,625, how many boxes were ordered?

94. Dimensions of a Patio   A contractor orders 8 cubic yards of premixed cement, all of which is to be used to pour a patio that will be 4 inches thick. If the length of the patio is specified to be twice the width, what will be the patio dimensions? (1 cubic yard = 27 cubic feet)

95. Constructing a Border around a Garden   A landscaper, who just completed a rectangular flower garden measuring 6 feet by 10 feet, orders 1 cubic yard of premixed cement, all of which is to be used to create a border of

uniform width around the garden. If the border is to have a depth of 3 inches, how wide will the border be?

96. Physics   An object is propelled vertically upward with an initial velocity of 20 meters per second. The distance s (in meters) of the object from the ground after t seconds is $s = -4.9t^2 + 20t$.
(a) When will the object be 15 meters above the ground?
(b) When will it strike the ground?
(c) Will the object reach a height of 100 meters?

97. Reducing the Size of a Candy Bar   A jumbo chocolate bar with a rectangular shape measures 12 centimeters in length, 7 centimeters in width, and 3 centimeters in thickness. Due to escalating costs of cocoa, management decides to reduce the volume of the bar by 10%. To accomplish this reduction, management decides that the new bar should have the same 3 centimeter thickness, but the length and width of each should be reduced an equal number of centimeters. What should be the dimensions of the new candy bar?

98. Reducing the Size of a Candy Bar   Rework Problem 97 if the reduction is to be 20%.

99. Constructing a Border around a Pool   A circular pool measures 10 feet across. One cubic yard of concrete is to be used to create a circular border of uniform width around the pool. If the border is to have a depth of 3 inches, how wide will the border be? (1 cubic yard = 27 cubic feet) See the illustration.

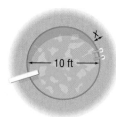

100. Constructing a Border around a Pool   Rework Problem 99 if the depth of the border is 4 inches.

101. Physics: Uniform Motion   A motorboat maintained a constant speed of 15 miles per hour relative to the water in going 10 miles upstream and then returning. The total time for the trip was 1.5 hours. Use this information to find the speed of the current.

102. Physics   An object is thrown down from the top of a building 1280 feet tall with an initial velocity of 32 feet per second. The distance s (in feet) of the object from the ground after t seconds is $s = 1280 - 32t - 16t^2$.
(a) When will the object strike ground?
(b) What is the height of the object after 4 seconds?

103. Show that the sum of the roots of a quadratic equation is $-b/a$.

104. Show that the product of the roots of a quadratic equation is c/a.

105. Find k such that the equation $kx^2 + x + k = 0$ has a repeated real solution.

106. Find k such that the equation $x^2 - kx + 4 = 0$ has a repeated real solution.

107. Show that the real solutions of the equation $ax^2 + bx + c = 0$ are the negatives of the real solutions of the equation $ax^2 - bx + c = 0$. Assume that $b^2 - 4ac \geq 0$.

108. Show that the real solutions of the equation $ax^2 + bx + c = 0$ are the reciprocals of the real solutions of the equation $cx^2 + bx + a = 0$. Assume that $b^2 - 4ac \geq 0$.

109. The sum of the consecutive integers $1, 2, 3, \ldots, n$ is given by the formula $\frac{1}{2}n(n + 1)$. How many consecutive integers, starting with 1, must be added to get a sum of 666?

110. Geometry   If a polygon of n sides has $\frac{1}{2}n(n - 3)$ diagonals, how many sides will a polygon with 65 diagonals have? Is there a polygon with 80 diagonals?

111. Computing Average Speed   In going from Chicago to Atlanta, a car averages 45 miles per hour, and in going from Atlanta to Miami, it averages 55 miles per hour. If Atlanta is halfway between Chicago and Miami, what is the average speed from Chicago to Miami? Discuss an intuitive solution. Write a paragraph defending your intuitive solution. Then solve the problem algebraically. Is your intuitive solution the same as the algebraic one? If not, find the flaw.

112. Speed of a Plane   On a recent flight from Phoenix to Kansas City, a distance of 919 nautical miles, the plane arrived 20 minutes early. On leaving the aircraft, I asked the captain, "What was our tail wind?" He replied, "I don't know, but our ground speed was 550 knots." How can you determine if enough information is provided to find the tail wind? If possible, find the tail wind. (1 knot = 1 nautical mile per hour)

113. Describe three ways you might solve a quadratic equation. State your preferred method; explain why you chose it.

114. Explain the benefits of evaluating the discriminant of a quadratic equation before attempting to solve it.

115. Make up three quadratic equations: one having two distinct solutions, one having no real solution, and one having exactly one real solution.

116. The word *quadratic* seems to imply four (*quad*), yet a quadratic equation is an equation that involves a polynomial of degree 2. Investigate the origin of the term *quadratic* as it is used in the expression *quadratic equation*. Write a brief essay on your findings.

PREPARING FOR THIS SECTION

Before getting started, review the following:

✓ Square Roots; Radicals (Review, Section 8, pp. 67–72)   ✓ Rational Exponents (Review, Section 9, pp. 74–77)

1.4   RADICAL EQUATIONS; EQUATIONS QUADRATIC IN FORM

> **OBJECTIVES**   1 Solve Radical Equations
> 2 Solve Equations Quadratic in Form

RADICAL EQUATIONS

1   When the variable in an equation occurs in a square root, cube root, and so on, that is, when it occurs in a radical, the equation is called a **radical equation.** Sometimes a suitable operation will change a radical equation to one that is linear or quadratic. A commonly used procedure is to isolate the most complicated radical on one side of the equation and then eliminate it by raising each side to a power equal to the index of the radical. Care must be taken, however, because apparent solutions that are not, in fact, solutions of the original equation may result. These are called **extraneous solutions.** Therefore, we need to check all answers when working with radical equations.

EXAMPLE 1   **Solving a Radical Equation**

Find the real solutions of the equation: $\sqrt[3]{2x - 4} - 2 = 0$

Solution   The equation contains a radical whose index is 3. We isolate it on the left side.

$$\sqrt[3]{2x - 4} - 2 = 0$$
$$\sqrt[3]{2x - 4} = 2$$

Now raise each side to the third power (the index of the radical is 3) and solve.

$$\left(\sqrt[3]{2x - 4}\right)^3 = 2^3 \qquad \text{Raise each side to the power 3.}$$
$$2x - 4 = 8 \qquad \text{Simplify.}$$
$$2x = 12 \qquad \text{Add 4 to both sides.}$$
$$x = 6 \qquad \text{Divide both sides by 2.}$$

Check: $\sqrt[3]{2(6) - 4} - 2 = \sqrt[3]{12 - 4} - 2 = \sqrt[3]{8} - 2 = 2 - 2 = 0.$ ∎

The solution set is $\{6\}$.   ∎

✎——— N O W   W O R K   P R O B L E M   **1**.

EXAMPLE 2   **Solving a Radical Equation**

Find the real solutions of the equation: $\sqrt{x - 1} = x - 7$

Solution   We square both sides since the index of a square root is 2.

$$\sqrt{x - 1} = x - 7$$
$$\left(\sqrt{x - 1}\right)^2 = (x - 7)^2 \qquad \text{Square both sides.}$$
$$x - 1 = x^2 - 14x + 49 \quad \text{Remove parentheses.}$$
$$x^2 - 15x + 50 = 0 \qquad \text{Put in standard form.}$$
$$(x - 10)(x - 5) = 0 \qquad \text{Factor.}$$
$$x = 10 \quad \text{or} \quad x = 5 \qquad \text{Apply the Zero-Product Property and solve.}$$

Check: $x = 10$: $\sqrt{x - 1} = \sqrt{10 - 1} = \sqrt{9} = 3$ and $x - 7 = 10 - 7 = 3$

$x = 5$: $\sqrt{x - 1} = \sqrt{5 - 1} = \sqrt{4} = 2$ and $x - 7 = 5 - 7 = -2$ ∎

The solution $x = 5$ is extraneous; the only solution of the equation is $x = 10$.   ∎

✎——— N O W   W O R K   P R O B L E M   **9**.

Sometimes, we need to raise each side to a power more than once in order to solve a radical equation algebraically.

EXAMPLE 3   **Solving a Radical Equation**

Find the real solutions of the equation: $\sqrt{2x + 3} - \sqrt{x + 2} = 2$

Solution   First, we choose to isolate the more complicated radical expression (in this case, $\sqrt{2x + 3}$) on the left side.

$$\sqrt{2x + 3} = \sqrt{x + 2} + 2$$

Now square both sides (the index of the radical is 2).

$$\left(\sqrt{2x + 3}\right)^2 = \left(\sqrt{x + 2} + 2\right)^2$$

$$2x + 3 = \left(\sqrt{x + 2}\right)^2 + 4\sqrt{x + 2} + 4 \quad \text{Remove parentheses.}$$

$$2x + 3 = x + 2 + 4\sqrt{x + 2} + 4 \quad \text{Simplify.}$$

$$2x + 3 = x + 6 + 4\sqrt{x + 2} \quad \text{Combine like terms.}$$

Because the equation still contains a radical, we isolate the remaining radical on the right side and again square both sides.

$$x - 3 = 4\sqrt{x + 2} \quad \text{Isolate the radical on the right side.}$$

$$(x - 3)^2 = 16(x + 2) \quad \text{Square both sides.}$$

$$x^2 - 6x + 9 = 16x + 32 \quad \text{Remove parentheses.}$$

$$x^2 - 22x - 23 = 0 \quad \text{Put in standard form.}$$

$$(x - 23)(x + 1) = 0 \quad \text{Factor.}$$

$$x = 23 \quad \text{or} \quad x = -1$$

The original equation appears to have the solution set $\{-1, 23\}$. However, we have not yet checked.

Check: $x = 23$: $\sqrt{2x + 3} - \sqrt{x + 2} = \sqrt{2(23) + 3} - \sqrt{23 + 2} = \sqrt{49} - \sqrt{25} = 7 - 5 = 2$

$x = -1$: $\sqrt{2x + 3} - \sqrt{x + 2} = \sqrt{2(-1) + 3} - \sqrt{-1 + 2} = \sqrt{1} - \sqrt{1} = 1 - 1 = 0$ ■

The equation has only one solution, 23; the solution -1 is extraneous. ■

NOW WORK PROBLEM **19**.

EQUATIONS QUADRATIC IN FORM

② The equation $x^4 + x^2 - 12 = 0$ is not quadratic in x, but it is quadratic in x^2. That is, if we let $u = x^2$, we get $u^2 + u - 12 = 0$, a quadratic equation. This equation can be solved for u and, in turn, by using $u = x^2$, we can find the solutions x of the original equation.

In general, if an appropriate substitution u transforms an equation into one of the form

$$au^2 + bu + c = 0 \qquad a \neq 0$$

then the original equation is called an **equation of the quadratic type** or an **equation quadratic in form.**

The difficulty of solving such an equation lies in the determination that the equation is, in fact, quadratic in form. After you are told an equation is quadratic in form, it is easy enough to see it, but some practice is needed to enable you to recognize them on your own.

EXAMPLE 4

Solving Equations That Are Quadratic in Form

Find the real solutions of the equation: $(x + 2)^2 + 11(x + 2) - 12 = 0$

Solution

For this equation, let $u = x + 2$. Then $u^2 = (x + 2)^2$, and the original equation,

$$(x + 2)^2 + 11(x + 2) - 12 = 0$$

becomes

$$u^2 + 11u - 12 = 0 \quad \text{Let } u = x + 2.$$

$$(u + 12)(u - 1) = 0 \quad \text{Factor.}$$

$$u = -12 \quad \text{or} \quad u = 1 \quad \text{Solve.}$$

But we want to solve for x. Because $u = x + 2$, we have

$$x + 2 = -12 \quad \text{or} \quad x + 2 = 1$$

$$x = -14 \qquad\qquad x = -1$$

Check: $x = -14$: $(-14 + 2)^2 + 11(-14 + 2) - 12$

$$= (-12)^2 + 11(-12) - 12 = 144 - 132 - 12 = 0$$

$x = -1$: $(-1 + 2)^2 + 11(-1 + 2) - 12 = 1 + 11 - 12 = 0$ ▬

The original equation has the solution set $\{-14, -1\}$. ■

EXAMPLE 5

Solving Equations That Are Quadratic in Form

Find the real solutions of the equation: $(x^2 - 1)^2 + (x^2 - 1) - 12 = 0$

Solution For the equation $(x^2 - 1)^2 + (x^2 - 1) - 12 = 0$, we let $u = x^2 - 1$ so that $u^2 = (x^2 - 1)^2$. Then the original equation

$$(x^2 - 1)^2 + (x^2 - 1) - 12 = 0$$

becomes

$$u^2 + u - 12 = 0 \quad \text{Let } u = x^2 - 1.$$

$$(u + 4)(u - 3) = 0 \quad \text{Factor.}$$

$$u = -4 \quad \text{or} \quad u = 3 \quad \text{Solve.}$$

But remember that we want to solve for x. Because $u = x^2 - 1$, we have

$$x^2 - 1 = -4 \quad \text{or} \quad x^2 - 1 = 3$$

$$x^2 = -3 \qquad\qquad x^2 = 4$$

The first of these has no real solution; the second has the solution set $\{-2, 2\}$.

Check: $x = -2$: $(4 - 1)^2 + (4 - 1) - 12 = 9 + 3 - 12 = 0$

$x = 2$: $(4 - 1)^2 + (4 - 1) - 12 = 9 + 3 - 12 = 0$ ▬

Thus, $\{-2, 2\}$ is the solution set of the original equation. ■

EXAMPLE 6

Solving Equations That Are Quadratic in Form

Find the real solutions of the equation: $x + 2\sqrt{x} - 3 = 0$

Solution For the equation $x + 2\sqrt{x} - 3 = 0$, let $u = \sqrt{x}$. Then $u^2 = x$, and the original equation,

$$x + 2\sqrt{x} - 3 = 0$$

becomes

$$u^2 + 2u - 3 = 0 \quad \text{Let } u = \sqrt{x}.$$

$$(u + 3)(u - 1) = 0 \quad \text{Factor.}$$

$$u = -3 \quad \text{or} \quad u = 1 \quad \text{Solve.}$$

Since $u = \sqrt{x}$, we have $\sqrt{x} = -3$ or $\sqrt{x} = 1$. The first of these, $\sqrt{x} = -3$ has no real solution, since the square root of a real number is never negative. The second, $\sqrt{x} = 1$, has the solution $x = 1$.

Check: $1 + 2\sqrt{1} - 3 = 1 + 2 - 3 = 0$

Thus, $x = 1$ is the only solution of the original equation.

ANOTHER METHOD FOR SOLVING EXAMPLE 6 WOULD BE TO TREAT IT AS A RADICAL EQUATION. SOLVE IT THIS WAY FOR PRACTICE.

The idea should now be clear. If an equation contains an expression and that same expression squared, make a substitution for the expression. You may get a quadratic equation.

NOW WORK PROBLEM 41.

1.4 EXERCISES

In Problems 1–30, find the real solutions of each equation.

1. $\sqrt{2t - 1} = 1$

2. $\sqrt{3t + 4} = 2$

3. $\sqrt{3t + 4} = -6$

4. $\sqrt{5t + 3} = -2$

5. $\sqrt[3]{1 - 2x} - 3 = 0$

6. $\sqrt[3]{1 - 2x} - 1 = 0$

7. $x = 8\sqrt{x}$

8. $x = 3\sqrt{x}$

9. $\sqrt{15 - 2x} = x$

10. $\sqrt{12 - x} = x$

11. $x = 2\sqrt{x - 1}$

12. $x = 2\sqrt{-x - 1}$

13. $\sqrt{x^2 - x - 4} = x + 2$

14. $\sqrt{3 - x + x^2} = x - 2$

15. $3 + \sqrt{3x + 1} = x$

16. $2 + \sqrt{12 - 2x} = x$

17. $\sqrt{2x + 3} - \sqrt{x + 1} = 1$

18. $\sqrt{3x + 7} + \sqrt{x + 2} = 1$

19. $\sqrt{3x + 1} - \sqrt{x - 1} = 2$

20. $\sqrt{3x - 5} - \sqrt{x + 7} = 2$

21. $\sqrt{3 - 2\sqrt{x}} = \sqrt{x}$

22. $\sqrt{10 + 3\sqrt{x}} = \sqrt{x}$

23. $(3x + 1)^{1/2} = 4$

24. $(3x - 5)^{1/2} = 2$

25. $(5x - 2)^{1/3} = 2$

26. $(2x + 1)^{1/3} = -1$

27. $(x^2 + 9)^{1/2} = 5$

28. $(x^2 - 16)^{1/2} = 9$

29. $x^{3/2} - 3x^{1/2} = 0$

30. $x^{3/4} - 9x^{1/4} = 0$

In Problems 31–62, find the real solutions of each equation.

31. $x^4 - 5x^2 + 4 = 0$

32. $x^4 - 10x^2 + 25 = 0$

33. $3x^4 - 2x^2 - 1 = 0$

34. $2x^4 - 5x^2 - 12 = 0$

35. $x^6 + 7x^3 - 8 = 0$

36. $x^6 - 7x^3 - 8 = 0$

37. $(x + 2)^2 + 7(x + 2) + 12 = 0$

38. $(2x + 5)^2 - (2x + 5) - 6 = 0$

39. $(3x + 4)^2 - 6(3x + 4) + 9 = 0$

40. $(2 - x)^2 + (2 - x) - 20 = 0$

41. $2(s + 1)^2 - 5(s + 1) = 3$

42. $3(1 - y)^2 + 5(1 - y) + 2 = 0$

43. $x - 4x\sqrt{x} = 0$

44. $x + 8\sqrt{x} = 0$

45. $x + \sqrt{x} = 20$

46. $x + \sqrt{x} = 6$

47. $t^{1/2} - 2t^{1/4} + 1 = 0$

48. $z^{1/2} - 4z^{1/4} + 4 = 0$

49. $4x^{1/2} - 9x^{1/4} + 4 = 0$

50. $x^{1/2} - 3x^{1/4} + 2 = 0$

51. $\sqrt[4]{5x^2 - 6} = x$

52. $\sqrt[4]{4 - 5x^2} = x$

53. $x^2 + 3x + \sqrt{x^2 + 3x} = 6$

54. $x^2 - 3x - \sqrt{x^2 - 3x} = 2$

55. $\dfrac{1}{(x + 1)^2} = \dfrac{1}{x + 1} + 2$

56. $\dfrac{1}{(x - 1)^2} + \dfrac{1}{x - 1} = 12$

57. $3x^{-2} - 7x^{-1} - 6 = 0$

58. $2x^{-2} - 3x^{-1} - 4 = 0$

59. $2x^{2/3} - 5x^{1/3} - 3 = 0$

60. $3x^{4/3} + 5x^{2/3} - 2 = 0$

61. $\left(\dfrac{v}{v + 1}\right)^2 + \dfrac{2v}{v + 1} = 8$

62. $\left(\dfrac{y}{y - 1}\right)^2 = 6\left(\dfrac{y}{y - 1}\right) + 7$

In Problems 63–68, find the real solutions of each equation. Use a calculator to express any solutions rounded to two decimal places.

63. $x - 4x^{1/2} + 2 = 0$

64. $x^{2/3} + 4x^{1/3} + 2 = 0$

65. $x^4 + \sqrt{3}x^2 - 3 = 0$

66. $x^4 + \sqrt{2}x^2 - 2 = 0$

67. $\pi(1 + t)^2 = \pi + 1 + t$

68. $\pi(1 + r)^2 = 2 + \pi(1 + r)$

69. If $k = \dfrac{x + 3}{x - 3}$ and $k^2 - k = 12$, find x.

70. If $k = \dfrac{x + 3}{x - 4}$ and $k^2 - 3k = 28$, find x.

71. **Physics: Using Sound to Measure Distance**   The distance to the surface of the water in a well can sometimes be found by dropping an object into the well and measuring the time elapsed until a sound is heard. If t_1 is the time (measured in seconds) that it takes for the object to strike the water, then t_1 will obey the equation $s = 16t_1^2$, where s is the distance (measured in feet). It follows that $t_1 = \sqrt{s}/4$. Suppose that t_2 is the time that it takes for the sound of the impact to reach your ears. Because sound waves are known to travel at a speed of approximately 1100 feet per second, the time t_2 to travel the distance s will be $t_2 = s/1100$. See the illustration.

Now $t_1 + t_2$ is the total time that elapses from the moment that the object is dropped to the moment that a sound is heard. We have the equation

$$\text{Total time elapsed} = \frac{\sqrt{s}}{4} + \frac{s}{1100}$$

Find the distance to the water's surface if the total time elapsed from dropping a rock to hearing it hit water is 4 seconds.

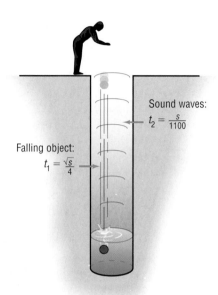

Sound waves: $t_2 = \frac{s}{1100}$

Falling object: $t_1 = \frac{\sqrt{s}}{4}$

72. Make up a radical equation that has no solution.

73. Make up a radical equation that has an extraneous solution.

74. Discuss the step in the solving process for radical equations that leads to the possibility of extraneous solutions. Why is there no such possibility for linear and quadratic equations?

PREPARING FOR THIS SECTION

Before getting started, review the following:

✓ Algebra Review (Review, Section 2, pp. 16–19)

1.5 SOLVING INEQUALITIES

OBJECTIVES
1. Use Interval Notation
2. Use Properties of Inequalities
3. Solve Inequalities
4. Solve Combined Inequalities

Suppose that a and b are two real numbers and $a < b$. We shall use the notation $a < x < b$ to mean that x is a number *between* a and b. The expression $a < x < b$ is equivalent to the two inequalities $a < x$ and $x < b$. Similarly, the expression $a \le x \le b$ is equivalent to the two inequalities $a \le x$ and $x \le b$. The remaining two possibilities, $a \le x < b$ and $a < x \le b$, are defined similarly.

Although it is acceptable to write $3 \ge x \ge 2$, it is preferable to reverse the inequality symbols and write instead $2 \le x \le 3$ so that, as you read from left to right, the values go from smaller to larger.

A statement such as $2 \le x \le 1$ is false because there is no number x for which $2 \le x$ and $x \le 1$. Finally, we never mix inequality symbols, as in $2 \le x \ge 3$.

INTERVALS

① Let a and b represent two real numbers with $a < b$.

> A **closed interval**, denoted by **[a, b]**, consists of all real numbers x for which $a \leq x \leq b$.
>
> An **open interval**, denoted by **(a, b)**, consists of all real numbers x for which $a < x < b$.
>
> The **half-open**, or **half-closed, intervals** are **(a, b]**, consisting of all real numbers x for which $a < x \leq b$, and **[a, b)**, consisting of all real numbers x for which $a \leq x < b$.

In each of these definitions, a is called the **left endpoint** and b the **right endpoint** of the interval.

The symbol ∞ (read as "infinity") is not a real number, but a notational device used to indicate unboundedness in the positive direction. The symbol $-\infty$ (read as "negative infinity") also is not a real number, but a notational device used to indicate unboundedness in the negative direction. Using the symbols ∞ and $-\infty$, we can define five other kinds of intervals:

$[a, \infty)$ Consists of all real numbers x for which $x \geq a$ $(a \leq x < \infty)$

(a, ∞) Consists of all real numbers x for which $x > a$ $(a < x < \infty)$

$(-\infty, a]$ Consists of all real numbers x for which $x \leq a$ $(-\infty < x \leq a)$

$(-\infty, a)$ Consists of all real numbers x for which $x < a$ $(-\infty < x < a)$

$(-\infty, \infty)$ Consists of all real numbers x $(-\infty < x < \infty)$

Note that ∞ and $-\infty$ are never included as endpoints, since neither is a real number.

Table 2 summarizes interval notation, corresponding inequality notation, and their graphs.

TABLE 2		
Interval	**Inequality**	**Graph**
The open interval (a, b)	$a < x < b$	
The closed interval $[a, b]$	$a \leq x \leq b$	
The half-open interval $[a, b)$	$a \leq x < b$	
The half-open interval $(a, b]$	$a < x \leq b$	
The interval $[a, \infty)$	$x \geq a$	
The interval (a, ∞)	$x > a$	
The interval $(-\infty, a]$	$x \leq a$	
The interval $(-\infty, a)$	$x < a$	
The interval $(-\infty, \infty)$	All real numbers	

EXAMPLE 1	**Writing Inequalities Using Interval Notation**

Write each inequality using interval notation.

(a) $1 \leq x \leq 3$ (b) $-4 < x < 0$ (c) $x > 5$ (d) $x \leq 1$

Solution (a) $1 \leq x \leq 3$ describes all numbers x between 1 and 3, inclusive. In interval notation, we write $[1, 3]$.

(b) In interval notation, $-4 < x < 0$ is written $(-4, 0)$.

(c) $x > 5$ consists of all numbers x greater than 5. In interval notation, we write $(5, \infty)$.

(d) In interval notation, $x \leq 1$ is written $(-\infty, 1]$.   ■

EXAMPLE 2	**Writing Intervals Using Inequality Notation**

Write each interval as an inequality involving x.

(a) $[1, 4)$ (b) $(2, \infty)$ (c) $[2, 3]$ (d) $(-\infty, -3]$

Solution (a) $[1, 4)$ consists of all numbers x for which $1 \leq x < 4$.

(b) $(2, \infty)$ consists of all numbers x for which $x > 2$ $(2 < x < \infty)$.

(c) $[2, 3]$ consists of all numbers x for which $2 \leq x \leq 3$.

(d) $(-\infty, -3]$ consists of all numbers x for which $x \leq -3$ $(-\infty < x \leq -3)$.   ■

 NOW WORK PROBLEMS **1**, **13**, AND **21**.

PROPERTIES OF INEQUALITIES

2   The product of two positive real numbers is positive, the product of two negative real numbers is positive, and the product of 0 and 0 is 0. For any real number a, the value of a^2 is 0 or positive; that is, a^2 is nonnegative. This is called the **nonnegative property.**

For any real number a, we have the following:

Nonnegative Property

$$a^2 \geq 0 \qquad\qquad \textbf{(1)}$$

If we add the same number to both sides of an inequality, we obtain an equivalent inequality. For example, since $3 < 5$, then $3 + 4 < 5 + 4$ or $7 < 9$. This is called the **addition property** of inequalities.

Addition Property of Inequalities

$$\text{If } a < b, \text{ then } a + c < b + c \qquad\qquad \textbf{(2a)}$$

$$\text{If } a > b, \text{ then } a + c > b + c \qquad\qquad \textbf{(2b)}$$

The addition property states that the sense, or direction, of an inequality remains unchanged if the same number is added to each side. Figure 6 illustrates the addition property (2a). In Figure 6(a), we see that a lies to the left of b. If c is positive, then $a + c$ and $b + c$ each lie c units to the right of a and b, respectively. Consequently, $a + c$ must lie to the left of $b + c$; that is, $a + c < b + c$. Figure 6(b) illustrates the situation if c is negative.

Figure 6

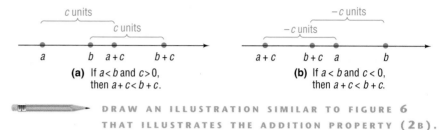

(a) If $a < b$ and $c > 0$, then $a + c < b + c$.

(b) If $a < b$ and $c < 0$, then $a + c < b + c$.

DRAW AN ILLUSTRATION SIMILAR TO FIGURE 6 THAT ILLUSTRATES THE ADDITION PROPERTY (2B).

EXAMPLE 3	**Addition Property of Inequalities**

(a) If $x < -5$, then $x + 5 < -5 + 5$ or $x + 5 < 0$.

(b) If $x > 2$, then $x + (-2) > 2 + (-2)$ or $x - 2 > 0$. ■

NOW WORK PROBLEM 29.

We will use two examples to arrive at our next property.

EXAMPLE 4	**Multiplying an Inequality by a Positive Number**

Express as an inequality the result of multiplying each side of the inequality $3 < 7$ by 2.

Solution   We begin with

$$3 < 7$$

Multiplying each side by 2 yields the numbers 6 and 14, so we have

$$6 < 14$$ ■

EXAMPLE 5	**Multiplying an Inequality by a Negative Number**

Express as an inequality the result of multiplying each side of the inequality $9 > 2$ by -4.

Solution   We begin with

$$9 > 2$$

Multiplying each side by -4 yields the numbers -36 and -8, so we have

$$-36 < -8$$ ■

Note that the effect of multiplying both sides of $9 > 2$ by the negative number -4 is that the direction of the inequality symbol is reversed.

Examples 4 and 5 illustrate the following general **multiplication properties** for inequalities:

Multiplication Properties for Inequalities

If $a < b$ and if $c > 0$, then $ac < bc$.

If $a < b$ and if $c < 0$, then $ac > bc$.   **(3a)**

If $a > b$ and if $c > 0$, then $ac > bc$.

If $a > b$ and if $c < 0$, then $ac < bc$.   **(3b)**

The multiplication properties state that the sense, or direction, of an inequality *remains the same* if each side is multiplied by a *positive* real number, whereas the direction is *reversed* if each side is multiplied by a *negative* real number.

EXAMPLE 6   **Multiplication Property of Inequalities**

(a) If $2x < 6$, then $\frac{1}{2}(2x) < \frac{1}{2}(6)$ or $x < 3$.

(b) If $\dfrac{x}{-3} > 12$, then $-3\left(\dfrac{x}{-3}\right) < -3(12)$ or $x < -36$.

(c) If $-4x > -8$, then $\dfrac{-4x}{-4} < \dfrac{-8}{-4}$ or $x < 2$.

(d) If $-x > 8$, then $(-1)(-x) < (-1)(8)$ or $x < -8$.   ∎

NOW WORK PROBLEM 35.

The **reciprocal property** states that the reciprocal of a positive real number is positive and that the reciprocal of a negative real number is negative.

Reciprocal Property for Inequalities

$$\text{If } a > 0, \text{ then } \frac{1}{a} > 0. \qquad \textbf{(4a)}$$

$$\text{If } a < 0, \text{ then } \frac{1}{a} < 0. \qquad \textbf{(4b)}$$

SOLVING INEQUALITIES

③ An **inequality in one variable** is a statement involving two expressions, at least one containing the variable, separated by one of the inequality symbols $<, \leq, >,$ or $\geq$. To **solve an inequality** means to find all values of the variable for which the statement is true. These values are called **solutions** of the inequality.

For example, the following are all inequalities involving one variable, x:

$$x + 5 < 8 \qquad 2x - 3 \geq 4 \qquad x^2 - 1 \leq 3 \qquad \frac{x + 1}{x - 2} > 0$$

Two inequalities having exactly the same solution set are called **equivalent inequalities.** As with equations, one method for solving an inequality is to replace it by a series of equivalent inequalities until an inequality with an obvious solution, such as $x < 3$, is obtained. We obtain equivalent inequalities by applying some of the same properties as those used to find equivalent equations. The addition property and the multiplication properties form the basis for the following procedures.

PROCEDURES THAT LEAVE THE INEQUALITY SYMBOL UNCHANGED

1. Simplify both sides of the inequality by combining like terms and eliminating parentheses:

$$\text{Replace} \quad (x + 2) + 6 > 2x + 5(x + 1)$$
$$\text{by} \quad x + 8 > 7x + 5$$

2. Add or subtract the same expression on both sides of the inequality:

$$\text{Replace} \quad 3x - 5 < 4$$
$$\text{by} \quad (3x - 5) + 5 < 4 + 5$$

3. Multiply or divide both sides of the inequality by the same positive expression:

$$\text{Replace} \quad 4x > 16 \quad \text{by} \quad \frac{4x}{4} > \frac{16}{4}$$

PROCEDURES THAT REVERSE THE SENSE OR DIRECTION OF THE INEQUALITY SYMBOL

1. Interchange the two sides of the inequality:

$$\text{Replace} \quad 3 < x \quad \text{by} \quad x > 3$$

2. Multiply or divide both sides of the inequality by the same *negative* expression:

$$\text{Replace} \quad -2x > 6 \quad \text{by} \quad \frac{-2x}{-2} < \frac{6}{-2}$$

As the examples that follow illustrate, we solve inequalities using many of the same steps that we would use to solve equations. In writing the solution of an inequality, we may use either set notation or interval notation, whichever is more convenient.

EXAMPLE 7   **Solving an Inequality**

Solve the inequality: $3 - 2x < 5$
Graph the solution set.

Solution

$$3 - 2x < 5$$

$$3 - 2x - 3 < 5 - 3 \qquad \text{Subtract 3 from both sides.}$$

$$-2x < 2 \qquad \text{Simplify.}$$

$$\frac{-2x}{-2} > \frac{2}{-2} \qquad \text{Divide both sides by } -2. \text{ (The sense of the inequality symbol is reversed.)}$$

$$x > -1 \qquad \text{Simplify.}$$

Figure 7
$x > -1$ or $(-1, \infty)$

The solution set is $\{x \mid x > -1\}$ or, using interval notation, all numbers in the interval $(-1, \infty)$. See Figure 7 for the graph. ∎

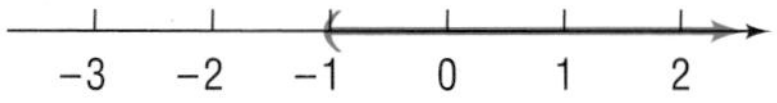

EXAMPLE 8	Solving an Inequality

Solve the inequality: $4x + 7 \geq 2x - 3$
Graph the solution set.

Solution

$$4x + 7 \geq 2x - 3$$

$$4x + 7 - 7 \geq 2x - 3 - 7 \qquad \text{Subtract 7 from both sides.}$$

$$4x \geq 2x - 10 \qquad \text{Simplify.}$$

$$4x - 2x \geq 2x - 10 - 2x \qquad \text{Subtract 2x from both sides.}$$

$$2x \geq -10 \qquad \text{Simplify.}$$

$$\frac{2x}{2} \geq \frac{-10}{2} \qquad \text{Divide both sides by 2. (The direction of the inequality symbol is unchanged.)}$$

$$x \geq -5 \qquad \text{Simplify.}$$

Figure 8
$x \geq -5$ or $[-5, \infty)$

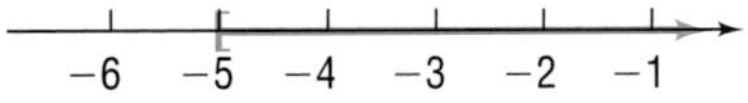

The solution set is $\{x \mid x \geq -5\}$ or, using interval notation, all numbers in the interval $[-5, \infty)$. See Figure 8 for the graph. ∎

 NOW WORK PROBLEM **43**.

EXAMPLE 9	Solving Combined Inequalities

④ Solve the inequality: $-5 < 3x - 2 < 1$
Graph the solution set.

Solution Recall that the inequality

$$-5 < 3x - 2 < 1$$

is equivalent to the two inequalities

$$-5 < 3x - 2 \quad \text{and} \quad 3x - 2 < 1$$

We will solve each of these inequalities separately.

$-5 < 3x - 2$		$3x - 2 < 1$
$-5 + 2 < 3x - 2 + 2$	Add 2 to both sides.	$3x - 2 + 2 < 1 + 2$
$-3 < 3x$	Simplify.	$3x < 3$
$\dfrac{-3}{3} < \dfrac{3x}{3}$	Divide both sides by 3.	$\dfrac{3x}{3} < \dfrac{3}{3}$
$-1 < x$	Simplify.	$x < 1$

The solution set of the original pair of inequalities consists of all x for which

$$-1 < x \quad \text{and} \quad x < 1$$

Figure 9
$-1 < x < 1$ or $(-1, 1)$

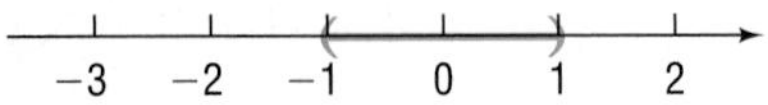

This may be written more compactly as $\{x \mid -1 < x < 1\}$. In interval notation, the solution is $(-1, 1)$. See Figure 9 for the graph. ∎

We observe in the preceding process that the two inequalities that we solved required exactly the same steps. A shortcut to solving the original inequality algebraically is to deal with the two inequalities at the same time, as follows:

$$-5 < \quad 3x - 2 \quad < 1$$

$$-5 + 2 < 3x - 2 + 2 < 1 + 2 \qquad \text{Add 2 to each part.}$$

$$-3 < \quad 3x \quad < 3 \qquad \text{Simplify.}$$

$$\frac{-3}{3} < \quad \frac{3x}{3} \quad < \frac{3}{3} \qquad \text{Divide each part by 3.}$$

$$-1 < \quad x \quad < 1 \qquad \text{Simplify.}$$

We use this shortcut in the next example.

EXAMPLE 10 **Solving Combined Inequalities**

Solve the inequality: $-1 \leq \dfrac{3 - 5x}{2} \leq 9$

Graph the solution set.

Solution
$$-1 \leq \quad \frac{3 - 5x}{2} \quad \leq 9$$

$$2(-1) \leq 2\left(\frac{3 - 5x}{2}\right) \leq 2(9) \qquad \begin{array}{l}\text{Multiply each part by 2 to remove}\\ \text{the denominator.}\end{array}$$

$$-2 \leq \quad 3 - 5x \quad \leq 18 \qquad \text{Simplify.}$$

$$-2 - 3 \leq 3 - 5x - 3 \leq 18 - 3 \qquad \begin{array}{l}\text{Subtract 3 from each part to isolate}\\ \text{the term containing } x.\end{array}$$

$$-5 \leq \quad -5x \quad \leq 15 \qquad \text{Simplify.}$$

$$\frac{-5}{-5} \geq \quad \frac{-5x}{-5} \quad \geq \frac{15}{-5} \qquad \begin{array}{l}\text{Divide each part by } -5 \text{ (change the sense}\\ \text{of each inequality symbol).}\end{array}$$

$$1 \geq \quad x \quad \geq -3 \qquad \text{Simplify.}$$

$$-3 \leq \quad x \quad \leq 1 \qquad \begin{array}{l}\text{Reverse the order so that the numbers get}\\ \text{larger as you read from left to right.}\end{array}$$

Figure 10
$-3 \leq x \leq 1$ or $[-3, 1]$

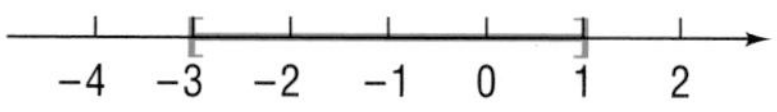

The solution set is $\{x \,|\, -3 \leq x \leq 1\}$, that is, all x in the interval $[-3, 1]$. Figure 10 illustrates the graph.   ∎

NOW WORK PROBLEM 63.

EXAMPLE 11 **Using the Reciprocal Property to Solve an Inequality**

Solve the inequality: $(4x - 1)^{-1} > 0$
Graph the solution set.

Solution   Since $(4x - 1)^{-1} = 1/(4x - 1)$ and since the Reciprocal Property states that when $1/a > 0$ then $a > 0$, we have

$$(4x - 1)^{-1} > 0$$

$$\frac{1}{4x - 1} > 0$$

$$4x - 1 > 0 \qquad \text{Reciprocal Property}$$

$$4x > 1$$

$$x > \frac{1}{4}$$

Figure 11
$x > \frac{1}{4}$ or $(\frac{1}{4}, \infty)$

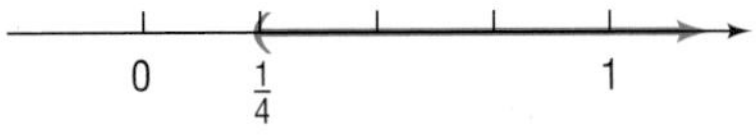

The solution set is $\{x | x > \frac{1}{4}\}$; that is, all x in the interval $(\frac{1}{4}, \infty)$. Figure 11 illustrates the graph.

NOW WORK PROBLEM 73.

EXAMPLE 12   **Creating Equivalent Inequalities**

If $-1 < x < 4$, find a and b so that $a < 2x + 1 < b$.

Solution   The idea here is to change the middle part of the combined inequality from x to $2x + 1$, using properties of inequalities.

$$-1 < \quad x \quad < 4$$

$$-2 < \quad 2x \quad < 8 \qquad \text{Multiply each part by 2.}$$

$$-1 < 2x + 1 < 9 \qquad \text{Add 1 to each part.}$$

Thus, $a = -1$ and $b = 9$.

NOW WORK PROBLEM 81.

Let's look at an applied problem involving inequalities.

EXAMPLE 13   **Physics: Ohm's Law**

In electricity, Ohm's law states that $E = IR$, where E is the voltage (in volts), I is the current (in amperes), and R is the resistance (in ohms). An air-conditioning unit is rated at a resistance of 10 ohms. If the voltage varies from 110 to 120 volts, inclusive, what corresponding range of current will the air conditioner draw?

Solution   The voltage lies between 110 and 120, inclusive, so

$$110 \leq \quad E \quad \leq 120$$

$$110 \leq \quad IR \quad \leq 120 \qquad \text{Ohm's law, } E = IR$$

$$110 \leq I(10) \leq 120 \qquad R = 10$$

$$\frac{110}{10} \leq \frac{I(10)}{10} \leq \frac{120}{10} \qquad \text{Divide each part by 10.}$$

$$11 \leq \quad I \quad \leq 12 \qquad \text{Simplify.}$$

The air conditioner will draw between 11 and 12 amperes of current, inclusive.

1.5 EXERCISES

In Problems 1–6, express the graph shown in color using interval notation. Also express each as an inequality involving x.

1.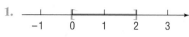
   -1 0 1 2 3

2. -1 0 1 2 3

3. -2 -1 0 1 2

4. -2 -1 0 1 2

5. -1 0 1 2 3

6. -1 0 1 2 3

In Problems 7–12, an inequality is given. Write the inequality obtained by:
(a) *Adding 3 to each side of the given inequality.*
(b) *Subtracting 5 from each side of the given inequality.*
(c) *Multiplying each side of the given inequality by 3.*
(d) *Multiplying each side of the given inequality by −2.*

7. $3 < 5$ 8. $2 > 1$ 9. $4 > -3$

10. $-3 > -5$ 11. $2x + 1 < 2$ 12. $1 - 2x > 5$

In Problems 13–20, write each inequality using interval notation, and illustrate each inequality using the real number line.

13. $0 \le x \le 4$ 14. $-1 < x < 5$ 15. $4 \le x < 6$ 16. $-2 < x < 0$

17. $x \ge 4$ 18. $x \le 5$ 19. $x < -4$ 20. $x > 1$

In Problems 21–28, write each interval as an inequality involving x, and illustrate each inequality using the real number line.

21. $[2, 5]$ 22. $(1, 2)$ 23. $(-3, -2)$ 24. $[0, 1)$

25. $[4, \infty)$ 26. $(-\infty, 2]$ 27. $(-\infty, -3)$ 28. $(-8, \infty)$

In Problems 29–42, fill in the blank with the correct inequality symbol

29. If $x < 5$, then $x - 5$ _____ 0.

30. If $x < -4$, then $x + 4$ _____ 0.

31. If $x > -4$, then $x + 4$ _____ 0.

32. If $x > 6$, then $x - 6$ _____ 0.

33. If $x \ge -4$, then $3x$ _____ -12.

34. If $x \le 3$, then $2x$ _____ 6.

35. If $x > 6$, then $-2x$ _____ -12.

36. If $x > -2$, then $-4x$ _____ 8.

37. If $x \ge 5$, then $-4x$ _____ -20.

38. If $x \le -4$, then $-3x$ _____ 12.

39. If $2x > 6$, then x _____ 3.

40. If $3x \le 12$, then x _____ 4.

41. If $-\frac{1}{2}x \le 3$, then x _____ -6.

42. If $-\frac{1}{4}x > 1$, then x _____ -4.

In Problems 43–78, solve each inequality. Express your answer using set notation or interval notation. Graph the solution set.

43. $x + 1 < 5$ 44. $x - 6 < 1$ 45. $1 - 2x \le 3$

46. $2 - 3x \le 5$ 47. $3x - 7 > 2$ 48. $2x + 5 > 1$

49. $3x - 1 \ge 3 + x$ 50. $2x - 2 \ge 3 + x$ 51. $-2(x + 3) < 8$

52. $-3(1 - x) < 12$ 53. $4 - 3(1 - x) \le 3$ 54. $8 - 4(2 - x) \le -2x$

55. $\frac{1}{2}(x - 4) > x + 8$ 56. $3x + 4 > \frac{1}{3}(x - 2)$ 57. $\frac{x}{2} \ge 1 - \frac{x}{4}$

58. $\frac{x}{3} \ge 2 + \frac{x}{6}$ 59. $0 \le 2x - 6 \le 4$ 60. $4 \le 2x + 2 \le 10$

61. $-5 \le 4 - 3x \le 2$ 62. $-3 \le 3 - 2x \le 9$ 63. $-3 < \dfrac{2x - 1}{4} < 0$

64. $0 < \dfrac{3x + 2}{2} < 4$ 65. $1 < 1 - \frac{1}{2}x < 4$ 66. $0 < 1 - \frac{1}{3}x < 1$

67. $(x + 2)(x - 3) > (x - 1)(x + 1)$ 68. $(x - 1)(x + 1) > (x - 3)(x + 4)$ 69. $x(4x + 3) \le (2x + 1)^2$

70. $x(9x - 5) \le (3x - 1)^2$ 71. $\frac{1}{2} \le \dfrac{x + 1}{3} < \frac{3}{4}$ 72. $\frac{1}{3} < \dfrac{x + 1}{2} \le \frac{2}{3}$

73. $(4x + 2)^{-1} < 0$

74. $(2x - 1)^{-1} > 0$

75. $0 < \dfrac{2}{x} < \dfrac{3}{5}$

76. $0 < \dfrac{4}{x} < \dfrac{2}{3}$

77. $0 < (2x - 4)^{-1} < \frac{1}{2}$

78. $0 < (3x + 6)^{-1} < \frac{1}{3}$

In Problems 79–88, find a and b.

79. If $-1 < x < 1$, then $a < x + 4 < b$.

80. If $-3 < x < 2$, then $a < x - 6 < b$.

81. If $2 < x < 3$, then $a < -4x < b$.

82. If $-4 < x < 0$, then $a < \frac{1}{2}x < b$.

83. If $0 < x < 4$, then $a < 2x + 3 < b$.

84. If $-3 < x < 3$, then $a < 1 - 2x < b$.

85. If $-3 < x < 0$, then $a < \dfrac{1}{x + 4} < b$.

86. If $2 < x < 4$, then $a < \dfrac{1}{x - 6} < b$.

87. If $6 < 3x < 12$, then $a < x^2 < b$.

88. If $0 < 2x < 6$, then $a < x^2 < b$.

89. What is the domain of the variable in the expression $\sqrt{3x + 6}$?

90. What is the domain of the variable in the expression $\sqrt{8 + 2x}$?

91. A young adult may be defined as someone older than 21, but less than 30 years of age. Express this statement using inequalities.

92. Middle-aged may be defined as being 40 or more and less than 60. Express the statement using inequalities.

93. **Life Expectancy** Metropolitan Life Insurance Co. reported that an average 25-year-old male in 1996 could expect to live at least 48.4 more years and an average 25-year-old female in 1996 could expect to live at least 54.7 more years.

(a) To what age can an average 25-year-old male expect to live? Express your answer as an inequality.
(b) To what age can an average 25-year-old female expect to live? Express your answer as an inequality.
(c) Who can expect to live longer, a male or a female? By how many years?

94. **General Chemistry** For a certain ideal gas, the volume V (in cubic centimeters) equals 20 times the temperature T (in degrees Celsius). If the temperature varies from $80°$ to $120°C$ inclusive, what is the corresponding range of the volume of the gas?

95. **Real Estate** A real estate agent agrees to sell a large apartment complex according to the following commission schedule: $45,000 plus 25% of the selling price in excess of $900,000. Assuming that the complex will sell at some price between $900,000 and $1,100,000 inclusive, over what range does the agent's commission vary? How does the commission vary as a percent of selling price?

96. **Sales Commission** A used car salesperson is paid a commission of $25 plus 40% of the selling price in excess of owner's cost. The owner claims that used cars typically sell for at least owner's cost plus $70 and at most owner's cost plus $300. For each sale made, over what range can the salesperson expect the commission to vary?

97. **Federal Tax Withholding** The percentage method of withholding for federal income tax (1998)* states that a single person whose weekly wages, after subtracting withholding allowances, are over $517, but not over $1105, shall have $69.90 plus 28% of the excess over $517 withheld. Over what range does the amount withheld vary if the weekly wages vary from $525 to $600 inclusive?

98. **Federal Tax Withholding** Rework Problem 97 if the weekly wages vary from $600 to $700 inclusive.

99. **Electricity Rates** Commonwealth Edison Company's summer charge for electricity is 10.494¢ per kilowatt-hour.† In addition, each monthly bill contains a customer charge of $9.36. If last summer's bills ranged from a low of $80.24 to a high of $271.80, over what range did usage vary (in kilowatt-hours)?

100. **Water Bills** The Village of Oak Lawn charges homeowners $21.60 per quarter-year plus $1.70 per 1000 gallons for water usage in excess of 12,000 gallons.‡ In 2000, one homeowner's quarterly bill ranged from a high of $65.75 to a low of $28.40. Over what range did water usage vary?

101. **Markup of a New Car** The markup over dealer's cost of a new car ranges from 12% to 18%. If the sticker price is $8800, over what range will the dealer's cost vary?

* *Source: Employer's Tax Guide*. Department of the Treasury, Internal Revenue Service, 1998.
† *Source:* Commonwealth Edison Co., Chicago, Illinois, 2000.
‡ *Source:* Village of Oak Lawn, Illinois, 2000.

102. IQ Tests A standard intelligence test has an average score of 100. According to statistical theory, of the people who take the test, the 2.5% with the highest scores will have scores of more than 1.96σ above the average, where σ (sigma, a number called the **standard deviation**) depends on the nature of the test. If $\sigma = 12$ for this test and there is (in principle) no upper limit to the score possible on the test, write the interval of possible test scores of the people in the top 2.5%.

103. Computing Grades In your Economics 101 class, you have scores of 68, 82, 87, and 89 on the first four of five tests. To get a grade of B, the average of the first five test scores must be greater than or equal to 80 and less than 90. Solve an inequality to find the range of the score that you need on the last test to get a B.

What do I need to get a B?

104. Computing Grades Repeat Problem 103 if the fifth test counts double.

105. Arithmetic Mean If $a < b$, show that $a < (a + b)/2 < b$. The number $(a + b)/2$ is called the **arithmetic mean** of a and b.

106. Refer to Problem 105. Show that the arithmetic mean of a and b is equidistant from a and b.

107. Geometric Mean If $0 < a < b$, show that $a < \sqrt{ab} < b$. The number $\sqrt{ab}$ is called the **geometric mean** of a and b.

108. Refer to Problems 105 and 107. Show that the geometric mean of a and b is less than the arithmetic mean of a and b.

109. Harmonic Mean For $0 < a < b$, let h be defined by

$$\frac{1}{h} = \frac{1}{2}\left(\frac{1}{a} + \frac{1}{b}\right)$$

Show that $a < h < b$. The number h is called the **harmonic mean** of a and b.

110. Refer to Problems 105, 107, and 109. Show that the harmonic mean of a and b equals the geometric mean squared divided by the arithmetic mean.

111. Make up an inequality that has no solution. Make up one that has exactly one solution.

112. The inequality $x^2 + 1 < -5$ has no solution. Explain why.

113. Do you prefer to use inequality notation or interval notation to express the solution to an inequality? Give your reasons. Are there particular circumstances when you prefer one to the other? Cite examples.

114. How would you explain to a fellow student the underlying reason for the multiplication properties for inequalities (page 127); that is, the sense or direction of an inequality remains the same if each side is multiplied by a positive real number, whereas the direction is reversed if each side is multiplied by a negative real number.

P R E P A R I N G   F O R   T H I S   S E C T I O N

Before getting started, review the following:

✓ Algebra Review (Review, Section 2, pp. 16–19)

1.6 EQUATIONS AND INEQUALITIES INVOLVING ABSOLUTE VALUE

OBJECTIVES   1   Solve Equations Involving Absolute Value
　　　　　　　 2   Solve Inequalities Involving Absolute Value

EQUATIONS INVOLVING ABSOLUTE VALUE

1   Recall that, on the real number line, the absolute value of a equals the distance from the origin to the point whose coordinate is a. For example, there are two points whose distance from the origin is 5 units, -5 and 5. So the equation $|x| = 5$ will have the solution set $\{-5, 5\}$. This leads to the following result:

Equations Involving Absolute Value

If a is a positive real number and if u is any algebraic expression, then

$$|u| = a \quad \text{is equivalent to} \quad u = a \quad \text{or} \quad u = -a \qquad (1)$$

| EXAMPLE 1 | Solving an Equation Involving Absolute Value |

Solve the equation: $|x + 4| = 13$

Solution This follows the form of equation (1), where $u = x + 4$. There are two possibilities.

$$x + 4 = 13 \quad \text{or} \quad x + 4 = -13$$
$$x = 9 \quad \text{or} \quad x = -17$$

The solution set is $\{-17, 9\}$. ■

NOW WORK PROBLEM 3.

② Let's look at an inequality involving absolute value.

| EXAMPLE 2 | Solving an Inequality Involving Absolute Value |

Solve the inequality: $|x| < 4$

Solution We are looking for all points whose coordinate x is a distance less than 4 units from the origin. See Figure 12 for an illustration. Because any x between -4 and 4 satisfies the condition $|x| < 4$, the solution set consists of all numbers x for which $-4 < x < 4$, that is, all x in the interval $(-4, 4)$. ■

Figure 12
$-4 < x < 4$ or $(-4, 4)$

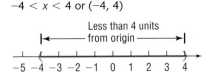

We are led to the following results:

Inequalities Involving Absolute Value

If a is a positive number and if u is an algebraic expression, then

> $|u| < a$ is equivalent to $-a < u < a$ **(2)**
>
> $|u| \leq a$ is equivalent to $-a \leq u \leq a$ **(3)**

In other words, $|u| < a$ is equivalent to $-a < u$ and $u < a$.

| EXAMPLE 3 | Solving an Inequality Involving Absolute Value |

Solve the inequality: $|2x + 4| \leq 3$
Graph the solution set.

Solution

$$|2x + 4| \leq 3 \qquad \text{This follows the form of statement (3);}$$
the expression $u = 2x + 4$ is inside the absolute value bars.

$$-3 \leq 2x + 4 \leq 3 \qquad \text{Apply statement (3).}$$

$$-3 - 4 \leq 2x + 4 - 4 \leq 3 - 4 \qquad \text{Subtract 4 from each part.}$$

$$-7 \leq 2x \leq -1 \qquad \text{Simplify.}$$

$$\begin{aligned} \frac{-7}{2} \le \quad & \frac{2x}{2} \quad \le \frac{-1}{2} && \text{Divide each part by 2.} \\ -\frac{7}{2} \le \quad & x \quad \le -\frac{1}{2} && \text{Simplify.} \end{aligned}$$

The solution set is $\left\{x \,\middle|\, -\frac{7}{2} \le x \le -\frac{1}{2}\right\}$, that is, all x in the interval $\left[-\frac{7}{2}, -\frac{1}{2}\right]$. See Figure 13. ■

Figure 13
$-\frac{7}{2} \le x \le -\frac{1}{2}$ or $\left[-\frac{7}{2}, -\frac{1}{2}\right]$

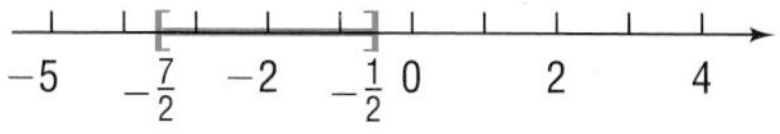

EXAMPLE 4 Solving an Inequality Involving Absolute Value

Solve the inequality: $|1 - 4x| < 5$
Graph the solution set.

Solution

$$\begin{aligned} |1 - 4x| &< 5 && \text{This expression follows the form of statement (2); the expression } u = 1 - 4x \text{ is inside the absolute value bars.} \\ -5 < 1 - 4x &< 5 && \text{Apply statement (2).} \\ -5 - 1 < 1 - 4x - 1 &< 5 - 1 && \text{Subtract 1 from each part.} \\ -6 < -4x &< 4 && \text{Simplify.} \\ \frac{-6}{-4} > \frac{-4x}{-4} &> \frac{4}{-4} && \text{Divide each part by } -4 \text{, which reverses the sense of the inequality symbols.} \\ \frac{3}{2} > x &> -1 && \text{Simplify.} \\ -1 < x &< \frac{3}{2} && \text{Rearrange the ordering.} \end{aligned}$$

The solution set is $\left\{x \,\middle|\, -1 < x < \frac{3}{2}\right\}$, that is, all x in the interval $\left(-1, \frac{3}{2}\right)$. See Figure 14. ■

Figure 14
$-1 < x < \frac{3}{2}$ or $\left(-1, \frac{3}{2}\right)$

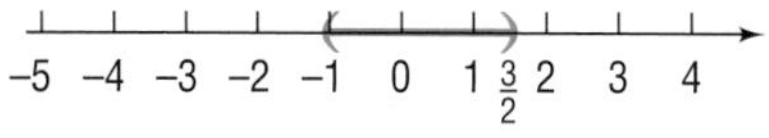

NOW WORK PROBLEM 29.

EXAMPLE 5 Solving an Inequality Involving Absolute Value

Solve the inequality: $|x| > 3$
Graph the solution set.

Solution We are looking for all points whose coordinate x is a distance greater than 3 units from the origin. Figure 15 illustrates the situation. We conclude that any x less than -3 or greater than 3 satisfies the condition $|x| > 3$. Consequently, the solution set consists of all numbers x for which $x < -3$ or $x > 3$, that is, all x in the intervals $(-\infty, -3)$ or $(3, \infty)$. ■

Figure 15
$x < -3$ or $x > 3$; $(-\infty, -3)$ or $(3, \infty)$

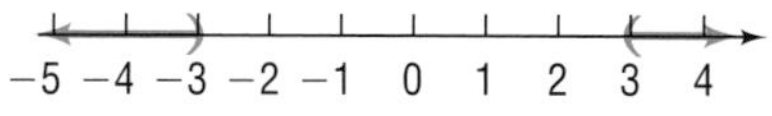

Inequalities Involving Absolute Value

If a is a positive number and u is an algebraic expression, then

$\lvert u\rvert > a$	is equivalent to	$u < -a$	or	$u > a$	**(4)**
$\lvert u\rvert \ge a$	is equivalent to	$u \le -a$	or	$u \ge a$	**(5)**

EXAMPLE 6	Solving an Inequality Involving Absolute Value

Solve the inequality: $|2x - 5| > 3$
Graph the solution set.

Solution

$$|2x - 5| > 3$$ This follows the form of statement (4); the expression $u = 2x - 5$ is inside the absolute value bars.

$2x - 5 < -3$	or	$2x - 5 > 3$	Apply statement (4).
$2x - 5 + 5 < -3 + 5$	or	$2x - 5 + 5 > 3 + 5$	Add 5 to each part.
$2x < 2$	or	$2x > 8$	Simplify.
$\dfrac{2x}{2} < \dfrac{2}{2}$	or	$\dfrac{2x}{2} > \dfrac{8}{2}$	Divide each part by 2.
$x < 1$	or	$x > 4$	Simplify.

Figure 16
$x < 1$ or $x > 4$; $(-\infty, 1)$ or $(4, \infty)$

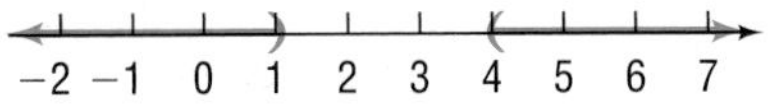

The solution set is $\{x \mid x < 1 \text{ or } x > 4\}$, that is, all x in the interval $(-\infty, 1)$ or $(4, \infty)$. See Figure 16. ∎

WARNING: A common error to be avoided is to attempt to write the solution $x < 1$ or $x > 4$ as $1 > x > 4$, which is incorrect, since there are no numbers x for which $1 > x$ and $x > 4$ ∎

NOW WORK PROBLEM **33**.

1.6 EXERCISES

In Problems 1–24, solve each equation.

1. $|2x| = 6$

2. $|3x| = 12$

3. $|2x + 3| = 5$

4. $|3x - 1| = 2$

5. $|1 - 4t| + 8 = 13$

6. $|1 - 2z| + 6 = 9$

7. $|-2x| = |8|$

8. $|-x| = |1|$

9. $-2|x| = 4$

10. $|3|x = 9$

11. $\frac{2}{3}|x| = 9$

12. $\frac{3}{4}|x| = 9$

13. $\left|\frac{x}{3} + \frac{2}{5}\right| = 2$

14. $\left|\frac{x}{2} - \frac{1}{3}\right| = 1$

15. $|u - 2| = -\frac{1}{2}$

16. $|2 - v| = -1$

17. $4 - |2x| = 3$

18. $5 - |\frac{1}{2}x| = 3$

19. $|x^2 - 9| = 0$

20. $|x^2 - 16| = 0$

21. $|x^2 - 2x| = 3$

22. $|x^2 + x| = 12$

23. $|x^2 + x - 1| = 1$

24. $|x^2 + 3x - 2| = 2$

In Problems 25–44, solve each inequality. Express your answer using set notation or interval notation.

25. $|2x| < 8$

26. $|3x| < 15$

27. $|3x| > 12$

28. $|2x| > 6$

29. $|x - 2| + 2 < 3$

30. $|x + 4| + 3 < 5$

31. $|3t - 2| \leq 4$

32. $|2u + 5| \leq 7$

33. $|x - 3| \geq 2$

34. $|x + 4| \geq 2$

35. $|1 - 4x| - 7 < -2$

36. $|1 - 2x| - 4 < -1$

37. $|1 - 2x| > 3$

38. $|2 - 3x| > 1$

39. $|-4x| + |-5| \leq 1$

40. $|-x| - |4| \leq 2$

41. $|-2x| > |-3|$

42. $|-x - 2| \geq 1$

43. $-|2x - 1| \geq -3$

44. $-|1 - 2x| \geq -3$

In Problems 45–50, find a and b.

45. If $|x - 1| < 3$, then $a < x + 4 < b$.

46. If $|x + 2| < 5$, then $a < x - 2 < b$.

47. If $|x + 4| \leq 2$, then $a \leq 2x - 3 \leq b$.

48. If $|x - 3| \leq 1$, then $a \leq 3x + 1 \leq b$.

49. If $|x - 2| \leq 7$, then $a \leq \dfrac{1}{x - 10} \leq b$.

50. If $|x + 1| \leq 3$, then $a \leq \dfrac{1}{x + 5} \leq b$.

51. If $b \neq 0$, prove that $\left|\dfrac{a}{b}\right| = \dfrac{|a|}{|b|}$.

52. Show that $a \leq |a|$.

53. Prove the *triangle inequality* $|a + b| \leq |a| + |b|$.
[**Hint:** Expand $|a + b|^2 = (a + b)^2$, and use the result of Problem 52.]

54. Prove that $|a - b| \geq |a| - |b|$.
[**Hint:** Apply the triangle inequality from Problem 53 to $|a| = |(a - b) + b|$.]

55. Express the fact that x differs from 3 by less than $\frac{1}{2}$ as an inequality involving an absolute value. Solve for x.

56. Express the fact that x differs from -4 by less than 1 as an inequality involving an absolute value. Solve for x.

57. Express the fact that x differs from -3 by more than 2 as an inequality involving an absolute value. Solve for x.

58. Express the fact that x differs from 2 by more than 3 as an inequality involving an absolute value. Solve for x.

59. Body Temperature "Normal" human body temperature is 98.6°F. If a temperature x that differs from normal by at least 1.5° is considered unhealthy, write the condition for an unhealthy temperature x as an inequality involving an absolute value, and solve for x.

60. Household Voltage In the United States, normal household voltage is 115 volts. However, it is not uncommon for actual voltage to differ from normal voltage by at most 5 volts. Express this situation as an inequality involving an absolute value. Use x as the actual voltage and solve for x.

61. If $a > 0$, show that the solution set of the inequality
$$x^2 < a$$
consists of all numbers x for which
$$-\sqrt{a} < x < \sqrt{a}$$

62. If $a > 0$, show that the solution set of the inequality
$$x^2 > a$$
consists of all numbers x for which
$$x < -\sqrt{a} \quad \text{or} \quad x > \sqrt{a}$$

In Problems 63–70, use the results found in Problems 61 and 62 to solve each inequality.

63. $x^2 < 1$

64. $x^2 < 4$

65. $x^2 \geq 9$

66. $x^2 \geq 1$

67. $x^2 \leq 16$

68. $x^2 \leq 9$

69. $x^2 > 4$

70. $x^2 > 16$

71. Solve $|3x - |2x + 1|| = 4$.

72. Solve $|x + |3x - 2|| = 2$.

---

73. The equation $|x| = -2$ has no solution. Explain why.

74. The inequality $|x| > -0.5$ has all real numbers as the solution. Explain why.

75. The inequality $|x| > 0$ has as solution set $\{x | x \neq 0\}$. Explain why.

Things To Know

Quadratic equation and quadratic formula (p. 111)

If $ax^2 + bx + c = 0$, $a \neq 0$, and if $b^2 - 4ac \geq 0$, then $x = \dfrac{-b \pm \sqrt{b^2 - 4ac}}{2a}$.

Discriminant (p. 111)

If $b^2 - 4ac > 0$, there are two distinct real solutions.

If $b^2 - 4ac = 0$, there is one repeated real solution.

If $b^2 - 4ac < 0$, there are no real solutions.

Interval notation (p. 125)

$[a, b]$	$\{x \mid a \leq x \leq b\}$	$[a, \infty)$	$\{x \mid x \geq a\}$
$[a, b)$	$\{x \mid a \leq x < b\}$	(a, ∞)	$\{x \mid x > a\}$
$(a, b]$	$\{x \mid a < x \leq b\}$	$(-\infty, a]$	$\{x \mid x \leq a\}$
(a, b)	$\{x \mid a < x < b\}$	$(-\infty, a)$	$\{x \mid x < a\}$
		$(-\infty, \infty)$	All real numbers

Properties of inequalities

Addition property (p. 126)   If $a < b$, then $a + c < b + c$.
If $a > b$, then $a + c > b + c$.

Multiplication properties (p. 127)   (a) If $a < b$ and if $c > 0$, then $ac < bc$.   (b) If $a > b$ and if $c > 0$, then $ac > bc$.
If $a < b$ and if $c < 0$, then $ac > bc$.   If $a > b$ and if $c < 0$, then $ac < bc$.

Reciprocal properties (p. 128)   If $a > 0$, then $\dfrac{1}{a} > 0$.   If $a < 0$, then $\dfrac{1}{a} < 0$.

Absolute value

If $|u| = a, a > 0$, then $u = -a$ or $u = a$. (p. 135)
If $|u| \leq a, a > 0$, then $-a \leq u \leq a$. (p. 136)
If $|u| \geq a, a > 0$, then $u \leq -a$ or $u \geq a$. (p. 137)

Objectives

You should be able to:

Solve an equation in one variable (p. 84)

Solve a linear equation (p. 88)

Solve equations that lead to linear equations. (p. 89)

Translate verbal descriptions into mathematical expressions (p. 95)

Set up applied problems (p.95)

Solve interest problems (p. 97)

Solve mixture problems (p. 99)

Solve uniform motion problems (p. 100)

Solve constant rate job problems (p. 101)

Solve a quadratic equation by factoring (p. 106)

Know how to complete the square (p. 108)

Solve a quadratic equation by completing the square (p. 109)

Solve a quadratic equation using the quadratic formula (p. 110)

Solve applied problems involving quadratic equations (p. 114)

Solve radical equations (p. 119)

Solve equations quadratic in form (p. 121)

Use interval notation (p. 125)

Use properties of inequalities (p. 126)

Solve inequalities (p. 128)

Solve combined inequalities (p. 130)

Solve equations involving absolute value (p. 135)

Solve inequalities involving absolute value (p. 136)

Fill-in-the-Blank Items

1. Two equations (or inequalities) that have precisely the same solution set are called _____.

2. An equation that is satisfied for every choice of the variable for which both sides are meaningful is called a(n) _____.

3. To complete the square of the expression $x^2 + 5x$, you would _____ the number _____.

4. The quantity $b^2 - 4ac$ is called the _____ of a quadratic equation. If it is _____, the equation has no real solution.

5. If $a < 0$, then $|a| =$ _____.

6. When a quadratic equation has a repeated solution, it is called a(n) _____ root or a root of _____ _____.

7. When an apparent solution does not satisfy the original equation, it is called a(n) _____ solution.

8. If each side of an inequality is multiplied by a(n) _____ number, then the sense of the inequality symbol is reversed.

9. The equation $|x^2| = 4$ has two real solutions, _____ and _____ .

True/False Items

T F **1.** Equations can have no solutions, one solution, or more than one solution.

T F **2.** Quadratic equations always have two real solutions.

T F **3.** If the discriminant of a quadratic equation is positive, then the equation has two solutions that are negatives of each other.

T F **4.** The square of any real number is always nonnegative.

T F **5.** The expression $x^2 + 1$ is positive for any real number x.

6. If $a < b$ and $c < 0$, determine whether the following statements are true or false.

T F   **(a)** $a \pm c < b \pm c$

T F   **(b)** $a \cdot c < b \cdot c$

T F   **(c)** $a/c > b/c$

7. If $a^2 < b^2, b \neq 0$, determine whether the following statements are true or false.

T F   **(a)** $a < b$

T F   **(b)** $a^2/b^2 < 1$

T F   **(c)** $b^2 - a^2 > 0$

Review Exercises

Blue problem numbers indicate the author's suggestions for use in a Practice Test.

In Problems 1–42, find all real solutions, if any, of each equation. (Where they appear, a, b, m, and n are positive constants.)

1. $2 - \dfrac{x}{3} = 8$

2. $\dfrac{x}{4} - 2 = 4$

3. $-2(5 - 3x) + 8 = 4 + 5x$

4. $(6 - 3x) - 2(1 + x) = 6x$

5. $\dfrac{3x}{4} - \dfrac{x}{3} = \dfrac{1}{12}$

6. $\dfrac{4 - 2x}{3} + \dfrac{1}{6} = 2x$

7. $\dfrac{x}{x - 1} = \dfrac{6}{5}, \quad x \neq 1$

8. $\dfrac{4x - 5}{3 - 7x} = 2, \quad x \neq \frac{3}{7}$

9. $x(1 - x) = 6$

10. $x(1 + x) = 6$

11. $\dfrac{1}{2}\left(x - \dfrac{1}{3}\right) = \dfrac{3}{4} - \dfrac{x}{6}$

12. $\dfrac{1 - 3x}{4} = \dfrac{x + 6}{3} + \dfrac{1}{2}$

13. $(x - 1)(2x + 3) = 3$

14. $x(2 - x) = 3(x - 4)$

15. $2x + 3 = 4x^2$

16. $1 + 6x = 4x^2$

17. $\sqrt[3]{x^2 - 1} = 2$

18. $\sqrt[3]{1 + x^3} = 3$

19. $x(x + 1) + 2 = 0$

20. $3x^2 - x + 1 = 0$

21. $x^4 - 5x^2 + 4 = 0$

22. $3x^4 + 4x^2 + 1 = 0$

23. $\sqrt{2x - 3} + x = 3$

24. $\sqrt{2x - 1} = x - 2$

25. $x^{3/2} + 5x^{1/2} = 0$

26. $x^{2/3} + x = 0$

27. $\sqrt{x + 1} + \sqrt{x - 1} = \sqrt{2x + 1}$

28. $\sqrt{2x - 1} - \sqrt{x - 5} = 3$

29. $2\sqrt[3]{x^2} - \sqrt[3]{x} = 1$

30. $4\sqrt[3]{x^2} = 1$

31. $x^{-6} - 7x^{-3} - 8 = 0$

32. $6x^{-1} - 5x^{-1/2} + 1 = 0$

33. $x^2 + m^2 = 2mx + (nx)^2$

34. $b^2x^2 + 2ax = x^2 + a^2$

35. $10a^2x^2 - 2abx - 36b^2 = 0$

36. $\dfrac{1}{x - m} + \dfrac{1}{x - n} = \dfrac{2}{x}, \quad x \neq 0, m, n$

37. $\sqrt{x^2 + 3x + 7} - \sqrt{x^2 - 3x + 9} + 2 = 0$

38. $\sqrt{x^2 + 3x + 7} - \sqrt{x^2 + 3x + 9} = 2$

39. $|2x + 3| = 7$

40. $|3x - 1| = 5$

41. $|2 - 3x| + 2 = 9$

42. $|1 - 2x| + 1 = 4$

In Problems 43–56, solve each inequality. Express your answer using set notation or interval notation. Graph the solution set.

43. $\dfrac{2x - 3}{5} + 2 \le \dfrac{x}{2}$

44. $\dfrac{5 - x}{3} \le 6x - 4$

45. $-9 \le \dfrac{2x + 3}{-4} \le 7$

46. $-4 < \dfrac{2x - 2}{3} < 6$

47. $6 > \dfrac{3 - 3x}{12} > 2$

48. $6 > \dfrac{5 - 3x}{2} \ge -3$

49. $|3x + 4| < \frac{1}{2}$

50. $|1 - 2x| < \frac{1}{3}$

51. $|2x - 5| \ge 9$

52. $|3x + 1| \ge 10$

53. $2 + |2 - 3x| \le 4$

54. $\dfrac{1}{2} + \left|\dfrac{2x - 1}{3}\right| \le 1$

55. $1 - |2 - 3x| < -4$

56. $1 - \left|\dfrac{2x - 1}{3}\right| < -2$

In Problems 57–60, what number should be added to complete the square of each expression.

57. $x^2 + 6x$

58. $x^2 - 10x$

59. $x^2 - \dfrac{4}{3}x$

60. $x^2 + \dfrac{4}{5}x$

---

61. Lightning and Thunder   A flash of lightning is seen, and the resulting thunderclap is heard 3 seconds later. If the speed of sound averages 1100 feet per second, how far away is the storm?

62. Physics: Intensity of Light   The intensity I (in candlepower) of a certain light source obeys the equation $I = 900/x^2$, where x is the distance (in meters) from the light. Over what range of distances can an object be placed from this light source so that the range of intensity of light is from 1600 to 3600 candlepower, inclusive?

63. Extent of Search and Rescue   A search plane has a cruising speed of 250 miles per hour and carries enough fuel for at most 5 hours of flying. If there is a wind that averages 30 miles per hour and the direction of the search is with the wind one way and against it the other, how far can the search plane travel before it has to turn back?

64. Extent of Search and Rescue   If the search plane described in Problem 63 is able to add a supplementary fuel tank that allows for an additional 2 hours of flying, how much farther can the plane extend its search?

65. Rescue at Sea   A life raft, set adrift from a sinking ship 150 miles offshore, travels directly toward a Coast Guard station at the rate of 5 miles per hour. At the time that the raft is set adrift, a rescue helicopter is dispatched from the Coast Guard station. If the helicopter's average speed is 90 miles per hour, how long will it take the helicopter to reach the life raft?

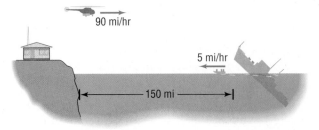

66. Physics: Uniform Motion   Two bees leave two locations 150 meters apart and fly, without stopping, back and forth between these two locations at average speeds of 3 meters per second and 5 meters per second, respectively. How long is it until the bees meet for the first time? How long is it until they meet for the second time?

67. Working Together to Get a Job Done   Clarissa and Shawna, working together, can paint the exterior of a house in 6 days. Clarissa by herself can complete this job in 5 days less than Shawna. How long will it take Clarissa to complete the job by herself?

68. Emptying a Tank   Two pumps of different sizes, working together, can empty a fuel tank in 5 hours. The larger pump can empty this tank in 4 hours less than the smaller. If the larger pump is out of order, how long will it take the smaller one to do the job alone?

69. Chemistry: Mixing Acids   A laboratory has 60 cubic centimeters (cm^3) of a solution that is 40% HCl acid. How many cubic centimeters of a 15% solution of HCl acid should be mixed with the 60 cm^3 of 40% acid to obtain a solution of 25% HCl? How much of the 25% solution is there?

70. Business: Blending Coffee   A coffee house has 20 pounds of a coffee that sells for $4 per pound. How many pounds of a coffee that sells for $8 per pound should be mixed with the 20 pounds of $4-per-pound coffee to obtain a blend that will sell for $5 per pound? How much of the $5-per-pound coffee is there to sell?

71. Chemistry: Salt Solutions   How much water should be added to 64 ounces of a 10% salt solution to make a 2% salt solution?

72. Chemistry: Salt Solutions   How much water must be evaporated from 64 ounces of a 2% salt solution to make a 10% salt solution?

73. Geometry   The hypotenuse of a right triangle measures 13 centimeters. Find the lengths of the legs if their sum is 17 centimeters.

74. Geometry   The diagonal of a rectangle measures 10 inches. If the length is 2 inches more than the width, find the dimensions of the rectangle.

75. Physics: Uniform Motion A man is walking at an average speed of 4 miles per hour alongside a railroad track. A freight train, going in the same direction at an average speed of 30 miles per hour, requires 5 seconds to pass the man. How long is the freight train? Give your answer in feet.

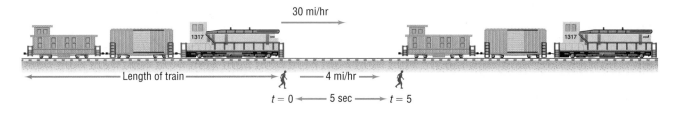

76. Framing a Painting An artist has 50 inches of oak trim to frame a painting. The frame is to have a border 3 inches wide surrounding the painting.
(a) If the painting is square, what are its dimensions? What are the dimensions of the frame?
(b) If the painting is rectangular with a length twice its width, what are the dimensions of the painting? What are the dimensions of the frame?

77. Using Two Pumps An 8 horsepower (hp) pump can fill a tank in 8 hours. A smaller, 3 hp pump fills the same tank in 12 hours. The pumps are used together to begin filling this tank. After four hours, the 8 hp pump breaks down. How long will it take the smaller pump to fill the tank?

78. Pleasing Proportions One formula stating the relationship between the length l and width w of a rectangle of "pleasing proportion" is $l^2 = w(l + w)$. How should a 4 foot by 8 foot sheet of plasterboard be cut so that the result is a rectangle of "pleasing proportion" with a width of 4 feet?

79. Business: Determining the Cost of a Charter A group of 20 senior citizens can charter a bus for a one-day excursion trip for $15 per person. The charter company agrees to reduce the price of each ticket by 10¢ for each additional passenger in excess of 20 who goes on the trip, up to a maximum of 44 passengers (the capacity of the bus).

If the final bill from the charter company was $482.40, how many seniors went on the trip, and how much did each pay?

80. Utilizing Copying Machines A new copying machine can do a certain job in 1 hour less than an older copier. Together they can do this job in 72 minutes. How long would it take the older copier by itself to do the job?

81. In a 100-meter race, Todd crosses the finish line 5 meters ahead of Scott. To even things up, Todd suggests to Scott that they race again, this time with Todd lining up 5 meters behind the start.
(a) Assuming that Todd and Scott run at the same pace as before, does the second race end in a tie?
(b) If not, who wins?
(c) By how many meters does he win?
(d) How far back should Todd start so that the race ends in a tie?
After running the race a second time, Scott, to even things up, suggests to Todd that he (Scott) line up 5 meters in front of the start.
(e) Assuming again that they run at the same pace as in the first race, does the third race result in a tie?
(f) If not, who wins?
(g) By how many meters?
(h) How far up should Scott start so that the race ends in a tie?

Project at Motorola

How Many Cellular Phones Can I Make?

Motorola uses several types of specialized robots to assemble electronic products accurately and quickly. The picture and figure on page 144 depict the operation of a turret-type machine for assembling surface mount technology (SMT) printed wiring boards. Resistors, capacitors, and other parts are picked from the feeder carriage by one of the vacuum nozzles on the rotating turret. The X–Y table, which holds a panel containing one or more boards, moves to the placement location for a given part, while the turret advances one position. The average time to pick and place one part is called the *tact time*, for example, 0.1 second per part. The actual throughput, measured in boards per hour or in placements per hour (PPH), depends on a number of factors, including the number and variety of parts on the panel, the time to load and unload a panel in the machine, and the time to visually inspect fiducial marks and calibrate the position of the panel with respect to the coordinate system of the machine. *(continued)*

Throughput T can be estimated with the following equation, where n is the number of boards per panel, p is the number of parts per board, C is the tact time, L is the load–unload time, and M is the mark reading time.

$$T = \frac{n}{Cnp + L + M} \quad \textbf{(1)}$$

1. Find an equation that relates throughput, T, with the number of parts per board, p, if there are 3 boards per panel, the load-unload time is 5 seconds, the mark reading is 1 second, and the tact time is 0.2 seconds.

2. For the equation found in Problem 1, determine the number of parts per board, p, if throughput T is to be 360 boards per hour (0.1 board per second).

3. For the equation found in Problem 1, determine the number of parts per board, p, if throughput T is to be 540 boards per hour (0.15 board per second).

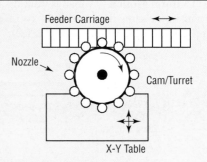

Feeder Carriage

Nozzle

Cam/Turret

X-Y Table

In Problems 4–6, use equation (1).
[**Hint:** Solve (1) for C first].

4. Determine the tact time, C, if the throughput is 216 boards per hour (0.06 board per second), there are 3 boards per panel, 100 parts per board, the mark reading is 1 second, and the load–unload time is 5 seconds.

5. Determine the tact time, C, if the throughput is 216 boards per hour (0.06 board per second), there are 3 boards per panel, 150 parts per board, the mark reading is 1 second, and the load–unload time is 5 seconds.

6. Determine the tact time, C, if the throughput is 216 boards per hour (0.06 board per second), there are 3 boards per panel, 200 parts per board, the mark reading is 1 second, and the load–unload time is 5 seconds.

7. Based on the results of Problems 4–6, as the number of parts per board increases, what happens to tact time if all other factors remain constant?

Graphs

Field Trip To Motorola

During the 1990s, mobile phones became a significant means of communication. Industry data indicate that in 1994 there were about 37.5 million mobile phones in use globally. In 1998, this figure rose to about 70 million mobile phones and is projected to rise to 102.5 million by 2002. Industry analysts base such predictions on a *mathematical model,* that is, a relationship (usually an equation) that involves one or more variables.

PREPARING FOR THIS SECTION

Before getting started, review the following:

✓ Real Numbers (Review, Section 1, pp. 2–13) ✓ Geometry Review (Review, Section 3, pp. 22–24)

✓ Algebra Review (Review, Section 2, pp. 16–21)

2.1 RECTANGULAR COORDINATES

OBJECTIVES **1** Use the Distance Formula
2 Use the Midpoint Formula

We locate a point on the real number line by assigning it a single real number, called the *coordinate of the point*. For work in a two-dimensional plane, we locate points by using two numbers.

We begin with two real number lines located in the same plane: one horizontal and the other vertical. We call the horizontal line the **x-axis;** the vertical line, the **y-axis;** and the point of intersection, the **origin O.** We assign coordinates to every point on these number lines as shown in Figure 1, using a convenient scale. In mathematics, we usually use the same scale on each axis; in applications, a different scale is often used on each axis.

The origin O has a value of 0 on both the x-axis and the y-axis. We follow the usual convention that points on the x-axis to the right of O are associated with positive real numbers, and those to the left of O are associated with negative real numbers. Those on the y-axis above O are associated with positive real numbers, and those below O are associated with negative real numbers. In Figure 1, the x-axis and y-axis are labeled as x and y, respectively, and we have used an arrow at the end of each axis to denote the positive direction.

Figure 1

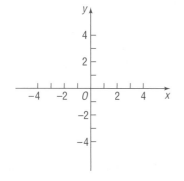

The coordinate system described here is called a **rectangular** or **Cartesian*** **coordinate system.** The plane formed by the *x*-axis and *y*-axis is sometimes called the **xy-plane,** and the *x*-axis and *y*-axis are referred to as the **coordinate axes.**

Any point *P* in the *xy*-plane can then be located by using an **ordered pair** (x, y) of real numbers. Let *x* denote the signed distance of *P* from the *y*-axis (*signed* in the sense that, if *P* is to the right of the *y*-axis, then $x > 0$, and if *P* is to the left of the *y*-axis, then $x < 0$); and let *y* denote the signed distance of *P* from the *x*-axis. The ordered pair (x, y), also called the **coordinates** of *P*, then gives us enough information to locate the point *P* in the plane.

For example, to locate the point whose coordinates are $(-3, 1)$, go 3 units along the *x*-axis to the left of *O* and then go straight up 1 unit. We **plot** this point by placing a dot at this location. See Figure 2, in which the points with coordinates $(-3, 1)$, $(-2, -3)$, $(3, -2)$, and $(3, 2)$ are plotted.

The origin has coordinates $(0, 0)$. Any point on the *x*-axis has coordinates of the form $(x, 0)$, and any point on the *y*-axis has coordinates of the form $(0, y)$.

If (x, y) are the coordinates of a point *P*, then *x* is called the **x-coordinate,** or **abscissa,** of *P* and *y* is the **y-coordinate,** or **ordinate,** of *P*. We identify the point *P* by its coordinates (x, y) by writing $P = (x, y)$, referring to it as "the point (x, y)," rather than "the point whose coordinates are (x, y)."

The coordinate axes divide the *xy*-plane into four sections, called **quadrants,** as shown in Figure 3. In quadrant I, both the *x*-coordinate and the *y*-coordinate of all points are positive; in quadrant II, *x* is negative and *y* is positive; in quadrant III, both *x* and *y* are negative; and in quadrant IV, *x* is positive and *y* is negative. Points on the coordinate axes belong to no quadrant.

Figure 2

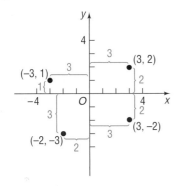

Figure 3

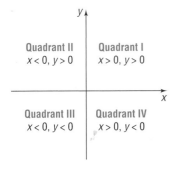

NOW WORK PROBLEM 1.

COMMENT: On a graphing calculator, you can set the scale on each axis. Once this has been done, you obtain the viewing rectangle. See Figure 4 for a typical **viewing rectangle.** You should now read Section 1, The Viewing Rectangle, in the Appendix.

Figure 4

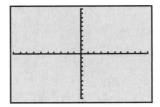

DISTANCE BETWEEN POINTS

1   If the same units of measurement, such as inches and centimeters, are used for both the *x*-axis and the *y*-axis, then all distances in the *xy*-plane can be measured using this unit of measurement.

*Named after René Descartes (1596–1650), a French mathematician, philosopher, and theologian.

| EXAMPLE 1 | **Finding the Distance between Two Points** |

Find the distance d between the points $(1, 3)$ and $(5, 6)$.

Figure 5

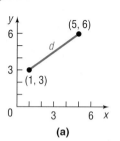

(a)

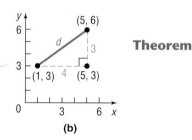

(b)

Solution First we plot the points $(1, 3)$ and $(5, 6)$ as shown in Figure 5(a). Then we draw a horizontal line from $(1, 3)$ to $(5, 3)$ and a vertical line from $(5, 3)$ to $(5, 6)$, forming a right triangle, as in Figure 5(b). One leg of the triangle is of length 4 and the other is of length 3. By the Pythagorean Theorem (Review, Section 3), the square of the distance d that we seek is

$$d^2 = 4^2 + 3^2 = 16 + 9 = 25$$

$$d = 5$$

The **distance formula** provides a straightforward method for computing the distance between two points.

Theorem **Distance Formula**

The distance between two points $P_1 = (x_1, y_1)$ and $P_2 = (x_2, y_2)$, denoted by $d(P_1, P_2)$, is

$$d(P_1, P_2) = \sqrt{(x_2 - x_1)^2 + (y_2 - y_1)^2} \qquad \textbf{(1)}$$

That is, to compute the distance between two points, find the difference of the x-coordinates, square it, and add this to the square of the difference of the y-coordinates. The square root of this sum is the distance. See Figure 6.

Figure 6

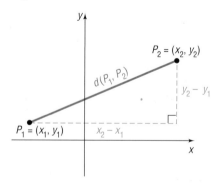

Proof of the Distance Formula Let (x_1, y_1) denote the coordinates of point P_1, and let (x_2, y_2) denote the coordinates of point P_2. Assume that the line joining P_1 and P_2 is neither horizontal nor vertical. Refer to Figure 7(a). The coordinates of P_3 are (x_2, y_1). The horizontal distance from P_1 to P_3 is the absolute value of the difference of the x-coordinates, $|x_2 - x_1|$. The vertical distance from P_3 to P_2 is the absolute value of the difference of the y-coordinates, $|y_2 - y_1|$. See Figure 7(b). The distance $d(P_1, P_2)$ that we seek is the length of the hypotenuse of the right triangle, so, by the Pythagorean Theorem, it follows that

$$[d(P_1, P_2)]^2 = |x_2 - x_1|^2 + |y_2 - y_1|^2$$

$$= (x_2 - x_1)^2 + (y_2 - y_1)^2$$

$$d(P_1, P_2) = \sqrt{(x_2 - x_1)^2 + (y_2 - y_1)^2}$$

Figure 7

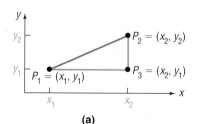

(a)

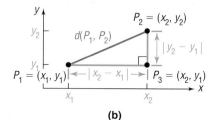

(b)

Now, if the line joining P_1 and P_2 is horizontal, then the y-coordinate of P_1 equals the y-coordinate of P_2; that is, $y_1 = y_2$. Refer to Figure 8(a). In this case, the distance formula (1) still works, because, for $y_1 = y_2$, it reduces to

$$d(P_1, P_2) = \sqrt{(x_2 - x_1)^2 + 0^2} = \sqrt{(x_2 - x_1)^2} = |x_2 - x_1|$$

Figure 8

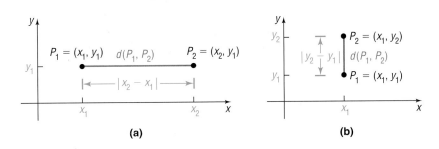

(a)   (b)

A similar argument holds if the line joining P_1 and P_2 is vertical. See Figure 8(b).

The distance formula is valid in all cases. ■

EXAMPLE 2 **Finding the Distance between Two Points**

Find the distance d between the points $(-4, 5)$ and $(3, 2)$.

Solution Using the distance formula (1), the solution is obtained as follows:

$$d = \sqrt{[3 - (-4)]^2 + (2 - 5)^2} = \sqrt{7^2 + (-3)^2}$$

$$= \sqrt{49 + 9} = \sqrt{58} \approx 7.62 \quad ■$$

✏ NOW WORK PROBLEMS **5** AND **9**.

The distance between two points $P_1 = (x_1, y_1)$ and $P_2 = (x_2, y_2)$ is never a negative number. Furthermore, the distance between two points is 0 only when the points are identical, that is, when $x_1 = x_2$ and $y_1 = y_2$. Also, because $(x_2 - x_1)^2 = (x_1 - x_2)^2$ and $(y_2 - y_1)^2 = (y_1 - y_2)^2$, it makes no difference whether the distance is computed from P_1 to P_2 or from P_2 to P_1; that is, $d(P_1, P_2) = d(P_2, P_1)$.

Rectangular coordinates enable us to translate geometry problems into algebra problems, and vice versa. The next example shows how algebra (the distance formula) can be used to solve geometry problems.

EXAMPLE 3 **Using Algebra to Solve Geometry Problems**

Consider the three points $A = (-2, 1)$, $B = (2, 3)$, and $C = (3, 1)$.

(a) Plot each point and form the triangle ABC.

(b) Find the length of each side of the triangle.

(c) Verify that the triangle is a right triangle.

(d) Find the area of the triangle.

Figure 9

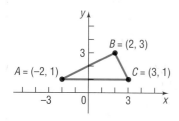

Solution

(a) Points A, B, and C and triangle ABC are plotted in Figure 9.

(b) $d(A, B) = \sqrt{[2 - (-2)]^2 + (3 - 1)^2} = \sqrt{16 + 4} = \sqrt{20} = 2\sqrt{5}$

$d(B, C) = \sqrt{(3 - 2)^2 + (1 - 3)^2} = \sqrt{1 + 4} = \sqrt{5}$

$d(A, C) = \sqrt{[3 - (-2)]^2 + (1 - 1)^2} = \sqrt{25 + 0} = 5$

(c) To show that the triangle is a right triangle, we need to show that the sum of the squares of the lengths of two of the sides equals the square of the length of the third side. (Why is this sufficient?) Looking at Figure 9, it seems reasonable to conjecture that the right angle is at vertex B. To verify, we check to see whether

$$[d(A, B)]^2 + [d(B, C)]^2 = [d(A, C)]^2$$

We find that

$$[d(A, B)]^2 + [d(B, C)]^2 = (2\sqrt{5})^2 + (\sqrt{5})^2$$
$$= 20 + 5 = 25 = [d(A, C)]^2$$

so it follows from the converse of the Pythagorean Theorem that triangle ABC is a right triangle.

(d) Because the right angle is at B, the sides AB and BC form the base and altitude of the triangle. Its area is therefore

$$\text{Area} = \frac{1}{2}(\text{Base})(\text{Altitude}) = \frac{1}{2}(2\sqrt{5})(\sqrt{5}) = 5 \text{ square units}$$

NOW WORK PROBLEM **19**.

MIDPOINT FORMULA

Figure 10

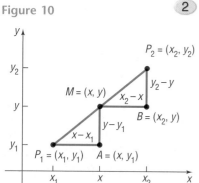

2

We now derive a formula for the coordinates of the **midpoint of a line segment.** Let $P_1 = (x_1, y_1)$ and $P_2 = (x_2, y_2)$ be the endpoints of a line segment, and let $M = (x, y)$ be the point on the line segment that is the same distance from P_1 as it is from P_2. See Figure 10. The triangles P_1AM and MBP_2 are congruent.* [Do you see why? Angle AP_1M = Angle BMP_2.† Angle P_1MA = Angle MP_2B, and $d(P_1, M) = d(M, P_2)$ is given. As a result, we have Angle–Side–Angle.] Hence, corresponding sides are equal in length. That is,

$$x - x_1 = x_2 - x \quad \text{and} \quad y - y_1 = y_2 - y$$
$$2x = x_1 + x_2 \qquad\qquad 2y = y_1 + y_2$$
$$x = \frac{x_1 + x_2}{2} \qquad\qquad y = \frac{y_1 + y_2}{2}$$

Theorem

Midpoint Formula

The midpoint $M = (x, y)$ of the line segment from $P_1 = (x_1, y_1)$ to $P_2 = (x_2, y_2)$ is

$$M = (x, y) = \left(\frac{x_1 + x_2}{2}, \frac{y_1 + y_2}{2} \right) \qquad\qquad (2)$$

---

* The following statement is a postulate from geometry. Two triangles are congruent if their sides are the same length (SSS), or if two sides and the included angle are the same (SAS), or if two angles and the included side are the same (ASA).

† Another postulate from geometry states that the transversal $\overline{P_1P_2}$ forms equal corresponding angles with the parallel lines $\overline{P_1A}$ and $\overline{MB}$.

To find the midpoint of a line segment, we average the x-coordinates and the y-coordinates of the endpoints.

| EXAMPLE 4 | **Finding the Midpoint of a Line Segment** |

Find the midpoint of the line segment from $P_1 = (-5, 3)$ to $P_2 = (3, 1)$. Plot the points P_1 and P_2 and their midpoint. Check your answer.

Solution   We apply the midpoint formula (2) using $x_1 = -5$, $x_2 = 3$, $y_1 = 3$, and $y_2 = 1$. Then the coordinates (x, y) of the midpoint M are

Figure 11

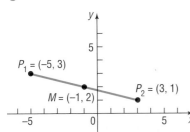

$$x = \frac{x_1 + x_2}{2} = \frac{-5 + 3}{2} = -1 \quad \text{and} \quad y = \frac{y_1 + y_2}{2} = \frac{3 + 1}{2} = 2$$

That is, $M = (-1, 2)$. See Figure 11.

Check: Because M is the midpoint, we check the answer by verifying that $d(P_1, M) = d(M, P_2)$.

$$d(P_1, M) = \sqrt{[-1 - (-5)]^2 + (2 - 3)^2} = \sqrt{16 + 1} = \sqrt{17}$$

$$d(M, P_2) = \sqrt{[3 - (-1)]^2 + (1 - 2)^2} = \sqrt{16 + 1} = \sqrt{17}$$ ◼ ◼

NOW WORK PROBLEM **29**.

2.1   EXERCISES

In Problems 1 and 2, plot each point in the xy-plane. Tell in which quadrant or on what coordinate axis each point lies.

1. (a) $A = (-3, 2)$
   (b) $B = (6, 0)$
   (c) $C = (-2, -2)$
   (d) $D = (6, 5)$
   (e) $E = (0, -3)$
   (f) $F = (6, -3)$

2. (a) $A = (1, 4)$
   (b) $B = (-3, -4)$
   (c) $C = (-3, 4)$
   (d) $D = (4, 1)$
   (e) $E = (0, 1)$
   (f) $F = (-3, 0)$

3. Plot the points $(2, 0)$, $(2, -3)$, $(2, 4)$, $(2, 1)$, and $(2, -1)$. Describe the set of all points of the form $(2, y)$, where y is a real number.

4. Plot the points $(0, 3)$, $(1, 3)$, $(-2, 3)$, $(5, 3)$, and $(-4, 3)$. Describe the set of all points of the form $(x, 3)$, where x is a real number.

In Problems 5–18, find the distance $d(P_1, P_2)$ between the points P_1 and P_2.

5.
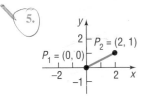

6.
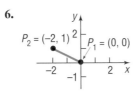

7.
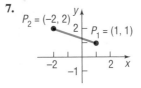

8.
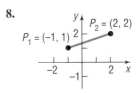

9. $P_1 = (3, -4);$ $P_2 = (5, 4)$

10. $P_1 = (-1, 0);$ $P_2 = (2, 4)$

11. $P_1 = (-3, 2);$ $P_2 = (6, 0)$

12. $P_1 = (2, -3);$ $P_2 = (4, 2)$

13. $P_1 = (4, -3);$ $P_2 = (6, 4)$

14. $P_1 = (-4, -3);$ $P_2 = (6, 2)$

15. $P_1 = (-0.2, 0.3);$ $P_2 = (2.3, 1.1)$

16. $P_1 = (1.2, 2.3);$ $P_2 = (-0.3, 1.1)$

17. $P_1 = (a, b);$ $P_2 = (0, 0)$

18. $P_1 = (a, a);$ $P_2 = (0, 0)$

In Problems 19–24, plot each point and form the triangle ABC. Verify that the triangle is a right triangle. Find its area.

19. $A = (-2, 5);$ $B = (1, 3);$ $C = (-1, 0)$

20. $A = (-2, 5);$ $B = (12, 3);$ $C = (10, -11)$

21. $A = (-5, 3);$ $B = (6, 0);$ $C = (5, 5)$

22. $A = (-6, 3);$ $B = (3, -5);$ $C = (-1, 5)$

23. $A = (4, -3);$ $B = (0, -3);$ $C = (4, 2)$

24. $A = (4, -3);$ $B = (4, 1);$ $C = (2, 1)$

25. Find all points having an x-coordinate of 2 whose distance from the point $(-2, -1)$ is 5.

26. Find all points having a y-coordinate of -3 whose distance from the point $(1, 2)$ is 13.

27. Find all points on the x-axis that are 5 units from the point $(4, -3)$.

28. Find all points on the y-axis that are 5 units from the point $(4, 4)$.

In Problems 29–38, find the midpoint of the line segment joining the points P_1 and P_2.

29. $P_1 = (5, -4);$ $P_2 = (3, 2)$

30. $P_1 = (-1, 0);$ $P_2 = (2, 4)$

31. $P_1 = (-3, 2);$ $P_2 = (6, 0)$

32. $P_1 = (2, -3);$ $P_2 = (4, 2)$

33. $P_1 = (4, -3);$ $P_2 = (6, 1)$

34. $P_1 = (-4, -3);$ $P_2 = (2, 2)$

35. $P_1 = (-0.2, 0.3);$ $P_2 = (2.3, 1.1)$

36. $P_1 = (1.2, 2.3);$ $P_2 = (-0.3, 1.1)$

37. $P_1 = (a, b);$ $P_2 = (0, 0)$

38. $P_1 = (a, a);$ $P_2 = (0, 0)$

39. The **medians** of a triangle are the line segments from each vertex to the midpoint of the opposite side (see the figure). Find the lengths of the medians of the triangle with vertices at $A = (0, 0)$, $B = (0, 6)$, and $C = (4, 4)$.

40. An **equilateral triangle** is one in which all three sides are of equal length. If two vertices of an equilateral triangle are $(0, 4)$ and $(0, 0)$, find the third vertex. How many of these triangles are possible?

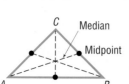

*In Problems 41–44, find the length of each side of the triangle determined by the three points P_1, P_2, and P_3. State whether the triangle is an isosceles triangle, a right triangle, neither of these, or both. (An **isosceles triangle** is one in which at least two of the sides are of equal length.)*

41. $P_1 = (2, 1);$ $P_2 = (-4, 1);$ $P_3 = (-4, -3)$

42. $P_1 = (-1, 4);$ $P_2 = (6, 2);$ $P_3 = (4, -5)$

43. $P_1 = (-2, -1);$ $P_2 = (0, 7);$ $P_3 = (3, 2)$

44. $P_1 = (7, 2);$ $P_2 = (-4, 0);$ $P_3 = (4, 6)$

In Problems 45–48, find the length of the line segment. Assume that the endpoints of each line segment have integer coordinates.

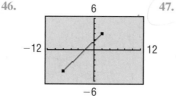

49. **Geometry** Find the midpoint of each diagonal of a square with side of length s. Draw the conclusion that the diagonals of a square intersect at their midpoints. [**Hint:** Use $(0, 0)$, $(0, s)$, $(s, 0)$, and (s, s) as the vertices of the square.]

50. **Geometry** Verify that the points $(0, 0)$, $(a, 0)$, and $(a/2, \sqrt{3}a/2)$ are the vertices of an equilateral triangle. Then show that the midpoints of the three sides are the vertices of a second equilateral triangle (refer to Problem 40.)

51. Baseball   A major league baseball "diamond" is actually a square, 90 feet on a side (see the figure). What is the distance directly from home plate to second base (the diagonal of the square)?

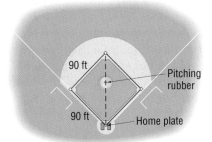

52. Little League Baseball   The layout of a Little League playing field is a square, 60 feet on a side.* How far is it directly from home plate to second base (the diagonal of the square)?

53. Baseball   Refer to Problem 51. Overlay a rectangular coordinate system on a major league baseball diamond so that the origin is at home plate, the positive *x*-axis lies in the direction from home plate to first base, and the positive *y*-axis lies in the direction from home plate to third base.
   (a) What are the coordinates of first base, second base, and third base? Use feet as the unit of measurement.
   (b) If the right fielder is located at $(310, 15)$, how far is it from the right fielder to second base?
   (c) If the center fielder is located at $(300, 300)$, how far is it from the center fielder to third base?

54. Little League Baseball   Refer to Problem 52. Overlay a rectangular coordinate system on a Little League baseball

* Source: Little League Baseball, Official Regulations and Playing Rules, 2000.

diamond so that the origin is at home plate, the positive *x*-axis lies in the direction from home plate to first base, and the positive *y*-axis lies in the direction from home plate to third base.
   (a) What are the coordinates of first base, second base, and third base? Use feet as the unit of measurement.
   (b) If the right fielder is located at $(180, 20)$, how far is it from the right fielder to second base?
   (c) If the center fielder is located at $(220, 220)$, how far is it from the center fielder to third base?

55. A Dodge Intrepid and a Mack truck leave an intersection at the same time. The Intrepid heads east at an average speed of 30 miles per hour, while the truck heads south at an average speed of 40 miles per hour. Find an expression for their distance apart *d* (in miles) at the end of *t* hours.

56. A hot-air balloon, headed due east at an average speed of 15 miles per hour and at a constant altitude of 100 feet, passes over an intersection (see the figure). Find an expression for its distance *d* (measured in feet) from the intersection *t* seconds later.

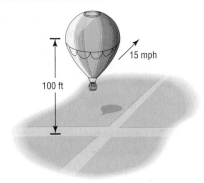

P R E P A R I N G   F O R   T H I S   S E C T I O N

Before getting started, review the following:

✓ Equations (Section 1.1, pp. 84–92)

2.2   GRAPHS OF EQUATIONS

OBJECTIVES
   1   Graph Equations by Plotting Points
   2   Find Intercepts from a Graph
   3   Find Intercepts from an Equation
   4   Test an Equation for Symmetry with Respect to the (a) *x*-axis, (b) *y*-axis, and (c) origin

An equation in two variables, say *x* and *y*, is a statement in which two expressions involving *x* and *y* are equal. The expressions are called the **sides** of the equation. Since an equation is a statement, it may be true or false, depending on the value of the variables. Any values of *x* and *y* that result in a true statement are said to **satisfy** the equation.

For example, the following are all equations in two variables x and y:

$$x^2 + y^2 = 5 \qquad 2x - y = 6 \qquad y = 2x + 5 \qquad x^2 = y$$

The first of these, $x^2 + y^2 = 5$, is satisfied for $x = 1$, $y = 2$, since $1^2 + 2^2 = 1 + 4 = 5$. Other choices of x and y also satisfy this equation. It is not satisfied for $x = 2$ and $y = 3$, since $2^2 + 3^2 = 4 + 9 = 13 \neq 5$.

The **graph of an equation** in two variables x and y consists of the set of points in the xy-plane whose coordinates (x, y) satisfy the equation.

Graphs play an important role in helping us to visualize the relationships that exist between two variable quantities. Figure 12 shows the monthly closing prices of Intel stock from August 31, 1999, through August 31, 2000. For example, the closing price on January 31, 2000, was about $50 per share.

Figure 12
Monthly closing prices of Intel stock,
8/31/99 to 8/31/00.

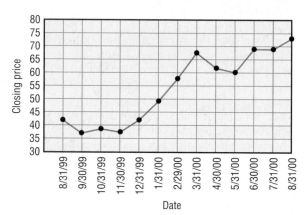

| EXAMPLE 1 | **Determining Whether a Point Is on the Graph of an Equation** |

Determine if the following points are on the graph of the equation $2x - y = 6$.

(a) $(2, 3)$ (b) $(2, -2)$

Solution   (a) For the point $(2, 3)$, we check to see if $x = 2$, $y = 3$ satisfies the equation $2x - y = 6$.

$$2x - y = 2(2) - 3 = 4 - 3 = 1 \neq 6$$

The equation is not satisfied, so the point $(2, 3)$ is not on the graph.

(b) For the point $(2, -2)$, we have

$$2x - y = 2(2) - (-2) = 4 + 2 = 6$$

The equation is satisfied, so the point $(2, -2)$ is on the graph.   ■

 NOW WORK PROBLEM **23**.

| EXAMPLE 2 | **Graphing an Equation by Plotting Points** |

Graph the equation: $y = 2x + 5$

Solution   We want to find all points (x, y) that satisfy the equation. To locate some of these points (and thus get an idea of the pattern of the graph), we assign some numbers to x and find corresponding values for y.

Figure 13
$y = 2x + 5$

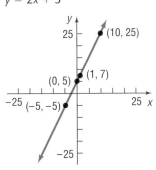

If	Then	Point on Graph
$x = 0$	$y = 2(0) + 5 = 5$	$(0, 5)$
$x = 1$	$y = 2(1) + 5 = 7$	$(1, 7)$
$x = -5$	$y = 2(-5) + 5 = -5$	$(-5, -5)$
$x = 10$	$y = 2(10) + 5 = 25$	$(10, 25)$

By plotting these points and then connecting them, we obtain the graph of the equation (a *line*), as shown in Figure 13.   ▬

EXAMPLE 3   **Graphing an Equation by Plotting Points**

Graph the equation: $y = x^2$

Solution   Table 1 provides several points on the graph. In Figure 14 we plot these points and connect them with a smooth curve to obtain the graph (a *parabola*).

	TABLE	1
x	$y = x^2$	(x, y)
-4	16	$(-4, 16)$
-3	9	$(-3, 9)$
-2	4	$(-2, 4)$
-1	1	$(-1, 1)$
0	0	$(0, 0)$
1	1	$(1, 1)$
2	4	$(2, 4)$
3	9	$(3, 9)$
4	16	$(4, 16)$

Figure 14
$y = x^2$

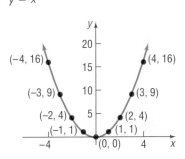

▬

The graphs of the equations shown in Figures 13 and 14 do not show all points. For example, in Figure 13, the point $(20, 45)$ is a part of the graph of $y = 2x + 5$, but it is not shown. Since the graph of $y = 2x + 5$ could be extended out as far as we please, we use arrows to indicate that the pattern shown continues. It is important when illustrating a graph to present enough of the graph so that any viewer of the illustration will "see" the rest of it as an obvious continuation of what is actually there. This is referred to as a **complete graph.**

So, one way to obtain a complete graph of an equation is to plot a sufficient number of points on the graph until a pattern becomes evident. Then these points are connected with a smooth curve following the suggested pattern. But how many points are sufficient? Sometimes knowledge about the equation tells us. For example, we will learn in the next section that, if an equation is of the form $y = mx + b$, then its graph is a line. In this case, only two points are needed to obtain the graph.

One purpose of this book is to investigate the properties of equations in order to decide whether a graph is complete. At first we shall graph equations by plotting a sufficient number of points. Shortly, we shall investigate various techniques that will enable us to graph an equation without plotting so many points. Other times we shall graph equations based solely on properties of the equation.

 COMMENT: Another way to obtain the graph of an equation is to use a graphing utility. Read Section 2, Using a Graphing Utility to Graph Equations, in the Appendix. ▬

EXAMPLE 4 **Graphing an Equation by Plotting Points**

Graph the equation: $y = x^3$

Solution We set up Table 2, listing several points on the graph. Figure 15 illustrates some of these points and the graph of $y = x^3$.

x	$y = x^3$	(x, y)
-3	-27	$(-3, -27)$
-2	-8	$(-2, -8)$
-1	-1	$(-1, -1)$
0	0	$(0, 0)$
1	1	$(1, 1)$
2	8	$(2, 8)$
3	27	$(3, 27)$

TABLE 2

Figure 15
$y = x^3$

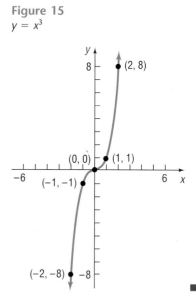

▬

EXAMPLE 5 **Graphing an Equation by Plotting Points**

Graph the equation: $x = y^2$

Solution We set up Table 3, listing several points on the graph. In this case, because of the form of the equation, we assign some numbers to y and find corresponding values of x. Figure 16 illustrates some of these points and the graph of $x = y^2$.

y	$x = y^2$	(x, y)
-3	9	$(9, -3)$
-2	4	$(4, -2)$
-1	1	$(1, -1)$
0	0	$(0, 0)$
1	1	$(1, 1)$
2	4	$(4, 2)$
3	9	$(9, 3)$
4	16	$(16, 4)$

TABLE 3

Figure 16
$x = y^2$

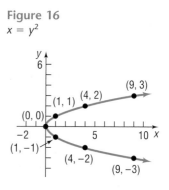

▬

Figure 17 $y = \sqrt{x}$

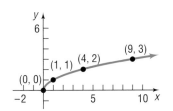

If we restrict y so that $y \geq 0$, the equation $x = y^2$, $y \geq 0$, may be written equivalently as $y = \sqrt{x}$. The portion of the graph of $x = y^2$ in quadrant I is therefore the graph of $y = \sqrt{x}$. See Figure 17.

COMMENT: To see the graph of the equation $x = y^2$ on a graphing calculator, you will need to graph two equations $Y_1 = \sqrt{x}$ and $Y_2 = -\sqrt{x}$. We discuss why in the next chapter. See Figure 18.

Figure 18 $x = y^2$

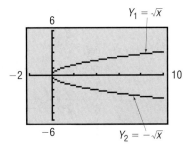

NOW WORK PROBLEM **45.**

We said earlier that we would discuss techniques that reduce the number of points required to graph an equation. Two such techniques involve finding *intercepts* and checking for *symmetry*.

INTERCEPTS

2 The points, if any, at which a graph crosses or touches the coordinate axes are called the **intercepts.** See Figure 19. The x-coordinate of a point at which the graph crosses or touches the x-axis is an ***x*-intercept,** and the y-coordinate of a point at which the graph crosses or touches the y-axis is a ***y*-intercept.**

Figure 19

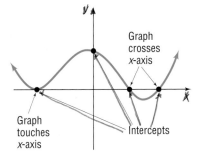

Graph crosses x-axis

Graph touches x-axis

Intercepts

| EXAMPLE 6 |

Finding Intercepts from a Graph

Find the intercepts of the graph in Figure 20. What are its x-intercepts? What are its y-intercepts?

Figure 20

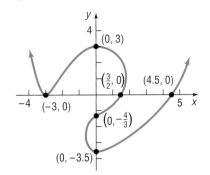

Solution   The intercepts of the graph are the points

$$(-3, 0), \quad (0, 3), \quad \left(\frac{3}{2}, 0\right), \quad \left(0, -\frac{4}{3}\right), \quad (0, -3.5), \quad (4.5, 0)$$

The x-intercepts are $-3, \frac{3}{2}$, and 4.5; the y-intercepts are $-3.5, -\frac{4}{3}$, and 3.

NOW WORK PROBLEM **11(a).**

③ The intercepts of the graph of an equation can be found by using the fact that points on the x-axis have y-coordinates equal to 0, and points on the y-axis have x-coordinates equal to 0.

PROCEDURE FOR FINDING INTERCEPTS

1. To find the x-intercept(s), if any, of the graph of an equation, let $y = 0$ in the equation and solve for x.
2. To find the y-intercept(s), if any, of the graph of an equation, let $x = 0$ in the equation and solve for y.

EXAMPLE 7

Finding Intercepts from an Equation

Find the x-intercept(s) and the y-intercept(s) of the graph of $y = x^2 - 4$. Graph $y = x^2 - 4$.

Solution   To find the x-intercept(s), we let $y = 0$ and obtain the equation

$$x^2 - 4 = 0$$

$$(x + 2)(x - 2) = 0 \quad \text{Factor.}$$

$$x + 2 = 0 \quad \text{or} \quad x - 2 = 0 \quad \text{Zero-Product Property.}$$

$$x = -2 \quad \text{or} \quad x = 2$$

The equation has the solution set $\{-2, 2\}$. The x-intercepts are -2 and 2.
To find the y-intercept(s), we let $x = 0$ and obtain the equation

$$y = -4$$

The y-intercept is -4.
Since $x^2 \geq 0$ for all x, we deduce from the equation $y = x^2 - 4$ that $y \geq -4$ for all x. This information, the intercepts, and the points from Table 4 enable us to graph $y = x^2 - 4$. See Figure 21.

TABLE	4	
x	$y = x^2 - 4$	(x, y)
-3	5	$(-3, 5)$
-1	-3	$(-1, -3)$
1	-3	$(1, -3)$
3	5	$(3, 5)$

Figure 21
$y = x^2 - 4$

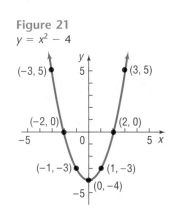

NOW WORK PROBLEM 33(list the intercepts).

COMMENT: For many equations, finding intercepts may not be so easy. In such cases, a graphing utility can be used. Read Section 3, Using a Graphing Utility to Locate Intercepts and Check for Symmetry in the Appendix, to find out how a graphing utility locates intercepts.

SYMMETRY

We have just seen the role that intercepts play in obtaining key points on the graph of an equation. Another helpful tool for graphing equations involves *symmetry,* particularly symmetry with respect to the *x*-axis, the *y*-axis, and the origin.

Figure 22
Symmetric with respect to the *x*-axis

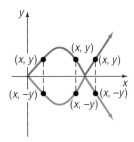

> A graph is said to be **symmetric with respect to the *x*-axis** if, for every point (x, y) on the graph, the point $(x, -y)$ is also on the graph.

Figure 22 illustrates the definition. Notice that, when a graph is symmetric with respect to the *x*-axis, the part of the graph above the *x*-axis is a reflection or mirror image of the part below it, and vice versa.

EXAMPLE 8   **Points Symmetric with Respect to the *x*-Axis**

If a graph is symmetric with respect to the *x*-axis and the point $(3, 2)$ is on the graph, then the point $(3, -2)$ is also on the graph.   ■

Figure 23
Symmetric with respect to the *y*-axis

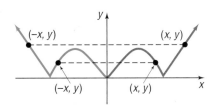

> A graph is said to be **symmetric with respect to the *y*-axis** if, for every point (x, y) on the graph, the point $(-x, y)$ is also on the graph.

Figure 23 illustrates the definition. Notice that, when a graph is symmetric with respect to the *y*-axis, the part of the graph to the right of the *y*-axis is a reflection of the part to the left of it, and vice versa.

EXAMPLE 9   **Points Symmetric with Respect to the *y*-Axis**

If a graph is symmetric with respect to the *y*-axis and the point $(5, 8)$ is on the graph, then the point $(-5, 8)$ is also on the graph.   ■

Figure 24
Symmetric with respect to the origin

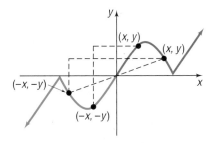

> A graph is said to be **symmetric with respect to the origin** if, for every point (x, y) on the graph, the point $(-x, -y)$ is also on the graph.

Figure 24 illustrates the definition. Notice that symmetry with respect to the origin may be viewed in two ways:

1. As a reflection about the *y*-axis, followed by a reflection about the *x*-axis.
2. As a projection along a line through the origin so that the distances from the origin are equal.

EXAMPLE 10   **Points Symmetric with Respect to the Origin**

If a graph is symmetric with respect to the origin and the point $(4, 2)$ is on the graph, then the point $(-4, -2)$ is also on the graph.   ■

NOW WORK PROBLEMS **1** AND **11(b)**.

④ When the graph of an equation is symmetric with respect to a coordinate axis or the origin, the number of points that you need to plot in order to see the pattern is reduced. For example, if the graph of an equation is symmetric with respect to the y-axis, then, once points to the right of the y-axis are plotted, an equal number of points on the graph can be obtained by reflecting them about the y-axis. Thus, before we graph an equation, we first want to determine whether it has any symmetry. The following tests are used for this purpose.

> ### TESTS FOR SYMMETRY
>
> To test the graph of an equation for symmetry with respect to the
>
> **x-Axis**   Replace y by $-y$ in the equation. If an equivalent equation results, the graph of the equation is symmetric with respect to the x-axis.
>
> **y-Axis**   Replace x by $-x$ in the equation. If an equivalent equation results, the graph of the equation is symmetric with respect to the y-axis.
>
> **Origin**   Replace x by $-x$ and y by $-y$ in the equation. If an equivalent equation results, the graph of the equation is symmetric with respect to the origin.

Let's look at an equation that we have already graphed to see how these tests are used.

EXAMPLE 11

Testing an Equation for Symmetry ($x = y^2$)

(a) To test the graph of the equation $x = y^2$ for symmetry with respect to the x-axis, we replace y by $-y$ in the equation, as follows:

$$x = y^2 \qquad \text{Original equation}$$
$$x = (-y)^2 \qquad \text{Replace } y \text{ by } -y.$$
$$x = y^2 \qquad \text{Simplify.}$$

When we replace y by $-y$, the result is the same equation. The graph is symmetric with respect to the x-axis.

(b) To test the graph of the equation $x = y^2$ for symmetry with respect to the y-axis, we replace x by $-x$ in the equation:

$$x = y^2 \qquad \text{Original equation}$$
$$-x = y^2 \qquad \text{Replace } x \text{ by } -x.$$

Because we arrive at the equation $-x = y^2$, which is not equivalent to the original equation, we conclude that the graph is not symmetric with respect to the y-axis.

(c) To test for symmetry with respect to the origin, we replace x by $-x$ and y by $-y$:

$$x = y^2 \qquad \text{Original equation}$$
$$-x = (-y)^2 \qquad \text{Replace } x \text{ by } -x \text{ and } y \text{ by } -y.$$
$$-x = y^2 \qquad \text{Simplify.}$$

The resulting equation, $-x = y^2$, is not equivalent to the original equation. We conclude that the graph is not symmetric with respect to the origin. ■

Figure 25(a) illustrates the graph of $x = y^2$. In forming a table of points on the graph of $x = y^2$, we can restrict ourselves to points whose y-coordinates are positive. Once these are plotted and connected, a reflection about the x-axis (because of the symmetry) provides the rest of the graph.

Figures 25(b) and (c) illustrate two other equations, $y = x^2$ and $y = x^3$, that we graphed earlier. Test each of these equations for symmetry to verify the conclusions stated in Figures 25(b) and (c). Notice how the existence of symmetry reduces the number of points that we need to plot.

Figure 25

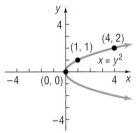

(a) Symmetry with respect to the x-axis

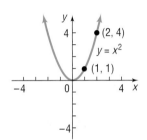

(b) Symmetry with respect to the y-axis

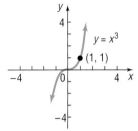

(c) Symmetry with respect to the origin

NOW WORK PROBLEM **33** (TEST FOR SYMMETRY).

EXAMPLE 12   **Graphing the Equation $y = 1/x$**

Graph the equation: $y = \dfrac{1}{x}$

Find any intercepts and check for symmetry first.

Solution   We check for intercepts first. If we let $x = 0$, we obtain a 0 denominator, which is not defined. We conclude that there is no y-intercept. If we let $y = 0$, we get the equation $1/x = 0$, which has no solution. We conclude that there is no x-intercept. The graph of $y = 1/x$ does not cross or touch the coordinate axes.

Next we check for symmetry:

x-Axis   Replacing y by $-y$ yields $-y = 1/x$, which is not equivalent to $y = 1/x$.

y-Axis   Replacing x by $-x$ yields $y = -1/x$, which is not equivalent to $y = 1/x$.

Origin   Replacing x by $-x$ and y by $-y$ yields $-y = -1/x$, which is equivalent to $y = 1/x$.

The graph is symmetric with respect to the origin.

Finally, we set up Table 5, listing several points on the graph. Because of the symmetry with respect to the origin, we use only positive values of x.

From Table 5 we infer that if x is a large and positive number then $y = 1/x$ is a positive number close to 0. We also infer that if x is a positive number close to 0 then $y = 1/x$ is a large and positive number. Armed with this information, we can graph the equation. Figure 26 illustrates some of these points and the graph of $y = 1/x$. Observe how the absence of intercepts and the existence of symmetry with respect to the origin were utilized.

TABLE 5		
x	$y = 1/x$	(x, y)
$\frac{1}{10}$	10	$\left(\frac{1}{10}, 10\right)$
$\frac{1}{3}$	3	$\left(\frac{1}{3}, 3\right)$
$\frac{1}{2}$	2	$\left(\frac{1}{2}, 2\right)$
1	1	$(1, 1)$
2	$\frac{1}{2}$	$\left(2, \frac{1}{2}\right)$
3	$\frac{1}{3}$	$\left(3, \frac{1}{3}\right)$
10	$\frac{1}{10}$	$\left(10, \frac{1}{10}\right)$

Figure 26

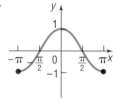

COMMENT: Refer to Example 3 in the Appendix, Section 3, for the graph of $y = 1/x$ using a graphing utility.

2.2 EXERCISES

In Problems 1–10, plot each point. Then plot the point that is symmetric to it with respect to (a) the x-axis; (b) the y-axis; (c) the origin.

1. $(3, 4)$ **2.** $(5, 3)$ **3.** $(-2, 1)$ **4.** $(4, -2)$ **5.** $(1, 1)$

6. $(-1, -1)$ **7.** $(-3, -4)$ **8.** $(4, 0)$ **9.** $(0, -3)$ **10.** $(-3, 0)$

In Problems 11–22, the graph of an equation is given.
  (a) *List the intercepts of the graph.*
  (b) *Based on the graph, tell whether the graph is symmetric with respect to the x-axis, the y-axis, and/or the origin.*

11.
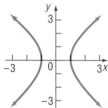

12.
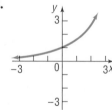

13.
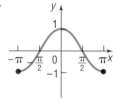

14.
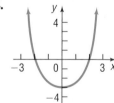

15.
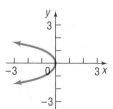

16.
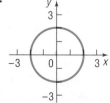

17.
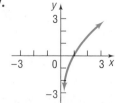

18.
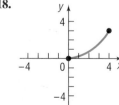

19.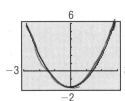

20.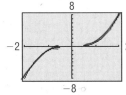

21.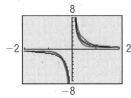

22.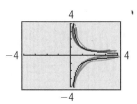

In Problems 23–28, tell whether the given points are on the graph of the equation.

23. Equation: $y = x^4 - \sqrt{x}$
Points: $(0,0); (1,1); (-1,0)$

24. Equation: $y = x^3 - 2\sqrt{x}$
Points: $(0,0); (1,1); (1,-1)$

25. Equation: $y^2 = x^2 + 9$
Points: $(0,3); (3,0); (-3,0)$

26. Equation: $y^3 = x + 1$
Points: $(1,2); (0,1); (-1,0)$

27. Equation: $x^2 + y^2 = 4$
Points: $(0,2); (-2,2); (\sqrt{2}, \sqrt{2})$

28. Equation: $x^2 + 4y^2 = 4$
Points: $(0,1); (2,0); (2, \frac{1}{2})$

In Problems 29–44, list the intercepts and test for symmetry.

29. $x^2 = y$

30. $y^2 = x$

31. $y = 3x$

32. $y = -5x$

33. $x^2 + y - 9 = 0$

34. $y^2 - x - 4 = 0$

35. $9x^2 + 4y^2 = 36$

36. $4x^2 + y^2 = 4$

37. $y = x^3 - 27$

38. $y = x^4 - 1$

39. $y = x^2 - 3x - 4$

40. $y = x^2 + 4$

41. $y = \dfrac{3x}{x^2 + 9}$

42. $y = \dfrac{x^2 - 4}{2x}$

43. $y = \dfrac{-x^3}{x^2 - 9}$

44. $y = \dfrac{x^4 + 1}{2x^5}$

In Problems 45–48, draw a quick sketch of each equation.

45. $y = x^3$

46. $x = y^2$

47. $y = \sqrt{x}$

48. $y = \dfrac{1}{x}$

49. If $(a, 2)$ is a point on the graph of $y = 3x + 5$, what is a?

50. If $(2, b)$ is a point on the graph of $y = x^2 + 4x$, what is b?

51. If (a, b) is a point on the graph of $2x + 3y = 6$, write an equation that relates a to b.

52. If $(2, 0)$ and $(0, 5)$ are points on the graph of $y = mx + b$, what are m and b?

In Problem 53, you may use a graphing utility, but it is not required.

53. (a) Graph $y = \sqrt{x^2}$, $y = x$, $y = |x|$, and $y = (\sqrt{x})^2$, noting which graphs are the same.
(b) Explain why the graphs of $y = \sqrt{x^2}$ and $y = |x|$ are the same.
(c) Explain why the graphs of $y = x$ and $y = (\sqrt{x})^2$ are not the same.
(d) Explain why the graphs of $y = \sqrt{x^2}$ and $y = x$ are not the same.

54. Make up an equation with the intercepts $(2, 0)$, $(4, 0)$, and $(0, 1)$. Compare your equation with a friend's equation. Comment on any similarities.

55. An equation is being tested for symmetry with respect to the x-axis, the y-axis, and the origin. Explain why, if two of these symmetries are present, the remaining one must also be present.

56. Draw a graph that contains the points $(-2, -1)$, $(0, 1)$, $(1, 3)$, and $(3, 5)$. Compare your graph with those of other students. Are most of the graphs almost straight lines? How many are "curved"? Discuss the various ways that these points might be connected.

2.3 LINES

OBJECTIVES
1. Calculate and Interpret the Slope of a Line
2. Graph Lines Given a Point and the Slope
3. Find the Equation of Vertical Lines
4. Use the Point–Slope Form of a Line; Identify Horizontal Lines
5. Find the Equation of a Line Given Two Points
6. Write the Equation of a Line in Slope–Intercept Form
7. Identify the Slope and y-Intercept of a Line from its Equation
8. Write the Equation of a Line in General Form

In this section we study a certain type of equation that contains two variables, called a *linear equation,* and its graph, a *line.*

SLOPE OF A LINE

1 Consider the staircase illustrated in Figure 27. Each step contains exactly the same horizontal **run** and the same vertical **rise.** The ratio of the rise to the run, called the *slope,* is a numerical measure of the steepness of the staircase. For example, if the run is increased and the rise remains the same, the staircase becomes less steep. If the run is kept the same, but the rise is increased, the staircase becomes more steep. This important characteristic of a line is best defined using rectangular coordinates.

Figure 27

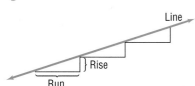

Let $P = (x_1, y_1)$ and $Q = (x_2, y_2)$ be two distinct points with $x_1 \neq x_2$. The **slope m** of the nonvertical line L containing P and Q is defined by the formula

$$m = \frac{y_2 - y_1}{x_2 - x_1} \qquad x_1 \neq x_2 \qquad \textbf{(1)}$$

If $x_1 = x_2$, L is a **vertical line** and the slope m of L is **undefined** (since this results in division by 0).

Figure 28(a) provides an illustration of the slope of a nonvertical line; Figure 28(b) illustrates a vertical line.

Figure 28

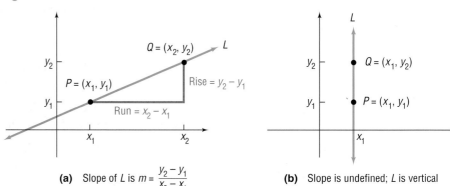

(a) Slope of L is $m = \dfrac{y_2 - y_1}{x_2 - x_1}$ **(b)** Slope is undefined; L is vertical

As Figure 28(a) illustrates, the slope m of a nonvertical line may be viewed as

$$m = \frac{y_2 - y_1}{x_2 - x_1} = \frac{\text{Rise}}{\text{Run}}$$

We can also express the slope m of a nonvertical line as

$$m = \frac{y_2 - y_1}{x_2 - x_1} = \frac{\text{Change in } y}{\text{Change in } x} = \frac{\Delta y}{\Delta x}$$

That is, the slope m of a nonvertical line L measures the amount y changes as x changes from x_1 to x_2. This is called the **average rate of change** of y with respect to x.

Two comments about computing the slope of a nonvertical line may prove helpful:

1. Any two distinct points on the line can be used to compute the slope of the line. (See Figure 29 for justification.)

Figure 29
Triangles ABC and PQR are similar (equal angles). Hence, ratios of corresponding sides are proportional so that

Slope using P and $Q = \dfrac{y_2 - y_1}{x_2 - x_1}$

$= $ Slope using A and $B = \dfrac{d(B, C)}{d(A, C)}$

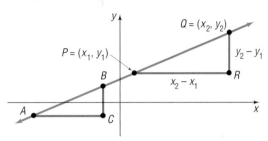

2. The slope of a line may be computed from $P = (x_1, y_1)$ to $Q = (x_2, y_2)$ or from Q to P, because

$$\frac{y_2 - y_1}{x_2 - x_1} = \frac{y_1 - y_2}{x_1 - x_2}$$

EXAMPLE 1

Finding and Interpreting the Slope of a Line Containing Two Points

The slope m of the line containing the points $(1, 2)$ and $(5, -3)$ may be computed as

$$m = \frac{-3 - 2}{5 - 1} = \frac{-5}{4} = -\frac{5}{4} \quad \text{or as} \quad m = \frac{2 - (-3)}{1 - 5} = \frac{5}{-4} = -\frac{5}{4}$$

For every 4-unit change in x, y will change by -5 units. That is, if x increases by 4 units, then y decreases by 5 units. The average rate of change of y with respect to x is $-\frac{5}{4}$. ■

NOW WORK PROBLEMS **1** AND **7.**

To get a better idea of the meaning of the slope m of a line L, consider the following example.

EXAMPLE 2

Finding the Slopes of Various Lines Containing the Same Point (2, 3)

Compute the slopes of the lines L_1, L_2, L_3, and L_4 containing the following pairs of points. Graph all four lines on the same set of coordinate axes.

L_1: $P = (2, 3)$ $Q_1 = (-1, -2)$

L_2: $P = (2, 3)$ $Q_2 = (3, -1)$

L_3: $P = (2, 3)$ $Q_3 = (5, 3)$

L_4: $P = (2, 3)$ $Q_4 = (2, 5)$

Solution Let $m_1, m_2, m_3,$ and m_4 denote the slopes of the lines $L_1, L_2, L_3,$ and L_4, respectively. Then

$$m_1 = \frac{-2 - 3}{-1 - 2} = \frac{-5}{-3} = \frac{5}{3}$$ A rise of 5 divided by a run of 3

$$m_2 = \frac{-1 - 3}{3 - 2} = \frac{-4}{1} = -4$$

$$m_3 = \frac{3 - 3}{5 - 2} = \frac{0}{3} = 0$$

m_4 is undefined

Figure 30

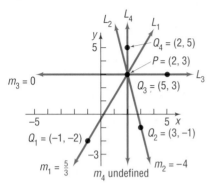

The graphs of these lines are given in Figure 30.

Figure 30 illustrates the following facts:

1. When the slope of a line is positive, the line slants upward from left to right (L_1).
2. When the slope of a line is negative, the line slants downward from left to right (L_2).
3. When the slope is 0, the line is horizontal (L_3).
4. When the slope is undefined, the line is vertical (L_4).

COMMENT: Now read Section 5, Square Screens, in the Appendix.

SEEING THE CONCEPT: On the same square screen, graph the following equations:

$Y_1 = 0$ Slope of line is 0.

$Y_2 = \frac{1}{4}x$ Slope of line is $\frac{1}{4}$.

$Y_3 = \frac{1}{2}x$ Slope of line is $\frac{1}{2}$.

$Y_4 = x$ Slope of line is 1.

$Y_5 = 2x$ Slope of line is 2.

$Y_6 = 6x$ Slope of line is 6.

Figure 31

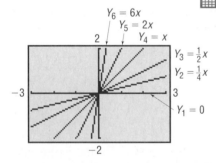

See Figure 31.

SEEING THE CONCEPT: On the same square screen, graph the following equations:

$Y_1 = 0$ Slope of line is 0.

$Y_2 = -\frac{1}{4}x$ Slope of line is $-\frac{1}{4}$.

$Y_3 = -\frac{1}{2}x$ Slope of line is $-\frac{1}{2}$.

$Y_4 = -x$ Slope of line is -1.

$Y_5 = -2x$ Slope of line is -2.

$Y_6 = -6x$ Slope of line is -6.

Figure 32

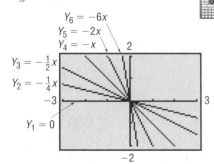

See Figure 32.

Figures 31 and 32 illustrate that the closer the line is to the vertical position, the greater the magnitude of the slope.

The next example illustrates how the slope of a line can be used to graph the line.

EXAMPLE 3 | Graphing a Line Given a Point and a Slope

Draw a graph of the line that contains the point $(3, 2)$ and has a slope of

(a) $\frac{3}{4}$ (b) $-\frac{4}{5}$

Solution (a) Slope = Rise/Run. The fact that the slope is $\frac{3}{4}$ means that for every horizontal movement (run) of 4 units to the right there will be a vertical movement (rise) of 3 units. If we start at the given point $(3, 2)$ and move 4 units to the right and 3 units up, we reach the point $(7, 5)$. By drawing the line through this point and the point $(3, 2)$, we have the graph. See Figure 33.

Figure 33 Slope $= \frac{3}{4}$

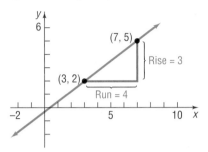

(b) The fact that the slope is

$$-\frac{4}{5} = \frac{-4}{5} = \frac{\text{Rise}}{\text{Run}}$$

means that for every horizontal movement of 5 units (run = 5) to the right there will be a corresponding vertical movement of -4 units (rise = -4, a downward movement of 4 units). If we start at the given point $(3, 2)$ and move 5 units to the right and then 4 units down, we arrive at the point $(8, -2)$. By drawing the line through these points, we have the graph. See Figure 34.

Alternatively, we can set

$$-\frac{4}{5} = \frac{4}{-5} = \frac{\text{Rise}}{\text{Run}}$$

Figure 34 Slope $= -\frac{4}{5}$

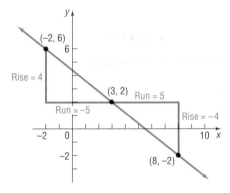

so that for every horizontal movement of -5 units (a movement to the left) there will be a corresponding vertical movement of 4 units (upward). This approach brings us to the point $(-2, 6)$, which is also on the graph shown in Figure 34. ∎

NOW WORK PROBLEMS **15** AND **21**.

EQUATIONS OF LINES

③ Now that we have discussed the slope of a line, we are ready to derive equations of lines. As we shall see, there are several forms of the equation of a line. Let's start with an example.

EXAMPLE 4 | Graphing a Line

Graph the equation: $x = 3$

Figure 35

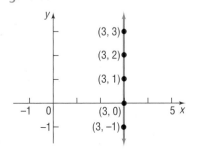

Solution We are looking for all points (x, y) in the plane for which $x = 3$. No matter what y-coordinate is used, the corresponding x-coordinate always equals 3. Consequently, the graph of the equation $x = 3$ is a vertical line with x-intercept 3 and undefined slope. See Figure 35. ∎

As suggested by Example 4, we have the following result:

Theorem   **Equation of a Vertical Line**

A vertical line is given by an equation of the form

$$x = a$$

where a is the x-intercept.

> **COMMENT:** To graph an equation using a graphing utility, we need to express the equation in the form $y =$ expression in x. But $x = 3$ cannot be put in this form. To overcome this, most graphing utilities have special ways for drawing vertical lines. LINE, PLOT, and VERT are among the more common ones. Consult your manual to determine the correct methodology for your graphing utility.

Figure 36

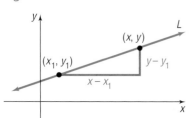

4   Now let L be a nonvertical line with slope m containing the point (x_1, y_1). See Figure 36. For any other point (x, y) on L, we have

$$m = \frac{y - y_1}{x - x_1} \quad \text{or} \quad y - y_1 = m(x - x_1)$$

Theorem   **Point–Slope Form of an Equation of a Line**

An equation of a nonvertical line of slope m that contains the point (x_1, y_1) is

$$y - y_1 = m(x - x_1) \tag{2}$$

EXAMPLE 5   **Using the Point–Slope Form of a Line**

Figure 37

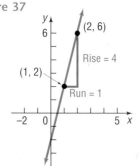

An equation of the line with slope 4 and containing the point $(1, 2)$ can be found by using the point–slope form with $m = 4$, $x_1 = 1$, and $y_1 = 2$.

$$y - y_1 = m(x - x_1) \qquad m = 4, \, x_1 = 1, \, y_1 = 2$$
$$y - 2 = 4(x - 1)$$

See Figure 37.

EXAMPLE 6   **Finding the Equation of a Horizontal Line**

Find an equation of the horizontal line containing the point $(3, 2)$.

Figure 38 $y = 2$ *Solution*

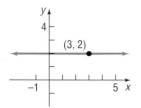

The slope of a horizontal line is 0. To get an equation, we use the point–slope form with $m = 0$, $x_1 = 3$, and $y_1 = 2$.

$$y - y_1 = m(x - x_1) \qquad m = 0, \, x_1 = 3, \, y_1 = 2$$
$$y - 2 = 0 \cdot (x - 3)$$
$$y - 2 = 0$$
$$y = 2$$

See Figure 38 for the graph.

As suggested by Example 6, we have the following result:

Theorem   **Equation of a Horizontal Line**

A horizontal line is given by an equation of the form

$$y = b$$

where b is the y-intercept.

EXAMPLE 7   **Finding an Equation of a Line Given Two Points**

⑤ Find an equation of the line L containing the points $(2, 3)$ and $(-4, 5)$. Graph the line L.

Solution   Since two points are given, we first compute the slope of the line.

$$m = \frac{5 - 3}{-4 - 2} = \frac{2}{-6} = -\frac{1}{3}$$

We use the point $(2, 3)$ and the fact that the slope $m = -\frac{1}{3}$ to get the point–slope form of the equation of the line.

$$y - 3 = -\frac{1}{3}(x - 2)$$

See Figure 39 for the graph.

Figure 39

$$y - 3 = -\frac{1}{3}(x - 2)$$

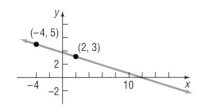

In the solution to Example 7, we could have used the other point, $(-4, 5)$, instead of the point $(2, 3)$. The equation that results, although it looks different, is equivalent to the equation that we obtained in the example. (Try it for yourself.)

NOW WORK PROBLEM **27**.

⑥ Another useful equation of a line is obtained when the slope m and y-intercept b are known. In this event, we know both the slope m of the line and a point $(0, b)$ on the line; we use the point-slope form, equation (2), to obtain the following equation:

$$y - b = m(x - 0) \quad \text{or} \quad y = mx + b$$

Theorem   **Slope–Intercept Form of an Equation of a Line**

An equation of a line L with slope m and y-intercept b is

$$y = mx + b \qquad \qquad \textbf{(3)}$$

Figure 40 $y = mx + 2$

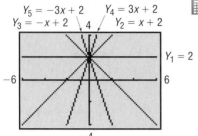

$Y_5 = -3x + 2$ $Y_4 = 3x + 2$
$Y_3 = -x + 2$ $Y_2 = x + 2$
$Y_1 = 2$

SEEING THE CONCEPT: To see the role that the slope m plays, graph the following lines on the same square screen.

$$Y_1 = 2$$
$$Y_2 = x + 2$$
$$Y_3 = -x + 2$$
$$Y_4 = 3x + 2$$
$$Y_5 = -3x + 2$$

See Figure 40. What do you conclude about the lines $y = mx + 2$? ■

Figure 41 $y = 2x + b$

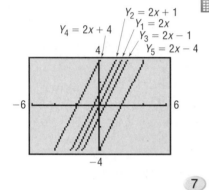

$Y_2 = 2x + 1$
$Y_1 = 2x$
$Y_4 = 2x + 4$
$Y_3 = 2x - 1$
$Y_5 = 2x - 4$

SEEING THE CONCEPT: To see the role of the y-intercept b, graph the following lines on the same square screen.

$$Y_1 = 2x$$
$$Y_2 = 2x + 1$$
$$Y_3 = 2x - 1$$
$$Y_4 = 2x + 4$$
$$Y_5 = 2x - 4$$

See Figure 41. What do you conclude about the lines $y = 2x + b$? ■

⑦ When the equation of a line is written in slope–intercept form, it is easy to find the slope m and y-intercept b of the line. For example, suppose that the equation of a line is

$$y = -2x + 3$$

Compare it to $y = mx + b$:

$$y = -2x + 3$$
$$\uparrow \qquad \uparrow$$
$$y = mx \; + \; b$$

The slope of this line is -2 and its y-intercept is 3.

EXAMPLE 8   Finding the Slope and y-Intercept

Find the slope m and y-intercept b of the equation $2x + 4y = 8$. Graph the equation.

Solution   To obtain the slope and y-intercept, we transform the equation into its slope–intercept form by solving the equation for y.

$$2x + 4y = 8$$
$$4y = -2x + 8$$
$$y = -\tfrac{1}{2}x + 2 \qquad \text{Solve for } y \text{ to obtain } y = mx + b.$$

The coefficient of x, $-\tfrac{1}{2}$, is the slope, and the y-intercept is 2.

We can graph the line in two ways:

Figure 42
$2x + 4y = 8$

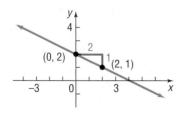

1. Use the fact that the y-intercept is 2 and the slope is $-\tfrac{1}{2}$. Then, starting at the point $(0, 2)$, go to the right 2 units and then down 1 unit to the point $(2, 1)$. See Figure 42.

2. Locate the intercepts. Because the y-intercept is 2, we know that one intercept is $(0, 2)$. To obtain the x-intercept, let $y = 0$ and solve for x. When $y = 0$, we have

$$2x + 4y = 8$$
$$2x + 4 \cdot 0 = 8 \qquad y = 0$$
$$2x = 8$$
$$x = 4$$

The intercepts are $(4, 0)$ and $(0, 2)$. See Figure 43.   ∎

Figure 43 $2x + 4y = 8$

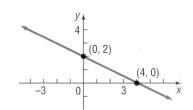

NOW WORK PROBLEM **49.**

⑧ The form of the equation of the line in Example 8, $2x + 4y = 8$, is called the *general form*.

> The equation of a line L is in **general form** when it is written as
>
> $$Ax + By = C \qquad\qquad (4)$$
>
> where A, B, and C are real numbers and A and B are not both 0.

Every line has an equation that is equivalent to an equation written in general form. For example, a vertical line whose equation is

$$x = a$$

can be written in the general form

$$1 \cdot x + 0 \cdot y = a \qquad A = 1, B = 0, C = a$$

A horizontal line whose equation is

$$y = b$$

can be written in the general form

$$0 \cdot x + 1 \cdot y = b \qquad A = 0, B = 1, C = b$$

Lines that are neither vertical nor horizontal have general equations of the form

$$Ax + By = C \qquad A \neq 0 \text{ and } B \neq 0$$

Because the equation of every line can be written in general form, any equation equivalent to (4) is called a **linear equation.**

NOW WORK PROBLEM **35.**

The next example illustrates a typical situation that requires the use of linear equations.

EXAMPLE 9   **Computing the Cost of Operating a Car**

The National Car Rental Company has determined that the cumulative cost of operating a vehicle is $0.41 per mile.

(a) Write an equation that relates the cumulative cost C, in dollars, of operating a car and the number x of miles it has been driven.

(b) What is the cost of operating a car with 1000 miles on it?

(c) What is the cost of operating a car with 2000 miles on it?

Solution (a) If x is the number of miles that the car has been driven, then the cumulative cost C, in dollars, is $0.41x$. An equation relating C and x is

$$C = 0.41x \qquad x \geq 0$$

The average cost per mile, $0.41, is the slope of the line $C = 0.41x$. In other words, the cost increases by $0.41 for each additional mile driven. See Figure 44.

(b) The cost of operating a car with 1000 miles on it is

$$C = 0.41x = 0.41(1000) = \$410$$

(c) The cost of operating a car with 2000 miles on it is

$$C = 0.41x = 0.41(2000) = \$820$$

Figure 44
$C = 0.41x$

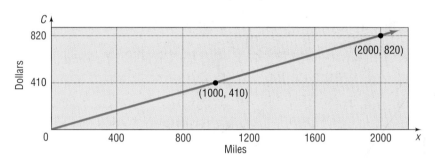

2.3 EXERCISES

In Problems 1–4, (a) find the slope of the line and (b) interpret the slope.

1.

2.
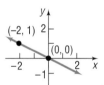

3.
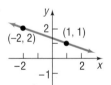

4.
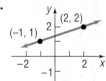

In Problems 5–12, plot each pair of points and determine the slope of the line containing them. Graph the line.

5. $(2, 3)$; $(4, 0)$ **6.** $(4, 2)$; $(3, 4)$ **7.** $(-2, 3)$; $(2, 1)$ **8.** $(-1, 1)$; $(2, 3)$

9. $(-3, -1)$; $(2, -1)$ **10.** $(4, 2)$; $(-5, 2)$ **11.** $(-1, 2)$; $(-1, -2)$ **12.** $(2, 0)$; $(2, 2)$

In Problems 13–20, graph the line containing the point P and having slope m.

13. $P = (1, 2)$; $m = 3$ **14.** $P = (2, 1)$; $m = 4$ **15.** $P = (2, 4)$; $m = -\frac{3}{4}$

16. $P = (1, 3)$; $m = -\frac{2}{5}$ **17.** $P = (-1, 3)$; $m = 0$ **18.** $P = (2, -4)$; $m = 0$

19. $P = (0, 3)$; slope undefined **20.** $P = (-2, 0)$; slope undefined

In Problems 21–26, the slope and a point on a line are given. Use this information to locate three additional points on the line. Answers may vary. [**Hint:** It is not necessary to find the equation of the line. See Example 3.]

21. Slope 4; point $(1, 2)$ **22.** Slope 2; point $(-2, 3)$ **23.** Slope $-\frac{3}{2}$; point $(2, -4)$

24. Slope $\frac{4}{3}$; point $(-3, 2)$ **25.** Slope -2; point $(-2, -3)$ **26.** Slope -1; point $(4, 1)$

In Problems 27–30, find an equation of each line.

27.

28.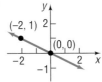

29.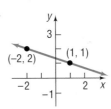

30.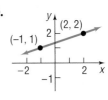

In Problems 31–42, find an equation for the line with the given properties. Express your answer using either the general form or the slope–intercept form of the equation of a line, whichever you prefer.

31. Slope $= 3$;   containing the point $(-2, 3)$

32. Slope $= 2$;   containing the point $(4, -3)$

33. Slope $= -\frac{2}{3}$;   containing the point $(1, -1)$

34. Slope $= \frac{1}{2}$;   containing the point $(3, 1)$

35. Containing the points $(1, 3)$ and $(-1, 2)$

36. Containing the points $(-3, 4)$ and $(2, 5)$

37. Slope $= -3$;   y-intercept $= 3$

38. Slope $= -2$;   y-intercept $= -2$

39. x-intercept $= 2$;   y-intercept $= -1$

40. x-intercept $= -4$;   y-intercept $= 4$

41. Slope undefined;   containing the point $(2, 4)$

42. Slope undefined;   containing the point $(3, 8)$

In Problems 43–62, find the slope and y-intercept of each line. Graph the line.

43. $y = 2x + 3$

44. $y = -3x + 4$

45. $\frac{1}{2}y = x - 1$

46. $\frac{1}{3}x + y = 2$

47. $y = \frac{1}{2}x + 2$

48. $y = 2x + \frac{1}{2}$

49. $x + 2y = 4$

50. $-x + 3y = 6$

51. $2x - 3y = 6$

52. $3x + 2y = 6$

53. $x + y = 1$

54. $x - y = 2$

55. $x = -4$

56. $y = -1$

57. $y = 5$

58. $x = 2$

59. $y - x = 0$

60. $x + y = 0$

61. $2y - 3x = 0$

62. $3x + 2y = 0$

---

63. Find an equation of the x-axis.

64. Find an equation of the y-axis.

65. **Measuring Temperature**   The relationship between Celsius (°C) and Fahrenheit (°F) degrees for measuring temperature is linear. Find an equation relating °C and °F if 0°C corresponds to 32°F and 100°C corresponds to 212°F. Use the equation to find the Celsius measure of 70°F.

66. **Measuring Temperature**   The Kelvin (K) scale for measuring temperature is obtained by adding 273 to the Celsius temperature.
    (a) Write an equation relating K and °C.
    (b) Write an equation relating K and °F (see Problem 65).

67. **Business: Computing Profit**   Each Sunday a newspaper agency sells x copies of a newspaper for $1.00 per copy. The cost to the agency of each newspaper is $0.50. The agency pays a fixed cost for storage, delivery, and so on, of $100 per Sunday.
    (a) Write an equation that relates the profit P, in dollars, to the number x of copies sold. Graph this equation.
    (b) What is the profit to the agency if 1000 copies are sold?
    (c) What is the profit to the agency if 5000 copies are sold?

68. **Business: Computing Profit**   Repeat Problem 67 if the cost to the agency is $0.45 per copy and the fixed cost is $125 per Sunday.

69. **Cost of Electricity**   In 2000, Florida Power and Light Company supplied electricity to residential customers for a monthly customer charge of $5.65 plus 6.543¢ per kilowatt-hour supplied in the month for the first 750 kilowatt-hours used.* Write an equation that relates the monthly charge C, in dollars, to the number x of kilowatt-hours used in the month. Graph this equation. What is the monthly charge for using 300 kilowatt-hours? For using 750 kilowatt-hours?

70. Show that an equation for a line with nonzero x- and y-intercepts can be written as

$$\frac{x}{a} + \frac{y}{b} = 1$$

where a is the x-intercept and b is the y-intercept. This is called the **intercept form** of the equation of a line.

Source: Florida Power and Light Co., Miami, Florida, 2000.

In Problems 71–74, match each graph with the correct equation:

(a) $y = x$;   (b) $y = 2x$;   (c) $y = \dfrac{x}{2}$;   (d) $y = 4x$.

71.
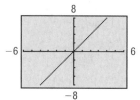

72.
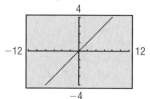

73.
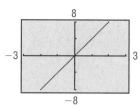

74.
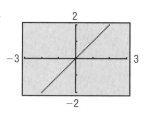

In Problems 75–78, write an equation of each line. Express your answer using either the general form or the slope–intercept form of the equation of a line, whichever you prefer.

75.
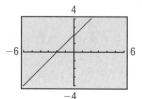

76.
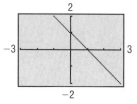

77.
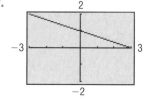

78.
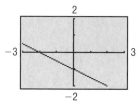

79. Which of the following equations might have the graph shown. (More than one answer is possible.)

(a) $2x + 3y = 6$ (e) $x - y = -1$

(b) $-2x + 3y = 6$ (f) $y = 3x - 5$

(c) $3x - 4y = -12$ (g) $y = 2x + 3$

(d) $x - y = 1$ (h) $y = -3x + 3$

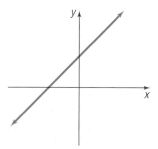

80. Which of the following equations might have the graph shown. (More than one answer is possible.)

(a) $2x + 3y = 6$ (e) $x - y = -1$

(b) $2x - 3y = 6$ (f) $y = -2x + 1$

(c) $3x + 4y = 12$ (g) $y = -\dfrac{1}{2}x + 10$

(d) $x - y = 1$ (h) $y = x + 4$

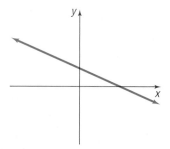

81. Which form of the equation of a line do you prefer to use? Justify your position with an example that shows that your choice is better than another. Have reasons.

82. Can every line be written in slope–intercept form? Explain.

83. Does every line have two distinct intercepts? Explain. Are there lines that have no intercepts? Explain.

84. What can you say about two lines that have equal slopes and equal y-intercepts?

85. What can you say about two lines with the same x-inter- cept and the same y-intercept? Assume that the x-inter- cept is not 0.

86. If two lines have the same slope, but different x-inter- cepts, can they have the same y-intercept?

87. If two lines have the same y-intercept, but different slopes, can they have the same x-intercept? What is the only way that this can happen?

88. The accepted symbol used to denote the slope of a line is the letter m. Investigate the origin of this symbolism. Begin by consulting a French dictionary and looking up the French word *monter*. Write a brief essay on your findings.

89. **Grade of a Road**   The term *grade* is used to describe the inclination of a road. How does this term relate to the notion of slope of a line? Is a 4% grade very steep? Investigate the grades of some mountainous roads and determine their slopes. Write a brief essay on your findings.

90. **Carpentry**   Carpenters use the term pitch to describe the steepness of staircases and roofs. How does pitch relate to slope? Investigate typical pitches used for stairs and for roofs. Write a brief essay on your findings.

PREPARING FOR THIS SECTION

Before getting started, review the following:

✓ For Objectives 5–7, The Distance Formula (Section 2.1, p. 148) and Completing the Square (Section 1.3, pp. 108–109)

2.4 | PARALLEL AND PERPENDICULAR LINES; CIRCLES

OBJECTIVES
1 Define Parallel Lines
2 Find Equations of Parallel Lines
3 Define Perpendicular Lines
4 Find Equations of Perpendicular Lines
5 Write the Standard Form of the Equation of a Circle
6 Graph a Circle
7 Find the Center and Radius of a Circle in General Form and Graph It

PARALLEL AND PERPENDICULAR LINES

1 When two lines (in the plane) do not intersect (that is, they have no points in common), they are said to be **parallel.** Look at Figure 45. There we have drawn two lines and have constructed two right triangles by drawing sides parallel to the coordinate axes. These lines are parallel if and only if the right triangles are similar. (Do you see why? Two angles are equal.) But the triangles are similar if and only if the ratios of corresponding sides are equal.

Figure 45
The lines are parallel if and only if their slopes are equal.

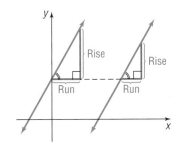

This suggests the following result:

Theorem

Criterion for Parallel Lines

Two nonvertical lines are parallel if and only if their slopes are equal and they have different y-intercepts.

■

The use of the words "if and only if" in the preceding theorem means that actually two statements are being made, one the converse of the other.

If two nonvertical lines are parallel, then their slopes are equal and they have different y-intercepts.

If two nonvertical lines have equal slopes and different y-intercepts, then they are parallel.

Refer to Seeing the Concept, Figure 41, $y = 2x + b$, on p. 170.

NOW WORK PROBLEM **1(a)**.

EXAMPLE 1

Showing That Two Lines Are Parallel

Show that the lines given by the following equations are parallel:

$$L_1:\quad 2x + 3y = 6 \qquad L_2:\quad 4x + 6y = 0$$

Figure 46
Parallel lines

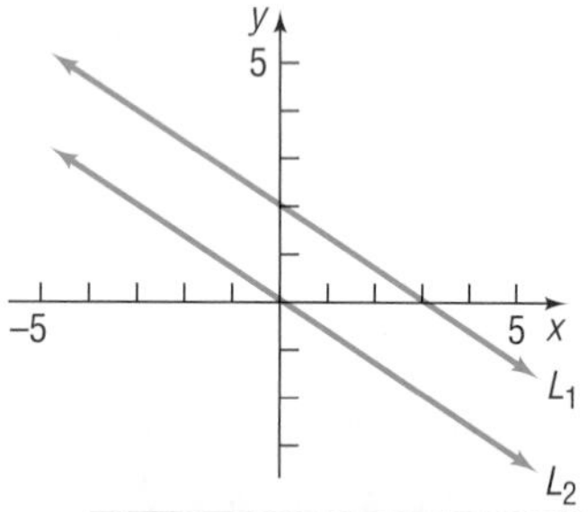

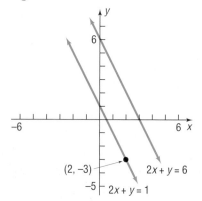

Solution To determine whether these lines have equal slopes, we write each equation in slope–intercept form:

$$L_1:\quad 2x + 3y = 6 \qquad\qquad L_2:\quad 4x + 6y = 0$$
$$3y = -2x + 6 \qquad\qquad\qquad 6y = -4x$$
$$y = -\tfrac{2}{3}x + 2 \qquad\qquad\qquad y = -\tfrac{2}{3}x$$
$$\text{Slope} = -\tfrac{2}{3} \qquad\qquad\qquad \text{Slope} = -\tfrac{2}{3}$$

Because these lines have the same slope, $-\tfrac{2}{3}$, but different y-intercepts, the lines are parallel. See Figure 46. ■

EXAMPLE 2

Finding a Line That Is Parallel to a Given Line

② Find an equation for the line that contains the point $(2, -3)$ and is parallel to the line $2x + y = 6$.

Solution The slope of the line that we seek equals the slope of the line $2x + y = 6$, since the two lines are to be parallel. We begin by writing the equation of the line $2x + y = 6$ in slope–intercept form.

Figure 47

$$2x + y = 6$$
$$y = -2x + 6$$

The slope is -2. Since the line that we seek contains the point $(2, -3)$, we use the point–slope form to obtain

$$y + 3 = -2(x - 2) \qquad \text{Point–slope form}$$
$$y + 3 = -2x + 4$$
$$y = -2x + 1 \qquad \text{Slope–intercept form}$$
$$2x + y = 1 \qquad \text{General form}$$

This line is parallel to the line $2x + y = 6$ and contains the point $(2, -3)$. See Figure 47. ■

NOW WORK PROBLEM **15**.

Figure 48
Perpendicular lines

③

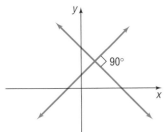

When two lines intersect at a right angle (90°), they are said to be **perpendicular.** See Figure 48.

The following result gives a condition, in terms of their slopes, for two lines to be perpendicular.

Theorem   Criterion for Perpendicular Lines

Two nonvertical lines are perpendicular if and only if the product of their slopes is −1.

Here, we shall prove the "only if" part of the statement:

> If two nonvertical lines are perpendicular, then the product of their slopes is −1.

In Problem 71, you are asked to prove the "if" part of the theorem; that is:

> If two nonvertical lines have slopes whose product is −1, then the lines are perpendicular.

Figure 49

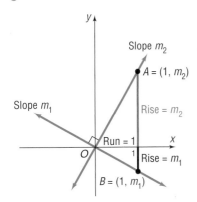

Proof   Let m_1 and m_2 denote the slopes of the two lines. There is no loss in generality (that is, neither the angle nor the slopes are affected) if we situate the lines so that they meet at the origin. See Figure 49. The point $A = (1, m_2)$ is on the line having slope m_2, and the point $B = (1, m_1)$ is on the line having slope m_1. (Do you see why this must be true?)

Suppose that the lines are perpendicular. Then triangle OAB is a right triangle. As a result of the Pythagorean Theorem, it follows that

$$\left[d(O, A)\right]^2 + \left[d(O, B)\right]^2 = \left[d(A, B)\right]^2 \tag{1}$$

By the distance formula, we can write each of these distances as

$$\left[d(O, A)\right]^2 = (1 - 0)^2 + (m_2 - 0)^2 = 1 + m_2^2$$

$$\left[d(O, B)\right]^2 = (1 - 0)^2 + (m_1 - 0)^2 = 1 + m_1^2$$

$$\left[d(A, B)\right]^2 = (1 - 1)^2 + (m_2 - m_1)^2 = m_2^2 - 2m_1m_2 + m_1^2$$

Using these facts in equation (1), we get

$$\left(1 + m_2^2\right) + \left(1 + m_1^2\right) = m_2^2 - 2m_1m_2 + m_1^2$$

which, upon simplification, can be written as

$$m_1m_2 = -1$$

If the lines are perpendicular, the product of their slopes is −1.   ■

You may find it easier to remember the condition for two nonvertical lines to be perpendicular by observing that the equality $m_1m_2 = -1$ means that m_1 and m_2 are negative reciprocals of each other; that is, $m_1 = -1/m_2$ and $m_2 = -1/m_1$.

EXAMPLE 3   **Finding the Slope of a Line Perpendicular to a Given Line**

If a line has slope $\frac{3}{2}$, any line having slope $-\frac{2}{3}$ is perpendicular to it.   ■

NOW WORK PROBLEM **1(b)**.

| EXAMPLE 4 | **Finding the Equation of a Line Perpendicular to a Given Line** |

 Find an equation of the line containing the point $(1, -2)$ that is perpendicular to the line $x + 3y = 6$. Graph the two lines.

Solution We first write the equation of the given line in slope–intercept form to find its slope.

$$x + 3y = 6$$
$$3y = -x + 6 \qquad \text{Proceed to solve for } y.$$
$$y = -\tfrac{1}{3}x + 2 \qquad \text{Place in the form } y = mx + b.$$

Figure 50

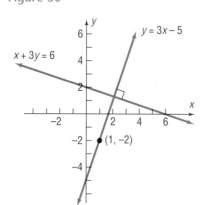

The given line has slope $-\tfrac{1}{3}$. Any line perpendicular to this line will have slope 3. Because we require the point $(1, -2)$ to be on this line with slope 3, we use the point–slope form of the equation of a line.

$$y - (-2) = 3(x - 1) \qquad \text{Point–slope form}$$

To obtain other forms of the equation we proceed as follows:

$$y + 2 = 3(x - 1)$$
$$y + 2 = 3x - 3$$
$$y = 3x - 5 \qquad \text{Slope–intercept form}$$
$$3x - y = 5 \qquad \text{General form}$$

Figure 50 shows the graphs. ■

NOW WORK PROBLEM **21**.

WARNING: Be sure to use a square screen when you graph perpendicular lines. Otherwise, the angle between the two lines will appear distorted. ■

CIRCLES

One advantage of a coordinate system is that it enables us to translate a geometric statement into an algebraic statement, and vice versa. Consider, for example, the following geometric statement that defines a circle.

> A **circle** is a set of points in the xy-plane that are a fixed distance r from a fixed point (h, k). The fixed distance r is called the **radius,** and the fixed point (h, k) is called the **center** of the circle.

Figure 51

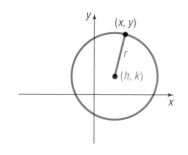

Figure 51 shows the graph of a circle. Is there an equation having this graph? If so, what is the equation? To find the equation, we let (x, y) represent the coordinates of any point on a circle with radius r and center (h, k). Then the distance between the points (x, y) and (h, k) must always equal r. That is, by the distance formula

$$\sqrt{(x - h)^2 + (y - k)^2} = r$$

or, equivalently,

$$(x - h)^2 + (y - k)^2 = r^2$$

The **standard form of an equation of a circle** with radius r and center (h, k) is

$$(x - h)^2 + (y - k)^2 = r^2 \qquad \textbf{(2)}$$

The standard form of an equation of a circle of radius r with center at the origin $(0, 0)$ is

$$x^2 + y^2 = r^2$$

If the radius $r = 1$, the circle whose center is at the origin is called the **unit circle** and has the equation

$$x^2 + y^2 = 1$$

See Figure 52.

Figure 52
Unit circle $x^2 + y^2 = 1$

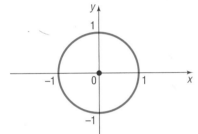

EXAMPLE 5	Writing the Standard Form of the Equation of a Circle

Write the standard form of the equation of the circle with radius 5 and center $(-3, 6)$.

Solution Using the form of equation (2) and substituting the values $r = 5$, $h = -3$, and $k = 6$, we have

$$(x - h)^2 + (y - k)^2 = r^2$$
$$(x + 3)^2 + (y - 6)^2 = 25 \qquad \blacksquare$$

→ NOW WORK PROBLEM 27.

⑥   The graph of any equation of the form of equation (2) is that of a circle with radius r and center (h, k).

EXAMPLE 6	Graphing a Circle

Graph the equation: $(x + 3)^2 + (y - 2)^2 = 16$

Solution The equation is of the form of equation (2), so its graph is a circle. To graph the equation, we first compare the given equation to the standard form of the equation of a circle. The comparison yields information about the circle.

$$(x + 3)^2 + (y - 2)^2 = 16$$
$$\left(x - (-3)\right)^2 + (y - 2)^2 = 4^2$$
$$\uparrow \qquad\qquad \uparrow \qquad \uparrow$$
$$(x - h)^2 + (y - k)^2 = r^2$$

Figure 53

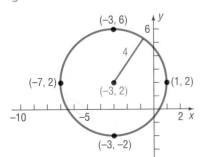

We see that $h = -3$, $k = 2$, and $r = 4$. The circle has center $(-3, 2)$ and a radius of 4 units. To graph this circle, we first plot the center $(-3, 2)$. Since the radius is 4, we can locate four points on the circle by plotting points 4 units to the left, to the right, up, and down from the center. These four points can then be used as guides to obtain the graph. See Figure 53. $\blacksquare$

→ NOW WORK PROBLEM 41.

If we eliminate the parentheses from the standard form of the equation of the circle given in Example 6, we get

$$(x + 3)^2 + (y - 2)^2 = 16$$
$$x^2 + 6x + 9 + y^2 - 4y + 4 = 16$$

which we find, upon simplifying, is equivalent to

$$x^2 + y^2 + 6x - 4y - 3 = 0$$

It can be shown that any equation of the form

$$x^2 + y^2 + ax + by + c = 0$$

has a graph that is a circle or a point, or it has no graph at all. For example, the graph of the equation $x^2 + y^2 = 0$ is the single point $(0, 0)$. The equation $x^2 + y^2 + 5 = 0$, or $x^2 + y^2 = -5$, has no graph, because sums of squares of real numbers are never negative.

When its graph is a circle, the equation

$$x^2 + y^2 + ax + by + c = 0$$

is referred to as the **general form of the equation of a circle.**

NOW WORK PROBLEM 33.

⑦ If an equation of a circle is in the general form, we use the method of completing the square to put the equation in standard form so that we can identify its center and radius.

EXAMPLE 7 Graphing a Circle Whose Equation Is in General Form

Graph the equation: $x^2 + y^2 + 4x - 6y + 12 = 0$

Solution We complete the square in both x and y to put the equation in standard form. Group the expressions involving x, group the expressions involving y, and put the constant on the right side of the equation. The result is

$$(x^2 + 4x) + (y^2 - 6y) = -12$$

Next, complete the square of each expression in parentheses. Remember that any number added on the left side of the equation must be added on the right.

$$(x^2 + 4x + 4) + (y^2 - 6y + 9) = -12 + 4 + 9$$

$$\left(\tfrac{4}{2}\right)^2 = 4 \qquad \left(\tfrac{-6}{2}\right)^2 = 9$$

$$(x + 2)^2 + (y - 3)^2 = 1 \qquad \text{Factor.}$$

We recognize this equation as the standard form of the equation of a circle with radius 1 and center $(-2, 3)$.

To graph the equation, we use the center $(-2, 3)$ and the radius 1. See Figure 54.

Figure 54

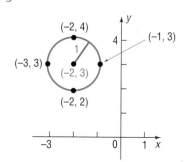

NOW WORK PROBLEM 43.

EXAMPLE 8	**Finding the General Equation of a Circle**

Find the general form of the equation of the circle whose center is $(1, -2)$ and whose graph contains the point $(4, -2)$.

Solution

To find the equation of a circle, we need to know its center and its radius. Here, we know that the center is $(1, -2)$. Since the point $(4, -2)$ is on the graph, the radius r will equal the distance from $(4, -2)$ to the center $(1, -2)$. See Figure 55. Then,

$$r = \sqrt{(4 - 1)^2 + [-2 - (-2)]^2}$$
$$= \sqrt{9} = 3$$

The standard form of the equation of the circle is

$$(x - 1)^2 + (y + 2)^2 = 9$$

Eliminating the parentheses and rearranging terms, we get the general form of the equation

$$x^2 + y^2 - 2x + 4y - 4 = 0 \qquad \blacksquare$$

Figure 55
$x^2 + y^2 - 2x + 4y - 4 = 0$

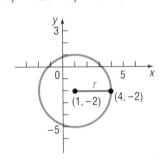

EXAMPLE 9	**Using a Graphing Utility to Graph a Circle**

Graph the equation: $x^2 + y^2 = 4$

Solution

This is the equation of a circle with center at the origin and radius 2. In order to graph this equation, we must first solve for y.

$$x^2 + y^2 = 4$$
$$y^2 = 4 - x^2 \qquad \text{Subtract } x^2 \text{ from each side}$$
$$y = \pm\sqrt{4 - x^2} \qquad \begin{array}{l}\text{Apply the Square Root}\\ \text{Property to solve for } y\end{array}$$

There are two equations to graph: first, we graph $Y_1 = \sqrt{4 - x^2}$ and then $Y_2 = -\sqrt{4 - x^2}$ on the same square screen. (Your circle will appear oval if you do not use a square screen.) See Figure 56. $\blacksquare$

Figure 56

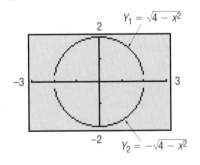

OVERVIEW

The preceding discussion about lines and circles dealt with two main types of problems that can be generalized as follows:

1. Given an equation, classify it and graph it.
2. Given a graph, or information about a graph, find its equation.

This text deals with both types of problems. We shall study various equations, classify them, and graph them. Although the second type of problem is usually more difficult to solve than the first, in many instances a graphing utility can be used to solve such problems.

2.4 EXERCISES

In Problems 1–10, the equation of a line L is given. Find the slope of a line that is (a) parallel to L and (b) perpendicular to L.

1. $y = 6x$

2. $y = -3x$

3. $y = -\frac{1}{2}x + 2$

4. $y = \frac{2}{3}x - 1$

5. $2x - 4y + 5 = 0$

6. $3x + y = 4$

7. $3x + 5y - 10 = 0$

8. $4x - 3y + 7 = 0$

9. $x = 7$

10. $y = 8$

In Problems 11–14, find an equation for the line L. Express your answer using either the general form or the slope–intercept form of the equation of a line, whichever you prefer.

11.

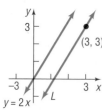

L is parallel to $y = 2x$

12.

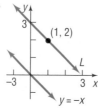

L is parallel to $y = -x$

13.

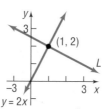

L is perpendicular to $y = 2x$

14.

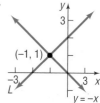

L is perpendicular to $y = -x$

In Problems 15–26, find an equation for the line with the given properties. Express your answer using either the general form or the slope–intercept form of the equation of a line, whichever you prefer.

15. Parallel to the line $y = 2x$; containing the point $(-1, 2)$

16. Parallel to the line $y = -3x$; containing the point $(-1, 2)$

17. Parallel to the line $2x - y = -2$; containing the point $(0, 0)$

18. Parallel to the line $x - 2y = -5$; containing the point $(0, 0)$

19. Parallel to the line $x = 5$; containing the point $(4, 2)$

20. Parallel to the line $y = 5$; containing the point $(4, 2)$

21. Perpendicular to the line $y = \frac{1}{2}x + 4$; containing the point $(1, -2)$

22. Perpendicular to the line $y = 2x - 3$; containing the point $(1, -2)$

23. Perpendicular to the line $2x + y = 2$; containing the point $(-3, 0)$

24. Perpendicular to the line $x - 2y = -5$; containing the point $(0, 4)$

25. Perpendicular to the line $x = 8$; containing the point $(3, 4)$

26. Perpendicular to the line $y = 8$; containing the point $(3, 4)$

In Problems 27–30, find the center and radius of each circle. Write the standard form of the equation.

27.

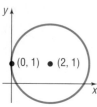

28.

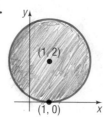

29.

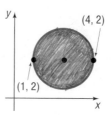

30.

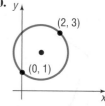

In Problems 31–38, write the standard form of the equation and the general form of the equation of each circle of radius r and center (h, k). Graph each circle.

31. $r = 2$; $(h, k) = (0, 0)$

32. $r = 3$; $(h, k) = (0, 0)$

33. $r = 1$; $(h, k) = (1, -1)$

34. $r = 2$; $(h, k) = (-2, 1)$

35. $r = 2$; $(h, k) = (0, 2)$

36. $r = 3$; $(h, k) = (1, 0)$

37. $r = 5$; $(h, k) = (4, -3)$

38. $r = 4$; $(h, k) = (2, -3)$

In Problems 39–48, find the center (h, k) and radius r of each circle. Graph each circle.

39. $x^2 + y^2 = 4$

40. $x^2 + (y - 1)^2 = 1$

41. $2(x - 3)^2 + 2y^2 = 8$

42. $3(x + 1)^2 + 3(y - 1)^2 = 6$

43. $x^2 + y^2 + 4x - 4y - 1 = 0$

44. $x^2 + y^2 - 6x + 2y + 9 = 0$

45. $x^2 + y^2 - x + 2y + 1 = 0$

46. $x^2 + y^2 + x + y - \frac{1}{2} = 0$

47. $2x^2 + 2y^2 - 12x + 8y - 24 = 0$

48. $2x^2 + 2y^2 + 8x + 7 = 0$

In Problems 49–54, find the general form of the equation of each circle.

49. Center at the origin and containing the point $(-3, 2)$

50. Center at the point $(1, 0)$ and containing the point $(-2, 3)$

51. Center at the point $(2, 3)$ and tangent to the x-axis

52. Center at the point $(-3, 1)$ and tangent to the y-axis

53. With endpoints of a diameter at the points $(1, 4)$ and $(-3, 2)$

54. With endpoints of a diameter at the points $(4, 3)$ and $(0, 1)$

55. Geometry   Use slopes to show that the triangle whose vertices are $(-2, 5)$, $(1, 3)$, and $(-1, 0)$ is a right triangle.

56. Geometry   Use slopes to show that the quadrilateral whose vertices are $(1, -1)$, $(4, 1)$, $(2, 2)$, and $(5, 4)$ is a parallelogram.

57. Geometry   Use slopes to show that the quadrilateral whose vertices are $(-1, 0)$, $(2, 3)$, $(1, -2)$, and $(4, 1)$ is a rectangle.

58. Geometry   Use slopes and the distance formula to show that the quadrilateral whose vertices are $(0, 0)$, $(1, 3)$, $(4, 2)$, and $(3, -1)$ is a square.

In Problems 59–62, match each graph with the correct equation.

(a) $(x - 3)^2 + (y + 3)^2 = 9$

(c) $(x - 1)^2 + (y + 2)^2 = 4$

(b) $(x + 1)^2 + (y - 2)^2 = 4$

(d) $(x + 3)^2 + (y - 3)^2 = 9$

59.

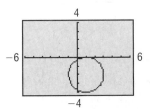

60.

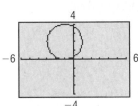

61.

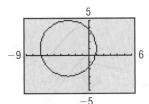

62.

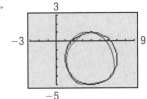

In Problems 63–66, find the standard form of the equation of each circle. Assume that the center has integer coordinates and that the radius is an integer.

63.

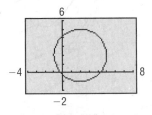

64.

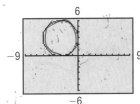

65.

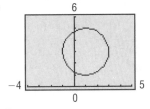

66.

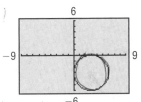

67. Which of the following equations might have the graph shown below. (More than one answer is possible.)
    (a) $(x - 2)^2 + (y + 3)^2 = 13$
    (b) $(x - 2)^2 + (y - 2)^2 = 8$
    (c) $(x - 2)^2 + (y - 3)^2 = 13$
    (d) $(x + 2) + (y - 2)^2 = 8$
    (e) $x^2 + y^2 - 4x - 9y = 0$
    (f) $x^2 + y^2 + 4x - 2y = 0$
    (g) $x^2 + y^2 - 9x - 4y = 0$
    (h) $x^2 + y^2 - 4x - 4y = 4$

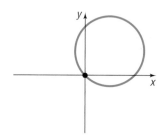

68. Which of the following equations might have the graph shown below. (More than one answer is possible.)
    (a) $(x - 2)^2 + y^2 = 3$ (e) $x^2 + y^2 + 10x + 16 = 0$
    (b) $(x + 2)^2 + y^2 = 3$ (f) $x^2 + y^2 + 10x - 2y = 1$
    (c) $x^2 + (y - 2)^2 = 3$ (g) $x^2 + y^2 + 9x + 10 = 0$
    (d) $(x + 2)^2 + y^2 = 4$ (h) $x^2 + y^2 - 9x - 10 = 0$

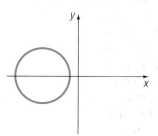

69. The figure below shows the graph of two parallel lines. Which of the following pairs of equations might have such a graph?
    (a) $x - 2y = 3$ (d) $x - y = -2$
        $x + 2y = 7$ $2x - 2y = -4$
    (b) $x + y = 2$ (e) $x + 2y = 2$
        $x + y = -1$ $x + 2y = -1$
    (c) $x - y = -2$
        $x - y = 1$

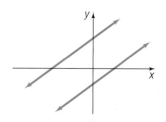

70. The figure below shows the graph of two perpendicular lines. Which of the following pairs of equations might have such a graph?
    (a) $y - 2x = 2$ (c) $2y - x = 2$ (e) $2x + y = -2$
        $y + 2x = -1$ $2y + x = -2$ $2y + x = -2$
    (b) $y - 2x = 0$ (d) $y - 2x = 2$
        $2y + x = 0$ $x + 2y = -1$

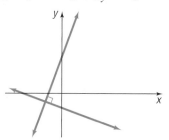

71. Prove that if two nonvertical lines have slopes whose product is -1 then the lines are perpendicular.
    [**Hint:** Refer to Figure 49, and use the converse of the Pythagorean Theorem.]

72. **Weather Satellites** Earth is represented on a map of a portion of the solar system so that its surface is the circle with equation $x^2 + y^2 + 2x + 4y - 4091 = 0$. A weather satellite circles 0.6 unit above Earth with the center of its circular orbit at the center of Earth. Find the equation for the orbit of the satellite on this map.

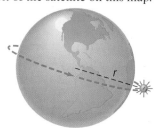

73. **Geometry** The **tangent line** to a circle may be defined as the line that intersects the circle in a single point, called the **point of tangency** (see the figure). If the equation of the circle is $x^2 + y^2 = r^2$ and the equation of the tangent line is $y = mx + b$, show that:
    (a) $r^2(1 + m^2) = b^2$
    [**Hint:** The quadratic equation $x^2 + (mx + b)^2 = r^2$ has exactly one solution.]
    (b) The point of tangency is $(-r^2m/b, r^2/b)$.
    (c) The tangent line is perpendicular to the line containing the center of the circle and the point of tangency.

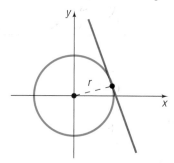

74. The **Greek method** for finding the equation of the tangent line to a circle used the fact that at any point on a circle the line containing the radius and the tangent line are perpendicular (see Problem 73). Use this method to find an equation of the tangent line to the circle $x^2 + y^2 = 9$ at the point $(1, 2\sqrt{2})$.

75. Use the Greek method described in Problem 74 to find an equation of the tangent line to the circle $x^2 + y^2 - 4x + 6y + 4 = 0$ at the point $(3, 2\sqrt{2} - 3)$.

76. Refer to Problem 73. The line $x - 2y + 4 = 0$ is tangent to a circle at $(0, 2)$. The line $y = 2x - 7$ is tangent to the same circle at $(3, -1)$. Find the center of the circle.

77. Find an equation of the line containing the centers of the two circles

$$x^2 + y^2 - 4x + 6y + 4 = 0$$
and $$x^2 + y^2 + 6x + 4y + 9 = 0$$

78. Show that the line containing the points (a, b) and (b, a) is perpendicular to the line $y = x$. Also show that the midpoint of (a, b) and (b, a) lies on the line $y = x$.

79. The equation $2x - y + C = 0$ defines a **family of lines,** one line for each value of C. On one set of coordinate axes, graph the members of the family when $C = -4$, $C = 0$, and $C = 2$. Can you draw a conclusion from the graph about each member of the family?

80. Rework Problem 79 for the family of lines $Cx + y + 4 = 0$.

81. If a circle of radius 2 is made to roll along the x-axis, what is an equation for the path of the center of the circle?

2.5 SCATTER DIAGRAMS; LINEAR CURVE FITTING

OBJECTIVES
1. Draw and Interpret Scatter Diagrams
2. Distinguish between Linear and Nonlinear Relations
3. Use a Graphing Utility to Find the Line of Best Fit

SCATTER DIAGRAMS

1. A **relation** is a correspondence between two sets. If x and y are two elements in these sets and if a relation exists between x and y, then we say that x **corresponds to** y or that y **depends on** x and write $x \rightarrow y$. We may also write $x \rightarrow y$ as the ordered pair (x, y). Here, y is referred to as the **dependent** variable and x is called the **independent** variable.

Often we are interested in specifying the type of relation (such as an equation) that might exist between two variables. The first step in finding this relation is to plot the ordered pairs using rectangular coordinates. The resulting graph is called a **scatter diagram.**

| EXAMPLE 1 | Drawing a Scatter Diagram |

The data listed in Table 6 represent the apparent temperature versus the relative humidity in a room whose actual temperature is 72° Fahrenheit.

TABLE 6

Relative Humidity (%), x	Apparent Temperature, y	(x, y)	Relative Humidity (%), x	Apparent Temperature, y	(x, y)
0	64	(0, 64)	60	72	(60, 72)
10	65	(10, 65)	70	73	(70, 73)
20	67	(20, 67)	80	74	(80, 74)
30	68	(30, 68)	90	75	(90, 75)
40	70	(40, 70)	100	76	(100, 76)
50	71	(50, 71)			

(a) Draw a scatter diagram by hand.

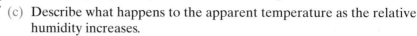 (b) Use a graphing utility to draw a scatter diagram.*

(c) Describe what happens to the apparent temperature as the relative humidity increases.

Solution (a) To draw a scatter diagram by hand, we plot the ordered pairs listed in Table 6, with the relative humidity as the x-coordinate and the apparent temperature as the y-coordinate. See Figure 57(a). Notice that the points in a scatter diagram are not connected.

(b) Figure 57(b) shows a scatter diagram using a graphing utility.

Figure 57

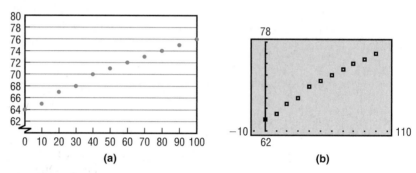

(a) (b)

(c) We see from the scatter diagrams that, as the relative humidity increases, the apparent temperature increases. ▬

✏ NOW WORK PROBLEM 7(a).

CURVE FITTING

② Scatter diagrams are used to help us see the type of relation that may exist between two variables. In this text, we will discuss a variety of different relations that may exist between two variables. For now, we concentrate on distinguishing between linear and nonlinear relations. See Figure 58.

Figure 58

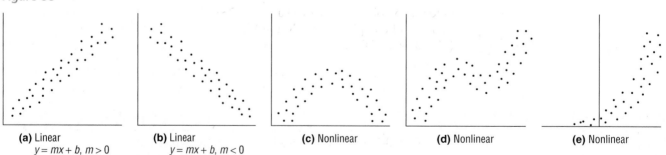

(a) Linear
$y = mx + b, m > 0$

(b) Linear
$y = mx + b, m < 0$

(c) Nonlinear

(d) Nonlinear

(e) Nonlinear

| EXAMPLE 2 | Distinguishing between Linear and Nonlinear Relations |

Determine whether the relation between the two variables in Figure 59 is linear or nonlinear.

*Consult your owner's manual for the proper keystrokes.

Figure 59

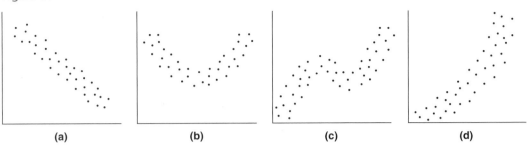

(a)	(b)	(c)	(d)

Solution   (a) Linear   (b) Nonlinear   (c) Nonlinear   (d) Nonlinear   ■

NOW WORK PROBLEM **1**.

In this section we will study data whose scatter diagrams imply that a linear relation exists between the two variables. Nonlinear data will be discussed in later chapters.

Suppose that the scatter diagram of a set of data appears to be linearly related as in Figure 58(a) or (b). We might wish to find an equation of a line that relates the two variables. One way to obtain an equation for such data is to draw a line through two points on the scatter diagram and estimate the equation of the line.

EXAMPLE 3   **Find an Equation for Linearly Related Data**

Using the data in Table 6 from Example 1, select two points from the data and find an equation of the line containing the points.

(a) Graph the line on the scatter diagram obtained in Example 1(a).

(b) Graph the line on the scatter diagram obtained in Example 1(b).

Solution   Select two points, say $(10, 65)$ and $(70, 73)$. (You should select your own two points and complete the solution.) The slope of the line joining the points $(10, 65)$ and $(70, 73)$ is

$$m = \frac{73 - 65}{70 - 10} = \frac{8}{60} = \frac{2}{15}$$

The equation of the line with slope $\frac{2}{15}$ and passing through $(10, 65)$ is found using the point–slope form with $m = \frac{2}{15}$, $x_1 = 10$, and $y_1 = 65$.

$$y - y_1 = m(x - x_1)$$

$$y - 65 = \frac{2}{15}(x - 10)$$

$$y = \frac{2}{15}x + \frac{191}{3}$$

(a) Figure 60(a) shows the scatter diagram with the graph of the line drawn by hand.

 (b) Figure 60(b) shows the scatter diagram with the graph of the line using a graphing utility.

Figure 60

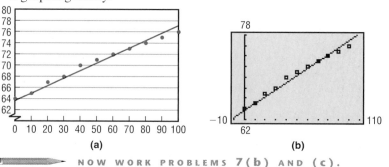

(a)   (b)

NOW WORK PROBLEMS 7(b) AND (c).

LINE OF BEST FIT

③ The line obtained in Example 3 depends on the selection of points, which will vary from person to person. So the line that we found might be different from the line that you found. Although the line that we found in Example 3 appears to "fit" the data well, there may be a line that "fits it better." Do you think your line fits the data better? Is there a line of *best fit*? As it turns out, there is a method for finding the line that best fits linearly related data (called the *line of best fit*).*

EXAMPLE 4

Finding the Line of Best Fit

Using the data in Table 6 from Example 1:

(a) Find the line of best fit using a graphing utility.
(b) Graph the line of best fit on the scatter diagram obtained in Example 1(b).
(c) Interpret the slope of the line of best fit.
(d) Use the line of best fit to predict the apparent temperature of a room whose actual temperature is 72°F and relative humidity is 45%.

Figure 61

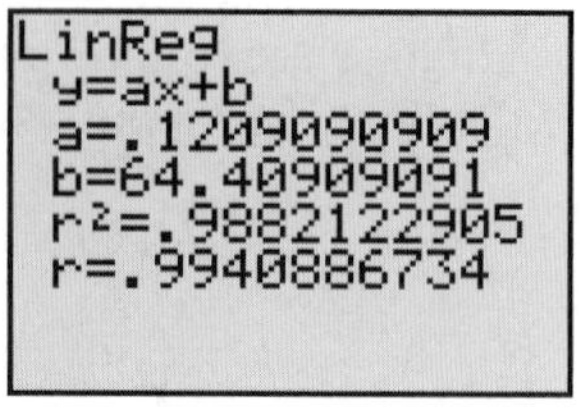

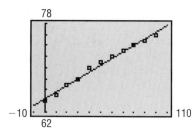

Solution

(a) Graphing utilities contain built-in programs that find the line of best fit for a collection of points in a scatter diagram. (Look in your owner's manual under Linear Regression or Line of Best Fit for details on how to execute the program.) Upon executing the LINear REGression program, we obtain the results shown in Figure 61. The output that the utility provides shows us the equation $y = ax + b$, where a is the slope of the line and b is the y-intercept. The line of best fit that relates relative humidity to apparent temperature may be expressed as the line $y = 0.121x + 64.409$.

(b) Figure 62 shows the graph of the line of best fit, along with the scatter diagram.

(c) The slope of the line of best fit is 0.121, which means that, for every 1% increase in the relative humidity, apparent room temperature increases 0.121°F.

(d) Letting $x = 45$ in the equation of the line of best fit, we obtain $y = 0.121(45) + 64.409 \approx 70°F$, which is the apparent temperature in the room.

Figure 62

NOW WORK PROBLEMS 7(d) AND (e).

*We shall not discuss in this book the underlying mathematics of lines of best fit. Most books in statistics and many in linear algebra discuss this topic.

Does the line of best fit appear to be a good fit? In other words, does the line appear to accurately describe the relation between temperature and relative humidity?

And just how "good" is this line of best fit? The answers are given by what is called the *correlation coefficient.* Look again at Figure 61. The last line of output is $r = 0.994$. This number, called the **correlation coefficient, r,** $-1 \leq r \leq 1$, is a measure of the strength of the *linear relation* that exists between two variables. The closer that $|r|$ is to 1, the more perfect the linear relationship is. If r is close to 0, there is little or no *linear* relationship between the variables. A negative value of r, $r < 0$, indicates that as x increases y decreases; a positive value of r, $r > 0$, indicates that as x increases y does also. The data given in Example 1, having a correlation coefficient of 0.994, are indicative of a strong linear relationship with positive slope.

2.5 EXERCISES

In Problems 1–6, examine the scatter diagram and determine whether the type of relation, if any, that may exist is linear or nonlinear.

1.
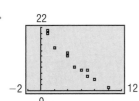

2.
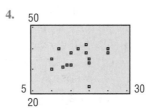

3.
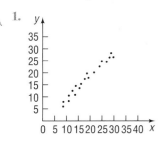

4.
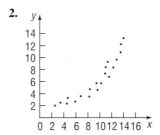

5.
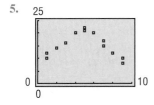

6.
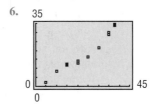

In Problems 7–14:
   (a) *Draw a scatter diagram by hand.*
   (b) *Select two points from the scatter diagram and find the equation of the line containing the points selected.**
   (c) *Graph the line found in part (b) on the scatter diagram.*
   (d) *Use a graphing utility to find the line of best fit.*
   (e) *Use a graphing utility to graph the line of best fit on the scatter diagram.*

7.

x	3	4	5	6	7	8	9
y	4	6	7	10	12	14	16

8.

x	3	5	7	9	11	13
y	0	2	3	6	9	11

* Answers will vary. We will use the first and last data points in the answer section.

9. | x | -2 | -1 | 0 | 1 | 2 |
|---|---|---|---|---|---|
| y | -4 | 0 | 1 | 4 | 5 |

10. | x | -2 | -1 | 0 | 1 | 2 |
|---|---|---|---|---|---|
| y | 7 | 6 | 3 | 2 | 0 |

11. | x | 20 | 30 | 40 | 50 | 60 |
|---|---|---|---|---|---|
| y | 100 | 95 | 91 | 83 | 70 |

12. | x | 5 | 10 | 15 | 20 | 25 |
|---|---|---|---|---|---|
| y | 2 | 4 | 7 | 11 | 18 |

13. | x | -20 | -17 | -15 | -14 | -10 |
|---|---|---|---|---|---|
| y | 100 | 120 | 118 | 130 | 140 |

14. | x | -30 | -27 | -25 | -20 | -14 |
|---|---|---|---|---|---|
| y | 10 | 12 | 13 | 13 | 18 |

---

15. **Consumption and Disposable Income** An economist wishes to estimate a line that relates personal consumption expenditures (C) and disposable income (I). Both C and I are in thousands of dollars. She interviews eight heads of households for families of size 3 and obtains the data below. Let I represent the independent variable and C the dependent variable.
   (a) Draw a scatter diagram by hand.
   (b) Find a line that fits the data.*
   (c) Interpret the slope. The slope of this line is called the **marginal propensity to consume.**
   (d) Predict the consumption of a family whose disposable income is $42,000.
   (e) Use a graphing utility to find the line of best fit to the data.

I (000)	C (000)
20	16
20	18
18	13
27	21
36	27
37	26
45	36
50	39

16. **Marginal Propensity to Save** The same economist as in Problem 15 wants to estimate a line that relates savings S and disposable income I. Let $S = I - C$ be the dependent variable and I the independent variable.
   (a) Draw a scatter diagram by hand.
   (b) Find a line that fits the data.
   (c) Interpret the slope. The slope of this line is called the **marginal propensity to save.**
   (d) Predict the savings of a family whose income is $42,000.
   (e) Use a graphing utility to find the line of best fit.

* Answers will vary. We will use the first and last data points in the answer section.

17. **Mortgage Qualification** The amount of money that a lending institution will allow you to borrow mainly depends on the interest rate and your annual income. The following data represent the annual income, I, required by a bank in order to lend L dollars at an interest rate of 7.5% for 30 years.

Annual Income, I ($)	Loan Amount, L ($)
15,000	44,600
20,000	59,500
25,000	74,500
30,000	89,400
35,000	104,300
40,000	119,200
45,000	134,100
50,000	149,000
55,000	163,900
60,000	178,800
65,000	193,700
70,000	208,600

Source: Information Please Almanac, 1999

Let I represent the independent variable and L the dependent variable.
   (a) Use a graphing utility to draw a scatter diagram of the data.
   (b) Use a graphing utility to find the line of best fit to the data.
   (c) Graph the line of best fit on the scatter diagram drawn in part (a).
   (d) Interpret the slope of the line of best fit.
   (e) Determine the loan amount that an individual would qualify for if her income is $42,000.

18. **Mortgage Qualification** The amount of money that a lending institution will allow you to borrow mainly depends on the interest rate and your annual income. The following data represent the annual income, I, required by a bank in order to lend L dollars at an interest rate of 8.5% for 30 years.

Annual Income, I ($)	Loan Amount, L ($)
15,000	40,600
20,000	54,100
25,000	67,700
30,000	81,200
35,000	94,800
40,000	108,300
45,000	121,900
50,000	135,400
55,000	149,000
60,000	162,500
65,000	176,100
70,000	189,600

Source: Information Please Almanac, 1999

Let I represent the independent variable and L the dependent variable.
(a) Use a graphing utility to draw a scatter diagram of the data.
(b) Use a graphing utility to find the line of best fit to the data.
(c) Graph the line of best fit on the scatter diagram drawn in part (a).
(d) Interpret the slope of the line of best fit.
(e) Determine the loan amount that an individual would qualify for if her income is $42,000.

19. Apparent Room Temperature   The following data represent the apparent temperature versus the relative humidity in a room whose actual temperature is 65° Fahrenheit.

Relative Humidity, h (%)	Apparent Temperature, T (°F)
0	59
10	60
20	61
30	61
40	62
50	63
60	64
70	65
80	65
90	66
100	67

Source: National Oceanic and Atmospheric Administration

Let h represent the independent variable and T the dependent variable.

(a) Use a graphing utility to draw a scatter diagram of the data.
(b) Use a graphing utility to find the line of best fit to the data.
(c) Graph the line of best fit on the scatter diagram drawn in part (a).
(d) Interpret the slope of the line of best fit.
(e) Determine the apparent temperature of a room whose actual temperature is 65°F if the relative humidity is 75%.

20. Apparent Room Temperature   The following data represent the apparent temperature versus the relative humidity in a room whose actual temperature is 75° Fahrenheit.

Relative Humidity, h (%)	Apparent Temperature, T (°F)
0	68
10	69
20	71
30	72
40	74
50	75
60	76
70	76
80	77
90	78
100	79

Source: National Oceanic and Atmospheric Administration

Let h represent the independent variable and let T be the dependent variable.

(a) Use a graphing utility to draw a scatter diagram of the data.
(b) Use a graphing utility to find the line of best fit to the data.
(c) Graph the line of best fit on the scatter diagram drawn in part (a).
(d) Interpret the slope of the line of best fit.
(e) Determine the apparent temperature of a room whose actual temperature is 75°F if the relative humidity is 75%.

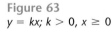

21. Average Miles per Car   The following data represent the average miles driven per car (in thousands) in the United States for the years 1985 to 1996. Let the year, x, represent the independent variable and average miles per car, M, represent the dependent variable.

(a) Use a graphing utility to draw a scatter diagram of the data.
(b) Use a graphing utility to find the line of best fit to the data.
(c) Graph the line of best fit on the scatter diagram drawn in part (a).
(d) Interpret the slope of the line of best fit.
(e) Predict the average number of miles driven per car in 1997.

Year, x	Average Miles per Car, M
1985	9.4
1986	9.5
1987	9.7
1988	10.0
1989	10.2
1990	10.3
1991	10.3
1992	10.6
1993	10.5
1994	10.8
1995	11.1
1996	11.3

Source: U.S. Federal Highway Administration

2.6   VARIATION

OBJECTIVES   **1** Construct a Model Using Direct Variation
2 Construct a Model Using Inverse Variation
3 Construct a Model Using Joint or Combined Variation

When a mathematical model is developed for a real-world problem, it often involves relationships between quantities that are expressed in terms of proportionality:

> Force is proportional to acceleration.
>
> For an ideal gas held at a constant temperature, pressure and volume are inversely proportional.
>
> The force of attraction between two heavenly bodies is inversely proportional to the square of the distance between them.
>
> Revenue is directly proportional to sales.

Each of these statements illustrates the idea of **variation,** or how one quantity varies in relation to another quantity. Quantities may vary *directly, inversely,* or *jointly.*

Figure 63
$y = kx; k > 0, x \geq 0$

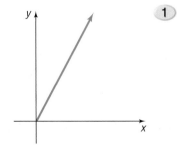

DIRECT VARIATION

1   Let x and y denote two quantities. Then y **varies directly** with x, or y is **directly proportional to** x, if there is a nonzero number k such that

$$y = kx$$

The number k is called the **constant of proportionality.**
The graph in Figure 63 illustrates the relationship between y and x if y varies directly with x and $k > 0$, $x \geq 0$. Note that the constant of proportionality is, in fact, the slope of the line.

If we know that two quantities vary directly, then knowing the value of each quantity in one instance enables us to write a formula that is true in all cases.

EXAMPLE 1

Chemistry: Gas Law

For a certain gas enclosed in a container of fixed volume, the pressure P (in newtons per square meter) varies directly with temperature T in Kelvins. If the pressure is found to be 20 newtons per square meter at a temperature of 60 K, find a formula that relates pressure P to temperature T. Then find the pressure P when $T = 120$ K.

Solution

Because P varies directly with T, we know that

$$P = kT$$

for some constant k. Because $P = 20$ when $T = 60$,

$$20 = k(60) \qquad \text{Substitute } P = 20, T = 60 \text{ in } P = kT.$$

$$k = \tfrac{1}{3} \qquad \text{Solve for } k.$$

Figure 64

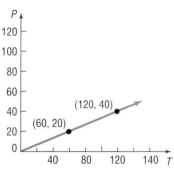

Thus, in all cases,

$$P = \tfrac{1}{3}T \qquad \text{Substitute } k = \tfrac{1}{3} \text{ in } P = kT.$$

In particular, when $T = 120$ K, we find

$$P = \tfrac{1}{3}(120) = 40 \text{ newtons per square meter}$$

Figure 64 illustrates the relationship between the pressure P and the temperature T.

NOW WORK PROBLEM **1**.

INVERSE VARIATION

Figure 65
$y = \frac{k}{x}; k > 0, x > 0$

(2)

Let x and y denote two quantities. Then y **varies inversely** with x, or y is **inversely proportional to** x, if there is a nonzero constant k such that

$$y = \frac{k}{x}$$

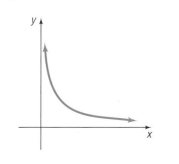

The graph in Figure 65 illustrates the relationship between y and x if y varies inversely with x and $k > 0$, $x > 0$.

	EXAMPLE 2	

Maximum Weight That Can Be Supported by a Piece of Pine

Figure 66

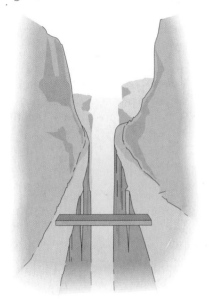

The maximum weight W that can be safely supported by a 2-inch by 4-inch piece of pine varies inversely with its length l. See Figure 66. Experiments indicate that the maximum weight that a 10-foot long pine 2-by-4 can support is 500 pounds. Write a general formula relating the maximum weight W (in pounds) to length l (in feet). Find the maximum weight W that can be safely supported by a length of 25 feet.

Solution Because W varies inversely with l, we know that

$$W = \frac{k}{l}$$

for some constant k. Because $W = 500$ when $l = 10$, we have

$$500 = \frac{k}{10}$$

$$k = 5000$$

Thus, in all cases,

$$W = \frac{5000}{l}$$

In particular, the maximum weight W that can be safely supported by a piece of pine 25 feet in length is

$$W = \frac{5000}{25} = 200 \text{ pounds}$$

Figure 67

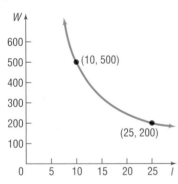

Figure 67 illustrates the relationship between the weight W and the length l. ▬

In direct or inverse variation, the quantities that vary may be raised to powers. For example, in the early seventeenth century, Johannes Kepler (1571–1630) discovered that the square of the period T of a planet varies directly with the cube of its mean distance a from the Sun. That is, $T^2 = ka^3$, where k is the constant of proportionality.

NOW WORK PROBLEM **25.**

JOINT VARIATION AND COMBINED VARIATION

③ When a variable quantity Q is proportional to the product of two or more other variables, we say that Q **varies jointly** with these quantities. Finally, combinations of direct and/or inverse variation may occur. This is usually referred to as **combined variation.**

Let's look at an example.

	EXAMPLE 3	

Loss of Heat through a Wall

The loss of heat through a wall varies jointly with the area of the wall and the difference between the inside and outside temperatures and varies inversely with the thickness of the wall. Write an equation that relates these quantities.

Solution   We begin by assigning symbols to represent the quantities:

$$L = \text{Heat loss} \qquad T = \text{Temperature difference}$$
$$A = \text{Area of wall} \qquad d = \text{Thickness of wall}$$

Then

$$L = k\,\frac{AT}{d}$$

where k is the constant of proportionality.   ∎

NOW WORK PROBLEM 17.

EXAMPLE 4   Force of the Wind on a Window

The force F of the wind on a flat surface positioned at a right angle to the direction of the wind varies jointly with the area A of the surface and the square of the speed v of the wind. A wind of 30 miles per hour blowing on a window measuring 4 feet by 5 feet has a force of 150 pounds. (See Figure 68.) What is the force on a window measuring 3 feet by 4 feet caused by a wind of 50 miles per hour?

Figure 68

Solution   Since F varies jointly with A and v^2, we have

$$F = kAv^2$$

where k is the constant of proportionality. We are told that $F = 150$ when $A = 4 \cdot 5 = 20$ and $v = 30$. Then, we have

$$150 = k(20)(900) \qquad F = kAv^2, F = 150, A = 20, v = 30$$

$$k = \frac{1}{120}$$

The general formula is therefore

$$F = \frac{1}{120}Av^2$$

For a wind of 50 miles per hour blowing on a window whose area is $A = 3 \cdot 4 = 12$ square feet, the force F is

$$F = \frac{1}{120}(12)(2500) = 250 \text{ pounds} \qquad ∎$$

2.6 EXERCISES

In Problems 1–12, write a general formula to describe each variation.

1. y varies directly with x; $\quad y = 2$ when $x = 10$
2. v varies directly with t; $\quad v = 16$ when $t = 2$
3. A varies directly with x^2; $\quad A = 4\pi$ when $x = 2$
4. V varies directly with x^3; $\quad V = 36\pi$ when $x = 3$
5. F varies inversely with d^2; $\quad F = 10$ when $d = 5$
6. y varies inversely with $\sqrt{x}$; $\quad y = 4$ when $x = 9$
7. z varies directly with the sum of the squares of x and y; $\quad z = 5$ when $x = 3$ and $y = 4$
8. T varies jointly with the cube root of x and the square of d; $\quad T = 18$ when $x = 8$ and $d = 3$
9. M varies directly with the square of d and inversely with the square root of x; $\quad M = 24$ when $x = 9$ and $d = 4$
10. z varies directly with the sum of the cube of x and the square of y; $\quad z = 1$ when $x = 2$ and $y = 3$
11. The square of T varies directly with the cube of a and inversely with the square of d; $\quad T = 2$ when $a = 2$ and $d = 4$
12. The cube of z varies directly with the sum of the squares of x and y; $\quad z = 2$ when $x = 9$ and $y = 4$

In Problems 13–20, write an equation that relates the quantities.

13. **Geometry** The volume V of a sphere varies directly with the cube of its radius r. The constant of proportionality is $4\pi/3$.

14. **Geometry** The square of the hypotenuse c of a right triangle varies directly with the sum of the squares of the legs a and b. The constant of proportionality is 1.

15. **Geometry** The area A of a triangle varies jointly with the lengths of the base b and the height h. The constant of proportionality is $\frac{1}{2}$.

16. **Geometry** The perimeter p of a rectangle varies directly with the sum of the lengths of its sides l and w. The constant of proportionality is 2.

17. **Geometry** The volume V of a right circular cylinder varies jointly with the square of its radius r and its height h. The constant of proportionality is π. (See the figure.)

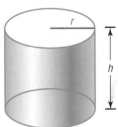

18. **Geometry** The volume V of a right circular cone varies jointly with the square of its radius r and its height h. The constant of proportionality is $\pi/3$. (See the figure.)

19. **Physics: Newton's Law** The force F (in newtons) of attraction between two bodies varies jointly with their masses m and M (in kilograms) and inversely with the square of the distance d (in meters) between them. The constant of proportionality is $G = 6.67 \times 10^{-11}$.

20. **Physics: Simple Pendulum** The *period* of a pendulum is the time required for one oscillation; the pendulum is usually referred to as *simple* when the angle made to the vertical is less than 5°. The period T of a simple pendulum (in seconds) varies directly with the square root of its length l (in feet). The constant of proportionality is $2\pi/\sqrt{32}$.

21. **Physics: Falling Objects** The distance s that an object falls is directly proportional to the square of the time t of the fall. If an object falls 16 feet in 1 second, how far will it fall in 3 seconds? How long will it take an object to fall 64 feet?

22. **Physics: Falling Objects** The velocity v of a falling object is directly proportional to the time t of the fall. If, after 2 seconds, the velocity of the object is 64 feet per second, what will its velocity be after 3 seconds?

23. **Physics: Stretching a Spring** The elongation E of a spring balance varies directly with the applied weight W (see the figure). If $E = 3$ when $W = 20$, find E when $W = 15$.

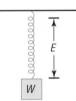

24. **Physics: Vibrating String** The rate of vibration of a string under constant tension varies inversely with the length of the string. If a string is 48 inches long and vibrates 256 times per second, what is the length of a string that vibrates 576 times per second?

25. **Weight of a Body** The weight of a body above the surface of Earth varies inversely with the square of the distance from the center of Earth. If a certain body weighs 55 pounds when it is 3960 miles from the center of Earth, how much will it weigh when it is 3965 miles from the center?

26. **Force of the Wind on a Window** The force exerted by the wind on a plane surface varies jointly with the area of the surface and the square of the velocity of the wind. If the force on an area of 20 square feet is 11 pounds when the wind velocity is 22 miles per hour, find the force on a surface area of 47.125 square feet when the wind velocity is 36.5 miles per hour.

27. **Horsepower** The horsepower (hp) that a shaft can safely transmit varies jointly with its speed (in revolutions per minute, rpm) and the cube of its diameter. If a shaft of a certain material 2 inches in diameter can transmit 36 hp at 75 rpm, what diameter must the shaft have in order to transmit 45 hp at 125 rpm?

28. **Weight of a Body** The weight of a body varies inversely with the square of its distance from the center of Earth. Assuming that the radius of Earth is 3960 miles, how much would a man weigh at an altitude of 1 mile above Earth's surface if he weighs 200 pounds on Earth's surface?

29. **Physics: Kinetic Energy** The kinetic energy K of a moving object varies jointly with its mass m and the square of its velocity v. If an object weighing 25 pounds and moving with a velocity of 100 feet per second has a kinetic energy of 400 foot-pounds, find its kinetic energy when the velocity is 150 feet per second.

30. **Electrical Resistance of a Wire** The electrical resistance of a wire varies directly with the length of the wire and inversely with the square of the diameter of the wire. If a wire 432 feet long and 4 millimeters in diameter has a resistance of 1.24 ohms, find the length of a wire of the same material whose resistance is 1.44 ohms and whose diameter is 3 millimeters.

31. **Measuring the Stress of Materials** The stress in the material of a pipe subject to internal pressure varies jointly with the internal pressure and the internal diameter of the pipe and inversely with the thickness of the pipe. The stress is 100 pounds per square inch when the diameter is 5 inches, the thickness is 0.75 inch, and the internal pressure is 25 pounds per square inch. Find the stress when the internal pressure is 40 pounds per square inch if the diameter is 8 inches and the thickness is 0.50 inch.

32. **Safe Load for a Beam** The maximum safe load for a horizontal rectangular beam varies jointly with the width of the beam and the square of the thickness of the beam and inversely with its length. If an 8-foot beam will support up to 750 pounds when the beam is 4 inches wide and 2 inches thick, what is the maximum safe load in a similar beam 10 feet long, 6 inches wide, and 2 inches thick?

33. **Resistance due to a Conductor** The resistance (in ohms) of a circular conductor varies directly with the length of the conductor and inversely with the square of the radius of the conductor. If 50 feet of wire with a radius of 6×10^{-3} inch has a resistance of 10 ohms, what would be the resistance of 100 feet of the same wire if the radius is increased to 7×10^{-3} inch?

34. **Chemistry: Gas Laws** The volume V of an ideal gas varies directly with the temperature T and inversely with the pressure P. Write an equation relating V, T, and P using k as the constant of proportionality. If a cylinder contains oxygen at a temperature of 300 K and a pressure of 15 atmospheres in a volume of 100 liters, what is the constant of proportionality k? If a piston is lowered into the cylinder, decreasing the volume occupied by the gas to 80 liters and raising the temperature to 310 K, what is the gas pressure?

In Problems 36–40, use the result obtained in Problem 35.

35. **Satellites in Orbit** The speed v required of a satellite to maintain a near-Earth circular orbit is directly proportional to the square root of the distance r of the satellite from the center of Earth.* The constant of proportionality is $\sqrt{g}$, where g is the acceleration of gravity for Earth. Write an equation that shows the relationship between v and r. (The radius of Earth is approximately 3960 miles.)

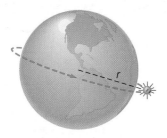

36. **Speed Required to Stay in Orbit** What speed is required to maintain a communications satellite in a circular orbit 500 miles above Earth's surface?

37. **Speed Required to Stay in Orbit** Find the speed of a satellite that moves in a circular orbit 140 miles above Earth's surface.

38. **Distance of a Satellite from Earth** Find the distance of a satellite from the surface of Earth as it moves around Earth in a circular orbit at a constant speed of 18,630 miles per hour.

39. **Distance of a Satellite from Earth** A weather satellite orbits Earth in a circle every 1.5 hours. How high is it above Earth?

40. **Distance and Speed of a Military Satellite** A military satellite orbits Earth every 2 hours. How high is this satellite and what is its speed?

---

*Near-Earth orbits are at least 100 miles above Earth's surface (out of Earth's atmosphere) and up to an altitude of approximately 15,000 miles. The effect of the gravitational attraction of other bodies is ignored. Although the acceleration of gravity at such altitudes is somewhat less than $g \approx 32$ feet per second per second $\approx 79{,}036$ miles per hour per hour, we shall ignore this discrepancy in our calculations.

In Problems 42–46, use the result obtained in Problem 41.

41. Physical Force The force F (in newtons) required to maintain an object in a circular path varies jointly with the mass m (in kilograms) of the object and the square of its speed v (in meters per second) and inversely with the radius r (in meters) of the circular path. The constant of proportionality is 1. Write an equation relating F, m, v, and r.

42. Physical Forces A motorcycle with mass 150 kilograms is driven at a constant speed of 120 kilometers per hour on a circular track with a radius of 100 meters. To keep the motorcycle from skidding, what frictional force must be exerted by the tires on the track?

43. Physical Forces If the speed of the motorcycle described in Problem 42 is increased by 10%, by how much is the frictional force of the tires increased?

44. Physical Forces If the radius of the track described in Problem 42 is cut in half, how much slower should the motorcycle be driven to maintain the same frictional force?

45. Physical Forces Judy is spinning a bucket of water in a horizontal plane at the end of a rope of length L (see the figure). If she triples the speed of the bucket, how many times as hard must she pull on the rope?

46. Physical Forces If Judy in Problem 45 doubles the length of the rope and maintains the same speed for the bucket, will she have to pull on the rope more or less? How much?

47. The formula on page 194 attributed to Johannes Kepler is one of the famous three Keplerian Laws of Planetary Motion. Go to the library and research these laws. Write a brief paper about these laws and Kepler's place in history.

48. Make up a real-world problem different from any in the text that you think involves two variables that vary directly. Exchange your problem with another student's to solve and critique.

49. Make up a real-world problem different from any in the text that you think involves two variables that vary inversely. Exchange your problem with another student's to solve and critique.

50. Make up a real-world problem different from any in the text that you think involves three variables that vary jointly. Exchange your problem with another student's to solve and critique.

CHAPTER REVIEW

Things To Know

Formulas

Distance formula (p. 148)	$d = \sqrt{(x_2 - x_1)^2 + (y_2 - y_1)^2}$
Midpoint formula (p.150)	$(x, y) = \left(\dfrac{x_1 + x_2}{2}, \dfrac{y_1 + y_2}{2} \right)$
Slope (p. 164)	$m = \dfrac{y_2 - y_1}{x_2 - x_1}$, if $x_1 \neq x_2$; undefined if $x_1 = x_2$
Parallel lines (p. 176)	Equal slopes $(m_1 = m_2)$ and different y-intercepts $(b_1 \neq b_2)$
Perpendicular lines (p. 177)	Product of slopes is -1 $(m_1 \cdot m_2 = -1)$
Direct variation (p.192)	$y = kx$
Inverse variation (p. 193)	$y = \dfrac{k}{x}$

Equations of Lines and Circles

Vertical line (p. 168)	$x = a$
Horizontal line (p. 169)	$y = b$

Point–slope form of the equation of a line (p. 168)	$y - y_1 = m(x - x_1)$; m is the slope of the line, (x_1, y_1) is a point on the line
Slope–intercept form of the equation of a line (p. 169)	$y = mx + b$; m is the slope of the line, b is the y-intercept
General form of the equation of a line (p. 171)	$Ax + By = C$; A, B not both 0
Standard form of the equation of a circle (p. 179)	$(x - h)^2 + (y - k)^2 = r^2$; r is the radius of the circle, (h, k) is the center of the circle
Equation of the unit circle (p. 179)	$x^2 + y^2 = 1$
General form of the equation of a circle (p. 180)	$x^2 + y^2 + ax + by + c = 0$

Objectives

You should be able to:

Use the distance formula (p. 147)

Use the midpoint formula (p. 150)

Graph equations by plotting points (p. 154)

Find intercepts from a graph (p. 157)

Find intercepts from an equation (p. 158)

Test an equation for symmetry with respect to the x-axis, the y-axis, and the origin (p. 160)

Calculate and interpret the slope of a line (p. 164)

Graph lines given a point and the slope (p. 167)

Find the equation of vertical lines (p. 167)

Use the point–slope form of a line; identify horizontal lines (p. 168)

Find the equation of a line given two points (p. 169)

Write the equation of a line in slope–intercept form (p.169)

Identify the slope and y-intercept of a line from its equation (p. 170)

Write the equation of a line in general form (p. 171)

Define parallel and perpendicular lines (p. 175 and p. 177)

Find equations of parallel lines (p. 176)

Find equations of perpendicular lines (p. 178)

Write the standard form of the equation of a circle (p. 178)

Graph a circle (p. 179)

Find the center and radius of a circle from an equation in general form and graph it (p. 180)

Draw and interpret scatter diagrams (p. 185)

Distinguish between linear and nonlinear relations (p. 186)

Use a graphing utility to find the line of best fit (p. 188)

Construct a model using direct variation (p. 192)

Construct a model using inverse variation (p. 193)

Construct a model using joint or combined variation (p. 194)

Fill-in-the-Blank Items

1. If (x, y) are the coordinates of a point P in the xy-plane, then x is called the _____ of P and y is the _____ of P.

2. If three distinct points P, Q, and R all lie on a line and if $d(P, Q) = d(Q, R)$, then Q is called the _____ of the line segment from P to R.

3. If for every point (x, y) on a graph the point $(-x, y)$ is also on the graph, then the graph is symmetric with respect to the _____ .

4. The set of points in the xy-plane that are a fixed distance from a fixed point is called a(n) _____ . The fixed distance is called the _____ ; the fixed point is called the _____ .

5. The slope of a vertical line is _____ ; the slope of a horizontal line is _____ .

6. Two nonvertical lines have slopes m_1 and m_2, respectively. The lines are parallel if _____ and the _____ are unequal; the lines are perpendicular if _____ .

7. If z varies jointly with x^2 and y^3 and inversely with $\sqrt{t}$, then $z = $ _____ , where k is the constant of proportionality.

True/False Items

T F   **1.**   The distance between two points is sometimes a negative number.

T F   **2.**   The graph of the equation $y = x^4 + x^2 + 1$ is symmetric with respect to the y-axis.

T F   **3.**   Vertical lines have undefined slope.

T F   **4.**   The slope of the line $2y = 3x + 5$ is 3.

T F   **5.**   Perpendicular lines have slopes that are reciprocals of one another.

T F   **6.**   The radius of the circle $x^2 + y^2 = 9$ is 3.

T F   **7.**   If y varies inversely with x, then as x increases in value y will decrease in value.

Review Exercises

Blue problem numbers indicate the author's suggestions for use in a Practice Test.

In Problems 1–8, test each equation for symmetry with respect to the x-axis, the y-axis, and the origin.

1. $2x = 3y^2$ **2.** $y = 5x$ **3.** $x^2 + 4y^2 = 16$ **4.** $9x^2 - y^2 = 9$

5. $y = x^4 + 2x^2 + 1$ **6.** $y = x^3 - x$ **7.** $x^2 + x + y^2 + 2y = 0$ **8.** $x^2 + 4x + y^2 - 2y = 0$

In Problems 9–18, find an equation of the line having the given characteristics. Express your answer using either the general form or the slope–intercept form of the equation of a line, whichever you prefer.

9. Slope $= -2$;   containing the point $(3, -1)$

10. Slope $= 0$;   containing the point $(-5, 4)$

11. Slope undefined;   containing the point $(-3, 4)$

12. x-intercept $= 2$;   containing the point $(4, -5)$

13. y-intercept $= -2$;   containing the point $(5, -3)$

14. Containing the points $(3, -4)$ and $(2, 1)$

15. Parallel to the line $2x - 3y = -4$;   containing the point $(-5, 3)$

16. Parallel to the line $x + y = 2$;   containing the point $(1, -3)$

17. Perpendicular to the line $x + y = 2$;   containing the point $(4, -3)$

18. Perpendicular to the line $3x - y = -4$;   containing the point $(-2, 4)$

In Problems 19–24, graph each line, labeling any intercepts.

19. $4x - 5y = -20$ **20.** $3x + 4y = 12$ **21.** $\dfrac{1}{2}x - \dfrac{1}{3}y = -\dfrac{1}{6}$

22. $-\dfrac{3}{4}x + \dfrac{1}{2}y = 0$ **23.** $\sqrt{2}x + \sqrt{3}y = \sqrt{6}$ **24.** $\dfrac{x}{3} + \dfrac{y}{4} = 1$

In Problems 25–30, find the center and radius of each circle. Graph each circle.

25. $x^2 + (y - 1)^2 = 4$ **26.** $(x + 2)^2 + y^2 = 9$ **27.** $x^2 + y^2 - 2x + 4y - 4 = 0$

28. $x^2 + y^2 + 4x - 4y - 1 = 0$ **29.** $3x^2 + 3y^2 - 6x + 12y = 0$ **30.** $2x^2 + 2y^2 - 4x = 0$

---

31. Find the slope of the line containing the points $(7, 4)$ and $(-3, 2)$. What is the distance between these points? What is their midpoint?

32. Show that the points $A = (3, 4)$, $B = (1, 1)$, and $C = (-2, 3)$ are the vertices of an isosceles triangle.

33. Show that the points $A = (-2, 0)$, $B = (-4, 4)$, and $C = (8, 5)$ are the vertices of a right triangle in two ways:

(a) By using the converse of the Pythagorean Theorem
(b) By using the slopes of the lines joining the vertices

34. The endpoints of the diameter of a circle are $(-3, 2)$ and $(5, -6)$. Find the center and radius of the circle. Write the general equation of this circle.

35. Show that the points $A = (2, 5)$, $B = (6, 1)$, and $C = (8, -1)$ lie on a line by using slopes.

36. Show that the points $A = (1, 5)$, $B = (2, 4)$, and $C = (-3, 5)$ lie on a circle with center $(-1, 2)$. What is the radius of this circle?

37. Geometry The area of an equilateral triangle varies directly with the square of the length of a side. If the area of the equilateral triangle whose sides are of length 1 centimeter is $\sqrt{3}/4$, find the length s of each side of an equilateral triangle whose area A is 16 square centimeters.

38. Vibrating Strings In a vibrating string, the pitch (measured in vibrations per second) varies directly with the square root of the tension of the string (measured in pounds). If a certain string vibrates 300 times per second under a tension of 9 pounds, find the tension required to cause the string to vibrate 400 times per second.

39. Kepler's Third Law of Planetary Motion Kepler's Third Law of Planetary Motion states that the square of the period T of revolution of a planet is proportional to the cube of its mean distance from the Sun. If the mean distance of Earth from the Sun is 93 million miles, what is the mean distance a of the planet Mercury from the Sun, given that Mercury has a "year" of 88 days?

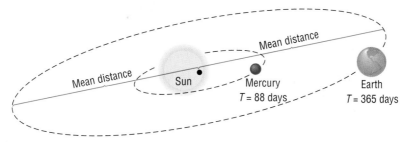

40. Use Problem 39 to find the mean distance of the planet Jupiter from the Sun, given that Jupiter circles the Sun every $5\sqrt{5}$ years.

41. Concentration of Carbon Monoxide in the Air The following data represent the average concentration of carbon monoxide in parts per million (ppm) in the air for 1987–1993.

Year	Concentration of Carbon Monoxide (ppm)
1987	6.69
1988	6.38
1989	6.34
1990	5.87
1991	5.55
1992	5.18
1993	4.88

Source: U.S. Environmental Protection Agency.

(a) Treating the year as the x-coordinate and the average level of carbon monoxide as the y-coordinate, draw a scatter diagram of the data.

(b) What is the slope of the line joining the points $(1987, 6.69)$ and $(1990, 5.87)$?

(c) Interpret this slope.

(d) What is the slope of the line joining the points $(1990, 5.87)$ and $(1993, 4.88)$?

(e) Interpret this slope.

(f) Use a graphing utility to find the slope of the line of best fit for these data.

(g) Interpret this slope.

(h) How do you explain the differences among the three slopes obtained?

(i) What is the trend in the data? In other words, as time passes, what is happening to the average level of carbon monoxide in the air? Why do you think this is happening?

42. Value of a Portfolio The following data represent the value of the Vanguard Index Trust-500 Portfolio for 1987–1995.

Year	Value (Dollars)
1987	54.26
1988	63.06
1989	82.84
1990	80.08
1991	104.28
1992	112.03
1993	123.11
1994	124.56
1995	171.20

Source: Vanguard Index Trust, Annual Report 1995.

(a) Treating the year as the x-coordinate and the value of the Vanguard Index Trust-500 Portfolio as the y-coordinate, draw a scatter diagram of the data.

(b) What is the slope of the line joining the points (1987, 54.26) and (1991, 104.28)?

(c) Interpret this slope.

(d) What is the slope of the line joining the points (1991, 104.28) and (1995, 171.20)?

(e) Interpret this slope.

 (f) Use a graphing utility to find the slope of the line of best fit for these data.

(g) Interpret this slope.

(h) If you were managing this trust, which of the three slopes would you use to convince someone to invest? Why?

(i) What is the trend in the data? In other words, as time passes, what is happening to the value of the Vanguard Index Trust-500 Portfolio?

43. Make up four problems that you might be asked to do given the two points $(-3, 4)$ and $(6, 1)$. Each problem should involve a different concept. Be sure that your directions are clearly stated.

44. Describe each of the following graphs in the xy-plane. Give justification.

(a) $x = 0$ (d) $xy = 0$

(b) $y = 0$ (e) $x^2 + y^2 = 0$

(c) $x + y = 0$

45. Suppose that you have a rectangular field that requires watering. Your watering system consists of an arm of variable length that rotates so that the watering pattern is a circle. Decide where to position the arm and what length it should be so that the entire field is watered most efficiently. When does it become desirable to use more than one arm? [**Hint:** Use a rectangular coordinate system positioned as shown in the figures below. Write equations for the circle(s) swept out by the watering arm(s).]

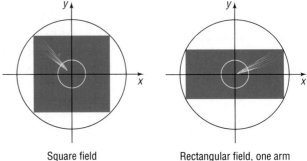

Square field   Rectangular field, one arm

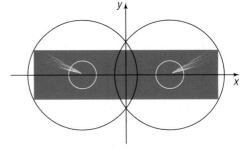

Rectangular field, two arms

Project at Motorola

During the 1990s mobile phones became a significant means of communication. Industry data indicated that in 1994 there were about 37.5 million mobile phones in use globally. In 1998, that figure rose to about 70 million mobile phones and is projected to rise to 102.5 million by 2002. Industry analysts base such predictions on a *mathematical model,* that is, a relationship (usually an equation) that involves one or more variables. One method of creating a mathematical model is to use the data available and construct an equation that describes the behavior of the data. The mathematical model is then used to forecast future values of one of the variables, in this case mobile phone usage.

1. Construct a scatter diagram of the following data treating year as the independent variable and mobile phone usage (in millions) as the dependent variable. Label each axis and create a title for the graph.

Year, t	Mobile phone usage, m (in millions)
1994	37.5
1998	70
2002	102.5

2. Using a graphing utility, find the line of best fit to the data.

3. Interpret the values of the slope and the y-intercept of the line of best fit.

4. Predict mobile phone usage for 2006.

5. Write a report to the director of strategic marketing that discusses your predictions. The report must address the adequacy of the model. That is, you must discuss whether you feel this model can be used to make predictions about future mobile phone usage. Consider factors such as the cost of phones and population growth when writing your report.

Functions and Their Graphs

3

Field Trip to Motorola

During the past decade the availability and usage of wireless Internet services has increased manyfold. The industry has developed a number of pricing proposals for such services. Marketing data have indicated that subscribers of wireless Internet services have tended to desire flat rate fee structures as compared with rates based totally on usage.

Interview at Motorola

Jocelyn Carter-Miller is Corporate Vice President and Chief Marketing Officer (CMO) for Motorola, Inc., an over $30 billion global provider of integrated communications and embedded electronics solutions. As CMO she has helped build the Motorola brand and image and has developed high-performance marketing organizations and processes. Jocelyn also heads motorola.com, the Motorola electronic commerce and information Web site. In this new role, she and her team have developed a strategy for serving Motorola's broad and diverse constituencies offering a full range of electronic services.

In her previous roles as Vice President–Latin American and Caribbean Operations and Director of European, Middle East and African Operations, Jocelyn headed international wireless data communications operations for Motorola, creating profitable opportunities through strategic alliances, value-added applications, and new product and service launches. She also developed skills in managing complex, high-risk ventures in countries like Brazil and Russia, setting standards for her company's practices in emerging markets.

Prior to her career at Motorola, Jocelyn served as Vice President, Marketing and Product Development for Mattel, where she broke new ground, driving record sales of Barbie and other toys using integrated product, entertainment, promotional, and licensing programs.

Jocelyn builds strong relationships and new opportunities through her involvement on outside boards and community organizations. She serves on the board of the Principal Financial Group and on the nonprofit boards of the Association of National Advertisers, the University of Chicago Women's Business Group Advisory Board, and the Smart School Charter Middle School.

Jocelyn holds a Master of Business Administration degree in marketing and finance from the University of Chicago and a Bachelor of Science degree in accounting from the University of Illinois, Urbana–Champaign and is a Certified Public Accountant. She is married to Edward Miller, President of Edventures, an educational reform development firm, and has two daughters, Alexis and Kimberly.

Jocelyn has won numerous awards, been featured in national publications, and regularly addresses business and community groups. She has co-authored with Melissa Giavagnoli the book *Networlding: Building Relationships and Opportunities for Success*, which was published in June 2000 by Jossey-Bass Publishers. Through their Web site Networlding.com, Jocelyn and Melissa facilitate meaningful connections and mutually beneficial opportunities for novice and expert Networlders alike.

PREPARING FOR THIS SECTION

Before getting started, review the following:

✓ Intervals (Section 1.5, pp. 125–126)

✓ Evaluating Algebraic Expressions, Domain of a Variable (Review, Section 2, pp. 19–21)

✓ Equations (Section 1.1, pp. 84–88)

✓ Intercepts (Section 2.2, pp. 157–158)

✓ Scatter Diagrams (Section 2.5, pp. 185–189)

✓ Solving Inequalities (Section 1.5, pp. 128–130)

3.1 FUNCTIONS

OBJECTIVES
1. Determine Whether a Relation Represents a Function
2. Find the Value of a Function
3. Find the Domain of a Function
4. Identify the Graph of a Function
5. Obtain Information from or about the Graph of a Function

1 A **relation** is a correspondence between two sets. If x and y are two elements in these sets and if a relation exists between x and y, then we say that x **corresponds** to y or that y **depends on** x, and we write $x \rightarrow y$. We may also write $x \rightarrow y$ as the ordered pair (x, y).

EXAMPLE 1	**An Example of a Relation**

Figure 1 depicts a relation between four individuals and their birthdays. The relation might be named "was born on." Then Katy corresponds to June 20, Dan corresponds to Sept 4, and so on. Using ordered pairs, this relation would be expressed as

$$\{(\text{Katy, June 20}), (\text{Dan, Sept 4}), (\text{Patrick, Dec 31}), (\text{Phoebe, Dec 31})\}$$

Figure 1

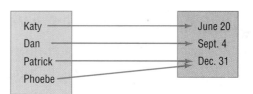

Often, we are interested in specifying the type of relation (such as an equation) that exists between the two variables. For example, the relation between the revenue R resulting from the sale of x items selling for $10 each may be expressed by the equation $R = 10x$. If we know how many items have been sold, then we can calculate the revenue by using the equation $R = 10x$. This equation is an example of a *function*.

As another example, suppose that an icicle falls off a building from a height of 64 feet above the ground. According to a law of physics, the distance s (in feet) of the icicle from the ground after t seconds is given (approximately) by the formula $s = 64 - 16t^2$. When $t = 0$ seconds, the icicle is $s = 64$ feet above the ground. After 1 second, the icicle is $s = 64 - 16(1)^2 = 48$ feet above the ground. After 2 seconds, the icicle strikes the ground. The formula $s = 64 - 16t^2$ provides a way of finding the distance s for any time t $(0 \le t \le 2)$. There is a correspondence between each time t in the interval $0 \le t \le 2$ and the distance s. We say that the distance s is a function of the time t because:

1. There is a correspondence between the set of times and the set of distances.

2. There is exactly one distance s obtained for any time t in the interval $0 \le t \le 2$.

Let's now look at the definition of a function.

Definition of Function

Let X and Y be two nonempty sets.* A **function** from X into Y is a relation that associates with each element of X exactly one element of Y.

Figure 2

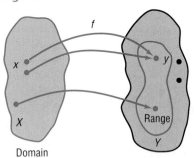

The set X is called the **domain** of the function. For each element x in X, the corresponding element y in Y is called the **value** of the function at x, or the image of x. The set of all images of the elements of the domain is called the **range** of the function. See Figure 2.

Since there may be some elements in Y that are not the image of any x in X, it follows that the range of a function is a subset of Y, as shown in Figure 2.

Not all relations between two sets are functions. The next example shows how to determine whether a relation is a function or not.

EXAMPLE 2

Determining Whether a Relation Represents a Function

Determine whether the following relations represent functions.

(a) See Figure 3. For this relation, the domain represents four individuals and the range represents their birthdays.

Figure 3

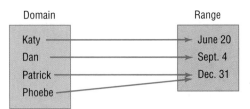

(b) See Figure 4. For this relation, the domain represents the employees of Sara's Pre-Owned Car Mart and the range represents their phone number(s).

Figure 4

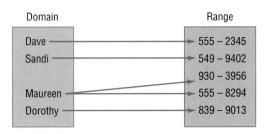

---

*The sets X and Y will usually be sets of real numbers. The sets X and Y can also be sets of complex numbers (discussed in Section 5.3), and then we have defined a complex function. In the broad definition (due to Lejeune Dirichlet), X and Y can be any two sets.

Solution   (a) The relation is a function because each element in the domain corresponds to a unique element in the range. Notice that more than one element in the domain can correspond to the same element in the range (Phoebe and Patrick were born on the same day of the year).

(b) The relation is not a function because each element in the domain does not correspond to a unique element in the range. Maureen has two telephone numbers; therefore, if Maureen is chosen from the domain, a unique telephone number cannot be assigned to her.   ■

 NOW WORK PROBLEM **1**.

We may think of a function as a set of ordered pairs (x, y) in which no two distinct pairs have the same first element. The set of all first elements x is the domain of the function, and the set of all second elements y is its range. Associated with each element x in the domain, there is a unique element y in the range.

EXAMPLE 3   **Determining Whether a Relation Represents a Function**

Determine whether each relation represents a function. For those that are functions, state the domain and range.

(a) $\{(1, 4), (2, 5), (3, 6), (4, 7)\}$
(b) $\{(1, 4), (2, 4), (3, 5), (6, 10)\}$
(c) $\{(-3, 9), (-2, 4), (0, 0), (1, 1), (-3, 8)\}$

Solution   (a) This relation is a function because there are no ordered pairs with the same first element and different second elements. The domain of this function is $\{1, 2, 3, 4\}$ and its range is $\{4, 5, 6, 7\}$.

(b) This relation is a function because there are no ordered pairs with the same first element and different second elements. The domain of this function is $\{1, 2, 3, 6\}$ and its range is $\{4, 5, 10\}$.

(c) This relation is not a function because there are two ordered pairs $(-3, 9)$ and $(-3, 8)$ that have the same first element, but different second elements.   ■

In Example 3(b), notice that 1 and 2 in the domain each have the same image in the range. This does not violate the definition of a function; two different first elements can have the same second element. A violation of the definition occurs when two ordered pairs have the same first element and different second elements, as in Example 3(c).

 NOW WORK PROBLEM **5**.

Example 2(a) demonstrates that a function may be defined by some correspondence between two sets. Examples 3(a) and 3(b) demonstrate that a function may be defined by a set of ordered pairs. A function may also be defined by an equation in two variables, usually denoted x and y.

EXAMPLE 4   **Example of a Function**

Consider the function defined by the equation

$$y = 2x - 5, \qquad 1 \leq x \leq 6$$

The domain $1 \le x \le 6$ specifies that the number x is restricted to the real numbers from 1 to 6, inclusive. The equation $y = 2x - 5$ specifies that the number x is to be multiplied by 2 and then 5 is to be subtracted from the result to get y. For example, if $x = \frac{3}{2}$, then $y = 2 \cdot \frac{3}{2} - 5 = -2$. ∎

FUNCTION NOTATION

Functions are often denoted by letters such as f, F, g, G, and others. If f is a function, then for each number x in its domain the corresponding image in the range is designated by the symbol $f(x)$, read as "f of x" or as "f at x." We refer to $f(x)$ as the **value of f at the number x;** $f(x)$ is the number that results when x is given and the function f is applied; $f(x)$ does *not* mean "f times x." For example, the function given in Example 4 may be written as $y = f(x) = 2x - 5$, $1 \le x \le 6$. Then $f(\frac{3}{2}) = -2$.

Figure 5 illustrates some other functions. Note that in every function illustrated, for each x in the domain there is one value in the range.

Figure 5

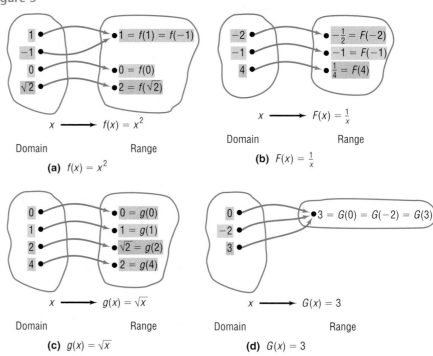

(a) $f(x) = x^2$

(b) $F(x) = \frac{1}{x}$

(c) $g(x) = \sqrt{x}$

(d) $G(x) = 3$

Sometimes it is helpful to think of a function f as a machine that receives as input a number from the domain, manipulates it, and outputs the value. See Figure 6.

Figure 6

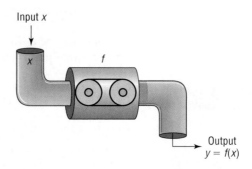

The restrictions on this input/output machine are as follows:

1. It only accepts numbers from the domain of the function.
2. For each input, there is exactly one output (which may be repeated for different inputs).

For a function $y = f(x)$, the variable x is called the **independent variable,** because it can be assigned any of the permissible numbers from the domain. The variable y is called the **dependent variable,** because its value depends on x.

Any symbol can be used to represent the independent and dependent variables. For example, if f is the *cube function*, then f can be given by $f(x) = x^3$ or $f(t) = t^3$ or $f(z) = z^3$. All three functions are the same. Each tells us to cube the independent variable. In practice, the symbols used for the independent and dependent variables are based on common usage, such as using C for cost in business.

The independent variable is also called the **argument** of the function. Thinking of the independent variable as an argument can sometimes make it easier to find the value of a function. For example, if f is the function defined by $f(x) = x^3$, then f tells us to cube the argument. Thus, $f(2)$ means to cube 2, $f(a)$ means to cube the number a, and $f(x + h)$ means to cube the quantity $x + h$.

EXAMPLE 5

Finding Values of a Function

For the function f defined by $f(x) = 2x^2 - 3x$, evaluate

(a) $f(3)$ (b) $f(x) + f(3)$ (c) $f(-x)$

(d) $-f(x)$ (e) $f(x + 3)$

Solution

(a) We substitute 3 for x in the equation for f to get

$$f(3) = 2(3)^2 - 3(3) = 18 - 9 = 9$$

(b) $f(x) + f(3) = (2x^2 - 3x) + (9) = 2x^2 - 3x + 9$

(c) We substitute $-x$ for x in the equation for f.

$$f(-x) = 2(-x)^2 - 3(-x) = 2x^2 + 3x$$

(d) $-f(x) = -(2x^2 - 3x) = -2x^2 + 3x$

(e) $f(x + 3) = 2(x + 3)^2 - 3(x + 3)$ Notice the use of parentheses here.

$$= 2(x^2 + 6x + 9) - 3x - 9$$
$$= 2x^2 + 12x + 18 - 3x - 9$$
$$= 2x^2 + 9x + 9$$

∎

Notice in this example that $f(x + 3) \neq f(x) + f(3)$ and $f(-x) \neq -f(x)$.

NOW WORK PROBLEM **13.**

Most calculators have special keys that enable you to find the value of certain commonly used functions. For example, you should be able to find the square function $f(x) = x^2$, the square root function $f(x) = \sqrt{x}$, the reciprocal function $f(x) = 1/x = x^{-1}$, and many others that will be discussed later in this book (such as $\ln x$ and $\log x$). Verify the results of Example 6, which follows, on your calculator.

| EXAMPLE 6 | **Finding Values of a Function on a Calculator** |

(a) $f(x) = x^2$; $f(1.234) = 1.522756$

(b) $F(x) = 1/x$; $F(1.234) = 0.8103727715$

(c) $g(x) = \sqrt{x}$; $g(1.234) = 1.110855526$ ■

 COMMENT Graphing calculators can be used to evaluate any function that you wish. Figure 7 shows the result obtained in Example 5(a) on a TI-83 graphing calculator with the function to be evaluated, $f(x) = 2x^2 - 3x$, in Y_1.*

Figure 7

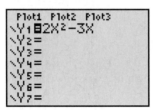 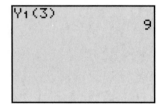

IMPLICIT FORM OF A FUNCTION

In general, when a function f is defined by an equation in x and y, we say that the function f is given **implicitly.** If it is possible to solve the equation for y in terms of x, then we write $y = f(x)$ and say that the function is given **explicitly.** For example,

Implicit Form	**Explicit Form**
$3x + y = 5$	$y = f(x) = -3x + 5$
$x^2 - y = 6$	$y = f(x) = x^2 - 6$
$xy = 4$	$y = f(x) = 4/x$

Not all equations in x and y define a function $y = f(x)$. If an equation is solved for y and two or more values of y can be obtained for a given x, then the equation does not define a function.

| EXAMPLE 7 | **Determining Whether an Equation Is a Function** |

Determine if the equation $x^2 + y^2 = 1$ is a function.

Solution To determine whether the equation $x^2 + y^2 = 1$, which defines the unit circle, is a function, we need to solve the equation for y.

$$x^2 + y^2 = 1$$
$$y^2 = 1 - x^2$$
$$y = \pm\sqrt{1 - x^2}$$

For values of x between -1 and 1, two values of y result. This means that the equation $x^2 + y^2 = 1$ does not define a function. ■

➤ NOW WORK PROBLEM 27.

---

*Consult your owner's manual for the required keystrokes.

COMMENT   The explicit form of a function is the form required by a graphing calculator. Now do you see why it is necessary to graph a circle in two "pieces"?   ■

We list next a summary of some important facts to remember about a function f.

SUMMARY OF IMPORTANT FACTS ABOUT FUNCTIONS

(a) To each x in the domain of f, there is exactly one image $f(x)$ in the range; however, an element in the range can result from more than one x in the domain.

(b) f is the symbol that we use to denote the function. It is symbolic of the equation that we use to get from an x in the domain to $f(x)$ in the range.

(c) If $y = f(x)$, then x is called the independent variable or argument of f, and y is called the dependent variable or the value of f at x.

DOMAIN OF A FUNCTION

3   Often the domain of a function f is not specified; instead, only the equation defining the function is given. In such cases, we agree that the domain of f is the largest set of real numbers for which the value $f(x)$ is a real number. The domain of a function f is the same as the domain of the variable x in the expression $f(x)$.

| EXAMPLE 8 | **Finding the Domain of a Function** |

Find the domain of each of the following functions:

(a) $f(x) = x^2 + 5x$ (b) $g(x) = \dfrac{3x}{x^2 - 4}$ (c) $h(t) = \sqrt{4 - 3t}$

Solution   (a) The function tells us to square a number and then add five times the number. Since these operations can be performed on any real number, we conclude that the domain of f is all real numbers.

(b) The function g tells us to divide $3x$ by $x^2 - 4$. Since division by 0 is not defined, the denominator $x^2 - 4$ can never be 0, so x can never equal -2 or 2. The domain of the function g is $\{x | x \neq -2, x \neq 2\}$.

(c) The function h tells us to take the square root of $4 - 3t$. But only nonnegative numbers have real square roots, so the expression under the square root must be nonnegative. This requires that

$$4 - 3t \geq 0$$

$$-3t \geq -4$$

$$t \leq \tfrac{4}{3}$$

The domain of h is $\{t \mid t \leq \tfrac{4}{3}\}$ or the interval $\left(-\infty, \tfrac{4}{3}\right]$.   ■

✏ **NOW WORK PROBLEM 37.**

If x is in the domain of a function f, we shall say that **f is defined at x,** or **$f(x)$ exists.** If x is not in the domain of f, we say that **f is not defined at x,** or **$f(x)$ does not exist.** For example, if $f(x) = x/(x^2 - 1)$, then $f(0)$ exists, but $f(1)$ and $f(-1)$ do not exist. (Do you see why?)

We have not said much about finding the range of a function. The reason is that when a function is defined by an equation it is often difficult to find the range.* Therefore, we shall usually be content to find just the domain of a function when only the rule for the function is given. We shall express the domain of a function using inequalities, interval notation, set notation, or words, whichever is most convenient.

THE GRAPH OF A FUNCTION

In applications, a graph often demonstrates more clearly the relationship between two variables than, say, an equation or table would. For example, Table 1 shows the price per share of Intel stock at the end of each month from 8/31/99 through 8/31/00. If we plot these data using the date as the *x*-coordinate and the price as the *y*-coordinate and then connect the points, we obtain Figure 8.

TABLE 1	
Date	**Closing Price ($)**
8/31/99	41.09
9/30/99	37.16
10/31/99	38.72
11/30/99	38.34
12/31/99	41.16
1/31/00	49.47
2/29/00	56.50
3/31/00	65.97
4/30/00	63.41
5/31/00	62.34
6/30/00	66.84
7/31/00	66.75
8/31/00	74.88

Courtesy of A.G. Edwards & Sons, Inc.

Figure 8
Monthly closing prices
of Intel stock 8/31/99
through 8/31/00

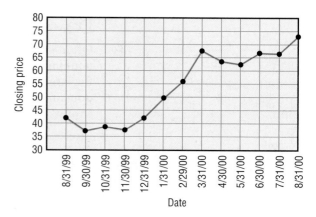

We can see from the graph that the price of the stock was rising rapidly from 11/30/99 through 3/31/00 and was falling slightly from 3/31/00 through 5/31/00. The graph also shows that the lowest price occurred at the end of September, 1999, whereas the highest occurred at the end of August, 2000. Equations and tables, on the other hand, usually require some calculations and interpretation before this kind of information can be "seen."

Look again at Figure 8. The graph shows that for each date on the horizontal axis there is only one price on the vertical axis. Thus, the graph represents a function, although the exact rule for getting from date to price is not given.

When a function is defined by an equation in *x* and *y*, the **graph** of the function is the graph of the equation, that is, the set of points (x, y) in the *xy*-plane that satisfies the equation.

 COMMENT When we select a viewing rectangle to graph a function, the values of Xmin, Xmax give the domain that we wish to view, while Ymin, Ymax give the range that we wish to view. These settings usually do not represent the actual domain and range of the function. ∎

④ Not every collection of points in the *xy*-plane represents the graph of a function. Remember, for a function, each number *x* in the domain has exactly one image *y* in the range. This means that the graph of a function cannot contain two points with the same *x*-coordinate and different *y*-coordinates. Therefore, the graph of a function must satisfy the following **vertical-line test.**

*In Section 6.1, we discuss a way to find the range of a certain class of functions.

Theorem   Vertical-line Test

A set of points in the xy-plane is the graph of a function if and only if every vertical line intersects the graph in at most one point. ∎

It follows that, if any vertical line intersects a graph at more than one point, the graph is not the graph of a function.

EXAMPLE 9   Identifying the Graph of a Function

Which of the graphs in Figure 9 are graphs of functions?

Figure 9

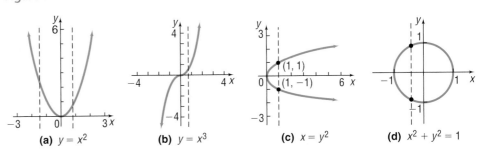

(a) $y = x^2$ (b) $y = x^3$ (c) $x = y^2$ (d) $x^2 + y^2 = 1$

Solution   The graphs in Figures 9(a) and 9(b) are graphs of functions, because every vertical line intersects each graph in at most one point. The graphs in Figures 9(c) and 9(d) are not graphs of functions, because there is a vertical line that intersects each graph in more than one point. ∎

NOW WORK PROBLEM 53.

⑤   If (x, y) is a point on the graph of a function f, then y is the value of f at x; that is, $y = f(x)$. The next example illustrates how to obtain information about a function if its graph is given.

EXAMPLE 10   Obtaining Information from the Graph of a Function

Figure 10

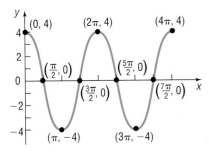

Let f be the function whose graph is given in Figure 10. (The graph of f might represent the distance that the bob of a pendulum is from its *at-rest* position. Negative values of y mean that the pendulum is to the left of the at-rest position, and positive values of y mean that the pendulum is to the right of the at-rest position.)

(a) What is $f(0)$, $f(3\pi/2)$, and $f(3\pi)$?

(b) What is the domain of f?

(c) What is the range of f?

(d) List the intercepts. (Recall that these are the points, if any, where the graph crosses or touches the coordinate axes.)

(e) How often does the line $y = 2$ intersect the graph?

(f) For what values of x does $f(x) = -4$?

Solution (a) Since $(0, 4)$ is on the graph of f, the y-coordinate 4 is the value of f at the x-coordinate 0; that is, $f(0) = 4$. In a similar way, we find that when $x = 3\pi/2$ then $y = 0$, so $f(3\pi/2) = 0$. When $x = 3\pi$, then $y = -4$, so $f(3\pi) = -4$.

(b) To determine the domain of f, we notice that the points on the graph of f will have x-coordinates between 0 and 4π, inclusive; and for each number x between 0 and 4π there is a point $(x, f(x))$ on the graph. The domain of f is $\{x \,|\, 0 \le x \le 4\pi\}$ or the interval $[0, 4\pi]$.

(c) The points on the graph all have y-coordinates between -4 and 4, inclusive; and for each such number y there is at least one number x in the domain. The range of f is $\{y \,|\, -4 \le y \le 4\}$ or the interval $[-4, 4]$.

(d) The intercepts are $(0, 4)$, $(\pi/2, 0)$, $(3\pi/2, 0)$, $(5\pi/2, 0)$, and $(7\pi/2, 0)$.

(e) Draw the horizontal line $y = 2$ on the graph in Figure 10. Then we find that it intersects the graph four times.

(f) Since $(\pi, -4)$ and $(3\pi, -4)$ are the only points on the graph for which $y = f(x) = -4$, we have $f(x) = -4$ when $x = \pi$ and $x = 3\pi$. ■

When the graph of a function is given, its domain may be viewed as the shadow created by the graph on the x-axis by vertical beams of light. Its range can be viewed as the shadow created by the graph on the y-axis by horizontal beams of light. Try this technique with the graph given in Figure 10.

NOW WORK PROBLEMS 47 AND 51.

EXAMPLE 11 **Obtaining Information about the Graph of a Function**

Consider the function: $f(x) = \dfrac{x}{x + 2}$

(a) Is the point $\left(1, \dfrac{1}{2}\right)$ on the graph of f?

(b) If $x = -1$, what is $f(x)$? What point is on the graph of f?

(c) If $f(x) = 2$, what is x? What point is on the graph of f?

Solution (a) When $x = 1$, then

$$f(x) = \frac{x}{x + 2}$$

$$f(1) = \frac{1}{1 + 2} = \frac{1}{3}$$

The point $\left(1, \dfrac{1}{3}\right)$ is on the graph of f; the point $\left(1, \dfrac{1}{2}\right)$ is not.

(b) If $x = -1$, then

$$f(x) = \frac{x}{x + 2}$$

$$f(-1) = \frac{-1}{-1 + 2} = -1$$

The point $(-1, -1)$ is on the graph of f.

(c) If $f(x) = 2$, then

$$f(x) = 2$$

$$\frac{x}{x + 2} = 2$$

$$x = 2(x + 2) \qquad \text{Multiply both sides by } x + 2$$

$$x = 2x + 4 \qquad \text{Remove parentheses}$$

$$x = -4 \qquad \text{Solve for } x$$

If $f(x) = 2$, then $x = -4$. The point $(-4, 2)$ is on the graph of f. ■

NOW WORK PROBLEM **63**.

APPLICATIONS

When we use functions in applications, the domain may be restricted by physical or geometric considerations. For example, the domain of the function f defined by $f(x) = x^2$ is the set of all real numbers. However, if f is used to obtain the area of a square when the length x of a side is known, then we must restrict the domain of f to the positive real numbers, since the length of a side can never be 0 or negative.

EXAMPLE 12   **Area of a Circle**

Express the area of a circle as a function of its radius.

Figure 11

Solution   See Figure 11. We know that the formula for the area A of a circle of radius r is $A = \pi r^2$. If we use r to represent the independent variable and A to represent the dependent variable, the function expressing this relationship is

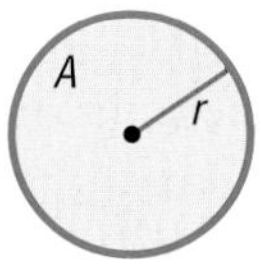

$$A(r) = \pi r^2$$

In this setting, the domain is $\{r \mid r > 0\}$. (Do you see why?) ■

EXAMPLE 13   **Construction Cost**

TABLE 2

Square Feet, x	Cost, C $
1500	165,000
1600	176,300
1700	183,000
1700	189,500
1830	201,700
1970	217,000
2050	220,000
2100	237,400

Sally, a builder of homes, is interested in finding a function that relates the cost C of building a house and x, the number of square feet of the house, so that she can easily provide estimates to clients for various-sized homes. The data that she obtained from constructing homes last year are listed in Table 2.

(a) Does the relation defined by the set of ordered pairs (x, C) represent a function?

(b) Draw a scatter diagram of the data.

(c) Select two points from the data and find an equation of the line containing the points.

(d) Interpret the slope.

(e) Express the relationship found in part (c) using function notation.

(f) What is the domain of the function?

(g) Use the function to find the cost to build a 2000 square foot house.

 (h) Using a graphing utility, find the line of best fit relating square feet and cost. Interpret the slope.

(i) Give some reasons to explain why the cost of building a 1700 square foot house might be $183,000 in one case and $189,500 in another.

Solution (a) No. Look at the ordered pairs (1700, 183,000) and (1700, 189,500). The first element 1700 has, corresponding to it, two second elements, 183,000 and 189,500, in violation of the definition of a function.

(b) See Figure 12.

Figure 12

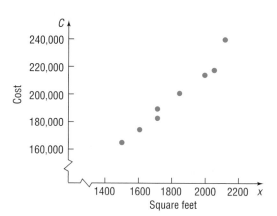

(c) We will choose the points (1500, 165,000) and (2050, 220,000). (You should select your own two points and complete the solution.) The slope of the line containing these points is

$$m = \frac{220,000 - 165,000}{2050 - 1500} = \frac{55,000}{550} = 100$$

The equation of the line containing these points is

$$y - 165,000 = 100(x - 1500)$$

$$y = 100x + 15,000$$

(d) The slope of the line, 100, represents the cost per square foot to build the house. For each additional square foot of house, the cost increases by $100.

(e) Using function notation, $C(x) = 100x + 15,000$.

(f) The domain is $\{x \mid x > 0\}$, since a house cannot have 0 or negative square feet.

(g) The cost to build a 2000 square foot house is

$$C(2000) = 100(2000) + 15,000 = \$215,000$$

(h) The line of best fit is $y = 112.09x - 3734.02$. The cost per square foot to build a house is approximately \$112.09. For each additional square foot of house being built, the cost increases by \$112.09.

(i) One explanation is that the fixtures used in one house are more expensive than those used in the other house.   ■

Observe in the solution to Example 13 that we used the symbol C in two ways: It is used to name the function, and it is used to symbolize the dependent variable. This double use is common in applications and should not cause any difficulty.

Refer to Example 13. Notice that the data in Table 2 do not represent a function since there is a first element paired with two different second elements $[(1700, 183,000), (1700, 189,500)]$. However, when we find a line that fits these data, we find a function that relates the data. **Curve fitting** is a process whereby we find a functional relationship between two or more variables even though the data, themselves, may not represent a function.

NOW WORK PROBLEM 89.

Examples 12 and 13 demonstrate that some functions are determined from data, while others are based on geometric, physical, or other relations.

SUMMARY
We list here some of the important vocabulary introduced in this section, with a brief description of each term.

Function
A relation between two sets of real numbers so that each number x in the first set, the domain, has corresponding to it exactly one number y in the second set.
A set of ordered pairs (x, y) or $(x, f(x))$ in which no first element is paired with two different second elements.
The range is the set of y values of the function for the x values in the domain.
A function f may be defined implicitly by an equation involving x and y or explicitly by writing $y = f(x)$.

Unspecified domain
If a function f is defined by an equation and no domain is specified, then the domain will be taken to be the largest set of real numbers for which the equation defines a real number.

Function notation
$y = f(x)$
f is a symbol for the function.
x is the independent variable or argument.
y is the dependent variable.
$f(x)$ is the value of the function at x, or the image of x.

Graph of a function
The collection of points (x, y) that satisfies the equation $y = f(x)$.
A collection of points is the graph of a function provided that every vertical line intersects the graph in at most one point (vertical-line test).

3.1 EXERCISES

In Problems 1–12, determine whether each relation represents a function. For each function, state the domain and range.

1.

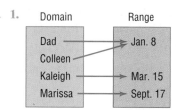

2.

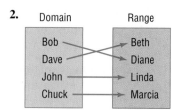

3.

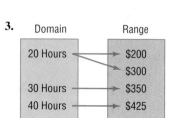

4.

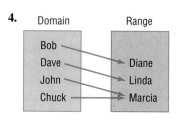

5. $\{(2, 6), (-3, 6), (4, 9), (2, 10)\}$

6. $\{(-2, 5), (-1, 3), (3, 7), (4, 12)\}$

7. $\{(1, 3), (2, 3), (3, 3), (4, 3)\}$

8. $\{(0, -2), (1, 3), (2, 3), (3, 7)\}$

9. $\{(-2, 4), (-2, 6), (0, 3), (3, 7)\}$

10. $\{(-4, 4), (-3, 3), (-2, 2), (-1, 1), (-4, 0)\}$

11. $\{(-2, 4), (-1, 1), (0, 0), (1, 1)\}$

12. $\{(-2, 16), (-1, 4), (0, 3), (1, 4)\}$

In Problems 13–20, find the following values for each function:
(a) $f(0)$ (b) $f(1)$ (c) $f(-1)$ (d) $f(-x)$ (e) $-f(x)$ (f) $f(x + 1)$ (g) $f(2x)$ (h) $f(x + h)$

13. $f(x) = 3x^2 + 2x - 4$

14. $f(x) = -2x^2 + x - 1$

15. $f(x) = \dfrac{x}{x^2 + 1}$

16. $f(x) = \dfrac{x^2 - 1}{x + 4}$

17. $f(x) = |x| + 4$

18. $f(x) = \sqrt{x^2 + x}$

19. $f(x) = \dfrac{2x + 1}{3x - 5}$

20. $f(x) = 1 - \dfrac{1}{(x + 2)^2}$

In Problems 21–32, determine whether the equation is a function.

21. $y = x^2$

22. $y = x^3$

23. $y = \dfrac{1}{x}$

24. $y = |x|$

25. $y^2 = 4 - x^2$

26. $y = \pm\sqrt{1 - 2x}$

27. $x = y^2$

28. $x + y^2 = 1$

29. $y = 2x^2 - 3x + 4$

30. $y = \dfrac{3x - 1}{x + 2}$

31. $2x^2 + 3y^2 = 1$

32. $x^2 - 4y^2 = 1$

In Problems 33–46, find the domain of each function.

33. $f(x) = -5x + 4$

34. $f(x) = x^2 + 2$

35. $f(x) = \dfrac{x}{x^2 + 1}$

36. $f(x) = \dfrac{x^2}{x^2 + 1}$

37. $g(x) = \dfrac{x}{x^2 - 16}$

38. $h(x) = \dfrac{2x}{x^2 - 4}$

39. $F(x) = \dfrac{x - 2}{x^3 + x}$

40. $G(x) = \dfrac{x + 4}{x^3 - 4x}$

41. $h(x) = \sqrt{3x - 12}$

42. $G(x) = \sqrt{1 - x}$

43. $f(x) = \dfrac{4}{\sqrt{x - 9}}$

44. $f(x) = \dfrac{x}{\sqrt{x - 4}}$

45. $p(x) = \sqrt{\dfrac{2}{x - 1}}$

46. $q(x) = \sqrt{-x - 2}$

47. Use the graph of the function f given below to answer parts (a)–(n).

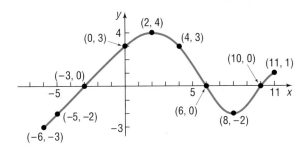

48. Use the graph of the function f given below to answer parts (a)–(n).

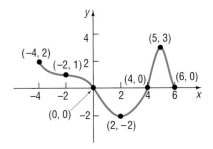

(a) Find $f(0)$ and $f(-6)$.

(b) Find $f(6)$ and $f(11)$.

(c) Is $f(3)$ positive or negative?

(d) Is $f(-4)$ positive or negative?

(e) For what numbers x is $f(x) = 0$?

(f) For what numbers x is $f(x) > 0$?

(g) What is the domain of f?

(h) What is the range of f?

(i) What are the x-intercepts?

(j) What are the y-intercepts?

(k) How often does the line $y = \frac{1}{2}$ intersect the graph?

(l) How often does the line $x = 5$ intersect the graph?

(m) For what values of x does $f(x) = 3$?

(n) For what values of x does $f(x) = -2$?

(a) Find $f(0)$ and $f(6)$.

(b) Find $f(2)$ and $f(-2)$.

(c) Is $f(3)$ positive or negative?

(d) Is $f(-1)$ positive or negative?

(e) For what numbers is $f(x) = 0$?

(f) For what numbers is $f(x) < 0$?

(g) What is the domain of f?

(h) What is the range of f?

(i) What are the x-intercepts?

(j) What are the y-intercepts?

(k) How often does the line $y = -1$ intersect the graph?

(l) How often does the line $x = 1$ intersect the graph?

(m) For what value of x does $f(x) = 3$?

(n) For what value of x does $f(x) = -2$?

In Problems 49–60, determine whether the graph is that of a function by using the vertical-line test. If it is, use the graph to find:
  (a) *Its domain and range*   (b) *The intercepts, if any*   (c) *Any symmetry with respect to the x-axis, the y-axis, or the origin*

49.

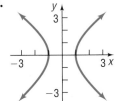

50.

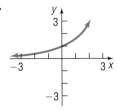

51.

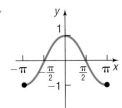

52.

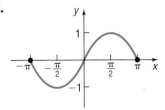

53.

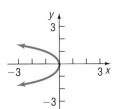

54.

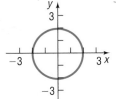

55.

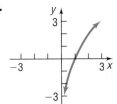

56.

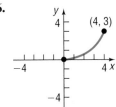

57.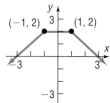

58.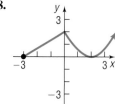

59.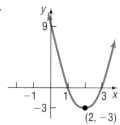

60.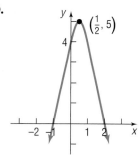

In Problems 61–66, answer the questions about the given function.

61. $f(x) = 2x^2 - x - 1$

   (a) Is the point $(-1, 2)$ on the graph of f?
   (b) If $x = -2$, what is $f(x)$? What point is on the graph of f?
   (c) If $f(x) = -1$, what is x? What point(s) are on the graph of f?
   (d) What is the domain of f?
   (e) List the x-intercepts, if any, of the graph of f.
   (f) List the y-intercept, if there is one, of the graph of f.

62. $f(x) = -3x^2 + 5x$

   (a) Is the point $(-1, 2)$ on the graph of f?
   (b) If $x = -2$, what is $f(x)$? What point is on the graph of f?
   (c) If $f(x) = -2$, what is x? What point(s) are on the graph of f?
   (d) What is the domain of f?
   (e) List the x-intercepts, if any, of the graph of f.
   (f) List the y-intercept, if there is one, of the graph of f.

63. $f(x) = \dfrac{x + 2}{x - 6}$

   (a) Is the point $(3, 14)$ on the graph of f?
   (b) If $x = 4$, what is $f(x)$? What point is on the graph of f?
   (c) If $f(x) = 2$, what is x? What point(s) are on the graph of f?
   (d) What is the domain of f?
   (e) List the x-intercepts, if any, of the graph of f.
   (f) List the y-intercept, if there is one, of the graph of f.

64. $f(x) = \dfrac{x^2 + 2}{x + 4}$

   (a) Is the point $\left(1, \frac{3}{5}\right)$ on the graph of f?
   (b) If $x = 0$, what is $f(x)$? What point is on the graph of f?
   (c) If $f(x) = \frac{1}{2}$, what is x? What point(s) are on the graph of f?
   (d) What is the domain of f?
   (e) List the x-intercepts, if any, of the graph of f.
   (f) List the y-intercept, if there is one, of the graph of f.

65. $f(x) = \dfrac{2x^2}{x^4 + 1}$

   (a) Is the point $(-1, 1)$ on the graph of f?
   (b) If $x = 2$, what is $f(x)$? What point is on the graph of f?
   (c) If $f(x) = 1$, what is x? What point(s) are on the graph of f?
   (d) What is the domain of f?
   (e) List the x-intercepts, if any, of the graph of f.
   (f) List the y-intercept, if there is one, of the graph of f.

66. $f(x) = \dfrac{2x}{x - 2}$

   (a) Is the point $\left(\frac{1}{2}, -\frac{2}{3}\right)$ on the graph of f?
   (b) If $x = 4$, what is $f(x)$? What point is on the graph of f?
   (c) If $f(x) = 1$, what is x? What point(s) are on the graph of f?
   (d) What is the domain of f?
   (e) List the x-intercepts, if any, of the graph of f.
   (f) List the y-intercept, if there is one, of the graph of f.

67. If $f(x) = 2x^3 + Ax^2 + 4x - 5$ and $f(2) = 5$, what is the value of A?

68. If $f(x) = 3x^2 - Bx + 4$ and $f(-1) = 12$, what is the value of B?

69. If $f(x) = (3x + 8)/(2x - A)$ and $f(0) = 2$, what is the value of A?

70. If $f(x) = (2x - B)/(3x + 4)$ and $f(2) = \frac{1}{2}$, what is the value of B?

71. If $f(x) = (2x - A)/(x - 3)$ and $f(4) = 0$, what is the value of A? Where is f not defined?

72. If $f(x) = (x - B)/(x - A)$, $f(2) = 0$, and $f(1)$ is undefined, what are the values of A and B?

73. Match each function with the graph that best describes the situation.
   (a) The cost of building a house as a function of its square footage
   (b) The height of an egg dropped from a 300-foot building as a function of time
   (c) The height of a human as a function of time
   (d) The demand for Big Macs as a function of price
   (e) The height of a child on a swing as a function of time

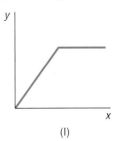   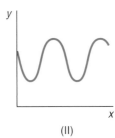   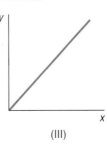   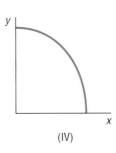   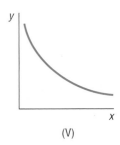

     (I)   (II)   (III)   (IV)   (V)

74. Match each function with the graph that best describes the situation.
   (a) The temperature of a bowl of soup as a function of time
   (b) The number of hours of daylight per day over a two-year period
   (c) The population of Florida as a function of time
   (d) The distance of a car traveling at a constant velocity as a function of time
   (e) The height of a golf ball hit with a 7-iron as a function of time

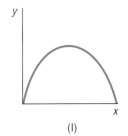   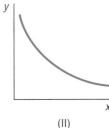   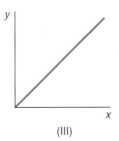   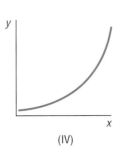   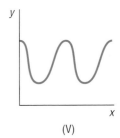

     (I)   (II)   (III)   (IV)   (V)

75. Consider the following scenario: Barbara decides to take a walk. She leaves home, walks 2 blocks in 5 minutes at a constant speed, and realizes that she forgot to lock the door. So Barbara runs home in 1 minute. While at her doorstep, it takes her 1 minute to find her keys and lock the door. Barbara walks 5 blocks in 15 minutes and then decides to jog home. It takes her 7 minutes to get home. Draw a graph of Barbara's distance from home (in blocks) as a function of time.

76. Consider the following scenario: Jayne enjoys riding her bicycle through the woods. At the forest preserve, she gets on her bicycle and rides up a 2000-foot incline in 10 minutes. She then travels down the incline in 3 minutes. The next 5000 feet is level terrain and she covers the distance in 20 minutes. She rests for 15 minutes. Jayne then travels 10,000 feet in 30 minutes. Draw a graph of Jayne's distance traveled (in feet) as a function of time.

77. The following sketch represents the distance d (in miles) that Kevin is from home as a function of time t (in hours). Answer the questions based on the graph. In parts (a)-(g),

how many hours elapsed and how far was Kevin from home during this time?

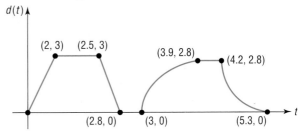

   (a) From $t = 0$ to $t = 2$
   (b) From $t = 2$ to $t = 2.5$
   (c) From $t = 2.5$ to $t = 2.8$
   (d) From $t = 2.8$ to $t = 3$
   (e) From $t = 3$ to $t = 3.9$
   (f) From $t = 3.9$ to $t = 4.2$
   (g) From $t = 4.2$ to $t = 5.3$
   (h) What is the farthest distance that Kevin is from home?
   (i) How many times did Kevin return home?

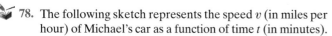

78. The following sketch represents the speed v (in miles per hour) of Michael's car as a function of time t (in minutes).

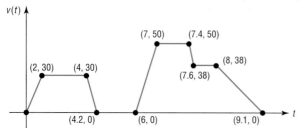

(a) Over what interval of time is Michael traveling fastest?
(b) Over what interval(s) of time is Michael's speed zero?
(c) What is Michael's speed between 0 and 2 minutes?
(d) What is Michael's speed between 4.2 and 6 minutes?
(e) What is Michael's speed between 7 and 7.4 minutes.
(f) When is Michael's speed constant?

79. **Effect of Gravity on Earth** If a rock falls from a height of 20 meters on Earth, the height H (in meters) after x seconds is approximately

$$H(x) = 20 - 4.9x^2$$

(a) What is the height of the rock when $x = 1$ second? $x = 1.1$ seconds? $x = 1.2$ seconds? $x = 1.3$ seconds?
(b) When is the height of the rock 15 meters? When is it 10 meters? When is it 5 meters?
(c) When does the rock strike the ground?

80. **Effect of Gravity on Jupiter** If a rock falls from a height of 20 meters on the planet Jupiter, its height H (in meters) after x seconds is approximately

$$H(x) = 20 - 13x^2$$

(a) What is the height of the rock when $x = 1$ second? $x = 1.1$ seconds? $x = 1.2$ seconds?
(b) When is the height of the rock 15 meters? When is it 10 meters? When is it 5 meters?
(c) When does the rock strike the ground?

81. **Motion of a Golf Ball** A golf ball is hit with an initial velocity of 130 feet per second at an inclination of 30° to the horizontal. In physics, it is established that the height h of the golf ball is given by the function

$$h(x) = \frac{-32x^2}{130^2} + x$$

where x is the horizontal distance that the golf ball has traveled.

(a) Determine the height of the golf ball after it has traveled 100 feet.
(b) 300 feet
(c) 500 feet
(d) How far was the golf ball hit?

82. **Cross-sectional Area** The cross-sectional area of a beam cut from a log with radius 1 foot is given by the function $A(x) = 4x\sqrt{1 - x^2}$, where x represents the length of half the base of the beam. See the figure. Determine the cross-sectional area of the beam if the length of half the base of the beam is as follows:
(a) One-third of a foot
(b) One-half of a foot
(c) Two-thirds of a foot

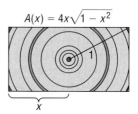

83. **Cost of Trans-Atlantic Travel** A Boeing 747 crosses the Atlantic Ocean (3000 miles) with an airspeed of 500 miles per hour. The cost C (in dollars) per passenger is given by

$$C(x) = 100 + \frac{x}{10} + \frac{36,000}{x}$$

where x is the ground speed (airspeed ± wind).
(a) What is the cost per passenger for quiescent (no wind) conditions?
(b) What is the cost per passenger with a head wind of 50 miles per hour?
(c) What is the cost per passenger with a tail wind of 100 miles per hour?
(d) What is the cost per passenger with a head wind of 100 miles per hour?
(e) Graph the function $C = C(x)$.
(f) As x varies from 400 to 600 miles per hour, how does the cost vary?

84. **Effect of Elevation on Weight** If an object weighs m pounds at sea level, then its weight W (in pounds) at a height of h miles above sea level is given approximately by

$$W(h) = m\left(\frac{4000}{4000 + h}\right)^2$$

(a) If Amy weighs 120 pounds at sea level, how much will she weigh on Pike's Peak, which is 14,110 feet above sea level?
(b) Use a graphing utility to graph the function $W = W(h)$. Use $m = 120$ pounds.
(c) Use the TRACE function to see how weight W varies as h changes from 0 to 5 miles.

(d) At what height will Amy weigh 119.5 pounds?

(e) Does your answer to part (d) seem reasonable?

85. Geometry   Express the area A of a rectangle as a function of the length x if the length is twice the width of the rectangle.

86. Geometry   Express the area A of an isosceles right triangle as a function of the length x of one of the two equal sides.

87. Express the gross salary G of a person who earns $10 per hour as a function of the number x of hours worked.

88. Tiffany, a commissioned salesperson, earns $100 base pay plus $10 per item sold. Express her gross salary G as a function of the number x of items sold.

89. Demand for Jeans   The marketing manager at Levi–Strauss wishes to find a function that relates the demand D of men's jeans and the price p of the jeans. The following data were obtained based on a price history of jeans.

Price ($/Pair), p	Demand (Pairs of Jeans Sold per Day), D
20	60
22	57
23	56
23	53
27	52
29	49
30	44

(a) Does the relation defined by the set of ordered pairs (p, D) represent a function?

(b) Draw a scatter diagram of the data.

(c) Select two points from the data and find an equation of the line containing the points. (The answer in the back of the book uses the first and last data points.)

(d) Interpret the slope.

(e) Express the relationship found in part (c) using function notation.

(f) What is the domain of the function?

(g) Predict how many jeans will be demanded if the price is $28 a pair?

(h) Using a graphing utility, find the line of best fit relating price and quantity demanded.

90. Advertising and Sales Revenue   A marketing firm wishes to find a function that relates the sales S of a product and the amount A spent on advertising the product. The data are obtained from past experience. Advertising and sales are measured in thousands of dollars.

Advertising Expenditures, A	Sales, S
20	335
22	339
22.5	338
24	343
24	341
27	350
28.3	351

(a) Does the relation defined by the set of ordered pairs (A, S) represent a function?

(b) Draw a scatter diagram of the data.

(c) Select two points from the data and find an equation of the line containing the points.

(d) Interpret the slope.

(e) Express the relationship found in part (c) using function notation.

(f) What is the domain of the function?

(g) Predict sales if advertising expenditures are $25,000.

(h) Using a graphing utility, find the line of best fit relating advertising expenditures and sales.

91. Distance and Time   Using the following data, find a function that relates the distance s, in miles, driven by a Ford Taurus, and the time t the Taurus has been driven.

Time (Hours), t	Distance (Miles), s
0	0
1	30
2	55
3	83
4	100
5	150
6	210
7	260
8	300

(a) Does the relation defined by the set of ordered pairs (t, s) represent a function?

(b) Draw a scatter diagram of the data.

(c) Select two points from the data and find an equation of the line containing the points. (The answer in the back of the book uses the first and last data points.)

(d) Interpret the slope.

(e) Express the relationship found in part (c) using function notation.

(f) What is the domain of the function?

(g) Predict the distance the car is driven after 11 hours.

(h) Using a graphing utility, find the line of best fit relating time and distance.

92. High School versus College GPA An administrator at Southern Illinois University wants to find a function that relates a student's college grade point average G to the high school grade point average x. She randomly selects 8 students and obtains the following data:

High School GPA, x	College GPA, G
2.73	2.43
2.92	2.97
3.45	3.63
3.78	3.81
2.56	2.83
2.98	2.81
3.67	3.45
3.10	2.93

(a) Does the relation defined by the set of ordered pairs (x, G) represent a function?
(b) Draw a scatter diagram of the data.
(c) Select two points from the data and find an equation of the line containing the points.
(d) Interpret the slope.
(e) Express the relationship found in part (c) using function notation.
(f) What is the domain of the function?
(g) Predict a student's college GPA if her high school GPA is 3.23.
(h) Using a graphing utility, find the line of best fit relating high school GPA and college GPA.

93. Some functions f have the property that $f(a + b) = f(a) + f(b)$ for all real numbers a and b. Which of the following functions have this property?
(a) $h(x) = 2x$ (b) $g(x) = x^2$
(c) $F(x) = 5x - 2$ (d) $G(x) = 1/x$

94. Draw the graph of a function whose domain is $\{x \mid -3 \le x \le 8, \ x \ne 5\}$ and whose range is $\{y \mid -1 \le y \le 2, y \ne 0\}$. What point(s) in the rectangle $-3 \le x \le 8, -1 \le y \le 2$ cannot be on the graph? Compare your graph with those of other students. What differences do you see?

95. Are the functions $f(x) = x - 1$ and $g(x) = (x^2 - 1)/(x + 1)$ the same? Explain.

96. Describe how you would proceed to find the domain and range of a function if you were given its graph. How would your strategy change if, instead, you were given the equation defining the function?

97. How many x-intercepts can the graph of a function have? How many y-intercepts can it have?

98. Is a graph that consists of a single point the graph of a function? If so, can you write the equation of such a function?

99. Is there a function whose graph is symmetric with respect to the x-axis? Explain.

100. Investigate when, historically, the use of the function notation $y = f(x)$ first appeared.

PREPARING FOR THIS SECTION

Before getting started, review the following:

✓ Intervals (Section 1.5, pp. 125–126)

✓ Point–slope Form of a Line (Section 2.3, p. 168)

✓ Slope of a Line (Section 2.3, pp. 164–167)

✓ Tests for Symmetry of an Equation (Section 2.2, pp. 159–162)

3.2   PROPERTIES OF FUNCTIONS

OBJECTIVES
1. Find the Average Rate of Change of a Function
2. Use a Graph to Determine Where a Function Is Increasing, Is Decreasing, or Is Constant
3. Use a Graph to Locate Local Maxima and Minima
4. Determine Even and Odd Functions from a Graph
5. Identify Even and Odd Functions from the Equation

AVERAGE RATE OF CHANGE

1. In Section 2.3 we said that the slope of a line could be interpreted as the average rate of change. Often we are interested in the rate at which functions

change. To find the average rate of change of a function between any two points on its graph, we calculate the slope of the line containing the two points.

| EXAMPLE 1 | **Finding the Average Rate of Change of a Function** |

The data in Table 3 represent the average cost of tuition and required fees (in dollars) at public four-year colleges.

TABLE 3	
Year	Average Cost of Tuition and Fees ($)
1991	2159
1992	2410
1993	2604
1994	2820
1995	2977
1996	3151
1997	3321

Source: U.S. Center for National Education Statistics

(a) Draw a scatter diagram of the data, treating the year as the independent variable.

(b) Draw a line through the points (1991, 2159) and (1992, 2410) on the scatter diagram found in part (a).

(c) Find the average rate of change of the cost of tuition and required fees from 1991 to 1992.

(d) Draw a line through the points (1995, 2977) and (1996, 3151) on the scatter diagram found in part (a).

(e) Find the average rate of change of the cost of tuition and required fees from 1995 to 1996.

(f) What is happening to the average rate of change as time passes?

Solution   (a) We plot the ordered pairs (1991, 2159), (1992, 2410), and so on, using rectangular coordinates. See Figure 13.

(b) Draw the line through (1991, 2159) and (1992, 2410). See Figure 14.

(c) The average rate of change is found by computing the slope of the line containing the points (1991, 2159) and (1992, 2410).

$$\text{Average rate of change} = \frac{2410 - 2159}{1992 - 1991} = \frac{251}{1} = \$251/\text{year}$$

(d) Draw the line through (1995, 2977) and (1996, 3151). See Figure 15.

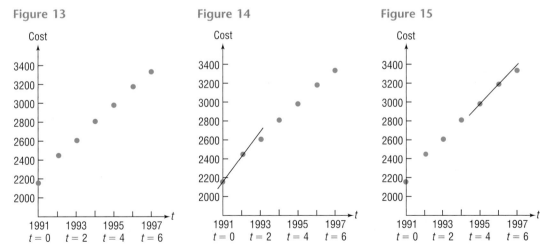

Figure 13   Figure 14   Figure 15

(e) The average rate of change is found by computing the slope of the line containing the points (1995, 2977) and (1996, 3151).

$$\text{Average rate of change} = \frac{3151 - 2977}{1996 - 1995} = \frac{174}{1} = \$174/\text{year}$$

(f) The cost of tuition and fees is increasing over time, but the rate of the increase is declining. ∎

If we know the function $C(t)$ that relates the year t to the cost C of tuition and fees, then the average rate of change from 1991 to 1992 may be expressed as

$$\text{Average rate of change} = \frac{C(1992) - C(1991)}{1992 - 1991} = \frac{251}{1} = \$251/\text{year}$$

 Expressions like this occur frequently in calculus.

> If c is in the domain of a function $y = f(x)$, the **average rate of change of f** from c to x is defined as
>
> $$\text{Average rate of change} = \frac{\Delta y}{\Delta x} = \frac{f(x) - f(c)}{x - c}, \qquad x \neq c \qquad \textbf{(1)}$$
>
> This expression is also called the **difference quotient** of f at c.*

The average rate of change of a function has an important geometric interpretation. Look at the graph of $y = f(x)$ in Figure 16. We have labeled two points on the graph: $(c, f(c))$ and $(x, f(x))$. The line containing these two points is called a **secant line;** its slope is

$$m_{\text{sec}} = \frac{f(x) - f(c)}{x - c}$$

Figure 16

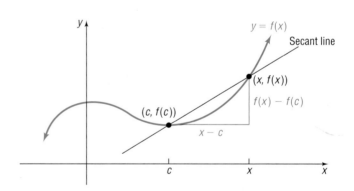

Theorem   **Slope of the Secant Line**

The average rate of change of a function equals the slope of the secant line containing two points on its graph.

EXAMPLE 2   **Finding the Average Rate of Change of a Function**

(a) Find the average rate of change of $f(x) = 2x^2 - 3x$ from 1 to x.

(b) Use this result to find the slope of the secant line containing $(1, f(1))$ and $(2, f(2))$.

(c) Find an equation of this secant line.

*See Problems 49–56 for an alternative definition of difference quotient.

Solution   (a) The average rate of change of f from 1 to x is

$$\frac{\Delta y}{\Delta x} = \frac{f(x) - f(1)}{x - 1} \qquad x \neq 1$$

$$= \frac{(2x^2 - 3x) - (-1)}{x - 1} \qquad f(x) = 2x^2 - 3x;\ f(1) = 2 \cdot 1^2 - 3(1) = -1$$

$$= \frac{2x^2 - 3x + 1}{x - 1} \qquad \text{Simplify}$$

$$= \frac{(2x - 1)(x - 1)}{x - 1} \qquad \text{Factor the numerator}$$

$$= 2x - 1 \qquad x \neq 1;\ \text{cancel } x - 1.$$

(b) The slope of the secant line containing $\big(1, f(1)\big)$ and $\big(2, f(2)\big)$ is the average rate of change of f from 1 to 2. Using $x = 2$ in part (a), we obtain $m_{\text{sec}} = 2(2) - 1 = 3$.

(c) Use the point–slope form to find the equation of the secant line.

$$y - y_1 = m_{\text{sec}}(x - x_1) \quad x_1 = 1,\ y_1 = f(1) = -1;\ m_{\text{sec}} = 3$$

$$y + 1 = 3(x - 1) \qquad \text{Point–slope form of the secant line}$$

$$y + 1 = 3x - 3$$

$$y = 3x - 4 \qquad \text{Slope–intercept form of the secant line} \quad \blacksquare$$

> ✏ **NOW WORK PROBLEM 29.**

⚠ INCREASING AND DECREASING FUNCTIONS

 Consider the graph given in Figure 17. If you look from left to right along the graph of the function, you will notice that parts of the graph are rising, parts are falling, and parts are horizontal. In such cases, the function is described as *increasing*, *decreasing*, or *constant*, respectively.

Figure 17

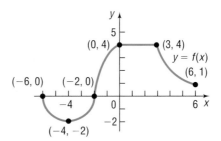

EXAMPLE 3   **Determining Where a Function Is Increasing, Decreasing, or Constant from Its Graph**

Where is the function in Figure 17 increasing? Where is it decreasing? Where is it constant?

Solution   To answer the question of where a function is increasing, where it is decreasing, and where it is constant, we use nonstrict inequalities involving the independent variable x, or we use open intervals* of x-coordinates. The graph

---

*The open interval (a, b) consists of all real numbers x for which $a < x < b$. Refer to Section 1.5, if necessary.

in Figure 17 is rising (increasing) from the point $(-4, -2)$ to the point $(0, 4)$, so we conclude that it is increasing on the open interval $(-4, 0)$ (or for $-4 < x < 0$). The graph is falling (decreasing) from the point $(-6, 0)$ to the point $(-4, -2)$ and from the point $(3, 4)$ to the point $(6, 1)$. We conclude that the graph is decreasing on the open intervals $(-6, -4)$ and $(3, 6)$ (or for $-6 < x < -4$ and $3 < x < 6$). The graph is constant on the open interval $(0, 3)$ (or $0 < x < 3$). ∎

More precise definitions follow:

A function f is **increasing** on an open interval I if, for any choice of x_1 and x_2 in I, with $x_1 < x_2$, we have $f(x_1) < f(x_2)$.

A function f is **decreasing** on an open interval I if, for any choice of x_1 and x_2 in I, with $x_1 < x_2$, we have $f(x_1) > f(x_2)$.

A function f is **constant** on an open interval I if, for all choices of x in I, the values $f(x)$ are equal.

Figure 18 illustrates the definitions. The graph of an increasing function goes up from left to right, the graph of a decreasing function goes down from left to right, and the graph of a constant function remains at a fixed height.

Figure 18

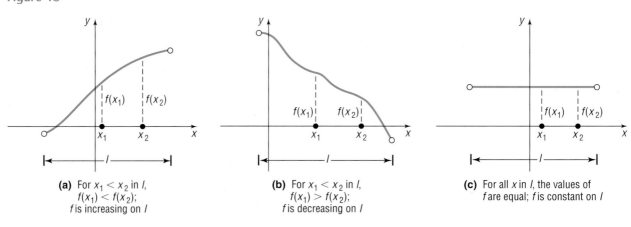

(a) For $x_1 < x_2$ in I, $f(x_1) < f(x_2)$; f is increasing on I

(b) For $x_1 < x_2$ in I, $f(x_1) > f(x_2)$; f is decreasing on I

(c) For all x in I, the values of f are equal; f is constant on I

   NOW WORK PROBLEMS **1**, **3**, AND **5**.

LOCAL MAXIMUM; LOCAL MINIMUM

③ When the graph of a function is increasing to the left of $x = c$ and decreasing to the right of $x = c$, then at c the value of f is largest. This value is called a *local maximum* of f.

When the graph of a function is decreasing to the left of $x = c$ and is increasing to the right of $x = c$, then at c the value of f is the smallest. This value is called a *local minimum* of f.

If f has a local maximum at c, then the value of f at c is greater than the values of f near c. If f has a local minimum at c, then the value of f at c is

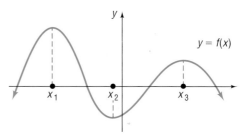

Figure 19
f has a local maximum at x_1 and x_3;
f has a local minimum at x_2

less than the values of f near c. The word *local* is used to suggest that it is only near c that the value $f(c)$ is largest or smallest. See Figure 19.

More precise definitions follow:

> A function f has a **local maximum at c** if there is an open interval I containing c so that, for all $x \neq c$ in I, $f(x) < f(c)$. We call $f(c)$ a **local maximum of f.**

> A function f has a **local minimum at c** if there is an open interval I containing c so that, for all $x \neq c$ in I, $f(x) > f(c)$. We call $f(c)$ a **local minimum of f.**

EXAMPLE 4   **Finding Local Maxima and Local Minima from a Graph**

Figure 20

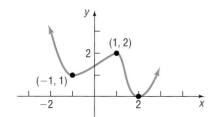

Figure 20 shows the graph of a function f.

(a) At what number(s), if any, does f have a local maximum?

(b) What are the local maxima?

(c) At what number(s), if any, does f have a local minimum?

(d) What are the local minima?

Solution   The domain of f is the set of real numbers.

(a) f has a local maximum at 1, since for all x close to 1, $x \neq 1$, we have $f(x) < f(1)$.

(b) The local maximum is $f(1) = 2$.

(c) f has a local minimum at -1 and at 2.

(d) The local minima are $f(-1) = 1$ and $f(2) = 0$.   ∎

⚠ If the graph of a function f is not given, then to locate the exact value at which f has a local maximum or a local minimum usually requires calculus.

 NOW WORK PROBLEMS 7 AND 9.

EVEN AND ODD FUNCTIONS

④ A function f is even if and only if whenever the point (x, y) is on the graph of f then the point $(-x, y)$ is also on the graph. Algebraically, we define an even function as follows:

A function f is **even** if for every number x in its domain the number $-x$ is also in the domain and

$$f(-x) = f(x)$$

A function f is odd if and only if whenever the point (x, y) is on the graph of f then the point $(-x, -y)$ is also on the graph. Algebraically, we define an odd function as follows:

A function f is **odd** if for every number x in its domain the number $-x$ is also in the domain and

$$f(-x) = -f(x)$$

Refer to Section 2.2, where the tests for symmetry are listed. The following results are then evident.

Theorem   A function is even if and only if its graph is symmetric with respect to the y-axis. A function is odd if and only if its graph is symmetric with respect to the origin.

EXAMPLE 5   ### Determining Even and Odd Functions from the Graph

Determine whether each graph given in Figure 21 is the graph of an even function, an odd function, or a function that is neither even nor odd.

Figure 21

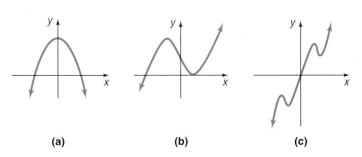

(a)   (b)   (c)

Solution   The graph in Figure 21(a) is that of an even function, because the graph is symmetric with respect to the y-axis. The function whose graph is given in Figure 21(b) is neither even nor odd, because the graph is neither symmetric with respect to the y-axis nor symmetric with respect to the origin. The function whose graph is given in Figure 21(c) is odd, because its graph is symmetric with respect to the origin.

NOW WORK PROBLEM 11.

⑤   In the next example, we use algebraic techniques to verify whether a given function is even, odd, or neither.

EXAMPLE 6	Identifying Even and Odd Functions Algebraically

Determine whether each of the following functions is even, odd, or neither. Then determine whether the graph is symmetric with respect to the y-axis or with respect to the origin.

(a) $f(x) = x^2 - 5$ (b) $g(x) = x^3 - 1$
(c) $h(x) = 5x^3 - x$ (d) $F(x) = |x|$

Solution (a) To determine whether f is even, odd, or neither, we replace x by $-x$ in $f(x) = x^2 - 5$. Then

$$f(-x) = (-x)^2 - 5 = x^2 - 5 = f(x)$$

Since $f(-x) = f(x)$, we conclude that f is an even function, and the graph is symmetric with respect to the y-axis.

(b) We replace x by $-x$ in $g(x) = x^3 - 1$. Then

$$g(-x) = (-x)^3 - 1 = -x^3 - 1$$

Since $g(-x) \neq g(x)$ and $g(-x) \neq -g(x) = -(x^3 - 1) = -x^3 + 1$, we conclude that g is neither even nor odd. The graph is not symmetric with respect to the y-axis nor is it symmetric with respect to the origin.

(c) We replace x by $-x$ in $h(x) = 5x^3 - x$. Then

$$h(-x) = 5(-x)^3 - (-x) = -5x^3 + x = -(5x^3 - x) = -h(x)$$

Since $h(-x) = -h(x)$, h is an odd function, and the graph of h is symmetric with respect to the origin.

(d) We replace x by $-x$ in $F(x) = |x|$. Then

$$F(-x) = |-x| = |-1| \cdot |x| = |x| = F(x)$$

Since $F(-x) = F(x)$, F is an even function, and the graph of F is symmetric with respect to the y-axis. ∎

NOW WORK PROBLEM 37.

3.2 EXERCISES

In Problems 1–10, use the graph of the function f given below.

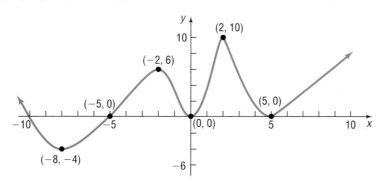

1. Is f increasing on the interval $(-8, -2)$?
2. Is f decreasing on the interval $(-8, -4)$?
3. Is f increasing on the interval $(2, 10)$?
4. Is f decreasing on the interval $(2, 5)$?
5. List the interval(s) on which f is increasing.
6. List the interval(s) on which f is decreasing.

7. Is there a local maximum at 2? If yes, what is it?
8. Is there a local maximum at 5? If yes, what is it?
9. List the numbers at which f has a local maximum. What are these local maxima?
10. List the numbers at which f has a local minimum. What are these local minima?

In Problems 11–20, the graph of a function is given. Use the graph to find:
   (a) *The intercepts, if any*
   (b) *Its domain and range*
   (c) *The intervals on which it is increasing, decreasing, or constant*
   (d) *Whether it is even, odd, or neither*

11.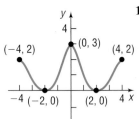

12.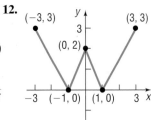

13.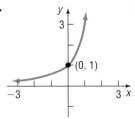

14.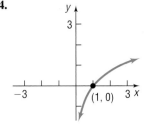

15.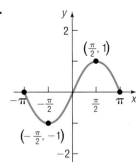

16.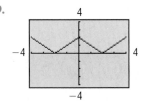

17.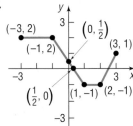

18.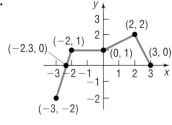

In Problems 19 and 20, assume that the entire graph is shown.

19.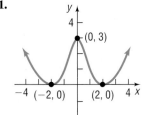

20.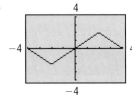

In Problems 21–24, the graph of a function f is given. Use the graph to find:
   (a) *The numbers, if any, at which f has a local maximum. What are these local maxima?*
   (b) *The numbers, if any, at which f has a local minimum. What are these local minima?*

21.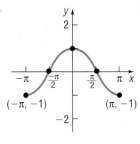

22.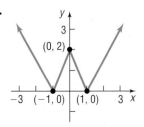

23.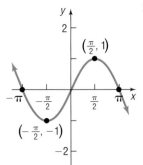

24.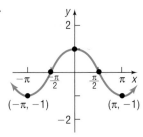

In Problems 25–36, (a) for each function find the average rate of change of f from 1 to x:

$$\frac{f(x) - f(1)}{x - 1}, \qquad x \neq 1$$

(b) Use the result from part (a) to compute the average rate of change from $x = 1$ to $x = 2$. Be sure to simplify.
(c) Find an equation of the secant line containing $(1, f(1))$ and $(2, f(2))$.

25. $f(x) = 5x$ **26.** $f(x) = -4x$ **27.** $f(x) = 1 - 3x$ **28.** $f(x) = x^2 + 1$

29. $f(x) = x^2 - 2x$ **30.** $f(x) = x - 2x^2$ **31.** $f(x) = x^3 - x$ **32.** $f(x) = x^3 + x$

33. $f(x) = \dfrac{2}{x + 1}$ **34.** $f(x) = \dfrac{1}{x^2}$ **35.** $f(x) = \sqrt{x}$ **36.** $f(x) = \sqrt{x + 3}$

In Problems 37–48, determine algebraically whether each function is even, odd, or neither.

37. $f(x) = 4x^3$ **38.** $f(x) = 2x^4 - x^2$ **39.** $g(x) = -3x^2 - 5$ **40.** $h(x) = 3x^3 + 5$

41. $F(x) = \sqrt[3]{x}$ **42.** $G(x) = \sqrt{x}$ **43.** $f(x) = x + |x|$ **44.** $f(x) = \sqrt[3]{2x^2 + 1}$

45. $g(x) = \dfrac{1}{x^2}$ **46.** $h(x) = \dfrac{x}{x^2 - 1}$ **47.** $h(x) = \dfrac{-x^3}{3x^2 - 9}$ **48.** $F(x) = \dfrac{2x}{|x|}$

Problems 49–56 require the following alternative definition:

The **difference quotient** of a function $y = f(x)$ at a number x in the domain of f is defined as the expression

$$\frac{f(x + h) - f(x)}{h}$$

where $h \neq 0$ is a nonzero real number.

Find the difference quotient of each of the following functions. Be sure to simplify your answer.

49. $f(x) = 2x + 5$ **50.** $f(x) = -3x + 2$ **51.** $f(x) = x^2 + 2x$ **52.** $f(x) = 2x^2 + x$

53. $f(x) = 2x^2 - 3x + 1$ **54.** $f(x) = -x^2 + 3x - 2$ **55.** $f(x) = 1/x$ **56.** $f(x) = 1/x^2$

57. Revenue from Selling Bikes   The following data represent the total revenue that would be received from selling x bicycles at Tunney's Bicycle Shop.

Number of Bicycles, x	Total Revenue, R (Dollars)
0	0
25	28,000
60	45,000
102	53,400
150	59,160
190	62,360
223	64,835
249	66,525

(a) Draw a scatter diagram of the data, treating the number of bicycles produced as the independent variable.
(b) Draw a line through the points $(0, 0)$ and $(25, 28{,}000)$ on the scatter diagram found in part (a).
(c) Find the average rate of change of revenue from 0 to 25 bicycles.
(d) Interpret the average rate of change found in part (c).
(e) Draw a line through the points $(190, 62{,}360)$ and $(223, 64{,}835)$ on the scatter diagram found in part (a).
(f) Find the average rate of change of revenue from 190 to 223 bicycles.
(g) Interpret the average rate of change found in part (f).

58. Cost of Manufacturing Bikes   The following data represent the monthly cost of producing bicycles at Tunney's Bicycle Shop.

Number of Bicycles, x	Total Cost of Production, C (Dollars)
0	24,000
25	27,750
60	31,500
102	35,250
150	39,000
190	42,750
223	46,500
249	50,250

(a) Draw a scatter diagram of the data, treating the number of bicycles produced as the independent variable.
(b) Draw a line through the points $(0, 24{,}000)$ and $(25, 27{,}750)$ on the scatter diagram found in part (a).
(c) Find the average rate of change of the cost from 0 to 25 bicycles.
(d) Interpret the average rate of change found in part (c).
(e) Draw a line through the points $(190, 42{,}750)$ and $(223, 46{,}500)$ on the scatter diagram found in part (a).
(f) Find the average rate of change of the cost from 190 to 223 bicycles.
(g) Interpret the average rate of change found in part (f).

59. Growth of Bacteria The following data represent the population of an unknown bacteria.

Time (Days)	Population
0	50
1	153
2	234
3	357
4	547
5	839
6	1280

(a) Draw a scatter diagram of the data, treating time as the independent variable.
(b) Draw a line through the points $(0, 50)$ and $(1, 153)$ on the scatter diagram found in part (a).

(c) Find the average rate of change of the population from day 0 to day 1.
(d) Interpret the average rate of change found in part (c).
(e) Draw a line through the points $(5, 839)$ and $(6, 1280)$ on the scatter diagram found in part (a).
(f) Find the average rate of change of the population from day 5 to day 6.
(g) Interpret the average rate of change found in part (f).
(h) What is happening to the average rate of change of the population as time passes?

60. Falling Objects Suppose that you drop a ball from a cliff 1000 feet high. You measure the distance s that the ball has fallen after time t using a motion detector and obtain the following data.

Time, t (Seconds)	Distance, s (Feet)
0	0
1	16
2	64
3	144
4	256
5	400
6	576
7	784

(a) Draw a scatter diagram of the data, treating time as the independent variable.
(b) Draw a line through the points $(0, 0)$ and $(2, 64)$.
(c) Find the average rate of change of the distance from 0 to 2 seconds; that is, find the slope of the line in part (b).
(d) Interpret the average rate of change found in part (c).
(e) Draw a line through the points $(5, 400)$ and $(7, 784)$.
(f) Find the average rate of change of the distance from 5 to 7 seconds; that is, find the slope of the line in part (e).
(g) Interpret the average rate of change found in part (f).
(h) What is happening to the average rate of change of the distance that the ball has fallen as time passes?

61. How many x-intercepts can a function defined on an interval have if it is increasing on that interval? Explain.

62. Can you think of a function that is both even and odd?

PREPARING FOR THIS SECTION

Before getting started, review the following:

✓ Graphs of Certain Equations (Section 2.2, Example 3, p. 155; Example 4, p. 156; Example 5, p. 156; Example 12, p. 161)

3.3 LIBRARY OF FUNCTIONS; PIECEWISE-DEFINED FUNCTIONS

OBJECTIVES 1 Graph the Functions Listed in the Library of Functions
2 Graph Piecewise-defined Functions

LIBRARY OF FUNCTIONS

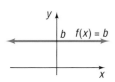

We now give names to some of the functions that we have encountered. In going through this list, pay special attention to the properties of each function, particularly to the shape of each graph. Knowing these graphs will lay the foundation for later graphing techniques.

Linear Functions

$$f(x) = mx + b \qquad m \text{ and } b \text{ are real numbers}$$

The domain of a **linear function** f consists of all real numbers. The graph of this function is a nonvertical line with slope m and y-intercept b. A linear function is increasing if $m > 0$, decreasing if $m < 0$, and constant if $m = 0$.

Constant Function

$$f(x) = b \qquad b \text{ is a real number}$$

Figure 22
Constant Function

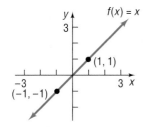

See Figure 22.

A **constant function** is a special linear function ($m = 0$). Its domain is the set of all real numbers; its range is the set consisting of a single number b. Its graph is a horizontal line whose y-intercept is b. The constant function is an even function whose graph is constant over its domain.

Identity Function

$$f(x) = x$$

Figure 23
Identity Function

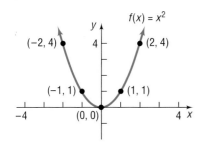

See Figure 23.

The **identity function** is also a special linear function. Its domain and range are the set of all real numbers. Its graph is a line whose slope is $m = 1$ and whose y-intercept is 0. The line consists of all points for which the x-coordinate equals the y-coordinate. The identity function is an odd function that is increasing over its domain. Note that the graph bisects quadrants I and III.

Figure 24
Square Function

Square Function

$$f(x) = x^2$$

See Figure 24.

The domain of the **square function** f is the set of all real numbers; its range is the set of nonnegative real numbers. The graph of this function is a parabola whose intercept is at $(0, 0)$. The square function is an even function that is decreasing on the interval $(-\infty, 0)$ and increasing on the interval $(0, \infty)$.

Figure 25
Cube Function

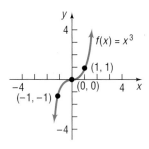

Cube Function

$$f(x) = x^3$$

See Figure 25.

The domain and range of the **cube function** are the set of all real numbers. The intercept of the graph is at $(0, 0)$. The cube function is odd and is increasing on the interval $(-\infty, \infty)$.

Figure 26
Square Root Function

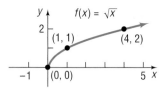

Square Root Function

$$f(x) = \sqrt{x}$$

See Figure 26.

The domain and range of the **square root function** are the set of nonnegative real numbers. The intercept of the graph is at $(0, 0)$. The square root function is neither even nor odd and is increasing on the interval $(0, \infty)$.

Figure 27
Reciprocal Function

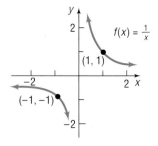

Reciprocal Function

$$f(x) = \frac{1}{x}$$

Refer to Example 12, p. 161, for a discussion of the equation $y = 1/x$. See Figure 27.

The domain and range of the **reciprocal function** are the set of all nonzero real numbers. The graph has no intercepts. The reciprocal function is decreasing on the intervals $(-\infty, 0)$ and $(0, \infty)$ and is an odd function.

Absolute Value Function

$$f(x) = |x|$$

Figure 28
Absolute Value Function

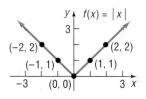

See Figure 28.

The domain of the **absolute value function** is the set of all real numbers; its range is the set of nonnegative real numbers. The intercept of the graph is at $(0, 0)$. If $x \geq 0$, then $f(x) = x$, and the graph of f is part of the line $y = x$; if $x < 0$, then $f(x) = -x$, and the graph of f is part of the line $y = -x$. The absolute value function is an even function; it is decreasing on the interval $(-\infty, 0)$ and increasing on the interval $(0, \infty)$.

SEEING THE CONCEPT: Graph $y = |x|$ on a square screen and compare what you see with Figure 28. Note that some graphing calculators use the symbols abs (x) for absolute value. If your utility has no built-in absolute value function, you can still graph $y = |x|$ by using the fact that $|x| = \sqrt{x^2}$.   ∎

The notation int(x) stands for the largest integer less than or equal to x. For example,

$$\text{int}(1) = 1 \quad \text{int}(2.5) = 2 \quad \text{int}\left(\tfrac{1}{2}\right) = 0 \quad \text{int}\left(-\tfrac{3}{4}\right) = -1 \quad \text{int}(\pi) = 3$$

This type of correspondence occurs frequently enough in mathematics that we give it a name.

Greatest Integer Function

$$f(x) = \text{int}(x) = \text{Greatest integer less than or equal to } x$$

NOTE: Some books use the notation $f(x) = [\![x]\!]$ instead of int(x).

We obtain the graph of $f(x) = \text{int}(x)$ by plotting several points. See Table 4. For values of $x, -1 \le x < 0$, the value of $f(x) = \text{int}(x)$ is -1; for values of $x, 0 \le x < 1$, the value of f is 0. See Figure 29 for the graph.

TABLE 4		
x	$y = f(x) =$ int(x)	(x, y)
-1	-1	$(-1, -1)$
$-\tfrac{1}{2}$	-1	$\left(-\tfrac{1}{2}, -1\right)$
$-\tfrac{1}{4}$	-1	$\left(-\tfrac{1}{4}, -1\right)$
0	0	$(0, 0)$
$\tfrac{1}{4}$	0	$\left(\tfrac{1}{4}, 0\right)$
$\tfrac{1}{2}$	0	$\left(\tfrac{1}{2}, 0\right)$
$\tfrac{3}{4}$	0	$\left(\tfrac{3}{4}, 0\right)$

Figure 29
Greatest Integer Function

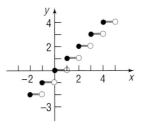

The domain of the **greatest integer function** is the set of all real numbers; its range is the set of integers. The y-intercept of the graph is 0. The x-intercepts lie in the interval $[0, 1)$. The greatest integer function is neither even nor odd. It is constant on every interval of the form $[k, k + 1)$, for k an integer. In Figure 29, we use a solid dot to indicate, for example, that at $x = 1$ the value of f is $f(1) = 1$; we use an open circle to illustrate that the function does not assume the value of 0 at $x = 1$.

From the graph of the greatest integer function, we can see why it is also called a **step function.** At $x = 0$, $x = \pm 1$, $x = \pm 2$, and so on, this function exhibits what is called a *discontinuity*; that is, at integer values, the graph suddenly "steps" from one value to another without taking on any of the intermediate values. For example, to the immediate left of $x = 3$, the y-coordinates are 2, and to the immediate right of $x = 3$, the y-coordinates are 3.

COMMENT: When graphing a function, you can choose either the **connected mode,** in which points plotted on the screen are connected, making the graph appear without any breaks, or the **dot mode,** in which only the points plotted appear. When graphing the greatest integer function with a graphing utility, it is necessary to be in the **dot mode.** This is to prevent the utility from "connecting the dots" when $f(x)$ changes from one integer value to the next. See Figure 30.

Figure 30
$f(x) = \text{int}(x)$

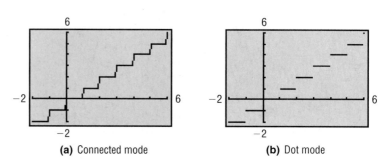

(a) Connected mode   **(b)** Dot mode

The functions that we have discussed so far are basic. Whenever you encounter one of them, you should see a mental picture of its graph. For example, if you encounter the function $f(x) = x^2$, you should see in your mind's eye a picture like Figure 24.

 NOW WORK PROBLEMS 1–8.

PIECEWISE-DEFINED FUNCTIONS

② Sometimes a function is defined differently on different parts of its domain. For example, the absolute value function $f(x) = |x|$ is actually defined by two equations: $f(x) = x$ if $x \geq 0$ and $f(x) = -x$ if $x < 0$. For convenience, we generally combine these equations into one expression as

$$f(x) = |x| = \begin{cases} x & \text{if } x \geq 0 \\ -x & \text{if } x < 0 \end{cases}$$

When functions are defined by more than one equation, they are called **piecewise-defined** functions.

Let's look at another example of a piecewise-defined function.

EXAMPLE 1

Analyzing a Piecewise-defined Function

The function f is defined as

$$f(x) = \begin{cases} -x + 1 & \text{if } -1 \leq x < 1 \\ 2 & \text{if } x = 1 \\ x^2 & \text{if } x > 1 \end{cases}$$

(a) Find $f(0)$, $f(1)$, and $f(2)$.   (b) Determine the domain of f.

(c) Graph f.   (d) Use the graph to find the range of f.

Solution   (a) To find $f(0)$, we observe that when $x = 0$ the equation for f is given by $f(x) = -x + 1$. So we have

$$f(0) = -0 + 1 = 1$$

When $x = 1$, the equation for f is $f(x) = 2$. Thus,

$$f(1) = 2$$

When $x = 2$, the equation for f is $f(x) = x^2$. So

$$f(2) = 2^2 = 4$$

Figure 31

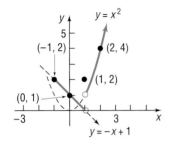

(b) To find the domain of f, we look at its definition. We conclude that the domain of f is $\{x \mid x \geq -1\}$, or the interval $[-1, \infty)$.

(c) To graph f, we graph "each piece." First we graph the line $y = -x + 1$ and keep only the part for which $-1 \leq x < 1$. Then we plot the point $(1, 2)$ because, when $x = 1$, $f(x) = 2$. Finally, we graph the parabola $y = x^2$ and keep only the part for which $x > 1$. See Figure 31.

(d) From the graph, we conclude that the range of f is $\{y \mid y > 0\}$, or the interval $(0, \infty)$.   ∎

NOW WORK PROBLEM **19.**

EXAMPLE 2

Cost of Electricity

In June 1999, Commonwealth Edison Company supplied electricity to residences for a monthly customer charge of \$8.02 plus 8.77¢ per kilowatt-hour (kWhr) for the first 400 kWhr supplied in the month and 6.574¢ per kWhr for all usage over 400 kWhr in the month.*

(a) What is the charge for using 300 kWhr in a month?

(b) What is the charge for using 700 kWhr in a month?

(c) If C is the monthly charge for x kWhr, express C as a function of x.

Solution

(a) For 300 kWhr, the charge is \$8.02 plus 8.77¢ $=$ \$0.0877 per kWhr. That is,

$$\text{Charge} = \$8.02 + \$0.0877(300) = \$34.33$$

(b) For 700 kWhr, the charge is \$8.02 plus 8.77¢ per kWhr for the first 400 kWhr plus 6.574¢ per kWhr for the 300 kWhr in excess of 400. That is,

$$\text{Charge} = \$8.02 + \$0.0877(400) + \$0.06574(300) = \$62.82$$

(c) If $0 \leq x \leq 400$, the monthly charge C (in dollars) can be found by multiplying x times \$0.0877 and adding the monthly customer charge of \$8.02. Thus, if $0 \leq x \leq 400$, then $C(x) = 0.0877x + 8.02$. For $x > 400$, the charge is $0.0877(400) + 8.02 + 0.06574(x - 400)$, since $x - 400$ equals the usage in excess of 400 kWhr, which costs \$0.06574 per kWhr. That is, if $x > 400$, then

$$C(x) = 0.0877(400) + 8.02 + 0.06574(x - 400)$$

$$= 43.10 + 0.06574(x - 400)$$

$$= 0.06574x + 16.80$$

*Source: Commonwealth Edison Co., Chicago, Illinois, 1999.

The rule for computing C follows two equations:

$$C(x) = \begin{cases} 0.0877x + 8.02 & \text{if } 0 \le x \le 400 \\ 0.06574x + 16.80 & \text{if } x > 400 \end{cases}$$

See Figure 32 for the graph.

Figure 32

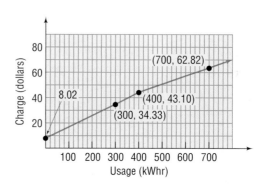

3.3 EXERCISES

In Problems 1–8, match each graph to the function listed whose graph most resembles the one given.

A. *Constant function*   B. *Linear function*   C. *Square function*
D. *Cube function*   E. *Square root function*   F. *Reciprocal function*
G. *Absolute value function*   H. *Greatest integer function*

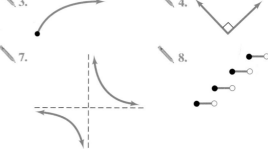

In Problems 9–14, sketch the graph of each function. Be sure to label three points on the graph.

9. $f(x) = x$ **10.** $f(x) = x^2$ **11.** $f(x) = x^3$
12. $f(x) = \sqrt{x}$ **13.** $f(x) = 1/x$ **14.** $f(x) = |x|$

15. If $f(x) = \begin{cases} x^2 & \text{if } x < 0 \\ 2 & \text{if } x = 0 \\ 2x + 1 & \text{if } x > 0 \end{cases}$

find: (a) $f(-2)$ (b) $f(0)$ (c) $f(2)$

16. If $f(x) = \begin{cases} x^3 & \text{if } x < 0 \\ 3x + 2 & \text{if } x \ge 0 \end{cases}$

find: (a) $f(-1)$ (b) $f(0)$ (c) $f(1)$

17. If $f(x) = \text{int}(2x)$, find: (a) $f(1.2)$ (b) $f(1.6)$ (c) $f(-1.8)$

18. If $f(x) = \text{int}(x/2)$, find: (a) $f(1.2)$ (b) $f(1.6)$ (c) $f(-1.8)$

In Problems 19–30:

(a) *Find the domain of each function.*   (b) *Locate any intercepts.*
(c) *Graph each function.*   (d) *Based on the graph, find the range.*

19. $f(x) = \begin{cases} 2x & \text{if } x \ne 0 \\ 1 & \text{if } x = 0 \end{cases}$

20. $f(x) = \begin{cases} 3x & \text{if } x \ne 0 \\ 4 & \text{if } x = 0 \end{cases}$

21. $f(x) = \begin{cases} -2x + 3 & x < 1 \\ 3x - 2 & x \ge 1 \end{cases}$

22. $f(x) = \begin{cases} x + 3 & x < -2 \\ -2x - 3 & x \ge -2 \end{cases}$

23. $f(x) = \begin{cases} x + 3 & -2 \le x < 1 \\ 5 & x = 1 \\ -x + 2 & x > 1 \end{cases}$

24. $f(x) = \begin{cases} 2x + 5 & -3 \le x < 0 \\ -3 & x = 0 \\ -5x & x > 0 \end{cases}$

25. $f(x) = \begin{cases} 1 + x & \text{if } x < 0 \\ x^2 & \text{if } x \ge 0 \end{cases}$

26. $f(x) = \begin{cases} 1/x & \text{if } x < 0 \\ \sqrt{x} & \text{if } x \ge 0 \end{cases}$

27. $f(x) = \begin{cases} |x| & \text{if } -2 \le x < 0 \\ 1 & \text{if } x = 0 \\ x^3 & \text{if } x > 0 \end{cases}$

28. $f(x) = \begin{cases} 3 + x & \text{if } -3 \le x < 0 \\ 3 & \text{if } x = 0 \\ \sqrt{x} & \text{if } x > 0 \end{cases}$

29. $f(x) = 2\,\text{int}(x)$

30. $f(x) = \text{int}(2x)$

In Problems 31–34, the graph of a piecewise-defined function is given. Write a definition for each function.

31.   **32.**   **33.**   **34.**

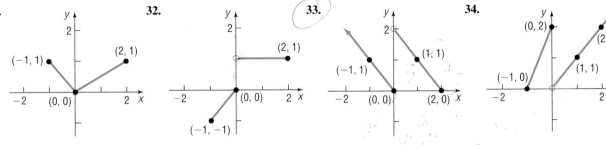

---

35. Cost of Natural Gas   In April 2000, the Peoples Gas Company had the following rate schedule* for natural gas usage in single-family residences:

Monthly service charge $9.45

Per therm service charge
1st 50 therms $0.36375/therm
Over 50 therms $0.11445/therm

Gas charge $0.3128/therm

(a) What is the charge for using 50 therms in a month?
(b) What is the charge for using 500 therms in a month?
(c) Construct a function that relates the monthly charge C for x therms of gas.
(d) Graph this function.

36. Cost of Natural Gas   In March 2000, Nicor Gas had the following rate schedule† for natural gas usage in single-family residences:

Monthly customer charge $6.45

Distribution charge
1st 20 therms $0.2012/therm
Next 30 therms $0.1117/therm
Over 50 therms $0.0374/therm

Gas supply charge $0.3209/therm

(a) What is the charge for using 40 therms in a month?
(b) What is the charge for using 202 therms in a month?

(c) Construct a function that gives the monthly charge C for x therms of gas.
(d) Graph this function.

37. Wind Chill   The wind chill factor represents the equivalent air temperature at a standard wind speed that would produce the same heat loss as the given temperature and wind speed. One formula for computing the equivalent temperature is

$$W = \begin{cases} t & 0 \le v < 1.79 \\ 33 - \dfrac{(10.45 + 10\sqrt{v} - v)(33 - t)}{22.04} & 1.79 \le v \le 20 \\ 33 - 1.5958(33 - t) & v > 20 \end{cases}$$

where v represents the wind speed (in meters per second) and t represents the air temperature (°C). Compute the wind chill for the following:
(a) An air temperature of 10°C and a wind speed of 1 meter per second (m/sec).
(b) An air temperature of 10°C and a wind speed of 5 m/sec.
(c) An air temperature of 10°C and a wind speed of 15 m/sec.
(d) An air temperature of 10°C and a wind speed of 25 m/sec.
(e) Explain the physical meaning of the equation corresponding to $0 \le v < 1.79$.
(f) Explain the physical meaning of the equation corresponding to $v > 20$.

---

* *Source:* The Peoples Gas Company, Chicago, Illinois.
† *Source:* Nicor Gas, Aurora, Illinois.

38. Wind Chill Redo Problem 37(a)–(d) for an air temperature of −10°C.

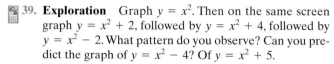

39. Exploration Graph $y = x^2$. Then on the same screen graph $y = x^2 + 2$, followed by $y = x^2 + 4$, followed by $y = x^2 - 2$. What pattern do you observe? Can you predict the graph of $y = x^2 - 4$? Of $y = x^2 + 5$.

40. Exploration Graph $y = x^2$. Then on the same screen graph $y = (x - 2)^2$, followed by $y = (x - 4)^2$, followed by $y = (x + 2)^2$. What pattern do you observe? Can you predict the graph of $y = (x + 4)^2$? Of $y = (x - 5)^2$?

41. Exploration Graph $y = |x|$. Then on the same screen graph $y = 2|x|$, followed by $y = 4|x|$, followed by $y = \frac{1}{2}|x|$. What pattern do you observe? Can you predict the graph of $y = \frac{1}{4}|x|$? Of $y = 5|x|$?

42. Exploration Graph $y = x^2$. Then on the same screen graph $y = -x^2$. What pattern do you observe? Now try $y = |x|$ and $y = -|x|$. What do you conclude?

43. Exploration Graph $y = \sqrt{x}$. Then on the same screen graph $y = \sqrt{-x}$. What pattern do you observe? Now try $y = 2x + 1$ and $y = 2(-x) + 1$. What do you conclude?

44. Exploration Graph $y = x^3$. Then on the same screen graph $y = (x - 1)^3 + 2$. Could you have predicted the result?

45. Exploration Graph $y = x^2$, $y = x^4$, and $y = x^6$ on the same screen. What do you notice is the same about each graph? What do you notice that is different?

46. Exploration Graph $y = x^3$, $y = x^5$, and $y = x^7$ on the same screen. What do you notice is the same about each graph? What do you notice that is different?

47. Consider the equation

$$y = \begin{cases} 1 & \text{if } x \text{ is rational} \\ 0 & \text{if } x \text{ is irrational} \end{cases}$$

Is this a function? What is its domain? What is its range? What is its y-intercept, if any? What are its x-intercepts, if any? Is it even, odd, or neither? How would you describe its graph?

48. Define some functions that pass through $(0, 0)$ and $(1, 1)$ and are increasing for $x \geq 0$. Begin your list with $y = \sqrt{x}$, $y = x$, and $y = x^2$. Can you propose a general result about such functions?

3.4 GRAPHING TECHNIQUES: TRANSFORMATIONS

OBJECTIVES 1 Graph Functions Using Horizontal and Vertical Shifts
2 Graph Functions Using Compressions and Stretches
3 Graph Functions Using Reflections about the x-Axis or y-Axis

At this stage, if you were asked to graph any of the functions defined by $y = x$, $y = x^2$, $y = x^3$, $y = \sqrt{x}$, $y = |x|$, or $y = 1/x$, your response should be, "Yes, I recognize these functions and know the general shapes of their graphs." (If this is not your answer, review the previous section, Figures 22 through 28.)

Sometimes we are asked to graph a function that is "almost" like one that we already know how to graph. In this section, we look at some of these functions and develop techniques for graphing them. Collectively, these techniques are referred to as **transformations.**

1 VERTICAL SHIFTS

EXAMPLE 1 | **Vertical Shift Up**

Use the graph of $f(x) = x^2$ to obtain the graph of $g(x) = x^2 + 3$.

Solution We begin by obtaining some points on the graphs of f and g. For example, when $x = 0$, then $y = f(0) = 0$ and $y = g(0) = 3$. When $x = 1$, then $y = f(1) = 1$ and $y = g(1) = 4$. Table 5 lists these and a few other points on each graph. We conclude that the graph of g is identical to that of f, except that it is shifted vertically up 3 units. See Figure 33.

TABLE 5

x	$y = f(x)$ $= x^2$	$y = g(x)$ $= x^2 + 3$
-2	4	7
-1	1	4
0	0	3
1	1	4
2	4	7

Figure 33

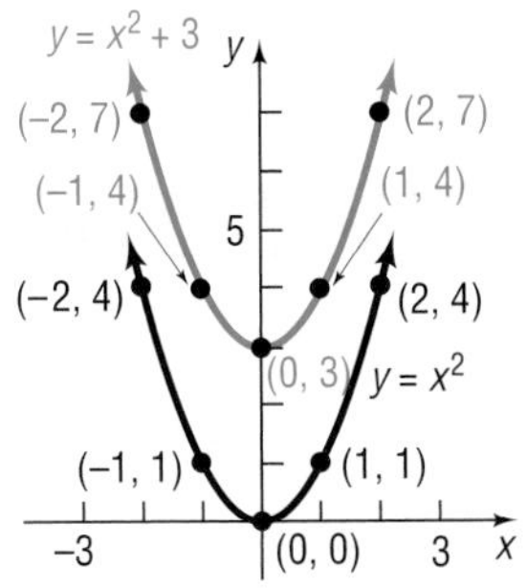

SEEING THE CONCEPT: On the same screen, graph each of the following functions:

$$Y_1 = x^2$$
$$Y_2 = x^2 + 1$$
$$Y_3 = x^2 + 2$$
$$Y_4 = x^2 - 1$$
$$Y_5 = x^2 - 2$$

Figure 34

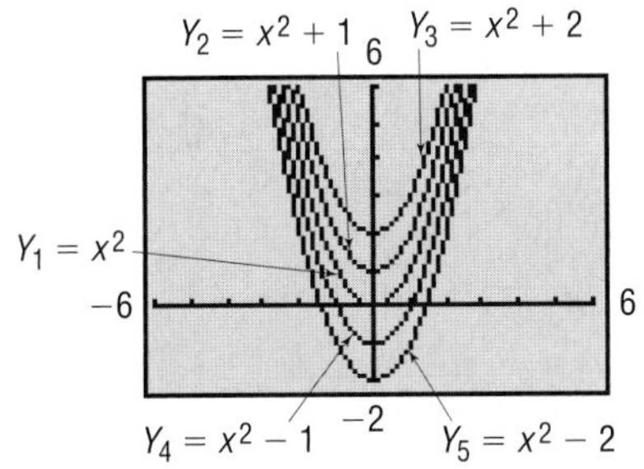

Figure 34 illustrates the graphs. You should have observed a general pattern. With $Y_1 = x^2$ on the screen, the graph of $Y_2 = x^2 + 1$ is identical to that of $Y_1 = x^2$, except that it is shifted vertically up 1 unit. Similarly, $Y_3 = x^2 + 2$ is identical to that of $Y_1 = x^2$, except that it is shifted vertically up 2 units. The graph of $Y_4 = x^2 - 1$ is identical to that of $Y_1 = x^2$, except that it is shifted vertically down 1 unit.

We are led to the following conclusion:

> If a real number k is added to the right side of a function $y = f(x)$, the graph of the new function $y = f(x) + k$ is the graph of f **shifted vertically up** (if $k > 0$) or **down** (if $k < 0$).

Let's look at another example.

EXAMPLE 2   **Vertical Shift Down**

Use the graph of $f(x) = x^2$ to obtain the graph of $g(x) = x^2 - 4$.

Solution   Table 6 lists some points on the graphs of f and g. Notice that each y-coordinate of g is 4 units less than the corresponding y-coordinate of f. The graph of g is identical to that of f, except that it is shifted down 4 units. See Figure 35.

TABLE 6

x	$y = f(x)$ $= x^2$	$y = g(x)$ $= x^2 - 4$
-2	4	0
-1	1	-3
0	0	-4
1	1	-3
2	4	0

Figure 35

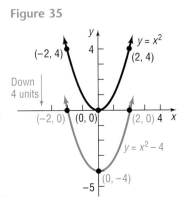

NOW WORK PROBLEM 29.

HORIZONTAL SHIFTS

| EXAMPLE 3 | Horizontal Shift to the Right |

Use the graph of $f(x) = x^2$ to obtain the graph of $g(x) = (x - 2)^2$.

Solution   The function $g(x) = (x - 2)^2$ is basically a square function. Table 7 lists some points on the graphs of f and g. Note that when $f(x) = 0$ then $x = 0$, and when $g(x) = 0$, then $x = 2$. Also, when $f(x) = 4$, then $x = -2$ or 2, and when $g(x) = 4$, then $x = 0$ or 4. We conclude that the graph of g is identical to that of f, except that it is shifted 2 units to the right. See Figure 36.

	TABLE 7	
x	$y = f(x)$ $= x^2$	$y = g(x)$ $= (x - 2)^2$
-2	4	16
0	0	4
2	4	0
4	16	4

Figure 36

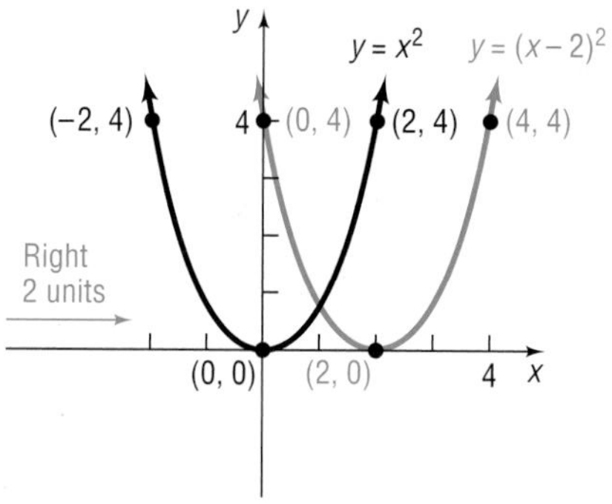

SEEING THE CONCEPT: On the same screen, graph each of the following functions:

$$Y_1 = x^2$$
$$Y_2 = (x - 1)^2$$
$$Y_3 = (x - 3)^2$$
$$Y_4 = (x + 2)^2$$

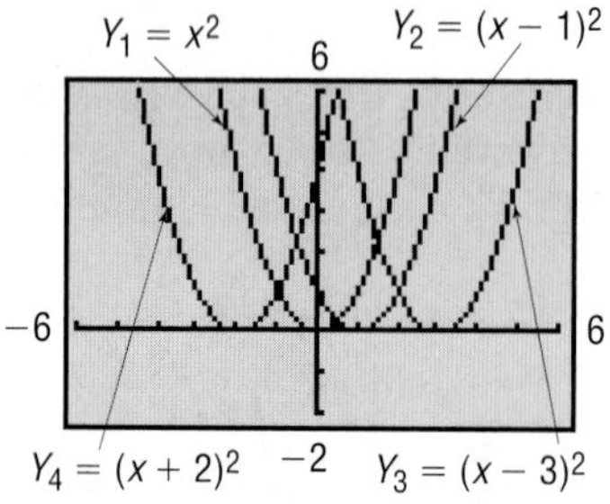

Figure 37 illustrates the graphs.

You should have observed the following pattern. With the graph of $Y_1 = x^2$ on the screen, the graph of $Y_2 = (x - 1)^2$ is identical to that of $Y = x^2$, except that it is shifted horizontally to the right 1 unit. Similarly, the graph of $Y_3 = (x - 3)^2$ is identical to that of $Y_1 = x^2$, except that it is shifted horizontally to the right 3 units. Finally, the graph of $Y_4 = (x + 2)^2$ is identical to that of $Y_1 = x^2$, except that it is shifted horizontally to the left 2 units.

Figure 37

We are led to the following conclusion.

> If the argument x of a function f is replaced by $x - h$, h a real number, the graph of the new function $g(x) = f(x - h)$ is the graph of f **shifted horizontally left** (if $h < 0$) or **right** (if $h > 0$).

EXAMPLE 4   **Horizontal Shift to the Left**

Use the graph of $f(x) = x^2$ to obtain the graph of $g(x) = (x + 4)^2$.

Solution   Again, the function $g(x) = (x + 4)^2$ is basically a square function. Its graph is the same as that of f, except that it is shifted horizontally 4 units to the left. (Do you see why? $(x + 4)^2 = [x - (-4)]^2$) See Figure 38.

Figure 38

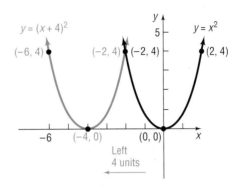

NOW WORK PROBLEM **33**.

Vertical and horizontal shifts are sometimes combined.

EXAMPLE 5   **Combining Vertical and Horizontal Shifts**

Graph the function: $f(x) = (x + 3)^2 - 5$

Solution   We graph f in steps. First, we note that the rule for f is basically a square function, so we begin with the graph of $y = x^2$ as shown in Figure 39(a). Next, to get the graph of $y = (x + 3)^2$, we shift the graph of $y = x^2$ horizontally 3 units to the left. See Figure 39(b). Finally, to get the graph of $y = (x + 3)^2 - 5$, we shift the graph of $y = (x + 3)^2$ vertically down 5 units. See Figure 39(c). Note the points plotted on each graph. Using key points can be helpful in keeping track of the transformation that has taken place.

Figure 39

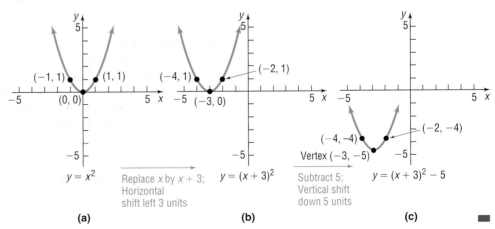

Check: Graph $Y_1 = f(x) = (x + 3)^2 - 5$ and compare the graph to Figure 39(c).

In Example 5, if the vertical shift had been done first, followed by the horizontal shift, the final graph would have been the same. Try it for yourself.

NOW WORK PROBLEM 35.

2   COMPRESSIONS AND STRETCHES

EXAMPLE 6   **Vertical Stretch**

Use the graph of $f(x) = |x|$ to obtain the graph of $g(x) = 2|x|$.

Solution   To see the relationship between the graphs of f and g, we form Table 8, listing points on each graph. For each x, the y-coordinate of a point on the graph of g is 2 times as large as the corresponding y-coordinate on the graph of f. The graph of $f(x) = |x|$ is vertically stretched by a factor of 2 [for example, from $(1, 1)$ to $(1, 2)$] to obtain the graph of $g(x) = 2|x|$. See Figure 40.

TABLE 8

x	$y = f(x)$ $= \|x\|$	$y = g(x)$ $= 2\|x\|$
-2	2	4
-1	1	2
0	0	0
1	1	2
2	2	4

Figure 40

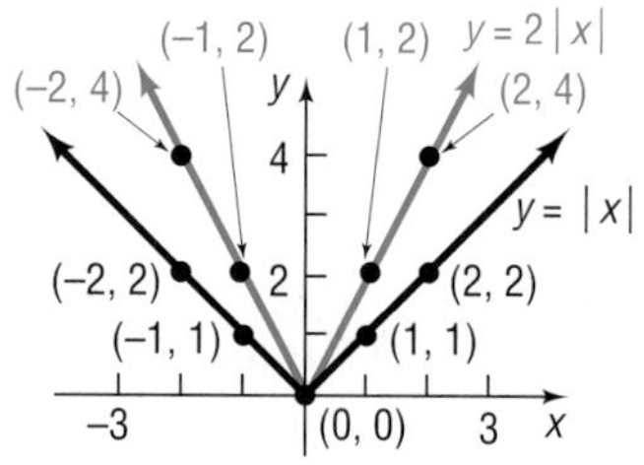

EXAMPLE 7   **Vertical Compression**

Use the graph of $f(x) = |x|$ to obtain the graph of $g(x) = \frac{1}{2}|x|$.

Solution   For each x, the y-coordinate of a point on the graph of g is $\frac{1}{2}$ as large as the corresponding y-coordinate on the graph of f. The graph of $f(x) = |x|$ is vertically compressed by a factor of $\frac{1}{2}$ [for example, from $(2, 2)$ to $(2, 1)$] to obtain the graph of $g(x) = \frac{1}{2}|x|$. See Table 9 and Figure 41.

TABLE 9

x	$y = f(x)$ $= \|x\|$	$y = g(x)$ $= \frac{1}{2}\|x\|$
-2	2	1
-1	1	$\frac{1}{2}$
0	0	0
1	1	$\frac{1}{2}$
2	2	1

Figure 41

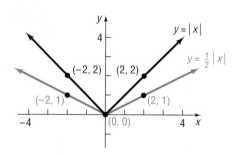

When the right side of a function $y = f(x)$ is multiplied by a positive number a, the graph of the new function $y = af(x)$ is obtained by multiplying each y-coordinate of $y = f(x)$ by a. A **vertical compression** results if $0 < a < 1$ and a **vertical stretch** occurs if $a > 1$.

NOW WORK PROBLEM **37**.

What happens if the argument x of a function $y = f(x)$ is multiplied by a positive number a, creating a new function $y = f(ax)$? To find the answer, we look at the following Exploration.

EXPLORATION   On the same screen, graph each of the following functions:

$$Y_1 = f(x) = x^2$$
$$Y_2 = f(2x) = (2x)^2$$
$$Y_3 = f\left(\frac{1}{2}x\right) = \left(\frac{1}{2}x\right)^2$$

RESULT   You should have obtained the graphs shown in Figure 42. The graph of $Y_2 = (2x)^2$ is the graph of $Y_1 = x^2$ compressed horizontally. Look at Table 10(a).

Figure 42

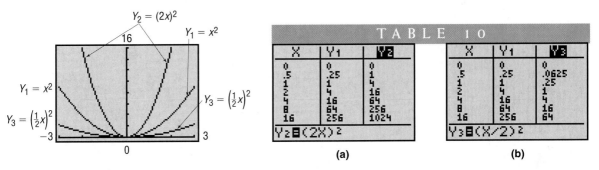

(a)   (b)

Notice that $(1, 1)$, $(2, 4)$, $(4, 16)$, and $(16, 256)$ are points on the graph of $Y_1 = x^2$. Also, $(0.5, 1)$, $(1, 4)$, $(2, 16)$, and $(8, 256)$ are points on the graph of $Y_2 = (2x)^2$. For each y-coordinate, the x-coordinate on the graph of Y_2 is $\frac{1}{2}$ the x-coordinate on Y_1. The graph of $Y_2 = (2x)^2$ is obtained by multiplying the x-coordinate of each point on the graph of $Y_1 = x^2$ by $\frac{1}{2}$. The graph of $Y_3 = (\frac{1}{2}x)^2$ is the graph of $Y_1 = x^2$ stretched horizontally. Look at Table 10(b). Notice that $(0.5, 0.25)$, $(1, 1)$, $(2, 4)$, and $(4, 16)$ are points on the graph of $Y_1 = x^2$. Also, $(1, 0.25)$, $(2, 1)$, $(4, 4)$, and $(8, 16)$ are points on the graph of $Y_3 = (\frac{1}{2}x)^2$. For each y-coordinate, the x-coordinate on the graph of Y_3 is 2 times the x-coordinate on Y_1. The graph of $Y_3 = (\frac{1}{2}x)^2$ is obtained by multiplying the x-coordinate of each point on the graph of $Y_1 = x^2$ by a factor of 2.

If the argument x of a function $y = f(x)$ is multiplied by a positive number a, the graph of the new function $y = f(ax)$ is obtained by multiplying each x-coordinate of $y = f(x)$ by $1/a$. A **horizontal compression** results if $a > 1$, and a **horizontal stretch** occurs if $0 < a < 1$.

Let's look at an example.

| EXAMPLE 8 | **Graphing Using Stretches and Compressions** |

The graph of $y = f(x)$ is given in Figure 43. Use this graph to find the graphs of:

(a) $y = 3f(x)$ (b) $y = f(3x)$

Figure 43

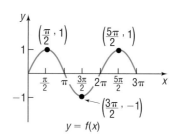

| **Solution** | (a) The graph of $y = 3f(x)$ is obtained by multiplying each y-coordinate of $y = f(x)$ by a factor of 3. See Figure 44(a). |
| | (b) The graph of $y = f(3x)$ is obtained from the graph of $y = f(x)$ by multiplying each x-coordinate of $y = f(x)$ by a factor of $\frac{1}{3}$. See Figure 44(b). |

Figure 44

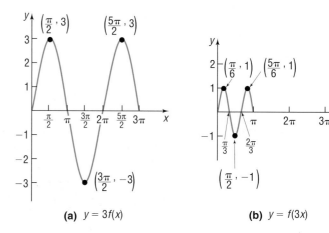

(a) $y = 3f(x)$ **(b)** $y = f(3x)$

NOW WORK PROBLEMS **59(e)** AND **(g)**.

---

  ③   **REFLECTIONS ABOUT THE x-AXIS AND THE y-AXIS**

| EXAMPLE 9 | **Reflection about the x-Axis** |

Graph the function: $f(x) = -x^2$

| **Solution** | We begin with the graph of $y = x^2$, as shown in Figure 45. For each point (x, y) on the graph of $y = x^2$, the point $(x, -y)$ is on the graph of $y = -x^2$, as indicated in Table 11. We can draw the graph of $y = -x^2$ by reflecting the graph of $y = x^2$ about the x-axis. See Figure 45. |

x	$y = x^2$	$y = -x^2$
-2	4	-4
-1	1	-1
0	0	0
1	1	-1
2	4	-4

TABLE 11

Figure 45

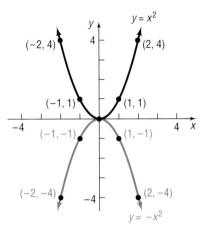

When the right side of the function $y = f(x)$ is multiplied by -1, the graph of the new function $y = -f(x)$ is the **reflection about the *x*-axis** of the graph of the function $y = f(x)$.

NOW WORK PROBLEM **41**.

| EXAMPLE 10 | **Reflection about the *y*-Axis** |

Graph the function: $f(x) = \sqrt{-x}$

Solution   First, notice that the domain of f consists of all real numbers x for which $-x \geq 0$ or, equivalently, $x \leq 0$. To get the graph of $f(x) = \sqrt{-x}$, we begin with the graph of $y = \sqrt{x}$, as shown in Figure 46. For each point (x, y) on the graph of $y = \sqrt{x}$, the point $(-x, y)$ is on the graph of $y = \sqrt{-x}$. We obtain the graph of $y = \sqrt{-x}$ by reflecting the graph of $y = \sqrt{x}$ about the y-axis. See Figure 46.

Figure 46

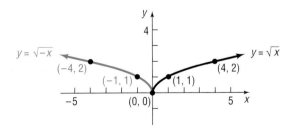

When the graph of the function $y = f(x)$ is known, the graph of the new function $y = f(-x)$ is the **reflection about the *y*-axis** of the graph of the function $y = f(x)$.

SUMMARY

Summary of Graphing Techniques

Table 12 summarizes the graphing procedures that we have just discussed.

TABLE 12		
To Graph:	**Draw the Graph of f and:**	**Functional Change to f(x)**
Vertical shifts $y = f(x) + k, \quad k > 0$ $y = f(x) - k, \quad k > 0$	Raise the graph of f by k units. Lower the graph of f by k units.	Add k to $f(x)$. Subtract k from $f(x)$.
Horizontal shifts $y = f(x + h), \quad h > 0$ $y = f(x - h), \quad h > 0$	Shift the graph of f to the left h units. Shift the graph of f to the right h units.	Replace x by $x + h$. Replace x by $x - h$.
Compressing or stretching $y = af(x), \quad a > 0$	Multiply each y-coordinate of $y = f(x)$ by a. Stretch the graph of f vertically if $a > 1$. Compress the graph of f vertically if $0 < a < 1$.	Multiply $f(x)$ by a.
$y = f(ax), \quad a > 0$	Multiply each x-coordinate of $y = f(x)$ by $\frac{1}{a}$. Stretch the graph of f horizontally if $0 < a < 1$. Compress the graph of f horizontally if $a > 1$.	Replace x by ax.
Reflection about the x-axis $y = -f(x)$	Reflect the graph of f about the x-axis.	Multiply $f(x)$ by -1.
Reflection about the y-axis $y = f(-x)$	Reflect the graph of f about the y-axis.	Replace x by $-x$.

The examples that follow combine some of the procedures outlined in this section to get the required graph.

EXAMPLE 11 **Determining the Function Obtained from a Series of Transformations**

Find the function that is finally graphed after the following three transformations are applied to the graph of $y = |x|$.

1. Shift left 2 units.   2. Shift up 3 units.   3. Reflect about the y-axis.

Solution
1. Shift left 2 units: Replace x by $x + 2$. $y = |x + 2|$
2. Shift up 3 units: Add 3. $y = |x + 2| + 3$
3. Reflect about the y-axis: Replace x by $-x$. $y = |-x + 2| + 3$ ∎

NOW WORK PROBLEM 21.

EXAMPLE 12 **Combining Graphing Procedures**

Graph the function: $f(x) = \dfrac{3}{x - 2} + 1$

Solution   We use the following steps to obtain the graph of f:

STEP 1: $y = \dfrac{1}{x}$ Reciprocal function.

STEP 2: $y = \dfrac{3}{x}$ Multiply by 3; vertical stretch of the graph of $y = \dfrac{1}{x}$ by a factor of 3.

STEP 3: $y = \dfrac{3}{x - 2}$ Replace x by $x - 2$; horizontal shift to the right 2 units.

STEP 4: $y = \dfrac{3}{x - 2} + 1$ Add 1; vertical shift up 1 unit.

See Figure 47.

Figure 47

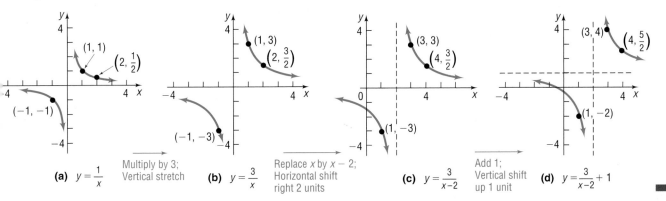

(a) $y = \dfrac{1}{x}$ — Multiply by 3; Vertical stretch

(b) $y = \dfrac{3}{x}$ — Replace x by $x - 2$; Horizontal shift right 2 units

(c) $y = \dfrac{3}{x-2}$ — Add 1; Vertical shift up 1 unit

(d) $y = \dfrac{3}{x-2} + 1$

Other orderings of the steps shown in Example 12 would also result in the graph of f. For example, try this one:

STEP 1: $y = \dfrac{1}{x}$ Reciprocal function

STEP 2: $y = \dfrac{1}{x - 2}$ Replace x by $x - 2$; horizontal shift to the right 2 units.

STEP 3: $y = \dfrac{3}{x - 2}$ Multiply by 3; vertical stretch of the graph of $y = \dfrac{1}{x - 2}$ by factor of 3.

STEP 4: $y = \dfrac{3}{x - 2} + 1$ Add 1; vertical shift up 1 unit.

NOW WORK PROBLEM **47.**

EXAMPLE 13   **Combining Graphing Procedures**

Graph the function: $f(x) = \sqrt{1 - x} + 2$

Solution   We use the following steps to get the graph of $y = \sqrt{1 - x} + 2$:

STEP 1: $y = \sqrt{x}$ Square root function

STEP 2: $y = \sqrt{x + 1}$ Replace x by $x + 1$; horizontal shift left 1 unit.

STEP 3: $y = \sqrt{-x + 1} = \sqrt{1 - x}$ Replace x by $-x$; reflect about y-axis.

STEP 4: $y = \sqrt{1 - x} + 2$ Add 2; vertical shift up 2 units.

See Figure 48.

Figure 48

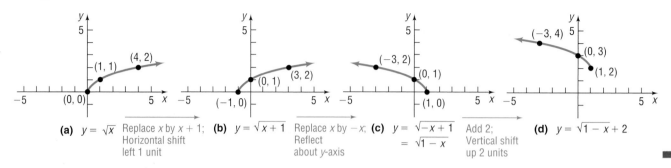

(a) $y = \sqrt{x}$ Replace x by $x + 1$;   **(b)** $y = \sqrt{x + 1}$ Replace x by $-x$;   **(c)** $y = \sqrt{-x + 1}$ Add 2;   **(d)** $y = \sqrt{1 - x} + 2$
           Horizontal shift   Reflect $= \sqrt{1 - x}$ Vertical shift
           left 1 unit   about y-axis   up 2 units

3.4 EXERCISES

In Problems 1–12, match each graph to one of the following functions.

A. $y = x^2 + 2$ B. $y = -x^2 + 2$ C. $y = |x| + 2$ D. $y = -|x| + 2$
E. $y = (x - 2)^2$ F. $y = -(x + 2)^2$ G. $y = |x - 2|$ H. $y = -|x + 2|$
I. $y = 2x^2$ J. $y = -2x^2$ K. $y = 2|x|$ L. $y = -2|x|$

1.

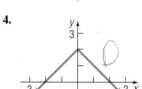

2.

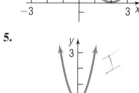

3.

4.

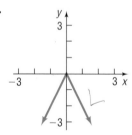

5.

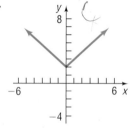

6.

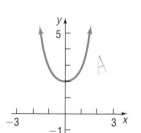

7.

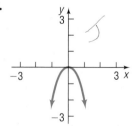

8.

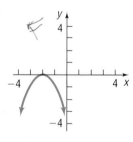

9.

10.

11.

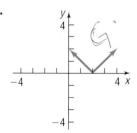

12.

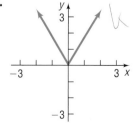

In Problems 13–20, write the function whose graph is the graph of $y = x^3$, but is:

13. Shifted to the right 4 units

14. Shifted to the left 4 units

15. Shifted up 4 units

16. Shifted down 4 units

17. Reflected about the y-axis

18. Reflected about the x-axis

19. Vertically stretched by a factor of 4

20. Horizontally stretched by a factor of 4

In Problems 21–24, find the function that is finally graphed after the following transformations are applied to the graph of $y = \sqrt{x}$.

21. (1) Shift up 2 units
    (2) Reflect about the x-axis
    (3) Reflect about the y-axis

22. (1) Reflect about the x-axis
    (2) Shift right 3 units
    (3) Shift down 2 units

23. (1) Reflect about the x-axis
    (2) Shift up 2 units
    (3) Shift left 3 units

24. (1) Shift up 2 units
    (2) Reflect about the y-axis
    (3) Shift left 3 units

25. If $(3, 0)$ is a point on the graph of $y = f(x)$, which of the following must be on the graph of $y = -f(x)$?
    (a) $(0, 3)$ (b) $(0, -3)$
    (c) $(3, 0)$ (d) $(-3, 0)$

26. If $(3, 0)$ is a point on the graph of $y = f(x)$, which of the following must be on the graph of $y = f(-x)$?
    (a) $(0, 3)$ (b) $(0, -3)$
    (c) $(3, 0)$ (d) $(-3, 0)$

27. If $(0, 3)$ is a point on the graph of $y = f(x)$, which of the following must be on the graph of $y = 2f(x)$?
    (a) $(0, 3)$ (b) $(0, 2)$
    (c) $(0, 6)$ (d) $(6, 0)$

28. If $(3, 0)$ is a point on the graph of $y = f(x)$, which of the following must be on the graph of $y = \frac{1}{2} f(x)$?
    (a) $(3, 0)$ (b) $(3/2, 0)$
    (c) $(0, 3/2)$ (d) $(1/2, 0)$

In Problems 29–58, graph each function using the techniques of shifting, compressing, stretching, and/or reflecting. Start with the graph of the basic function (for example, $y = x^2$) and show all stages.

29. $f(x) = x^2 - 1$

30. $f(x) = x^2 + 4$

31. $g(x) = x^3 + 1$

32. $g(x) = x^3 - 1$

33. $h(x) = \sqrt{x - 2}$

34. $h(x) = \sqrt{x + 1}$

35. $f(x) = (x - 1)^3 + 2$

36. $f(x) = (x + 2)^3 - 3$

37. $g(x) = 4\sqrt{x}$

38. $g(x) = \frac{1}{2}\sqrt{x}$

39. $h(x) = \dfrac{1}{2x}$

40. $h(x) = \dfrac{4}{x}$

41. $f(x) = -|x|$

42. $f(x) = -\sqrt{x}$

43. $g(x) = |-x|$

44. $g(x) = (-x)^3$

45. $h(x) = -x^3 + 2$

46. $h(x) = \dfrac{1}{-x}$

47. $f(x) = 2(x + 1)^2 - 3$

48. $f(x) = 3(x - 2)^2 + 1$

49. $g(x) = \sqrt{x - 2} + 1$

50. $g(x) = |x + 1| - 3$

51. $h(x) = \sqrt{-x} - 2$

52. $h(x) = \dfrac{4}{x} + 2$

53. $f(x) = -(x + 1)^3 - 1$

54. $f(x) = -4\sqrt{x - 1}$

55. $g(x) = 2|1 - x|$

56. $g(x) = 4\sqrt{2 - x}$

57. $h(x) = 2\,\text{int}(x - 1)$

58. $h(x) = \text{int}(-x)$

In Problems 59–64, the graph of a function f is illustrated. Use the graph of f as the first step toward graphing each of the following functions:

(a) $F(x) = f(x) + 3$ (b) $G(x) = f(x + 2)$ (c) $P(x) = -f(x)$ (d) $H(x) = f(x + 1) - 2$
(e) $Q(x) = \frac{1}{2}f(x)$ (f) $g(x) = f(-x)$ (g) $h(x) = f(2x)$

59.

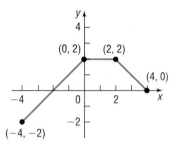

60.

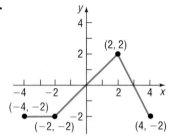

61.

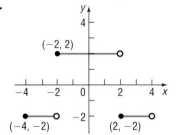

62.

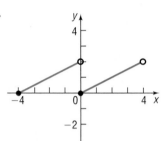

63.

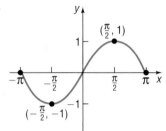

64.

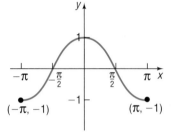

65. Exploration
(a) Use a graphing utility to graph $y = x + 1$ and $y = |x + 1|$.
(b) Graph $y = 4 - x^2$ and $y = |4 - x^2|$.
(c) Graph $y = x^3 + x$ and $y = |x^3 + x|$.
(d) What do you conclude about the relationship between the graphs of $y = f(x)$ and $y = |f(x)|$?

66. Exploration
(a) Use a graphing utility to graph $y = x + 1$ and $y = |x| + 1$.
(b) Graph $y = 4 - x^2$ and $y = 4 - |x|^2$.
(c) Graph $y = x^3 + x$ and $y = |x|^3 + |x|$.
(d) What do you conclude about the relationship between the graphs of $y = f(x)$ and $y = f(|x|)$?

67. The graph of a function f is illustrated in the figure.
(a) Draw the graph of $y = |f(x)|$.
(b) Draw the graph of $y = f(|x|)$.

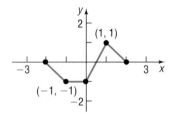

68. The graph of a function f is illustrated in the figure.
(a) Draw the graph of $y = |f(x)|$.
(b) Draw the graph of $y = f(|x|)$.

In Problems 69–74, complete the square of each quadratic expression. Then graph each function using the technique of shifting. (If necessary, refer to Section 1.3 to review completing the square.)

69. $f(x) = x^2 + 2x$

70. $f(x) = x^2 - 6x$

71. $f(x) = x^2 - 8x + 1$

72. $f(x) = x^2 + 4x + 2$

73. $f(x) = x^2 + x + 1$

74. $f(x) = x^2 - x + 1$

75. The equation $y = (x - c)^2$ defines a *family of parabolas*, one parabola for each value of c. On one set of coordinate axes, graph the members of the family for $c = 0$, $c = 3$, and $c = -2$.

76. Repeat Problem 75 for the family of parabolas $y = x^2 + c$.

77. **Temperature Measurements**   The relationship between the Celsius (°C) and Fahrenheit (°F) scales for measuring temperature is given by the equation

$$F = \frac{9}{5}C + 32$$

The relationship between the Celsius (°C) and Kelvin (K) scales is $K = C + 273$. Graph the equation $F = \frac{9}{5}C + 32$ using degrees Fahrenheit on the y-axis and degrees Celsius on the x-axis. Use the techniques introduced in this section to obtain the graph showing the relationship between Kelvin and Fahrenheit temperatures.

78. **Period of a Pendulum**   The period T (in seconds) of a simple pendulum is a function of its length l (in feet) defined by the equation

$$T = 2\pi\sqrt{\frac{l}{g}}$$

where $g \approx 32.2$ feet per second per second is the acceleration of gravity.

(a) Use a graphing utility to graph the function $T = T(l)$.

(b) Now graph the functions $T = T(l + 1), T = T(l + 2)$, and $T = T(l + 3)$.

(c) Discuss how adding to the length l changes the period T.

(d) Now graph the functions $T = T(2l), T = T(3l)$, and $T = T(4l)$.

(e) Discuss how multiplying the length l by factors of 2, 3, and 4 changes the period T.

79. **Cigar Company Profits**   The daily profits of a cigar company from selling x cigars are given by

$$p(x) = -0.05x^2 + 100x - 2000$$

The government wishes to impose a tax on cigars (sometimes called a *sin tax*) that gives the company the option of either paying a flat tax of $10,000 per day or a tax of 10% on profits. As chief financial officer of the company, you need to decide which tax is the better option for the company.

(a) On the same screen, graph $Y_1 = p(x) - 10{,}000$ and $Y_2 = (1 - 0.10)p(x)$.

(b) Based on the graph, which option would you select? Why?

(c) Using the terminology learned in this section, describe each graph in terms of the graph of $p(x)$.

(d) Suppose that the government offered the options of a flat tax of $4800 or a tax of 10% on profits. Which would you select? Why?

3.5 OPERATIONS ON FUNCTIONS; COMPOSITE FUNCTIONS

OBJECTIVES   **1**   Form the Sum, Difference, Product, and Quotient of Two Functions

2   Form the Composite Function and Find Its Domain

1   In this section, we introduce some operations on functions. We shall see that functions, like numbers, can be added, subtracted, multiplied, and divided. For example, if $f(x) = x^2 + 9$ and $g(x) = 3x + 5$, then

$$f(x) + g(x) = (x^2 + 9) + (3x + 5) = x^2 + 3x + 14$$

The new function $y = x^2 + 3x + 14$ is called the *sum function $f + g$*. Similarly,

$$f(x) \cdot g(x) = (x^2 + 9)(3x + 5) = 3x^3 + 5x^2 + 27x + 45$$

The new function $y = 3x^3 + 5x^2 + 27x + 45$ is called the *product function $f \cdot g$*.

The general definitions are given next.

If f and g are functions:

The **sum** $f + g$ is the function defined by

$$(f + g)(x) = f(x) + g(x)$$

The domain of $f + g$ consists of the numbers x that are in the domains of both f and g.

The **difference** $f - g$ is the function defined by

$$(f - g)(x) = f(x) - g(x)$$

The domain of $f - g$ consists of the numbers x that are in the domains of both f and g.

The **product** $f \cdot g$ is the function defined by

$$(f \cdot g)(x) = f(x) \cdot g(x)$$

The domain of $f \cdot g$ consists of the numbers x that are in the domains of both f and g.

The **quotient** f/g is the function defined by

$$\left(\frac{f}{g}\right)(x) = \frac{f(x)}{g(x)}, \qquad g(x) \neq 0$$

The domain of f/g consists of the numbers x for which $g(x) \neq 0$ that are in the domains of both f and g.

EXAMPLE 1   Operations on Functions

Let f and g be two functions defined as

$$f(x) = \sqrt{x + 2} \quad \text{and} \quad g(x) = \sqrt{3 - x}$$

Find the following, and determine the domain in each case.

(a) $(f + g)(x)$ (b) $(f - g)(x)$ (c) $(f \cdot g)(x)$ (d) $(f/g)(x)$

Solution (a) $(f + g)(x) = f(x) + g(x) = \sqrt{x + 2} + \sqrt{3 - x}$

(b) $(f - g)(x) = f(x) - g(x) = \sqrt{x + 2} - \sqrt{3 - x}$

(c) $(f \cdot g)(x) = f(x) \cdot g(x) = (\sqrt{x + 2})(\sqrt{3 - x}) = \sqrt{(x + 2)(3 - x)}$

(d) $\left(\frac{f}{g}\right)(x) = \frac{f(x)}{g(x)} = \frac{\sqrt{x + 2}}{\sqrt{3 - x}} = \frac{\sqrt{(x + 2)(3 - x)}}{3 - x}$

The domain of f consists of all numbers x for which $x \geq -2$; the domain of g consists of all numbers x for which $x \leq 3$. The numbers x common to both these domains are those for which $-2 \leq x \leq 3$. As a result, the numbers x for which $-2 \leq x \leq 3$, or the interval $[-2, 3]$ comprise the domain of the sum function $f + g$, the difference function $f - g$, and the product function $f \cdot g$. For the quotient function f/g, we must exclude from this set the number 3, because the denominator, g, has the value 0 when $x = 3$. The domain of f/g consists of all x for which $-2 \leq x < 3$, or the interval $[-2, 3)$. ▬

NOW WORK PROBLEM **1.**

   In calculus, it is sometimes helpful to view a complicated function as the sum, difference, product, or quotient of simpler functions. For example,

$$F(x) = x^2 + \sqrt{x} \text{ is the sum of } f(x) = x^2 \text{ and } g(x) = \sqrt{x}.$$
$$H(x) = (x^2 - 1)/(x^2 + 1) \text{ is the quotient of } f(x) = x^2 - 1 \text{ and}$$
$$g(x) = x^2 + 1.$$

COMPOSITE FUNCTIONS

Consider the function $y = (2x + 3)^2$. If we write $y = f(u) = u^2$ and $u = g(x) = 2x + 3$, then, by a substitution process, we can obtain the original function: $y = f(u) = f(g(x)) = (2x + 3)^2$. This process is called **composition.**

In general, suppose that f and g are two functions and that x is a number in the domain of g. By evaluating g at x, we get $g(x)$. If $g(x)$ is in the domain of f, then we may evaluate f at $g(x)$ and thereby obtain the expression $f(g(x))$. The correspondence from x to $f(g(x))$ is called a *composite function $f \circ g$.*

> Given two functions f and g, the **composite function,** denoted by $f \circ g$ (read as "f composed with g"), is defined by
>
> $$(f \circ g)(x) = f(g(x))$$
>
> The domain of $f \circ g$ is the set of all numbers x in the domain of g such that $g(x)$ is in the domain of f.

Look carefully at Figure 49. Only those x's in the domain of g for which $g(x)$ is in the domain of f can be in the domain of $f \circ g$. The reason is that if $g(x)$ is not in the domain of f then $f(g(x))$ is not defined. Because of this, the domain of $f \circ g$ is a subset of the domain of g; the range of $f \circ g$ is a subset of the range of f.

Figure 49

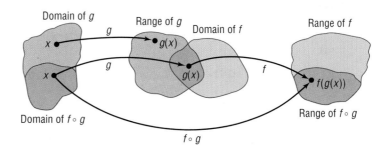

Figure 50 provides a second illustration of the definition. Notice that the "inside" function g in $f(g(x))$ is done first.

Figure 50

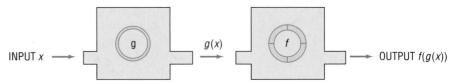

Let's look at some examples.

EXAMPLE 2 Evaluating a Composite Function

Suppose that $f(x) = 2x^2 - 3$ and $g(x) = 4x$. Find:

(a) $(f \circ g)(1)$ (b) $(g \circ f)(1)$ (c) $(f \circ f)(-2)$ (d) $(g \circ g)(-1)$

Solution   (a) $(f \circ g)(1) = f(g(1)) = f(4) = 2 \cdot 16 - 3 = 29$

$$g(x) = 4x \qquad f(x) = 2x^2 - 3$$
$$g(1) = 4$$

(b) $(g \circ f)(1) = g(f(1)) = g(-1) = 4 \cdot (-1) = -4$

$$f(x) = 2x^2 - 3 \qquad g(x) = 4x$$
$$f(1) = -1$$

(c) $(f \circ f)(-2) = f(f(-2)) = f(5) = 2 \cdot 25 - 3 = 47$

$$f(-2) = 5$$

(d) $(g \circ g)(-1) = g(g(-1)) = g(-4) = 4 \cdot (-4) = -16$

$$g(-1) = -4$$

NOW WORK PROBLEM **13**.

Look back at Figure 49. In determining the domain of the composite function $(f \circ g)(x) = f(g(x))$, keep the following two thoughts in mind about the input x.

1. $g(x)$ must be defined so that any x not in the domain of g must be excluded.
2. $f(g(x))$ must be defined so that any x for which $g(x)$ is not in the domain of f must be excluded.

EXAMPLE 3 Finding the Domain of $f \circ g$

Find the domain of $(f \circ g)(x)$ if $f(x) = \dfrac{1}{x + 2}$ and $g(x) = \dfrac{4}{x - 1}$.

Solution   For $(f \circ g)(x) = f(g(x))$, we first note that the domain of g is $\{x \mid x \neq 1\}$, so we exclude 1 from the domain of $f \circ g$. Next, we note that the domain of f is $\{x \mid x \neq -2\}$. This means $g(x)$ cannot equal -2, so we solve the equation

$$\frac{4}{x-1} = -2 \qquad \qquad g(x) = -2$$

$$4 = -2(x-1)$$

$$4 = -2x + 2$$

$$2x = -2$$

$$x = -1$$

We also exclude -1 from the domain of $f \circ g$. The domain of $f \circ g$ is $\{x \,|\, x \neq -1, x \neq 1\}$.

Check: For $x = 1, g(x) = \dfrac{4}{x-1}$ is not defined, so $(f \circ g)(x) = f(g(x))$ is not defined.
      For $x = -1, g(-1) = 4/-2 = -2$, and $(f \circ g)(-1) = f(g(-1)) = f(-2)$ is not defined.   ▪ ■

✏━━━━━▶   **N O W   W O R K   P R O B L E M   2 3 .**

EXAMPLE 4   **Finding a Composite Function**

Suppose that $f(x) = \dfrac{1}{x+2}$ and $g(x) = \dfrac{4}{x-1}$. Find the following composite functions, and then find the domain of each composite function.

(a) $f \circ g$ (b) $g \circ f$ (c) $f \circ f$ (d) $g \circ g$

Solution   The domain of f is $\{x \,|\, x \neq -2\}$ and the domain of g is $\{x \,|\, x \neq 1\}$.

(a) $(f \circ g)(x) = f(g(x)) = f\left(\dfrac{4}{x-1}\right) = \dfrac{1}{\dfrac{4}{x-1} + 2} = \dfrac{x-1}{4 + 2(x-1)} = \dfrac{x-1}{2x+2}$

$$\uparrow \qquad\qquad\qquad\qquad\qquad \uparrow$$
$$f(x) = \dfrac{1}{x+2} \qquad \text{Multiply by } \dfrac{x-1}{x-1}.$$

The domain of $f \circ g$ consists of those x in the domain of $g(x \neq 1)$ for which $g(x) = 4/(x-1) \neq -2$ or, equivalently, $x \neq -1$. Thus, the domain of $f \circ g$ is $\{x \,|\, x \neq -1, x \neq 1\}$. (See Example 3.)

(b) $(g \circ f)(x) = g(f(x)) = g\left(\dfrac{1}{x+2}\right) = \dfrac{4}{\dfrac{1}{x+2} - 1} = \dfrac{4(x+2)}{1 - (x+2)} = \dfrac{4(x+2)}{-x-1}$

$$\uparrow$$
$$g(x) = \dfrac{4}{x-1}$$

The domain of $g \circ f$ consists of those x in the domain of $f(x \neq -2)$ for which

$$f(x) = \dfrac{1}{x+2} \neq 1 \qquad \dfrac{1}{x+2} = 1$$
$$1 = x + 2$$
or, equivalently, $x = -1$

$$x \neq -1$$

The domain of $g \circ f$ is $\{x \,|\, x \neq -1, x \neq -2\}$.

(c) $(f \circ f)(x) = f(f(x)) = f\left(\dfrac{1}{x+2}\right) = \dfrac{1}{\dfrac{1}{x+2}+2} = \dfrac{x+2}{1+2(x+2)} = \dfrac{x+2}{2x+5}$

$\uparrow$

$f(x) = \dfrac{1}{x+2}$

The domain of $f \circ f$ consists of those x in the domain of $f (x \neq -2)$ for which

$$f(x) = \frac{1}{x+2} \neq -2 \qquad \frac{1}{x+2} = -2$$
$$1 = -2(x+2)$$
$$1 = -2x - 4$$
$$2x = -5$$
$$x = -\tfrac{5}{2}$$

or, equivalently,

$$x \neq -\tfrac{5}{2}$$

The domain of $f \circ f$ is $\left\{x \mid x \neq -\tfrac{5}{2},\, x \neq -2\right\}$.

(d) $(g \circ g)(x) = g(g(x)) = g\left(\dfrac{4}{x-1}\right) = \dfrac{4}{\dfrac{4}{x-1}-1} = \dfrac{4(x-1)}{4-(x-1)} = \dfrac{4(x-1)}{-x+5}$

$\uparrow$

$g(x) = \dfrac{4}{x-1}$

The domain of $g \circ g$ consists of those x in the domain of $g (x \neq 1)$ for which

$$g(x) = \frac{4}{x-1} \neq 1 \qquad \frac{4}{x-1} = 1$$
$$4 = x - 1$$
$$x = 5$$

or, equivalently,

$$x \neq 5$$

The domain of $g \circ g$ is $\{x \mid x \neq 1,\, x \neq 5\}$. ■

NOW WORK PROBLEMS 35 AND 37.

Examples 4(a) and 4(b) illustrate that, in general, $f \circ g \neq g \circ f$. However, sometimes $f \circ g$ does equal $g \circ f$, as shown in the next example.

EXAMPLE 5 — **Showing That Two Composite Functions Are Equal**

If $f(x) = 3x - 4$ and $g(x) = \tfrac{1}{3}(x+4)$, show that
$$(f \circ g)(x) = (g \circ f)(x) = x$$
for every x.

Solution

$(f \circ g)(x) = f(g(x))$

$\qquad = f\left(\dfrac{x+4}{3}\right)$ $g(x) = \dfrac{1}{3}(x+4) = \dfrac{x+4}{3}$.

$\qquad = 3\left(\dfrac{x+4}{3}\right) - 4$ Substitute $g(x)$ into the rule for f, $f(x) = 3x - 4$.

$\qquad = x + 4 - 4 = x$

$$(g \circ f)(x) = g(f(x))$$

$$= g(3x - 4) \qquad f(x) = 3x - 4.$$

$$= \tfrac{1}{3}[(3x - 4) + 4] \quad \text{Substitute } f(x) \text{ into the rule for } g,$$
$$g(x) = \tfrac{1}{3}(x + 4).$$

$$= \tfrac{1}{3}(3x) = x$$

Thus, $(f \circ g)(x) = (g \circ f)(x) = x$. ∎

In Section 6.1, we shall see that there is an important relationship between functions f and g for which $(f \circ g)(x) = (g \circ f)(x) = x$.

 NOW WORK PROBLEM 47.

CALCULUS APPLICATION

Some techniques in calculus require that we be able to determine the components of a composite function. For example, the function $H(x) = \sqrt{x + 1}$ is the composition of the functions f and g, where $f(x) = \sqrt{x}$ and $g(x) = x + 1$, because $H(x) = (f \circ g)(x) = f(g(x)) = f(x + 1) = \sqrt{x + 1}$.

EXAMPLE 6

Finding the Components of a Composite Function

Find functions f and g such that $f \circ g = H$ if $H(x) = (x^2 + 1)^{50}$.

Solution   The function H takes $x^2 + 1$ and raises it to the power 50. A natural way to decompose H is to raise the function $g(x) = x^2 + 1$ to the power 50. If we let $f(x) = x^{50}$ and $g(x) = x^2 + 1$, then

$$(f \circ g)(x) = f(g(x))$$
$$= f(x^2 + 1)$$
$$= (x^2 + 1)^{50} = H(x)$$

See Figure 51.

Figure 51

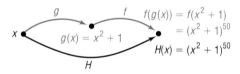

Other functions f and g may be found for which $f \circ g = H$ in Example 6. For example, if $f(x) = x^2$ and $g(x) = (x^2 + 1)^{25}$, then

$$(f \circ g)(x) = f(g(x)) = f((x^2 + 1)^{25}) = [(x^2 + 1)^{25}]^2 = (x^2 + 1)^{50}$$

Although the functions f and g found as a solution to Example 6 are not unique, there is usually a "natural" selection for f and g that comes to mind first.

EXAMPLE 7

Finding the Components of a Composite Function

Find functions f and g such that $f \circ g = H$ if $H(x) = 1/(x + 1)$.

Solution   Here H is the reciprocal of $g(x) = x + 1$. If we let $f(x) = 1/x$ and $g(x) = x + 1$, we find that

$$(f \circ g)(x) = f(g(x)) = f(x + 1) = \frac{1}{x + 1} = H(x)$$ ∎

3.5 EXERCISES

In Problems 1–10, for the given functions f and g, find the following functions and state the domain of each.

(a) $f + g$ (b) $f - g$ (c) $f \cdot g$ (d) f/g

1. $f(x) = 3x + 4$; $g(x) = 2x - 3$
2. $f(x) = 2x + 1$; $g(x) = 3x - 2$
3. $f(x) = x - 1$; $g(x) = 2x^2$
4. $f(x) = 2x^2 + 3$; $g(x) = 4x^3 + 1$
5. $f(x) = \sqrt{x}$; $g(x) = 3x - 5$
6. $f(x) = |x|$; $g(x) = x$
7. $f(x) = 1 + \dfrac{1}{x}$; $g(x) = \dfrac{1}{x}$
8. $f(x) = \sqrt{x - 2}$; $g(x) = \sqrt{4 - x}$
9. $f(x) = \dfrac{2x + 3}{3x - 2}$; $g(x) = \dfrac{4x}{3x - 2}$
10. $f(x) = \sqrt{x + 1}$; $g(x) = \dfrac{2}{x}$
11. Given $f(x) = 3x + 1$ and $(f + g)(x) = 6 - \frac{1}{2}x$, find the function g.
12. Given $f(x) = 1/x$ and $(f/g)(x) = (x + 1)/(x^2 - x)$, find the function g.

In Problems 13–22, for the given functions f and g, find

(a) $(f \circ g)(4)$ (b) $(g \circ f)(2)$ (c) $(f \circ f)(1)$ (d) $(g \circ g)(0)$

13. $f(x) = 2x$; $g(x) = 3x^2 + 1$
14. $f(x) = 3x + 2$; $g(x) = 2x^2 - 1$
15. $f(x) = 4x^2 - 3$; $g(x) = 3 - \frac{1}{2}x^2$
16. $f(x) = 2x^2$; $g(x) = 1 - 3x^2$
17. $f(x) = \sqrt{x}$; $g(x) = 2x$
18. $f(x) = \sqrt{x + 1}$; $g(x) = 3x$
19. $f(x) = |x|$; $g(x) = \dfrac{1}{x^2 + 1}$
20. $f(x) = |x - 2|$; $g(x) = \dfrac{3}{x^2 + 2}$
21. $f(x) = \dfrac{3}{x + 1}$; $g(x) = \sqrt[3]{x}$
22. $f(x) = x^{3/2}$; $g(x) = \dfrac{2}{x + 1}$

In Problems 23–30, find the domain of the composite function $f \circ g$.

23. $f(x) = \dfrac{3}{x - 1}$; $g(x) = \dfrac{2}{x}$
24. $f(x) = \dfrac{1}{x + 3}$; $g(x) = \dfrac{-2}{x}$
25. $f(x) = \dfrac{x}{x - 1}$; $g(x) = \dfrac{-4}{x}$
26. $f(x) = \dfrac{x}{x + 3}$; $g(x) = \dfrac{2}{x}$
27. $f(x) = \sqrt{x}$; $g(x) = 2x + 3$
28. $f(x) = x - 2$; $g(x) = \sqrt{1 - x}$
29. $f(x) = x^2 + 1$; $g(x) = \sqrt{x - 1}$
30. $f(x) = x^2 + 4$; $g(x) = \sqrt{x - 2}$

In Problems 31–46, for the given functions f and g, find:

(a) $f \circ g$ (b) $g \circ f$ (c) $f \circ f$ (d) $g \circ g$

State the domain of each composite function.

31. $f(x) = 2x + 3$; $g(x) = 3x$
32. $f(x) = -x$; $g(x) = 2x - 4$
33. $f(x) = 3x + 1$; $g(x) = x^2$
34. $f(x) = x + 1$; $g(x) = x^2 + 4$
35. $f(x) = x^2$; $g(x) = x^2 + 4$
36. $f(x) = x^2 + 1$; $g(x) = 2x^2 + 3$
37. $f(x) = \dfrac{3}{x - 1}$; $g(x) = \dfrac{2}{x}$
38. $f(x) = \dfrac{1}{x + 3}$; $g(x) = \dfrac{-2}{x}$
39. $f(x) = \dfrac{x}{x - 1}$; $g(x) = \dfrac{-4}{x}$
40. $f(x) = \dfrac{x}{x + 3}$; $g(x) = \dfrac{2}{x}$
41. $f(x) = \sqrt{x}$; $g(x) = 2x + 3$
42. $f(x) = \sqrt{x - 2}$; $g(x) = 1 - 2x$
43. $f(x) = x^2 + 1$; $g(x) = \sqrt{x - 1}$
44. $f(x) = x^2 + 4$; $g(x) = \sqrt{x - 2}$
45. $f(x) = ax + b$; $g(x) = cx + d$
46. $f(x) = \dfrac{ax + b}{cx + d}$; $g(x) = mx$

In Problems 47–54, show that $(f \circ g)(x) = (g \circ f)(x) = x.$

47. $f(x) = 2x; \quad g(x) = \frac{1}{2}x$

48. $f(x) = 4x; \quad g(x) = \frac{1}{4}x$

49. $f(x) = x^3; \quad g(x) = \sqrt[3]{x}$

50. $f(x) = x + 5; \quad g(x) = x - 5$

51. $f(x) = 2x - 6; \quad g(x) = \frac{1}{2}(x + 6)$

52. $f(x) = 4 - 3x; \quad g(x) = \frac{1}{3}(4 - x)$

53. $f(x) = ax + b; \quad g(x) = \frac{1}{a}(x - b), \quad a \neq 0$

54. $f(x) = \frac{1}{x}; \quad g(x) = \frac{1}{x}$

In Problems 55–60, find functions f and g so that f $\circ$ g = H.

55. $H(x) = (2x + 3)^4$

56. $H(x) = (1 + x^2)^3$

57. $H(x) = \sqrt{x^2 + 1}$

58. $H(x) = \sqrt{1 - x^2}$

59. $H(x) = |2x + 1|$

60. $H(x) = |2x^2 + 3|$

---

61. If $f(x) = 2x^3 - 3x^2 + 4x - 1$ and $g(x) = 2$, find $(f \circ g)(x)$ and $(g \circ f)(x)$.

62. If $f(x) = x/(x - 1)$, find $(f \circ f)(x)$.

63. If $f(x) = 2x^2 + 5$ and $g(x) = 3x + a$, find a so that the graph of $f \circ g$ crosses the y-axis at 23.

64. If $f(x) = 3x^2 - 7$ and $g(x) = 2x + a$, find a so that the graph of $f \circ g$ crosses the y-axis at 68.

65. Surface Area of a Balloon   The surface area S (in square meters) of a hot-air balloon is given by

$$S(r) = 4\pi r^2$$

where r is the radius of the balloon (in meters). If the radius r is increasing with time t (in seconds) according to the formula $r(t) = \frac{2}{3}t^3, t \geq 0$, find the surface area S of the balloon as a function of the time t.

66. Volume of a Balloon   The volume V (in cubic meters) of the hot-air balloon described in Problem 65 is given by $V(r) = \frac{4}{3}\pi r^3$. If the radius r is the same function of t as in Problem 65, find the volume V as a function of the time t.

67. Automobile Production   The number N of cars produced at a certain factory in 1 day after t hours of operation is given by $N(t) = 100t - 5t^2, 0 \leq t \leq 10$. If the cost C (in dollars) of producing N cars is $C(N) = 15{,}000 + 8000N$, find the cost C as a function of the time t of operation of the factory.

68. Environmental Concerns   The spread of oil leaking from a tanker is in the shape of a circle. If the radius r (in feet) of the spread after t hours is $r(t) = 200\sqrt{t}$, find the area A of the oil slick as a function of the time t.

69. Production Cost   The price p of a certain product and the quantity x sold obey the demand equation

$$p = -\frac{1}{4}x + 100, \qquad 0 \leq x \leq 400$$

Suppose that the cost C of producing x units is

$$C = \frac{\sqrt{x}}{25} + 600$$

Assuming that all items produced are sold, find the cost C as a function of the price p.
[**Hint:** Solve for x in the demand equation and then form the composite.]

70. Cost of a Commodity   The price p of a certain commodity and the quantity x sold obey the demand equation

$$p = -\frac{1}{5}x + 200, \qquad 0 \leq x \leq 1000$$

Suppose that the cost C of producing x units is

$$C = \frac{\sqrt{x}}{10} + 400$$

Assuming that all items produced are sold, find the cost C as a function of the price p.

71. Volume of a Cylinder   The volume V of a right circular cylinder of height h and radius r is $V = \pi r^2 h$. If the height is twice the radius, express the volume V as a function of r.

72. Volume of a Cone   The volume V of a right circular cone is $V = \frac{1}{3}\pi r^2 h$. If the height is twice the radius, express the volume V as a function of r.

73. If f and g are odd functions, show that the composite function $f \circ g$ is also odd.

74. If f is an odd function and g is an even function, show that the composite functions $f \circ g$ and $g \circ f$ are also even.

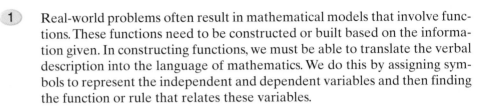

3.6 MATHEMATICAL MODELS: CONSTRUCTING FUNCTIONS

OBJECTIVE ① Construct and Analyze Functions

① Real-world problems often result in mathematical models that involve functions. These functions need to be constructed or built based on the information given. In constructing functions, we must be able to translate the verbal description into the language of mathematics. We do this by assigning symbols to represent the independent and dependent variables and then finding the function or rule that relates these variables.

EXAMPLE 1   **Area of a Rectangle with Fixed Perimeter**

The perimeter of a rectangle is 50 feet. Express its area A as a function of the length x of a side.

Solution   Consult Figure 52. If the length of the rectangle is x and if w is its width, then the sum of the lengths of the sides is the perimeter, 50.

Figure 52

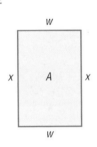

$$x + w + x + w = 50$$
$$2x + 2w = 50$$
$$x + w = 25$$
$$w = 25 - x$$

The area A is length times width, so

$$A = xw = x(25 - x)$$

The area A as a function of x is

$$A(x) = x(25 - x)$$ ∎

Note that we use the symbol A as the dependent variable and also as the name of the function that relates the length x to the area A. As we mentioned earlier, this double usage is common in applications and should cause no difficulties.

EXAMPLE 2   **Economics: Demand Equations**

In economics, revenue R is defined as the amount of money received from the sale of a product and is equal to the unit selling price p of the product times the number x of units actually sold. That is,

$$R = xp$$

In economics, the Law of Demand states that p and x are related: As one increases, the other decreases. Suppose that p and x are related by the following **demand equation:**

$$p = -\tfrac{1}{10}x + 20, \qquad 0 \le x \le 200$$

Express the revenue R as a function of the number x of units sold.

Solution   Since $R = xp$ and $p = -\frac{1}{10}x + 20$, it follows that

$$R(x) = xp = x\left(-\tfrac{1}{10}x + 20\right) = -\tfrac{1}{10}x^2 + 20x$$ ∎

NOW WORK PROBLEM **3**.

EXAMPLE 3

Finding the Distance from the Origin to a Point on a Graph

Let $P = (x, y)$ be a point on the graph of $y = x^2 - 1$.

(a) Express the distance d from P to the origin O as a function of x.

(b) What is d if $x = 0$?

(c) What is d if $x = 1$?

(d) What is d if $x = \sqrt{2}/2$?

(e) Use a graphing utility to graph the function $d = d(x)$, $x \geq 0$. Rounded to two decimal places, find the value(s) of x at which d has a local minimum. [This gives the point(s) on the graph of $y = x^2 - 1$ closest to the origin.]

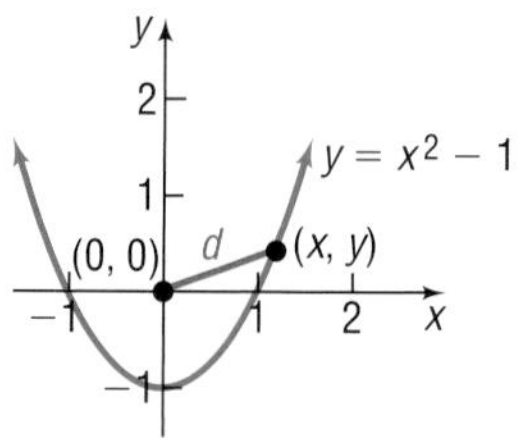

Solution   (a) Figure 53 illustrates the graph. The distance d from P to O is

Figure 53

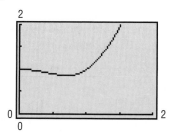

$$d = \sqrt{(x - 0)^2 + (y - 0)^2} = \sqrt{x^2 + y^2}$$

Since P is a point on the graph of $y = x^2 - 1$, we have

$$d(x) = \sqrt{x^2 + (x^2 - 1)^2} = \sqrt{x^4 - x^2 + 1}$$

We have expressed the distance d as a function of x.

(b) If $x = 0$, the distance d is

$$d(0) = \sqrt{1} = 1$$

(c) If $x = 1$, the distance d is

$$d(1) = \sqrt{1 - 1 + 1} = 1$$

Figure 54

(d) If $x = \sqrt{2}/2$, the distance d is

$$d\left(\frac{\sqrt{2}}{2}\right) = \sqrt{\left(\frac{\sqrt{2}}{2}\right)^4 - \left(\frac{\sqrt{2}}{2}\right)^2 + 1} = \sqrt{\frac{1}{4} - \frac{1}{2} + 1} = \frac{\sqrt{3}}{2}$$

(e) Figure 54 shows the graph of $Y_1 = \sqrt{x^4 - x^2 + 1}$. Using the MINIMUM feature on a graphing utility, we find that when $x \approx 0.71$ the value of d is smallest ($d \approx 0.87$ rounded to two decimal places is a local minimum). By symmetry, it follows that when $x \approx -0.71$, the value of d is also a local minimum.   ∎

NOW WORK PROBLEM **9**.

EXAMPLE 4

Filling a Swimming Pool

A rectangular swimming pool 20 meters long and 10 meters wide is 4 meters deep at one end and 1 meter deep at the other. Figure 55 illustrates a cross-sectional view of the pool. Water is being pumped into the pool to a height of 3 meters at the deep end.

Figure 55

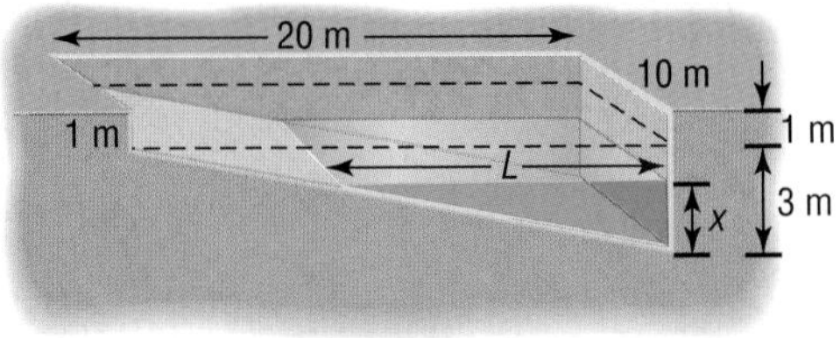

(a) Find a function that expresses the volume V of water in the pool as a function of the height x of the water at the deep end.

(b) Find the volume when the height is 1 meter.

(c) Find the volume when the height is 2 meters.

(d) Use a graphing utility to graph the function $V = V(x)$. At what height is the volume 20 cubic meters? 100 cubic meters?

Solution (a) Let L denote the distance (in meters) measured at water level from the deep end to the short end. Notice that L and x form the sides of a triangle that is similar to the triangle whose sides are 20 meters by 3 meters. Thus, L and x are related by the equation

$$\frac{L}{x} = \frac{20}{3} \quad \text{or} \quad L = \frac{20x}{3}, \qquad 0 \le x \le 3$$

The volume V of water in the pool at any time is

$$V = \left(\begin{array}{c} \text{cross-sectional} \\ \text{triangular area} \end{array}\right)(\text{width}) = (\tfrac{1}{2}Lx)(10) \quad \text{cubic meters}$$

Since $L = 20x/3$, we have

$$V(x) = \left(\frac{1}{2} \cdot \frac{20x}{3} \cdot x\right)(10) = \frac{100}{3}x^2 \quad \text{cubic meters}$$

(b) When the height x of the water is 1 meter, the volume $V = V(x)$ is

$$V(1) = \frac{100}{3} \cdot 1^2 = 33\frac{1}{3} \text{ cubic meters}$$

Figure 56

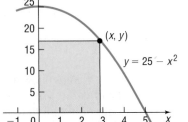

(c) When the height x of the water is 2 meters, the volume $V = V(x)$ is

$$V(2) = \frac{100}{3} \cdot 2^2 = \frac{400}{3} = 133\frac{1}{3} \text{ cubic meters}$$

(d) See Figure 56. Use TRACE. When $x \approx 0.77$ meter, the volume is 20 cubic meters. When $x \approx 1.73$ meters, the volume is 100 cubic meters. ■

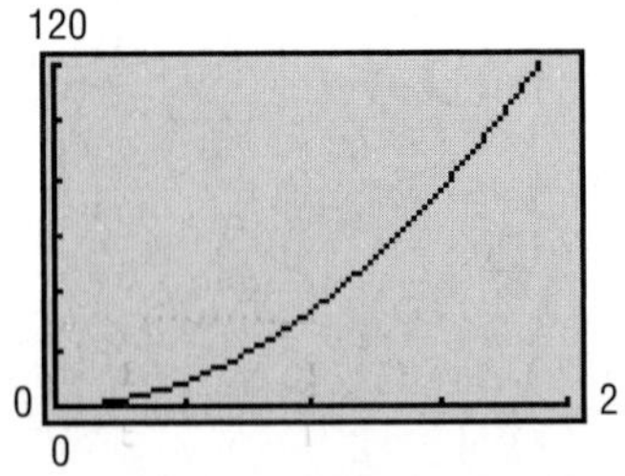

EXAMPLE 5   **Area of a Rectangle**

Figure 57

A rectangle has one corner on the graph of $y = 25 - x^2$, another at the origin, a third on the positive y-axis, and the fourth on the positive x-axis. See Figure 57.

(a) Express the area A of the rectangle as a function of x.

(b) What is the domain of A?

(c) Graph $A = A(x)$.

(d) For what value of x is the area largest?

of installation as a function of the distance x (in miles) from the connection box to the point where the cable installation turns off the road. Give the domain.

(b) Compute the cost if $x = 1$ mile.

(c) Compute the cost if $x = 3$ miles.

(d) Graph the function $C = C(x)$. Use TRACE to see how the cost C varies as x changes from 0 to 5.

(e) What value of x results in the least cost?

36. Time Required to Go from an Island to a Town An island is 2 miles from the nearest point P on a straight shoreline. A town is 12 miles down the shore from P. See the illustration.

(a) If a person can row a boat at an average speed of 3 miles per hour and the same person can walk 5 miles per hour, express the time T that it takes to go from the island to town as a function of the distance x from P to where the person lands the boat.

(b) What is the domain of T?

(c) How long will it take to travel from the island to town if the person lands the boat 4 miles from P?

(d) How long will it take if the person lands the boat 8 miles from P?

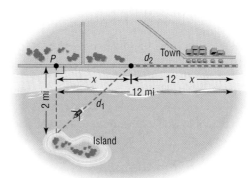

37. Water is poured into a container in the shape of a right circular cone with radius 4 feet and height 16 feet (see the figure). Express the volume V of the water in the cone as a function of the height h of the water.
[**Hint:** The volume V of a cone of radius r and height h is $V = \frac{1}{3}\pi r^2 h$.]

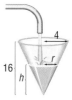

38. Federal Income Tax Two 1999 Tax Rate Schedules are given in the accompanying table. If x equals the amount on Form 1040, line 37, and y equals the tax due, construct a function f for each schedule.

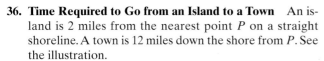

1999 TAX RATE SCHEDULES

SCHEDULE X—IF YOUR FILING STATUS IS SINGLE				SCHEDULE Y-1—USE IF YOUR FILING STATUS IS MARRIED FILING JOINTLY OR QUALIFYING WIDOW(ER)			
If the amount on Form 1040, line 37, is: Over—	But not over—	Enter on Form 1040, line 38	of the amount over—	If the amount on Form 1040, line 37, is: Over—	But not over—	Enter on Form 1040, line 38	of the amount over—
$0	$25,750	_____15%	$0	$0	$43,050	_____15%	$0
25,750	62,450	$3,862.50 + 28%	25,750	43,050	104,050	$6,457.50 + 28%	43,050
62,450	130,250	14,138.50 + 31%	62,450	104,050	158,550	23,537.50 + 31%	104,050
130,250	283,150	35,156.50 + 36%	130,250	158,550	283,150	40,432.50 + 36%	158,550
283,150		90,200.50 + 39.6%	283,150	283,150		85,288.50 + 39.6%	283,150

C H A P T E R   R E V I E W

Library of Functions

Linear function (p. 235)

$f(x) = mx + b$

Graph is a line with slope m and y-intercept b.

Constant function (p. 235)

$f(x) = b$

Graph is a horizontal line with y-interc
See Figure 22.

Identity function (p. 235)

$f(x) = x$

Graph is a line with slope 1 and y-intercept 0. See Figure 23.

Cube function (p. 236)

$f(x) = x^3$

See Figure 25.

Reciprocal function (p. 236)

$f(x) = 1/x$

See Figure 27.

Square function (p. 235)

$f(x) = x^2$

Graph is a parabola with intercept at $(0, 0)$. See Figure 24.

Square root function (p. 236)

$f(x) = \sqrt{x}$

See Figure 26.

Absolute value function (p. 236)

$f(x) = |x|$

See Figure 28.

Greatest integer function (p. 237)

$f(x) = \text{int}(x)$

See Figure 29.

Things To Know

Function (p. 206)	A relation between two sets of real numbers so that each number x in the first set, the domain, has corresponding to it exactly one number y in the second set. The range is the set of y values of the function for the x values in the domain.
	x is the independent variable; y is the dependent variable.
	A function f may be defined implicitly by an equation involving x and y or explicitly by writing $y = f(x)$.
	A function can also be characterized as a set of ordered pairs (x, y) or $(x, f(x))$ in which no first element is paired with two different second elements.
Function notation (p. 208)	$y = f(x)$
	f is a symbol for the function.
	x is the argument, or independent variable.
	y is the dependent variable.
	$f(x)$ is the value of the function at x, or the image of x.
Domain (p. 211)	If unspecified, the domain of a function f is the largest set of real numbers for which $f(x)$ is a real number.
Vertical-line test (p. 213)	A set of points in the plane is the graph of a function if and only if every vertical line intersects the graph in at most one point.
Average rate of change of a function (p. 226)	The average rate of change of f from c to x is $$\frac{\Delta y}{\Delta x} = \frac{f(x) - f(c)}{x - c}, \qquad x \neq c$$
Increasing function (p. 228)	A function f is increasing on an open interval I if, for any choice of x_1 and x_2 in I, with $x_1 < x_2$, we have $f(x_1) < f(x_2)$.
Decreasing function (p. 228)	A function f is decreasing on an open interval I if, for any choice of x_1 and x_2 in I, with $x_1 < x_2$, we have $f(x_1) > f(x_2)$.
Constant function (p. 228)	A function f is constant on an interval I if, for all choices of x in I, the values of $f(x)$ are equal.
Local maximum (p. 229)	A function f has a local maximum at c if there is an open interval I containing c so that, for all $x \neq c$ in $I, f(x) < f(c)$.
Local minimum (p. 229)	A function f has a local minimum at c if there is an open interval I containing c so that, for all $x \neq c$ in $I, f(x) > f(c)$.
Even function f (p. 230)	$f(-x) = f(x)$ for every x in the domain ($-x$ must also be in the domain).
Odd function f (p. 230)	$f(-x) = -f(x)$ for every x in the domain ($-x$ must also be in the domain).
Difference quotient of f (p. 233)	$\dfrac{f(x + h) - f(x)}{h}, \quad h \neq 0$

Objectives

You should be able to:

Determine whether a relation represents a function (p. 205)

Find the value of a function (p. 209)

Find the domain of a function (p. 211)

Identify the graph of a function (p. 212)

Obtain information from or about the graph of a function (p. 213)

Find the average rate of a change of a function (p. 224)

Use a graph to determine where a function is increasing, is decreasing, or is constant (p. 227)

Use a graph to locate local maxima and minima (p. 228)

Determine even and odd functions from a graph (p. 229)

Identify even or odd functions from the equation (p. 230)

Graph the functions listed in the library of functions (p. 235)

Graph piecewise-defined functions (p. 238)

Graph functions using horizontal and vertical shifts (p. 242)

Graph functions using compressions and stretches (p. 246)

Graph functions using reflections about the x-axis or y-axis (p. 248)

Form the sum, difference, product, and quotient of two functions (p. 255)

Form the composite function and find its domain (p. 257)

Construct and analyze functions (p. 264)

Fill-in-the-Blank Items

1. If f is a function defined by the equation $y = f(x)$, then x is called the _____ variable and y is the _____ variable.

2. A set of points in the xy-plane is the graph of a function if and only if every _____ line intersects the graph in at most one point.

3. The average rate of change of a function equals the _____ of the secant line containing two points on its graph.

4. A(n) _____ function f is one for which $f(-x) = f(x)$ for every x in the domain of f; a(n) _____ function f is one for which $f(-x) = -f(x)$ for every x in the domain of f.

5. Suppose that the graph of a function f is known. Then the graph of $y = f(x - 2)$ may be obtained by a(n) _____ shift of the graph of f to the _____ a distance of 2 units.

6. If $f(x) = x + 1$ and $g(x) = x^3$, then _____ $= (x + 1)^3$.

True/False Items

T F **1.** Every relation is a function.

T F **2.** Vertical lines intersect the graph of a function in no more than one point.

T F **3.** The y-intercept of the graph of the function $y = f(x)$, whose domain is all real numbers, is $f(0)$.

T F **4.** A function f is decreasing on an open interval I if, for any choice of x_1 and x_2 in I, with $x_1 < x_2$, we have $f(x_1) < f(x_2)$.

T F **5.** Even functions have graphs that are symmetric with respect to the origin.

T F **6.** The graph of $y = f(-x)$ is the reflection about the y-axis of the graph of $y = f(x)$.

T F **7.** $f(g(x)) = f(x) \cdot g(x)$.

T F **8.** The domain of the composite function $(f \circ g)(x)$ is the same as that of $g(x)$.

Review Exercises

Blue problem numbers indicate the author's suggestions for use in a Practice Test.

1. Given that f is a linear function, $f(4) = -5$ and $f(0) = 3$, write the equation that defines f.

2. Given that g is a linear function with slope $= -4$ and $g(-2) = 2$, write the equation that defines g.

3. A function f is defined by $f(x) = \dfrac{Ax + 5}{6x - 2}$. If $f(1) = 4$, find A.

4. A function g is defined by $g(x) = \dfrac{A}{x} + \dfrac{8}{x^2}$. If $g(-1) = 0$, find A.

5. Tell which of the following graphs are graphs of functions.

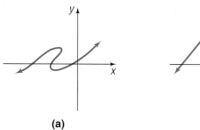

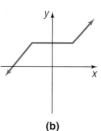

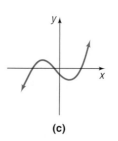

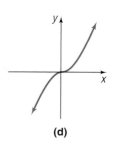

(a)   (b)   (c)   (d)

6. Use the graph of the function f shown to find:

(a) The domain and range of f

(b) $f(-1)$

(c) The intercepts of f

(d) The intervals on which f is increasing, decreasing, or constant

(e) Whether the function is even, odd, or neither

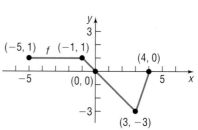

In Problems 7–12, find the following for each function:
    (a) $f(-x)$ (b) $-f(x)$ (c) $f(x + 2)$ (d) $f(x - 2)$ (e) $f(2x)$

7. $f(x) = \dfrac{3x}{x^2 - 4}$

8. $f(x) = \dfrac{x^2}{x + 2}$

9. $f(x) = \sqrt{x^2 - 4}$

10. $f(x) = |x^2 - 4|$

11. $f(x) = \dfrac{x^2 - 4}{x^2}$

12. $f(x) = \dfrac{x^3}{x^2 - 4}$

In Problems 13–20, find the domain of each function.

13. $f(x) = \dfrac{x}{x^2 - 9}$

14. $f(x) = \dfrac{3x^2}{x - 2}$

15. $f(x) = \sqrt{2 - x}$

16. $f(x) = \sqrt{x + 2}$

17. $h(x) = \dfrac{\sqrt{x}}{|x|}$

18. $g(x) = \dfrac{|x|}{x}$

19. $f(x) = \dfrac{x}{x^2 + 2x - 3}$

20. $F(x) = \dfrac{1}{x^2 - 3x - 4}$

In Problems 21–24:
    (a) *Find the domain of each function.*   (b) *Locate any intercepts.*
    (c) *Graph each function.*   (d) *Based on the graph, find the range.*

21. $f(x) = \begin{cases} 3x & -2 < x \le 1 \\ x + 1 & x > 1 \end{cases}$

22. $f(x) = \begin{cases} x - 1 & -3 < x < 0 \\ 3x - 1 & x \ge 0 \end{cases}$

23. $f(x) = \begin{cases} x & -4 \le x < 0 \\ 1 & x = 0 \\ 3x & x > 0 \end{cases}$

24. $f(x) = \begin{cases} x^2 & -2 \le x \le 2 \\ 2x - 1 & x > 2 \end{cases}$

In Problems 25–28, find the average rate of change from 2 to x for each function f. Be sure to simplify.

25. $f(x) = 2 - 5x$

26. $f(x) = 2x^2 + 7$

27. $f(x) = 3x - 4x^2$

28. $f(x) = x^2 - 3x + 2$

In Problems 29–36, determine (algebraically) whether the given function is even, odd, or neither.

29. $f(x) = x^3 - 4x$

30. $g(x) = \dfrac{4 + x^2}{1 + x^4}$

31. $h(x) = \dfrac{1}{x^4} + \dfrac{1}{x^2} + 1$

32. $F(x) = \sqrt{1 - x^3}$

33. $G(x) = 1 - x + x^3$

34. $H(x) = 1 + x + x^2$

35. $f(x) = \dfrac{x}{1 + x^2}$

36. $g(x) = \dfrac{1 + x^2}{x^3}$

In Problems 37–48, graph each function using the techniques of shifting, compressing or stretching, and reflections. Identify any intercepts on the graph. State the domain and, based on the graph, find the range.

37. $F(x) = |x| - 4$

38. $f(x) = |x| + 4$

39. $g(x) = -2|x|$

40. $g(x) = \frac{1}{2}|x|$

41. $h(x) = \sqrt{x - 1}$

42. $h(x) = \sqrt{x} - 1$

43. $f(x) = \sqrt{1 - x}$

44. $f(x) = -\sqrt{x + 3}$

45. $h(x) = (x - 1)^2 + 2$

46. $h(x) = (x + 2)^2 - 3$

47. $g(x) = 3(x - 1)^3 + 1$

48. $g(x) = -2(x + 2)^3 - 8$

In Problems 49–54, for the given functions f and g find:

(a) $(f \circ g)(2)$ (b) $(g \circ f)(-2)$ (c) $(f \circ f)(4)$ (d) $(g \circ g)(-1)$

49. $f(x) = 3x - 5$; $g(x) = 1 - 2x^2$

50. $f(x) = 4 - x$; $g(x) = 1 + x^2$

51. $f(x) = \sqrt{x + 2}$; $g(x) = 2x^2 + 1$

52. $f(x) = 1 - 3x^2$; $g(x) = \sqrt{4 - x}$

53. $f(x) = \dfrac{1}{x^2 + 4}$; $g(x) = 3x - 2$

54. $f(x) = \dfrac{2}{1 + 2x^2}$; $g(x) = 3x$

In Problems 55–60, find f ∘ g, g ∘ f, f ∘ f, and g ∘ g for each pair of functions. State the domain of each.

55. $f(x) = 2 - x$; $g(x) = 3x + 1$

56. $f(x) = 2x - 1$; $g(x) = 2x + 1$

57. $f(x) = 3x^2 + x + 1$; $g(x) = |3x|$

58. $f(x) = \sqrt{3x}$; $g(x) = 1 + x$

59. $f(x) = \dfrac{x + 1}{x - 1}$; $g(x) = \dfrac{1}{x}$

60. $f(x) = \sqrt{x - 3}$; $g(x) = 3x$

61. For the graph of the function f shown below, draw the graph of:
   (a) $y = f(-x)$ (b) $y = -f(x)$
   (c) $y = f(x + 2)$ (d) $y = f(x) + 2$
   (e) $y = 2f(x)$ (f) $y = f(3x)$

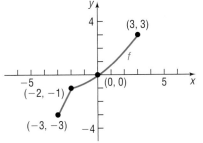

62. For the graph of the function g shown below, draw the graph of:
   (a) $y = g(-x)$ (b) $y = -g(x)$
   (c) $y = g(x + 2)$ (d) $y = g(x) + 2$
   (e) $y = 2g(x)$ (f) $y = g(3x)$

63. Temperature Conversion   The temperature T of the air is approximately a linear function of the altitude h for altitudes within 10,000 meters of the surface of Earth. If the surface temperature is 30°C and the temperature at 10,000 meters is 5°C, find the function $T = T(h)$.

64. Speed as a Function of Time   The speed v (in feet per second) of a car is a linear function of the time t (in seconds) for $10 \le t \le 30$. If after each second the speed of the car has increased by 5 feet per second and if after 20 seconds the speed is 80 feet per second, how fast is the car going after 30 seconds? Find the function $v = v(t)$.

65. Strength of a Beam   The strength of a rectangular wooden beam is proportional to the product of the width and the cube of its depth (see the figure). If the beam is to be cut from a log in the shape of a cylinder of radius 3 feet, express the strength S of the beam as a function of the width x. What is the domain of S?

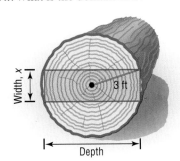

66. Productivity versus Earnings   The following data represent the average hourly earnings and productivity (output per hour) of production workers for the years 1986–1995. Let productivity x be the independent variable and average hourly earnings y be the dependent variable.

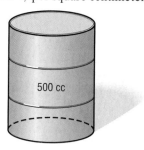

Productivity	Average Hourly Earnings
94.2	8.76
94.1	8.98
94.6	9.28
95.3	9.66
96.1	10.01
96.7	10.32
100	10.57
100.2	10.83
100.7	11.12
100.8	11.44

Source: Bureau of Labor Statistics.

(a) Draw a scatter diagram of the data.
(b) Draw a line from the point $(94.2, 8.76)$ to $(96.7, 10.32)$ on the scatter diagram found in part (a).
(c) Find the average rate of change of hourly earnings for productivity from 94.2 to 96.7.
(d) Interpret the average rate of change found in part (c).
(e) Draw a line from the point $(96.7, 10.32)$ to $(100.8, 11.44)$ on the scatter diagram found in part (a).
(f) Find the average rate of change of hourly earnings for productivity from 96.7 to 100.8.
(g) Interpret the average rate of change found in part (f).
(h) What is happening to the average rate of change of hourly earnings as productivity increases?

67. Speed of a Parachutist   The following data represent the distance that a parachutist has fallen over time.

Time (Seconds)	Distance (Meters)
0	0
5	112.5
10	490
15	1102.5
20	1960

(a) Draw a scatter diagram of the data.
(b) Draw a line from the point $(0, 0)$ to $(5, 112.5)$.
(c) Find the average rate of change of distance from 0 to 5 seconds.
(d) Interpret the average rate of change found in part (c).
(e) Draw a line from the points $(15, 1102.5)$ to $(20, 1960)$.
(f) Find the average rate of change of distance from 15 to 20 seconds.
(g) Interpret the average rate of change found in part (f).

(h) What is happening to the average rate of change of distance as time passes?

68. Material Needed to Make a Drum   A steel drum in the shape of a right circular cylinder is required to have a volume of 100 cubic feet.
(a) Express the amount A of material required to make the drum as a function of the radius r of the cylinder.
(b) How much material is required if the drum is of radius 3 feet?
(c) Of radius 4 feet?
(d) Of radius 5 feet?
(e) Graph $A = A(r)$. For what value of r is A smallest?

69. Cost of a Drum   A drum in the shape of a right circular cylinder is required to have a volume of 500 cubic centimeters. The top and bottom are made of material that costs 6¢ per square centimeter; the sides are made of material that costs 4¢ per square centimeter.

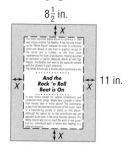

500 cc

(a) Express the total cost C of the material as a function of the radius r of the cylinder.
(b) What is the cost if the radius is 4 cm?
(c) What is the cost if the radius is 8 cm?
(d) Graph $C = C(r)$. For what value of r is the cost C least?

70. Page Design   A page with dimensions of $8\frac{1}{2}$ inches by 11 inches has a border of uniform width x surrounding the printed matter of the page, as shown in the figure.

$8\frac{1}{2}$ in.

x

And the
Rock 'n Roll
Beat is On

x x 11 in.

x

(a) Write a formula for the area A of the printed part of the page as a function of the width x of the border.
(b) Give the domain and range of A.
(c) Find the area of the printed page for borders of widths 1 inch, 1.2 inches, and 1.5 inches.
(d) Graph the function $A = A(x)$.
(e) Use TRACE to determine what margin should be used to obtain an area of 70 square inches and of 50 square inches.

71. Spheres The volume V of a sphere of radius r is $V = \frac{4}{3}\pi r^3$; the surface area S of this sphere is $S = 4\pi r^2$. Express the volume V as a function of the surface area S. If the surface area doubles, how does the volume change?

72. Constructing a Closed Box A closed box with a square base is required to have a volume of 10 cubic feet.
   (a) Express the amount A of material used to make such a box as a function of the length x of a side of the square base.

   (b) How much material is required for a base 1 foot by 1 foot?
   (c) How much material is required for a base 2 feet by 2 feet?
   (d) Graph $A = A(x)$. For what value of x is A smallest?

Project at Motorola

During the past decade the availability and usage of wireless Internet services have increased. The industry has developed a number of pricing proposals for such services. Marketing data have indicated that subscribers of wireless Internet services have tended to desire flat fee rate structures as compared with rates based totally on usage. The Computer Resource Department of Indigo Media (hypothetical) has entered into a contractual agreement for wireless Internet services. As a part of the contractual agreement, employees are able to sign up for their own wireless services. Three pricing options are available:

Silver Plan: $20/month for up to 200 K-bytes of service plus $0.16 for each additional K-byte of service

Gold Plan: $50/month for up to 1000 K-bytes of service plus $0.08 for each additional K-byte of service

Platinum Plan: $100/month for up to 3000 K-bytes of service plus $0.04 for each additional K-byte of service

You have been requested to write a report that answers the following questions in order to aid employees in choosing the appropriate pricing plan.

(a) If C is the monthly charge for x K-bytes of service, express C as a function of x for each of the three plans.

(b) Graph each of the three functions found in part (a).

(c) For how many K-bytes of service is the Silver Plan the best pricing option? When is the Gold Plan best? When is the Platinum Plan best? Explain your reasoning.

(d) Write a report that summarizes your findings.

Polynomial and Rational Functions

Field Trip to Motorola

First-, second-, and third-degree polynomial functions are widely used at Motorola to model manufacturing processes and product design features, based on the measurements made in factories and testing labs. Equipment cycle time characterization is one example. By performing a set of designed experiments, engineers can develop empirical models to predict the assembly time for a given product on a given type of machine. Trade-off analyses, for example, to optimize production costs, are frequently performed using ratios of functions that express the costs and benefits of a particular course of action.

Interview at Motorola

Tom Tirpak works in the Virtual Design and Manufacturing Group at the Motorola Advanced Technology Center in Schaumburg, Illinois. Since joining Motorola in 1991, Tom and his team have developed methods and software for improving the cycle time, quality, and cost of electronics manufacturing and product design operations. In cooperation with Motorola University, he has taught classes on Surface Mount Technology (SMT), Manufacturing Optimization, and Factory Physics. Tom is a member of several professional societies, including the Institute for Operations Research and Management Science (INFORMS) and Tau Beta Pi.

Tom received his bachelor's and master's degrees in General Engineering and a doctorate in Electrical and Computer Engineering from the University of Illinois at Urbana–Champaign, specializing in Robotics and Control Theory. Math concepts that he uses most frequently on the job include fitting curves for measured data, identifying input and output variables for a model, constructing functions to express the benefit of certain decisions, determining the statistical validity of a proposed model, and solving systems of inequalities for an optimum solution.

PREPARING FOR THIS SECTION

Before getting started, review the following concepts:

✓ Intercepts (Section 2.2, pp. 157–158)

✓ Solving Quadratic Equations (Section 1.3, pp. 105–108, 110–114)

✓ Completing the Square (Section 1.3, pp. 108–110)

✓ Graphing Techniques: Transformations (Section 3.4, pp. 242–252)

4.1 QUADRATIC FUNCTIONS AND MODELS

OBJECTIVES

1. Graph a Quadratic Function Using Transformations
2. Identify the Vertex and Axis of Symmetry of a Quadratic Function
3. Graph a Quadratic Function Using Its Vertex, Axis, and Intercepts
4. Use the Maximum or Minimum Value of a Quadratic Function to Solve Applied Problems
5. Use a Graphing Utility to Find the Quadratic Function of Best Fit to Data

QUADRATIC FUNCTIONS

A *quadratic function* is a function that is defined by a second-degree polynomial in one variable.

A **quadratic function** is a function of the form

$$f(x) = ax^2 + bx + c \tag{1}$$

where a, b, and c are real numbers and $a \neq 0$. The domain of a quadratic function consists of all real numbers.

Many applications require a knowledge of quadratic functions. For example, suppose that Texas Instruments collects the data shown in Table 1,

TABLE 1	
Price per Calculator, p (Dollars)	Number of Calculators, x
60	11,100
65	10,115
70	9,652
75	8,731
80	8,087
85	7,205
90	6,439

which relate the number of calculators sold at the price p per calculator. Since the price of a product determines the quantity that will be purchased, we treat price as the independent variable. The relationship between the number x of calculators sold and the price p per calculator may be approximated by the linear equation

$$x = 21,000 - 150p$$

Then the revenue R derived from selling x calculators at the price p per calculator is

$$R = xp$$
$$R(p) = (21,000 - 150p)p$$
$$= -150p^2 + 21,000p$$

So the revenue R is a quadratic function of the price p. Figure 1 illustrates the graph of this revenue function, whose domain is $0 \le p \le 140$, since both x and p must be nonnegative.

A second situation in which a quadratic function appears involves the motion of a projectile. Based on Newton's second law of motion (force equals mass times acceleration, $F = ma$), it can be shown that, ignoring air resistance, the path of a projectile propelled upward at an inclination to the horizontal is the graph of a quadratic function. See Figure 2 for an illustration.

Figure 1
Graph of a revenue function: $R = -150p^2 + 21,000p$

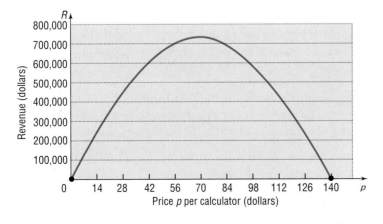

Figure 2
Path of a cannonball

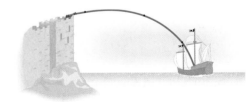

GRAPHING QUADRATIC FUNCTIONS

We know how to graph the quadratic function $f(x) = x^2$. Figure 3 shows the graph of three functions of the form $f(x) = ax^2$, $a > 0$, for $a = 1$, $a = \frac{1}{2}$, and $a = 3$. Notice that the larger the value of a, the "narrower" the graph, and the smaller the value of a, the "wider" the graph.

Figure 4 shows the graphs of $f(x) = ax^2$ for $a < 0$. Notice that these graphs are reflections about the x-axis of the graphs in Figure 3. Based on the results of these two figures, we can draw some general conclusions about the graph of $f(x) = ax^2$. First, as $|a|$ increases, the graph becomes *narrower* (a vertical stretch), and as $|a|$ gets closer to zero, the graph gets *wider* (a vertical compression). Second, if a is positive, then the graph opens *up*, and if a is negative, the graph opens *down*.

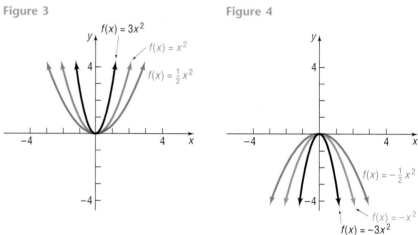

Figure 3

Figure 4

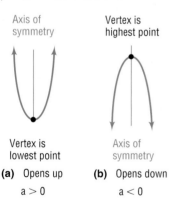

Figure 5
Graphs of a quadratic function,
$f(x) = ax^2 + bx + c, a \neq 0$

(a) Opens up
$a > 0$

(b) Opens down
$a < 0$

The graphs in Figures 3 and 4 are typical of the graphs of all quadratic functions, which we call **parabolas**.* Refer to Figure 5, where two parabolas are pictured. The one on the left **opens up** and has a lowest point; the one on the right **opens down** and has a highest point. The lowest or highest point of a parabola is called the **vertex**. The vertical line passing through the vertex in each parabola in Figure 5 is called the **axis of symmetry** (sometimes abbreviated to **axis**) of the parabola. Because the parabola is symmetric about its axis, the axis of symmetry of a parabola can be used to find additional points on the parabola.

The parabolas shown in Figure 5 are the graphs of a quadratic function $f(x) = ax^2 + bx + c, a \neq 0$. Notice that the coordinate axes are not included in the figure. Depending on the values of a, b, and c, the axes could be placed anywhere. The important fact is that, except possibly for compression or stretching, the shape of the graph of a quadratic function will look like one of the parabolas in Figure 5.

① In the following example, we use techniques from Section 3.4 to graph a quadratic function $f(x) = ax^2 + bx + c, a \neq 0$. In so doing, we shall complete the square and write the function f in the form $f(x) = a(x - h)^2 + k$.

---

* We shall study parabolas using a geometric definition later in this book.

| **EXAMPLE 1** | **Graphing a Quadratic Function Using Trasformations** |

Graph the function $f(x) = 2x^2 + 8x + 5$. Find the vertex and axis of symmetry.

Solution We begin by completing the square on the right side.

$$f(x) = 2x^2 + 8x + 5$$

$$= 2(x^2 + 4x) + 5 \qquad \text{Factor out the 2 from } 2x^2 + 8x.$$

$$= 2(x^2 + 4x + 4) + 5 - 8 \qquad \text{Complete the square of } 2(x^2 + 4x).$$
$$\text{Notice that the factor of 2 requires}$$
$$= 2(x + 2)^2 - 3 \qquad \text{that 8 be added and subtracted.} \qquad \textbf{(2)}$$

The graph of f can be obtained in three stages, as shown in Figure 6. Now compare this graph to the graph in Figure 5(a). The graph of $f(x) = 2x^2 + 8x + 5$ is a parabola that opens up and has its vertex (lowest point) at $(-2, -3)$. Its axis of symmetry is the line $x = -2$.

Figure 6

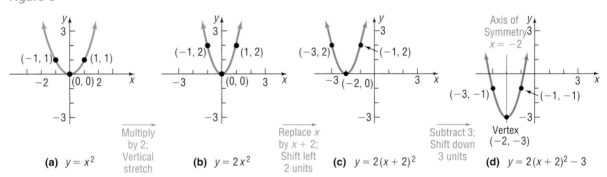

(a) $y = x^2$ Multiply by 2; Vertical stretch   **(b)** $y = 2x^2$ Replace x by $x + 2$; Shift left 2 units   **(c)** $y = 2(x + 2)^2$ Subtract 3; Shift down 3 units   Vertex $(-2, -3)$ **(d)** $y = 2(x + 2)^2 - 3$

Check: Graph $f(x) = 2x^2 + 8x + 5$ and use the MINIMUM command to locate its vertex.

NOW WORK PROBLEM **17**.

The method used in Example 1 can be used to graph any quadratic function $f(x) = ax^2 + bx + c, a \neq 0$, as follows:

$$f(x) = ax^2 + bx + c$$

$$= a\left(x^2 + \frac{b}{a}x\right) + c \qquad \text{Factor out } a \text{ from } ax^2 + bx.$$

$$= a\left(x^2 + \frac{b}{a}x + \frac{b^2}{4a^2}\right) + c - a\left(\frac{b^2}{4a^2}\right) \qquad \begin{array}{l}\text{Complete the square by adding}\\ \text{and subtracting } a(b^2/4a^2).\\ \text{Look closely at this step!}\end{array}$$

$$= a\left(x + \frac{b}{2a}\right)^2 + c - \frac{b^2}{4a}$$

$$= a\left(x + \frac{b}{2a}\right)^2 + \frac{4ac - b^2}{4a} \qquad c - \frac{b^2}{4a} = c \cdot \frac{4a}{4a} - \frac{b^2}{4a} = \frac{4ac - b^2}{4a}$$

Based on these results, we conclude the following:

If $h = -b/2a$ and $k = (4ac - b^2)/4a$, then

$$f(x) = ax^2 + bx + c = a(x - h)^2 + k \qquad \textbf{(3)}$$

The graph of f is the parabola $y = ax^2$ shifted horizontally h units and vertically k units. As a result, the vertex is at (h, k), and the graph opens up if $a > 0$ and down if $a < 0$. The axis of symmetry is the vertical line $x = h$.

For example, compare equation (3) with equation (2) of Example 1.

$$f(x) = 2(x + 2)^2 - 3$$
$$= a(x - h)^2 + k$$

We conclude that $a = 2$, so the graph opens up. Also, we find that $h = -2$ and $k = -3$, so its vertex is at $(-2, -3)$.

② It is not required to complete the square to obtain the vertex. In almost every case, it is easier to obtain the vertex of a quadratic function f by remembering that its x-coordinate is $h = -b/2a$. The y-coordinate can then be found by evaluating f at $-b/2a$.

We summarize these remarks as follows:

> **CHARACTERISTICS OF THE GRAPH OF A QUADRATIC FUNCTION**
>
> $$f(x) = ax^2 + bx + c$$
>
> $$\text{Vertex} = \left(\frac{-b}{2a}, f\left(\frac{-b}{2a} \right) \right) \qquad \text{Axis of symmetry: the line } x = \frac{-b}{2a} \qquad \textbf{(4)}$$
>
> Parabola opens up if $a > 0$; the vertex is a minimum point.
> Parabola opens down if $a < 0$; the vertex is a maximum point.

EXAMPLE 2 **Locating the Vertex without Graphing**

Without graphing, locate the vertex and axis of symmetry of the parabola defined by $f(x) = -3x^2 + 6x + 1$. Does it open up or down?

Solution For this quadratic function, $a = -3$, $b = 6$, and $c = 1$. The x-coordinate of the vertex is

$$h = \frac{-b}{2a} = \frac{-6}{-6} = 1$$

The y-coordinate of the vertex is therefore

$$k = f\left(\frac{-b}{2a} \right) = f(1) = -3 + 6 + 1 = 4$$

The vertex is located at the point $(1, 4)$. The axis of symmetry is the line $x = 1$. Finally, because $a = -3 < 0$, the parabola opens down. ∎

③ The information we gathered in Example 2, together with the location of the intercepts, usually provides enough information to graph the quadratic function $f(x) = ax^2 + bx + c$, $a \neq 0$. The y-intercept is the value of f at $x = 0$, that is, $f(0) = c$.

The x-intercepts, if there are any, are found by solving the quadratic equation

$$f(x) = ax^2 + bx + c = 0$$

This equation has two, one, or no real solutions, depending on whether the discriminant $b^2 - 4ac$ is positive, 0, or negative. Depending on the value of the discriminant, the graph of f has x-intercepts, as follows:

THE x-INTERCEPTS OF A QUADRATIC FUNCTION

1. If the discriminant $b^2 - 4ac > 0$, the graph of $f(x) = ax^2 + bx + c$ has two distinct x-intercepts and so will cross the x-axis in two places.
2. If the discriminant $b^2 - 4ac = 0$, the graph of $f(x) = ax^2 + bx + c$ has one x-intercept and touches the x-axis at its vertex.
3. If the discriminant $b^2 - 4ac < 0$, the graph of $f(x) = ax^2 + bx + c$ has no x-intercept and so will not cross or touch the x-axis.

Figure 7 illustrates these possibilities for parabolas that open up.

Figure 7
$f(x) = ax^2 + bx + c$, $a > 0$

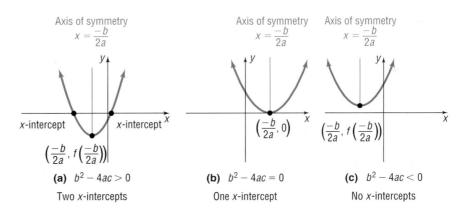

(a) $b^2 - 4ac > 0$
Two x-intercepts

(b) $b^2 - 4ac = 0$
One x-intercept

(c) $b^2 - 4ac < 0$
No x-intercepts

EXAMPLE 3

Graphing a Quadratic Function Using Its Vertex, Axis, and Intercepts

Use the information from Example 2 and the locations of the intercepts to graph $f(x) = -3x^2 + 6x + 1$.

Solution   In Example 2, we found the vertex to be at $(1, 4)$ and the axis of symmetry to be $x = 1$. The y-intercept is found by letting $x = 0$. The y-intercept is $f(0) = 1$. The x-intercepts are found by solving the equation $f(x) = 0$. This results in the equation

$$-3x^2 + 6x + 1 = 0 \quad a = -3, b = 6, c = 1$$

The discriminant $b^2 - 4ac = (6)^2 - 4(-3)(1) = 36 + 12 = 48 > 0$, so the equation has two real solutions and the graph has two x-intercepts. Using the quadratic formula, we find that

$$x = \frac{-b + \sqrt{b^2 - 4ac}}{2a} = \frac{-6 + \sqrt{48}}{-6} = \frac{-6 + 4\sqrt{3}}{-6} \approx -0.15$$

Figure 8 $f(x) = -3x^2 + 6x + 1$

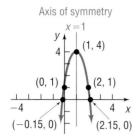

and

$$x = \frac{-b - \sqrt{b^2 - 4ac}}{2a} = \frac{-6 - \sqrt{48}}{-6} = \frac{-6 - 4\sqrt{3}}{-6} \approx 2.15$$

The x-intercepts are approximately -0.15 and 2.15.

The graph is illustrated in Figure 8. Notice how we used the y-intercept and the axis of symmetry, $x = 1$, to obtain the additional point $(2, 1)$ on the graph. ∎

 GRAPH THE FUNCTION IN EXAMPLE 3 USING THE METHOD PRESENTED IN EXAMPLE 1.

 WHICH OF THE TWO METHODS DO YOU PREFER? GIVE REASONS.

Check: Graph $f(x) = -3x^2 + 6x + 1$. Use ROOT or ZERO to locate the two x-intercepts and use MAXIMUM to locate the vertex. ∎

 NOW WORK PROBLEM 25.

If the graph of a quadratic function has only one x-intercept or none, it is usually necessary to plot an additional point to obtain the graph.

EXAMPLE 4

Graphing a Quadratic Function Using Its Vertex, Axis, and Intercepts

Graph $f(x) = x^2 - 6x + 9$ by determining whether the graph opens up or down. Find its vertex, axis of symmetry, y-intercept, and x-intercepts, if any.

Figure 9

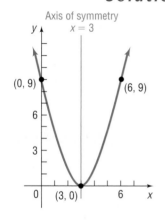

Solution For $f(x) = x^2 - 6x + 9$, we have $a = 1$, $b = -6$, and $c = 9$. Since $a = 1 > 0$, the parabola opens up. The x-coordinate of the vertex is

$$h = \frac{-b}{2a} = \frac{-(-6)}{2(1)} = 3$$

The y-coordinate of the vertex is

$$k = f(3) = (3)^2 - 6(3) + 9 = 0$$

So the vertex is at $(3, 0)$. The axis of symmetry is the line $x = 3$. The y-intercept is $f(0) = 9$. Since the vertex $(3, 0)$ lies on the x-axis, the graph touches the x-axis at the x-intercept. By using the axis of symmetry and the y-intercept at $(0, 9)$, we can locate the additional point $(6, 9)$ on the graph. See Figure 9. ∎

 GRAPH THE FUNCTION IN EXAMPLE 4 USING THE METHOD PRESENTED IN EXAMPLE 1.

 WHICH OF THE TWO METHODS DO YOU PREFER? GIVE REASONS.

 NOW WORK PROBLEM 31.

EXAMPLE 5

Graphing a Quadratic Function Using Its Vertex, Axis, and Intercepts

Graph $f(x) = 2x^2 + x + 1$ by determining whether the graph opens up or down. Find its vertex, axis of symmetry, y-intercept, and x-intercepts, if any.

Solution   For $f(x) = 2x^2 + x + 1$, we have $a = 2, b = 1$, and $c = 1$. Since $a = 2 > 0$, the parabola opens up. The x-coordinate of the vertex is

$$h = \frac{-b}{2a} = -\frac{1}{4}$$

The y-coordinate of the vertex is

$$k = f\left(-\frac{1}{4}\right) = 2\left(\frac{1}{16}\right) + \left(-\frac{1}{4}\right) + 1 = \frac{7}{8}$$

So the vertex is at $\left(-\frac{1}{4}, \frac{7}{8}\right)$. The axis of symmetry is the line $x = -\frac{1}{4}$. The y–intercept is $f(0) = 1$. The x-intercept(s), if any, obey the equation $2x^2 + x + 1 = 0$. Since the discriminant $b^2 - 4ac = (1)^2 - 4(2)(1) = -7 < 0$, this equation has no real solutions, and therefore the graph has no x-intercepts. We use the point $(0, 1)$ and the axis of symmetry $x = -\frac{1}{4}$ to locate the additional point $\left(-\frac{1}{2}, 1\right)$ on the graph. See Figure 10. ∎

Figure 10

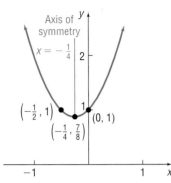

 NOW WORK PROBLEM 33.

SUMMARY

> **Steps for Graphing a Quadratic Function $f(x) = ax^2 + bx + c, a \neq 0$**
>
> **Option 1**
>
> **STEP 1:** Complete the square in x to write the quadratic function in the form $f(x) = a(x - h)^2 + k$.
>
> **STEP 2:** Graph the function in stages using transformations.
>
> **Option 2**
>
> **STEP 1:** Determine the vertex, $\left(\dfrac{-b}{2a}, f\left(\dfrac{-b}{2a}\right)\right)$.
>
> **STEP 2:** Determine the axis of symmetry, $x = \dfrac{-b}{2a}$.
>
> **STEP 3:** Determine the y-intercept, $f(0)$.
>
> **STEP 4:** a. If $b^2 - 4ac > 0$, then the graph of the quadratic function has two x-intercepts, which are found by solving the equation $ax^2 + bx + c = 0$.
> b. If $b^2 - 4ac = 0$, the vertex is the x-intercept.
> c. If $b^2 - 4ac < 0$, there are no x-intercepts.
>
> **STEP 5:** Determine an additional point if $b^2 - 4ac \leq 0$ by using the y-intercept and the axis of symmetry.
>
> **STEP 6:** Plot the points and draw the graph.

QUADRATIC MODELS

④ When a mathematical model leads to a quadratic function, the characteristics of this quadratic function can provide important information about the model. For example, for a quadratic revenue function, we can find the maximum revenue; for a quadratic cost function, we can find the minimum cost.

To see why, recall that the graph of a quadratic function $f(x) = ax^2 + bx + c$ is a parabola with vertex at $(-b/2a, f(-b/2a))$. This vertex is the highest point on the graph if $a < 0$ and the lowest point on the graph if $a > 0$. If the vertex is the highest point $(a < 0)$, then $f(-b/2a)$ is the **maximum value** of f. If the vertex is the lowest point $(a > 0)$, then $f(-b/2a)$ is the **minimum value** of f.

EXAMPLE 6

Finding the Maximum or Minimum Value of a Quadratic Function

Determine whether the quadratic function

$$f(x) = x^2 - 4x + 7$$

has a maximum or minimum value. Then find the maximum or minimum value.

Solution   We compare $f(x) = x^2 - 4x + 7$ to $f(x) = ax^2 + bx + c$. We conclude that $a = 1, b = -4$, and $c = 7$. Since $a > 0$, the graph of f opens up, so the vertex is a minimum point. The minimum value occurs at

$$x = \frac{-b}{2a} \underset{\substack{\uparrow \\ a = 1, b = -4}}{=} \frac{-(-4)}{2(1)} = \frac{4}{2} = 2$$

The minimum value is

$$f\left(\frac{-b}{2a}\right) = f(2) = 2^2 - 4(2) + 7 = 4 - 8 + 7 = 3 \qquad \blacksquare$$

NOW WORK PROBLEM 43.

EXAMPLE 7

Maximizing Revenue

The marketing department at Texas Instruments has found that, when certain calculators are sold at a price of p dollars per unit, the revenue R (in dollars) as a function of the price p is

$$R(p) = -150p^2 + 21{,}000p$$

What unit price should be established in order to maximize revenue? If this price is charged, what is the maximum revenue?

Solution   The revenue R is

$$R(p) = -150p^2 + 21{,}000p \qquad R(p) = ap^2 + bp + c$$

The function R is a quadratic function with $a = -150, b = 21{,}000$, and $c = 0$. Because $a < 0$, the vertex is the highest point of the parabola. The revenue R is therefore a maximum when the price p is

$$p = \frac{-b}{2a} = \frac{-21{,}000}{2(-150)} = \frac{-21{,}000}{-300} = \$70.00$$

The maximum revenue R is

$$R(70) = -150(70)^2 + 21{,}000(70) = \$735{,}000$$

See Figure 11 for an illustration.

Figure 11
$R(p) = -150p^2 + 21,000p$

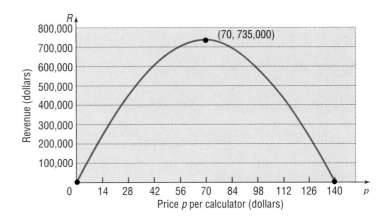

 NOW WORK PROBLEM 51.

| EXAMPLE 8 | **Maximizing the Area Enclosed by a Fence** |

A farmer has 2000 yards of fence to enclose a rectangular field. What are the dimensions of the rectangle that encloses the most area?

Solution Figure 12 illustrates the situation. The available fence represents the perimeter of the rectangle. If x is the length and w is the width, then

Figure 12

$$2x + 2w = 2000 \qquad \textbf{(5)}$$

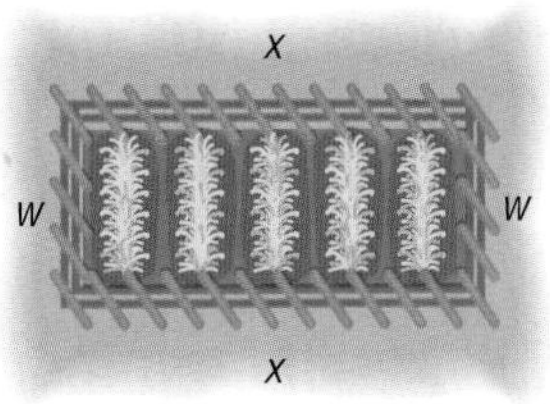

The area A of the rectangle is

$$A = xw$$

To express A in terms of a single variable, we solve equation (5) for w and substitute the result in $A = xw$. Then A involves only the variable x. [You could also solve equation (5) for x and express A in terms of w alone. Try it!]

$$2x + 2w = 2000 \qquad \text{Equation (5)}$$

$$2w = 2000 - 2x \qquad \text{Solve for } w.$$

$$w = \frac{2000 - 2x}{2} = 1000 - x$$

Then the area A is

$$A = xw = x(1000 - x) = -x^2 + 1000x$$

Now, A is a quadratic function of x.

$$A(x) = -x^2 + 1000x \quad a = -1, b = 1000, c = 0$$

Since $a < 0$, the vertex is a maximum point on the graph of A. The maximum value occurs at

$$x = \frac{-b}{2a} = \frac{-1000}{2(-1)} = 500$$

The maximum value of A is

$$A\left(\frac{-b}{2a}\right) = A(500) = -500^2 + 1000(500)$$

$$= -250,000 + 500,000 = 250,000$$

The maximum area that can be enclosed by 2000 yards of fence is 250,000 square yards. ∎

NOW WORK PROBLEM **57**.

EXAMPLE 9	**Analyzing the Motion of a Projectile**

A projectile is fired from a cliff 500 feet above the water at an inclination of 45° to the horizontal, with a muzzle velocity of 400 feet per second. In physics, it is established that the height h of the projectile above the water is given by

$$h(x) = \frac{-32x^2}{(400)^2} + x + 500$$

where x is the horizontal distance of the projectile from the base of the cliff. See Figure 13.

Figure 13

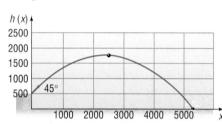

(a) Find the maximum height of the projectile.

(b) How far from the base of the cliff will the projectile strike the water?

Solution

(a) The height of the projectile is given by a quadratic function.

$$h(x) = \frac{-32x^2}{(400)^2} + x + 500 = \frac{-1}{5000}x^2 + x + 500$$

We are looking for the maximum value of h. Since the maximum value is obtained at the vertex, we compute

$$x = \frac{-b}{2a} = \frac{-1}{2(-1/5000)} = \frac{5000}{2} = 2500$$

The maximum height of the projectile is

$$h(2500) = \frac{-1}{5000}(2500)^2 + 2500 + 500$$

$$= -1250 + 2500 + 500 = 1750 \text{ ft}$$

(b) The projectile will strike the water when the height is zero. To find the distance x traveled, we need to solve the equation

$$h(x) = \frac{-1}{5000}x^2 + x + 500 = 0$$

We use the quadratic formula with

$$b^2 - 4ac = 1 - 4\left(\frac{-1}{5000}\right)(500) = 1.4$$

$$x = \frac{-1 \pm \sqrt{1.4}}{2(-1/5000)} \approx \begin{cases} -458 \\ 5458 \end{cases}$$

We discard the negative solution and find that the projectile will strike the water at a distance of about 5458 feet from the base of the cliff. ∎

SEEING THE CONCEPT: Graph

$$h(x) = \frac{-1}{5000}x^2 + x + 500, \qquad 0 \le x \le 5500$$

Use MAXIMUM to find the maximum height of the projectile, and use ROOT or ZERO to find the distance from the base of the cliff to where the projectile strikes the water. Compare your results with those obtained in the text. TRACE the path of the projectile. How far from the base of the cliff is the projectile when its height is 1000 ft? 1500 ft?

NOW WORK PROBLEM **61**.

EXAMPLE 10 The Golden Gate Bridge

The Golden Gate Bridge, a suspension bridge, spans the entrance to San Francisco Bay. Its 746-foot-tall towers are 4200 feet apart. The bridge is suspended from two huge cables more than 3 feet in diameter; the 90-foot-wide roadway is 220 feet above the water. The cables are parabolic in shape and touch the road surface at the center of the bridge. Find the height of the cable at a distance of 1000 feet from the center.

Solution   We begin by choosing the placement of the coordinate axes so that the x-axis coincides with the road surface and the origin coincides with the center of the bridge. As a result, the twin towers will be vertical (height $746 - 220 = 526$ feet above the road) and located 2100 feet from the center. Also, the cable, which has the shape of a parabola, will extend from the towers, open up, and have its vertex at $(0, 0)$. As illustrated in Figure 14, the choice of placement of the axes enables us to identify the equation of the parabola as $y = ax^2$, $a > 0$. We can also see that the points $(-2100, 526)$ and $(2100, 526)$ are on the graph.

Figure 14

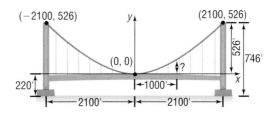

Based on these facts, we can find the value of a in $y = ax^2$.

$$y = ax^2$$

$$526 = a(2100)^2 \qquad y = 526; x = 2100$$

$$a = \frac{526}{(2100)^2}$$

The equation of the parabola is therefore

$$y = \frac{526}{(2100)^2}x^2$$

The height of the cable when $x = 1000$ is

$$y = \frac{526}{(2100)^2}(1000)^2 \approx 119.3 \text{ feet}$$

The cable is 119.3 feet high at a distance of 1000 feet from the center of the bridge. ∎

NOW WORK PROBLEM 63.

FITTING A QUADRATIC FUNCTION TO DATA

⑤ In Section 2.5 we found the line of best fit for data that appeared to be linearly related. It was noted that data may also follow a nonlinear relation. Figures 15(a) and (b) show scatter diagrams of data that follow a quadratic relation.

Figure 15

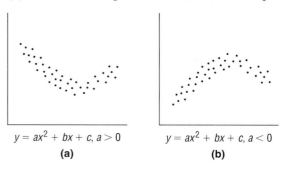

$y = ax^2 + bx + c, a > 0$ $y = ax^2 + bx + c, a < 0$
(a)   (b)

EXAMPLE 11

Fitting a Quadratic Function to Data

A farmer collected the data given in Table 2, which shows crop yields Y for various amounts of fertilizer used, x.

(a) Draw a scatter diagram of the data. Comment on the type of relation that may exist between the two variables.

(b) The quadratic function of best fit to these data is

$$Y(x) = -0.0171x^2 + 1.0765x + 3.8939$$

Use this function to determine the optimal amount of fertilizer to apply.

(c) Use the function to predict crop yield when the optimal amount of fertilizer is applied.

(d) Use a graphing utility to verify that the function given in part (b) is the quadratic function of best fit.

(e) With a graphing utility, draw a scatter diagram of the data and then graph the quadratic function of best fit on the scatter diagram.

Solution

(a) Figure 16 shows the scatter diagram. It appears that the data follow a quadratic relation, with $a < 0$.

TABLE 2

Plot	Fertilizer, x (Pounds/100 ft²)	Yield (Bushels)
1	0	4
2	0	6
3	5	10
4	5	7
5	10	12
6	10	10
7	15	15
8	15	17
9	20	18
10	20	21
11	25	20
12	25	21
13	30	21
14	30	22
15	35	21
16	35	20
17	40	19
18	40	19

Figure 16

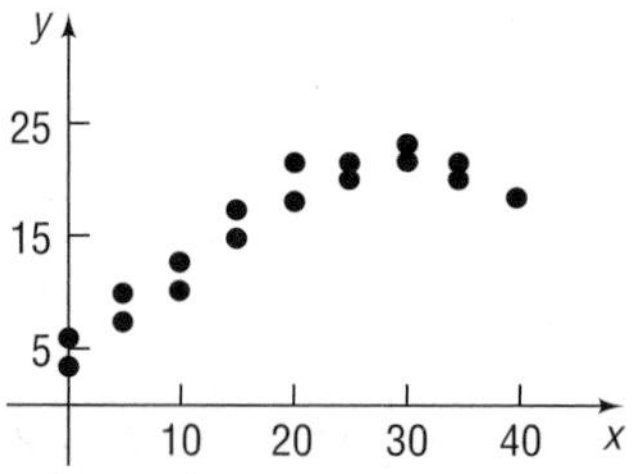

(b) Based on the quadratic function of best fit, the optimal amount of fertilizer to apply is

$$h = \frac{-b}{2a} = \frac{-1.0765}{2(-0.0171)} \approx 31.5 \text{ pounds of fertilizer per 100 square feet}$$

(c) We evaluate the function $Y(x)$ for $x = 31.5$.

$$Y(31.5) = -0.0171(31.5)^2 + 1.0765(31.5) + 3.8939 \approx 20.8 \text{ bushels}$$

If we apply 31.5 pounds of fertilizer per 100 square feet, the crop yield will be 20.8 bushels according to the quadratic function of best fit.

 (d) Upon executing the QUADratic REGression program, we obtain the results shown in Figure 17. The output of the utility shows us the equation $y = ax^2 + bx + c$. The quadratic function of best fit is $Y(x) = -0.0171x^2 + 1.0765x + 3.8939$, where x represents the amount of fertilizer used and Y represents crop yield.

(e) Figure 18 shows the graph of the quadratic function found in part (b) drawn on the scatter diagram.

Figure 17

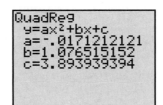

Figure 18

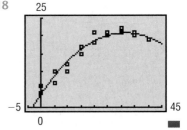

Look again at Figure 17. Notice that the output given by the graphing calculator does not include r, the correlation coefficient. Recall that the correlation coefficient is a measure of the strength of a **linear** relation that exists between two variables. The graphing calculator does not provide an indication of how well the function fits the data in terms of r since the function cannot be expressed as a linear function.

NOW WORK PROBLEM **71**.

4.1   EXERCISES

In Problems 1–8, match each graph to one the following functions without using a graphing utility.

A. $f(x) = x^2 - 1$ B. $f(x) = -x^2 - 1$ C. $f(x) = x^2 - 2x + 1$ D. $f(x) = x^2 + 2x + 1$

E. $f(x) = x^2 - 2x + 2$ F. $f(x) = x^2 + 2x$ G. $f(x) = x^2 - 2x$ G. $f(x) = x^2 + 2x + 2$

1.

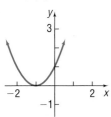

2.

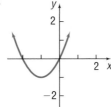

3.

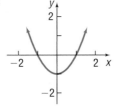

4.

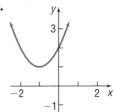

5.

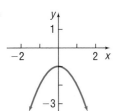

6.

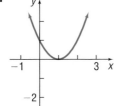

7.

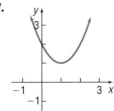

8.

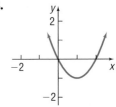

In Problems 9–24, graph the function f by starting with the graph of $y = x^2$ and using transformations (shifting, compressing, stretching, and/or reflection).
[**Hint:** If necessary, write f in the form $f(x) = a(x - h)^2 + k$.]

9. $f(x) = \dfrac{1}{4}x^2$

10. $f(x) = 2x^2$

11. $f(x) = \dfrac{1}{4}x^2 - 2$

12. $f(x) = 2x^2 - 3$

13. $f(x) = \dfrac{1}{4}x^2 + 2$

14. $f(x) = 2x^2 + 4$

15. $f(x) = \dfrac{1}{4}x^2 + 1$

16. $f(x) = -2x^2 - 2$

17. $f(x) = x^2 + 4x + 2$

18. $f(x) = x^2 - 6x - 1$

19. $f(x) = 2x^2 - 4x + 1$

20. $f(x) = 3x^2 + 6x$

21. $f(x) = -x^2 - 2x$

22. $f(x) = -2x^2 + 6x + 2$

23. $f(x) = \dfrac{1}{2}x^2 + x - 1$

24. $f(x) = \dfrac{2}{3}x^2 + \dfrac{4}{3}x - 1$

In Problems 25–40, graph each quadratic function by determining whether its graph opens up or down and by finding its vertex, axis of symmetry, y-intercept, and x-intercepts, if any.

25. $f(x) = -x^2 - 6x$

26. $f(x) = -x^2 + 4x$

27. $f(x) = 2x^2 - 8x$

28. $f(x) = 3x^2 + 18x$

29. $f(x) = x^2 + 2x - 8$

30. $f(x) = x^2 - 2x - 3$

31. $f(x) = x^2 + 2x + 1$

32. $f(x) = x^2 + 6x + 9$

33. $f(x) = 2x^2 - x + 2$

34. $f(x) = 4x^2 - 2x + 1$

35. $f(x) = -2x^2 + 2x - 3$

36. $f(x) = -3x^2 + 3x - 2$

37. $f(x) = 3x^2 + 6x + 2$

38. $f(x) = 2x^2 + 5x + 3$

39. $f(x) = -4x^2 - 6x + 2$

40. $f(x) = 3x^2 - 8x + 2$

In Problems 41–48, determine, without graphing, whether the given quadratic function has a maximum value or a minimum value and then find the value.

41. $f(x) = 2x^2 + 12x$

42. $f(x) = -2x^2 + 12x$

43. $f(x) = 2x^2 + 12x - 3$

44. $f(x) = 4x^2 - 8x + 3$

45. $f(x) = -x^2 + 10x - 4$

46. $f(x) = -2x^2 + 8x + 3$

47. $f(x) = -3x^2 + 12x + 1$

48. $f(x) = 4x^2 - 4x$

Answer Problems 49 and 50 using the following: A quadratic function of the form $f(x) = ax^2 + bx + c$ with $b^2 - 4ac > 0$ may also be written in the form $f(x) = a(x - r_1)(x - r_2)$, where r_1 and r_2 are the x-intercepts of the graph of the quadratic function.

49. (a) Find a quadratic function whose x-intercepts are -3 and 1 with $a = 1$; $a = 2$; $a = -2$; $a = 5$.
(b) How does the value of a affect the intercepts?
(c) How does the value of a affect the axis of symmetry?
(d) How does the value of a affect the vertex?
(e) Compare the x-coordinate of the vertex with the midpoint of the x-intercepts. What might you conclude?

50. (a) Find a quadratic function whose x-intercepts are -5 and 3 with $a = 1$; $a = 2$; $a = -2$; $a = 5$.
(b) How does the value of a affect the intercepts?
(c) How does the value of a affect the axis of symmetry?
(d) How does the value of a affect the vertex?
(e) Compare the x-coordinate of the vertex with the midpoint of the x-intercepts. What might you conclude?

51. Maximizing Revenue Suppose that the manufacturer of a gas clothes dryer has found that, when the unit price is p dollars, the revenue R (in dollars) is

$$R(p) = -4p^2 + 4000p$$

What unit price should be established for the dryer to maximize revenue? What is the maximum revenue?

52. Maximizing Revenue The John Deere company has found that the revenue from sales of heavy-duty tractors is a function of the unit price p that it charges. If the revenue R is

$$R(p) = -\dfrac{1}{2}p^2 + 1900p$$

what unit price p should be charged to maximize revenue? What is the maximum revenue?

53. Demand Equation The price p and the quantity x sold of a certain product obey the demand equation

$$p = -\dfrac{1}{6}x + 100, \qquad 0 \le x \le 600$$

(a) Express the revenue R as a function of x. (Remember, $R = xp$.)
(b) What is the revenue if 200 units are sold?
(c) What quantity x maximizes revenue? What is the maximum revenue?
(d) What price should the company charge to maximize revenue?

54. Demand Equation The price p and the quantity x sold of a certain product obey the demand equation

$$p = -\dfrac{1}{3}x + 100, \qquad 0 \le x \le 300$$

(a) Express the revenue R as a function of x.
(b) What is the revenue if 100 units are sold?
(c) What quantity x maximizes revenue? What is the maximum revenue?
(d) What price should the company charge to maximize revenue?

55. Demand Equation The price p and the quantity x sold of a certain product obey the demand equation

$$x = -5p + 100, \qquad 0 \le p \le 20$$

(a) Express the revenue R as a function of x.
(b) What is the revenue if 15 units are sold?
(c) What quantity x maximizes revenue? What is the maximum revenue?
(d) What price should the company charge to maximize revenue?

56. Demand Equation   The price p and the quantity x sold of a certain product obey the demand equation

$$x = -20p + 500, \qquad 0 \le p \le 25$$

(a) Express the revenue R as a function of x.
(b) What is the revenue if 20 units are sold?
(c) What quantity x maximizes revenue? What is the maximum revenue?
(d) What price should the company charge to maximize revenue?

57. Enclosing a Rectangular Field   David has available 400 yards of fencing and wishes to enclose a rectangular area.
(a) Express the area A of the rectangle as a function of the width x of the rectangle.
(b) For what value of x is the area largest?
(c) What is the maximum area?

58. Enclosing a Rectangular Field   Beth has 3000 feet of fencing available to enclose a rectangular field.
(a) Express the area A of the rectangle as a function of x, where x is the length of the rectangle.
(b) For what value of x is the area largest?
(c) What is the maximum area?

59. Enclosing the Most Area with a Fence   A farmer with 4000 meters of fencing wants to enclose a rectangular plot that borders on a river. If the farmer does not fence the side along the river, what is the largest area that can be enclosed? (See the figure.)

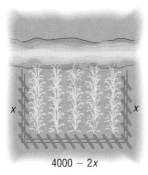

$$4000 - 2x$$

60. Enclosing the Most Area with a Fence   A farmer with 2000 meters of fencing wants to enclose a rectangular plot that borders on a straight highway. If the farmer does not fence the side along the highway, what is the largest area that can be enclosed?

61. Analyzing the Motion of a Projectile   A projectile is fired from a cliff 200 feet above the water at an inclination of 45° to the horizontal, with a muzzle velocity of 50 feet per second. The height h of the projectile above the water is given by

$$h(x) = \frac{-32x^2}{(50)^2} + x + 200$$

where x is the horizontal distance of the projectile from the base of the cliff.
(a) How far from the base of the cliff is the height of the projectile a maximum?

(b) Find the maximum height of the projectile.
(c) How far from the base of the cliff will the projectile strike the water?
(d) Using a graphing utility, graph the function h, $0 \le x \le 200$.
(e) When the height of the projectile is 100 feet above the water, how far is it from the cliff?

62. Analyzing the Motion of a Projectile   A projectile is fired at an inclination of 45° to the horizontal, with a muzzle velocity of 100 feet per second. The height h of the projectile is given by

$$h(x) = \frac{-32x^2}{(100)^2} + x$$

where x is the horizontal distance of the projectile from the firing point.
(a) How far from the firing point is the height of the projectile a maximum?
(b) Find the maximum height of the projectile.
(c) How far from the firing point will the projectile strike the ground?
(d) Using a graphing utility, graph the function h, $0 \le x \le 350$.
(e) When the height of the projectile is 50 feet above the ground, how far has it traveled horizontally?

63. Suspension Bridge   A suspension bridge with weight uniformly distributed along its length has twin towers that extend 75 meters above the road surface and are 400 meters apart. The cables are parabolic in shape and are suspended from the tops of the towers. The cables touch the road surface at the center of the bridge. Find the height of the cables at a point 100 meters from the center. (Assume that the road is level.)

64. Architecture   A parabolic arch has a span of 120 feet and a maximum height of 25 feet. Choose suitable rectangular coordinate axes and find the equation of the parabola. Then calculate the height of the arch at points 10 feet, 20 feet, and 40 feet from the center.

65. Constructing Rain Gutters   A rain gutter is to be made of aluminum sheets that are 12 inches wide by turning up the edges 90°. What depth will provide maximum cross-sectional area and hence allow the most water to flow?

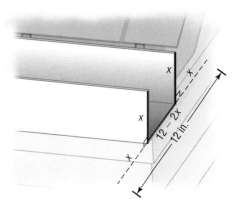

66. Norman Windows A Norman window has the shape of a rectangle surmounted by a semicircle of diameter equal to the width of the rectangle (see the figure). If the perimeter of the window is 20 feet, what dimensions will admit the most light (maximize the area)?

[**Hint:** Circumference of a circle $= 2\pi r$; area of a circle $= \pi r^2$, where r is the radius of the circle.]

67. Constructing a Stadium A track and field playing area is in the shape of a rectangle with semicircles at each end (see the figure). The inside perimeter of the track is to be 1500 meters. What should the dimensions of the rectangle be so that the area of the rectangle is a maximum?

68. Architecture A special window has the shape of a rectangle surmounted by an equilateral triangle (see the figure). If the perimeter of the window is 16 feet, what dimensions will admit the most light?

[**Hint:** Area of an equilateral triangle $= (\sqrt{3}/4)x^2$, where x is the length of a side of the triangle.]

69. Life Cycle Hypothesis An individual's income varies with his or her age. The following table shows the median income I of individuals of different age groups within the United States for 1995. For each age group, let the class midpoint represent the independent variable, x. For the class "65 years and older," we will assume that the class midpoint is 69.5.

Age	Class Midpoint, x	Median Income, I
15–24 years	19.5	$20,979
25–34 years	29.5	$34,701
35–44 years	39.5	$43,465
45–54 years	49.5	$48,058
55–64 years	59.5	$38,077
65 years and older	69.5	$19,096

Source: U.S. Census Bureau

(a) Draw a scatter diagram of the data. Comment on the type of relation that may exist between the two variables.

(b) The quadratic function of best fit to these data is

$$I(x) = -42.6x^2 + 3806x - 38,526$$

Use this function to determine the age at which an individual can expect to earn the most income.

(c) Use the function to predict the peak income earned.

(d) Use a graphing utility to verify that the function given in part (b) is the quadratic function of best fit.

(e) With a graphing utility, draw a scatter diagram of the data and then graph the quadratic function of best fit on the scatter diagram.

70. Life Cycle Hypothesis An individual's income varies with his or her age. The following table shows the median income I of individuals of different age groups within the United States for 1996. For each age group, the class midpoint represents the independent variable, x. For the age group "65 years and older," we will assume that the class midpoint is 69.5.

Age	Class Midpoint, x	Median Income, I
15–24 years	19.5	$21,438
25–34 years	29.5	$35,888
35–44 years	39.5	$44,420
45–54 years	49.5	$50,472
55–64 years	59.5	$39,815
65 years and older	69.5	$19,448

Source: U.S. Census Bureau

(a) Draw a scatter diagram of the data. Comment on the type of relation that may exist between the two variables.

(b) The quadratic function of best fit to these data is

$$I(x) = -44.8x^2 + 4009x - 41392$$

Use this function to determine the age at which an individual can expect to earn the most income.

(c) Use the function to predict the peak income earned.

(d) Use a graphing utility to verify that the function given in part (b) is the quadratic function of best fit.

(e) With a graphing utility, draw a scatter diagram of the data and then graph the quadratic function of best fit on the scatter diagram.

71. Price of Crude Oil   The following data represent the imports of crude oil (1000 barrels per day) for the years 1980–1997.

Year, x	Imports, I	Year, x	Imports, I
1980	5263	1989	5843
1981	4396	1990	5894
1982	3488	1991	5782
1983	3329	1992	6083
1984	3426	1993	6787
1985	3201	1994	7063
1986	4178	1995	7230
1987	4674	1996	7508
1988	5107	1997	7996

Source: U.S. Energy Information Administration

(a) Draw a scatter diagram of the data. Comment on the type of relation that may exist between the two variables.

(b) The quadratic function of best fit to these data is

$$I(x) = 18.04x^2 - 71,495.6x + 70,831,298$$

Use this function to determine the year in which imports of crude oil were lowest.

(c) Use the function found in part (b) to predict the number of barrels of imported crude oil in 1998.

(d) Use a graphing utility to verify that the function given in part (b) is the quadratic function of best fit.

(e) With a graphing utility, draw a scatter diagram of the data and then graph the quadratic function of best fit on the scatter diagram.

72. Miles per Gallon   An engineer collects data showing the speed s of a Ford Taurus and its average miles per gallon, M. See the table.

Speed, s	Miles per Gallon, M
30	18
35	20
40	23
40	25
45	25
50	28
55	30
60	29
65	26
65	25
70	25

(a) Draw a scatter diagram of the data. Comment on the type of relation that may exist between the two variables.

(b) The quadratic function of best fit to these data is

$$M(s) = -0.018s^2 + 1.93s - 25.34$$

Use this function to determine the speed that maximizes miles per gallon.

(c) Use the function to predict miles per gallon for a speed of 63 miles per hour.

(d) Use a graphing utility to verify that the function given in part (b) is the quadratic function of best fit.

(e) With a graphing utility, draw a scatter diagram of the data and then graph the quadratic function of best fit on the scatter diagram.

73. Height of a Ball   A shot-putter throws a ball at an inclination of 45° to the horizontal. The following data represent the height of the ball h at the instant that it has traveled x feet horizontally.

Distance, x	Height, h
20	25
40	40
60	55
80	65
100	71
120	77
140	77
160	75
180	71
200	64

(a) Draw a scatter diagram of the data. Comment on the type of relation that may exist between the two variables.

(b) The quadratic function of best fit to these data is

$$h(x) = -0.0037x^2 + 1.03x + 5.7$$

Use this function to determine how far the ball will travel before it reaches its maximum height.

(c) Use the function to find the maximum height of the ball.

(d) Use a graphing utility to verify that the function given in part (b) is the quadratic function of best fit.

(e) With a graphing utility, draw a scatter diagram of the data and then graph the quadratic function of best fit on the scatter diagram.

74. Chemical Reactions A self-catalytic chemical reaction results in the formation of a compound that causes the formation ratio to increase. If the reaction rate V is given by

$$V(x) = kx(a - x), \qquad 0 \le x \le a$$

where k is a positive constant, a is the initial amount of the compound, and x is the variable amount of the compound, for what value of x is the reaction rate a maximum?

75. Calculus: Simpson's Rule The figure shows the graph of $y = ax^2 + bx + c$. Suppose that the points $(-h, y_0)$, $(0, y_1)$, and (h, y_2) are on the graph. It can be shown that the area enclosed by the parabola, the x-axis, and the lines $x = -h$ and $x = h$ is

$$\text{Area} = \frac{h}{3}(2ah^2 + 6c)$$

Show that this area may also be given by

$$\text{Area} = \frac{h}{3}(y_0 + 4y_1 + y_2)$$

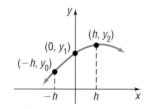

76. Use the result obtained in Problem 75 to find the area enclosed by $f(x) = -5x^2 + 8$, the x-axis, and the lines $x = -1$ and $x = 1$.

77. Use the result obtained in Problem 75 to find the area enclosed by $f(x) = 2x^2 + 8$, the x-axis, and the lines $x = -2$ and $x = 2$.

78. Use the result obtained in Problem 75 to find the area enclosed by $f(x) = x^2 + 3x + 5$, the x-axis, and the lines $x = -4$ and $x = 4$.

79. Use the result obtained in Problem 75 to find the area enclosed by $f(x) = -x^2 + x + 4$, the x-axis, and the lines $x = -1$ and $x = 1$.

80. A rectangle has one vertex on the line $y = 10 - x$, $x > 0$, another at the origin, one on the positive x-axis, and one on the positive y-axis. Find the largest area A that can be enclosed by the rectangle.

81. Let $f(x) = ax^2 + bx + c$, where a, b, and c are odd integers. If x is an integer, show that $f(x)$ must be an odd integer.

[**Hint:** x is either an even integer or an odd integer.]

82. Make up a quadratic function that opens down and has only one x-intercept. Compare yours with others in the class. What are the similarities? What are the differences?

83. On one set of coordinate axes, graph the family of parabolas $f(x) = x^2 + 2x + c$ for $c = -3$, $c = 0$, and $c = 1$. Describe the characteristics of a member of this family.

84. On one set of coordinate axes, graph the family of parabolas $f(x) = x^2 + bx + 1$ for $b = -4$, $b = 0$, and $b = 4$. Describe the general characteristics of this family.

Before getting started, review the following concepts:

✓ Polynomials (Review, Section 5, pp. 37–45)

✓ Intercepts (Section 2.2, pp. 157–158)

✓ Graphing Techniques: Transformations (Section 3.4, pp. 242–252)

4.2 POLYNOMIAL FUNCTIONS

OBJECTIVES
1. Identify Polynomials and Their Degree
2. Graph Polynomial Functions Using Transformations
3. Identify the Zeros of a Polynomial and Their Multiplicity
4. Analyze the Graph of a Polynomial Function

1   *Polynomial functions* are among the simplest expressions in algebra. They are easy to evaluate: only addition and repeated multiplication are required. Because of this, they are often used to approximate other, more complicated functions. In this section, we investigate characteristics of this important class of function.

> A **polynomial function** is a function of the form
>
> $$f(x) = a_n x^n + a_{n-1} x^{n-1} + \cdots + a_1 x + a_0 \qquad \textbf{(1)}$$
>
> where $a_n, a_{n-1}, \ldots, a_1, a_0$ are real numbers and n is a nonnegative integer. The domain consists of all real numbers.

A polynomial function is a function whose rule is given by a polynomial in one variable. The **degree** of a polynomial function is the degree of the polynomial in one variable, that is, the largest power of x that appears.

EXAMPLE 1

Identifying Polynomial Functions

Determine which of the following are polynomial functions. For those that are, state the degree; for those that are not, tell why not.

(a) $f(x) = 2 - 3x^4$ (b) $g(x) = \sqrt{x}$

(c) $h(x) = \dfrac{x^2 - 2}{x^3 - 1}$ (d) $F(x) = 0$

(e) $G(x) = 8$ (f) $H(x) = -2x^3(x - 1)^2$

Solution   (a) f is a polynomial function of degree 4.

(b) g is not a polynomial function. The variable x is raised to the $\frac{1}{2}$ power, which is not a nonnegative integer.

(c) h is not a polynomial function. It is the ratio of two polynomials, and the polynomial in the denominator is of positive degree.

(d) F is the zero polynomial function; it is not assigned a degree.

(e) G is a nonzero constant function, a polynomial function of degree 0 since $G(x) = 8 = 8x^0$.

(f) $H(x) = -2x^3(x - 1)^2 = -2x^3(x^2 - 2x + 1) = -2x^5 + 4x^4 - 2x^3$. So H is a polynomial function of degree 5. Do you see how to find the degree of H without multiplying out?   ∎

NOW WORK PROBLEMS **1** AND **5**.

We have already discussed in detail polynomial functions of degrees 0, 1, and 2. See Table 3 (on page 302) for a summary of the characteristics of the graphs of these polynomial functions.

One of the objectives of this section is to analyze the graph of a polynomial function. If you take a course in calculus, you will learn that the graph of every polynomial function is both smooth and continuous. By **smooth,** we mean that the graph contains no sharp corners or cusps; by **continuous,** we mean that the graph has no gaps or holes and can be drawn without lifting pencil from paper. See Figures 19(a) and (b).

		TABLE 3		
Degree	**Form**		**Name**	**Graph**
No degree	$f(x) = 0$		Zero function	The x-axis
0	$f(x) = a_0, \quad a_0 \neq 0$		Constant function	Horizontal line with y-intercept a_0
1	$f(x) = a_1 x + a_0, \quad a_1 \neq 0$		Linear function	Nonvertical, nonhorizontal line with slope a_1 and y-intercept a_0
2	$f(x) = a_2 x^2 + a_1 x + a_0, \quad a_2 \neq 0$		Quadratic function	Parabola: Graph opens up if $a_2 > 0$; graph opens down if $a_2 < 0$

Figure 19

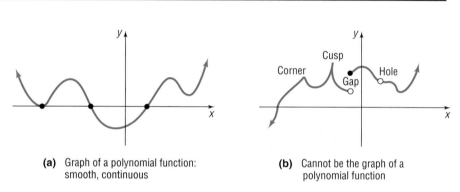

(a) Graph of a polynomial function: smooth, continuous

(b) Cannot be the graph of a polynomial function

We begin the analysis of the graph of a polynomial function by discussing *power functions,* a special kind of polynomial function.

POWER FUNCTION

A **power function of degree n** is a function of the form

$$f(x) = ax^n \qquad (2)$$

where a is a real number, $a \neq 0$, and $n > 0$ is an integer.

Thus, a power function is a function that is defined by a single monomial.

The graph of a power function of degree 1, $f(x) = ax$, is a straight line, with slope a, that passes through the origin. The graph of a power function of degree 2, $f(x) = ax^2$, is a parabola, with vertex at the origin, that opens up if $a > 0$ and down if $a < 0$.

If we know how to graph a power function of the form $f(x) = x^n$, then a compression or stretch and, perhaps, a reflection about the x-axis will enable us to obtain the graph of $g(x) = ax^n$. Consequently, we shall concentrate on graphing power functions of the form $f(x) = x^n$.

We begin with power functions of even degree of the form $f(x) = x^n$, $n \geq 2$ and n even. The domain of f is the set of all real numbers, and the range is the set of nonnegative real numbers. Such a power function is an even function (do you see why?), so its graph is symmetric with respect to the y-axis. Its graph always contains the origin and the points $(-1, 1)$ and $(1, 1)$.

If $n = 2$, the graph is the familiar parabola $y = x^2$ that opens up, with vertex at the origin. If $n \geq 4$, the graph of $f(x) = x^n$, n even, will be closer to the x-axis than the parabola $y = x^2$ if $-1 < x < 1$ and farther from the

Figure 20

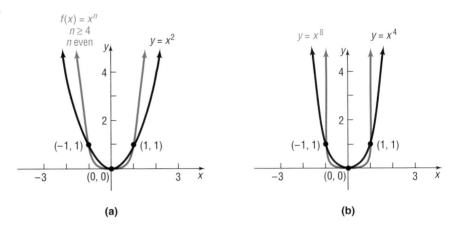

(a)   (b)

x-axis than the parabola $y = x^2$ if $x < -1$ or if $x > 1$. Figure 20(a) illustrates this conclusion. Figure 20(b) shows the graphs of $y = x^4$ and $y = x^8$ for comparison.

From Figure 20, we can see that as n increases the graph of $f(x) = x^n$, $n \geq 2$ and n even, tends to flatten out near the origin and to increase very rapidly when x is far from 0. For large n, it may appear that the graph coincides with the x-axis near the origin, but it does not; the graph actually touches the x-axis only at the origin (see Table 4). Also, for large n, it may appear that for $x < -1$ or for $x > 1$ the graph is vertical, but it is not; it is only increasing very rapidly in these intervals. If the graphs were enlarged many times, these distinctions would be clear.

TABLE 4			
	$x = 0.1$	$x = 0.3$	$x = 0.5$
$f(x) = x^8$	10^{-8}	0.0000656	0.0039063
$f(x) = x^{20}$	10^{-20}	$3.487 \cdot 10^{-11}$	0.000001
$f(x) = x^{40}$	10^{-40}	$1.216 \cdot 10^{-21}$	$9.095 \cdot 10^{-13}$

SEEING THE CONCEPT: Graph $Y_1 = x^4$, $Y_2 = x^8$, and $Y_3 = x^{12}$ using the viewing rectangle $-2 \leq x \leq 2, -4 \leq y \leq 16$. Then graph each again using the viewing rectangle $-1 \leq x \leq 1, 0 \leq y \leq 1$. See Figure 21. TRACE along one of the graphs to confirm that for x close to 0 the graph is above the x-axis and that for $x > 0$ the graph is increasing.

Figure 21

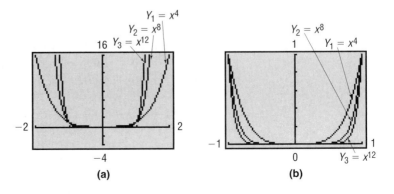

(a)   (b)

> ### CHARACTERISTICS OF POWER FUNCTIONS, $y = x^n$, n IS AN EVEN INTEGER
>
> 1. The graph is symmetric with respect to the y-axis.
> 2. The domain is the set of all real numbers. The range is the set of non-negative real numbers.
> 3. The graph always contains the points $(0, 0)$, $(1, 1)$, and $(-1, 1)$.
> 4. As the exponent n increases in magnitude, the graph becomes more vertical when $x < -1$ or $x > 1$; but for x near the origin, the graph tends to flatten out and lie closer to the x-axis.

Now we consider power functions of odd degree of the form $f(x) = x^n$, $n \geq 3$ and n odd. The domain and range of f are the set of real numbers. Such a power function is an odd function (do you see why?), so its graph is symmetric with respect to the origin. Its graph always contains the origin and the points $(-1, -1)$ and $(1, 1)$.

The graph of $f(x) = x^n$ when $n = 3$ has been shown several times and is repeated in Figure 22. If $n \geq 5$, the graph of $f(x) = x^n$, n odd, will be closer to the x-axis than that of $y = x^3$ if $-1 < x < 1$ and farther from the x-axis than that of $y = x^3$ if $x < -1$ or if $x > 1$. Figure 22 also illustrates this conclusion.

Figure 23 shows the graph of $y = x^5$ and the graph of $y = x^9$ for further comparison.

Figure 22

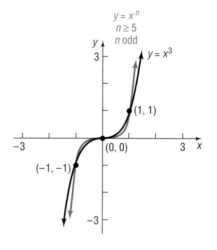

Figure 23

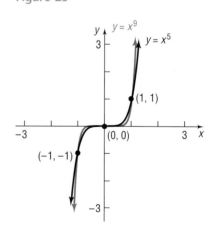

It appears that each graph coincides with the x-axis near the origin, but it does not; each graph actually touches the x-axis only at the origin. Also, it appears that as x increases the graph becomes vertical, but it does not; each graph is increasing very rapidly.

 SEEING THE CONCEPT: Graph $Y_1 = x^3$, $Y_2 = x^7$, and $Y_3 = x^{11}$ using the viewing rectangle $-2 \leq x \leq 2$, $-16 \leq y \leq 16$. Then graph each again using the viewing rectangle $-1 \leq x \leq 1$, $0 \leq y \leq 1$. See Figure 24. TRACE along one of the graphs to confirm that the graph is increasing and only touches the x-axis at the origin.

Figure 24

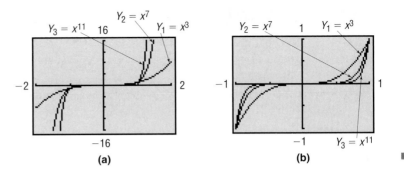

(a)   (b)

To summarize:

> ### CHARACTERISTICS OF POWER FUNCTIONS, $y = x^n$, n IS AN ODD INTEGER
>
> 1. The graph is symmetric with respect to the origin.
> 2. The domain and range are the set of all real numbers.
> 3. The graph always contains the points $(0, 0)$, $(1, 1)$, and $(-1, -1)$.
> 4. As the exponent n increases in magnitude, the graph becomes more vertical when $x < -1$ or $x > 1$; but for x near the origin, the graph tends to flatten out and lie closer to the x-axis.

2 The methods of shifting, compression, stretching, and reflection studied in Section 3.4, when used with the facts just presented, will enable us to graph polynomial functions that are transformations of power functions.

EXAMPLE 2

Graphing Polynomial Functions Using Transformations

Graph: $f(x) = 1 - x^5$

Solution   Figure 25 shows the required stages.

Figure 25

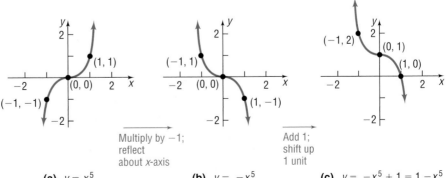

(a) $y = x^5$ (b) $y = -x^5$ (c) $y = -x^5 + 1 = 1 - x^5$

EXAMPLE 3

Graphing Polynomial Functions Using Transformations

Graph: $f(x) = \frac{1}{2}(x - 1)^4$

Solution Figure 26 shows the required stages.

Figure 26

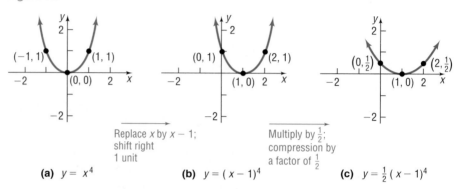

Replace x by $x - 1$; shift right 1 unit

Multiply by $\frac{1}{2}$; compression by a factor of $\frac{1}{2}$

(a) $y = x^4$ **(b)** $y = (x - 1)^4$ **(c)** $y = \frac{1}{2}(x - 1)^4$

NOW WORK PROBLEM **13** AND **17**.

GRAPHING OTHER POLYNOMIALS

To graph most polynomials of degree 3 or higher requires advanced techniques. However, if we can locate the x-intercepts of the graph, then algebraic techniques can be used to obtain the graph.

Figure 27 shows the graph of a polynomial function with four x-intercepts. Notice that at the x-intercepts the graph must either cross the x-axis or touch the x-axis. Consequently, between consecutive x-intercepts the graph is either above the x-axis or below the x-axis. We will make use of this characteristic of the graph of a polynomial shortly.

Figure 27

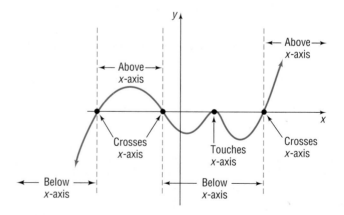

3   If a polynomial function f is factored completely, it is easy to solve the equation $f(x) = 0$ and locate the x-intercepts of the graph. For example, if $f(x) = (x - 1)^2(x + 3)$, then the solutions of the equation

$$f(x) = (x - 1)^2(x + 3) = 0$$

are identified as 1 and -3. Based on this result, we make the following observations:

> If f is a polynomial function and r is a real number for which $f(r) = 0$, then r is called a (real) **zero of f,** or **root of f.** If r is a (real) zero of f, then
>
> (a) r is an x-intercept of the graph of f.
> (b) $(x - r)$ is a factor of f.

EXAMPLE 4 **Finding a Polynomial from Its Zeros**

(a) Find a polynomial of degree 3 whose zeros are -3, 2, and 5.

 (b) Graph the polynomial found in part (a) to verify your result.

Solution (a) If r is a zero of a polynomial f, then $x - r$ is a factor of f. This means that $x - (-3) = x + 3$, $x - 2$, and $x - 5$ are factors of f. As a result, any polynomial of the form

$$f(x) = a(x + 3)(x - 2)(x - 5)$$

where a is any nonzero real number, qualifies.

(b) The value of a causes a stretch, compression, or reflection, but does not affect the x-intercepts. We choose to graph f with $a = 1$.

$$f(x) = (x + 3)(x - 2)(x - 5) = x^3 - 4x^2 - 11x + 30$$

Figure 28 shows the graph of f. Notice that the x-intercepts are -3, 2, and 5. ∎

Figure 28

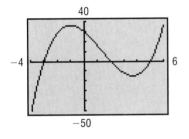

SEEING THE CONCEPT: Graph the function found in Example 4 for $a = 2$ and $a = -1$. Does the value of a affect the zeros of f? How does the value of a affect the graph of f?

NOW WORK PROBLEM **27.**

If the same factor $x - r$ occurs more than once, then r is called a **repeated, or multiple, zero of f.** More precisely, we have the following definition.

> If $(x - r)^m$ is a factor of a polynomial f and $(x - r)^{m+1}$ is not a factor of f, then r is called a **zero of multiplicity m of f.**

EXAMPLE 5 **Identifying Zeros and Their Multiplicities**

For the polynomial

$$f(x) = 5(x - 2)(x + 3)^2\left(x - \frac{1}{2}\right)^4$$

2 is a zero of multiplicity 1.

-3 is a zero of multiplicity 2.

$\frac{1}{2}$ is a zero of multiplicity 4. ∎

In Example 5 notice that, if you add the multiplicities $(1 + 2 + 4 = 7)$, you obtain the degree of the polynomial.

Suppose that it is possible to factor completely a polynomial function and, as a result, locate all the x-intercepts of its graph (the real zeros of the function). As mentioned earlier, these x-intercepts then divide the x-axis into open intervals and, on each such interval, the graph of the polynomial will be either above or below the x-axis. Let's look at an example.

EXAMPLE 6

Graphing a Polynomial Using Its x-Intercepts

For the polynomial: $f(x) = x^2(x - 2)$

(a) Find the x- and y-intercepts of the graph of f.

(b) Use the x-intercepts to find the intervals on which the graph of f is above the x-axis and the intervals on which the graph of f is below the x-axis.

(c) Locate other points on the graph and connect all the points plotted with a smooth, continuous curve.

Solution
(a) The y-intercept is $f(0) = 0^2(0 - 2) = 0$. The x-intercepts satisfy the equation

$$f(x) = x^2(x - 2) = 0$$

from which we find

$$x^2 = 0 \quad \text{or} \quad x - 2 = 0$$
$$x = 0 \qquad \qquad x = 2$$

The x-intercepts are 0 and 2.

(b) The two x-intercepts divide the x-axis into three intervals:

$$(-\infty, 0) \qquad (0, 2) \qquad (2, \infty)$$

Since the graph of f crosses or touches the x-axis only at $x = 0$ and $x = 2$, it follows that the graph of f is either above the x-axis $[f(x) > 0]$ or below the x-axis $[f(x) < 0]$ on each of these three intervals. To see where the graph lies, we only need to pick a number in each interval, evaluate f there, and see whether the value is positive (above the x-axis) or negative (below the x- axis). See Table 5.

(c) In constructing Table 5, we obtained three additional points on the graph: $(-1, -3), (1, -1),$ and $(3, 9)$. Figure 29 illustrates these points, the intercepts, and a smooth, continuous curve (the graph of f) connecting them.

TABLE 5

Interval	$(-\infty, 0)$	$(0, 2)$	$(2, \infty)$
Number Chosen	-1	1	3
Value of f	$f(-1) = -3$	$f(1) = -1$	$f(3) = 9$
Location of Graph	Below x-axis	Below x-axis	Above x-axis
Point on Graph	$(-1, -3)$	$(1, -1)$	$(3, 9)$

Figure 29

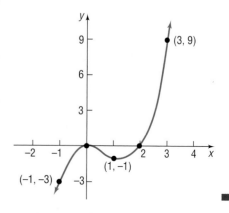

Look again at Table 5. Since the graph of f is below the x-axis on both sides of 0, the graph of f *touches* the x-axis at $x = 0$, a *zero of multiplicity 2.* Since the graph of f is below the x-axis for $x < 2$ and above the x-axis for $x > 2$, the graph of f *crosses* the x-axis at $x = 2$, a *zero of multiplicity 1.*
This suggests the following result:

IF r IS A ZERO OF EVEN MULTIPLICITY

Sign of $f(x)$ does not change from one side to the other side of r.

Graph **touches** x-axis at r.

IF r IS A ZERO OF ODD MULTIPLICITY

Sign of $f(x)$ changes from one side to the other side of r.

Graph **crosses** x-axis at r.

> NOW WORK PROBLEMS 33(a) AND (b).

 Look again at Figure 29. We cannot be sure just how low the graph actually goes between $x = 0$ and $x = 2$. But we do know that somewhere in the interval $(0, 2)$ the graph of f must change direction (from decreasing to increasing). The points at which a graph changes direction are called **turning points.** In calculus, such points are called **local maxima** or **local minima,** and techniques for locating them are given. So we shall not ask for the location of turning points in our graphs. Instead, we will use the following result from calculus, which tells us the maximum number of turning points that the graph of a polynomial function can have.

Theorem   If f is a polynomial function of degree n, then f has at most $n - 1$ turning points.

For example, the graph of $f(x) = x^2(x - 2)$ shown in Figure 29 is the graph of a polynomial of degree 3 and has $3 - 1 = 2$ turning points: one at $(0, 0)$ and the other somewhere between $x = 0$ and $x = 2$.

Figure 30

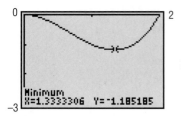

EXPLORATION   A graphing utility can be used to locate the turning points of a graph. Graph $y = x^2(x - 2)$. Use MINIMUM to find the location of the turning point for $0 < x < 2$. See Figure 30.

 One last remark about Figure 29. Notice that the graph of $f(x) = x^2(x - 2)$ looks somewhat like the graph of $y = x^3$. In fact, for very large values of x, either positive or negative, there is little difference. To see for yourself, use your calculator to compare the values of $f(x) = x^2(x - 2)$ and $y = x^3$ for $x = -100,000$ and $x = 100,000$. The behavior of the graph of a function for large values of x, either positive or negative, is referred to as its **end behavior.**

Theorem   **End Behavior**

For large values of x, either positive or negative, the graph of the polynomial

$$f(x) = a_n x^n + a_{n-1} x^{n-1} + \cdots + a_1 x + a_0$$

resembles the graph of the power function

$$y = a_n x^n$$

Look back at Figures 20 and 22. Based on the above theorem and the previous discussion on power functions, the end behavior of a polynomial can only be of four types. See Figure 31.

Figure 31
End Behavior

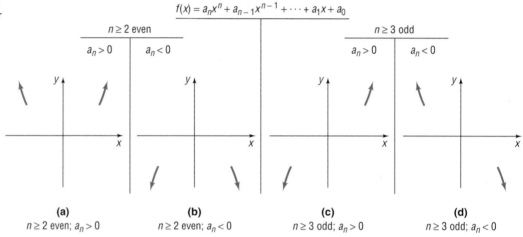

$f(x) = a_n x^n + a_{n-1} x^{n-1} + \cdots + a_1 x + a_0$

$n \geq 2$ even $n \geq 3$ odd

$a_n > 0$ $a_n < 0$ $a_n > 0$ $a_n < 0$

(a)   **(b)**   **(c)**   **(d)**
$n \geq 2$ even; $a_n > 0$ $n \geq 2$ even; $a_n < 0$ $n \geq 3$ odd; $a_n > 0$ $n \geq 3$ odd; $a_n < 0$

NOW WORK PROBLEM 33(c).

SUMMARY

Graph of a Polynomial Function $f(x) = a_n x^n + a_{n-1} x^{n-1} + \cdots + a_1 x + a_0, \quad a_n \neq 0$

Degree of the polynomial f: n
Maximum number of turning points: $n - 1$
At a zero of even multiplicity: The graph of f touches the x-axis.
At a zero of odd multiplicity: The graph of f crosses the x-axis.
Between zeros, the graph of f is either above or below the x-axis.
End behavior: For large $|x|$, the graph of f behaves like the graph of $y = a_n x^n$.

EXAMPLE 7   **Analyzing the Graph of a Polynomial Function**

For the polynomial: $f(x) = x^3 + x^2 - 12x$

(a) Find the x- and y-intercepts of the graph of f.
(b) Determine whether the graph crosses or touches the x-axis at each x-intercept.
(c) End behavior: find the power function that the graph of f resembles for large values of x.
(d) Determine the maximum number of turning points on the graph of f.

(e) Use the x-intercepts to find the intervals on which the graph of f is above the x-axis and the intervals on which the graph is below the x-axis.

(f) Put all the information together and connect the points with a smooth, continuous curve to obtain the graph of f.

Solution   (a) The y-intercept is $f(0) = 0$. To find the x-intercepts, if any, we factor f.

$$f(x) = x^3 + x^2 - 12x = x(x^2 + x - 12) = x(x + 4)(x - 3)$$

Solving the equation $f(x) = x(x + 4)(x - 3) = 0$, we find that the x-intercepts, or zeros of f, are $-4, 0$, and 3.

(b) Since each zero of f is of multiplicity 1, the graph of f will cross the x-axis at each x-intercept.

(c) End behavior: the graph of f resembles that of the power function $y = x^3$ for large values of $|x|$.

(d) The graph of f will contain at most two turning points.

(e) The three x-intercepts divide the x-axis into four intervals:

$$(-\infty, -4) \qquad (-4, 0) \qquad (0, 3) \qquad (3, \infty)$$

To determine the location of the graph of $f(x)$ in each interval, we create Table 6.

(f) The graph of f is given in Figure 32.

Figure 32

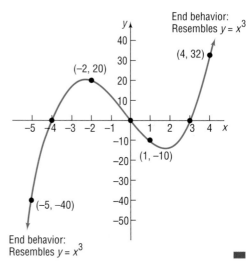

End behavior:
Resembles $y = x^3$

	TABLE 6			
	−4	0	3	→x
Interval	$(-\infty, -4)$	$(-4, 0)$	$(0, 3)$	$(3, \infty)$
Number Chosen	-5	-2	1	4
Value of f	$f(-5) = -40$	$f(-2) = 20$	$f(1) = -10$	$f(4) = 32$
Location of Graph	Below x-axis	Above x-axis	Below x-axis	Above x-axis
Point on Graph	$(-5, -40)$	$(-2, 20)$	$(1, -10)$	$(4, 32)$

 EXPLORATION   Graph $y = x^3 + x^2 - 12x$. Compare what you see with Figure 32. Use MAXIMUM/MINIMUM to locate the two turning points.   ■

NOW WORK PROBLEM **47**.

EXAMPLE 8   **Analyzing the Graph of a Polynomial Function**

Follow the instructions of Example 7 for the following polynomial:

$$f(x) = x^2(x - 4)(x + 1)$$

Solution   (a) The y-intercept is $f(0) = 0$. The x-intercepts satisfy the equation

$$f(x) = x^2(x - 4)(x + 1) = 0$$

So
$$x^2 = 0 \quad \text{or} \quad x - 4 = 0 \quad \text{or} \quad x + 1 = 0$$
$$x = 0 \qquad\qquad x = 4 \qquad\qquad x = -1$$

The x-intercepts are $-1, 0,$ and 4.

(b) The intercept 0 is a zero of multiplicity 2, so the graph of f will touch the x-axis at 0; 4 and -1 are zeros of multiplicity 1, so the graph of f will cross the x-axis at 4 and -1.

(c) End behavior: the graph of f resembles that of the power function $y = x^4$ for large values of $|x|$.

(d) The graph of f will contain at most three turning points.

(e) The three x-intercepts divide the x-axis into four intervals:
$$(-\infty, -1) \qquad (-1, 0) \qquad (0, 4) \qquad (4, \infty)$$

To determine the location of the graph of $f(x)$ in each interval, we create Table 7.

(f) The graph of f is given in Figure 33.

TABLE 7

Interval	$(-\infty, -1)$	$(-1, 0)$	$(0, 4)$	$(4, \infty)$
Number Chosen	-2	$-\frac{1}{2}$	2	5
Value of f	$f(-2) = 24$	$f\left(-\frac{1}{2}\right) = -\frac{9}{16}$	$f(2) = -24$	$f(5) = 150$
Location of Graph	Above x-axis	Below x-axis	Below x-axis	Above x-axis
Point on Graph	$(-2, 24)$	$\left(-\frac{1}{2}, -\frac{9}{16}\right)$	$(2, -24)$	$(5, 150)$

Figure 33

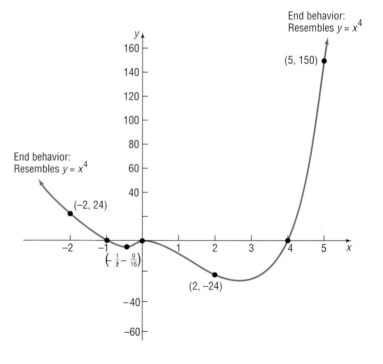

 EXPLORATION   Graph $y = x^2(x - 4)(x + 1)$. Compare what you see with Figure 33. Use MAXIMUM/MINIMUM to locate the two turning points besides $(0, 0)$.   ■

NOW WORK PROBLEM **57.**

EXAMPLE 9   **Using a Graphing Utility to Graph a Polynomial Function**

Graph: $f(x) = \pi x^4 - \sqrt{7}x^3 + 5x - 1$

Solution   We begin by making the following observations:

1. The y-intercept is $f(0) = -1$.

2. End behavior: the graph will behave like $y = \pi x^4$ for large values of $|x|$.

3. The graph will have no more than four x-intercepts and no more than three turning points.

These observations help us set the viewing rectangle for our first attempt at a complete graph. See Figure 34(a). Figure 34(b) shows the complete graph.
   Using ROOT or ZERO, we find the x-intercepts to be -1.01 and 0.20, correct to two decimal places. Using MINIMUM, we find that the only turning point is located at $(-0.57, -3.02)$, correct to two decimal places.

Figure 34

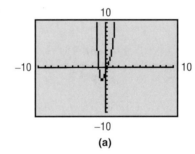

(a)

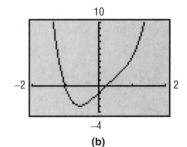

(b)   ■

SUMMARY

To analyze the graph of a polynomial function $y = f(x)$, follow these steps:

STEPS FOR ANALYZING THE GRAPH OF A POLYNOMIAL

STEP 1:   (a) Find the x-intercepts, if any, by solving the equation $f(x) = 0$.
       (b) Find the y-intercept by letting $x = 0$ and finding the value of $f(0)$.

STEP 2:   Determine whether the graph of f crosses or touches the x-axis at each x-intercept.

STEP 3:   End behavior: find the power function that the graph of f resembles for large values of x.

STEP 4:   Determine the maximum number of turning points on the graph of f.

STEP 5:   Use the x-intercept(s) to find the intervals on which the graph of f is above the x-axis and the intervals on which the graph is below the x-axis.

STEP 6:   Plot the points obtained in Steps 1 and 5, and use the remaining information to connect them with a smooth, continuous curve.

4.2   EXERCISES

In Problems 1–12, determine which functions are polynomial functions. For those that are, state the degree. For those that are not, tell why not.

1. $f(x) = 4x + x^3$

2. $f(x) = 5x^2 + 4x^4$

3. $g(x) = \dfrac{1 - x^2}{2}$

4. $h(x) = 3 - \dfrac{1}{2}x$

5. $f(x) = 1 - \dfrac{1}{x}$

6. $f(x) = x(x - 1)$

7. $g(x) = x^{3/2} - x^2 + 2$

8. $h(x) = \sqrt{x}(\sqrt{x} - 1)$

9. $F(x) = 5x^4 - \pi x^3 + \dfrac{1}{2}$

10. $F(x) = \dfrac{x^2 - 5}{x^3}$

11. $G(x) = 2(x - 1)^2(x^2 + 1)$

12. $G(x) = -3x^2(x + 2)^3$

In Problems 13–26, use transformations of the graph of $y = x^4$ or $y = x^5$ to graph each function.

13. $f(x) = (x + 1)^4$

14. $f(x) = (x - 2)^5$

15. $f(x) = x^5 - 3$

16. $f(x) = x^4 + 2$

17. $f(x) = \dfrac{1}{2}x^4$

18. $f(x) = 3x^5$

19. $f(x) = -x^5$

20. $f(x) = -x^4$

21. $f(x) = (x - 1)^5 + 2$

22. $f(x) = (x + 2)^4 - 3$

23. $f(x) = 2(x + 1)^4 + 1$

24. $f(x) = \dfrac{1}{2}(x - 1)^5 - 2$

25. $f(x) = 4 - (x - 2)^5$

26. $f(x) = 3 - (x + 2)^4$

In Problems 27–32, form a polynomial whose zeros and degree are given.

27. Zeros: $-1, 1, 3$;   degree 3

28. Zeros: $-2, 2, 3$;   degree 3

29. Zeros: $-3, 0, 4$;   degree 3

30. Zeros: $-4, 0, 2$;   degree 3

31. Zeros: $-4, -1, 2, 3$;   degree 4

32. Zeros: $-3, -1, 2, 5$;   degree 4

In Problems 33–44, for each polynomial function: (a) List each real zero and its multiplicity. (b) Determine whether the graph crosses or touches the x-axis at each x-intercept. (c) Find the power function that the graph of f resembles for large values of |x|.

33. $f(x) = 3(x - 7)(x + 3)^2$

34. $f(x) = 4(x + 4)(x + 3)^3$

35. $f(x) = 4(x^2 + 1)(x - 2)^3$

36. $f(x) = 2(x - 3)(x + 4)^3$

37. $f(x) = -2\left(x + \dfrac{1}{2}\right)^2(x^2 + 4)^2$

38. $f(x) = \left(x - \dfrac{1}{3}\right)^2(x - 1)^3$

39. $f(x) = (x - 5)^3(x + 4)^2$

40. $f(x) = (x + \sqrt{3})^2(x - 2)^4$

41. $f(x) = 3(x^2 + 8)(x^2 + 9)^2$

42. $f(x) = -2(x^2 + 3)^3$

43. $f(x) = -2x^2(x^2 - 2)$

44. $f(x) = 4x(x^2 - 3)$

In Problems 45–68, for each polynomial function f:
  (a) Find the x- and y-intercepts of f.
  (b) Determine whether the graph of f crosses or touches the x-axis at each x-intercept.
  (c) End behavior: find the power function that the graph of f resembles for large values of |x|.
  (d) Determine the maximum number of turning points on the graph of f.
  (e) Use the x-intercept(s) to find the intervals on which the graph of f is above and below the x-axis.
  (f) Plot the points obtained in parts (a) and (e), and use the remaining information to connect them with a smooth, continuous curve.

45. $f(x) = (x - 1)^2$

46. $f(x) = (x - 2)^3$

47. $f(x) = x^2(x - 3)$

48. $f(x) = x(x + 2)^2$

49. $f(x) = 6x^3(x + 4)$

50. $f(x) = 5x(x - 1)^3$

51. $f(x) = -4x^2(x + 2)$

52. $f(x) = -\dfrac{1}{2}x^3(x + 4)$

53. $f(x) = x(x - 2)(x + 4)$

54. $f(x) = x(x + 4)(x - 3)$　　　　**55.** $f(x) = 4x - x^3$　　　　**56.** $f(x) = x - x^3$

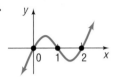 **57.** $f(x) = x^2(x - 2)(x + 2)$　　**58.** $f(x) = x^2(x - 3)(x + 4)$　　**59.** $f(x) = x^2(x - 2)^2$

60. $f(x) = x^3(x - 3)$　　　　　**61.** $f(x) = x^2(x - 3)(x + 1)$　　**62.** $f(x) = x^2(x - 3)(x - 1)$

63. $f(x) = (x + 2)^2(x - 4)^2$　　**64.** $f(x) = (x - 2)^2(x + 2)(x + 4)$　　**65.** $f(x) = x^2(x - 2)(x^2 + 3)$

66. $f(x) = x^2(x^2 + 1)(x + 4)$　　**67.** $f(x) = -x^2(x^2 - 1)(x + 1)$　　**68.** $f(x) = -x^2(x^2 - 4)(x - 5)$

In Problems 69–72, decide which of the polynomial functions in the list might have the given graph. (More than one answer may be possible.)

69.

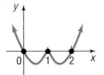

(a) $y = -4x(x - 1)(x - 2)$
(b) $y = x^2(x - 1)^2(x - 2)$
(c) $y = 3x(x - 1)(x - 2)$
(d) $y = x(x - 1)^2(x - 2)^2$
(e) $y = x^3(x - 1)(x - 2)$
(f) $y = -x(1 - x)(x - 2)$

70.

(a) $y = 2x^3(x - 1)(x - 2)^2$
(b) $y = x^2(x - 1)(x - 2)$
(c) $y = x^3(x - 1)^2(x - 2)$
(d) $y = x^2(x - 1)^2(x - 2)^2$
(e) $y = 5x(x - 1)^2(x - 2)$
(f) $y = -2x(x - 1)^2(2 - x)$

71.

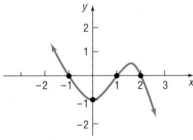

(a) $y = \dfrac{1}{2}(x^2 - 1)(x - 2)$

(b) $y = -\dfrac{1}{2}(x^2 + 1)(x - 2)$

(c) $y = (x^2 - 1)\left(1 - \dfrac{x}{2}\right)$

(d) $y = -\dfrac{1}{2}(x^2 - 1)^2(x - 2)$

(e) $y = \left(x^2 + \dfrac{1}{2}\right)(x^2 - 1)(2 - x)$

(f) $y = -(x - 1)(x - 2)(x + 1)$

72.

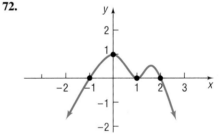

(a) $y = -\dfrac{1}{2}(x^2 - 1)(x - 2)(x + 1)$

(b) $y = -\dfrac{1}{2}(x^2 + 1)(x - 2)(x + 1)$

(c) $y = -\dfrac{1}{2}(x + 1)^2(x - 1)(x - 2)$

(d) $y = (x - 1)^2(x + 1)\left(1 - \dfrac{x}{2}\right)$

(e) $y = -(x - 1)^2(x - 2)(x + 1)$

(f) $y = -\left(x^2 + \dfrac{1}{2}\right)(x - 1)^2(x + 1)(x - 2)$

In Problems 73–86, graph each polynomial function. Approximate the x-intercepts, if any, and the turning points rounded to two decimal places.

73. $f(x) = x^3 + 0.2x^2 - 1.5876x - 0.31752$　　　　**74.** $f(x) = x^3 - 0.8x^2 - 4.6656x + 3.73248$

75. $f(x) = x^3 + 2.56x^2 - 3.31x + 0.89$　　　　　**76.** $f(x) = x^3 - 2.91x^2 - 7.668x - 3.8151$

77. $f(x) = x^4 - 2.5x^2 + 0.5625$　　　　　　　**78.** $f(x) = x^4 - 18.5x^2 + 50.2619$

79. $f(x) = x^4 + 0.65x^3 - 16.6319x^2 + 14.209335x - 3.1264785$

80. $f(x) = x^4 + 3.45x^3 - 11.6639x^2 - 5.864241x - 0.69257738$

81. $f(x) = \pi x^3 + \sqrt{2}x^2 - x - 2$

82. $f(x) = -2x^3 + \pi x^2 + \sqrt{3}x + 1$

83. $f(x) = 2x^4 - \pi x^3 + \sqrt{5}x - 4$

84. $f(x) = -1.2x^4 + 0.5x^2 - \sqrt{3}x + 2$

85. $f(x) = -2x^5 - \sqrt{2}x^2 - x - \sqrt{2}$

86. $f(x) = \pi x^5 + \pi x^4 + \sqrt{3}x + 1$

---

87. **Motor Vehicle Thefts** The following data represent the number of motor vehicle thefts (in thousands) in the United States for the years 1987–1997, where 1 represents 1987, 2 represents 1988, and so on.

Year, x	Motor Vehicle Thefts, T
1987, 1	1289
1988, 2	1433
1989, 3	1565
1990, 4	1636
1991, 5	1662
1992, 6	1611
1993, 7	1563
1994, 8	1539
1995, 9	1472
1996, 10	1394
1997, 11	1354

Source: U.S. Federal Bureau of Investigation

(a) Draw a scatter diagram of the data. Comment on the type of relation that may exist between the two variables.
(b) The cubic function of best fit to these data is

$$T(x) = 1.52x^3 - 39.81x^2 + 282.29x + 1035.5$$

Use this function to predict the number of motor vehicle thefts in 1994.
(c) Use a graphing utility to verify that the function given in part (b) is the cubic function of best fit.
(d) With a graphing utility, draw a scatter diagram of the data and then graph the cubic function of best fit on the scatter diagram.
(e) Do you think that the function given in part (b) will be useful in predicting the number of motor vehicle thefts in 1999?

88. **Larceny Thefts** The following data represent the number of larceny thefts (in thousands) in the United States

for the years 1983–1993, where 1 represents 1983, 2 represents 1984, and so on.

Year, x	Larceny Thefts, L
1983, 1	6713
1984, 2	6592
1985, 3	6926
1986, 4	7257
1987, 5	7500
1988, 6	7706
1989, 7	7872
1990, 8	7946
1991, 9	8142
1992, 10	7915
1993, 11	7821

Source: U.S. Federal Bureau of Investigation

(a) Draw a scatter diagram of the data. Comment on the type of relation that may exist between the two variables.
(b) The cubic function of best fit to these data is

$$L(x) = -4.37x^3 + 59.29x^2 - 14.02x + 6578$$

Use this function to predict the number of larceny thefts in 1994.
(c) Use a graphing utility to verify that the function given in part (b) is the cubic function of best fit.
(d) With a graphing utility, draw a scatter diagram of the data and then graph the cubic function of best fit on the scatter diagram.
(e) Do you think the function found in part (b) will be useful in predicting the number of larceny thefts in 1999?

89. **Cost of Manufacturing** The following data represent the cost C (in thousands of dollars) of manufacturing Chevy Cavaliers and the number x of Cavaliers produced.

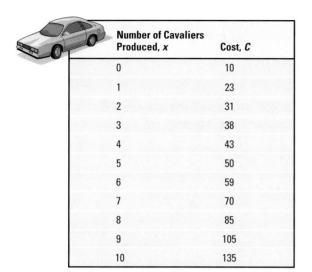

Number of Cavaliers Produced, x	Cost, C
0	10
1	23
2	31
3	38
4	43
5	50
6	59
7	70
8	85
9	105
10	135

(a) Draw a scatter diagram of the data. Comment on the type of relation that may exist between the two variables.

(b) Find the average rate of change in cost from four to five Cavaliers.

(c) What is the average rate of change in cost from eight to nine Cavaliers?

(d) The cubic function of best fit to these data is

$$C(x) = 0.2x^3 - 2.3x^2 + 14.3x + 10.2$$

Use this function to predict the cost of manufacturing eleven Cavaliers.

(e) Use a graphing utility to verify that the function given in part (d) is the cubic function of best fit.

(f) With a graphing utility, draw a scatter diagram of the data and then graph the cubic function of best fit on the scatter diagram.

(g) Interpret the y-intercept.

90. Cost of Printing   The following data represent the weekly cost C (in thousands of dollars) of printing textbooks and the number x (in thousands of units) of texts printed.

(a) Draw a scatter diagram of the data. Comment on the type of relation that may exist between the two variables.

(b) Find the average rate of change in cost from 10,000 to 13,000 textbooks.

(c) What is the average rate of change in cost from 18,000 to 20,000 textbooks?

(d) The cubic function of best fit to these data is

$$C(x) = 0.015x^3 - 0.595x^2 + 9.15x + 98.43$$

Use this function to predict the cost of printing 22,000 texts per week.

(e) Use a graphing utility to verify that the function given in part (d) is the cubic function of best fit.

(f) With a graphing utility, draw a scatter diagram of the data and then graph the cubic function of best fit on the scatter diagram.

(g) Interpret the y-intercept.

Number of Textbooks, x	Cost, C
0	100
5	128.1
10	144
13	153.5
17	161.2
18	162.6
20	166.3
23	178.9
25	190.2
27	221.8

91. Can the graph of a polynomial function have no y-intercept? Can it have no x-intercepts? Explain.

92. Write a few paragraphs that provide a general strategy for graphing a polynomial function. Be sure to mention the following: degree, intercepts, end behavior, and turning points.

93. Make up a polynomial that has the following characteristics: crosses the x-axis at -1 and 4, touches the x-axis at 0 and 2, and is above the x-axis between 0 and 2. Give your polynomial to a fellow classmate and ask for a written critique of your polynomial.

94. Make up two polynomials, not of the same degree, with the following characteristics: crosses the x-axis at -2, touches the x-axis at 1, and is above the x-axis between -2 and 1. Give your polynomials to a fellow classmate and ask for a written critique of your polynomials.

95. The graph of a polynomial function is always smooth and continuous. Name a function studied earlier that is smooth and not continuous. Name one that is continuous, but not smooth.

96. Which of the following statements are true regarding the graph of the cubic polynomial $f(x) = x^3 + bx^2 + cx + d$? (Give reasons for your conclusions.)

(a) It intersects the y-axis in one and only one point.

(b) It intersects the x-axis in at most three points.

(c) It intersects the x-axis at least once.

(d) For x very large, it behaves like the graph of $y = x^3$.

(e) It is symmetric with respect to the origin.

(f) It passes through the origin.

PREPARING FOR THIS SECTION

Before getting started, review the following concepts:

✓ Rational Expressions (Review, Section 7, pp. 55–64)

✓ Polynomial Division (Review, Section 5, pp. 43–45)

✓ Graph of $f(x) = 1/x$ (Section 2.2, Example 12, pp. 161–162)

✓ Graphing Techniques: Transformations (Section 3.4, pp. 242–252)

4.3 RATIONAL FUNCTIONS I

OBJECTIVES
1. Find the Domain of a Rational Function
2. Determine the Vertical Asymptotes of a Rational Function
3. Determine the Horizontal or Oblique Asymptotes of a Rational Function

Ratios of integers are called *rational numbers*. Similarly, ratios of polynomial functions are called *rational functions*.

> A **rational function** is a function of the form
>
> $$R(x) = \frac{p(x)}{q(x)}$$
>
> where p and q are polynomial functions and q is not the zero polynomial. The domain consists of all real numbers except those for which the denominator q is 0.

EXAMPLE 1

Finding the Domain of a Rational Function

1. (a) The domain of $R(x) = \dfrac{2x^2 - 4}{x + 5}$ consists of all real numbers x except -5, that is, $\{x \mid x \neq -5\}$.

(b) The domain of $R(x) = \dfrac{1}{x^2 - 4}$ consists of all real numbers x except -2 and 2, $\{x \mid x \neq -2, x \neq 2\}$.

(c) The domain of $R(x) = \dfrac{x^3}{x^2 + 1}$ consists of all real numbers.

(d) The domain of $R(x) = \dfrac{-x^2 + 2}{3}$ consists of all real numbers.

(e) The domain of $R(x) = \dfrac{x^2 - 1}{x - 1}$ consists of all real numbers x except 1, that is $\{x \mid x \neq 1\}$. ∎

It is important to observe that the functions

$$R(x) = \frac{x^2 - 1}{x - 1} \quad \text{and} \quad f(x) = x + 1$$

are not equal, since the domain of R is $\{x \mid x \neq 1\}$ and the domain of f is all real numbers.

NOW WORK PROBLEM 3.

If $R(x) = p(x)/q(x)$ is a rational function and if p and q have no common factors, then the rational function R is said to be in **lowest terms.** For a rational function $R(x) = p(x)/q(x)$ in lowest terms, the zeros, if any, of the numerator are the x-intercepts of the graph of R and so will play a major role in the graph of R. The zeros of the denominator of R [that is, the numbers x, if any, for which $q(x) = 0$], although not in the domain of R, also play a major role in the graph of R. We will discuss this role shortly.

We have already discussed the characteristics of the rational function $f(x) = 1/x$. (Refer to Example 12, page 161). The next rational function that we take up is $H(x) = 1/x^2$.

EXAMPLE 2

Graphing $y = \dfrac{1}{x^2}$

Analyze the graph: $H(x) = \dfrac{1}{x^2}$

Solution   The domain of $H(x) = 1/x^2$ consists of all real numbers x except 0. The graph has no y-intercept, because x can never equal 0. The graph has no x-intercept because the equation $H(x) = 0$ has no solution. Therefore, the graph of H will not cross either of the coordinate axes. Because

$$H(-x) = \frac{1}{(-x)^2} = \frac{1}{x^2} = H(x)$$

H is an even function, so its graph is symmetric with respect to the y-axis.

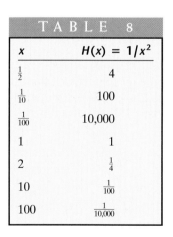

TABLE 8	
x	$H(x) = 1/x^2$
$\frac{1}{2}$	4
$\frac{1}{10}$	100
$\frac{1}{100}$	10,000
1	1
2	$\frac{1}{4}$
10	$\frac{1}{100}$
100	$\frac{1}{10,000}$

Table 8 shows the behavior of $H(x) = 1/x^2$ for selected positive numbers x (we will use symmetry to obtain the graph of H when $x < 0$). From the first three rows of Table 8, we see that, as the values of x approach (get closer to) 0, the values of $H(x)$ become larger and larger positive numbers. When this happens, we say that H is **unbounded in the positive direction.** We symbolize this by writing $H \to \infty$ (read as "H **approaches infinity**"). In calculus, **limits** are used to convey these ideas. There we use the symbolism $\lim_{x \to 0} H(x) = \infty$, read "the limit of $H(x)$ as x approaches zero equals infinity," to mean that $H(x) \to \infty$ as $x \to 0$.

Look at the last four rows of Table 8. As $x \to \infty$, the values of $H(x)$ approach 0 (the end behavior of the graph). In calculus, this is symbolized by writing $\lim_{x \to \infty} H(x) = 0$. Figure 35 shows the graph. Notice the use of red dashed lines to convey the ideas discussed above.

Figure 35

$H(x) = \dfrac{1}{x^2}$

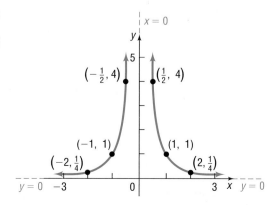

Sometimes transformations (shifting, compressing, stretching, and reflection) can be used to graph a rational function.

EXAMPLE 3

Using Transformations to Graph a Rational Function

Graph the rational function: $R(x) = \dfrac{1}{(x-2)^2} + 1$

Solution   First, we take note of the fact that the domain of R consists of all real numbers except $x = 2$. To graph R, we start with the graph of $y = 1/x^2$. See Figure 36 for the steps.

Figure 36

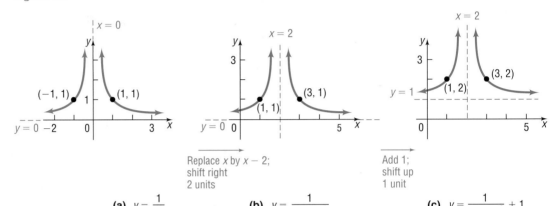

(a) $y = \dfrac{1}{x^2}$

Replace x by $x - 2$;
shift right
2 units

(b) $y = \dfrac{1}{(x-2)^2}$

Add 1;
shift up
1 unit

(c) $y = \dfrac{1}{(x-2)^2} + 1$

NOW WORK PROBLEM 21.

ASYMPTOTES

In Figure 36(c), notice that as the values of x become more negative, that is, as x becomes **unbounded in the negative direction** ($x \rightarrow -\infty$, read as "x **approaches negative infinity**"), the values $R(x)$ approach 1. In fact, we can conclude the following from Figure 36(c):

1. As $x \rightarrow -\infty$, the values $R(x)$ approach 1. $\left[\lim\limits_{x \to -\infty} R(x) = 1\right]$

2. As x approaches 2, the values $R(x) \rightarrow \infty$. $\left[\lim\limits_{x \to 2} R(x) = \infty\right]$

3. As $x \rightarrow \infty$, the values $R(x)$ approach 1. $\left[\lim\limits_{x \to \infty} R(x) = 1\right]$

This behavior of the graph is depicted by the vertical line $x = 2$ and the horizontal line $y = 1$. These lines are called *asymptotes* of the graph, which we define as follows:

Let R denote a function.

> If, as $x \rightarrow -\infty$ or as $x \rightarrow \infty$, the values of $R(x)$ approach some fixed number, L, then the line $y = L$ is a **horizontal asymptote** of the graph of R.

> If, as x approaches some number c, the values $|R(x)| \rightarrow \infty$, then the line $x = c$ is a **vertical asymptote** of the graph of R. The graph of R never intersects a vertical asymptote.

Figure 37

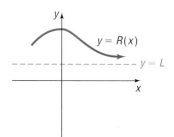

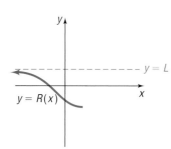

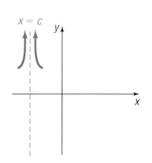

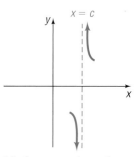

(a) End behavior:
As $x \to \infty$, the values of
$R(x)$ approach L. $[\lim\limits_{x \to \infty} R(x) = L]$.
That is, the points on the graph
of R are getting closer to
the line $y = L$; $y = L$ is a
horizontal asymptote.

(b) End behavior:
As $x \to -\infty$, the values
of $R(x)$ approach L.
$[\lim\limits_{x \to -\infty} R(x) = L]$. That is, the
points on the graph of R
are getting closer to the line
$y = L$; $y = L$ is a horizontal
asymptote.

(c) As x approaches c, the
values of $|R(x)| \to \infty$,
$[\lim\limits_{x \to c^-} R(x) = \infty$,
$\lim\limits_{x \to c^+} R(x) = \infty]$. That is,
the points on the graph
of R are getting closer to
the line $x = c$; $x = c$ is a
vertical asymptote.

(d) As x approaches c, the
values of $|R(x)| \to \infty$,
$[\lim\limits_{x \to c^-} R(x) = -\infty$;
$\lim\limits_{x \to c^+} R(x) = \infty]$. That is,
the points on the graph
of R are getting closer to
the line $x = c$; $x = c$ is a
vertical asymptote.

Figure 38
Oblique asymptote

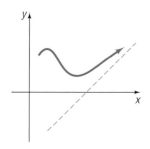

Even though the asymptotes of a function are not part of the graph of
the function, they provide information about how the graph looks. Figure 37
illustrates some of the possibilities.

A horizontal asymptote, when it occurs, describes a certain behavior of
the graph as $x \to \infty$ or as $x \to -\infty$, that is, its end behavior. The graph of a
function may intersect a horizontal asymptote.

A vertical asymptote, when it occurs, describes a certain behavior of the
graph when x is close to some number c. The graph of the function will never
intersect a vertical asymptote.

If an asymptote is neither horizontal nor vertical, it is called **oblique.** Figure 38 shows an oblique asymptote. An oblique asymptote, when it occurs,
describes the end behavior of the graph. The graph of a function may intersect an oblique asymptote.

FINDING ASYMPTOTES

② The vertical asymptotes, if any, of a rational function $R(x) = p(x)/q(x)$, in
lowest terms, are located at the zeros of the denominator of $q(x)$. Suppose
that r is a zero so $x - r$ is a factor. Now, as x approaches r, symbolized as
$x \to r$, the values of $x - r$ approach 0, causing the ratio to become un-
bounded, that is, causing $|R(x)| \to \infty$. Based on the definition, we conclude
that the line $x = r$ is a vertical asymptote.

Theorem

Locating Vertical Asymptotes

A rational function $R(x) = p(x)/q(x)$, *in lowest terms*, will have a vertical asymptote $x = r$ if r is a real zero of the *denominator q*. That is, if
$x - r$ is a factor of the denominator q of a rational function
$R(x) = p(x)/q(x)$, in lowest terms, then R will have the vertical asymptote $x = r$.

◼

WARNING: If a rational function is not in lowest terms, an application of this theorem may result in an incorrect listing of vertical asymptotes.   ◼

EXAMPLE 4	**Finding Vertical Asymptotes**

Find the vertical asymptotes, if any, of the graph of each rational function.

(a) $R(x) = \dfrac{x}{x^2 - 4}$ (b) $F(x) = \dfrac{x + 3}{x - 1}$

(c) $H(x) = \dfrac{x^2}{x^2 + 1}$ (d) $G(x) = \dfrac{x^2 - 9}{x^2 + 4x - 21}$

Solution (a) R is in lowest terms and the zeros of the denominator $x^2 - 4$ are -2 and 2. Hence, the lines $x = -2$ and $x = 2$ are the vertical asymptotes of the graph of R.

(b) F is in lowest terms and the only zero of the denominator is 1. Hence, the line $x = 1$ is the only vertical asymptote of the graph of F.

(c) H is in lowest terms and the denominator has no real zeros. Hence, the graph of H has no vertical asymptotes.

(d) Factor $G(x)$ to determine if it is in lowest terms.

$$G(x) = \frac{x^2 - 9}{x^2 + 4x - 21} = \frac{(x + 3)(x - 3)}{(x + 7)(x - 3)} = \frac{x + 3}{x + 7}, \qquad x \neq 3$$

The only zero of the denominator of $G(x)$ in lowest terms is -7. Hence, the line $x = -7$ is the only vertical asymptote of the graph of G. ■

As Example 4 points out, rational functions can have no vertical asymptotes, one vertical asymptote, or more than one vertical asymptote. However, the graph of a rational function will never intersect any of its vertical asymptotes. (Do you know why?)

 EXPLORATION   Graph each of the following rational functions:

$$R(x) = \frac{1}{x - 1} \qquad R(x) = \frac{1}{(x - 1)^2} \qquad R(x) = \frac{1}{(x - 1)^3} \qquad R(x) = \frac{1}{(x - 1)^4}$$

Each has the vertical asymptote $x = 1$. What happens to the value of $R(x)$ as x approaches 1 from the right side of the vertical asymptote; that is, what is $\lim_{x \to 1^+} R(x)$?

What happens to the value of $R(x)$ as x approaches 1 from the left side of the vertical asymptote; that is, what is $\lim_{x \to 1^-} R(x)$? How does the multiplicity of the zero in the denominator affect the graph of R? ■

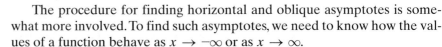

NOW WORK PROBLEM 35.
(FIND THE VERTICAL ASYMPTOTES, IF ANY.)

③   The procedure for finding horizontal and oblique asymptotes is somewhat more involved. To find such asymptotes, we need to know how the values of a function behave as $x \to -\infty$ or as $x \to \infty$.

If a rational function $R(x)$ is **proper,** that is, if the degree of the numerator is less than the degree of the denominator, then as $x \to -\infty$ or as $x \to \infty$, the values of $R(x)$ approach 0. Consequently, the line $y = 0$ (the x-axis) is a horizontal asymptote of the graph.

Theorem   If a rational function is proper, the line $y = 0$ is a horizontal asymptote of its graph.

■

| EXAMPLE 5 | **Finding Horizontal Asymptotes** |

Find the horizontal asymptotes, if any, of the graph of

$$R(x) = \frac{x - 12}{4x^2 + x + 1}$$

Solution The rational function R is proper, since the degree of the numerator, 1, is less than the degree of the denominator, 2. We conclude that the line $y = 0$ is a horizontal asymptote of the graph of R. ■

To see why $y = 0$ is a horizontal asymptote of the function R in Example 5, we need to investigate the behavior of R as $x \to -\infty$ and $x \to \infty$. When $|x|$ is unbounded, the numerator of R, which is $x - 12$, can be approximated by the power function $y = x$, while the denominator of R, which is $4x^2 + x + 1$, can be approximated by the power function $y = 4x^2$. Applying these ideas to $R(x)$, we find that

$$R(x) = \frac{x - 12}{4x^2 + x + 1} \approx \frac{x}{4x^2} = \frac{1}{4x} \to 0$$

For |x| unbounded   As $x \to -\infty$ or $x \to \infty$

This shows that the line $y = 0$ is a horizontal asymptote of the graph of R.

If a rational function $R(x) = p(x)/q(x)$ is **improper,** that is, if the degree of the numerator is greater than or equal to the degree of the denominator, we must use long division to write the rational function as the sum of a polynomial $f(x)$ plus a proper rational function $r(x)/q(x)$. That is, we write

$$R(x) = \frac{p(x)}{q(x)} = f(x) + \frac{r(x)}{q(x)}$$

where $f(x)$ is a polynomial and $r(x)/q(x)$ is a proper rational function. Since $r(x)/q(x)$ is proper, then $r(x)/q(x) \to 0$ as $x \to -\infty$ or as $x \to \infty$. As a result,

$$R(x) = \frac{p(x)}{q(x)} \to f(x), \qquad \text{as } x \to -\infty \text{ or as } x \to \infty$$

The possibilities are listed next.

1. If $f(x) = b$, a constant, then the line $y = b$ is a horizontal asymptote of the graph of R.
2. If $f(x) = ax + b$, $a \neq 0$, then the line $y = ax + b$ is an oblique asymptote of the graph of R.
3. In all other cases, the graph of R approaches the graph of f, and there are no horizontal or oblique asymptotes.

The following examples demonstrate these conclusions.

| EXAMPLE 6 | **Finding Horizontal or Oblique Asymptotes** |

Find the horizontal or oblique asymptotes, if any, of the graph of

$$H(x) = \frac{3x^4 - x^2}{x^3 - x^2 + 1}$$

Solution The rational function H is improper, since the degree of the numerator, 4, is larger than the degree of the denominator, 3. To find any horizontal or oblique asymptotes, we use long division.

$$
\begin{array}{r}
3x + 3 \\
x^3 - x^2 + 1 \overline{)\,3x^4 - x^2 } \\
\underline{3x^4 - 3x^3 + 3x} \\
3x^3 - x^2 - 3x \\
\underline{3x^3 - 3x^2 + 3} \\
2x^2 - 3x - 3
\end{array}
$$

As a result,

$$
H(x) = \frac{3x^4 - x^2}{x^3 - x^2 + 1} = 3x + 3 + \frac{2x^2 - 3x - 3}{x^3 - x^2 + 1}
$$

Then, as $x \to -\infty$ or as $x \to \infty$,

$$
\frac{2x^2 - 3x - 3}{x^3 - x^2 + 1} \approx \frac{2x^2}{x^3} = \frac{2}{x} \to 0
$$

So, as $x \to -\infty$ or as $x \to \infty$, we have $H(x) \to 3x + 3$. We conclude that the graph of the rational function H has an oblique asymptote $y = 3x + 3$. ■

EXAMPLE 7   **Finding Horizontal or Oblique Asymptotes**

Find the horizontal or oblique asymptotes, if any, of the graph of

$$
R(x) = \frac{8x^2 - x + 2}{4x^2 - 1}
$$

Solution The rational function R is improper, since the degree of the numerator, 2, equals the degree of the denominator, 2. To find any horizontal or oblique asymptotes, we use long division.

$$
\begin{array}{r}
2 \\
4x^2 - 1 \overline{)\,8x^2 - x + 2} \\
\underline{8x^2 - 2} \\
-x + 4
\end{array}
$$

As a result,

$$
R(x) = \frac{8x^2 - x + 2}{4x^2 - 1} = 2 + \frac{-x + 4}{4x^2 - 1}
$$

Then, as $x \to -\infty$ or as $x \to \infty$,

$$
\frac{-x + 4}{4x^2 - 1} \approx \frac{-x}{4x^2} = \frac{-1}{4x} \to 0
$$

So, as $x \to -\infty$ or as $x \to \infty$, we have $R(x) \to 2$. We conclude that $y = 2$ is a horizontal asymptote of the graph. ■

In Example 7, we note that the quotient 2 obtained by long division is the quotient of the leading coefficients of the numerator polynomial and the denominator polynomial $\left(\frac{8}{4}\right)$. This means that we can avoid the long division process for rational functions whose numerator and denominator *are of the same degree* and conclude that the quotient of the leading coefficients will give us the horizontal asymptote.

NOW WORK PROBLEM 31.

EXAMPLE 8   **Finding Horizontal or Oblique Asymptotes**

Find the horizontal or oblique asymptotes, if any, of the graph of

$$G(x) = \frac{2x^5 - x^3 + 2}{x^3 - 1}$$

Solution   The rational function G is improper, since the degree of the numerator, 5, is larger than the degree of the denominator, 3. To find any horizontal or oblique asymptotes, we use long division.

$$
\begin{array}{r}
2x^2 - 1 \\
x^3 - 1 \overline{\smash{\big)}\ 2x^5 - x^3 \qquad\ + 2} \\
\underline{2x^5 \qquad\quad - 2x^2} \\
-x^3 + 2x^2 + 2 \\
\underline{-x^3 \qquad\ + 1} \\
2x^2 + 1
\end{array}
$$

As a result,

$$G(x) = \frac{2x^5 - x^3 + 2}{x^3 - 1} = 2x^2 - 1 + \frac{2x^2 + 1}{x^3 - 1}$$

Then, as $x \to -\infty$ or as $x \to \infty$,

$$\frac{2x^2 + 1}{x^3 - 1} \approx \frac{2x^2}{x^3} = \frac{2}{x} \to 0$$

So, as $x \to -\infty$ or as $x \to \infty$, we have $G(x) \to 2x^2 - 1$. We conclude that, for large values of $|x|$, the graph of G approaches the graph of $y = 2x^2 - 1$. That is, the graph of G will look like the graph of $y = 2x^2 - 1$ as $x \to -\infty$ or $x \to \infty$. Since $y = 2x^2 - 1$ is not a linear function, G has no horizontal or oblique asymptotes.   ■

We now summarize the procedure for finding horizontal and oblique asymptotes.

SUMMARY

FINDING HORIZONTAL AND OBLIQUE ASYMPTOTES OF A RATIONAL FUNCTION R

Consider the rational function

$$R(x) = \frac{p(x)}{q(x)} = \frac{a_n x^n + a_{n-1} x^{n-1} + \cdots + a_1 x + a_0}{b_m x^m + b_{m-1} x^{m-1} + \cdots + b_1 x + b_0}$$

in which the degree of the numerator is n and the degree of the denominator is m.

1. If $n < m$, then R is a proper rational function, and the graph of R will have the horizontal asymptote $y = 0$ (the x-axis).
2. If $n \geq m$, then R is improper. Here long division is used.
   (a) If $n = m$, the quotient obtained will be a number $L(= a_n/b_m)$, and the line $y = L(= a_n/b_m)$ is a horizontal asymptote.
   (b) If $n = m + 1$, the quotient obtained is of the form $ax + b$ (a polynomial of degree 1), and the line $y = ax + b$ is an oblique asymptote.
   (c) If $n > m + 1$, the quotient obtained is a polynomial of degree 2 or higher, and R has neither a horizontal nor an oblique asymptote. In this case, for x unbounded, the graph of R will behave like the graph of the quotient.

NOTE: The graph of a rational function either has one horizontal or one oblique asymptote or else has no horizontal and no oblique asymptote.

4.3   EXERCISES

In Problems 1–12, find the domain of each rational function.

1. $R(x) = \dfrac{4x}{x - 3}$

2. $R(x) = \dfrac{5x^2}{3 + x}$

3. $H(x) = \dfrac{-4x^2}{(x - 2)(x + 4)}$

4. $G(x) = \dfrac{6}{(x + 3)(4 - x)}$

5. $F(x) = \dfrac{3x(x - 1)}{2x^2 - 5x - 3}$

6. $Q(x) = \dfrac{-x(1 - x)}{3x^2 + 5x - 2}$

7. $R(x) = \dfrac{x}{x^3 - 8}$

8. $R(x) = \dfrac{x}{x^4 - 1}$

9. $H(x) = \dfrac{3x^2 + x}{x^2 + 4}$

10. $G(x) = \dfrac{x - 3}{x^4 + 1}$

11. $R(x) = \dfrac{3(x^2 - x - 6)}{4(x^2 - 9)}$

12. $F(x) = \dfrac{-2(x^2 - 4)}{3(x^2 + 4x + 4)}$

In Problems 13–18, use the graph shown to find:
   (a) *The domain and range of each function*   (b) *The intercepts, if any*   (c) *Horizontal asymptotes, if any*
   (d) *Vertical asymptotes, if any*   (e) *Oblique asymptotes, if any*

13.

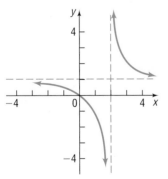

14.

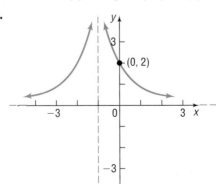

15.

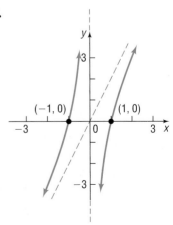

16.

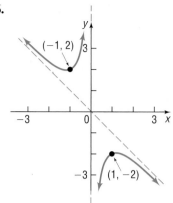

17.

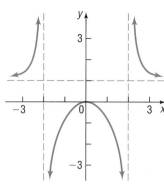

18.

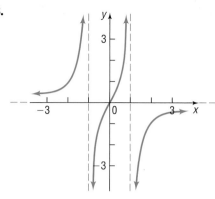

In Problems 19–30, graph each rational function using transformations.

19. $F(x) = 2 + \dfrac{1}{x}$

20. $Q(x) = 3 + \dfrac{1}{x^2}$

21. $R(x) = \dfrac{1}{(x-1)^2}$

22. $R(x) = \dfrac{3}{x}$

23. $H(x) = \dfrac{-2}{x+1}$

24. $G(x) = \dfrac{2}{(x+2)^2}$

25. $R(x) = \dfrac{-1}{x^2 + 4x + 4}$

26. $R(x) = \dfrac{1}{x-1} + 1$

27. $G(x) = 1 + \dfrac{2}{(x-3)^2}$

28. $F(x) = 2 - \dfrac{1}{x+1}$

29. $R(x) = \dfrac{x^2 - 4}{x^2}$

30. $R(x) = \dfrac{x-4}{x}$

In Problems 31–42, find the vertical, horizontal, and oblique asymptotes, if any, of each rational function.

31. $R(x) = \dfrac{3x}{x+4}$

32. $R(x) = \dfrac{3x+5}{x-6}$

33. $H(x) = \dfrac{x^4 + 2x^2 + 1}{x^2 - x + 1}$

34. $G(x) = \dfrac{-x^2 + 1}{x+5}$

35. $T(x) = \dfrac{x^3}{x^4 - 1}$

36. $P(x) = \dfrac{4x^5}{x^3 - 1}$

37. $Q(x) = \dfrac{5 - x^2}{3x^4}$

38. $F(x) = \dfrac{-2x^2 + 1}{2x^3 + 4x^2}$

39. $R(x) = \dfrac{3x^4 + 4}{x^3 + 3x}$

40. $R(x) = \dfrac{6x^2 + x + 12}{3x^2 - 5x - 2}$

41. $G(x) = \dfrac{x^3 - 1}{x - x^2}$

42. $F(x) = \dfrac{x-1}{x - x^3}$

---

43. Gravity In physics, it is established that the acceleration due to gravity, g, at a height h meters above sea level is given by

$$g(h) = \frac{3.99 \times 10^{14}}{(6.374 \times 10^6 + h)^2}$$

where 6.374×10^6 is the radius of Earth in meters.
  (a) What is the acceleration due to gravity at sea level?
  (b) The Sears Tower in Chicago, Illinois, is 443 meters tall. What is the acceleration due to gravity at the top of the Sears Tower?
  (c) The peak of Mount Everest is 8848 meters above sea level. What is the acceleration due to gravity on the peak of Mount Everest?
  (d) Find the horizontal asymptote of $g(h)$.
  (e) Using your graphing utility, graph $g(h)$.
  (f) Solve $g(h) = 0$. How do you interpret your answer?

44. Population Model A rare species of insect was discovered in the Amazon Rain Forest. In order to protect the species, environmentalists declare the insect endangered and transplant the insects into a protected area. The population of the insect t months after being transplanted is given by P.

$$P(t) = \frac{50(1 + 0.5t)}{(2 + 0.01t)}$$

  (a) How many insects were discovered? In other words, what was the population when $t = 0$?
  (b) What will the population be after 5 years?
  (c) Determine the horizontal asymptote of $P(t)$. What is the largest population that the protected area can sustain?
  (d) Using your graphing utility, graph $P(t)$.
  (e) TRACE $P(t)$ for large values of t to verify your answer to (c).

45. If the graph of a rational function R has the vertical asymptote $x = 4$, then the factor $x - 4$ must be present in the denominator of R. Explain why.

46. If the graph of a rational function R has the horizontal asymptote $y = 2$, then the degree of the numerator of R equals the degree of the denominator of R. Explain why.

47. Can the graph of a rational function have both a horizontal and an oblique asymptote? Explain.

48. Make up a rational function that has $y = 2x + 1$ as an oblique asymptote. Explain the methodology that you used.

PREPARING FOR THIS SECTION

Before getting started, review the following concept:

✓ Intercepts (Section 2.2, pp. 157–158)   ✓ Symmetry (Section 2.2, pp. 159–162)

4.4 RATIONAL FUNCTIONS II: ANALYZING GRAPHS

OBJECTIVES ① Analyze the Graph of a Rational Function
② Solve Applied Problems Involving Rational Functions

GRAPHING RATIONAL FUNCTIONS

① We commented earlier that calculus provides the tools required to graph a polynomial function accurately. The same holds true for rational functions. However, we can gather together quite a bit of information about their graphs to get an idea of the general shape and position of the graph.

In the examples that follow, we will analyze the graph of a rational function $R(x) = p(x)/q(x)$ by applying the following steps:

> **ANALYZING THE GRAPH OF A RATIONAL FUNCTION**
>
> STEP 1: Find the domain of the rational function.
>
> STEP 2: Locate the intercepts, if any, of the graph. The x-intercepts, if any, of $R(x) = p(x)/q(x)$, in lowest terms, are numbers in the domain that satisfy the equation $p(x) = 0$. The y-intercept, if there is one, is $R(0)$.
>
> STEP 3: Test for symmetry. Replace x by $-x$ in $R(x)$. If $R(-x) = R(x)$, there is symmetry with respect to the y-axis; if $R(-x) = -R(x)$, there is symmetry with respect to the origin.
>
> STEP 4: Write R in lowest terms and find the real zeros of the denominator. With R in lowest terms, each zero will give rise to a vertical asymptote.
>
> STEP 5: Locate the horizontal or oblique asymptotes, if any, using the procedure given earlier. Determine points, if any, at which the graph of R intersects these asymptotes.
>
> STEP 6: Determine where the graph is above the x-axis and where the graph is below the x-axis, using the zeros of the numerator and the denominator to divide the x-axis into intervals.
>
> STEP 7: Graph the asymptotes, if any, found in Steps 4 and 5. Plot the points found in Steps 2, 5, and 6. Use all the information to connect the points and graph R.

EXAMPLE 1   **Analyzing the Graph of a Rational Function**

Analyze the graph of the rational function: $R(x) = \dfrac{x - 1}{x^2 - 4}$

Solution   First, we factor both the numerator and the denominator of R.

$$R(x) = \frac{x - 1}{(x + 2)(x - 2)}$$

R is in lowest terms.

STEP 1: The domain of R is $\{x \mid x \neq -2, x \neq 2\}$.

STEP 2: We locate the x-intercepts by finding the zeros of the numerator. By inspection, 1 is the only x-intercept. The y-intercept is $R(0) = \frac{1}{4}$.

STEP 3: Because

$$R(-x) = \frac{-x - 1}{x^2 - 4}$$

we conclude that R is neither even nor odd. There is no symmetry with respect to the y-axis or the origin.

STEP 4: We locate the vertical asymptotes by factoring the denominator: $x^2 - 4 = (x + 2)(x - 2)$. Since R is in lowest terms, the graph of R has two vertical asymptotes: the lines $x = -2$ and $x = 2$.

STEP 5: The degree of the numerator is less than the degree of the denominator, so R is proper and the line $y = 0$ (the x-axis) is a horizontal asymptote of the graph. To determine if the graph of R intersects the horizontal asymptote, we solve the equation $R(x) = 0$.

$$\frac{x - 1}{x^2 - 4} = 0$$

$$x - 1 = 0$$

$$x = 1$$

The only solution is $x = 1$, so the graph of R intersects the horizontal asymptote at $(1, 0)$.

STEP 6: The zero of the numerator, 1, and the zeros of the denominator, -2 and 2, divide the x-axis into four intervals:

$$(-\infty, -2) \qquad (-2, 1) \qquad (1, 2) \qquad (2, \infty)$$

Now construct Table 9.

TABLE 9				
		-2	1	2
Interval	$(-\infty, -2)$	$(-2, 1)$	$(1, 2)$	$(2, \infty)$
Number Chosen	-3	0	$\frac{3}{2}$	3
Value of R	$R(-3) = -0.8$	$R(0) = \frac{1}{4}$	$R(\frac{3}{2}) = -\frac{2}{7}$	$R(3) = 0.4$
Location of Graph	Below x-axis	Above x-axis	Below x-axis	Above x-axis
Point on Graph	$(-3, -0.8)$	$(0, \frac{1}{4})$	$(\frac{3}{2}, -\frac{2}{7})$	$(3, 0.4)$

STEP 7: We begin by graphing the asymptotes and plotting the points found in Steps 2, 5, and 6. See Figure 39(a). Next, we determine the behavior of the graph near the asymptotes. Since the x-axis is a horizontal asymptote and the graph lies below the x-axis for $-\infty < x < -2$, we can sketch a portion of the graph by placing a small arrow to the far left and under the x-axis. Since the line $x = -2$ is a vertical asymptote and the graph lies below the x-axis for $-\infty < x < -2$, we continue the sketch with an arrow placed well below the x-axis and approaching the line $x = -2$ on the left. Similar explanations account for the positions of the other portions of the graph. In particular, note how we use the facts that the graph lies above the x-axis for $-2 < x < 1$ and below the x-axis for $1 < x < 2$ and $(1, 0)$ is an x-intercept to draw the conclusion that the graph crosses the x-axis at $(1, 0)$. Figure 39(b) shows the complete graph.

Figure 39

$$R(x) = \frac{x - 1}{x^2 - 4}$$

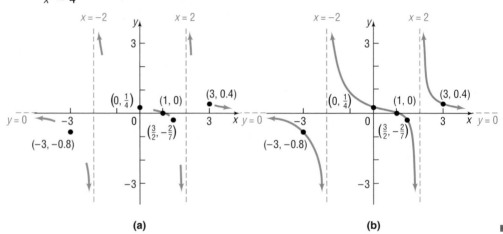

(a)　　　　　　　　(b)

EXPLORATION Graph $R(x) = (x - 1)/(x^2 - 4)$.

SOLUTION The analysis just completed in Example 1 helps us to set the viewing rectangle to obtain a complete graph. Figure 40(a) shows the graph of $R(x) = \dfrac{x - 1}{x^2 - 4}$ in connected mode, and Figure 40(b) shows it in dot mode. Notice in Figure 40(a) that the graph has vertical lines at $x = -2$ and $x = 2$. This is due to the fact that, when the graphing utility is in connected mode, it will "connect the dots" between consecutive pixels. We know that the graph

Figure 40

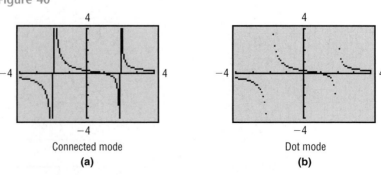

Connected mode　　　　Dot mode
(a)　　　　　　　　(b)

of R does not cross the lines $x = -2$ and $x = 2$, since R is not defined at $x = -2$ or $x = 2$. So, when graphing rational functions, dot mode should be used to avoid extraneous vertical lines that are not part of the graph.

NOW WORK PROBLEM **1**.

| EXAMPLE 2 | **Analyzing the Graph of a Rational Function** |

Analyze the graph of the rational function: $R(x) = \dfrac{x^2 - 1}{x}$

Solution   STEP 1: The domain of R is $\{x \mid x \neq 0\}$.

STEP 2: The graph has two x-intercepts: -1 and 1. There is no y-intercept, since x cannot equal 0.

STEP 3: Since $R(-x) = -R(x)$, the function is odd and the graph is symmetric with respect to the origin.

STEP 4: R is in lowest terms, so the graph of $R(x)$ has the line $x = 0$ (the y-axis) as a vertical asymptote.

STEP 5: The rational function R is improper, since the degree of the numerator, 2, is larger than the degree of the denominator, 1. To find any horizontal or oblique asymptotes, we use long division.

$$
\begin{array}{r}
x \phantom{{}-1} \\
x \overline{)\, x^2 - 1} \\
\underline{x^2 \phantom{{}- 1}} \\
-1
\end{array}
$$

The quotient is x, so the line $y = x$ is an oblique asymptote of the graph. To determine whether the graph of R intersects the asymptote $y = x$, we solve the equation $R(x) = x$.

$$
R(x) = \frac{x^2 - 1}{x} = x
$$
$$
x^2 - 1 = x^2
$$
$$
-1 = 0 \qquad \text{Impossible}
$$

We conclude that the equation $(x^2 - 1)/x = x$ has no solution, so the graph of $R(x)$ does not intersect the line $y = x$.

STEP 6: The zeros of the numerator are -1 and 1; the denominator has the zero 0. We divide the x-axis into four intervals:

$$(-\infty, -1) \qquad (-1, 0) \qquad (0, 1) \qquad (1, \infty)$$

Now we construct Table 10.

TABLE 10				
		-1	0	1 $\rightarrow x$
Interval	$(-\infty, -1)$	$(-1, 0)$	$(0, 1)$	$(1, \infty)$
Number Chosen	-2	$-\frac{1}{2}$	$\frac{1}{2}$	2
Value of R	$R(-2) = -\frac{3}{2}$	$R(-\frac{1}{2}) = \frac{3}{2}$	$R(\frac{1}{2}) = -\frac{3}{2}$	$R(2) = \frac{3}{2}$
Location of Graph	Below x-axis	Above x-axis	Below x-axis	Above x-axis
Point on Graph	$(-2, -\frac{3}{2})$	$(-\frac{1}{2}, \frac{3}{2})$	$(\frac{1}{2}, -\frac{3}{2})$	$(2, \frac{3}{2})$

STEP 7:   Figure 41(a) shows a partial graph using the facts that we have gathered. The complete graph is given in Figure 41(b).

Figure 41
$$R(x) = \frac{x^2 - 1}{x}$$

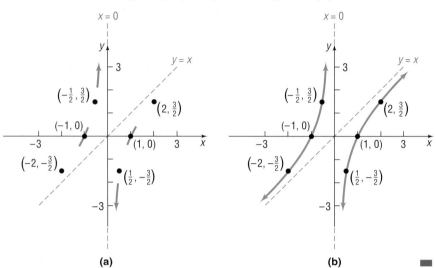

(a)   (b)

SEEING THE CONCEPT: Graph $R(x) = (x^2 - 1)/x$ and compare what you see with Figure 41(b). Could you have predicted from the graph that $y = x$ is an oblique asymptote? Graph $y = x$ and ZOOM-OUT. What do you observe?

NOW WORK PROBLEM 9.

EXAMPLE 3   **Analyzing the Graph of a Rational Function**

Analyze the graph of the rational function: $R(x) = \dfrac{x^4 + 1}{x^2}$

Solution   STEP 1:   The domain of R is $\{x \mid x \neq 0\}$.

STEP 2:   The graph has no x-intercepts and no y-intercepts.

STEP 3:   Since $R(-x) = R(x)$, the function is even and the graph is symmetric with respect to the y-axis.

STEP 4: R is in lowest terms, so the graph of $R(x)$ has the line $x = 0$ (the y-axis) as a vertical asymptote.

STEP 5:   The rational function R is improper. To find any horizontal or oblique asymptotes, we use long division.

$$\begin{array}{r} x^2 \\ x^2 \overline{) x^4 + 1} \\ \underline{x^4 } \\ 1 \end{array}$$

The quotient is x^2, so the graph has no horizontal or oblique asymptotes. However, the graph of R will approach the graph of $y = x^2$ as $x \to -\infty$ and as $x \to \infty$.

STEP 6:   The numerator has no zeros, and the denominator has one zero at 0. We divide the x-axis into the two intervals

$$(-\infty, 0) \qquad (0, \infty)$$

Now we construct Table 11.

STEP 7:   Figure 42 shows the graph.

TABLE 11		
	|—————•—————→ x \ 0	
Interval	$(-\infty, 0)$	$(0, \infty)$
Number Chosen	-1	1
Value of R	$R(-1) = 2$	$R(1) = 2$
Location of Graph	Above x-axis	Above x-axis
Point on Graph	$(-1, 2)$	$(1, 2)$

Figure 42

$R(x) = \dfrac{x^4 + 1}{x^2}$

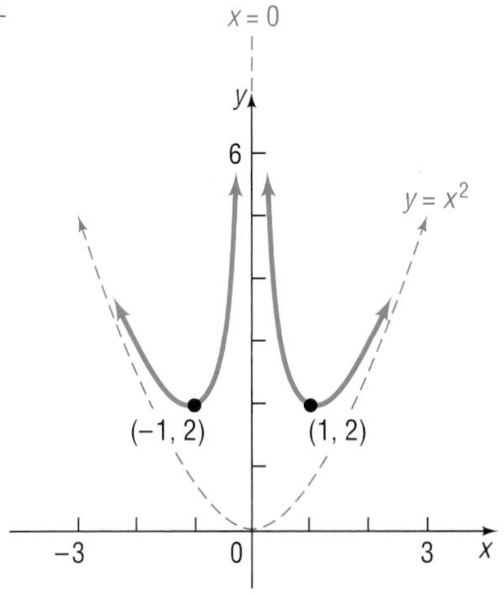

SEEING THE CONCEPT: Graph $R(x) = (x^4 + 1)/x^2$ and compare what you see with Figure 42. Use MINIMUM to find the two turning points. Enter $y = x^2$ and ZOOM-OUT. What do you see?

NOW WORK PROBLEM **7.**

EXAMPLE 4   **Analyzing the Graph of a Rational Function**

Analyze the graph of the rational function: $R(x) = \dfrac{3x^2 - 3x}{x^2 + x - 12}$

Solution   We factor R to get

$$R(x) = \frac{3x(x - 1)}{(x + 4)(x - 3)}$$

R is in lowest terms.

STEP 1: The domain of R is $\{x \mid x \neq -4, x \neq 3\}$.

STEP 2: The graph has two x-intercepts: 0 and 1. The y-intercept is $R(0) = 0$.

STEP 3: There is no symmetry with respect to the y-axis or the origin.

STEP 4: Since R is in lowest terms, the graph of R has two vertical asymptotes: $x = -4$ and $x = 3$.

STEP 5: Since the degree of the numerator equals the degree of the denominator, the graph has a horizontal asymptote. To find it, we either use long division or form the quotient of the leading coefficient of the numerator, 3, and the leading coefficient of the denominator, 1. Thus, the graph of R has the horizontal asymptote $y = 3$. To find out whether the graph of R intersects the asymptote, we solve the equation $R(x) = 3$.

$$R(x) = \frac{3x^2 - 3x}{x^2 + x - 12} = 3$$
$$3x^2 - 3x = 3x^2 + 3x - 36$$
$$-6x = -36$$
$$x = 6$$

The graph intersects the line $y = 3$ only at $x = 6$, and $(6, 3)$ is a point on the graph of R.

STEP 6: The zeros of the numerator, 0 and 1, and the zeros of the denominator, -4 and 3, divide the x-axis into five intervals:

$$(-\infty, -4) \qquad (-4, 0) \qquad (0, 1) \qquad (1, 3) \qquad (3, \infty)$$

Now we construct Table 12.

TABLE 12					
		-4	0	1	3
Interval	$(-\infty, -4)$	$(-4, 0)$	$(0, 1)$	$(1, 3)$	$(3, \infty)$
Number Chosen	-5	-2	$\frac{1}{2}$	2	4
Value of R	$R(-5) = 11.25$	$R(-2) = -1.8$	$R(\frac{1}{2}) = \frac{1}{15}$	$R(2) = -1$	$R(4) = 4.5$
Location of Graph	Above x-axis	Below x-axis	Above x-axis	Below x-axis	Above x-axis
Point on Graph	$(-5, 11.25)$	$(-2, -1.8)$	$(\frac{1}{2}, \frac{1}{15})$	$(2, -1)$	$(4, 4.5)$

STEP 7: Figure 43(a) shows a partial graph. Notice that we have not yet used the fact that the line $y = 3$ is a horizontal asymptote, because we do

Figure 43

$$R(x) = \frac{3x^2 - 3x}{x^2 + x - 12}$$

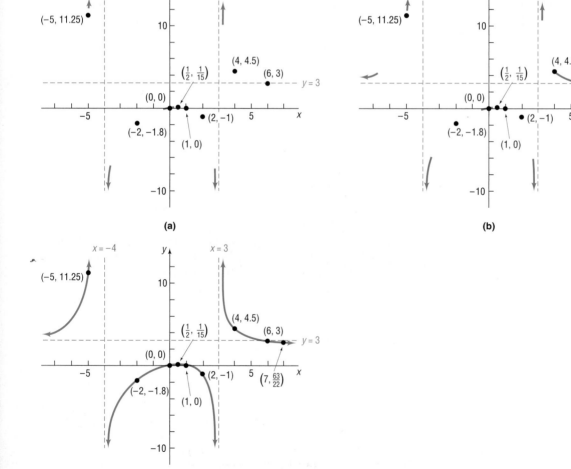

not know yet whether the graph of R crosses or touches the line $y = 3$ at $(6, 3)$. To see whether the graph, in fact, crosses or touches the line $y = 3$, we plot an additional point to the right of $(6, 3)$. We use $x = 7$ to find $R(7) = \frac{63}{22} < 3$. The graph crosses $y = 3$ at $x = 6$. Because $(6, 3)$ is the only point where the graph of R intersects the asymptote $y = 3$, the graph must approach the line $y = 3$ from above as $x \to -\infty$ and approach the line $y = 3$ from below as $x \to \infty$. See Figure 43(b). The completed graph is shown in Figure 43(c).   ■

EXPLORATION   Graph $R(x) = \dfrac{3x^2 - 3x}{x^2 + x - 12}$.

SOLUTION   Figure 44 shows the graph in connected mode, and Figure 45(a) shows it in dot mode. Neither graph displays well the behavior between the two x-intercepts, 0 and 1. Nor do they display well the fact that the graph crosses the horizontal asymptote at $(6, 3)$. To see these parts better, we graph R for $-1 \le x \le 2$, Figure 45(b), and for $4 \le x \le 60$, Figure 46(b).

The new graphs reflect the behavior produced by the analysis. Furthermore, we observe two turning points, one between 0 and 1 and the other to the right of 4. Rounded to two decimal places, these turning points are $(0.52, 0.07)$ and $(11.48, 2.75)$.

Figure 44

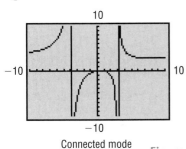

Connected mode

Figure 45

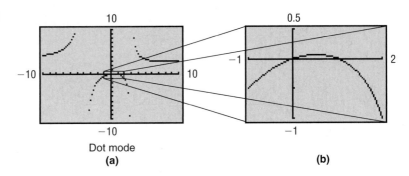

Dot mode
(a)   (b)

Figure 46

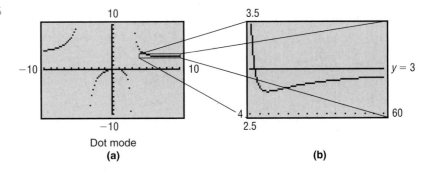

Dot mode
(a)   (b)

EXAMPLE 5   **Analyzing the Graph of a Rational Function with a Hole**

Analyze the graph of the rational function: $R(x) = \dfrac{2x^2 - 5x + 2}{x^2 - 4}$

Solution   We factor R and obtain

$$R(x) = \frac{(2x - 1)(x - 2)}{(x + 2)(x - 2)}$$

In lowest terms,

$$R(x) = \frac{2x - 1}{x + 2}, \qquad x \neq -2$$

STEP 1: The domain of R is $\{x \mid x \neq -2, x \neq 2\}$.

STEP 2: The graph has one x-intercept: $1/2$. The y-intercept is $R(0) = -1/2$.

STEP 3: Because

$$R(-x) = \frac{2x^2 + 5x + 2}{x^2 - 4}$$

we conclude that R is neither even nor odd. Thus, there is no symmetry with respect to the y-axis or the origin.

STEP 4: The graph has one vertical asymptote, $x = -2$, since $x + 2$ is the only factor of the denominator of $R(x)$ *in lowest terms*. However, the rational function is undefined at both $x = 2$ and $x = -2$.

STEP 5: Since the degree of the numerator equals the degree of the denominator, the graph has a horizontal asymptote. To find it, we either use long division or form the quotient of the leading coefficient of the numerator, 2, and the leading coefficient of the denominator, 1. Thus, the graph of R has the horizontal asymptote $y = 2$. To find out whether the graph of R intersects the asymptote, we solve the equation $R(x) = 2$.

$$R(x) = \frac{2x - 1}{x + 2} = 2$$

$$2x - 1 = 2(x + 2)$$

$$2x - 1 = 2x + 4$$

$$-1 = 4 \qquad \text{Impossible}$$

The graph does not intersect the line $y = 2$.

STEP 6: The zeros of the numerator and denominator, $-2, 1/2,$ and 2, divide the x-axis into four intervals:

$$(-\infty, -2) \qquad (-2, 1/2) \qquad (1/2, 2) \qquad (2, \infty)$$

Now we construct Table 13.

TABLE	1 3			
	-2		**1/2**	**2** → x
Interval	$(-\infty, -2)$	$(-2, 1/2)$	$(1/2, 2)$	$(2, \infty)$
Number Chosen	-3	-1	1	3
Value of R	$R(-3) = 7$	$R(-1) = -3$	$R(1) = \frac{1}{3}$	$R(3) = 1$
Location of Graph	Above x-axis	Below x-axis	Above x-axis	Above x-axis
Point on Graph	$(-3, 7)$	$(-1, -3)$	$(1, \frac{1}{3})$	$(3, 1)$

STEP 7: See Figure 47. Notice the vertical asymptote $x = -2$ and the hole at the point $(2, 3/4)$. R is not defined at -2 and 2.

Figure 47

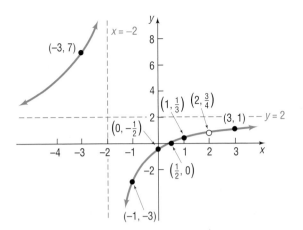

As Example 5 shows, the zeros of the denominator of a rational function give rise to either vertical asymptotes or holes on the graph.

 EXPLORATION   Graph $R(x) = \dfrac{2x^2 - 5x + 2}{x^2 - 4}$. Do you see the hole at $(2, 3/4)$? TRACE along the graph. Did you obtain an ERROR at $x = 2$? Are you convinced that an algebraic analysis of a rational function is required in order to accurately interpret the graph obtained with a graphing utility?

NOW WORK PROBLEM **27.**

We now discuss the problem of finding a rational function from its graph.

EXAMPLE 6	**Constructing a Rational Function from Its Graph**

Make up a rational function that might have the graph shown in Figure 48.

Figure 48

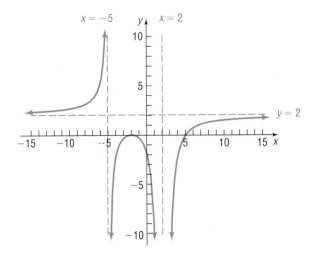

Solution   The numerator of a rational function $R(x) = p(x)/q(x)$ in lowest terms determines the x-intercepts of its graph. The graph shown in Figure 48 has x-intercepts -2 (even multiplicity; graph touches the x-axis) and 5 (odd multiplicity; graph crosses the x-axis). So, one possibility for the numerator is $p(x) = (x + 2)^2(x - 5)$. The denominator of a rational function in lowest terms determines the vertical asymptotes of its graph. The vertical asymptotes of the graph are $x = -5$ and $x = 2$. Since $R(x)$ approaches ∞ from the left of $x = -5$ and $R(x)$ approaches $-\infty$ from the right of $x = -5$, we know that $(x + 5)$ is a factor of odd multiplicity in $q(x)$. Also, $R(x)$ approaches $-\infty$ from both sides of $x = 2$, so $(x - 2)$ is a factor of even multiplicity in $q(x)$. A possibility for the denominator is $q(x) = (x + 5)(x - 2)^2$.

So far we have $R(x) = \dfrac{(x + 2)^2(x - 5)}{(x + 5)(x - 2)^2}$. However, the horizontal asymptote of the graph given in Figure 48 is $y = 2$, so we know that the degree of the numerator must equal the degree in the denominator and the quotient of leading coefficients must be $2/1$. This leads to

$$R(x) = \frac{2(x + 2)^2(x - 5)}{(x + 5)(x - 2)^2}$$

Figure 49

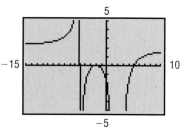

 Check: Figure 49 shows the graph of R on a graphing utility. Since Figure 49 looks similar to Figure 48, we have found a rational function R for the graph in Figure 48.

NOW WORK PROBLEM 39.

(2) **APPLICATION**

EXAMPLE 7   **Finding the Least Cost of a Can**

Reynolds Metal Company manufactures aluminum cans in the shape of a cylinder with a capacity of 500 cubic centimeters $\left(\frac{1}{2}\right.$ liter$)$. The top and bottom of the can are made of a special aluminum alloy that costs 0.05¢ per square centimeter. The sides of the can are made of material that costs 0.02¢ per square centimeter.

(a) Express the cost of material for the can as a function of the radius r of the can.

(b) Use a graphing utility to graph the function $C = C(r)$.

(c) What value of r will result in the least cost?

(d) What is this least cost?

Solution (a) Figure 50 illustrates the situation. Notice that the material required to produce a cylindrical can of height h and radius r consists of a rectangle of area $2\pi rh$ and two circles, each of area πr^2. The total cost C (in cents) of manufacturing the can is therefore

Figure 50

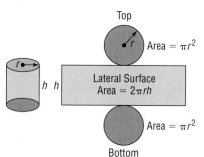

$$C = \text{Cost of top and bottom} + \text{Cost of side}$$

$$= \underbrace{2(\pi r^2)}_{\substack{\text{Total area}\\\text{of top and}\\\text{bottom}}} \quad \underbrace{(0.05)}_{\substack{\text{Cost/unit}\\\text{area}}} \quad + \quad \underbrace{(2\pi rh)}_{\substack{\text{Total}\\\text{area of}\\\text{side}}} \quad \underbrace{(0.02)}_{\substack{\text{Cost/unit}\\\text{area}}}$$

$$= 0.10\pi r^2 + 0.04\pi rh$$

But we have the additional restriction that the height h and radius r must be chosen so that the volume V of the can is 500 cubic centimeters. Since $V = \pi r^2 h$, we have

$$500 = \pi r^2 h \quad \text{or} \quad h = \frac{500}{\pi r^2}$$

Substituting this expression for h, the cost C, in cents, as a function of the radius r is

Figure 51

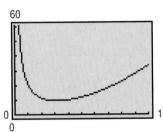

$$C(r) = 0.10\pi r^2 + 0.04\pi r \frac{500}{\pi r^2} = 0.10\pi r^2 + \frac{20}{r} = \frac{0.10\pi r^3 + 20}{r}$$

(b) See Figure 51 for the graph of $C(r)$.

(c) Using the MINIMUM command, the cost is least for a radius of about 3.17 centimeters.

(d) The least cost is $C(3.17) \approx 9.47¢$.

4.4   EXERCISES

In Problems 1–38, follow Steps 1 through 7 on page 328 to analyze the graph of each function.

1. $R(x) = \dfrac{x + 1}{x(x + 4)}$

2. $R(x) = \dfrac{x}{(x - 1)(x + 2)}$

3. $R(x) = \dfrac{3x + 3}{2x + 4}$

4. $R(x) = \dfrac{2x + 4}{x - 1}$

5. $R(x) = \dfrac{3}{x^2 - 4}$

6. $R(x) = \dfrac{6}{x^2 - x - 6}$

7. $P(x) = \dfrac{x^4 + x^2 + 1}{x^2 - 1}$

8. $Q(x) = \dfrac{x^4 - 1}{x^2 - 4}$

9. $H(x) = \dfrac{x^3 - 1}{x^2 - 9}$

10. $G(x) = \dfrac{x^3 + 1}{x^2 + 2x}$

11. $R(x) = \dfrac{x^2}{x^2 + x - 6}$

12. $R(x) = \dfrac{x^2 + x - 12}{x^2 - 4}$

13. $G(x) = \dfrac{x}{x^2 - 4}$

14. $G(x) = \dfrac{3x}{x^2 - 1}$

15. $R(x) = \dfrac{3}{(x - 1)(x^2 - 4)}$

16. $R(x) = \dfrac{-4}{(x + 1)(x^2 - 9)}$

17. $H(x) = 4\dfrac{x^2 - 1}{x^4 - 16}$

18. $H(x) = \dfrac{x^2 + 4}{x^4 - 1}$

19. $F(x) = \dfrac{x^2 - 3x - 4}{x + 2}$

20. $F(x) = \dfrac{x^2 + 3x + 2}{x - 1}$

21. $R(x) = \dfrac{x^2 + x - 12}{x - 4}$

22. $R(x) = \dfrac{x^2 - x - 12}{x + 5}$

23. $F(x) = \dfrac{x^2 + x - 12}{x + 2}$

24. $G(x) = \dfrac{x^2 - x - 12}{x + 1}$

25. $R(x) = \dfrac{x(x - 1)^2}{(x + 3)^3}$

26. $R(x) = \dfrac{(x - 1)(x + 2)(x - 3)}{x(x - 4)^2}$

27. $R(x) = \dfrac{x^2 + x - 12}{x^2 - x - 6}$

28. $R(x) = \dfrac{x^2 + 3x - 10}{x^2 + 8x + 15}$

29. $R(x) = \dfrac{6x^2 - 7x - 3}{2x^2 - 7x + 6}$

30. $R(x) = \dfrac{8x^2 + 26x + 15}{2x^2 - x - 15}$

31. $R(x) = \dfrac{x^2 + 5x + 6}{x + 3}$

32. $R(x) = \dfrac{x^2 + x - 30}{x + 6}$

33. $f(x) = x + \dfrac{1}{x}$

34. $f(x) = 2x + \dfrac{9}{x}$

35. $f(x) = x^2 + \dfrac{1}{x}$

36. $f(x) = 2x^2 + \dfrac{9}{x}$

37. $f(x) = x + \dfrac{1}{x^3}$

38. $f(x) = 2x + \dfrac{9}{x^3}$

In Problems 39–42, make up a rational function that might have the given graph. (More than one answer might be possible.)

39.

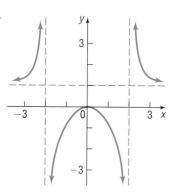

40.

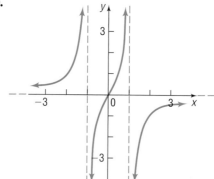

41.

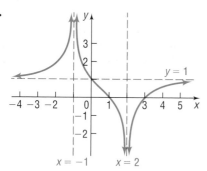

42.

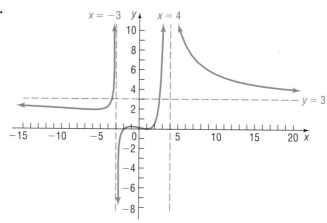

43. Drug Concentration   The concentration C of a certain drug in a patient's bloodstream t hours after injection is given by

$$C(t) = \frac{t}{2t^2 + 1}$$

(a) Find the horizontal asymptote of $C(t)$. What happens to the concentration of the drug as t increases?
(b) Using your graphing utility, graph $C(t)$.
(c) Determine the time at which the concentration is highest.

44. Drug Concentration   The concentration C of a certain drug in a patient's bloodstream t minutes after injection is given by

$$C(t) = \frac{50t}{(t^2 + 25)}$$

(a) Find the horizontal asymptote of $C(t)$. What happens to the concentration of the drug as t increases?
(b) Using your graphing utility, graph $C(t)$.
(c) Determine the time at which the concentration is highest.

45. Average Cost   In Problem 89, Exercise 4.2, the cost function C (in thousands of dollars) for manufacturing x Chevy Cavaliers was found to be

$$C(x) = 0.2x^3 - 2.3x^2 + 14.3x + 10.2$$

Economists define the **average cost function** as

$$\overline{C}(x) = \frac{C(x)}{x}$$

(a) Find the average cost function.
(b) What is the average cost of producing six Cavaliers per hour?
(c) What is the average cost of producing nine Cavaliers per hour?
(d) Using your graphing utility, graph the average cost function.
(e) Using your graphing utility, find the number of Cavaliers that should be produced per hour to minimize average cost.
(f) What is the minimum average cost?

46. Average Cost   In Problem 90, Exercise 4.2, the cost function C (in thousands of dollars) for printing x textbooks (in thousands of units) was found to be

$$C(x) = 0.015x^3 - 0.595x^2 + 9.15x + 98.43$$

(a) Find the average cost function (refer to Problem 45).

(b) What is the average cost of printing 13 thousand textbooks per week?

(c) What is the average cost of printing 25 thousand textbooks per week?

(d) Using your graphing utility, graph the average cost function.

(e) Using your graphing utility, find the number of textbooks that should be printed to minimize average cost.

(f) What is the minimum average cost?

47. Minimizing Surface Area   United Parcel Service has contracted you to design a closed box with a square base that has a volume of 10,000 cubic inches. See the illustration.

(a) Find a function for the surface area of the box.

(b) Using a graphing utility, graph the function found in part (a).

(c) What is the minimum amount of cardboard that can be used to construct the box?

(d) What are the dimensions of the box that minimize surface area?

(e) Why might UPS be interested in designing a box that minimizes surface area?

48. Minimizing Surface Area   United Parcel Service has contracted you to design a closed box with a square base that has a volume of 5000 cubic inches. See the illustration.

(a) Find a function for the surface area of the box.

(b) Using a graphing utility, graph the function found in part (a).

(c) What is the minimum amount of cardboard that can be used to construct the box?

(d) What are the dimensions of the box that minimize surface area?

(e) Why might UPS be interested in designing a box that minimizes surface area?

49. Cost of a Can   A can in the shape of a right circular cylinder is required to have a volume of 500 cubic centimeters. The top and bottom are made of material that costs 6¢ per square centimeter, while the sides are made of material that costs 4¢ per square centimeter.

(a) Express the total cost C of the material as a function of the radius r of the cylinder. (Refer to Figure 50.)

(b) Graph $C = C(r)$. For what value of r is the cost C least?

50. Material Needed to Make a Drum   A steel drum in the shape of a right circular cylinder is required to have a volume of 100 cubic feet.

(a) Express the amount A of material required to make the drum as a function of the radius r of the cylinder.

(b) How much material is required if the drum is of radius 3 feet?

(c) Of radius 4 feet?

(d) Of radius 5 feet?

(e) Graph $A = A(r)$. For what value of r is A smallest?

51. Graph each of the following functions:

$$y = \frac{x^2 - 1}{x - 1} \qquad y = \frac{x^3 - 1}{x - 1}$$

$$y = \frac{x^4 - 1}{x - 1} \qquad y = \frac{x^5 - 1}{x - 1}$$

Is $x = 1$ a vertical asymptote? Why not? What is happening for $x = 1$? What do you conjecture about $y = \dfrac{x^n - 1}{x - 1}$, $n \geq 1$ an integer, for $x = 1$?

52. Graph each of the following functions:

$$y = \frac{x^2}{x - 1} \qquad y = \frac{x^4}{x - 1}$$

$$y = \frac{x^6}{x - 1} \qquad y = \frac{x^8}{x - 1}$$

What similarities do you see? What differences?

In Problems 53–58, graph each function and use MINIMUM to obtain the minimum value, rounded to two decimal places.

53. $f(x) = x + \dfrac{1}{x}, \quad x > 0$

54. $f(x) = 2x + \dfrac{9}{x}, \quad x > 0$

55. $f(x) = x^2 + \dfrac{1}{x}, \quad x > 0$

56. $f(x) = 2x^2 + \dfrac{9}{x}, \quad x > 0$

57. $f(x) = x + \dfrac{1}{x^3}, \quad x > 0$

58. $f(x) = 2x + \dfrac{9}{x^3}, \quad x > 0$

59. Write a few paragraphs that provide a general strategy for graphing a rational function. Be sure to mention the following: proper, improper, intercepts, and asymptotes.

60. Make up a rational function that has the following characteristics: crosses the x-axis at 2; touches the x-axis at -1; one vertical asymptote at -5 and another at 6; and one horizontal asymptote, $y = 3$. Compare yours to a fellow classmate's. How do they differ? What are the similarities?

61. Make up a rational function that has the following characteristics: crosses the x-axis at 3; touches the x-axis at -2; one vertical asymptote, $x = 1$; and one horizontal asymptote, $y = 2$. Give your rational function to a fellow classmate and ask for a written critique of your rational function.

PREPARING FOR THIS SECTION

Before getting started, review the following concept:

✓ Solving Inequalities (Section 1.5, pp. 128–132)

4.5 POLYNOMIAL AND RATIONAL INEQUALITIES

OBJECTIVES
1. Solve Polynomial Inequalities
2. Solve Rational Inequalities

1 In this section we solve inequalities that involve polynomials of degree 2 and higher, as well as some that involve rational expressions. To solve such inequalities, we use the information obtained in the previous three sections about the graph of polynomial and rational functions. The general idea follows:

Suppose that the polynomial or rational inequality is in one of the forms

$$f(x) < 0 \qquad f(x) > 0 \qquad f(x) \le 0 \qquad f(x) \ge 0$$

Locate the zeros of f if f is a polynomial function, and locate the zeros of the numerator and the denominator if f is a rational function. If we use these zeros to divide the real number line into intervals, then we know that on each interval the graph of f is either above the x-axis $\big[f(x) > 0\big]$ or below the x-axis $\big[f(x) < 0\big]$. In other words, we have found the solution of the inequality.

The following steps provide more detail.

> ### STEPS FOR SOLVING POLYNOMIAL AND RATIONAL INEQUALITIES
>
> **STEP 1:** Write the inequality so that a polynomial or rational expression f is on the left side and zero is on the right side in one of the following forms:
>
> $$f(x) > 0 \qquad f(x) \ge 0 \qquad f(x) < 0 \qquad f(x) \le 0$$
>
> For rational expressions, be sure that the left side is written as a single quotient.
>
> **STEP 2:** Determine the numbers at which the expression f on the left side equals zero and, if the expression is rational, the numbers at which the expression f on the left side is undefined.
>
> **STEP 3:** Use the numbers found in Step 2 to separate the real number line into intervals.

STEP 4: Select a number in each interval and evaluate f at the number.
(a) If the value of f is positive, then $f(x) > 0$ for all numbers x in the interval.
(b) If the value of f is negative, then $f(x) < 0$ for all numbers x in the interval.
If the inequality is not strict, include the solutions of $f(x) = 0$ in the solution set.

EXAMPLE 1

Solving a Polynomial Inequality

Solve the inequality $x^2 \leq 4x + 12$, and graph the solution set.

Solution STEP 1: Rearrange the inequality so that 0 is on the right side.

$$x^2 \leq 4x + 12$$

$$x^2 - 4x - 12 \leq 0 \quad \text{Subtract } 4x + 12 \text{ from both sides of the inequality.}$$

This inequality is equivalent to the one that we wish to solve.

STEP 2: Find the zeros of $f(x) = x^2 - 4x - 12$ by solving the equation $x^2 - 4x - 12 = 0$.

$$x^2 - 4x - 12 = 0$$

$$(x + 2)(x - 6) = 0 \quad \text{Factor.}$$

$$x = -2 \quad \text{or} \quad x = 6$$

STEP 3: We use the zeros of f to separate the real number line into three intervals:

$$(-\infty, -2) \qquad (-2, 6) \qquad (6, \infty)$$

STEP 4: We choose a number in each interval and evaluate $f(x) = x^2 - 4x - 12$ to determine if $f(x)$ is positive or negative. See Table 14.

TABLE 14

Interval	$(-\infty, -2)$	$(-2, 6)$	$(6, \infty)$
Number Chosen	-3	0	7
Value of f	$f(-3) = 9$	$f(0) = -12$	$f(7) = 9$
Conclusion	Positive	Negative	Positive

Based on Table 14, we know that $f(x) < 0$ for all x in the interval $(-2, 6)$, that is, for all x such that $-2 < x < 6$. However, because the original inequality is not strict, numbers x that satisfy the equation $f(x) = x^2 - 4x - 12 = 0$ are also solutions of the inequality $x^2 \leq 4x + 12$. Thus, we include -2 and 6. The solution set of the given inequality is $\{x \mid -2 \leq x \leq 6\}$ or, using interval notation, $[-2, 6]$.

Figure 52
$-2 \leq x \leq 6; [-2, 6]$

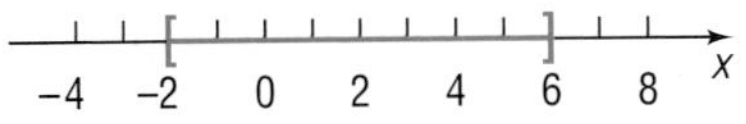

Figure 52 shows the graph of the solution set.

NOW WORK PROBLEMS **3** AND **7**.

| EXAMPLE 2 | **Solving a Polynomial Inequality** |

Solve the inequality $x^4 > x$, and graph the solution set.

Solution STEP 1: Rearrange the inequality so that 0 is on the right side.

$$x^4 > x$$

$$x^4 - x > 0 \quad \text{Subtract } x \text{ from both sides of the inequality.}$$

This inequality is equivalent to the one that we wish to solve.

STEP 2: Find the zeros of $f(x) = x^4 - x$ by solving $x^4 - x = 0$.

$$x^4 - x = 0$$

$$x(x^3 - 1) = 0 \quad \text{Factor out } x.$$

$$x(x - 1)(x^2 + x + 1) = 0 \quad \text{Factor the difference of two cubes.}$$

$$x = 0 \quad \text{or} \quad x - 1 = 0 \quad \text{or} \quad x^2 + x + 1 = 0 \quad \text{Set each factor equal to zero and solve.}$$

$$x = 0 \quad \text{or} \quad x = 1$$

The equation $x^2 + x + 1 = 0$ has no real solutions. (Do you see why?)

STEP 3: We use the zeros to separate the real number line into three intervals:

$$(-\infty, 0) \qquad (0, 1) \qquad (1, \infty)$$

STEP 4: We choose a number in each interval and evaluate $f(x) = x^4 - x$ to determine if $f(x)$ is positive or negative. See Table 15.

<table>
<tr><th colspan="4">T A B L E   1 5</th></tr>
<tr><td colspan="4"></td></tr>
<tr><td>Interval</td><td>$(-\infty, 0)$</td><td>$(0, 1)$</td><td>$(1, \infty)$</td></tr>
<tr><td>Number Chosen</td><td>-1</td><td>$\frac{1}{2}$</td><td>2</td></tr>
<tr><td>Value of f</td><td>$f(-1) = 2$</td><td>$f(\frac{1}{2}) = -\frac{7}{16}$</td><td>$f(2) = 14$</td></tr>
<tr><td>Conclusion</td><td>Positive</td><td>Negative</td><td>Positive</td></tr>
</table>

Based on Table 15, we know that $f(x) > 0$ for all x in the intervals $(-\infty, 0)$ or $(1, \infty)$, that is, for all numbers x for which $x < 0$ or $x > 1$. Because the original inequality is strict, the solution set of the given inequality is $\{x \mid x < 0 \text{ or } x > 1\}$ or, using interval notation, $(-\infty, 0)$ or $(1, \infty)$.

Figure 53
$x < 0$ or $x > 1$; $(-\infty, 0)$ or $(1, \infty)$

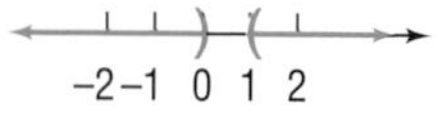

Figure 53 shows the graph of the solution set.

NOW WORK PROBLEM **19**.

(2) Let's solve a rational inequality.

EXAMPLE 3 ### Solving a Rational Inequality

Solve the inequality $\dfrac{(x + 3)(2 - x)}{(x - 1)^2} > 0$, and graph the solution set.

Solution **STEP 1:** The domain of the variable x is $\{x \,|\, x \neq 1\}$. The inequality is already in a form with 0 on the right side.

STEP 2: Let $f(x) = \dfrac{(x + 3)(2 - x)}{(x - 1)^2}$. The zeros of the numerator of f are -3 and 2; the zero of the denominator is 1.

STEP 3: We use the zeros found in Step 2 to separate the real number line into four intervals:

$$(-\infty, -3) \qquad (-3, 1) \qquad (1, 2) \qquad (2, \infty)$$

STEP 4: We choose a number in each interval and evaluate $f(x) = \dfrac{(x + 3)(2 - x)}{(x - 1)^2}$ to determine if $f(x)$ is positive or negative. See Table 16.

TABLE 16

	-3		1		2	

Interval	$(-\infty, -3)$	$(-3, 1)$	$(1, 2)$	$(2, \infty)$
Number Chosen	-4	0	$\frac{3}{2}$	3
Value of f	$f(-4) = -\frac{6}{25}$	$f(0) = 6$	$f(\frac{3}{2}) = 9$	$f(3) = -\frac{3}{2}$
Conclusion	Negative	Positive	Positive	Negative

Based on Table 16, we know that $f(x) > 0$ for all x in the intervals $(-3, 1)$ or $(1, 2)$, that is, for all x such that $-3 < x < 1$ or $1 < x < 2$. Because the original inequality is strict, the solution set of the given inequality is $\{x \,|\, -3 < x < 2, x \neq 1\}$ or, using interval notation, $(-3, 1)$ or $(1, 2)$. Figure 54 shows the graph of the solution set. Notice the hole at $x = 1$ to indicate that 1 is to be excluded.

Figure 54
$-3 < x < 2, x \neq 1; (-3, 2), x \neq 1$

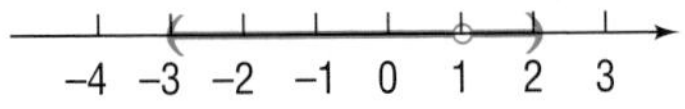

$-4 \;-3 \;-2 \;-1 \;\;0 \;\;1 \;\;2 \;\;3$

✏️ **NOW WORK PROBLEMS 29.**

EXAMPLE 4 ### Solving a Rational Inequality

Solve the inequality $\dfrac{4x + 5}{x + 2} \geq 3$, and graph the solution set.

Solution **STEP 1:** The domain of the variable x is $\{x \,|\, x \neq -2\}$. We rearrange terms so that 0 is on the right side.

$$\frac{4x + 5}{x + 2} - 3 \geq 0 \qquad \text{Subtract 3 from both sides of the inequality.}$$

STEP 2:   Let $f(x) = \dfrac{4x + 5}{x + 2} - 3$. To find the zeros of the numerator and the denominator, we must express f as a quotient.

$$f(x) = \frac{4x + 5}{x + 2} - 3 \qquad \text{Least Common Denominator: } x + 2$$

$$= \frac{4x + 5}{x + 2} - 3 \cdot \frac{x + 2}{x + 2} \qquad \text{Multiply } -3 \text{ by } \frac{x + 2}{x + 2}.$$

$$= \frac{4x + 5 - 3x - 6}{x + 2} \qquad \text{Write as a single quotient.}$$

$$= \frac{x - 1}{x + 2} \qquad \text{Combine like terms.}$$

The zero of the numerator of f is 1, and the zero of the denominator is -2.

STEP 3:   We use the zeros found in Step 2 to separate the real number line into three intervals:

$$(-\infty, -2) \qquad (-2, 1) \qquad (1, \infty)$$

STEP 4:   We choose a number in each interval and evaluate $f(x) = \dfrac{4x + 5}{x + 2} - 3$ to determine if $f(x)$ is positive or negative. See Table 17.

TABLE 17

Interval	$(-\infty, -2)$	$(-2, 1)$	$(1, \infty)$
Number Chosen	-3	0	2
Value of f	$f(-3) = 4$	$f(0) = -\frac{1}{2}$	$f(2) = \frac{1}{4}$
Conclusion	Positive	Negative	Positive

Based on Table 17, we know that $f(x) > 0$ for all x in the intervals $(-\infty, -2)$ or $(1, \infty)$, that is, for all x such that $x < -2$ or $x > 1$. Because the original inequality is not strict, numbers x that satisfy the equation $f(x) = \dfrac{x - 1}{x + 2} = 0$ are also solutions of the inequality. Since $\dfrac{x - 1}{x + 2} = 0$ only if $x = 1$, we conclude that the solution set is $\{x \mid x < -2 \text{ or } x \geq 1\}$ or, using interval notation, $(-\infty, -2)$ or $[1, \infty)$.

Figure 55
$x < -2$ or $x \geq 1$;
$(-\infty, -2)$ or $[1, \infty)$

Figure 55 shows the graph of the solution set. ∎

NOW WORK PROBLEM 37.

4.5 EXERCISES

In Problems 1–46, solve each inequality.

1. $(x - 5)(x + 2) < 0$

2. $(x - 5)(x + 2) > 0$

3. $x^2 - 4x \geq 0$

4. $x^2 + 8x \geq 0$

5. $x^2 - 9 < 0$

6. $x^2 - 1 < 0$

7. $x^2 + x \geq 2$

8. $x^2 + 7x \leq -12$

9. $2x^2 \leq 5x + 3$

10. $6x^2 \leq 6 + 5x$

11. $x(x - 7) > 8$

12. $x(x + 1) > 20$

13. $4x^2 + 9 < 6x$

14. $25x^2 + 16 < 40x$

15. $6(x^2 - 1) > 5x$

16. $2(2x^2 - 3x) > -9$

17. $(x - 1)(x^2 + x + 4) \geq 0$

18. $(x + 2)(x^2 - x + 1) \geq 0$

19. $(x - 1)(x - 2)(x - 3) \leq 0$

20. $(x + 1)(x + 2)(x + 3) \leq 0$

21. $x^3 - 2x^2 - 3x > 0$

22. $x^3 + 2x^2 - 3x > 0$

23. $x^4 > x^2$

24. $x^4 < 4x^2$

25. $x^3 \geq 4x^2$

26. $x^3 \leq 9x^2$

27. $x^4 > 1$

28. $x^3 > 1$

29. $\dfrac{x + 1}{x - 1} > 0$

30. $\dfrac{x - 3}{x + 1} > 0$

31. $\dfrac{(x - 1)(x + 1)}{x} \leq 0$

32. $\dfrac{(x - 3)(x + 2)}{x - 1} \leq 0$

33. $\dfrac{(x - 2)^2}{x^2 - 1} \geq 0$

34. $\dfrac{(x + 5)^2}{x^2 - 4} \geq 0$

35. $6x - 5 < \dfrac{6}{x}$

36. $x + \dfrac{12}{x} < 7$

37. $\dfrac{x + 4}{x - 2} \leq 1$

38. $\dfrac{x + 2}{x - 4} \geq 1$

39. $\dfrac{3x - 5}{x + 2} \leq 2$

40. $\dfrac{x - 4}{2x + 4} \geq 1$

41. $\dfrac{1}{x - 2} < \dfrac{2}{3x - 9}$

42. $\dfrac{5}{x - 3} > \dfrac{3}{x + 1}$

43. $\dfrac{2x + 5}{x + 1} > \dfrac{x + 1}{x - 1}$

44. $\dfrac{1}{x + 2} > \dfrac{3}{x + 1}$

45. $\dfrac{x^2(3 + x)(x + 4)}{(x + 5)(x - 1)} \geq 0$

46. $\dfrac{x(x^2 + 1)(x - 2)}{(x - 1)(x + 1)} \geq 0$

---

47. For what positive numbers will the cube of a number exceed four times its square?

48. For what positive numbers will the square of a number exceed twice the number?

49. What is the domain of the function $f(x) = \sqrt{x^2 - 16}$?

50. What is the domain of the function $f(x) = \sqrt{x^3 - 3x^2}$?

51. What is the domain of the function $f(x) = \sqrt{\dfrac{x - 2}{x + 4}}$?

52. What is the domain of the function $f(x) = \sqrt{\dfrac{x - 1}{x + 4}}$?

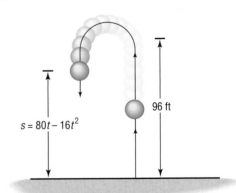

$s = 80t - 16t^2$

96 ft

53. Physics   A ball is thrown vertically upward with an initial velocity of 80 feet per second. The distance s (in feet) of the ball from the ground after t seconds is $s = 80t - 16t^2$. For what time interval is the ball more than 96 feet above the ground? (See the figure.)

54. Physics   A ball is thrown vertically upward with an initial velocity of 96 feet per second. The distance s (in feet) of the ball from the ground after t seconds is $s = 96t - 16t^2$. For what time interval is the ball more than 112 feet above the ground?

55. Business   The monthly revenue achieved by selling x wristwatches is calculated to be $x(40 - 0.2x)$ dollars. The wholesale cost of each watch is \$32. How many watches must be sold each month to achieve a profit (revenue − cost) of at least \$50?

56. Business   The monthly revenue achieved by selling x boxes of candy is calculated to be $x(5 - 0.05x)$ dollars. The wholesale cost of each box of candy is \$1.50. How many boxes must be sold each month to achieve a profit of at least \$60?

57. Find k such that the equation $x^2 + kx + 1 = 0$ has no real solution.

58. Find k such that the equation $kx^2 + 2x + 1 = 0$ has two distinct real solutions.

 59. Make up an inequality that has no solution. Make up one that has exactly one solution.

60. The inequality $x^2 + 1 < -5$ has no solution. Explain why.

C H A P T E R   R E V I E W

Things To Know

Quadratic function (pp. 286–287)

$f(x) = ax^2 + bx + c$

Graph is a parabola that opens up if $a > 0$ and opens down if $a < 0$.

Vertex: $\left(\dfrac{-b}{2a}, f\left(\dfrac{-b}{2a}\right)\right)$

Axis of symmetry: $x = \dfrac{-b}{2a}$

y-intercept: $f(0)$

x-intercept(s):   If any, found by solving the equation $ax^2 + bx + c = 0$.

Power function (p. 302)

$f(x) = x^n,\quad n \geq 2$ even (p. 304)

Even function
Passes through $(-1, 1), (0, 0), (1, 1)$
Opens up

$f(x) = x^n,\quad n \geq 3$ odd (p. 305)

Odd function
Passes through $(-1, -1), (0, 0), (1, 1)$
Increasing

Polynomial function (pp. 301, 309–310)

$f(x) = a_n x^n + a_{n-1} x^{n-1}$
$\quad + \cdots + a_1 x + a_0,\quad a_n \neq 0$

Domain: all real numbers
At most $n - 1$ turning points
End behavior: Behaves like $y = a_n x^n$ for large $|x|$

Rational function (p. 318)

$R(x) = \dfrac{p(x)}{q(x)}$

p, q are polynomial functions. Domain: $\{x \mid q(x) \neq 0\}$

Objectives

You should be able to:

Graph a quadratic function using transformations (p. 284)

Identify the vertex and axis of symmetry of a quadratic function (p. 286)

Graph a quadratic function using its vertex, axis, and intercepts (p. 286)

Use the maximum or minimum value of a quadratic function to solve applied problems (p. 289)

Use a graphing utility to find the quadratic function of best fit to data (p. 294)

Identify polynomials and their degree (p. 301)

Graph polynomial functions using transformations (p. 305)

Identify the zeros of a polynomial function and their multiplicity (p. 306)

Analyze the graph of a polynomial function (p. 308)

Find the domain of a rational function (p. 318)

Determine the vertical asymptotes of a rational function (p. 321)

Determine the horizontal or oblique asymptotes of a rational function (p. 322)

Analyze the graph of a rational function (p. 328)

Solve applied problems involving rational functions (p. 338)

Solve polynomial inequalities (p. 342)

Solve rational inequalities (p. 345)

Fill-in-the-Blank Items

1. The graph of a quadratic function is called a(n) _____.

2. The function $f(x) = ax^2 + bx + c, a \neq 0$, has a minimum value if _____. The minimum value occurs at $x =$ _____.

3. A number r for which $f(r) = 0$ is called a(n) _____ of the function f.

4. The line _____ is a horizontal asymptote of $R(x) = \dfrac{x^3 - 1}{x^3 + 1}$.

5. The line _____ is a vertical asymptote of $R(x) = \dfrac{x^3 - 1}{x^3 + 1}$.

True/False Items

T F   **1.** The graph of $f(x) = 2x^2 + 3x - 4$ opens up.

T F   **2.** The minimum value of $f(x) = -x^2 + 4x + 5$ is $f(2)$.

T F   **3.** The graph of $R(x) = \dfrac{x^2}{x - 1}$ has exactly one vertical asymptote.

T F   **4.** The graph of $f(x) = x^2(x - 3)(x + 4)$ has exactly three x-intercepts.

T F   **5.** The graph of a polynomial function sometimes has a hole.

T F   **6.** The graph of a rational function sometimes has a hole.

Review Exercises

Blue problem numbers indicate the author's suggestions for use in a Practice Test.

In Problems 1–10, graph each quadratic function by determining whether its graph opens up or down and by finding its vertex, axis of symmetry, y-intercept, and x-intercepts, if any.

1. $f(x) = (x - 2)^2 + 2$ **2.** $f(x) = (x + 1)^2 - 4$ **3.** $f(x) = \frac{1}{4}x^2 - 16$ **4.** $f(x) = -\frac{1}{3}x^2 + 2$

5. $f(x) = -4x^2 + 4x$ **6.** $f(x) = 9x^2 - 6x + 3$ **7.** $f(x) = \frac{9}{2}x^2 + 3x + 1$ **8.** $f(x) = -x^2 + x + \frac{1}{2}$

9. $f(x) = 3x^2 + 4x - 1$ **10.** $f(x) = -2x^2 - x + 4$

In Problems 11–16, graph each function using transformations (shifting, compressing, stretching, and reflection).

11. $f(x) = (x + 2)^3$ **12.** $f(x) = -x^3 + 3$ **13.** $f(x) = -(x - 1)^4$

14. $f(x) = (x - 1)^4 - 2$ **15.** $f(x) = (x - 1)^4 + 2$ **16.** $f(x) = (1 - x)^3$

In Problems 17–22, determine whether the given quadratic function has a maximum value or a minimum value, and then find the value.

17. $f(x) = 3x^2 - 6x + 4$ **18.** $f(x) = 2x^2 + 8x + 5$ **19.** $f(x) = -x^2 + 8x - 4$

20. $f(x) = -x^2 - 10x - 3$ **21.** $f(x) = -3x^2 + 12x + 4$ **22.** $f(x) = -2x^2 + 4$

In Problems 23–30:
  (a) *Find the x- and y-intercepts of each polynomial function f.*
  (b) *Determine whether the graph of f touches or crosses the x-axis at each x-intercept.*
  (c) *End behavior: find the power function that the graph of f resembles for large values of $|x|$.*
  (d) *Determine the maximum number of turning points of the graph of f.*
  (e) *Use the x-intercept(s) to find the intervals on which the graph of f is above and below the x-axis.*
  (f) *Plot the points obtained in parts (a) and (e), and use the remaining information to connect them with a smooth curve.*

23. $f(x) = x(x + 2)(x + 4)$ **24.** $f(x) = x(x - 2)(x - 4)$ **25.** $f(x) = (x - 2)^2(x + 4)$

26. $f(x) = (x - 2)(x + 4)^2$ **27.** $f(x) = -2x^3 + 4x^2$ **28.** $f(x) = -4x^3 + 4x$

29. $f(x) = (x - 1)^2(x + 3)(x + 1)$ **30.** $f(x) = (x - 4)(x + 2)^2(x - 2)$

In Problems 31–42, discuss each rational function following the seven steps outlined in Section 4.4.

31. $R(x) = \dfrac{2x - 6}{x}$

32. $R(x) = \dfrac{4 - x}{x}$

33. $H(x) = \dfrac{x + 2}{x(x - 2)}$

34. $H(x) = \dfrac{x}{x^2 - 1}$

35. $R(x) = \dfrac{x^2 + x - 6}{x^2 - x - 6}$

36. $R(x) = \dfrac{x^2 - 6x + 9}{x^2}$

37. $F(x) = \dfrac{x^3}{x^2 - 4}$

38. $F(x) = \dfrac{3x^3}{(x - 1)^2}$

39. $R(x) = \dfrac{2x^4}{(x - 1)^2}$

40. $R(x) = \dfrac{x^4}{x^2 - 9}$

41. $G(x) = \dfrac{x^2 - 4}{x^2 - x - 2}$

42. $F(x) = \dfrac{(x - 1)^2}{x^2 - 1}$

In Problems 43–52, solve each inequality.

43. $2x^2 + 5x - 12 < 0$

44. $3x^2 - 2x - 1 \geq 0$

45. $\dfrac{6}{x + 3} \geq 1$

46. $\dfrac{-2}{1 - 3x} < 1$

47. $\dfrac{2x - 6}{1 - x} < 2$

48. $\dfrac{3 - 2x}{2x + 5} \geq 2$

49. $\dfrac{(x - 2)(x - 1)}{x - 3} \geq 0$

50. $\dfrac{x + 1}{x(x - 5)} \leq 0$

51. $\dfrac{x^2 - 8x + 12}{x^2 - 16} > 0$

52. $\dfrac{x(x^2 + x - 2)}{x^2 + 9x + 20} \leq 0$

---

53. Find the point on the line $y = x$ that is closest to the point $(3, 1)$.

[**Hint:** Find the minimum value of the function $f(x) = d^2$, where d is the distance from $(3, 1)$ to a point on the line.]

54. Find the point on the line $y = x + 1$ that is closest to the point $(4, 1)$.

55. Landscaping   A landscape engineer has 200 feet of border to enclose a rectangular pond. What dimensions will result in the largest pond?

56. Geometry   Find the length and width of a rectangle whose perimeter is 20 feet and whose area is 16 square feet.

57. Enclosing the Most Area with a Fence   A farmer with 10,000 meters of fencing wants to enclose a rectangular field and then divide it into two plots with a fence parallel to one of the sides (see the figure). What is the largest area that can be enclosed?

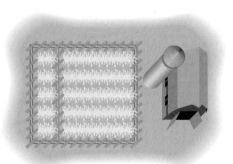

58. A rectangle has one vertex on the line $y = 8 - 2x, x > 0$, another at the origin, one on the positive x-axis, and one on the positive y-axis. Find the largest area A that can be enclosed by the rectangle.

59. Architecture   A special window in the shape of a rectangle with semicircles at each end is to be constructed so that the outside dimensions are 100 feet in length. See

the illustration. Find the dimensions of the rectangle that maximizes its area.

60. Parabolic Arch Bridges   A horizontal bridge is in the shape of a parabolic arch. Given the information shown in the figure, what is the height h of the arch 2 feet from shore?

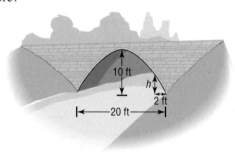

61. AIDS Cases in the United States   The following data represent the cumulative number of reported AIDS cases in the United States for 1990–1997.

Year, t	Number of AIDS Cases, A
1990, 1	193,878
1991, 2	251,638
1992, 3	326,648
1993, 4	399,613
1994, 5	457,280
1995, 6	528,215
1996, 7	594,760
1997, 8	653,253

Source: U.S. Center for Disease Control and Prevention.

(a) Draw a scatter diagram of the data.

(b) The cubic function of best fit to these data is

$$A(t) = -212t^3 + 2429t^2 + 59{,}569t + 130{,}003$$

Use this function to predict the cumulative number of AIDS cases reported in the United States in 2000.

(c) Use a graphing utility to verify that the function given in part (b) is the cubic function of best fit.

(d) With a graphing utility, draw a scatter diagram of the data and then graph the cubic function of best fit on the scatter diagram.

(e) Do you think the function found in part (b) will be useful in predicting the number of AIDS cases in 2005?

62. Gravity on the Moon Neil Armstrong wishes to estimate the acceleration due to gravity on the Moon. He throws a ball (with a special chip that measures the distance to the ground in 2-second intervals) straight down into a crater that is known to be 1000 feet deep. The ball records the following data:

Time (seconds)	0	2	4	6	8	10	12	14	16
Distance (feet)	1000	969	917	844	749	633	496	337	157

(a) Draw a scatter diagram using time as the independent variable and distance as the dependent variable.

(b) The quadratic function of best fit to these data is

$$s(t) = -2.7t^2 - 10t + 1000$$

Use this function to determine the initial velocity of the ball.

(c) Use this function to predict how long it will take for the ball to strike the ground.

(d) Use a graphing utility to verify that the function given in part (b) is the quadratic function of best fit.

(e) With a graphing utility, draw a scatter diagram of the data and then graph the quadratic function of best fit on the scatter diagram.

(f) According to physics theory,

$$s(t) = -\frac{1}{2}gt^2 + v_0 t + s_0$$

where $s(t)$ is the height of the object at time t, g is acceleration due to gravity, v_0 is the initial velocity of the object (this number will be negative if the object is thrown down), and s_0 is the initial height of the object. What is the acceleration due to gravity on the Moon? (Compare your answer with the fact that the acceleration due to gravity on the Moon is approximately 5.33 feet/sec².)

63. Construct a polynomial function with the following characteristics: degree 6; four real zeros, one of multiplicity 3; y-intercept 3; behaves like $y = -5x^6$ for large values of $|x|$. Is this polynomial unique? Compare your polynomial with those of other students. What terms will be the same as everyone else's? Add some more characteristics, such as symmetry or naming the real zeros. How does this modify the polynomial?

64. Construct a rational function with the following characteristics: three real zeros, one of multiplicity 2; y-intercept 1; vertical asymptotes $x = -2$ and $x = 3$; oblique asymptote $y = 2x + 1$. Is this rational function unique? Compare yours with those of other students. What will be the same as everyone else's? Add some more characteristics, such as symmetry or naming the real zeros. How does this modify the rational function?

65. The illustration shows the graph of a polynomial function.
(a) Is the degree of the polynomial even or odd?
(b) Is the leading coefficient positive or negative?
(c) Is the function even, odd, or neither?
(d) Why is x^2 necessarily a factor of the polynomial?
(e) What is the minimum degree of the polynomial?
(f) Formulate five different polynomials whose graphs could look like the one shown. Compare yours to those of other students. What similarities do you see? What differences?

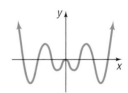

Project at Motorola

How many cellular phones can I make?

Motorola uses several types of specialized robots to assemble electronic products accurately and quickly. Resistors, capacitors, and other parts are picked from the feeder carriage by one of the vacuum nozzles on the rotating turret. The X–Y table, which holds a panel containing one or more boards, moves to the placement location for a given part, while the turret advances one position. The average time to pick and place one part is called the *tact time*, for example, 0.1 sec per part. The actual throughput, measured in boards per hour or in placements per hour (PPH), depends on a number of factors, including the number and variety of parts on the panel, the time to load–unload a panel in the machine, and the time to visually inspect *fiducial marks* and calibrate the position of the panel with respect to the coordinate system of the machine.

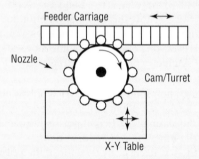

Feeder Carriage

Nozzle

Cam/Turret

X-Y Table

Throughput can be estimated with the following equation, where n is the number of boards per panel, p is the number of parts per board, C is the tact time, L is the load–unload time, and M is the mark reading time.

$$TH(n) = \frac{n}{Cnp + L + M}$$

$$f(n) = \frac{L + M}{Cpn^2 + (L + M)n}$$

As shown for three different machines in the graph, throughput is highly dependent on the number of placements per board. Manufacturing engineers work with product designers to select the best number of boards per panel, that is, the one that results in the highest possible throughput.

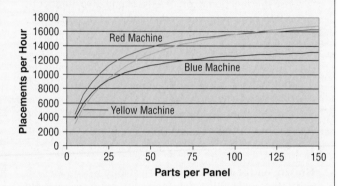

1. Write a rational function for TH in terms of the variable n, where the tact time is 0.2 sec per part, the load–unload time is 5 sec, the mark reading time is 1 sec, and there are 50 parts per board.

2. Graph the function $Y_1 = TH(n)$, and determine the minimum number of boards per panel, such that the throughput is at least 300 boards per hour.

3. If it were possible to fit 10 boards per panel, what would the throughput be? What would be the highest possible throughput if there were no limit on the number of boards per panel?

4. The function $f(n)$ represents the percent increase in throughput for a 1-unit increase in the number of boards per panel. Graph the function $Y_1 = f(n)$ for the values of tact time and other factors given in question 1.

5. Is $f(n)$ a proper rational function?

6. At what value of n is it not possible to increase throughput by more than 2% when one more board is added to the panel?

7. Assuming that all machines (blue, red, and yellow) cost the same to operate, which one is the best choice for a panel with one board? With three boards? Why?

The Zeros of a Polynomial Function

5

Field Trip to Motorola

Complex numbers may appear on the surface to be an abstract concept with little practical value. However, in the field of electrical engineering they are a powerful tool that aid in the design of new products. Some examples in the area of communications include cellular phones, pagers, two-way radios, cordless telephones, police and fire dispatch systems, and satellites. All these products contain electrical circuits that must operate according to detailed design specifications. Engineers use circuit analysis techniques based on complex numbers to gain understanding into circuit operation in order to optimize their designs. Software tools (circuit simulators) are also available that automate this analytical process and reduce the design time. Without complex numbers the design process would be a much more difficult task.

PREPARING FOR THIS SECTION

Before getting started, review the following concepts:

✓ Polynomial Division (Review, Section 5, pp. 43–45)

5.1 SYNTHETIC DIVISION

OBJECTIVE ① Use Synthetic Division

To find the quotient as well as the remainder when a polynomial function f of degree 1 or higher is divided by $g(x) = x - c$, a shortened version of long division, called **synthetic division,** makes the task simpler.

① To see how synthetic division works, we will use long division to divide the polynomial $f(x) = 2x^3 - x^2 + 3$ by $g(x) = x - 3$.

$$
\begin{array}{r}
2x^2 + 5x + 15 \qquad \leftarrow \text{Quotient} \\
x - 3 \overline{)\, 2x^3 - x^2 \qquad\quad + 3} \\
\underline{2x^3 - 6x^2} \\
5x^2 \\
\underline{5x^2 - 15x} \\
15x + 3 \\
\underline{15x - 45} \\
48 \qquad \leftarrow \text{Remainder}
\end{array}
$$

Check: (Divisor) · (Quotient) + Remainder

$$
= (x - 3)(2x^2 + 5x + 15) + 48
$$

$$
= 2x^3 + 5x^2 + 15x - 6x^2 - 15x - 45 + 48
$$

$$
= 2x^3 - x^2 + 3
$$

The process of synthetic division arises from rewriting the long division in a more compact form, using simpler notation. For example, in the long division above, the terms in blue are not really necessary because they are identical to the terms directly above them. With these terms removed, we have

$$
\begin{array}{r}
2x^2 + 5x\ \ + 15 \\
x - 3 \overline{)\ 2x^3 - \ x^2 \qquad\ \ + 3} \\
\underline{-\ 6x^2} \\
5x^2 \\
\underline{-\ 15x} \\
15x \\
\underline{-\ 45} \\
48
\end{array}
$$

Most of the x's that appear in this process can also be removed, provided that we are careful about positioning each coefficient. In this regard, we will need to use 0 as the coefficient of x in the dividend, because that power of x is missing. Now we have

$$
\begin{array}{r}
2x^2 + 5x + 15 \\
x - 3 \overline{)\ 2\ \ -1 \qquad 0 \qquad 3} \\
\underline{-\ 6} \\
5 \\
\underline{-\ 15} \\
15 \\
\underline{-\ 45} \\
48
\end{array}
$$

We can make this display more compact by moving the lines up until the numbers in color align horizontally.

$$
\begin{array}{lr}
2x^2 + 5x + 15 & \text{Row 1} \\
x - 3 \overline{)\ 2\quad -1\qquad 0\qquad 3} & \text{Row 2} \\
\underline{\quad -6\ -15\ -45} & \text{Row 3} \\
\bigcirc\quad\ 5\quad\ 15\quad 48 & \text{Row 4}
\end{array}
$$

Now, if we place the leading coefficient of the quotient, 2, in the circled position, the first three numbers in row 4 are precisely the coefficients of the quotient, and the last number in row 4 is the remainder. Thus, row 1 is not really needed, so we can compress the process to three rows, where the bottom row contains the coefficients of the quotient and the remainder.

$$
\begin{array}{lr}
x - 3 \overline{)\ 2\quad -1\qquad 0\qquad 3} & \text{Row 1} \\
\underline{\quad\ -6\ -15\ -45} & \text{Row 2 (subtract)} \\
2\quad\ 5\quad\ 15\quad 48 & \text{Row 3}
\end{array}
$$

Recall that the entries in row 3 are obtained by subtracting the entries in row 2 from those in row 1. Rather than subtracting the entries in row 2, we can change the sign of each entry and add. With this modification, our display will look like this:

$$
\begin{array}{lr}
x - 3 \overline{)\ 2\quad -1\qquad 0\qquad 3} & \text{Row 1} \\
\underline{\quad\ \ 6\quad 15\quad 45} & \text{Row 2 (add)} \\
2\quad\ 5\quad\ 15\quad 48 & \text{Row 3}
\end{array}
$$

Notice that the entries in row 2 are three times the prior entries in row 3. Our last modification to the display replaces the $x - 3$ by 3. The entries in row 3 give the quotient and the remainder, as shown next.

$$
\begin{array}{r|rrrr}
3) & 2 & -1 & 0 & 3 \\
   &   & 6 & 15 & 45 \\
\hline
   & 2 & 5 & 15 & 48
\end{array}
\quad
\begin{array}{l}
\text{Row 1} \\
\text{Row 2 (add)} \\
\text{Row 3}
\end{array}
$$

Quotient   Remainder

$$2x^2 + 5x + 15 \quad\quad 48$$

Let's go through an example step by step.

EXAMPLE 1

Using Synthetic Division to Find the Quotient and Remainder

Use synthetic division to find the quotient $q(x)$ and remainder R when

$$f(x) = x^3 - 4x^2 - 5 \quad \text{is divided by} \quad g(x) = x - 3$$

Solution   **STEP 1:** Write the dividend in descending powers of x. Then copy the coefficients, remembering to insert a 0 for any missing powers of x.

$$
\begin{array}{rrrr}
1 & -4 & 0 & -5
\end{array}
\quad \text{Row 1}
$$

STEP 2: Insert the usual division symbol. In synthetic division, the divisor is of the form $x - c$, and c is the number placed to the left of the division symbol. Here, since the divisor is $x - 3$, we insert 3 to the left of the division symbol.

$$
\begin{array}{r|rrrr}
3) & 1 & -4 & 0 & -5
\end{array}
\quad \text{Row 1}
$$

STEP 3: Bring the 1 down two rows, and enter it in row 3.

$$
\begin{array}{r|rrrr}
3) & 1 & -4 & 0 & -5 \\
   & \downarrow & & & \\
\hline
   & 1 & & &
\end{array}
\quad
\begin{array}{l}
\text{Row 1} \\
\text{Row 2} \\
\text{Row 3}
\end{array}
$$

STEP 4: Multiply the latest entry in row 3 by 3, and place the result in row 2, one column over to the right.

$$
\begin{array}{r|rrrr}
3) & 1 & -4 & 0 & -5 \\
   &   & 3 &   &   \\
\hline
   & 1 &   &   &
\end{array}
\quad
\begin{array}{l}
\text{Row 1} \\
\text{Row 2} \\
\text{Row 3}
\end{array}
$$

STEP 5: Add the entry in row 2 to the entry above it in row 1, and enter the sum in row 3.

$$
\begin{array}{r|rrrr}
3) & 1 & -4 & 0 & -5 \\
   &   & 3 &   &   \\
\hline
   & 1 & -1 &   &
\end{array}
\quad
\begin{array}{l}
\text{Row 1} \\
\text{Row 2} \\
\text{Row 3}
\end{array}
$$

STEP 6: Repeat Steps 4 and 5 until no more entries are available in row 1.

$$
\begin{array}{r|rrrr}
3) & 1 & -4 & 0 & -5 \\
   &   & 3 & -3 & -9 \\
\hline
   & 1 & -1 & -3 & -14
\end{array}
\quad
\begin{array}{l}
\text{Row 1} \\
\text{Row 2} \\
\text{Row 3}
\end{array}
$$

STEP 7: The final entry in row 3, the -14, is the remainder; the other entries in row 3 the 1, -1, and -3, are the coefficients (in descending order) of a polynomial whose degree is 1 less than that of the dividend. This is the quotient. Thus,

$$q(x) = x^2 - x - 3 \qquad R = -14$$

Check: (Divisor)(Quotient) + Remainder

$$
\begin{aligned}
&= (x - 3)(x^2 - x - 3) + (-14)\\
&= (x^3 - x^2 - 3x - 3x^2 + 3x + 9) + (-14)\\
&= x^3 - 4x^2 - 5 = \text{Dividend}
\end{aligned}
$$

Let's do an example in which all seven steps are combined.

EXAMPLE 2

Using Synthetic Division to Verify a Factor

Use synthetic division to show that $x + 3$ is a factor of

$$f(x) = 2x^5 + 5x^4 - 2x^3 + 2x^2 - 2x + 3$$

Solution The divisor is $x + 3 = x - (-3)$, so we place -3 to the left of the division symbol. Then the row 3 entries will be multiplied by -3, entered in row 2, and added to row 1.

$-3)$	2	5	-2	2	-2	3	Row 1
		-6	3	-3	3	-3	Row 2
	2	-1	1	-1	1	0	Row 3

Because the remainder is 0, we have

(Divisor)(Quotient) + Remainder

$$= (x + 3)(2x^4 - x^3 + x^2 - x + 1) = 2x^5 + 5x^4 - 2x^3 + 2x^2 - 2x + 3$$

As we see, $x + 3$ is a factor of $2x^5 + 5x^4 - 2x^3 + 2x^2 - 2x + 3$.

As Example 2 illustrates, the remainder after division gives information about whether the divisor is, or is not, a factor. We shall have more to say about this in the next section.

NOW WORK PROBLEMS **3** AND **13**.

5.1 EXERCISES

In Problems 1–12, use synthetic division to find the quotient $q(x)$ and remainder R when $f(x)$ is divided by $g(x)$.

1. $f(x) = x^3 - x^2 + 2x + 4$; $g(x) = x - 2$

2. $f(x) = x^3 + 2x^2 - 3x + 1$; $g(x) = x + 1$

3. $f(x) = 3x^3 + 2x^2 - x + 3$; $g(x) = x - 3$

4. $f(x) = -4x^3 + 2x^2 - x + 1$; $g(x) = x + 2$

5. $f(x) = x^5 - 4x^3 + x$; $g(x) = x + 3$

6. $f(x) = x^4 + x^2 + 2$; $g(x) = x - 2$

7. $f(x) = 4x^6 - 3x^4 + x^2 + 5$; $g(x) = x - 1$

8. $f(x) = x^5 + 5x^3 - 10$; $g(x) = x + 1$

9. $f(x) = 0.1x^3 + 0.2x$; $g(x) = x + 1.1$

10. $f(x) = 0.1x^2 - 0.2$; $g(x) = x + 2.1$

11. $f(x) = x^5 - 1$; $g(x) = x - 1$

12. $f(x) = x^5 + 1$; $g(x) = x + 1$

In Problems 13–22, use synthetic division to determine whether $x - c$ is a factor of $f(x)$.

13. $f(x) = 4x^3 - 3x^2 - 8x + 4;\quad x - 2$

14. $f(x) = -4x^3 + 5x^2 + 8;\quad x + 3$

15. $f(x) = 3x^4 - 6x^3 - 5x + 10;\quad x - 2$

16. $f(x) = 4x^4 - 15x^2 - 4;\quad x - 2$

17. $f(x) = 3x^6 + 82x^3 + 27;\quad x + 3$

18. $f(x) = 2x^6 - 18x^4 + x^2 - 9;\quad x + 3$

19. $f(x) = 4x^6 - 64x^4 + x^2 - 15;\quad x + 4$

20. $f(x) = x^6 - 16x^4 + x^2 - 16;\quad x + 4$

21. $f(x) = 2x^4 - x^3 + 2x - 1;\quad x - \dfrac{1}{2}$

22. $f(x) = 3x^4 + x^3 - 3x + 1;\quad x + \dfrac{1}{3}$

23. Find the sum of $a, b, c,$ and d if

$$\frac{x^3 - 2x^2 + 3x + 5}{x + 2} = ax^2 + bx + c + \frac{d}{x + 2}$$

24. When dividing a polynomial by $x - c$, do you prefer to use long division or synthetic division? Does the value of c make a difference to you in choosing? Give reasons

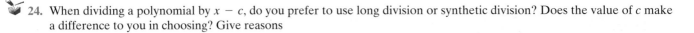

PREPARING FOR THIS SECTION

Before getting started, review the following concepts:

✓ Classification of Numbers; Real Numbers
(Review, Section 1, pp. 2–4)

✓ Polynomial Functions (Section 4.2, pp. 300–313)

✓ Factoring Polynomials (Review, Section 6, pp. 43–54)

✓ Polynomial Division
(Review, Section 5, pp. 43–45)

✓ Quadratic Formula (Section 1.3, pp. 110–113)

5.2 THE REAL ZEROS OF A POLYNOMIAL FUNCTION

OBJECTIVES

1. Use the Remainder and Factor Theorems
2. Use Descartes' Rule of Signs to Determine the Number of Positive and the Number of Negative Real Zeros of a Polynomial Function
3. Use the Rational Zeros Theorem to List the Potential Rational Zeros of a Polynomial Function
4. Find the Real Zeros of a Polynomial Function
5. Solve Polynomial Equations
6. Use the Theorem for Bounds on Zeros
7. Use the Intermediate Value Theorem

In this section, we discuss techniques that can be used to find the real zeros of a polynomial function. Recall that if r is a real zero of a polynomial function f then $f(r) = 0$, r is an x-intercept of the graph of f, and r is a solution of the equation $f(x) = 0$. For polynomial and rational functions, we have seen the importance of the zeros for graphing. In most cases, however, the zeros of a polynomial function are difficult to find using algebraic methods. No nice formulas like the quadratic formula are available to help us find zeros for polynomials of degree 3 or higher. Formulas do exist for solving any third- or fourth-degree polynomial equation, but they are somewhat complicated. No general formulas exist for polynomial equations of degree 5 or higher. Refer to the Historical Feature at the end of this section for more information.

REMAINDER AND FACTOR THEOREMS

① When we divide one polynomial (the dividend) by another (the divisor), we obtain a quotient polynomial and a remainder, the remainder being either the zero polynomial or a polynomial whose degree is less than the degree of the divisor. To check our work, we verify that

$$(\text{Quotient})(\text{Divisor}) + \text{Remainder} = \text{Dividend}$$

This checking routine is the basis for a famous theorem called the **division algorithm*** for **polynomials,** which we now state without proof.

Theorem

Division Algorithm for Polynomials

If $f(x)$ and $g(x)$ denote polynomial functions and if $g(x)$ is not the zero polynomial, then there are unique polynomial functions $q(x)$ and $r(x)$ such that

$$\frac{f(x)}{g(x)} = q(x) + \frac{r(x)}{g(x)} \quad \text{or} \quad \underset{\underset{\text{dividend}}{\uparrow}}{f(x)} = \underset{\underset{\text{quotient}}{\uparrow}}{q(x)}\,\underset{\underset{\text{divisor}}{\uparrow}}{g(x)} + \underset{\underset{\text{remainder}}{\uparrow}}{r(x)} \quad \textbf{(1)}$$

where $r(x)$ is either the zero polynomial or a polynomial of degree less than that of $g(x)$.

In equation (1), $f(x)$ is the **dividend,** $g(x)$ is the **divisor,** $q(x)$ is the **quotient,** and $r(x)$ is the **remainder.**

If the divisor $g(x)$ is a first-degree polynomial of the form

$$g(x) = x - c, \qquad c \text{ a real number}$$

then the remainder $r(x)$ is either the zero polynomial or a polynomial of degree 0. As a result, for such divisors, the remainder is some number, say R, and we may write

$$f(x) = (x - c)q(x) + R \qquad \textbf{(2)}$$

This equation is an identity in x and is true for all real numbers x. Suppose that $x = c$. Then equation (2) becomes

$$f(c) = (c - c)q(c) + R$$
$$f(c) = R$$

Substitute $f(c)$ for R in equation (2) to obtain

$$f(x) = (x - c)q(x) + f(c) \qquad \textbf{(3)}$$

We have now proved the **Remainder Theorem.**

Remainder Theorem

Let f be a polynomial function. If $f(x)$ is divided by $x - c$, then the remainder is $f(c)$.

---

* A systematic process in which certain steps are repeated a finite number of times is called an **algorithm.** For example, long division is an algorithm.

EXAMPLE 1

Using the Remainder Theorem

Find the remainder if $f(x) = x^3 - 4x^2 - 5$ is divided by

(a) $x - 3$ (b) $x + 2$

Solution (a) We could use long division or synthetic division, but it is easier to use the Remainder Theorem, which says that the remainder is $f(3)$.

$$f(3) = (3)^3 - 4(3)^2 - 5 = 27 - 36 - 5 = -14$$

The remainder is -14.

(b) To find the remainder when $f(x)$ is divided by $x + 2 = x - (-2)$, we evaluate $f(-2)$.

$$f(-2) = (-2)^3 - 4(-2)^2 - 5 = -8 - 16 - 5 = -29$$

The remainder is -29.   ■

 Compare the method used in Example 1(a) above with the method used in Example 1 on page 356 (synthetic division). Which method do you prefer? Give reasons.   ▨

 COMMENT: A graphing utility provides another way to find the value of a function, using the eVALUEate feature. Consult your manual for details. Then check the results of Example 1.   ▨

An important and useful consequence of the Remainder Theorem is the **Factor Theorem.**

Factor Theorem   Let f be a polynomial function. Then $x - c$ is a factor of $f(x)$ if and only if $f(c) = 0$.

     ■

The Factor Theorem actually consists of two separate statements:

> 1. If $f(c) = 0$, then $x - c$ is a factor of $f(x)$.
> 2. If $x - c$ is a factor of $f(x)$, then $f(c) = 0$.

The proof requires two parts.

Proof

1. Suppose that $f(c) = 0$. Then, by equation (3), we have

$$f(x) = (x - c)q(x)$$

for some polynomial $q(x)$. That is, $x - c$ is a factor of $f(x)$.

2. Suppose that $x - c$ is a factor of $f(x)$. Then there is a polynomial function q such that

$$f(x) = (x - c)q(x)$$

Replacing x by c, we find that

$$f(c) = (c - c)q(c) = 0 \cdot q(c) = 0$$

This completes the proof.   ■

One use of the Factor Theorem is to determine whether a polynomial has a particular factor.

EXAMPLE 2

Using the Factor Theorem

Use the Factor Theorem to determine whether the function $f(x) = 2x^3 - x^2 + 2x - 3$ has the factor

(a) $x - 1$　　　　　　　　　　　　(b) $x + 3$

Solution　The Factor Theorem states that if $f(c) = 0$ then $x - c$ is a factor.

(a) Because $x - 1$ is of the form $x - c$ with $c = 1$, we find the value of $f(1)$. We choose to use substitution.

$$f(1) = 2(1)^3 - (1)^2 + 2(1) - 3 = 2 - 1 + 2 - 3 = 0$$

By the Factor Theorem, $x - 1$ is a factor of $f(x)$.

(b) To test the factor $x + 3$, we first need to write it in the form $x - c$. Since $x + 3 = x - (-3)$, we find the value of $f(-3)$. We choose to use synthetic division.

$$
\begin{array}{r}
-3\overline{)\,2 \quad -1 \quad\ 2 \quad\ -3\,} \\
\underline{-6 \quad 21 \quad -69} \\
2 \quad -7 \quad 23 \quad -72
\end{array}
$$

Because $f(-3) = -72 \neq 0$, we conclude from the Factor Theorem that $x - (-3) = x + 3$ is not a factor of $f(x)$. ∎

NOW WORK PROBLEM **1.**

In Example 2(a), we found that $x - 1$ was a factor of f. To write f in factored form, we use long division or synthetic division. Using synthetic division, we find that

$$
\begin{array}{r}
1\overline{)\,2 \quad -1 \quad 2 \quad -3\,} \\
\underline{2 \quad\ 1 \quad\ 3} \\
2 \quad\ 1 \quad 3 \quad\ 0
\end{array}
$$

The quotient is $q(x) = 2x^2 + x + 3$ with a remainder of 0, as expected. We can write f in factored form as

$$f(x) = 2x^3 - x^2 + 2x - 3 = (x - 1)(2x^2 + x + 3)$$

THE NUMBER AND LOCATION OF REAL ZEROS

The next theorem concerns the number of real zeros that a polynomial function may have. In counting the zeros of a polynomial, we count each zero as many times as its multiplicity.

Theorem　**Number of Real Zeros**

A polynomial function cannot have more real zeros than its degree.

Proof The proof is based on the Factor Theorem. If r is a zero of a polynomial function f, then $f(r) = 0$ and, hence, $x - r$ is a factor of $f(x)$. Each zero corresponds to a factor of degree 1. Because f cannot have more first-degree factors than its degree, the result follows. ∎

Descartes' Rule of Signs provides information about the number and location of the real zeros of a polynomial function written in standard form (descending powers of x). It requires that we count the number of variations in sign of the coefficients of $f(x)$ and $f(-x)$.

For example, the following polynomial function has two variations in the signs of coefficients.

$$f(x) = -3x^7 + 4x^4 + 3x^2 - 2x - 1$$
$$= -3x^7 + 0x^6 + 0x^5 + 4x^4 + 0x^3 + 3x^2 - 2x - 1$$

$-$ to $+$ $+$ to $-$

Notice that we ignored the zero coefficients in $0x^6, 0x^5$ and $0x^3$ in counting the number of variations in the sign of $f(x)$. Replacing x by $-x$, we get

$$f(-x) = -3(-x)^7 + 4(-x)^4 + 3(-x)^2 - 2(-x) - 1$$
$$= 3x^7 + 4x^4 + 3x^2 + 2x - 1$$

$+$ to $-$

which has one variation in sign.

Theorem

Descartes' Rule of Signs

Let f denote a polynomial function written in standard form.

The number of positive real zeros of f either equals the number of variations in the sign of the nonzero coefficients of $f(x)$ or else equals that number less an even integer.

The number of negative real zeros of f either equals the number of variations in the sign of the nonzero coefficients of $f(-x)$ or else equals that number less an even integer.

∎

We shall not prove Descartes' Rule of Signs. Let's see how it is used.

EXAMPLE 3

Using the Number of Real Zeros Theorem and Descartes' Rule of Signs

Discuss the real zeros of $f(x) = 3x^6 - 4x^4 + 3x^3 + 2x^2 - x - 3$.

Solution Because the polynomial is of degree 6, by the Number of Real Zeros Theorem there are at most six real zeros. Since there are three variations in sign of the nonzero coefficients of $f(x)$, by Descartes' Rule of Signs we expect either three or one positive real zeros. To continue, we look at $f(-x)$.

$$f(-x) = 3x^6 - 4x^4 - 3x^3 + 2x^2 + x - 3$$

There are three variations in sign, so we expect either three (or one) negative real zeros. Equivalently, we now know that the graph of f has either three or one positive x-intercepts and three or one negative x-intercepts. ∎

NOW WORK PROBLEM **11**.

RATIONAL ZEROS THEOREM

③   The next result, called the **Rational Zeros Theorem,** provides information about the rational zeros of a polynomial *with integer coefficients.*

Theorem

Rational Zeros Theorem

Let f be a polynomial function of degree 1 or higher of the form

$$f(x) = a_n x^n + a_{n-1} x^{n-1} + \cdots + a_1 x + a_0, \qquad a_n \neq 0, a_0 \neq 0$$

where each coefficient is an integer. If p/q, in lowest terms, is a rational zero of f, then p must be a factor of a_0, and q must be a factor of a_n.

EXAMPLE 4

Listing Potential Rational Zeros

List the potential rational zeros of

$$f(x) = 2x^3 + 11x^2 - 7x - 6$$

Solution   Because f has integer coefficients, we may use the Rational Zeros Theorem. First, we list all the integers p that are factors of the constant term $a_0 = -6$ and all the integers q that are factors of the leading coefficient $a_3 = 2$.

$$p: \quad \pm 1, \pm 2, \pm 3, \pm 6 \quad \text{Factors of } -6$$

$$q: \quad \pm 1, \pm 2 \qquad\qquad \text{Factors of } 2$$

Now we form all possible ratios p/q.

$$\frac{p}{q}: \quad \pm 1, \pm 2, \pm 3, \pm 6, \pm \frac{1}{2}, \pm \frac{3}{2}$$

If f has a rational zero, it will be found in this list, which contains 12 possibilities.

━━━━━━▶ NOW WORK PROBLEM **23.**

Be sure that you understand what the Rational Zeros Theorem says: For a polynomial with integer coefficients, *if* there is a rational zero, it is one of those listed. It may be the case that the function does not have any rational zeros.

④   Long division, synthetic division, or substitution can be used to test each potential rational zero to determine whether it is indeed a zero. To make the work easier, the integers are usually tested first. Let's continue this example.

| EXAMPLE 5 | **Finding the Rational Zeros of a Polynomial Function** |

Continue working with Example 4 to find the rational zeros of

$$f(x) = 2x^3 + 11x^2 - 7x - 6$$

Write f in factored form.

Solution　We gather all the information that we can about the zeros.

STEP 1: There are at most three real zeros.

STEP 2: By Descartes' Rule of Signs, there is one positive real zero. Also, because

$$f(-x) = -2x^3 + 11x^2 + 7x - 6$$

there are two negative zeros or no negative zeros.

STEP 3: Now we use the list of potential rational zeros obtained in Example 4: $\pm 1, \pm 2, \pm 3, \pm 6, \pm \frac{1}{2}, \pm \frac{3}{2}$. We choose to test the potential rational zero 1 using substitution.

$$f(1) = 2(1)^3 + 11(1)^2 - 7(1) - 6 = 2 + 11 - 7 - 6 = 0$$

Since $f(1) = 0$, 1 is a zero and $x - 1$ is a factor of f. We can use long division or synthetic division to factor f.

$$f(x) = 2x^3 + 11x^2 - 7x - 6$$
$$= (x - 1)(2x^2 + 13x + 6)$$

Now any solution of the equation $2x^2 + 13x + 6 = 0$ will be a zero of f. Because of this, we call the equation $2x^2 + 13x + 6 = 0$ a **depressed equation** of f. Since the degree of the depressed equation of f is less than that of the original polynomial, we work with the depressed equation to find the zeros of f.

STEP 4: The depressed equation $2x^2 + 13x + 6 = 0$ is a quadratic equation with discriminant $b^2 - 4ac = 169 - 48 = 121 > 0$. The equation has two real solutions, which can be found by factoring.

$$2x^2 + 13x + 6 = (2x + 1)(x + 6) = 0$$

$$2x + 1 = 0 \quad \text{or} \quad x + 6 = 0$$

$$x = -\tfrac{1}{2} \qquad\qquad x = -6$$

The zeros of f are $-6, -\frac{1}{2}$, and 1.

　　We use the Factor Theorem to factor f. Each zero gives rise to a factor, so $x - (-6) = x + 6$, $x - \left(-\frac{1}{2}\right) = x + \frac{1}{2}$, and $x - 1$ are factors of f. Since the leading coefficient of f is 2 and f is of degree 3, we have

$$f(x) = 2x^3 + 11x^2 - 7x - 6 = 2(x + 6)(x + \tfrac{1}{2})(x - 1)$$

　　Notice that the three zeros of f found in Example 5 are among those given in the list of potential rational zeros in Example 4. ■

To obtain information about the real zeros of a polynomial function, follow these steps:

STEPS FOR FINDING THE REAL ZEROS OF A POLYNOMIAL FUNCTION

STEP 1:   Use the degree of the polynomial to determine the maximum number of zeros.

STEP 2:   Use Descartes' Rule of Signs to determine the possible number of positive zeros and negative zeros.

STEP 3:   (a) If the polynomial has integer coefficients, use the Rational Zeros Theorem to identify those rational numbers that potentially can be zeros.
(b) Use substitution, synthetic division, or long division to test each potential rational zero.
(c) Each time that a zero (and thus a factor) is found, repeat Step 3 on the depressed equation.

STEP 4:   In attempting to find the zeros, remember to use (if possible) the factoring techniques that you already know (special products, factoring by grouping, and so on).

EXAMPLE 6

Finding the Real Zeros of a Polynomial Function

Find the real zeros of: $f(x) = x^5 - 5x^4 + 12x^3 - 24x^2 + 32x - 16$
Write f in factored form.

Solution   We gather all the information that we can about the zeros.

STEP 1:   There are at most five real zeros.

STEP 2:   By Descartes' Rule of Signs, there are five, three, or one positive zeros. Because

$$f(-x) = -x^5 - 5x^4 - 12x^3 - 24x^2 - 32x - 16$$

there are no negative zeros.

STEP 3:   Because the leading coefficient $a_5 = 1$ and there are no negative zeros, the potential rational zeros are the integers $1, 2, 4, 8,$ and $16,$ the positive factors of the constant term, $16.$ We test the potential rational zero 1 first, using synthetic division.

$$
\begin{array}{r|rrrrrr}
1) & 1 & -5 & 12 & -24 & 32 & -16 \\
  &   & 1 & -4 & 8 & -16 & 16 \\
\hline
  & 1 & -4 & 8 & -16 & 16 & 0
\end{array}
$$

The remainder is $f(1) = 0,$ so 1 is a zero and $x - 1$ is a factor of $f.$ Using the entries in the bottom row of the synthetic division, we can begin to factor $f.$

$$f(x) = x^5 - 5x^4 + 12x^3 - 24x^2 + 32x - 16$$
$$= (x - 1)(x^4 - 4x^3 + 8x^2 - 16x + 16)$$

We now work with the first depressed equation:

$$q_1(x) = x^4 - 4x^3 + 8x^2 - 16x + 16 = 0$$

REPEAT STEP 3:   The potential rational zeros of q_1 are still $1, 2, 4, 8,$ and $16.$ We test 1 first, since it may be a repeated zero.

$$
\begin{array}{r|rrrrr}
1) & 1 & -4 & 8 & -16 & 16 \\
  &   & 1 & -3 & 5 & -11 \\
\hline
  & 1 & -3 & 5 & -11 & 5
\end{array}
$$

Since the remainder is 5, 1 is not a repeated zero. We try 2 next.

$$2)\overline{\begin{array}{rrrrr} 1 & -4 & 8 & -16 & 16 \\ & 2 & -4 & 8 & -16 \\ \hline 1 & -2 & 4 & -8 & 0 \end{array}}$$

The remainder is $f(2) = 0$, so 2 is a zero and $x - 2$ is a factor of f. Again using the bottom row, we find

$$f(x) = x^5 - 5x^4 + 12x^3 - 24x^2 + 32x - 16$$
$$= (x - 1)(x - 2)(x^3 - 2x^2 + 4x - 8)$$

The remaining zeros satisfy the new depressed equation

$$q_2(x) = x^3 - 2x^2 + 4x - 8 = 0$$

Notice that $q_2(x)$ can be factored using grouping. [Alternatively, you could repeat Step 3 and check the potential rational zero 2]. Then

$$\begin{aligned} x^3 - 2x^2 + 4x - 8 &= 0 \\ x^2(x - 2) + 4(x - 2) &= 0 \\ (x^2 + 4)(x - 2) &= 0 \\ x^2 + 4 = 0 \quad \text{or} \quad x - 2 &= 0 \\ x &= 2 \end{aligned}$$

Since $x^2 + 4 = 0$ has no real solutions, the real zeros of f are 1 and 2, the latter being a zero of multiplicity 2. The factored form of f is

$$f(x) = x^5 - 5x^4 + 12x^3 - 24x^2 + 32x - 16$$
$$= (x - 1)(x - 2)^2(x^2 + 4)$$

■

NOW WORK PROBLEM 35.

EXAMPLE 7   **Solving a Polynomial Equation**

⑤   Solve the equation: $x^5 - 5x^4 + 12x^3 - 24x^2 + 32x - 16 = 0$

Solution   The solutions of this equation are the zeros of the polynomial function

$$f(x) = x^5 - 5x^4 + 12x^3 - 24x^2 + 32x - 16$$

Using the result of Example 6, the real zeros of f are 1 and 2. These are the real solutions of the equation

$$x^5 - 5x^4 + 12x^3 - 24x^2 + 32x - 16 = 0$$

■

NOW WORK PROBLEM 47.

In Example 6, the quadratic factor $x^2 + 4$ that appears in the factored form of f is called *irreducible*, because the polynomial $x^2 + 4$ cannot be factored over the real numbers. In general, we say that a quadratic factor $ax^2 + bx + c$ is **irreducible** if it cannot be factored over the real numbers, that is, if it is prime over the real numbers.

Refer back to Examples 5 and 6. The polynomial function of Example 5 has three real zeros, and its factored form contains three linear factors. The polynomial function of Example 6 has two distinct real zeros, and its factored form contains two distinct linear factors and one irreducible quadratic factor.

Theorem   Every polynomial function (with real coefficients) can be uniquely factored into a product of linear factors and/or irreducible quadratic factors.

◼

We shall prove this result in Section 5.4, and, in fact, we shall draw several additional conclusions about the zeros of a polynomial function. One conclusion is worth noting now. If a polynomial (with real coefficients) is of odd degree, then it must contain at least one linear factor. (Do you see why?) This means that it must have at least one real zero.

Corollary   A polynomial function (with real coefficients) of odd degree has at least one real zero.

◼

BOUNDS ON ZEROS

⑥   The search for the real zeros of a polynomial function can be reduced somewhat if *bounds* on the zeros are found. A number M is a **bound** on the zeros of a polynomial if every zero lies between $-M$ and M, inclusive. That is, M is a bound to the zeros of a polynomial f if

$$-M \le \text{any zero of } f \le M$$

Theorem   **Bounds on Zeros**

Let f denote a polynomial function whose leading coefficient is 1.

$$f(x) = x^n + a_{n-1}x^{n-1} + \cdots + a_1 x + a_0$$

A bound M on the zeros of f is the smaller of the two numbers

$$\boxed{\text{Max}\{1, |a_0| + |a_1| + \cdots + |a_{n-1}|\}, \quad 1 + \text{Max}\{|a_0|, |a_1|, \ldots, |a_{n-1}|\}}\ \textbf{(4)}$$

where Max{ } means "choose the largest entry in { }."

◼

An example will help to make the theorem clear.

EXAMPLE 8   **Using the Theorem for Finding Bounds on Zeros**

Find a bound to the zeros of each polynomial.

(a) $f(x) = x^5 + 3x^3 - 9x^2 + 5$ (b) $g(x) = 4x^5 - 2x^3 + 2x^2 + 1$

Solution   (a) The leading coefficient of f is 1.

$$f(x) = x^5 + 3x^3 - 9x^2 + 5 \qquad a_4 = 0, a_3 = 3, a_2 = -9, a_1 = 0, a_0 = 5$$

We evaluate the two expressions in (4).

$$\text{Max}\{1, |a_0| + |a_1| + \cdots + |a_{n-1}|\} = \text{Max}\{1, |5| + |0| + |-9| + |3| + |0|\}$$
$$= \text{Max}\{1, 17\} = 17$$

$$1 + \text{Max}\{|a_0|, |a_1|, \ldots, |a_{n-1}|\} = 1 + \text{Max}\{|5|, |0|, |-9|, |3|, |0|\}$$
$$= 1 + 9 = 10$$

The smaller of the two numbers, 10, is the bound. Every zero of f lies between -10 and 10.

(b) First we write g so that it is the product of a constant times a polynomial whose leading coefficient is 1.

$$g(x) = 4x^5 - 2x^3 + 2x^2 + 1 = 4\left(x^5 - \tfrac{1}{2}x^3 + \tfrac{1}{2}x^2 + \tfrac{1}{4}\right)$$

Next we evaluate the two expressions in (4) with $a_4 = 0$, $a_3 = -\tfrac{1}{2}$, $a_2 = \tfrac{1}{2}$, $a_1 = 0$, and $a_0 = \tfrac{1}{4}$.

$$\text{Max}\{1, |a_0| + |a_1| + \cdots + |a_{n-1}|\} = \text{Max}\{1, |\tfrac{1}{4}| + |0| + |\tfrac{1}{2}| + |-\tfrac{1}{2}| + |0|\}$$
$$= \text{Max}\{1, \tfrac{5}{4}\} = \tfrac{5}{4}$$

$$1 + \text{Max}\{|a_0|, |a_1|, \ldots, |a_{n-1}|\} = 1 + \text{Max}\{|\tfrac{1}{4}|, |0|, |\tfrac{1}{2}|, |-\tfrac{1}{2}|, |0|\}$$
$$= 1 + \tfrac{1}{2} = \tfrac{3}{2}$$

The smaller of the two numbers, $\tfrac{5}{4}$, is the bound. Every zero of g lies between $-\tfrac{5}{4}$ and $\tfrac{5}{4}$. ∎

 COMMENT: The bounds on the zeros of a polynomial provide good choices for setting Xmin and Xmax of the viewing rectangle. With these choices, all the x-intercepts of the graph can be seen. ∎

 NOW WORK PROBLEM 71.

INTERMEDIATE VALUE THEOREM

7   The next result, called the **Intermediate Value Theorem,** is based on the fact that the graph of a polynomial function is continuous; that is, it contains no "holes" or "gaps."

Theorem   **Intermediate Value Theorem**

Let f denote a polynomial function. If $a < b$ and if $f(a)$ and $f(b)$ are of opposite sign, then there is at least one zero of f between a and b. ∎

Although the proof of this result requires advanced methods in calculus, it is easy to "see" why the result is true. Look at Figure 1.

Figure 1
If $f(a) < 0$ and $f(b) > 0$, there is a zero between a and b.

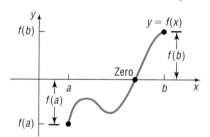

EXAMPLE 9   **Using the Intermediate Value Theorem to Locate Zeros**

Show that $f(x) = x^5 - x^3 - 1$ has a zero between 1 and 2.

Solution   We evaluate f at 1 and at 2.

$$f(1) = -1 \quad \text{and} \quad f(2) = 23$$

Because $f(1) < 0$ and $f(2) > 0$, it follows from the Intermediate Value Theorem that f has a zero between 1 and 2. ∎

NOW WORK PROBLEM 79.

Let's look at the polynomial f of Example 9 more closely. Based on Descartes' Rule of Signs, f has exactly one positive real zero. Based on the Rational Zeros Theorem, 1 is the only potential positive rational zero. Since $f(1) \neq 0$, we conclude that the zero between 1 and 2 is irrational. We can use the Intermediate Value Theorem to approximate it. The steps to use follow:

APPROXIMATING THE ZEROS OF A POLYNOMIAL FUNCTION

STEP 1: Find two consecutive integers a and $a + 1$ such that f has a zero between them.

STEP 2: Divide the interval $[a, a + 1]$ into 10 equal subintervals.

STEP 3: Evaluate f at each endpoint of the subintervals until the Intermediate Value Theorem applies; this interval then contains a zero.

STEP 4: Repeat the process starting at Step 2 until the desired accuracy is achieved.

| EXAMPLE 10 | **Approximating the Zeros of a Polynomial Function** |

Find the positive zero of $f(x) = x^5 - x^3 - 1$ correct to two decimal places.

Solution   From Example 9 we know that the positive zero is between 1 and 2. We divide the interval $[1, 2]$ into 10 equal subintervals: $[1, 1.1]$, $[1.1, 1.2]$, $[1.2, 1.3]$, $[1.3, 1.4]$, $[1.4, 1.5]$, $[1.5, 1.6]$, $[1.6, 1.7]$, $[1.7, 1.8]$, $[1.8, 1.9]$, $[1.9, 2]$. Now, we find the value of f at each endpoint until the Intermediate Value Theorem applies.

$$f(x) = x^5 - x^3 - 1$$

$$f(1.0) = -1 \qquad\qquad f(1.2) = -0.23968$$

$$f(1.1) = -0.72049 \qquad f(1.3) = 0.51593$$

We can stop here and conclude that the zero is between 1.2 and 1.3. Now we divide the interval $[1.2, 1.3]$ into 10 equal subintervals and proceed to evaluate f at each endpoint.

$$f(1.20) = -0.23968 \qquad f(1.23) \approx -0.0455613$$

$$f(1.21) \approx -0.1778185 \qquad f(1.24) \approx 0.025001$$

$$f(1.22) \approx -0.1131398$$

We conclude that the zero lies between 1.23 and 1.24, and so, correct to two decimal places, the zero is 1.23.   ■

Figure 2

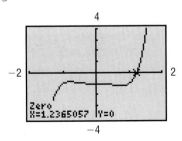

EXPLORATION   We continue to examine the polynomial f. The Theorem on Bounds of Zeros tells us that every zero is between -2 and 2. If we graph f using $-2 \leq x \leq 2$, we see that f has exactly one x-intercept. See Figure 2. Using ZERO or ROOT, we find this zero to be 1.24 rounded to two decimal places.   ■

There are many other numerical techniques for approximating the zeros of a polynomial. The one outlined in Example 10 (a variation of the *bisection method*) has the advantages that it will always work, that it can be programmed rather easily on a computer, and each time it is used another decimal place of accuracy is achieved. See Problem 109 for the bisection method, which places the zero in a succession of intervals, with each new interval being half the length of the preceding one.

HISTORICAL FEATURE

Formulas for the solution of third- and fourth-degree polynomial equations exist, and, while not very practical, they do have an interesting history.

In the 1500s in Italy, mathematical contests were a popular pastime, and persons possessing methods for solving problems kept them secret. (Solutions that were published were already common knowledge.) Niccolo of Brescia (1499–1557), commonly referred to as Tartaglia ("the stammerer"), had the secret for solving cubic (third-degree) equations, which gave him a decided advantage in the contests. Girolamo Cardano (1501–1576) found out that Tartaglia had the secret, and, being interested in cubics, he requested it from Tartaglia. The reluctant Tartaglia hesitated for some time, but finally, swearing Cardano to secrecy with midnight oaths by candlelight, told him the secret. Cardano then pub-

lished the solution in his book *Ars Magna* (1545), giving Tartaglia the credit but rather compromising the secrecy. Tartaglia exploded into bitter recriminations, and each wrote pamphlets that reflected on the other's mathematics, moral character, and ancestry.

The quartic (fourth-degree) equation was solved by Cardano's student Lodovico Ferrari, and this solution also was included, with credit and this time with permission, in the *Ars Magna*.

Attempts were made to solve the fifth-degree equation in similar ways, all of which failed. In the early 1800s, P. Ruffini, Niels Abel, and Evariste Galois all found ways to show that it is not possible to solve fifth-degree equations by formula, but the proofs required the introduction of new methods. Galois' methods eventually developed into a large part of modern algebra.

HISTORICAL PROBLEMS

Problems 1–8, develop the Tartaglia–Cardano solution of the cubic equation and show why it is not altogether practical.

1. Show that the general cubic equation $y^3 + by^2 + cy + d = 0$ can be transformed into an equation of the form $x^3 + px + q = 0$ by using the substitution $y = x - b/3$.

2. In the equation $x^3 + px + q = 0$, replace x by $H + K$. Let $3HK = -p$, and show that $H^3 + K^3 = -q$.
   [**Hint:** $3H^2K + 3HK^2 = 3HKx$.]

3. Based on Problem 2, we have the two equations
   $$3HK = -p \quad \text{and} \quad H^3 + K^3 = -q$$
   Solve for K in $3HK = -p$ and substitute into $H^3 + K^3 = -q$. Then show that
   $$H = \sqrt[3]{\frac{-q}{2} + \sqrt{\frac{q^2}{4} + \frac{p^3}{27}}}$$
   [**Hint:** Look for an equation that is quadratic in form.]

4. Use the solution for H from Problem 3 and the equation $H^3 + K^3 = -q$ to show that
   $$K = \sqrt[3]{\frac{-q}{2} - \sqrt{\frac{q^2}{4} + \frac{p^3}{27}}}$$

5. Use the results from Problems 2–4 to show that the solution of $x^3 + px + q = 0$ is
   $$x = \sqrt[3]{\frac{-q}{2} + \sqrt{\frac{q^2}{4} + \frac{p^3}{27}}} + \sqrt[3]{\frac{-q}{2} - \sqrt{\frac{q^2}{4} + \frac{p^3}{27}}}$$

6. Use the result of Problem 5 to solve the equation $x^3 - 6x - 9 = 0$.

7. Use a calculator and the result of Problem 5 to solve the equation $x^3 + 3x - 14 = 0$.

8. Use the methods of this chapter to solve the equation $x^3 + 3x - 14 = 0$.

5.2 EXERCISES

In Problems 1–10, use the Factor Theorem to determine whether $x - c$ is a factor of $f(x)$.

1. $f(x) = 4x^3 - 3x^2 - 8x + 4; \quad x - 2$

2. $f(x) = -4x^3 + 5x^2 + 8; \quad x + 3$

3. $f(x) = 3x^4 - 6x^3 - 5x + 10; \quad x - 2$

4. $f(x) = 4x^4 - 15x^2 - 4; \quad x - 2$

5. $f(x) = 3x^6 + 82x^3 + 27; \quad x + 3$

6. $f(x) = 2x^6 - 18x^4 + x^2 - 9; \quad x + 3$

7. $f(x) = 4x^6 - 64x^4 + x^2 - 15; \quad x + 4$

8. $f(x) = x^6 - 16x^4 + x^2 - 16; \quad x + 4$

9. $f(x) = 2x^4 - x^3 + 2x - 1; \quad x - \frac{1}{2}$

10. $f(x) = 3x^4 + x^3 - 3x + 1; \quad x + \frac{1}{3}$

In Problems 11–22, tell the maximum number of zeros that each polynomial function may have. Then use Descartes' Rule of Signs to determine how many positive and how many negative zeros each polynomial function may have. Do not attempt to find the zeros.

11. $f(x) = -4x^7 + x^3 - x^2 + 2$

12. $f(x) = 5x^4 + 2x^2 - 6x - 5$

13. $f(x) = 2x^6 - 3x^2 - x + 1$

14. $f(x) = -3x^5 + 4x^4 + 2$

15. $f(x) = 3x^3 - 2x^2 + x + 2$

16. $f(x) = -x^3 - x^2 + x + 1$

17. $f(x) = -x^4 + x^2 - 1$

18. $f(x) = x^4 + 5x^3 - 2$

19. $f(x) = x^5 + x^4 + x^2 + x + 1$

20. $f(x) = x^5 - x^4 + x^3 - x^2 + x - 1$

21. $f(x) = x^6 - 1$

22. $f(x) = x^6 + 1$

In Problems 23–34, list the potential rational zeros of each polynomial function. Do not attempt to find the zeros.

23. $f(x) = 3x^4 - 3x^3 + x^2 - x + 1$

24. $f(x) = x^5 - x^4 + 2x^2 + 3$

25. $f(x) = x^5 - 6x^2 + 9x - 3$

26. $f(x) = 2x^5 - x^4 - x^2 + 1$

27. $f(x) = -4x^3 - x^2 + x + 2$

28. $f(x) = 6x^4 - x^2 + 2$

29. $f(x) = 6x^4 - x^2 + 9$

30. $f(x) = -4x^3 + x^2 + x + 6$

31. $f(x) = 2x^5 - x^3 + 2x^2 + 12$

32. $f(x) = 3x^5 - x^2 + 2x + 18$

33. $f(x) = 6x^4 + 2x^3 - x^2 + 20$

34. $f(x) = -6x^3 - x^2 + x + 10$

In Problems 35–46, use Descartes' Rule of Signs and the Rational Zeros Theorem to find all the real zeros of each polynomial function. Use the zeros to factor f over the real numbers.

35. $f(x) = x^3 + 2x^2 - 5x - 6$

36. $f(x) = x^3 + 8x^2 + 11x - 20$

37. $f(x) = 2x^3 - x^2 + 2x - 1$

38. $f(x) = 2x^3 + x^2 + 2x + 1$

39. $f(x) = x^4 + x^2 - 2$

40. $f(x) = x^4 - 3x^2 - 4$

41. $f(x) = 4x^4 + 7x^2 - 2$

42. $f(x) = 4x^4 + 15x^2 - 4$

43. $f(x) = x^4 + x^3 - 3x^2 - x + 2$

44. $f(x) = x^4 - x^3 - 6x^2 + 4x + 8$

45. $f(x) = 4x^5 - 8x^4 - x + 2$

46. $f(x) = 4x^5 + 12x^4 - x - 3$

In Problems 47–58, solve each equation in the real number system.

47. $x^4 - x^3 + 2x^2 - 4x - 8 = 0$

48. $2x^3 + 3x^2 + 2x + 3 = 0$

49. $3x^3 + 4x^2 - 7x + 2 = 0$

50. $2x^3 - 3x^2 - 3x - 5 = 0$

51. $3x^3 - x^2 - 15x + 5 = 0$

52. $2x^3 - 11x^2 + 10x + 8 = 0$

53. $x^4 + 4x^3 + 2x^2 - x + 6 = 0$

54. $x^4 - 2x^3 + 10x^2 - 18x + 9 = 0$

55. $x^3 - \frac{2}{3}x^2 + \frac{8}{3}x + 1 = 0$

56. $x^3 + \frac{3}{2}x^2 + 3x - 2 = 0$

57. $2x^4 - 19x^3 + 57x^2 - 64x + 20 = 0$

58. $2x^4 + x^3 - 24x^2 + 20x + 16 = 0$

*In Problems 59–70, find the intercepts of each polynomial function f(x). Find the intervals x for which the graph of f is above the x-axis and below the x-axis. Obtain several other points on the graph, and connect them with a smooth curve. [**Hint:** Use the factored form of f (see Problems 35–46).]*

59. $f(x) = x^3 + 2x^2 - 5x - 6$

60. $f(x) = x^3 + 8x^2 + 11x - 20$

61. $f(x) = 2x^3 - x^2 + 2x - 1$

62. $f(x) = 2x^3 + x^2 + 2x + 1$

63. $f(x) = x^4 + x^2 - 2$

64. $f(x) = x^4 - 3x^2 - 4$

65. $f(x) = 4x^4 + 7x^2 - 2$

66. $f(x) = 4x^4 + 15x^2 - 4$

67. $f(x) = x^4 + x^3 - 3x^2 - x + 2$

68. $f(x) = x^4 - x^3 - 6x^2 + 4x + 8$

69. $f(x) = 4x^5 - 8x^4 - x + 2$

70. $f(x) = 4x^5 + 12x^4 - x - 3$

In Problems 71–78, find a bound on the real zeros of each polynomial function.

71. $f(x) = x^4 - 3x^2 - 4$

72. $f(x) = x^4 - 5x^2 - 36$

73. $f(x) = x^4 + x^3 - x - 1$

74. $f(x) = x^4 - x^3 + x - 1$

75. $f(x) = 3x^4 + 3x^3 - x^2 - 12x - 12$

76. $f(x) = 3x^4 - 3x^3 - 5x^2 + 27x - 36$

77. $f(x) = 4x^5 - x^4 + 2x^3 - 2x^2 + x - 1$

78. $f(x) = 4x^5 + x^4 + x^3 + x^2 - 2x - 2$

In Problems 79–84, use the Intermediate Value Theorem to show that each polynomial function has a zero in the given interval.

79. $f(x) = 8x^4 - 2x^2 + 5x - 1;$ $[0, 1]$

80. $f(x) = x^4 + 8x^3 - x^2 + 2;$ $[-1, 0]$

81. $f(x) = 2x^3 + 6x^2 - 8x + 2;$ $[-5, -4]$

82. $f(x) = 3x^3 - 10x + 9;$ $[-3, -2]$

83. $f(x) = x^5 - x^4 + 7x^3 - 7x^2 - 18x + 18;$ $[1.4, 1.5]$

84. $f(x) = x^5 - 3x^4 - 2x^3 + 6x^2 + x + 2;$ $[1.7, 1.8]$

In Problems 85–88, each equation has a solution r in the interval indicated. Use the method of Example 10 to approximate this solution correct to two decimal places.

85. $8x^4 - 2x^2 + 5x - 1 = 0;$ $0 \le r \le 1$

86. $x^4 + 8x^3 - x^2 + 2 = 0;$ $-1 \le r \le 0$

87. $2x^3 + 6x^2 - 8x + 2 = 0;$ $-5 \le r \le -4$

88. $3x^3 - 10x + 9 = 0;$ $-3 \le r \le -2$

In Problems 89–92, each polynomial function has exactly one positive zero. Use the method of Example 10 to approximate the zero correct to two decimal places.

89. $f(x) = x^3 + x^2 + x - 4$

90. $f(x) = 2x^4 + x^2 - 1$

91. $f(x) = 2x^4 - 3x^3 - 4x^2 - 8$

92. $f(x) = 3x^3 - 2x^2 - 20$

---

93. Find k such that $f(x) = x^3 - kx^2 + kx + 2$ has the factor $x - 2$.

94. Find k such that $f(x) = x^4 - kx^3 + kx^2 + 1$ has the factor $x + 2$.

95. What is the remainder when
$f(x) = 2x^{20} - 8x^{10} + x - 2$ is divided by $x - 1$?

96. What is the remainder when
$f(x) = -3x^{17} + x^9 - x^5 + 2x$ is divided by $x + 1$?

97. Use the Factor Theorem to prove that $x - c$ is a factor of $x^n - c^n$ for any positive integer n.

98. Use the Factor Theorem to prove that $x + c$ is a factor of $x^n + c^n$ if $n \geq 1$ is an odd integer.

99. One solution of the equation $x^3 - 8x^2 + 16x - 3 = 0$ is 3. Find the sum of the remaining solutions.

100. One solution of the equation $x^3 + 5x^2 + 5x - 2 = 0$ is -2. Find the sum of the remaining solutions.

101. Is $\frac{1}{3}$ a zero of $f(x) = 2x^3 + 3x^2 - 6x + 7$? Explain.

102. Is $\frac{1}{3}$ a zero of $f(x) = 4x^3 - 5x^2 - 3x + 1$? Explain.

103. Is $\frac{3}{5}$ a zero of $f(x) = 2x^6 - 5x^4 + x^3 - x + 1$? Explain.

104. Is $\frac{2}{3}$ a zero of $f(x) = x^7 + 6x^5 - x^4 + x + 2$? Explain.

105. What is the length of the edge of a cube if, after a slice 1 inch thick is cut from one side, the volume remaining is 294 cubic inches?

106. What is the length of the edge of a cube if its volume could be doubled by an increase of 6 centimeters in one edge, an increase of 12 centimeters in a second edge, and a decrease of 4 centimeters in the third edge?

107. Let $f(x)$ be a polynomial function whose coefficients are integers. Suppose that r is a real zero of f and that the leading coefficient of f is 1. Use the Rational Zeros Theorem to show that r is either an integer or an irrational number.

108. Prove the Rational Zeros Theorem.
[**Hint:** Let p/q, where p and q have no common factors except 1 and -1, be a zero of the polynomial function $f(x) = a_n x^n + a_{n-1} x^{n-1} + \cdots + a_1 x + a_0$, whose coefficients are all integers. Show that $a_n p^n + a_{n-1} p^{n-1} q + \cdots + a_1 p q^{n-1} + a_0 q^n = 0$. Now, because p is a factor of the first n terms of this equation, p must also be a factor of the term $a_0 q^n$. Since p is not a factor of q (why?), p must be a factor of a_0. Similarly, q must be a factor of a_n.]

109. **Bisection Method for Approximating Zeros of a Function f** We begin with two consecutive integers, a and $a + 1$, such that $f(a)$ and $f(a + 1)$ are of opposite sign. Evaluate f at the midpoint m_1 of a and $a + 1$. If $f(m_1) = 0$, then m_1 is the zero of f, and we are finished. Otherwise, $f(m_1)$ is of opposite sign to either $f(a)$ or $f(a + 1)$. Suppose that it is $f(a)$ and $f(m_1)$ that are of opposite sign. Now evaluate f at the midpoint m_2 of a and m_1. Repeat this process until the desired degree of accuracy is obtained. Note that each iteration places the zero in an interval whose length is half that of the previous interval. Use the bisection method to solve Problems 85–92.

PREPARING FOR THIS SECTION

Before getting started on this section, review the following:

✓ Classification of Numbers; (Review, Section 1, pp. 2–6)

✓ Rationalizing Square Roots (Review, Section 8, pp. 69–70)

✓ Quadratic Formula (Section 1.3, pp. 110–113)

5.3 COMPLEX NUMBERS; QUADRATIC EQUATIONS WITH A NEGATIVE DISCRIMINANT

OBJECTIVES
1. Add, Subtract, Multiply, and Divide Complex Numbers
2. Solve Quadratic Equations with a Negative Discriminant

One property of a real number is that its square is nonnegative. For example, there is no real number x for which

$$x^2 = -1$$

To remedy this situation, we introduce a number called the **imaginary unit,** which we denote by i and whose square is -1;

$$i^2 = -1$$

This should not surprise you. If our universe were to consist only of integers, there would be no number x for which $2x = 1$. This unfortunate circumstance was remedied by introducing numbers such as $\frac{1}{2}$ and $\frac{2}{3}$, the *rational numbers.* If our universe were to consist only of rational numbers, there would be no x whose square equals 2. That is, there would be no number x for which $x^2 = 2$. To remedy this, we introduced numbers such as $\sqrt{2}$ and $\sqrt[3]{5}$, the *irrational numbers.* The *real numbers,* you will recall, consist of the rational numbers and the irrational numbers. Now, if our universe were to consist only of real numbers, then there would be no number x whose square is -1. To remedy this, we introduce a number i, whose square is -1.

In the progression outlined, each time that we encountered a situation that was unsuitable, we introduced a new number system to remedy this situation. And each new number system contained the earlier number system as a subset. The number system that results from introducing the number i is called the **complex number system.**

> **Complex numbers** are numbers of the form $a + bi$, where a and b are real numbers. The real number a is called the **real part** of the number $a + bi$; the real number b is called the **imaginary part** of $a + bi$.

For example, the complex number $-5 + 6i$ has the real part -5 and the imaginary part 6.

When a complex number is written in the form $a + bi$, where a and b are real numbers, we say it is in **standard form.** However, if the imaginary part of a complex number is negative, such as in the complex number $3 + (-2)i$, we agree to write it instead in the form $3 - 2i$.

Also, the complex number $a + 0i$ is usually written merely as a. This serves to remind us that the real numbers are a subset of the complex numbers. The complex number $0 + bi$ is usually written as bi. Sometimes the complex number bi is called a **pure imaginary number.**

Equality, addition, subtraction, and multiplication of complex numbers are defined so as to preserve the familiar rules of algebra for real numbers. Two complex numbers are equal if and only if their real parts are equal and their imaginary parts are equal. That is,

Equality of Complex Numbers

$$a + bi = c + di \qquad \text{if and only if } a = c \text{ and } b = d \qquad \textbf{(1)}$$

Two complex numbers are added by forming the complex number whose real part is the sum of the real parts and whose imaginary part is the sum of the imaginary parts. That is,

Sum of Complex Numbers

$$(a + bi) + (c + di) = (a + c) + (b + d)i \qquad \textbf{(2)}$$

To subtract two complex numbers, we use this rule:

Difference of Complex Numbers

$$(a + bi) - (c + di) = (a - c) + (b - d)i \qquad \textbf{(3)}$$

EXAMPLE 1

Adding and Subtracting Complex Numbers

(a) $(3 + 5i) + (-2 + 3i) = [3 + (-2)] + (5 + 3)i = 1 + 8i$

(b) $(6 + 4i) - (3 + 6i) = (6 - 3) + (4 - 6)i = 3 + (-2)i = 3 - 2i$ ■

NOW WORK PROBLEM **5**.

Products of complex numbers are calculated as illustrated in Example 2.

EXAMPLE 2

Multiplying Complex Numbers

$$(5 + 3i) \cdot (2 + 7i) = 5 \cdot (2 + 7i) + 3i(2 + 7i) = 10 + 35i + 6i + 21i^2$$

        ↑   ↑
  Distributive property   Distributive property

$$= 10 + 41i + 21(-1)$$

↑
$i^2 = -1$

$$= -11 + 41i$$ ■

Based on the procedure of Example 2, we define the **product** of two complex numbers by the following formula:

Product of Complex Numbers

$$(a + bi) \cdot (c + di) = (ac - bd) + (ad + bc)i \qquad \textbf{(4)}$$

Do not bother to memorize formula (4). Instead, whenever it is necessary to multiply two complex numbers, follow the usual rules for multiplying two binomials, as in Example 2, remembering that $i^2 = -1$. For example,

$$(2i)(2i) = 4i^2 = -4$$

$$(2 + i)(1 - i) = 2 - 2i + i - i^2 = 3 - i$$

NOW WORK PROBLEM **11**.

Algebraic properties for addition and multiplication, such as the commutative, associative, and distributive properties, hold for complex numbers. The property that every nonzero complex number has a multiplicative inverse, or reciprocal, requires a closer look.

CONJUGATES

If $z = a + bi$ is a complex number, then its **conjugate,** denoted by $\bar{z}$, is defined as

$$\bar{z} = \overline{a + bi} = a - bi$$

For example, $\overline{2 + 3i} = 2 - 3i$ and $\overline{-6 - 2i} = -6 + 2i$.

EXAMPLE 3	Multiplying a Complex Number by Its Conjugate

Find the product of the complex number $z = 3 + 4i$ and its conjugate $\bar{z}$.

Solution   Since $\bar{z} = 3 - 4i$, we have

$$z\bar{z} = (3 + 4i)(3 - 4i) = 9 - 12i + 12i - 16i^2 = 9 + 16 = 25 \qquad ■$$

The result obtained in Example 3 has an important generalization.

Theorem   The product of a complex number and its conjugate is a nonnegative real number. That is, if $z = a + bi$, then

$$z\bar{z} = a^2 + b^2 \tag{5}$$

■

Proof   If $z = a + bi$, then

$$z\bar{z} = (a + bi)(a - bi) = a^2 - (bi)^2 = a^2 - b^2 i^2 = a^2 + b^2 \qquad ■$$

To express the reciprocal of a nonzero complex number z in standard form, multiply the numerator and denominator of $\dfrac{1}{z}$ by $\bar{z}$. That is, if $z = a + bi$ is a nonzero complex number, then

$$\frac{1}{a + bi} = \frac{1}{z} = \frac{1}{z} \cdot \frac{\bar{z}}{\bar{z}} = \frac{\bar{z}}{z\bar{z}} \underset{\uparrow}{=} \frac{a - bi}{a^2 + b^2}$$

$$\text{Use (5).}$$

$$= \frac{a}{a^2 + b^2} - \frac{b}{a^2 + b^2} i$$

EXAMPLE 4	Writing the Reciprocal of a Complex Number in Standard Form

Write $\dfrac{1}{3 + 4i}$ in standard form $a + bi$; that is, find the reciprocal of $3 + 4i$.

Solution   The idea is to multiply the numerator and denominator by the conjugate of $3 + 4i$, that is, the complex number $3 - 4i$. The result is

$$\frac{1}{3 + 4i} = \frac{1}{3 + 4i} \cdot \frac{3 - 4i}{3 - 4i} = \frac{3 - 4i}{9 + 16} = \frac{3}{25} - \frac{4}{25} i \qquad ■$$

To express the quotient of two complex numbers in standard form, we multiply the numerator and denominator of the quotient by the conjugate of the denominator.

EXAMPLE 5

Writing the Quotient of Complex Numbers in Standard Form

Write each of the following in standard form.

(a) $\dfrac{1 + 4i}{5 - 12i}$

(b) $\dfrac{2 - 3i}{4 - 3i}$

Solution (a) $\dfrac{1 + 4i}{5 - 12i} = \dfrac{1 + 4i}{5 - 12i} \cdot \dfrac{5 + 12i}{5 + 12i} = \dfrac{5 + 12i + 20i + 48i^2}{25 + 144}$

$$= \dfrac{-43 + 32i}{169} = -\dfrac{43}{169} + \dfrac{32}{169}i$$

(b) $\dfrac{2 - 3i}{4 - 3i} = \dfrac{2 - 3i}{4 - 3i} \cdot \dfrac{4 + 3i}{4 + 3i} = \dfrac{8 + 6i - 12i - 9i^2}{16 + 9}$

$$= \dfrac{17 - 6i}{25} = \dfrac{17}{25} - \dfrac{6}{25}i$$

NOW WORK PROBLEM 19.

EXAMPLE 6

Writing Other Expressions in Standard Form

If $z = 2 - 3i$ and $w = 5 + 2i$, write each of the following expressions in standard form.

(a) $\dfrac{z}{w}$

(b) $\overline{z + w}$

(c) $z + \bar{z}$

Solution (a) $\dfrac{z}{w} = \dfrac{z \cdot \bar{w}}{w \cdot \bar{w}} = \dfrac{(2 - 3i)(5 - 2i)}{(5 + 2i)(5 - 2i)} = \dfrac{10 - 4i - 15i + 6i^2}{25 + 4}$

$$= \dfrac{4 - 19i}{29} = \dfrac{4}{29} - \dfrac{19}{29}i$$

(b) $\overline{z + w} = \overline{(2 - 3i) + (5 + 2i)} = \overline{7 - i} = 7 + i$

(c) $z + \bar{z} = (2 - 3i) + (2 + 3i) = 4$

The conjugate of a complex number has certain general properties that we shall find useful later.

For a real number $a = a + 0i$, the conjugate is $\bar{a} = \overline{a + 0i} = a - 0i = a$. That is,

Theorem　　The conjugate of a real number is the real number itself.

Other properties of the conjugate that are direct consequences of the definition are given next. In each statement, z and w represent complex numbers.

Theorem   The conjugate of the conjugate of a complex number is the complex number itself.

$$(\bar{\bar{z}}) = z \tag{6}$$

The conjugate of the sum of two complex numbers equals the sum of their conjugates.

$$\overline{z + w} = \bar{z} + \bar{w} \tag{7}$$

The conjugate of the product of two complex numbers equals the product of their conjugates.

$$\overline{z \cdot w} = \bar{z} \cdot \bar{w} \tag{8}$$

We leave the proofs of equations (6), (7), and (8) as exercises.

POWERS OF i

The **powers of i** follow a pattern that is useful to know.

$$i^1 = i \qquad\qquad i^5 = i^4 \cdot i = 1 \cdot i = i$$
$$i^2 = -1 \qquad\qquad i^6 = i^4 \cdot i^2 = -1$$
$$i^3 = i^2 \cdot i = -i \qquad\qquad i^7 = i^4 \cdot i^3 = -i$$
$$i^4 = i^2 \cdot i^2 = (-1)(-1) = 1 \qquad\qquad i^8 = i^4 \cdot i^4 = 1$$

And so on. The powers of i repeat with every fourth power.

EXAMPLE 7   **Evaluating Powers of i**

(a) $i^{27} = i^{24} \cdot i^3 = \left(i^4\right)^6 \cdot i^3 = 1^6 \cdot i^3 = -i$

(b) $i^{101} = i^{100} \cdot i^1 = \left(i^4\right)^{25} \cdot i = 1^{25} \cdot i = i$

EXAMPLE 8   **Writing the Power of a Complex Number in Standard Form**

Write $(2 + i)^3$ in standard form.

Solution   We use the special product formula for $(x + a)^3$.

$$(x + a)^3 = x^3 + 3ax^2 + 3a^2x + a^3$$

Using this special product formula,

$$(2 + i)^3 = 2^3 + 3 \cdot i \cdot 2^2 + 3 \cdot i^2 \cdot 2 + i^3$$
$$= 8 + 12i + 6(-1) + (-i)$$
$$= 2 + 11i$$

NOW WORK PROBLEM 33.

QUADRATIC EQUATIONS WITH A NEGATIVE DISCRIMINANT

② Quadratic equations with a negative discriminant have no real number solution. However, if we extend our number system to allow complex numbers, quadratic equations will always have a solution. Since the solution to a quadratic equation involves the square root of the discriminant, we begin with a discussion of square roots of negative numbers.

If N is a positive real number, we define the **principal square root of** $-N$, denoted by $\sqrt{-N}$, as

$$\sqrt{-N} = \sqrt{N}\,i$$

where i is the imaginary unit and $i^2 = -1$.

WARNING: In writing $\sqrt{-N} = \sqrt{N}\,i$, be sure to place i outside the $\sqrt{}$ symbol. ∎

EXAMPLE 9 **Evaluating the Square Root of a Negative Number**

(a) $\sqrt{-1} = \sqrt{1}\,i = i$ (b) $\sqrt{-4} = \sqrt{4}\,i = 2i$

(c) $\sqrt{-8} = \sqrt{8}\,i = 2\sqrt{2}\,i$ ∎

EXAMPLE 10 **Solving Equations**

Solve each equation in the complex number system.

(a) $x^2 = 4$ (b) $x^2 = -9$

Solution (a) $x^2 = 4$
$x = \pm\sqrt{4} = \pm 2$
The equation has two solutions, -2 and 2.

(b) $x^2 = -9$
$x = \pm\sqrt{-9} = \pm\sqrt{9}\,i = \pm 3i$
The equation has two solutions, $-3i$ and $3i$. ∎

NOW WORK PROBLEMS 41 AND 45.

WARNING: When working with square roots of negative numbers, do not set the square root of a product equal to the product of the square roots (which can be done with positive numbers). To see why, look at this calculation: We know that $\sqrt{100} = 10$. However, it is also true that $100 = (-25)(-4)$, so

$$10 = \sqrt{100} = \sqrt{(-25)(-4)} \ne \sqrt{-25}\,\sqrt{-4} = (\sqrt{25}i)(\sqrt{4}i) = (5i)(2i) = 10i^2 = -10$$

$\uparrow$
Here is the error. ∎

Because we have defined the square root of a negative number, we can now restate the quadratic formula without restriction.

Theorem   In the complex number system, the solutions of the quadratic equation $ax^2 + bx + c = 0$, where a, b, and c are real numbers and $a \neq 0$, are given by the formula

$$x = \frac{-b \pm \sqrt{b^2 - 4ac}}{2a} \qquad (9)$$

| EXAMPLE 11 | **Solving Quadratic Equations in the Complex Number System** |

Solve the equation $x^2 - 4x + 8 = 0$ in the complex number system.

Solution   Here $a = 1$, $b = -4$, $c = 8$, and $b^2 - 4ac = 16 - 4(1)(8) = -16$. Using equation (9), we find that

$$x = \frac{-(-4) \pm \sqrt{-16}}{2(1)} = \frac{4 \pm \sqrt{16}\,i}{2} = \frac{4 \pm 4i}{2} = 2 \pm 2i$$

The equation has the solution set $\{2 - 2i, 2 + 2i\}$.   ∎

Check:
$$2 + 2i\colon \quad (2 + 2i)^2 - 4(2 + 2i) + 8 = 4 + 8i + 4i^2 - 8 - 8i + 8$$
$$= 4 - 4 = 0$$
$$2 - 2i\colon \quad (2 - 2i)^2 - 4(2 - 2i) + 8 = 4 - 8i + 4i^2 - 8 + 8i + 8$$
$$= 4 - 4 = 0 \qquad ∎$$

✏️ **NOW WORK PROBLEM 51.**

The discriminant $b^2 - 4ac$ of a quadratic equation still serves as a way to determine the character of the solutions.

> **CHARACTER OF THE SOLUTIONS OF A QUADRATIC EQUATION**
>
> In the complex number system, consider a quadratic equation $ax^2 + bx + c = 0$ with real coefficients.
>
> 1. If $b^2 - 4ac > 0$, the equation has two unequal real solutions.
> 2. If $b^2 - 4ac = 0$, the equation has a repeated real solution, a double root.
> 3. If $b^2 - 4ac < 0$, the equation has two complex solutions that are not real. The solutions are conjugates of each other.

The third conclusion in the display is a consequence of the fact that if $b^2 - 4ac = -N < 0$ then, by the quadratic formula, the solutions are

$$x = \frac{-b + \sqrt{b^2 - 4ac}}{2a} = \frac{-b + \sqrt{-N}}{2a} = \frac{-b + \sqrt{N}\,i}{2a} = \frac{-b}{2a} + \frac{\sqrt{N}}{2a}\,i$$

and

$$x = \frac{-b - \sqrt{b^2 - 4ac}}{2a} = \frac{-b - \sqrt{-N}}{2a} = \frac{-b - \sqrt{N}i}{2a} = \frac{-b}{2a} - \frac{\sqrt{N}}{2a}i$$

which are conjugates of each other.

EXAMPLE 12

Determining the Character of the Solution of a Quadratic Equation

Without solving, determine the character of the solution of each equation.

(a) $3x^2 + 4x + 5 = 0$ (b) $2x^2 + 4x + 1 = 0$

(c) $9x^2 - 6x + 1 = 0$

Solution

(a) Here $a = 3$, $b = 4$, and $c = 5$, so $b^2 - 4ac = 16 - 4(3)(5) = -44$. The solutions are two complex numbers that are not real and are conjugates of each other.

(b) Here $a = 2$, $b = 4$, and $c = 1$, so $b^2 - 4ac = 16 - 8 = 8$. The solutions are two unequal real numbers.

(c) Here $a = 9$, $b = -6$, and $c = 1$, so $b^2 - 4ac = 36 - 4(9)(1) = 0$. The solution is a repeated real number, that is, a double root. ∎

NOW WORK PROBLEM 65.

5.3 EXERCISES

In Problems 1–38, write each expression in the standard form a + bi.

1. $(2 - 3i) + (6 + 8i)$
2. $(4 + 5i) + (-8 + 2i)$
3. $(-3 + 2i) - (4 - 4i)$
4. $(3 - 4i) - (-3 - 4i)$
5. $(2 - 5i) - (8 + 6i)$
6. $(-8 + 4i) - (2 - 2i)$
7. $3(2 - 6i)$
8. $-4(2 + 8i)$
9. $2i(2 - 3i)$
10. $3i(-3 + 4i)$
11. $(3 - 4i)(2 + i)$
12. $(5 + 3i)(2 - i)$
13. $(-6 + i)(-6 - i)$
14. $(-3 + i)(3 + i)$
15. $\dfrac{10}{3 - 4i}$
16. $\dfrac{13}{5 - 12i}$
17. $\dfrac{2 + i}{i}$
18. $\dfrac{2 - i}{-2i}$
19. $\dfrac{6 - i}{1 + i}$
20. $\dfrac{2 + 3i}{1 - i}$
21. $\left(\dfrac{1}{2} + \dfrac{\sqrt{3}}{2}i\right)^2$
22. $\left(\dfrac{\sqrt{3}}{2} - \dfrac{1}{2}i\right)^2$
23. $(1 + i)^2$
24. $(1 - i)^2$
25. i^{23}
26. i^{14}
27. i^{-15}
28. i^{-23}
29. $i^6 - 5$
30. $4 + i^3$
31. $6i^3 - 4i^5$
32. $4i^3 - 2i^2 + 1$
33. $(1 + i)^3$
34. $(3i)^4 + 1$
35. $i^7(1 + i^2)$
36. $2i^4(1 + i^2)$
37. $i^6 + i^4 + i^2 + 1$
38. $i^7 + i^5 + i^3 + i$

In Problems 39–44, perform the indicated operations and express your answer in the form a + bi.

39. $\sqrt{-4}$
40. $\sqrt{-9}$
41. $\sqrt{-25}$
42. $\sqrt{-64}$
43. $\sqrt{(3 + 4i)(4i - 3)}$
44. $\sqrt{(4 + 3i)(3i - 4)}$

In Problems 45–64, solve each equation in the complex number system.

45. $x^2 + 4 = 0$
46. $x^2 - 4 = 0$
47. $x^2 - 16 = 0$
48. $x^2 + 25 = 0$
49. $x^2 - 6x + 13 = 0$
50. $x^2 + 4x + 8 = 0$
51. $x^2 - 6x + 10 = 0$
52. $x^2 - 2x + 5 = 0$
53. $8x^2 - 4x + 1 = 0$
54. $10x^2 + 6x + 1 = 0$
55. $5x^2 + 1 = 2x$
56. $13x^2 + 1 = 6x$
57. $x^2 + x + 1 = 0$
58. $x^2 - x + 1 = 0$
59. $x^3 - 8 = 0$
60. $x^3 + 27 = 0$
61. $x^4 = 16$
62. $x^4 = 1$
63. $x^4 + 13x^2 + 36 = 0$
64. $x^4 + 3x^2 - 4 = 0$

In Problems 65–70, without solving, determine the character of the solutions of each equation in the complex number system.

65. $3x^2 - 3x + 4 = 0$ **66.** $2x^2 - 4x + 1 = 0$ **67.** $2x^2 + 3x = 4$

68. $x^2 + 6 = 2x$ **69.** $9x^2 - 12x + 4 = 0$ **70.** $4x^2 + 12x + 9 = 0$

71. $2 + 3i$ is a solution of a quadratic equation with real coefficients. Find the other solution.

72. $4 - i$ is a solution of a quadratic equation with real coefficients. Find the other solution.

In Problems 73–76, $z = 3 - 4i$ and $w = 8 + 3i$. Write each expression in the standard form $a + bi$.

73. $z + \bar{z}$ **74.** $w - \bar{w}$ **75.** $z\bar{z}$ **76.** $\overline{z - w}$

77. Use $z = a + bi$ to show that $z + \bar{z} = 2a$ and $z - \bar{z} = 2bi$.

78. Use $z = a + bi$ to show that $\bar{\bar{z}} = z$.

79. Use $z = a + bi$ and $w = c + di$ to show that $\overline{z + w} = \bar{z} + \bar{w}$.

80. Use $z = a + bi$ and $w = c + di$ to show that $\overline{z \cdot w} = \bar{z} \cdot \bar{w}$.

81. Explain to a friend how you would add two complex numbers and how you would multiply two complex numbers. Explain any differences in the two explanations.

82. Write a brief paragraph that compares the method used to rationalize the denominator of a rational expression and the method used to write the quotient of two complex numbers in standard form.

5.4 COMPLEX ZEROS; FUNDAMENTAL THEOREM OF ALGEBRA

OBJECTIVES **1** Utilize the Conjugate Pairs Theorem to Find the Complex Zeros of a Polynomial

2 Find a Polynomial Function with Specified Zeros

3 Find the Complex Zeros of a Polynomial

In Section 5.2 we found the **real** zeros of a polynomial function. In this section we will find the **complex** zeros of a polynomial function. Finding the complex zeros of a function requires finding all zeros of the form $a + bi$. These zeros will be real if $b = 0$.

A variable in the complex number system is referred to as a **complex variable.** A **complex polynomial function** f of degree n is a function of the form

$$f(x) = a_n x^n + a_{n-1} x^{n-1} + \cdots + a_1 x + a_0 \qquad (1)$$

where $a_n, a_{n-1}, \ldots, a_1, a_0$ are complex numbers, $a_n \neq 0$, n is a nonnegative integer, and x is a complex variable. As before, a_n is called the **leading coefficient** of f. A complex number r is called a (complex) **zero** of f if $f(r) = 0$.

We have learned that some quadratic equations have no real solutions, but that in the complex number system every quadratic equation has a solution, either real or complex. The next result, proved by Karl Friedrich Gauss (1777–1855) when he was 22 years old,* gives an extension to complex polynomials. In fact, this result is so important and useful that it has become known as the **Fundamental Theorem of Algebra.**

---

*In all, Gauss gave four different proofs of this theorem, the first one in 1799 being the subject of his doctoral dissertation.

Fundamental Theorem of Algebra	Every complex polynomial function $f(x)$ of degree $n \geq 1$ has at least one complex zero.

We shall not prove this result, as the proof is beyond the scope of this book. However, using the Fundamental Theorem of Algebra and the Factor Theorem, we can prove the following result:

Theorem	Every complex polynomial function $f(x)$ of degree $n \geq 1$ can be factored into n linear factors (not necessarily distinct) of the form

$$f(x) = a_n(x - r_1)(x - r_2) \cdot \ldots \cdot (x - r_n) \tag{2}$$

where $a_n, r_1, r_2, \ldots, r_n$ are complex numbers. That is, every complex polynomial function of degree $n \geq 1$ has exactly n (not necessarily distinct) zeros.

Proof Let

$$f(x) = a_n x^n + a_{n-1} x^{n-1} + \cdots + a_1 x + a_0$$

By the Fundamental Theorem of Algebra, f has at least one zero, say r_1. Then, by the Factor Theorem, $x - r_1$ is a factor, and

$$f(x) = (x - r_1)q_1(x)$$

where $q_1(x)$ is a complex polynomial of degree $n - 1$ whose leading coefficient is a_n. Again by the Fundamental Theorem of Algebra, the complex polynomial $q_1(x)$ has at least one zero, say r_2. By the Factor Theorem, $q_1(x)$ has the factor $x - r_2$, so

$$q_1(x) = (x - r_2)q_2(x)$$

where $q_2(x)$ is a complex polynomial of degree $n - 2$ whose leading coefficient is a_n. Consequently,

$$f(x) = (x - r_1)(x - r_2)q_2(x)$$

Repeating this argument n times, we finally arrive at

$$f(x) = (x - r_1)(x - r_2) \cdot \ldots \cdot (x - r_n)q_n(x)$$

where $q_n(x)$ is a complex polynomial of degree $n - n = 0$ whose leading coefficient is a_n. Thus, $q_n(x) = a_n x^0 = a_n$, and so

$$f(x) = a_n(x - r_1)(x - r_2) \cdot \ldots \cdot (x - r_n)$$

We conclude that every complex polynomial function $f(x)$ of degree $n \geq 1$ has exactly n (not necessarily distinct) zeros. ∎

COMPLEX ZEROS OF POLYNOMIALS WITH REAL COEFFICIENTS

① We can use the Fundamental Theorem of Algebra to obtain valuable information about the complex zeros of polynomials whose coefficients are real numbers.

Conjugate Pairs Theorem

Let $f(x)$ be a polynomial whose coefficients are real numbers. If $r = a + bi$ is a zero of f, then the complex conjugate $\bar{r} = a - bi$ is also a zero of f.

In other words, for polynomials whose coefficients are real numbers, the zeros occur in conjugate pairs.

Proof   Let

$$f(x) = a_n x^n + a_{n-1} x^{n-1} + \cdots + a_1 x + a_0$$

where $a_n, a_{n-1}, \ldots, a_1, a_0$ are real numbers and $a_n \neq 0$. If $r = a + bi$ is a zero of f, then $f(r) = f(a + bi) = 0$, so

$$a_n r^n + a_{n-1} r^{n-1} + \cdots + a_1 r + a_0 = 0$$

We take the conjugate of both sides to get

$$\overline{a_n r^n + a_{n-1} r^{n-1} + \cdots + a_1 r + a_0} = \bar{0}$$

$$\overline{a_n r^n} + \overline{a_{n-1} r^{n-1}} + \cdots + \overline{a_1 r} + \overline{a_0} = \bar{0} \qquad \text{The conjugate of a sum equals the sum of the conjugates (see Section 5.3).}$$

$$\overline{a_n}(\bar{r})^n + \overline{a_{n-1}}(\bar{r})^{n-1} + \cdots + \bar{a}_1 \bar{r} + \bar{a}_0 = \bar{0} \qquad \text{The conjugate of a product equals the product of the conjugates.}$$

$$a_n(\bar{r})^n + a_{n-1}(\bar{r})^{n-1} + \cdots + a_1 \bar{r} + a_0 = 0 \qquad \text{The conjugate of a real number equals the real number.}$$

This last equation states that $f(\bar{r}) = 0$; that is, $\bar{r} = a - bi$ is a zero of f.   ∎

The value of this result should be clear. Once we know that, say, $3 + 4i$ is a zero of a polynomial with real coefficients, then we know that $3 - 4i$ is also a zero. This result has an important corollary.

Corollary

A polynomial f of odd degree with real coefficients has at least one real zero.

Proof   Because complex zeros occur as conjugate pairs in a polynomial with real coefficients, there will always be an even number of zeros that are not real numbers. Consequently, since f is of odd degree, one of its zeros has to be a real number.   ∎

For example, the polynomial $f(x) = x^5 - 3x^4 + 4x^3 - 5$ has at least one zero that is a real number, since f is of degree 5 (odd) and has real coefficients.

EXAMPLE 1

Using the Conjugate Pairs Theorem

A polynomial f of degree 5 whose coefficients are real numbers has the zeros $1, 5i$, and $1 + i$. Find the remaining two zeros.

Solution   Since complex zeros appear as conjugate pairs, it follows that $-5i$, the conjugate of $5i$, and $1 - i$, the conjugate of $1 + i$, are the two remaining zeros.   ∎

NOW WORK PROBLEM 1.

| EXAMPLE 2 | **Finding a Polynomial Function Whose Zeros Are Given** |

2 Find a polynomial f of degree 4 whose coefficients are real numbers and that has the zeros 1, 1, and $-4 + i$.

Solution   Since $-4 + i$ is a zero, by the Conjugate Pairs Theorem, $-4 - i$ must also be a zero of f. Because of the Factor Theorem, if $f(c) = 0$, then $x - c$ is a factor of $f(x)$. So we can now write f as

$$f(x) = a(x - 1)(x - 1)[x - (-4 + i)][x - (-4 - i)]$$

where a is any real number. If we let $a = 1$, we obtain

$$
\begin{aligned}
f(x) &= (x - 1)(x - 1)[x - (-4 + i)][x - (-4 - i)] \\
&= (x^2 - 2x + 1)[x^2 - (-4 + i)x - (-4 - i)x + (-4 + i)(-4 - i)] \\
&= (x^2 - 2x + 1)(x^2 + 4x - ix + 4x + ix + 16 + 4i - 4i - i^2) \\
&= (x^2 - 2x + 1)(x^2 + 8x + 17) \\
&= (x^4 + 8x^3 + 17x^2 - 2x^3 - 16x^2 - 34x + x^2 + 8x + 17) \\
&= x^4 + 6x^3 + 2x^2 - 26x + 17
\end{aligned}
$$

EXPLORATION   Graph the function f found in Example 2 for $a = 1$. Does the value of a affect the zeros of f? How does the value of a affect the graph of f?

SOLUTION   A quick analysis of the polynomial f tells us what to expect:

At most three turning points.
For large $|x|$, the graph will behave like $y = x^4$.
A repeated real zero at 1 so that the graph will touch the x-axis at 1.
The only x-intercept is at 1; the y-intercept is 17.

Figure 3 shows the complete graph. (Do you see why? The graph has exactly three turning points.) The value of a causes a stretch or compression; a reflection also occurs if $a < 0$. The zeros are not affected.

Now we can prove the theorem we conjectured earlier in Section 5.2.

Figure 3

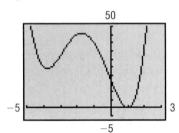

| Theorem | Every polynomial function with real coefficients can be uniquely factored over the real numbers into a product of linear factors and/or irreducible quadratic factors. |

Proof   Every complex polynomial f of degree n has exactly n zeros and can be factored into a product of n linear factors. If its coefficients are real, then those zeros that are complex numbers will always occur as conjugate pairs. As a result, if $r = a + bi$ is a complex zero, then so is $\bar{r} = a - bi$. Consequently, when the linear factors $x - r$ and $x - \bar{r}$ of f are multiplied, we have

$$(x - r)(x - \bar{r}) = x^2 - (r + \bar{r})x + r\bar{r} = x^2 - 2ax + a^2 + b^2$$

This second-degree polynomial has real coefficients and is irreducible (over the real numbers). Thus, the factors of f are either linear or irreducible quadratic factors.

| EXAMPLE 3 | **Finding the Complex Zeros of a Polynomial** |

③ Find the complex zeros of the polynomial function

$$f(x) = 3x^4 + 5x^3 + 25x^2 + 45x - 18$$

Write f in factored form.

Solution STEP 1: The degree of f is 4. So f will have four complex zeros.

STEP 2: Descartes' Rule of Signs provides information about the real zeros. For this polynomial, there is one positive real zero. Because

$$f(-x) = 3x^4 - 5x^3 + 25x^2 - 45x - 18$$

there are three or one negative real zeros.

STEP 3: The Rational Zeros Theorem provides information about the potential rational zeros of polynomials with integer coefficients. For this polynomial (which has integer coefficients), the potential rational zeros are

$$\pm\frac{1}{3}, \quad \pm\frac{2}{3}, \quad \pm1, \quad \pm2, \quad \pm3, \quad \pm6, \quad \pm9, \quad \pm18$$

We test 1 first:
$$1\overline{)3 \quad 5 \quad 25 \quad 45 \quad -18}$$
$$ 3 \quad 8 \quad 33 \quad 78$$
$$\overline{3 \quad 8 \quad 33 \quad 78 \quad 60}$$

We test -1:
$$-1\overline{)3 \quad\ 5 \quad 25 \quad\ 45 \quad -18}$$
$$ -3 \quad -2 \quad -23 \quad -22$$
$$\overline{3 \quad\ 2 \quad 23 \quad\ 22 \quad -40}$$

We test 2:
$$2\overline{)3 \quad\ 5 \quad 25 \quad\ 45 \quad -18}$$
$$ 6 \quad 22 \quad 94 \quad 278$$
$$\overline{3 \quad 11 \quad 47 \quad 139 \quad 260}$$

We test -2:
$$-2\overline{)3 \quad\ 5 \quad 25 \quad\ 45 \quad -18}$$
$$ -6 \quad\ 2 \quad -54 \quad 18$$
$$\overline{3 \quad -1 \quad 27 \quad -9 \quad\ 0}$$

Since, $f(-2) = 0$, then -2 is a zero and $x + 2$ is a factor of f. The depressed equation is

$$3x^3 - x^2 + 27x - 9 = 0$$

STEP 4: We factor by grouping.

$$3x^3 - x^2 + 27x - 9 = 0$$
$$x^2(3x - 1) + 9(3x - 1) = 0$$
$$(x^2 + 9)(3x - 1) = 0$$

Factor x^2 from $3x^3 - x^2$ and 9 from $27x - 9$

$$x^2 + 9 = 0 \qquad\qquad 3x - 1 = 0$$

Factor out the common factor $3x - 1$

$$x^2 = -9 \qquad\qquad x = \frac{1}{3}$$

Apply Zero-Product Property

$$x = -3i \qquad x = 3i \qquad x = \frac{1}{3}$$

The four complex zeros of f are $\left\{-3i, 3i, -2, \dfrac{1}{3}\right\}$.

The factored form of f is

$$f(x) = 3x^4 + 5x^3 + 25x^2 + 45x - 18$$

$$= 3(x + 3i)(x - 3i)(x + 2)\left(x - \dfrac{1}{3}\right) \quad \blacksquare$$

NOW WORK PROBLEM **27.**

5.4 EXERCISES

In Problems 1–10, information is given about a polynomial $f(x)$ whose coefficients are real numbers. Find the remaining zeros of f.

1. Degree 3;   zeros: $3, 4 - i$

2. Degree 3;   zeros: $4, 3 + i$

3. Degree 4;   zeros: $i, 1 + i$

4. Degree 4;   zeros: $1, 2, 2 + i$

5. Degree 5;   zeros: $1, i, 2i$

6. Degree 5;   zeros: $0, 1, 2, i$

7. Degree 4;   zeros: $i, 2, -2$

8. Degree 4;   zeros: $2 - i, -i$

9. Degree 6;   zeros: $2, 2 + i, -3 - i, 0$

10. Degree 6;   zeros: $i, 3 - 2i, -2 + i$

In Problems 11–16, form a polynomial $f(x)$ with real coefficients having the given degree and zeros.

11. Degree 4;   zeros: $3 + 2i$; 4, multiplicity 2

12. Degree 4;   zeros: $i, 1 + 2i$

13. Degree 5;   zeros: 2, multiplicity 1; $-i$; $1 + i$

14. Degree 6;   zeros: $i, 4 - i; 2 + i$

15. Degree 4;   zeros: 3, multiplicity 2; $-i$

16. Degree 5;   zeros: 1, multiplicity 3; $1 + i$

In Problems 17–24, use the given zero to find the remaining zeros of each function.

17. $f(x) = x^3 - 4x^2 + 4x - 16$;   zero: $2i$

18. $g(x) = x^3 + 3x^2 + 25x + 75$;   zero: $-5i$

19. $f(x) = 2x^4 + 5x^3 + 5x^2 + 20x - 12$;   zero: $-2i$

20. $h(x) = 3x^4 + 5x^3 + 25x^2 + 45x - 18$;   zero: $3i$

21. $h(x) = x^4 - 9x^3 + 21x^2 + 21x - 130$;   zero: $3 - 2i$

22. $f(x) = x^4 - 7x^3 + 14x^2 - 38x - 60$;   zero: $1 + 3i$

23. $h(x) = 3x^5 + 2x^4 + 15x^3 + 10x^2 - 528x - 352$;   zero: $-4i$

24. $g(x) = 2x^5 - 3x^4 - 5x^3 - 15x^2 - 207x + 108$;   zero: $3i$

In Problems 25–34, find the complex zeros of each polynomial function. Write f in factored form.

25. $f(x) = x^3 - 1$

26. $f(x) = x^4 - 1$

27. $f(x) = x^3 - 8x^2 + 25x - 26$

28. $f(x) = x^3 + 13x^2 + 57x + 85$

29. $f(x) = x^4 + 5x^2 + 4$

30. $f(x) = x^4 + 13x^2 + 36$

31. $f(x) = x^4 + 2x^3 + 22x^2 + 50x - 75$

32. $f(x) = x^4 + 3x^3 - 19x^2 + 27x - 252$

33. $f(x) = 3x^4 - x^3 - 9x^2 + 159x - 52$

34. $f(x) = 2x^4 + x^3 - 35x^2 - 113x + 65$

In Problems 35 and 36, tell why the facts given are contradictory.

35. $f(x)$ is a polynomial of degree 3 whose coefficients are real numbers; its zeros are $4 + i, 4 - i$, and $2 + i$.

36. $f(x)$ is a polynomial of degree 3 whose coefficients are real numbers; its zeros are $2, i$, and $3 + i$.

37. $f(x)$ is a polynomial of degree 4 whose coefficients are real numbers; three of its zeros are $2, 1 + 2i$, and $1 - 2i$. Explain why the remaining zero must be a real number.

38. $f(x)$ is a polynomial of degree 4 whose coefficients are real numbers; two of its zeros are -3 and $4 - i$. Explain why one of the remaining zeros must be a real number. Write down one of the missing zeros.

CHAPTER REVIEW

Things To Know

Zeros of a polynomial f (p. 358)	Numbers for which $f(x) = 0$; these are the x-intercepts of the graph of f.
Remainder Theorem (p. 359)	If a polynomial $f(x)$ is divided by $x - c$, then the remainder is $f(c)$.
Factor Theorem (p. 360)	$x - c$ is a factor of a polynomial $f(x)$ if and only if $f(c) = 0$.
Descartes' Rule of Signs (p. 362)	Let f denote a polynomial function. The number of positive zeros of f either equals the number of variations in sign of the nonzero coefficients of $f(x)$ or else equals that number less some even integer. The number of negative zeros of f either equals the number of variations in sign of the nonzero coefficients of $f(-x)$ or else equals that number less some even integer.
Rational Zeros Theorem (p. 363)	Let f be a polynomial function of degree 1 or higher of the form $$f(x) = a_n x^n + a_{n-1} x^{n-1} + \cdots + a_1 x + a_0, \qquad a_n \neq 0, a_0 \neq 0$$ where each coefficient is an integer. If p/q, in lowest terms, is a rational zero of f, then p must be a factor of a_0 and q must be a factor of a_n.
Intermediate Value Theorem (p. 368)	Let f be a polynomial function. If $a < b$ and $f(a)$ and $f(b)$ are of opposite sign, then there is at least one zero of f between a and b.
Quadratic equation and quadratic formula (p. 379)	If $ax^2 + bx + c = 0$, $a \neq 0$, then $x = \dfrac{-b \pm \sqrt{b^2 - 4ac}}{2a}$.
Discriminant (p. 379)	If $b^2 - 4ac > 0$, there are two distinct real solutions. If $b^2 - 4ac = 0$, there is one repeated real solution. If $b^2 - 4ac < 0$, there are two distinct complex solutions that are not real; the solutions are conjugates of each other.
Fundamental Theorem of Algebra (p. 382)	Every complex polynomial function $f(x)$ of degree $n \geq 1$ has at least one complex zero.
Conjugate Pairs Theorem (p. 383)	Let $f(x)$ be a polynomial whose coefficients are real numbers. If $r = a + bi$ is a zero of f, then its complex conjugate $\bar{r} = a - bi$ is also a zero of f.

Objectives

You should be able to:

Use synthetic division (p. 354)

Use the Remainder and Factor Theorems (p. 359)

Use Descartes' Rule of Signs (p. 362)

Use the Rational Zeros Theorem (p. 363)

Find the real zeros of a polynomial function (p. 363)

Solve polynomial equations (p. 366)

Use the Theorem for Bounds on Zeros (p. 367)

Use the Intermediate Value Theorem (p. 368)

Add, subtract, multiply and divide complex numbers (p. 373)

Solve quadratic equations with a negative discriminant (p. 378)

Utilize the conjugate pairs theorem (p. 382)

Find a polynomial function with specified zeros (p. 384)

Find the complex zeros of a polynomial (p. 385)

Fill-in-the-Blank Items

1. In the process of polynomial division, (Divisor)(Quotient) + _____ = _____.

2. When a polynomial function f is divided by $x - c$, the remainder is _____.

3. A polynomial function f has the factor $x - c$ if and only if _____.

4. The polynomial function $f(x) = x^5 - 2x^3 + x^2 - x + 1$ has at most _____ real zeros.

5. The possible rational zeros of $f(x) = 2x^5 - x^3 + x^2 - x + 1$ are _____.

6. In the complex number $5 + 2i$, the number 5 is called the _____ part; the number 2 is called the _____ part; the number i is called the _____ _____.

7. If $3 + 4i$ is a zero of a polynomial of degree 5 with real coefficients, then so is _____.

8. The equation $|x^2| = 4$ has four complex solutions: _____, _____, _____, and _____.

9. If a function f whose domain is all real numbers is even and 4 is a zero of f, then _____ is also a zero.

True/False Items

T F **1.** Every polynomial of degree 3 with real coefficients has exactly three real zeros.

T F **2.** If $2 - 3i$ is a zero of a polynomial with real coefficients, then so is $-2 + 3i$.

T F **3.** If f is a polynomial function of degree 4 and if $f(2) = 5$, then

$$\frac{f(x)}{x - 2} = p(x) + \frac{5}{x - 2}$$

where $p(x)$ is a polynomial of degree 3.

T F **4.** The conjugate of $2 + 5i$ is $-2 - 5i$.

T F **5.** A polynomial of degree n with real coefficients has exactly n complex zeros. At most n of them are real numbers.

Review Exercises

Blue problem numbers indicate the author's suggestions for use in a Practice Test.

In Problems 1–4, use synthetic division to find the quotient $q(x)$ and remainder R when $f(x)$ is divided by $g(x)$.

1 $f(x) = 8x^3 - 3x^2 + x + 4$; $g(x) = x - 1$

2. $f(x) = 2x^3 + 8x^2 - 5x + 5$; $g(x) = x - 2$

3. $f(x) = x^4 - 2x^3 + x - 1$; $g(x) = x + 2$

4. $f(x) = x^4 - x^2 + 3x$; $g(x) = x + 1$

5. Find the value of $f(x) = 12x^6 - 8x^4 + 1$ at $x = 4$.

6. Find the value of $f(x) = -16x^3 + 18x^2 - x + 2$ at $x = -2$.

In Problems 7 and 8, use Descartes' Rule of Signs to determine how many positive and negative zeros each polynomial function may have. Do not attempt to find the zeros.

7. $f(x) = 12x^8 - x^7 + 8x^4 - 2x^3 + x + 3$

8. $f(x) = -6x^5 + x^4 + 5x^3 + x + 1$

9. List all the potential rational zeros of $f(x) = 12x^8 - x^7 + 6x^4 - x^3 + x - 3$.

10. List all the potential rational zeros of $f(x) = -6x^5 + x^4 + 2x^3 - x + 1$.

In Problems 11–16, use Descartes' Rule of Signs and the Rational Zeros Theorem to find all the real zeros of each polynomial function. Use the zeros to factor f over the real numbers.

11. $f(x) = x^3 - 3x^2 - 6x + 8$

12. $f(x) = x^3 - x^2 - 10x - 8$

13. $f(x) = 4x^3 + 4x^2 - 7x + 2$

14. $f(x) = 4x^3 - 4x^2 - 7x - 2$

15. $f(x) = x^4 - 4x^3 + 9x^2 - 20x + 20$

16. $f(x) = x^4 + 6x^3 + 11x^2 + 12x + 18$

In Problems 17–20, solve each equation in the real number system.

17. $2x^4 + 2x^3 - 11x^2 + x - 6 = 0$

18. $3x^4 + 3x^3 - 17x^2 + x - 6 = 0$

19. $2x^4 + 7x^3 + x^2 - 7x - 3 = 0$

20. $2x^4 + 7x^3 - 5x^2 - 28x - 12 = 0$

In Problems 21–30, find the complex of zeros of each polynomial function f(x). Write f in factored form.

21. $f(x) = x^3 - 3x^2 - 6x + 8$

22. $f(x) = x^3 - x^2 - 10x - 8$

23. $f(x) = 4x^3 + 4x^2 - 7x + 2$

24. $f(x) = 4x^3 - 4x^2 - 7x - 2$

25. $f(x) = x^4 - 4x^3 + 9x^2 - 20x + 20$

26. $f(x) = x^4 + 6x^3 + 11x^2 + 12x + 18$

27. $f(x) = 2x^4 + 2x^3 - 11x^2 + x - 6$

28. $f(x) = 3x^4 + 3x^3 - 17x^2 + x - 6$

29. $f(x) = 2x^4 + 7x^3 + x^2 - 7x - 3$

30. $f(x) = 2x^4 + 7x^3 - 5x^2 - 28x - 12$

In Problems 31–34, find a bound to the zeros of each polynomial function.

31. $f(x) = x^3 - x^2 - 4x + 2$

32. $f(x) = x^3 + x^2 - 10x - 5$

33. $f(x) = 2x^3 - 7x^2 - 10x + 35$

34. $f(x) = 3x^3 - 7x^2 - 6x + 14$

In Problems 35–38, use the Intermediate Value Theorem to show that each polynomial has a zero in the given interval.

35. $f(x) = 3x^3 - x - 1;$ $[0, 1]$

36. $f(x) = 2x^3 - x^2 - 3;$ $[1, 2]$

37. $f(x) = 8x^4 - 4x^3 - 2x - 1;$ $[0, 1]$

38. $f(x) = 3x^4 + 4x^3 - 8x - 2;$ $[1, 2]$

In Problems 39–42, each polynomial has exactly one positive zero. Approximate the zero rounded to two decimal places.

39. $f(x) = x^3 - x - 2$

40. $f(x) = 2x^3 - x^2 - 3$

41. $f(x) = 8x^4 - 4x^3 - 2x - 1$

42. $f(x) = 3x^4 + 4x^3 - 8x - 2$

In Problems 43–52, use the complex number system and write each expression in the standard form a + bi.

43. $(6 + 3i) - (2 - 4i)$

44. $(8 - 3i) + (-6 + 2i)$

45. $4(3 - i) + 3(-5 + 2i)$

46. $2(1 + i) - 3(2 - 3i)$

47. $\dfrac{3}{3 + i}$

48. $\dfrac{4}{2 - i}$

49. i^{50}

50. i^{29}

51. $(2 + 3i)^3$

52. $(3 - 2i)^3$

In Problems 53–56, information is given about a complex polynomial f(x) whose coefficients are real numbers. Find the remaining zeros of f.

53. Degree 3; zeros: $4 + i, 6$

54. Degree 3; zeros: $3 + 4i, 5$

55. Degree 4; zeros: $i, 1 + i$

56. Degree 4; zeros: $1, 2, 1 + i$

In Problems 57–70, solve each equation in the complex number system.

57. $x^2 + x + 1 = 0$

58. $x^2 - x + 1 = 0$

59. $2x^2 + x - 2 = 0$

60. $3x^2 - 2x - 1 = 0$

61. $x^2 + 3 = x$

62. $2x^2 + 1 = 2x$

63. $x(1 - x) = 6$

64. $x(1 + x) = 2$

65. $x^4 + 2x^2 - 8 = 0$

66. $x^4 + 8x^2 - 9 = 0$

67. $x^3 - x^2 - 8x + 12 = 0$

68. $x^3 - 3x^2 - 4x + 12 = 0$

69. $3x^4 - 4x^3 + 4x^2 - 4x + 1 = 0$

70. $x^4 + 4x^3 + 2x^2 - 8x - 8 = 0$

Alternating Current (AC) Circuit Analysis

AC circuit analysis techniques are used on electrical circuits that are energized with sinusoidal sources. A common example of an AC source is your electric power at home. Its effective amplitude is 120 volts and its frequency (f) is 60 cycles/second (60 Hz). The figure shows an electric circuit with sinusoidal voltage source V_s with amplitude 10 volts and frequency $5000/\pi$ Hz. It is connected to two complex load impedances $Z_1 = R_1 - X_1 i$ and $Z_2 = R_2 - X_2 i$ that are connected in series with each other. Each load contains a resistor (R) connected to a capacitor (C) or an inductor (L).

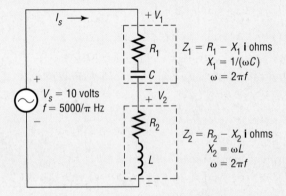

Resistors are electronic components designed to resist or limit electric current flow. The greater the resistance is, the more effectively it limits current. Electric power is lost or dissipated as heat in resistors. Capacitors and inductors limit current also, and their ability to do this is measured as reactance. Electric power is not lost but is stored in ideal capacitors and inductors. The capacitive and inductive reactance values X_1 and X_2 are given by $X_1 = 1/(\omega C)$ and $X_2 = \omega L$, where $\omega = 2\pi f$, C is capacitance in farads (F), and L is inductance in henrys (H). Reactance is unique to resistance in that it is independent of the frequency of operation.

1. Given that resistors R_1 and R_2 are each 5 ohms, capacitor C is 1 F, and inductor L is 0.0015 H, calculate X_1 and X_2 for the given frequency of operation. Write out Z_1 and Z_2 in standard complex form. Impedance, Z_t, is formed by adding Z_1 and Z_2 as shown in the figure below. What is Z_t? What two circuit elements (resistors, capacitors, inductors) do you think Z_t contains?

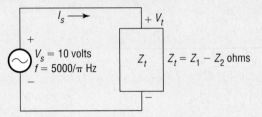

2. The current flow (I_s) is sinusoidal and its frequency is the same as the voltage source. Calculate I_s using the formula $I_s = V_s/Z_t$ (amps).

3. As I_s flows through Z_1 a voltage drop V_1 is created given by $V_1 = I_s Z_1$. Calculate the voltage across Z_1.

4. Find the voltage drop V_2 across $Z_2 (V_2 = I_s Z_2)$.

5. Does $V_s = V_1 + V_2$? Do you think this is reasonable?

6. Find the voltage drop V_t across $Z_t (V_t = I_s Z_t)$. Does $V_t = V_1 + V_2$? Is this a reasonable result?

7. The power dissipated in load Z_1 in watts is given by $P_1 = (1/2) \operatorname{Re}\{V_1 I_{s*}\}$, where I_{s*} is the conjugate of I_s and $\operatorname{Re}\{V_1 I_{s*}\}$, is the real part of $V_1 I_{s*}$. Calculate P_1.

8. The power dissipated in load Z_2 in watts is given by $P_2 = (1/2) \operatorname{Re}\{V_2 I_{s*}\}$, Calculate P_2.

9. Calculate P_t, the power dissipated in load Z_t.

10. Does $P_t = P_1 + P_2$? Do you think this is reasonable?

11. What do you think will happen to Z_t, I_s, and P_t as frequency goes to zero? To infinity?

Exponential and Logarithmic Functions

Field Trip to Motorola

Many physical and chemical processes exhibit exponential or logarithmic behavior. These functions are the cornerstones of science and engineering. For example, chemical reactions often can be modeled by an exponential function. Metal fatigue, which is very complex when we introduce temperature and the rate of cycling, can be described with logarithmic functions

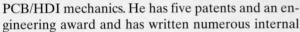

Interview at Motorola

Andrew Skipor is a Principal Staff Engineer with the Design Reliability Group in Motorola Lab's Motorola Advanced Technology Center. He has been with Motorola over 16 years. His interests include development of methods to evaluate package reliability, numerical simulation, experimental methods such as Moiré interferometry, nonlinear solder mechanics, fatigue, and PCB/HDI mechanics. He has five patents and an engineering award and has written numerous internal and external publications. Skipor is the Chairman of the Motorola– IEEE/CPMT Graduate Student Fellowship for Research in Electronic Packaging. He received his Ph.D. in engineering mechanics from the University of Illinois at Chicago, supported in part by the Motorola Distinguished Scholar Program. Throughout his career at Motorola, he has used advanced topics in mathematics that directly depend on a sound foundation in algebra, trigonometry, and analytic geometry.

PREPARING FOR THIS SECTION

Before getting started, review the following:

✓ Functions (Section 3.1, pp. 204–217)

✓ Increasing/Decreasing Functions (Section 3.2, pp. 227–228)

6.1   ONE-TO-ONE FUNCTIONS; INVERSE FUNCTIONS

OBJECTIVES   1   Determine the Inverse of a Function
2   Obtain the Graph of the Inverse Function from the Graph of the Function
3   Find the Inverse Function f^{-1}

1   In Section 3.1 we said that a function f can be thought of as a machine that receives as input a number, say x, from the domain, manipulates it, and outputs the value $f(x)$. The **inverse of f** receives as input a number $f(x)$, manipulates it, and outputs the value x.

EXAMPLE 1   **Finding the Inverse of a Function**

Find the inverse of the following functions.

(a) Let the domain of the function represent the employees of Yolanda's Preowned Car Mart and let the range represent their base salaries.

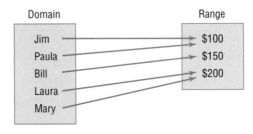

(b) Let the domain of the function represent the employees of Yolanda's Preowned Car Mart and let the range represent their spouse's names.

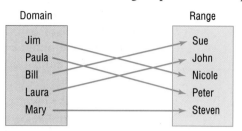

Solution   (a) The elements in the domain represent inputs to the function, and the elements in the range represent the outputs. To find the inverse, interchange the elements in the domain with the elements in the range. For example, the function receives as input Bill and outputs $150. So the inverse receives an input $150 and outputs Bill. The inverse of the given function takes the form

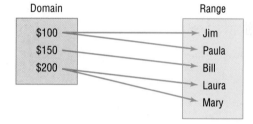

(b) The inverse of the given function is

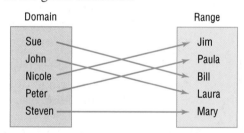

Notice that the inverse found in Example 1(b) is a function, since each element in the domain corresponds to a unique element in the range. The inverse found in Example 1(a) is not a function, since each element in the domain does not correspond to a unique element in the range.

If the function f is a set of ordered pairs (x, y), then the inverse of f is the set of ordered pairs (y, x).

EXAMPLE 2   **Finding the Inverse of a Function**

Find the inverse of the following functions:

(a) $\{(-3, -27), (-2, -8), (-1, -1), (0, 0), (1, 1), (2, 8), (3, 27)\}$
(b) $\{(-3, 9), (-2, 4), (-1, 1), (0, 0), (1, 1), (2, 4), (3, 9)\}$

Solution   (a) The inverse of the given function is found by interchanging the entries in each ordered pair and so is given by

$$\{(-27, -3), (-8, -2), (-1, -1), (0, 0), (1, 1), (8, 2), (27, 3)\}$$

(b) The inverse of the given function is

$$\{(9, -3), (4, -2), (1, -1), (0, 0), (1, 1), (4, 2), (9, 3)\}$$ ▄

The inverse obtained in the solution to Example 2(a) is a function, but the inverse obtained in Example 2(b) is not a function. So, sometimes the inverse of a function is a function, and sometimes it is not.

> When the inverse of a function f is itself a function, then f is said to be a **one-to-one function.** That is, f is **one-to-one** if, for any choice of elements x_1 and x_2 in the domain of f, if $x_1 \neq x_2$, then the corresponding values $f(x_1)$ and $f(x_2)$ are unequal, $f(x_1) \neq f(x_2)$.

In other words, if a function f is one-to-one, then for each x in the domain of f there is exactly one y in the range, and no y in the range is the image of more than one x in the domain. See Figure 1.

Figure 1

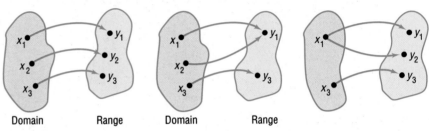

Domain   **Range**   **Domain**   **Range**

(a) One-to-one function: Each x in the domain has one and only one image in the range

(b) Not a one-to-one function: y_1 is the image of both x_1 and x_2

(c) Not a function: x_1 has two images, y_1 and y_2

✏ NOW WORK PROBLEMS **1** AND **5**.

Figure 2
$f(x_1) = f(x_2) = h$,
and $x_1 \neq x_2$;
f is not a
one-to-one function.

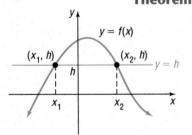

If the graph of a function f is known, there is a simple test, called the **horizontal-line test,** to determine whether f is one-to-one.

Theorem   Horizontal-line Test

> If every horizontal line intersects the graph of a function f in at most one point, then f is one-to-one.

▄

The reason that this test works can be seen in Figure 2, where the horizontal line $y = h$ intersects the graph at two distinct points, (x_1, h) and (x_2, h). Since h is the image of both x_1 and x_2, $x_1 \neq x_2$, f is not one-to-one.

EXAMPLE 3   **Using the Horizontal-line Test**

For each function, use the graph to determine whether the function is one-to-one.

(a) $f(x) = x^2$ (b) $g(x) = x^3$

Solution   (a) Figure 3(a) illustrates the horizontal-line test for $f(x) = x^2$. The horizontal line $y = 1$ intersects the graph of f twice, at $(1, 1)$ and at $(-1, 1)$, so f is not one-to-one.

(b) Figure 3(b) illustrates the horizontal-line test for $g(x) = x^3$. Because every horizontal line will intersect the graph of g exactly once, it follows that g is one-to-one.

Figure 3

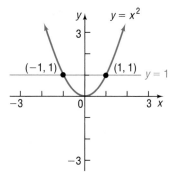

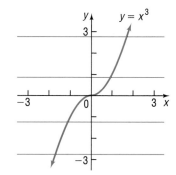

(a) A horizontal line intersects the graph twice; thus, f is not one-to-one

(b) Every horizontal line intersects the graph exactly once; thus, g is one-to-one

NOW WORK PROBLEM **9**.

Let's look more closely at the one-to-one function $g(x) = x^3$. This function is an increasing function. Because an increasing (or decreasing) function will always have different y values for unequal x values, it follows that a function that is increasing (decreasing) over its domain is also a one-to-one function.

Theorem

A function that is increasing over its domain is a one-to-one function.
A function that is decreasing over its domain is a one-to-one function.

INVERSE FUNCTION OF $y = f(x)$

Figure 4

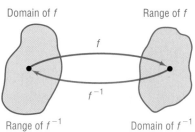

If f is a one-to-one function, its inverse is a function. Then, to each x in the domain of f, there is exactly one y in the range (because f is a function); and to each y in the range of f, there is exactly one x in the domain (because f is one-to-one). The correspondence from the range of f back to the domain of f is called the **inverse function of f** and is denoted by the symbol f^{-1}. Figure 4 illustrates this definition.

WARNING: Be careful! f^{-1} is a symbol for the inverse function of f. The -1 used in f^{-1} is not an exponent. That is, f^{-1} does *not* mean the reciprocal of f; $f^{-1}(x)$ is not equal to $\dfrac{1}{f(x)}$.

Two facts are now apparent about a function f and its inverse f^{-1}.

Domain of f = Range of f^{-1} Range of f = Domain of f^{-1}

Look again at Figure 4 to visualize the relationship. If we start with x, apply f, and then apply f^{-1}, we get x back again. If we start with x, apply f^{-1}, and then apply f, we get the number x back again. To put it simply, what f does, f^{-1} undoes, and vice versa.

$$\boxed{\text{Input } x} \xrightarrow{\text{Apply } f} \boxed{f(x)} \xrightarrow{\text{Apply } f^{-1}} \boxed{f^{-1}(f(x)) = x}$$

$$\boxed{\text{Input } x} \xrightarrow{\text{Apply } f^{-1}} \boxed{f^{-1}(x)} \xrightarrow{\text{Apply } f} \boxed{f(f^{-1}(x)) = x}$$

In other words,

$$f^{-1}(f(x)) = x \quad \text{and} \quad f(f^{-1}(x)) = x$$

For example, the function $f(x) = 2x$ multiplies the argument x by 2. The inverse function f^{-1} undoes whatever f does. So the inverse function of f is $f^{-1}(x) = \frac{1}{2}x$, which divides the argument by 2. We can verify this by showing that

$$f^{-1}(f(x)) = f^{-1}(2x) = \frac{1}{2}(2x) = x \quad \text{and} \quad f(f^{-1}(x)) = f\left(\frac{1}{2}x\right) = 2\left(\frac{1}{2}x\right) = x$$

See Figure 5.

Figure 5

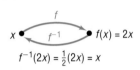

$f^{-1}(2x) = \frac{1}{2}(2x) = x$

EXAMPLE 4

Verifying Inverse Functions

(a) We verify that the inverse of $g(x) = x^3$ is $g^{-1}(x) = \sqrt[3]{x}$ by showing that

$$g^{-1}(g(x)) = g^{-1}(x^3) = \sqrt[3]{x^3} = x$$

and

$$g(g^{-1}(x)) = g(\sqrt[3]{x}) = (\sqrt[3]{x})^3 = x$$

(b) We verify that the inverse of $h(x) = 3x$ is $h^{-1}(x) = \frac{1}{3}x$ by showing that

$$h^{-1}(h(x)) = h^{-1}(3x) = \frac{1}{3}(3x) = x$$

and

$$h(h^{-1}(x)) = h\left(\frac{1}{3}x\right) = 3\left(\frac{1}{3}x\right) = x$$

(c) We verify that the inverse of $f(x) = 2x + 3$ is $f^{-1}(x) = \frac{1}{2}(x - 3)$ by showing that

$$f^{-1}(f(x)) = f^{-1}(2x + 3) = \frac{1}{2}[(2x + 3) - 3] = \frac{1}{2}(2x) = x$$

and

$$f(f^{-1}(x)) = f\left(\frac{1}{2}(x - 3)\right) = 2\left[\frac{1}{2}(x - 3)\right] + 3 = (x - 3) + 3 = x \quad \blacksquare$$

 NOW WORK PROBLEM **21**.

EXPLORATION: Simultaneously graph $Y_1 = x$, $Y_2 = x^3$, and $Y_3 = \sqrt[3]{x}$ on a square screen, using the viewing rectangle $-3 \le x \le 3, -2 \le y \le 2$. What do you observe about the graphs of $Y_2 = x^3$, its inverse $Y_3 = \sqrt[3]{x}$, and the line $Y_1 = x$?

Repeat this experiment by simultaneously graphing $Y_1 = x$, $Y_2 = 2x + 3$, and $Y_3 = \frac{1}{2}(x - 3)$, using the viewing rectangle $-6 \le x \le 3, -8 \le y \le 4$. Do you see the symmetry of the graph of Y_2 and its inverse Y_3 with respect to the line $Y_1 = x$? $\blacksquare$

GEOMETRIC INTERPRETATION

Figure 6

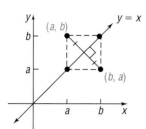

Suppose that (a, b) is a point on the graph of the one-to-one function f defined by $y = f(x)$. Then $b = f(a)$. This means that $a = f^{-1}(b)$, so (b, a) is a point on the graph of the inverse function f^{-1}. The relationship between the point (a, b) on f and the point (b, a) on f^{-1} is shown in Figure 6. The line segment containing (a, b) and (b, a) is perpendicular to the line $y = x$ and is bisected by the line $y = x$. (Do you see why?) It follows that the point (b, a) on f^{-1} is the reflection about the line $y = x$ of the point (a, b) on f.

Theorem

The graph of a function f and the graph of its inverse f^{-1} are symmetric with respect to the line $y = x$.

Figure 7 illustrates this result. Notice that, once the graph of f is known, the graph of f^{-1} may be obtained by reflecting the graph of f about the line $y = x$.

Figure 7

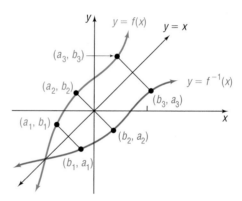

EXAMPLE 5   **Graphing the Inverse Function**

The graph in Figure 8(a) is that of a one-to-one function $y = f(x)$. Draw the graph of its inverse.

Figure 8

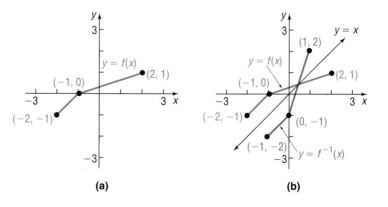

(a)   (b)

Solution   We begin by adding the graph of $y = x$ to Figure 8(a). Since the points $(-2, -1), (-1, 0)$, and $(2, 1)$ are on the graph of f, we know that the points $(-1, -2), (0, -1)$, and $(1, 2)$ must be on the graph of f^{-1}. Keeping in mind that the graph of f^{-1} is the reflection about the line $y = x$ of the graph of f, we can draw f^{-1}. See Figure 8(b).   ■

➤ NOW WORK PROBLEM 15.

3 ### FINDING THE INVERSE FUNCTION f^{-1}

The fact that the graph of a one-to-one function f and its inverse function f^{-1} are symmetric with respect to the line $y = x$ tells us more. It says that we can obtain f^{-1} by interchanging the roles of x and y in f. Look again at Figure 7. If f is defined by the equation

$$y = f(x)$$

then f^{-1} is defined by the equation

$$x = f(y)$$

The equation $x = f(y)$ defines f^{-1} *implicitly*. If we can solve this equation for y, we will have the *explicit* form of f^{-1}, that is,

$$y = f^{-1}(x)$$

Let's use this procedure to find the inverse of $f(x) = 2x + 3$. (Since f is a linear function and is increasing, we know that f is one-to-one and so has an inverse function.)

EXAMPLE 6   ### Finding the Inverse Function f^{-1}

Find the inverse of $f(x) = 2x + 3$. Also find the domain and range of f and f^{-1}. Graph f and f^{-1} on the same coordinate axes.

Solution   In the equation $y = 2x + 3$, interchange the variables x and y. The result,

$$x = 2y + 3$$

is an equation that defines the inverse f^{-1} implicitly. To find the explicit form, we solve for y.

$$2y + 3 = x$$
$$2y = x - 3$$
$$y = \tfrac{1}{2}(x - 3)$$

Figure 9

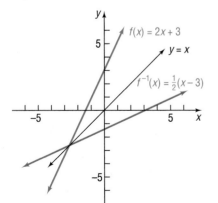

The explicit form of the inverse f^{-1} is therefore

$$f^{-1}(x) = \tfrac{1}{2}(x - 3)$$

which we verified in Example 4(c).

Next we find

$$\text{Domain } f = \text{Range } f^{-1} = (-\infty, \infty)$$
$$\text{Range } f = \text{Domain } f^{-1} = (-\infty, \infty)$$

The graphs of $f(x) = 2x + 3$ and its inverse $f^{-1}(x) = \tfrac{1}{2}(x - 3)$ are shown in Figure 9. Note the symmetry of the graphs with respect to the line $y = x$.   ■

We outline next the steps to follow for finding the inverse of a one-to-one function.

> **PROCEDURE FOR FINDING THE INVERSE OF A ONE-TO-ONE FUNCTION**
>
> STEP 1:   In $y = f(x)$, interchange the variables x and y to obtain
> $$x = f(y)$$
> This equation defines the inverse function f^{-1} implicitly.
>
> STEP 2:   If possible, solve the implicit equation for y in terms of x to obtain the explicit form of f^{-1}.
> $$y = f^{-1}(x)$$
>
> STEP 3:   Check the result by showing that
> $$f^{-1}(f(x)) = x \quad \text{and} \quad f(f^{-1}(x)) = x$$

EXAMPLE 7   **Finding the Inverse Function**

The function

$$f(x) = \frac{2x + 1}{x - 1}, \qquad x \neq 1$$

is one-to-one. Find its inverse and check the result.

Solution   STEP 1:   Interchange the variables x and y in

$$y = \frac{2x + 1}{x - 1}$$

to obtain

$$x = \frac{2y + 1}{y - 1}$$

STEP 2:   Solve for y.

$$x = \frac{2y + 1}{y - 1}$$

$$x(y - 1) = 2y + 1 \qquad \text{Multiply both sides by } y - 1.$$

$$xy - x = 2y + 1 \qquad \text{Apply the distributive property.}$$

$$xy - 2y = x + 1 \qquad \text{Subtract } 2y \text{ from both sides; add } x \text{ to both sides.}$$

$$(x - 2)y = x + 1 \qquad \text{Factor the left side.}$$

$$y = \frac{x + 1}{x - 2} \qquad \text{Divide by } x - 2.$$

The inverse is

$$f^{-1}(x) = \frac{x + 1}{x - 2}, \qquad x \neq 2 \qquad \text{Replace } y \text{ by } f^{-1}(x).$$

STEP 3: Check:

$$f^{-1}(f(x)) = f^{-1}\left(\frac{2x+1}{x-1}\right) = \frac{\dfrac{2x+1}{x-1}+1}{\dfrac{2x+1}{x-1}-2} = \frac{2x+1+x-1}{2x+1-2(x-1)} = \frac{3x}{3} = x$$

$$f(f^{-1}(x)) = f\left(\frac{x+1}{x-2}\right) = \frac{2\left(\dfrac{x+1}{x-2}\right)+1}{\dfrac{x+1}{x-2}-1} = \frac{2(x+1)+x-2}{x+1-(x-2)} = \frac{3x}{3} = x$$

EXPLORATION   In Example 7, we found that, if $f(x) = (2x+1)/(x-1)$, then $f^{-1}(x) = (x+1)/(x-2)$. Compare the vertical and horizontal asymptotes of f and f^{-1}. What did you find? Are you surprised?

NOW WORK PROBLEM 33.

We said in Chapter 3 that finding the range of a function f is not easy. However, if f is one-to-one, we can find its range by finding the domain of the inverse function f^{-1}.

EXAMPLE 8 **Finding the Range of a Function**

Find the domain and range of

$$f(x) = \frac{2x+1}{x-1}$$

Solution The domain of f is $\{x \mid x \neq 1\}$. To find the range of f, we first find the inverse f^{-1}. Based on Example 7, we have

$$f^{-1}(x) = \frac{x+1}{x-2}$$

The domain of f^{-1} is $\{x \mid x \neq 2\}$, so the range of f is $\{y \mid y \neq 2\}$.

NOW WORK PROBLEM 47.

If a function is not one-to-one, then its inverse is not a function. Sometimes, though, an appropriate restriction on the domain of such a function will yield a new function that is one-to-one. Then its inverse is a function. Let's look at an example of this common practice.

EXAMPLE 9 **Finding the Inverse of a Domain-restricted Function**

Find the inverse of $y = f(x) = x^2$ if $x \geq 0$.

Solution The function $y = x^2$ is not one-to-one. [Refer to Example 3(a).] However, if we restrict this function to only that part of its domain for which $x \geq 0$, as indicated, we have a new function that is increasing and therefore is one-to-one. As a result, the function defined by $y = f(x) = x^2$, $x \geq 0$, has an inverse function, f^{-1}.

We follow the steps given previously to find f^{-1}.

STEP 1: In the equation $y = x^2$, $x \geq 0$, interchange the variables x and y. The result is

$$x = y^2, \qquad y \geq 0$$

This equation defines (implicitly) the inverse function.

STEP 2: We solve for y to get the explicit form of the inverse. Since $y \geq 0$, only one solution for y is obtained.

$$y = \sqrt{x}$$

So

$$f^{-1}(x) = \sqrt{x}$$

STEP 3: Check: $f^{-1}(f(x)) = f^{-1}(x^2) = \sqrt{x^2} = |x| = x$, since $x \geq 0$
$$f(f^{-1}(x)) = f(\sqrt{x}) = (\sqrt{x})^2 = x.$$

Figure 10 illustrates the graphs of $f(x) = x^2$, $x \geq 0$, and $f^{-1}(x) = \sqrt{x}$.

Figure 10

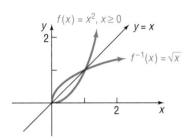

SUMMARY

1. If a function f is one-to-one, then it has an inverse function f^{-1}.
2. Domain f = Range f^{-1}; Range f = Domain f^{-1}.
3. To verify that f^{-1} is the inverse of f, show that $f^{-1}(f(x)) = x$ and $f(f^{-1}(x)) = x$.
4. The graphs of f and f^{-1} are symmetric with respect to the line $y = x$.
5. To find the range of a one-to-one function f, find the domain of the inverse function f^{-1}.

6.1   EXERCISES

In Problems 1–8, (a) find the inverse and (b) determine whether the inverse is a function.

1.

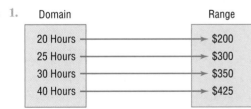

2.

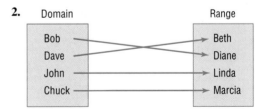

3.

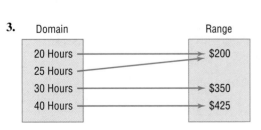

4.

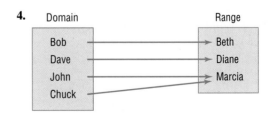

5. $\{(2, 6), (-3, 6), (4, 9), (1, 10)\}$

6. $\{(-2, 5), (-1, 3), (3, 7), (4, 12)\}$

7. $\{(0, 0), (1, 1), (2, 16), (3, 81)\}$

8. $\{(1, 2), (2, 8), (3, 18), (4, 32)\}$

In Problems 9–14, the graph of a function f is given. Use the horizontal-line test to determine whether f is one-to-one.

9.

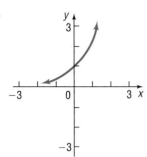

10.

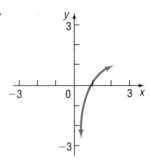

11.

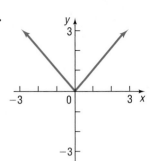

12.

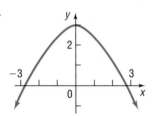

13.

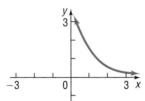

14.

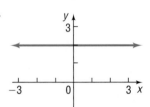

In Problems 15–20, the graph of a one-to-one function f is given. Draw the graph of the inverse function f^{-1}. For convenience (and as a hint), the graph of $y = x$ is also given.

15.

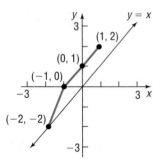

16.

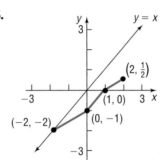

17.

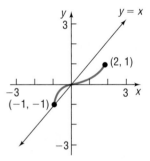

18.

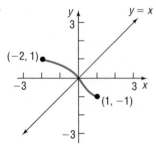

19.

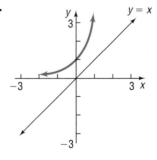

20.

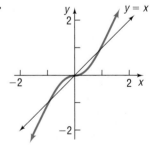

In Problems 21–30, verify that the functions f and g are inverses of each other by showing that $f(g(x)) = x$ and $g(f(x)) = x$.

21. $f(x) = 3x + 4$; $g(x) = \frac{1}{3}(x - 4)$

22. $f(x) = 3 - 2x$; $g(x) = -\frac{1}{2}(x - 3)$

23. $f(x) = 4x - 8$; $g(x) = \frac{x}{4} + 2$

24. $f(x) = 2x + 6$; $g(x) = \frac{1}{2}x - 3$

25. $f(x) = x^3 - 8$; $g(x) = \sqrt[3]{x + 8}$

26. $f(x) = (x - 2)^2, x \geq 2$; $g(x) = \sqrt{x} + 2$

27. $f(x) = \frac{1}{x}$; $g(x) = \frac{1}{x}$

28. $f(x) = x$; $g(x) = x$

29. $f(x) = \frac{2x + 3}{x + 4}$; $g(x) = \frac{4x - 3}{2 - x}$

30. $f(x) = \frac{x - 5}{2x + 3}$; $g(x) = \frac{3x + 5}{1 - 2x}$

In Problems 31–42, the function f is one-to-one. Find its inverse and check your answer. State the domain and range of f and f^{-1}. Graph f, f^{-1}, and y = x on the same coordinate axes.

31. $f(x) = 3x$

32. $f(x) = -4x$

33. $f(x) = 4x + 2$

34. $f(x) = 1 - 3x$

35. $f(x) = x^3 - 1$

36. $f(x) = x^3 + 1$

37. $f(x) = x^2 + 4, \quad x \geq 0$

38. $f(x) = x^2 + 9, \quad x \geq 0$

39. $f(x) = \dfrac{4}{x}$

40. $f(x) = -\dfrac{3}{x}$

41. $f(x) = \dfrac{1}{x - 2}$

42. $f(x) = \dfrac{4}{x + 2}$

In Problems 43–54, the function f is one-to-one. Find its inverse and check your answer. State the domain of f and find its range using f^{-1}.

43. $f(x) = \dfrac{2}{3 + x}$

44. $f(x) = \dfrac{4}{2 - x}$

45. $f(x) = \dfrac{3x}{x + 2}$

46. $f(x) = \dfrac{-2x}{x - 1}$

47. $f(x) = \dfrac{2x}{3x - 1}$

48. $f(x) = \dfrac{3x + 1}{-x}$

49. $f(x) = \dfrac{3x + 4}{2x - 3}$

50. $f(x) = \dfrac{2x - 3}{x + 4}$

51. $f(x) = \dfrac{2x + 3}{x + 2}$

52. $f(x) = \dfrac{-3x - 4}{x - 2}$

53. $f(x) = \dfrac{x^2 - 4}{2x^2}, \quad x > 0$

54. $f(x) = \dfrac{x^2 + 3}{3x^2}, \quad x > 0$

---

55. Find the inverse of the linear function
$$f(x) = mx + b, \quad m \neq 0$$

56. Find the inverse of the function
$$f(x) = \sqrt{r^2 - x^2}, \quad 0 \leq x \leq r$$

57. A function f has an inverse function. If the graph of f lies in quadrant I, in which quadrant does the graph of f^{-1} lie?

58. A function f has an inverse function. If the graph of f lies in quadrant II, in which quadrant does the graph of f^{-1} lie?

59. The function $f(x) = |x|$ is not one-to-one. Find a suitable restriction on the domain of f so that the new function that results is one-to-one. Then find the inverse of f.

60. The function $f(x) = x^4$ is not one-to-one. Find a suitable restriction on the domain of f so that the new function that results is one-to-one. Then find the inverse of f.

61. Temperature Conversion To convert from x degrees Celsius to y degrees Fahrenheit, we use the formula $y = f(x) = \frac{9}{5}x + 32$. To convert from x degrees Fahrenheit to y degrees Celsius, we use the formula $y = g(x) = \frac{5}{9}(x - 32)$. Show that f and g are inverse functions.

62. Demand for Corn The demand for corn obeys the equation $p(x) = 300 - 50x$, where p is the price per bushel (in dollars) and x is the number of bushels produced, in millions. Express the production amount x as a function of the price p.

63. Period of a Pendulum The period T (in seconds) of a simple pendulum is a function of its length l (in feet), given by $T(l) = 2\pi\sqrt{l/g}$, where $g \approx 32.2$ feet per sec-

ond per second is the acceleration of gravity. Express the length l as a function of the period T.

64. The given function f is one-to-one.
$$f(x) = \dfrac{ax + b}{cx + d}$$
   (a) Find the domain of f.
   (b) Find f^{-1}.
   (c) Find the range of f.
   (d) If $c \neq 0$, under what conditions on a, b, c, and d is $f = f^{-1}$?

65. Can an even function be one-to-one? Explain.

66. Is every odd function one-to-one? Explain.

67. If the graph of a function and its inverse intersect, where must this necessarily occur? Can they intersect anywhere else? Must they intersect?

68. Can a one-to-one function and its inverse be equal? What must be true about the graph of f for this to happen? Give some examples to support your conclusion.

69. Draw the graph of a one-to-one function that contains the points $(-2, -3)$, $(0, 0)$, and $(1, 5)$. Now draw the graph of its inverse. Compare your graph to those of other students. Discuss any similarities. What differences do you see?

70. Give an example of a function whose domain is the set of real numbers and that is neither increasing nor decreasing on its domain, but is one-to-one.
   [**Hint:** Use a piecewise-defined function.]

Before getting started, review the following:

✓ Integer Exponents (Review, Section 4, pp. 28–35)

✓ Graphing Techniques: Transformations
(Section 3.4, pp. 242–252)

✓ Rational Exponents (Review, Section 9, pp. 74–77)

✓ Equations (Section 1.1, pp. 84–92)

6.2   EXPONENTIAL FUNCTIONS

OBJECTIVES   1   Evaluate Exponential Functions
                 2   Graph Exponential Functions
                 3   Define the Number e
                 4   Solve Exponential Equations

1   In the Review, Section 9, we give a definition for raising a real number a to a rational power. Based on that discussion, we gave meaning to expressions of the form

$$a^r$$

where the base a is a positive real number and the exponent r is a rational number.

But what is the meaning of a^x, where the base a is a positive real number and the exponent x is an irrational number? Although a rigorous definition requires methods discussed in calculus, the basis for the definition is easy to follow: Select a rational number r that is formed by truncating (removing) all but a finite number of digits from the irrational number x. Then it is reasonable to expect that

$$a^x \approx a^r$$

For example, take the irrational number $\pi = 3.14159\ldots$. Then, an approximation to a^π is

$$a^\pi \approx a^{3.14}$$

where the digits after the hundredths position have been removed from the value for π. A better approximation would be

$$a^\pi \approx a^{3.14159}$$

where the digits after the hundred-thousandths position have been removed. Continuing in this way, we can obtain approximations to a^π to any desired degree of accuracy.

Most calculators have an $\boxed{x^y}$ key or a caret key $\boxed{\wedge}$ for working with exponents. To evaluate expressions of the form a^x, enter the base a, then press the $\boxed{x^y}$ key (or the $\boxed{\wedge}$ key), enter the exponent x, and press $\boxed{=}$ (or $\boxed{\text{enter}}$).

EXAMPLE 1   **Using a Calculator to Evaluate Powers of 2**

Using a calculator, evaluate:

(a) $2^{1.4}$ (b) $2^{1.41}$ (c) $2^{1.414}$ (d) $2^{1.4142}$ (e) $2^{\sqrt{2}}$

Solution   (a) $2^{1.4} \approx 2.639015822$ (b) $2^{1.41} \approx 2.657371628$
(c) $2^{1.414} \approx 2.66474965$ (d) $2^{1.4142} \approx 2.665119089$
(e) $2^{\sqrt{2}} \approx 2.665144143$ ■

━━━━━━━ NOW WORK PROBLEM **1**.

It can be shown that the familiar laws for rational exponents hold for real exponents.

Theorem   **Laws of Exponents**

If s, t, a, and b are real numbers, with $a > 0$ and $b > 0$, then

$$a^s \cdot a^t = a^{s+t} \qquad \left(a^s\right)^t = a^{st} \qquad (ab)^s = a^s \cdot b^s$$

$$1^s = 1 \qquad a^{-s} = \frac{1}{a^s} = \left(\frac{1}{a}\right)^s \qquad a^0 = 1 \qquad \textbf{(1)}$$

■

We are now ready for the following definition:

An **exponential function** is a function of the form

$$f(x) = a^x$$

where a is a positive real number $(a > 0)$ and $a \neq 1$. The domain of f is the set of all real numbers.

We exclude the base $a = 1$, because this function is simply the constant function $f(x) = 1^x = 1$. We also need to exclude the bases that are negative, because, otherwise, we would have to exclude many values of x from the domain, such as $x = \frac{1}{2}$ and $x = \frac{3}{4}$. [Recall that $(-2)^{1/2} = \sqrt{-2}$, $(-3)^{3/4} = \sqrt[4]{(-3)^3} = \sqrt[4]{-27}$, and so on, are not defined in the system of real numbers.]

GRAPHS OF EXPONENTIAL FUNCTIONS

② First, we graph the exponential function $f(x) = 2^x$.

━━

EXAMPLE 2   **Graphing an Exponential Function**

Graph the exponential function: $f(x) = 2^x$

Solution   The domain of $f(x) = 2^x$ consists of all real numbers. We begin by locating some points on the graph of $f(x) = 2^x$, as listed in Table 1 (page 406).

Since $2^x > 0$ for all x, the range of f is the interval $(0, \infty)$. From this, we conclude that the graph has no x-intercepts, and, in fact, the graph will lie above the x-axis. As Table 1 indicates, the y-intercept is 1. Table 1 also indicates that as $x \rightarrow -\infty$ the value of $f(x) = 2^x$ gets closer and closer to 0. We conclude that the x-axis is a horizontal asymptote to the graph as $x \rightarrow -\infty$. This gives us the end behavior of the graph for x large and negative.

To determine the end behavior for x large and positive, look again at Table 1. As $x \rightarrow \infty$, $f(x) = 2^x$ grows very quickly, causing the graph of

$f(x) = 2^x$ to rise very rapidly. It is apparent that f is an increasing function and hence is one-to-one.

Using all this information, we plot some of the points from Table 1 and connect them with a smooth, continuous curve, as shown in Figure 11.

x	$f(x) = 2^x$
−10	$2^{-10} \approx 0.00098$
−3	$2^{-3} = \frac{1}{8}$
−2	$2^{-2} = \frac{1}{4}$
−1	$2^{-1} = \frac{1}{2}$
0	$2^0 = 1$
1	$2^1 = 2$
2	$2^2 = 4$
3	$2^3 = 8$
10	$2^{10} = 1024$

TABLE 1

Figure 11
$y = 2^x$

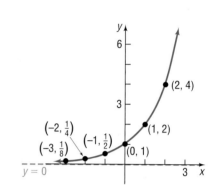

As we shall see, graphs that look like the one in Figure 11 occur very frequently in a variety of situations. For example, look at the graph in Figure 12, which illustrates the closing price of a share of Dell Computer stock.

Figure 12

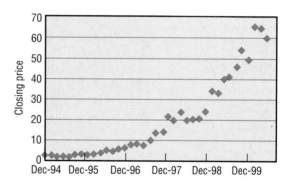

Investors might conclude from this graph that the price of Dell Computer is *behaving exponentially*; that is, the graph exhibits rapid, or exponential, growth. We shall have more to say about situations that lead to exponential growth later in this chapter. For now, we continue to seek properties of the exponential functions.

The graph of $f(x) = 2^x$ in Figure 11 is typical of all exponential functions that have a base larger than 1. Such functions are increasing functions and hence are one-to-one. Their graphs lie above the x-axis, pass through the point $(0, 1)$, and thereafter rise rapidly as $x \to \infty$. As $x \to -\infty$, the x-axis ($y = 0$) is a horizontal asymptote. There are no vertical asymptotes. Finally, the graphs are smooth and continuous, with no corners or gaps.

Figure 13 illustrates the graphs of two more exponential functions whose bases are larger than 1. Notice that for the larger base the graph is steeper when $x > 0$ and is closer to the x-axis when $x < 0$.

Figure 13

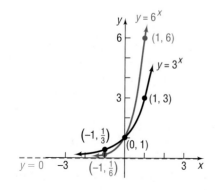

![SEEING THE CONCEPT icon] **SEEING THE CONCEPT**   Graph $y = 2^x$ and compare what you see to Figure 11. Clear the screen and graph $y = 3^x$ and $y = 6^x$ and compare what you see to Figure 13. Clear the screen and graph $y = 10^x$ and $y = 100^x$. What viewing rectangle seems to work best? ▄

The following display summarizes the information that we have about $f(x) = a^x, a > 1$.

Facts about the Graph of an Exponential Function $f(x) = a^x,\ a > 1$

1. The domain is all real numbers; the range is the set of positive real numbers.
2. There are no x-intercepts; the y-intercept is 1.
3. The x-axis ($y = 0$) is a horizontal asymptote as $x \to -\infty$.
4. $f(x) = a^x, a > 1$, is an increasing function and is one-to-one.
5. The graph of f contains the points $(0, 1)$, $(1, a)$, and $(-1, 1/a)$.
6. The graph of f is smooth and continuous, with no corners or gaps. See Figure 14.

Figure 14
$f(x) = a^x, a > 1$

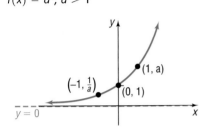

Now we consider $f(x) = a^x$ when $0 < a < 1$.

EXAMPLE 3   Graphing an Exponential Function

Graph the exponential function: $f(x) = \left(\frac{1}{2}\right)^x$

Solution   The domain of $f(x) = \left(\frac{1}{2}\right)^x$ consists of all real numbers. As before, we locate some points on the graph by creating Table 2. Since $\left(\frac{1}{2}\right)^x > 0$ for all x, the range of f is the interval $(0, \infty)$. The graph lies above the x-axis and so has no x-intercepts. The y-intercept is 1. As $x \to -\infty$, $f(x) = \left(\frac{1}{2}\right)^x$ grows very quickly. As $x \to \infty$, the values of $f(x)$ approach 0. The x-axis ($y = 0$) is a horizontal asymptote as $x \to \infty$. It is apparent that f is a decreasing function and hence is one-to-one. Figure 15 illustrates the graph.

TABLE 2	
x	$f(x) = \left(\frac{1}{2}\right)^x$
-10	$\left(\frac{1}{2}\right)^{-10} = 1024$
-3	$\left(\frac{1}{2}\right)^{-3} = 8$
-2	$\left(\frac{1}{2}\right)^{-2} = 4$
-1	$\left(\frac{1}{2}\right)^{-1} = 2$
0	$\left(\frac{1}{2}\right)^{0} = 1$
1	$\left(\frac{1}{2}\right)^{1} = \frac{1}{2}$
2	$\left(\frac{1}{2}\right)^{2} = \frac{1}{4}$
3	$\left(\frac{1}{2}\right)^{3} = \frac{1}{8}$
10	$\left(\frac{1}{2}\right)^{10} \approx 0.00098$

Figure 15
$y = \left(\dfrac{1}{2}\right)^x$

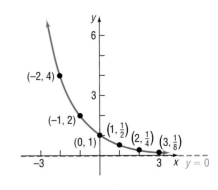

▄

We could have obtained the graph of $y = \left(\frac{1}{2}\right)^x$ from the graph of $y = 2^x$ using transformations. If $f(x) = 2^x$, then $f(-x) = 2^{-x} = 1/2^x = \left(\frac{1}{2}\right)^x$. The graph of $y = \left(\frac{1}{2}\right)^x = 2^{-x}$ is a reflection about the y-axis of the graph of $y = 2^x$. See Figures 16(a) and (b).

Figure 16

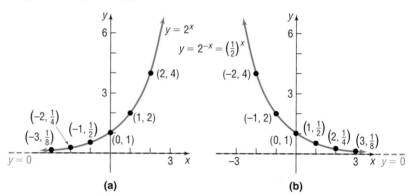

(a)   (b)

SEEING THE CONCEPT   Using a graphing utility, simultaneously graph

(a) $Y_1 = 3^x$, $Y_2 = \left(\dfrac{1}{3}\right)^x$ (b) $Y_1 = 6^x$, $Y_2 = \left(\dfrac{1}{6}\right)^x$

Figure 17

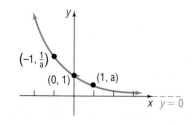

Conclude that the graph of $Y_2 = \left(\dfrac{1}{a}\right)^x$, for $a > 0$, is the reflection about the y-axis of the graph of $Y_1 = a^x$.   ▬

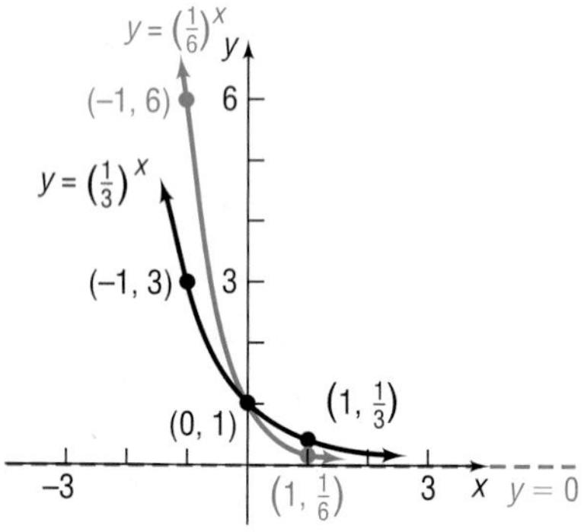

The graph of $f(x) = \left(\frac{1}{2}\right)^x$ in Figure 15 is typical of all exponential functions that have a base between 0 and 1. Such functions are decreasing and one-to-one. Their graphs lie above the x-axis and pass through the point $(0, 1)$. The graphs rise rapidly as $x \to -\infty$. As $x \to \infty$, the x-axis is a horizontal asymptote. There are no vertical asymptotes. Finally, the graphs are smooth and continuous, with no corners or gaps.

Figure 17 illustrates the graphs of two more exponential functions whose bases are between 0 and 1. Notice that the choice of a base closer to 0 results in a graph that is steeper when $x < 0$ and closer to the x-axis when $x > 0$.

SEEING THE CONCEPT   Graph $y = \left(\frac{1}{2}\right)^x$ and compare what you see to Figure 15. Clear the screen and graph $y = \left(\frac{1}{3}\right)^x$ and $y = \left(\frac{1}{6}\right)^x$ and compare what you see to Figure 17. Clear the screen and graph $y = \left(\frac{1}{10}\right)^x$ and $y = \left(\frac{1}{100}\right)^x$. What viewing rectangle seems to work best?   ▬

The following display summarizes the information that we have about the function $f(x) = a^x, 0 < a < 1$.

Figure 18
$f(x) = a^x, 0 < a < 1$

> ### Facts about the Graph of an Exponential Function $f(x) = a^x, 0 < a < 1$
>
> 1. The domain is all real numbers; the range is the set of positive real numbers.
> 2. There are no x-intercepts; the y-intercept is 1.
> 3. The x-axis ($y = 0$) is a horizontal asymptote as $x \to \infty$.
> 4. $f(x) = a^x, 0 < a < 1$, is a decreasing function and is one-to-one.
> 5. The graph of f contains the points $(0, 1)$, $(1, a)$, and $(-1, 1/a)$.
> 6. The graph of f is smooth and continuous, with no corners or gaps. See Figure 18.

| EXAMPLE 4 | **Graphing Exponential Functions Using Transformations** |

Graph $f(x) = 2^{-x} - 3$ and determine the domain, range, and horizontal asymptote of f.

Solution   We begin with the graph of $y = 2^x$. Figure 19 shows the various steps.

Figure 19

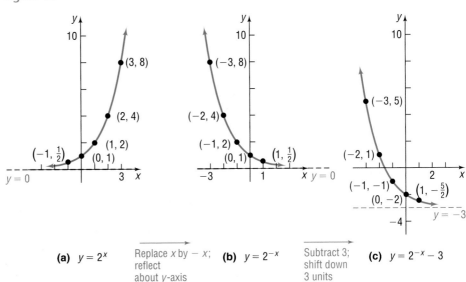

(a) $y = 2^x$ Replace x by $-x$; **(b)** $y = 2^{-x}$ Subtract 3; **(c)** $y = 2^{-x} - 3$
reflect   shift down
about y-axis   3 units

As Figure 19(c) illustrates, the domain of $f(x) = 2^{-x} - 3$ is the interval $(-\infty, \infty)$ and the range is the interval $(-3, \infty)$. The horizontal asymptote of f is the line $y = -3$.   ∎

NOW WORK PROBLEM **19**.

THE BASE e

③   As we shall see shortly, many problems that occur in nature require the use of an exponential function whose base is a certain irrational number, symbolized by the letter e.

   Let's look now at one way of arriving at this important number e.

The **number e** is defined as the number that the expression

$$\left(1 + \frac{1}{n}\right)^n \qquad \textbf{(2)}$$

approaches as $n \to \infty$. In calculus, this is expressed using limit notation as

$$e = \lim_{n \to \infty} \left(1 + \frac{1}{n}\right)^n$$

Table 3 illustrates what happens to the defining expression (2) as n takes on increasingly large values. The last number in the last column in the table is correct to nine decimal places and is the same as the entry given for e on your calculator (if expressed correctly to nine decimal places).

T A B L E 3			
n	$\dfrac{1}{n}$	$1 + \dfrac{1}{n}$	$\left(1 + \dfrac{1}{n}\right)^n$
1	1	2	2
2	0.5	1.5	2.25
5	0.2	1.2	2.48832
10	0.1	1.1	2.59374246
100	0.01	1.01	2.704813829
1,000	0.001	1.001	2.716923932
10,000	0.0001	1.0001	2.718145927
100,000	0.00001	1.00001	2.718268237
1,000,000	0.000001	1.000001	2.718280469
1,000,000,000	10^{-9}	$1 + 10^{-9}$	2.718281827

The exponential function $f(x) = e^x$, whose base is the number e, occurs with such frequency in applications that it is usually referred to as *the* exponential function. Indeed, most calculators have the key* $\boxed{e^x}$ or $\boxed{\exp(x)}$, which may be used to evaluate the exponential function for a given value of x.

Now use your calculator to approximate e^x for $x = -2$, $x = -1$, $x = 0$, $x = 1$, and $x = 2$, as we have done to create Table 4. The graph of the exponential function $f(x) = e^x$ is given in Figure 20. Since $2 < e < 3$, the graph of $y = e^x$ lies between the graphs of $y = 2^x$ and $y = 3^x$. Do you see why? (Refer to Figures 11 and 13.)

Figure 20
$y = e^x$

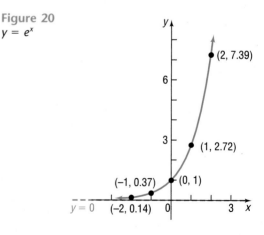

TABLE 4	
x	e^x
-2	0.14
-1	0.37
0	1
1	2.72
2	7.39

*If your calculator does not have this key but does have a $\boxed{\text{SHIFT}}$ key (or $\boxed{2^{\text{nd}}}$ key) and an $\boxed{\ln}$ key, you can display the number e as follows:

Keystrokes: $\boxed{1}$ $\boxed{\text{SHIFT}}$ $\boxed{\ln}$

Display: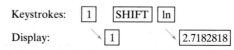

The reason this works will become clear in Section 6.3.

SEEING THE CONCEPT   Graph $Y_1 = e^x$ and compare what you see to Figure 20. Use eVALUEate or TABLE to verify the points on the graph shown in Figure 20. Now graph $Y_2 = 2^x$ and $Y_3 = 3^x$ on the same screen as $Y_1 = e^x$. Notice that the graph of $Y_1 = e^x$ lies between these two graphs.   ∎

EXAMPLE 5

Graphing Exponential Functions Using Transformations

Graph $f(x) = -e^{x-3}$ and determine the domain, range, and horizontal asymptote of f.

Solution   We begin with the graph of $y = e^x$. Figure 21 shows the various steps.

Figure 21

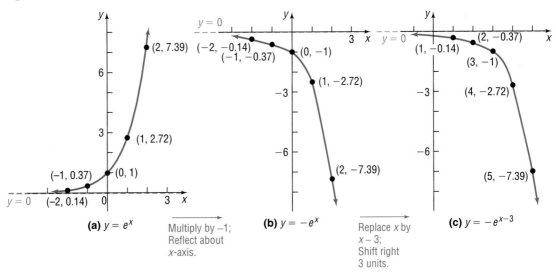

(a) $y = e^x$ →   Multiply by -1; Reflect about x-axis.   (b) $y = -e^x$ →   Replace x by $x - 3$; Shift right 3 units.   (c) $y = -e^{x-3}$

As Figure 21(c) illustrates, the domain of $f(x) = -e^{x-3}$ is the interval $(-\infty, \infty)$ and the range is the interval $(-\infty, 0)$. The horizontal asymptote is the line $y = 0$.   ∎

NOW WORK PROBLEM **27.**

EXPONENTIAL EQUATIONS

④ Equations that involve terms of the form $a^x, a > 0, a \neq 1$, are often referred to as **exponential equations.** Such equations can sometimes be solved by appropriately applying the Laws of Exponents and property (3).

$$\text{If } a^u = a^v, \quad \text{then } u = v \qquad (3)$$

Property (3) is a consequence of the fact that exponential functions are one-to-one. To use property (3), each side of the equality must be written with the same base.

EXAMPLE 6

Solving an Exponential Equation

Solve: $3^{x+1} = 81$

Solution   Since $81 = 3^4$, we can write the equation as

$$3^{x+1} = 81 = 3^4$$

Now we have the same base, 3, on each side, so we can apply property (3) to obtain

$$x + 1 = 4$$

$$x = 3$$ ∎

✎ NOW WORK PROBLEM 35.

EXAMPLE 7

Solving an Exponential Equation

Solve: $e^{-x^2} = \left(e^x\right)^2 \cdot \dfrac{1}{e^3}$

Solution   We use Laws of Exponents first to get the base e on the right side.

$$\left(e^x\right)^2 \cdot \frac{1}{e^3} = e^{2x} \cdot e^{-3} = e^{2x-3}$$

As a result,

$$e^{-x^2} = e^{2x-3}$$

$$-x^2 = 2x - 3 \qquad \text{Apply property (3).}$$

$$x^2 + 2x - 3 = 0 \qquad \text{Place the quadratic equation in standard form.}$$

$$(x + 3)(x - 1) = 0 \qquad \text{Factor.}$$

$$x = -3 \quad \text{or} \quad x = 1 \qquad \text{Use the Zero-Product Property.}$$

The solution set is $\{-3, 1\}$.   ∎

APPLICATION

Many applications involve the exponential functions. Let's look at one.

EXAMPLE 8

Exponential Probability

Between 9:00 PM and 10:00 PM cars arrive at Burger King's drive-thru at the rate of 12 cars per hour (0.2 car per minute). The following formula from probability can be used to determine the probability that a car will arrive within t minutes of 9:00 PM.

$$F(t) = 1 - e^{-0.2t}$$

(a) Determine the probability that a car will arrive within 5 minutes of 9 PM (that is, before 9:05 PM).

(b) Determine the probability that a car will arrive within 30 minutes of 9 PM (before 9:30 PM).

(c) What value does F approach as t becomes unbounded in the positive direction?

 (d) Graph $F(t) = 1 - e^{-0.2t}, t > 0$. Use eVALUEate or TABLE to compare the values of F at $t = 5$ [part (a)] and at $t = 30$ [part (b)].

(e) Within how many minutes of 9 PM will the probability of a car arriving equal 50%? [**Hint:** Use TRACE or TABLE].

Solution   (a) The probability that a car will arrive within 5 minutes is found by evaluating $F(t)$ at $t = 5$.

$$F(5) = 1 - e^{-0.2(5)} \approx 0.63212$$
$$\uparrow$$
$$\text{Use a calculator}$$

We conclude that there is a 63% probability that a car will arrive within 5 minutes.

(b) The probability that a car will arrive within 30 minutes is found by evaluating $F(t)$ at $t = 30$.

$$F(30) = 1 - e^{-0.2(30)} \approx 0.9975$$
$$\uparrow$$
$$\text{Use a calculator}$$

There is a 99.75% probability that a car will arrive within 30 minutes.

(c) As time passes, the probability that a car will arrive increases. The value that F approaches can be found by letting $t \to \infty$. Since $e^{-0.2t} = 1/e^{0.2t}$, it follows that $e^{-0.2t} \to 0$ as $t \to \infty$. Thus, F approaches 1 as t gets large.

(d) See Figure 22 for the graph of F.

(e) Within 3.5 minutes of 9 PM, the probability of a car arriving equals 50%. ▬

Figure 22

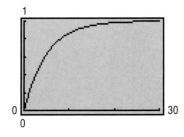

SUMMARY

Properties of the Exponential Function

$f(x) = a^x, \quad a > 1$ Domain: the interval $(-\infty, \infty)$; Range: the interval $(0, \infty)$;
x-intercepts: none; y-intercept: 1;
horizontal asymptote: x-axis as $x \to -\infty$;
increasing; one-to-one; smooth; continuous
See Figure 14 for a typical graph.

$f(x) = a^x, \quad 0 < a < 1$ Domain: the interval $(-\infty, \infty)$; Range: the interval $(0, \infty)$;
x-intercepts: none; y-intercept: 1;
horizontal asymptote: x-axis as $x \to \infty$;
decreasing; one-to-one; smooth; continuous
See Figure 18 for a typical graph.

If $a^u = a^v$, then $u = v$.

6.2 EXERCISES

In Problems 1–10, approximate each number using a calculator. Express your answer rounded to three decimal places.

1. (a) $3^{2.2}$ (b) $3^{2.23}$ (c) $3^{2.236}$ (d) $3^{\sqrt{5}}$

2. (a) $5^{1.7}$ (b) $5^{1.73}$ (c) $5^{1.732}$ (d) $5^{\sqrt{3}}$

3. (a) $2^{3.14}$ (b) $2^{3.141}$ (c) $2^{3.1415}$ (d) 2^{π}

4. (a) $2^{2.7}$ (b) $2^{2.71}$ (c) $2^{2.718}$ (d) 2^{e}

5. (a) $3.1^{2.7}$ (b) $3.14^{2.71}$ (c) $3.141^{2.718}$ (d) π^{e}

6. (a) $2.7^{3.1}$ (b) $2.71^{3.14}$ (c) $2.718^{3.141}$ (d) e^{π}

7. $e^{1.2}$ 8. $e^{-1.3}$ 9. $e^{-0.85}$ 10. $e^{2.1}$

In Problems 11–18, the graph of an exponential function is given. Match each graph to one of the following functions.

A. $y = 3^x$ B. $y = 3^{-x}$ C. $y = -3^x$ D. $y = -3^{-x}$

E. $y = 3^x - 1$ F. $y = 3^{x-1}$ G. $y = 3^{1-x}$ H. $y = 1 - 3^x$

11.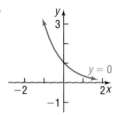

12.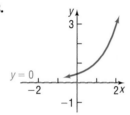

13.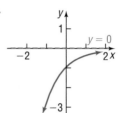

14.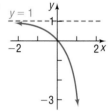

15.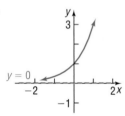

16.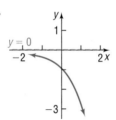

17.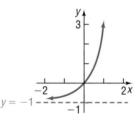

18.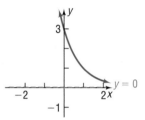

In Problems 19–26, use transformations to graph each function. Determine the domain, range, and horizontal asymptote of each function.

19. $f(x) = 2^x + 1$ 20. $f(x) = 2^{x+2}$ 21. $f(x) = 3^{-x} - 2$ 22. $f(x) = -3^x + 1$

23. $f(x) = 2 + 3(4^x)$ 24. $f(x) = 1 - 3(2^x)$ 25. $f(x) = 2 + 3^{x/2}$ 26. $f(x) = 1 - 2^{-x/3}$

In Problems 27–34, begin with the graph of $y = e^x$ (Figure 20) and use transformations to graph each function. Determine the domain, range, and horizontal asymptote of each function.

27. $f(x) = e^{-x}$ 28. $f(x) = -e^x$ 29. $f(x) = e^{x+2}$ 30. $f(x) = e^x - 1$

31. $f(x) = 5 - e^{-x}$ 32. $f(x) = 9 - 3e^{-x}$ 33. $f(x) = 2 - e^{-x/2}$ 34. $f(x) = 7 - 3e^{2x}$

In Problems 35–48, solve each equation.

35. $2^{2x+1} = 4$ 36. $5^{1-2x} = \frac{1}{5}$ 37. $3^{x^3} = 9^x$ 38. $4^{x^2} = 2^x$

39. $8^{x^2-2x} = \frac{1}{2}$ 40. $9^{-x} = \frac{1}{3}$ 41. $2^x \cdot 8^{-x} = 4^x$ 42. $\left(\frac{1}{2}\right)^{1-x} = 4$

43. $\left(\frac{1}{5}\right)^{2-x} = 25$ 44. $4^x - 2^x = 0$ 45. $4^x = 8$ 46. $9^{2x} = 27$

47. $e^{x^2} = (e^{3x}) \cdot \dfrac{1}{e^2}$ 48. $(e^4)^x \cdot e^{x^2} = e^{12}$

49. If $4^x = 7$, what does 4^{-2x} equal?   50. If $2^x = 3$, what does 4^{-x} equal?

51. If $3^{-x} = 2$, what does 3^{2x} equal?   52. If $5^{-x} = 3$, what does 5^{3x} equal?

In Problems 53–56, graph each function f. Based on the graph, state the domain, range, and intercepts, if any, of f.

53. $f(x) = \begin{cases} e^{-x} & \text{if } x < 0 \\ e^x & \text{if } x \geq 0 \end{cases}$

54. $f(x) = \begin{cases} e^x & \text{if } x < 0 \\ e^{-x} & \text{if } x \geq 0 \end{cases}$

55. $f(x) = \begin{cases} -e^x & \text{if } x < 0 \\ -e^{-x} & \text{if } x \geq 0 \end{cases}$

56. $f(x) = \begin{cases} -e^{-x} & \text{if } x < 0 \\ -e^x & \text{if } x \geq 0 \end{cases}$

---

57. Optics   If a single pane of glass obliterates 3% of the light passing through it, then the percent p of light that passes through n successive panes is given approximately by the function

$$p(n) = 100e^{-0.03n}$$

(a) What percent of light will pass through 10 panes?
(b) What percent of light will pass through 25 panes?

58. Atmospheric Pressure   The atmospheric pressure p on a balloon or plane decreases with increasing height. This pressure, measured in millimeters of mercury, is related to the number of kilometers h above sea level by the function

$$p(h) = 760e^{-0.145h}$$

(a) Find the atmospheric pressure at a height of 2 kilometers (over 1 mile).
(b) What is it at a height of 10 kilometers (over 30,000 feet)?

59. Space Satellites   The number of watts w provided by a space satellite's power supply after a period of d days is given by the function

$$w(d) = 50e^{-0.004d}$$

(a) How much power will be available after 30 days?
(b) How much power will be available after 1 year (365 days)?

60. Healing of Wounds   The normal healing of wounds can be modeled by an exponential function. If A_0 represents the original area of the wound and if A equals the area of the wound after n days, then the function

$$A(n) = A_0e^{-0.35n}$$

describes the area of a wound on the nth day following an injury when no infection is present to retard the healing. Suppose that a wound initially had an area of 100 square millimeters.

(a) If healing is taking place, how large will the area of the wound be after 3 days?
(b) How large will it be after 10 days?

61. Drug Medication   The function

$$D(h) = 5e^{-0.4h}$$

can be used to find the number of milligrams D of a certain drug that is in a patient's bloodstream h hours after the drug has been administered. How many milligrams will be present after 1 hour? After 6 hours?

62. Spreading of Rumors   A model for the number of people N in a college community who have heard a certain rumor is

$$N = P\left(1 - e^{-0.15d}\right)$$

where P is the total population of the community and d is the number of days that have elapsed since the rumor began. In a community of 1000 students, how many students will have heard the rumor after 3 days?

63. Exponential Probability   Between 12:00 PM and 1:00 PM, cars arrive at Citibank's drive-thru at the rate of 6 cars per hour (0.1 car per minute). The following formula from probability can be used to determine the probability that a car will arrive within t minutes of 12:00 PM:

$$F(t) = 1 - e^{-0.1t}$$

(a) Determine the probability that a car will arrive within 10 minutes of 12:00 PM (that is, before 12:10 PM).
(b) Determine the probability that a car will arrive within 40 minutes of 12:00 PM (before 12:40 PM).
(c) What value does F approach as t becomes unbounded in the positive direction?
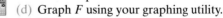 (d) Graph F using your graphing utility.
(e) Using TRACE, determine how many minutes are needed for the probability to reach 50%?

64. Exponential Probability   Between 5:00 PM and 6:00 PM, cars arrive at Jiffy Lube at the rate of 9 cars per hour (0.15 car per minute). The following formula from probability can be used to determine the probability that a car will arrive within t minutes of 5:00 PM:

$$F(t) = 1 - e^{-0.15t}$$

(a) Determine the probability that a car will arrive within 15 minutes of 5:00 PM (that is, before 5:15 PM).
(b) Determine the probability that a car will arrive within 30 minutes of 5:00 PM (before 5:30 PM).
(c) What value does F approach as t becomes unbounded in the positive direction?
(d) Graph F using your graphing utility.
(e) Using TRACE, determine how many minutes are needed for the probability to reach 60%?

65. Poisson Probability Between 5:00 PM and 6:00 PM, cars arrive at McDonald's drive-thru at the rate of 20 cars per hour. The following formula from probability can be used to determine the probability that x cars will arrive between 5:00 PM and 6:00 PM.

$$P(x) = \frac{20^x e^{-20}}{x!}$$

where

$$x! = x \cdot (x-1) \cdot (x-2) \cdots \cdots 3 \cdot 2 \cdot 1$$

(a) Determine the probability that $x = 15$ cars will arrive between 5:00 PM and 6:00 PM.

(b) Determine the probability that $x = 20$ cars will arrive between 5:00 PM and 6:00 PM.

66. Poisson Probability People enter a line for the *Demon Roller Coaster* at the rate of 4 per minute. The following formula from probability can be used to determine the probability that x people will arrive within the next minute.

$$P(x) = \frac{4^x e^{-4}}{x!}$$

where

$$x! = x \cdot (x-1) \cdot (x-2) \cdots \cdots 3 \cdot 2 \cdot 1$$

(a) Determine the probability that $x = 5$ people will arrive within the next minute.

(b) Determine the probability that $x = 8$ people will arrive within the next minute.

67. Relative Humidity The relative humidity is the ratio (expressed as a percent) of the amount of water vapor in the air to the maximum amount that the air can hold at a specific temperature. The relative humidity, R, is found using the following formula:

$$R = 10^{\left(\frac{2345}{T} - \frac{2345}{D} + 2\right)}$$

where T is the air temperature (in Kelvins) and D is the dew point temperature (in Kelvins).

[**NOTE:** Kelvins are found by adding 273 to Celsius degrees.]

(a) Determine the relative humidity if the air temperature is 10° Celsius and the dew point temperature is 5° Celsius.

(b) Determine the relative humidity if the air temperature is 20° Celsius and the dew point temperature is 15° Celsius.

(c) What is the relative humidity if the air temperature and the dew point temperature are the same?

68. Learning Curve Suppose that a student has 500 vocabulary words to learn. If the student learns 15 words after 5 minutes, the function

$$L(t) = 500(1 - e^{-0.0061t})$$

approximates the number of words L that the student will learn after t minutes.

(a) How many words will the student learn after 30 minutes?

(b) How many words will the student learn after 60 minutes?

69. Alternating Current in a *RL* Circuit The equation governing the amount of current I (in amperes) after time t (in seconds) in a single *RL* circuit consisting of a resistance R (in ohms), an inductance L (in henrys), and an electromotive force E (in volts) is

$$I = \frac{E}{R}\left[1 - e^{-(R/L)t}\right]$$

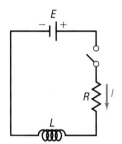

(a) If $E = 120$ volts, $R = 10$ ohms, and $L = 5$ henrys, how much current I_1 is flowing after 0.3 second? After 0.5 second? After 1 second?

(b) What is the maximum current?

(c) Graph this function $I = I_1(t)$, measuring I along the y-axis and t along the x-axis.

(d) If $E = 120$ volts, $R = 5$ ohms, and $L = 10$ henrys, how much current I_2 is flowing after 0.3 second? After 0.5 second? After 1 second?

(e) What is the maximum current?

(f) Graph this function $I = I_2(t)$ on the same coordinate axes as $I_1(t)$.

70. Alternating Current in a *RC* Circuit The equation governing the amount of current I (in amperes) after time t (in microseconds) in a single *RC* circuit consisting of a resistance R (in ohms), a capacitance C (in microfarads), and an electromotive force E (in volts) is

$$I = \frac{E}{R} e^{-t/(RC)}$$

(a) If $E = 120$ volts, $R = 2000$ ohms, and $C = 1.0$ microfarad, how much current I_1 is flowing initially ($t = 0$)? After 1000 microseconds? After 3000 microseconds?

(b) What is the maximum current?

(c) Graph this function $I = I_1(t)$, measuring I along the y-axis and t along the x-axis.

(d) If $E = 120$ volts, $R = 1000$ ohms, and $C = 2.0$ microfarads, how much current I_2 is flowing initially? After 1000 microseconds? After 3000 microseconds?

(e) What is the maximum current?

(f) Graph this function $I = I_2(t)$ on the same coordinate axes as $I_1(t)$.

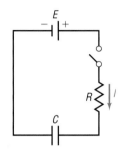

71. Another Formula for e Use a calculator to compute the values of

$$2 + \frac{1}{2!} + \frac{1}{3!} + \cdots + \frac{1}{n!}$$

for $n = 4, 6, 8$, and 10. Compare each result with e.
[**Hint:** $1! = 1, 2! = 2 \cdot 1, 3! = 3 \cdot 2 \cdot 1$,
$n! = n(n - 1) \cdot \cdots \cdot (3)(2)(1)]$

72. Another Formula for e Use a calculator to compute the various values of the expression. Compare the values to e.

$$2 + \cfrac{1}{1 + \cfrac{1}{2 + \cfrac{2}{3 + \cfrac{3}{4 + \cfrac{4}{\text{etc.}}}}}}$$

73. Difference Quotient If $f(x) = a^x$, show that

$$\frac{f(x + h) - f(x)}{h} = a^x \left(\frac{a^h - 1}{h} \right)$$

74. If $f(x) = a^x$, show that $f(A + B) = f(A) \cdot f(B)$.

75. If $f(x) = a^x$, show that $f(-x) = \dfrac{1}{f(x)}$.

76. If $f(x) = a^x$, show that $f(\alpha x) = \left[f(x) \right]^\alpha$.

77. The *Challenger* Disaster* After the *Challenger* disaster in 1986, a study of the 23 launches that preceded the fatal flight was made. A mathematical model was developed involving the relationship between the Fahrenheit temperature x around the O-rings and the number y of eroded or leaky primary O-rings. The model stated that

$$y = \frac{6}{1 + e^{-(5.085 - 0.1156x)}}$$

where the number 6 indicates the 6 primary O-rings on the spacecraft.

(a) What is the predicted number of eroded or leaky primary O-rings at a temperature of 100°F?

(b) What is the predicted number of eroded or leaky primary O-rings at a temperature of 60°F?

(c) What is the predicted number of eroded or leaky primary O-rings at a temperature of 30°F?

(d) Graph the equation and TRACE. At what temperature is the predicted number of eroded or leaky O-rings 1? 3? 5?

78. Historical Problem Pierre de Fermat (1601–1665) conjectured that the function

$$f(x) = 2^{(2^x)} + 1$$

for $x = 1, 2, 3, \ldots$, would always have a value equal to a prime number. But Leonhard Euler (1707–1783) showed that this formula fails for $x = 5$. Use a calculator to determine the prime numbers produced by f for $x = 1, 2, 3, 4$. Then show that $f(5) = 641 \times 6,700,417$, which is not prime.

79. The bacteria in a 4-liter container double every minute. After 60 minutes the container is full. How long did it take to fill half the container?

80. Explain in your own words what the number e is. Provide at least two applications that require the use of this number.

81. Do you think that there is a power function that increases more rapidly than an exponential function whose base is greater than 1? Explain.

* Linda Tappin, "Analyzing Data Relating to the *Challenger* Disaster," *Mathematics Teacher*, Vol. 87, No. 6, September 1994, pp. 423–426.

PREPARING FOR THIS SECTION

Before getting started, review the following:

✓ Solving Inequalities (Section 1.5, pp. 128–132)   ✓ Polynomial and Rational Inequalities
                                                         (Section 4.5, pp. 342–346)

6.3 LOGARITHMIC FUNCTIONS

OBJECTIVES   **1** Change Exponential Expressions to Logarithmic Expressions
               2 Change Logarithmic Expressions to Exponential Expressions
               3 Evaluate Logarithmic Functions
               4 Determine the Domain of a Logarithmic Function
               5 Graph Logarithmic Functions
               6 Solve Logarithmic Equations

Recall that a one-to-one function $y = f(x)$ has an inverse function that is defined implicitly by the equation $x = f(y)$. In particular, the exponential function $y = f(x) = a^x$, $a > 0$, $a \neq 1$, is one-to-one and hence has an inverse function that is defined implicitly by the equation

$$x = a^y, \qquad a > 0, \quad a \neq 1$$

This inverse function is so important that it is given a name, the *logarithmic function*.

> The **logarithmic function to the base a,** where $a > 0$ and $a \neq 1$, is denoted by $y = \log_a x$ (read as "y is the logarithm to the base a of x") and is defined by
>
> $$y = \log_a x \quad \text{if and only if} \quad x = a^y$$
>
> The domain of the logarithmic function $y = \log_a x$ is $x > 0$.

A *logarithm* is merely a name for a certain exponent.

EXAMPLE 1   **Relating Logarithms to Exponents**

(a) If $y = \log_3 x$, then $x = 3^y$. For example, $2 = \log_3 9$ is equivalent to $9 = 3^2$.
(b) If $y = \log_5 x$, then $x = 5^y$. For example, $-1 = \log_5\left(\frac{1}{5}\right)$ is equivalent to $\frac{1}{5} = 5^{-1}$.   ∎

EXAMPLE 2   **Changing Exponential Expressions to Logarithmic Expressions**

1   Change each exponential expression to an equivalent expression involving a logarithm.

(a) $1.2^3 = m$ (b) $e^b = 9$ (c) $a^4 = 24$

Solution We use the fact that $y = \log_a x$ and $x = a^y, a > 0, a \neq 1$, are equivalent.

(a) If $1.2^3 = m$, then $3 = \log_{1.2} m$. (b) If $e^b = 9$, then $b = \log_e 9$.

(c) If $a^4 = 24$, then $4 = \log_a 24$. ∎

➤ NOW WORK PROBLEM **1.**

| EXAMPLE 3 | **Changing Logarithmic Expressions to Exponential Expressions** |

② Change each logarithmic expression to an equivalent expression involving an exponent.

(a) $\log_a 4 = 5$ (b) $\log_e b = -3$ (c) $\log_3 5 = c$

Solution (a) If $\log_a 4 = 5$, then $a^5 = 4$. (b) If $\log_e b = -3$, then $e^{-3} = b$.

(c) If $\log_3 5 = c$, then $3^c = 5$. ∎

➤ NOW WORK PROBLEM **13.**

③ To find the exact value of a logarithm, we write the logarithm in exponential notation and use the fact that if $a^u = a^v$ then $u = v$.

| EXAMPLE 4 | **Finding the Exact Value of a Logarithmic Expression** |

Find the exact value of

(a) $\log_2 16$ (b) $\log_3 \dfrac{1}{27}$

Solution (a)

$$y = \log_2 16$$
$$2^y = 16 \qquad \text{Change to exponential form.}$$
$$2^y = 2^4 \qquad 16 = 2^4$$
$$y = 4 \qquad \text{Equate exponents.}$$

Therefore, $\log_2 16 = 4$.

(b)

$$y = \log_3 \dfrac{1}{27}$$
$$3^y = \dfrac{1}{27} \qquad \text{Change to exponential form.}$$
$$3^y = 3^{-3} \qquad \dfrac{1}{27} = \dfrac{1}{3^3} = 3^{-3}$$
$$y = -3 \qquad \text{Equate exponents.}$$

Therefore, $\log_3 \dfrac{1}{27} = -3$. ∎

➤ NOW WORK PROBLEM **25.**

DOMAIN OF A LOGARITHMIC FUNCTION

④ The logarithmic function $y = \log_a x$ has been defined as the inverse of the exponential function $y = a^x$. That is, if $f(x) = a^x$, then $f^{-1}(x) = \log_a x$. Based on the discussion given in Section 6.1 on inverse functions, we know that, for a function f and its inverse f^{-1},

$$\text{Domain } f^{-1} = \text{Range } f \quad \text{and} \quad \text{Range } f^{-1} = \text{Domain } f$$

Consequently, it follows that

> Domain of logarithmic function = Range of exponential function = $(0, \infty)$
>
> Range of logarithmic function = Domain of exponential function = $(-\infty, \infty)$

In the next box, we summarize some properties of the logarithmic function:

> $y = \log_a x$ (defining equation: $x = a^y$)
>
> Domain: $0 < x < \infty$ Range: $-\infty < y < \infty$

The domain of a logarithmic function consists of the *positive* real numbers, so the argument of a logarithmic function must be greater than zero.

EXAMPLE 5   **Finding the Domain of a Logarithmic Function**

Find the domain of each logarithmic function.

(a) $F(x) = \log_2 (1 - x)$ (b) $g(x) = \log_5 \left(\dfrac{1 + x}{1 - x} \right)$

(c) $h(x) = \log_{1/2} |x|$

Solution   (a) The domain of F consists of all x for which $1 - x > 0$, that is, all $x < 1$, or using interval notation, $(-\infty, 1)$.

(b) The domain of g is restricted to

$$\frac{1 + x}{1 - x} > 0$$

Solving this inequality, we find that the domain of g consists of all x between -1 and 1, that is, $-1 < x < 1$, or using interval notation, $(-1, 1)$.

(c) Since $|x| > 0$, provided that $x \neq 0$, the domain of h consists of all nonzero real numbers, or using interval notation, $(-\infty, 0)$ or $(0, \infty)$. ∎

NOW WORK PROBLEMS **39** AND **45**.

GRAPHS OF LOGARITHMIC FUNCTIONS

5   Since exponential functions and logarithmic functions are inverses of each other, the graph of a logarithmic function $y = \log_a x$ is the reflection about the line $y = x$ of the graph of the exponential function $y = a^x$, as shown in Figure 23.

Figure 23

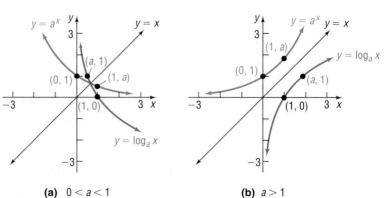

(a) $0 < a < 1$ **(b)** $a > 1$

Facts about the Graph of a Logarithmic Function $f(x) = \log_a x$

1. The domain is the set of positive real numbers; the range is all real numbers.
2. The x-intercept of the graph is 1. There is no y-intercept.
3. The y-axis ($x = 0$) is a vertical asymptote of the graph.
4. A logarithmic function is decreasing if $0 < a < 1$ and increasing if $a > 1$.
5. The graph of f contains the points $(1, 0)$, $(a, 1)$, and $(1/a, -1)$.
6. The graph is smooth and continuous, with no corners or gaps.

If the base of a logarithmic function is the number e, then we have the **natural logarithm function.** This function occurs so frequently in applications that it is given a special symbol, **ln** (from the Latin, *logarithmus naturalis*). Thus,

$$y = \log_e x = \ln x \quad \text{if and only if} \quad x = e^y \qquad \textbf{(1)}$$

Since $y = \ln x$ and the exponential function $y = e^x$ are inverse functions, we can obtain the graph of $y = \ln x$ by reflecting the graph of $y = e^x$ about the line $y = x$. See Figure 24.

Figure 24

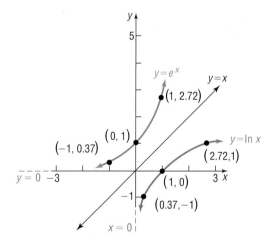

Using a calculator with an [ln] key, we can obtain other points on the graph of $f(x) = \ln x$. See Table 5.

SEEING THE CONCEPT   Graph $Y_1 = e^x$ and $Y_2 = \ln x$ on the same square screen. Use eVALUEate to verify the points on the graph given in Figure 24. Do you see the symmetry of the two graphs with respect to the line $y = x$?

TABLE 5

x	$\ln x$
$\frac{1}{2}$	-0.69
2	0.69
3	1.10

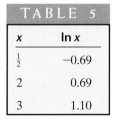

EXAMPLE 6

Graphing Logarithmic Functions Using Transformations

Graph $f(x) = -\ln(x + 2)$ by starting with the graph of $y = \ln x$. Determine the domain, range, and vertical asymptote of f.

Solution   The domain of f consists of all x for which

$$x + 2 > 0 \quad \text{or} \quad x > -2$$

To obtain the graph of $y = -\ln(x + 2)$, we use the steps illustrated in Figure 25.

Figure 25

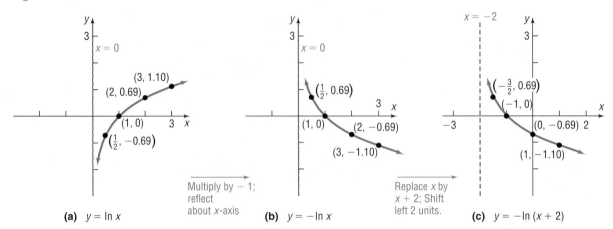

(a) $y = \ln x$

Multiply by -1; reflect about x-axis

(b) $y = -\ln x$

Replace x by $x + 2$; Shift left 2 units.

(c) $y = -\ln(x + 2)$

The range of $f(x) = -\ln(x + 2)$ is the interval $(-\infty, \infty)$, and the vertical asymptote is $x = -2$. [Do you see why? The original asymptote $(x = 0)$ is shifted to the left 2 units.]

EXAMPLE 7　**Graphing Logarithmic Functions Using Transformations**

Graph $f(x) = \ln(1 - x)$. Determine the domain, range, and vertical asymptote of f.

Solution　The domain of f consists of all x for which

$$1 - x > 0 \quad \text{or} \quad x < 1$$

To obtain the graph of $y = \ln(1 - x)$, we use the steps illustrated in Figure 26.

Figure 26

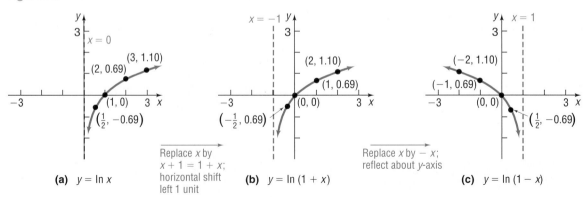

(a) $y = \ln x$

Replace x by $x + 1 = 1 + x$; horizontal shift left 1 unit

(b) $y = \ln(1 + x)$

Replace x by $-x$; reflect about y-axis

(c) $y = \ln(1 - x)$

The range of $f(x) = \ln(1 - x)$ is the interval $(-\infty, \infty)$, and the vertical asymptote is $x = 1$.

NOW WORK PROBLEM **61**.

LOGARITHMIC EQUATIONS

⑥ Equations that contain logarithms are called **logarithmic equations.** Care must be taken when solving logarithmic equations algebraically. Be sure to check each apparent solution in the original equation and discard any that are extraneous. In the expression $\log_a M$, remember that a and M are positive and $a \neq 1$.

Some logarithmic equations can be solved by changing from a logarithmic expression to an exponential expression.

EXAMPLE 8 **Solving a Logarithmic Equation**

Solve: (a) $\log_3(4x - 7) = 2$ (b) $\log_x 64 = 2$

Solution (a) We can obtain an exact solution by changing the logarithm to exponential form.

$$\log_3(4x - 7) = 2$$
$$4x - 7 = 3^2 \qquad \text{Change to exponential form.}$$
$$4x - 7 = 9$$
$$4x = 16$$
$$x = 4$$

(b) We can obtain an exact solution by changing the logarithm to exponential form.

$$\log_x 64 = 2$$
$$x^2 = 64 \qquad \text{Change to exponential form.}$$
$$x = \pm\sqrt{64} = \pm 8$$

The base of a logarithm is always positive. As a result, we discard -8; the only solution is 8. ∎

EXAMPLE 9 **Using Logarithms to Solve Exponential Equations**

Solve: $e^{2x} = 5$

Solution We can obtain an exact solution by changing the exponential equation to logarithmic form.

$$e^{2x} = 5$$
$$\ln 5 = 2x \qquad \text{Change to a logarithmic expression using (1).}$$
$$x = \frac{\ln 5}{2} \approx 0.805 \qquad ∎$$

✎━━━━━ NOW WORK PROBLEMS **73** AND **85**.

EXAMPLE 10 **Alcohol and Driving**

The concentration of alcohol in a person's blood is measurable. Recent medical research suggests that the risk R (given as a percent) of having an accident while driving a car can be modeled by the equation

$$R = 6e^{kx}$$

where x is the variable concentration of alcohol in the blood and k is a constant.

(a) Suppose that a concentration of alcohol in the blood of 0.04 results in a 10% risk ($R = 10$) of an accident. Find the constant k in the equation.

(b) Using this value of k, what is the risk if the concentration is 0.17?

(c) Using the same value of k, what concentration of alcohol corresponds to a risk of 100%?

(d) If the law asserts that anyone with a risk of having an accident of 20% or more should not have driving privileges, at what concentration of alcohol in the blood should a driver be arrested and charged with a DUI (Driving Under the Influence)?

Solution   (a) For a concentration of alcohol in the blood of 0.04 and a risk of 10%, we let $x = 0.04$ and $R = 10$ in the equation and solve for k.

$$R = 6e^{kx}$$

$$10 = 6e^{k(0.04)}$$

$$\frac{10}{6} = e^{0.04k} \qquad \text{Divide both sides by 6.}$$

$$0.04k = \ln \frac{10}{6} \qquad \text{Change to a logarithmic expression.}$$

$$k = \frac{\ln (10/6)}{0.04} \qquad \text{Solve for } k.$$

$$k \approx 12.77$$

(b) Using $k = 12.77$ and $x = 0.17$ in the equation, we find the risk R to be

$$R = 6e^{kx} = 6e^{(12.77)(0.17)} = 52.6$$

For a concentration of alcohol in the blood of 0.17, the risk of an accident is about 52.6%.

(c) Using $k = 12.77$ and $R = 100$ in the equation, we find the concentration x of alcohol in the blood to be

$$R = 6e^{kx}$$

$$100 = 6e^{12.77x}$$

$$\frac{100}{6} = e^{12.77x} \qquad \text{Divide both sides by 6.}$$

$$12.77x = \ln \frac{100}{6} \qquad \text{Change to a logarithmic expression.}$$

$$x = \frac{\ln (100/6)}{12.77} \qquad \text{Solve for } x.$$

$$x \approx 0.22$$

For a concentration of alcohol in the blood of 0.22, the risk of an accident is 100%.

(d) Using $k = 12.77$ and $R = 20$ in the equation, we find the concentration x of alcohol in the blood to be

$$R = 6e^{kx}$$
$$20 = 6e^{12.77x}$$
$$\frac{20}{6} = e^{12.77x}$$
$$12.77x = \ln\frac{20}{6}$$
$$x = \frac{\ln(20/6)}{12.77}$$
$$x \approx 0.094$$

A driver with a concentration of alcohol in the blood of 0.094 or more should be arrested and charged with DUI. ■

[**NOTE:** Most states use 0.08 or 0.10 as the blood alcohol content at which a DUI citation is given.]

SUMMARY

Properties of the Logarithmic Function

$f(x) = \log_a x, \quad a > 1$
$\left(y = \log_a x \text{ means } x = a^y\right)$

Domain: the interval $(0, \infty)$; Range: the interval $(-\infty, \infty)$;
x-intercept: 1; y-intercept: none; vertical asymptote: $x = 0$ (y-axis);
increasing; one-to-one

See Figure 27(a) for a typical graph.

$f(x) = \log_a x, \quad 0 < a < 1$
$\left(y = \log_a x \text{ means } x = a^y\right)$

Domain: the interval $(0, \infty)$; Range: the interval $(-\infty, \infty)$;
x-intercept: 1; y-intercept: none; vertical asymptote: $x = 0$ (y-axis);
decreasing; one-to-one

See Figure 27(b) for a typical graph.

Figure 27

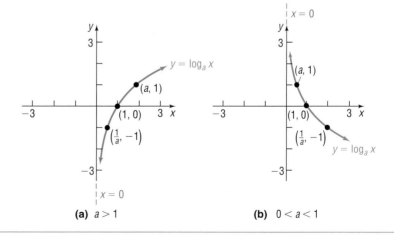

(a) $a > 1$　　　　　　**(b)** $0 < a < 1$

6.3 EXERCISES

In Problems 1–12, change each exponential expression to an equivalent expression involving a logarithm.

1. $9 = 3^2$ **2.** $16 = 4^2$ **3.** $a^2 = 1.6$ **4.** $a^3 = 2.1$

5. $1.1^2 = M$ **6.** $2.2^3 = N$ **7.** $2^x = 7.2$ **8.** $3^x = 4.6$

9. $x^{\sqrt{2}} = \pi$ **10.** $x^\pi = e$ **11.** $e^x = 8$ **12.** $e^{2.2} = M$

In Problems 13–24, change each logarithmic expression to an equivalent expression involving an exponent.

13. $\log_2 8 = 3$

14. $\log_3\left(\frac{1}{9}\right) = -2$

15. $\log_a 3 = 6$

16. $\log_b 4 = 2$

17. $\log_3 2 = x$

18. $\log_2 6 = x$

19. $\log_2 M = 1.3$

20. $\log_3 N = 2.1$

21. $\log_{\sqrt{2}} \pi = x$

22. $\log_\pi x = \frac{1}{2}$

23. $\ln 4 = x$

24. $\ln x = 4$

In Problems 25–36, find the exact value of each logarithm without using a calculator.

25. $\log_2 1$

26. $\log_8 8$

27. $\log_5 25$

28. $\log_3\left(\frac{1}{9}\right)$

29. $\log_{1/2} 16$

30. $\log_{1/3} 9$

31. $\log_{10} \sqrt{10}$

32. $\log_5 \sqrt[3]{25}$

33. $\log_{\sqrt{2}} 4$

34. $\log_{\sqrt{3}} 9$

35. $\ln \sqrt{e}$

36. $\ln e^3$

In Problems 37–46, find the domain of each function.

37. $f(x) = \ln(x - 3)$

38. $g(x) = \ln(x - 1)$

39. $F(x) = \log_2 x^2$

40. $H(x) = \log_5 x^3$

41. $h(x) = \log_{1/2}(x^2 - 2x + 1)$

42. $G(x) = \log_{1/2}(x^2 - 1)$

43. $f(x) = \ln\left(\dfrac{1}{x + 1}\right)$

44. $g(x) = \ln\left(\dfrac{1}{x - 5}\right)$

45. $g(x) = \log_5\left(\dfrac{x + 1}{x}\right)$

46. $h(x) = \log_3\left(\dfrac{x}{x - 1}\right)$

In Problems 47–50, use a calculator to evaluate each expression. Round your answer to three decimal places.

47. $\ln\dfrac{5}{3}$

48. $\dfrac{\ln 5}{3}$

49. $\dfrac{\ln(10/3)}{0.04}$

50. $\dfrac{\ln(2/3)}{-0.1}$

51. Find a so that the graph of $f(x) = \log_a x$ contains the point $(2, 2)$.

52. Find a so that the graph of $f(x) = \log_a x$ contains the point $\left(\frac{1}{2}, -4\right)$.

In Problems 53–60, the graph of a logarithmic function is given. Match each graph to one of the following functions:

A. $y = \log_3 x$

B. $y = \log_3(-x)$

C. $y = -\log_3 x$

D. $y = -\log_3(-x)$

E. $y = \log_3 x - 1$

F. $y = \log_3(x - 1)$

G. $y = \log_3(1 - x)$

H. $y = 1 - \log_3 x$

53.

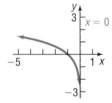

54.

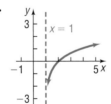

55.

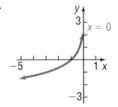

56.

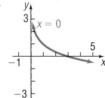

57.

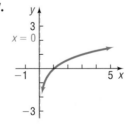

58.

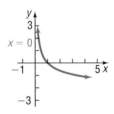

59.

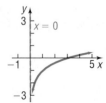

60.

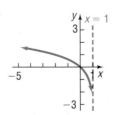

In Problems 61–72, use transformations to graph each function. Determine the domain, range, and vertical asymptote of each function.

61. $f(x) = \ln(x + 4)$

62. $f(x) = \ln(x - 3)$

63. $f(x) = \ln(-x)$

64. $f(x) = -\ln(-x)$

65. $g(x) = \ln(2x)$

66. $h(x) = \ln(\frac{1}{2}x)$

67. $f(x) = 3 \ln x$

68. $f(x) = -2 \ln x$

69. $g(x) = \ln(3 - x)$

70. $h(x) = \ln(4 - x)$

71. $f(x) = -\ln(x - 1)$

72. $f(x) = 2 - \ln x$

In Problems 73–92, solve each equation.

73. $\log_3 x = 2$

74. $\log_5 x = 3$

75. $\log_2(2x + 1) = 3$

76. $\log_3(3x - 2) = 2$

77. $\log_x 4 = 2$

78. $\log_x(\frac{1}{8}) = 3$

79. $\ln e^x = 5$

80. $\ln e^{-2x} = 8$

81. $\log_4 64 = x$

82. $\log_5 625 = x$

83. $\log_3 243 = 2x + 1$

84. $\log_6 36 = 5x + 3$

85. $e^{3x} = 10$

86. $e^{-2x} = \frac{1}{3}$

87. $e^{2x+5} = 8$

88. $e^{-2x+1} = 13$

89. $\log_3(x^2 + 1) = 2$

90. $\log_5(x^2 + x + 4) = 2$

91. $\log_2 8^x = -3$

92. $\log_3 3^x = -1$

In Problems 93–96, graph each function f. Based on the graph, state the domain, range, and intercepts, if any, of f.

93. $f(x) = \begin{cases} \ln(-x) & \text{if } x < 0 \\ \ln x & \text{if } x > 0 \end{cases}$

94. $f(x) = \begin{cases} \ln(-x) & \text{if } x \le -1 \\ -\ln(-x) & \text{if } -1 < x < 0 \end{cases}$

95. $f(x) = \begin{cases} -\ln x & \text{if } 0 < x < 1 \\ \ln x & \text{if } x \ge 1 \end{cases}$

96. $f(x) = \begin{cases} \ln x & \text{if } 0 < x < 1 \\ -\ln x & \text{if } x \ge 1 \end{cases}$

97. Optics If a single pane of glass obliterates 10% of the light passing through it, then the percent P of light that passes through n successive panes is given approximately by the equation

$$P = 100e^{-0.1n}$$

(a) How many panes are necessary to block at least 50% of the light?

(b) How many panes are necessary to block at least 75% of the light?

98. Chemistry The pH of a chemical solution is given by the formula

$$pH = -\log_{10}[H^+]$$

where $[H^+]$ is the concentration of hydrogen ions in moles per liter. Values of pH range from 0 (acidic) to 14 (alkaline).

(a) Find the pH of a 1-liter container of water with 0.0000001 mole of hydrogen ion.

(b) Find the hydrogen ion concentration of a mildly acidic solution with a pH of 4.2.

99. Space Satellites The number of watts w provided by a space satellite's power supply after d days is given by the formula

$$w = 50e^{-0.004d}$$

(a) How long will it take for the available power to drop to 30 watts?

(b) How long will it take for the available power to drop to only 5 watts?

100. Healing of Wounds The normal healing of wounds can be modeled by an exponential function. If A_0 represents the original area of the wound and if A equals the area of the wound after n days, then the formula

$$A = A_0 e^{-0.35n}$$

describes the area of a wound on the nth day following an injury when no infection is present to retard the healing. Suppose that a wound initially had an area of 100 square millimeters.

(a) If healing is taking place, how many days will pass before the wound is one-half its original size?

(b) How long before the wound is 10% of its original size?

101. Exponential Probability Between 12:00 PM and 1:00 PM, cars arrive at Citibank's drive-thru at the rate of 6 cars per hour (0.1 car per minute). The following formula from probability can be used to determine the probability that a car will arrive within t minutes of 12:00 PM.

$$F(t) = 1 - e^{-0.1t}$$

(a) Determine how many minutes are needed for the probability to reach 50%.
(b) Determine how many minutes are needed for the probability to reach 80%.
 (c) Is it possible for the probability to equal 100%? Explain.

102. Exponential Probability Between 5:00 PM and 6:00 PM, cars arrive at Jiffy Lube at the rate of 9 cars per hour (0.15 car per minute). The following formula from probability can be used to determine the probability that a car will arrive within t minutes of 5:00 PM.

$$F(t) = 1 - e^{-0.15t}$$

(a) Determine how many minutes are needed for the probability to reach 50%.
(b) Determine how many minutes are needed for the probability to reach 80%.

103. Drug Medication The formula

$$D = 5e^{-0.4h}$$

can be used to find the number of milligrams D of a certain drug that is in a patient's bloodstream h hours after the drug has been administered. When the number of milligrams reaches 2, the drug is to be administered again. What is the time between injections?

104. Spreading of Rumors A model for the number of people N in a college community who have heard a certain rumor is

$$N = P(1 - e^{-0.15d})$$

where P is the total population of the community and d is the number of days that have elapsed since the rumor began. In a community of 1000 students, how many days will elapse before 450 students have heard the rumor?

105. Current in a *RL* Circuit The equation governing the amount of current I (in amperes) after time t (in seconds) in a simple *RL* circuit consisting of a resistance R (in ohms), an inductance L (in henrys), and an electromotive force E (in volts) is

$$I = \frac{E}{R}\left[1 - e^{-(R/L)t}\right]$$

If $E = 12$ volts, $R = 10$ ohms, and $L = 5$ henrys, how long does it take to obtain a current of 0.5 ampere? Of 1.0 ampere? Graph the equation.

106. Learning Curve Psychologists sometimes use the function

$$L(t) = A(1 - e^{-kt})$$

to measure the amount L learned at time t. The number A represents the amount to be learned, and the number k measures the rate of learning. Suppose that a student has an amount A of 200 vocabulary words to learn. A psychologist determines that the student learned 20 vocabulary words after 5 minutes.
(a) Determine the rate of learning k.
(b) Approximately how many words will the student have learned after 10 minutes?
(c) After 15 minutes?
(d) How long does it take for the student to learn 180 words?

107. Alcohol and Driving The concentration of alcohol in a person's blood is measurable. Suppose that the risk R (given as a percent) of having an accident while driving a car can be modeled by the equation

$$R = 3e^{kx}$$

where x is the variable concentration of alcohol in the blood and k is a constant.
(a) Suppose that a concentration of alcohol in the blood of 0.06 results in a 10% risk ($R = 10$) of an accident. Find the constant k in the equation.
(b) Using this value of k, what is the risk if the concentration is 0.17?
(c) Using the same value of k, what concentration of alcohol corresponds to a risk of 100%?
(d) If the law asserts that anyone with a risk of having an accident of 15% or more should not have driving privileges, at what concentration of alcohol in the blood should a driver be arrested and charged with a DUI?
(e) Compare this situation with that of Example 10. If you were a lawmaker, which situation would you support? Give your reasons.

108. Is there any function of the form $y = x^\alpha, 0 < \alpha < 1$, that increases more slowly than a logarithmic function whose base is greater than 1? Explain.

109. Critical Thinking In buying a new car, one consideration might be how well the price of the car holds up over time. Different makes of cars have different depreciation rates. One way to compute a depreciation rate for a car is given here. Suppose that the current prices of a certain Mercedes automobile are as follows:

	Age in Years				
New	1	2	3	4	5
$38,000	$36,600	$32,400	$28,750	$25,400	$21,200

Use the formula New $= Old(e^{Rt})$ to find R, the annual depreciation rate, for a specific time t. When might be the best time to trade in the car? Consult the NADA ("blue") book and compare two like models that you are interested in. Which has the better depreciation rate?

P R E P A R I N G   F O R   T H I S   S E C T I O N

Before getting started, review the following:

✓ Inverse Functions (Section 6.1, pp. 395–396)

6.4 | PROPERTIES OF LOGARITHMS; EXPONENTIAL AND LOGARITHMIC MODELS

OBJECTIVES   **1**   Work with the Properties of Logarithms
2   Write a Logarithmic Expression as a Sum or Difference of Logarithms
3   Write a Logarithmic Expression as a Single Logarithm
4   Evaluate Logarithms Whose Base Is Neither 10 nor e
5   Work with Exponential and Logarithmic Models

1   Logarithms have some very useful properties that can be derived directly from the definition and the laws of exponents.

EXAMPLE 1   **Establishing Properties of Logarithms**

(a) Show that $\log_a 1 = 0$.   (b) Show that $\log_a a = 1$.

Solution   (a) This fact was established when we graphed $y = \log_a x$ (see Figure 23). To show the result algebraically, let $y = \log_a 1$. Then

$$y = \log_a 1$$
$$a^y = 1 \qquad \text{Change to an exponent.}$$
$$a^y = a^0 \qquad a^0 = 1$$
$$y = 0 \qquad \text{Solve for } y.$$
$$\log_a 1 = 0 \qquad y = \log_a 1$$

(b) Let $y = \log_a a$. Then

$$y = \log_a a$$
$$a^y = a \qquad \text{Change to an exponent.}$$
$$a^y = a^1 \qquad a^1 = a$$
$$y = 1 \qquad \text{Solve for } y.$$
$$\log_a a = 1 \qquad y = \log_a a$$

■

To summarize:

$$\log_a 1 = 0 \qquad \log_a a = 1$$

Theorem | **Properties of Logarithms**

In the properties given next, M and a are positive real numbers, with $a \neq 1$, and r is any real number.

The number $\log_a M$ is the exponent to which a must be raised to obtain M. That is,

$$a^{\log_a M} = M \tag{1}$$

The logarithm to the base a of a raised to a power equals that power. That is,

$$\log_a a^r = r \tag{2}$$

The proof uses the fact that $y = a^x$ and $y = \log_a x$ are inverses.

Proof of Property (1) For inverse functions,

$$f\left(f^{-1}(x)\right) = x$$

Using $f(x) = a^x$ and $f^{-1}(x) = \log_a x$, we find

$$f\left(f^{-1}(x)\right) = a^{\log_a x} = x$$

Now let $x = M$ to obtain $a^{\log_a M} = M$.

Proof of Property (2) For inverse functions,

$$f^{-1}\left(f(x)\right) = x$$

Using $f(x) = a^x$ and $f^{-1}(x) = \log_a x$, we find

$$f^{-1}\left(f(x)\right) = \log_a a^x = x$$

Now let $x = r$ to obtain $\log_a a^r = r$.

EXAMPLE 2 | **Using Properties (1) and (2)**

(a) $2^{\log_2 \pi} = \pi$ (b) $\log_{0.2} 0.2^{-\sqrt{2}} = -\sqrt{2}$ (c) $\ln e^{kt} = kt$

NOW WORK PROBLEM 3.

Other useful properties of logarithms are given next.

Theorem | **Properties of Logarithms**

In the following properties, M, N, and a are positive real numbers, with $a \neq 1$, and r is any real number.

The Log of a Product Equals the Sum of the Logs

$$\log_a(MN) = \log_a M + \log_a N \tag{3}$$

The Log of a Quotient Equals the Difference of the Logs

$$\log_a\left(\frac{M}{N}\right) = \log_a M - \log_a N \qquad (4)$$

The Log of a Power Equals the Product of the Power and the Log

$$\log_a M^r = r \log_a M \qquad (5)$$

We shall derive properties (3) and (5) and leave the derivation of property (4) as an exercise (see Problem 87).

Proof of Property (3)   Let $A = \log_a M$ and let $B = \log_a N$. These expressions are equivalent to the exponential expressions

$$a^A = M \quad \text{and} \quad a^B = N$$

Now

$$\log_a(MN) = \log_a(a^A a^B) = \log_a a^{A+B} \qquad \text{Law of Exponents}$$

$$= A + B \qquad \text{Property (2) of logarithms}$$

$$= \log_a M + \log_a N \qquad \blacksquare$$

Proof of Property (5)   Let $A = \log_a M$. This expression is equivalent to

$$a^A = M$$

Now

$$\log_a M^r = \log_a(a^A)^r = \log_a a^{rA} \qquad \text{Law of Exponents}$$

$$= rA \qquad \text{Property (2) of logarithms}$$

$$= r \log_a M \qquad \blacksquare$$

NOW WORK PROBLEM **7**.

Logarithms can be used to transform products into sums, quotients into differences, and powers into factors. Such transformations prove useful in certain types of calculus problems.

EXAMPLE 3

Writing a Logarithmic Expression as a Sum of Logarithms

Write $\log_a(x\sqrt{x^2 + 1})$ as a sum of logarithms. Express all powers as factors.

Solution $\log_a(x\sqrt{x^2 + 1}) = \log_a x + \log_a \sqrt{x^2 + 1} \qquad \text{Property (3)}$

$$= \log_a x + \log_a(x^2 + 1)^{1/2}$$

$$= \log_a x + \tfrac{1}{2}\log_a(x^2 + 1) \qquad \text{Property (5)} \qquad \blacksquare$$

EXAMPLE 4 — Writing a Logarithmic Expression as a Difference of Logarithms

Write

$$\ln \frac{x^2}{(x-1)^3}$$

as a difference of logarithms. Express all powers as factors.

Solution

$$\ln \frac{x^2}{(x-1)^3} = \ln x^2 - \ln(x-1)^3 = 2\ln x - 3\ln(x-1)$$

$\quad\quad\quad\quad\quad\quad\quad\uparrow\quad\quad\quad\quad\quad\quad\quad\quad\quad\uparrow$

$\quad\quad\quad\quad\quad\quad$ Property (4) $\quad\quad\quad\quad\quad\quad$ Property (5)

EXAMPLE 5 — Writing a Logarithmic Expression as a Sum and Difference of Logarithms

Write

$$\log_a \frac{x^3\sqrt{x^2+1}}{(x+1)^4}$$

as a sum and difference of logarithms. Express all powers as factors.

Solution

$$\log_a \frac{x^3\sqrt{x^2+1}}{(x+1)^4} = \log_a(x^3\sqrt{x^2+1}) - \log_a(x+1)^4 \qquad \text{Property (4)}$$

$$= \log_a x^3 + \log_a \sqrt{x^2+1} - \log_a(x+1)^4 \qquad \text{Property (3)}$$

$$= \log_a x^3 + \log_a(x^2+1)^{1/2} - \log_a(x+1)^4$$

$$= 3\log_a x + \tfrac{1}{2}\log_a(x^2+1) - 4\log_a(x+1) \qquad \text{Property (5)}$$

NOW WORK PROBLEM **35**.

③ Another use of properties (3) through (5) is to write sums and/or differences of logarithms with the same base as a single logarithm. This skill will be needed to solve certain logarithmic equations discussed in the next section.

EXAMPLE 6 — Writing Expressions as a Single Logarithm

Write each of the following as a single logarithm.

(a) $\log_a 7 + 4\log_a 3$ \qquad\qquad (b) $\tfrac{2}{3}\ln 8 - \ln(3^4 - 8)$

(c) $\log_a x + \log_a 9 + \log_a(x^2+1) - \log_a 5$

Solution

(a) $\log_a 7 + 4\log_a 3 = \log_a 7 + \log_a 3^4 \qquad \text{Property (5)}$

$\qquad\qquad\qquad\qquad = \log_a 7 + \log_a 81$

$\qquad\qquad\qquad\qquad = \log_a(7 \cdot 81) \qquad \text{Property (3)}$

$\qquad\qquad\qquad\qquad = \log_a 567$

(b) $\tfrac{2}{3}\ln 8 - \ln(3^4 - 8) = \ln 8^{2/3} - \ln(81 - 8) \qquad \text{Property (5)}$

$\qquad\qquad\qquad\qquad\qquad = \ln 4 - \ln 73$

$\qquad\qquad\qquad\qquad\qquad = \ln\left(\tfrac{4}{73}\right) \qquad \text{Property (4)}$

(c) $\log_a x + \log_a 9 + \log_a(x^2 + 1) - \log_a 5 = \log_a(9x) + \log_a(x^2 + 1) - \log_a 5$

$$= \log_a[9x(x^2 + 1)] - \log_a 5$$

$$= \log_a\left[\frac{9x(x^2 + 1)}{5}\right]$$ ∎

WARNING: A common error made by some students is to express the logarithm of a sum as the sum of logarithms.

$$\log_a(M + N) \quad \text{is not equal to} \quad \log_a M + \log_a N$$

Correct statement $\log_a(MN) = \log_a M + \log_a N$ Property (3)

Another common error is to express the difference of logarithms as the quotient of logarithms.

$$\log_a M - \log_a N \quad \text{is not equal to} \quad \frac{\log_a M}{\log_a N}$$

Correct statement $\log_a M - \log_a N = \log_a\left(\dfrac{M}{N}\right)$ Property (4)

A third common error is to express a logarithm raised to a power as the product of the power times the logarithm.

$$(\log_a M)^r \quad \text{is not equal to} \quad r \log_a M$$

Correct statement $\log_a M^r = r \log_a M$ Property (5) ∎

✎━━━━ **NOW WORK PROBLEM 41.**

Two other properties of logarithms that we need to know are consequences of the fact that the logarithmic function $y = \log_a x$ is one-to-one.

Theorem   In the following properties, M, N, and a are positive real numbers, with $a \neq 1$.

> If $M = N$, then $\log_a M = \log_a N$. **(6)**
>
> If $\log_a M = \log_a N$, then $M = N$. **(7)**

∎

When property (6) is used, we start with the equation $M = N$ and say "take the logarithm of both sides" to obtain $\log_a M = \log_a N$.

Properties (6) and (7) are useful for solving *exponential and logarithmic equations*, a topic discussed in the next section.

USING A CALCULATOR TO EVALUATE LOGARITHMS WITH BASES OTHER THAN 10 OR e

④ Logarithms to the base 10, called **common logarithms,** were used to facilitate arithmetic computations before the widespread use of calculators. (See the Historical Feature at the end of this section.) Natural logarithms, that is, logarithms whose base is the number e, remain very important because they arise frequently in the study of natural phenomena.

Common logarithms are usually abbreviated by writing **log,** with the base understood to be 10, just as natural logarithms are abbreviated by **ln,** with the base understood to be e.

Most calculators have both $\boxed{\log}$ and $\boxed{\ln}$ keys to calculate the common logarithm and natural logarithm of a number. Let's look at an example to see how to approximate logarithms having a base other than 10 or e.

EXAMPLE 7

Approximating Logarithms Whose Base Is Neither 10 nor e

Approximate $\log_2 7$.
Round the answer to four decimal places.

Solution   Let $y = \log_2 7$. Then $2^y = 7$, so

$$2^y = 7$$
$$\ln 2^y = \ln 7 \qquad \text{Property (6)}$$
$$y \ln 2 = \ln 7 \qquad \text{Property (5)}$$
$$y = \frac{\ln 7}{\ln 2} \qquad \text{Solve for } y.$$
$$y \approx 2.8074 \qquad \begin{array}{l}\text{Use calculator (}\boxed{\ln}\text{ key) and}\\\text{round to four decimal places.}\end{array}$$

Example 7 shows how to approximate a logarithm whose base is 2 by changing to logarithms involving the base e. In general, we use the **Change-of-Base Formula.**

Theorem

Change-of-Base Formula

If $a \neq 1, b \neq 1$, and M are positive real numbers, then

$$\log_a M = \frac{\log_b M}{\log_b a} \qquad (8)$$

Proof   We derive this formula as follows: Let $y = \log_a M$. Then

$$a^y = M$$
$$\log_b a^y = \log_b M \qquad \text{Property (6)}$$
$$y \log_b a = \log_b M \qquad \text{Property (5)}$$
$$y = \frac{\log_b M}{\log_b a} \qquad \text{Solve for } y.$$
$$\log_a M = \frac{\log_b M}{\log_b a} \qquad y = \log_a M \qquad \blacksquare$$

Since calculators have only keys for $\boxed{\log}$ and $\boxed{\ln}$, in practice, the Change-of-Base Formula uses either $b = 10$ or $b = e$. Thus,

$$\log_a M = \frac{\log M}{\log a} \quad \text{and} \quad \log_a M = \frac{\ln M}{\ln a} \qquad (9)$$

EXAMPLE 8

Using the Change-of-Base Formula

Approximate: (a) $\log_5 89$ (b) $\log_{\sqrt{2}} \sqrt{5}$
Round answers to four decimal places.

Solution (a) $\log_5 89 = \dfrac{\log 89}{\log 5} \approx \dfrac{1.949390007}{0.6989700043} \approx 2.7889$

or

$\log_5 89 = \dfrac{\ln 89}{\ln 5} \approx \dfrac{4.48863637}{1.609437912} \approx 2.7889$

(b) $\log_{\sqrt{2}} \sqrt{5} = \dfrac{\log \sqrt{5}}{\log \sqrt{2}} = \dfrac{\frac{1}{2}\log 5}{\frac{1}{2}\log 2} \approx 2.3219$

or

$\log_{\sqrt{2}} \sqrt{5} = \dfrac{\ln \sqrt{5}}{\ln \sqrt{2}} = \dfrac{\frac{1}{2}\ln 5}{\frac{1}{2}\ln 2} \approx 2.3219$ ■

COMMENT: To graph logarithmic functions when the base is different from e or 10 requires the Change-of-Base Formula. For example, to graph $y = \log_2 x$, we would instead graph $y = (\ln x)/(\ln 2)$. Try it.   ■

 NOW WORK PROBLEMS **11** AND **51**.

EXPONENTIAL AND LOGARITHMIC MODELS

⑤ When placed in a scatter diagram, data sometimes indicate a relation between the variables that is exponential or logarithmic. Figure 28 shows typical scatter diagrams for data that are exponential or logarithmic.

Figure 28

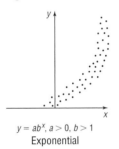

$y = ab^x, a > 0, b > 1$
Exponential

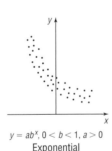

$y = ab^x, 0 < b < 1, a > 0$
Exponential

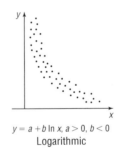

$y = a + b \ln x, a > 0, b < 0$
Logarithmic

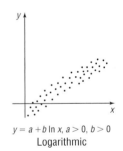

$y = a + b \ln x, a > 0, b > 0$
Logarithmic

EXAMPLE 9

Fitting a Curve to an Exponential Model

Beth is interested in finding a function that explains the closing price (adjusted for stock splits) of Dell Computer stock at the end of each year. She obtains the data in the table:

Year	Closing Price
1989	0.0573
1990	0.1927
1991	0.2669
1992	0.75
1993	0.3535
1994	0.6406
1995	1.082
1996	3.3203
1997	10.5
1998	36.5938
1999	51

Source: NASDAQ

(a) Draw a scatter diagram using year as the independent variable. Use $x = 1$ for 1989, $x = 2$ for 1990, and so on.

(b) The exponential function of best fit to the data is found to be

$$y = 0.03028(1.8905)^x$$

Express the function of best fit in the form $A = A_0 e^{kt}$, where $x = t$.

(c) Use the solution to part (b) to predict the closing price of Dell Computer stock at the end of the year 2000.

(d) Use a graphing utility to verify the exponential function of best fit.

(e) Use a graphing utility to draw a scatter diagram of the data and then graph the exponential function of best fit on it.

Solution (a) Letting 1 represent 1989, 2 represent 1990, and so on, we obtain the scatter diagram shown in Figure 29.

Figure 29

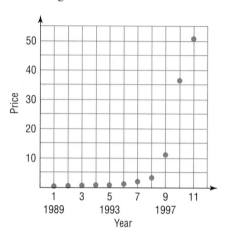

(b) To express $y = ab^x$ in the form $A = A_0 e^{kt}$, where $x = t$, we proceed as follows:

$$ab^x = A_0 e^{kt}, \qquad x = t$$

We set the coefficients equal to each other and the exponential expressions equal to each other.

$$a = A_0, \qquad b^x = e^{kt}, \qquad x = t$$
$$b^x = (e^k)^t, \qquad x = t$$
$$b = e^k$$

Since $y = ab^x = 0.03028(1.8905)^x$, we find that $a = 0.03028$ and $b = 1.8905$. Thus,

$$a = A_0 = 0.03028 \quad \text{and} \quad b = 1.8905 = e^k$$
$$k = \ln(1.8905) = 0.6368$$

As a result, $A = A_0 e^{kt} = 0.03028 e^{0.6368t}$.

(c) Let $t = 12$ (the year 2000) in the function found in part (b). The predicted closing price of Dell Computer stock at the end of the year 2000 is

$$A = 0.03028 e^{0.6368(12)} = \$63.08$$

Figure 30

 (d) A graphing utility fits the data to an exponential model of the form $y = ab^x$ by using the EXPonential REGression option.* See Figure 30. Thus, $y = ab^x = 0.03028(1.8905)^x$.

---

* If your utility does not have such an option but does have a LINear REGression option, you can transform the exponential model to a linear model by the following technique:

$y = ab^x$	Exponential form
$\ln y = \ln(ab^x)$	Take natural logs of each side.
$\ln y = \ln a + \ln b^x$	Property of logarithms
$\ln y = \ln a + (\ln b)x$	Property of logarithms
$Y = \ln a + (\ln b)x$	

Now apply the LINear REGression techniques discussed in Chapter 2. Be careful though. The dependent variable is now $Y = \ln y$, while the independent variable remains x. Once the line of best fit is obtained, you can find a and b and obtain the exponential curve $y = ab^x$.

Figure 31

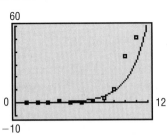

(e) Figure 31 shows the graph of $y = 0.03028(1.8905)^x$ on a scatter diagram of the data. ▬

➤ NOW WORK PROBLEM **89**.

There are relations between variables that do not follow an exponential model, but, instead, follow a logarithmic model in which the independent variable is related to the dependent variable by a logarithmic function.

| EXAMPLE 10 | **Fitting a Curve to a Logarithmic Model** |

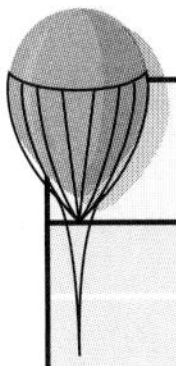

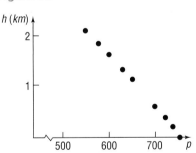

Atmospheric Pressure, p	Height, h
760	0
740	0.184
725	0.328
700	0.565
650	1.079
630	1.291
600	1.634
580	1.862
550	2.235

Jodi, a meteorologist, is interested in finding a function that explains the relation between the height of a weather balloon (in kilometers) and the atmospheric pressure (measured in millimeters of mercury) on the balloon. She collects the data in the table to the left.

(a) Draw a scatter diagram of the data with atmospheric pressure as the independent variable.

(b) The logarithmic function of best fit to the data is found to be

$$h = 45.8 - 6.9 \ln p$$

where h is the height of the weather balloon and p is the atmospheric pressure. Use the logarithmic function of best fit to predict the height of the balloon if the atmospheric pressure is 560 millimeters of mercury.

(c) Use a graphing utility to verify the logarithmic function of best fit.

(d) Use a graphing utility to draw a scatter diagram of the data, and then graph the logarithmic function of best fit on it.

Solution (a) See Figure 32.

Figure 32

h (km)

(b) Using the given function, Jodi predicts the height of the weather balloon when the atmospheric pressure is 560 to be

$$h = 45.8 - 6.9 \ln 560 \approx 2.14 \text{ kilometers}$$

(c) A graphing utility fits the data to a logarithmic model of the form $y = a + b \ln x$ by using the Logarithm REGression option. See Figure 33. Notice that $|r|$ is close to 1, indicating a good fit.

(d) Figure 34 shows the graph of $h = 45.8 - 6.9 \ln p$ on the scatter diagram.

Figure 33

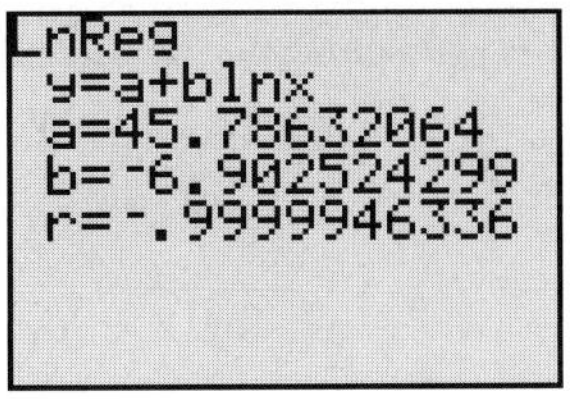

Figure 34

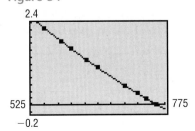

SUMMARY

Properties of Logarithms

In the list that follows, $a > 0$, $a \neq 1$, and $b > 0$, $b \neq 1$; also, $M > 0$ and $N > 0$.

Definition	$y = \log_a x$ means $x = a^y$
Properties of logarithms	$\log_a 1 = 0;\quad \log_a a = 1$
	$a^{\log_a M} = M;\quad \log_a a^r = r$
	$\log_a(MN) = \log_a M + \log_a N$
	$\log_a\left(\dfrac{M}{N}\right) = \log_a M - \log_a N$
	$\log_a M^r = r \log_a M$
	If $M = N$, then $\log_a M = \log_a N$.
	If $\log_a M = \log_a N$, then $M = N$.
Change-of-Base Formula	$\log_a M = \dfrac{\log_b M}{\log_b a}$

HISTORICAL FEATURE

John Napier
(1550–1617)

Logarithms were invented about 1590 by John Napier (1550–1617) and Jobst Bürgi (1552–1632), working independently. Napier, whose work had the greater influence, was a Scottish lord, a secretive man whose neighbors were inclined to believe him to be in league with the devil. His approach to logarithms was very different from ours; it was based on the relationship between arithmetic and geometric sequences, discussed in a later chapter, and not on the inverse function relationship of logarithms to exponential functions (described in Section 6.3). Napier's tables, published in 1614, listed what would now be called *natural logarithms* of sines and were rather difficult to use. A London professor, Henry Briggs, became interested in the tables and visited Napier. In their conversations, they developed the idea of common logarithms, which were published in 1617. Their importance for calculation was immediately recognized, and by 1650 they were being printed as far away as China. They remained an important calculation tool until the advent of the inexpensive handheld calculator about 1972, which has decreased their calculational, but not their theoretical, importance.

A side effect of the invention of logarithms was the popularization of the decimal system of notation for real numbers.

6.4   EXERCISES

In Problems 1–16, use properties of logarithms to find the exact value of each expression. Do not use a calculator.

1. $\log_3 3^{71}$

2. $\log_2 2^{-13}$

3. $\ln e^{-4}$

4. $\ln e^{\sqrt{2}}$

5. $2^{\log_2 7}$

6. $e^{\ln 8}$

7. $\log_8 2 + \log_8 4$

8. $\log_6 9 + \log_6 4$

9. $\log_6 18 - \log_6 3$

10. $\log_8 16 - \log_8 2$

11. $\log_2 6 \cdot \log_6 4$

12. $\log_3 8 \cdot \log_8 9$

13. $3^{\log_3 5 - \log_3 4}$

14. $5^{\log_5 6 + \log_5 7}$

15. $e^{\log_{e^2} 16}$

16. $e^{\log_{e^2} 9}$

In Problems 17–24, suppose that $\ln 2 = a$ *and* $\ln 3 = b$. *Use properties of logarithms to write each logarithm in terms of a and b.*

17. $\ln 6$

18. $\ln \frac{2}{3}$

19. $\ln 1.5$

20. $\ln 0.5$

21. $\ln 8$

22. $\ln 27$

23. $\ln \sqrt[5]{6}$

24. $\ln \sqrt[4]{\frac{2}{3}}$

In Problems 25–40, write each expression as a sum and/or difference of logarithms. Express powers as factors.

25. $\log_a(u^2 v^3)$

26. $\log_2\left(\dfrac{a}{b^2}\right)$

27. $\log \dfrac{1}{M^3}$

28. $\log(10u^2)$

29. $\log_5 \sqrt{\dfrac{a^3}{b}}$

30. $\log_6\left(\dfrac{ab^4}{\sqrt[3]{c^2}}\right)$

31. $\ln(x^2 \sqrt{1-x})$

32. $\ln(x\sqrt{1+x^2})$

33. $\log_2\left(\dfrac{x^3}{x-3}\right)$

34. $\log_5\left(\dfrac{\sqrt[3]{x^2+1}}{x^2-1}\right)$

35. $\log\left[\dfrac{x(x+2)}{(x+3)^2}\right]$

36. $\log\left[\dfrac{x^3 \sqrt{x+1}}{(x-2)^2}\right]$

37. $\ln\left[\dfrac{x^2-x-2}{(x+4)^2}\right]^{1/3}$

38. $\ln\left[\dfrac{(x-4)^2}{x^2-1}\right]^{2/3}$

39. $\ln \dfrac{5x\sqrt{1-3x}}{(x-4)^3}$

40. $\ln\left[\dfrac{5x^2 \sqrt[3]{1-x}}{4(x+1)^2}\right]$

In Problems 41–50, write each expression as a single logarithm.

41. $3\log_5 u + 4\log_5 v$

42. $\log_3 u^2 - \log_3 v$

43. $\log_{1/2} \sqrt{x} - \log_{1/2} x^3$

44. $\log_2\left(\dfrac{1}{x}\right) + \log_2\left(\dfrac{1}{x^2}\right)$

45. $\ln\left(\dfrac{x}{x-1}\right) + \ln\left(\dfrac{x+1}{x}\right) - \ln(x^2-1)$

46. $\log\left(\dfrac{x^2+2x-3}{x^2-4}\right) - \log\left(\dfrac{x^2+7x+6}{x+2}\right)$

47. $8\log_2 \sqrt{3x-2} - \log_2\left(\dfrac{4}{x}\right) + \log_2 4$

48. $21\log_3 \sqrt[3]{x} + \log_3(9x^2) - \log_3 25$

49. $2\log_a(5x^3) - \frac{1}{2}\log_a(2x+3)$

50. $\frac{1}{3}\log(x^3+1) + \frac{1}{2}\log(x^2+1)$

In Problems 51–58, use the Change-of-Base Formula and a calculator to approximate each logarithm. Round your answer to three decimal places.

51. $\log_3 21$

52. $\log_5 18$

53. $\log_{1/3} 71$

54. $\log_{1/2} 15$

55. $\log_{\sqrt{2}} 7$

56. $\log_{\sqrt{5}} 8$

57. $\log_\pi e$

58. $\log_\pi \sqrt{2}$

In Problems 59–62, find the exact value of each expression. Do not use a calculator.

59. $\log_2 3 \cdot \log_3 4 \cdot \log_4 5 \cdot \log_5 6 \cdot \log_6 7 \cdot \log_7 8$

60. $\log_2 4 \cdot \log_4 6 \cdot \log_6 8$

61. $\log_2 3 \cdot \log_3 4 \cdot \ldots \cdot \log_n(n+1) \cdot \log_{n+1} 2$

62. $\log_2 2 \cdot \log_2 4 \cdot \log_2 8 \cdot \ldots \cdot \log_2 2^n$

In Problems 63–68, graph each function using a graphing utility and the Change-of-Base Formula.

63. $y = \log_4 x$

64. $y = \log_5 x$

65. $y = \log_2(x+2)$

66. $y = \log_4(x-3)$

67. $y = \log_{x-1}(x+1)$

68. $y = \log_{x+2}(x-2)$

In Problems 69–78, express y as a function of x. The constant C is a positive number.

69. $\ln y = \ln x + \ln C$

70. $\ln y = \ln(x+C)$

71. $\ln y = \ln x + \ln(x+1) + \ln C$

72. $\ln y = 2\ln x - \ln(x+1) + \ln C$

73. $\ln y = 3x + \ln C$

74. $\ln y = -2x + \ln C$

75. $\ln(y - 3) = -4x + \ln C$

76. $\ln(y + 4) = 5x + \ln C$

77. $3 \ln y = \frac{1}{2} \ln(2x + 1) - \frac{1}{3} \ln(x + 4) + \ln C$

78. $2 \ln y = -\frac{1}{2} \ln x + \frac{1}{3} \ln(x^2 + 1) + \ln C$

79. Show that $\log_a\left(x + \sqrt{x^2 - 1}\right) + \log_a\left(x - \sqrt{x^2 - 1}\right) = 0$.

80. Show that $\log_a\left(\sqrt{x} + \sqrt{x - 1}\right) + \log_a\left(\sqrt{x} - \sqrt{x - 1}\right) = 0$.

81. Show that $\ln\left(1 + e^{2x}\right) = 2x + \ln\left(1 + e^{-2x}\right)$.

82. If $f(x) = \log_a x$, show that $\dfrac{f(x + h) - f(x)}{h} = \log_a\left(1 + \dfrac{h}{x}\right)^{1/h}$, $h \neq 0$.

83. If $f(x) = \log_a x$, show that $-f(x) = \log_{1/a} x$.

84. If $f(x) = \log_a x$, show that $f(AB) = f(A) + f(B)$.

85. If $f(x) = \log_a x$, show that $f(1/x) = -f(x)$.

86. If $f(x) = \log_a x$, show that $f(x^\alpha) = \alpha f(x)$.

87. Show that $\log_a(M/N) = \log_a M - \log_a N$, where a, M, and N are positive real numbers, with $a \neq 1$.

88. Show that $\log_a(1/N) = -\log_a N$, where a and N are positive real numbers, with $a \neq 1$.

---

89. Chemistry A chemist has a 100-gram sample of a radioactive material. He records the amount of radioactive material every week for 6 weeks and obtains the following data:

Week	Weight (in Grams)
0	100.0
1	88.3
2	75.9
3	69.4
4	59.1
5	51.8
6	45.5

(a) Draw a scatter diagram with week as the independent variable.

(b) The exponential function of best fit to the data is found to be

$$y = 100(0.88)^x$$

Express the function of best fit in the form $A = A_0 e^{kt}$, where $x = t$.

(c) Use the function in part (b) to estimate the time it takes until 50 grams of material is left. (Since 50 grams is one-half of the original amount, this time is referred to as the **half-life** of the radioactive material).

(d) Use the function in part (b) to predict how much radioactive material will be left after 50 weeks.

(e) Use a graphing utility to verify the exponential function of best fit.

(f) Use a graphing utility to draw a scatter diagram of the data and then graph the exponential function of best fit on it.

90. Chemistry A chemist has a 1000-gram sample of a radioactive material. She records the amount of radioactive material remaining in the sample every day for a week and obtains the following data:

Day	Weight (in Grams)
0	1000.0
1	897.1
2	802.5
3	719.8
4	651.1
5	583.4
6	521.7
7	468.3

(a) Draw a scatter diagram with day as the independent variable.

(b) The exponential function of best fit to the data is found to be

$$y = 999(0.898)^x$$

Express the function of best fit in the form $A = A_0 e^{kt}$, where $x = t$.

(c) Use the function in part (b) to estimate the time it takes until 500 grams of material is left. (Since 500 grams is one-half of the original amount, this time is referred to as the **half-life** of the radioactive material.)

(d) Use the function in part (b) to predict how much radioactive material is left after 20 days.

(e) Use a graphing utility to verify the exponential function of best fit.

(f) Use a graphing utility to draw a scatter diagram of the data and then graph the exponential function of best fit on it.

91. Economics and Marketing   A store manager collected the following data regarding price and quantity demanded of shoes:

Price ($/Unit)	Quantity Demanded
79	10
67	20
54	30
46	40
38	50
31	60

(a) Draw a scatter diagram with quantity demanded on the x-axis and price on the y-axis.

(b) The exponential function of best fit to the data is found to be

$$y = 96(0.98)^x$$

Express the function of best fit in the form $y = A_0 e^{kt}$, where $x = t$.

(c) Use the function in part (b) to predict the quantity demanded if the price is $60.

(d) Use a graphing utility to verify the exponential function of best fit.

(e) Use a graphing utility to draw a scatter diagram of the data and then graph the exponential function of best fit on it.

92. Economics and Marketing   A store manager collected the following data regarding price and quantity supplied of dresses:

Price ($/Unit)	Quantity Supplied
25	10
32	20
40	30
46	40
60	50
74	60

(a) Draw a scatter diagram with quantity supplied on the x-axis and price on the y-axis.

(b) The exponential function of best fit to the data is found to be

$$y = 20.524(1.022)^x$$

Express the function of best fit in the form $A = A_0 e^{kx}$, where $x = t$.

(c) Use the function in part (b) to predict the quantity supplied if the price is $45.

(d) Use a graphing utility to verify the exponential function of best fit.

(e) Use a graphing utility to draw a scatter diagram of the data and then graph the exponential function of best fit on it.

93. Economics and Marketing   The following data represent the price and quantity demanded in 1997 for IBM personal computers at Best Buy.

Price ($/Computer)	Quantity Demanded
2300	152
2000	159
1700	164
1500	171
1300	176
1200	180
1000	189

(a) Draw a scatter diagram of the data with price as the dependent variable.

(b) The logarithmic function of best fit to the data is found to be

$$y = 32,741 - 6071 \ln x$$

Use it to predict the number of IBM personal computers that would be demanded if the price were $1650.

(c) Use a graphing utility to verify the logarithmic function of best fit.

(d) Use a graphing utility to draw a scatter diagram of the data and then graph the logarithmic function of best fit on it.

94. Find the domain of $f(x) = \log_a x^2$ and the domain of $g(x) = 2 \log_a x$. Since $\log_a x^2 = 2 \log_a x$, how do you reconcile the fact that the domains are not equal? Write a brief explanation.

6.5 LOGARITHMIC AND EXPONENTIAL EQUATIONS

OBJECTIVES **1** Solve Logarithmic Equations Using the Properties of Logarithms
 2 Solve Exponential Equations
 3 Solve Logarithmic and Exponential Equations Using a Graphing Utility

LOGARITHMIC EQUATIONS

1 In Section 6.3 we solved logarithmic equations by changing a logarithm to exponential form. Often, however, some manipulation of the equation (usually using the properties of logarithms) is required before we can change to exponential form.

Our practice will be to solve equations, whenever possible, by finding exact solutions using algebraic methods. When algebraic methods cannot be used, approximate solutions will be obtained using a graphing utility. The reader is encouraged to pay particular attention to the form of equations for which exact solutions are possible.

EXAMPLE 1 **Solving a Logarithmic Equation**

Solve: $2 \log_5 x = \log_5 9$

Solution Because each logarithm has the same base, 5, we can obtain an exact solution as follows:

$$2 \log_5 x = \log_5 9$$
$$\log_5 x^2 = \log_5 9 \qquad \log_a M^r = r \log_a M$$
$$x^2 = 9 \qquad \text{If } \log_a M = \log_a N, \text{ then } M = N$$
$$x = 3 \quad \text{or} \quad \cancel{x = -3} \qquad \text{Recall that logarithms of negative numbers are not defined, so, in the expression } 2 \log_5 x, x \text{ must be positive.}$$
$$\text{Therefore, } -3 \text{ is extraneous and we discard it.}$$

The equation has only one solution, 3. ∎

NOW WORK PROBLEM **5**.

EXAMPLE 2 **Solving a Logarithmic Equation**

Solve: $\log_4 (x + 3) + \log_4 (2 - x) = 1$

Solution To obtain an exact solution, we need to express the left side as a single logarithm. Then we will change the expression to exponential form.

$$\log_4 (x + 3) + \log_4 (2 - x) = 1$$
$$\log_4 [(x + 3)(2 - x)] = 1 \qquad \log_a M + \log_a N = \log_a (MN)$$
$$(x + 3)(2 - x) = 4^1 = 4 \qquad \text{Change to an exponential expression.}$$
$$-x^2 - x + 6 = 4 \qquad \text{Simplify.}$$
$$x^2 + x - 2 = 0 \qquad \text{Place the quadratic equation in standard form.}$$
$$(x + 2)(x - 1) = 0 \qquad \text{Factor.}$$
$$x = -2 \quad \text{or} \quad x = 1 \qquad \text{Zero-Product Property}$$

Since the arguments of each logarithmic expression in the equation are positive for both $x = -2$ and $x = 1$, neither is extraneous. The solution set is $\{-2, 1\}$.   ∎

NOW WORK PROBLEM 9.

EXPONENTIAL EQUATIONS

② In Sections 6.2 and 6.3, we solved certain exponential equations by expressing each side of the equation with the same base. However, many exponential equations cannot be rewritten so that each side has the same base. In such cases, properties of logarithms along with algebraic techniques can sometimes be used to obtain a solution.

EXAMPLE 3

Solving an Exponential Equation

Solve: $4^x - 2^x - 12 = 0$

Solution   We note that $4^x = \left(2^2\right)^x = 2^{2x} = \left(2^x\right)^2$, so the equation is actually quadratic in form, and we can rewrite it as

$$\left(2^x\right)^2 - 2^x - 12 = 0 \quad \text{Let } u = 2^x; \text{ then } u^2 - u - 12 = 0.$$

Now we can factor as usual.

$$\left(2^x - 4\right)\left(2^x + 3\right) = 0 \qquad (u - 4)(u + 3) = 0$$

$$2^x - 4 = 0 \quad \text{or} \quad 2^x + 3 = 0 \qquad u - 4 = 0 \quad \text{or} \quad u + 3 = 0$$

$$2^x = 4 \qquad\qquad 2^x = -3 \qquad u = 2^x = 4 \qquad u = 2^x = -3$$

The equation on the left has the solution $x = 2$, since $2^x = 4 = 2^2$; the equation on the right has no solution, since $2^x > 0$ for all x. The only solution is 2.   ∎

In Example 3, we were able to write the exponential expression using the same base after utilizing some algebra, obtaining an exact solution to the equation. When this is not possible, logarithms can sometimes be used to obtain the solution.

EXAMPLE 4

Solving an Exponential Equation

Solve: $2^x = 5$

Solution   We write the exponential equation as the equivalent logarithmic equation.

$$2^x = 5$$

$$x = \log_2 5 = \frac{\ln 5}{\ln 2}$$
$$\uparrow$$
Change-of-Base Formula (9), Section 6.4

Alternatively, we can solve the equation $2^x = 5$ by taking the logarithm of each side (refer to Property (6), Section 6.4). Taking the natural logarithm,

$$2^x = 5$$

$$\ln 2^x = \ln 5 \qquad \text{If } M = N, \log_a M = \log_a N$$

$$x \ln 2 = \ln 5 \qquad \log_a M^r = r\log_a M$$

$$x = \frac{\ln 5}{\ln 2}$$

Using a calculator, the solution, rounded to three decimal places, is

$$x = \frac{\ln 5}{\ln 2} \approx 2.322$$

NOW WORK PROBLEM **17**.

EXAMPLE 5 **Solving an Exponential Equation**

Solve: $8 \cdot 3^x = 5$

Solution
$$8 \cdot 3^x = 5$$

$$3^x = \tfrac{5}{8} \qquad \text{Solve for } 3^x.$$

$$x = \log_3\left(\tfrac{5}{8}\right) = \frac{\ln \tfrac{5}{8}}{\ln 3} \qquad \text{Solve for } x.$$

The solution, rounded to three decimal places, is

$$x = \frac{\ln\left(\tfrac{5}{8}\right)}{\ln 3} \approx -0.428$$

EXAMPLE 6 **Solving an Exponential Equation**

Solve: $5^{x-2} = 3^{3x+2}$

Solution Because the bases are different, we first apply Property (6), Section 6.4 (taking the natural logarithm of each side) and then use appropriate properties of logarithms. The result is an equation in x that we can solve.

$$5^{x-2} = 3^{3x+2}$$

$$\ln 5^{x-2} = \ln 3^{3x+2} \qquad \text{If } M = N, \log_a M = \log_a N$$

$$(x-2)\ln 5 = (3x+2)\ln 3 \qquad \log_a M^r = r\log_a M$$

$$x \ln 5 - 2\ln 5 = 3x \ln 3 + 2\ln 3 \qquad \text{Distribute.}$$

$$x \ln 5 - 3x \ln 3 = 2\ln 3 + 2\ln 5 \qquad \text{Place terms involving } x \text{ on the left.}$$

$$(\ln 5 - 3\ln 3)x = 2(\ln 3 + \ln 5) \qquad \text{Factor.}$$

$$x = \frac{2(\ln 3 + \ln 5)}{\ln 5 - 3\ln 3} \approx -3.212$$

NOW WORK PROBLEM **25**.

GRAPHING UTILITY SOLUTIONS

△ ③ The techniques introduced in this section apply only to certain types of logarithmic and exponential equations. Solutions for other types are usually studied in calculus, using numerical methods. However, we can use a graphing utility to approximate the solution.

EXAMPLE 7 | **Solving Equations Using a Graphing Utility**

Solve: $\log_3 x + \log_4 x = 4$
Express the solution(s) rounded to two decimal places.

Figure 35

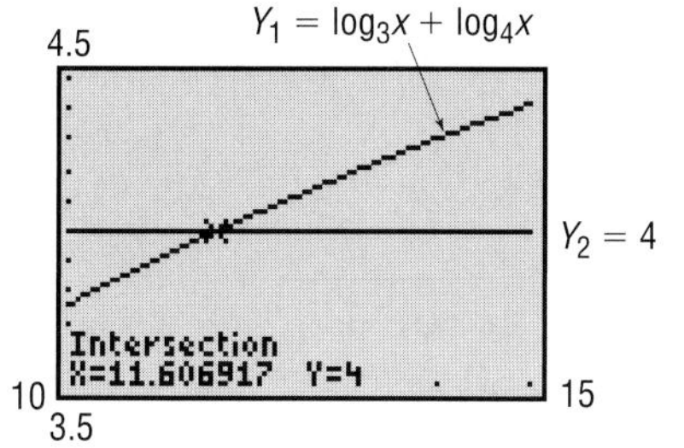

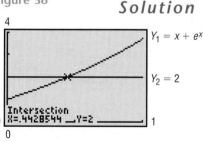

Solution The solution is found by graphing

$$Y_1 = \log_3 x + \log_4 x = \frac{\log x}{\log 3} + \frac{\log x}{\log 4} \quad \text{and} \quad Y_2 = 4$$

(Remember that you must use the Change-of-Base Formula to graph Y_1.) Y_1 is an increasing function (do you know why?), and so there is only one point of intersection for Y_1 and Y_2. Figure 35 shows the graphs of Y_1 and Y_2. Using the INTERSECT command, the solution is 11.61, rounded to two decimal places. ■

Can you discover an algebraic solution to Example 7?
[**Hint:** Factor $\log x$ from Y_1.]

EXAMPLE 8 | **Solving Equations Using a Graphing Utility**

Solve: $x + e^x = 2$
Express the solution(s) rounded to two decimal places.

Figure 36

Solution The solution is found by graphing $Y_1 = x + e^x$ and $Y_2 = 2$. Y_1 is an increasing function (do you know why?), and so there is only one point of intersection for Y_1 and Y_2. Figure 36 shows the graphs of Y_1 and Y_2. Using the INTERSECT command, the solution is 0.44 rounded to two decimal places. ■

6.5   EXERCISES

In Problems 1–44, solve each equation.

1. $\log_4(x + 2) = \log_4 8$

2. $\log_5(2x + 3) = \log_5 3$

3. $\frac{1}{2}\log_3 x = 2\log_3 2$

4. $-2\log_4 x = \log_4 9$

5. $2\log_5 x = 3\log_5 4$

6. $3\log_2 x = -\log_2 27$

7. $3\log_2(x - 1) + \log_2 4 = 5$

8. $2\log_3(x + 4) - \log_3 9 = 2$

9. $\log x + \log(x + 15) = 2$

10. $\log_4 x + \log_4(x - 3) = 1$

11. $\ln x + \ln(x + 2) = 4$

12. $\ln(x + 1) - \ln x = 2$

13. $2^{2x} + 2^x - 12 = 0$

14. $3^{2x} + 3^x - 2 = 0$

15. $3^{2x} + 3^{x+1} - 4 = 0$

16. $2^{2x} + 2^{x+2} - 12 = 0$

17. $2^x = 10$

18. $3^x = 14$

19. $8^{-x} = 1.2$

20. $2^{-x} = 1.5$

21. $3^{1-2x} = 4^x$

22. $2^{x+1} = 5^{1-2x}$

23. $\left(\frac{3}{5}\right)^x = 7^{1-x}$

24. $\left(\frac{4}{3}\right)^{1-x} = 5^x$

25. $1.2^x = (0.5)^{-x}$

26. $(0.3)^{1+x} = 1.7^{2x-1}$

27. $\pi^{1-x} = e^x$

28. $e^{x+3} = \pi^x$

29. $5(2^{3x}) = 8$

30. $0.3(4^{0.2x}) = 0.2$

31. $\log_a(x-1) - \log_a(x+6) = \log_a(x-2) - \log_a(x+3)$

32. $\log_a x + \log_a(x-2) = \log_a(x+4)$

33. $\log_{1/3}(x^2+x) - \log_{1/3}(x^2-x) = -1$

34. $\log_4(x^2-9) - \log_4(x+3) = 3$

35. $\log_2(x+1) - \log_4 x = 1$
  [**Hint:** Change $\log_4 x$ to base 2.]

36. $\log_2(3x+2) - \log_4 x = 3$

37. $\log_{16} x + \log_4 x + \log_2 x = 7$

38. $\log_9 x + 3\log_3 x = 14$

39. $\left(\sqrt[3]{2}\right)^{2-x} = 2^{x^2}$

40. $\log_2 x^{\log_2 x} = 4$

41. $\dfrac{e^x + e^{-x}}{2} = 1$

42. $\dfrac{e^x + e^{-x}}{2} = 3$

43. $\dfrac{e^x - e^{-x}}{2} = 2$

44. $\dfrac{e^x - e^{-x}}{2} = -2$

  [**Hint:** Multiply each side by e^x.]

In Problems 45–60, use a graphing utility to solve each equation. Express your answer rounded to two decimal places.

45. $\log_5 x + \log_3 x = 1$

46. $\log_2 x + \log_6 x = 3$

47. $\log_5(x+1) - \log_4(x-2) = 1$

48. $\log_2(x-1) - \log_6(x+2) = 2$

49. $e^x = -x$

50. $e^{2x} = x + 2$

51. $e^x = x^2$

52. $e^x = x^3$

53. $\ln x = -x$

54. $\ln(2x) = -x + 2$

55. $\ln x = x^3 - 1$

56. $\ln x = -x^2$

57. $e^x + \ln x = 4$

58. $e^x - \ln x = 4$

59. $e^{-x} = \ln x$

60. $e^{-x} = -\ln x$

PREPARING FOR THIS SECTION

Before getting started, review the following:

✓ Simple Interest (Section 1.2, pp. 97–99)

6.6 COMPOUND INTEREST

OBJECTIVES
1. Determine the Future Value of a Lump Sum of Money
2. Calculate Effective Rates of Return
3. Determine the Present Value of a Lump Sum of Money
4. Determine the Time Required to Double or Triple a Lump Sum of Money

1. Interest is money paid for the use of money. The total amount borrowed (whether by an individual from a bank in the form of a loan or by a bank from an individual in the form of a savings account) is called the **principal.** The **rate of interest,** expressed as a percent, is the amount charged for the use of the principal for a given period of time, usually on a yearly (that is, per annum) basis.

Simple Interest Formula

If a principal of P dollars is borrowed for a period of t years at a per annum interest rate r, expressed as a decimal, the interest I charged is

$$I = Prt \qquad (1)$$

Interest charged according to formula (1) is called **simple interest.**

In working with problems involving interest, we define the term **payment period** as follows:

Annually	Once per year	Monthly	12 times per year
Semiannually	Twice per year	Daily	365 times per year*
Quarterly	Four times per year		

When the interest due at the end of a payment period is added to the principal so that the interest computed at the end of the next payment period is based on this new principal amount (old principal + interest), the interest is said to have been **compounded. Compound interest** is interest paid on principal and previously earned interest.

EXAMPLE 1   **Computing Compound Interest**

A credit union pays interest of 8% per annum compounded quarterly on a certain savings plan. If $1000 is deposited in such a plan and the interest is left to accumulate, how much is in the account after 1 year?

Solution   We use the simple interest formula, $I = Prt$. The principal P is $1000 and the rate of interest is $8\% = 0.08$. After the first quarter of a year, the time t is $\frac{1}{4}$ year, so the interest earned is

$$I = Prt = (\$1000)(0.08)(\tfrac{1}{4}) = \$20$$

The new principal is $P + I = \$1000 + \$20 = \$1020$. At the end of the second quarter, the interest on this principal is

$$I = (\$1020)(0.08)(\tfrac{1}{4}) = \$20.40$$

At the end of the third quarter, the interest on the new principal of $\$1020 + \$20.40 = \$1040.40$ is

$$I = (\$1040.40)(0.08)(\tfrac{1}{4}) = \$20.81$$

Finally, after the fourth quarter, the interest is

$$I = (\$1061.21)(0.08)(\tfrac{1}{4}) = \$21.22$$

After 1 year the account contains $\$1061.21 + \$21.22 = \$1082.43$. ∎

The pattern of the calculations performed in Example 1 leads to a general formula for compound interest. To fix our ideas, let P represent the principal to be invested at a per annum interest rate r that is compounded n times per year, so the time of each compounding period is $1/n$ years. (For computing purposes, r is expressed as a decimal.) The interest earned after each compounding period is the principal $P \times$ rate $r \times$ time $1/n = P \cdot r/n$. The amount A after one compounding period is

$$A = P + P\left(\frac{r}{n}\right) = P\left(1 + \frac{r}{n}\right)$$

---

*Most banks use a 360-day "year." Why do you think they do?

After two compounding periods, the amount A, based on the new principal $P(1 + r/n)$, is

$$A = \underbrace{P\left(1 + \frac{r}{n}\right)}_{\substack{\text{New} \\ \text{principal}}} + \underbrace{P\left(1 + \frac{r}{n}\right)\left(\frac{r}{n}\right)}_{\substack{\text{Interest on} \\ \text{new principal}}} = P\left(1 + \frac{r}{n}\right)\left(1 + \frac{r}{n}\right) = P\left(1 + \frac{r}{n}\right)^2$$

After three compounding periods,

$$A = P\left(1 + \frac{r}{n}\right)^2 + P\left(1 + \frac{r}{n}\right)^2\left(\frac{r}{n}\right) = P\left(1 + \frac{r}{n}\right)^2\left(1 + \frac{r}{n}\right) = P\left(1 + \frac{r}{n}\right)^3$$

Continuing this way, after n compounding periods (1 year),

$$A = P\left(1 + \frac{r}{n}\right)^n$$

Because t years will contain $n \cdot t$ compounding periods, after t years we have

$$A = P\left(1 + \frac{r}{n}\right)^{nt}$$

Theorem

Compound Interest Formula

The amount A after t years due to a principal P invested at an annual interest rate r compounded n times per year is

$$A = P\left(1 + \frac{r}{n}\right)^{nt} \tag{2}$$

For example, to rework Example 1, we would use $P = \$1000$, $r = 0.08$, $n = 4$, (quarterly compounding) and $t = 1$ year to obtain

$$A = 1000\left(1 + \frac{0.08}{4}\right)^4 = \$1082.43$$

In Equation (2), the amount A is typically referred to as the **accumulated value or future value** of the account, while P is called the **present value.**

 EXPLORATION   To see the effects of compounding interest monthly on an initial deposit of $1, graph $Y_1 = \left(1 + \frac{r}{12}\right)^{12x}$ with $r = 0.06$ and $r = 0.12$ for $0 \le x \le 30$. What is the future value of $1 in 30 years when the interest rate per annum is $r = 0.06$ (6%)? What is the future value of $1 in 30 years when the interest rate per annum is $r = 0.12$ (12%)? Does doubling the interest rate double the future value?

NOTE: In using your calculator, be sure to use stored values, rather than approximations, in order to avoid round-off errors. At the final step, round money to the nearest cent.

NOW WORK PROBLEM 1.

EXAMPLE 2

Comparing Investments Using Different Compounding Periods

Investing $1000 at an annual rate of 10% compounded annually, quarterly, monthly, and daily will yield the following amounts after 1 year:

Annual compounding: $A = P(1 + r)$
$$= (\$1000)(1 + 0.10) = \$1100.00$$

Quarterly compounding: $A = P\left(1 + \dfrac{r}{4}\right)^4$
$$= (\$1000)(1 + 0.025)^4 = \$1103.81$$

Monthly compounding: $A = P\left(1 + \dfrac{r}{12}\right)^{12}$
$$= (\$1000)(1 + 0.00833)^{12} = \$1104.71$$

Daily compounding: $A = P\left(1 + \dfrac{r}{365}\right)^{365}$
$$= (\$1000)(1 + 0.000274)^{365} = \$1105.16$$

From Example 2 we can see that the effect of compounding more frequently is that the amount after 1 year is higher: $1000 compounded 4 times a year at 10% results in $1103.81; $1000 compounded 12 times a year at 10% results in $1104.71; and $1000 compounded 365 times a year at 10% results in $1105.16. This leads to the following question: What would happen to the amount after 1 year if the number of times that the interest is compounded were increased without bound?

Let's find the answer. Suppose that P is the principal, r is the per annum interest rate, and n is the number of times that the interest is compounded each year. The amount after 1 year is

$$A = P\left(1 + \frac{r}{n}\right)^n$$

Rewrite this expression as follows:

$$A = P\left(1 + \frac{r}{n}\right)^n = P\left(1 + \frac{1}{n/r}\right)^n = P\left[\left(1 + \frac{1}{n/r}\right)^{n/r}\right]^r = P\left[\left(1 + \frac{1}{h}\right)^h\right]^r \quad (3)$$

$$\underset{h = \frac{n}{r}}{\uparrow}$$

Now suppose that the number n of times that the interest is compounded per year gets larger and larger; that is, suppose that $n \rightarrow \infty$. Then $h = n/r \rightarrow \infty$, and the expression in brackets equals e. [Refer to equation (2), p. 409.] That is, $A \rightarrow Pe^r$.

Table 6 compares $(1 + r/n)^n$, for large values of n, to e^r for $r = 0.05$, $r = 0.10, r = 0.15$, and $r = 1$. The larger that n gets, the closer $(1 + r/n)^n$ gets to e^r. No matter how frequent the compounding, the amount after 1 year has the definite ceiling Pe^r.

TABLE 6			
$\left(1 + \frac{r}{n}\right)^n$			
$n = 100$	$n = 1000$	$n = 10,000$	e^r
$r = 0.05$ 1.0512580	1.0512698	1.051271	1.0512711
$r = 0.10$ 1.1051157	1.1051654	1.1051704	1.1051709
$r = 0.15$ 1.1617037	1.1618212	1.1618329	1.1618342
$r = 1$ 2.7048138	2.7169239	2.7181459	2.7182818

When interest is compounded so that the amount after 1 year is Pe^r, we say the interest is **compounded continuously.**

Theorem   **Continuous Compounding**

The amount A after t years due to a principal P invested at an annual interest rate r compounded continuously is

$$A = Pe^{rt} \qquad\qquad\qquad (4)$$

EXAMPLE 3   **Using Continuous Compounding**

The amount A that results from investing a principal P of $1000 at an annual rate r of 10% compounded continuously for a time t of 1 year is

$$A = \$1000e^{0.10} = (\$1000)(1.10517) = \$1105.17$$

 NOW WORK PROBLEM **9**.

2   The **effective rate of interest** is the equivalent annual simple rate of interest that would yield the same amount as compounding after 1 year. For example, based on Example 3, a principal of $1000 will result in $1105.17 at a rate of 10% compounded continuously. To get this same amount using a simple rate of interest would require that interest of $1105.17 − $1000.00 = $105.17 be earned on the principal. Since $105.17 is 10.517% of $1000, a simple rate of interest of 10.517% is needed to equal 10% compounded continuously. The effective rate of interest of 10% compounded continuously is 10.517%.

Based on the results of Examples 2 and 3, we find the following comparisons:

	Annual Rate	Effective Rate
Annual compounding	10%	10%
Quarterly compounding	10%	10.381%
Monthly compounding	10%	10.471%
Daily compounding	10%	10.516%
Continuous compounding	10%	10.517%

NOW WORK PROBLEM **21**.

EXAMPLE 4   **Computing the Value of an IRA**

On January 2, 2000, $2000 is placed in an Individual Retirement Account (IRA) that will pay interest of 10% per annum compounded continuously. What will the IRA be worth on January 1, 2020?

Solution   The amount A after 20 years is

$$A = Pe^{rt} = \$2000e^{(0.10)(20)} = \$14{,}778.11$$

 EXPLORATION   How long will it be until $A = \$4000$? $6000? [**Hint:** Graph $Y_1 = 2000e^{0.1x}$ and $Y_2 = 4000$. Use INTERSECT to find x.]

③ When people engaged in finance speak of the "time value of money," they are usually referring to the *present value* of money. The **present value** of A dollars to be received at a future date is the principal that you would need to invest now so that it would grow to A dollars in the specified time period. The present value of money to be received at a future date is always less than the amount to be received, since the amount to be received will equal the present value (money invested now) *plus* the interest accrued over the time period.

We use the compound interest formula (2) to get a formula for present value. If P is the present value of A dollars to be received after t years at a per annum interest rate r compounded n times per year, then, by formula (2),

$$A = P\left(1 + \frac{r}{n}\right)^{nt}$$

To solve for P, we divide both sides by $(1 + r/n)^{nt}$. The result is

$$\frac{A}{(1 + r/n)^{nt}} = P \quad \text{or} \quad P = A\left(1 + \frac{r}{n}\right)^{-nt}$$

Time is money

Theorem

Present Value Formulas

The present value P of A dollars to be received after t years, assuming a per annum interest rate r compounded n times per year, is

$$P = A\left(1 + \frac{r}{n}\right)^{-nt} \tag{5}$$

If the interest is compounded continuously, then

$$P = Ae^{-rt} \tag{6}$$

To prove (6), solve formula (4) for P.

EXAMPLE 5

Computing the Value of a Zero-Coupon Bond

A zero-coupon (noninterest-bearing) bond can be redeemed in 10 years for $1000. How much should you be willing to pay for it now if you want a return of

(a) 8% compounded monthly?   (b) 7% compounded continuously?

Solution   (a) We are seeking the present value of $1000. We use formula (5) with $A = \$1000, n = 12, r = 0.08$, and $t = 10$.

$$P = A\left(1 + \frac{r}{n}\right)^{-nt}$$

$$= \$1000\left(1 + \frac{0.08}{12}\right)^{-12(10)}$$

$$= \$450.52$$

For a return of 8% compounded monthly, you should pay $450.52 for the bond.

(b) Here we use formula (6) with $A = \$1000$, $r = 0.07$, and $t = 10$.

$$P = Ae^{-rt}$$
$$= \$1000e^{-(0.07)(10)}$$
$$= \$496.59$$

For a return of 7% compounded continuously, you should pay $496.59 for the bond. ■

NOW WORK PROBLEM 11.

EXAMPLE 6	**Rate of Interest Required to Double an Investment**

④   What annual rate of interest compounded annually should you seek if you want to double your investment in 5 years?

Solution   If P is the principal and we want P to double, the amount A will be $2P$. We use the compound interest formula with $n = 1$ and $t = 5$ to find r.

$$2P = P(1 + r)^5$$
$$2 = (1 + r)^5$$
$$1 + r = \sqrt[5]{2}$$
$$r = \sqrt[5]{2} - 1 = 1.148698 - 1 = 0.148698$$

The annual rate of interest needed to double the principal in 5 years is 14.87%. ■

NOW WORK PROBLEM 23.

EXAMPLE 7	**Doubling and Tripling Time for an Investment**

(a) How long will it take for an investment to double in value if it earns 5% compounded continuously?

(b) How long will it take to triple at this rate?

Solution   (a) If P is the initial investment and we want P to double, the amount A will be $2P$. We use formula (4) for continuously compounded interest with $r = 0.05$. Then

$$A = Pe^{rt}$$
$$2P = Pe^{0.05t}$$
$$2 = e^{0.05t}$$
$$0.05t = \ln 2$$
$$t = \frac{\ln 2}{0.05} = 13.86$$

It will take about 14 years to double the investment.

(b) To triple the investment, we set $A = 3P$ in formula (4).

$$A = Pe^{rt}$$

$$3P = Pe^{0.05t}$$

$$3 = e^{0.05t}$$

$$0.05t = \ln 3$$

$$t = \frac{\ln 3}{0.05} = 21.97$$

It will take about 22 years to triple the investment.

NOW WORK PROBLEM **29.**

6.6 EXERCISES

In Problems 1–10, find the amount that results from each investment.

1. $100 invested at 4% compounded quarterly after a period of 2 years

2. $50 invested at 6% compounded monthly after a period of 3 years

3. $500 invested at 8% compounded quarterly after a period of $2\frac{1}{2}$ years

4. $300 invested at 12% compounded monthly after a period of $1\frac{1}{2}$ years

5. $600 invested at 5% compounded daily after a period of 3 years

6. $700 invested at 6% compounded daily after a period of 2 years

7. $10 invested at 11% compounded continuously after a period of 2 years

8. $40 invested at 7% compounded continuously after a period of 3 years

9. $100 invested at 10% compounded continuously after a period of $2\frac{1}{4}$ years

10. $100 invested at 12% compounded continuously after a period of $3\frac{3}{4}$ years

In Problems 11–20, find the principal needed now to get each amount; that is, find the present value.

11. To get $100 after 2 years at 6% compounded monthly

12. To get $75 after 3 years at 8% compounded quarterly

13. To get $1000 after $2\frac{1}{2}$ years at 6% compounded daily

14. To get $800 after $3\frac{1}{2}$ years at 7% compounded monthly

15. To get $600 after 2 years at 4% compounded quarterly

16. To get $300 after 4 years at 3% compounded daily

17. To get $80 after $3\frac{1}{4}$ years at 9% compounded continuously

18. To get $800 after $2\frac{1}{2}$ years at 8% compounded continuously

19. To get $400 after 1 year at 10% compounded continuously

20. To get $1000 after 1 year at 12% compounded continuously

21. Find the effective rate of interest for $5\frac{1}{4}$% compounded quarterly.

22. What interest rate compounded quarterly will give an effective interest rate of 7%?

23. What rate of interest compounded annually is required to double an investment in 3 years?

24. What rate of interest compounded annually is required to double an investment in 10 years?

In Problems 25–28, which of the two rates would yield the larger amount in 1 year?
[**Hint:** Start with a principal of $10,000 in each instance.]

25. 6% compounded quarterly or $6\frac{1}{4}$% compounded annually

26. 9% compounded quarterly or $9\frac{1}{4}$% compounded annually

27. 9% compounded monthly or 8.8% compounded daily

28. 8% compounded semiannually or 7.9% compounded daily

29. How long does it take for an investment to double in value if it is invested at 8% per annum compounded monthly? Compounded continuously?

30. How long does it take for an investment to double in value if it is invested at 10% per annum compounded monthly? Compounded continuously?

31. If Tanisha has $100 to invest at 8% per annum compounded monthly, how long will it be before she has $150? If the compounding is continuous, how long will it be?

32. If Angela has $100 to invest at 10% per annum compounded monthly, how long will it be before she has $175? If the compounding is continuous, how long will it be?

33. How many years will it take for an initial investment of $10,000 to grow to $25,000? Assume a rate of interest of 6% compounded continuously.

34. How many years will it take for an initial investment of $25,000 to grow to $80,000? Assume a rate of interest of 7% compounded continuously.

35. What will a $90,000 house cost 5 years from now if the inflation rate over that period averages 3% compounded annually?

36. Sears charges 1.25% per month on the unpaid balance for customers with charge accounts (interest is compounded monthly). A customer charges $200 and does not pay her bill for 6 months. What is the bill at that time?

37. Jerome will be buying a used car for $15,000 in 3 years. How much money should he ask his parents for now so that, if he invests it at 5% compounded continuously, he will have enough to buy the car?

38. John will require $3000 in 6 months to pay off a loan that has no prepayment privileges. If he has the $3000 now, how much of it should he save in an account paying 3% compounded monthly so that in 6 months he will have exactly $3000?

39. George is contemplating the purchase of 100 shares of a stock selling for $15 per share. The stock pays no dividends. The history of the stock indicates that it should grow at an annual rate of 15% per year. How much will the 100 shares of stock be worth in 5 years?

40. Tracy is contemplating the purchase of 100 shares of a stock selling for $15 per share. The stock pays no dividends. Her broker says that the stock will be worth $20 per share in 2 years. What is the annual rate of return on this investment?

41. A business purchased for $650,000 in 1994 is sold in 1997 for $850,000. What is the annual rate of return for this investment?

42. Tanya has just inherited a diamond ring appraised at $5000. If diamonds have appreciated in value at an an-
nual rate of 8%, what was the value of the ring 10 years ago when the ring was purchased?

43. Jim places $1000 in a bank account that pays 5.6% compounded continuously. After 1 year, will he have enough money to buy a computer system that costs $1060? If another bank will pay Jim 5.9% compounded monthly, is this a better deal?

44. On January 1, Kim places $1000 in a certificate of deposit that pays 6.8% compounded continuously and matures in 3 months. Then Kim places the $1000 and the interest in a passbook account that pays 5.25% compounded monthly. How much does Kim have in the passbook account on May 1?

45. Will invests $2000 in a bond trust that pays 9% interest compounded semiannually. His friend Henry invests $2000 in a certificate of deposit (CD) that pays $8\frac{1}{2}$% compounded continuously. Who has more money after 20 years, Will or Henry?

46. Suppose that April has access to an investment that will pay 10% interest compounded continuously. Which is better: To be given $1000 now so that she can take advantage of this investment opportunity or to be given $1325 after 3 years?

47. Colleen and Bill have just purchased a house for $150,000, with the seller holding a second mortgage of $50,000. They promise to pay the seller $50,000 plus all accrued interest 5 years from now. The seller offers them three interest options on the second mortgage:
(a) Simple interest at 12% per annum
(b) $11\frac{1}{2}$% interest compounded monthly
(c) $11\frac{1}{4}$% interest compounded continuously
Which option is best; that is, which results in the least interest on the loan?

48. The First National Bank advertises that it pays interest on savings accounts at the rate of 4.25% compounded daily. Find the effective rate if the bank uses (a) 360 days or (b) 365 days in determining the daily rate.

Problems 49–52 involve zero-coupon bonds. A zero-coupon bond is a bond that is sold now at a discount and will pay its face value at the time when it matures; no interest payments are made.

49. A zero-coupon bond can be redeemed in 20 years for $10,000. How much should you be willing to pay for it now if you want a return of:
(a) 10% compounded monthly?
(b) 10% compounded continuously?

50. A child's grandparents are considering buying a $40,000 face value zero-coupon bond at birth so that she will have enough money for her college education 17 years later. If they want a rate of return of 8% compounded annually, what should they pay for the bond?

51. How much should a $10,000 face value zero-coupon bond, maturing in 10 years, be sold for now if its rate of return is to be 8% compounded annually?

52. If Pat pays $12,485.52 for a $25,000 face value zero-coupon bond that matures in 8 years, what is his annual rate of return?

53. Time to Double or Triple an Investment The formula

$$t = \frac{\ln m}{n \ln\left(1 + \frac{r}{n}\right)}$$

can be used to find the number of years t required to multiply an investment m times when r is the per annum interest rate compounded n times a year.
(a) How many years will it take to double the value of an IRA that compounds annually at the rate of 12%?
(b) How many years will it take to triple the value of a savings account that compounds quarterly at an annual rate of 6%?
(c) Give a derivation of this formula.

54. Time to Reach an Investment Goal   The formula

$$t = \frac{\ln A - \ln P}{r}$$

can be used to find the number of years t required for an investment P to grow to a value A when compounded continuously at an annual rate r.

(a) How long will it take to increase an initial investment of \$1000 to \$8000 at an annual rate of 10%?

(b) What annual rate is required to increase the value of a \$2000 IRA to \$30,000 in 35 years?

(c) Give a derivation of this formula.

55. Explain in your own words what the term *compound interest* means. What does *continuous compounding* mean?

56. Explain in your own words the meaning of *present value*.

57. Critical Thinking   You have just contracted to buy a house and will seek financing in the amount of \$100,000. You go to several banks. Bank 1 will lend you \$100,000 at the rate of 8.75% amortized over 30 years with a loan origination fee of 1.75%. Bank 2 will lend you \$100,000 at the

rate of 8.375% amortized over 15 years with a loan origination fee of 1.5%. Bank 3 will lend you \$100,000 at the rate of 9.125% amortized over 30 years with no loan origination fee. Bank 4 will lend you \$100,000 at the rate of 8.625% amortized over 15 years with no loan origination fee. Which loan would you take? Why? Be sure to have sound reasons for your choice. Use the information in the table to assist you. If the amount of the monthly payment does not matter to you, which loan would you take? Again, have sound reasons for your choice. Compare your final decision with others in the class. Discuss.

	Monthly Payment	Loan Origination Fee
Bank 1	\$786.70	\$1,750.00
Bank 2	\$977.42	\$1,500.00
Bank 3	\$813.63	\$0.00
Bank 4	\$990.68	\$0.00

6.7   GROWTH AND DECAY; NEWTON'S LAW; LOGISTIC MODELS

OBJECTIVES
1. Find Equations of Populations That Obey the Law of Uninhibited Growth
2. Find Equations of Populations That Obey the Law of Decay
3. Use Newton's Law of Cooling
4. Use Logistic Growth Models

1 Many natural phenomena have been found to follow the law that an amount A varies with time t according to

$$A(t) = A_0 e^{kt} \tag{1}$$

Here $A_0 = A(0)$ is the original amount $(t = 0)$ and $k \neq 0$ is a constant.

If $k > 0$, then equation (1) states that the amount A is increasing over time; if $k < 0$, the amount A is decreasing over time. In either case, when an amount A varies over time according to equation (1), it is said to follow the **exponential law** or the **law of uninhibited growth** $(k > 0)$ **or decay** $(k < 0)$. See Figure 37.

Figure 37

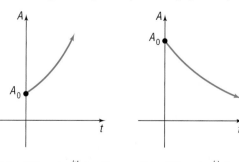

(a) $A(t) = A_0 e^{kt}, \, k > 0$ (b) $A(t) = A_0 e^{kt}, \, k < 0$

For example, we saw in Section 6.6 that continuously compounded interest follows the law of uninhibited growth. In this section we shall look at three additional phenomena that follow the exponential law.

UNINHIBITED GROWTH

Cell division is a process in the growth of many living organisms, such as amoebas, plants, and human skin cells. Based on an ideal situation in which no cells die and no by-products are produced, the number of cells present at a given time follows the law of uninhibited growth. Actually, however, after enough time has passed, growth at an exponential rate will cease due to the influence of factors such as lack of living space and dwindling food supply. The law of uninhibited growth accurately reflects only the early stages of the cell division process.

The cell division process begins with a culture containing N_0 cells. Each cell in the culture grows for a certain period of time and then divides into two identical cells. We assume that the time needed for each cell to divide in two is constant and does not change as the number of cells increases. These new cells then grow, and eventually each divides in two, and so on.

A model that gives the number N of cells in the culture after a time t has passed (in the early stages of growth) is given next.

Uninhibited Growth of Cells

$$N(t) = N_0 e^{kt}, \qquad k > 0 \qquad \text{(2)}$$

Here $N_0 = N(0)$ is the initial number of cells and k is a positive constant that represents the growth rate of the cells.

In using formula (2) to model the growth of cells, we are using a function that yields positive real numbers, even though we are counting the number of cells, which must be an integer. This is a common practice in many applications.

EXAMPLE 1   ## Uninhibited Growth

A certain culture of cells increases according to the equation

$$N(t) = 75e^{0.32t}$$

where t is time in hours.

(a) How many cells are in the culture initially?

(b) What is the growth rate of the culture?

(c) How many cells are in the culture after 4 hours?

Solution   (a) The initial number of cells is obtained when $t = 0$.

$$N_0 = N(0) = 75e^{0.32(0)} = 75$$

There are 75 cells present initially.

(b) Compare $N(t) = 75e^{0.32t}$ to $N(t) = 75e^{kt}$. The value of k, 0.32, indicates a growth rate of 32% per hour.

(c) After 4 hours, $t = 4$, we have

$$N(4) = 75e^{0.32(4)} \approx 269.75$$

There are about 270 cells after 4 hours.   ■

EXAMPLE 2   ## Bacterial Growth

A colony of bacteria increases according to the law of uninhibited growth.

(a) If the number of bacteria doubles in 3 hours, find k in formula (2) and express N as a function of t.

(b) How long will it take for the size of the colony to triple?

(c) How long does it take for the population to double a second time (that is, increase four times).

Solution   (a) Using formula (2), the number N of cells at a time t is

$$N(t) = N_0 e^{kt}$$

where N_0 is the initial number of bacteria present and k is a positive number. We first seek the number k. The number of cells doubles in 3 hours, so we have

$$N(3) = 2N_0$$

But $N(3) = N_0 e^{k(3)}$, so

$$N_0 e^{k(3)} = 2N_0$$

$$e^{3k} = 2$$

$$3k = \ln 2 \qquad \text{Write the exponential equation as a logarithm.}$$

$$k = \tfrac{1}{3}\ln 2 \approx 0.2310$$

Formula (2) for this growth process is therefore

$$N(t) = N_0 e^{0.2310t}$$

(b) The time t needed for the size of the colony to triple requires that $N(t) = 3N_0$. We substitute $3N_0$ for $N(t)$ to get

$$3N_0 = N_0 e^{0.2310t}$$

$$3 = e^{0.2310t}$$

$$0.2310t = \ln 3$$

$$t = \frac{\ln 3}{0.2310} \approx 4.756 \text{ hours}$$

It will take about 4.756 hours for the size of the colony to triple.

(c) If a population doubles in 3 hours, it will double a second time in 3 more hours, for a total time of 6 hours.   ■

✏️ NOW WORK PROBLEM 1.

RADIOACTIVE DECAY

② Radioactive materials follow the law of uninhibited decay. The amount A of a radioactive material present at time t is given by the following model:

Uninhibited Radioactive Decay

$$A(t) = A_0 e^{kt}, \qquad k < 0 \tag{3}$$

Here $A_0 = A(0)$ is the original amount of radioactive material and k is a negative number that represents the rate of decay.

All radioactive substances have a specific **half-life,** which is the time required for half of the radioactive substance to decay. In **carbon dating,** we use the fact that all living organisms contain two kinds of carbon, carbon 12 (a stable carbon) and carbon 14 (a radioactive carbon, with a half-life of 5600 years). While an organism is living, the ratio of carbon 12 to carbon 14 is constant. But when an organism dies, the original amount of carbon 12 present remains unchanged, whereas the amount of carbon 14 begins to decrease. This change in the amount of carbon 14 present relative to the amount of carbon 12 present makes it possible to calculate when an organism died.

EXAMPLE 3 ### Estimating the Age of Ancient Tools

Traces of burned wood along with ancient stone tools in an archeological dig in Chile were found to contain approximately 1.67% of the original amount of carbon 14. If the half-life of carbon 14 is 5600 years, approximately when was the tree cut and burned?

Solution Using formula (3), the amount A of carbon 14 present at time t is

$$A(t) = A_0 e^{kt}$$

where A_0 is the original amount of carbon 14 present and k is a negative number. We first seek the number k. To find it, we use the fact that after 5600 years, half of the original amount of carbon 14 remains, so $A(5600) = \frac{1}{2} A_0$. Thus,

$$\frac{1}{2} A_0 = A_0 e^{k(5600)}$$

$$\frac{1}{2} = e^{5600k}$$

$$5600k = \ln \frac{1}{2}$$

$$k = \frac{\ln \frac{1}{2}}{5600} \approx -0.000124$$

Formula (3) therefore becomes

$$A(t) = A_0 e^{-0.000124t}$$

If the amount A of carbon 14 now present is 1.67% of the original amount, it follows that

$$0.0167 A_0 = A_0 e^{-0.000124t}$$

$$0.0167 = e^{-0.000124t}$$

$$-0.000124t = \ln 0.0167$$

$$t = \frac{\ln 0.0167}{-0.000124} \approx 33{,}000 \text{ years}$$

The tree was cut and burned about 33,000 years ago. Some archeologists use this conclusion to argue that humans lived in the Americas 33,000 years ago, much earlier than is generally accepted. ■

NOW WORK PROBLEM 3.

NEWTON'S LAW OF COOLING

③ **Newton's Law of Cooling*** states that the temperature of a heated object decreases exponentially over time toward the temperature of the surrounding medium.

Newton's Law of Cooling

The temperature u of a heated object at a given time t can be modeled by the following function:

$$u(t) = T + (u_0 - T)e^{kt}, \quad k < 0 \qquad \textbf{(4)}$$

where T is the constant temperature of the surrounding medium, u_0 is the initial temperature of the heated object, and k is a negative constant.

EXAMPLE 4

Using Newton's Law of Cooling

An object is heated to 100°C (degrees Celsius) and is then allowed to cool in a room whose air temperature is 30°C.

(a) If the temperature of the object is 80°C after 5 minutes, when will its temperature be 50°C?

(b) Determine the elapsed time before the temperature of the object is 35°C.

(c) What do you notice about $u(t)$, the temperature, as t, time, passes?

Solution   (a) Using formula (4) with $T = 30$ and $u_0 = 100$, the temperature (in degrees Celsius) of the object at time t (in minutes) is

$$u(t) = 30 + (100 - 30)e^{kt} = 30 + 70e^{kt} \qquad \textbf{(5)}$$

where k is a negative constant. To find k, we use the fact that $u = 80$ when $t = 5$. Then

$$u(t) = 30 + 70e^{kt}$$
$$80 = 30 + 70e^{k(5)} \qquad t = 5; u(5) = 80$$
$$50 = 70e^{5k}$$
$$e^{5k} = \frac{50}{70}$$
$$5k = \ln\frac{5}{7}$$
$$k = \frac{\ln\frac{5}{7}}{5} \approx -0.0673$$

Formula (5) therefore becomes

$$u(t) = 30 + 70e^{-0.0673t} \qquad \textbf{(6)}$$

*Named after Sir Isaac Newton (1642–1727), one of the cofounders of calculus.

We want to find t when $u = 50°C$, so

$$50 = 30 + 70e^{-0.0673t}$$

$$20 = 70e^{-0.0673t}$$

$$e^{-0.0673t} = \frac{20}{70}$$

$$-0.0673t = \ln\frac{2}{7}$$

$$t = \frac{\ln\frac{2}{7}}{-0.0673} \approx 18.6 \text{ minutes}$$

The temperature of the object will be 50°C after about 18.6 minutes.

(b) If $u = 35°C$, then, based on equation (6), we have

$$35 = 30 + 70e^{-0.0673t}$$

$$5 = 70e^{-0.0673t}$$

$$e^{-0.0673t} = \frac{5}{70}$$

$$-0.0673t = \ln\frac{5}{70}$$

$$t = \frac{\ln\frac{5}{70}}{-0.0673} \approx 39.2 \text{ minutes}$$

The object will reach a temperature of 35°C after about 39.2 minutes.

(c) Refer to equation (6). As time passes, the value of t increases, the value of $e^{-0.0673t}$ approaches zero, and the value of $u(t)$ approaches 30°C. ■

NOW WORK PROBLEM 13.

LOGISTIC MODELS

④   The exponential growth model $A(t) = A_0 e^{kt}$, $k > 0$, assumes uninhibited growth, meaning that the value of the function grows without limit. Earlier we stated that cell division could be modeled using this function, assuming that no cells die and no by-products are produced. However, cell division would eventually be limited by factors such as living space and food supply. The **logistic growth model** is an exponential function that can model situations where the growth of the dependent variable is limited.

Other situations that lead to a logistic growth model include population growth and the sales of a product due to advertising. See Problems 21 through 24. The logistic growth model is given next.

Logistic Growth Model

$$P(t) = \frac{c}{1 + ae^{-bt}}$$

Here a, b, and c are constants with $c > 0$ and $b > 0$.

The number c is called the **carrying capacity** because the value $P(t)$ approaches c as t approaches infinity; that is, $\lim_{t \to \infty} P(t) = c$.

| EXAMPLE 5 | **Fruit Fly Population** |

Fruit flies are placed in a half-pint milk bottle with a banana (for food) and yeast plants (for food and to provide a stimulus to lay eggs). Suppose that the fruit fly population after t days is given by

$$P(t) = \frac{230}{1 + 56.5e^{-0.37t}}$$

(a) What is the carrying capacity of the half-pint bottle? That is, what is $P(t)$ as $t \to \infty$?

(b) How many fruit flies were initially placed in the half-pint bottle?

(c) When will the population of fruit flies be 180?

 (d) Using a graphing utility, graph $P(t)$.

Solution   (a) As $t \to \infty$, $e^{-0.37t} \to 0$ and $P(t) \to 230/1$. The carrying capacity of the half-pint bottle is 230 fruit flies.

(b) To find the initial number of fruit flies in the half-pint bottle, we evaluate $P(0)$.

$$P(0) = \frac{230}{1 + 56.5e^{-0.37(0)}}$$
$$= \frac{230}{1 + 56.5}$$
$$= 4$$

So initially there were four fruit flies in the half-pint bottle.

(c) To determine when the population of fruit flies will be 180, we solve the equation

$$\frac{230}{1 + 56.5e^{-0.37t}} = 180$$

$230 = 180(1 + 56.5e^{-0.37t})$	
$1.2778 = 1 + 56.5e^{-0.37t}$	Divide both sides by 180.
$0.2778 = 56.5e^{-0.37t}$	Subtract 1 from both sides.
$0.0049 = e^{-0.37t}$	Divide both sides by 56.5.
$\ln(0.0049) = -0.37t$	Rewrite as a logarithmic expression.
$t \approx 14.4$ days	Divide both sides by -0.37.

It will take approximately 14.4 days for the population to reach 180 fruit flies.

(d) See Figure 38 for the graph of $P(t)$.   ■

Figure 38

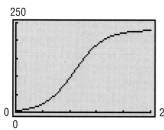

 EXPLORATION   On the same viewing rectangle, graph $Y_1 = \dfrac{500}{1 + 24e^{-0.03t}}$ and $Y_2 = \dfrac{500}{1 + 24e^{-0.08t}}$. What effect does b have on the logistic growth function?   ■

6.7 EXERCISES

1. **Growth of an Insect Population** The size P of a certain insect population at time t (in days) obeys the function $P(t) = 500e^{0.02t}$. After how many days will the population reach 1000? 2000?

2. **Growth of Bacteria** The number N of bacteria present in a culture at time t (in hours) obeys the function $N(t) = 1000e^{0.01t}$. After how many hours will the population equal 1500? 2000?

3. **Radioactive Decay** Strontium 90 is a radioactive material that decays according to the function $A(t) = A_0e^{-0.0244t}$, where A_0 is the initial amount present and A is the amount present at time t (in years).
   (a) What is the half-life of strontium 90?
   (b) Determine how long it takes for 100 grams of strontium 90 to decay to 10 grams.

4. **Radioactive Decay** Iodine 131 is a radioactive material that decays according to the function $A(t) = A_0e^{-0.087t}$, where A_0 is the initial amount present and A is the amount present at time t (in days).
   (a) What is the half-life of iodine 131?
   (b) Determine how long it takes for 100 grams of iodine 131 to decay to 10 grams.

5. **Growth of a Colony of Mosquitoes** The population of a colony of mosquitoes obeys the law of uninhibited growth. If there are 1000 mosquitoes initially and there are 1800 after 1 day, what is the size of the colony after 3 days? How long is it until there are 10,000 mosquitoes?

6. **Bacterial Growth** A culture of bacteria obeys the law of uninhibited growth. If 500 bacteria are present initially and there are 800 after 1 hour, how many will be present in the culture after 5 hours? How long is it until there are 20,000 bacteria?

7. **Population Growth** The population of a southern city follows the exponential law. If the population doubled in size over an 18-month period and the current population is 10,000, what will the population be 2 years from now?

8. **Population Growth** The population of a midwestern city follows the exponential law. If the population decreased from 900,000 to 800,000 from 1993 to 1995, what will the population be in 1997?

9. **Radioactive Decay** The half-life of radium is 1690 years. If 10 grams is present now, how much will be present in 50 years?

10. **Radioactive Decay** The half-life of radioactive potassium is 1.3 billion years. If 10 grams is present now, how much will be present in 100 years? In 1000 years?

11. **Estimating the Age of a Tree** A piece of charcoal is found to contain 30% of the carbon 14 that it originally had. When did the tree from which the charcoal came die? Use 5600 years as the half-life of carbon 14.

12. **Estimating the Age of a Fossil** A fossilized leaf contains 70% of its normal amount of carbon 14. How old is the fossil?

13. **Cooling Time of a Pizza** A pizza baked at 450°F is removed from the oven at 5:00 PM into a room that is a constant 70°F. After 5 minutes, the pizza is at 300°F.
    (a) At what time can you begin eating the pizza if you want its temperature to be 135°F?
    (b) Determine the time that needs to elapse before the pizza is 160°F.
    (c) What do you notice about the temperature as time passes?

14. **Newton's Law of Cooling** A thermometer reading 72°F is placed in a refrigerator where the temperature is a constant 38°F.
    (a) If the thermometer reads 60°F after 2 minutes, what will it read after 7 minutes?
    (b) How long will it take before the thermometer reads 39°F?
    (c) Determine the time needed to elapse before the thermometer reads 45°F.
    (d) What do you notice about the temperature as time passes?

15. **Newton's Law** A thermometer reading 8°C is brought into a room with a constant temperature of 35°C. If the thermometer reads 15°C after 3 minutes, what will it read after being in the room for 5 minutes? For 10 minutes?
    [**Hint:** You need to construct a formula similar to equation (4).]

16. **Thawing Time of a Steak** A frozen steak has a temperature of 28°F. It is placed in a room with a constant temperature of 70°F. After 10 minutes, the temperature of the steak has risen to 35°F. What will the temperature of the steak be after 30 minutes? How long will it take the steak to thaw to a temperature of 45°F? [See the hint given for Problem 15.]

17. **Decomposition of Salt in Water** Salt (NaCl) decomposes in water into sodium (NA^+) and chloride (Cl^-) ions according to the law of uninhibited decay. If the initial amount of salt is 25 kilograms and, after 10 hours, 15 kilograms of salt is left, how much salt is left after 1 day? How long does it take until $\frac{1}{2}$ kilogram of salt is left?

18. **Voltage of a Conductor** The voltage of a certain conductor decreases over time according to the law of uninhibited decay. If the initial voltage is 40 volts, and 2 seconds later it is 10 volts, what is the voltage after 5 seconds?

19. **Radioactivity from Chernobyl** After the release of radioactive material into the atmosphere from a nuclear power plant at Chernobyl (Ukraine) in 1986, the hay in Austria was contaminated by iodine 131 (half-life 8 days). If it is all right to feed the hay to cows when 10% of the iodine 131 remains, how long do the farmers need to wait to use this hay?

20. **Pig Roasts** The hotel Bora-Bora is having a pig roast. At noon, the chef put the pig in a large earthen oven. The pig's original temperature was 75°F. At 2:00 PM the chef checked the pig's temperature and was upset because it had reached only 100°F. If the oven's temperature remains a constant 325°F, at what time may the hotel serve its guests, assuming that pork is done when it reaches 175°F?

21. **Proportion of the Population That Owns a VCR** The logistic growth model

$$P(t) = \frac{0.9}{1 + 6e^{-0.32t}}$$

relates the proportion of U.S. households that own a VCR to the year. Let $t = 0$ represent 1984, $t = 1$ represent 1985, and so on.
(a) What proportion of the U.S. households owned a VCR in 1984?
(b) Determine the maximum proportion of households that will own a VCR.
(c) When will 0.8 (80%) of U.S. households own a VCR?

22. **Market Penetration of Intel's Coprocessor** The logistic growth model

$$P(t) = \frac{0.90}{1 + 3.5e^{-0.339t}}$$

relates the proportion of new personal computers sold at Best Buy that have Intel's latest coprocessor t months after it has been introduced.
(a) What proportion of new personal computers sold at Best Buy will have Intel's latest coprocessor when it is first introduced (that is, at $t = 0$)?
(b) Determine the maximum proportion of new personal computers sold at Best Buy that will have Intel's latest coprocessor.
(c) When will 0.75 (75%) of new personal computers sold at Best Buy have Intel's latest coprocessor?

23. **Population of a Bacteria Culture** The logistic growth model

$$P(t) = \frac{1000}{1 + 32.33e^{-0.439t}}$$

represents the population of a bacterium after t hours.
(a) What is the carrying capacity of the environment?
(b) What was the initial amount of bacteria in the population?
(c) When will the amount of bacteria be 800?

24. **Population of a Endangered Species** Often environmentalists will capture an endangered species and transport the species to a controlled environment where the species can produce offspring and regenerate its population. Suppose that six American bald eagles are captured, transported to Montana, and set free. Based on experience, the environmentalists expect the population to grow according to the model

$$P(t) = \frac{500}{1 + 83.33e^{-0.162t}}$$

where $P(t)$ is the population after t years.
(a) What is the carrying capacity of the environment?
(b) What is the predicted population of the American bald eagle in 20 years?
(c) When will the population be 300?

| **6.8** | LOGARITHMIC SCALES |

OBJECTIVES 1 Work with Decibels (Loudness of Sound)
 2 Work with the Richter Scale (Earthquake Magnitude)

Common logarithms often appear in the measurement of quantities, because they provide a way to scale down positive numbers that vary from very small to very large. For example, if a certain quantity can take on values from $0.0000000001 = 10^{-10}$ to $10,000,000,000 = 10^{10}$, the common logarithms of such numbers would be between -10 and 10, respectively.

LOUDNESS OF SOUND

 Our first application utilizes a logarithmic scale to measure the loudness of a sound. Physicists define the **intensity of a sound wave** as the amount of energy that the wave transmits through a given area. For example, the least intense sound that a human ear can detect at a frequency of 100 hertz is about 10^{-12} watt per square meter. The **loudness** $L(x)$, measured in **decibels** (named in honor of Alexander Graham Bell), of a sound of intensity x (measured in watts per square meter) is defined as

$$L(x) = 10 \log \frac{x}{I_0} \qquad \textbf{(1)}$$

where $I_0 = 10^{-12}$ watt per square meter is the least intense sound that a human ear can detect. If we let $x = I_0$ in equation (1), we get

$$L(I_0) = 10 \log \frac{I_0}{I_0} = 10 \log 1 = 0$$

At the threshold of human hearing, the loudness is 0 decibels. Table 7 gives the loudness of some common sounds.

TABLE 7
LOUDNESS OF COMMON SOUNDS (IN DECIBELS)

Decibels	Common Sound	Result	
140	Shotgun blast, jet 100 feet away at takeoff	Pain	
130	Motor test chamber	Human ear pain threshold	
120	Firecrackers, severe thunder, pneumatic jackhammer, hockey crowd	Uncomfortably loud	
110	Amplified rock music		
100	Textile loom, subway train, elevated train, farm tractor, power lawn mower, newspaper press	Loud	
90	Heavy city traffic, noisy factory		
80	Diesel truck going 40 mi/hr 50 feet away, crowded restaurant, garbage disposal, average factory, vacuum cleaner	Moderately loud	
70	Passenger car going 50 mi/hr 50 feet away		
60	Quiet typewriter, singing birds, window air conditioner, quiet automobile	Quiet	
50	Normal conversation, average office		
40	Household refrigerator, quiet office	Very quiet	
30	Average home, dripping faucet, whisper 5 feet away		
20	Light rainfall, rustle of leaves	Average person's threshold of hearing	
10	Whisper across room	Just audible	
0		Threshold for acute hearing	

Note that a decibel is not a linear unit like the meter. For example, a noise level of 10 decibels is 10 times as great as a noise level of 0 decibels. [If $L(x) = 10$, then $x = 10I_0$.] A noise level of 20 decibels is 100 times as great as a noise level of 0 decibels. [If $L(x) = 20$, then $x = 100I_0$.] A noise level of 30 decibels is 1000 times as great as a noise level of 0 decibels, and so on.

EXAMPLE 1 — Finding the Intensity of a Sound

Use Table 7 to find the intensity of the sound of a dripping faucet.

Solution From Table 7 we see that the loudness of the sound of a dripping faucet is 30 decibels. By equation (1), its intensity x may be found as follows:

$$L(x) = 10 \log\left(\frac{x}{I_0}\right) \qquad \text{Equation (1)}$$

$$30 = 10 \log\left(\frac{x}{I_0}\right) \qquad L(x) = 30$$

$$3 = \log\left(\frac{x}{I_0}\right) \qquad \text{Divide by 10.}$$

$$\frac{x}{I_0} = 10^3 \qquad \text{Write in exponential form.}$$

$$x = 1000I_0$$

where $I_0 = 10^{-12}$ watt per square meter. The intensity of the sound of a dripping faucet is 1000 times as great as a noise level of 0 decibels; that is, such a sound has an intensity of $1000 \cdot 10^{-12} = 10^{-9}$ watt per square meter. ■

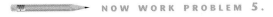 NOW WORK PROBLEM 5.

EXAMPLE 2 — Finding the Loudness of a Sound

Use Table 7 to determine the loudness of a subway train if it is known that this sound is 10 times as intense as the sound due to heavy city traffic.

Solution The sound due to heavy city traffic has a loudness of 90 decibels. Its intensity, therefore, is the value of x in the equation

$$90 = 10 \log\left(\frac{x}{I_0}\right)$$

A sound 10 times as intense as x has loudness $L(10x)$. The loudness of the subway train is

$$L(10x) = 10 \log\left(\frac{10x}{I_0}\right) \qquad \text{Substitute } 10x \text{ for } x.$$

$$= 10 \log\left(10 \cdot \frac{x}{I_0}\right)$$

$$= 10\left[\log 10 + \log\left(\frac{x}{I_0}\right)\right] \qquad \text{Log of product} = \text{sum of logs}$$

$$= 10 \log 10 + 10 \log\left(\frac{x}{I_0}\right)$$

$$= 10 + 90 = 100 \text{ decibels} \qquad \log 10 = 1;\ 90 = 10 \log(x/I_0) \quad ■$$

MAGNITUDE OF AN EARTHQUAKE

2 Our second application uses a logarithmic scale to measure the magnitude of an earthquake.

The **Richter scale**[*] is one way of converting seismographic readings into numbers that provide an easy reference for measuring the magnitude M of an earthquake. All earthquakes are compared to a **zero-level earthquake** whose seismographic reading measures 0.001 millimeter at a distance of 100 kilometers from the epicenter. An earthquake whose seismographic reading measures x millimeters has **magnitude** $M(x)$, given by

$$M(x) = \log \frac{x}{x_0} = \log \frac{x}{10^{-3}} \qquad (2)$$

where $x_0 = 10^{-3}$ is the reading of a zero-level earthquake the same distance from its epicenter.

EXAMPLE 3 | **Finding the Magnitude of an Earthquake**

What is the magnitude of an earthquake whose seismographic reading is 0.1 millimeter at a distance of 100 kilometers from its epicenter?

Solution If $x = 0.1$, the magnitude $M(x)$ of this earthquake is

$$M(0.1) = \log \frac{x}{x_0} = \log\left(\frac{0.1}{10^{-3}}\right) = \log \frac{10^{-1}}{10^{-3}} = \log 10^2 = 2$$

This earthquake measures 2.0 on the Richter scale. ∎

NOW WORK PROBLEM 7.

Based on formula (2), we define the **intensity of an earthquake** as the ratio of x to x_0. For example, the intensity of the earthquake described in Example 3 is $\frac{0.1}{0.001} = 10^2 = 100$. That is, it is 100 times as intense as a zero-level earthquake.

EXAMPLE 4 | **Comparing the Intensity of Two Earthquakes**

The devastating San Francisco earthquake of 1906 measured 6.9 on the Richter scale. How did the intensity of that earthquake compare to the Papua, New Guinea, earthquake of 1988, which measured 6.7 on the Richter scale?

Figure 39

---

[*]Named after the American scientist, C. F. Richter, who devised it in 1935.

Solution   Let x_1 and x_2 denote the seismographic readings, respectively, of the 1906 San Francisco earthquake and the 1988 Papua, New Guinea, earthquake. Then, based on formula (2),

$$6.9 = \log \frac{x_1}{x_0} \qquad 6.7 = \log \frac{x_2}{x_0}$$

Consequently,

$$\frac{x_1}{x_0} = 10^{6.9} \qquad \frac{x_2}{x_0} = 10^{6.7}$$

The 1906 San Francisco earthquake was $10^{6.9}$ times as intense as a zero-level earthquake. The Papua, New Guinea, earthquake was $10^{6.7}$ times as intense as a zero-level earthquake. Form the ratio x_1/x_2 to compare the two earthquakes,

$$\frac{x_1}{x_2} = \frac{10^{6.9} x_0}{10^{6.7} x_0} = 10^{0.2} \approx 1.58$$

$$x_1 \approx 1.58 x_2$$

The San Francisco earthquake was 1.58 times as intense as the Papua, New Guinea, earthquake.   ■

   Example 4 demonstrates that the relative intensity of two earthquakes can be found by raising 10 to a power equal to the difference of their readings on the Richter scale.

6.8   EXERCISES

1. **Loudness of a Dishwasher**   Find the loudness of a dishwasher that operates at an intensity of 10^{-5} watt per square meter. Express your answer in decibels.

2. **Loudness of a Diesel Engine**   Find the loudness of a diesel engine that operates at an intensity of 10^{-3} watt per square meter. Express your answer in decibels.

3. **Loudness of a Jet Engine**   With engines at full throttle, a Boeing 727 jetliner produces noise at an intensity of 0.15 watt per square meter. Find the loudness of the engines in decibels.

4. **Loudness of a Whisper**   A whisper produces noise at an intensity of $10^{-9.8}$ watt per square meter. What is the loudness of a whisper in decibels?

5. **Intensity of a Sound at the Threshold of Pain**   For humans, the threshold of pain due to sound averages 130 decibels. What is the intensity of such a sound in watts per square meter?

6. **Comparing Sounds**   If one sound is 50 times as intense as another, what is the difference in the loudness of the two sounds? Express your answer in decibels.

7. **Magnitude of an Earthquake**   Find the magnitude of an earthquake whose seismographic reading is 10.0 mil-

limeters at a distance of 100 kilometers from its epicenter.

8. **Magnitude of an Earthquake**   Find the magnitude of an earthquake whose seismographic reading is 1210 millimeters at a distance of 100 kilometers from its epicenter.

9. **Comparing Earthquakes**   The Mexico City earthquake of 1985 registered 8.1 on the Richter scale. What would a seismograph 100 kilometers from the epicenter have measured for this earthquake? How does this earthquake compare in intensity to the 1906 San Francisco earthquake, which registered 6.9 on the Richter scale?

10. **Comparing Earthquakes**   Two earthquakes differ by 1.0 when measured on the Richter scale. How would the seismographic readings differ at a distance of 100 kilometers from the epicenter? How do their intensities compare?

11. **NBA Finals 1997**   In game 5 of the NBA Finals between the Chicago Bulls and the Utah Jazz at the Delta Center, the crowd noise measured 110 decibels. NBA guidelines say sound levels are not to exceed 95 decibels. Compute the ratio of the intensities of these two sounds to determine by how much the crowd noise exceeded guidelines.

CHAPTER REVIEW

Things To Know

One-to-one function f (p. 394)	A function whose inverse is itself a function. For any choice of elements x_1, x_2 in the domain of f, if $x_1 \neq x_2$, then $f(x_1) \neq f(x_2)$
Horizontal-line test (p. 394)	If every horizontal line intersects the graph of a function f in at most one point, then f is one-to-one.
Inverse function f^{-1} of f (pp. 395-396)	Domain of f = Range of f^{-1}; Range of f = Domain of f^{-1}. $f^{-1}(f(x)) = x$ and $f(f^{-1}(x)) = x$. Graphs of f and f^{-1} are symmetric with respect to the line $y = x$.

Properties of the exponential function (pp. 407 and 408)

$f(x) = a^x$, $a > 1$ Domain: the interval $(-\infty, \infty)$; Range: the interval $(0, \infty)$; x-intercepts: none; y-intercept: 1; horizontal asymptote: x-axis ($y = 0$) as $x \to -\infty$; increasing; one-to-one; smooth; continuous
See Figure 14 for a typical graph.

$f(x) = a^x$, $0 < a < 1$ Domain: the interval $(-\infty, \infty)$; Range: the interval $(0, \infty)$; x-intercepts: none; y-intercept: 1; horizontal asymptote: x-axis ($y = 0$) as $x \to \infty$; decreasing; one-to-one; smooth; continuous
See Figure 18 for a typical graph.

Number e (p. 409)   Value approached by the expression $\left(1 + \dfrac{1}{n}\right)^n$ as $n \to \infty$; that is, $\displaystyle\lim_{n \to \infty}\left(1 + \dfrac{1}{n}\right)^n = e$.

Property of exponents (p. 411)   If $a^u = a^v$, then $u = v$.

Properties of the logarithmic function (p. 421)

$f(x) = \log_a x$, $a > 1$ Domain: the interval $(0, \infty)$; Range: the interval $(-\infty, \infty)$;
$(y = \log_a x$ means $x = a^y)$ x-intercept: 1; y-intercept: none; vertical asymptote: $x = 0$ (y-axis); increasing; one-to-one; smooth; continuous
See Figure 27(a) for a typical graph.

$f(x) = \log_a x$, $0 < a < 1$ Domain: the interval $(0, \infty)$; Range: the interval $(-\infty, \infty)$;
$(y = \log_a x$ means $x = a^y)$ x-intercept: 1; y-intercept: none; vertical asymptote: $x = 0$ (y-axis); decreasing; one-to-one; smooth; continuous
See Figure 27(b) for a typical graph.

Natural logarithm (p. 421) $y = \ln x$ means $x = e^y$.

Properties of logarithms (pp. 429–431) $\log_a 1 = 0$ $\log_a a = 1$ $a^{\log_a M} = M$ $\log_a a^r = r$

$\log_a(MN) = \log_a M + \log_a N$ $\log_a\left(\dfrac{M}{N}\right) = \log_a M - \log_a N$
$\log_a M^r = r \log_a M$

(p. 433)   If $M = N$, then $\log_a M = \log_a N$.
If $\log_a M = \log_a N$, then $M = N$.

Formulas

Change-of-Base Formula (p. 434)	$\log_a M = \dfrac{\log_b M}{\log_b a}$
Compound Interest Formula (p. 448)	$A = P\left(1 + \dfrac{r}{n}\right)^{nt}$
Continuous compounding (p. 450)	$A = Pe^{rt}$

Present Value Formulas (p. 451) $P = A\left(1 + \dfrac{r}{n}\right)^{-nt}$ or $P = Ae^{-rt}$

Growth and Decay (p. 455) $A(t) = A_0 e^{kt}$

Logistic Growth Model (p. 460) $P(t) = \dfrac{c}{1 + ae^{-bt}}$

Objectives

You should be able to:

Determine the inverse of a function (p. 392)

Obtain the graph of the inverse function from the graph of a function (p. 397)

Find the inverse function f^{-1} (p. 398)

Evaluate exponential functions (p. 404)

Graph exponential functions (p. 405)

Define the number e (p. 409)

Solve exponential equations (p. 411)

Change exponential expressions to logarithmic expressions (p. 418)

Change logarithmic expressions to exponential expressions (p. 419)

Evaluate logarithmic functions (p. 419)

Determine the domain of a logarithmic function (p. 419)

Graph logarithmic functions (p. 420)

Solve logarithmic equations (p. 423)

Work with the properties of logarithms (p. 429)

Write a logarithmic expression as a sum or difference of logarithms (p. 431)

Write a logarithmic expression as a single logarithm (p. 432)

Evaluate logarithms whose base is neither 10 nor e (p. 433)

Work with exponential and logarithmic models (p. 435)

Solve logarithmic equations using the properties of logarithms (p. 442)

Solve exponential equations (p. 443)

Solve logarithmic and exponential equations using a graphing utility (p. 445)

Determine the future value of a lump sum of money (p. 446)

Calculate effective rates of return (p. 450)

Determine the present value of a lump sum of money (p. 451)

Determine the time required to double or triple a lump sum of money (p. 452)

Find equations of populations that obey the law of uninhibited growth (p. 455)

Find equations of populations that obey the law of decay (p. 457)

Use Newton's Law of Cooling (p. 459)

Use logistic growth models (p. 460)

Work with decibels (p. 464)

Work with the Richter scale (p. 466)

Fill-in-the-Blank Items

1. If every horizontal line intersects the graph of a function f at no more than one point, then f is a(n) _____ function.

2. If f^{-1} denotes the inverse of a function f, then the graphs of f and f^{-1} are symmetric with respect to the line _____.

3. The graph of every exponential function $f(x) = a^x, a > 0, a \neq 1$, passes through the three points _____.

4. If the graph of an exponential function $f(x) = a^x, a > 0, a \neq 1$, is decreasing, then its base must be less than _____.

5. If $3^x = 3^4$, then $x = $ _____.

6. The logarithm of a product equals the _____ of the logarithms.

7. For every base, the logarithm of _____ equals 0.

8. If $\log_8 M = \log_5 7/\log_5 8$, then $M = $ _____.

9. The domain of the logarithmic function $f(x) = \log_a x$ consists of _____.

10. The graph of every logarithmic function $f(x) = \log_a x, a > 0, a \neq 1$, passes through the three points _____.

11. If the graph of a logarithmic function $f(x) = \log_a x, a > 0, a \neq 1$, is increasing, then its base must be larger than _____.

12. If $\log_3 x = \log_3 7$, then $x = $ _____.

True/False Items

T F **1.** If f and g are inverse functions, then the domain of f is the same as the domain of g.

T F **2.** If f and g are inverse functions, then their graphs are symmetric with respect to the line $y = x$.

T F **3.** The graph of every exponential function $f(x) = a^x, a > 0, a \neq 1$, will contain the points $(0, 1)$ and $(1, a)$.

T F **4.** The graphs of $y = 3^{-x}$ and $y = \left(\frac{1}{3}\right)^x$ are identical.

T F **5.** The present value of $1000 to be received after 2 years at 10% per annum compounded continuously is approximately $1205.

T F **6.** If $y = \log_a x$, then $y = a^x$.

T F **7.** The graph of every logarithmic function $f(x) = \log_a x, a > 0, a \neq 1$, will contain the points $(1, 0)$ and $(a, 1)$.

T F **8.** $a^{\log_M a} = M$, where $a > 0, a \neq 1, M > 0$.

T F **9.** $\log_a(M + N) = \log_a M + \log_a N$, where $a > 0, a \neq 1, M > 0, N > 0$.

T F **10.** $\log_a M - \log_a N = \log_a(M/N)$, where $a > 0, a \neq 1, M > 0, N > 0$.

Review Exercises

Blue problem numbers indicate the author's suggestions for use in a Practice Test.

In Problems 1–6, the function f is one-to-one. Find the inverse of each function and check your answer. Find the domain and range of f and f^{-1}.

1. $f(x) = \dfrac{2x + 3}{5x - 2}$

2. $f(x) = \dfrac{2 - x}{3 + x}$

3. $f(x) = \dfrac{1}{x - 1}$

4. $f(x) = \sqrt{x - 2}$

5. $f(x) = \dfrac{3}{x^{1/3}}$

6. $f(x) = x^{1/3} + 1$

In Problems 7–12, find the exact value of each expression. Do not use a calculator.

7. $\log_2\left(\frac{1}{8}\right)$

8. $\log_3 81$

9. $\ln e^{\sqrt{2}}$

10. $e^{\ln 0.1}$

11. $2^{\log_2 0.4}$

12. $\log_2 2^{\sqrt{3}}$

In Problems 13–18, write each expression as the sum and/or difference of logarithms. Express powers as factors.

13. $\log_3\left(\dfrac{uv^2}{w}\right)$

14. $\log_2\left(a^2 \sqrt{b}\right)^4$

15. $\log\left(x^2 \sqrt{x^3 + 1}\right)$

16. $\log_5\left(\dfrac{x^2 + 2x + 1}{x^2}\right)$

17. $\ln\left(\dfrac{x\sqrt[3]{x^2 + 1}}{x - 3}\right)$

18. $\ln\left(\dfrac{2x + 3}{x^2 - 3x + 2}\right)^2$

In Problems 19–24, write each expression as a single logarithm.

19. $3\log_4 x^2 + \dfrac{1}{2}\log_4 \sqrt{x}$

20. $-2\log_3\left(\dfrac{1}{x}\right) + \dfrac{1}{3}\log_3 \sqrt{x}$

21. $\ln\left(\dfrac{x - 1}{x}\right) + \ln\left(\dfrac{x}{x + 1}\right) - \ln(x^2 - 1)$

22. $\log(x^2 - 9) - \log(x^2 + 7x + 12)$

23. $2\log 2 + 3\log x - \dfrac{1}{2}\left[\log(x + 3) + \log(x - 2)\right]$

24. $\dfrac{1}{2}\ln(x^2 + 1) - 4\ln\dfrac{1}{2} - \dfrac{1}{2}\left[\ln(x - 4) + \ln x\right]$

In Problems 25 and 26, use the Change-of-Base Formula and a calculator to evaluate each logarithm. Round your answer to three decimal places.

25. $\log_4 19$

26. $\log_2 21$

In Problems 27–32, find y as a function of x. The constant C is a positive number.

27. $\ln y = 2x^2 + \ln C$

28. $\ln(y - 3) = \ln 2x^2 + \ln C$

29. $\ln(y - 3) + \ln(y + 3) = x + C$

30. $\ln(y - 1) + \ln(y + 1) = -x + C$

31. $e^{y+C} = x^2 + 4$

32. $e^{3y-C} = (x + 4)^2$

In Problems 33–42, use transformations to graph each function. Determine the domain, range, and any asymptotes.

33. $f(x) = 2^{x-3}$

34. $f(x) = -2^x + 3$

35. $f(x) = \dfrac{1}{2}(3^{-x})$

36. $f(x) = 1 + 3^{2x}$

37. $f(x) = 1 - e^x$

38. $f(x) = 3e^x$

39. $f(x) = 3 + \ln x$

40. $f(x) = \frac{1}{2}\ln x$

41. $f(x) = 3 - e^{-x}$

42. $f(x) = 4 - \ln(-x)$

In Problems 43–62, solve each equation.

43. $4^{1-2x} = 2$

44. $8^{6+3x} = 4$

45. $3^{x^2+x} = \sqrt{3}$

46. $4^{x-x^2} = \frac{1}{2}$

47. $\log_x 64 = -3$

48. $\log_{\sqrt{2}} x = -6$

49. $5^x = 3^{x+2}$

50. $5^{x+2} = 7^{x-2}$

51. $9^{2x} = 27^{3x-4}$

52. $25^{2x} = 5^{x^2-12}$

53. $\log_3 \sqrt{x - 2} = 2$

54. $2^{x+1} \cdot 8^{-x} = 4$

55. $8 = 4^{x^2} \cdot 2^{5x}$

56. $2^x \cdot 5 = 10^x$

57. $\log_6(x + 3) + \log_6(x + 4) = 1$

58. $\log_{10}(7x - 12) = 2 \log_{10} x$

59. $e^{1-x} = 5$

60. $e^{1-2x} = 4$

61. $2^{3x} = 3^{2x+1}$

62. $2^{x^3} = 3^{x^2}$

In Problems 63–66, use the following result: If x is the atmospheric pressure (measured in millimeters of mercury), then the formula for the altitude h(x) (measured in meters above sea level) is

$$h(x) = (30T + 8000) \log\left(\frac{P_0}{x}\right)$$

where T is the temperature (in degrees Celsius) and P_0 is the atmospheric pressure at sea level, which is approximately 760 millimeters of mercury.

63. Finding the Altitude of an Airplane At what height is a Piper Cub whose instruments record an outside temperature of 0°C and a barometric pressure of 300 millimeters of mercury?

64. Finding the Height of a Mountain How high is a mountain if instruments placed on its peak record a temperature of 5°C and a barometric pressure of 500 millimeters of mercury?

65. Atmospheric Pressure Outside an Airplane What is the atmospheric pressure outside a Boeing 737 flying at an altitude of 10,000 meters if the outside air temperature is −100°C?

66. Atmospheric Pressure at High Altitudes What is the atmospheric pressure (in millimeters of mercury) on Mt. Everest, which has an altitude of approximately 8900 meters, if the air temperature is 5°C?

67. Amplifying Sound An amplifier's power output P (in watts) is related to its decibel voltage gain d by the formula $P = 25e^{0.1d}$.

(a) Find the power output for a decibel voltage gain of 4 decibels.

(b) For a power output of 50 watts, what is the decibel voltage gain?

68. Limiting Magnitude of a Telescope A telescope is limited in its usefulness by the brightness of the star it is aimed at and by the diameter of its lens. One measure of a star's brightness is its *magnitude*: the dimmer the star, the larger its magnitude. A formula for the limiting magnitude L of a telescope, that is, the magnitude of the dimmest star that it can be used to view, is given by

$$L = 9 + 5.1 \log d$$

where d is the diameter (in inches) of the lens.
(a) What is the limiting magnitude of a 3.5-inch telescope?
(b) What diameter is required to view a star of magnitude 14?

69. Product Demand The demand for a new product increases rapidly at first and then levels off. The percent P of actual purchases of this product after it has been on the market t months is

$$P = 90 - 80\left(\frac{3}{4}\right)^t$$

(a) What is the percent of purchases of the product after 5 months?
(b) What is the percent of purchases of the product after 10 months?
(c) What is the maximum percent of purchases of the product?
(d) How many months does it take before 40% of purchases occur?
(e) How many months before 70% of purchases occur?

70. Disseminating Information A survey of a certain community of 10,000 residents shows that the number of residents N who have heard a piece of information after m months is given by the formula

$$m = 55.3 - 6 \ln(10,000 - N)$$

How many months will it take for half of the citizens to learn about a community program of free blood pressure readings?

71. Salvage Value The number of years n for a piece of machinery to depreciate to a known salvage value can be found using the formula

$$n = \frac{\log s - \log i}{\log(1 - d)}$$

where s is the salvage value of the machinery, i is its initial value, and d is the annual rate of depreciation.
(a) How many years will it take for a piece of machinery to decline in value from $90,000 to $10,000 if the annual rate of depreciation is 0.20 (20%)?
(b) How many years will it take for a piece of machinery to lose half of its value if the annual rate of depreciation is 15%?

72. Funding a College Education A child's grandparents purchase a $10,000 bond fund that matures in 18 years to be used for her college education. The bond fund pays 4% interest compounded semiannually. How much will the bond fund be worth at maturity?

73. Funding a College Education A child's grandparents wish to purchase a bond fund that matures in 18 years to be used for her college education. The bond fund pays 4% interest compounded semiannually. How much should they purchase so that the bond fund will be worth $85,000 at maturity?

74. Funding an IRA First Colonial Bankshares Corporation advertised the following IRA investment plans.

Target IRA Plans

For each $5000 Maturity Value Desired	
Deposit:	At a Term of:
$620.17	20 Years
$1045.02	15 Years
$1760.92	10 Years
$2967.26	5 Years

(a) Assuming continuous compounding, what was the annual rate of interest they offered?
(b) First Colonial Bankshares claims that $4000 invested today will have a value of over $32,000 in 20 years. Use the answer found in part (a) to find the actual value of $4000 in 20 years. Assume continuous compounding.

75. Loudness of a Garbage Disposal Find the loudness of a garbage disposal unit that operates at an intensity of 10^{-4} watt per square meter. Express your answer in decibels.

76. Comparing Earthquakes On September 9, 1985, the western suburbs of Chicago experienced a mild earthquake that registered 3.0 on the Richter scale. How did this earthquake compare in intensity to the great San Francisco earthquake of 1906, which registered 6.9 on the Richter scale?

77. Estimating the Date on Which a Prehistoric Man Died The bones of a prehistoric man found in the desert of New Mexico contain approximately 5% of the original amount of carbon 14. If the half-life of carbon 14 is 5600 years, approximately how long ago did the man die?

78. Temperature of a Skillet A skillet is removed from an oven whose temperature is 450°F and placed in a room whose temperature is 70°F. After 5 minutes, the temperature of the skillet is 400°F. How long will it be until its temperature is 150°F?

79. Biology A certain bacteria initially increases according to the law of uninhibited growth. A biologist collects the following data for this bacteria:

Time (Hours)	Population
0	1000
1	1415
2	2000
3	2828
4	4000
5	5656
6	8000

(a) Draw a scatter diagram.
(b) The exponential function of best fit to the data is found to be

$$y = 1000(\sqrt{2})^x$$

Express the function of best fit in the form $N = N_0 e^{kt}$.
(c) Use the solution to part (b) to predict the population at $t = 7$ hours.
(d) Use a graphing utility to verify the exponential function of best fit.
(e) Use a graphing utility to draw a scatter diagram of the data and then graph the exponential function of best fit on it.

80. Finance The following data represent the amount of money an investor has in an investment account each year for 10 years. She wishes to determine the effective rate of return on her investment.

Year	Value of Account
1985	$10,000
1986	$10,573
1987	$11,260
1988	$11,733
1989	$12,424
1990	$13,269
1991	$13,968
1992	$14,823
1993	$15,297
1994	$16,539

(a) Draw a scatter diagram with the number of years after the initial investment as the independent variable and the value of the account as the dependent variable.
(b) The exponential function of best fit to the data is found to be

$$y = 10,014(1.057)^x$$

Express the function of best fit in the form $A = A_0 e^{kt}$, where $x = t$.
(c) Use the solution to part (b) to estimate the value of the account in the year 2020.
(d) Use a graphing utility to verify the exponential function of best fit.
(e) Use a graphing utility to draw a scatter diagram of the data and then graph the exponential function of best fit on it.

81. World Population According to the U.S. Census Bureau, the growth rate of the world's population in 1997 was $k = 1.33\% = 0.0133$. The population of the world in 1997 was 5,840,445,216. Letting $t = 0$ represent 1997, use the uninhibited growth model to predict the world's population in the year 2002.

82. Radioactive Decay The half-life of radioactive cobalt is 5.27 years. If 100 grams of radioactive cobalt is present now, how much will be present in 20 years? In 40 years?

83. Logistic Growth The logistic growth model

$$P(t) = \frac{0.8}{1 + 1.67e^{-0.16t}}$$

represents the proportion of new computers sold that utilize the Microsoft Windows 98 operating system. Let $t = 0$ represent 1998, $t = 1$ represent 1999, and so on.
(a) What proportion of new computers sold in 1998 utilized Windows 98?
(b) Determine the maximum proportion of new computers sold that will utilize Windows 98.
(c) Using a graphing utility, graph $P(t)$.
(d) When will 75% of new computers sold utilize Windows 98?

Project at Motorola

Thermal Fatigue of Solder Interconnects

What happens to an electronic package when it is subjected to repeated temperature changes? This question is important if you want your electronic product to be reliable. Every time you use your cell phone, pager, computer, or start your car, the electronics inside begin to warm up. When materials warm up, they expand. However, different materials expand at different rates. For example, the glass–epoxy laminate, called a printed circuit board, has a coefficient of thermal expansion (CTE) of 12 to 15 $\times$ 10^{-6}/°C, while the silicon IC located inside an electronic package has a CTE equal to 2.5 $\times$ 10^{-6}/°C. These differences will induce inelastic deformation to the solder interconnect. The solder interconnect makes the electrical connection from the PCB to the electronic package and is usually made of a low-melting-point (183°C) alloy comprised of tin and lead. After using your portable product or computer or turning the car engine off, the electronics will cool off. These temperature cycles result in repeated expansion and contraction of the material used to make the electronic assemblies. The greater the temperature change or the greater the difference in CTE between materials, the greater will be the inelastic strain imparted to the solder joint. A decrease in strain will increase fatigue life. Let's look at a typical situation.

Example of an electronic package (top view) soldered to a PCB used to evaluate fatigue life.

EXPERIMENTAL FATIGUE DATA	
Solder Joint Strain, εp	**Fatigue Cycles, Nf**
0.01	10,000
0.035	1000
0.1	100
0.4	10
1.5	1

Using the experimental fatigue life data given in the table, answer the following:

1. Draw a scatter diagram of the data with solder joint strain as the independent variable.

2. Let $X = \ln(\varepsilon p)$. Draw a scatter diagram of the transformed data with X as the independent variable and fatigue cycles as the dependent variable. What happens to the shape of the scatter diagram?

3. Let $Y = \ln(Nf)$. Draw a scatter diagram of the transformed data with X as the independent variable and Y as the dependent variable. What type of relation would best describe the data?

4. Find the line of best fit to the transformed data using a graphing utility. Graph the line of best fit on the scatter diagram drawn in problem 3.

5. Write the equation from problem 4 in the form $Nf = e^{b}(\varepsilon p)^{m}$.

   [**Hint:** The line of best fit is $\ln(Nf) = m\ln(\varepsilon p) + b$. Think of a property of logarithms that will eliminate the natural logarithm.]

6. If the solder joint strain is 0.02, what is the expected fatigue life? What is the inelastic solder joint strain if the fatigue life is 3000 cycles?

7. Rewrite the function from problem 5 so that the inelastic solder joint strain is a function of the fatigue life. Compute the solder joint strain when the fatigue life equals 3000 cycles

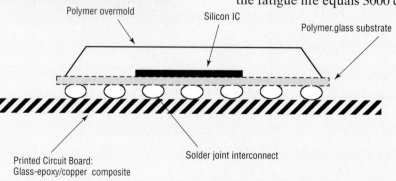

Polymer overmold

Silicon IC

Polymer.glass substrate

Printed Circuit Board:
Glass-epoxy/copper composite

Solder joint interconnect

The Conics

Field Trip to Motorola

One of the many activities that NASA (National Aeronautic and Space Administration) is involved in is placing antennas in orbit. These antennas are usually in the shape of a paraboloid of revolution. In transporting objects of any kind into orbit, cost is a major factor. One factor that reduces the cost is to keep the weight as small as possible. A second factor is to keep the size of the object as small as possible.

As a result, a parabolic reflector is manufactured by NASA for use on the space shuttle by making the antenna out of material that can be folded up. Once in orbit, the folded antenna is deployed and then opened up to assume the form of a parabolic reflector. One problem associated with this method of deployment is to be sure that when the antenna is opened up it aligns properly with the axis of the parabola.

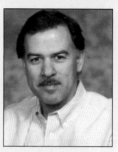
7.1 PRELIMINARIES

OBJECTIVE (1) Know the Names of the Conics

(1) The word *conic* derives from the word *cone*, which is a geometric figure that can be constructed in the following way: Let *a* and *g* be two distinct lines that intersect at a point *V*. Keep the line *a* fixed. Now rotate the line *g* about *a* while maintaining the same angle between *a* and *g*. The collection of points swept out (generated) by the line *g* is called a (**right circular**) **cone.** See Figure 1. The fixed line *a* is called the **axis** of the cone; the point *V* is called its **vertex;** the lines that pass through *V* and make the same angle with *a* as *g* are called **generators** of the cone. Each generator is a line that lies entirely on the cone. The cone consists of two parts, called **nappes,** that intersect at the vertex.

Figure 1

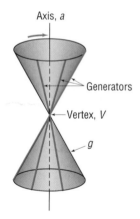

Axis, *a*

Generators

Vertex, *V*

g

    Conics, an abbreviation for **conic sections,** are curves that result from the intersection of a (right circular) cone and a plane. The conics we shall study arise when the plane does not contain the vertex, as shown in Figure 2. These conics are **circles** when the plane is perpendicular to the axis of the cone and intersects each generator; **ellipses** when the plane is tilted slightly so that it intersects each generator, but intersects only one nappe of the cone; **parabolas** when the plane is tilted farther so that it is parallel to one (and only one) generator and intersects only one nappe of the cone; and **hyperbolas** when the plane intersects both nappes.

    If the plane does contain the vertex, the intersection of the plane and the cone is a point, a line, or a pair of intersecting lines. These are usually called **degenerate conics.**

Figure 2

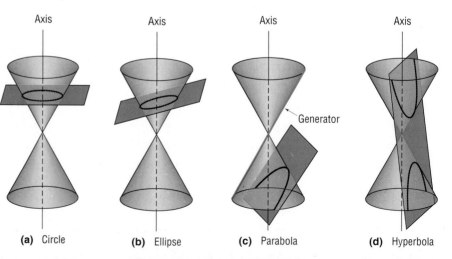

(a) Circle   (b) Ellipse   (c) Parabola   (d) Hyperbola

PREPARING FOR THIS SECTION

Before getting started, review the following:

✓ Distance Formula (Section 2.1, p. 148)

✓ Symmetry (Section 2.2, pp. 159–161)

✓ Completing the Square (Section 1.3, pp. 108–110)

✓ Graphing Techniques: Transformations
  (Section 3.4, pp. 242–252)

7.2 THE PARABOLA

OBJECTIVES
1. Find the Equation of a Parabola
2. Graph Parabolas
3. Discuss the Equation of a Parabola
4. Work with Parabolas with Vertex at (h, k)
5. Solve Applied Problems Involving Parabolas

We stated earlier (Section 4.1) that the graph of a quadratic function is a parabola. In this section, we begin with a geometric definition of a parabola and use it to obtain an equation.

> A **parabola** is defined as the collection of all points P in the plane that are the same distance from a fixed point F as they are from a fixed line D. The point F is called the **focus** of the parabola, and the line D is its **directrix.** As a result, a parabola is the set of points P for which

$$d(F, P) = d(P, D) \qquad \textbf{(1)}$$

1   Figure 3 shows a parabola. The line through the focus F and perpendicular to the directrix D is called the **axis of symmetry** of the parabola. The point of intersection of the parabola with its axis of symmetry is called the **vertex** V.

Figure 3

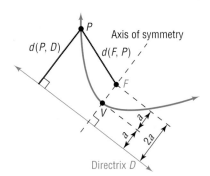

Because the vertex V lies on the parabola, it must satisfy equation (1): $d(F, V) = d(V, D)$. The vertex is midway between the focus and the directrix. We shall let a equal the distance $d(F, V)$ from F to V. Now we are ready to derive an equation for a parabola. To do this, we use a rectangular system of coordinates positioned so that the vertex V, focus F, and directrix D of the parabola are conveniently located. If we choose to locate the vertex V at the origin $(0, 0)$, then we can conveniently position the focus F on either the x-axis or the y-axis.

First, we consider the case where the focus F is on the positive x-axis, as shown in Figure 4. Because the distance from F to V is a, the coordinates of F will be $(a, 0)$ with $a > 0$. Similarly, because the distance from V to the directrix D is also a and because D must be perpendicular to the x-axis (since the x-axis is the axis of symmetry), the equation of the directrix D must be $x = -a$. Now, if $P = (x, y)$ is any point on the parabola, then P must obey equation (1):

$$d(F, P) = d(P, D)$$

Figure 4
$y^2 = 4ax$

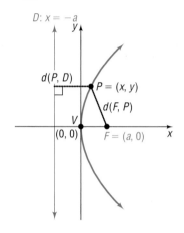

So we have

$$\sqrt{(x - a)^2 + y^2} = |x + a| \qquad \text{Use the distance formula.}$$

$$(x - a)^2 + y^2 = (x + a)^2 \qquad \text{Square both sides.}$$

$$x^2 - 2ax + a^2 + y^2 = x^2 + 2ax + a^2 \qquad \text{Remove parentheses.}$$

$$y^2 = 4ax \qquad \text{Simplify.}$$

Theorem

Equation of a Parabola; Vertex at $(0, 0)$, Focus at $(a, 0)$, $a > 0$

The equation of a parabola with vertex at $(0, 0)$, focus at $(a, 0)$, and directrix $x = -a$, $a > 0$, is

$$y^2 = 4ax \qquad (2)$$

EXAMPLE 1

Finding the Equation of a Parabola and Graphing It

Find an equation of the parabola with vertex at $(0, 0)$ and focus at $(3, 0)$. Graph the equation.

Solution   The distance from the vertex $(0, 0)$ to the focus $(3, 0)$ is $a = 3$. Based on equation (2), the equation of this parabola is

$$y^2 = 4ax$$

$$y^2 = 12x \qquad a = 3$$

Figure 5
$y^2 = 12x$

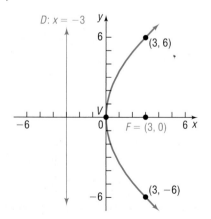

To graph this parabola, it is helpful to plot the two points on the graph above and below the focus. To locate them, we let $x = 3$. Then

$$y^2 = 12x = 12(3) = 36$$

$$y = \pm 6 \qquad \text{Solve for } y.$$

The points on the parabola above and below the focus are $(3, 6)$ and $(3, -6)$. These points help in graphing the parabola because they determine the "opening." See Figure 5.

In general, the points on a parabola $y^2 = 4ax$ that lie above and below the focus $(a, 0)$ are each at a distance $2a$ from the focus. This follows from the fact that if $x = a$ then $y^2 = 4ax = 4a^2$ so $y = \pm 2a$. The line segment joining these two points is called the **latus rectum;** its length is $4a$.

 COMMENT   To graph the parabola $y^2 = 12x$ discussed in Example 1, we need to graph the two functions $Y_1 = \sqrt{12x}$ and $Y_2 = -\sqrt{12x}$. Do this and compare what you see with Figure 5.   ■

⟶   N O W   W O R K   P R O B L E M   **9** .

By reversing the steps we used to obtain equation (2), it follows that the graph of an equation of the form of equation (2), $y^2 = 4ax$, is a parabola; its vertex is at $(0, 0)$, its focus is at $(a, 0)$, its directrix is the line $x = -a$, and its axis of symmetry is the x-axis.

③   For the remainder of this section, the direction "Discuss the equation" will mean to find the vertex, focus, and directrix of the parabola and graph it.

EXAMPLE 2   Discussing the Equation of a Parabola

Discuss the equation: $y^2 = 8x$

Solution   The equation $y^2 = 8x$ is of the form $y^2 = 4ax$, where $4a = 8$ so that $a = 2$. Consequently, the graph of the equation is a parabola with vertex at $(0, 0)$ and focus on the positive x-axis at $(2, 0)$. The directrix is the vertical line $x = -2$. The two points defining the latus rectum are obtained by letting $x = 2$. Then $y^2 = 16$ so $y = \pm 4$. See Figure 6.   ■

Figure 6
$y^2 = 8x$

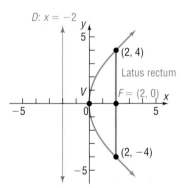

Recall that we arrived at equation (2) after placing the focus on the positive x-axis. If the focus is placed on the negative x-axis, positive y-axis, or negative y-axis, a different form of the equation for the parabola results. The four forms of the equation of a parabola with vertex at $(0, 0)$ and focus on a coordinate axis a distance a from $(0, 0)$ are given in Table 1, and their graphs are given in Figure 7. Notice that each graph is symmetric with respect to its axis of symmetry.

TABLE 1 EQUATIONS OF A PARABOLA: VERTEX AT (0,0); FOCUS ON AXIS; $a > 0$				
Vertex	**Focus**	**Directrix**	**Equation**	**Description**
$(0, 0)$	$(a, 0)$	$x = -a$	$y^2 = 4ax$	Parabola, axis of symmetry is the x-axis, opens to right
$(0, 0)$	$(-a, 0)$	$x = a$	$y^2 = -4ax$	Parabola, axis of symmetry is the x-axis, opens to left
$(0, 0)$	$(0, a)$	$y = -a$	$x^2 = 4ay$	Parabola, axis of symmetry is the y-axis, opens up
$(0, 0)$	$(0, -a)$	$y = a$	$x^2 = -4ay$	Parabola, axis of symmetry is the y-axis, opens down

Figure 7

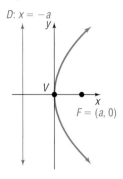

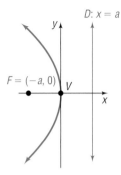

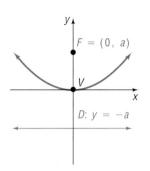

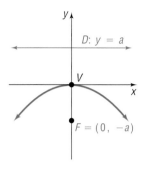

(a) $y^2 = 4ax$ **(b)** $y^2 = -4ax$ **(c)** $x^2 = 4ay$ **(d)** $x^2 = -4ay$

EXAMPLE 3

Discussing the Equation of a Parabola

Discuss the equation: $x^2 = -12y$

Figure 8
$x^2 = -12y$

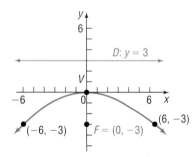

Solution   The equation $x^2 = -12y$ is of the form $x^2 = -4ay$, with $a = 3$. Consequently, the graph of the equation is a parabola with vertex at $(0, 0)$, focus at $(0, -3)$ and directrix the line $y = 3$. The parabola opens down, and its axis of symmetry is the y-axis. To obtain the points defining the latus rectum, let $y = -3$. Then $x^2 = 36$ so $x = \pm 6$. See Figure 8.   ∎

> NOW WORK PROBLEM 27.

EXAMPLE 4

Finding the Equation of a Parabola

Find the equation of the parabola with focus at $(0, 4)$ and directrix the line $y = -4$. Graph the equation.

Figure 9
$x^2 = 16y$

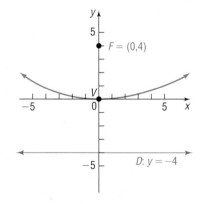

Solution   A parabola whose focus is at $(0, 4)$ and whose directrix is the horizontal line $y = -4$ will have its vertex at $(0, 0)$. (Do you see why? The vertex is midway between the focus and the directrix.) Since the focus is on the positive y-axis at $(0, 4)$, the equation of this parabola is of the form $x^2 = 4ay$, with $a = 4$; that is,

$$x^2 = 4ay = 4(4)y = 16y$$
$$\uparrow$$
$$a = 4$$

Figure 9 shows the graph of $x^2 = 16y$.   ∎

EXAMPLE 5	**Finding the Equation of a Parabola**

Find the equation of a parabola with vertex at $(0, 0)$ if its axis of symmetry is the x-axis and its graph contains the point $\left(-\frac{1}{2}, 2\right)$. Find its focus and directrix, and graph the equation.

Solution   The vertex is at the origin, the axis of symmetry is the x-axis, and the graph contains a point in the second quadrant, so the parabola opens to the left. We see from Table 1 that the form of the equation is

$$y^2 = -4ax$$

Because the point $\left(-\frac{1}{2}, 2\right)$ is on the parabola, the coordinates $x = -\frac{1}{2}$, $y = 2$ must satisfy the equation. Substituting $x = -\frac{1}{2}$ and $y = 2$ into the equation, we find that

$$4 = -4a\left(-\frac{1}{2}\right)$$
$$a = 2$$

The equation of the parabola is

$$y^2 = -4(2)x = -8x$$

The focus is at $(-2, 0)$ and the directrix is the line $x = 2$. Letting $x = -2$, we find $y^2 = 16$ so $y = \pm 4$. The points $(-2, 4)$ and $(-2, -4)$ define the latus rectum. See Figure 10. ■

Figure 10
$y^2 = -8x$

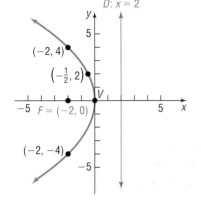

━━── NOW WORK PROBLEM **19.**

VERTEX AT (h, k)

④   If a parabola with vertex at the origin and axis of symmetry along a coordinate axis is shifted horizontally h units and then vertically k units, the result is a parabola with vertex at (h, k) and axis of symmetry parallel to a coordinate axis. The equations of such parabolas have the same forms as those in Table 1, but with x replaced by $x - h$ (the horizontal shift) and y replaced by $y - k$ (the vertical shift). Table 2 gives the forms of the equations of such parabolas. Figure 11(a)–(d) illustrates the graphs for $h > 0$, $k > 0$.

TABLE 2
PARABOLAS WITH VERTEX AT (h, k); AXIS OF SYMMETRY PARALLEL TO A COORDINATE AXIS, $a > 0$

Vertex	Focus	Directrix	Equation	Description
(h, k)	$(h + a, k)$	$x = h - a$	$(y - k)^2 = 4a(x - h)$	Parabola, axis of symmetry parallel to x-axis, opens to right
(h, k)	$(h - a, k)$	$x = h + a$	$(y - k)^2 = -4a(x - h)$	Parabola, axis of symmetry parallel to x-axis, opens to left
(h, k)	$(h, k + a)$	$y = k - a$	$(x - h)^2 = 4a(y - k)$	Parabola, axis of symmetry parallel to y-axis, opens up
(h, k)	$(h, k - a)$	$y = k + a$	$(x - h)^2 = -4a(y - k)$	Parabola, axis of symmetry parallel to y-axis, opens down

Figure 11

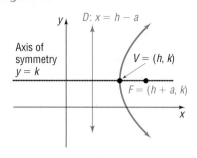

(a) $(y - k)^2 = 4a(x - h)$

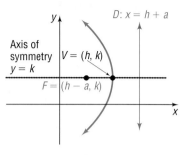

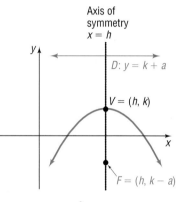

(b) $(y - k)^2 = -4a(x - h)$

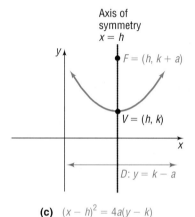

(c) $(x - h)^2 = 4a(y - k)$

Axis of symmetry $x = h$

$D: y = k + a$

$V = (h, k)$

$F = (h, k - a)$

(d) $(x - h)^2 = -4a(y - k)$

EXAMPLE 6 — Finding the Equation of a Parabola, Vertex Not at Origin

Find an equation of the parabola with vertex at $(-2, 3)$ and focus at $(0, 3)$. Graph the equation.

Figure 12
$(y - 3)^2 = 8(x + 2)$

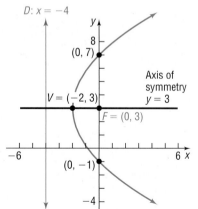

Solution The vertex $(-2, 3)$ and focus $(0, 3)$ both lie on the horizontal line $y = 3$ (the axis of symmetry). The distance a from the vertex $(-2, 3)$ to the focus $(0, 3)$ is $a = 2$. Also, because the focus lies to the right of the vertex, we know that the parabola opens to the right. Consequently, the form of the equation is

$$(y - k)^2 = 4a(x - h)$$

where $(h, k) = (-2, 3)$ and $a = 2$. Therefore, the equation is

$$(y - 3)^2 = 4 \cdot 2[x - (-2)]$$
$$(y - 3)^2 = 8(x + 2)$$

If $x = 0$ then $(y - 3)^2 = 16$. Then, $y - 3 = \pm 4$ so $y = -1$ or $y = 7$. The points $(0, -1)$ and $(0, 7)$ define the latus rectum; the line $x = -4$ is the directrix. See Figure 12. ■

NOW WORK PROBLEM **17.**

Polynomial equations define parabolas whenever they involve two variables that are quadratic in one variable and linear in the other. To discuss this type of equation, we first complete the square of the variable that is quadratic.

EXAMPLE 7	**Discussing the Equation of a Parabola**

Discuss the equation: $x^2 + 4x - 4y = 0$

Solution   To discuss the equation $x^2 + 4x - 4y = 0$, we complete the square involving the variable x.

$$x^2 + 4x - 4y = 0$$

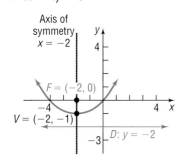

Figure 13
$x^2 + 4x - 4y = 0$

Axis of
symmetry,
$x = -2$

$F = (-2, 0)$

$V = (-2, -1)$

$D: y = -2$

$$x^2 + 4x = 4y \qquad \text{Isolate the terms involving } x \text{ on the left side.}$$

$$x^2 + 4x + 4 = 4y + 4 \qquad \text{Complete the square on the left side.}$$

$$(x + 2)^2 = 4(y + 1) \qquad \text{Factor.}$$

This equation is of the form $(x - h)^2 = 4a(y - k)$, with $h = -2$, $k = -1$, and $a = 1$. The graph is a parabola with vertex at $(h, k) = (-2, -1)$ that opens up. The focus is at $(-2, 0)$, and the directrix is the line $y = -2$. See Figure 13.   ■

⎯ NOW WORK PROBLEM 35.

⑤   Parabolas find their way into many applications. For example, as we discussed in Section 4.1, suspension bridges have cables in the shape of a parabola. Another property of parabolas that is used in applications is their reflecting property.

REFLECTING PROPERTY

Suppose that a mirror is shaped like a **paraboloid of revolution,** a surface formed by rotating a parabola about its axis of symmetry. If a light (or any other emitting source) is placed at the focus of the parabola, all the rays emanating from the light will reflect off the mirror in lines parallel to the axis of symmetry. This principle is used in the design of searchlights, flashlights, certain automobile headlights, and other such devices. See Figure 14.

Conversely, suppose that rays of light (or other signals) emanate from a distant source so that they are essentially parallel. When these rays strike the surface of a parabolic mirror whose axis of symmetry is parallel to these rays, they are reflected to a single point at the focus. This principle is used in the design of some solar energy devices, satellite dishes, and the mirrors used in some types of telescopes. See Figure 15.

Figure 14
Searchlight

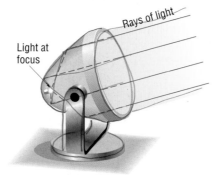

Rays of light

Light at
focus

Figure 15
Telescope

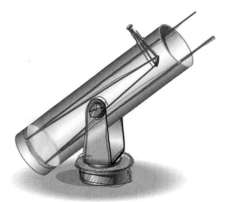

| EXAMPLE 8 | **Satellite Dish** |

A satellite dish is shaped like a paraboloid of revolution. The signals that emanate from a satellite strike the surface of the dish and are reflected to a single point, where the receiver is located. If the dish is 8 feet across at its opening and is 3 feet deep at its center, at what position should the receiver be placed?

Solution   Figure 16(a) shows the satellite dish. We draw the parabola used to form the dish on a rectangular coordinate system so that the vertex of the parabola is at the origin and its focus is on the positive *y*-axis. See Figure 16(b).

Figure 16

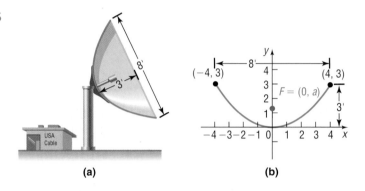

(a)   (b)

The form of the equation of the parabola is

$$x^2 = 4ay$$

and its focus is at $(0, a)$. Since $(4, 3)$ is a point on the graph, we have

$$4^2 = 4a(3)$$

$$a = \frac{4}{3}$$

The receiver should be located $1\frac{1}{3}$ feet from the base of the dish, along its axis of symmetry.   ■

NOW WORK PROBLEM **51**.

7.2 EXERCISES

In Problems 1–8, the graph of a parabola is given. Match each graph to its equation.

A. $y^2 = 4x$ C. $y^2 = -4x$ E. $(y - 1)^2 = 4(x - 1)$ G. $(y - 1)^2 = -4(x - 1)$
B. $x^2 = 4y$ D. $x^2 = -4y$ F. $(x + 1)^2 = 4(y + 1)$ H. $(x + 1)^2 = -4(y + 1)$

1.   **2.**   **3.**   **4.**

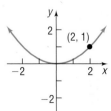   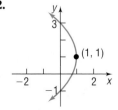   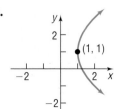   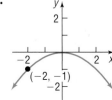

5.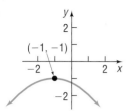

6.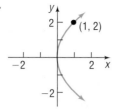

7.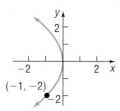

8.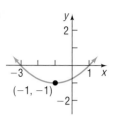

In Problems 9–24, find the equation of the parabola described. Find the two points that define the latus rectum, and graph the equation.

9. Focus at $(4, 0)$;   vertex at $(0, 0)$

10. Focus at $(0, 2)$;   vertex at $(0, 0)$

11. Focus at $(0, -3)$;   vertex at $(0, 0)$

12. Focus at $(-4, 0)$;   vertex at $(0, 0)$

13. Focus at $(-2, 0)$;   directrix the line $x = 2$

14. Focus at $(0, -1)$;   directrix the line $y = 1$

15. Directrix the line $y = -\frac{1}{2}$;   vertex at $(0, 0)$

16. Directrix the line $x = -\frac{1}{2}$;   vertex at $(0, 0)$

17. Vertex at $(2, -3)$;   focus at $(2, -5)$

18. Vertex at $(4, -2)$;   focus at $(6, -2)$

19. Vertex at $(0, 0)$;   axis of symmetry the y-axis;   containing the point $(2, 3)$

20. Vertex at $(0, 0)$;   axis of symmetry the x-axis;   containing the point $(2, 3)$

21. Focus at $(-3, 4)$;   directrix the line $y = 2$

22. Focus at $(2, 4)$;   directrix the line $x = -4$

23. Focus at $(-3, -2)$;   directrix the line $x = 1$

24. Focus at $(-4, 4)$;   directrix the line $y = -2$

In Problems 25–42, find the vertex, focus, and directrix of each parabola. Graph the equation.

25. $x^2 = 4y$

26. $y^2 = 8x$

27. $y^2 = -16x$

28. $x^2 = -4y$

29. $(y - 2)^2 = 8(x + 1)$

30. $(x + 4)^2 = 16(y + 2)$

31. $(x - 3)^2 = -(y + 1)$

32. $(y + 1)^2 = -4(x - 2)$

33. $(y + 3)^2 = 8(x - 2)$

34. $(x - 2)^2 = 4(y - 3)$

35. $y^2 - 4y + 4x + 4 = 0$

36. $x^2 + 6x - 4y + 1 = 0$

37. $x^2 + 8x = 4y - 8$

38. $y^2 - 2y = 8x - 1$

39. $y^2 + 2y - x = 0$

40. $x^2 - 4x = 2y$

41. $x^2 - 4x = y + 4$

42. $y^2 + 12y = -x + 1$

In Problems 43–50, write an equation for each parabola.

43.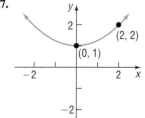

44.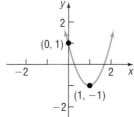

45.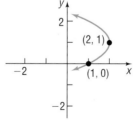

46.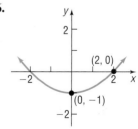

47.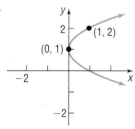

48.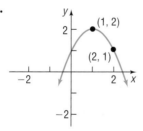

49.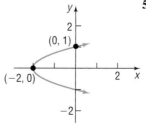

50.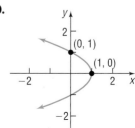

51. **Satellite Dish** A satellite dish is shaped like a paraboloid of revolution. The signals that emanate from a satellite strike the surface of the dish and are reflected to a single point, where the receiver is located. If the dish is 10 feet across at its opening and is 4 feet deep at its center, at what position should the receiver be placed?

52. **Constructing a TV Dish** A cable TV receiving dish is in the shape of a paraboloid of revolution. Find the location of the receiver, which is placed at the focus, if the dish is 6 feet across at its opening and 2 feet deep.

53. **Constructing a Flashlight** The reflector of a flashlight is in the shape of a paraboloid of revolution. Its diameter is 4 inches and its depth is 1 inch. How far from the vertex should the light bulb be placed so that the rays will be reflected parallel to the axis?

54. **Constructing a Headlight** A sealed-beam headlight is in the shape of a paraboloid of revolution. The bulb, which is placed at the focus, is 1 inch from the vertex. If the depth is to be 2 inches, what is the diameter of the headlight at its opening?

55. **Suspension Bridge** The cables of a suspension bridge are in the shape of a parabola, as shown in the figure. The towers supporting the cable are 600 feet apart and 80 feet high. If the cables touch the road surface midway between the towers, what is the height of the cable at a point 150 feet from the center of the bridge?

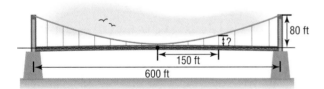

56. **Suspension Bridge** The cables of a suspension bridge are in the shape of a parabola. The towers supporting the cable are 400 feet apart and 100 feet high. If the cables are at a height of 10 feet midway between the towers, what is the height of the cable at a point 50 feet from the center of the bridge?

57. **Searchlight** A searchlight is shaped like a paraboloid of revolution. If the light source is located 2 feet from the base along the axis of symmetry and the opening is 5 feet across, how deep should the searchlight be?

58. **Searchlight** A searchlight is shaped like a paraboloid of revolution. If the light source is located 2 feet from the base along the axis of symmetry and the depth of the searchlight is 4 feet, what should the width of the opening be?

59. **Solar Heat** A mirror is shaped like a paraboloid of revolution and will be used to concentrate the rays of the sun at its focus, creating a heat source. If the mirror is 20 feet across at its opening and is 6 feet deep, where will the heat source be concentrated?

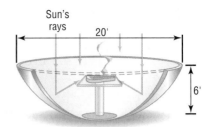

60. **Reflecting Telescope** A reflecting telescope contains a mirror shaped like a paraboloid of revolution. If the mirror is 4 inches across at its opening and is 3 feet deep, where will the collected light be concentrated?

61. **Parabolic Arch Bridge** A bridge is built in the shape of a parabolic arch. The bridge has a span of 120 feet and a maximum height of 25 feet. See the illustration. Choose a suitable rectangular coordinate system and find the height of the arch at distances of 10, 30, and 50 feet from the center.

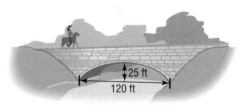

62. **Parabolic Arch Bridge** A bridge is to be built in the shape of a parabolic arch and is to have a span of 100 feet. The height of the arch a distance of 40 feet from the center is to be 10 feet. Find the height of the arch at its center.

63. Show that an equation of the form
$$Ax^2 + Ey = 0, \qquad A \neq 0, E \neq 0$$
is the equation of a parabola with vertex at $(0, 0)$ and axis of symmetry the y-axis. Find its focus and directrix.

64. Show that an equation of the form
$$Cy^2 + Dx = 0, \qquad C \neq 0, D \neq 0$$
is the equation of a parabola with vertex at $(0, 0)$ and axis of symmetry the x-axis. Find its focus and directrix.

65. Show that the graph of an equation of the form
$$Ax^2 + Dx + Ey + F = 0, \qquad A \neq 0$$
   (a) Is a parabola if $E \neq 0$.
   (b) Is a vertical line if $E = 0$ and $D^2 - 4AF = 0$.
   (c) Is two vertical lines if $E = 0$ and $D^2 - 4AF > 0$.
   (d) Contains no points if $E = 0$ and $D^2 - 4AF < 0$.

66. Show that the graph of an equation of the form
$$Cy^2 + Dx + Ey + F = 0, \qquad C \neq 0$$
   (a) Is a parabola if $D \neq 0$.
   (b) Is a horizontal line if $D = 0$ and $E^2 - 4CF = 0$.
   (c) Is two horizontal lines if $D = 0$ and $E^2 - 4CF > 0$.
   (d) Contains no points if $D = 0$ and $E^2 - 4CF < 0$.

PREPARING   FOR   THIS   SECTION

Before getting started, review the following:

✓ Distance Formula (Section 2.1, p. 148)

✓ Completing the Square (Section 1.3, pp. 108–110)

✓ Intercepts (Section 2.2, pp. 157–158)

✓ Symmetry (Section 2.2, pp. 159–161)

✓ Circles (Section 2.4, pp. 178–181)

✓ Graphing Techniques: Transformations
  (Section 3.4, pp. 242–252)

7.3   THE ELLIPSE

OBJECTIVES   **1**   Find the Equation of an Ellipse

2   Graph Ellipses

3   Discuss the Equation of an Ellipse

4   Work with Ellipses with Center at (h, k)

5   Solve Applied Problems Involving Ellipses

> An **ellipse** is the collection of all points in the plane the sum of whose distances from two fixed points, called the **foci**, is a constant.

Figure 17

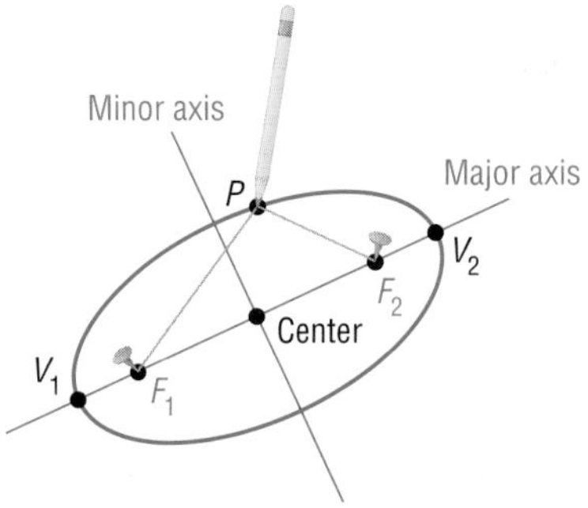

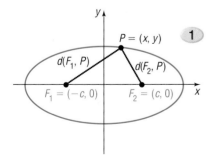

The definition actually contains within it a physical means for drawing an ellipse. Find a piece of string (the length of this string is the constant referred to in the definition). Then take two thumbtacks (the foci) and stick them on a piece of cardboard so that the distance between them is less than the length of the string. Now attach the ends of the string to the thumbtacks and, using the point of a pencil, pull the string taut. See Figure 17. Keeping the string taut, rotate the pencil around the two thumbtacks. The pencil traces out an ellipse, as shown in Figure 17.

In Figure 17, the foci are labeled F_1 and F_2. The line containing the foci is called the **major axis.** The midpoint of the line segment joining the foci is called the **center** of the ellipse. The line through the center and perpendicular to the major axis is called the **minor axis.**

The two points of intersection of the ellipse and the major axis are the **vertices,** V_1 and V_2, of the ellipse. The distance from one vertex to the other is called the **length of the major axis.** The ellipse is symmetric with respect to its major axis and with respect to its minor axis.

With these ideas in mind, we are now ready to find the equation of an ellipse in a rectangular coordinate system. First, we place the center of the ellipse at the origin. Second, we position the ellipse so that its major axis coincides with a coordinate axis. Suppose that the major axis coincides with the x-axis, as shown in Figure 18. If c is the distance from the center to a focus, then one focus will be at $F_1 = (-c, 0)$ and the other at $F_2 = (c, 0)$. As we shall see, it is convenient to let $2a$ denote the constant distance referred to in the definition. Then, if $P = (x, y)$ is any point on the ellipse, we have

Figure 18
$$d(F_1, P) + d(F_2, P) = 2a$$

$$d(F_1, P) + d(F_2, P) = 2a \qquad \text{Sum of the distances from } P \text{ to the foci equals a constant, } 2a.$$

$$\sqrt{(x + c)^2 + y^2} + \sqrt{(x - c)^2 + y^2} = 2a \qquad \text{Use the distance formula.}$$

$$\sqrt{(x + c)^2 + y^2} = 2a - \sqrt{(x - c)^2 + y^2} \qquad \text{Isolate one radical.}$$

$$(x + c)^2 + y^2 = 4a^2 - 4a\sqrt{(x - c)^2 + y^2}$$ Square both sides.
$$+ (x - c)^2 + y^2$$

$$x^2 + 2cx + c^2 + y^2 = 4a^2 - 4a\sqrt{(x - c)^2 + y^2}$$ Remove parentheses.
$$+ x^2 - 2cx + c^2 + y^2$$

$$4cx - 4a^2 = -4a\sqrt{(x - c)^2 + y^2}$$ Simplify; Isolate the radical.

$$cx - a^2 = -a\sqrt{(x - c)^2 + y^2}$$ Divide each side by 4.

$$(cx - a^2)^2 = a^2\left[(x - c)^2 + y^2\right]$$ Square both sides again.

$$c^2x^2 - 2a^2cx + a^4 = a^2(x^2 - 2cx + c^2 + y^2)$$ Remove parentheses.

$$(c^2 - a^2)x^2 - a^2y^2 = a^2c^2 - a^4$$ Rearrange the terms.

$$(a^2 - c^2)x^2 + a^2y^2 = a^2(a^2 - c^2)$$ Multiply each side by -1; factor a^2 on the right side.   **(1)**

To obtain points on the ellipse off the x-axis, it must be that $a > c$. To see why, look again at Figure 18.

$$d(F_1, P) + d(F_2, P) > d(F_1, F_2)$$ The sum of the lengths of two sides of a triangle is greater than the length of the third side.

$$2a > 2c$$ $d(F_1, P) + d(F_2, P) = 2a; d(F_1, F_2) = 2c.$

$$a > c$$

Since $a > c$, we also have $a^2 > c^2$, so $a^2 - c^2 > 0$. Let $b^2 = a^2 - c^2, b > 0$. Then $a > b$ and equation (1) can be written as

$$b^2x^2 + a^2y^2 = a^2b^2$$

$$\frac{x^2}{a^2} + \frac{y^2}{b^2} = 1$$ Divide each side by a^2b^2.

Theorem   **Equation of an Ellipse; Center at (0, 0); Foci at ($\pm c$, 0); Major Axis along the x-Axis**

An equation of the ellipse with center at $(0, 0)$ and foci at $(-c, 0)$ and $(c, 0)$ is

$$\frac{x^2}{a^2} + \frac{y^2}{b^2} = 1, \quad \text{where } a > b > 0 \text{ and } b^2 = a^2 - c^2 \quad \textbf{(2)}$$

The major axis is the x-axis. The vertices are at $(-a, 0)$ and $(a, 0)$.

Figure 19

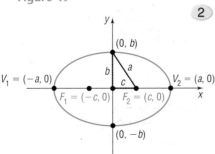

As you can verify, the ellipse defined by equation (2) is symmetric with respect to the x-axis, y-axis, and origin.

To find the vertices of the ellipse defined by equation (2), let $y = 0$. The vertices satisfy the equation $x^2/a^2 = 1$, the solutions of which are $x = \pm a$. Consequently, the vertices of the ellipse given by equation (2) are $V_1 = (-a, 0)$ and $V_2 = (a, 0)$. The y-intercepts of the ellipse, found by letting $x = 0$, have coordinates $(0, -b)$ and $(0, b)$. These four intercepts, $(a, 0)$, $(-a, 0)$, $(0, b)$, and $(0, -b)$, are used to graph the ellipse. See Figure 19.

Notice in Figure 19 the right triangle formed with the points $(0, 0)$, $(c, 0)$, and $(0, b)$. Because $b^2 = a^2 - c^2$ (or $b^2 + c^2 = a^2$), the distance from the focus at $(c, 0)$ to the point $(0, b)$ is a.

| EXAMPLE 1 | Finding an Equation of an Ellipse |

Find an equation of the ellipse with center at the origin, one focus at $(3, 0)$, and a vertex at $(-4, 0)$. Graph the equation.

Solution The ellipse has its center at the origin and, since the given focus and vertex lie on the x-axis, the major axis is the x-axis. The distance from the center, $(0, 0)$, to one of the foci, $(3, 0)$, is $c = 3$. The distance from the center, $(0, 0)$, to one of the vertices, $(-4, 0)$, is $a = 4$. From equation (2), it follows that

$$b^2 = a^2 - c^2 = 16 - 9 = 7$$

so an equation of the ellipse is

$$\frac{x^2}{16} + \frac{y^2}{7} = 1$$

Figure 20 shows the graph.   ■

Figure 20

$\dfrac{x^2}{16} + \dfrac{y^2}{7} = 1$

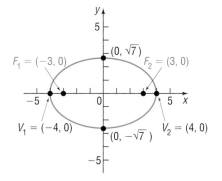

Notice in Figure 20 how we used the intercepts of the equation to graph the ellipse. Following this practice will make it easier for you to obtain an accurate graph of an ellipse.

COMMENT The intercepts of the ellipse also provide information about how to set the viewing rectangle. To graph the ellipse

$$\frac{x^2}{16} + \frac{y^2}{7} = 1$$

discussed in Example 1, we would set the viewing rectangle using a square screen that includes the intercepts, perhaps $-4.5 \leq x \leq 4.5$, $-3 \leq y \leq 3$. Then we would proceed to solve the equation for y:

$$\frac{x^2}{16} + \frac{y^2}{7} = 1$$

$$\frac{y^2}{7} = 1 - \frac{x^2}{16} \qquad \text{Subtract } \frac{x^2}{16} \text{ from each side.}$$

$$y^2 = 7\left(1 - \frac{x^2}{16}\right) \qquad \text{Multiply both sides by 7.}$$

$$y = \pm\sqrt{7\left(1 - \frac{x^2}{16}\right)} \qquad \text{Take the square root of each side.}$$

Figure 21

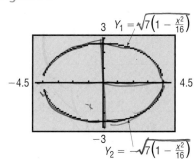

$3 \quad Y_1 = \sqrt{7\left(1 - \frac{x^2}{16}\right)}$

$-4.5 \qquad\qquad 4.5$

-3

$Y_2 = -\sqrt{7\left(1 - \frac{x^2}{16}\right)}$

Now graph the two functions

$$Y_1 = \sqrt{7\left(1 - \frac{x^2}{16}\right)} \quad \text{and} \quad Y_2 = -\sqrt{7\left(1 - \frac{x^2}{16}\right)}$$

Figure 21 shows the result.   ■

NOW WORK PROBLEM 15.

An equation of the form of equation (2), with $a > b$, is the equation of an ellipse with center at the origin, foci on the x-axis at $(-c, 0)$ and $(c, 0)$, where $c^2 = a^2 - b^2$, and major axis along the x-axis.

3   For the remainder of this section, the direction "Discuss the equation" will mean to find the center, major axis, foci, and vertices of the ellipse and graph it.

EXAMPLE 2   **Discussing the Equation of an Ellipse**

Discuss the equation: $\dfrac{x^2}{25} + \dfrac{y^2}{9} = 1$

Solution   The given equation is of the form of equation (2), with $a^2 = 25$ and $b^2 = 9$. The equation is that of an ellipse with center $(0, 0)$ and major axis along the x-axis. The vertices are at $(\pm a, 0) = (\pm 5, 0)$. Because $b^2 = a^2 - c^2$, we find that

$$c^2 = a^2 - b^2 = 25 - 9 = 16$$

The foci are at $(\pm c, 0) = (\pm 4, 0)$. Figure 22 shows the graph.

Figure 22

$\dfrac{x^2}{25} + \dfrac{y^2}{9} = 1$

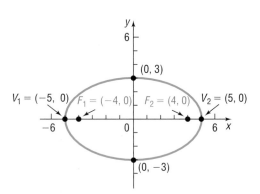

NOW WORK PROBLEM **5.**

If the major axis of an ellipse with center at $(0, 0)$ lies on the y-axis, then the foci are at $(0, -c)$ and $(0, c)$. Using the same steps as before, the definition of an ellipse leads to the following result:

Theorem   **Equation of an Ellipse; Center at (0, 0); Foci at (0, ± c); Major Axis along the y-Axis**

An equation of the ellipse with center at $(0, 0)$ and foci at $(0, -c)$ and $(0, c)$ is

$$\frac{x^2}{b^2} + \frac{y^2}{a^2} = 1, \qquad \text{where } a > b > 0 \text{ and } b^2 = a^2 - c^2 \qquad \textbf{(3)}$$

The major axis is the y-axis; the vertices are at $(0, -a)$ and $(0, a)$.

Figure 23

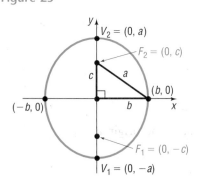

Figure 23 illustrates the graph of such an ellipse. Again, notice the right triangle with the points at $(0, 0)$, $(b, 0)$, and $(0, c)$.

Look closely at equations (2) and (3). Although they may look alike, there is a difference! In equation (2), the larger number, a^2, is in the denominator of the x^2-term, so the major axis of the ellipse is along the x-axis. In equation (3), the larger number, a^2, is in the denominator of the y^2-term, so the major axis is along the y-axis.

EXAMPLE 3   **Discussing the Equation of an Ellipse**

Discuss the equation: $9x^2 + y^2 = 9$

Solution   To put the equation in proper form, we divide each side by 9.

$$x^2 + \frac{y^2}{9} = 1$$

Figure 24
$$x^2 + \frac{y^2}{9} = 1$$

The larger number, 9, is in the denominator of the y^2-term so, based on equation (3), this is the equation of an ellipse with center at the origin and major axis along the y-axis. Also, we conclude that $a^2 = 9$, $b^2 = 1$, and $c^2 = a^2 - b^2 = 9 - 1 = 8$. The vertices are at $(0, \pm a) = (0, \pm 3)$, and the foci are at $(0, \pm c) = (0, \pm 2\sqrt{2})$. The graph is given in Figure 24.   ∎

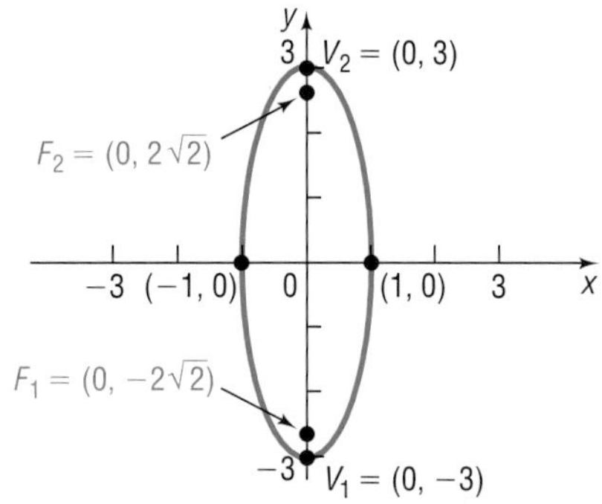

NOW WORK PROBLEM **9**.

EXAMPLE 4   **Finding an Equation of an Ellipse**

Find an equation of the ellipse having one focus at $(0, 2)$ and vertices at $(0, -3)$ and $(0, 3)$. Graph the equation.

Solution   Because the vertices are at $(0, -3)$ and $(0, 3)$, the center of this ellipse is at their midpoint, the origin. Also, its major axis lies on the y-axis. The distance from the center, $(0, 0)$, to one of the foci, $(0, 2)$, is $c = 2$. The distance from the center, $(0, 0)$, to one of the vertices, $(0, 3)$, is $a = 3$. So $b^2 = a^2 - c^2 = 9 - 4 = 5$. The form of the equation of this ellipse is given by equation (3).

Figure 25
$$\frac{x^2}{5} + \frac{y^2}{9} = 1$$

$$\frac{x^2}{b^2} + \frac{y^2}{a^2} = 1$$

$$\frac{x^2}{5} + \frac{y^2}{9} = 1$$

Figure 25 shows the graph.   ∎

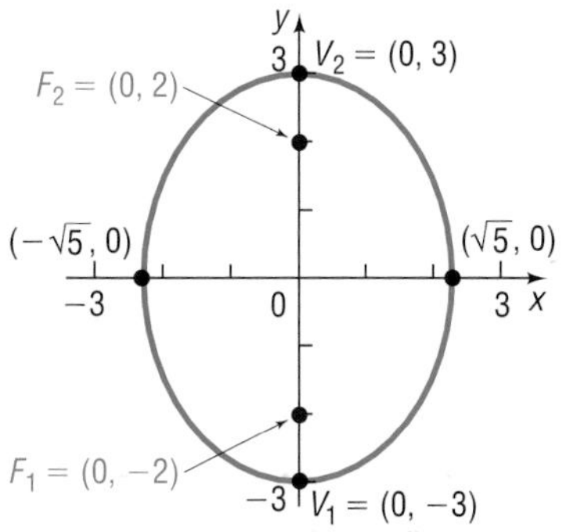

NOW WORK PROBLEM **17**.

The circle may be considered a special kind of ellipse. To see why, let $a = b$ in equation (2) or (3). Then

$$\frac{x^2}{a^2} + \frac{y^2}{a^2} = 1$$

$$x^2 + y^2 = a^2$$

This is the equation of a circle with center at the origin and radius a. The value of c is

$$c^2 = a^2 - b^2 = 0$$

We conclude that the closer the two foci of an ellipse are to the center, the more the ellipse will look like a circle.

CENTER AT (h, k)

④ If an ellipse with center at the origin and major axis coinciding with a coordinate axis is shifted horizontally h units and then vertically k units, the result is an ellipse with center at (h, k) and major axis parallel to a coordinate axis. The equations of such ellipses have the same forms as those given in equations (2) and (3), except that x is replaced by $x - h$ (the horizontal shift) and y is replaced by $y - k$ (the vertical shift). Table 3 gives the forms of the equations of such ellipses, and Figure 26 shows their graphs.

TABLE 3
ELLIPSES WITH CENTER AT (h, k) AND MAJOR AXIS PARALLEL TO A COORDINATE AXIS

Center	Major Axis	Foci	Vertices	Equation
(h, k)	Parallel to x-axis	$(h + c, k)$	$(h + a, k)$	$\dfrac{(x - h)^2}{a^2} + \dfrac{(y - k)^2}{b^2} = 1,$
		$(h - c, k)$	$(h - a, k)$	$a > b$ and $b^2 = a^2 - c^2$
(h, k)	Parallel to y-axis	$(h, k + c)$	$(h, k + a)$	$\dfrac{(x - h)^2}{b^2} + \dfrac{(y - k)^2}{a^2} = 1,$
		$(h, k - c)$	$(h, k - a)$	$a > b$ and $b^2 = a^2 - c^2$

Figure 26

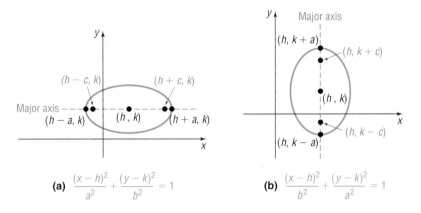

(a) $\dfrac{(x - h)^2}{a^2} + \dfrac{(y - k)^2}{b^2} = 1$

(b) $\dfrac{(x - h)^2}{b^2} + \dfrac{(y - k)^2}{a^2} = 1$

EXAMPLE 5 | **Finding an Equation of an Ellipse, Center Not at the Origin**

Find an equation for the ellipse with center at $(2, -3)$, one focus at $(3, -3)$, and one vertex at $(5, -3)$. Graph the equation.

Solution The center is at $(h, k) = (2, -3)$, so $h = 2$ and $k = -3$. Since the center, focus, and vertex all lie on the line $y = -3$, the major axis is parallel to the x-axis. The distance from the center $(2, -3)$ to a focus $(3, -3)$ is $c = 1$; the distance from the center $(2, -3)$ to a vertex $(5, -3)$ is $a = 3$. Then, $b^2 = a^2 - c^2 = 9 - 1 = 8$. The form of the equation is

$$\frac{(x - h)^2}{a^2} + \frac{(y - k)^2}{b^2} = 1, \qquad \text{where } h = 2, k = -3, a = 3, b = 2\sqrt{2}$$

$$\frac{(x - 2)^2}{9} + \frac{(y + 3)^2}{8} = 1$$

To graph the equation, we use the center $(h, k) = (2, -3)$ to locate the vertices. The major axis is parallel to the x-axis, so the vertices are $a = 3$ units left and right of the center $(2, -3)$. Therefore, the vertices are

$$V_1 = (2 - 3, -3) = (-1, -3) \quad \text{and} \quad V_2 = (2 + 3, -3) = (5, -3)$$

Figure 27

$$\frac{(x - 2)^2}{9} + \frac{(y + 3)^2}{8} = 1$$

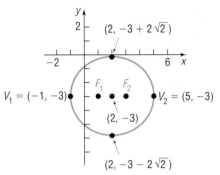

Since $c = 1$ and the major axis is parallel to the x-axis, the foci are 1 unit left and right of the center. Therefore, the foci are

$$F_1 = (2 - 1, -3) = (1, -3) \quad \text{and} \quad F_2 = (2 + 1, -3) = (3, -3)$$

Finally, we use the value of $b = 2\sqrt{2}$ to find the two points above and below the center.

$$\left(2, -3 - 2\sqrt{2}\right) \quad \text{and} \quad \left(2, -3 + 2\sqrt{2}\right)$$

Figure 27 shows the graph.

NOW WORK PROBLEM **41.**

EXAMPLE 6   **Discussing the Equation of an Ellipse**

Discuss the equation: $4x^2 + y^2 - 8x + 4y + 4 = 0$

Solution   We proceed to complete the squares in x and in y.

$$4x^2 + y^2 - 8x + 4y + 4 = 0$$

$$4x^2 - 8x + y^2 + 4y = -4 \qquad \text{Group like variables; place the constant on the right side.}$$

$$4\left(x^2 - 2x\right) + \left(y^2 + 4y\right) = -4 \qquad \text{Factor out 4.}$$

$$4\left(x^2 - 2x + 1\right) + \left(y^2 + 4y + 4\right) = -4 + 4 + 4 \qquad \text{Complete each square.}$$

$$4(x - 1)^2 + (y + 2)^2 = 4$$

$$(x - 1)^2 + \frac{(y + 2)^2}{4} = 1 \qquad \text{Divide each side by 4.}$$

Figure 28

$$(x - 1)^2 + \frac{(y + 2)^2}{4} = 1$$

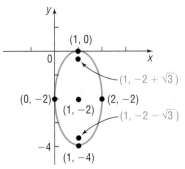

This is the equation of an ellipse with center at $(1, -2)$ and major axis parallel to the y-axis. Since $a^2 = 4$ and $b^2 = 1$, we have $c^2 = a^2 - b^2 = 4 - 1 = 3$. The vertices are at $(h, k \pm a) = (1, -2 \pm 2)$ or $(1, 0)$ and $(1, -4)$. The foci are at $(h, k \pm c) = \left(1, -2 \pm \sqrt{3}\right)$ or $\left(1, -2 - \sqrt{3}\right)$ and $\left(1, -2 + \sqrt{3}\right)$. Figure 28 shows the graph.

NOW WORK PROBLEM **33.**

APPLICATIONS

(5) Ellipses are found in many applications in science and engineering. For example, the orbits of the planets around the Sun are elliptical, with the Sun's position at a focus. See Figure 29.

Figure 29

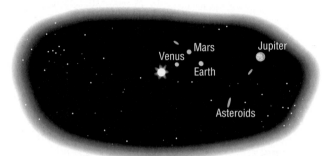

Stone and concrete bridges are often shaped as semielliptical arches. Elliptical gears are used in machinery when a variable rate of motion is required.

Ellipses also have an interesting reflection property. If a source of light (or sound) is placed at one focus, the waves transmitted by the source will reflect off the ellipse and concentrate at the other focus. This is the principle behind *whispering galleries*, which are rooms designed with elliptical ceilings. A person standing at one focus of the ellipse can whisper and be heard by a person standing at the other focus, because all the sound waves that reach the ceiling are reflected to the other person.

EXAMPLE 7

Whispering Galleries

Figure 30 shows the specifications for an elliptical ceiling in a hall designed to be a whispering gallery. In a whispering gallery, a person standing at one focus of the ellipse can whisper and be heard by another person standing at the other focus, because all the sound waves that reach the ceiling from one focus are reflected to the other focus. Where are the foci located in the hall?

Solution   We set up a rectangular coordinate system so that the center of the ellipse is at the origin and the major axis is along the x-axis. See Figure 31. The equation of the ellipse is

$$\frac{x^2}{a^2} + \frac{y^2}{b^2} = 1$$

where $a = 25$ and $b = 20$.

Figure 30

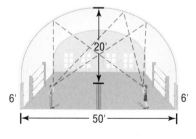

Figure 31

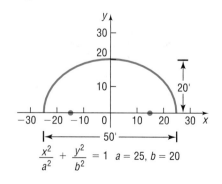

$$\frac{x^2}{a^2} + \frac{y^2}{b^2} = 1 \quad a = 25, b = 20$$

Then, since

$$c^2 = a^2 - b^2 = 25^2 - 20^2 = 625 - 400 = 225$$

we have $c = 15$. The foci are located 15 feet from the center of the ellipse along the major axis.   ■

7.3   EXERCISES

In Problems 1–4, the graph of an ellipse is given. Match each graph to its equation.

A. $\dfrac{x^2}{4} + y^2 = 1$ B. $x^2 + \dfrac{y^2}{4} = 1$ C. $\dfrac{x^2}{16} + \dfrac{y^2}{4} = 1$ D. $\dfrac{x^2}{4} + \dfrac{y^2}{16} = 1$

1.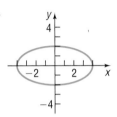   **2.**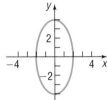   **3.**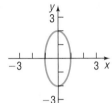   **4.**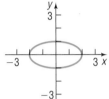

In Problems 5–14, find the vertices and foci of each ellipse. Graph each equation.

5. $\dfrac{x^2}{25} + \dfrac{y^2}{4} = 1$ **6.** $\dfrac{x^2}{9} + \dfrac{y^2}{4} = 1$ **7.** $\dfrac{x^2}{9} + \dfrac{y^2}{25} = 1$ **8.** $x^2 + \dfrac{y^2}{16} = 1$

9. $4x^2 + y^2 = 16$ **10.** $x^2 + 9y^2 = 18$ **11.** $4y^2 + x^2 = 8$ **12.** $4y^2 + 9x^2 = 36$

13. $x^2 + y^2 = 16$ **14.** $x^2 + y^2 = 4$

In Problems 15–24, find an equation for each ellipse. Graph the equation.

15. Center at $(0,0)$;   focus at $(3,0)$;   vertex at $(5,0)$
16. Center at $(0,0)$;   focus at $(-1,0)$;   vertex at $(3,0)$
17. Center at $(0,0)$;   focus at $(0,-4)$;   vertex at $(0,5)$
18. Center at $(0,0)$;   focus at $(0,1)$;   vertex at $(0,-2)$
19. Foci at $(\pm 2,0)$;   length of the major axis is 6
20. Focus at $(0,-4)$;   vertices at $(0,\pm 8)$
21. Foci at $(0,\pm 3)$; x-intercepts are ± 2
22. Foci at $(0,\pm 2)$;   length of the major axis is 8
23. Center at $(0,0)$;   vertex at $(0,4)$; $b = 1$
24. Vertices at $(\pm 5,0)$; $c = 2$

In Problems 25–28, write an equation for each ellipse.

25.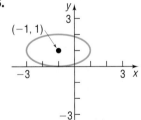   **26.**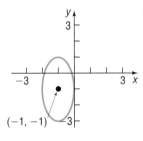   **27.**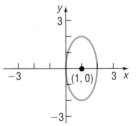   **28.**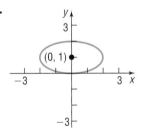

In Problems 29–40, find the center, foci, and vertices of each ellipse. Graph each equation.

29. $\dfrac{(x-3)^2}{4} + \dfrac{(y+1)^2}{9} = 1$ **30.** $\dfrac{(x+4)^2}{9} + \dfrac{(y+2)^2}{4} = 1$

31. $(x+5)^2 + 4(y-4)^2 = 16$ **32.** $9(x-3)^2 + (y+2)^2 = 18$

33. $x^2 + 4x + 4y^2 - 8y + 4 = 0$ **34.** $x^2 + 3y^2 - 12y + 9 = 0$

35. $2x^2 + 3y^2 - 8x + 6y + 5 = 0$ **36.** $4x^2 + 3y^2 + 8x - 6y = 5$

37. $9x^2 + 4y^2 - 18x + 16y - 11 = 0$ **38.** $x^2 + 9y^2 + 6x - 18y + 9 = 0$

39. $4x^2 + y^2 + 4y = 0$ **40.** $9x^2 + y^2 - 18x = 0$

In Problems 41–50, find an equation for each ellipse. Graph the equation.

41. Center at $(2, -2)$;   vertex at $(7, -2)$;   focus at $(4, -2)$

42. Center at $(-3, 1)$;   vertex at $(-3, 3)$;   focus at $(-3, 0)$

43. Vertices at $(4, 3)$ and $(4, 9)$;   focus at $(4, 8)$

44. Foci at $(1, 2)$ and $(-3, 2)$;   vertex at $(-4, 2)$

45. Foci at $(5, 1)$ and $(-1, 1)$;   length of the major axis is 8

46. Vertices at $(2, 5)$ and $(2, -1)$; $c = 2$

47. Center at $(1, 2)$;   focus at $(4, 2)$;   contains the point $(1, 3)$

48. Center at $(1, 2)$;   focus at $(1, 4)$;   contains the point $(2, 2)$

49. Center at $(1, 2)$;   vertex at $(4, 2)$;   contains the point $(1, 3)$

50. Center at $(1, 2)$;   vertex at $(1, 4)$;   contains the point $(2, 2)$

In Problems 51–54, graph each function.
[**Hint:** Notice that each function is half an ellipse.]

51. $f(x) = \sqrt{16 - 4x^2}$

52. $f(x) = \sqrt{9 - 9x^2}$

53. $f(x) = -\sqrt{64 - 16x^2}$ **54.** $f(x) = -\sqrt{4 - 4x^2}$

---

55. Semielliptical Arch Bridge   An arch in the shape of the upper half of an ellipse is used to support a bridge that is to span a river 20 meters wide. The center of the arch is 6 meters above the center of the river (see the figure). Write an equation for the ellipse in which the *x*-axis coincides with the water level and the *y*-axis passes through the center of the arch.

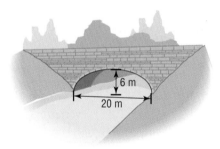

56. Semielliptical Arch Bridge   The arch of a bridge is a semiellipse with a horizontal major axis. The span is 30 feet, and the top of the arch is 10 feet above the major axis. The roadway is horizontal and is 2 feet above the top of the arch. Find the vertical distance from the roadway to the arch at 5-foot intervals along the roadway.

57. Whispering Gallery   A hall 100 feet in length is to be designed as a whispering gallery. If the foci are located 25 feet from the center, how high will the ceiling be at the center?

58. Whispering Gallery   Jim, standing at one focus of a whispering gallery, is 6 feet from the nearest wall. His friend is standing at the other focus, 100 feet away. What is the length of this whispering gallery? How high is its elliptical ceiling at the center?

59. Semielliptical Arch Bridge   A bridge is built in the shape of a semielliptical arch. The bridge has a span of 120 feet and a maximum height of 25 feet. Choose a suitable rectangular coordinate system and find the height of the arch at distances of 10, 30, and 50 feet from the center.

60. Semielliptical Arch Bridge   A bridge is built in the shape of a semielliptical arch and is to have a span of 100 feet. The height of the arch, at a distance of 40 feet from the center, is to be 10 feet. Find the height of the arch at its center.

61. Semielliptical Arch   An arch in the form of half an ellipse is 40 feet wide and 15 feet high at the center. Find the height of the arch at intervals of 10 feet along its width.

62. Semielliptical Arch Bridge   An arch for a bridge over a highway is in the form of half an ellipse. The top of the arch is 20 feet above the ground level (the major axis). The highway has four lanes, each 12 feet wide; a center safety strip 8 feet wide; and two side strips, each 4 feet wide. What should the span of the bridge be (the length of its major axis) if the height 28 feet from the center is to be 13 feet?

---

*In Problems 63–66, use the fact that the orbit of a planet about the Sun is an ellipse, with the Sun at one focus. The **aphelion** of a planet is its greatest distance from the Sun, and the **perihelion** is its shortest distance. The **mean distance** of a planet from the Sun is the length of the semimajor axis of the elliptical orbit. See the illustration.*

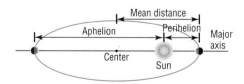

63. Earth   The mean distance of Earth from the Sun is 93 million miles. If the aphelion of Earth is 94.5 million miles, what is the perihelion? Write an equation for the orbit of Earth around the Sun.

64. Mars   The mean distance of Mars from the Sun is 142 million miles. If the perihelion of Mars is 128.5 million miles, what is the aphelion? Write an equation for the orbit of Mars about the Sun.

65. Jupiter   The aphelion of Jupiter is 507 million miles. If the distance from the Sun to the center of its elliptical orbit is 23.2 million miles, what is the perihelion? What is the mean distance? Write an equation for the orbit of Jupiter around the Sun.

66. Pluto   The perihelion of Pluto is 4551 million miles, and the distance of the Sun from the center of its elliptical orbit is 897.5 million miles. Find the aphelion of Pluto. What is the mean distance of Pluto from the Sun? Write an equation for the orbit of Pluto about the Sun.

67. Racetrack Design   Consult the figure. A racetrack is in the shape of an ellipse, 100 feet long and 50 feet wide. What is the width 10 feet from the side?

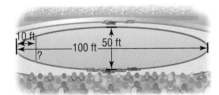

68. Racetrack Design   A racetrack is in the shape of an ellipse 80 feet long and 40 feet wide. What is the width 10 feet from the side?

69. Show that an equation of the form

$$Ax^2 + Cy^2 + F = 0, \qquad A \neq 0, C \neq 0, F \neq 0$$

where A and C are of the same sign and F is of opposite sign,
(a) Is the equation of an ellipse with center at $(0, 0)$ if $A \neq C$.
(b) Is the equation of a circle with center $(0, 0)$ if $A = C$.

70. Show that the graph of an equation of the form

$$Ax^2 + Cy^2 + Dx + Ey + F = 0, \qquad A \neq 0, C \neq 0$$

where A and C are of the same sign,
(a) Is an ellipse if $(D^2/4A) + (E^2/4C) - F$ is the same sign as A.
(b) Is a point if $(D^2/4A) + (E^2/4C) - F = 0$.
(c) Contains no points if $(D^2/4A) + (E^2/4C) - F$ is of opposite sign to A.

71. The **eccentricity** e of an ellipse is defined as the number c/a, where a and c are the numbers given in equation (2). Because $a > c$, it follows that $e < 1$. Write a brief paragraph about the general shape of each of the following ellipses. Be sure to justify your conclusions.
(a) Eccentricity close to 0
(b) Eccentricity = 0.5
(c) Eccentricity close to 1

P R E P A R I N G   F O R   T H I S   S E C T I O N

Before getting started, review the following:

✓ Distance Formula (Section 2.1, p. 148)

✓ Completing the Square (Section 1.3, pp. 108–110)

✓ Symmetry (Section 2.2, pp. 159–161)

✓ Asymptotes (Section 4.3, pp. 320–321)

✓ Graphing Techniques: Transformations (Section 3.4, pp. 242–252)

7.4   THE HYPERBOLA

OBJECTIVES
1. Find the Equation of a Hyperbola
2. Graph Hyperbolas
3. Discuss the Equation of a Hyperbola
4. Find the Asymptotes of a Hyperbola
5. Work with Hyperbolas with Center at (h, k)
6. Solve Applied Problems Involving Hyperbolas

A **hyperbola** is the collection of all points in the plane the difference of whose distances from two fixed points, called the **foci**, is a constant.

Figure 32

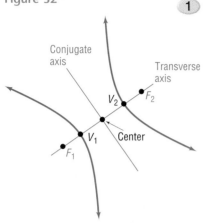

Figure 33
$d(F_1, P) - d(F_2, P) = \pm 2a$

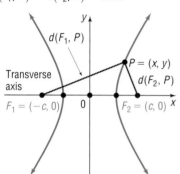

Figure 32 illustrates a hyperbola with foci F_1 and F_2. The line containing the foci is called the **transverse axis.** The midpoint of the line segment joining the foci is called the **center** of the hyperbola. The line through the center and perpendicular to the transverse axis is called the **conjugate axis.** The hyperbola consists of two separate curves, called **branches,** that are symmetric with respect to the transverse axis, conjugate axis, and center. The two points of intersection of the hyperbola and the transverse axis are the **vertices,** V_1 and V_2, of the hyperbola.

With these ideas in mind, we are now ready to find the equation of a hyperbola in a rectangular coordinate system. First, we place the center at the origin. Next, we position the hyperbola so that its transverse axis coincides with a coordinate axis. Suppose that the transverse axis coincides with the x-axis, as shown in Figure 33.

If c is the distance from the center to a focus, then one focus will be at $F_1 = (-c, 0)$ and the other at $F_2 = (c, 0)$. Now we let the constant difference of the distances from any point $P = (x, y)$ on the hyperbola to the foci F_1 and F_2 be denoted by $\pm 2a$. (If P is on the right branch, the $+$ sign is used; if P is on the left branch, the $-$ sign is used.) The coordinates of P must satisfy the equation

$$d(F_1, P) - d(F_2, P) = \pm 2a \qquad \text{Difference of the distances from } P \text{ to the foci equals } \pm 2a.$$

$$\sqrt{(x + c)^2 + y^2} - \sqrt{(x - c)^2 + y^2} = \pm 2a \qquad \text{Use the distance formula.}$$

$$\sqrt{(x + c)^2 + y^2} = \pm 2a + \sqrt{(x - c)^2 + y^2} \qquad \text{Isolate one radical.}$$

$$(x + c)^2 + y^2 = 4a^2 \pm 4a\sqrt{(x - c)^2 + y^2} \qquad \text{Square both sides.}$$
$$+ (x - c)^2 + y^2$$

Next, we remove the parentheses.

$$x^2 + 2cx + c^2 + y^2 = 4a^2 \pm 4a\sqrt{(x - c)^2 + y^2} + x^2 - 2cx + c^2 + y^2$$

$$4cx - 4a^2 = \pm 4a\sqrt{(x - c)^2 + y^2} \qquad \text{Simplify; Isolate the radical.}$$

$$cx - a^2 = \pm a\sqrt{(x - c)^2 + y^2} \qquad \text{Divide each side by 4.}$$

$$(cx - a^2)^2 = a^2[(x - c)^2 + y^2] \qquad \text{Square both sides.}$$

$$c^2x^2 - 2ca^2x + a^4 = a^2(x^2 - 2cx + c^2 + y^2) \qquad \text{Simplify.}$$

$$c^2x^2 + a^4 = a^2x^2 + a^2c^2 + a^2y^2 \qquad \text{Remove parentheses.}$$

$$(c^2 - a^2)x^2 - a^2y^2 = a^2c^2 - a^4 \qquad \text{Rearrange terms.}$$

$$(c^2 - a^2)x^2 - a^2y^2 = a^2(c^2 - a^2) \qquad \text{Factor } a^2 \text{ on the right side. (1)}$$

To obtain points on the hyperbola off the x-axis, it must be that $a < c$. To see why, look again at Figure 33.

$$d(F_1, P) < d(F_2, P) + d(F_1, F_2) \qquad \text{Use triangle } F_1 PF_2.$$

$$d(F_1, P) - d(F_2, P) < d(F_1, F_2) \qquad \begin{array}{l} P \text{ is on the right branch, so} \\ d(F_1, P) - d(F_2, P) = 2a. \end{array}$$

$$2a < 2c$$

$$a < c$$

Since $a < c$, we also have $a^2 < c^2$, so $c^2 - a^2 > 0$. Let $b^2 = c^2 - a^2, b > 0$. Then equation (1) can be written as

$$b^2x^2 - a^2y^2 = a^2b^2$$

$$\frac{x^2}{a^2} - \frac{y^2}{b^2} = 1 \qquad \text{Divide each side by } a^2b^2.$$

To find the vertices of the hyperbola defined by this equation, let $y = 0$. The vertices satisfy the equation $x^2/a^2 = 1$, the solutions of which are $x = \pm a$. Consequently, the vertices of the hyperbola are $V_1 = (-a, 0)$ and $V_2 = (a, 0)$. Notice that the distance from the center $(0, 0)$ to either vertex is a.

Theorem

Equation of a Hyperbola; Center at (0, 0); Foci at ($\pm$ c, 0); Vertices at ($\pm$ a, 0); Transverse Axis along the x-Axis

An equation of the hyperbola with center at $(0, 0)$, foci at $(-c, 0)$ and $(c, 0)$, and vertices at $(-a, 0)$ and $(a, 0)$ is

$$\frac{x^2}{a^2} - \frac{y^2}{b^2} = 1, \qquad \text{where } b^2 = c^2 - a^2 \qquad \textbf{(2)}$$

The transverse axis is the x-axis.

② As you can verify, the hyperbola defined by equation (2) is symmetric with respect to the x-axis, y-axis, and origin. To find the y-intercepts, if any, let $x = 0$ in equation (2). This results in the equation $y^2/b^2 = -1$, which has no real solution. We conclude that the hyperbola defined by equation (2) has no y-intercepts. In fact, since $x^2/a^2 - 1 = y^2/b^2 \ge 0$, it follows that $x^2/a^2 \ge 1$. There are no points on the graph for $-a < x < a$. See Figure 34.

Figure 34

$$\frac{x^2}{a^2} - \frac{y^2}{b^2} = 1, \quad b^2 = c^2 - a^2$$

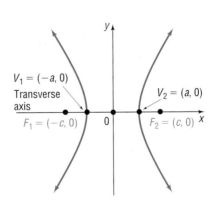

| EXAMPLE 1 | Finding and Graphing an Equation of a Hyperbola |

Find an equation of the hyperbola with center at the origin, one focus at $(3, 0)$, and one vertex at $(-2, 0)$. Graph the equation.

Solution The hyperbola has its center at the origin, and the transverse axis coincides with the x-axis. One focus is at $(c, 0) = (3, 0)$, so $c = 3$. One vertex is at $(-a, 0) = (-2, 0)$, so $a = 2$. From equation (2), it follows that $b^2 = c^2 - a^2 = 9 - 4 = 5$, so an equation of the hyperbola is

$$\frac{x^2}{4} - \frac{y^2}{5} = 1$$

To graph a hyperbola, it is helpful to locate and plot other points on the graph. For example, to find the points above and below the foci, we let $x = \pm 3$. Then

$$\frac{x^2}{4} - \frac{y^2}{5} = 1$$

$$\frac{(\pm 3)^2}{4} - \frac{y^2}{5} = 1 \qquad x = \pm 3$$

$$\frac{9}{4} - \frac{y^2}{5} = 1$$

$$\frac{y^2}{5} = \frac{5}{4}$$

$$y^2 = \frac{25}{4}$$

$$y = \pm \frac{5}{2}$$

Figure 35
$$\frac{x^2}{4} - \frac{y^2}{5} = 1$$

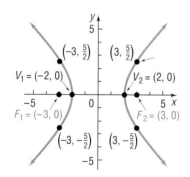

The points above and below the foci are $\left(\pm 3, \frac{5}{2}\right)$ and $\left(\pm 3, -\frac{5}{2}\right)$. These points help because they determine the "opening" of the hyperbola. See Figure 35. ■

COMMENT To graph the hyperbola $(x^2/4) - (y^2/5) = 1$ discussed in Example 1, we need to graph the two functions $Y_1 = \sqrt{5}\sqrt{(x^2/4) - 1}$ and $Y_2 = -\sqrt{5}\sqrt{(x^2/4) - 1}$. Do this and compare what you see with Figure 35. ■

NOW WORK PROBLEM 5.

An equation of the form of equation (2) is the equation of a hyperbola with center at the origin; foci on the x-axis at $(-c, 0)$ and $(c, 0)$, where $c^2 = a^2 + b^2$; and transverse axis along the x-axis.

③ For the remainder of this section, the direction "Discuss the equation" will mean to find the center, transverse axis, vertices, and foci of the hyperbola and graph it.

| EXAMPLE 2 | Discussing the Equation of a Hyperbola |

Discuss the equation: $\dfrac{x^2}{16} - \dfrac{y^2}{4} = 1$

Solution   The given equation is of the form of equation (2), with $a^2 = 16$ and $b^2 = 4$. The graph of the equation is a hyperbola with center at $(0, 0)$ and transverse axis along the x-axis. Also, we know that $c^2 = a^2 + b^2 = 16 + 4 = 20$. The vertices are at $(\pm a, 0) = (\pm 4, 0)$, and the foci are at $(\pm c, 0) = (\pm 2\sqrt{5}, 0)$.

To locate the points on the graph above and below the foci, we let $x = \pm 2\sqrt{5}$. Then

$$\frac{x^2}{16} - \frac{y^2}{4} = 1$$

$$\frac{(\pm 2\sqrt{5})^2}{16} - \frac{y^2}{4} = 1 \qquad x = \pm 2\sqrt{5}$$

$$\frac{20}{16} - \frac{y^2}{4} = 1$$

$$\frac{5}{4} - \frac{y^2}{4} = 1$$

$$\frac{y^2}{4} = \frac{1}{4}$$

$$y = \pm 1$$

The points above and below the foci are $\left(\pm 2\sqrt{5}, 1\right)$ and $\left(\pm 2\sqrt{5}, -1\right)$. See Figure 36.

Figure 36
$$\frac{x^2}{16} - \frac{y^2}{4} = 1$$

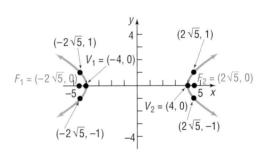

NOW WORK PROBLEM **15.**

The next result gives the form of the equation of a hyperbola with center at the origin and transverse axis along the y-axis.

Theorem   **Equation of a Hyperbola; Center at (0, 0); Foci at (0, ± c); Vertices at (0, ± a); Transverse Axis along the y-Axis**

An equation of the hyperbola with center at $(0, 0)$, foci at $(0, -c)$ and $(0, c)$, and vertices at $(0, -a)$ and $(0, a)$ is

$$\frac{y^2}{a^2} - \frac{x^2}{b^2} = 1, \qquad \text{where } b^2 = c^2 - a^2 \qquad \textbf{(3)}$$

The transverse axis is the y-axis.

Figure 37
$$\frac{y^2}{a^2} - \frac{x^2}{b^2} = 1, \quad b^2 = c^2 - a^2$$

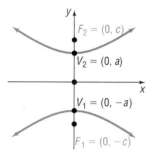

Figure 37 shows the graph of a typical hyperbola defined by equation (3).

An equation of the form of equation (2), $\frac{x^2}{a^2} - \frac{y^2}{b^2} = 1$, is the equation of a hyperbola with center at the origin, foci on the x-axis at $(-c, 0)$ and $(c, 0)$, where $c^2 = a^2 + b^2$, and transverse axis along the x-axis.

An equation of the form of equation (3), $\frac{y^2}{a^2} - \frac{x^2}{b^2} = 1$, is the equation of a hyperbola with center at the origin, foci on the y-axis at $(0, -c)$ and $(0, c)$, where $c^2 = a^2 + b^2$, and transverse axis along the y-axis.

Notice the difference in the forms of equations (2) and (3). When the y^2-term is subtracted from the x^2-term, the transverse axis is the x-axis. When the x^2-term is subtracted from the y^2-term, the transverse axis is the y-axis.

EXAMPLE 3   **Discussing the Equation of a Hyperbola**

Discuss the equation: $y^2 - 4x^2 = 4$

Solution   To put the equation in proper form, we divide each side by 4:

$$\frac{y^2}{4} - x^2 = 1$$

Since the x^2-term is subtracted from the y^2-term, the equation is that of a hyperbola with center at the origin and transverse axis along the y-axis. Also, comparing the above equation to equation (3), we find $a^2 = 4$, $b^2 = 1$, and $c^2 = a^2 + b^2 = 5$. The vertices are at $(0, \pm a) = (0, \pm 2)$, and the foci are at $(0, \pm c) = (0, \pm\sqrt{5})$.

To locate other points on the graph, we let $x = \pm 2$. Then

Figure 38
$$\frac{y^2}{4} - x^2 = 1$$

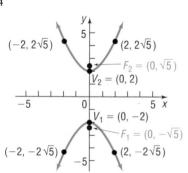

$$y^2 - 4x^2 = 4$$
$$y^2 - 4(\pm 2)^2 = 4 \qquad x = \pm 2$$
$$y^2 - 16 = 4$$
$$y^2 = 20$$
$$y = \pm 2\sqrt{5}$$

Four other points on the graph are $(\pm 2, 2\sqrt{5})$ and $(\pm 2, -2\sqrt{5})$. See Figure 38.   ∎

EXAMPLE 4   **Finding an Equation of a Hyperbola**

Find an equation of the hyperbola having one vertex at $(0, 2)$ and foci at $(0, -3)$ and $(0, 3)$.

Solution   Since the foci are at $(0, -3)$ and $(0, 3)$, the center of the hyperbola is at their midpoint, the origin. Also, the transverse axis is along the y-axis. The given information also reveals that $c = 3$, $a = 2$, and $b^2 = c^2 - a^2 = 9 - 4 = 5$. The form of the equation of the hyperbola is given by equation (3):

Figure 39
$$\frac{y^2}{4} - \frac{x^2}{5} = 1$$

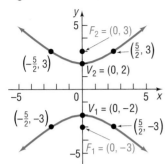

$$\frac{y^2}{a^2} - \frac{x^2}{b^2} = 1$$

$$\frac{y^2}{4} - \frac{x^2}{5} = 1$$

Let $y = 3$ to obtain points on the graph across from the foci. See Figure 39. ■

NOW WORK PROBLEM **9**.

Look at the equations of the hyperbolas in Examples 1 and 4. For the hyperbola in Example 1, $a^2 = 9$ and $b^2 = 7$, so $a > b$; for the hyperbola in Example 4, $a^2 = 4$ and $b^2 = 5$, so $a < b$. We conclude that, for hyperbolas, there are no requirements involving the relative sizes of a and b. Contrast this situation to the case of an ellipse, in which the relative sizes of a and b dictate which axis is the major axis. Hyperbolas have another feature to distinguish them from ellipses and parabolas: Hyperbolas have asymptotes.

ASYMPTOTES

4   Recall from Section 4.3 that a horizontal or oblique asymptote of a graph is a line with the property that the distance from the line to points on the graph approaches 0 as $x \to -\infty$ or as $x \to \infty$. The asymptotes provide information about the end behavior of the graph of a hyperbola.

Theorem

Asymptotes of a Hyperbola

The hyperbola $\dfrac{x^2}{a^2} - \dfrac{y^2}{b^2} = 1$ has the two oblique asymptotes

$$y = \frac{b}{a}x \quad \text{and} \quad y = -\frac{b}{a}x \qquad \textbf{(4)}$$

■

Proof   We begin by solving for y in the equation of the hyperbola.

$$\frac{x^2}{a^2} - \frac{y^2}{b^2} = 1$$

$$\frac{y^2}{b^2} = \frac{x^2}{a^2} - 1$$

$$y^2 = b^2\left(\frac{x^2}{a^2} - 1\right)$$

Since $x \neq 0$, we can rearrange the right side in the form

$$y^2 = \frac{b^2 x^2}{a^2}\left(1 - \frac{a^2}{x^2}\right)$$

$$y = \pm\frac{bx}{a}\sqrt{1 - \frac{a^2}{x^2}}$$

Now, as $x \to -\infty$ or as $x \to \infty$, the term a^2/x^2 approaches 0, so the expression under the radical approaches 1. Thus, as $x \to -\infty$ or as $x \to \infty$, the value of y approaches $\pm bx/a$; that is, the graph of the hyperbola approaches the lines

$$y = -\frac{b}{a}x \quad \text{and} \quad y = \frac{b}{a}x$$

These lines are oblique asymptotes of the hyperbola. ■

The asymptotes of a hyperbola are not part of the hyperbola, but they do serve as a guide for graphing a hyperbola. For example, suppose that we want to graph the equation

$$\frac{x^2}{a^2} - \frac{y^2}{b^2} = 1$$

We begin by plotting the vertices $(-a, 0)$ and $(a, 0)$. Then we plot the points $(0, -b)$ and $(0, b)$ and use these four points to construct a rectangle, as shown in Figure 40. The diagonals of this rectangle have slopes b/a and $-b/a$, and their extensions are the asymptotes $y = (b/a)x$ and $y = -(b/a)x$ of the hyperbola. If we graph the asymptotes, we can use them to establish the "opening" of the hyperbola and avoid plotting other points.

Figure 40

$$\frac{x^2}{a^2} - \frac{y^2}{b^2} = 1$$

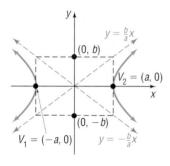

Theorem **Asymptotes of a Hyperbola**

The hyperbola $\dfrac{y^2}{a^2} - \dfrac{x^2}{b^2} = 1$ has the two oblique asymptotes

$$y = \frac{a}{b}x \quad \text{and} \quad y = -\frac{a}{b}x \qquad \textbf{(5)}$$

■

You are asked to prove this result in Problem 60.

For the remainder of this section, the direction "Discuss the equation" will mean to find the center, transverse axis, vertices, foci, and asymptotes of the hyperbola and graph it.

EXAMPLE 5 **Discussing the Equation of a Hyperbola**

Figure 41

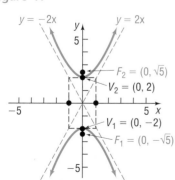

Discuss the equation: $\dfrac{y^2}{4} - x^2 = 1$

Solution Since the x^2-term is subtracted from the y^2-term, the equation is of the form of equation (3) and is a hyperbola with center at the origin and transverse axis along the y-axis. Also, comparing this equation to equation (3), we find that $a^2 = 4$, $b^2 = 1$, and $c^2 = a^2 + b^2 = 5$. The vertices are at $(0, \pm a) = (0, \pm 2)$, and the foci are at $(0, \pm c) = (0, \pm\sqrt{5})$. Using equation (5), the asymptotes are the lines $y = \dfrac{a}{b}x = 2x$ and $y = -\dfrac{a}{b}x = -2x$. Form the rectangle containing the points $(0, \pm a) = (0, \pm 2)$ and $(\pm b, 0) = (\pm 1, 0)$. The extensions of the diagonals of this rectangle are the asymptotes. Now graph the rectangle, the asymptotes, and the hyperbola. See Figure 41. ■

| EXAMPLE 8 | **Discussing the Equation of a Hyperbola** |

Discuss the equation: $-x^2 + 4y^2 - 2x - 16y + 11 = 0$

Solution We complete the squares in x and in y.

$$-x^2 + 4y^2 - 2x - 16y + 11 = 0$$

$$-(x^2 + 2x) + 4(y^2 - 4y) = -11 \qquad \text{Group terms.}$$

$$-(x^2 + 2x + 1) + 4(y^2 - 4y + 4) = -11 - 1 + 16 \qquad \text{Complete each square.}$$

$$-(x + 1)^2 + 4(y - 2)^2 = 4$$

$$(y - 2)^2 - \frac{(x + 1)^2}{4} = 1 \qquad \text{Divide by 4.}$$

Figure 45

$$(y - 2)^2 - \frac{(x + 1)^2}{4} = 1$$

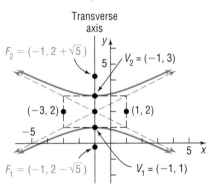

This is the equation of a hyperbola with center at $(-1, 2)$ and transverse axis parallel to the y-axis. Also, $a^2 = 1$ and $b^2 = 4$, so $c^2 = a^2 + b^2 = 5$. Since the transverse axis is parallel to the y-axis, the vertices and foci are located a and c units above and below the center, respectively. The vertices are at $(h, k \pm a) = (-1, 2 \pm 1)$, or $(-1, 1)$ and $(-1, 3)$. The foci are at $(h, k \pm c) = (-1, 2 \pm \sqrt{5})$. The asymptotes are $y - 2 = \frac{1}{2}(x + 1)$ and $y - 2 = -\frac{1}{2}(x + 1)$. Figure 45 shows the graph. ∎

NOW WORK PROBLEM **41**.

APPLICATIONS

Figure 46 ⑥

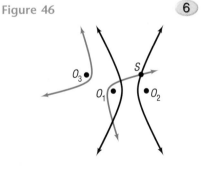

See Figure 46. Suppose that a gun is fired from an unknown source S. An observer at O_1 hears the report (sound of gun shot) 1 second after another observer at O_2. Because sound travels at about 1100 feet per second, it follows that the point S must be 1100 feet closer to O_2 than to O_1. S lies on one branch of a hyperbola with foci at O_1 and O_2. (Do you see why? The difference of the distances from S to O_1 and from S to O_2 is the constant 1100.) If a third observer at O_3 hears the same report 2 seconds after O_1 hears it, then S will lie on a branch of a second hyperbola with foci at O_1 and O_3. The intersection of the two hyperbolas will pinpoint the location of S.

LORAN

Figure 47

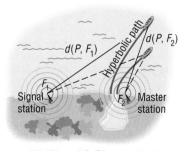

$d(P, F_1) - d(P, F_2) = \text{constant}$

In the LOng RAnge Navigation system (LORAN), a master radio sending station and a secondary sending station emit signals that can be received by a ship at sea. See Figure 47. Because a ship monitoring the two signals will usually be nearer to one of the two stations, there will be a difference in the distance that the two signals travel, which will register as a slight time difference between the signals. As long as the time difference remains constant, the difference of the two distances will also be constant. If the ship follows a path corresponding to the fixed time difference, it will follow the path of a hyperbola whose foci are located at the positions of the two sending stations. So for each time difference a different hyperbolic path results, each bringing the ship to a different shore location. Navigation charts show the various hyperbolic paths corresponding to different time differences.

EXAMPLE 9   LORAN

Two LORAN stations are positioned 250 miles apart along a straight shore.

(a) A ship records a time difference of 0.00054 second between the LORAN signals. Set up an appropriate rectangular coordinate system to determine where the ship would reach shore if it were to follow the hyperbola corresponding to this time difference.

(b) If the ship wants to enter a harbor located between the two stations 25 miles from the master station, what time difference should it be looking for?

(c) If the ship is 80 miles offshore when the desired time difference is obtained, what is the approximate location of the ship?

[**NOTE:**   The speed of each radio signal is 186,000 miles per second.]

Solution

Figure 48

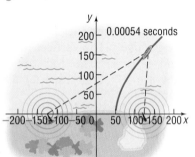

(a) We set up a rectangular coordinate system so that the two stations lie on the x-axis and the origin is midway between them. See Figure 48. The ship lies on a hyperbola whose foci are the locations of the two stations. The reason for this is that the constant time difference of the signals from each station results in a constant difference in the distance of the ship from each station. Since the time difference is 0.00054 second and the speed of the signal is 186,000 miles per second, the difference of the distances from the ship to each station (foci) is

$$\text{Distance} = \text{Speed} \times \text{Time} = 186{,}000 \times 0.00054 \approx 100 \text{ miles}$$

The difference of the distances from the ship to each station, 100, equals $2a$, so $a = 50$ and the vertex of the corresponding hyperbola is at $(50, 0)$. Since the focus is at $(125, 0)$, following this hyperbola the ship would reach shore 75 miles from the master station.

(b) To reach shore 25 miles from the master station, the ship would follow a hyperbola with vertex at $(100, 0)$. For this hyperbola, $a = 100$, so the constant difference of the distances from the ship to each station is $2a = 200$. The time difference that the ship should look for is

$$\text{Time} = \frac{\text{Distance}}{\text{Speed}} = \frac{200}{186{,}000} = 0.001075 \text{ second}$$

(c) To find the approximate location of the ship, we need to find the equation of the hyperbola with vertex at $(100, 0)$ and a focus at $(125, 0)$. The form of the equation of this hyperbola is

$$\frac{x^2}{a^2} - \frac{y^2}{b^2} = 1$$

where $a = 100$. Since $c = 125$, we have

$$b^2 = c^2 - a^2 = 125^2 - 100^2 = 5625$$

The equation of the hyperbola is

$$\frac{x^2}{100^2} - \frac{y^2}{5625} = 1$$

Figure 49

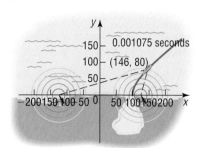

Since the ship is 80 miles from shore, we use $y = 80$ in the equation and solve for x.

$$\frac{x^2}{100^2} - \frac{80^2}{5625} = 1$$

$$\frac{x^2}{100^2} = 1 + \frac{80^2}{5625} \approx 2.14$$

$$x^2 \approx 100^2(2.14)$$

$$x \approx 146$$

The ship is at the position $(146, 80)$. See Figure 49.

NOW WORK PROBLEM **53**.

7.4   EXERCISES

In Problems 1–4, the graph of a hyperbola is given. Match each graph to its equation.

A. $\dfrac{x^2}{4} - y^2 = 1$ B. $x^2 - \dfrac{y^2}{4} = 1$ C. $\dfrac{y^2}{4} - x^2 = 1$ D. $y^2 - \dfrac{x^2}{4} = 1$

1.

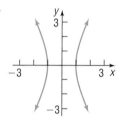

2.

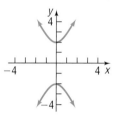

3.

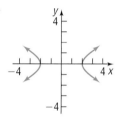

4.

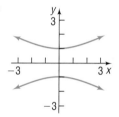

In Problems 5–14, find an equation for the hyperbola described. Graph the equation.

5. Center at $(0, 0)$;   focus at $(3, 0)$;   vertex at $(1, 0)$
6. Center at $(0, 0)$;   focus at $(0, 5)$;   vertex at $(0, 3)$
7. Center at $(0, 0)$;   focus at $(0, -6)$;   vertex at $(0, 4)$
8. Center at $(0, 0)$;   focus at $(-3, 0)$;   vertex at $(2, 0)$
9. Foci at $(-5, 0)$ and $(5, 0)$;   vertex at $(3, 0)$
10. Focus at $(0, 6)$;   vertices at $(0, -2)$ and $(0, 2)$
11. Vertices at $(0, -6)$ and $(0, 6)$;   asymptote the line $y = 2x$
12. Vertices at $(-4, 0)$ and $(4, 0)$;   asymptote the line $y = 2x$
13. Foci at $(-4, 0)$ and $(4, 0)$;   asymptote the line $y = -x$
14. Foci at $(0, -2)$ and $(0, 2)$;   asymptote the line $y = -x$

In Problems 15–22, find the center, transverse axis, vertices, foci, and asymptotes. Graph each equation.

15. $\dfrac{x^2}{25} - \dfrac{y^2}{9} = 1$ **16.** $\dfrac{y^2}{16} - \dfrac{x^2}{4} = 1$ **17.** $4x^2 - y^2 = 16$ **18.** $y^2 - 4x^2 = 16$

19. $y^2 - 9x^2 = 9$ **20.** $x^2 - y^2 = 4$ **21.** $y^2 - x^2 = 25$ **22.** $2x^2 - y^2 = 4$

In Problems 23–26, write an equation for each hyperbola.

23.

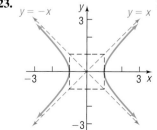

24.

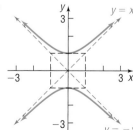

25.

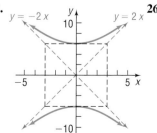

26.

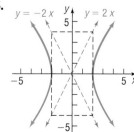

In Problems 27–34, find an equation for the hyperbola described. Graph the equation.

27. Center at $(4, -1)$;   focus at $(7, -1)$;   vertex at $(6, -1)$

28. Center at $(-3, 1)$;   focus at $(-3, 6)$;   vertex at $(-3, 4)$

29. Center at $(-3, -4)$;   focus at $(-3, -8)$;   vertex at $(-3, -2)$

30. Center at $(1, 4)$;   focus at $(-2, 4)$;   vertex at $(0, 4)$

31. Foci at $(3, 7)$ and $(7, 7)$;   vertex at $(6, 7)$

32. Focus at $(-4, 0)$;   vertices at $(-4, 4)$ and $(-4, 2)$

33. Vertices at $(-1, -1)$ and $(3, -1)$;   asymptote the line $y + 1 = \dfrac{3}{2}(x - 1)$

34. Vertices at $(1, -3)$ and $(1, 1)$;   asymptote the line $y + 1 = \dfrac{3}{2}(x - 1)$

In Problems 35–48, find the center, transverse axis, vertices, foci, and asymptotes. Graph each equation.

35. $\dfrac{(x - 2)^2}{4} - \dfrac{(y + 3)^2}{9} = 1$

36. $\dfrac{(y + 3)^2}{4} - \dfrac{(x - 2)^2}{9} = 1$

37. $(y - 2)^2 - 4(x + 2)^2 = 4$

38. $(x + 4)^2 - 9(y - 3)^2 = 9$

39. $(x + 1)^2 - (y + 2)^2 = 4$

40. $(y - 3)^2 - (x + 2)^2 = 4$

41. $x^2 - y^2 - 2x - 2y - 1 = 0$

42. $y^2 - x^2 - 4y + 4x - 1 = 0$

43. $y^2 - 4x^2 - 4y - 8x - 4 = 0$

44. $2x^2 - y^2 + 4x + 4y - 4 = 0$

45. $4x^2 - y^2 - 24x - 4y + 16 = 0$

46. $2y^2 - x^2 + 2x + 8y + 3 = 0$

47. $y^2 - 4x^2 - 16x - 2y - 19 = 0$

48. $x^2 - 3y^2 + 8x - 6y + 4 = 0$

In Problems 49–52, graph each function.
[**Hint:** Notice that each function is half a hyperbola.]

49. $f(x) = \sqrt{16 + 4x^2}$

50. $f(x) = -\sqrt{9 + 9x^2}$

51. $f(x) = -\sqrt{-25 + x^2}$

52. $f(x) = \sqrt{-1 + x^2}$

53. LORAN   Two LORAN stations are positioned 200 miles apart along a straight shore.
 (a) A ship records a time difference of 0.00038 second between the LORAN signals. Set up an appropriate rectangular coordinate system to determine where the ship would reach shore if it were to follow the hyperbola corresponding to this time difference.
 (b) If the ship wants to enter a harbor located between the two stations 20 miles from the master station, what time difference should it be looking for?
 (c) If the ship is 50 miles offshore when the desired time difference is obtained, what is the approximate location of the ship?
 [**Note:** The speed of each radio signal is 186,000 miles per second.]

54. LORAN   Two LORAN stations are positioned 100 miles apart along a straight shore.
 (a) A ship records a time difference of 0.00032 second between the LORAN signals. Set up an appropriate rectangular coordinate system to determine where the ship would reach shore if it were to follow the hyperbola corresponding to this time difference.
 (b) If the ship wants to enter a harbor located between the two stations 10 miles from the master station, what time difference should it be looking for?
 (c) If the ship is 20 miles offshore when the desired time difference is obtained, what is the approximate location of the ship?
 [**Note:** The speed of each radio signal is 186,000 miles per second.]

55. Calibrating Instruments   In a test of their recording devices, a team of seismologists positioned two of the devices 2000 feet apart, with the device at point A to the west of the device at point B. At a point between the devices and 200 feet from point B, a small amount of explosive was detonated and a note made of the time at which the sound reached each device. A second explosion is to be carried out at a point directly north of point B.
 (a) How far north should the site of the second explosion be chosen so that the measured time difference recorded by the devices for the second detonation is the same as that recorded for the first detonation?
 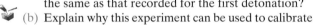
 (b) Explain why this experiment can be used to calibrate the instruments.

56. Explain in your own words the LORAN system of navigation.

57. The **eccentricity** e of a hyperbola is defined as the number c/a, where a and c are the numbers given in equation (2). Because $c > a$, it follows that $e > 1$. Describe the general shape of a hyperbola whose eccentricity is close to 1. What is the shape if e is very large?

58. A hyperbola for which $a = b$ is called an **equilateral hyperbola.** Find the eccentricity e of an equilateral hyperbola.
 [**Note:** The eccentricity of a hyperbola is defined in Problem 57.]

59. Two hyperbolas that have the same set of asymptotes are called **conjugate.** Show that the hyperbolas

$$\frac{x^2}{4} - y^2 = 1 \quad \text{and} \quad y^2 - \frac{x^2}{4} = 1$$

are conjugate. Graph each hyperbola on the same set of coordinate axes.

60. Prove that the hyperbola

$$\frac{y^2}{a^2} - \frac{x^2}{b^2} = 1$$

has the two oblique asymptotes

$$y = \frac{a}{b}x \quad \text{and} \quad y = -\frac{a}{b}x$$

61. Show that the graph of an equation of the form

$$Ax^2 + Cy^2 + F = 0, \qquad A \neq 0, C \neq 0, F \neq 0$$

where A and C are of opposite sign, is a hyperbola with center at $(0, 0)$.

62. Show that the graph of an equation of the form

$$Ax^2 + Cy^2 + Dx + Ey + F = 0, \qquad A \neq 0, C \neq 0$$

where A and C are of opposite sign,
(a) Is a hyperbola if $(D^2/4A) + (E^2/4C) - F \neq 0$.
(b) Is two intersecting lines if

$$(D^2/4A) + (E^2/4C) - F = 0$$

CHAPTER REVIEW

Things To Know

Equations

Parabola	See Tables 1 and 2 (pages 479 and 481).
Ellipse	See Table 3 (page 492).
Hyperbola	See Table 4 (page 505).

Objectives

You should be able to:

Know the names of the conics (p. 476)

Find the equation of a parabola (p. 477)

Graph parabolas (p. 478)

Discuss the equation of a parabola (p. 479)

Work with parabolas with vertex at (h, k) (p. 481)

Solve applied problems involving parabolas (p. 483)

Find the equation of an ellipse (p. 487)

Graph ellipses (p. 488)

Discuss the equation of an ellipse (p. 490)

Work with ellipses with center at (h, k) (p. 492)

Solve applied problems involving ellipses (p. 494)

Find the equation of a hyperbola (p. 498)

Graph hyperbolas (p. 499)

Discuss the equation of a hyperbola (p. 500)

Find the asymptotes of a hyperbola (p. 503)

Work with hyperbolas with center at (h, k) (p. 505)

Solve applied problems involving hyperbolas (p. 507)

Fill-in-the-Blank Items

1. A(n) _____ is the collection of all points in the plane such that the distance from each point to a fixed point equals its distance to a fixed line.

2. A(n) _____ is the collection of all points in the plane the sum of whose distances from two fixed points is a constant.

3. A(n) _____ is the collection of all points in the plane the difference of whose distances from two fixed points is a constant.

4. For an ellipse, the foci lie on the _____ axis; for a hyperbola, the foci lie on the _____ axis.

5. For the ellipse $\dfrac{x^2}{9} + \dfrac{y^2}{16} = 1$, the major axis is along the _____.

6. The equations of the asymptotes of the hyperbola $\dfrac{y^2}{9} - \dfrac{x^2}{4} = 1$ are _____ and _____.

True/False Items

T F **1.** On a parabola, the distance from any point to the focus equals the distance from that point to the directrix.

T F **2.** The foci of an ellipse lie on its minor axis.

T F **3.** The foci of a hyperbola lie on its transverse axis.

T F **4.** Hyperbolas always have asymptotes, and ellipses never have asymptotes.

T F **5.** A hyperbola never intersects its conjugate axis.

T F **6.** A hyperbola always intersects its transverse axis.

Review Exercises

Blue problem numbers indicate the author's suggestions for use in a Practice Test.

In Problems 1–20, identify each equation. If it is a parabola, gives its vertex, focus, and directrix; if it is an ellipse, give its center, vertices, and foci; if it is a hyperbola, give its center, vertices, foci, and asymptotes.

1. $y^2 = -16x$

2. $16x^2 = y$

3. $\dfrac{x^2}{25} - y^2 = 1$

4. $\dfrac{y^2}{25} - x^2 = 1$

5. $\dfrac{y^2}{25} + \dfrac{x^2}{16} = 1$

6. $\dfrac{x^2}{9} + \dfrac{y^2}{16} = 1$

7. $x^2 + 4y = 4$

8. $3y^2 - x^2 = 9$

9. $4x^2 - y^2 = 8$

10. $9x^2 + 4y^2 = 36$

11. $x^2 - 4x = 2y$

12. $2y^2 - 4y = x - 2$

13. $y^2 - 4y - 4x^2 + 8x = 4$

14. $4x^2 + y^2 + 8x - 4y + 4 = 0$

15. $4x^2 + 9y^2 - 16x - 18y = 11$

16. $4x^2 + 9y^2 - 16x + 18y = 11$

17. $4x^2 - 16x + 16y + 32 = 0$

18. $4y^2 + 3x - 16y + 19 = 0$

19. $9x^2 + 4y^2 - 18x + 8y = 23$

20. $x^2 - y^2 - 2x - 2y = 1$

In Problems 21–36, obtain an equation of the conic described. Graph the equation.

21. Parabola;   focus at $(-2, 0)$;   directrix the line $x = 2$

22. Ellipse;   center at $(0, 0)$;   focus at $(0, 3)$; vertex at $(0, 5)$

23. Hyperbola;   center at $(0, 0)$;   focus at $(0, 4)$; vertex at $(0, -2)$

24. Parabola;   vertex at $(0, 0)$;   directrix the line $y = -3$

25. Ellipse;   foci at $(-3, 0)$ and $(3, 0)$;   vertex at $(4, 0)$

26. Hyperbola;   vertices at $(-2, 0)$ and $(2, 0)$; focus at $(4, 0)$

27. Parabola;   vertex at $(2, -3)$;   focus at $(2, -4)$

28. Ellipse;   center at $(-1, 2)$;   focus at $(0, 2)$; vertex at $(2, 2)$

29. Hyperbola;   center at $(-2, -3)$;   focus at $(-4, -3)$; vertex at $(-3, -3)$

30. Parabola;   focus at $(3, 6)$;   directrix the line $y = 8$

31. Ellipse;   foci at $(-4, 2)$ and $(-4, 8)$;   vertex at $(-4, 10)$

32. Hyperbola;   vertices at $(-3, 3)$ and $(5, 3)$; focus at $(7, 3)$

33. Center at $(-1, 2)$; $a = 3$; $c = 4$;   transverse axis parallel to the x-axis

34. Center at $(4, -2)$; $a = 1$; $c = 4$;   transverse axis parallel to the y-axis

35. Vertices at $(0, 1)$ and $(6, 1)$;   asymptote the line $3y + 2x - 9 = 0$

36. Vertices at $(4, 0)$ and $(4, 4)$;   asymptote the line $y + 2x - 10 = 0$

37. Find an equation of the hyperbola whose foci are the vertices of the ellipse $4x^2 + 9y^2 = 36$ and whose vertices are the foci of this ellipse.

38. Find an equation of the ellipse whose foci are the vertices of the hyperbola $x^2 - 4y^2 = 16$ and whose vertices are the foci of this hyperbola.

39. Describe the collection of points in a plane so that the distance from each point to the point $(3, 0)$ is three-fourths of its distance from the line $x = \frac{16}{3}$.

40. Describe the collection of points in a plane so that the distance from each point to the point $(5, 0)$ is five-fourths of its distance from the line $x = \frac{16}{5}$.

41. **Mirrors**   A mirror is shaped like a paraboloid of revolution. If a light source is located 1 foot from the base along the axis of symmetry and the opening is 2 feet across, how deep should the mirror be?

42. **Parabolic Arch Bridge**   A bridge is built in the shape of a parabolic arch. The bridge has a span of 60 feet and a maximum height of 20 feet. Find the height of the arch at distances of 5, 10, and 20 feet from the center.

43. **Semielliptical Arch Bridge**   A bridge is built in the shape of a semielliptical arch. The bridge has a span of 60 feet and a maximum height of 20 feet. Find the height of the arch at distances of 5, 10, and 20 feet from the center.

44. **Whispering Galleries**   The figure shows the specifications for an elliptical ceiling in a hall designed to be a whispering gallery. Where in the hall are the foci located?

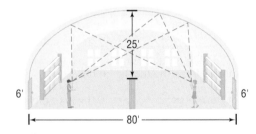

45. **LORAN**   Two LORAN stations are positioned 150 miles apart along a straight shore.
   (a) A ship records a time difference of 0.00032 second between the LORAN signals. Set up an appropriate rectangular coordinate system to determine where the ship would reach shore if it were to follow the hyperbola corresponding to this time difference.
   (b) If the ship wants to enter a harbor located between the two stations 15 miles from the master station, what time difference should it be looking for?

(c) If the ship is 20 miles offshore when the desired time difference is obtained, what is the approximate location of the ship?

[**Note:** The speed of each radio signal is 186,000 miles per second.]

46. Use the definition of a parabola and the idea of the eccentricity of an ellipse and hyperbola to construct a uni-

fying definition for all three conics. [Refer to Problem 71 in Exercises 7.3 and Problem 57 in Exercises 7.4.]

47. Formulate a strategy for discussing and graphing an equation of the form

$$Ax^2 + Cy^2 + Dx + Ey + F = 0$$

Project at Motorola

Distorted Deployable Space Reflector Antennas

A certain parabolic reflector antenna is to be used in a space application. To avoid excessive costs transporting the antenna into orbit, NASA makes these antennas from a very thin metallic cloth material attached on a frame that is deployed while in space to open up and take the form of a parabolic umbrella. In the correct configuration, the axis of the antenna aligns with the z-axis. To ensure that the deployed antenna is of the right shape, the following scheme was devised. Eight small, light-reflecting targets are placed on the cloth with the coordinates given in Table 1 (when the antenna is correctly deployed). Four of them are on the x–z plane and four are on the y–z plane. Tension wires are used to adjust the position of these targets when distorted, that is, when the actual values of x, y, and z do not equal the correct values. When in the cargo bay of the space shuttle, a laser placed on the z-axis and at a distance $L = 10$ meters (m) from the vertex performs measurements and records its distance from the targets and their angle θ (see the figure). Table 1 contains the correct values of x, y, and z for the parabolic reflector.

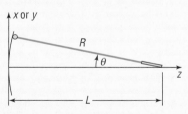

TABLE 1

Target	x	y (m)	z (m)
T_1	0	−2	0.5
T_2	0	−1	0.125
T_3	0	1	0.125
T_4	0	2	0.5
T_5	−2	0	0.5
T_6	−1	0	0.125
T_7	1	0	0.125
T_8	2	0	0.5

The laser measurements of the targets on the antenna immediately after deployment are tabulated in Table 2.

TABLE 2

Target	x (m)	R (m)	θ (deg)
T_1	0	9.551	−11.78
T_2	0	9.948	−5.65
T_3	0	9.928	5.90
T_4	0	9.708	11.89

Target	y (m)	R (m)	θ (deg)
T_5	0	9.722	−11.99
T_6	0	9.917	−5.85
T_7	0	9.925	5.78
T_8	0	9.551	11.78

In Problems 1–3, use any of the target points from Table 1 and the fact that the equation of the parabolic reflector (assuming vertex at the origin) is $z = 4ax^2 + 4ay^2$.

1. Find the focal distance of the undistorted parabolic reflector.

2. Use the tabulated measurements of the target points (Table 2) to convert the R and θ coordinates of the targets into Cartesian coordinates (z and x or y, as appropriate). That is, find the actual values of x, y, and z.

4. Find how much the targets T_1 through T_4 have to be moved in the y and z directions and the targets T_5 through T_8 in the x and z directions in order to correct the distortions of the antenna.

NOTE: In actuality, instead of adjusting the tension wires, the laser measurements are used to determine how much the feed antenna at the focus of the parabola needs to change its electronic distribution so as to correct for the distortions (phase conjugation).

Systems of Equations and Inequalities

Field Trip to Motorola

Digital cell phones communicate using the language of computers, with voice and data sent as groups of information bits (zeros and ones) over the air. Wireless communication is very difficult; the signals are reflected off buildings, absorbed by trees and rain, and scattered by street signs before being received by the antenna on the phone. Because of these impairments, it is difficult to receive all the information bits correctly; some will have invariably flipped from 0 to 1 or vice versa. On a voice call, the effect is digital distortion, which may make the call sound like it is coming from under water, for example. On a data call, the effect changes the data present in the transferred files. One primary reason digital phones are so powerful is that Motorola phone designers can use *error control coding* to ensure that phone calls are clear and error free.

PREPARING FOR THIS SECTION

Before getting started on this section, review the following:

✓ Linear Equations (Section 1.1, pp. 88–89)   ✓ Lines (Section 2.3, pp. 163–172)

8.1 SYSTEMS OF LINEAR EQUATIONS: TWO EQUATIONS CONTAINING TWO VARIABLES

OBJECTIVES
1. Solve Systems of Equations by Substitution
2. Solve Systems of Equations by Elimination
3. Identify Inconsistent Systems
4. Express the Solutions of a System of Dependent Equations

We begin with an example.

EXAMPLE 1   **Movie Theater Ticket Sales**

A movie theater sells tickets for $8.00 each, with seniors receiving a discount of $2.00. One evening the theater took in $3580 in revenue. If x represents the number of tickets sold at $8.00 and y the number of tickets sold at the discounted price of $6.00, write an equation that relates these variables.

Solution   Each nondiscounted ticket brings in $8.00, so x tickets will bring in $8x$ dollars. Similarly, y discounted tickets bring in $6y$ dollars. Since the total brought in is $3580, we must have

$$8x + 6y = 3580$$

∎

In Example 1, suppose that we also know that 525 tickets were sold that evening. Then we have another equation relating the variables x and y.

$$x + y = 525$$

The two equations

$$8x + 6y = 3580$$

$$x + y = 525$$

form a *system* of equations.

In general, a **system of equations** is a collection of two or more equations, each containing one or more variables. Example 2 gives some samples of systems of equations.

EXAMPLE 2

Examples of Systems of Equations

(a) $\begin{cases} 2x + y = 5 & \text{(1)} \\ -4x + 6y = -2 & \text{(2)} \end{cases}$ Two equations containing two variables, x and y

(b) $\begin{cases} x + y^2 = 5 & \text{(1)} \\ 2x + y = 4 & \text{(2)} \end{cases}$ Two equations containing two variables, x and y

(c) $\begin{cases} x + y + z = 6 & \text{(1)} \\ 3x - 2y + 4z = 9 & \text{(2)} \\ x - y - z = 0 & \text{(3)} \end{cases}$ Three equations containing three variables, x, y, and z

(d) $\begin{cases} x + y + z = 5 & \text{(1)} \\ x - y = 2 & \text{(2)} \end{cases}$ Two equations containing three variables, x, y, and z

(e) $\begin{cases} x + y + z = 6 & \text{(1)} \\ 2x + 2z = 4 & \text{(2)} \\ y + z = 2 & \text{(3)} \\ x = 4 & \text{(4)} \end{cases}$ Four equations containing three variables, x, y, and z ■

We use a brace, as shown, to remind us that we are dealing with a system of equations. We also will find it convenient to number each equation in the system.

A **solution** of a system of equations consists of values for the variables that are solutions of each equation of the system. To **solve** a system of equations means to find all solutions of the system.

For example, $x = 2$, $y = 1$ is a solution of the system in Example 2(a), because

$$\begin{cases} 2x + y = 5 & \text{(1)} \\ -4x + 6y = -2 & \text{(2)} \end{cases} \qquad \begin{cases} 2(2) + 1 = 4 + 1 = 5 \\ -4(2) + 6(1) = -8 + 6 = -2 \end{cases}$$

A solution of the system in Example 2(b) is $x = 1$, $y = 2$, because

$$\begin{cases} x + y^2 = 5 & \text{(1)} \\ 2x + y = 4 & \text{(2)} \end{cases} \qquad \begin{cases} 1 + 2^2 = 1 + 4 = 5 \\ 2(1) + 2 = 2 + 2 = 4 \end{cases}$$

Another solution of the system in Example 2(b) is $x = \frac{11}{4}$, $y = -\frac{3}{2}$, which you can check for yourself.

A solution of the system in Example 2(c) is $x = 3$, $y = 2$, $z = 1$, because

$$\begin{cases} x + y + z = 6 & \text{(1)} \\ 3x - 2y + 4z = 9 & \text{(2)} \\ x - y - z = 0 & \text{(3)} \end{cases} \qquad \begin{cases} 3 + 2 + 1 = 6 & \text{(1)} \\ 3(3) - 2(2) + 4(1) = 9 - 4 + 4 = 9 & \text{(2)} \\ 3 - 2 - 1 = 0 & \text{(3)} \end{cases}$$

Note that $x = 3, y = 3, z = 0$ is not a solution of the system in Example 2(c).

$$\begin{cases} x + y + z = 6 & (1) \\ 3x + 2y + 4z = 9 & (2) \\ x - y - z = 0 & (3) \end{cases} \qquad \begin{cases} 3 + 3 + 0 = 6 & (1) \\ 3(3) - 2(3) + 4(0) = 3 \ne 9 & (2) \\ 3 - 3 - 0 = 0 & (3) \end{cases}$$

Although these values satisfy equations (1) and (3), they do not satisfy equation (2). Any solution of the system must satisfy *each* equation of the system.

> **NOW WORK PROBLEM 3.**

When a system of equations has at least one solution, it is said to be **consistent;** otherwise, it is called **inconsistent.**

An equation in n variables is said to be **linear** if it is equivalent to an equation of the form

$$a_1 x_1 + a_2 x_2 + \cdots + a_n x_n = b$$

where $x_1, x_2, \ldots, x_n$ are n distinct variables, $a_1, a_2, \ldots, a_n, b$ are constants, and at least one of the a's is not 0.

Some examples of linear equations are

$$2x + 3y = 2 \qquad 5x - 2y + 3z = 10 \qquad 8x + 8y - 2z + 5w = 0$$

If each equation in a system of equations is linear, then we have a **system of linear equations.** The systems in Examples 2(a), (c), (d), and (e) are linear, whereas the system in Example 2(b) is nonlinear. In this chapter we shall solve linear systems in Sections 8.1–8.4. We discuss nonlinear systems in Section 8.7.

TWO LINEAR EQUATIONS CONTAINING TWO VARIABLES

We can view the problem of solving a system of two linear equations containing two variables as a geometry problem. The graph of each equation in such a system is a line. So, a system of two equations containing two variables represents a pair of lines. The lines either (1) intersect or (2) are parallel or (3) are **coincident** (that is, identical).

1. If the lines intersect, then the system of equations has one solution, given by the point of intersection. The system is **consistent** and the equations are **independent.**
2. If the lines are parallel, then the system of equations has no solution, because the lines never intersect. The system is **inconsistent.**
3. If the lines are coincident, then the system of equations has infinitely many solutions, represented by the totality of points on the line. The system is **consistent** and the equations are **dependent.**

Figure 1 illustrates these conclusions.

Figure 1

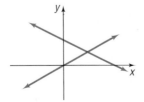

(a) Intersecting lines; system has one solution

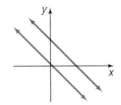

(b) Parallel lines; system has no solution

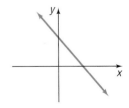

(c) Coincident lines; system has infinitely many solutions

| EXAMPLE 3 | Graphing a System of Linear Equations |

Graph the system: $\begin{cases} 2x + y = 5 & (1) \\ -4x + 6y = 12 & (2) \end{cases}$

Solution   Equation (1) in slope–intercept form is $y = -2x + 5$, which has slope -2 and y-intercept 5. Equation (2) in slope–intercept form is $y = \frac{2}{3}x + 2$, which has slope $\frac{2}{3}$ and y-intercept 2. Figure 2 shows their graph.   ■

Figure 2

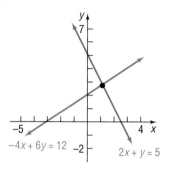

From the graph in Figure 2, we see that the lines intersect, so the system given in Example 3 is consistent. We can also use the graph as a means of approximating the solution. For this system, the solution appears to be close to the point $(1, 3)$. The actual solution, which you should verify, is $\left(\frac{9}{8}, \frac{11}{4}\right)$. To obtain exact solutions, we use algebraic methods. The first algebraic method we take up is the *method of substitution*.

SEEING THE CONCEPT   Graph the lines $2x + y = 5$ $(Y_1 = -2x + 5)$ and $-4x + 6y = 12$ $(Y_2 = \frac{2}{3}x + 2)$ and compare what you see with Figure 2. Use IN-TERSECT to verify that the point of intersection is $(1.125, 2.75)$.   ■

METHOD OF SUBSTITUTION

①   We illustrate the **method of substitution** by solving the system given in Example 3.

| EXAMPLE 4 | Solving a System of Linear Equations by Substitution |

Solve: $\begin{cases} 2x + y = 5 & (1) \\ -4x + 6y = 12 & (2) \end{cases}$

Solution   We solve the first equation for y, obtaining

$$y = -2x + 5$$

We substitute this result for y in the second equation. The result is an equation containing just the variable x, which we can then solve.

$$-4x + 6y = 12 \qquad (2)$$

$$-4x + 6(-2x + 5) = 12 \qquad \text{Substitute } y = -2x + 5.$$

$$-4x - 12x + 30 = 12 \qquad \text{Remove parentheses.}$$

$$-16x = -18 \qquad \text{Combine like terms.}$$

$$x = \frac{-18}{-16} = \frac{9}{8} \qquad \text{Solve for } x.$$

Once we know that $x = \frac{9}{8}$, we can easily find the value of y by **back-substitution**, that is, by substituting $\frac{9}{8}$ for x in one of the previous equations. We use the first one.

$$2x + y = 5 \tag{1}$$

$$2\left(\frac{9}{8}\right) + y = 5 \qquad \text{Substitute } x = 9/8.$$

$$\frac{9}{4} + y = 5 \qquad \text{Simplify.}$$

$$y = 5 - \frac{9}{4} = \frac{20}{4} - \frac{9}{4} = \frac{11}{4} \qquad \text{Solve for } y.$$

The solution of the system is $x = \frac{9}{8} = 1.125$, $y = \frac{11}{4} = 2.75$.

$$\text{Check:} \begin{cases} 2x + y = 5 & 2\left(\frac{9}{8}\right) + \frac{11}{4} = \frac{9}{4} + \frac{11}{4} = \frac{20}{4} = 5 \\ -4x + 6y = 12 & -4\left(\frac{9}{8}\right) + 6\left(\frac{11}{4}\right) = -\frac{9}{2} + \frac{33}{2} = \frac{24}{2} = 12 \end{cases}$$

The method used to solve the system in Example 4 is called **substitution.** The steps to be used are outlined next.

STEPS FOR SOLVING BY SUBSTITUTION

STEP 1: Pick one of the equations and solve for one of the variables in terms of the remaining variables.

STEP 2: Substitute the result in the remaining equations.

STEP 3: If one equation in one variable results, solve this equation. Otherwise, repeat Steps 1 and 2 until a single equation with one variable remains.

STEP 4: Find the values of the remaining variables by back-substitution.

STEP 5: Check the solution found.

EXAMPLE 5 — Solving a System of Linear Equations by Substitution

Solve: $\begin{cases} 3x - 2y = 5 & (1) \\ 5x - y = 6 & (2) \end{cases}$

Solution **STEP 1:** After looking at the two equations, we conclude that it is easiest to solve for the variable y in equation (2).

$$5x - y = 6 \tag{2}$$

$$-y = 6 - 5x \qquad \text{Subtract } 5x \text{ from both sides.}$$

$$y = 5x - 6 \qquad \text{Multiply both sides by } -1.$$

STEP 2: We substitute this result into equation (1) and simplify.

$$3x - 2y = 5 \qquad (1)$$

$$3x - 2(5x - 6) = 5 \qquad \text{Substitute } y = 5x - 6 \text{ in (1).}$$

STEP 3: $\qquad\qquad\qquad -7x + 12 = 5 \qquad$ Remove parentheses and collect like terms.

$$-7x = -7 \qquad \text{Subtract 12 from both sides.}$$

$$x = 1 \qquad \text{Divide both sides by } -7.$$

STEP 4: Knowing $x = 1$, we can find y from the equation

$$y = 5x - 6 = 5(1) - 6 = -1$$
$$\underset{x \,=\, 1}{\uparrow}$$

STEP 5: Check: $\begin{cases} 3(1) - 2(-1) = 3 + 2 = 5 \\ 5(1) - (-1) \ = 5 + 1 = 6 \end{cases}$

The solution of the system is $x = 1$, $y = -1$.

NOW USE SUBSTITUTION TO WORK PROBLEM **11.**

METHOD OF ELIMINATION

② A second method for solving a system of linear equations is the **method of elimination.** This method is usually preferred over substitution if substitution leads to fractions or if the system contains more than two variables. Elimination also provides the necessary motivation for solving systems using matrices (the subject of Section 8.3).

The idea behind the method of elimination is to replace the original system of equations by an equivalent system so that adding two of the equations eliminates a variable. The rules for obtaining equivalent equations are the same as those studied earlier. However, we may also interchange any two equations of the system and/or replace any equation in the system by the sum (or difference) of that equation and any other equation in the system.

RULES FOR OBTAINING AN EQUIVALENT SYSTEM OF EQUATIONS

1. Interchange any two equations of the system.

2. Multiply (or divide) each side of an equation by the same nonzero constant.

3. Replace any equation in the system by the sum (or difference) of that equation and a nonzero multiple of any other equation in the system.

An example will give you the idea. As you work through the example, pay particular attention to the pattern being followed.

| EXAMPLE 6 | Solving a System of Linear Equations by Elimination |

Solve: $\begin{cases} 2x + 3y = 1 & (1) \\ -x + y = -3 & (2) \end{cases}$

Solution We multiply each side of equation (2) by 2 so that the coefficients of x in the two equations are negatives of one another. The result is the equivalent system

$$\begin{cases} 2x + 3y = 1 & (1) \\ -2x + 2y = -6 & (2) \end{cases}$$

If we add the equations of this system, we obtain an equation containing just the variable y, which we can solve.

$$\begin{array}{rl}
\begin{cases} 2x + 3y = 1 & (1) \\ -2x + 2y = -6 & (2) \end{cases} & \\
 5y = -5 & \text{Add (1) and (2).} \\
 y = -1 &
\end{array}$$

We back-substitute by using this value for y in equation (1) and simplify to get

$$\begin{array}{rl}
2x + 3y = 1 & (1) \\
2x + 3(-1) = 1 & \text{Substitute } y = -1. \\
2x = 4 & \text{Simplify.} \\
x = 2 & \text{Solve for } x.
\end{array}$$

The solution of the original system is $x = 2$, $y = -1$. We leave it to you to check the solution. ■

The procedure used in Example 6 is called the *method of elimination.* Notice the pattern of the solution. First, we eliminated the variable x from the second equation. Then we back-substituted; that is, we substituted the value found for y back into the first equation to find x.

✏ NOW USE ELIMINATION TO WORK PROBLEM **11.**

Let's return to the movie theater example (Example 1).

| EXAMPLE 7 | Movie Theater Ticket Sales |

A movie theater sells tickets for $8.00 each, with seniors receiving a discount of $2.00. One evening the theater sold 525 tickets and took in $3580 in revenue. How many of each type of ticket was sold?

Solution If x represents the number of tickets sold at $8.00 and y the number of tickets sold at the discounted price of $6.00, then the given information results in the system of equations

$$\begin{cases} 8x + 6y = 3580 \\ x + y = 525 \end{cases}$$

We use elimination and multiply the second equation by -6 and then add the equations.

$$\begin{cases} 8x + 6y = 3580 \\ -6x - 6y = -3150 \end{cases}$$
$$\underline{\hspace{6cm}}$$
$$2x = 430 \qquad \text{Add the equations.}$$
$$x = 215$$

Since $x + y = 525$, then $y = 525 - x = 525 - 215 = 310$: 215 nondiscounted tickets and 310 senior discount tickets were sold. ▬

━━━▶ **NOW WORK PROBLEM 43.**

③ The previous examples dealt with consistent systems of equations that had a unique solution. The next two examples deal with the other possibilities that may occur, the first being a system that has no solution.

EXAMPLE 8   **An Inconsistent System of Linear Equations**

Solve: $\begin{cases} 2x + y = 5 & (1) \\ 4x + 2y = 8 & (2) \end{cases}$

Solution   We choose to use the method of substitution and solve equation (1) for y.

$$2x + y = 5 \qquad (1)$$
$$y = -2x + 5 \qquad \text{Subtract } 2x \text{ from each side.}$$

Now substitute $y = -2x + 5$ for y in equation (2) and solve for x.

$$4x + 2y = 8 \qquad (2)$$
$$4x + 2(-2x + 5) = 8 \qquad \text{Substitute } y = -2x + 5.$$
$$4x - 4x + 10 = 8 \qquad \text{Remove parentheses.}$$
$$0 \cdot x = -2 \qquad \text{Subtract 10 from both sides.}$$

Figure 3

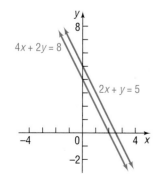

This equation has no solution. We conclude that the system itself has no solution and is therefore inconsistent. ▬

Figure 3 illustrates the pair of lines whose equations form the system in Example 8. Notice that the graphs of the two equations are lines, each with slope -2; one has a y-intercept of 5, the other a y-intercept of 4. The lines are parallel and have no point of intersection. This geometric statement is equivalent to the algebraic statement that the system has no solution.

SEEING THE CONCEPT   Graph the lines $2x + y = 5$ $(Y_1 = -2x + 5)$ and $4x + 2y = 8$ $(Y_2 = -2x + 4)$ and compare what you see with Figure 3. How can you be sure that the lines are parallel? ▬

④   The next example is an illustration of a system of dependent equations.

| EXAMPLE 9 | ### Solving a System of Dependent Equations |

Solve: $\begin{cases} 2x + y = 4 & (1) \\ -6x - 3y = -12 & (2) \end{cases}$

Solution   We choose to use the method of elimination:

$$\begin{cases} 2x + y = 4 & (1) \\ -6x - 3y = -12 & (2) \end{cases}$$

$$\begin{cases} 6x + 3y = 12 & (1) \text{ Multiply each side of equation (1) by 3.} \\ -6x - 3y = -12 & (2) \end{cases}$$

$$\begin{cases} 6x + 3y = 12 & (1) \\ 0 = 0 & (2) \text{ Replace equation (2) by the sum} \\ & \text{of equations (1) and (2).} \end{cases}$$

The original system is equivalent to a system containing one equation, so the equations are dependent. This means that any values of x and y for which $6x + 3y = 12$ or, equivalently, $2x + y = 4$ are solutions. For example, $x = 2, y = 0; x = 0, y = 4; x = -1, y = 6; x = 3, y = -2;$ and so on, are solutions. There are, in fact, infinitely many values of x and y for which $2x + y = 4$, so the original system has infinitely many solutions. We will write the solutions of the original system either as

$$y = 4 - 2x$$

where x can be any real number, or as

$$x = 2 - \frac{1}{2}y$$

where y can be any real number.   ∎

Figure 4

$\begin{cases} 2x + y = 4 \\ -6x - 3y = -12 \end{cases}$

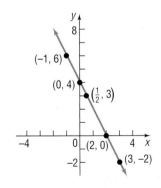

Figure 4 illustrates the situation presented in Example 9. Notice that the graphs of the two equations are lines, each with slope -2 and each with y-intercept 4. The lines are coincident. Notice also that equation (2) in the original system is just -3 times equation (1), indicating that the two equations are dependent.

For the system in Example 9, we can write down some of the infinite number of solutions by assigning values to x and then finding $y = 4 - 2x$.

If $x = -1$, then $y = 6$.

If $x = 0$, then $y = 4$.

If $x = 2$, then $y = 0$.

The pairs (x, y) are points on the line in Figure 4.

 SEEING THE CONCEPT   Graph the lines $2x + y = 4$ $(Y_1 = -2x + 4)$ and $-6x - 3y = -12$ $(Y_2 = -2x + 4)$ and compare what you see with Figure 4. How can you be sure that the lines are coincident?   ∎

NOW WORK PROBLEMS **17** AND **21.**

8.1 EXERCISES

In Problems 1–8, verify that the values of the variables listed are solutions of the system of equations.

1. $\begin{cases} 2x - y = 5 \\ 5x + 2y = 8 \end{cases}$

$x = 2, y = -1$

2. $\begin{cases} 3x + 2y = 2 \\ x - 7y = -30 \end{cases}$

$x = -2, y = 4$

3. $\begin{cases} 3x - 4y = 4 \\ \frac{1}{2}x - 3y = -\frac{1}{2} \end{cases}$

$x = 2, y = \frac{1}{2}$

4. $\begin{cases} 2x + \frac{1}{2}y = 0 \\ 3x - 4y = -\frac{19}{2} \end{cases}$

$x = -\frac{1}{2}, y = 2$

5. $\begin{cases} x^2 - y^2 = 3 \\ xy = 2 \end{cases}$

$x = 2, y = 1$

6. $\begin{cases} x^2 - y^2 = 3 \\ xy = 2 \end{cases}$

$x = -2, y = -1$

7. $\begin{cases} \dfrac{x}{1+x} + 3y = 6 \\ x + 9y^2 = 36 \end{cases}$

$x = 0, y = 2$

8. $\begin{cases} \dfrac{x}{x-1} + y = 5 \\ 3x - y = 3 \end{cases}$

$x = 2, y = 3$

In Problems 9–32, solve each system of equations. If the system has no solution, say that it is inconsistent. Use either substitution or elimination.

9. $\begin{cases} x + y = 8 \\ x - y = 4 \end{cases}$

10. $\begin{cases} x + 2y = 5 \\ x + y = 3 \end{cases}$

11. $\begin{cases} 5x - y = 13 \\ 2x + 3y = 12 \end{cases}$

12. $\begin{cases} x + 3y = 5 \\ 2x - 3y = -8 \end{cases}$

13. $\begin{cases} 3x = 24 \\ x + 2y = 0 \end{cases}$

14. $\begin{cases} 4x + 5y = -3 \\ -2y = -4 \end{cases}$

15. $\begin{cases} 3x - 6y = 2 \\ 5x + 4y = 1 \end{cases}$

16. $\begin{cases} 2x + 4y = \frac{2}{3} \\ 3x - 5y = -10 \end{cases}$

17. $\begin{cases} 2x + y = 1 \\ 4x + 2y = 3 \end{cases}$

18. $\begin{cases} x - y = 5 \\ -3x + 3y = 2 \end{cases}$

19. $\begin{cases} 2x - y = 0 \\ 3x + 2y = 7 \end{cases}$

20. $\begin{cases} 3x + 3y = -1 \\ 4x + y = \frac{8}{3} \end{cases}$

21. $\begin{cases} x + 2y = 4 \\ 2x + 4y = 8 \end{cases}$

22. $\begin{cases} 3x - y = 7 \\ 9x - 3y = 21 \end{cases}$

23. $\begin{cases} 2x - 3y = -1 \\ 10x + y = 11 \end{cases}$

24. $\begin{cases} 3x - 2y = 0 \\ 5x + 10y = 4 \end{cases}$

25. $\begin{cases} 2x + 3y = 6 \\ x - y = \frac{1}{2} \end{cases}$

26. $\begin{cases} \frac{1}{2}x + y = -2 \\ x - 2y = 8 \end{cases}$

27. $\begin{cases} \frac{1}{2}x + \frac{1}{3}y = 3 \\ \frac{1}{4}x - \frac{2}{3}y = -1 \end{cases}$

28. $\begin{cases} \frac{1}{3}x - \frac{3}{2}y = -5 \\ \frac{3}{4}x + \frac{1}{3}y = 11 \end{cases}$

29. $\begin{cases} 3x - 5y = 3 \\ 15x + 5y = 21 \end{cases}$

30. $\begin{cases} 2x - y = -1 \\ x + \frac{1}{2}y = \frac{3}{2} \end{cases}$

31. $\begin{cases} \dfrac{1}{x} + \dfrac{1}{y} = 8 \\ \dfrac{3}{x} - \dfrac{5}{y} = 0 \end{cases}$

[**Hint:** Let $u = 1/x$ and $v = 1/y$, and solve for u and v. Then $x = 1/u$ and $y = 1/y$.]

32. $\begin{cases} \dfrac{4}{x} - \dfrac{3}{y} = 0 \\ \dfrac{6}{x} + \dfrac{3}{2y} = 2 \end{cases}$

In Problems 33–38, use a graphing utility to solve each system of equations. Express the solution rounded to two decimal places.

33. $\begin{cases} y = \sqrt{2}x - 20\sqrt{7} \\ y = -0.1x + 20 \end{cases}$

34. $\begin{cases} y = -\sqrt{2}x + 100 \\ y = 0.2x + \sqrt{19} \end{cases}$

35. $\begin{cases} \sqrt{2}x + \sqrt{3}y + \sqrt{6} = 0 \\ \sqrt{3}x - \sqrt{2}y + 60 = 0 \end{cases}$

36. $\begin{cases} \sqrt{5}x - \sqrt{6}y + 60 = 0 \\ 0.2x + 0.3y + \sqrt{5} = 0 \end{cases}$

37. $\begin{cases} \sqrt{3}x + \sqrt{2}y = \sqrt{0.3} \\ 100x - 95y = 20 \end{cases}$

38. $\begin{cases} \sqrt{6}x - \sqrt{5}y + \sqrt{1.1} = 0 \\ y = -0.2x + 0.1 \end{cases}$

39. Supply and Demand Suppose that the quantity supplied Q_s and quantity demanded Q_d of *T*-shirts at a concert is given by the following equations:

$$Q_s = -200 + 50p$$

$$Q_d = 1000 - 25p$$

where p is the price. The equilibrium price of a market is defined as the price at which quantity supplied equals quantity demanded $(Q_s = Q_d)$. Find the equilibrium price for *T*-shirts at this concert. What is the equilibrium quantity?

40. Supply and Demand Suppose that the quantity supplied Q_s and quantity demanded Q_d of hot dogs at a baseball game is given by the following equations:

$$Q_s = -2000 + 3000p$$

$$Q_d = 10,000 - 1000p$$

where p is the price. The equilibrium price of a market is defined as the price at which quantity supplied equals quantity demanded $(Q_s = Q_d)$. Find the equilibrium price for hot dogs at the baseball game. What is the equilibrium quantity?

41. The perimeter of a rectangular floor is 90 feet. Find the dimensions of the floor if the length is twice the width.

42. The length of fence required to enclose a rectangular field is 3000 meters. What are the dimensions of the field if it is known that the difference between its length and width is 50 meters?

43. Cost of Fast Food Four large cheeseburgers and two chocolate shakes cost a total of $7.90. Two shakes cost 15¢ more than one cheeseburger. What is the cost of a cheeseburger? A shake?

44. Movie Theater Tickets A movie theater charges $9.00 for adults and $7.00 for senior citizens. On a day when 325 people paid an admission, the total receipts were $2495. How many who paid were adults? How many were seniors?

45. Mixing Nuts A store sells cashews for $5.00 per pound and peanuts for $1.50 per pound. The manager decides to mix 30 pounds of peanuts with some cashews and sell the mixture for $3.00 per pound. How many pounds of cashews should be mixed with the peanuts so that the mixture will produce the same revenue as would selling the nuts separately?

46. Financial Planning A recently retired couple need $12,000 per year to supplement their Social Security. They have $150,000 to invest to obtain this income. They have decided on two investment options: AA bonds yielding 10% per annum and a Bank Certificate yielding 5%.
(a) How much should be invested in each to realize exactly $12,000?
(b) If, after two years, the couple requires $14,000 per year in income, how should they reallocate their investment to achieve the new amount?

47. Computing Wind Speed With a tail wind, a small Piper aircraft can fly 600 miles in 3 hours. Against this same wind, the Piper can fly the same distance in 4 hours. Find the average wind speed and the average airspeed of the Piper.

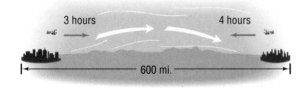

48. Computing Wind Speed The average airspeed of a single-engine aircraft is 150 miles per hour. If the aircraft flew the same distance in 2 hours with the wind as it flew in 3 hours against the wind, what was the wind speed?

49. Restaurant Management A restaurant manager wants to purchase 200 sets of dishes. One design costs

$25 per set, while another costs $45 per set. If she only has $7400 to spend, how many of each design should be ordered?

50. Cost of Fast Food One group of people purchased 10 hot dogs and 5 soft drinks at a cost of $12.50. A second bought 7 hot dogs and 4 soft drinks at a cost of $9.00. What is the cost of a single hot dog? A single soft drink?

We paid $12.50.   We paid $9.00.
How much is one hot dog?   How much is one hot dog?
How much is one cola?   How much is one cola?

51. Computing a Refund The grocery store we use does not mark prices on its goods. My wife went to this store, bought three 1-pound packages of bacon and two cartons of eggs, and paid a total of $7.45. Not knowing that she went to the store, I also went to the same store, purchased two 1-pound packages of bacon and three cartons of eggs, and paid a total of $6.45. Now we want to return two 1-pound packages of bacon and two cartons of eggs. How much will be refunded?

52. Finding the Current of a Stream Pamela requires 3 hours to swim 15 miles downstream on the Illinois River. The return trip upstream takes 5 hours. Find Pamela's average speed in still water. How fast is the current? (Assume that Pamela's speed is the same in each direction.)

53. Pharmacy A doctor's prescription calls for a daily intake of liquid containing 40 mg of vitamin C and 30 mg of vitamin D. Your pharmacy stocks two liquids that can be used: one contains 20% vitamin C and 30% vitamin D, the other 40% vitamin C and 20% vitamin D. How many milligrams of each liquid should be mixed to fill the prescription?

54. Pharmacy A doctor's prescription calls for the creation of pills that contain 12 units of vitamin B_{12} and 12 units of vitamin E. Your pharmacy stocks two powders that can be used to make these pills: one contains 20% vitamin B_{12} and 30% vitamin E, the other 40% vitamin B_{12} and 20% vitamin E. How many units of each powder should be mixed in each pill?

*The point at which a company's profits equal zero is called the company's **break-even point**. For Problems 55 and 56, let R represent a company's revenue, let C represent the company's costs, and let x represent the number of units produced and sold each day. Find the firm's break-even point; that is, find x so that R = C.*

55. $R = 8x$
$C = 4.5x + 17,500$

56. $R = 12x$
$C = 10x + 15,000$

57. Solve: $\begin{cases} y = m_1 x + b_1 \\ y = m_2 x + b_2 \end{cases}$
where $m_1 \neq m_2$.

58. Solve: $\begin{cases} y = m_1 x + b_1 \\ y = m_2 x + b_2 \end{cases}$
where $m_1 = m_2 = m$ and $b_1 \neq b_2$.

59. Solve: $\begin{cases} y = m_1 x + b_1 \\ y = m_2 x + b_2 \end{cases}$
where $m_1 = m_2 = m$ and $b_1 = b_2 = b$.

60. Make up a system of two linear equations containing two variables that has
(a) No solution
(b) Exactly one solution
(c) Infinitely many solutions
Give the three systems to a friend to solve and critique.

61. Write a brief paragraph outlining your strategy for solving a system of two linear equations containing two variables.

62. Do you prefer the method of substitution or the method of elimination for solving a system of two linear equations containing two variables? Give reasons.

8.2   SYSTEMS OF LINEAR EQUATIONS: THREE EQUATIONS CONTAINING THREE VARIABLES

OBJECTIVES
1. Solve Systems of Three Equations Containing Three Variables
2. Identify Inconsistent Systems
3. Express the Solutions of a System of Dependent Equations

Just as with a system of two linear equations containing two variables, a system of three linear equations containing three variables also has either (1) exactly one solution (a consistent system with independent equations), or (2) no solution (an inconsistent system), or (3) infinitely many solutions (a consistent system with dependent equations).

We can view the problem of solving a system of three linear equations containing three variables as a geometry problem. The graph of each equation in such a system is a plane in space. A system of three linear equations containing three variables represents three planes in space. Figure 5 illustrates some of the possibilities.

Figure 5

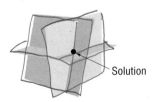

(a) Consistent system; one solution

(b) Consistent system; infinite number of solutions

(c) Inconsistent system; no solution

Recall that a **solution** to a system of equations consists of values for the variables that are solutions of each equation of the system. For example, $x = 3, y = -1, z = -5$ is a solution to the system of equations

$$\begin{cases} x + y + z = -3 & (1) \quad 3 + (-1) + (-5) = -3 \\ 2x - 3y + 6z = -21 & (2) \quad 2(3) - 3(-1) + 6(-5) = 6 + 3 - 30 = -21 \\ -3x + 5y = -14 & (3) \quad -3(3) + 5(-1) = -9 - 5 = -14 \end{cases}$$

because these values of the variables are solutions of each equation.

Typically, when solving a system of three linear equations containing three variables, we use the method of elimination. Recall that the idea behind the method of elimination is to form equivalent equations so that adding two of the equations eliminates a variable. We list the rules for obtaining an equivalent system of equations as a reminder.

RULES FOR OBTAINING AN EQUIVALENT SYSTEM OF EQUATIONS

1. Interchange any two equations of the system.
2. Multiply (or divide) each side of an equation by the same nonzero constant.
3. Replace any equation in the system by the sum (or difference) of that equation and a nonzero multiple of any other equation in the system.

---

① Let's see how elimination works on a system of three equations containing three variables.

EXAMPLE 1   **Solving a System of Three Linear Equations Containing Three Variables**

Use the method of elimination to solve the system of equations.

$$\begin{cases} x + y - z = -1 & (1) \\ 4x - 3y + 2z = 16 & (2) \\ 2x - 2y - 3z = 5 & (3) \end{cases}$$

Solution   For a system of three equations, we attempt to eliminate one variable at a time, using pairs of equations. Our plan of attack on this system will be to first eliminate the variable x from equations (2) and (3). Next, we will eliminate the variable y from equation (3), leaving only the variable z. Back-substitution can then be used to obtain the values of y and then x.

We begin by multiplying each side of equation (1) by -4 and adding the result to equation (2). (Do you see why? The coefficients of x are now opposites of each other.) We also multiply equation (1) by -2 and add the result to equation (3). Notice that these two procedures result in the removal of the x-variable from equations (2) and (3).

$$\begin{array}{ll} x + y - z = -1 & (1) \\ 4x - 3y + 2z = 16 & (2) \end{array} \quad \text{Multiply by } -4$$

$$\begin{array}{ll} -4x - 4y + 4z = 4 & (1) \\ 4x - 3y + 2z = 16 & (2) \\ \hline -7y + 6z = 20 & \text{Add.} \end{array}$$

$$\begin{array}{ll} x + y - z = -1 & (1) \\ 2x - 2y - 3z = 5 & (3) \end{array} \quad \text{Multiply by } -2$$

$$\begin{array}{ll} -2x - 2y + 2z = 2 & (1) \\ 2x - 2y - 3z = 5 & (3) \\ \hline -4y - z = 7 & \text{Add.} \end{array}$$

$$\begin{cases} x + y - z = -1 & (1) \\ -7y + 6z = 20 & (2) \\ -4y - z = 7 & (3) \end{cases}$$

We now concentrate on equations (2) and (3), treating them as a system of two equations containing two variables. We multiply each side of equation (2) by 4 and each side of equation (3) by -7 and add these equations. The result is the new equation (3).

$$
\begin{array}{ll}
-7y + 6z = 20 & \text{Multiply by 4.} \\
-4y - z = 7 & \text{Multiply by } -7.
\end{array}
\quad
\begin{array}{r}
-28y + 24z = 80 \\
28y + 7z = -49 \quad \text{Add.} \\
\hline
31z = 31 \longrightarrow
\end{array}
\quad
\begin{cases}
x + y - z = -1 & (1) \\
-7y + 6z = 20 & (2) \\
31z = 31 & (3)
\end{cases}
$$

We now solve equation (3) for z by dividing both sides of the equation by 31.

$$
\begin{cases}
x + y - z = -1 & (1) \\
 -7y + 6z = 20 & (2) \\
 z = 1 & (3)
\end{cases}
$$

Back-substitute $z = 1$ in equation (2) and solve for y.

$$-7y + 6z = 20 \quad (2)$$

$$-7y + 6(1) = 20 \quad \text{\small $z = 1$}$$

$$-7y = 14 \quad \text{\small Subtract 6 from both sides of the equation.}$$

$$y = -2 \quad \text{\small Divide both sides of the equation by } -7.$$

Finally, we back-substitute $y = -2$ and $z = 1$ in equation (1) and solve for x.

$$x + y - z = -1 \quad (1)$$

$$x + (-2) - 1 = -1 \quad \text{\small $y = -2$ and $z = 1$.}$$

$$x - 3 = -1 \quad \text{\small Simplify.}$$

$$x = 2 \quad \text{\small Add 3 to both sides.}$$

The solution of the original system is $x = 2$, $y = -2$, $z = 1$. You should verify this solution. ■

Look back over the solution given in Example 1. Note the pattern of removing one of the variables from two of the equations, followed by solving this system of two equations and two unknowns. Although which variables to remove is your choice, the methodology remains the same for all systems.

◖▭▭▭▭━ **N O W   W O R K   P R O B L E M   3 .**

② The previous example was a consistent system that had a unique solution. The next two examples deal with the other possibilities that may occur.

EXAMPLE 2   ### An Inconsistent System of Linear Equations

Solve:
$$
\begin{cases}
2x + y - z = 2 & (1) \\
x + 2y - z = -9 & (2) \\
x - 4y + z = 1 & (3)
\end{cases}
$$

Solution By adding equations (1) and (3) and equations (2) and (3), we can eliminate the variable z from equations (1) and (2).

$$
\begin{array}{rl}
2x + y - z = & 2 \quad (1) \\
x - 4y + z = & 1 \quad (3) \\
\hline
3x - 3y = & 3 \quad \text{Add}
\end{array}
$$

$$
\begin{array}{rl}
x + 2y - z = & -9 \quad (2) \\
x - 4y + z = & 1 \quad (3) \\
\hline
2x - 2y = & -8 \quad \text{Add}
\end{array}
$$

$$
\begin{cases}
3x - 3y = & 3 \quad (1) \\
2x - 2y = & -8 \quad (2) \\
x - 4y + z = & 1 \quad (3)
\end{cases}
$$

We work with equations (1) and (2), treating them as a system of two equations containing two variables. Multiply each side of equation (1) by $\frac{1}{3}$ and multiply each side of equation (2) by $-\frac{1}{2}$. Then add.

$$
\begin{cases}
3x - 3y = 3 & (1) \text{ Multiply by } \frac{1}{3} \\
2x - 2y = -8 & (2) \text{ Multiply by } -\frac{1}{2}
\end{cases}
\qquad
\begin{cases}
x - y = 1 \quad (1) \\
-x + y = 4 \quad (2) \\
\hline
 0 = 5 \quad \text{Add}
\end{cases}
$$

The system is inconsistent. ∎

③ Now let's look at a system of dependent equations.

EXAMPLE 3　　**Solving a System of Dependent Equations**

Solve:
$$
\begin{cases}
x - 2y - z = 8 & (1) \\
2x - 3y + z = 23 & (2) \\
4x - 5y + 5z = 53 & (3)
\end{cases}
$$

Solution Multiply each side of equation (1) by -2 and add the result to equation (2). Also, multiply each side of equation (1) by -4 and add the result to equation (3).

$$
\begin{array}{l}
x - 2y - z = 8 \quad (1) \text{ Multiply by } -2 \\
2x - 3y + z = 23 \quad (2)
\end{array}
\qquad
\begin{array}{rl}
-2x + 4y + 2z = & -16 \quad (1) \\
2x - 3y + z = & 23 \quad (2) \\
\hline
y + 3z = & 7 \quad \text{Add.}
\end{array}
$$

$$
\begin{array}{l}
x - 2y - z = 8 \quad (1) \text{ Multiply by } -4 \\
4x - 5y + 5z = 53 \quad (3)
\end{array}
\qquad
\begin{array}{rl}
-4x + 8y + 4z = & -32 \quad (1) \\
4x - 5y + 5z = & 53 \quad (2) \\
\hline
3y + 9z = & 21 \quad \text{Add.}
\end{array}
$$

$$
\begin{cases}
x - 2y - z = 8 & (1) \\
y + 3z = 7 & (2) \\
3y + 9z = 21 & (3)
\end{cases}
$$

Treat equations (2) and (3) as a system of two equations containing two variables, and eliminate the y variable by multiplying both sides of equation (2) by -3 and adding the result to equation (3).

$$
\begin{array}{l}
y + 3z = 7 \quad \text{Multiply by } -3. \\
3y + 9z = 21
\end{array}
\qquad
\begin{array}{rl}
-3y - 9z = & -21 \\
3y + 9z = & 21 \quad \text{Add.} \\
\hline
0 = & 0
\end{array}
\qquad
\begin{cases}
x - 2y - z = 8 & (1) \\
y + 3z = 7 & (2) \\
0 = 0 & (3)
\end{cases}
$$

The original system is equivalent to a system containing two equations, so the equations are dependent and the system has infinitely many solutions. If we let z represent any real number, then, solving equation (2) for y, we determine that $y = -3z + 7$. Substitute this expression into equation (1) to determine x in terms of z.

$$x - 2y - z = 8 \qquad \text{(1)}$$
$$x - 2(-3z + 7) - z = 8 \qquad y = -3z + 7$$
$$x + 6z - 14 - z = 8 \qquad \text{Remove parentheses.}$$
$$x + 5z = 22 \qquad \text{Combine like terms.}$$
$$x = -5z + 22 \qquad \text{Solve for } x.$$

We will write the solution to the system as

$$\begin{cases} x = -5z + 22 \\ y = -3z + \ 7 \end{cases}$$

where z can be any real number.

To find specific solutions to the system, choose any value of z and use the equations $x = -5z + 22$ and $y = -3z + 7$ to determine x and y. For example, if $z = 0$, then $x = 22$ and $y = 7$, and if $z = 1$, then $x = 17$ and $y = 4$. ∎

NOW WORK PROBLEM 7.

Two points in the Cartesian plane determine a unique line. Given three noncollinear points, we can find the (unique) quadratic function whose graph contains these three points.

EXAMPLE 4

Curve Fitting

Find real numbers $a, b,$ and c so that the graph of the quadratic function $y = ax^2 + bx + c$ contains the points $(-1, -4), (1, 6),$ and $(3, 0)$.

Solution   We require that the three points satisfy the equation $y = ax^2 + bx + c$.

For the point $(-1, -4)$ $-4 = a(-1)^2 + b(-1) + c$ $-4 = a - b + c$

For the point $(1, 6)$ $6 = a(1)^2 + b(1) + c$ $6 = a + b + c$

For the point $(3, 0)$ $0 = a(3)^2 + b(3) + c$ $0 = 9a + 3b + c$

We wish to determine $a, b,$ and c so that each equation is satisfied. That is, we want to solve the following system of three equations containing three variables:

$$\begin{cases} a - b + c = -4 & \text{(1)} \\ a + b + c = \ \ 6 & \text{(2)} \\ 9a + 3b + c = \ \ 0 & \text{(3)} \end{cases}$$

Solving this system of equations, we obtain $a = -2, b = 5,$ and $c = 3$. So the quadratic function whose graph contains the points $(-1, -4), (1, 6),$ and $(3, 0)$ is

$$y = -2x^2 + 5x + 3 \qquad y = ax^2 + bx + c, \quad a = -2, b = 5, c = 3$$

Figure 6 shows the graph of the function along with the three points. ∎

Figure 6

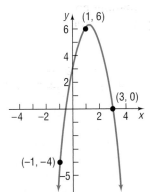

8.2 EXERCISES

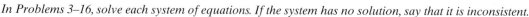

In Problems 1 and 2, verify that the values of the variables listed are solutions of the system of equations.

1. $\begin{cases} 3x + 3y + 2z = 4 \\ x - y - z = 0 \\ 2y - 3z = -8 \end{cases}$
$x = 1, y = -1, z = 2$

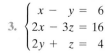

2. $\begin{cases} 4x - z = 7 \\ 8x + 5y - z = 0 \\ -x - y + 5z = 6 \end{cases}$
$x = 2, y = -3, z = 1$

In Problems 3–16, solve each system of equations. If the system has no solution, say that it is inconsistent.

3. $\begin{cases} x - y = 6 \\ 2x - 3z = 16 \\ 2y + z = 4 \end{cases}$

4. $\begin{cases} 2x + y = -4 \\ -2y + 4z = 0 \\ 3x - 2z = -11 \end{cases}$

5. $\begin{cases} x - 2y + 3z = 7 \\ 2x + y + z = 4 \\ -3x + 2y - 2z = -10 \end{cases}$

6. $\begin{cases} 2x + y - 3z = 0 \\ -2x + 2y + z = -7 \\ 3x - 4y - 3z = 7 \end{cases}$

7. $\begin{cases} x - y - z = 1 \\ 2x + 3y + z = 2 \\ 3x + 2y = 0 \end{cases}$

8. $\begin{cases} 2x - 3y - z = 0 \\ -x + 2y + z = 5 \\ 3x - 4y - z = 1 \end{cases}$

9. $\begin{cases} x - y - z = 1 \\ -x + 2y - 3z = -4 \\ 3x - 2y - 7z = 0 \end{cases}$

10. $\begin{cases} 2x - 3y - z = 0 \\ 3x + 2y + 2z = 2 \\ x + 5y + 3z = 2 \end{cases}$

11. $\begin{cases} 2x - 2y + 3z = 6 \\ 4x - 3y + 2z = 0 \\ -2x + 3y - 7z = 1 \end{cases}$

12. $\begin{cases} 3x - 2y + 2z = 6 \\ 7x - 3y + 2z = -1 \\ 2x - 3y + 4z = 0 \end{cases}$

13. $\begin{cases} x + y - z = 6 \\ 3x - 2y + z = -5 \\ x + 3y - 2z = 14 \end{cases}$

14. $\begin{cases} x - y + z = -4 \\ 2x - 3y + 4z = -15 \\ 5x + y - 2z = 12 \end{cases}$

15. $\begin{cases} x + 2y - z = -3 \\ 2x - 4y + z = -7 \\ -2x + 2y - 3z = 4 \end{cases}$

16. $\begin{cases} x + 4y - 3z = -8 \\ 3x - y + 3z = 12 \\ x + y + 6z = 1 \end{cases}$

---

17. **Curve Fitting** Find real numbers $a, b,$ and c so that the graph of the function $y = ax^2 + bx + c$ contains the points $(-1, 4), (2, 3),$ and $(0, 1)$.

18. **Curve Fitting** Find real numbers $a, b,$ and c so that the graph of the function $y = ax^2 + bx + c$ contains the points $(-1, -2), (1, -4),$ and $(2, 4)$.

19. **Electricity: Kirchhoff's Rules** An application of Kirchhoff's Rules to the circuit shown results in the following system of equations:

$$\begin{cases} I_2 = I_1 + I_3 \\ 5 - 3I_1 - 5I_2 = 0 \\ 10 - 5I_2 - 7I_3 = 0 \end{cases}$$

Find the currents $I_1, I_2,$ and I_3.*

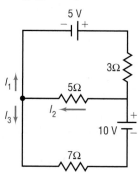

20. **Electricity: Kirchhoff's Rules** An application of Kirchhoff's Rules to the circuit shown results in the following system of equations:

$$\begin{cases} I_3 = I_1 + I_2 \\ 8 = 4I_3 + 6I_2 \\ 8I_1 = 4 + 6I_2 \end{cases}$$

Find the currents $I_1, I_2,$ and I_3.[†]

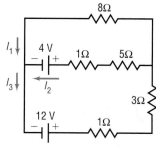

21. **Theater Revenues** A Broadway theater has 500 seats, divided into orchestra, main, and balcony seating. Orchestra seats sell for $50, main seats for $35, and balcony seats for $25. If all the seats are sold, the gross revenue to the theater is $17,100. If all the main and balcony seats are sold, but only half the orchestra seats are sold, the gross revenue is $14,600. How many are there of each kind of seat?

*Source: Based on Raymond Serway, *Physics*, 3rd ed. (Philadelphia: Saunders, 1990), Prob. 26, p. 790.
[†]Source: Ibid., Prob. 27, p. 790.

22. **Theater Revenues**   A movie theater charges $8.00 for adults, $4.50 for children, and $6.00 for senior citizens. One day the theater sold 405 tickets and collected $2320 in receipts. There were twice as many children's tickets sold as adult tickets. How many adults, children, and senior citizens went to the theater that day?

23. **Nutrition**   A dietitian wishes a patient to have a meal that has 66 grams of protein, 94.5 grams of carbohydrates, and 910 milligrams of calcium. The hospital food service tells the dietitian that the dinner for today is chicken, corn, and 2% milk. Each serving of chicken has 30 grams of protein, 35 grams of carbohydrates, and 200 milligrams of calcium. Each serving of corn has 3 grams of protein, 16 grams of carbohydrates, and 10 milligrams of calcium. Each glass of 2% milk has 9 grams of protein, 13 grams of carbohydrates, and 300 milligrams of calcium. How many servings of each food should the dietitian provide for the patient?

24. **Investments**   Kelly has $20,000 to invest. As her financial planner, you recommend that she diversify into three investments: Treasury bills that yield 5% simple interest, Treasury bonds that yield 7% simple interest, and corporate bonds that yield 10% simple interest. Kelly wishes to earn $1390 per year in income. Also, Kelly wants her investment in Treasury bills to be $3000 more than her investment in corporate bonds. How much money should Kelly place in each investment?

25. **Prices of Fast Food**   One group of customers bought 8 deluxe hamburgers, 6 orders of large fries, and 6 large colas for $26.10. A second group ordered 10 deluxe ham-

burgers, 6 large fries, and 8 large colas and paid $31.60. Is there sufficient information to determine the price of each food item? If not, construct a table showing the various possibilities. Assume that the hamburgers cost between $1.75 and $2.25, the fries between $0.75 and $1.00, and the colas between $0.60 and $0.90.

26. **Prices of Fast Food**   Use the information given in Problem 25, and suppose that a third group purchased 3 deluxe hamburgers, 2 large fries, and 4 large colas for $10.95. Now is there sufficient information to determine the price of each food item?

27. **Painting a House**   Three painters, Beth, Bill, and Edie, working together, can paint the exterior of a home in 10 hours. Bill and Edie together have painted a similar house in 15 hours. One day, all three worked on this same kind of house for 4 hours, after which Edie left. Beth and Bill required 8 more hours to finish. Assuming no gain or loss in efficiency, how long should it take each person to complete such a job alone?

P R E P A R I N G   F O R   T H I S   S E C T I O N

Before getting started on this section, review the following:

✓ Linear Equations (Section 1.1, pp. 88–89)   ✓ Lines (Section 2.3, pp. 163–172)

8.3 SYSTEMS OF LINEAR EQUATIONS: MATRICES

OBJECTIVES ① Write the Augmented Matrix of a System of Linear Equations
② Write the System from the Augmented Matrix
③ Perform Row Operations on a Matrix
④ Solve Systems of Linear Equations Using Matrices

The systematic approach of the method of elimination for solving a system of linear equations provides another method of solution that involves a simplified notation.

Consider the following system of linear equations:

$$\begin{cases} x + 4y = 14 \\ 3x - 2y = 0 \end{cases}$$

If we choose not to write the symbols used for the variables, we can represent this system as

$$\begin{bmatrix} 1 & 4 & | & 14 \\ 3 & -2 & | & 0 \end{bmatrix}$$

where it is understood that the first column represents the coefficients of the variable x, the second column the coefficients of y, and the third column the constants on the right side of the equal signs. The vertical line serves as a reminder of the equal signs. The large square brackets are the traditional symbols used to denote a *matrix* in algebra.

A **matrix** is defined as a rectangular array of numbers,

$$\begin{array}{c} \\ \text{Row 1} \\ \text{Row 2} \\ \vdots \\ \text{Row } i \\ \vdots \\ \text{Row } m \end{array} \begin{array}{cccccc} \text{Column 1} & \text{Column 2} & & \text{Column } j & & \text{Column } n \\ \begin{bmatrix} a_{11} & a_{12} & \cdots & a_{1j} & \cdots & a_{1n} \\ a_{21} & a_{22} & \cdots & a_{2j} & \cdots & a_{2n} \\ \vdots & \vdots & & \vdots & & \vdots \\ a_{i1} & a_{i2} & \cdots & a_{ij} & \cdots & a_{in} \\ \vdots & \vdots & & \vdots & & \vdots \\ a_{m1} & a_{m2} & \cdots & a_{mj} & \cdots & a_{mn} \end{bmatrix} \end{array} \quad (1)$$

Each number a_{ij} of the matrix has two indexes: the **row index** i and the **column index** j. The matrix shown in display (1) has m rows and n columns. The numbers a_{ij} are usually referred to as the **entries** of the matrix. For example, a_{23} refers to the entry in the second row, third column.

Now we will use matrix notation to represent a system of linear equations. The matrices used to represent systems of linear equations are called **augmented matrices.** In writing the augmented matrix of a system, the variables of each equation must be on the left side of the equal sign and the constants on the right side. A variable that does not appear in an equation has a coefficient of 0.

EXAMPLE 1

Writing the Augmented Matrix of a System of Linear Equations

Write the augmented matrix of each system of equations.

(a) $\begin{cases} 3x - 4y = -6 & (1) \\ 2x - 3y = -5 & (2) \end{cases}$

(b) $\begin{cases} 2x - y + z = 0 & (1) \\ x + z - 1 = 0 & (2) \\ x + 2y - 8 = 0 & (3) \end{cases}$

Solution   (a) The augmented matrix is

$$\begin{bmatrix} 3 & -4 & | & -6 \\ 2 & -3 & | & -5 \end{bmatrix}$$

(b) Care must be taken that the system be written so that the coefficients of all variables are present (if any variable is missing, its coefficient is 0).

Also, all constants must be to the right of the equal sign. We need to re-arrange the given system as follows:

$$\begin{cases} 2x - y + z = 0 & (1) \\ x + z - 1 = 0 & (2) \\ x + 2y - 8 = 0 & (3) \end{cases}$$

$$\begin{cases} 2x - y + z = 0 & (1) \\ x + 0 \cdot y + z = 1 & (2) \\ x + 2y + 0 \cdot z = 8 & (3) \end{cases}$$

The augmented matrix is

$$\begin{bmatrix} 2 & -1 & 1 & | & 0 \\ 1 & 0 & 1 & | & 1 \\ 1 & 2 & 0 & | & 8 \end{bmatrix}$$ ▬

If we do not include the constants to the right of the equal sign, that is, to the right of the vertical bar in the augmented matrix of a system of equations, the resulting matrix is called the **coefficient matrix** of the system. For the systems discussed in Example 1, the coefficient matrices are

$$\begin{bmatrix} 3 & -4 \\ 2 & -3 \end{bmatrix} \quad \text{and} \quad \begin{bmatrix} 2 & -1 & 1 \\ 1 & 0 & 1 \\ 1 & 2 & 0 \end{bmatrix}$$

✏ NOW WORK PROBLEM **3**.

EXAMPLE 2

Writing the System of Linear Equations from the Augmented Matrix

② Write the system of linear equations corresponding to each augmented matrix.

(a) $$\begin{bmatrix} 5 & 2 & | & 13 \\ -3 & 1 & | & -10 \end{bmatrix}$$

(b) $$\begin{bmatrix} 3 & -1 & -1 & | & 7 \\ 2 & 0 & 2 & | & 8 \\ 0 & 1 & 1 & | & 0 \end{bmatrix}$$

Solution   (a) The matrix has two rows and so represents a system of two equations. The two columns to the left of the vertical bar indicate that the system has two variables. If x and y are used to denote these variables, the system of equations is

$$\begin{cases} 5x + 2y = 13 & (1) \\ -3x + y = -10 & (2) \end{cases}$$

(b) Since the augmented matrix has three rows, it represents a system of three equations. Since there are three columns to the left of the vertical bar, the system contains three variables. If x, y, and z are the three variables, the system is

$$\begin{cases} 3x - y - z = 7 & (1) \\ 2x + 2z = 8 & (2) \\ y + z = 0 & (3) \end{cases}$$ ▬

Row Operations on a Matrix

3 **Row operations** on a matrix are used to solve systems of equations when the system is written as an augmented matrix. There are three basic row operations.

> ### Row Operations
>
> 1. Interchange any two rows.
> 2. Replace a row by a nonzero multiple of that row.
> 3. Replace a row by the sum of that row and a constant nonzero multiple of some other row.

These three row operations correspond to the three rules given earlier for obtaining an equivalent system of equations. When a row operation is performed on a matrix, the resulting matrix represents a system of equations equivalent to the system represented by the original matrix.

For example, consider the augmented matrix

$$\begin{bmatrix} 1 & 2 & | & 3 \\ 4 & -1 & | & 2 \end{bmatrix}$$

Suppose that we want to apply a row operation to this matrix that results in a matrix whose entry in row 2, column 1 is a 0. The row operation to use is

Multiply each entry in row 1 by -4 and add the result
to the corresponding entries in row 2. **(2)**

If we use R_2 to represent the new entries in row 2 and we use r_1 and r_2 to represent the original entries in rows 1 and 2, respectively, then we can represent the row operation in statement (2) by

$$R_2 = -4r_1 + r_2$$

Then

$$\begin{bmatrix} 1 & 2 & | & 3 \\ 4 & -1 & | & 2 \end{bmatrix} \xrightarrow[\substack{\uparrow \\ R_2 = -4r_1 + r_2}]{} \begin{bmatrix} 1 & 2 & | & 3 \\ -4(1) + 4 & -4(2) + (-1) & | & -4(3) + 2 \end{bmatrix} = \begin{bmatrix} 1 & 2 & | & 3 \\ 0 & -9 & | & -10 \end{bmatrix}$$

As desired, we now have the entry 0 in row 2, column 1.

EXAMPLE 3 **Applying a Row Operation to an Augmented Matrix**

Apply the row operation $R_2 = -3r_1 + r_2$ to the augmented matrix

$$\begin{bmatrix} 1 & -2 & | & 2 \\ 3 & -5 & | & 9 \end{bmatrix}$$

Solution The row operation $R_2 = -3r_1 + r_2$ tells us that the entries in row 2 are to be replaced by the entries obtained after multiplying each entry in row 1 by -3 and adding the result to the corresponding entries in row 2. Thus,

$$\begin{bmatrix} 1 & -2 & | & 2 \\ 3 & -5 & | & 9 \end{bmatrix} \xrightarrow[\substack{\uparrow \\ R_2 = -3r_1 + r_2}]{} \begin{bmatrix} 1 & -2 & | & 2 \\ -3(1) + 3 & (-3)(-2) + (-5) & | & -3(2) + 9 \end{bmatrix} = \begin{bmatrix} 1 & -2 & | & 2 \\ 0 & 1 & | & 3 \end{bmatrix}$$

NOW WORK PROBLEM **13.**

EXAMPLE 4 Finding a Particular Row Operation

Using the matrix

$$\begin{bmatrix} 1 & -2 & | & 2 \\ 0 & 1 & | & 3 \end{bmatrix}$$

find a row operation that will result in this matrix having a 0 in row 1, column 2.

Solution We want a 0 in row 1, column 2. This result can be accomplished by multiplying row 2 by 2 and adding the result to row 1. That is, we apply the row operation $R_1 = 2r_2 + r_1$.

$$\begin{bmatrix} 1 & -2 & | & 2 \\ 0 & 1 & | & 3 \end{bmatrix} \underset{\substack{\uparrow \\ R_1 = 2r_2 + r_1}}{\rightarrow} \begin{bmatrix} 2(0) + 1 & 2(1) + (-2) & | & 2(3) + 2 \\ 0 & 1 & | & 3 \end{bmatrix} = \begin{bmatrix} 1 & 0 & | & 8 \\ 0 & 1 & | & 3 \end{bmatrix}$$ ∎

A word about the notation that we have introduced. A row operation such as $R_1 = 2r_2 + r_1$ changes the entries in row 1. Note also that for this type of row operation we change the entries in a given row by multiplying the entries in some other row by an appropriate nonzero number and adding the results to the original entries of the row to be changed.

SOLVING A SYSTEM OF LINEAR EQUATIONS USING MATRICES

④ To solve a system of linear equations using matrices, we use row operations on the augmented matrix of the system to obtain a matrix that is in *row echelon form*.

A matrix is in **row echelon form** when

1. The entry in row 1, column 1 is a 1, and 0's appear below it.
2. The first nonzero entry in each row after the first row is a 1, 0's appear below it, and it appears to the right of the first nonzero entry in any row above.
3. Any rows that contain all 0's to the left of the vertical bar appear at the bottom.

For example, for a system of three equations containing three variables with a unique solution, the augmented matrix is in row echelon form if it is of the form

$$\begin{bmatrix} 1 & a & b & | & d \\ 0 & 1 & c & | & e \\ 0 & 0 & 1 & | & f \end{bmatrix}$$

where a, b, c, d, e, and f are real numbers. The last row of the augmented matrix states that $z = f$. We can then determine the value of y using back-substitution with $z = f$, since row 2 represents the equation $y + cz = e$. Finally, x is determined using back-substitution again.

Two advantages of solving a system of equations by writing the augmented matrix in row echelon form are the following:

1. The process is algorithmic; that is, it consists of repetitive steps that can be programmed on a computer.
2. The process works on any system of linear equations, no matter how many equations or variables are present.

The next example shows how to write a matrix in row echelon form.

EXAMPLE 5

Solving a System of Linear Equations Using Matrices (Row Echelon Form)

Solve: $\begin{cases} 2x + 2y = 6 & (1) \\ x + y + z = 1 & (2) \\ 3x + 4y - z = 13 & (3) \end{cases}$

Solution First, we write the augmented matrix that represents this system.

$$\begin{bmatrix} 2 & 2 & 0 & | & 6 \\ 1 & 1 & 1 & | & 1 \\ 3 & 4 & -1 & | & 13 \end{bmatrix}$$

The first step requires getting the entry 1 in row 1, column 1. An interchange of rows 1 and 2 is the easiest way to do this. [Note that this is equivalent to interchanging equations (1) and (2) of the system.]

$$\begin{bmatrix} 1 & 1 & 1 & | & 1 \\ 2 & 2 & 0 & | & 6 \\ 3 & 4 & -1 & | & 13 \end{bmatrix}$$

Next, we want a 0 in row 2, column 1 and a 0 in row 3, column 1. We use the row operations $R_2 = -2r_1 + r_2$ and $R_3 = -3r_1 + r_3$ to accomplish this. Notice that row 1 is unchanged using these row operations. Also, do you see that performing these row operations simultaneously is the same as doing one followed by the other?

$$\begin{bmatrix} 1 & 1 & 1 & | & 1 \\ 2 & 2 & 0 & | & 6 \\ 3 & 4 & -1 & | & 13 \end{bmatrix} \rightarrow \begin{bmatrix} 1 & 1 & 1 & | & 1 \\ 0 & 0 & -2 & | & 4 \\ 0 & 1 & -4 & | & 10 \end{bmatrix}$$
$$\uparrow$$
$$R_2 = -2r_1 + r_2$$
$$R_3 = -3r_1 + r_3$$

Now we want the entry 1 in row 2, column 2. Interchanging rows 2 and 3 will accomplish this.

$$\begin{bmatrix} 1 & 1 & 1 & | & 1 \\ 0 & 0 & -2 & | & 4 \\ 0 & 1 & -4 & | & 10 \end{bmatrix} \rightarrow \begin{bmatrix} 1 & 1 & 1 & | & 1 \\ 0 & 1 & -4 & | & 10 \\ 0 & 0 & -2 & | & 4 \end{bmatrix}$$

Finally, we want a 1 in row 3, column 3. To obtain it, we use the row operation $R_3 = -\frac{1}{2}r_3$. The result is

$$\begin{bmatrix} 1 & 1 & 1 & | & 1 \\ 0 & 1 & -4 & | & 10 \\ 0 & 0 & -2 & | & 4 \end{bmatrix} \rightarrow \begin{bmatrix} 1 & 1 & 1 & | & 1 \\ 0 & 1 & -4 & | & 10 \\ 0 & 0 & 1 & | & -2 \end{bmatrix}$$
$$\uparrow$$
$$R_3 = -\frac{1}{2}r_3$$

This matrix is the row echelon form of the augmented matrix. The third row of this matrix represents the equation $z = -2$. Using $z = -2$, we back-substitute into the equation $y - 4z = 10$ (from the second row) and obtain

$$y - 4z = 10$$
$$y - 4(-2) = 10 \quad z = -2$$
$$y = 2 \quad \text{Solve for } y.$$

Finally, we back-substitute $y = 2$ and $z = -2$ into the equation $x + y + z = 1$ (from the first row) and obtain

$$x + y + z = 1$$
$$x + 2 + (-2) = 1 \quad y = 2, z = -2$$
$$x = 1 \quad \text{Solve for } x.$$

The solution of the system is $x = 1$, $y = 2$, $z = -2$. ∎

The steps that we used to solve the system of linear equations in Example 5 can be summarized as follows:

MATRIX METHOD FOR SOLVING A SYSTEM OF LINEAR EQUATIONS (ROW ECHELON FORM)

STEP 1:   Write the augmented matrix that represents the system.

STEP 2:   Perform row operations that place the entry 1 in row 1, column 1.

STEP 3:   Perform row operations that leave the entry 1 in row 1, column 1 unchanged, while causing 0's to appear below it in column 1.

STEP 4:   Perform row operations that place the entry 1 in row 2, column 2, but leave the entries in columns to the left unchanged. If it is impossible to place a 1 in row 2, column 2, then proceed to place a 1 in row 2, column 3. Once a 1 is in place, perform row operations to place 0's under it.
[If any rows are obtained that contain only 0's on the left side of the vertical bar, place such rows at the bottom of the matrix.]

STEP 5:   Now repeat Step 4, placing a 1 in the next row, but one column to the right. Continue until the bottom row or the vertical bar is reached.

STEP 6:   The matrix that results is the row echelon form of the augmented matrix. Analyze the system of equations corresponding to it to solve the original system.

EXAMPLE 6   **Solving a System of Linear Equations Using Matrices (Row Echelon Form)**

Solve:
$$\begin{cases} x - y + z = 8 & (1) \\ 2x + 3y - z = -2 & (2) \\ 3x - 2y - 9z = 9 & (3) \end{cases}$$

Solution   STEP 1: The augmented matrix of the system is

$$\begin{bmatrix} 1 & -1 & 1 & | & 8 \\ 2 & 3 & -1 & | & -2 \\ 3 & -2 & -9 & | & 9 \end{bmatrix}$$

STEP 2: Because the entry 1 is already present in row 1, column 1, we can go to step 3.

STEP 3: Perform the row operations $R_2 = -2r_1 + r_2$ and $R_3 = -3r_1 + r_3$. Each of these leaves the entry 1 in row 1, column 1 unchanged, while causing 0's to appear under it.

$$\begin{bmatrix} 1 & -1 & 1 & | & 8 \\ 2 & 3 & -1 & | & -2 \\ 3 & -2 & -9 & | & 9 \end{bmatrix} \rightarrow \begin{bmatrix} 1 & -1 & 1 & | & 8 \\ 0 & 5 & -3 & | & -18 \\ 0 & 1 & -12 & | & -15 \end{bmatrix}$$
$$\begin{array}{l} R_2 = -2r_1 + r_2 \\ R_3 = -3r_1 + r_3 \end{array}$$

STEP 4: The easiest way to obtain the entry 1 in row 2, column 2 without altering column 1 is to interchange rows 2 and 3 (another way would be to multiply row 2 by $\frac{1}{5}$, but this introduces fractions).

$$\begin{bmatrix} 1 & -1 & 1 & | & 8 \\ 0 & 1 & -12 & | & -15 \\ 0 & 5 & -3 & | & -18 \end{bmatrix}$$

To get 0's under the 1 in row 2, column 2, perform the row operation $R_3 = -5r_2 + r_3$.

$$\begin{bmatrix} 1 & -1 & 1 & | & 8 \\ 0 & 1 & -12 & | & -15 \\ 0 & 5 & -3 & | & -18 \end{bmatrix} \rightarrow \begin{bmatrix} 1 & -1 & 1 & | & 8 \\ 0 & 1 & -12 & | & -15 \\ 0 & 0 & 57 & | & 57 \end{bmatrix}$$
$$R_3 = -5r_2 + r_3$$

STEP 5: Continuing, we obtain a 1 in row 3, column 3 by using $R_3 = \frac{1}{57}r_3$.

$$\begin{bmatrix} 1 & -1 & 1 & | & 8 \\ 0 & 1 & -12 & | & -15 \\ 0 & 0 & 57 & | & 57 \end{bmatrix} \rightarrow \begin{bmatrix} 1 & -1 & 1 & | & 8 \\ 0 & 1 & -12 & | & -15 \\ 0 & 0 & 1 & | & 1 \end{bmatrix}$$
$$R_3 = \frac{1}{57}r_3$$

STEP 6: The matrix on the right is the row echelon form of the augmented matrix. The system of equations represented by the matrix in row echelon form is

$$\begin{cases} x - y + z = 8 & (1) \\ y - 12z = -15 & (2) \\ z = 1 & (3) \end{cases}$$

Using $z = 1$, we back-substitute to get

$$\begin{cases} x - y + 1 = 8 & (1) \\ y - 12(1) = -15 & (2) \end{cases} \underset{\text{Simplify.}}{\longrightarrow} \begin{cases} x - y = 7 & (1) \\ y = -3 & (2) \end{cases}$$

We get $y = -3$, and back-substituting into $x - y = 7$, we find that $x = 4$. The solution of the system is $x = 4$, $y = -3$, $z = 1$. ∎

Sometimes it is advantageous to write a matrix in **reduced row echelon form.** In this form, row operations are used to obtain entries that are 0 above (as well as below) the leading 1 in a row. For example, the row echelon form obtained in Example 6 is

$$\begin{bmatrix} 1 & -1 & 1 & | & 8 \\ 0 & 1 & -12 & | & -15 \\ 0 & 0 & 1 & | & 1 \end{bmatrix}$$

To write this matrix in reduced row echelon form, we proceed as follows:

$$\begin{bmatrix} 1 & -1 & 1 & | & 8 \\ 0 & 1 & -12 & | & -15 \\ 0 & 0 & 1 & | & 1 \end{bmatrix} \rightarrow \underset{\substack{\uparrow \\ R_1 = r_2 + r_1}}{\begin{bmatrix} 1 & 0 & -11 & | & -7 \\ 0 & 1 & -12 & | & -15 \\ 0 & 0 & 1 & | & 1 \end{bmatrix}} \rightarrow \underset{\substack{\uparrow \\ R_1 = 11r_3 + r_1 \\ R_2 = 12r_3 + r_2}}{\begin{bmatrix} 1 & 0 & 0 & | & 4 \\ 0 & 1 & 0 & | & -3 \\ 0 & 0 & 1 & | & 1 \end{bmatrix}}$$

The matrix is now written in reduced row echelon form. The advantage of writing the matrix in this form is that the solution to the system, $x = 4$, $y = -3$, $z = 1$, is readily found, without the need to back-substitute. Another advantage will be seen in Section 8.5, where the inverse of a matrix is discussed.

 **NOW WORK PROBLEMS 33 AND 43.**

The matrix method for solving a system of linear equations also identifies systems that have infinitely many solutions and systems that are inconsistent. Let's see how.

EXAMPLE 7

Solving a System of Linear Equations Using Matrices

Solve: $\begin{cases} 6x - y - z = 4 & (1) \\ -12x + 2y + 2z = -8 & (2) \\ 5x + y - z = 3 & (3) \end{cases}$

Solution   We start with the augmented matrix of the system.

$$\begin{bmatrix} 6 & -1 & -1 & | & 4 \\ -12 & 2 & 2 & | & -8 \\ 5 & 1 & -1 & | & 3 \end{bmatrix} \rightarrow \underset{\substack{\uparrow \\ R_1 = -1r_3 + r_1}}{\begin{bmatrix} 1 & -2 & 0 & | & 1 \\ -12 & 2 & 2 & | & -8 \\ 5 & 1 & -1 & | & 3 \end{bmatrix}} \rightarrow \underset{\substack{\uparrow \\ R_2 = 12r_1 + r_2 \\ R_3 = -5r_1 + r_3}}{\begin{bmatrix} 1 & -2 & 0 & | & 1 \\ 0 & -22 & 2 & | & 4 \\ 0 & 11 & -1 & | & -2 \end{bmatrix}}$$

Obtaining a 1 in row 2, column 2 without altering column 1 can be accomplished only by $R_2 = -\frac{1}{22}r_2$ or by $R_3 = -\frac{1}{11}r_3$ and interchanging rows or by $R_2 = \frac{23}{11}r_3 + r_2$. We shall use the first of these.

$$\begin{bmatrix} 1 & -2 & 0 & | & 1 \\ 0 & -22 & 2 & | & 4 \\ 0 & 11 & -1 & | & -2 \end{bmatrix} \rightarrow \underset{\substack{\uparrow \\ R_2 = -\frac{1}{22}r_2}}{\begin{bmatrix} 1 & -2 & 0 & | & 1 \\ 0 & 1 & -\frac{1}{11} & | & -\frac{2}{11} \\ 0 & 11 & -1 & | & -2 \end{bmatrix}} \rightarrow \underset{\substack{\uparrow \\ R_3 = -11r_2 + r_3}}{\begin{bmatrix} 1 & -2 & 0 & | & 1 \\ 0 & 1 & -\frac{1}{11} & | & -\frac{2}{11} \\ 0 & 0 & 0 & | & 0 \end{bmatrix}}$$

This matrix is in row echelon form. Because the bottom row consists entirely of 0's, the system actually consists of only two equations.

$$\begin{cases} x - 2y \quad\quad = 1 & (1) \\ \quad\quad y - \frac{1}{11}z = -\frac{2}{11} & (2) \end{cases}$$

We back-substitute the solution for y from the second equation, $y = \frac{1}{11}z - \frac{2}{11}$, into the first equation to get

$$x = 2y + 1 = 2\left(\frac{1}{11}z - \frac{2}{11}\right) + 1 = \frac{2}{11}z + \frac{7}{11}$$

The original system is equivalent to the system

$$\begin{cases} x = \frac{2}{11}z + \frac{7}{11} & (1) \\ y = \frac{1}{11}z - \frac{2}{11} & (2) \end{cases}$$

where z can be any real number.

Let's look at the situation. The original system of three equations is equivalent to a system containing two equations. This means that any values of x, y, z that satisfy both

$$x = \frac{2}{11}z + \frac{7}{11} \quad \text{and} \quad y = \frac{1}{11}z - \frac{2}{11}$$

will be solutions. For example, $z = 0, x = \frac{7}{11}, y = -\frac{2}{11}$; $z = 1, x = \frac{9}{11}, y = -\frac{1}{11}$; and $z = -1, x = \frac{5}{11}, y = -\frac{3}{11}$ are some of the solutions of the original system. There are, in fact, infinitely many values of $x, y,$ and z for which the two equations are satisfied. That is, the original system has infinitely many solutions. We will write the solution of the original system as

$$\begin{cases} x = \frac{2}{11}z + \frac{7}{11} \\ y = \frac{1}{11}z - \frac{2}{11} \end{cases}$$

where z can be any real number. ■

We can also find the solution by writing the augmented matrix in reduced row echelon form. Starting with the echelon form, we have

$$\begin{bmatrix} 1 & -2 & 0 & | & 1 \\ 0 & 1 & -\frac{1}{11} & | & -\frac{2}{11} \\ 0 & 0 & 0 & | & 0 \end{bmatrix} \rightarrow \begin{bmatrix} 1 & 0 & -\frac{2}{11} & | & \frac{7}{11} \\ 0 & 1 & -\frac{1}{11} & | & -\frac{2}{11} \\ 0 & 0 & 0 & | & 0 \end{bmatrix}$$
$$\uparrow$$
$$R_1 = 2r_2 + r_1$$

The matrix on the right is in reduced row echelon form. The corresponding system of equations is

$$\begin{cases} x - \frac{2}{11}z = \frac{7}{11} & (1) \\ y - \frac{1}{11}z = -\frac{2}{11} & (2) \end{cases}$$

or, equivalently,

$$\begin{cases} x = \frac{2}{11}z + \frac{7}{11} & (1) \\ y = \frac{1}{11}z - \frac{2}{11} & (2) \end{cases}$$

where z can be any real number.

━ NOW WORK PROBLEM **49**.

EXAMPLE 8	**Solving a System of Linear Equations Using Matrices**

Solve: $\begin{cases} x + y + z = 6 \\ 2x - y - z = 3 \\ x + 2y + 2z = 0 \end{cases}$

Solution   We proceed as follows, beginning with the augmented matrix.

$$\begin{bmatrix} 1 & 1 & 1 & | & 6 \\ 2 & -1 & -1 & | & 3 \\ 1 & 2 & 2 & | & 0 \end{bmatrix} \rightarrow \begin{bmatrix} 1 & 1 & 1 & | & 6 \\ 0 & -3 & -3 & | & -9 \\ 0 & 1 & 1 & | & -6 \end{bmatrix} \rightarrow \begin{bmatrix} 1 & 1 & 1 & | & 6 \\ 0 & 1 & 1 & | & -6 \\ 0 & -3 & -3 & | & -9 \end{bmatrix} \rightarrow \begin{bmatrix} 1 & 1 & 1 & | & 6 \\ 0 & 1 & 1 & | & -6 \\ 0 & 0 & 0 & | & -27 \end{bmatrix}$$

$\uparrow$
$R_2 = -2r_1 + r_2$
$R_3 = -1r_1 + r_3$

$\uparrow$
Interchange rows 2 and 3.

$\uparrow$
$R_3 = 3r_2 + r_3$

This matrix is in row echelon form. The bottom row is equivalent to the equation

$$0x + 0y + 0z = -27$$

which has no solution. Hence, the original system is inconsistent. ∎

 **NOW WORK PROBLEM 23.**

The matrix method is especially effective for systems of equations for which the number of equations and the number of variables are unequal. Here, too, such a system is either inconsistent or consistent. If it is consistent, it will have either exactly one solution or infinitely many solutions.

Let's look at a system of four equations containing three variables.

EXAMPLE 9	**Solving a System of Linear Equations Using Matrices**

Solve: $\begin{cases} x - 2y + z = 0 & (1) \\ 2x + 2y - 3z = -3 & (2) \\ y - z = -1 & (3) \\ -x + 4y + 2z = 13 & (4) \end{cases}$

Solution   We proceed as follows, beginning with the augmented matrix.

$$\begin{bmatrix} 1 & -2 & 1 & | & 0 \\ 2 & 2 & -3 & | & -3 \\ 0 & 1 & -1 & | & -1 \\ -1 & 4 & 2 & | & 13 \end{bmatrix} \rightarrow \begin{bmatrix} 1 & -2 & 1 & | & 0 \\ 0 & 6 & -5 & | & -3 \\ 0 & 1 & -1 & | & -1 \\ 0 & 2 & 3 & | & 13 \end{bmatrix} \rightarrow \begin{bmatrix} 1 & -2 & 1 & | & 0 \\ 0 & 1 & -1 & | & -1 \\ 0 & 6 & -5 & | & -3 \\ 0 & 2 & 3 & | & 13 \end{bmatrix}$$

$\uparrow$
$R_2 = -2r_1 + r_2$
$R_4 = r_1 + r_4$

$\uparrow$
Interchange rows 2 and 3.

$$\rightarrow \begin{bmatrix} 1 & -2 & 1 & | & 0 \\ 0 & 1 & -1 & | & -1 \\ 0 & 0 & 1 & | & 3 \\ 0 & 0 & 5 & | & 15 \end{bmatrix} \rightarrow \begin{bmatrix} 1 & -2 & 1 & | & 0 \\ 0 & 1 & -1 & | & -1 \\ 0 & 0 & 1 & | & 3 \\ 0 & 0 & 0 & | & 0 \end{bmatrix}$$

$\uparrow$
$R_3 = -6r_2 + r_3$
$R_4 = -2r_2 + r_4$

$\uparrow$
$R_4 = -5r_3 + r_4$

We could stop here, since the matrix is in row echelon form, and back-substitute $z = 3$ to find x and y. Or we can continue to obtain the reduced row echelon form.

$$\rightarrow \begin{bmatrix} 1 & 0 & -1 & | & -2 \\ 0 & 1 & -1 & | & -1 \\ 0 & 0 & 1 & | & 3 \\ 0 & 0 & 0 & | & 0 \end{bmatrix} \rightarrow \begin{bmatrix} 1 & 0 & 0 & | & 1 \\ 0 & 1 & 0 & | & 2 \\ 0 & 0 & 1 & | & 3 \\ 0 & 0 & 0 & | & 0 \end{bmatrix}$$

$\uparrow$
$R_1 = 2r_2 + r_1$

$\uparrow$
$R_1 = r_3 + r_1$
$R_2 = r_3 + r_2$

The matrix is now in reduced row echelon form, and we can see that the solution is $x = 1$, $y = 2$, $z = 3$.

NOW WORK PROBLEM 65.

EXAMPLE 10 | **Nutrition**

A dietitian at Cook County Hospital wants a patient to have a meal that has 65 grams of protein, 95 grams of carbohydrates, and 905 milligrams of calcium. The hospital food service tells the dietitian that the dinner for today is chicken *a la king*, baked potatoes, and 2% milk. Each serving of chicken *a la king* has 30 grams of protein, 35 grams of carbohydrates, and 200 milligrams of calcium. Each serving of baked potatoes contains 4 grams of protein, 33 grams of carbohydrates, and 10 milligrams of calcium. Each glass of 2% milk contains 9 grams of protein, 13 grams of carbohydrates, and 300 milligrams of calcium. How many servings of each food should the dietitian provide for the patient?

Solution Let c, p, and m represent the number of servings of chicken *a la king*, baked potatoes, and milk, respectively. The dietitian wants the patient to have 65 grams of protein. Each serving of chicken *a la king* has 30 grams of protein, so c servings will have $30c$ grams of protein. Each serving of baked potatoes contains 4 grams of protein, so p potatoes will have $4p$ grams of protein. Finally, each glass of milk has 9 grams of protein, so m glasses of milk will have $9m$ grams of protein. The same logic will result in equations for carbohydrates and calcium, and we have the following system of equations:

$$\begin{cases} 30c + 4p + 9m = 65 & \text{protein equation} \\ 35c + 33p + 13m = 95 & \text{carbohydrate equation} \\ 200c + 10p + 300m = 905 & \text{calcium equation} \end{cases}$$

We begin with the augmented matrix and proceed as follows:

$$\begin{bmatrix} 30 & 4 & 9 & | & 65 \\ 35 & 33 & 13 & | & 95 \\ 200 & 10 & 300 & | & 905 \end{bmatrix} \rightarrow \begin{bmatrix} 1 & \frac{2}{15} & \frac{3}{10} & | & \frac{13}{6} \\ 35 & 33 & 13 & | & 95 \\ 200 & 10 & 300 & | & 905 \end{bmatrix} \rightarrow \begin{bmatrix} 1 & \frac{2}{15} & \frac{3}{10} & | & \frac{13}{6} \\ 0 & \frac{85}{3} & \frac{5}{2} & | & \frac{115}{6} \\ 0 & -\frac{50}{3} & 240 & | & \frac{1415}{3} \end{bmatrix}$$

$\uparrow$
$R_1 = \left(\frac{1}{30}\right)r_1$

$\uparrow$
$R_2 = -35r_1 + r_2$
$R_3 = -200r_1 + r_3$

$$\rightarrow \begin{bmatrix} 1 & \frac{2}{15} & \frac{3}{10} & | & \frac{13}{6} \\ 0 & 1 & \frac{3}{34} & | & \frac{23}{34} \\ 0 & -\frac{50}{3} & 240 & | & \frac{1415}{3} \end{bmatrix} \rightarrow \begin{bmatrix} 1 & \frac{2}{15} & \frac{3}{10} & | & \frac{13}{6} \\ 0 & 1 & \frac{3}{34} & | & \frac{23}{34} \\ 0 & 0 & \frac{4105}{17} & | & \frac{8210}{17} \end{bmatrix} \rightarrow \begin{bmatrix} 1 & \frac{2}{15} & \frac{3}{10} & | & \frac{13}{6} \\ 0 & 1 & \frac{3}{34} & | & \frac{23}{34} \\ 0 & 0 & 1 & | & 2 \end{bmatrix}$$

$\uparrow$
$R_2 = \left(\frac{3}{85}\right)r_2$

$\uparrow$
$R_3 = \left(\frac{50}{3}\right)r_2 + r_3$

$\uparrow$
$R_3 = \left(\frac{17}{4105}\right)r_3$

The matrix is now in echelon form. The final matrix represents the system

$$\begin{cases} c + \frac{2}{15}p + \frac{3}{10}m = \frac{13}{6} & (1) \\ p + \frac{3}{34}m = \frac{23}{34} & (2) \\ m = 2 & (3) \end{cases}$$

From (3), we determine that 2 glasses of milk should be served. Back-substitute $m = 2$ into equation (2) to find that $p = \frac{1}{2}$, so $\frac{1}{2}$ of a baked potato should be served. Back-substitute these values into equation (1) and find that $c = 1.5$, so 1.5 servings of chicken *a la king* should be given to the patient to meet the dietary requirements. ■

COMMENT: Most graphing utilities have the capability to put an augmented matrix into row echelon form (ref) and also reduced row echelon form (rref). See Appendix 7 for a discussion. ■

8.3 EXERCISES

In Problems 1–12, write the augmented matrix of the given system of equations.

1. $\begin{cases} x - 5y = 5 \\ 4x + 3y = 6 \end{cases}$

2. $\begin{cases} 3x + 4y = 7 \\ 4x - 2y = 5 \end{cases}$

3. $\begin{cases} 2x + 3y - 6 = 0 \\ 4x - 6y + 2 = 0 \end{cases}$

4. $\begin{cases} 9x - y = 0 \\ 3x - y - 4 = 0 \end{cases}$

5. $\begin{cases} 0.01x - 0.03y = 0.06 \\ 0.13x + 0.10y = 0.20 \end{cases}$

6. $\begin{cases} \frac{4}{3}x - \frac{3}{2}y = \frac{3}{4} \\ -\frac{1}{4}x + \frac{1}{3}y = \frac{2}{3} \end{cases}$

7. $\begin{cases} x - y + z = 10 \\ 3x + 3y = 5 \\ x + y + 2z = 2 \end{cases}$

8. $\begin{cases} 5x - y - z = 0 \\ x + y = 5 \\ 2x - 3z = 2 \end{cases}$

9. $\begin{cases} x + y - z = 2 \\ 3x - 2y = 2 \\ 5x + 3y - z = 1 \end{cases}$

10. $\begin{cases} 2x + 3y - 4z = 0 \\ x - 5z + 2 = 0 \\ x + 2y - 3z = -2 \end{cases}$

11. $\begin{cases} x - y - z = 10 \\ 2x + y + 2z = -1 \\ -3x + 4y = 5 \\ 4x - 5y + z = 0 \end{cases}$

12. $\begin{cases} x - y + 2z - w = 5 \\ x + 3y - 4z + 2w = 2 \\ 3x - y - 5z - w = -1 \end{cases}$

In Problems 13–20, perform each row operation on the given augmented matrix.

13. $\left[\begin{array}{cc|c} 1 & -3 & -2 \\ 2 & -5 & 5 \end{array}\right]$ (a) $R_2 = -2r_1 + r_2$

14. $\left[\begin{array}{cc|c} 1 & -3 & -3 \\ 2 & -5 & -4 \end{array}\right]$ (a) $R_2 = -2r_1 + r_2$

15. $\left[\begin{array}{ccc|c} 1 & -3 & 4 & 3 \\ 2 & -5 & 6 & 6 \\ -3 & 3 & 4 & 6 \end{array}\right]$ (a) $R_2 = -2r_1 + r_2$ (b) $R_3 = 3r_1 + r_3$

16. $\left[\begin{array}{ccc|c} 1 & -3 & 3 & -5 \\ 2 & -5 & -3 & -5 \\ -3 & -2 & 4 & 6 \end{array}\right]$ (a) $R_2 = -2r_1 + r_2$ (b) $R_3 = 3r_1 + r_3$

17. $\left[\begin{array}{ccc|c} 1 & -3 & 2 & -6 \\ 2 & -5 & 3 & -4 \\ -3 & -6 & 4 & 6 \end{array}\right]$ (a) $R_2 = -2r_1 + r_2$ (b) $R_3 = 3r_1 + r_3$

18. $\left[\begin{array}{ccc|c} 1 & -3 & -4 & -6 \\ 2 & -5 & 6 & -6 \\ -3 & 1 & 4 & 6 \end{array}\right]$ (a) $R_2 = -2r_1 + r_2$ (b) $R_3 = 3r_1 + r_3$

19. $\left[\begin{array}{ccc|c} 1 & -3 & 1 & -2 \\ 2 & -5 & 6 & -2 \\ -3 & 1 & 4 & 6 \end{array}\right]$ (a) $R_2 = -2r_1 + r_2$ (b) $R_3 = 3r_1 + r_3$

20. $\left[\begin{array}{ccc|c} 1 & -3 & -1 & 2 \\ 2 & -5 & 2 & 6 \\ -3 & -6 & 4 & 6 \end{array}\right]$ (a) $R_2 = -2r_1 + r_2$ (b) $R_3 = 3r_1 + r_3$

In Problems 21–32, the reduced row echelon form of a system of linear equations is given. Write the system of equations corresponding to the given matrix. Use x, y; or x, y, z; or x_1, x_2, x_3, x_4 as variables. Determine whether the system is consistent or inconsistent. If it is consistent, give the solution.

21. $\left[\begin{array}{cc|c} 1 & 0 & 5 \\ 0 & 1 & -1 \end{array}\right]$

22. $\left[\begin{array}{cc|c} 1 & 0 & -4 \\ 0 & 1 & 0 \end{array}\right]$

23. $\left[\begin{array}{ccc|c} 1 & 0 & 0 & 1 \\ 0 & 1 & 0 & 2 \\ 0 & 0 & 0 & 3 \end{array}\right]$

24. $\left[\begin{array}{ccc|c} 1 & 0 & 0 & 0 \\ 0 & 1 & 0 & 0 \\ 0 & 0 & 0 & 2 \end{array}\right]$

25. $\left[\begin{array}{ccc|c} 1 & 0 & 2 & -1 \\ 0 & 1 & -4 & -2 \\ 0 & 0 & 0 & 0 \end{array}\right]$

26. $\left[\begin{array}{ccc|c} 1 & 0 & 4 & 4 \\ 0 & 1 & 3 & 2 \\ 0 & 0 & 0 & 0 \end{array}\right]$

27. $\begin{bmatrix} 1 & 0 & 0 & 0 & | & 1 \\ 0 & 1 & 0 & 1 & | & 2 \\ 0 & 0 & 1 & 2 & | & 3 \end{bmatrix}$

28. $\begin{bmatrix} 1 & 0 & 0 & 0 & | & 1 \\ 0 & 1 & 0 & 2 & | & 2 \\ 0 & 0 & 1 & 3 & | & 0 \end{bmatrix}$

29. $\begin{bmatrix} 1 & 0 & 0 & 4 & | & 2 \\ 0 & 1 & 1 & 3 & | & 3 \\ 0 & 0 & 0 & 0 & | & 0 \end{bmatrix}$

30. $\begin{bmatrix} 1 & 0 & 0 & 0 & | & 1 \\ 0 & 1 & 0 & 0 & | & 2 \\ 0 & 0 & 1 & 2 & | & 3 \end{bmatrix}$

31. $\begin{bmatrix} 1 & 0 & 0 & 1 & | & -2 \\ 0 & 1 & 0 & 2 & | & 2 \\ 0 & 0 & 1 & -1 & | & 0 \\ 0 & 0 & 0 & 0 & | & 0 \end{bmatrix}$

32. $\begin{bmatrix} 1 & 0 & 0 & 0 & | & 1 \\ 0 & 1 & 0 & 0 & | & 2 \\ 0 & 0 & 1 & 0 & | & 3 \\ 0 & 0 & 0 & 1 & | & 0 \end{bmatrix}$

In Problems 33–68, solve each system of equations using matrices (row operations). If the system has no solution, say that it is inconsistent.

33. $\begin{cases} x + y = 8 \\ x - y = 4 \end{cases}$

34. $\begin{cases} x + 2y = 5 \\ x + y = 3 \end{cases}$

35. $\begin{cases} 2x - 4y = -2 \\ 3x + 2y = 3 \end{cases}$

36. $\begin{cases} 3x + 3y = 3 \\ 4x + 2y = \frac{8}{3} \end{cases}$

37. $\begin{cases} x + 2y = 4 \\ 2x + 4y = 8 \end{cases}$

38. $\begin{cases} 3x - y = 7 \\ 9x - 3y = 21 \end{cases}$

39. $\begin{cases} 2x + 3y = 6 \\ x - y = \frac{1}{2} \end{cases}$

40. $\begin{cases} \frac{1}{2}x + y = -2 \\ x - 2y = 8 \end{cases}$

41. $\begin{cases} 3x - 5y = 3 \\ 15x + 5y = 21 \end{cases}$

42. $\begin{cases} 2x - y = -1 \\ x + \frac{1}{2}y = \frac{3}{2} \end{cases}$

43. $\begin{cases} x - y = 6 \\ 2x - 3z = 16 \\ 2y + z = 4 \end{cases}$

44. $\begin{cases} 2x + y = -4 \\ -2y + 4z = 0 \\ 3x - 2z = -11 \end{cases}$

45. $\begin{cases} x - 2y + 3z = 7 \\ 2x + y + z = 4 \\ -3x + 2y - 2z = -10 \end{cases}$

46. $\begin{cases} 2x + y - 3z = 0 \\ -2x + 2y + z = -7 \\ 3x - 4y - 3z = 7 \end{cases}$

47. $\begin{cases} 2x - 2y - 2z = 2 \\ 2x + 3y + z = 2 \\ 3x + 2y = 0 \end{cases}$

48. $\begin{cases} 2x - 3y - z = 0 \\ -x + 2y + z = 5 \\ 3x - 4y - z = 1 \end{cases}$

49. $\begin{cases} -x + y + z = -1 \\ -x + 2y - 3z = -4 \\ 3x - 2y - 7z = 0 \end{cases}$

50. $\begin{cases} 2x - 3y - z = 0 \\ 3x + 2y + 2z = 2 \\ x + 5y + 3z = 2 \end{cases}$

51. $\begin{cases} 2x - 2y + 3z = 6 \\ 4x - 3y + 2z = 0 \\ -2x + 3y - 7z = 1 \end{cases}$

52. $\begin{cases} 3x - 2y + 2z = 6 \\ 7x - 3y + 2z = -1 \\ 2x - 3y + 4z = 0 \end{cases}$

53. $\begin{cases} x + y - z = 6 \\ 3x - 2y + z = -5 \\ x + 3y - 2z = 14 \end{cases}$

54. $\begin{cases} x - y + z = -4 \\ 2x - 3y + 4z = -15 \\ 5x + y - 2z = 12 \end{cases}$

55. $\begin{cases} x + 2y - z = -3 \\ 2x - 4y + z = -7 \\ -2x + 2y - 3z = 4 \end{cases}$

56. $\begin{cases} x + 4y - 3z = -8 \\ 3x - y + 3z = 12 \\ x + y + 6z = 1 \end{cases}$

57. $\begin{cases} 3x + y - z = \frac{2}{3} \\ 2x - y + z = 1 \\ 4x + 2y = \frac{8}{3} \end{cases}$

58. $\begin{cases} x + y = 1 \\ 2x - y + z = 1 \\ x + 2y + z = \frac{8}{3} \end{cases}$

59. $\begin{cases} x + y + z + w = 4 \\ 2x - y + z = 0 \\ 3x + 2y + z - w = 6 \\ x - 2y - 2z + 2w = -1 \end{cases}$

60. $\begin{cases} x + y + z + w = 4 \\ -x + 2y + z = 0 \\ 2x + 3y + z - w = 6 \\ -2x + y - 2z + 2w = -1 \end{cases}$

61. $\begin{cases} x + 2y + z = 1 \\ 2x - y + 2z = 2 \\ 3x + y + 3z = 3 \end{cases}$

62. $\begin{cases} x + 2y - z = 3 \\ 2x - y + 2z = 6 \\ x - 3y + 3z = 4 \end{cases}$

63. $\begin{cases} x - y + z = 5 \\ 3x + 2y - 2z = 0 \end{cases}$

64. $\begin{cases} 2x + y - z = 4 \\ -x + y + 3z = 1 \end{cases}$

65. $\begin{cases} 2x + 3y - z = 3 \\ x - y - z = 0 \\ -x + y + z = 0 \\ x + y + 3z = 5 \end{cases}$

66. $\begin{cases} x - 3y + z = 1 \\ 2x - y - 4z = 0 \\ x - 3y + 2z = 1 \\ x - 2y = 5 \end{cases}$

67. $\begin{cases} 4x + y + z - w = 4 \\ x - y + 2z + 3w = 3 \end{cases}$

68. $\begin{cases} -4x + y = 5 \\ 2x - y + z - w = 5 \\ z + w = 4 \end{cases}$

69. **Curve Fitting**   Find the function $y = ax^2 + bx + c$ whose graph contains the points $(1, 2)$, $(-2, -7)$, and $(2, -3)$.

70. **Curve Fitting**   Find the function $y = ax^2 + bx + c$ whose graph contains the points $(1, -1)$, $(3, -1)$, and $(-2, 14)$.

71. **Curve Fitting**   Find the function $f(x) = ax^3 + bx^2 + cx + d$ for which $f(-3) = -112, f(-1) = -2, f(1) = 4$, and $f(2) = 13$.

72. **Curve Fitting**   Find the function $f(x) = ax^3 + bx^2 + cx + d$ for which $f(-2) = -10, f(-1) = 3, f(1) = 5$, and $f(3) = 15$.

73. **Nutrition**   A dietitian at Palos Community Hospital wants a patient to have a meal that has 78 grams of protein, 59 grams of carbohydrates, and 75 milligrams of vitamin A. The hospital food service tells the dietitian that the dinner for today is salmon steak, baked eggs, and acorn squash. Each serving of salmon steak has 30 grams of protein, 20 grams of carbohydrates, and 2 milligrams of vitamin A. Each serving of baked eggs contains 15 grams of protein, 2 grams of carbohydrates, and 20 milligrams of vitamin A. Each serving of acorn squash contains 3 grams of protein, 25 grams of carbohydrates, and 32 milligrams of vitamin A. How many servings of each food should the dietitian provide for the patient?

74. **Nutrition**   A dietitian at General Hospital wants a patient to have a meal that has 47 grams of protein, 58 grams of carbohydrates, and 630 milligrams of calcium. The hospital food service tells the dietitian that the dinner for today is pork chops, corn on the cob, and 2% milk. Each serving of pork chops has 23 grams of protein, 0 grams of carbohydrates, and 10 milligrams of calcium. Each serving of corn on the cob contains 3 grams of protein, 16 grams of carbohydrates, and 10 milligrams of calcium. Each glass of 2% milk contains 9 grams of protein, 13 grams of carbohydrates, and 300 milligrams of calcium. How many servings of each food should the dietitian provide for the patient?

75. **Financial Planning**   Carletta has $10,000 to invest. As her financial consultant, you recommend that she invest in Treasury bills that yield 6%, Treasury bonds that yield 7%, and corporate bonds that yield 8%. Carletta wants to have an annual income of $680, and the amount invested in corporate bonds must be half that invested in Treasury bills. Find the amount in each investment.

76. **Financial Planning**   John has $20,000 to invest. As his financial consultant, you recommend that he invest in Treasury bills that yield 5%, Treasury bonds that yield 7%, and corporate bonds that yield 9%. John wants to have an annual income of $1280, and the amount invested in Treasury bills must be two times the amount invested in corporate bonds. Find the amount in each investment.

77. **Production**   To manufacture an automobile requires painting, drying, and polishing. Epsilon Motor Company produces three types of cars: the Delta, the Beta, and the Sigma. Each Delta requires 10 hours for painting, 3 hours for drying, and 2 hours for polishing. A Beta requires 16 hours of painting, 5 hours of drying, and 3 hours of polishing, while a Sigma requires 8 hours for painting, 2 hours for drying, and 1 hour for polishing. If the company has 240 hours for painting, 69 hours for drying, and 41 hours for polishing per month, how many of each type of car are produced?

78. **Production**   A Florida juice company completes the preparation of its products by sterilizing, filling, and labeling bottles. Each case of orange juice requires 9 minutes for sterilizing, 6 minutes for filling, and 1 minute for labeling. Each case of grapefruit juice requires 10 minutes for sterilizing, 4 minutes for filling, and 2 minutes for labeling. Each case of tomato juice requires 12 minutes for sterilizing, 4 minutes for filling, and 1 minute for labeling. If the company runs the sterilizing machine for 398 minutes, the filling machine for 164 minutes, and the labeling machine for 58 minutes, how many cases of each type of juice are prepared?

79. **Electricity: Kirchhoff's Rules**   An application of Kirchhoff's Rules to the circuit shown results in the following system of equations:

$$\begin{cases} -4 + 8 - 2I_2 = 0 \\ 8 = 5I_4 + I_1 \\ 4 = 3I_3 + I_1 \\ I_3 + I_4 = I_1 \end{cases}$$

Find the currents I_1, I_2, I_3, and I_4.*

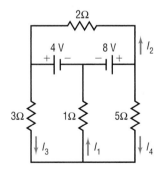

80. **Electricity: Kirchhoff's Rules**   An application of Kirchhoff's Rules to the circuit shown on page 548, results in the following system of equations:

$$\begin{cases} I_1 = I_3 + I_2 \\ 24 - 6I_1 - 3I_3 = 0 \\ 12 + 24 - 6I_1 - 6I_2 = 0 \end{cases}$$

*Source: Based on Raymond Serway, *Physics,* 3rd ed. (Philadelphia: Saunders, 1990), Prob. 34, p. 79.

Find the currents I_1, I_2, and I_3.[†]

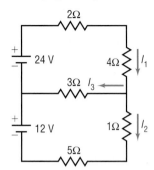

81. Financial Planning   Three retired couples each require an additional annual income of $2000 per year. As their financial consultant, you recommend that they invest some money in Treasury bills that yield 7%, some money in corporate bonds that yield 9%, and some money in junk bonds that yield 11%. Prepare a table for each couple showing the various ways that their goals can be achieved:
  (a) If the first couple has $20,000 to invest.
  (b) If the second couple has $25,000 to invest.
  (c) If the third couple has $30,000 to invest.
  (d) What advice would you give each couple regarding the amount to invest and the choices available?
  [**Hint:** Higher yields generally carry more risk.]

82. Financial Planning   A young couple has $25,000 to invest. As their financial consultant, you recommend that they invest some money in Treasury bills that yield 7%, some money in corporate bonds that yield 9%, and some money in junk bonds that yield 11%. Prepare a table showing the various ways this couple can achieve the following goals:
  (a) The couple wants $1500 per year in income.

[†] *Source:* Ibid., Prob. 38, p. 791.

  (b) The couple wants $2000 per year in income.
  (c) The couple wants $2500 per year in income.
  (d) What advice would you give this couple regarding the income that they require and the choices available?
  [**Hint:** Higher yields generally carry more risk.]

83. Pharmacy   A doctor's prescription calls for a daily intake of liquid containing 40 mg of vitamin C and 30 mg of vitamin D. Your pharmacy stocks three liquids that can be used: one contains 20% vitamin C and 30% vitamin D; a second, 40% vitamin C and 20% vitamin D; and a third, 30% vitamin C and 50% vitamin D. Create a table showing the possible combinations that could be used to fill the prescription.

84. Pharmacy   A doctor's prescription calls for the creation of pills that contain 12 units of vitamin B_{12} and 12 units of vitamin E. Your pharmacy stocks three powders that can be used to make these pills: one contains 20% vitamin B_{12} and 30% vitamin E; a second, 40% vitamin B_{12} and 20% vitamin E; and a third, 30% vitamin B_{12} and 40% vitamin E. Create a table showing the possible combinations of each powder that could be mixed in each pill.

85. Write a brief paragraph or two that outline your strategy for solving a system of linear equations using matrices.

86. When solving a system of linear equations using matrices, do you prefer to place the augmented matrix in row echelon form or in reduced row echelon form? Give reasons for your choice.

87. Make up a system of three linear equations containing three variables that has:
  (a) No solution
  (b) Exactly one solution
  (c) Infinitely many solutions
  Give the three systems to a friend to solve and critique.

8.4   SYSTEMS OF LINEAR EQUATIONS: DETERMINANTS

OBJECTIVES   **1** Evaluate 2 by 2 Determinants
  2 Use Cramer's Rule to Solve a System of Two Equations, Two Variables
  3 Evaluate 3 by 3 Determinants
  4 Use Cramer's Rule to Solve a System of Three Equations, Three Variables
  5 Know Properties of Determinants

1   In the preceding section, we described a method of using matrices to solve a system of linear equations. This section deals with yet another method for solving systems of linear equations; however, it can be used only when the number of equations equals the number of variables. Although the method will work for any system (provided that the number of equations equals the number of variables), it is most often used for systems of two equations containing two variables or three equations containing three variables. This method, called *Cramer's Rule,* is based on the concept of a *determinant.*

2 BY 2 DETERMINANTS

If $a, b, c,$ and d are four real numbers, the symbol

$$D = \begin{vmatrix} a & b \\ c & d \end{vmatrix}$$

is called a **2 by 2 determinant.** Its value is the number $ad - bc$; that is,

$$D = \begin{vmatrix} a & b \\ c & d \end{vmatrix} = ad - bc \qquad (1)$$

A device that may be helpful for remembering the value of a 2 by 2 determinant is the following:

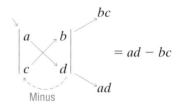

EXAMPLE 1   Evaluating a 2 × 2 Determinant

$$\begin{vmatrix} 3 & -2 \\ 6 & 1 \end{vmatrix} = (3)(1) - (6)(-2) = 3 - (-12) = 15$$

NOW WORK PROBLEM 3.

CRAMER'S RULE

② Let's now see the role that a 2 by 2 determinant plays in the solution of a system of two equations containing two variables. Consider the system

$$\begin{cases} ax + by = s & (1) \\ cx + dy = t & (2) \end{cases} \qquad (2)$$

We shall use the method of elimination to solve this system.

Provided $d \neq 0$ and $b \neq 0$, this system is equivalent to the system

$$\begin{cases} adx + bdy = sd & \text{(1) Multiply by } d. \\ bcx + bdy = tb & \text{(2) Multiply by } b. \end{cases}$$

On subtracting the second equation from the first equation, we get

$$\begin{cases} (ad - bc)x + 0 \cdot y = sd - tb & (1) \\ bcx \quad\;\; + bdy = tb & (2) \end{cases}$$

Now, the first equation can be rewritten using determinant notation.

$$\begin{vmatrix} a & b \\ c & d \end{vmatrix} x = \begin{vmatrix} s & b \\ t & d \end{vmatrix}$$

If $D = \begin{vmatrix} a & b \\ c & d \end{vmatrix} = ad - bc \neq 0$, we can solve for x to get

$$x = \frac{\begin{vmatrix} s & b \\ t & d \end{vmatrix}}{\begin{vmatrix} a & b \\ c & d \end{vmatrix}} = \frac{\begin{vmatrix} s & b \\ t & d \end{vmatrix}}{D} \qquad (3)$$

Return now to the original system (2). Provided that $a \neq 0$ and $c \neq 0$, the system is equivalent to

$$\begin{cases} acx + bcy = cs & \text{(1) Multiply by } c. \\ acx + ady = at & \text{(2) Multiply by } a. \end{cases}$$

On subtracting the first equation from the second equation, we get

$$\begin{cases} acx + bcy = cs & \text{(1)} \\ 0 \cdot x + (ad - bc)y = at - cs & \text{(2)} \end{cases}$$

The second equation can now be rewritten using determinant notation.

$$\begin{vmatrix} a & b \\ c & d \end{vmatrix} y = \begin{vmatrix} a & s \\ c & t \end{vmatrix}$$

If $D = \begin{vmatrix} a & b \\ c & d \end{vmatrix} = ad - bc \neq 0$, we can solve for y to get

$$y = \frac{\begin{vmatrix} a & s \\ c & t \end{vmatrix}}{\begin{vmatrix} a & b \\ c & d \end{vmatrix}} = \frac{\begin{vmatrix} a & s \\ c & t \end{vmatrix}}{D} \qquad (4)$$

Equations (3) and (4) lead us to the following result, called **Cramer's Rule.**

Theorem

Cramer's Rule for Two Equations Containing Two Variables

The solution to the system of equations

$$\begin{cases} ax + by = s & \text{(1)} \\ cx + dy = t & \text{(2)} \end{cases} \qquad (5)$$

is given by

$$x = \frac{\begin{vmatrix} s & b \\ t & d \end{vmatrix}}{\begin{vmatrix} a & b \\ c & d \end{vmatrix}}, \qquad y = \frac{\begin{vmatrix} a & s \\ c & t \end{vmatrix}}{\begin{vmatrix} a & b \\ c & d \end{vmatrix}} \qquad (6)$$

provided that

$$D = \begin{vmatrix} a & b \\ c & d \end{vmatrix} = ad - bc \neq 0$$

In the derivation given for Cramer's Rule above, we assumed that none of the numbers a, b, c, and d was 0. In Problem 58 you will be asked to complete the proof under the less stringent conditions that $D = ad - bc \neq 0$.

Now look carefully at the pattern in Cramer's Rule. The denominator in the solution (6) is the determinant of the coefficients of the variables.

$$\begin{cases} ax + by = s \\ cx + dy = t \end{cases} \qquad D = \begin{vmatrix} a & b \\ c & d \end{vmatrix}$$

In the solution for x, the numerator is the determinant, denoted by D_x, formed by replacing the entries in the first column (the coefficients of x) of D by the constants on the right side of the equal sign.

$$D_x = \begin{vmatrix} s & b \\ t & d \end{vmatrix}$$

In the solution for y, the numerator is the determinant, denoted by D_y, formed by replacing the entries in the second column (the coefficients of y) of D by the constants on the right side of the equal sign.

$$D_y = \begin{vmatrix} a & s \\ c & t \end{vmatrix}$$

Cramer's Rule then states that, if $D \neq 0$,

$$x = \frac{D_x}{D}, \qquad y = \frac{D_y}{D} \qquad \qquad \textbf{(7)}$$

EXAMPLE 2   **Solving a System of Linear Equations Using Determinants**

Use Cramer's Rule, if applicable, to solve the system

$$\begin{cases} 3x - 2y = 4 & (1) \\ 6x + y = 13 & (2) \end{cases}$$

Solution   The determinant D of the coefficients of the variables is

$$D = \begin{vmatrix} 3 & -2 \\ 6 & 1 \end{vmatrix} = (3)(1) - (6)(-2) = 15$$

Because $D \neq 0$, Cramer's Rule (7) can be used.

$$x = \frac{D_x}{D} = \frac{\begin{vmatrix} 4 & -2 \\ 13 & 1 \end{vmatrix}}{15} = \frac{30}{15} = 2 \qquad y = \frac{D_y}{D} = \frac{\begin{vmatrix} 3 & 4 \\ 6 & 13 \end{vmatrix}}{15} = \frac{15}{15} = 1$$

The solution is $x = 2$, $y = 1$.   ∎

In attempting to use Cramer's Rule, if the determinant D of the coefficients of the variables is found to equal 0 (so that Cramer's Rule is not applicable), then the system is either inconsistent or has infinitely many solutions.

NOW WORK PROBLEM **11**.

3 BY 3 DETERMINANTS

③ To use Cramer's Rule to solve a system of three equations containing three variables, we need to define a 3 by 3 determinant.

A **3 by 3 determinant** is symbolized by

$$\begin{vmatrix} a_{11} & a_{12} & a_{13} \\ a_{21} & a_{22} & a_{23} \\ a_{31} & a_{32} & a_{33} \end{vmatrix} \tag{8}$$

in which $a_{11}, a_{12}, \ldots,$ are real numbers.

As with matrices, we use a double subscript to identify an entry by indicating its row and column numbers. For example, the entry a_{23} is in row 2, column 3.

The value of a 3 by 3 determinant may be defined in terms of 2 by 2 determinants by the following formula:

$$\begin{vmatrix} a_{11} & a_{12} & a_{13} \\ a_{21} & a_{22} & a_{23} \\ a_{31} & a_{32} & a_{33} \end{vmatrix} = a_{11}\begin{vmatrix} a_{22} & a_{23} \\ a_{32} & a_{33} \end{vmatrix} \overset{\text{Minus}}{\underset{\downarrow}{-}} a_{12}\begin{vmatrix} a_{21} & a_{23} \\ a_{31} & a_{33} \end{vmatrix} + a_{13}\begin{vmatrix} a_{21} & a_{22} \\ a_{31} & a_{32} \end{vmatrix} \tag{9}$$

2 by 2 determinant left after removing row and column containing a_{11}	2 by 2 determinant left after removing row and column containing a_{12}	2 by 2 determinant left after removing row and column containing a_{13}

The 2 by 2 determinants shown in formula (9) are called **minors** of the 3 by 3 determinant. For an n by n determinant, the **minor** M_{ij} of element a_{ij} is the determinant resulting from removing the ith row and jth column.

EXAMPLE 3

Finding Minors of a 3 by 3 Determinant

For the determinant $A = \begin{vmatrix} 2 & -1 & 3 \\ -2 & 5 & 1 \\ 0 & 6 & -9 \end{vmatrix}$, find: (a) M_{12} (b) M_{23}

Solution   (a) M_{12} is the determinant that results from removing the first row and second column from A.

$$A = \begin{vmatrix} 2 & -1 & 3 \\ -2 & 5 & 1 \\ 0 & 6 & -9 \end{vmatrix} \qquad M_{12} = \begin{vmatrix} -2 & 1 \\ 0 & -9 \end{vmatrix} = (-2)(-9) - (0)(1) = 18$$

(b) M_{23} is the determinant that results from removing the second row and third column from A.

$$A = \begin{vmatrix} 2 & -1 & 3 \\ -2 & 5 & 1 \\ 0 & 6 & -9 \end{vmatrix} \qquad M_{23} = \begin{vmatrix} 2 & -1 \\ 0 & 6 \end{vmatrix} = (2)(6) - (0)(-1) = 12$$

■

Referring back to formula (9), we see that each element a_{ij} is multiplied by its minor, but sometimes this term is added and other times subtracted. To determine whether to add or subtract a term, we must consider the *cofactor*.

For an n by n determinant A, the **cofactor** of element a_{ij}, denoted by A_{ij}, is given by

$$A_{ij} = (-1)^{i+j} M_{ij}$$

where M_{ij} is the minor of element a_{ij}.

The exponent of $(-1)^{i+j}$ is the sum of the row and column of the element a_{ij}; so if $i + j$ is even, $(-1)^{i+j}$ will equal 1, and if $i + j$ is odd, $(-1)^{i+j}$ will equal -1.

To find the value of a determinant, multiply each element in any row or column by its cofactor and sum the results. This process is referred to as *expanding across a row or column*. For example, the value of the 3 by 3 determinant in formula (9) was found by expanding across row 1.

If we choose to expand down column 2, we obtain

$$\begin{vmatrix} a_{11} & a_{12} & a_{13} \\ a_{21} & a_{22} & a_{23} \\ a_{31} & a_{32} & a_{33} \end{vmatrix} = (-1)^{1+2} a_{12} \begin{vmatrix} a_{21} & a_{23} \\ a_{31} & a_{33} \end{vmatrix} + (-1)^{2+2} a_{22} \begin{vmatrix} a_{11} & a_{13} \\ a_{31} & a_{33} \end{vmatrix} + (-1)^{3+2} a_{32} \begin{vmatrix} a_{11} & a_{13} \\ a_{21} & a_{23} \end{vmatrix}$$

↑
Expand down column 2.

If we choose to expand across row 3, we obtain

$$\begin{vmatrix} a_{11} & a_{12} & a_{13} \\ a_{21} & a_{22} & a_{23} \\ a_{31} & a_{32} & a_{33} \end{vmatrix} = (-1)^{3+1} a_{31} \begin{vmatrix} a_{12} & a_{13} \\ a_{22} & a_{23} \end{vmatrix} + (-1)^{3+2} a_{32} \begin{vmatrix} a_{11} & a_{13} \\ a_{21} & a_{23} \end{vmatrix} + (-1)^{3+3} a_{33} \begin{vmatrix} a_{11} & a_{12} \\ a_{21} & a_{22} \end{vmatrix}$$

↑
Expand across row 3.

It can be shown that the value of a determinant does not depend on the choice of the row or column used in the expansion. However, expanding across a row or column that has an element equal to 0 reduces the amount of work needed to compute the value of the determinant.

EXAMPLE 4 **Evaluating a 3 × 3 Determinant**

Find the value of the 3 by 3 determinant: $\begin{vmatrix} 3 & 4 & -1 \\ 4 & 6 & 2 \\ 8 & -2 & 3 \end{vmatrix}$

Solution We choose to expand across row 1.

$$\begin{vmatrix} 3 & 4 & -1 \\ 4 & 6 & 2 \\ 8 & -2 & 3 \end{vmatrix} = (-1)^{1+1} 3 \begin{vmatrix} 6 & 2 \\ -2 & 3 \end{vmatrix} + (-1)^{1+2} 4 \begin{vmatrix} 4 & 2 \\ 8 & 3 \end{vmatrix} + (-1)^{1+3} (-1) \begin{vmatrix} 4 & 6 \\ 8 & -2 \end{vmatrix}$$

$$= 3(18 + 4) - 4(12 - 16) + (-1)(-8 - 48)$$

$$= 3(22) - 4(-4) + (-1)(-56)$$

$$= 66 + 16 + 56 = 138$$

We could also find the value of the 3 by 3 determinant in Example 4 by expanding down column 3.

$$\begin{vmatrix} 3 & 4 & -1 \\ 4 & 6 & 2 \\ 8 & -2 & 3 \end{vmatrix} = (-1)^{1+3}(-1)\begin{vmatrix} 4 & 6 \\ 8 & -2 \end{vmatrix} + (-1)^{2+3}2\begin{vmatrix} 3 & 4 \\ 8 & -2 \end{vmatrix} + (-1)^{3+3}3\begin{vmatrix} 3 & 4 \\ 4 & 6 \end{vmatrix}$$

$$= -1(-8 - 48) - 2(-6 - 32) + 3(18 - 16)$$

$$= 56 + 76 + 6 = 138$$

COMMENT: A graphing utility can be used to evaluate determinants. Check your manual to see how. Then check the answer obtained in Example 4. ∎

 NOW WORK PROBLEM 7.

SYSTEMS OF THREE EQUATIONS CONTAINING THREE VARIABLES

④ Consider the following system of three equations containing three variables.

$$\begin{cases} a_{11}x + a_{12}y + a_{13}z = c_1 \\ a_{21}x + a_{22}y + a_{23}z = c_2 \\ a_{31}x + a_{32}y + a_{33}z = c_3 \end{cases} \qquad \textbf{(10)}$$

It can be shown that if the determinant D of the coefficients of the variables is not 0, that is, if

$$D = \begin{vmatrix} a_{11} & a_{12} & a_{13} \\ a_{21} & a_{22} & a_{23} \\ a_{31} & a_{32} & a_{33} \end{vmatrix} \neq 0$$

then the unique solution of system (10) is given by

Cramer's Rule for Three Equations Containing Three Variables

$$x = \frac{D_x}{D} \qquad y = \frac{D_y}{D} \qquad z = \frac{D_z}{D}$$

where

$$D_x = \begin{vmatrix} c_1 & a_{12} & a_{13} \\ c_2 & a_{22} & a_{23} \\ c_3 & a_{32} & a_{33} \end{vmatrix} \qquad D_y = \begin{vmatrix} a_{11} & c_1 & a_{13} \\ a_{21} & c_2 & a_{23} \\ a_{31} & c_3 & a_{33} \end{vmatrix} \qquad D_z = \begin{vmatrix} a_{11} & a_{12} & c_1 \\ a_{21} & a_{22} & c_2 \\ a_{31} & a_{32} & c_3 \end{vmatrix}$$

The similarity of this pattern and the pattern observed earlier for a system of two equations containing two variables should be apparent.

EXAMPLE 5   ### Using Cramer's Rule

Use Cramer's Rule, if applicable, to solve the following system:

$$\begin{cases} 2x + y - z = 3 & (1) \\ -x + 2y + 4z = -3 & (2) \\ x - 2y - 3z = 4 & (3) \end{cases}$$

Solution   The value of the determinant D of the coefficients of the variables is

$$D = \begin{vmatrix} 2 & 1 & -1 \\ -1 & 2 & 4 \\ 1 & -2 & -3 \end{vmatrix} = (-1)^{1+1}2 \begin{vmatrix} 2 & 4 \\ -2 & -3 \end{vmatrix} + (-1)^{1+2}1 \begin{vmatrix} -1 & 4 \\ 1 & -3 \end{vmatrix} + (-1)^{1+3}(-1) \begin{vmatrix} -1 & 2 \\ 1 & -2 \end{vmatrix}$$

$$= 2(2) - 1(-1) + (-1)(0)$$

$$= 4 + 1 = 5$$

Because $D \neq 0$, we proceed to find the values of D_x, D_y, D_z.

$$D_x = \begin{vmatrix} 3 & 1 & -1 \\ -3 & 2 & 4 \\ 4 & -2 & -3 \end{vmatrix} = (-1)^{1+1}3 \begin{vmatrix} 2 & 4 \\ -2 & -3 \end{vmatrix} + (-1)^{1+2}1 \begin{vmatrix} -3 & 4 \\ 4 & -3 \end{vmatrix} + (-1)^{1+3}(-1) \begin{vmatrix} -3 & 2 \\ 4 & -2 \end{vmatrix}$$

$$= 3(2) - 1(-7) + (-1)(-2) = 15$$

$$D_y = \begin{vmatrix} 2 & 3 & -1 \\ -1 & -3 & 4 \\ 1 & 4 & -3 \end{vmatrix} = (-1)^{1+1}2 \begin{vmatrix} -3 & 4 \\ 4 & -3 \end{vmatrix} + (-1)^{1+2}3 \begin{vmatrix} -1 & 4 \\ 1 & -3 \end{vmatrix} + (-1)^{1+3}(-1) \begin{vmatrix} -1 & -3 \\ 1 & 4 \end{vmatrix}$$

$$= 2(-7) - 3(-1) + (-1)(-1)$$

$$= -14 + 3 + 1 = -10$$

$$D_z = \begin{vmatrix} 2 & 1 & 3 \\ -1 & 2 & -3 \\ 1 & -2 & 4 \end{vmatrix} = (-1)^{1+1}2 \begin{vmatrix} 2 & -3 \\ -2 & 4 \end{vmatrix} + (-1)^{1+2}1 \begin{vmatrix} -1 & -3 \\ 1 & 4 \end{vmatrix} + (-1)^{1+3}3 \begin{vmatrix} -1 & 2 \\ 1 & -2 \end{vmatrix}$$

$$= 2(2) - 1(-1) + 3(0) = 5$$

As a result,

$$x = \frac{D_x}{D} = \frac{15}{5} = 3 \qquad y = \frac{D_y}{D} = \frac{-10}{5} = -2 \qquad z = \frac{D_z}{D} = \frac{5}{5} = 1$$

The solution is $x = 3$, $y = -2$, $z = 1$.   ∎

If the determinant of the coefficients of the variables of a system of three linear equations containing three variables is 0, then Cramer's Rule is not applicable. In such a case, the system is either inconsistent or has infinitely many solutions.

NOW WORK PROBLEM 29.

PROPERTIES OF DETERMINANTS

⑤   Determinants have several properties that are sometimes helpful for obtaining their value. We list some of them here.

Theorem   The value of a determinant changes sign if any two rows (or any two columns) are interchanged.   **(11)**   ∎

Proof for 2 by 2 Determinants

$$\begin{vmatrix} a & b \\ c & d \end{vmatrix} = ad - bc \quad \text{and} \quad \begin{vmatrix} c & d \\ a & b \end{vmatrix} = bc - ad = -(ad - bc)$$ ∎

EXAMPLE 6   Demonstrating Theorem (11)

$$\begin{vmatrix} 3 & 4 \\ 1 & 2 \end{vmatrix} = 6 - 4 = 2 \qquad \begin{vmatrix} 1 & 2 \\ 3 & 4 \end{vmatrix} = 4 - 6 = -2$$

  ■

Theorem   If all the entries in any row (or any column) equal 0, the value of the determinant is 0.   **(12)**

  ■

Proof   Merely expand across the row (or down the column) containing the 0's.   ■

Theorem   If any two rows (or any two columns) of a determinant have corresponding entries that are equal, the value of the determinant is 0.   **(13)**

  ■

     You are asked to prove this result for a 3 by 3 determinant in which the entries in column 1 equal the entries in column 3 in Problem 61.

EXAMPLE 7   Demonstrating Theorem (13)

$$\begin{vmatrix} 1 & 2 & 3 \\ 1 & 2 & 3 \\ 4 & 5 & 6 \end{vmatrix} = (-1)^{1+1}1\begin{vmatrix} 2 & 3 \\ 5 & 6 \end{vmatrix} + (-1)^{1+2}2\begin{vmatrix} 1 & 3 \\ 4 & 6 \end{vmatrix} + (-1)^{1+3}3\begin{vmatrix} 1 & 2 \\ 4 & 5 \end{vmatrix}$$

$$= 1(-3) - 2(-6) + 3(-3)$$

$$= -3 + 12 - 9 = 0$$

  ■

Theorem   If any row (or any column) of a determinant is multiplied by a nonzero number k, the value of the determinant is also changed by a factor of k.   **(14)**

  ■

     You are asked to prove this result for a 3 by 3 determinant using row 2 in Problem 60.

EXAMPLE 8   Demonstrating Theorem (14)

$$\begin{vmatrix} 1 & 2 \\ 4 & 6 \end{vmatrix} = 6 - 8 = -2$$

$$\begin{vmatrix} k & 2k \\ 4 & 6 \end{vmatrix} = 6k - 8k = -2k = k(-2) = k\begin{vmatrix} 1 & 2 \\ 4 & 6 \end{vmatrix}$$

  ■

Theorem   If the entries of any row (or any column) of a determinant are multiplied by a nonzero number k and the result is added to the corresponding entries of another row (or column), the value of the determinant remains unchanged.   **(15)**

  ■

     In Problem 62, you are asked to prove this result for a 3 by 3 determinant using rows 1 and 2.

| EXAMPLE 9 | **Demonstrating Theorem (15)** |

$$\begin{vmatrix} 3 & 4 \\ 5 & 2 \end{vmatrix} = \begin{vmatrix} -7 & 0 \\ 5 & 2 \end{vmatrix} = -14$$

Multiply row 2 by −2 and add to row 1. ∎

8.4 EXERCISES

In Problems 1–10, find the value of each determinant.

1. $\begin{vmatrix} 3 & 1 \\ 4 & 2 \end{vmatrix}$
2. $\begin{vmatrix} 6 & 1 \\ 5 & 2 \end{vmatrix}$
3. $\begin{vmatrix} 6 & 4 \\ -1 & 3 \end{vmatrix}$
4. $\begin{vmatrix} 8 & -3 \\ 4 & 2 \end{vmatrix}$

5. $\begin{vmatrix} -3 & -1 \\ 4 & 2 \end{vmatrix}$
6. $\begin{vmatrix} -4 & 2 \\ -5 & 3 \end{vmatrix}$
7. $\begin{vmatrix} 3 & 4 & 2 \\ 1 & -1 & 5 \\ 1 & 2 & -2 \end{vmatrix}$
8. $\begin{vmatrix} 1 & 3 & -2 \\ 6 & 1 & -5 \\ 8 & 2 & 3 \end{vmatrix}$

9. $\begin{vmatrix} 4 & -1 & 2 \\ 6 & -1 & 0 \\ 1 & -3 & 4 \end{vmatrix}$
10. $\begin{vmatrix} 3 & -9 & 4 \\ 1 & 4 & 0 \\ 8 & -3 & 1 \end{vmatrix}$

In Problems 11–40, solve each system of equations using Cramer's Rule if it is applicable. If Cramer's Rule is not applicable, say so.

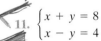

11. $\begin{cases} x + y = 8 \\ x - y = 4 \end{cases}$
12. $\begin{cases} x + 2y = 5 \\ x - y = 3 \end{cases}$
13. $\begin{cases} 5x - y = 13 \\ 2x + 3y = 12 \end{cases}$
14. $\begin{cases} x + 3y = 5 \\ 2x - 3y = -8 \end{cases}$

15. $\begin{cases} 3x = 24 \\ x + 2y = 0 \end{cases}$
16. $\begin{cases} 4x + 5y = -3 \\ -2y = -4 \end{cases}$
17. $\begin{cases} 3x - 6y = 24 \\ 5x + 4y = 12 \end{cases}$
18. $\begin{cases} 2x + 4y = 16 \\ 3x - 5y = -9 \end{cases}$

19. $\begin{cases} 3x - 2y = 4 \\ 6x - 4y = 0 \end{cases}$
20. $\begin{cases} -x + 2y = 5 \\ 4x - 8y = 6 \end{cases}$
21. $\begin{cases} 2x - 4y = -2 \\ 3x + 2y = 3 \end{cases}$
22. $\begin{cases} 3x + 3y = 3 \\ 4x + 2y = \frac{8}{3} \end{cases}$

23. $\begin{cases} 2x - 3y = -1 \\ 10x + 10y = 5 \end{cases}$
24. $\begin{cases} 3x - 2y = 0 \\ 5x + 10y = 4 \end{cases}$
25. $\begin{cases} 2x + 3y = 6 \\ x - y = \frac{1}{2} \end{cases}$
26. $\begin{cases} \frac{1}{2}x + y = -2 \\ x - 2y = 8 \end{cases}$

27. $\begin{cases} 3x - 5y = 3 \\ 15x + 5y = 21 \end{cases}$
28. $\begin{cases} 2x - y = -1 \\ x + \frac{1}{2}y = \frac{3}{2} \end{cases}$
29. $\begin{cases} x + y - z = 6 \\ 3x - 2y + z = -5 \\ x + 3y - 2z = 14 \end{cases}$
30. $\begin{cases} x - y + z = -4 \\ 2x - 3y + 4z = -15 \\ 5x + y - 2z = 12 \end{cases}$

31. $\begin{cases} x + 2y - z = -3 \\ 2x - 4y + z = -7 \\ -2x + 2y - 3z = 4 \end{cases}$
32. $\begin{cases} x + 4y - 3z = -8 \\ 3x - y + 3z = 12 \\ x + y + 6z = 1 \end{cases}$
33. $\begin{cases} x - 2y + 3z = 1 \\ 3x + y - 2z = 0 \\ 2x - 4y + 6z = 2 \end{cases}$
34. $\begin{cases} x - y + 2z = 5 \\ 3x + 2y = 4 \\ -2x + 2y - 4z = -10 \end{cases}$

35. $\begin{cases} x + 2y - z = 0 \\ 2x - 4y + z = 0 \\ -2x + 2y - 3z = 0 \end{cases}$
36. $\begin{cases} x + 4y - 3z = 0 \\ 3x - y + 3z = 0 \\ x + y + 6z = 0 \end{cases}$
37. $\begin{cases} x - 2y + 3z = 0 \\ 3x + y - 2z = 0 \\ 2x - 4y + 6z = 0 \end{cases}$
38. $\begin{cases} x - y + 2z = 0 \\ 3x + 2y = 0 \\ -2x + 2y - 4z = 0 \end{cases}$

39. $\begin{cases} \dfrac{1}{x} + \dfrac{1}{y} = 8 \\ \dfrac{3}{x} - \dfrac{5}{y} = 0 \end{cases}$
40. $\begin{cases} \dfrac{4}{x} - \dfrac{3}{y} = 0 \\ \dfrac{6}{x} + \dfrac{3}{2y} = 2 \end{cases}$

[**Hint:** Let $u = 1/x$ and $v = 1/y$, and solve for u and v.]

In Problems 41–46, solve for x.

41. $\begin{vmatrix} x & x \\ 4 & 3 \end{vmatrix} = 5$

42. $\begin{vmatrix} x & 1 \\ 3 & x \end{vmatrix} = -2$

43. $\begin{vmatrix} x & 1 & 1 \\ 4 & 3 & 2 \\ -1 & 2 & 5 \end{vmatrix} = 2$

44. $\begin{vmatrix} 3 & 2 & 4 \\ 1 & x & 5 \\ 0 & 1 & -2 \end{vmatrix} = 0$

45. $\begin{vmatrix} x & 2 & 3 \\ 1 & x & 0 \\ 6 & 1 & -2 \end{vmatrix} = 7$

46. $\begin{vmatrix} x & 1 & 2 \\ 1 & x & 3 \\ 0 & 1 & 2 \end{vmatrix} = -4x$

In Problems 47–54, use properties of determinants to find the value of each determinant if it is known that

$$\begin{vmatrix} x & y & z \\ u & v & w \\ 1 & 2 & 3 \end{vmatrix} = 4$$

47. $\begin{vmatrix} 1 & 2 & 3 \\ u & v & w \\ x & y & z \end{vmatrix}$

48. $\begin{vmatrix} x & y & z \\ u & v & w \\ 2 & 4 & 6 \end{vmatrix}$

49. $\begin{vmatrix} x & y & z \\ -3 & -6 & -9 \\ u & v & w \end{vmatrix}$

50. $\begin{vmatrix} 1 & 2 & 3 \\ x-u & y-v & z-w \\ u & v & w \end{vmatrix}$

51. $\begin{vmatrix} 1 & 2 & 3 \\ x-3 & y-6 & z-9 \\ 2u & 2v & 2w \end{vmatrix}$

52. $\begin{vmatrix} x & y & z-x \\ u & v & w-u \\ 1 & 2 & 2 \end{vmatrix}$

53. $\begin{vmatrix} 1 & 2 & 3 \\ 2x & 2y & 2z \\ u-1 & v-2 & w-3 \end{vmatrix}$

54. $\begin{vmatrix} x+3 & y+6 & z+9 \\ 3u-1 & 3v-2 & 3w-3 \\ 1 & 2 & 3 \end{vmatrix}$

---

55. Geometry: Equation of a Line An equation of the line containing the two points (x_1, y_1) and (x_2, y_2) may be expressed as the determinant

$$\begin{vmatrix} x & y & 1 \\ x_1 & y_1 & 1 \\ x_2 & y_2 & 1 \end{vmatrix} = 0$$

Prove this result by expanding the determinant and comparing the result to the two-point form of the equation of a line.

56. Geometry: Collinear Points Using the result obtained in Problem 55, show that three distinct points (x_1, y_1), (x_2, y_2), and (x_3, y_3) are collinear (lie on the same line) if and only if

$$\begin{vmatrix} x_1 & y_1 & 1 \\ x_2 & y_2 & 1 \\ x_3 & y_3 & 1 \end{vmatrix} = 0$$

57. Show that $\begin{vmatrix} x^2 & x & 1 \\ y^2 & y & 1 \\ z^2 & z & 1 \end{vmatrix} = (y-z)(x-y)(x-z)$.

58. Complete the proof of Cramer's Rule for two equations containing two variables.

[**Hint:** In system (5), page 550, if $a = 0$, then $b \neq 0$ and $c \neq 0$, since $D = -bc \neq 0$. Now show that equation (6) provides a solution of the system when $a = 0$. There are then three remaining cases: $b = 0, c = 0$, and $d = 0$.]

59. Interchange columns 1 and 3 of a 3 by 3 determinant. Show that the value of the new determinant is -1 times the value of the original determinant.

60. Multiply each entry in row 2 of a 3 by 3 determinant by the number $k, k \neq 0$. Show that the value of the new determinant is k times the value of the original determinant.

61. Prove that a 3 by 3 determinant in which the entries in column 1 equal those in column 3 has the value 0.

62. Prove that, if row 2 of a 3 by 3 determinant is multiplied by $k, k \neq 0$, and the result is added to the entries in row 1, there is no change in the value of the determinant.

8.5   MATRIX ALGEBRA

OBJECTIVES **1** Find the Sum and Difference of Two Matrices
2 Find Scalar Multiples of a Matrix
3 Find the Product of Two Matrices
4 Find the Inverse of a Matrix
5 Solve Systems of Equations Using Inverse Matrices

In Section 8.3, we defined a matrix as a rectangular array of real numbers and used an augmented matrix to represent a system of linear equations. There is, however, a branch of mathematics, called **linear algebra,** that deals with matrices in such a way that an algebra of matrices is permitted. In this section, we provide a survey of how this **matrix algebra** is developed.

Before getting started, we restate the definition of a matrix.

A **matrix** is defined as a rectangular array of numbers:

$$
\begin{array}{c}
\begin{array}{cccccc}
\text{Column 1} & \text{Column 2} & & \text{Column } j & & \text{Column } n
\end{array} \\
\begin{array}{c}
\text{Row 1} \\
\text{Row 2} \\
\vdots \\
\text{Row } i \\
\vdots \\
\text{Row } m
\end{array}
\begin{bmatrix}
a_{11} & a_{12} & \cdots & a_{1j} & \cdots & a_{1n} \\
a_{21} & a_{22} & \cdots & a_{2j} & \cdots & a_{2n} \\
\vdots & \vdots & & \vdots & & \vdots \\
a_{i1} & a_{i2} & \cdots & a_{ij} & \cdots & a_{in} \\
\vdots & \vdots & & \vdots & & \vdots \\
a_{m1} & a_{m2} & \cdots & a_{mj} & \cdots & a_{mn}
\end{bmatrix}
\end{array}
$$

Each number a_{ij} of the matrix has two indexes: the **row index** i and the **column index** j. The matrix shown above has m rows and n columns. The numbers a_{ij} are usually referred to as the **entries** of the matrix. For example, a_{23} refers to the entry in the second row, third column.

Let's begin with an example that illustrates how matrices can be used to conveniently represent an array of information.

EXAMPLE 1   Arranging Data in a Matrix

In a survey of 900 people, the following information was obtained:

200 males	Thought federal defense spending was too high
150 males	Thought federal defense spending was too low
45 males	Had no opinion
315 females	Thought federal defense spending was too high
125 females	Thought federal defense spending was too low
65 females	Had no opinion

We can arrange these data in a rectangular array as follows:

	Too High	Too Low	No Opinion
Male	200	150	45
Female	315	125	65

or as the matrix

$$\begin{bmatrix} 200 & 150 & 45 \\ 315 & 125 & 65 \end{bmatrix}$$

This matrix has two rows (representing males and females) and three columns (representing "too high," "too low," and "no opinion"). ■

The matrix we developed in Example 1 has 2 rows and 3 columns. In general, a matrix with m rows and n columns is called an **m by n matrix.** The matrix we developed in Example 1 is a 2 by 3 matrix and contains $2 \cdot 3 = 6$ entries. An m by n matrix will contain $m \cdot n$ entries.

If an m by n matrix has the same number of rows as columns, that is, if $m = n$, then the matrix is referred to as a **square matrix.**

EXAMPLE 2

Examples of Matrices

(a) $\begin{bmatrix} 5 & 0 \\ -6 & 1 \end{bmatrix}$ A 2 by 2 square matrix

(b) $\begin{bmatrix} 1 & 0 & 3 \end{bmatrix}$ A 1 by 3 matrix

(c) $\begin{bmatrix} 6 & -2 & 4 \\ 4 & 3 & 5 \\ 8 & 0 & 1 \end{bmatrix}$ A 3 by 3 square matrix

■

THE SUM AND DIFFERENCE OF TWO MATRICES

① We begin our discussion of matrix algebra by first defining what is meant by two matrices being equal and then defining the operations of addition and subtraction. It is important to note that these definitions require each matrix to have the same number of rows *and* the same number of columns.

We usually represent matrices by capital letters, such as A, B, C, and so on.

Two m by n matrices A and B are said to be **equal,** written as

$$A = B$$

provided that each entry a_{ij} in A is equal to the corresponding entry b_{ij} in B.

For example,

$$\begin{bmatrix} 2 & 1 \\ 0.5 & -1 \end{bmatrix} = \begin{bmatrix} \sqrt{4} & 1 \\ \frac{1}{2} & -1 \end{bmatrix} \quad \text{and} \quad \begin{bmatrix} 3 & 2 & 1 \\ 0 & 1 & -2 \end{bmatrix} = \begin{bmatrix} \sqrt{9} & \sqrt{4} & 1 \\ 0 & 1 & \sqrt[3]{-8} \end{bmatrix}$$

$$\begin{bmatrix} 4 & 1 \\ 6 & 1 \end{bmatrix} \neq \begin{bmatrix} 4 & 0 \\ 6 & 1 \end{bmatrix}$$ Because the entries in row 1, column 2 are not equal

$$\begin{bmatrix} 4 & 1 & 2 \\ 6 & 1 & 2 \end{bmatrix} \neq \begin{bmatrix} 4 & 1 & 2 & 3 \\ 6 & 1 & 2 & 4 \end{bmatrix}$$ Because the matrix on the left is 2 by 3 and the matrix on the right is 2 by 4

Suppose that A and B represent two m by n matrices. We define their **sum** $A + B$ to be the m by n matrix formed by adding the corresponding entries a_{ij} of A and b_{ij} of B. The **difference $A - B$** is defined as the m by n matrix formed by subtracting the entries b_{ij} in B from the corresponding entries a_{ij} in A. Addition and subtraction of matrices are allowed only for matrices having the same number m of rows and the same number n of columns. For example, a 2 by 3 matrix and a 2 by 4 matrix cannot be added or subtracted.

| EXAMPLE 3 | **Adding and Subtracting Matrices** |

Suppose that

$$A = \begin{bmatrix} 2 & 4 & 8 & -3 \\ 0 & 1 & 2 & 3 \end{bmatrix} \quad \text{and} \quad B = \begin{bmatrix} -3 & 4 & 0 & 1 \\ 6 & 8 & 2 & 0 \end{bmatrix}$$

Find: (a) $A + B$ (b) $A - B$

Solution   First, we observe that both A and B have 2 rows and 4 columns so it is possible to find their sum and their difference.

(a) $A + B = \begin{bmatrix} 2 & 4 & 8 & -3 \\ 0 & 1 & 2 & 3 \end{bmatrix} + \begin{bmatrix} -3 & 4 & 0 & 1 \\ 6 & 8 & 2 & 0 \end{bmatrix}$

$= \begin{bmatrix} 2 + (-3) & 4 + 4 & 8 + 0 & -3 + 1 \\ 0 + 6 & 1 + 8 & 2 + 2 & 3 + 0 \end{bmatrix}$ Add corresponding entries.

$= \begin{bmatrix} -1 & 8 & 8 & -2 \\ 6 & 9 & 4 & 3 \end{bmatrix}$

(b) $A - B = \begin{bmatrix} 2 & 4 & 8 & -3 \\ 0 & 1 & 2 & 3 \end{bmatrix} - \begin{bmatrix} -3 & 4 & 0 & 1 \\ 6 & 8 & 2 & 0 \end{bmatrix}$

$= \begin{bmatrix} 2 - (-3) & 4 - 4 & 8 - 0 & -3 - 1 \\ 0 - 6 & 1 - 8 & 2 - 2 & 3 - 0 \end{bmatrix}$ Subtract corresponding entries.

$= \begin{bmatrix} 5 & 0 & 8 & -4 \\ -6 & -7 & 0 & 3 \end{bmatrix}$ ■

Figure 7

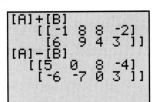

SEEING THE CONCEPT   Graphing utilities can make the sometimes tedious process of matrix algebra easy. In fact, most graphing calculators can handle matrices as large as 9 by 9, some even larger ones. Enter the matrices into a graphing utility. Name them $[A]$ and $[B]$. Figure 7 shows the results of adding and subtracting $[A]$ and $[B]$.   ■

NOW WORK PROBLEM **1**.

Many of the algebraic properties of sums of real numbers are also true for sums of matrices. Suppose that A, B, and C are m by n matrices. Then matrix addition is **commutative.** That is,

Commutative Property

$$A + B = B + A$$

Matrix addition is also **associative.** That is,

Associative Property

$$(A + B) + C = A + (B + C)$$

Although we shall not prove these results, the proofs, as the following example illustrates, are based on the commutative and associative properties for real numbers.

EXAMPLE 4   **Demonstrating the Commutative Property**

$$\begin{bmatrix} 2 & 3 & -1 \\ 4 & 0 & 7 \end{bmatrix} + \begin{bmatrix} -1 & 2 & 1 \\ 5 & -3 & 4 \end{bmatrix} = \begin{bmatrix} 2 + (-1) & 3 + 2 & -1 + 1 \\ 4 + 5 & 0 + (-3) & 7 + 4 \end{bmatrix}$$

$$= \begin{bmatrix} -1 + 2 & 2 + 3 & 1 + (-1) \\ 5 + 4 & -3 + 0 & 4 + 7 \end{bmatrix}$$

$$= \begin{bmatrix} -1 & 2 & 1 \\ 5 & -3 & 4 \end{bmatrix} + \begin{bmatrix} 2 & 3 & -1 \\ 4 & 0 & 7 \end{bmatrix}$$ ∎

A matrix whose entries are all equal to 0 is called a **zero matrix.** Each of the following matrices is a zero matrix.

$$\begin{bmatrix} 0 & 0 \\ 0 & 0 \end{bmatrix}$$ 2 by 2 square zero matrix $$\begin{bmatrix} 0 & 0 & 0 \\ 0 & 0 & 0 \end{bmatrix}$$ 2 by 3 zero matrix $$\begin{bmatrix} 0 & 0 & 0 \end{bmatrix}$$ 1 by 3 zero matrix

Zero matrices have properties similar to the real number 0. If A is an m by n matrix and 0 is an m by n zero matrix, then

$$A + 0 = A$$

In other words, the zero matrix is the additive identity in matrix algebra.

SCALAR MULTIPLES OF A MATRIX

(2)   We can multiply a matrix by a real number. If k is a real number and A is an m by n matrix, the matrix kA is the m by n matrix formed by multiplying each entry a_{ij} in A by k. The number k is sometimes referred to as a **scalar,** and the matrix kA is called a **scalar multiple** of A.

EXAMPLE 5   **Operations Using Matrices**

Suppose that

$$A = \begin{bmatrix} 3 & 1 & 5 \\ -2 & 0 & 6 \end{bmatrix} \qquad B = \begin{bmatrix} 4 & 1 & 0 \\ 8 & 1 & -3 \end{bmatrix} \qquad C = \begin{bmatrix} 9 & 0 \\ -3 & 6 \end{bmatrix}$$

Find: (a) $4A$ (b) $\frac{1}{3}C$ (c) $3A - 2B$

Solution   (a) $4A = 4\begin{bmatrix} 3 & 1 & 5 \\ -2 & 0 & 6 \end{bmatrix} = \begin{bmatrix} 4 \cdot 3 & 4 \cdot 1 & 4 \cdot 5 \\ 4(-2) & 4 \cdot 0 & 4 \cdot 6 \end{bmatrix} = \begin{bmatrix} 12 & 4 & 20 \\ -8 & 0 & 24 \end{bmatrix}$

(b) $\frac{1}{3}C = \frac{1}{3}\begin{bmatrix} 9 & 0 \\ -3 & 6 \end{bmatrix} = \begin{bmatrix} \frac{1}{3} \cdot 9 & \frac{1}{3} \cdot 0 \\ \frac{1}{3}(-3) & \frac{1}{3} \cdot 6 \end{bmatrix} = \begin{bmatrix} 3 & 0 \\ -1 & 2 \end{bmatrix}$

(c) $3A - 2B = 3\begin{bmatrix} 3 & 1 & 5 \\ -2 & 0 & 6 \end{bmatrix} - 2\begin{bmatrix} 4 & 1 & 0 \\ 8 & 1 & -3 \end{bmatrix}$

$= \begin{bmatrix} 3 \cdot 3 & 3 \cdot 1 & 3 \cdot 5 \\ 3(-2) & 3 \cdot 0 & 3 \cdot 6 \end{bmatrix} - \begin{bmatrix} 2 \cdot 4 & 2 \cdot 1 & 2 \cdot 0 \\ 2 \cdot 8 & 2 \cdot 1 & 2(-3) \end{bmatrix}$

$= \begin{bmatrix} 9 & 3 & 15 \\ -6 & 0 & 18 \end{bmatrix} - \begin{bmatrix} 8 & 2 & 0 \\ 16 & 2 & -6 \end{bmatrix}$

$= \begin{bmatrix} 9 - 8 & 3 - 2 & 15 - 0 \\ -6 - 16 & 0 - 2 & 18 - (-6) \end{bmatrix}$

$= \begin{bmatrix} 1 & 1 & 15 \\ -22 & -2 & 24 \end{bmatrix}$ ■

Check: Enter the matrices $[A], [B]$, and $[C]$ into a graphing utility. Then find $4A$, $\frac{1}{3}C$, and $3A - 2B$. ■

NOW WORK PROBLEM 5.

We list next some of the algebraic properties of scalar multiplication. Let h and k be real numbers, and let A and B be m by n matrices. Then

Properties of Scalar Multiplication

$$k(hA) = (kh)A$$
$$(k + h)A = kA + hA$$
$$k(A + B) = kA + kB$$

THE PRODUCT OF TWO MATRICES

③ Unlike the straightforward definition for adding two matrices, the definition for multiplying two matrices is not what we might expect. In preparation for this definition, we need the following definitions:

A **row vector** R is a 1 by n matrix

$$R = \begin{bmatrix} r_1 & r_2 & \cdots & r_n \end{bmatrix}$$

A **column vector** C is an n by 1 matrix

$$C = \begin{bmatrix} c_1 \\ c_2 \\ \vdots \\ c_n \end{bmatrix}$$

The **product** RC of R times C is defined as the number

$$RC = \begin{bmatrix} r_1 & r_2 & \cdots & r_n \end{bmatrix} \begin{bmatrix} c_1 \\ c_2 \\ \vdots \\ c_n \end{bmatrix} = r_1 c_1 + r_2 c_2 + \cdots + r_n c_n$$

Notice that a row vector and a column vector can be multiplied only if they contain the same number of entries.

| EXAMPLE 6 | **The Product of a Row Vector by a Column Vector** |

If $R = \begin{bmatrix} 3 & -5 & 2 \end{bmatrix}$ and $C = \begin{bmatrix} 3 \\ 4 \\ -5 \end{bmatrix}$, then

$$RC = \begin{bmatrix} 3 & -5 & 2 \end{bmatrix} \begin{bmatrix} 3 \\ 4 \\ -5 \end{bmatrix} = 3 \cdot 3 + (-5)4 + 2(-5)$$

$$= 9 - 20 - 10 = -21 \qquad \blacksquare$$

Let's look at an application of the product of a row vector by a column vector.

| EXAMPLE 7 | **Using Matrices to Compute Revenue** |

A clothing store sells men's shirts for $25, silk ties for $8, and wool suits for $300. Last month, the store had sales consisting of 100 shirts, 200 ties, and 50 suits. What was the total revenue due to these sales?

Solution   We set up a row vector R to represent the prices of each item and a column vector C to represent the corresponding number of items sold.
Then

$$
\begin{array}{cc}
\begin{array}{c} \text{Prices} \\ \text{Shirts Ties Suits} \end{array} & \begin{array}{c} \text{Number} \\ \text{sold} \end{array} \\
R = \begin{bmatrix} 25 & 8 & 300 \end{bmatrix} & C = \begin{bmatrix} 100 \\ 200 \\ 50 \end{bmatrix} \begin{array}{l} \text{Shirts} \\ \text{Ties} \\ \text{Suits} \end{array}
\end{array}
$$

The total revenue obtained is the product RC. That is,

$$RC = \begin{bmatrix} 25 & 8 & 300 \end{bmatrix} \begin{bmatrix} 100 \\ 200 \\ 50 \end{bmatrix}$$

$$= \underbrace{25 \cdot 100}_{\text{Shirt revenue}} + \underbrace{8 \cdot 200}_{\text{Tie revenue}} + \underbrace{300 \cdot 50}_{\text{Suit revenue}} = \underbrace{\$19{,}100}_{\text{Total revenue}} \qquad \blacksquare$$

The definition for multiplying two matrices is based on the definition of a row vector times a column vector.

Let A denote an m by r matrix, and let B denote an r by n matrix. The **product** AB is defined as the m by n matrix whose entry in row i, column j is the product of the ith row of A and the jth column of B.

The definition of the product AB of two matrices A and B, in this order, requires that the number of columns of A equal the number of rows of B; otherwise, no product is defined.

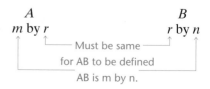

An example will help to clarify the definition.

| EXAMPLE 8 | **Multiplying Two Matrices** |

Find the product AB if

$$A = \begin{bmatrix} 2 & 4 & -1 \\ 5 & 8 & 0 \end{bmatrix} \quad \text{and} \quad B = \begin{bmatrix} 2 & 5 & 1 & 4 \\ 4 & 8 & 0 & 6 \\ -3 & 1 & -2 & -1 \end{bmatrix}$$

Solution   First, we note that A is 2 by 3 and B is 3 by 4, so the product AB is defined and will be a 2 by 4 matrix.

Suppose that we want the entry in row 2, column 3 of AB. To find it, we find the product of the row vector from row 2 of A and the column vector from column 3 of B.

Column 3 of B
Row 2 of A

$$\begin{bmatrix} 5 & 8 & 0 \end{bmatrix} \begin{bmatrix} 1 \\ 0 \\ -2 \end{bmatrix} = 5 \cdot 1 + 8 \cdot 0 + 0(-2) = 5$$

So far we have

Column 3
↓

$$AB = \begin{bmatrix} \underline{\quad} & \underline{\quad} & 5 & \underline{\quad} \\ \underline{\quad} & \underline{\quad} & & \underline{\quad} \end{bmatrix} \quad \leftarrow \text{Row 2}$$

Now, to find the entry in row 1, column 4 of AB, we find the product of row 1 of A and column 4 of B.

Column 4 of B
Row 1 of A

$$\begin{bmatrix} 2 & 4 & -1 \end{bmatrix} \begin{bmatrix} 4 \\ 6 \\ -1 \end{bmatrix} = 2 \cdot 4 + 4 \cdot 6 + (-1)(-1) = 33$$

Continuing in this fashion, we find AB.

$$AB = \begin{bmatrix} 2 & 4 & -1 \\ 5 & 8 & 0 \end{bmatrix} \begin{bmatrix} 2 & 5 & 1 & 4 \\ 4 & 8 & 0 & 6 \\ -3 & 1 & -2 & -1 \end{bmatrix}$$

$$= \begin{bmatrix} \text{Row 1 of } A & \text{Row 1 of } A & \text{Row 1 of } A & \text{Row 1 of } A \\ \text{times} & \text{times} & \text{times} & \text{times} \\ \text{column 1 of } B & \text{column 2 of } B & \text{column 3 of } B & \text{column 4 of } B \\ & & & \\ \text{Row 2 of } A & \text{Row 2 of } A & \text{Row 2 of } A & \text{Row 2 of } A \\ \text{times} & \text{times} & \text{times} & \text{times} \\ \text{column 1 of } B & \text{column 2 of } B & \text{column 3 of } B & \text{column 4 of } B \end{bmatrix}$$

$$= \begin{bmatrix} 2 \cdot 2 + 4 \cdot 4 + (-1)(-3) & 2 \cdot 5 + 4 \cdot 8 + (-1)1 & 2 \cdot 1 + 4 \cdot 0 + (-1)(-2) & 33 \text{ (from earlier)} \\ 5 \cdot 2 + 8 \cdot 4 + 0(-3) & 5 \cdot 5 + 8 \cdot 8 + 0 \cdot 1 & 5 \text{ (from earlier)} & 5 \cdot 4 + 8 \cdot 6 + 0(-1) \end{bmatrix}$$

$$= \begin{bmatrix} 23 & 41 & 4 & 33 \\ 42 & 89 & 5 & 68 \end{bmatrix}$$

Check: Enter the matrices A and B. Then find AB. (See what happens if you try to find BA.)

NOW WORK PROBLEM 17.

Notice that for the matrices given in Example 8 the product BA is not defined, because B is 3 by 4 and A is 2 by 3.

Another result that can occur when multiplying two matrices is illustrated in the next example.

EXAMPLE 9   **Multiplying Two Matrices**

If

$$A = \begin{bmatrix} 2 & 1 & 3 \\ 1 & -1 & 0 \end{bmatrix} \quad \text{and} \quad B = \begin{bmatrix} 1 & 0 \\ 2 & 1 \\ 3 & 2 \end{bmatrix}$$

find: (a) AB (b) BA

Solution   (a) $AB = \underset{\text{2 by 3}}{\begin{bmatrix} 2 & 1 & 3 \\ 1 & -1 & 0 \end{bmatrix}} \underset{\text{3 by 2}}{\begin{bmatrix} 1 & 0 \\ 2 & 1 \\ 3 & 2 \end{bmatrix}} = \underset{\text{2 by 2}}{\begin{bmatrix} 13 & 7 \\ -1 & -1 \end{bmatrix}}$

(b) $BA = \underset{\text{3 by 2}}{\begin{bmatrix} 1 & 0 \\ 2 & 1 \\ 3 & 2 \end{bmatrix}} \underset{\text{2 by 3}}{\begin{bmatrix} 2 & 1 & 3 \\ 1 & -1 & 0 \end{bmatrix}} = \underset{\text{3 by 3}}{\begin{bmatrix} 2 & 1 & 3 \\ 5 & 1 & 6 \\ 8 & 1 & 9 \end{bmatrix}}$

Notice in Example 9 that AB is 2 by 2 and BA is 3 by 3. It is possible for both AB and BA to be defined, yet be unequal. In fact, even if A and B are both n by n matrices, so that AB and BA are each defined and n by n, AB and BA will usually be unequal.

EXAMPLE 10   **Multiplying Two Square Matrices**

If

$$A = \begin{bmatrix} 2 & 1 \\ 0 & 4 \end{bmatrix} \quad \text{and} \quad B = \begin{bmatrix} -3 & 1 \\ 1 & 2 \end{bmatrix}$$

find: (a) AB (b) BA

Solution   (a) $AB = \begin{bmatrix} 2 & 1 \\ 0 & 4 \end{bmatrix} \begin{bmatrix} -3 & 1 \\ 1 & 2 \end{bmatrix} = \begin{bmatrix} -5 & 4 \\ 4 & 8 \end{bmatrix}$

(b) $BA = \begin{bmatrix} -3 & 1 \\ 1 & 2 \end{bmatrix} \begin{bmatrix} 2 & 1 \\ 0 & 4 \end{bmatrix} = \begin{bmatrix} -6 & 1 \\ 2 & 9 \end{bmatrix}$

The preceding examples demonstrate that an important property of real numbers, the commutative property of multiplication, is not shared by matrices. In general:

Theorem   Matrix multiplication is not commutative.

NOW WORK PROBLEMS **7** AND **9.**

Next we give two of the properties of real numbers that are shared by matrices. Assuming that each product and sum is defined, we have the following:

Associative Property

$$A(BC) = (AB)C$$

Distributive Property

$$A(B + C) = AB + AC$$

THE IDENTITY MATRIX

For an n by n square matrix, the entries located in row i, column i, $1 \le i \le n$, are called the **diagonal entries.** An n by n square matrix whose diagonal entries are 1's, while all other entries are 0's, is called the **identity matrix I_n.** For example,

$$I_2 = \begin{bmatrix} 1 & 0 \\ 0 & 1 \end{bmatrix} \qquad I_3 = \begin{bmatrix} 1 & 0 & 0 \\ 0 & 1 & 0 \\ 0 & 0 & 1 \end{bmatrix}$$

and so on.

EXAMPLE 11   **Multiplication with an Identity Matrix**

Let

$$A = \begin{bmatrix} -1 & 2 & 0 \\ 0 & 1 & 3 \end{bmatrix} \quad \text{and} \quad B = \begin{bmatrix} 3 & 2 \\ 4 & 6 \\ 5 & 2 \end{bmatrix}$$

Find: (a) AI_3 (b) $I_2 A$ (c) BI_2

Solution   (a) $AI_3 = \begin{bmatrix} -1 & 2 & 0 \\ 0 & 1 & 3 \end{bmatrix} \begin{bmatrix} 1 & 0 & 0 \\ 0 & 1 & 0 \\ 0 & 0 & 1 \end{bmatrix} = \begin{bmatrix} -1 & 2 & 0 \\ 0 & 1 & 3 \end{bmatrix} = A$

(b) $I_2 A = \begin{bmatrix} 1 & 0 \\ 0 & 1 \end{bmatrix} \begin{bmatrix} -1 & 2 & 0 \\ 0 & 1 & 3 \end{bmatrix} = \begin{bmatrix} -1 & 2 & 0 \\ 0 & 1 & 3 \end{bmatrix} = A$

(c) $BI_2 = \begin{bmatrix} 3 & 2 \\ 4 & 6 \\ 5 & 2 \end{bmatrix} \begin{bmatrix} 1 & 0 \\ 0 & 1 \end{bmatrix} = \begin{bmatrix} 3 & 2 \\ 4 & 6 \\ 5 & 2 \end{bmatrix} = B$

Example 11 demonstrates the following property:

Identity Property

If A is an m by n matrix, then

$$I_m A = A \quad \text{and} \quad AI_n = A$$

If A is an n by n square matrix, then $AI_n = I_n A = A$.

An identity matrix has properties analogous to those of the real number 1. In other words, the identity matrix is a multiplicative identity in matrix algebra.

THE INVERSE OF A MATRIX

④ Let A be a square n by n matrix. If there exists an n by n matrix A^{-1}, read "A inverse," for which

$$AA^{-1} = A^{-1}A = I_n$$

then A^{-1} is called the **inverse** of the matrix A.

As we shall soon see, not every square matrix has an inverse. When a matrix A does have an inverse A^{-1}, then A is said to be **nonsingular.** If a matrix A has no inverse, it is called **singular.** *

EXAMPLE 12   **Multiplying a Matrix by Its Inverse**

Show that the inverse of

$$A = \begin{bmatrix} 3 & 1 \\ 2 & 1 \end{bmatrix} \quad \text{is} \quad A^{-1} = \begin{bmatrix} 1 & -1 \\ -2 & 3 \end{bmatrix}$$

Solution   We need to show that $AA^{-1} = A^{-1}A = I_2$.

$$AA^{-1} = \begin{bmatrix} 3 & 1 \\ 2 & 1 \end{bmatrix}\begin{bmatrix} 1 & -1 \\ -2 & 3 \end{bmatrix} = \begin{bmatrix} 1 & 0 \\ 0 & 1 \end{bmatrix} = I_2$$

$$A^{-1}A = \begin{bmatrix} 1 & -1 \\ -2 & 3 \end{bmatrix}\begin{bmatrix} 3 & 1 \\ 2 & 1 \end{bmatrix} = \begin{bmatrix} 1 & 0 \\ 0 & 1 \end{bmatrix} = I_2$$

■

We now show one way to find the inverse of

$$A = \begin{bmatrix} 3 & 1 \\ 2 & 1 \end{bmatrix}$$

---

* If the determinant of A is zero, then A is singular. (Refer to Section 8.4.)

Suppose that A^{-1} is given by

$$A^{-1} = \begin{bmatrix} x & y \\ z & w \end{bmatrix} \qquad \textbf{(1)}$$

where x, y, z, and w are four variables. Based on the definition of an inverse, if, indeed, A has an inverse, we have

$$AA^{-1} = I_2$$

$$\begin{bmatrix} 3 & 1 \\ 2 & 1 \end{bmatrix}\begin{bmatrix} x & y \\ z & w \end{bmatrix} = \begin{bmatrix} 1 & 0 \\ 0 & 1 \end{bmatrix}$$

$$\begin{bmatrix} 3x + z & 3y + w \\ 2x + z & 2y + w \end{bmatrix} = \begin{bmatrix} 1 & 0 \\ 0 & 1 \end{bmatrix}$$

Because corresponding entries must be equal, it follows that this matrix equation is equivalent to four ordinary equations.

$$\begin{cases} 3x + z = 1 \\ 2x + z = 0 \end{cases} \qquad \begin{cases} 3y + w = 0 \\ 2y + w = 1 \end{cases}$$

The augmented matrix of each system is

$$\left[\begin{array}{cc|c} 3 & 1 & 1 \\ 2 & 1 & 0 \end{array}\right] \qquad \left[\begin{array}{cc|c} 3 & 1 & 0 \\ 2 & 1 & 1 \end{array}\right] \qquad \textbf{(2)}$$

The usual procedure would be to transform each augmented matrix into reduced row echelon form. Notice, though, that the left sides of the augmented matrices are equal, so the same row operations (see Section 8.3) can be used to reduce each side. We find it more efficient to combine the two augmented matrices (2) into a single matrix, as shown next, and then transform it into reduced row echelon form.

$$\left[\begin{array}{cc|cc} 3 & 1 & 1 & 0 \\ 2 & 1 & 0 & 1 \end{array}\right]$$

Now we attempt to transform the left side into an identity matrix.

$$\left[\begin{array}{cc|cc} 3 & 1 & 1 & 0 \\ 2 & 1 & 0 & 1 \end{array}\right] \underset{\underset{R_1 = -1r_2 + r_1}{\uparrow}}{\rightarrow} \left[\begin{array}{cc|cc} 1 & 0 & 1 & -1 \\ 2 & 1 & 0 & 1 \end{array}\right]$$

$$\underset{\underset{R_2 = -2r_1 + r_2}{\uparrow}}{\rightarrow} \left[\begin{array}{cc|cc} 1 & 0 & 1 & -1 \\ 0 & 1 & -2 & 3 \end{array}\right] \qquad \textbf{(3)}$$

Matrix (3) is in reduced row echelon form. Now we reverse the earlier step of combining the two augmented matrices in (2) and write the single matrix (3) as two augmented matrices.

$$\left[\begin{array}{cc|c} 1 & 0 & 1 \\ 0 & 1 & -2 \end{array}\right] \quad \text{and} \quad \left[\begin{array}{cc|c} 1 & 0 & -1 \\ 0 & 1 & 3 \end{array}\right]$$

We conclude from these matrices that $x = 1$, $z = -2$, and $y = -1$, $w = 3$. Substituting these values into matrix (1), we find that

$$A^{-1} = \begin{bmatrix} 1 & -1 \\ -2 & 3 \end{bmatrix}$$

Notice in display (3) that the 2 by 2 matrix to the right of the vertical bar is, in fact, the inverse of A. Also notice that the identity matrix I_2 is the matrix that appears to the left of the vertical bar. These observations and the procedures followed above will work in general.

> ## PROCEDURE FOR FINDING THE INVERSE OF A NONSINGULAR MATRIX
>
> To find the inverse of an n by n nonsingular matrix A, proceed as follows:
>
> STEP 1: Form the matrix $[A \,|\, I_n]$.
>
> STEP 2: Transform the matrix $[A \,|\, I_n]$ into reduced row echelon form.
>
> STEP 3: The reduced row echelon form of $[A \,|\, I_n]$ will contain the identity matrix I_n on the left of the vertical bar; the n by n matrix on the right of the vertical bar is the inverse of A.

In other words, if A is nonsingular, we begin with the matrix $[A \,|\, I_n]$ and, after transforming it into reduced row echelon form, we end up with the matrix $[I_n \,|\, A^{-1}]$.

Let's look at another example.

EXAMPLE 13   Finding the Inverse of a Matrix

The matrix
$$A = \begin{bmatrix} 1 & 1 & 0 \\ -1 & 3 & 4 \\ 0 & 4 & 3 \end{bmatrix}$$

is nonsingular. Find its inverse.

Solution First, we form the matrix

$$[A \,|\, I_3] = \left[\begin{array}{rrr|rrr} 1 & 1 & 0 & 1 & 0 & 0 \\ -1 & 3 & 4 & 0 & 1 & 0 \\ 0 & 4 & 3 & 0 & 0 & 1 \end{array}\right]$$

Next, we use row operations to transform $[A \,|\, I_3]$ into reduced row echelon form.

$$\left[\begin{array}{rrr|rrr} 1 & 1 & 0 & 1 & 0 & 0 \\ -1 & 3 & 4 & 0 & 1 & 0 \\ 0 & 4 & 3 & 0 & 0 & 1 \end{array}\right] \rightarrow \underset{\underset{R_2 = r_1 + r_2}{\uparrow}}{\left[\begin{array}{rrr|rrr} 1 & 1 & 0 & 1 & 0 & 0 \\ 0 & 4 & 4 & 1 & 1 & 0 \\ 0 & 4 & 3 & 0 & 0 & 1 \end{array}\right]} \rightarrow \underset{\underset{R_2 = \frac{1}{4} r_2}{\uparrow}}{\left[\begin{array}{rrr|rrr} 1 & 1 & 0 & 1 & 0 & 0 \\ 0 & 1 & 1 & \frac{1}{4} & \frac{1}{4} & 0 \\ 0 & 4 & 3 & 0 & 0 & 1 \end{array}\right]}$$

$$\rightarrow \underset{\underset{\substack{R_1 = -1r_2 + r_1 \\ R_3 = -4r_2 + r_3}}{\uparrow}}{\left[\begin{array}{rrr|rrr} 1 & 0 & -1 & \frac{3}{4} & -\frac{1}{4} & 0 \\ 0 & 1 & 1 & \frac{1}{4} & \frac{1}{4} & 0 \\ 0 & 0 & -1 & -1 & -1 & 1 \end{array}\right]} \rightarrow \underset{\underset{R_3 = -1r_3}{\uparrow}}{\left[\begin{array}{rrr|rrr} 1 & 0 & -1 & \frac{3}{4} & -\frac{1}{4} & 0 \\ 0 & 1 & 1 & \frac{1}{4} & \frac{1}{4} & 0 \\ 0 & 0 & 1 & 1 & 1 & -1 \end{array}\right]}$$

$$\rightarrow \underset{\underset{\substack{R_1 = r_3 + r_1 \\ R_2 = -1r_3 + r_2}}{\uparrow}}{\left[\begin{array}{rrr|rrr} 1 & 0 & 0 & \frac{7}{4} & \frac{3}{4} & -1 \\ 0 & 1 & 0 & -\frac{3}{4} & -\frac{3}{4} & 1 \\ 0 & 0 & 1 & 1 & 1 & -1 \end{array}\right]}$$

The matrix $\left[A \mid I_3\right]$ is now in reduced row echelon form, and the identity matrix I_3 is on the left of the vertical bar. Hence, the inverse of A is

$$A^{-1} = \begin{bmatrix} \frac{7}{4} & \frac{3}{4} & -1 \\ -\frac{3}{4} & -\frac{3}{4} & 1 \\ 1 & 1 & -1 \end{bmatrix}$$

Figure 8

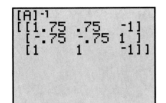

You can (and should) verify that this is the correct inverse by showing that $AA^{-1} = A^{-1}A = I_3$.

Check: Enter the matrix A into a graphing utility. Figure 8 shows A^{-1}.

─ NOW WORK PROBLEM **23**.

If transforming the matrix $\left[A \mid I_n\right]$ into reduced row echelon form does not result in the identity matrix I_n to the left of the vertical bar, then A is singular and has no inverse. The next example demonstrates such a matrix.

EXAMPLE 14   **Showing That a Matrix Has No Inverse**

Show that the following matrix has no inverse.

$$A = \begin{bmatrix} 4 & 6 \\ 2 & 3 \end{bmatrix}$$

Solution   Proceeding as in Example 13, we form the matrix

$$\left[A \mid I_2\right] = \begin{bmatrix} 4 & 6 & 1 & 0 \\ 2 & 3 & 0 & 1 \end{bmatrix}$$

Then we use row operations to transform $\left[A \mid I_2\right]$ into reduced row echelon form.

$$\left[A \mid I_2\right] = \begin{bmatrix} 4 & 6 & 1 & 0 \\ 2 & 3 & 0 & 1 \end{bmatrix} \underset{\underset{R_1 = \frac{1}{4}r_1}{\uparrow}}{\rightarrow} \begin{bmatrix} 1 & \frac{3}{2} & \frac{1}{4} & 0 \\ 2 & 3 & 0 & 1 \end{bmatrix} \underset{\underset{R_2 = -2r_1 + r_2}{\uparrow}}{\rightarrow} \begin{bmatrix} 1 & \frac{3}{2} & \frac{1}{4} & 0 \\ 0 & 0 & -\frac{1}{2} & 1 \end{bmatrix}$$

The matrix $\left[A \mid I_2\right]$ is sufficiently reduced for us to see that the identity matrix cannot appear to the left of the vertical bar. We conclude that A is singular and so has no inverse.

Check: Enter the matrix A. Try to find its inverse. What happens?

 ─ NOW WORK PROBLEM **51**.

SOLVING SYSTEMS OF LINEAR EQUATIONS

⑤ Inverse matrices can be used to solve systems of equations in which the number of equations is the same as the number of variables.

EXAMPLE 15   **Using the Inverse Matrix to Solve a System of Linear Equations**

Solve the system of equations: $\begin{cases} x + y = 3 \\ -x + 3y + 4z = -3 \\ 4y + 3z = 2 \end{cases}$

Solution If we let

$$A = \begin{bmatrix} 1 & 1 & 0 \\ -1 & 3 & 4 \\ 0 & 4 & 3 \end{bmatrix} \qquad X = \begin{bmatrix} x \\ y \\ z \end{bmatrix} \qquad B = \begin{bmatrix} 3 \\ -3 \\ 2 \end{bmatrix}$$

then the original system of equations can be written compactly as the matrix equation

$$AX = B \qquad \qquad \textbf{(4)}$$

We know from Example 13 that the matrix A has the inverse A^{-1}, so we multiply each side of equation (4) by A^{-1}.

$$AX = B$$
$$A^{-1}(AX) = A^{-1}B$$
$$(A^{-1}A)X = A^{-1}B \qquad \text{Associative property of multiplication}$$
$$I_3 X = A^{-1}B \qquad \text{Definition of inverse matrix}$$
$$X = A^{-1}B \qquad \text{Property of identity matrix} \qquad \textbf{(5)}$$

Now we use (5) to find $X = \begin{bmatrix} x \\ y \\ z \end{bmatrix}$.

$$X = \begin{bmatrix} x \\ y \\ z \end{bmatrix} = A^{-1}B = \begin{bmatrix} \frac{7}{4} & \frac{3}{4} & -1 \\ -\frac{3}{4} & -\frac{3}{4} & 1 \\ 1 & 1 & -1 \end{bmatrix} \begin{bmatrix} 3 \\ -3 \\ 2 \end{bmatrix} = \begin{bmatrix} 1 \\ 2 \\ -2 \end{bmatrix}$$

$\uparrow$ Example 13

Thus, $x = 1$, $y = 2$, $z = -2$. ■

The method used in Example 15 to solve a system of equations is particularly useful when it is necessary to solve several systems of equations in which the constants appearing to the right of the equal signs change, while the coefficients of the variables on the left side remain the same. See Problems 31–50 for some illustrations. Be careful; this method can only be used if the inverse exists. If it does not exist, row reduction must be used, since the system is either inconsistent or dependent.

HISTORICAL FEATURE

*Arthur Cayley
(1821–1895)*

Matrices were invented in 1857 by Arthur Cayley (1821–1895) as a way of efficiently computing the result of substituting one linear system into another (see Historical Problem 2). The resulting system had incredible richness, in the sense that a very wide variety of mathematical systems could be mimicked by the matrices. Cayley and his friend James J. Sylvester (1814–1897) spent much of the rest of their lives elaborating the theory. The torch was then passed to Georg Frobenius (1849–1917), whose deep investigations established a central place for matrices in modern mathematics. In 1924, rather to the surprise of physicists, it was found that matrices (with complex numbers in them) were exactly the right tool for describing the behavior of atomic systems. Today, matrices are used in a wide variety of applications.

HISTORICAL PROBLEMS

1. *Matrices and Complex Numbers*   Frobenius emphasized in his research how matrices could be used to mimic other mathematical systems. Here, we mimic the behavior of complex numbers using matrices. Mathematicians call such a relationship an isomorphism.

Complex number $\longleftrightarrow$ Matrix

$$a + bi \longleftrightarrow \begin{bmatrix} a & b \\ -b & a \end{bmatrix}$$

Note that the complex number can be read off the top line of the matrix.

$$2 + 3i \longleftrightarrow \begin{bmatrix} 2 & 3 \\ -3 & 2 \end{bmatrix} \text{ and } \begin{bmatrix} 4 & -2 \\ 2 & 4 \end{bmatrix} \longleftrightarrow 4 - 2i$$

(a) Find the matrices corresponding to $2 - 5i$ and $1 + 3i$.

(b) Multiply the two matrices.

(c) Find the corresponding complex number for the matrix found in part (b).

(d) Multiply $2 - 5i$ by $1 + 3i$. The result should be the same as that found in part (c).

The process also works for addition and subtraction. Try it for yourself.

2. *Cayley's Definition of Matrix Multiplication*   Cayley invented matrix multiplication to simplify the following problem:

$$\begin{cases} u = ar + bs \\ v = cr + ds \end{cases} \qquad \begin{cases} x = ku + lv \\ y = mu + nv \end{cases}$$

(a) Find x and y in terms of r and s by substituting u and v from the first system of equations into the second system of equations.

(b) Use the result of part (a) to find the 2 by 2 matrix A in

$$\begin{bmatrix} x \\ y \end{bmatrix} = A \begin{bmatrix} r \\ s \end{bmatrix}$$

(c) Now look at the following way to do it. Write the equations in matrix form.

$$\begin{bmatrix} u \\ v \end{bmatrix} = \begin{bmatrix} a & b \\ c & d \end{bmatrix} \begin{bmatrix} r \\ s \end{bmatrix} \qquad \begin{bmatrix} x \\ y \end{bmatrix} = \begin{bmatrix} k & l \\ m & n \end{bmatrix} \begin{bmatrix} u \\ v \end{bmatrix}$$

So

$$\begin{bmatrix} x \\ y \end{bmatrix} = \begin{bmatrix} k & l \\ m & n \end{bmatrix} \begin{bmatrix} a & b \\ c & d \end{bmatrix} \begin{bmatrix} r \\ s \end{bmatrix}$$

Do you see how Cayley defined matrix multiplication?

8.5 EXERCISES

In Problems 1–16, use the following matrices to compute the given expression.

$$A = \begin{bmatrix} 0 & 3 & -5 \\ 1 & 2 & 6 \end{bmatrix} \qquad B = \begin{bmatrix} 4 & 1 & 0 \\ -2 & 3 & -2 \end{bmatrix} \qquad C = \begin{bmatrix} 4 & 1 \\ 6 & 2 \\ -2 & 3 \end{bmatrix}$$

1. $A + B$

2. $A - B$

3. $4A$

4. $-3B$

5. $3A - 2B$

6. $2A + 4B$

7. AC

8. BC

9. CA

10. CB

11. $C(A + B)$

12. $(A + B)C$

13. $AC - 3I_2$

14. $CA + 5I_3$

15. $CA - CB$

16. $AC + BC$

In Problems 17–20, compute each product.

17. $\begin{bmatrix} 2 & -2 \\ 1 & 0 \end{bmatrix} \begin{bmatrix} 2 & 1 & 4 & 6 \\ 3 & -1 & 3 & 2 \end{bmatrix}$

18. $\begin{bmatrix} 4 & 1 \\ 2 & 1 \end{bmatrix} \begin{bmatrix} -6 & 6 & 1 & 0 \\ 2 & 5 & 4 & -1 \end{bmatrix}$

19. $\begin{bmatrix} 1 & 0 & 1 \\ 2 & 4 & 1 \\ 3 & 6 & 1 \end{bmatrix} \begin{bmatrix} 1 & 3 \\ 6 & 2 \\ 8 & -1 \end{bmatrix}$

20. $\begin{bmatrix} 4 & -2 & 3 \\ 0 & 1 & 2 \\ -1 & 0 & 1 \end{bmatrix} \begin{bmatrix} 2 & 6 \\ 1 & -1 \\ 0 & 2 \end{bmatrix}$

In Problems 21–30, each matrix is nonsingular. Find the inverse of each matrix. Be sure to check your answer.

21. $\begin{bmatrix} 2 & 1 \\ 1 & 1 \end{bmatrix}$

22. $\begin{bmatrix} 3 & -1 \\ -2 & 1 \end{bmatrix}$

23. $\begin{bmatrix} 6 & 5 \\ 2 & 2 \end{bmatrix}$

24. $\begin{bmatrix} -4 & 1 \\ 6 & -2 \end{bmatrix}$

25. $\begin{bmatrix} 2 & 1 \\ a & a \end{bmatrix}, \quad a \neq 0$

26. $\begin{bmatrix} b & 3 \\ b & 2 \end{bmatrix}, \quad b \neq 0$

27. $\begin{bmatrix} 1 & -1 & 1 \\ 0 & -2 & 1 \\ -2 & -3 & 0 \end{bmatrix}$

28. $\begin{bmatrix} 1 & 0 & 2 \\ -1 & 2 & 3 \\ 1 & -1 & 0 \end{bmatrix}$

29. $\begin{bmatrix} 1 & 1 & 1 \\ 3 & 2 & -1 \\ 3 & 1 & 2 \end{bmatrix}$

30. $\begin{bmatrix} 3 & 3 & 1 \\ 1 & 2 & 1 \\ 2 & -1 & 1 \end{bmatrix}$

In Problems 31–50, use the inverses found in Problems 21–30 to solve each system of equations.

31. $\begin{cases} 2x + y = 8 \\ x + y = 5 \end{cases}$

32. $\begin{cases} 3x - y = 8 \\ -2x + y = 4 \end{cases}$

33. $\begin{cases} 2x + y = 0 \\ x + y = 5 \end{cases}$

34. $\begin{cases} 3x - y = 4 \\ -2x + y = 5 \end{cases}$

35. $\begin{cases} 6x + 5y = 7 \\ 2x + 2y = 2 \end{cases}$

36. $\begin{cases} -4x + y = 0 \\ 6x - 2y = 14 \end{cases}$

37. $\begin{cases} 6x + 5y = 13 \\ 2x + 2y = 5 \end{cases}$

38. $\begin{cases} -4x + y = 5 \\ 6x - 2y = -9 \end{cases}$

39. $\begin{cases} 2x + y = -3 \\ ax + ay = -a \end{cases}, \quad a \neq 0$

40. $\begin{cases} bx + 3y = 2b + 3 \\ bx + 2y = 2b + 2 \end{cases}, \quad b \neq 0$

41. $\begin{cases} 2x + y = 7/a \\ ax + ay = 5 \end{cases}, \quad a \neq 0$

42. $\begin{cases} bx + 3y = 14 \\ bx + 2y = 10 \end{cases}, \quad b \neq 0$

43. $\begin{cases} x - y + z = 0 \\ -2y + z = -1 \\ -2x - 3y = -5 \end{cases}$

44. $\begin{cases} x + 2z = 6 \\ -x + 2y + 3z = -5 \\ x - y = 6 \end{cases}$

45. $\begin{cases} x - y + z = 2 \\ -2y + z = 2 \\ -2x - 3y = \frac{1}{2} \end{cases}$

46. $\begin{cases} x + 2z = 2 \\ -x + 2y + 3z = -\frac{3}{2} \\ x - y = 2 \end{cases}$

47. $\begin{cases} x + y + z = 9 \\ 3x + 2y - z = 8 \\ 3x + y + 2z = 1 \end{cases}$

48. $\begin{cases} 3x + 3y + z = 8 \\ x + 2y + z = 5 \\ 2x - y + z = 4 \end{cases}$

49. $\begin{cases} x + y + z = 2 \\ 3x + 2y - z = \frac{7}{3} \\ 3x + y + 2z = \frac{10}{3} \end{cases}$

50. $\begin{cases} 3x + 3y + z = 1 \\ x + 2y + z = 0 \\ 2x - y + z = 4 \end{cases}$

In Problems 51–56, show that each matrix has no inverse.

51. $\begin{bmatrix} 4 & 2 \\ 2 & 1 \end{bmatrix}$

52. $\begin{bmatrix} -3 & \frac{1}{2} \\ 6 & -1 \end{bmatrix}$

53. $\begin{bmatrix} 15 & 3 \\ 10 & 2 \end{bmatrix}$

54. $\begin{bmatrix} -3 & 0 \\ 4 & 0 \end{bmatrix}$

55. $\begin{bmatrix} -3 & 1 & -1 \\ 1 & -4 & -7 \\ 1 & 2 & 5 \end{bmatrix}$

56. $\begin{bmatrix} 1 & 1 & -3 \\ 2 & -4 & 1 \\ -5 & 7 & 1 \end{bmatrix}$

In Problems 57–60, use a graphing utility to find the inverse, if it exists, of each matrix. Round answers to two decimal places.

57. $\begin{bmatrix} 25 & 61 & -12 \\ 18 & -2 & 4 \\ 8 & 35 & 21 \end{bmatrix}$

58. $\begin{bmatrix} 18 & -3 & 4 \\ 6 & -20 & 14 \\ 10 & 25 & -15 \end{bmatrix}$

59. $\begin{bmatrix} 44 & 21 & 18 & 6 \\ -2 & 10 & 15 & 5 \\ 21 & 12 & -12 & 4 \\ -8 & -16 & 4 & 9 \end{bmatrix}$

60. $\begin{bmatrix} 16 & 22 & -3 & 5 \\ 21 & -17 & 4 & 8 \\ 2 & 8 & 27 & 20 \\ 5 & 15 & -3 & -10 \end{bmatrix}$

In Problems 61–64, use the idea behind Example 15 with a graphing utility to solve the following systems of equations. Round answers to two decimal places.

61. $\begin{cases} 25x + 61y - 12z = 10 \\ 18x - 12y + 7y = -9 \\ 3x + 4y - z = 12 \end{cases}$

62. $\begin{cases} 25x + 61y - 12z = 15 \\ 18x - 12y + 7z = -3 \\ 3x + 4y - z = 12 \end{cases}$

63. $\begin{cases} 25x + 61y - 12z = 21 \\ 18x - 12y + 7z = 7 \\ 3x + 4y - z = -2 \end{cases}$

64. $\begin{cases} 25x + 61y - 12z = 25 \\ 18x - 12y + 7z = 10 \\ 3x + 4y - z = -4 \end{cases}$

65. **Computing the Cost of Production**   The Acme Steel Company is a producer of stainless steel and aluminum containers. On a certain day, the following stainless steel containers were manufactured: 500 with 10-gallon capacity, 350 with 5-gallon capacity, and 400 with 1-gallon capacity. On the same day, the following aluminum containers were manufactured: 700 with 10-gallon capacity, 500 with 5-gallon capacity, and 850 with 1-gallon capacity.
    (a) Find a 2 by 3 matrix representing the above data. Find a 3 by 2 matrix to represent the same data.
    (b) If the amount of material used in the 10-gallon containers is 15 pounds, the amount used in the 5-gallon containers is 8 pounds, and the amount used in the 1-gallon containers is 3 pounds, find a 3 by 1 matrix representing the amount of material.
    (c) Multiply the 2 by 3 matrix found in part (a) and the 3 by 1 matrix found in part (b) to get a 2 by 1 matrix showing the day's usage of material.
    (d) If stainless steel costs Acme $0.10 per pound and aluminum costs $0.05 per pound, find a 1 by 2 matrix representing cost.
    (e) Multiply the matrices found in parts (c) and (d) to determine the total cost of the day's production.

66. **Computing Profit**   Rizza Ford has two locations, one in the city and the other in the suburbs. In January, the city location sold 400 subcompacts, 250 intermediate-size cars, and 50 station wagons; in February, it sold 350 subcompacts, 100 intermediates, and 30 station wagons. At the suburban location in January, 450 subcompacts, 200 intermediates, and 140 station wagons were sold. In February, the suburban location sold 350 subcompacts, 300 intermediates, and 100 station wagons.
    (a) Find 2 by 3 matrices that summarize the sales data for each location for January and February (one matrix for each month).
    (b) Use matrix addition to obtain total sales for the two-month period.
    (c) The profit on each kind of car is $100 per subcompact, $150 per intermediate, and $200 per station wagon. Find a 3 by 1 matrix representing this profit.
    (d) Multiply the matrices found in parts (b) and (c) to get a 2 by 1 matrix showing the profit at each location.

67. Consider the 2 by 2 square matrix

$$A = \begin{bmatrix} a & b \\ c & d \end{bmatrix}$$

If $D = ad - bc \neq 0$, show that A is nonsingular and that

$$A^{-1} = \frac{1}{D} \begin{bmatrix} d & -b \\ -c & a \end{bmatrix}$$

68. Make up a situation different from any found in the text that can be represented by a matrix.

PREPARING FOR THIS SECTION

Before getting started, review the following:

✓ Identity (Section 1.1, p. 85)

✓ Proper and Improper Rational Functions (Section 4.3, pp. 322–323)

✓ Factoring Polynomials (Review, Section 6, pp. 47–54)

✓ Fundamental Theorem of Algebra (Section 5.4, p. 382)

8.6   PARTIAL FRACTION DECOMPOSITION

OBJECTIVES   ① Decompose P/Q, Where Q Has Only Nonrepeated Linear Factors
② Decompose P/Q, Where Q Has Repeated Linear Factors
③ Decompose P/Q, where Q Has Only Nonrepeated Irreducible Quadratic Factors
④ Decompose P/Q, Where Q Has Repeated Irreducible Quadratic Factors

Consider the problem of adding two fractions:

$$\frac{3}{x + 4} \quad \text{and} \quad \frac{2}{x - 3}$$

The result is

$$\frac{3}{x+4} + \frac{2}{x-3} = \frac{3(x-3) + 2(x+4)}{(x+4)(x-3)} = \frac{5x-1}{x^2+x-12}$$

The reverse procedure, of starting with the rational expression $(5x-1)/(x^2+x-12)$ and writing it as the sum (or difference) of the two simpler fractions $3/(x+4)$ and $2/(x-3)$, is referred to as **partial fraction decomposition,** and the two simpler fractions are called **partial fractions.** Decomposing a rational expression into a sum of partial fractions is important in solving certain types of calculus problems. This section presents a systematic way to decompose rational expressions.

We begin by recalling that a rational expression is the ratio of two polynomials, say, P and $Q \neq 0$. We assume that P and Q have no common factors. Recall also that a rational expression P/Q is called **proper** if the degree of the polynomial in the numerator is less than the degree of the polynomial in the denominator. Otherwise, the rational expression is termed **improper.**

Because any improper rational expression can be reduced by long division to a mixed form consisting of the sum of a polynomial and a proper rational expression, we shall restrict the discussion that follows to proper rational expressions.

The partial fraction decomposition of the rational expression P/Q depends on the factors of the denominator Q. Recall (from Section 5.2) that any polynomial whose coefficients are real numbers can be factored (over the real numbers) into products of linear and/or irreducible quadratic factors. Thus, the denominator Q of the rational expression P/Q will contain only factors of one or both of the following types:

1. *Linear factors* of the form $x - a$, where a is a real number.

2. *Irreducible quadratic factors* of the form $ax^2 + bx + c$, where a, b, and c are real numbers, $a \neq 0$, and $b^2 - 4ac < 0$ (which guarantees that $ax^2 + bx + c$ cannot be written as the product of two linear factors with real coefficients).

As it turns out, there are four cases to be examined. We begin with the case for which Q has only nonrepeated linear factors.

CASE 1:   *Q* has only nonrepeated linear factors.

Under the assumption that Q has only nonrepeated linear factors, the polynomial Q has the form

$$Q(x) = (x - a_1)(x - a_2) \cdots \cdots (x - a_n)$$

where none of the numbers $a_1, a_2, \ldots, a_n$ are equal. In this case, the partial fraction decomposition of P/Q is of the form

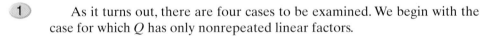

$$\frac{P(x)}{Q(x)} = \frac{A_1}{x - a_1} + \frac{A_2}{x - a_2} + \cdots + \frac{A_n}{x - a_n} \qquad \textbf{(1)}$$

where the numbers $A_1, A_2, \ldots, A_n$ are to be determined.

We show how to find these numbers in the example that follows.

EXAMPLE 1   **Nonrepeated Linear Factors**

Write the partial fraction decomposition of $\dfrac{x}{x^2 - 5x + 6}$.

Solution   First, we factor the denominator,

$$x^2 - 5x + 6 = (x - 2)(x - 3)$$

and conclude that the denominator contains only nonrepeated linear factors. Then we decompose the rational expression according to equation (1):

$$\frac{x}{x^2 - 5x + 6} = \frac{A}{x - 2} + \frac{B}{x - 3} \qquad \textbf{(2)}$$

where A and B are to be determined. To find A and B, we clear the fractions by multiplying each side by $(x - 2)(x - 3) = x^2 - 5x + 6$. The result is

$$x = A(x - 3) + B(x - 2) \qquad \textbf{(3)}$$

or

$$x = (A + B)x + (-3A - 2B)$$

This equation is an identity in x. We equate the coefficients of like powers of x to get

$$\begin{cases} 1 = \quad A + \ B \quad \text{Equate coefficients of } x: 1x = (A + B)x. \\ 0 = -3A - 2B \quad \text{Equate coefficients of } x^0, \text{ the constants: } 0x^0 = (-3A - 2B)x^0. \end{cases}$$

This system of two equations containing two variables, A and B, can be solved using whatever method you wish. Solving it, we get

$$A = -2 \qquad B = 3$$

From equation (2), the partial fraction decomposition is

$$\frac{x}{x^2 - 5x + 6} = \frac{-2}{x - 2} + \frac{3}{x - 3} \qquad \blacksquare$$

Check: The decomposition can be checked by adding the fractions.

$$\frac{-2}{x - 2} + \frac{3}{x - 3} = \frac{-2(x - 3) + 3(x - 2)}{(x - 2)(x - 3)} = \frac{x}{(x - 2)(x - 3)}$$

$$= \frac{x}{x^2 - 5x + 6} \qquad \blacksquare$$

NOW WORK PROBLEM **9**.

The numbers to be found in the partial fraction decomposition can sometimes be found more readily by using suitable choices for x (which may include complex numbers) in the identity obtained after fractions have been cleared. In Example 1, the identity after clearing fractions, equation (3), is

$$x = A(x - 3) + B(x - 2)$$

If we let $x = 2$ in this expression, the term containing B drops out, leaving $2 = A(-1)$, or $A = -2$. Similarly, if we let $x = 3$, the term containing A drops out, leaving $3 = B$. As before, $A = -2$ and $B = 3$.

We use this method in the next example.

② **CASE 2:   Q has repeated linear factors.**

If the polynomial Q has a repeated factor, say $(x - a)^n$, $n \geq 2$ an integer, then, in the partial fraction decomposition of P/Q, we allow for the terms

$$\frac{A_1}{x - a} + \frac{A_2}{(x - a)^2} + \cdots + \frac{A_n}{(x - a)^n}$$

where the numbers $A_1, A_2, \ldots, A_n$ are to be determined.

EXAMPLE 2

Repeated Linear Factors

Write the partial fraction decomposition of $\dfrac{x + 2}{x^3 - 2x^2 + x}$.

Solution   First, we factor the denominator,

$$x^3 - 2x^2 + x = x(x^2 - 2x + 1) = x(x - 1)^2$$

and find that the denominator has the nonrepeated linear factor x and the repeated linear factor $(x - 1)^2$. By Case 1, we must allow for the term A/x in the decomposition; and, by Case 2, we must allow for the terms $B/(x - 1) + C/(x - 1)^2$ in the decomposition.

We write

$$\frac{x + 2}{x^3 - 2x^2 + x} = \frac{A}{x} + \frac{B}{x - 1} + \frac{C}{(x - 1)^2} \tag{4}$$

Again, we clear fractions by multiplying each side by $x^3 - 2x^2 + x = x(x - 1)^2$. The result is the identity

$$x + 2 = A(x - 1)^2 + Bx(x - 1) + Cx \tag{5}$$

If we let $x = 0$ in this expression, the terms containing B and C drop out, leaving $2 = A(-1)^2$, or $A = 2$. Similarly, if we let $x = 1$, the terms containing A and B drop out, leaving $3 = C$. Then, equation (5) becomes

$$x + 2 = 2(x - 1)^2 + Bx(x - 1) + 3x$$

Now let $x = 2$ (any choice other than 0 or 1 will work as well). The result is

$$4 = 2(1)^2 + B(2)(1) + 3(2)$$

$$2B = 4 - 2 - 6 = -4$$

$$B = -2$$

We have $A = 2$, $B = -2$, and $C = 3$.

From equation (4), the partial fraction decomposition is

$$\frac{x + 2}{x^3 - 2x^2 + x} = \frac{2}{x} + \frac{-2}{x - 1} + \frac{3}{(x - 1)^2}$$

∎

EXAMPLE 3   Repeated Linear Factors

Write the partial fraction decomposition of $\dfrac{x^3 - 8}{x^2(x - 1)^3}$.

Solution   The denominator contains the repeated linear factor x^2 and the repeated linear factor $(x - 1)^3$. The partial fraction decomposition takes the form

$$\frac{x^3 - 8}{x^2(x - 1)^3} = \frac{A}{x} + \frac{B}{x^2} + \frac{C}{x - 1} + \frac{D}{(x - 1)^2} + \frac{E}{(x - 1)^3} \qquad \textbf{(6)}$$

As before, we clear fractions and obtain the identity

$$x^3 - 8 = Ax(x - 1)^3 + B(x - 1)^3 + Cx^2(x - 1)^2 + Dx^2(x - 1) + Ex^2 \qquad \textbf{(7)}$$

Let $x = 0$. (Do you see why this choice was made?) Then

$$-8 = B(-1)$$

$$B = 8$$

Now let $x = 1$ in equation (7). Then

$$-7 = E$$

Use $B = 8$ and $E = -7$ in equation (7) and collect like terms.

$$x^3 - 8 = Ax(x - 1)^3 + 8(x - 1)^3$$
$$+ Cx^2(x - 1)^2 + Dx^2(x - 1) - 7x^2$$
$$x^3 - 8 - 8(x^3 - 3x^2 + 3x - 1) + 7x^2 = Ax(x - 1)^3 + Cx^2(x - 1)^2 + Dx^2(x - 1)$$
$$-7x^3 + 31x^2 - 24x = x(x - 1)\left[A(x - 1)^2 + Cx(x - 1) + Dx\right]$$
$$x(x - 1)(-7x + 24) = x(x - 1)\left[A(x - 1)^2 + Cx(x - 1) + Dx\right]$$
$$-7x + 24 = A(x - 1)^2 + Cx(x - 1) + Dx \qquad \textbf{(8)}$$

We now work with equation (8). Let $x = 0$. Then

$$24 = A$$

Now let $x = 1$ in equation (8). Then

$$17 = D$$

Use $A = 24$ and $D = 17$ in equation (8) and collect like terms.

$$-7x + 24 = 24(x - 1)^2 + Cx(x - 1) + 17x$$
$$-24x^2 + 48x - 24 - 17x - 7x + 24 = Cx(x - 1)$$
$$-24x^2 + 24x = Cx(x - 1)$$
$$-24x(x - 1) = Cx(x - 1)$$
$$-24 = C$$

We now know all the numbers A, B, C, D, and E, so, from equation (6), we have the decomposition

$$\frac{x^3 - 8}{x^2(x - 1)^3} = \frac{24}{x} + \frac{8}{x^2} + \frac{-24}{x - 1} + \frac{17}{(x - 1)^2} + \frac{-7}{(x - 1)^3}$$

The method employed in Example 3, although somewhat tedious, is still preferable to solving the system of five equations containing five variables that the expansion of equation (6) leads to.

NOW WORK PROBLEM 15.

3 The final two cases involve irreducible quadratic factors. A quadratic factor is irreducible if it cannot be factored into linear factors with real coefficients. A quadratic expression $ax^2 + bx + c$ is irreducible whenever $b^2 - 4ac < 0$. For example, $x^2 + x + 1$ and $x^2 + 4$ are irreducible.

> **CASE 3: Q contains a nonrepeated irreducible quadratic factor.**
>
> If Q contains a nonrepeated irreducible quadratic factor of the form $ax^2 + bx + c$, then, in the partial fraction decomposition of P/Q, allow for the term
>
> $$\frac{Ax + B}{ax^2 + bx + c}$$

where the numbers A and B are to be determined.

EXAMPLE 4 **Nonrepeated Irreducible Quadratic Factor**

Write the partial fraction decomposition of $\dfrac{3x - 5}{x^3 - 1}$.

Solution We factor the denominator,

$$x^3 - 1 = (x - 1)(x^2 + x + 1)$$

and find that it has a nonrepeated linear factor $x - 1$ and a nonrepeated irreducible quadratic factor $x^2 + x + 1$. We allow for the term $A/(x - 1)$ by Case 1, and we allow for the term $(Bx + C)/(x^2 + x + 1)$ by Case 3. We write

$$\frac{3x - 5}{x^3 - 1} = \frac{A}{x - 1} + \frac{Bx + C}{x^2 + x + 1} \tag{9}$$

We clear fractions by multiplying each side of equation (9) by $x^3 - 1 = (x - 1)(x^2 + x + 1)$ to obtain the identity

$$3x - 5 = A(x^2 + x + 1) + (Bx + C)(x - 1) \tag{10}$$

Now let $x = 1$. Then equation (10) gives $-2 = A(3)$, or $A = -\frac{2}{3}$. We use this value of A in equation (10) and simplify.

$$3x - 5 = -\frac{2}{3}(x^2 + x + 1) + (Bx + C)(x - 1)$$

$$3(3x - 5) = -2(x^2 + x + 1) + 3(Bx + C)(x - 1) \quad \text{Multiply each side by 3.}$$

$$9x - 15 = -2x^2 - 2x - 2 + 3(Bx + C)(x - 1)$$

$$2x^2 + 11x - 13 = 3(Bx + C)(x - 1) \quad \text{Collect terms.}$$

$$(2x + 13)(x - 1) = 3(Bx + C)(x - 1) \quad \text{Factor the left side.}$$

$$2x + 13 = 3Bx + 3C$$

$$2 = 3B \quad \text{and} \quad 13 = 3C \quad \text{Equate coefficients.}$$

$$B = \frac{2}{3} \qquad C = \frac{13}{3}$$

From equation (9), we see that

$$\frac{3x - 5}{x^3 - 1} = \frac{-\frac{2}{3}}{x - 1} + \frac{\frac{2}{3}x + \frac{13}{3}}{x^2 + x + 1}$$

∎

NOW WORK PROBLEM 17.

④ **CASE 4: Q contains repeated irreducible quadratic factors.**

If the polynomial Q contains a repeated irreducible quadratic factor $(ax^2 + bx + c)^n$, $n \geq 2$, n an integer, then, in the partial fraction decomposition of P/Q, allow for the terms

$$\frac{A_1 x + B_1}{ax^2 + bx + c} + \frac{A_2 x + B_2}{(ax^2 + bx + c)^2} + \cdots + \frac{A_n x + B_n}{(ax^2 + bx + c)^n}$$

where the numbers $A_1, B_1, A_2, B_2, \ldots, A_n, B_n$ are to be determined.

EXAMPLE 5 **Repeated Irreducible Quadratic Factor**

Write the partial fraction decomposition of $\dfrac{x^3 + x^2}{(x^2 + 4)^2}$.

Solution The denominator contains the repeated irreducible quadratic factor $(x^2 + 4)^2$, so we write

$$\frac{x^3 + x^2}{(x^2 + 4)^2} = \frac{Ax + B}{x^2 + 4} + \frac{Cx + D}{(x^2 + 4)^2} \tag{11}$$

We clear fractions to obtain

$$x^3 + x^2 = (Ax + B)(x^2 + 4) + Cx + D$$

Collecting like terms yields

$$x^3 + x^2 = Ax^3 + Bx^2 + (4A + C)x + D + 4B$$

Equating coefficients, we arrive at the system

$$\begin{cases} A = 1 \\ B = 1 \\ 4A + C = 0 \\ D + 4B = 0 \end{cases}$$

The solution is $A = 1$, $B = 1$, $C = -4$, $D = -4$. From equation (11),

$$\frac{x^3 + x^2}{(x^2 + 4)^2} = \frac{x + 1}{x^2 + 4} + \frac{-4x - 4}{(x^2 + 4)^2}$$

∎

NOW WORK PROBLEM 31.

8.6 EXERCISES

In Problems 1–8, tell whether the given rational expression is proper or improper. If improper, rewrite it as the sum of a polynomial and a proper rational expression.

1. $\dfrac{x}{x^2 - 1}$

2. $\dfrac{5x + 2}{x^3 - 1}$

3. $\dfrac{x^2 + 5}{x^2 - 4}$

4. $\dfrac{3x^2 - 2}{x^2 - 1}$

5. $\dfrac{5x^3 + 2x - 1}{x^2 - 4}$

6. $\dfrac{3x^4 + x^2 - 2}{x^3 + 8}$

7. $\dfrac{x(x - 1)}{(x + 4)(x - 3)}$

8. $\dfrac{2x(x^2 + 4)}{x^2 + 1}$

In Problems 9–42, write the partial fraction decomposition of each rational expression.

9. $\dfrac{4}{x(x - 1)}$

10. $\dfrac{3x}{(x + 2)(x - 1)}$

11. $\dfrac{1}{x(x^2 + 1)}$

12. $\dfrac{1}{(x + 1)(x^2 + 4)}$

13. $\dfrac{x}{(x - 1)(x - 2)}$

14. $\dfrac{3x}{(x + 2)(x - 4)}$

15. $\dfrac{x^2}{(x - 1)^2(x + 1)}$

16. $\dfrac{x + 1}{x^2(x - 2)}$

17. $\dfrac{1}{x^3 - 8}$

18. $\dfrac{2x + 4}{x^3 - 1}$

19. $\dfrac{x^2}{(x - 1)^2(x + 1)^2}$

20. $\dfrac{x + 1}{x^2(x - 2)^2}$

21. $\dfrac{x - 3}{(x + 2)(x + 1)^2}$

22. $\dfrac{x^2 + x}{(x + 2)(x - 1)^2}$

23. $\dfrac{x + 4}{x^2(x^2 + 4)}$

24. $\dfrac{10x^2 + 2x}{(x - 1)^2(x^2 + 2)}$

25. $\dfrac{x^2 + 2x + 3}{(x + 1)(x^2 + 2x + 4)}$

26. $\dfrac{x^2 - 11x - 18}{x(x^2 + 3x + 3)}$

27. $\dfrac{x}{(3x - 2)(2x + 1)}$

28. $\dfrac{1}{(2x + 3)(4x - 1)}$

29. $\dfrac{x}{x^2 + 2x - 3}$

30. $\dfrac{x^2 - x - 8}{(x + 1)(x^2 + 5x + 6)}$

31. $\dfrac{x^2 + 2x + 3}{(x^2 + 4)^2}$

32. $\dfrac{x^3 + 1}{(x^2 + 16)^2}$

33. $\dfrac{7x + 3}{x^3 - 2x^2 - 3x}$

34. $\dfrac{x^5 + 1}{x^6 - x^4}$

35. $\dfrac{x^2}{x^3 - 4x^2 + 5x - 2}$

36. $\dfrac{x^2 + 1}{x^3 + x^2 - 5x + 3}$

37. $\dfrac{x^3}{(x^2 + 16)^3}$

38. $\dfrac{x^2}{(x^2 + 4)^3}$

39. $\dfrac{4}{2x^2 - 5x - 3}$

40. $\dfrac{4x}{2x^2 + 3x - 2}$

41. $\dfrac{2x + 3}{x^4 - 9x^2}$

42. $\dfrac{x^2 + 9}{x^4 - 2x^2 - 8}$

PREPARING FOR THIS SECTION

Before getting started, review the following:

✓ Lines (Section 2.3, pp. 163–172)

✓ Circles (Section 2.4, pp. 178–181)

✓ Parabolas (Section 7.2, pp. 477–484)

✓ Ellipses (Section 7.3, pp. 487–495)

✓ Hyperbolas (Section 7.4, pp. 497–509)

8.7 SYSTEMS OF NONLINEAR EQUATIONS

OBJECTIVES
1. Solve a System of Nonlinear Equations Using Substitution
2. Solve a System of Nonlinear Equations Using Elimination

1. There is no general methodology for solving a system of nonlinear equations. There are times when substitution is best; other times, elimination is best; and there are times when neither of these methods works. Experience and a certain degree of imagination are your allies here.

Before we begin, two comments are in order.

1. If the system contains two variables and if the equations in the system are easy to graph, then graph them. By graphing each equation in the

system, you can get an idea of how many solutions a system has and approximately where they are located.

2. Extraneous solutions can creep in when solving nonlinear systems, so it is imperative that all apparent solutions be checked.

EXAMPLE 1 Solving a System of Nonlinear Equations

Solve the following system of equations:

$$\begin{cases} 3x - y = -2 & \text{(1) A line} \\ 2x^2 - y = 0 & \text{(2) A parabola} \end{cases}$$

Solution Using Substitution

Figure 9

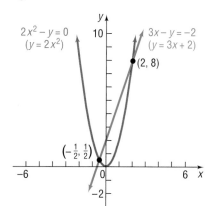

First, we notice that the system contains two variables and that we know how to graph each equation. In Figure 9, we see that the system apparently has two solutions.

We will use substitution to solve the system. Equation (1) is easily solved for y.

$$3x - y = -2$$
$$y = 3x + 2$$

We substitute this expression for y in equation (2). The result is an equation containing just the variable x, which we can then solve.

$$2x^2 - y = 0 \qquad \text{(2)}$$
$$2x^2 - (3x + 2) = 0 \qquad \text{Substitute } y = 3x + 2.$$
$$2x^2 - 3x - 2 = 0 \qquad \text{Remove parentheses.}$$
$$(2x + 1)(x - 2) = 0 \qquad \text{Factor.}$$
$$2x + 1 = 0 \quad \text{or} \quad x - 2 = 0 \qquad \text{Apply the Zero-Product Property.}$$
$$x = -\frac{1}{2} \quad \text{or} \quad x = 2$$

Using these values for x in $y = 3x + 2$, we find

$$y = 3\left(-\frac{1}{2}\right) + 2 = \frac{1}{2} \quad \text{or} \quad y = 3(2) + 2 = 8$$

The apparent solutions are $x = -\frac{1}{2}$, $y = \frac{1}{2}$ and $x = 2$, $y = 8$.
Check: For $x = -\frac{1}{2}$, $y = \frac{1}{2}$:

$$\begin{cases} 3\left(-\frac{1}{2}\right) - \frac{1}{2} = -\frac{3}{2} - \frac{1}{2} = -2 & \text{(1)} \\ 2\left(-\frac{1}{2}\right)^2 - \frac{1}{2} = 2\left(\frac{1}{4}\right) - \frac{1}{2} = 0 & \text{(2)} \end{cases}$$

For $x = 2$, $y = 8$:

$$\begin{cases} 3(2) - 8 = 6 - 8 = -2 & \text{(1)} \\ 2(2)^2 - 8 = 2(4) - 8 = 0 & \text{(2)} \end{cases}$$

Each solution checks. Now we know that the graphs in Figure 9 intersect at $\left(-\frac{1}{2}, \frac{1}{2}\right)$ and at $(2, 8)$. ∎

Check: Graph $3x - y = -2$ $(Y_1 = 3x + 2)$ and $2x^2 - y = 0$ $(Y_2 = 2x^2)$ and compare what you see with Figure 9. Use INTERSECT (twice) to find the two points of intersection. ∎

 NOW WORK PROBLEM 11 USING SUBSTITUTION.

② Our next example illustrates how the method of elimination works for nonlinear systems.

EXAMPLE 2

Solving a System of Nonlinear Equations

Solve: $\begin{cases} x^2 + y^2 = 13 & \text{(1) A circle} \\ x^2 - y = 7 & \text{(2) A parabola} \end{cases}$

Solution Using Elimination

Figure 10

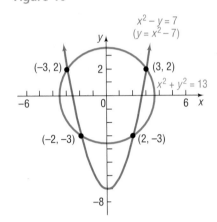

First, we graph each equation, as shown in Figure 10. Based on the graph, we expect four solutions. By subtracting equation (2) from equation (1), the variable x can be eliminated.

$$\begin{cases} x^2 + y^2 = 13 \\ \underline{x^2 - y = 7} \qquad \text{Subtract.} \\ y^2 + y = 6 \end{cases}$$

This quadratic equation in y is easily solved by factoring.

$$y^2 + y - 6 = 0$$
$$(y + 3)(y - 2) = 0$$
$$y = -3 \quad \text{or} \quad y = 2$$

We use these values for y in equation (2) to find x.

If $y = 2$, then $x^2 = y + 7 = 9$ so $x = 3$ or -3.

If $y = -3$, then $x^2 = y + 7 = 4$ so $x = 2$ or -2.

We have four solutions: $x = 3$, $y = 2$; $x = -3$, $y = 2$; $x = 2$, $y = -3$; and $x = -2$, $y = -3$.

You should verify that, in fact, these four solutions also satisfy equation (1), so all four are solutions of the system. The four points, $(3, 2)$, $(-3, 2)$, $(2, -3)$, and $(-2, -3)$, are the points of intersection of the graphs. Look again at Figure 10. ∎

 Check: Graph $x^2 + y^2 = 13$ and $x^2 - y = 7$. (Remember that to graph $x^2 + y^2 = 13$ requires two functions: $Y_1 = \sqrt{13 - x^2}$ and $Y_2 = -\sqrt{13 - x^2}$.) Compare what you see with Figure 10. Use INTERSECT to find the four points of intersection. ∎

 NOW WORK PROBLEM 9 USING ELIMINATION.

EXAMPLE 3

Solving a System of Nonlinear Equations

Solve: $\begin{cases} x^2 - y^2 = 1 & \text{(1)} \\ x^3 - y^2 = x & \text{(2)} \end{cases}$

Solution Using Elimination

Because the second equation is not easy to graph, we omit the graphing step. We use elimination, subtracting equation (2) from equation (1), to eliminate the variable y.

$$\begin{cases} x^2 - y^2 = 1 \qquad\qquad \text{(1)} \\ \underline{x^3 - y^2 = x} \qquad\qquad \text{(2)} \\ x^2 - x^3 = 1 - x \qquad \text{Subtract.} \end{cases}$$

$$x^3 - x^2 - x + 1 = 0$$
$$x^2(x - 1) - (x - 1) = 0 \qquad \text{Factor by grouping.}$$
$$(x^2 - 1)(x - 1) = 0$$
$$(x + 1)(x - 1)^2 = 0 \qquad x^2 - 1 = (x + 1)(x - 1)$$
$$x + 1 = 0 \quad \text{or} \quad (x - 1)^2 = 0$$
$$x = -1 \quad \text{or} \qquad\quad x = 1$$

We now use equation (1) to get y.

If $x = 1$, then $1 - y^2 = 1$ so $y = 0$.

If $x = -1$, then $1 - y^2 = 1$ so $y = 0$.

There are two apparent solutions: $x = 1$, $y = 0$ and $x = -1$, $y = 0$. Because each of these solutions also satisfies equation (2), the system has two solutions: $x = 1$, $y = 0$ and $x = -1$, $y = 0$. ∎

 SEEING THE CONCEPT   Graph $x^2 - y^2 = 1$ and $x^3 - y^2 = x$. (You will need to graph four functions:

$$Y_1 = \sqrt{x^2 - 1},\, Y_2 = -\sqrt{x^2 - 1},\, Y_3 = \sqrt{x^3 - x},\, \text{and } Y_4 = -\sqrt{x^3 - x}).$$

See Figure 11. The graphs appear to intersect twice, so the system apparently has two solutions. Using INTERSECT, the solutions to the system of equations are $(-1, 0)$ and $(1, 0)$.

Figure 11

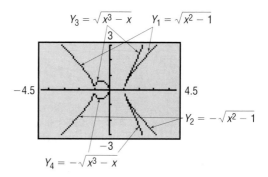

$Y_3 = \sqrt{x^3 - x}$ $Y_1 = \sqrt{x^2 - 1}$

$Y_2 = -\sqrt{x^2 - 1}$

$Y_4 = -\sqrt{x^3 - x}$

EXAMPLE 4

Solving a System of Nonlinear Equations

Solve: $\begin{cases} x^2 + x + y^2 - 3y + 2 = 0 & (1) \\ x + 1 + \dfrac{y^2 - y}{x} = 0 & (2) \end{cases}$

Solution Using Elimination   First, we multiply equation (2) by x to eliminate the fraction. The result is an equivalent system because x cannot be 0 [look at equation (2) to see why].

$$\begin{cases} x^2 + x + y^2 - 3y + 2 = 0 & (1) \\ x^2 + x + y^2 - y = 0 & (2) \end{cases}$$

Now subtract equation (2) from equation (1) to eliminate x. The result is

$$-2y + 2 = 0$$

$$y = 1 \quad \text{Solve for } y.$$

To find x, we back-substitute $y = 1$ in equation (1).

$$x^2 + x + 1 - 3 + 2 = 0$$

$$x^2 + x = 0$$

$$x(x + 1) = 0$$

$$x = 0 \quad \text{or} \quad x = -1$$

Because x cannot be 0, the value $x = 0$ is extraneous, and we discard it. We proceed to check the solution $x = -1$, $y = 1$.

Check: $\begin{cases} (-1)^2 + (-1) + 1^2 - 3(1) + 2 = 1 - 1 + 1 - 3 + 2 = 0 & (1) \\ -1 + 1 + \dfrac{1^2 - 1}{-1} = 0 + \dfrac{0}{-1} = 0 & (2) \end{cases}$

The only solution to the system is $x = -1$, $y = 1$ ▬ ■

NOW WORK PROBLEMS **25** AND **45**.

EXAMPLE 5   Solving a System of Nonlinear Equations

Solve: $\begin{cases} x^2 - y^2 = 4 & \text{(1) A hyperbola} \\ \quad\quad y = x^2 & \text{(2) A parabola} \end{cases}$

Figure 12

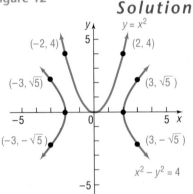

Solution Either substitution or elimination can be used here. We use substitution and replace x^2 by y in equation (1). The result is

$$y - y^2 = 4$$
$$y^2 - y + 4 = 0$$

This is a quadratic equation whose discriminant is $(-1)^2 - 4 \cdot 1 \cdot 4 = 1 - 4 \cdot 4 = -15 < 0$. The equation has no real solutions, so the system is inconsistent. The graphs of these two equations do not intersect. See Figure 12. ▬

EXAMPLE 6   Solving a System of Nonlinear Equations

Solve: $\begin{cases} 3xy - 2y^2 = -2 & (1) \\ 9x^2 + 4y^2 = 10 & (2) \end{cases}$

Solution We multiply equation (1) by 2 and add the result to equation (2) to eliminate the y^2 terms.

$$\begin{array}{ll} \begin{cases} 6xy - 4y^2 = -4 & (1) \\ 9x^2 + 4y^2 = \ \ 10 & (2) \text{ Add.} \end{cases} \\ \hline \quad 9x^2 + 6xy = \ \ 6 \\ \quad 3x^2 + 2xy = \ \ 2 & \text{Divide each side by 3.} \end{array}$$

Since $x \neq 0$ (do you see why?), we can solve for y in this equation to get

$$y = \frac{2 - 3x^2}{2x}, \qquad x \neq 0 \tag{1}$$

Now substitute for y in equation (2) of the system.

$$9x^2 + 4y^2 = 10 \qquad (2)$$
$$9x^2 + 4\left(\frac{2 - 3x^2}{2x}\right)^2 = 10 \qquad \text{Substitute } y = \frac{2 - 3x^2}{2x}.$$
$$9x^2 + \frac{4 - 12x^2 + 9x^4}{x^2} = 10$$
$$9x^4 + 4 - 12x^2 + 9x^4 = 10x^2 \qquad \text{Multiply both sides by } x^2.$$
$$18x^4 - 22x^2 + 4 = 0 \qquad \text{Subtract } 10x^2 \text{ from both sides.}$$
$$9x^4 - 11x^2 + 2 = 0 \qquad \text{Divide both sides by 2.}$$

This quadratic equation (in x^2) can be factored.

$$\left(9x^2 - 2\right)\left(x^2 - 1\right) = 0$$

$$9x^2 - 2 = 0 \quad \text{or} \quad x^2 - 1 = 0$$

$$x^2 = \frac{2}{9} \quad \text{or} \quad x^2 = 1$$

$$x = \pm\sqrt{\frac{2}{9}} = \pm\frac{\sqrt{2}}{3} \quad \text{or} \quad x = \pm1$$

To find y, we use equation (1):

If $x = \dfrac{\sqrt{2}}{3}$: $\quad y = \dfrac{2 - 3x^2}{2x} = \dfrac{2 - \frac{2}{3}}{2(\sqrt{2}/3)} = \dfrac{4}{2\sqrt{2}} = \sqrt{2}$

If $x = -\dfrac{\sqrt{2}}{3}$: $\quad y = \dfrac{2 - 3x^2}{2x} = \dfrac{2 - \frac{2}{3}}{2(-\sqrt{2}/3)} = \dfrac{4}{-2\sqrt{2}} = -\sqrt{2}$

If $x = 1$: $\quad y = \dfrac{2 - 3x^2}{2x} = \dfrac{2 - 3}{2} = -\dfrac{1}{2}$

If $x = -1$: $\quad y = \dfrac{2 - 3x^2}{2x} = \dfrac{2 - 3}{-2} = \dfrac{1}{2}$

The system has four solutions. Check them for yourself. ∎

NOW WORK PROBLEM **43**.

EXAMPLE 7	**Running Long Distance Races**

In a 50-mile race, the winner crosses the finish line 1 mile ahead of the second-place runner and 4 miles ahead of the third-place runner. Assuming that each runner maintains a constant speed throughout the race, by how many miles does the second-place runner beat the third-place runner?

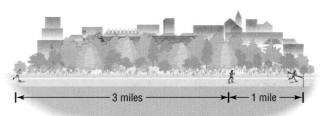

|← 3 miles →|←— 1 mile —→|

Solution   Let v_1, v_2, v_3 denote the speeds of the first-, second-, and third-place runners, respectively. Let t_1 and t_2 denote the times (in hours) required for the first-place runner and second-place runner to finish the race. Then we have the system of equations

$$\begin{cases} 50 = v_1 t_1 & \text{(1) First-place runner goes 50 miles in } t_1 \text{ hours.} \\ 49 = v_2 t_1 & \text{(2) Second-place runner goes 49 miles in } t_1 \text{ hours.} \\ 46 = v_3 t_1 & \text{(3) Third-place runner goes 46 miles in } t_1 \text{ hours.} \\ 50 = v_2 t_2 & \text{(4) Second-place runner goes 50 miles in } t_2 \text{ hours.} \end{cases}$$

We seek the distance of the third-place runner from the finish at time t_2. At time t_2, the third-place runner has gone a distance of $v_3 t_2$ miles, so the distance remaining is $50 - v_3 t_2$. Now

$$50 - v_3 t_2 = 50 - v_3 \left(t_1 \cdot \frac{t_2}{t_1} \right)$$

$$= 50 - (v_3 t_1) \cdot \frac{t_2}{t_1}$$

$$= 50 - 46 \cdot \frac{50/v_2}{50/v_1} \qquad \left\{ \begin{array}{l} \text{From (3), } v_3 t_1 = 46; \\ \text{from (4), } t_2 = 50/v_2; \\ \text{from (1), } t_1 = 50/v_1. \end{array} \right.$$

$$= 50 - 46 \cdot \frac{v_1}{v_2}$$

$$= 50 - 46 \cdot \frac{50}{49} \qquad \text{Form the quotient of (1) and (2).}$$

$$\approx 3.06 \text{ miles} \qquad\qquad ■$$

HISTORICAL FEATURE

In the beginning of this section, we said that imagination and experience are important in solving simultaneous nonlinear equations. Indeed, these kinds of problems lead into some of the deepest and most difficult parts of modern mathematics. Look again at the graphs in Examples 1 and 2 of this section (Figures 9 and 10). We see that Example 1 has two solutions, and Example 2 has four solutions. We might conjecture that the number of solutions is equal to the product of the degrees of the equations involved. This conjecture was indeed made by Etienne Bezout (1730–1783), but working out the details took about 150 years. It turns out that, to arrive at the correct number of intersections, we must count not only the complex number intersections, but also those intersections that, in a certain sense, lie at infinity. For example, a parabola and a line lying on the axis of the parabola intersect at the vertex and at infinity. This topic is part of the study of algebraic geometry.

HISTORICAL PROBLEMS

A papyrus dating back to 1950 BC contains the following problem: "A given surface area of 100 units of area shall be represented as the sum of two squares whose sides are to each other as $1 \div \frac{3}{4}$." Solve for the sides by solving the system of equations

$$\begin{cases} x^2 + y^2 = 100 \\ x = \frac{3}{4} y \end{cases}$$

8.7 EXERCISES

In Problems 1–20, graph each equation of the system. Then solve the system to find the points of intersection.

1. $\begin{cases} y = x^2 + 1 \\ y = x + 1 \end{cases}$

2. $\begin{cases} y = x^2 + 1 \\ y = 4x + 1 \end{cases}$

3. $\begin{cases} y = \sqrt{36 - x^2} \\ y = 8 - x \end{cases}$

4. $\begin{cases} y = \sqrt{4 - x^2} \\ y = 2x + 4 \end{cases}$

5. $\begin{cases} y = \sqrt{x} \\ y = 2 - x \end{cases}$

6. $\begin{cases} y = \sqrt{x} \\ y = 6 - x \end{cases}$

7. $\begin{cases} x = 2y \\ x = y^2 - 2y \end{cases}$

8. $\begin{cases} y = x - 1 \\ y = x^2 - 6x + 9 \end{cases}$

9. $\begin{cases} x^2 + y^2 = 4 \\ x^2 + 2x + y^2 = 0 \end{cases}$

10. $\begin{cases} x^2 + y^2 = 8 \\ x^2 + y^2 + 4y = 0 \end{cases}$

11. $\begin{cases} y = 3x - 5 \\ x^2 + y^2 = 5 \end{cases}$

12. $\begin{cases} x^2 + y^2 = 10 \\ y = x + 2 \end{cases}$

13. $\begin{cases} x^2 + y^2 = 4 \\ y^2 - x = 4 \end{cases}$

14. $\begin{cases} x^2 + y^2 = 16 \\ x^2 - 2y = 8 \end{cases}$

15. $\begin{cases} xy = 4 \\ x^2 + y^2 = 8 \end{cases}$

16. $\begin{cases} x^2 = y \\ xy = 1 \end{cases}$

17. $\begin{cases} x^2 + y^2 = 4 \\ y = x^2 - 9 \end{cases}$

18. $\begin{cases} xy = 1 \\ y = 2x + 1 \end{cases}$

19. $\begin{cases} y = x^2 - 4 \\ y = 6x - 13 \end{cases}$

20. $\begin{cases} x^2 + y^2 = 10 \\ xy = 3 \end{cases}$

In Problems 21–50, solve each system. Use any method you wish.

21. $\begin{cases} 2x^2 + y^2 = 18 \\ xy = 4 \end{cases}$

22. $\begin{cases} x^2 - y^2 = 21 \\ x + y = 7 \end{cases}$

23. $\begin{cases} y = 2x + 1 \\ 2x^2 + y^2 = 1 \end{cases}$

24. $\begin{cases} x^2 - 4y^2 = 16 \\ 2y - x = 2 \end{cases}$

25. $\begin{cases} x + y + 1 = 0 \\ x^2 + y^2 + 6y - x = -5 \end{cases}$

26. $\begin{cases} 2x^2 - xy + y^2 = 8 \\ xy = 4 \end{cases}$

27. $\begin{cases} 4x^2 - 3xy + 9y^2 = 15 \\ 2x + 3y = 5 \end{cases}$

28. $\begin{cases} 2y^2 - 3xy + 6y + 2x + 4 = 0 \\ 2x - 3y + 4 = 0 \end{cases}$

29. $\begin{cases} x^2 - 4y^2 + 7 = 0 \\ 3x^2 + y^2 = 31 \end{cases}$

30. $\begin{cases} 3x^2 - 2y^2 + 5 = 0 \\ 2x^2 - y^2 + 2 = 0 \end{cases}$

31. $\begin{cases} 7x^2 - 3y^2 + 5 = 0 \\ 3x^2 + 5y^2 = 12 \end{cases}$

32. $\begin{cases} x^2 - 3y^2 + 1 = 0 \\ 2x^2 - 7y^2 + 5 = 0 \end{cases}$

33. $\begin{cases} x^2 + 2xy = 10 \\ 3x^2 - xy = 2 \end{cases}$

34. $\begin{cases} 5xy + 13y^2 + 36 = 0 \\ xy + 7y^2 = 6 \end{cases}$

35. $\begin{cases} 2x^2 + y^2 = 2 \\ x^2 - 2y^2 + 8 = 0 \end{cases}$

36. $\begin{cases} y^2 - x^2 + 4 = 0 \\ 2x^2 + 3y^2 = 6 \end{cases}$

37. $\begin{cases} x^2 + 2y^2 = 16 \\ 4x^2 - y^2 = 24 \end{cases}$

38. $\begin{cases} 4x^2 + 3y^2 = 4 \\ 2x^2 - 6y^2 = -3 \end{cases}$

39. $\begin{cases} \dfrac{5}{x^2} - \dfrac{2}{y^2} + 3 = 0 \\ \dfrac{3}{x^2} + \dfrac{1}{y^2} = 7 \end{cases}$

40. $\begin{cases} \dfrac{2}{x^2} - \dfrac{3}{y^2} + 1 = 0 \\ \dfrac{6}{x^2} - \dfrac{7}{y^2} + 2 = 0 \end{cases}$

41. $\begin{cases} \dfrac{1}{x^4} + \dfrac{6}{y^4} = 6 \\ \dfrac{2}{x^4} - \dfrac{2}{y^4} = 19 \end{cases}$

42. $\begin{cases} \dfrac{1}{x^4} - \dfrac{1}{y^4} = 1 \\ \dfrac{1}{x^4} + \dfrac{1}{y^4} = 4 \end{cases}$

43. $\begin{cases} x^2 - 3xy + 2y^2 = 0 \\ x^2 + xy = 6 \end{cases}$

44. $\begin{cases} x^2 - xy - 2y^2 = 0 \\ xy + x + 6 = 0 \end{cases}$

45. $\begin{cases} y^2 + y + x^2 - x - 2 = 0 \\ y + 1 + \dfrac{x - 2}{y} = 0 \end{cases}$

46. $\begin{cases} x^3 - 2x^2 + y^2 + 3y - 4 = 0 \\ x - 2 + \dfrac{y^2 - y}{x^2} = 0 \end{cases}$

47. $\begin{cases} \log_x y = 3 \\ \log_x(4y) = 5 \end{cases}$

48. $\begin{cases} \log_x(2y) = 3 \\ \log_x(4y) = 2 \end{cases}$

49. $\begin{cases} \ln x = 4 \ln y \\ \log_3 x = 2 + 2 \log_3 y \end{cases}$

50. $\begin{cases} \ln x = 5 \ln y \\ \log_2 x = 3 + 2 \log_2 y \end{cases}$

In Problems 51–56, graph each equation and find the point(s) of intersection, if any.

51. The line $x + 2y = 0$ and the circle $(x - 1)^2 + (y - 1)^2 = 5$

52. The line $x + 2y + 6 = 0$ and the circle $(x + 1)^2 + (y + 1)^2 = 5$

53. The circle $(x - 1)^2 + (y + 2)^2 = 4$ and the parabola $y^2 + 4y - x + 1 = 0$

54. The circle $(x + 2)^2 + (y - 1)^2 = 4$ and the parabola $y^2 - 2y - x - 5 = 0$

55. The graph of $y = \dfrac{4}{x - 3}$ and the circle $x^2 - 6x + y^2 + 1 = 0$

56. The graph of $y = \dfrac{4}{x + 2}$ and the circle $x^2 + 4x + y^2 - 4 = 0$

In Problems 57–64, use a graphing utility to solve each system of equations. Express the solution(s) rounded to two decimal places.

57. $\begin{cases} y = x^{2/3} \\ y = e^{-x} \end{cases}$

58. $\begin{cases} y = x^{3/2} \\ y = e^{-x} \end{cases}$

59. $\begin{cases} x^2 + y^3 = 2 \\ x^3 y = 4 \end{cases}$

60. $\begin{cases} x^3 + y^2 = 2 \\ x^2 y = 4 \end{cases}$

61. $\begin{cases} x^4 + y^4 = 12 \\ xy^2 = 2 \end{cases}$ **62.** $\begin{cases} x^4 + y^4 = 6 \\ xy = 1 \end{cases}$ **63.** $\begin{cases} xy = 2 \\ y = \ln x \end{cases}$ **64.** $\begin{cases} x^2 + y^2 = 4 \\ y = \ln x \end{cases}$

65. The difference of two numbers is 2 and the sum of their squares is 10. Find the numbers.

66. The sum of two numbers is 7 and the difference of their squares is 21. Find the numbers.

67. The product of two numbers is 4 and the sum of their squares is 8. Find the numbers.

68. The product of two numbers is 10 and the difference of their squares is 21. Find the numbers.

69. The difference of two numbers is the same as their product, and the sum of their reciprocals is 5. Find the numbers.

70. The sum of two numbers is the same as their product, and the difference of their reciprocals is 3. Find the numbers.

71. The ratio of a to b is $\frac{2}{3}$. The sum of a and b is 10. What is the ratio of $a + b$ to $b - a$?

72. The ratio of a to b is $\frac{4}{3}$. The sum of a and b is 14. What is the ratio of $a - b$ to $a + b$?

73. Geometry   The perimeter of a rectangle is 16 inches and its area is 15 square inches. What are its dimensions?

74. Geometry   An area of 52 square feet is to be enclosed by two squares whose sides are in the ratio of $2:3$. Find the sides of the squares.

75. Geometry   Two circles have perimeters that add up to 12π centimeters and areas that add up to 20π square centimeters. Find the radius of each circle.

76. Geometry   The altitude of an isosceles triangle drawn to its base is 3 centimeters, and its perimeter is 18 centimeters. Find the length of its base.

77. The Tortoise and the Hare   In a 21-meter race between a tortoise and a hare, the tortoise leaves 9 minutes before the hare. The hare, by running at an average speed of 0.5 meter per hour faster than the tortoise, crosses the finish line 3 minutes before the tortoise. What are the average speeds of the tortoise and the hare?

78. Running a Race   In a 1-mile race, the winner crosses the finish line 10 feet ahead of the second-place runner and 20 feet ahead of the third-place runner. Assuming that each runner maintains a constant speed throughout the race, by how many feet does the second-place runner beat the third-place runner?

79. Constructing a Box   A rectangular piece of cardboard, whose area is 216 square centimeters, is made into an open box by cutting a 2-centimeter square from each corner and turning up the sides. See the figure. If the box is to have a volume of 224 cubic centimeters, what size cardboard should you start with?

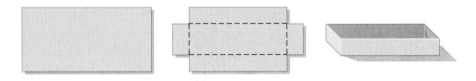

80. Constructing a Cylindrical Tube   A rectangular piece of cardboard, whose area is 216 square centimeters, is made into a cylindrical tube by joining together two sides of the rectangle. See the figure. If the tube is to have a volume of 224 cubic centimeters, what size cardboard should you start with?

81. Fencing   A farmer has 300 feet of fence available to enclose a 4500 square foot region in the shape of adjoining squares, with sides of length x and y. See the figure. Find x and y.

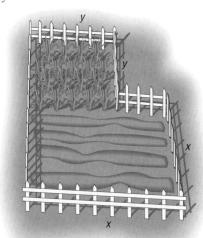

82. Bending Wire   A wire 60 feet long is cut into two pieces. Is it possible to bend one piece into the shape of a square and the other into the shape of a circle so that the total area enclosed by the two pieces is 100 square feet? If this is possible, find the length of the side of the square and the radius of the circle.

83. Geometry   Find formulas for the length l and width w of a rectangle in terms of its area A and perimeter P.

84. Geometry   Find formulas for the base b and one of the equal sides l of an isosceles triangle in terms of its altitude h and perimeter P.

85. Descartes's Method of Equal Roots   Descartes's method for finding tangents depends on the idea that, for many

graphs, the tangent line at a given point is the *unique* line that intersects the graph at that point only. We will apply his method to find an equation of the tangent line to the parabola $y = x^2$ at the point $(2, 4)$. See the figure. First, we know that the equation of the tangent line must be in the form $y = mx + b$. Using the fact that the point $(2, 4)$ is on the line, we can solve for b in terms of m and get the equation $y = mx + (4 - 2m)$. Now we want $(2, 4)$ to be the *unique* solution to the system

$$\begin{cases} y = x^2 \\ y = mx + 4 - 2m \end{cases}$$

From this system, we get $x^2 = mx + 4 - 2m$ or $x^2 - mx + (2m - 4) = 0$. By using the quadratic formula, we get

$$x = \frac{m \pm \sqrt{m^2 - 4(2m - 4)}}{2}$$

To obtain a unique solution for x, the two roots must be equal; in other words, the discriminant $m^2 - 4(2m - 4)$ must be 0. Complete the work to get m, and write an equation of the tangent line.

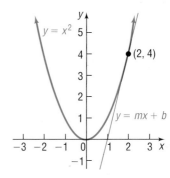

In Problems 86–92, use Descartes's method from Problem 85 to find the equation of the line tangent to each graph at the given point.

86. $x^2 + y^2 = 10$;   at $(1, 3)$

87. $y = x^2 + 2$;   at $(1, 3)$

88. $x^2 + y = 5$;   at $(-2, 1)$

89. $2x^2 + 3y^2 = 14$;   at $(1, 2)$

90. $3x^2 + y^2 = 7$;   at $(-1, 2)$

91. $x^2 - y^2 = 3$;   at $(2, 1)$

92. $2y^2 - x^2 = 14$;   at $(2, 3)$

93. If r_1 and r_2 are two solutions of a quadratic equation $ax^2 + bx + c = 0$, then it can be shown that

$$r_1 + r_2 = -\frac{b}{a} \quad \text{and} \quad r_1 r_2 = \frac{c}{a}$$

Solve this system of equations for r_1 and r_2.

94. A circle and a line intersect at most twice. A circle and a parabola intersect at most four times. Deduce that a circle and the graph of a polynomial of degree 3 intersect at most six times. What do you conjecture about a polynomial of degree 4? What about a polynomial of degree n? Can you explain your conclusions using an algebraic argument?

95. Suppose that you are the manager of a sheet metal shop. A customer asks you to manufacture 10,000 boxes, each box being open on top. The boxes are required to have a square base and a 9-cubic-foot capacity. You construct the boxes by cutting out a square from each corner of a square piece of sheet metal and folding along the edges.
(a) What are the dimensions of the square to be cut if the area of the square piece of sheet metal is 100 square feet?
(b) Could you make the box using a smaller piece of sheet metal? Make a list of the dimensions of the box for various pieces of sheet metal.

PREPARING FOR THIS SECTION

Before getting started, review the following:

✓ Linear Equations (Section 1.1, pp. 88–89) ✓ Lines (Section 2.3, pp. 163–172)

8.8 SYSTEMS OF INEQUALITIES

OBJECTIVES ① Graph an Inequality

 ② Graph a System of Linear Inequalities

In Chapter 1, we discussed inequalities in one variable. In this section, we discuss inequalities in two variables.

EXAMPLE 1   **Samples of Inequalities in Two Variables**

(a) $3x + y \leq 6$ (b) $x^2 + y^2 < 4$ (c) $y^2 \leq x$ ■

①   An inequality in two variables x and y is **satisfied** by an ordered pair (a, b) if, when x is replaced by a and y by b, a true statement results. A **graph of an inequality in two variables** x and y consists of all points (x, y) whose coordinates satisfy the inequality.

Let's look at an example.

EXAMPLE 2   **Graphing a Linear Inequality**

Graph the linear inequality: $3x + y \leq 6$

Solution   We begin with the associated problem of the graph of the linear equation

$$3x + y = 6$$

formed by replacing (for now) the $\leq$ symbol with an $=$ sign. The graph of the linear equation is a line. See Figure 13(a). This line is part of the graph of the inequality that we seek because the inequality is nonstrict. (Do you see why? We are seeking points for which $3x + y$ is less than *or equal to* 6.)

Now let's test a few randomly selected points to see whether they belong to the graph of the inequality.

	$3x + y \leq 6$	Conclusion
$(4, -1)$	$3(4) + (-1) = 11 > 6$	Does not belong to graph
$(5, 5)$	$3(5) + 5 = 20 > 6$	Does not belong to graph
$(-1, 2)$	$3(-1) + 2 = -1 \leq 6$	Belongs to graph
$(-2, -2)$	$3(-2) + (-2) = -8 \leq 6$	Belongs to graph

Look again at Figure 13(a). Notice that the two points that belong to the graph both lie on the same side of the line, and the two points that do not belong to the graph lie on the opposite side. As it turns out, this is always the case. The graph we seek consists of all points that lie on the same side of the line as do $(-1, 2)$ and $(-2, -2)$. The graph we seek is the shaded region in Figure 13(b).

Figure 13

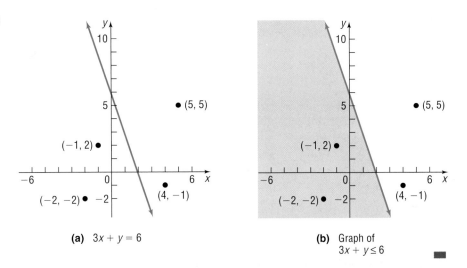

(a) $3x + y = 6$

(b) Graph of $3x + y \le 6$

The graph of any inequality in two variables may be obtained in a like way. First, the equation corresponding to the inequality is graphed, using dashes if the inequality is strict and solid marks if it is nonstrict. This graph, in almost every case, will separate the xy-plane into two or more regions. In each region, either all points satisfy the inequality or no points satisfy the inequality. The use of a single test point in each region is all that is required to determine whether the points of that region are part of the graph. The steps to follow are given next.

> ### STEPS FOR GRAPHING AN INEQUALITY
>
> **STEP 1:**   Replace the inequality symbol by an equal sign and graph the resulting equation. If the inequality is strict, use dashes; if it is nonstrict, use a solid mark. This graph separates the xy-plane into two or more regions.
>
> **STEP 2:**   In each of the regions, select a test point P.
>
> (a) If the coordinates of P satisfy the inequality, then so do all the points in that region. Indicate this by shading the region.
>
> (b) If the coordinates of P do not satisfy the inequality, then none of the points in that region do.

EXAMPLE 3   **Graphing an Inequality**

Figure 14

Graph: $x^2 + y^2 \le 4$

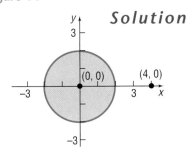

Solution   First, we graph the equation $x^2 + y^2 = 4$, a circle of radius 2, center at the origin. A solid circle will be used because the inequality is not strict. We use two test points, one inside the circle, the other outside.

Inside	$(0, 0)$: $\quad x^2 + y^2 = 0^2 + 0^2 = 0 \le 4$	Belongs to the graph
Outside	$(4, 0)$: $\quad x^2 + y^2 = 4^2 + 0^2 = 16 > 4$	Does not belong to graph

All the points inside and on the circle satisfy the inequality. See Figure 14.

NOW WORK PROBLEM 7.

Linear inequalities are inequalities in one of the forms

$$Ax + By < C \qquad Ax + By > C \qquad Ax + By \le C \qquad Ax + By \ge C$$

where A and B are not both zero.

Figure 15

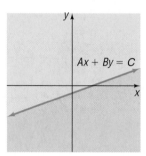

The graph of the corresponding equation of a linear inequality is a line, which separates the xy-plane into two regions, called **half-planes.** See Figure 15.

As shown, $Ax + By = C$ is the equation of the boundary line and it divides the plane into two half-planes: one for which $Ax + By < C$ and the other for which $Ax + By > C$. Because of this, for linear inequalities, only one test point is required.

EXAMPLE 4

Graphing Linear Inequalities

Graph: (a) $y < 2$ (b) $y \ge 2x$

Solution

(a) The graph of the equation $y = 2$ is a horizontal line and is not part of the graph of the inequality. Since $(0, 0)$ satisfies the inequality, the graph consists of the half-plane below the line $y = 2$. See Figure 16.

(b) The graph of the equation $y = 2x$ is a line and is part of the graph of the inequality. Using $(3, 0)$ as a test point, we find it does not satisfy the inequality $[0 < 2 \cdot 3]$. Points in the half-plane on the opposite side of $y = 2x$ satisfy the inequality. See Figure 17. ∎

Figure 16

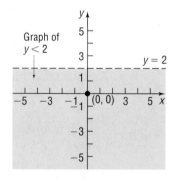

Figure 17

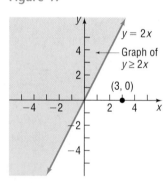

COMMENT: A graphing utility can be used to graph inequalities. To see how, read Section 6 in the Appendix. ∎

NOW WORK PROBLEM 3.

SYSTEMS OF INEQUALITIES IN TWO VARIABLES

② The **graph of a system of inequalities** in two variables x and y is the set of all points (x, y) that simultaneously satisfy *each* of the inequalities in the system. The graph of a system of inequalities can be obtained by graphing each inequality individually and then determining where, if at all, they intersect.

| EXAMPLE 5 | **Graphing a System of Linear Inequalities** |

Graph the system: $\begin{cases} x + y \geq 2 \\ 2x - y \leq 4 \end{cases}$

Solution   First, we graph the inequality $x + y \geq 2$ as the shaded region in Figure 18(a). Next, we graph the inequality $2x - y \leq 4$ as the shaded region in Figure 18(b). Now, superimpose the two graphs, as shown in Figure 18(c). The points that are in both shaded regions [the overlapping, darker region in Figure 18(c)] are the solutions we seek to the system, because they simultaneously satisfy each linear inequality.

Figure 18

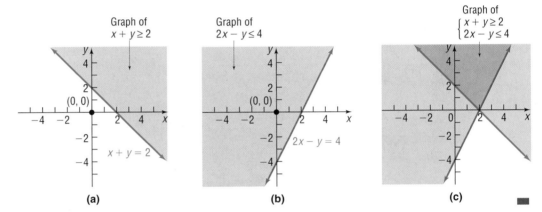

(a)   (b)   (c)

| EXAMPLE 6 | **Graphing a System of Linear Inequalities** |

Graph the system: $\begin{cases} x + y \leq 2 \\ x + y \geq 0 \end{cases}$

Solution   See Figure 19. The overlapping, darker shaded region between the two boundary lines is the graph of the system.

Figure 19

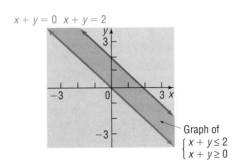

NOW WORK PROBLEM **25.**

EXAMPLE 7

Graphing a System of Linear Inequalities

Graph the system: $\begin{cases} 2x - y \geq 2 \\ 2x - y \geq 0 \end{cases}$

Solution See Figure 20. The overlapping, darker shaded region is the graph of the system. Note that the graph of the system is identical to the graph of the single inequality $2x - y \geq 2$.

Figure 20

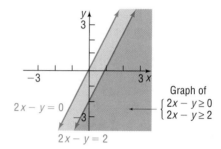

EXAMPLE 8

Graphing a System of Linear Inequalities

Graph the system: $\begin{cases} x + 2y \leq 2 \\ x + 2y \geq 6 \end{cases}$

Solution See Figure 21. Because no overlapping region results, there are no points in the xy-plane that simultaneously satisfy each inequality. Hence, the system has no solution.

Figure 21

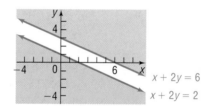

EXAMPLE 9

Graphing a System of Inequalities

Graph the system: $\begin{cases} y \geq x^2 - 4 \\ x + y \leq 2 \end{cases}$

Solution Figure 22 shows that the graph of the system consists of the region enclosed by the graphs of the parabola $y = x^2 - 4$ and the line $x + y = 2$. The points of intersection of the two equations are found by solving the system of equations

$$\begin{cases} y = x^2 - 4 \\ x + y = 2 \end{cases}$$

Figure 22

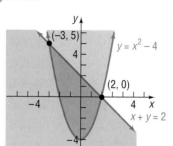

Using substitution, we find

$$x + (x^2 - 4) = 2$$
$$x^2 + x - 6 = 0$$
$$(x + 3)(x - 2) = 0$$
$$x = -3 \qquad x = 2$$

The two points of intersection are $(-3, 5)$ and $(2, 0)$. ■

▬▬▬▬ NOW WORK PROBLEM **19**.

EXAMPLE 10

Graphing a System of Four Linear Inequalities

Figure 23

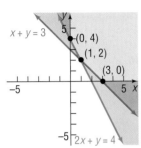

Graph the system:
$$\begin{cases} x + y \geq 3 \\ 2x + y \geq 4 \\ x \geq 0 \\ y \geq 0 \end{cases}$$

Solution The two inequalities $x \geq 0$ and $y \geq 0$ require the graph of the system to be in quadrant I. We concentrate on the remaining two inequalities. The intersection of the graphs of these two inequalities and quadrant I, shown in gray in Figure 23, is the graph of the system. ■

EXAMPLE 11

Financial Planning

A retired couple has up to $25,000 to invest. As their financial adviser, you recommend that they place at least $15,000 in Treasury bills yielding 6% and at most $5000 in corporate bonds yielding 9%.

(a) Using x to denote the amount of money invested in Treasury bills and y the amount invested in corporate bonds, write a system of linear inequalities that describes the possible amounts of each investment. We shall assume that x and y are in thousands of dollars.

(b) Graph the system.

Solution

Figure 24

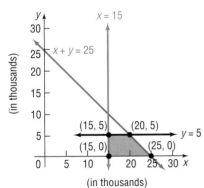

(a) The system of linear inequalities is

$$\begin{cases} x \geq 0 & \text{\small x and y are nonnegative variables since they represent} \\ y \geq 0 & \text{\small money invested in thousands of dollars.} \\ x + y \leq 25 & \text{\small The total of the two investments, $x + y$, cannot exceed \$25,000.} \\ x \geq 15 & \text{\small At least \$15,000 in Treasury bills} \\ y \leq 5 & \text{\small At most \$5000 in corporate bonds.} \end{cases}$$

(b) See the shaded region in Figure 24. Note that the inequalities $x \geq 0$ and $y \geq 0$ again require that the graph of the system be in quadrant I. ■

The graph of the system of linear inequalities in Figure 24 is said to be **bounded,** because it can be contained within some circle of sufficiently large radius. A graph that cannot be contained in any circle is said to be **unbounded.** For example, the graph of the system of linear inequalities in Figure 23 is unbounded, since it extends indefinitely in a particular direction.

Notice in Figures 23 and 24 that those points belonging to the graph that are also points of intersection of boundary lines have been plotted. Such points are referred to as **vertices** or **corner points** of the graph. The system graphed in Figure 23 has three corner points: $(0, 4)$, $(1, 2)$, and $(3, 0)$. The system graphed in Figure 24 has four corner points: $(15, 0)$, $(25, 0)$, $(20, 5)$, and $(15, 5)$.

These ideas will be used in the next section in developing a method for solving linear programming problems, an important application of linear inequalities.

▬▬▬▬ NOW WORK PROBLEM **33**.

8.8 EXERCISES

In Problems 1–12, graph each inequality.

1. $x \geq 0$

2. $y \geq 0$

3. $x \geq 4$

4. $y \leq 2$

5. $2x + y \geq 6$

6. $3x + 2y \leq 6$

7. $x^2 + y^2 > 1$

8. $x^2 + y^2 \leq 9$

9. $y \leq x^2 - 1$

10. $y > x^2 + 2$

11. $xy \geq 4$

12. $xy \leq 1$

In Problems 13–30, graph each system of inequalities.

13. $\begin{cases} x + y \leq 2 \\ 2x + y \geq 4 \end{cases}$

14. $\begin{cases} 3x - y \geq 6 \\ x + 2y \leq 2 \end{cases}$

15. $\begin{cases} 2x - y \leq 4 \\ 3x + 2y \geq -6 \end{cases}$

16. $\begin{cases} 4x - 5y \leq 0 \\ 2x - y \geq 2 \end{cases}$

17. $\begin{cases} 2x - 3y \leq 0 \\ 3x + 2y \leq 6 \end{cases}$

18. $\begin{cases} 4x - y \geq 2 \\ x + 2y \geq 2 \end{cases}$

19. $\begin{cases} x^2 + y^2 \leq 9 \\ x + y \geq 3 \end{cases}$

20. $\begin{cases} x^2 + y^2 \geq 9 \\ x + y \leq 3 \end{cases}$

21. $\begin{cases} y \geq x^2 - 4 \\ y \leq x - 2 \end{cases}$

22. $\begin{cases} y^2 \leq x \\ y \geq x \end{cases}$

23. $\begin{cases} xy \geq 4 \\ y \geq x^2 + 1 \end{cases}$

24. $\begin{cases} y + x^2 \leq 1 \\ y \geq x^2 - 1 \end{cases}$

25. $\begin{cases} x - 2y \leq 6 \\ 2x - 4y \geq 0 \end{cases}$

26. $\begin{cases} x + 4y \leq 8 \\ x + 4y \geq 4 \end{cases}$

27. $\begin{cases} 2x + y \geq -2 \\ 2x + y \geq 2 \end{cases}$

28. $\begin{cases} x - 4y \leq 4 \\ x - 4y \geq 0 \end{cases}$

29. $\begin{cases} 2x + 3y \geq 6 \\ 2x + 3y \leq 0 \end{cases}$

30. $\begin{cases} 2x + y \geq 0 \\ 2x + y \geq 2 \end{cases}$

In Problems 31–40, graph each system of linear inequalities. Tell whether the graph is bounded or unbounded, and label the corner points.

31. $\begin{cases} x \geq 0 \\ y \geq 0 \\ 2x + y \leq 6 \\ x + 2y \leq 6 \end{cases}$

32. $\begin{cases} x \geq 0 \\ y \geq 0 \\ x + y \geq 4 \\ 2x + 3y \geq 6 \end{cases}$

33. $\begin{cases} x \geq 0 \\ y \geq 0 \\ x + y \geq 2 \\ 2x + y \geq 4 \end{cases}$

34. $\begin{cases} x \geq 0 \\ y \geq 0 \\ 3x + y \leq 6 \\ 2x + y \leq 2 \end{cases}$

35. $\begin{cases} x \geq 0 \\ y \geq 0 \\ x + y \geq 2 \\ 2x + 3y \leq 12 \\ 3x + y \leq 12 \end{cases}$

36. $\begin{cases} x \geq 0 \\ y \geq 0 \\ x + y \geq 2 \\ x + y \leq 10 \\ 2x + y \leq 3 \end{cases}$

37. $\begin{cases} x \geq 0 \\ y \geq 0 \\ x + y \geq 2 \\ x + y \leq 8 \\ 2x + y \leq 10 \end{cases}$

38. $\begin{cases} x \geq 0 \\ y \geq 0 \\ x + y \geq 2 \\ x + y \leq 8 \\ x + 2y \geq 1 \end{cases}$

39. $\begin{cases} x \geq 0 \\ y \geq 0 \\ x + 2y \geq 1 \\ x + 2y \leq 10 \end{cases}$

40. $\begin{cases} x \geq 0 \\ y \geq 0 \\ x + 2y \geq 1 \\ x + 2y \leq 10 \\ x + y \geq 2 \\ x + y \leq 8 \end{cases}$

In Problems 41–44, write a system of linear inequalities that has the given graph.

41.

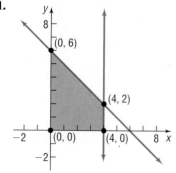

42.

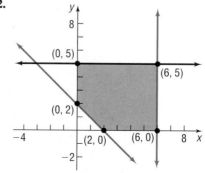

43.

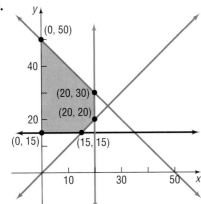

44.

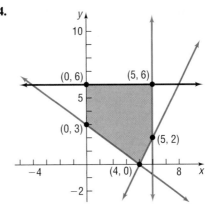

45. Financial Planning A retired couple has up to $50,000 to invest. As their financial adviser, you recommend that they place at least $35,000 in Treasury bills yielding 7% and at most $10,000 in corporate bonds yielding 10%.
(a) Using x to denote the amount of money invested in Treasury bills and y the amount invested in corporate bonds, write a system of linear inequalities that describes the possible amounts of each investment.
(b) Graph the system and label the corner points.

46. Manufacturing Trucks Mike's Toy Truck Company manufacturers two models of toy trucks, a standard model and a deluxe model. Each standard model requires 2 hours for painting and 3 hours for detail work; each deluxe model requires 3 hours for painting and 4 hours for detail work. Two painters and three detail workers are employed by the company, and each works 40 hours per week.
(a) Using x to denote the number of standard model trucks and y to denote the number of deluxe model trucks, write a system of linear inequalities that describes the possible number of each model of truck that can be manufactured in a week.
(b) Graph the system and label the corner points.

47. Blending Coffee Bill's Coffee House, a store that specializes in coffee, has available 75 pounds of A grade coffee and 120 pounds of B grade coffee. These will be blended into 1 pound packages as follows: An economy blend that contains 4 ounces of A grade coffee and 12 ounces of B grade coffee and a superior blend that contains 8 ounces of A grade coffee and 8 ounces of B grade coffee.
(a) Using x to denote the number of packages of the economy blend and y to denote the number of packages of the superior blend, write a system of linear inequalities that describes the possible number of packages of each kind of blend.
(b) Graph the system and label the corner points.

48. Mixed Nuts Nola's Nuts, a store that specializes in selling nuts, has available 90 pounds of cashews and 120 pounds of peanuts. These are to be mixed in 12 ounce packages as follows: a lower priced package containing 8 ounces of peanuts and 4 ounces of cashews and a quality package containing 6 ounces of peanuts and 6 ounces of cashews.
(a) Use x to denote the number of lower priced packages and use y to denote the number of quality packages. Write a system of linear inequalities that describes the possible number of each kind of package.
(b) Graph the system and label the corner points.

49. Transporting Goods A small truck can carry no more than 1600 pounds of cargo or 150 cubic feet of cargo. A printer weighs 20 pounds and occupies 3 cubic feet of space. A microwave oven weighs 30 pounds and occupies 2 cubic feet of space.
(a) Using x to represent the number of microwave ovens and y to represent the number of printers, write a system of linear inequalities that describes the number of ovens and printers that can be hauled by the truck.
(b) Graph the system and label the corner points.

8.9 LINEAR PROGRAMMING

OBJECTIVES 1 Set up a Linear Programming Problem
2 Solve a Linear Programming Problem

1 Historically, linear programming evolved as a technique for solving problems involving resource allocation of goods and materials for the U.S. Air Force during World War II. Today, linear programming techniques are used to solve a wide variety of problems, such as optimizing airline scheduling and establishing telephone lines. Although most practical linear programming problems involve systems of several hundred linear inequalities containing several hundred variables, we will limit our discussion to problems containing only two variables, because we can solve such problems using graphing techniques.*

We begin by returning to Example 11 of the previous section.

EXAMPLE 1   **Financial Planning**

A retired couple has up to $25,000 to invest. As their financial adviser, you recommend that they place at least $15,000 in Treasury bills yielding 6% and at most $5000 in corporate bonds yielding 9%. How much money should be placed in each investment so that income is maximized?   ■

The problem given here is typical of a *linear programming problem*. The problem requires that a certain linear expression, the income, be maximized. If I represents income, x the amount invested in Treasury bills at 6%, and y the amount invested in corporate bonds at 9%, then

$$I = 0.06x + 0.09y$$

We shall assume, as before, that I, x, and y are in thousands of dollars.

The linear expression $I = 0.06x + 0.09y$ is called the **objective function.** Furthermore, the problem requires that the maximum income be achieved under certain conditions or **constraints,** each of which is a linear inequality involving the variables. (See Example 11 in Section 8.8.) The linear programming problem given in Example 1 may be restated as

Maximize $I = 0.06x + 0.09y$

subject to the conditions that

$$x \geq 0, \qquad y \geq 0$$
$$x + y \leq 25$$
$$x \geq 15$$
$$y \leq 5$$

---

*The **simplex method** is a way to solve linear programming problems involving many inequalities and variables. This method was developed by George Dantzig in 1946 and is particularly well suited for computerization. In 1984, Narendra Karmarkar of Bell Laboratories discovered a way of solving large linear programming problems that improves on the simplex method.

In general, every linear programming problem has two components:

1. A linear objective function that is to be maximized or minimized.
2. A collection of linear inequalities that must be satisfied simultaneously.

> A **linear programming problem** in two variables x and y consists of maximizing (or minimizing) a linear objective function
>
> $$z = Ax + By, \qquad A \text{ and } B \text{ are real numbers, not both } 0$$
>
> subject to certain conditions, or constraints, expressible as linear inequalities in x and y.

To maximize (or minimize) the quantity $z = Ax + By$, we need to identify points (x, y) that make the expression for z the largest (or smallest) possible. But not all points (x, y) are eligible; only those that also satisfy each linear inequality (constraint) can be used. We refer to each point (x, y) that satisfies the system of linear inequalities (the constraints) as a **feasible point.** In a linear programming problem, we seek the feasible point(s) that maximizes (or minimizes) the objective function.

Let's look again at the linear programming problem in Example 1.

EXAMPLE 2   Analyzing a Linear Programming Problem

Consider the linear programming problem

$$\text{Maximize} \qquad I = 0.06x + 0.09y$$

subject to the conditions that

$$x \geq 0, \qquad y \geq 0$$

$$x + y \leq 25$$

$$x \geq 15$$

$$y \leq 5$$

Graph the constraints. Then graph the objective function for $I = 0, 0.9, 1.35, 1.65,$ and 1.8.

Solution   Figure 25 shows the graph of the constraints. We superimpose on this graph the graph of the objective function for the given values of I.

For $I = 0$, the objective function is the line $0 = 0.06x + 0.09y$.

For $I = 0.9$, the objective function is the line $0.9 = 0.06x + 0.09y$.

For $I = 1.35$, the objective function is the line $1.35 = 0.06x + 0.09y$.

For $I = 1.65$, the objective function is the line $1.65 = 0.06x + 0.09y$.

For $I = 1.8$, the objective function is the line $1.8 = 0.06x + 0.09y$.   ■

Figure 25

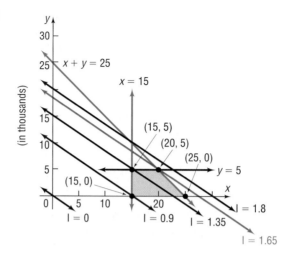

A **solution** to a linear programming problem consists of a feasible point that maximizes (or minimizes) the objective function, together with the corresponding value of the objective function.

2 One condition for a linear programming problem in two variables to have a solution is that the graph of the feasible points be bounded. (Refer to page 597.)

If none of the feasible points maximizes (or minimizes) the objective function or if there are no feasible points, then the linear programming problem has no solution.

Consider the linear programming problem stated in Example 2, and look again at Figure 25. The feasible points are the points that lie in the shaded region. For example, $(20, 3)$ is a feasible point, as are $(15, 5)$, $(20, 5)$, $(18, 4)$, and so on. To find the solution of the problem requires that we find a feasible point (x, y) that makes $I = 0.06x + 0.09y$ as large as possible. Notice that as I increases in value from $I = 0$ to $I = 0.9$ to $I = 1.35$ to $I = 1.65$ to $I = 1.8$, we obtain a collection of parallel lines. Furthermore, notice that the largest value of I that can be obtained using feasible points is $I = 1.65$, which corresponds to the line $1.65 = 0.06x + 0.09y$. Any larger value of I results in a line that does not pass through any feasible points. Finally, notice that the feasible point that yields $I = 1.65$ is the point $(20, 5)$, a corner point. These observations form the basis of the following result, which we state without proof.

Theorem

Location of the Solution of a Linear Programming Problem

If a linear programming problem has a solution, it is located at a corner point of the graph of the feasible points.

If a linear programming problem has multiple solutions, at least one of them is located at a corner point of the graph of the feasible points.

In either case, the corresponding value of the objective function is unique.

We shall not consider here linear programming problems that have no solution. As a result, we can outline the procedure for solving a linear programming problem as follows:

PROCEDURE FOR SOLVING A LINEAR PROGRAMMING PROBLEM

STEP 1:   Write an expression for the quantity to be maximized (or minimized). This expression is the objective function.

STEP 2:   Write all the constraints as a system of linear inequalities and graph the system.

STEP 3:   List the corner points of the graph of the feasible points.

STEP 4:   List the corresponding values of the objective function at each corner point. The largest (or smallest) of these is the solution.

EXAMPLE 3

Solving a Minimum Linear Programming Problem

Minimize the expression

$$z = 2x + 3y$$

subject to the constraints

$$y \le 5, \qquad x \le 6 \qquad x + y \ge 2, \qquad x \ge 0, \qquad y \ge 0$$

Solution   The objective function is $z = 2x + 3y$. We seek the smallest value of z that can occur if x and y are solutions of the system of linear inequalities

$$\begin{cases} y \le 5 \\ x \le 6 \\ x + y \ge 2 \\ x \ge 0 \\ y \ge 0 \end{cases}$$

The graph of this system (the feasible points) is shown as the shaded region in Figure 26. We have also plotted the corner points. Table 1 lists the corner points and the corresponding values of the objective function. From the table, we can see that the minimum value of z is 4, and it occurs at the point $(2, 0)$.

Figure 26

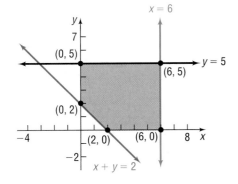

TABLE 1	
Corner Point	**Value of the Objective Function**
(x, y)	$z = 2x + 3y$
$(0, 2)$	$z = 2(0) + 3(2) = 6$
$(0, 5)$	$z = 2(0) + 3(5) = 15$
$(6, 5)$	$z = 2(6) + 3(5) = 27$
$(6, 0)$	$z = 2(6) + 3(0) = 12$
$(2, 0)$	$z = 2(2) + 3(0) = 4$

NOW WORK PROBLEMS **3** AND **9**.

EXAMPLE 4 **Maximizing Profit**

At the end of every month, after filling orders for its regular customers, a coffee company has some pure Colombian coffee and some special-blend coffee remaining. The practice of the company has been to package a mixture of the two coffees into 1-pound packages as follows: a low-grade mixture containing 4 ounces of Colombian coffee and 12 ounces of special-blend coffee and a high-grade mixture containing 8 ounces of Colombian and 8 ounces of special-blend coffee. A profit of $0.30 per package is made on the low-grade mixture, whereas a profit of $0.40 per package is made on the high-grade mixture. This month, 120 pounds of special-blend coffee and 100 pounds of pure Colombian coffee remain. How many packages of each mixture should be prepared to achieve a maximum profit? Assume that all packages prepared can be sold.

Solution We begin by assigning symbols for the two variables.

$$x = \text{Number of packages of the low–grade mixture}$$

$$y = \text{Number of packages of the high–grade mixture}$$

If P denotes the profit, then

$$P = \$0.30x + \$0.40y$$

This expression is the objective function. We seek to maximize P subject to certain constraints on x and y. Because x and y represent numbers of packages, the only meaningful values for x and y are nonnegative integers. Thus, we have the two constraints

$$x \geq 0, \qquad y \geq 0 \quad \text{Nonnegative constraints}$$

We also have only so much of each type of coffee available. For example, the total amount of Colombian coffee used in the two mixtures cannot exceed 100 pounds, or 1600 ounces. Because we use 4 ounces in each low-grade package and 8 ounces in each high-grade package, we are led to the constraint

$$4x + 8y \leq 1600 \quad \text{Colombian coffee constraint}$$

Similarly, the supply of 120 pounds, or 1920 ounces, of special-blend coffee leads to the constraint

$$12x + 8y \leq 1920 \quad \text{Special-blend coffee constraint}$$

The linear programming problem may be stated as

$$\text{Maximize} \qquad P = 0.3x + 0.4y$$

subject to the constraints

$$x \geq 0, \qquad y \geq 0, \qquad 4x + 8y \leq 1600, \qquad 12x + 8y \leq 1920$$

The graph of the constraints (the feasible points) is illustrated in Figure 27. We list the corner points and evaluate the objective function at each. In Table 2, we can see that the maximum profit, $84, is achieved with 40 packages of the low-grade mixture and 180 packages of the high-grade mixture.

Figure 27

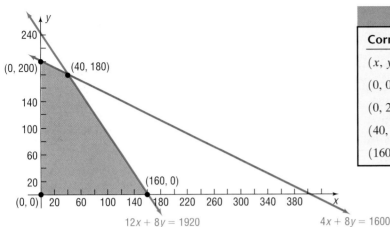

12x + 8y = 1920   4x + 8y = 1600

TABLE 2	
Corner Point	Value of Profit
(x, y)	$P = 0.3x + 0.4y$
$(0, 0)$	$P = 0$
$(0, 200)$	$P = 0.3(0) + 0.4(200) = \$80$
$(40, 180)$	$P = 0.3(40) + 0.4(180) = \$84$
$(160, 0)$	$P = 0.3(160) + 0.4(0) = \$48$

NOW WORK PROBLEM **17.**

8.9   EXERCISES

In Problems 1–6, find the maximum and minimum value of the given objective function of a linear programming problem. The figure illustrates the graph of the feasible points.

1. $z = x + y$

2. $z = 2x + 3y$

3. $z = x + 10y$

4. $z = 10x + y$

5. $z = 5x + 7y$

6. $z = 7x + 5y$

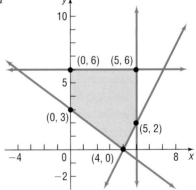

In Problems 7–16, solve each linear programming problem.

7. Maximize $z = 2x + y$ subject to $x \geq 0,\ y \geq 0,\ x + y \leq 6,\quad x + y \geq 1$

8. Maximize $z = x + 3y$ subject to $x \geq 0,\ y \geq 0,\ x + y \geq 3,\quad x \leq 5,\quad y \leq 7$

9. Minimize $z = 2x + 5y$ subject to $x \geq 0,\ y \geq 0,\ x + y \geq 2,\quad x \leq 5,\quad y \leq 3$

10. Minimize $z = 3x + 4y$ subject to $x \geq 0,\ y \geq 0,\ 2x + 3y \geq 6, x + y \leq 8$

11. Maximize $z = 3x + 5y$ subject to $x \geq 0,\ y \geq 0,\ x + y \geq 2,\quad 2x + 3y \leq 12,\quad 3x + 2y \leq 12$

12. Maximize $z = 5x + 3y$ subject to $x \geq 0,\ y \geq 0,\ x + y \geq 2,\quad x + y \leq 8,\quad 2x + y \leq 10$

13. Minimize $z = 5x + 4y$ subject to $x \geq 0,\ y \geq 0,\ x + y \geq 2,\quad 2x + 3y \leq 12,\quad 3x + y \leq 12$

14. Minimize $z = 2x + 3y$ subject to $x \geq 0,\ y \geq 0,\ x + y \geq 3,\quad x + y \leq 9,\quad x + 3y \geq 6$

15. Maximize $z = 5x + 2y$ subject to $x \geq 0,\ y \geq 0,\ x + y \leq 10,\quad 2x + y \geq 10,\quad x + 2y \geq 10$

16. Maximize $z = 2x + 4y$ subject to $x \geq 0,\ y \geq 0,\ 2x + y \geq 4,\quad x + y \leq 9$

17. Maximizing Profit   A manufacturer of skis produces two types: downhill and cross-country. Use the following table to determine how many of each kind of ski should be produced to achieve a maximum profit. What is the maximum profit? What would the maximum profit be if the maximum time available for manufacturing is increased to 48 hours?

	Downhill	Cross-country	Maximum Time Available
Manufacturing time per ski	2 hours	1 hour	40 hours
Finishing time per ski	1 hour	1 hour	32 hours
Profit per ski	$70	$50	

18. Farm Management A farmer has 70 acres of land available for planting either soybeans or wheat. The cost of preparing the soil, the workdays required, and the expected profit per acre planted for each type of crop are given in the following table:

	Soybeans	Wheat
Preparation cost per acre	$60	$30
Workdays required per acre	3	4
Profit per acre	$180	$100

The farmer cannot spend more than $1800 in preparation costs nor more than a total of 120 workdays. How many acres of each crop should be planted in order to maximize the profit? What is the maximum profit? What is the maximum profit if the farmer is willing to spend no more than $2400 on preparation?

19. Farm Management A small farm in Illinois has 100 acres of land available on which to grow corn and soybeans. The following table shows the cultivation cost per acre, the labor cost per acre, and the expected profit per acre. The column on the right shows the amount of money available for each of these expenses. Find the number of acres of each crop that should be planted in order to maximize profit.

	Soybeans	Corn	Money Available
Cultivation cost per acre	$40	$60	$1800
Labor cost per acre	$60	$60	$2400
Profit per acre	$200	$250	

20. Dietary Requirements A certain diet requires at least 60 units of carbohydrates, 45 units of protein, and 30 units of fat each day. Each ounce of Supplement A provides 5 units of carbohydrates, 3 units of protein, and 4 units of fat. Each ounce of Supplement B provides 2 units of carbohydrates, 2 units of protein, and 1 unit of fat. If Supplement A costs $1.50 per ounce and Supplement B costs $1.00 per ounce, how many ounces of each supplement should be taken daily to minimize the cost of the diet?

21. Production Scheduling In a factory, machine 1 produces 8-inch pliers at the rate of 60 units per hour and 6-inch pliers at the rate of 70 units per hour. Machine 2 produces 8-inch pliers at the rate of 40 units per hour and 6-inch pliers at the rate of 20 units per hour. It costs $50 per hour to operate machine 1, and machine 2 costs $30 per hour to operate. The production schedule requires that at least 240 units of 8-inch pliers and at least 140 units of 6-inch pliers be produced during each 10-hour day. Which combination of machines will cost the least money to operate?

22. Farm Management An owner of a fruit orchard hires a crew of workers to prune at least 25 of his 50 fruit trees. Each newer tree requires one hour to prune, while each older tree needs one-and-a-half hours. The crew contracts to work for at least 30 hours and charge $15 for each newer tree and $20 for each older tree. To minimize his cost, how many of each kind of tree will the orchard owner have pruned? What will be the cost?

23. Managing a Meat Market A meat market combines ground beef and ground pork in a single package for meat loaf. The ground beef is 75% lean (75% beef, 25% fat) and costs the market $0.75 per pound. The ground pork is 60% lean and costs the market $0.45 per pound. The meat loaf must be at least 70% lean. If the market wants to use at least 50 lb of its available pork, but no more than 200 lb of its available ground beef, how much ground beef should be mixed with ground pork so that the cost is minimized?

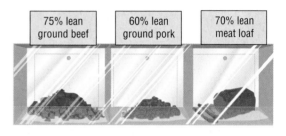

24. Return on Investment An investment broker is instructed by her client to invest up to $20,000, some in a junk bond yielding 9% per annum and some in Treasury bills yielding 7% per annum. The client wants to invest at least $8000 in T-bills and no more than $12,000 in the junk bond.
(a) How much should the broker recommend that the client place in each investment to maximize income if the client insists that the amount invested in T-bills must equal or exceed the amount placed in junk bonds?
(b) How much should the broker recommend that the client place in each investment to maximize income if the client insists that the amount invested in T-bills must not exceed the amount placed in junk bonds?

25. Maximizing Profit on Ice Skates A factory manufactures two kinds of ice skates: racing skates and figure skates. The racing skates require 6 work-hours in the fabrication department, whereas the figure skates require 4 work-hours there. The racing skates require 1 work-hour in the finishing department, whereas the figure skates require 2 work-hours there. The fabricating department has available at most 120 work-hours per day, and the finishing department has no more than 40 work-hours per day available. If the profit on each racing skate is $10 and the profit on each figure skate is $12, how many of each should be manufactured each day to maximize profit? (Assume that all skates made are sold.)

26. Financial Planning A retired couple has up to $50,000 to place in fixed-income securities. Their financial adviser suggests two securities to them: one is an AAA bond that yields 8% per annum; the other is a Certificate of Deposit (CD) that yields 4%. After careful consideration of the alternatives, the couple decides to place at most $20,000 in the AAA bond and at least $15,000 in the CD. They also instruct the financial adviser to place at least as much in the CD as in the AAA bond. How should the financial adviser proceed to maximize the return on their investment?

27. **Product Design** An entrepreneur is having a design group produce at least six samples of a new kind of fastener that he wants to market. It costs $9.00 to produce each metal fastener and $4.00 to produce each plastic fastener. He wants to have at least two of each version of the fastener and needs to have all the samples 24 hours from now. It takes 4 hours to produce each metal sample and 2 hours to produce each plastic sample. To minimize the cost of the samples, how many of each kind should the entrepreneur order? What will be the cost of the samples?

28. **Animal Nutrition** Kevin's dog Amadeus likes two kinds of canned dog food. "Gourmet Dog" costs 40 cents a can and has 20 units of a vitamin complex; the calorie content is 75 calories. "Chow Hound" costs 32 cents a can and has 35 units of vitamins and 50 calories. Kevin likes Amadeus to have at least 1175 units of vitamins a month and at least 2375 calories during the same time period. Kevin has space to store only 60 cans of dog food at a time. How much of each kind of dog food should Kevin buy each month in order to minimize his cost?

29. **Airline Revenue** An airline has two classes of service: first class and coach. Management's experience has been that each aircraft should have at least 8 but no more than 16 first-class seats and at least 80 but not more than 120 coach seats.

(a) If management decides that the ratio of first class to coach seats should never exceed $1 : 12$, with how many of each type of seat should an aircraft be configured to maximize revenue?

(b) If management decides that the ratio of first class to coach seats should never exceed $1 : 8$, with how many of each type of seat should an aircraft be configured to maximize revenue?

 (c) If you were management, what would you do?
[**Hint:** Assume that the airline charges $C for a coach seat and $F for a first-class seat; $C > 0, F > C.$]

30. **Minimizing Cost** A farm that specializes in raising frying chickens supplements the regular chicken feed with four vitamins. The owner wants the supplemental food to contain at least 50 units of vitamin I, 90 units of vitamin II, 60 units of vitamin III, and 100 units of vitamin IV per 100 ounces of feed. Two supplements are available: supplement A, which contains 5 units of vitamin I, 25 units of vitamin II, 10 units of vitamin III, and 35 units of vitamin IV per ounce, and supplement B, which contains 25 units of vitamin I, 10 units of vitamin II, 10 units of vitamin III, and 20 units of vitamin IV per ounce. If supplement A costs $0.06 per ounce and supplement B costs $0.08 per ounce, how much of each supplement should the manager of the farm buy to add to each 100 ounces of feed in order to keep the total cost at a minimum, while still meeting the owner's vitamin specifications?

31. Explain in your own words what a linear programming problem is and how it can be solved.

Things To Know

Systems of equations (pp. 517–518)

Systems with no solutions are inconsistent.   Systems with a solution are consistent.

Consistent systems of linear equations have either a unique solution or an infinite number of solutions.

Matrix (pp. 534 and 559)	Rectangular array of numbers, called entries
m by n matrix (p. 560)	Matrix with m rows and n columns
Identity matrix I (p. 567)	Square matrix whose diagonal entries are 1's, while all other entries are 0's
Inverse of a matrix (p. 568)	A^{-1} is the inverse of A if $AA^{-1} = A^{-1}A = I$
Nonsingular matrix (p. 568)	A square matrix that has an inverse

Linear programming problems (p. 601)

Maximize (or minimize) a linear objective function, $z = Ax + By$, subject to certain conditions, or constraints, expressible as linear inequalities in x and y.

Feasible point (p. 601)

A point (x, y) that satisfies the constraints of a linear programming problem

Location of solution (p. 602)

If a linear programming problem has a solution, it is located at a corner point of the graph of the feasible points.

If a linear programming problem has multiple solutions, at least one of them is located at a corner point of the graph of the feasible points.

In either case, the corresponding value of the objective function is unique.

Objectives:

You should be able to:

Solve systems of equations by substitution (p. 519)

Solve systems of equations by elimination (p. 521)

Identify inconsistent systems (pp. 523 and 529)

Express the solution of a system of dependent equations (pp. 524 and 530)

Solve a system of three equations containing three variables (p. 528)

Write the augmented matrix of a system of linear equation (p. 534)

Write the system from the augmented matrix (p. 535)

Perform row operations on a matrix (p. 536)

Solve systems of linear equations using matrices (p. 537)

Evaluate 2 by 2 determinants (p. 548)

Use Cramer's Rule to solve a system of equations (pp. 549 and 554)

Evaluate 3 by 3 determinants (p. 552)

Know properties of determinants (p. 555)

Find the sum and difference of two matrices (p. 560)

Find scalar multiples of a matrix (p. 562)

Find the product of two matrices (p. 563)

Find the inverse of a matrix (p. 568)

Solve a system of equations using inverse matrices (p. 571)

Decompose P/Q, where Q has only nonrepeated linear factors (p. 576)

Decompose P/Q, where Q has repeated linear factors (p. 578)

Decompose P/Q, where Q has nonrepeated irreducible quadratic factors (p. 580)

Decompose P/Q, where Q has repeated irreducible quadratic factors (p. 581)

Solve a system of nonlinear equations using substitution (p. 582).

Solve a system of nonlinear equations using elimination (p. 584).

Graph an inequality (p. 592)

Graph a system of linear inequalities (p. 594)

Set up and solve a linear programming problem (pp. 600 and 602)

Fill-in-the-Blank Items

1. If a system of equations has no solution, it is said to be _____.

2. An m by n rectangular array of numbers is called a(n) _____.

3. Cramer's Rule uses _____ to solve a system of linear equations.

4. The matrix used to represent a system of linear equations is called a(n) _____ matrix.

5. A matrix B, for which $AB = I_n$, the identity matrix, is called the _____ of A.

6. A matrix that has the same number of rows as columns is called a(n) _____ matrix.

7. In the algebra of matrices, the matrix that has properties similar to the number 1 is called the _____ matrix.

8. A rational function is called _____ if the degree of its numerator is less than the degree of its denominator.

9. The graph of a linear inequality is called a(n) _____

10. A linear programming problem requires that a linear expression, called the _____, be maximized or minimized.

11. Each point that satisfies the constraints of a linear programming problem is called a(n) _____.

True/False Items

T F	**1.**	A system of two linear equations containing two variables always has at least one solution.
T F	**2.**	The augmented matrix of a system of two equations containing three variables has two rows and four columns.
T F	**3.**	A 3 by 3 determinant can never equal 0.
T F	**4.**	A consistent system of equations will have exactly one solution.
T F	**5.**	Every square matrix has an inverse.
T F	**6.**	Matrix multiplication is commutative.
T F	**7.**	Any pair of matrices can be multiplied.
T F	**8.**	The factors of the denominator of a rational expression are used to arrive at the partial fraction decomposition.
T F	**9.**	The graph of a linear inequality is a half-plane.
T F	**10.**	The graph of a system of linear inequalities is sometimes unbounded.
T F	**11.**	If a linear programming problem has a solution, it is located at a corner point of the graph of the feasible points.

Review Exercises

Blue problem numbers indicate the author's suggestions for use in a Practice Test.

In Problems 1–20, solve each system of linear equations using the method of substitution or the method of elimination. If the system has no solution, say that it is inconsistent.

1. $\begin{cases} 2x - y = 5 \\ 5x + 2y = 8 \end{cases}$
2. $\begin{cases} 2x + 3y = 2 \\ 7x - y = 3 \end{cases}$
3. $\begin{cases} 3x - 4y = 4 \\ x - 3y = \frac{1}{2} \end{cases}$
4. $\begin{cases} 2x + y = 0 \\ 5x - 4y = -\frac{13}{2} \end{cases}$

5. $\begin{cases} x - 2y - 4 = 0 \\ 3x + 2y - 4 = 0 \end{cases}$
6. $\begin{cases} x - 3y + 5 = 0 \\ 2x + 3y - 5 = 0 \end{cases}$
7. $\begin{cases} y = 2x - 5 \\ x = 3y + 4 \end{cases}$
8. $\begin{cases} x = 5y + 2 \\ y = 5x + 2 \end{cases}$

9. $\begin{cases} x - y + 4 = 0 \\ \frac{1}{2}x + \frac{1}{6}y + \frac{2}{5} = 0 \end{cases}$
10. $\begin{cases} x + \frac{1}{4}y = 2 \\ y + 4x + 2 = 0 \end{cases}$
11. $\begin{cases} x - 2y - 8 = 0 \\ 2x + 2y - 10 = 0 \end{cases}$
12. $\begin{cases} x - 3y + \frac{7}{2} = 0 \\ \frac{1}{2}x + 3y - 5 = 0 \end{cases}$

13. $\begin{cases} y - 2x = 11 \\ 2y - 3y = 18 \end{cases}$
14. $\begin{cases} 3x - 4y - 12 = 0 \\ 5x + 2y + 6 = 0 \end{cases}$
15. $\begin{cases} 2x + 3y - 13 = 0 \\ 3x - 2y = 0 \end{cases}$
16. $\begin{cases} 4x + 5y = 21 \\ 5x + 6y = 42 \end{cases}$

17. $\begin{cases} 3x - 2y = 8 \\ x - \frac{2}{3}y = 12 \end{cases}$
18. $\begin{cases} 2x + 5y = 10 \\ 4x + 10y = 15 \end{cases}$
19. $\begin{cases} x + 2y - z = 6 \\ 2x - y + 3z = -13 \\ 3x - 2y + 3z = -16 \end{cases}$
20. $\begin{cases} x + 5y - z = 2 \\ 2x + y + z = 7 \\ x - y + 2z = 11 \end{cases}$

In Problems 21–28, use the following matrices to compute each expression.

$$A = \begin{bmatrix} 1 & 0 \\ 2 & 4 \\ -1 & 2 \end{bmatrix} \qquad B = \begin{bmatrix} 4 & -3 & 0 \\ 1 & 1 & -2 \end{bmatrix} \qquad C = \begin{bmatrix} 3 & -4 \\ 1 & 5 \\ 5 & -2 \end{bmatrix}$$

21. $A + C$
22. $A - C$
23. $6A$
24. $-4B$
25. AB
26. BA
27. CB
28. BC

In Problems 29–34, find the inverse of each matrix algebraically, if there is one. If there is not an inverse, say that the matrix is singular.

29. $\begin{bmatrix} 4 & 6 \\ 1 & 3 \end{bmatrix}$
30. $\begin{bmatrix} -3 & 2 \\ 1 & -2 \end{bmatrix}$
31. $\begin{bmatrix} 1 & 3 & 3 \\ 1 & 2 & 1 \\ 1 & -1 & 2 \end{bmatrix}$

32. $\begin{bmatrix} 3 & 1 & 2 \\ 3 & 2 & -1 \\ 1 & 1 & 1 \end{bmatrix}$
33. $\begin{bmatrix} 4 & -8 \\ -1 & 2 \end{bmatrix}$
34. $\begin{bmatrix} -3 & 1 \\ -6 & 2 \end{bmatrix}$

In Problems 35–44, solve each system of equations using matrices. If the system has no solution, say that it is inconsistent.

35. $\begin{cases} 3x - 2y = 1 \\ 10x + 10y = 5 \end{cases}$

36. $\begin{cases} 3x + 2y = 6 \\ x - y = -\frac{1}{2} \end{cases}$

37. $\begin{cases} 5x + 6y - 3z = 6 \\ 4x - 7y - 2z = -3 \\ 3x + y - 7z = 1 \end{cases}$

38. $\begin{cases} 2x + y + z = 5 \\ 4x - y - 3z = 1 \\ 8x + y - z = 5 \end{cases}$

39. $\begin{cases} x - 2z = 1 \\ 2x + 3y = -3 \\ 4x - 3y - 4z = 3 \end{cases}$

40. $\begin{cases} x + 2y - z = 2 \\ 2x - 2y + z = -1 \\ 6x + 4y + 3z = 5 \end{cases}$

41. $\begin{cases} x - y + z = 0 \\ x - y - 5z - 6 = 0 \\ 2x - 2y + z - 1 = 0 \end{cases}$

42. $\begin{cases} 4x - 3y + 5z = 0 \\ 2x + 4y - 3z = 0 \\ 6x + 2y + z = 0 \end{cases}$

43. $\begin{cases} x - y - z - t = 1 \\ 2x + y + z + 2t = 3 \\ x - 2y - 2z - 3t = 0 \\ 3x - 4y + z + 5t = -3 \end{cases}$

44. $\begin{cases} x - 3y + 3z - t = 4 \\ x + 2y - z = -3 \\ x + 3z + 2t = 3 \\ x + y + 5z = 6 \end{cases}$

In Problems 45–50, find the value of each determinant.

45. $\begin{vmatrix} 3 & 4 \\ 1 & 3 \end{vmatrix}$

46. $\begin{vmatrix} -4 & 0 \\ 1 & 3 \end{vmatrix}$

47. $\begin{vmatrix} 1 & 4 & 0 \\ -1 & 2 & 6 \\ 4 & 1 & 3 \end{vmatrix}$

48. $\begin{vmatrix} 2 & 3 & 10 \\ 0 & 1 & 5 \\ -1 & 2 & 3 \end{vmatrix}$

49. $\begin{vmatrix} 2 & 1 & -3 \\ 5 & 0 & 1 \\ 2 & 6 & 0 \end{vmatrix}$

50. $\begin{vmatrix} -2 & 1 & 0 \\ 1 & 2 & 3 \\ -1 & 4 & 2 \end{vmatrix}$

In Problems 51–56, use Cramer's Rule, if applicable, to solve each system.

51. $\begin{cases} x - 2y = 4 \\ 3x + 2y = 4 \end{cases}$

52. $\begin{cases} x - 3y = -5 \\ 2x + 3y = 5 \end{cases}$

53. $\begin{cases} 2x + 3y - 13 = 0 \\ 3x - 2y = 0 \end{cases}$

54. $\begin{cases} 3x - 4y - 12 = 0 \\ 5x + 2y + 6 = 0 \end{cases}$

55. $\begin{cases} x + 2y - z = 6 \\ 2x - y + 3z = -13 \\ 3x - 2y + 3z = -16 \end{cases}$

56. $\begin{cases} x - y + z = 8 \\ 2x + 3y - z = -2 \\ 3x - y - 9z = 9 \end{cases}$

In Problems 57–66, write the partial fraction decomposition of each rational expression.

57. $\dfrac{6}{x(x-4)}$

58. $\dfrac{x}{(x+2)(x-3)}$

59. $\dfrac{x-4}{x^2(x-1)}$

60. $\dfrac{2x-6}{(x-2)^2(x-1)}$

61. $\dfrac{x}{(x^2+9)(x+1)}$

62. $\dfrac{3x}{(x-2)(x^2+1)}$

63. $\dfrac{x^3}{(x^2+4)^2}$

64. $\dfrac{x^3+1}{(x^2+16)^2}$

65. $\dfrac{x^2}{(x^2+1)(x^2-1)}$

66. $\dfrac{4}{(x^2+4)(x^2-1)}$

In Problems 67–76, solve each system of nonlinear equations.

67. $\begin{cases} 2x + y + 3 = 0 \\ x^2 + y^2 = 5 \end{cases}$

68. $\begin{cases} x^2 + y^2 = 16 \\ 2x - y^2 = -8 \end{cases}$

69. $\begin{cases} 2xy + y^2 = 10 \\ 3y^2 - xy = 2 \end{cases}$

70. $\begin{cases} 3x^2 - y^2 = 1 \\ 7x^2 - 2y^2 - 5 = 0 \end{cases}$

71. $\begin{cases} x^2 + y^2 = 6y \\ x^2 = 3y \end{cases}$

72. $\begin{cases} 2x^2 + y^2 = 9 \\ x^2 + y^2 = 9 \end{cases}$

73. $\begin{cases} 3x^2 + 4xy + 5y^2 = 8 \\ x^2 + 3xy + 2y^2 = 0 \end{cases}$

74. $\begin{cases} 3x^2 + 2xy - 2y^2 = 6 \\ xy - 2y^2 + 4 = 0 \end{cases}$

75. $\begin{cases} x^2 - 3x + y^2 + y = -2 \\ \dfrac{x^2 - x}{y} + y + 1 = 0 \end{cases}$

76. $\begin{cases} x^2 + x + y^2 = y + 2 \\ x + 1 = \dfrac{2-y}{x} \end{cases}$

In Problems 77–82, graph each system of inequalities. Tell whether the graph is bounded or unbounded, and label the corner points.

77. $\begin{cases} -2x + y \le 2 \\ x + y \ge 2 \end{cases}$

78. $\begin{cases} x - 2y \le 6 \\ 2x + y \ge 2 \end{cases}$

79. $\begin{cases} x \ge 0 \\ y \ge 0 \\ x + y \le 4 \\ 2x + 3y \le 6 \end{cases}$

80. $\begin{cases} x \ge 0 \\ y \ge 0 \\ 3x + y \ge 6 \\ 2x + y \ge 2 \end{cases}$

81. $\begin{cases} x \ge 0 \\ y \ge 0 \\ 2x + y \le 8 \\ x + 2y \ge 2 \end{cases}$

82. $\begin{cases} x \ge 0 \\ y \ge 0 \\ 3x + y \le 9 \\ 2x + 3y \ge 6 \end{cases}$

In Problems 83–86, graph each system of inequalities.

83. $\begin{cases} x^2 + y^2 \le 16 \\ x + y \ge 2 \end{cases}$

84. $\begin{cases} y^2 \le x - 1 \\ x - y \le 3 \end{cases}$

85. $\begin{cases} y \le x^2 \\ xy \le 4 \end{cases}$

86. $\begin{cases} x^2 + y^2 \ge 1 \\ x^2 + y^2 \le 4 \end{cases}$

In Problems 87–92, solve each linear programming problem.

87. Maximize $z = 3x + 4y$ subject to $x \ge 0$, $y \ge 0$, $3x + 2y \ge 6$, $x + y \le 8$

88. Maximize $z = 2x + 4y$ subject to $x \ge 0$, $y \ge 0$, $x + y \le 6$, $x \ge 2$

89. Minimize $z = 3x + 5y$ subject to $x \ge 0$, $y \ge 0$, $x + y \ge 1$, $3x + 2y \le 12$, $x + 3y \le 12$

90. Minimize $z = 3x + y$ subject to $x \ge 0$, $y \ge 0$, $x \le 8$, $y \le 6$, $2x + y \ge 4$

91. Maximize $z = 5x + 4y$ subject to $x \ge 0$, $y \ge 0$, $x + 2y \ge 2$, $3x + 4y \le 12$, $y \ge x$

92. Maximize $z = 4x + 5y$ subject to $x \ge 0$, $y \ge 0$, $2x + 3y \ge 6$, $x \ge y$, $2x + y \le 12$

---

93. Find A so that the system of equations has infinitely many solutions.

$$\begin{cases} 2x + 5y = 5 \\ 4x + 10y = A \end{cases}$$

94. Find A so that the system in Problem 93 is inconsistent.

95. Curve Fitting Find real numbers b and c so that the graph of the parabola $y = x^2 + bx + c$ contains the points $(1, 2)$ and $(-1, 3)$.

96. Curve Fitting Find real numbers b and c so that the graph of the parabola $y = x^2 + bx + c$ contains the points $(1, 3)$ and $(3, 5)$.

97. Curve Fitting Find the quadratic function $y = ax^2 + bx + c$ whose graph contains the three points $(0, 1)$, $(1, 0)$, and $(-2, 1)$.

98. Curve Fitting Find the general equation of the circle that contains the three points $(0, 1)$, $(1, 0)$, and $(-2, 1)$.
[**Hint:** The general equation of a circle is $x^2 + y^2 + Dx + Ey + F = 0$.]

99. Blending Coffee A coffee distributor is blending a new coffee that will cost \$3.90 per pound. It will consist of a blend of \$3.00-per-pound coffee and \$6.00-per-pound

coffee. What amounts of each type of coffee should be mixed to achieve the desired blend?
[**Hint:** Assume that the weight of the blended coffee is 100 pounds].

$3.00/lb $3.90/lb $6.00/lb

100. Farming A 1000-acre farm in Illinois is used to raise corn and soybeans. The cost per acre for raising corn is \$65, and the cost per acre for soybeans is \$45. If \$54,325 has been budgeted for costs and all the acreage is to be used, how many acres should be allocated for each crop?

101. Cookie Orders A cookie company makes three kinds of cookies, oatmeal raisin, chocolate chip, and shortbread, packaged in small, medium, and large boxes. The small box contains 1 dozen oatmeal raisin and 1 dozen chocolate chip; the medium box has 2 dozen oatmeal raisin, 1 dozen chocolate chip, and 1 dozen shortbread; the large box contains 2 dozen oatmeal raisin, 2 dozen chocolate chip, and 3 dozen shortbread. If you require exactly 15 dozen oatmeal raisin, 10 dozen chocolate chip, and 11 dozen shortbread, how many of each size box should you buy?

102. **Mixed Nuts**   A store that specializes in selling nuts has available 72 pounds of cashews and 120 pounds of peanuts. These are to be mixed in 12-ounce packages as follows: a lower-priced package containing 8 ounces of peanuts and 4 ounces of cashews and a quality package containing 6 ounces of peanuts and 6 ounces of cashews.
    (a) Use x to denote the number of lower-priced packages and use y to denote the number of quality packages. Write a system of linear inequalities that describes the possible number of each kind of package.
    (b) Graph the system and label the corner points.

103. A small rectangular lot has a perimeter of 68 feet. If its diagonal is 26 feet, what are the dimensions of the lot?

104. The area of a rectangular window is 4 square feet. If the diagonal measures $2\sqrt{2}$ feet, what are the dimensions of the window?

105. **Geometry**   A certain right triangle has a perimeter of 14 inches. If the hypotenuse is 6 inches long, what are the lengths of the legs?

106. **Geometry**   A certain isosceles triangle has a perimeter of 18 inches. If the altitude is 6 inches, what is the length of the base?

107. **Building a Fence**   How much fence is required to enclose a 5000 square foot region in the shape of adjoining squares whose sides are in the ratio of 1 : 2?
    [**Hint:** See the illustration in Problem 81, page 591.]

108. **Calculating Allowances**   Katy, Mike, Danny, and Colleen agreed to do yard work at home for $45 to be split among them. After they finished, their father determined that Mike deserves twice what Katy gets, Katy and Colleen deserve the same amount, and Danny deserves half of what Katy gets. How much does each one receive?

109. **Determining the Speed of the Current of the Aguarico River**   On a recent trip to the Cuyabeno Wildlife Reserve in the Amazon region of Ecuador, a 100-kilometer trip by speedboat was taken down the Aguarico River from Chiritza to the Flotel Orellana. As I watched the Amazon unfold, I wondered how fast the speedboat was going and how fast the current of the white-water Aguarico River was. I timed the trip downstream at 2.5 hours and the return trip at 3 hours. What were the two speeds?

110. **Finding the Speed of the Jet Stream**   On a flight between Midway Airport in Chicago and Ft. Lauderdale, Florida, a Boeing 737 jet maintains an airspeed of 475 miles per hour. If the trip from Chicago to Ft. Lauderdale takes 2 hours, 30 minutes and the return flight takes 2 hours, 50 minutes, what is the speed of the jet stream? (Assume that the speed of the jet stream remains constant at the various altitudes of the plane.)

111. **Constant Rate Jobs**   If Bruce and Bryce work together for 1 hour and 20 minutes, they will finish a certain job. If Bryce and Marty work together for 1 hour and 36 minutes, the same job can be finished. If Marty and Bruce work together, they can complete this job in 2 hours and 40 minutes. How long will it take each of them working alone to finish the job?

112. **Maximizing Profit on Figurines**   A factory manufactures two kinds of ceramic figurines: a dancing girl and a mermaid, each requiring three processes—molding, painting, and glazing. The daily labor available for molding is no more than 90 work-hours, labor available for painting does not exceed 120 work-hours, and labor available for glazing is no more than 60 work-hours. The dancing girl requires 3 work-hours for molding, 6 work-hours for painting, and 2 work-hours for glazing. The mermaid requires 3 work-hours for molding, 4 work-hours for painting, and 3 work-hours for glazing. If the profit on each figurine is $25 for dancing girls and $30 for mermaids, how many of each should be produced each day to maximize profit? If management decides to produce the number of each figurine that maximizes profit, determine which of these processes has excess work-hours assigned to it.

113. **Minimizing Production Cost**   A factory produces gasoline engines and diesel engines. Each week the factory is obligated to deliver at least 20 gasoline engines and at least 15 diesel engines. Due to physical limitations, however, the factory cannot make more than 60 gasoline engines nor more than 40 diesel engines. Finally, to prevent layoffs, a total of at least 50 engines must be produced. If gasoline engines cost $450 each to produce and diesel engines cost $550 each to produce, how many of each should be produced per week to minimize the cost? What is the excess capacity of the factory; that is, how many of each kind of engine is being produced in excess of the number that the factory is obligated to deliver?

114. Describe four ways of solving a system of three linear equations containing three variables. Which method do you prefer? Why?

Error Control Coding

Digital cell phones communicate using the language of computers, with voice and data sent as groups of information bits (zeros and ones) over the air. Wireless communication is very difficult; the signals are reflected off buildings, absorbed by trees and rain, and scattered by street signs before being received by the antenna on the phone. Because of these impairments, it is difficult to receive all the information bits correctly; some will have invariably flipped from 0 to 1, or vice versa. On a voice call, the effect is digital distortion, which may make the call sound like it is coming from under water, for example. On a data call, the effect changes the data present in the transferred files. One primary reason that digital phones are so powerful is that Motorola phone designers can use *error control coding* to ensure that your phone call is clear and error free.

In practice, error control coding adds extra bits (called *parity*) to the information bits so that errors can be corrected. The process of *encoding* the information by adding these extra bits and *decoding* to detect and correct any errors is called **error control coding.** Let's see how using matrix algebra to perform encoding and decoding makes your phones calls clear and error free.

1. A set of information bits plus parity bits forms a *codeword,* and the set of all possible codewords forms a *code.* Let us encode 4 information bits with 3 parity bits to form 7-bit codewords (such as 0000 000, 0001 110, etc). This is known as a $(7, 4)$ code. How many codewords are in a $(7, 4)$ code? (**Hint:** The number of codewords depends only on the number of information bits.)

2. Matrix algebra provides a succinct representation of a code by describing an encoder for the code. The *generator matrix G* encodes an information vector u into codeword v by performing the matrix multiplication $v = uG$. Use the matrix G below and vector-matrix multiplication to list all the codewords of this $(7, 4)$ code. Note that the algebra must be performed mod 2 so that the codewords are composed of zeros and ones (that is, vector 1001 is encoded to codeword 1001 001, not 1001 221).

$$G = \begin{bmatrix} 1 & 0 & 0 & 0 & 1 & 1 & 1 \\ 0 & 1 & 0 & 0 & 0 & 1 & 1 \\ 0 & 0 & 1 & 0 & 1 & 0 & 1 \\ 0 & 0 & 0 & 1 & 1 & 1 & 0 \end{bmatrix}$$

3. A code that can be generated with matrix multiplication is a linear code. A linear code always contains the all-zero vector (0000 000) and has the property that the sum (mod 2) of any two codewords equals another codeword. Pick two nonzero codewords of the code and show that the sum is another codeword.

4. The generator matrix G has a corresponding *parity check* matrix H. The parity check matrix has the property that $vH = 0$ for any codeword v. Explain why the following H satisfies this property. (**Hint:** Note that the codeword $v = uG$ for some information vector u, and perform the matrix multiplication GH).

$$H = \begin{bmatrix} 1 & 1 & 1 \\ 0 & 1 & 1 \\ 1 & 0 & 1 \\ 1 & 1 & 0 \\ 1 & 0 & 0 \\ 0 & 1 & 0 \\ 0 & 0 & 1 \end{bmatrix}$$

5. The code described by G and H is used to correct single-bit errors; for example, if the codeword 0000 000 is sent and the word 1000 000 is received, the decoder will correct the error and produce 0000 000. The decoder multiplies the received matrix r with the parity check matrix H to determine an error pattern, which is added to the received matrix to produce the original codeword. Use matrix multiplication and the following table to produce the codeword associated with the received matrix $r = [0101\ 000]$. (**Hint:** Your answer should appear in the codeword list from question 2.)

rH	Error pattern
[000]	0000 000
[001]	0000 001
[010]	0000 010
[011]	0100 000
[100]	0000 100
[101]	0010 000
[110]	0001 000
[111]	1000 000

Sequences; Induction; The Binomial Theorem

Field Trip to Motorola

How would you like to design products that allow you to talk on a cellular phone from anywhere in the world or surf the web while riding in a car? Motorola designs and manufactures products that do these things and more. Probability of error is a performance measure that is important to the design of wireless communication systems and products. Many parts of a wireless system are designed to decrease the probability of error. This is what allows you to hear someone clearly on your cellular phone or surf the web more quickly. In the chapter project, you will learn how speech signals are coded for transmission and error control.

PREPARING FOR THIS SECTION

Before getting started, review the following concept:

✓ Evaluating Functions (Section 3.1, p. 209)

✓ Integer Exponents (Review, Section 4, pp. 28–35)

✓ Compound Interest (Section 6.6, pp. 446–453)

9.1 SEQUENCES

OBJECTIVES
1. Write the First Several Terms of a Sequence
2. Write the Terms of a Sequence Defined by a Recursive Formula
3. Use Summation Notation
4. Find the Sum of a Sequence

A **sequence** is a function whose domain is the set of positive integers.

Because a sequence is a function, it will have a graph. In Figure 1(a), you will recognize the graph of the function $f(x) = 1/x$, $x > 0$. If all the points on this graph were removed except those whose x-coordinates are positive integers, that is, if all points were removed except $(1, 1), (2, \frac{1}{2}), (3, \frac{1}{3})$, and so on, the remaining points would be the graph of the sequence $f(n) = 1/n$, as shown in Figure 1(b).

Figure 1

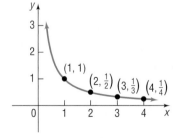

(a) $f(x) = \frac{1}{x}$, $x > 0$

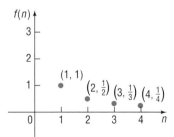

(b) $f(n) = \frac{1}{n}$, n a positive integer

A sequence is usually represented by listing its values in order. For example, the sequence whose graph is given in Figure 1(b) might be represented as

$$f(1), f(2), f(3), f(4), \dots \quad \text{or} \quad 1, \frac{1}{2}, \frac{1}{3}, \frac{1}{4}, \dots$$

The list never ends, as the ellipsis indicates. The numbers in this ordered list are called the **terms** of the sequence.

In dealing with sequences, we usually use subscripted letters, such as a_1, to represent the first term, a_2 for the second term, a_3 for the third term, and so on. For the sequence $f(n) = 1/n$, we write

$$a_1 = f(1) = 1, \quad a_2 = f(2) = \frac{1}{2}, \quad a_3 = f(3) = \frac{1}{3}, \quad a_4 = f(4) = \frac{1}{4}, \dots, \quad a_n = f(n) = \frac{1}{n}, \dots$$

In other words, we usually do not use the traditional function notation $f(n)$ for sequences. For this particular sequence, we have a rule for the nth term, which is $a_n = \frac{1}{n}$, so it is easy to find any term of the sequence.

When a formula for the nth term (sometimes called the **general term**) of a sequence is known, rather than write out the terms of the sequence, we usually represent the entire sequence by placing braces around the formula for the nth term. For example, the sequence whose nth term is $b_n = \left(\frac{1}{2}\right)^n$ may be represented as

$$\{b_n\} = \left\{\left(\frac{1}{2}\right)^n\right\}$$

or by

$$b_1 = \frac{1}{2}, \quad b_2 = \frac{1}{4}, \quad b_3 = \frac{1}{8}, \quad \dots, \quad b_n = \left(\frac{1}{2}\right)^n, \quad \dots$$

EXAMPLE 1 — Writing the First Several Terms of a Sequence

Write down the first six terms of the following sequence and graph it.

$$\{a_n\} = \left\{\frac{n-1}{n}\right\}$$

Figure 2

Solution   The first six terms of the sequence are

$$a_1 = 0, \quad a_2 = \frac{1}{2}, \quad a_3 = \frac{2}{3}, \quad a_4 = \frac{3}{4}, \quad a_5 = \frac{4}{5}, \quad a_6 = \frac{5}{6}$$

See Figure 2 for the graph.

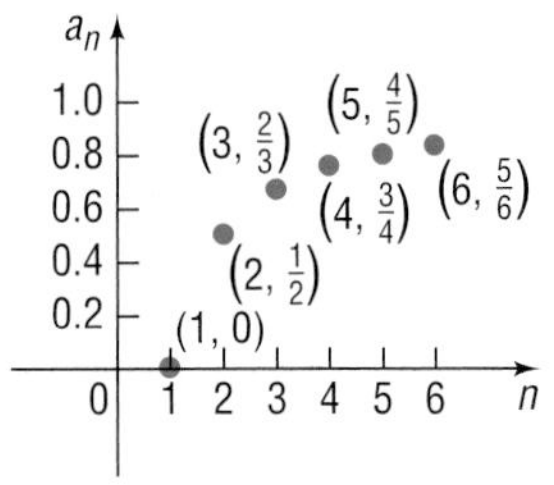

COMMENT: Graphing utilities can be used to write the terms of a sequence and graph them. Figure 3 shows the sequence given in Example 1 generated on a TI-83 graphing calculator. We can see the first few terms of the sequence on the viewing window. You need to press the right arrow key to scroll right in order to see the remaining terms of the sequence. Figure 4 shows a graph of the sequence. Notice that the first term of the sequence is not visible since it lies on the x-axis. TRACEing the graph will allow you to see the terms of the sequence. The TABLE feature can also be used to generate the terms of the sequence. See Table 1.

Figure 3

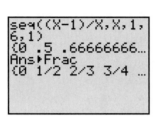

Figure 4

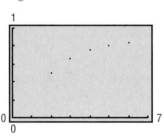

T A B L E 1

n	u(n)
1	0
2	.5
3	.66667
4	.75
5	.8
6	.83333
7	.85714

u(n)◻(n−1)/n

EXAMPLE 2

Writing the First Several Terms of a Sequence

Write down the first six terms of the following sequence and graph it.

$$\{b_n\} = \left\{(-1)^{n-1}\left(\frac{2}{n}\right)\right\}$$

Solution The first six terms of the sequence are

$$b_1 = 2, \quad b_2 = -1, \quad b_3 = \frac{2}{3}, \quad b_4 = -\frac{1}{2}, \quad b_5 = \frac{2}{5}, \quad b_6 = -\frac{1}{3}$$

See Figure 5 for the graph.

Figure 5

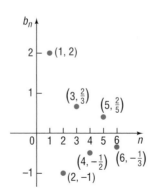

EXAMPLE 3

Writing the First Several Terms of a Sequence

Write down the first six terms of the following sequence and graph it.

$$\{c_n\} = \begin{cases} n & \text{if } n \text{ is even} \\ \dfrac{1}{n} & \text{if } n \text{ is odd} \end{cases}$$

Solution   The first six terms of the sequence are

$$c_1 = 1, \quad c_2 = 2, \quad c_3 = \frac{1}{3}, \quad c_4 = 4, \quad c_5 = \frac{1}{5}, \quad c_6 = 6$$

See Figure 6 for the graph.

Figure 6

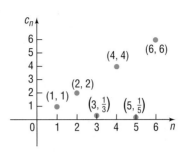

NOW WORK PROBLEMS 3 AND 5.

Sometimes a sequence is indicated by an observed pattern in the first few terms that makes it possible to infer the makeup of the nth term. In the example that follows, a sufficient number of terms of the sequence is given so that a natural choice for the nth term is suggested.

EXAMPLE 4   Determining a Sequence from a Pattern

(a) $e, \dfrac{e^2}{2}, \dfrac{e^3}{3}, \dfrac{e^4}{4}, \ldots$ $\qquad$ $a_n = \dfrac{e^n}{n}$

(b) $1, \dfrac{1}{3}, \dfrac{1}{9}, \dfrac{1}{27}, \ldots$ $\qquad$ $b_n = \dfrac{1}{3^{n-1}}$

(c) $1, 3, 5, 7, \ldots$ $\qquad$ $c_n = 2n - 1$

(d) $1, 4, 9, 16, 25, \ldots$ $\qquad$ $d_n = n^2$

(e) $1, -\dfrac{1}{2}, \dfrac{1}{3}, -\dfrac{1}{4}, \dfrac{1}{5}, \ldots$ $\qquad$ $e_n = (-1)^{n+1}\left(\dfrac{1}{n}\right)$

Notice in the sequence $\{e_n\}$ in Example 4(e) that the signs of the terms **alternate.** When this occurs, we use factors such as $(-1)^{n+1}$, which equals 1 if n is odd and -1 if n is even, or $(-1)^n$, which equals -1 if n is odd and 1 if n is even.

NOW WORK PROBLEM 13.

THE FACTORIAL SYMBOL

If $n \geq 0$ is an integer, the **factorial symbol** $n!$ is defined as follows:

$$0! = 1 \qquad 1! = 1$$
$$n! = n(n - 1) \cdot \ldots \cdot 3 \cdot 2 \cdot 1 \qquad \text{if } n \geq 2$$

TABLE 2						
n	0	1	2	3	4	5	6
$n!$	1	1	2	6	24	120	720

For example, $2! = 2 \cdot 1 = 2$, $\quad 3! = 3 \cdot 2 \cdot 1 = 6$, $\quad 4! = 4 \cdot 3 \cdot 2 \cdot 1 = 24$, and so on. Table 2 lists the values of $n!$ for $0 \leq n \leq 6$.

Because

$$n! = n\underbrace{(n-1)(n-2)\cdot\ldots\cdot 3\cdot 2\cdot 1}_{(n-1)!}$$

we can use the formula

$$n! = n(n-1)!$$

to find successive factorials. For example, because $6! = 720$, we have

$$7! = 7\cdot 6! = 7(720) = 5040$$

and

$$8! = 8\cdot 7! = 8(5040) = 40{,}320$$

 COMMENT: Your calculator has a factorial key. Use it to see how fast factorials increase in value. Find the value of 69!. What happens when you try to find 70!? In fact, 70! is larger than 10^{100} (a **googol**), the largest number most calculators can display.   ■

RECURSIVE FORMULAS

2 A second way of defining a sequence is to assign a value to the first (or the first few) term(s) and specify the nth term by a formula or equation that involves one or more of the terms preceding it. Sequences defined this way are said to be defined **recursively,** and the rule or formula is called a **recursive formula.**

EXAMPLE 5 **Writing the Terms of a Recursively Defined Sequence**

Write down the first five terms of the following recursively defined sequence.

$$s_1 = 1, \qquad s_n = 4s_{n-1}$$

Solution The first term is given as $s_1 = 1$. To get the second term, we use $n = 2$ in the formula to get $s_2 = 4s_1 = 4\cdot 1 = 4$. To get the third term, we use $n = 3$ in the formula to get $s_3 = 4s_2 = 4\cdot 4 = 16$. To get a new term requires that we know the value of the preceding term. The first five terms are

$$s_1 = 1$$
$$s_2 = 4\cdot 1 = 4$$
$$s_3 = 4\cdot 4 = 16$$
$$s_4 = 4\cdot 16 = 64$$
$$s_5 = 4\cdot 64 = 256$$
■

EXAMPLE 6 **Writing the Terms of a Recursively Defined Sequence**

Write down the first five terms of the following recursively defined sequence.

$$f_1 = 1, \qquad f_n = nf_{n-1}$$

Solution   Here

$$f_1 = 1$$
$$f_2 = 2f_1 = 2 \cdot 1 = 2$$
$$f_3 = 3f_2 = 3 \cdot 2 = 6$$
$$f_4 = 4f_3 = 4 \cdot 6 = 24$$
$$f_5 = 5f_4 = 5 \cdot 24 = 120$$ ■

You should recognize the nth term of the sequence in Example 6 as $n!$.

EXAMPLE 7   **Writing the Terms of a Recursively Defined Sequence**

Write down the first five terms of the following recursively defined sequence.

$$u_1 = 1, \qquad u_2 = 1, \qquad u_{n+2} = u_n + u_{n+1}$$

Solution   We are given the first two terms. To get the third term requires that we know each of the previous two terms. Thus,

$$u_1 = 1$$
$$u_2 = 1$$
$$u_3 = u_1 + u_2 = 1 + 1 = 2$$
$$u_4 = u_2 + u_3 = 1 + 2 = 3$$
$$u_5 = u_3 + u_4 = 2 + 3 = 5$$ ■

The sequence defined in Example 7 is called a **Fibonacci sequence,** and the terms of this sequence are called **Fibonacci numbers.** These numbers appear in a wide variety of applications (see Problems 71–74).

✏ NOW WORK PROBLEMS **21** AND **29**.

SUMMATION NOTATION

③ It is often important to be able to find the sum of the first n terms of a sequence $\{a_n\}$, that is,

$$a_1 + a_2 + a_3 + \cdots + a_n \tag{1}$$

Rather than write down all these terms, we introduce a more concise way to express the sum, called **summation notation.** Using summation notation, we would write the sum (1) as

$$a_1 + a_2 + a_3 + \cdots + a_n = \sum_{k=1}^{n} a_k$$

The symbol Σ (a stylized version of the Greek letter sigma, which is an S in our alphabet) is simply an instruction to sum, or add up, the terms. The integer k is called the **index** of the sum; it tells you where to start the sum and where to end it. The expression

$$\sum_{k=1}^{n} a_k \tag{2}$$

is an instruction to add the terms a_k of the sequence $\{a_n\}$ from $k = 1$ through $k = n$. We read expression (2) as "the sum of a_k from $k = 1$ to $k = n$."

EXAMPLE 8   **Expanding Summation Notation**

Write out each sum.

(a) $\displaystyle\sum_{k=1}^{n} \frac{1}{k}$

(b) $\displaystyle\sum_{k=1}^{n} k!$

Solution   (a) $\displaystyle\sum_{k=1}^{n} \frac{1}{k} = 1 + \frac{1}{2} + \frac{1}{3} + \cdots + \frac{1}{n}$ (b) $\displaystyle\sum_{k=1}^{n} k! = 1! + 2! + \cdots + n!$ ∎

EXAMPLE 9   **Writing a Sum in Summation Notation**

Express each sum using summation notation.

(a) $1^2 + 2^2 + 3^2 + \cdots + 9^2$

(b) $1 + \dfrac{1}{2} + \dfrac{1}{4} + \dfrac{1}{8} + \cdots + \dfrac{1}{2^{n-1}}$

Solution   (a) The sum $1^2 + 2^2 + 3^2 + \cdots + 9^2$ has 9 terms, each of the form k^2, and starts at $k = 1$ and ends at $k = 9$:

$$1^2 + 2^2 + 3^2 + \cdots + 9^2 = \sum_{k=1}^{9} k^2$$

(b) The sum

$$1 + \frac{1}{2} + \frac{1}{4} + \frac{1}{8} + \cdots + \frac{1}{2^{n-1}}$$

has n terms, each of the form $\dfrac{1}{2^{k-1}}$, and starts at $k = 1$ and ends at $k = n$:

$$1 + \frac{1}{2} + \frac{1}{4} + \frac{1}{8} + \cdots + \frac{1}{2^{n-1}} = \sum_{k=1}^{n} \frac{1}{2^{k-1}}$$ ∎

The index of summation need not always begin at 1 or end at n; for example, we could have expressed the sum in Example 9(b) as

$$\sum_{k=0}^{n-1} \frac{1}{2^k} = 1 + \frac{1}{2} + \frac{1}{4} + \cdots + \frac{1}{2^{n-1}}$$

Letters other than k may be used as the index. For example,

$$\sum_{j=1}^{n} j! \quad \text{and} \quad \sum_{i=1}^{n} i!$$

each represent the same sum as the one given in Example 8(b).

✏️━━━ **NOW WORK PROBLEMS 49 AND 59.**

ADDING THE FIRST n TERMS OF A SEQUENCE

④ Next, we list some properties of sequences using summation notation. These properties are useful for adding the terms of a sequence.

Theorem　**Properties of Sequences**

If $\{a_n\}$ and $\{b_n\}$ are two sequences and c is a real number, then:

1. $\displaystyle\sum_{k=1}^{n} c = cn$

2. $\displaystyle\sum_{k=1}^{n} (ca_k) = c \sum_{k=1}^{n} a_k$

3. $\displaystyle\sum_{k=1}^{n} (a_k + b_k) = \sum_{k=1}^{n} a_k + \sum_{k=1}^{n} b_k$

4. $\displaystyle\sum_{k=1}^{n} (a_k - b_k) = \sum_{k=1}^{n} a_k - \sum_{k=1}^{n} b_k$

5. $\displaystyle\sum_{k=1}^{n} a_k = \sum_{k=1}^{j} a_k + \sum_{k=j+1}^{n} a_k,$　where $1 < j < n$

6. $\displaystyle\sum_{k=1}^{n} k = 1 + 2 + 3 + \cdots + n = \frac{n(n + 1)}{2}$

7. $\displaystyle\sum_{k=1}^{n} k^2 = 1^2 + 2^2 + 3^2 + \cdots + n^2 = \frac{n(n + 1)(2n + 1)}{6}$

8. $\displaystyle\sum_{k=1}^{n} k^3 = 1^3 + 2^3 + 3^3 + \cdots + n^3 = \left(\frac{n(n + 1)}{2}\right)^2$

We shall not prove these properties. The proofs of 1 through 5 are based on properties of real numbers; the proofs of 7 and 8 require mathematical induction, which is discussed in Section 9.4. See Problem 75 for a derivation of 6.

EXAMPLE 10　**Finding the Sum of a Sequence**

Find the sum of each sequence.

(a) $\displaystyle\sum_{k=1}^{5} (3k)$　(b) $\displaystyle\sum_{k=1}^{3} (k^3 + 1)$　(c) $\displaystyle\sum_{k=1}^{4} (k^2 - 7k + 2)$

Solution　(a) $\displaystyle\sum_{k=1}^{5} (3k) = 3 \sum_{k=1}^{5} k = 3\left(\frac{5(5 + 1)}{2}\right) = 3(15) = 45$

　　　　　　　　↑　　　↑
　　　　Property 2　Property 6

(b) $\displaystyle\sum_{k=1}^{3} (k^3 + 1) = \sum_{k=1}^{3} k^3 + \sum_{k=1}^{3} 1$　　　Property 3

$$= \left(\frac{3(3 + 1)}{2}\right)^2 + 1(3)$$　　Properties 1, 8

$$= 36 + 3$$

$$= 39$$

(c) $\displaystyle\sum_{k=1}^{4}(k^2 - 7k + 2) = \sum_{k=1}^{4} k^2 - \sum_{k=1}^{4}(7k) + \sum_{k=1}^{4} 2$ Properties 3, 4

$\displaystyle = \sum_{k=1}^{4} k^2 - 7\sum_{k=1}^{4} k + \sum_{k=1}^{4} 2$ Property 2

$\displaystyle = \frac{4(4 + 1)(2 \cdot 4 + 1)}{6} - 7\left(\frac{4(4 + 1)}{2}\right) + 2(4)$

Properties 1, 6, 7

$= 30 - 70 + 8$

$= -32$

NOW WORK PROBLEM **39**.

9.1 EXERCISES

In Problems 1–12, write down the first five terms of each sequence.

1. $\{n\}$

2. $\{n^2 + 1\}$

3. $\left\{\dfrac{n}{n + 2}\right\}$

4. $\left\{\dfrac{2n + 1}{2n}\right\}$

5. $\{(-1)^{n+1}n^2\}$

6. $\left\{(-1)^{n-1}\left(\dfrac{n}{2n - 1}\right)\right\}$

7. $\left\{\dfrac{2^n}{3^n + 1}\right\}$

8. $\left\{\left(\dfrac{4}{3}\right)^n\right\}$

9. $\left\{\dfrac{(-1)^n}{(n + 1)(n + 2)}\right\}$

10. $\left\{\dfrac{3^n}{n}\right\}$

11. $\left\{\dfrac{n}{e^n}\right\}$

12. $\left\{\dfrac{n^2}{2^n}\right\}$

In Problems 13–20, the given pattern continues. Write down the nth term of each sequence suggested by the pattern.

13. $\dfrac{1}{2}, \dfrac{2}{3}, \dfrac{3}{4}, \dfrac{4}{5}, \cdots$

14. $\dfrac{1}{1 \cdot 2}, \dfrac{1}{2 \cdot 3}, \dfrac{1}{3 \cdot 4}, \dfrac{1}{4 \cdot 5}, \cdots$

15. $1, \dfrac{1}{2}, \dfrac{1}{4}, \dfrac{1}{8}, \cdots$

16. $\dfrac{2}{3}, \dfrac{4}{9}, \dfrac{8}{27}, \dfrac{16}{81}, \cdots$

17. $1, -1, 1, -1, 1, -1, \ldots$

18. $1, \dfrac{1}{2}, 3, \dfrac{1}{4}, 5, \dfrac{1}{6}, 7, \dfrac{1}{8}, \cdots$

19. $1, -2, 3, -4, 5, -6, \ldots$

20. $2, -4, 6, -8, 10, \ldots$

In Problems 21–34, a sequence is defined recursively. Write the first five terms.

21. $a_1 = 2;\quad a_n = 3 + a_{n-1}$

22. $a_1 = 3;\quad a_n = 4 - a_{n-1}$

23. $a_1 = -2;\quad a_n = n + a_{n-1}$

24. $a_1 = 1;\quad a_n = n - a_{n-1}$

25. $a_1 = 5;\quad a_n = 2a_{n-1}$

26. $a_1 = 2;\quad a_n = -a_{n-1}$

27. $a_1 = 3;\quad a_n = \dfrac{a_{n-1}}{n}$

28. $a_1 = -2;\quad a_n = n + 3a_{n-1}$

29. $a_1 = 1;\quad a_2 = 2;\quad a_n = a_{n-1} \cdot a_{n-2}$

30. $a_1 = -1;\quad a_2 = 1;\quad a_n = a_{n-2} + na_{n-1}$

31. $a_1 = A;\quad a_n = a_{n-1} + d$

32. $a_1 = A;\quad a_n = ra_{n-1},\quad r \neq 0$

33. $a_1 = \sqrt{2};\quad a_n = \sqrt{2 + a_{n-1}}$

34. $a_1 = \sqrt{2};\quad a_n = \sqrt{\dfrac{a_{n-1}}{2}}$

In Problems 35–46, find the sum of each sequence.

35. $\displaystyle\sum_{k=1}^{10} 5$

36. $\displaystyle\sum_{k=1}^{20} 8$

37. $\displaystyle\sum_{k=1}^{6} k$

38. $\displaystyle\sum_{k=1}^{4} (-k)$

39. $\displaystyle\sum_{k=1}^{5} (5k + 3)$

40. $\displaystyle\sum_{k=1}^{6} (3k - 7)$

41. $\displaystyle\sum_{k=1}^{3} (k^2 + 4)$

42. $\displaystyle\sum_{k=0}^{4} (k^2 - 4)$

43. $\displaystyle\sum_{k=1}^{6} (-1)^k 2^k$

44. $\displaystyle\sum_{k=1}^{4} (-1)^k 3^k$

45. $\displaystyle\sum_{k=1}^{4} (k^3 - 1)$

46. $\displaystyle\sum_{k=0}^{3} (k^3 + 2)$

In Problems 47–56, write out each sum.

47. $\sum_{k=1}^{n} (k + 2)$

48. $\sum_{k=1}^{n} (2k + 1)$

49. $\sum_{k=1}^{n} \frac{k^2}{\ }$

50. $\sum_{k=1}^{n} (k + 1)^2$

51. $\sum_{k=0}^{n} \frac{1}{3^k}$

52. $\sum_{k=0}^{n} \left(\frac{3}{2}\right)^k$

53. $\sum_{k=0}^{n-1} \frac{1}{3^{k+1}}$

54. $\sum_{k=0}^{n-1} (2k + 1)$

55. $\sum_{k=2}^{n} (-1)^k \ln k$

56. $\sum_{k=3}^{n} (-1)^{k+1} 2^k$

In Problems 57–66, express each sum using summation notation.

57. $1 + 2 + 3 + \cdots + 20$

58. $1^3 + 2^3 + 3^3 + \cdots$ ^{3}

59. $\frac{1}{2} + \frac{2}{3} + \frac{3}{4} + \cdots + \frac{13}{13 + 1}$

60. $1 + 3 + 5 + 7 + \cdots$ $(12) - 1]$

61. $1 - \frac{1}{3} + \frac{1}{9} - \frac{1}{27} + \cdots + (-1)^6\left(\frac{1}{3^6}\right)$

62. $\frac{2}{3} - \frac{4}{9} + \frac{8}{27} - \cdots + (-1)$ $\left(\frac{2}{\ }\right)^{11}$

63. $3 + \frac{3^2}{2} + \frac{3^3}{3} + \cdots + \frac{3^n}{n}$

64. $\frac{1}{e} + \frac{2}{e^2} + \frac{3}{e^3} + \cdots + \frac{n}{e^n}$

65. $a + (a + d) + (a + 2d) + \cdots + (a + nd)$

66. $a + ar + ar^2 + \cdots + ar^{n-1}$

67. **Credit Card Debt** John has a balance of \$3000 on his Discover card that charges 1% interest per month on any unpaid balance. John can afford to pay \$100 toward the balance each month. His balance each month after making a \$100 payment is given by the recursively defined sequence

$$B_0 = \$3000, \qquad B_n = 1.01B_{n-1} - 100$$

Determine John's balance after making the first payment. That is, determine B_1.

68. **Car Loans** Phil bought a car by taking out a loan for \$18,500 at 0.5% interest per month. Phil's normal monthly payment is \$434.47 per month, but he decides that he can afford to pay \$100 extra toward the balance each month. His balance each month is given by the recursively defined sequence

$$B_0 = \$18{,}500, \qquad B_n = 1.005B_{n-1} - 534.47$$

Determine Phil's balance after making the first payment. That is, determine B_1.

69. **Trout Population** A pond currently has 2000 trout in it. A fish hatchery decides to add an additional 20 trout each month. In addition, it is known that the trout population is growing 3% per month. The size of the population after n months is given by the recursively defined sequence

$$p_0 = 2000, \qquad p_n = 1.03p_{n-1} + 20$$

How many trout are in the pond after two months? That is, what is p_2?

70. **Environmental Control** The Environmental Protection Agency (EPA) determines that Maple Lake has 250 tons of pollutants as a result of industrial waste and that 10% of the pollutant present is neutralized by solar oxidation every year. The EPA imposes new pollution control laws that result in 15 tons of new pollutant entering the lake each year. The amount of pollut… years is given by the recursively de… he lake after n … uence

$$p_0 = 250, \qquad p_n = 0.9p_{n-1} + $$

Determine the amount of pollutant in the … years. That is, determine p_2. … two

71. **Growth of a Rabbit Colony** A colony of rabb… with one pair of mature rabbits, which will produ… of offspring (one male, one female) each month. A… that all rabbits mature in 1 month and produce a pa… offspring (one male, one female) after 2 months. If no ra… bits ever die, how many pairs of mature rabbits are ther… after 7 months?
[**Hint:** A Fibonacci sequence models this colony. Do you see why?]

72. **Fibonacci Sequence** Let

$$u_n = \frac{(1 + \sqrt{5})^n - (1 - \sqrt{5})^n}{2^n\sqrt{5}}$$

define the nth term of a sequence.
(a) Show that $u_1 = 1$ and $u_2 = 1$.
(b) Show that $u_{n+2} = u_{n+1} + u_n$.
(c) Draw the conclusion that $\{u_n\}$ is a Fibonacci sequence.

73. Pascal's Triangle Divide the triangular array shown (called Pascal's triangle) using diagonal lines as indicated. Find the sum of the numbers in each of these diagonal rows. Do you recognize this sequence?

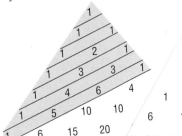

74. Fibonacci Sequence Use th   ult of Problem 72 to do the following problems: the Fibonacci sequence.

(a) Write the first 10 ter

(b) Compute the rati   for the first 10 terms.

(c) As n gets larger   it number does the ratio approach? Thi whose   is referred to as the **golden**
ratio. Rect ing   hose sides are in this ratio were
considere de   to the eye by the Greeks. For example, en ratio.   de of the Parthenon was constructed
using

(d) Compute the ratio $\dfrac{u_n}{u_{n+1}}$ for the first 10 terms.

(e) As n gets large, what number does the ratio approach? This number is also referred to as the **golden ratio.** This ratio is believed to have been used in the construction of the Great Pyramid in Egypt. The ratio equals the sum of the areas of the four face triangles divided by the total surface area of the Great Pyramid.

75. Show that

$$1 + 2 + \cdots + (n - 1) + n = \frac{n(n + 1)}{2}$$

[**Hint:** Let

$$S = 1 + 2 + \cdots + (n - 1) + n$$
$$S = n + (n - 1) + (n - 2) + \cdots + 1$$

Add these equations. Then

$$2S = \underbrace{[1 + n] + [2 + (n - 1)] + \cdots + [n + 1]}_{n \text{ terms in brackets}}$$

Now complete the derivation].

76. Investigate various applications that lead to a Fibonacci sequence, such as art, architecture, or financial markets. Write an essay on these applications.

9.2   THMETIC SEQUENCES

OBJECTIVES
1. Determine If a Sequence Is Arithmetic
2. Find a Formula for an Arithmetic Sequence
3. Find the Sum of an Arithmetic Sequence

1. When the difference between successive terms of a sequence is always the same number, the sequence is called **arithmetic.** An **arithmetic sequence*** may be defined recursively as $a_1 = a$, $a_n - a_{n-1} = d$, or as

$$a_1 = a, \qquad a_n = a_{n-1} + d \tag{1}$$

where $a = a_1$ and d are real numbers. The number a is the first term, and the number d is called the **common difference.**

The terms of an arithmetic sequence with first term a and common difference d follow the pattern

$$a, \quad a + d, \quad a + 2d, \quad a + 3d, \ldots$$

EXAMPLE 1   **Determining If a Sequence Is Arithmetic**

The sequence

$$4, \quad 7, \quad 10, \quad 13, \ldots$$

is arithmetic since the difference of successive terms is 3. The first term is 4, and the common difference is 3.   ▬

* Sometimes called an **arithmetic progression.**

EXAMPLE 2

Determining If a Sequence Is Arithmetic

Show that the following sequence is arithmetic. Find the first term and the common difference.

$$\{s_n\} = \{3n + 5\}$$

Solution   The first term is $s_1 = 3 \cdot 1 + 5 = 8$. The nth and $(n - 1)$st terms of the sequence $\{s_n\}$ are

$$s_n = 3n + 5 \quad \text{and} \quad s_{n-1} = 3(n - 1) + 5 = 3n + 2$$

Their difference is

$$s_n - s_{n-1} = (3n + 5) - (3n + 2) = 5 - 2 = 3$$

Since the difference of two successive terms is constant, the sequence is arithmetic and the common difference is 3.   ■

EXAMPLE 3

Determining If a Sequence Is Arithmetic

Show that the sequence $\{t_n\} = \{4 - n\}$ is arithmetic. Find the first term and the common difference.

Solution   The first term is $t_1 = 4 - 1 = 3$. The nth and $(n - 1)$st terms are

$$t_n = 4 - n \quad \text{and} \quad t_{n-1} = 4 - (n - 1) = 5 - n$$

Their difference is

$$t_n - t_{n-1} = (4 - n) - (5 - n) = 4 - 5 = -1$$

The difference of two successive terms is constant. $\{t_n\}$ is an arithmetic sequence whose common difference is -1.   ■

NOW WORK PROBLEM **3**.

②   Suppose that a is the first term of an arithmetic sequence whose common difference is d. We seek a formula for the nth term, a_n. To see the pattern, we write down the first few terms.

$$a_1 = a$$
$$a_2 = a_1 + d = a + 1 \cdot d$$
$$a_3 = a_2 + d = (a + d) + d = a + 2 \cdot d$$
$$a_4 = a_3 + d = (a + 2 \cdot d) + d = a + 3 \cdot d$$
$$a_5 = a_4 + d = (a + 3 \cdot d) + d = a + 4 \cdot d$$
$$\vdots$$
$$a_n = a_{n-1} + d = \left[a + (n - 2)d\right] + d = a + (n - 1)d$$

We are led to the following result:

Theorem

nth Term of an Arithmetic Sequence

For an arithmetic sequence $\{a_n\}$ whose first term is a and whose common difference is d, the nth term is determined by the formula

$$a_n = a + (n - 1)d \qquad (2)$$

EXAMPLE 4   Finding a Particular Term of an Arithmetic Sequence

Find the thirteenth term of the arithmetic sequence: $2, 6, 10, 14, 18, \ldots$

Solution   The first term of this arithmetic sequence is $a = 2$, and the common difference is 4. By formula (2), the nth term is

$$a_n = 2 + (n - 1)4$$

Hence, the thirteenth term is

$$a_{13} = 2 + 12 \cdot 4 = 50$$

 EXPLORATION   Use a graphing utility to find the thirteenth term of the sequence given in Example 4. Use it to find the twentieth and fiftieth terms.

EXAMPLE 5   Finding a Recursive Formula for an Arithmetic Sequence

The eighth term of an arithmetic sequence is 75, and the twentieth term is 39. Find the first term and the common difference. Give a recursive formula for the sequence.

Solution   By formula (2), we know that $a_n = a + (n - 1)d$. As a result,

$$\begin{cases} a_8 = a + 7d = 75 \\ a_{20} = a + 19d = 39 \end{cases}$$

This is a system of two linear equations containing two variables, a and d, which we can solve by elimination. Subtracting the second equation from the first equation, we get

$$-12d = 36$$

$$d = -3$$

With $d = -3$, we find that $a = 75 - 7d = 75 - 7(-3) = 96$. The first term is $a = 96$, and the common difference is $d = -3$. A recursive formula for this sequence is found using formula (1).

$$a_1 = 96, \qquad a_n = a_{n-1} - 3$$

Based on formula (2), a formula for the nth term of the sequence $\{a_n\}$ in Example 5 is

$$a_n = a + (n - 1)d = 96 + (n - 1)(-3) = 99 - 3n$$

 NOW WORK PROBLEMS **19** AND **25**.

 EXPLORATION   Graph the recursive formula from Example 5, $a_1 = 96$, $a_n = a_{n-1} - 3$, using a graphing utility. Conclude that the graph of the recursive formula behaves like the graph of a linear function. How is d, the common difference, related to m, the slope of a line?

ADDING THE FIRST n TERMS
OF AN ARITHMETIC SEQUENCE

③ The next result gives a formula for finding the sum of the first n terms of an arithmetic sequence.

Theorem

Sum of *n* Terms of an Arithmetic Sequence

Let $\{a_n\}$ be an arithmetic sequence with first term a and common difference d. The sum S_n of the first n terms of $\{a_n\}$ is

$$S_n = \frac{n}{2}\big[2a + (n-1)d\big] = \frac{n}{2}(a + a_n) \qquad \textbf{(3)}$$

Proof

$$
\begin{aligned}
S_n &= a_1 + a_2 + a_3 + \cdots + a_n & &\text{Sum of first } n \text{ terms} \\
&= a + (a+d) + (a+2d) + \cdots + [a+(n-1)d] & &\text{Formula (2)} \\
&= \underbrace{(a + a + \cdots + a)}_{n\text{ terms}} + [d + 2d + \cdots + (n-1)d] & &\text{Rearrange terms} \\
&= na + d[1 + 2 + \cdots + (n-1)] & & \\
&= na + d\left[\frac{(n-1)n}{2}\right] & &\text{Property 6, Section 9.1} \\
&= na + \frac{n}{2}(n-1)d & & \\
&= \frac{n}{2}[2a + (n-1)d] & &\text{Factor out } n/2 & &\textbf{(4)} \\
&= \frac{n}{2}[a + a + (n-1)d] & & \\
&= \frac{n}{2}(a + a_n) & &\text{Formula (2)} & &\textbf{(5)}
\end{aligned}
$$

Formula (3) provides two ways to find the sum of the first n terms of an arithmetic sequence. Notice that formula (4) involves the first term and common difference, whereas formula (5) involves the first term and the nth term. Use whichever form is easier.

EXAMPLE 6

Finding the Sum of *n* Terms of an Arithmetic Sequence

Find the sum S_n of the first n terms of the sequence $\{3n + 5\}$; that is, find

$$8 + 11 + 14 + \cdots + (3n + 5)$$

Solution   The sequence $\{3n + 5\}$ is an arithmetic sequence with first term $a = 8$ and the nth term $(3n + 5)$. To find the sum S_n, we use formula (3), as given in (5).

$$S_n = \frac{n}{2}(a + a_n) = \frac{n}{2}\big[8 + (3n + 5)\big] = \frac{n}{2}(3n + 13) \qquad \blacksquare$$

 NOW WORK PROBLEM 33.

EXAMPLE 7

Using a Graphing Utility to Find the Sum of 20 Terms of an Arithmetic Sequence

Use a graphing utility to find the sum S_n of the first 20 terms of the sequence $\{9.5n + 2.6\}$.

Solution   Figure 7 shows the results obtained using a TI-83 graphing calculator.

Figure 7

```
sum(seq(9.5n+2.6
,n,1,20,1)
              2047
```

The sum of the first 20 terms of the sequence $\{9.5n + 2.6\}$ is 2047. ■

   WORK EXAMPLE **7** USING FORMULA **(3)**.
   NOW WORK PROBLEM **41**.

EXAMPLE 8	**Creating a Floor Design**

A ceramic tile floor is designed in the shape of a trapezoid 20 feet wide at the base and 10 feet wide at the top. See Figure 8. The tiles, 12 inches by 12 inches, are to be placed so that each successive row contains one less tile than the preceding row. How many tiles will be required?

Figure 8

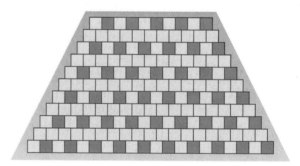

Solution   The bottom row requires 20 tiles and the top row, 10 tiles. Since each successive row requires one less tile, the total number of tiles required is

$$S = 20 + 19 + 18 + \cdots + 11 + 10$$

This is the sum of an arithmetic sequence; the common difference is -1. The number of terms to be added is $n = 11$, with the first term $a = 20$ and the last term $a_{11} = 10$. The sum S is

$$S = \frac{n}{2}(a + a_{11}) = \frac{11}{2}(20 + 10) = 165$$

In all, 165 tiles will be required. ■

9.2   EXERCISES

In Problems 1–10, an arithmetic sequence is given. Find the common difference and write out the first four terms.

1. $\{n + 4\}$ **2.** $\{n - 5\}$ **3.** $\{2n - 5\}$ **4.** $\{3n + 1\}$ **5.** $\{6 - 2n\}$

6. $\{4 - 2n\}$ **7.** $\left\{\dfrac{1}{2} - \dfrac{1}{3}n\right\}$ **8.** $\left\{\dfrac{2}{3} + \dfrac{n}{4}\right\}$ **9.** $\{\ln 3^n\}$ **10.** $\{e^{\ln n}\}$

In Problems 11–18, find the nth term of the arithmetic sequence whose initial term a and common difference d are given. What is the fifth term?

11. $a = 2; \quad d = 3$ **12.** $a = -2; \quad d = 4$ **13.** $a = 5; \quad d = -3$ **14.** $a = 6; \quad d = -2$

15. $a = 0; \quad d = \frac{1}{2}$ **16.** $a = 1; \quad d = -\frac{1}{3}$ **17.** $a = \sqrt{2}; \quad d = \sqrt{2}$ **18.** $a = 0; \quad d = \pi$

In Problems 19–24, find the indicated term in each arithmetic sequence.

19. 12th term of $2, 4, 6, \ldots$ **20.** 8th term of $-1, 1, 3, \ldots$

21. 10th term of $1, -2, -5, \ldots$ **22.** 9th term of $5, 0, -5, \ldots$

23. 8th term of $a, a + b, a + 2b, \ldots$ **24.** 7th term of $2\sqrt{5}, 4\sqrt{5}, 6\sqrt{5}, \ldots$

In Problems 25–32, find the first term and the common difference of the arithmetic sequence described. Give a recursive formula for the sequence.

25. 8th term is 8;   20th term is 44   **26.** 4th term is 3;   20th term is 35

27. 9th term is -5;   15th term is 31   **28.** 8th term is 4;   18th term is -96

29. 15th term is 0;   40th term is -50 **30.** 5th term is -2;   13th term is 30

31. 14th term is -1;   18th term is -9 **32.** 12th term is 4;   18th term is 28

In Problems 33–40, find the sum.

33. $1 + 3 + 5 + \cdots + (2n - 1)$ **34.** $2 + 4 + 6 + \cdots + 2n$

35. $7 + 12 + 17 + \cdots + (2 + 5n)$ **36.** $-1 + 3 + 7 + \cdots + (4n - 5)$

37. $2 + 4 + 6 + \cdots + 70$ **38.** $1 + 3 + 5 + \cdots + 59$

39. $5 + 9 + 13 + \cdots + 49$ **40.** $2 + 5 + 8 + \cdots + 41$

For Problems 41–46, use a graphing utility to find the sum of each sequence.

41. $\{3.45n + 4.12\}, \quad n = 20$ **42.** $\{2.67n - 1.23\}, \quad n = 25$

43. $2.8 + 5.2 + 7.6 + \cdots + 36.4$ **44.** $5.4 + 7.3 + 9.2 + \cdots + 32$

45. $4.9 + 7.48 + 10.06 + \cdots + 66.82$ **46.** $3.71 + 6.9 + 10.09 + \cdots + 80.27$

---

47. Find x so that $x + 3, 2x + 1,$ and $5x + 2$ are consecutive terms of an arithmetic sequence.

48. Find x so that $2x, 3x + 2,$ and $5x + 3$ are consecutive terms of an arithmetic sequence.

49. Drury Lane Theater The Drury Lane Theater has 25 seats in the first row and 30 rows in all. Each successive row contains one additional seat. How many seats are in the theater?

50. Football Stadium The corner section of a football stadium has 15 seats in the first row and 40 rows in all. Each successive row contains two additional seats. How many seats are in this section?

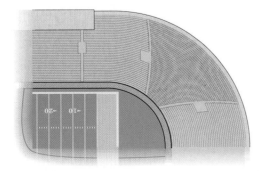

51. Creating a Mosaic A mosaic is designed in the shape of an equilateral triangle, 20 feet on each side. Each tile in the mosaic is in the shape of an equilateral triangle, 12 inches to a side. The tiles are to alternate in color as shown in the illustration. How many tiles of each color will be required?

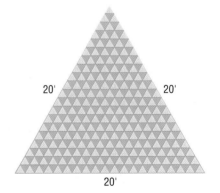

20'   20'   20'

52. Constructing a Brick Staircase A brick staircase has a total of 30 steps. The bottom step requires 100 bricks. Each successive step requires two less bricks than the prior step.
(a) How many bricks are required for the top step?
(b) How many bricks are required to build the staircase?

53. Stadium Construction How many rows are in the corner section of a stadium containing 2040 seats if the first row has 10 seats and each successive row has 4 additional seats?

54. Salary Suppose that you just received a job offer with a starting salary of $35,000 per year and a guaranteed raise of $1400 per year. How many years will it take before your aggregate salary is $280,000?
[**Hint:** Your aggregate salary after two years is $35,000 + ($35,000 + $1400).]

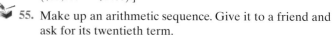 **55.** Make up an arithmetic sequence. Give it to a friend and ask for its twentieth term.

9.3 GEOMETRIC SEQUENCES; GEOMETRIC SERIES

OBJECTIVES
1. Determine If a Sequence Is Geometric
2. Find a Formula for a Geometric Sequence
3. Find the Sum of a Geometric Sequence
4. Find the Sum of a Geometric Series
5. Solve Annuity Problems

1 When the ratio of successive terms of a sequence is always the same nonzero number, the sequence is called **geometric. A geometric sequence*** may be defined recursively as $a_1 = a$, $a_n/a_{n-1} = r$, or as

$$a_1 = a, \qquad a_n = ra_{n-1} \qquad \textbf{(1)}$$

where $a_1 = a$ and $r \neq 0$ are real numbers. The number a is the first term, and the nonzero number r is called the **common ratio.**

The terms of a geometric sequence with first term a and common ratio r follow the pattern

$$a, \quad ar, \quad ar^2, \quad ar^3, \ldots$$

EXAMPLE 1 | **Determining If a Sequence Is Geometric**

The sequence

$$2, 6, 18, 54, 162, \ldots$$

is geometric since the ratio of successive terms is 3 $\left(\frac{6}{2} = \frac{18}{6} = \cdots = 3\right)$. The first term is 2, and the common ratio is 3. ∎

EXAMPLE 2 | **Determining If a Sequence Is Geometric**

Show that the following sequence is geometric. Find the first term and the common ratio.

$$\{s_n\} = 2^{-n}$$

Solution The first term is $s_1 = 2^{-1} = \frac{1}{2}$. The nth and $(n-1)$st terms of the sequence $\{s_n\}$ are

$$s_n = 2^{-n} \quad \text{and} \quad s_{n-1} = 2^{-(n-1)}$$

---

*Sometimes called a **geometric progression.**

Their ratio is

$$\frac{s_n}{s_{n-1}} = \frac{2^{-n}}{2^{-(n-1)}} = 2^{-n+(n-1)} = 2^{-1} = \frac{1}{2}$$

Because the ratio of successive terms is a nonzero constant, the sequence $\{s_n\}$ is geometric with common ratio $\frac{1}{2}$. ∎

EXAMPLE 3   **Determining If a Sequence Is Geometric**

Show that the following sequence is geometric. Find the first term and the common ratio.

$$\{t_n\} = \{4^n\}$$

Solution   The first term is $t_1 = 4^1 = 4$. The nth and $(n-1)$st terms are

$$t_n = 4^n \quad \text{and} \quad t_{n-1} = 4^{n-1}$$

Their ratio is

$$\frac{t_n}{t_{n-1}} = \frac{4^n}{4^{n-1}} = 4^{n-(n-1)} = 4$$

The sequence, $\{t_n\}$ is a geometric sequence with common ratio 4. ∎

NOW WORK PROBLEM 3.

②   Suppose that a is the first term of a geometric sequence with common ratio $r \neq 0$. We seek a formula for the nth term a_n. To see the pattern, we write down the first few terms:

$$a_1 = 1a = ar^0$$
$$a_2 = ra_1 = ar^1$$
$$a_3 = ra_2 = r(ar) = ar^2$$
$$a_4 = ra_3 = r(ar^2) = ar^3$$
$$a_5 = ra_4 = r(ar^3) = ar^4$$
$$\vdots$$
$$a_n = ra_{n-1} = r(ar^{n-2}) = ar^{n-1}$$

We are led to the following result:

Theorem   **nth Term of a Geometric Sequence**

For a geometric sequence $\{a_n\}$ whose first term is a and whose common ratio is r, the nth term is determined by the formula

$$a_n = ar^{n-1}, \qquad r \neq 0 \qquad \textbf{(2)}$$

∎

EXAMPLE 4   **Finding a Particular Term of a Geometric Sequence**

(a) Find the ninth term of the geometric sequence: $10, 9, \frac{81}{10}, \frac{729}{100} \ldots$

(b) Find a recursive formula for this sequence.

Solution   (a) The first term of this geometric sequence is $a = 10$ and the common ratio is $9/10$. (Use $9/10$, or $\frac{81/10}{9} = \frac{9}{10}$, or any two successive terms.) By formula (2), the nth term is

$$a_n = 10\left(\frac{9}{10}\right)^{n-1}$$

The ninth term is

$$a_9 = 10\left(\frac{9}{10}\right)^{9-1} = 10\left(\frac{9}{10}\right)^8 = 4.3046721$$

(b) The first term in the sequence is 10 and the common ratio is $r = 9/10$. Using formula (1), the recursive formula is $a_1 = 10$, $a_n = \frac{9}{10}a_{n-1}$. ∎

 EXPLORATION   Use a graphing utility to find the ninth term of the sequence given in Example 4. Use it to find the twentieth and fiftieth terms. Now use a graphing utility to graph the recursive formula found in Example 4(b). Conclude that the graph of the recursive formula behaves like the graph of an exponential function. How is r, the common ratio, related to a, the base of the exponential function $y = a^x$? ∎

 NOW WORK PROBLEMS **25** AND **33**.

ADDING THE FIRST n TERMS OF A GEOMETRIC SEQUENCE

③ The next result gives us a formula for finding the sum of the first n terms of a geometric sequence.

Theorem

Sum of n Terms of a Geometric Sequence

Let $\{a_n\}$ be a geometric sequence with first term a and common ratio r where $r \neq 0$, $r \neq 1$. The sum S_n of the first n terms of $\{a_n\}$ is

$$S_n = a\frac{1 - r^n}{1 - r}, \qquad r \neq 0, 1 \tag{3}$$

■

Proof   The sum S_n of the first n terms of $\{a_n\} = \{ar^{n-1}\}$ is

$$S_n = a + ar + \cdots + ar^{n-1} \tag{4}$$

Multiply each side by r to obtain

$$rS_n = ar + ar^2 + \cdots + ar^n \tag{5}$$

Now, subtract (5) from (4). The result is

$$S_n - rS_n = a - ar^n$$
$$(1 - r)S_n = a(1 - r^n)$$

Since $r \neq 1$, we can solve for S_n.

$$S_n = a\frac{1 - r^n}{1 - r}$$ ■

| **EXAMPLE 5** | **Finding the Sum of n Terms of a Geometric Sequence** |

Find the sum S_n of the first n terms of the sequence $\left\{\left(\frac{1}{2}\right)^n\right\}$; that is, find

$$\frac{1}{2} + \frac{1}{4} + \frac{1}{8} + \cdots + \left(\frac{1}{2}\right)^n$$

Solution   The sequence $\left\{\left(\frac{1}{2}\right)^n\right\}$ is a geometric sequence with $a = \frac{1}{2}$ and $r = \frac{1}{2}$. The sum S_n that we seek is the sum of the first n terms of the sequence, so we use formula (3) to get

$$S_n = \sum_{k=1}^{n} \left(\frac{1}{2}\right)^k = \frac{1}{2} + \frac{1}{4} + \frac{1}{8} + \cdots + \left(\frac{1}{2}\right)^n$$

$$= \frac{1}{2}\left[\frac{1 - \left(\frac{1}{2}\right)^n}{1 - \frac{1}{2}}\right] \qquad \text{Formula (3)}$$

$$= \frac{1}{2}\left[\frac{1 - \left(\frac{1}{2}\right)^n}{\frac{1}{2}}\right]$$

$$= 1 - \left(\frac{1}{2}\right)^n \qquad\qquad ▬$$

NOW WORK PROBLEM 39.

| **EXAMPLE 6** | **Using a Graphing Utility to Find the Sum of a Geometric Sequence** |

 Use a graphing utility to find the sum S_n of the first 15 terms of the sequence $\left\{\left(\frac{1}{3}\right)^n\right\}$; that is, find

$$\frac{1}{3} + \frac{1}{9} + \frac{1}{27} + \cdots + \left(\frac{1}{3}\right)^{15}$$

Solution   Figure 9 shows the result obtained using a TI-83 graphing calculator.

Figure 9

The sum of the first 15 terms of the sequence $\left\{\left(\frac{1}{3}\right)^n\right\}$ is 0.4999999652.   ▬

NOW WORK PROBLEM 45.

GEOMETRIC SERIES

An infinite sum of the form

$$a + ar + ar^2 + \cdots + ar^{n-1} + \cdots$$

with first term a and common ratio r, is called an **infinite geometric series** and is denoted by

$$\sum_{k=1}^{\infty} ar^{k-1}$$

④ Based on formula (3), the sum S_n of the first n terms of a geometric series is

$$S_n = a\,\frac{1 - r^n}{1 - r} = \frac{a}{1 - r} - \frac{ar^n}{1 - r} \tag{6}$$

If this finite sum S_n approaches a number L as $n \to \infty$, then we call L the **sum of the infinite geometric series,** and we write

$$L = \sum_{k=1}^{\infty} ar^{k-1}$$

Theorem   **Sum of an Infinite Geometric Series**

If $|r| < 1$, the sum of the infinite geometric series $\displaystyle\sum_{k=1}^{\infty} ar^{k-1}$ is

$$\sum_{k=1}^{\infty} ar^{k-1} = \frac{a}{1 - r} \tag{7}$$

■

Intuitive Proof   Since $|r| < 1$, it follows that $|r^n|$ approaches 0 as $n \to \infty$. Then, based on formula (6), the sum S_n approaches $\dfrac{a}{1 - r}$ as $n \to \infty$. ■

EXAMPLE 7   **Finding the Sum of a Geometric Series**

Find the sum of the geometric series: $2 + \frac{4}{3} + \frac{8}{9} + \cdots$

Solution   The first term is $a = 2$, and the common ratio is

$$r = \frac{\frac{4}{3}}{2} = \frac{4}{6} = \frac{2}{3}$$

Since $|r| < 1$, we use formula (7) to find that

$$2 + \frac{4}{3} + \frac{8}{9} + \cdots = \frac{2}{1 - \frac{2}{3}} = 6$$

■

NOW WORK PROBLEM **51.**

EXPLORATION   Use a graphing utility to graph $U_n = 2\left(\frac{2}{3}\right)^{n-1}$ in sequence mode. TRACE the graph for large values of n. What happens to the value of U_n as n increases without bound? What can you conclude about $\displaystyle\sum_{n=1}^{\infty} 2\left(\frac{2}{3}\right)^{(n-1)}$?

EXAMPLE 8   **Repeating Decimals**

Show that the repeating decimal $0.999\ldots$ equals 1.

Solution

$$0.999\ldots = \frac{9}{10} + \frac{9}{100} + \frac{9}{1000} + \cdots$$

The decimal, $0.999\ldots$ is a geometric series with first term $\frac{9}{10}$ and common ratio $\frac{1}{10}$. Using formula (7), we find

$$0.999\ldots = \frac{\frac{9}{10}}{1 - \frac{1}{10}} = \frac{\frac{9}{10}}{\frac{9}{10}} = 1$$

EXAMPLE 9   **Pendulum Swings**

Figure 10

Initially, a pendulum swings through an arc of 18 inches. See Figure 10. On each successive swing, the length of the arc is 0.98 of the previous length.

(a) What is the length of the arc after 10 swings?

(b) On which swing is the length of the arc first less than 12 inches?

(c) After 15 swings, what total distance will the pendulum have swung?

(d) When it stops, what total distance will the pendulum have swung?

Solution   (a) The length of the first swing is 18 inches.
The length of the second swing is $0.98(18)$ inches.
The length of the third swing is $0.98(0.98)(18) = 0.98^2(18)$ inches.
The length of the arc of the tenth swing is

$$(0.98)^9(18) = 15.007 \text{ inches}$$

(b) The length of the arc of the nth swing is $(0.98)^{n-1}(18)$. For this to be exactly 12 inches requires that

$$(0.98)^{n-1}(18) = 12$$

$$(0.98)^{n-1} = \frac{12}{18} = \frac{2}{3}$$

$$n - 1 = \log_{0.98}\left(\frac{2}{3}\right)$$

$$n = 1 + \frac{\ln\left(\frac{2}{3}\right)}{\ln 0.98} \approx 1 + 20.07 \approx 21.07$$

The length of the arc of the pendulum exceeds 12 inches on the twenty-first swing and is first less than 12 inches on the twenty-second swing.

(c) After 15 swings, the pendulum will have swung the following total distance L:

$$L = \underset{\text{1st}}{18} + \underset{\text{2nd}}{0.98(18)} + \underset{\text{3rd}}{(0.98)^2(18)} + \underset{\text{4th}}{(0.98)^3(18)} + \cdots + \underset{\text{15th}}{(0.98)^{14}(18)}$$

This is the sum of a geometric sequence. The common ratio is 0.98; the first term is 18. The sum has 15 terms, so

$$L = 18 \frac{1 - 0.98^{15}}{1 - 0.98} \approx 18(13.07) \approx 235.29 \text{ inches}$$

The pendulum will have swung through 235.29 inches after 15 swings.

(d) When the pendulum stops, it will have swung the following total distance T:

$$T = 18 + 0.98(18) + (0.98)^2(18) + (0.98)^3(18) + \cdots$$

This is the sum of a geometric series. The common ratio is $r = 0.98$; the first term is $a = 18$. The sum is

$$T = \frac{a}{1 - r} = \frac{18}{1 - 0.98} = 900$$

The pendulum will have swung a total of 900 inches when it finally stops. ∎

ANNUITIES

In Section 6.6 we developed the compound interest formula that gives the future value when a fixed amount of money is deposited in an account that pays interest compounded periodically. Often, though, money is invested in small amounts at periodic intervals. An **annuity** is a sequence of equal periodic deposits. The periodic deposits may be made annually, quarterly, monthly, or daily.

When deposits are made at the same time that the interest is credited, the annuity is called **ordinary.** We will only deal with ordinary annuities here. The **amount of an annuity** is the sum of all deposits made plus all interest paid.

Suppose that the interest an account earns is i percent per payment period (expressed as a decimal). For example, if an account pays 12% compounded monthly (12 times a year) then $i = 0.12/12 = 0.01$. If an account pays 8% compounded quarterly (4 times a year) then $i = 0.08/4 = 0.02$. To develop a formula for the amount of an annuity, suppose that $\$P$ is deposited each payment period for n payment periods in an account that earns i percent per payment period. When the last deposit is made at the nth payment period, the first deposit of $\$P$ has earned interest compounded for $n - 1$ payment periods, the second deposit of $\$P$ has earned interest compounded for $n - 2$ payment periods, and so on. Table 3 shows the value of each deposit after n deposits have been made.

TABLE 3					
Deposit	1	2	3	...	$n - 1$	n
Amount	$P(1 + i)^{n-1}$	$P(1 + i)^{n-2}$	$P(1 + i)^{n-3}$	...	$P(1 + i)$	P

The amount A of the annuity is the sum of the amounts shown in Table 3; that is,

$$A = P(1 + i)^{n-1} + P(1 + i)^{n-2} + \cdots + P(1 + i) + P$$
$$= P\left[1 + (1 + i) + \ldots + (1 + i)^{n-1}\right]$$

The expression in brackets is the sum of a geometric sequence with n terms and a common ratio of $(1 + i)$. As a result,

$$A = P\left[1 + (1 + i) + \cdots + (1 + i)^{n-2} + (1 + i)^{n-1}\right]$$

$$= P\,\frac{1 - (1 + i)^n}{1 - (1 + i)} = P\,\frac{1 - (1 + i)^n}{-i} = P\,\frac{(1 + i)^n - 1}{i}$$

We have established the following result:

Theorem   **Amount of an Annuity**

If P represents the deposit in dollars made at each payment period for an annuity at i percent interest per payment period, the amount A of the annuity after n payment periods is

$$A = P\,\frac{(1 + i)^n - 1}{i} \qquad \textbf{(8)}$$

EXAMPLE 10   **Determining the Amount of an Annuity**

To save for retirement, Brett decides to place $2000 into an Individual Retirement Account (IRA) each year for the next 30 years. What will the value of the IRA be when Brett retires in 30 years if the rate of return of the IRA is assumed to be 10% per annum compounded annually?

Solution   This is an ordinary annuity with 30 annual deposits of $P = \$2000$. The rate of interest per payment period is $i = 0.10/1 = 0.10$. The number of payment periods is $n = 30$. The amount A of the annuity in 30 years is

$$A = 2000\left\{\frac{(1 + 0.10)^{30} - 1}{0.10}\right\} = \$2000(164.49402) = \$328,988.05 \quad \blacksquare$$

EXAMPLE 11   **Determining the Amount of an Annuity**

To save for his daughter's college education, Mr. McGowen decides to put $50 aside every month in a credit union account paying 10% interest compounded monthly. If he begins this savings program when his daughter is 3 years old, how much will he have saved by the time his daughter is 18 years old?

Solution   When his daughter is 18 years old, Mr. McGowen will have made his 180th payment (15 years $\times$ 12 payments per year). This is an annuity with $P = \$50$, $n = 180$, and $i = \frac{0.10}{12}$. The amount A saved is

$$A = 50\left[\frac{\left(1 + \dfrac{0.10}{12}\right)^{180} - 1}{\dfrac{0.10}{12}}\right] = \$50(414.4703) = \$20,723.52 \quad \blacksquare$$

HISTORICAL FEATURE

Fibonacci

Sequences are among the oldest objects of mathematical investigation, having been studied for over 3500 years. After the initial steps, however, little progress was made until about 1600.

Arithmetic and geometric sequences appear in the Rhind papyrus, a mathematical text containing 85 problems copied around 1650 BC by the Egyptian scribe Ahmes from an earlier work (see Historical Problem 1). Fibonacci (AD 1220) wrote about problems similar to those found in the Rhind papyrus, leading one to suspect that Fibonacci may have had material available that is now lost. This material would have been in the non-Euclidean Greek tradition of Heron (about AD 75) and Diophantus (about AD 250). One problem, again modified slightly, is still with us in the familiar puzzle rhyme "As I was going to St. Ives ..." (see Historical Problem 2).

The Rhind papyrus indicates that the Egyptians knew how to add up the terms of an arithmetic or geometric sequence, as did the Babylonians. The rule for summing up a geometric sequence is found in Euclid's *Elements* (book IX, 35, 36), where, like all Euclid's algebra, it is presented in a geometric form.

Investigations of other kinds of sequences began in the 1500s, when algebra became sufficiently developed to handle the more complicated problems. The development of calculus in the 1600s added a powerful new tool, especially for finding the sum of infinite series, and the subject continues to flourish today.

HISTORICAL PROBLEMS

1. *Arithmetic sequence problem from the Rhind papyrus (statement modified slightly for clarity)* One hundred loaves of bread are to be divided among five people so that the amounts that they receive form an arithmetic sequence. The first two together receive one-seventh of what the last three receive. How many does each receive? [*Partial answer:* First person receives $1\frac{2}{3}$ loaves.]

2. The following old English children's rhyme resembles one of the Rhind papyrus problems.

    As I was going to St. Ives
    I met a man with seven wives
    Each wife had seven sacks
    Each sack had seven cats
    Each cat had seven kits [kittens]
    Kits, cats, sacks, wives
    How many were going to St. Ives?

    (a) Assuming that the speaker and the cat fanciers met by traveling in opposite directions, what is the answer?

    (b) How many kittens are being transported?

    (c) Kits, cats, sacks, wives; how many?

    [**Hint:** It is easier to include the man, find the sum with the formula, and then subtract 1 for the man.]

9.3 EXERCISES

In Problems 1–10, a geometric sequence is given. Find the common ratio and write out the first four terms.

1. $\{3^n\}$

2. $\{(-5)^n\}$

3. $\left\{-3\left(\frac{1}{2}\right)^n\right\}$

4. $\left\{\left(\frac{5}{2}\right)^n\right\}$

5. $\left\{\frac{2^{n-1}}{4}\right\}$

6. $\left\{\frac{3^n}{9}\right\}$

7. $\{2^{n/3}\}$

8. $\{3^{2n}\}$

9. $\left\{\frac{3^{n-1}}{2^n}\right\}$

10. $\left\{\frac{2^n}{3^{n-1}}\right\}$

In Problems 11–24, determine whether the given sequence is arithmetic, geometric, or neither. If the sequence is arithmetic, find the common difference; if it is geometric, find the common ratio.

11. $\{n+2\}$

12. $\{2n-5\}$

13. $\{4n^2\}$

14. $\{5n^2+1\}$

15. $\{3-\frac{2}{3}n\}$

16. $\{8-\frac{3}{4}n\}$

17. $1, 3, 6, 10, \ldots$

18. $2, 4, 6, 8, \ldots$

19. $\left\{\left(\frac{2}{3}\right)^n\right\}$

20. $\left\{\left(\frac{5}{4}\right)^n\right\}$

21. $-1, -2, -4, -8, \ldots$

22. $1, 1, 2, 3, 5, 8, \ldots$

23. $\{3^{n/2}\}$

24. $\{(-1)^n\}$

In Problems 25–32, find the fifth term and the nth term of the geometric sequence whose initial term a and common ratio r are given.

25. $a = 2$; $r = 3$
26. $a = -2$; $r = 4$
27. $a = 5$; $r = -1$
28. $a = 6$; $r = -2$

29. $a = 0$; $r = \frac{1}{2}$
30. $a = 1$; $r = -\frac{1}{3}$
31. $a = \sqrt{2}$; $r = \sqrt{2}$
32. $a = 0$; $r = 1/\pi$

In Problems 33–38, find the indicated term of each geometric sequence.

33. 7th term of $1, \frac{1}{2}, \frac{1}{4}, \ldots$
34. 8th term of $1, 3, 9, \ldots$
35. 9th term of $1, -1, 1, \ldots$

36. 10th term of $-1, 2, -4, \ldots$
37. 8th term of $0.4, 0.04, 0.004, \ldots$
38. 7th term of $0.1, 1.0, 10.0, \ldots$

In Problems 39–44, find the sum.

39. $\dfrac{1}{4} + \dfrac{2}{4} + \dfrac{2^2}{4} + \dfrac{2^3}{4} + \cdots + \dfrac{2^{n-1}}{4}$
40. $\dfrac{3}{9} + \dfrac{3^2}{9} + \dfrac{3^3}{9} + \cdots + \dfrac{3^n}{9}$
41. $\displaystyle\sum_{k=1}^{n} \left(\tfrac{2}{3}\right)^k$

42. $\displaystyle\sum_{k=1}^{n} 4 \cdot 3^{k-1}$
43. $-1 - 2 - 4 - 8 - \cdots - \left(2^{n-1}\right)$
44. $2 + \dfrac{6}{5} + \dfrac{18}{25} + \cdots + 2\left(\dfrac{3}{5}\right)^n$

For Problems 45–50, use a graphing utility to find the sum of each geometric sequence.

45. $\dfrac{1}{4} + \dfrac{2}{4} + \dfrac{2^2}{4} + \dfrac{2^3}{4} + \cdots + \dfrac{2^{14}}{4}$
46. $\dfrac{3}{9} + \dfrac{3^2}{9} + \dfrac{3^3}{9} + \cdots + \dfrac{3^{15}}{9}$
47. $\displaystyle\sum_{n=1}^{15} \left(\tfrac{2}{3}\right)^n$

48. $\displaystyle\sum_{n=1}^{15} 4 \cdot 3^{n-1}$
49. $-1 - 2 - 4 - 8 - \cdots - 2^{14}$
50. $2 + \dfrac{6}{5} + \dfrac{18}{25} + \cdots + 2\left(\dfrac{3}{5}\right)^{15}$

In Problems 51–60, find the sum of each infinite geometric series.

51. $1 + \frac{1}{3} + \frac{1}{9} + \cdots$
52. $2 + \frac{4}{3} + \frac{8}{9} + \cdots$
53. $8 + 4 + 2 + \cdots$
54. $6 + 2 + \frac{2}{3} + \cdots$

55. $2 - \frac{1}{2} + \frac{1}{8} - \frac{1}{32} + \cdots$
56. $1 - \frac{3}{4} + \frac{9}{16} - \frac{27}{64} + \cdots$
57. $\displaystyle\sum_{k=1}^{\infty} 5\left(\tfrac{1}{4}\right)^{k-1}$
58. $\displaystyle\sum_{k=1}^{\infty} 8\left(\tfrac{1}{3}\right)^{k-1}$

59. $\displaystyle\sum_{k=1}^{\infty} 6\left(-\tfrac{2}{3}\right)^{k-1}$
60. $\displaystyle\sum_{k=1}^{\infty} 4\left(-\tfrac{1}{2}\right)^{k-1}$

61. Find x so that x, $x + 2$, and $x + 3$ are consecutive terms of a geometric sequence.

62. Find x so that $x - 1$, x, and $x + 2$ are consecutive terms of a geometric sequence.

63. Salary Increases   Suppose that you have just been hired at an annual salary of $18,000 and expect to receive annual increases of 5%. What will your salary be when you begin your fifth year?

64. Equipment Depreciation   A new piece of equipment cost a company $15,000. Each year, for tax purposes, the company depreciates the value by 15%. What value should the company give the equipment after 5 years?

65. Pendulum Swings   Initially, a pendulum swings through an arc of 2 feet. On each successive swing, the length of the arc is 0.9 of the previous length.
(a) What is the length of the arc after 10 swings?
(b) On which swing is the length of the arc first less than 1 foot?
(c) After 15 swings, what total length will the pendulum have swung?
(d) When it stops, what total length will the pendulum have swung?

66. Bouncing Balls   A ball is dropped from a height of 30 feet. Each time it strikes the ground, it bounces up to 0.8 of the previous height.

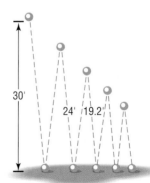

(a) What height will the ball bounce up to after it strikes the ground for the third time?
(b) How high will it bounce after it strikes the ground for the nth time?
(c) How many times does the ball need to strike the ground before its bounce is less than 6 inches?
(d) What total distance does the ball travel before it stops bouncing?

67. Retirement Christine contributes $100 each month to her 401(k). What will be the value of Christine's 401(k) in 30 years if the per annum rate of return is assumed to be 12% compounded monthly?

68. Saving for a Home Jolene wants to purchase a new home. Suppose that she invests $400 per month into a mutual fund. If the per annum rate of return of the mutual fund is assumed to be 10% compounded monthly, how much will Jolene have for a down payment after 3 years?

69. Tax Sheltered Annuity Don contributes $500 at the end of each quarter to a Tax Sheltered Annuity (TSA). What will the value of the TSA be in 20 years if the per annum rate of return is assumed to be 8% compounded quarterly?

70. Retirement Ray, planning on retiring in 15 years, contributes $1000 to an Individual Retirement Account (IRA) semiannually. What will the value of the IRA be when Ray retires if the per annum rate of return is assumed to be 10% compounded semiannually?

71. Sinking Fund Scott and Alice want to purchase a vacation home in 10 years and need $50,000 for a down payment. How much should they place in a savings account each month if the per annum rate of return is assumed to be 6% compounded monthly?

72. Sinking Fund For a child born in 1996, a 4-year college education at a public university is projected to be $150,000. Assuming an 8% per annum rate of return compounded monthly, how much must be contributed to a college fund every month in order to have $150,000 in 18 years when the child begins college.

73. Critical Thinking You are interviewing for a job and receive two offers:

> *A*: $20,000 to start, with guaranteed annual increases of 6% for the first 5 years
> *B*: $22,000 to start, with guaranteed annual increases of 3% for the first 5 years

Which offer is best if your goal is to be making as much as possible after 5 years? Which is best if your goal is to make as much money as possible over the contract (5 years)?

74. Critical Thinking Which of the following choices, *A* or *B*, results in more money?

> *A*: To receive $1000 on day 1, $999 on day 2, $998 on day 3, with the process to end after 1000 days
> *B*: To receive $1 on day 1, $2 on day 2, $4 on day 3, for 19 days

75. Critical Thinking You have just signed a 7-year professional football league contract with a beginning salary of $2,000,000 per year. Management gives you the following options with regard to your salary over the 7 years.

> **1.** A bonus of $100,000 each year

> **2.** An annual increase of 4.5% per year beginning after 1 year
> **3.** An annual increase of $95,000 per year beginning after 1 year

Which option provides the most money over the 7-year period? Which the least? Which would you choose? Why?

76. A Rich Man's Promise A rich man promises to give you $1000 on September 1, 2001. Each day thereafter he will give you $\frac{9}{10}$ of what he gave you the previous day. What is the first date on which the amount you receive is less than 1¢? How much have you received when this happens?

77. Grains of Wheat on a Chess Board In an old fable, a commoner who had just saved the king's life was told he could ask the king for any just reward. Being a shrewd man, the commoner said, "A simple wish, sire. Place one grain of wheat on the first square of a chessboard, two grains on the second square, four grains on the third square, continuing until you have filled the board. This is all I seek." Compute the total number of grains needed to do this to see why the request, seemingly simple, could not be granted. (A chessboard consists of 8 × 8 = 64 squares.)

78. Look at the figure below. What fraction of the square is eventually shaded if the indicated shading process continues indefinitely?

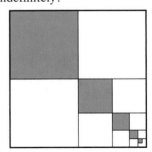

79. Multiplier Suppose that, throughout the U.S. economy, individuals spend 90% of every additional dollar that they earn. Economists would say that an individual's **marginal propensity to consume** is 0.90. For example, if Jane earns an additional dollar, she will spend $0.9(1) = \$0.90$ of it. The individual that earns $0.90 (from Jane) will spend 90% of it or $0.81. This process of spending continues and results in an infinite geometric series as follows:

$$1, 0.90, 0.90^2, 0.90^3, 0.90^4, \ldots$$

The sum of this infinite geometric series is called the **multiplier**. What is the multiplier if individuals spend 90% of every additional dollar that they earn?

80. Multiplier   Refer to Problem 79. Suppose that the marginal propensity to consume throughout the U.S. economy is 0.95. What is the multiplier for the U.S. economy?

81. Stock Price   One method of pricing a stock is to discount the stream of future dividends of the stock. Suppose that a stock pays $P per year in dividends and, historically, the dividend has been increased $i\%$ per year. If you desire an annual rate of return of $r\%$, this method of pricing a stock states that the price that you should pay is the present value of an infinite stream of payments:

$$\text{Price} = P + P\,\frac{1+i}{1+r} + P\left(\frac{1+i}{1+r}\right)^{2} + P\left(\frac{1+i}{1+r}\right)^{3} + \cdots$$

The price of the stock is the sum of an infinite geometric series. Suppose that a stock pays an annual dividend of $4.00 and, historically, the dividend has been increased

3% per year. You desire an annual rate of return of 9%. What is the most you should pay for the stock?

82. Stock Price   Refer to Problem 81. Suppose that a stock pays an annual dividend of $2.50 and, historically, the dividend has increased 4% per year. You desire an annual rate of return of 11%. What is the most that you should pay for the stock?

83. Can a sequence be both arithmetic and geometric? Give reasons for your answer.

84. Make up a geometric sequence. Give it to a friend and ask for its 20th term.

85. Make up two infinite geometric series, one that has a sum and one that does not. Give them to a friend and ask for the sum of each series.

9.4   MATHEMATICAL INDUCTION

OBJECTIVE   (1)   Prove Statements Using Mathematical Induction

(1)   *Mathematical induction* is a method for proving that statements involving natural numbers are true for all natural numbers.* For example, the statement "$2n$ is always an even integer" can be proved true for all natural numbers n by using mathematical induction. Also, the statement "the sum of the first n positive odd integers equals n^2," that is,

$$1 + 3 + 5 + \cdots + (2n - 1) = n^2 \tag{1}$$

can be proved for all natural numbers n by using mathematical induction.

Before stating the method of mathematical induction, let's try to gain a sense of the power of the method. We shall use the statement in equation (1) for this purpose by restating it for various values of $n = 1, 2, 3, \ldots$.

$n = 1$ The sum of the first positive odd integer is 1^2; $1 = 1^2$.

$n = 2$ The sum of the first 2 positive odd integers is 2^2; $1 + 3 = 4 = 2^2$.

$n = 3$ The sum of the first 3 positive odd integers is 3^2; $1 + 3 + 5 = 9 = 3^2$.

$n = 4$ The sum of the first 4 positive odd integers is 4^2; $1 + 3 + 5 + 7 = 16 = 4^2$.

Although from this pattern we might conjecture that statement (1) is true for any choice of n, can we really be sure that it does not fail for some choice of n? The method of proof by mathematical induction will, in fact, prove that the statement is true for all n.

---

* Recall that the natural numbers are the numbers $1, 2, 3, 4, \ldots$. In other words, the terms *natural numbers* and *positive integers* are synonymous.

Theorem   **The Principle of Mathematical Induction**

Suppose that the following two conditions are satisfied with regard to a statement about natural numbers:

CONDITION I: The statement is true for the natural number 1.

CONDITION II: If the statement is true for some natural number k, it is also true for the next natural number $k + 1$.

Then the statement is true for all natural numbers.

Figure 11

We shall not prove this principle. However, we can provide a physical interpretation that will help us to see why the principle works. Think of a collection of natural numbers obeying a statement as a collection of infinitely many dominoes (see Figure 11).

Now, suppose that we are told two facts:

1. The first domino is pushed over.
2. If one domino falls over, say the kth domino, then so will the next one, the $(k + 1)$st domino.

Is it safe to conclude that *all* the dominoes fall over? The answer is yes, because if the first one falls (Condition I), then the second one does also (by Condition II); and if the second one falls, then so does the third (by Condition II); and so on.

Now let's prove some statements about natural numbers using mathematical induction.

EXAMPLE 1   **Using Mathematical Induction**

Show that the following statement is true for all natural numbers n.

$$1 + 3 + 5 + \cdots + (2n - 1) = n^2 \qquad \textbf{(2)}$$

Solution   We need to show first that statement (2) holds for $n = 1$. Because $1 = 1^2$, statement (2) is true for $n = 1$. Condition I holds.

Next, we need to show that Condition II holds. Suppose that we know for some k that

$$1 + 3 + \cdots + (2k - 1) = k^2 \qquad \textbf{(3)}$$

We wish to show that, based on equation (3), statement (2) holds for $k + 1$. We look at the sum of the first $k + 1$ positive odd integers to determine whether this sum equals $(k + 1)^2$.

$$1 + 3 + \cdots + (2k - 1) + (2k + 1) = \underbrace{[1 + 3 + \cdots + (2k - 1)]}_{= \, k^2 \text{ by equation (3)}} + (2k + 1)$$
$$= k^2 + (2k + 1)$$
$$= k^2 + 2k + 1 = (k + 1)^2$$

Conditions I and II are satisfied; by the Principle of Mathematical Induction, statement (2) is true for all natural numbers n.

EXAMPLE 2	**Using Mathematical Induction**

Show that the following statement is true for all natural numbers n.

$$2^n > n$$

Solution   First, we show that the statement $2^n > n$ holds when $n = 1$. Because $2^1 = 2 > 1$, the inequality is true for $n = 1$. Condition I holds.

Next, we assume, for some natural number k, that $2^k > k$. We wish to show that the formula holds for $k + 1$; that is, we wish to show that $2^{k+1} > k + 1$. Now,

$$2^{k+1} = 2 \cdot 2^k > 2 \cdot k = k + k \geq k + 1$$

$$\uparrow \qquad\qquad\qquad \uparrow$$
$$\text{We know that} \qquad\quad k \geq 1$$
$$2^k > k.$$

If $2^k > k$, then $2^{k+1} > k + 1$, so Condition II of the Principle of Mathematical Induction is satisfied. The statement $2^n > n$ is true for all natural numbers n.   ■

EXAMPLE 3	**Using Mathematical Induction**

Show that the following formula is true for all natural numbers n.

$$1 + 2 + 3 + \cdots + n = \frac{n(n + 1)}{2} \tag{4}$$

Solution   First, we show that formula (4) is true when $n = 1$. Because

$$\frac{1(1 + 1)}{2} = \frac{1(2)}{2} = 1$$

Condition I of the Principle of Mathematical Induction holds.

Next, we assume that formula (4) holds for some k, and we determine whether the formula then holds for $k + 1$. We assume that

$$1 + 2 + 3 + \cdots + k = \frac{k(k + 1)}{2} \quad \text{for some } k \tag{5}$$

Now we need to show that

$$1 + 2 + 3 + \cdots + k + (k + 1) = \frac{(k + 1)(k + 1 + 1)}{2} = \frac{(k + 1)(k + 2)}{2}$$

We do this as follows:

$$1 + 2 + 3 + \cdots + k + (k + 1) = \underbrace{[1 + 2 + 3 + \cdots + k]}_{= \frac{k(k+1)}{2} \quad \text{by equation (5)}} + (k + 1)$$

$$= \frac{k(k + 1)}{2} + (k + 1)$$

$$= \frac{k^2 + k + 2k + 2}{2}$$

$$= \frac{k^2 + 3k + 2}{2} = \frac{(k + 1)(k + 2)}{2}$$

Condition II also holds. As a result, formula (4) is true for all natural numbers n.   ■

NOW WORK PROBLEM **1**.

| EXAMPLE 4 | **Using Mathematical Induction** |

Show that $3^n - 1$ is divisible by 2 for all natural numbers n.

Solution   First, we show that the statement is true when $n = 1$. Because $3^1 - 1 = 3 - 1 = 2$ is divisible by 2, the statement is true when $n = 1$. Condition I is satisfied.

Next, we assume that the statement holds for some k, and we determine whether the statement then holds for $k + 1$. We assume that $3^k - 1$ is divisible by 2 for some k. We need to show that $3^{k+1} - 1$ is divisible by 2. Now

$$3^{k+1} - 1 = 3^{k+1} - 3^k + 3^k - 1 \qquad \text{Subtract and add } 3^k$$
$$= 3^k(3 - 1) + (3^k - 1) = 3^k \cdot 2 + (3^k - 1)$$

Because $3^k \cdot 2$ is divisible by 2 and $3^k - 1$ is divisible by 2, it follows that $3^k \cdot 2 + (3^k - 1) = 3^{k+1} - 1$ is divisible by 2. Condition II is also satisfied. As a result, the statement "$3^n - 1$ is divisible by 2" is true for all natural numbers n. ∎

WARNING: The conclusion that a statement involving natural numbers is true for all natural numbers is made only after *both* Conditions I and II of the Principle of Mathematical Induction have been satisfied. Problem 27 demonstrates a statement for which only Condition I holds, but the statement is not true for all natural numbers. Problem 28 demonstrates a statement for which only Condition II holds, but the statement is *not* true for any natural number. ∎

9.4 EXERCISES

In Problems 1–26, use the Principle of Mathematical Induction to show that the given statement is true for all natural numbers n.

1. $2 + 4 + 6 + \cdots + 2n = n(n + 1)$

2. $1 + 5 + 9 + \cdots + (4n - 3) = n(2n - 1)$

3. $3 + 4 + 5 + \cdots + (n + 2) = \frac{1}{2}n(n + 5)$

4. $3 + 5 + 7 + \cdots + (2n + 1) = n(n + 2)$

5. $2 + 5 + 8 + \cdots + (3n - 1) = \frac{1}{2}n(3n + 1)$

6. $1 + 4 + 7 + \cdots + (3n - 2) = \frac{1}{2}n(3n - 1)$

7. $1 + 2 + 2^2 + \cdots + 2^{n-1} = 2^n - 1$

8. $1 + 3 + 3^2 + \cdots + 3^{n-1} = \frac{1}{2}(3^n - 1)$

9. $1 + 4 + 4^2 + \cdots + 4^{n-1} = \frac{1}{3}(4^n - 1)$

10. $1 + 5 + 5^2 + \cdots + 5^{n-1} = \frac{1}{4}(5^n - 1)$

11. $\dfrac{1}{1 \cdot 2} + \dfrac{1}{2 \cdot 3} + \dfrac{1}{3 \cdot 4} + \cdots + \dfrac{1}{n(n + 1)} = \dfrac{n}{n + 1}$

12. $\dfrac{1}{1 \cdot 3} + \dfrac{1}{3 \cdot 5} + \dfrac{1}{5 \cdot 7} + \cdots + \dfrac{1}{(2n - 1)(2n + 1)} = \dfrac{n}{2n + 1}$

13. $1^2 + 2^2 + 3^2 + \cdots + n^2 = \frac{1}{6}n(n + 1)(2n + 1)$

14. $1^3 + 2^3 + 3^3 + \cdots + n^3 = \frac{1}{4}n^2(n + 1)^2$

15. $4 + 3 + 2 + \cdots + (5 - n) = \frac{1}{2}n(9 - n)$

16. $-2 - 3 - 4 - \cdots - (n + 1) = -\frac{1}{2}n(n + 3)$

17. $1 \cdot 2 + 2 \cdot 3 + 3 \cdot 4 + \cdots + n(n + 1) = \frac{1}{3}n(n + 1)(n + 2)$

18. $1 \cdot 2 + 3 \cdot 4 + 5 \cdot 6 + \cdots + (2n - 1)(2n) = \frac{1}{3}n(n + 1)(4n - 1)$

19. $n^2 + n$ is divisible by 2.

20. $n^3 + 2n$ is divisible by 3.

21. $n^2 - n + 2$ is divisible by 2.

22. $n(n + 1)(n + 2)$ is divisible by 6.

23. If $x > 1$, then $x^n > 1$.

24. If $0 < x < 1$, then $0 < x^n < 1$.

25. $a - b$ is a factor of $a^n - b^n$.
    [**Hint:** $a^{k+1} - b^{k+1} = a(a^k - b^k) + b^k(a - b)$]

26. $a + b$ is a factor of $a^{2n+1} + b^{2n+1}$.

27. Show that the statement "$n^2 - n + 41$ is a prime number" is true for $n = 1$, but is not true for $n = 41$.

28. Show that the formula

$$2 + 4 + 6 + \cdots + 2n = n^2 + n + 2$$

obeys Condition II of the Principle of Mathematical Induction. That is, show that if the formula is true for some k it is also true for $k + 1$. Then show that the formula is false for $n = 1$ (or for any other choice of n).

29. Use mathematical induction to prove that if $r \neq 1$ then

$$a + ar + ar^2 + \cdots + ar^{n-1} = a\frac{1 - r^n}{1 - r}$$

30. Use mathematical induction to prove that

$$a + (a + d) + (a + 2d) + \cdots$$

$$+ \left[a + (n - 1)d\right] = na + d\frac{n(n - 1)}{2}$$

31. Extended Principle of Mathematical Induction   The Extended Principle of Mathematical Induction states that if Conditions I and II hold, that is,

(I)   A statement is true for a natural number j.

(II) If the statement is true for some natural number $k \geq j$, then it is also true for the next natural number $k + 1$.

then the statement is true for all natural numbers $\geq j$.

   Use the Extended Principle of Mathematical Induction to show that the number of diagonals in a convex polygon of n sides is $\frac{1}{2}n(n - 3)$.
[**Hint:** Begin by showing that the result is true when $n = 4$ (Condition I).]

32. Geometry   Use the Extended Principle of Mathematical Induction to show that the sum of the interior angles of a convex polygon of n sides equals $(n - 2) \cdot 180°$.

33. How would you explain to a friend the Principle of Mathematical Induction?

9.5 | THE BINOMIAL THEOREM

OBJECTIVES   **1** Evaluate a Binomial Coefficient

   2 Expand a Binomial

Formulas have been given for expanding $(x + a)^n$ for $n = 2$ and $n = 3$. The *Binomial Theorem** is a formula for the expansion of $(x + a)^n$ for any positive integer n. If $n = 1, 2, 3$, and 4, the expansion of $(x + a)^n$ is straightforward.

$$(x + a)^1 = x + a$$
Two terms, beginning with x^1 and ending with a^1

$$(x + a)^2 = x^2 + 2ax + a^2$$
Three terms, beginning with x^2 and ending with a^2

$$(x + a)^3 = x^3 + 3ax^2 + 3a^2x + a^3$$
Four terms, beginning with x^3 and ending with a^3

$$(x + a)^4 = x^4 + 4ax^3 + 6a^2x^2 + 4a^3x + a^4$$
Five terms, beginning with x^4 and ending with a^4

Notice that each expansion of $(x + a)^n$ begins with x^n and ends with a^n. As you read from left to right, the powers of x are decreasing, while the powers of a are increasing. Also, the number of terms that appears equals $n + 1$. Notice, too, that the degree of each monomial in the expansion equals n. For example, in the expansion of $(x + a)^3$, each monomial $(x^3, 3ax^2, 3a^2x, a^3)$ is of degree 3. As a result, we might conjecture that the expansion of $(x + a)^n$ would look like this:

$$(x + a)^n = x^n + \_ax^{n-1} + \_a^2x^{n-2} + \cdots + \_a^{n-1}x + a^n$$

---

*The name *binomial* is derived from the fact that $x + a$ is a binomial, that is, it contains two terms.

where the blanks are numbers to be found. This is, in fact, the case, as we shall see shortly.

First, we need to introduce a symbol.

THE SYMBOL $\dbinom{n}{j}$

① We define the symbol $\dbinom{n}{j}$, read "n taken j at a time," as follows:

If j and n are integers with $0 \leq j \leq n$, the symbol $\dbinom{n}{j}$ is defined as

$$\binom{n}{j} = \frac{n!}{j!(n-j)!} \tag{1}$$

COMMENT: On a graphing calculator, the symbol $\dbinom{n}{j}$ may be denoted by the key $\boxed{\text{nCr}}$. ∎

EXAMPLE 1

Evaluating $\dbinom{n}{j}$

Find:

(a) $\dbinom{3}{1}$ (b) $\dbinom{4}{2}$ (c) $\dbinom{8}{7}$ (d) $\dbinom{65}{15}$

Solution

(a) $\dbinom{3}{1} = \dfrac{3!}{1!(3-1)!} = \dfrac{3!}{1!2!} = \dfrac{3 \cdot 2 \cdot 1}{1(2 \cdot 1)} = \dfrac{6}{2} = 3$

(b) $\dbinom{4}{2} = \dfrac{4!}{2!(4-2)!} = \dfrac{4!}{2!2!} = \dfrac{4 \cdot 3 \cdot 2 \cdot 1}{(2 \cdot 1)(2 \cdot 1)} = \dfrac{24}{4} = 6$

(c) $\dbinom{8}{7} = \dfrac{8!}{7!(8-7)!} = \dfrac{8!}{7!1!} = \dfrac{8 \cdot 7!}{7! \cdot 1!} = \dfrac{8}{1} = 8$

$$\underset{8! \,=\, 8 \cdot 7!}{\uparrow}$$

Figure 12

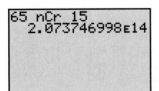
```
65 nCr 15
    2.073746998E14
```

(d) Figure 12 shows the solution using a TI-83 graphing calculator:
$$\binom{65}{15} = 2.073746998 \times 10^{14}.$$ ∎

NOW WORK PROBLEM 1.

Four useful formulas involving the symbol $\dbinom{n}{j}$ are

$$\binom{n}{0} = 1 \qquad \binom{n}{1} = n \qquad \binom{n}{n-1} = n \qquad \binom{n}{n} = 1$$

Proof

$$\binom{n}{0} = \frac{n!}{0!(n-0)!} = \frac{n!}{0!n!} = \frac{1}{1} = 1$$

$$\binom{n}{1} = \frac{n!}{1!(n-1)!} = \frac{n!}{(n-1)!} = \frac{n(n-1)!}{(n-1)!} = n$$

You are asked to show the remaining two formulas in Problem 41.   ■

Suppose that we arrange the various values of the symbol $\binom{n}{j}$ in a triangular display, as shown next and in Figure 13.

$$\binom{0}{0}$$

$$\binom{1}{0} \quad \binom{1}{1}$$

$$\binom{2}{0} \quad \binom{2}{1} \quad \binom{2}{2}$$

$$\binom{3}{0} \quad \binom{3}{1} \quad \binom{3}{2} \quad \binom{3}{3}$$

$$\binom{4}{0} \quad \binom{4}{1} \quad \binom{4}{2} \quad \binom{4}{3} \quad \binom{4}{4}$$

$$\binom{5}{0} \quad \binom{5}{1} \quad \binom{5}{2} \quad \binom{5}{3} \quad \binom{5}{4} \quad \binom{5}{5}$$

Figure 13
Pascal triangle

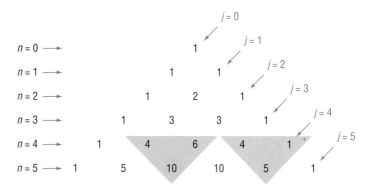

This display is called the **Pascal triangle**, named after Blaise Pascal (1623–1662), a French mathematician.

The Pascal triangle has 1's down the sides. To get any other entry, merely add the two nearest entries in the row above it. The shaded triangles in Figure 13 illustrate this feature of the Pascal triangle. Based on this feature, the row corresponding to $n = 6$ is found as follows:

$$
\begin{array}{cccccccc}
n = 5 \rightarrow & & 1 & 5 & 10 & 10 & 5 & 1 \\
n = 6 \rightarrow & 1 & 6 & 15 & 20 & 15 & 6 & 1
\end{array}
$$

Later we shall prove that this addition always works (see the theorem on page 652).

Although the Pascal triangle provides an interesting and organized display of the symbol $\binom{n}{j}$, in practice it is not all that helpful. For example, if you wanted to know the value of $\binom{12}{5}$, you would need to produce 12 rows of the triangle before seeing the answer. It is much faster to use the definition (1).

BINOMIAL THEOREM

(2) Now we are ready to state the **Binomial Theorem**. A proof is given at the end of this section.

Theorem

Binomial Theorem

Let x and a be real numbers. For any positive integer n, we have

$$(x + a)^n = \binom{n}{0}x^n + \binom{n}{1}ax^{n-1} + \cdots + \binom{n}{j}a^j x^{n-j} + \cdots + \binom{n}{n}a^n$$

$$= \sum_{j=0}^{n} \binom{n}{j} x^{n-j} a^j \qquad \text{(2)}$$

Now you know why we needed to introduce the symbol $\binom{n}{j}$; these symbols are the numerical coefficients that appear in the expansion of $(x + a)^n$. Because of this, the symbol $\binom{n}{j}$ is called the **binomial coefficient.**

EXAMPLE 2

Expanding a Binomial

Use the Binomial Theorem to expand $(x + 2)^5$.

Solution   In the Binomial Theorem, let $a = 2$ and $n = 5$. Then

$$(x + 2)^5 = \binom{5}{0}x^5 + \binom{5}{1}2x^4 + \binom{5}{2}2^2 x^3 + \binom{5}{3}2^3 x^2 + \binom{5}{4}2^4 x + \binom{5}{5}2^5$$

Use equation (2).

$$= 1 \cdot x^5 + 5 \cdot 2x^4 + 10 \cdot 4x^3 + 10 \cdot 8x^2 + 5 \cdot 16x + 1 \cdot 32$$

Use row $n = 5$ of the Pascal triangle or formula (1) for $\binom{n}{j}$.

$$= x^5 + 10x^4 + 40x^3 + 80x^2 + 80x + 32$$

EXAMPLE 3

Expanding a Binomial

Expand $(2y - 3)^4$ using the Binomial Theorem.

Solution   First, we rewrite the expression $(2y - 3)^4$ as $[2y + (-3)]^4$. Now we use the Binomial Theorem with $n = 4$, $x = 2y$, and $a = -3$.

$$[2y + (-3)]^4 = \binom{4}{0}(2y)^4 + \binom{4}{1}(-3)(2y)^3 + \binom{4}{2}(-3)^2(2y)^2$$

$$+ \binom{4}{3}(-3)^3(2y) + \binom{4}{4}(-3)^4$$

$$= 1 \cdot 16y^4 + 4(-3)8y^3 + 6 \cdot 9 \cdot 4y^2 + 4(-27)2y + 1 \cdot 81$$
↑
Use row $n = 4$ of the Pascal triangle or formula (1) for $\binom{n}{j}$.

$$= 16y^4 - 96y^3 + 216y^2 - 216y + 81$$

In this expansion, note that the signs alternate due to the fact that $a = -3 < 0$. ■

NOW WORK PROBLEM **17**.

EXAMPLE 4

Finding a Particular Coefficient in a Binomial Expansion

Find the coefficient of y^8 in the expansion of $(2y + 3)^{10}$.

Solution   We write out the expansion using the Binomial Theorem.

$$(2y + 3)^{10} = \binom{10}{0}(2y)^{10} + \binom{10}{1}(2y)^9(3)^1 + \binom{10}{2}(2y)^8(3)^2 + \binom{10}{3}(2y)^7(3)^3$$

$$+ \binom{10}{4}(2y)^6(3)^4 + \cdots + \binom{10}{9}(2y)(3)^9 + \binom{10}{10}(3)^{10}$$

From the third term in the expansion, the coefficient of y^8 is

$$\binom{10}{2}(2)^8(3)^2 = \frac{10!}{2!8!} \cdot 2^8 \cdot 9 = \frac{10 \cdot 9 \cdot 8!}{2 \cdot 8!} \cdot 2^8 \cdot 9 = 103,680 \quad ■$$

As this solution demonstrates, we can use the Binomial Theorem to find a particular term in an expansion without writing the entire expansion.

Based on the expansion of $(x + a)^n$, the term containing x^j is

$$\binom{n}{n - j}a^{n-j}x^j \tag{3}$$

For example, we can solve Example 4 by using formula (3) with $n = 10$, $a = 3$, $x = 2y$, and $j = 8$. Then the term containing y^8 is

$$\binom{10}{10 - 8}3^{10-8}(2y)^8 = \binom{10}{2} \cdot 3^2 \cdot 2^8 \cdot y^8 = \frac{10!}{2!8!} \cdot 9 \cdot 2^8 y^8$$

$$= \frac{10 \cdot 9 \cdot 8!}{2 \cdot 8!} \cdot 9 \cdot 2^8 y^8 = 103,680 y^8$$

| EXAMPLE 5 | **Finding a Particular Term in a Binomial Expansion** |

Find the sixth term in the expansion of $(x + 2)^9$.

Solution A   We expand using the Binomial Theorem until the sixth term is reached.

$$(x + 2)^9 = \binom{9}{0}x^9 + \binom{9}{1}x^8 \cdot 2 + \binom{9}{2}x^7 \cdot 2^2 + \binom{9}{3}x^6 \cdot 2^3 + \binom{9}{4}x^5 \cdot 2^4$$

$$+ \binom{9}{5}x^4 \cdot 2^5 + \cdots$$

The sixth term is

$$\binom{9}{5}x^4 \cdot 2^5 = \frac{9!}{5!4!} \cdot x^4 \cdot 32 = 4032x^4$$

Solution B   The sixth term in the expansion of $(x + 2)^9$, which has 10 terms total, contains x^4. (Do you see why?) By formula (3), the sixth term is

$$\binom{9}{9 - 4}2^{9-4}x^4 = \binom{9}{5}2^5x^4 = \frac{9!}{5!4!} \cdot 32x^4 = 4032x^4 \quad ∎$$

NOW WORK PROBLEMS **25** AND **31**.

Next we show that the *triangular addition* feature of the Pascal triangle illustrated in Figure 13 always works.

Theorem   If n and j are integers with $1 \le j \le n$, then

$$\boxed{\binom{n}{j - 1} + \binom{n}{j} = \binom{n + 1}{j}} \tag{4}$$

∎

Proof

$$\binom{n}{j - 1} + \binom{n}{j} = \frac{n!}{(j - 1)![n - (j - 1)]!} + \frac{n!}{j!(n - j)!}$$

$$= \frac{n!}{(j - 1)!(n - j + 1)!} + \frac{n!}{j!(n - j)!}$$

$$= \frac{jn!}{j(j - 1)!(n - j + 1)!} + \frac{(n - j + 1)n!}{j!(n - j + 1)(n - j)!} \qquad \text{Multiply the first term by } j/j \text{ and the second term by } (n - j + 1)/(n - j + 1).$$

$$= \frac{jn!}{j!(n - j + 1)!} + \frac{(n - j + 1)n!}{j!(n - j + 1)!} \qquad \text{Now the denominators are equal.}$$

$$= \frac{jn! + (n - j + 1)n!}{j!(n - j + 1)!}$$

$$= \frac{n!(j + n - j + 1)}{j!(n - j + 1)!}$$

$$= \frac{n!(n + 1)}{j!(n - j + 1)!} = \frac{(n + 1)!}{j![(n + 1) - j]!} = \binom{n + 1}{j} \qquad ∎$$

Proof of the Binomial Theorem　We use Mathematical Induction to prove the Binomial Theorem. First, we show that formula (2) is true for $n = 1$.

$$(x + a)^1 = x + a = \binom{1}{0}x^1 + \binom{1}{1}a^1$$

Next we suppose that formula (2) is true for some k. That is, we assume that

$$(x + a)^k = \binom{k}{0}x^k + \binom{k}{1}ax^{k-1} + \cdots + \binom{k}{j-1}a^{j-1}x^{k-j+1} + \binom{k}{j}a^j x^{k-j} + \cdots + \binom{k}{k}a^k \quad \textbf{(5)}$$

Now we calculate $(x + a)^{k+1}$.

$$(x + a)^{k+1} = (x + a)(x + a)^k = x(x + a)^k + a(x + a)^k$$

Use equation (5) →

$$= x\left[\binom{k}{0}x^k + \binom{k}{1}ax^{k-1} + \cdots + \binom{k}{j-1}a^{j-1}x^{k-j+1} + \binom{k}{j}a^j x^{k-j} + \cdots + \binom{k}{k}a^k\right]$$

$$+ a\left[\binom{k}{0}x^k + \binom{k}{1}ax^{k-1} + \cdots + \binom{k}{j-1}a^{j-1}x^{k-j+1} + \binom{k}{j}a^j x^{k-j} + \cdots + \binom{k}{k-1}a^{k-1}x + \binom{k}{k}a^k\right]$$

$$= \binom{k}{0}x^{k+1} + \binom{k}{1}ax^k + \cdots + \binom{k}{j-1}a^{j-1}x^{k-j+2} + \binom{k}{j}a^j x^{k-j+1} + \cdots + \binom{k}{k}a^k x$$

$$+ \binom{k}{0}ax^k + \binom{k}{1}a^2 x^{k-1} + \cdots + \binom{k}{j-1}a^j x^{k-j+1} + \binom{k}{j}a^{j+1}x^{k-j} + \cdots + \binom{k}{k-1}a^k x + \binom{k}{k}a^{k+1}$$

$$= \binom{k}{0}x^{k+1} + \left[\binom{k}{1} + \binom{k}{0}\right]ax^k + \cdots + \left[\binom{k}{j} + \binom{k}{j-1}\right]a^j x^{k-j+1}$$

$$+ \cdots + \left[\binom{k}{k} + \binom{k}{k-1}\right]a^k x + \binom{k}{k}a^{k+1}$$

Because

$$\binom{k}{0} = 1 = \binom{k+1}{0}, \qquad \underset{\underset{(4)}{\uparrow}}{\binom{k}{1} + \binom{k}{0} = \binom{k+1}{1}}, \cdots,$$

$$\underset{\underset{(4)}{\uparrow}}{\binom{k}{j} + \binom{k}{j-1} = \binom{k+1}{j}}, \cdots, \binom{k}{k} = 1 = \binom{k+1}{k+1}$$

we have

$$(x + a)^{k+1} = \binom{k+1}{0}x^{k+1} + \binom{k+1}{1}ax^k + \cdots + \binom{k+1}{j}a^j x^{k-j+1} + \cdots + \binom{k+1}{k+1}a^{k+1}$$

Conditions I and II of the Principle of Mathematical Induction are satisfied, so formula (2) is true for all n. ∎

HISTORICAL FEATURE

Omar Khayyám (1050–1123)

The case $n = 2$ of the Binomial Theorem, $(a + b)^2$, was known to Euclid in 300 BC, but the general law seems to have been discovered by the Persian mathematician and astronomer Omar Khayyám (1050–1123), who is also well known as the author of the *Rubáiyát*, a collection of four-line poems making observations on the human condition. Omar Khayyám did not state the Binomial Theorem explicitly, but he claimed to have a method for extracting third, fourth, fifth roots, and so on. A little study shows that one must know the Binomial Theorem to create such a method.

The heart of the Binomial Theorem is the formula for the numerical coefficients, and, as we saw, they can be written out in a symmetric triangular form. The Pascal triangle appears first in the books of Yang Hui (about 1270) and Chu Shih-chieh (1303). Pascal's name is attached to the triangle because of the many applications he made of it, especially to counting and probability. In establishing these results, he was one of the earliest users of mathematical induction.

Many people worked on the proof of the Binomial Theorem, which was finally completed for all n (including complex numbers) by Niels Abel (1802–1829).

9.5 EXERCISES

In Problems 1–12, evaluate each expression.

1. $\dbinom{5}{3}$

2. $\dbinom{7}{3}$

3. $\dbinom{7}{5}$

4. $\dbinom{9}{7}$

5. $\dbinom{50}{49}$

6. $\dbinom{100}{98}$

7. $\dbinom{1000}{1000}$

8. $\dbinom{1000}{0}$

9. $\dbinom{55}{23}$

10. $\dbinom{60}{20}$

11. $\dbinom{47}{25}$

12. $\dbinom{37}{19}$

In Problems 13–24, expand each expression using the Binomial Theorem.

13. $(x + 1)^5$

14. $(x - 1)^5$

15. $(x - 2)^6$

16. $(x + 3)^5$

17. $(3x + 1)^4$

18. $(2x + 3)^5$

19. $(x^2 + y^2)^5$

20. $(x^2 - y^2)^6$

21. $(\sqrt{x} + \sqrt{2})^6$

22. $(\sqrt{x} - \sqrt{3})^4$

23. $(ax + by)^5$

24. $(ax - by)^4$

In Problems 25–38, use the Binomial Theorem to find the indicated coefficient or term.

25. The coefficient of x^6 in the expansion of $(x + 3)^{10}$

26. The coefficient of x^3 in the expansion of $(x - 3)^{10}$

27. The coefficient of x^7 in the expansion of $(2x - 1)^{12}$

28. The coefficient of x^3 in the expansion of $(2x + 1)^{12}$

29. The coefficient of x^7 in the expansion of $(2x + 3)^9$

30. The coefficient of x^2 in the expansion of $(2x - 3)^9$

31. The fifth term in the expansion of $(x + 3)^7$

32. The third term in the expansion of $(x - 3)^7$

33. The third term in the expansion of $(3x - 2)^9$

34. The sixth term in the expansion of $(3x + 2)^8$

35. The coefficient of x^0 in the expansion of $\left(x^2 + \dfrac{1}{x}\right)^{12}$

36. The coefficient of x^0 in the expansion of $\left(x - \dfrac{1}{x^2}\right)^9$

37. The coefficient of x^4 in the expansion of $\left(x - \dfrac{2}{\sqrt{x}}\right)^{10}$

38. The coefficient of x^2 in the expansion of $\left(\sqrt{x} + \dfrac{3}{\sqrt{x}}\right)^8$

39. Use the Binomial Theorem to find the numerical value of $(1.001)^5$ correct to five decimal places.
    [**Hint:** $(1.001)^5 = (1 + 10^{-3})^5$]

40. Use the Binomial Theorem to find the numerical value of $(0.998)^6$ correct to five decimal places.

41. Show that $\dbinom{n}{n-1} = n$ and $\dbinom{n}{n} = 1$.

42. Show that if n and j are integers with $0 \le j \le n$ then

$$\binom{n}{j} = \binom{n}{n-j}$$

Conclude that the Pascal triangle is symmetric with respect to a vertical line drawn from the topmost entry.

43. If n is a positive integer, show that

$$\binom{n}{0} + \binom{n}{1} + \cdots + \binom{n}{n} = 2^n$$

[**Hint:** $2^n = (1 + 1)^n$; now use the Binomial Theorem.]

44. If n is a positive integer, show that

$$\binom{n}{0} - \binom{n}{1} + \binom{n}{2} - \cdots + (-1)^n \binom{n}{n} = 0$$

45. $\dbinom{5}{0}\left(\dfrac{1}{4}\right)^5 + \dbinom{5}{1}\left(\dfrac{1}{4}\right)^4\left(\dfrac{3}{4}\right) + \dbinom{5}{2}\left(\dfrac{1}{4}\right)^3\left(\dfrac{3}{4}\right)^2$

$+ \dbinom{5}{3}\left(\dfrac{1}{4}\right)^2\left(\dfrac{3}{4}\right)^3 + \dbinom{5}{4}\left(\dfrac{1}{4}\right)\left(\dfrac{3}{4}\right)^4 + \dbinom{5}{5}\left(\dfrac{3}{4}\right)^5 = ?$

46. Stirling's formula An approximation for $n!$, when n is large, is given by

$$n! \approx \sqrt{2n\pi}\left(\frac{n}{e}\right)^n\left(1 + \frac{1}{12n-1}\right)$$

Calculate 12!, 20!, and 25! on your calculator. Then use Stirling's formula to approximate 12!, 20!, and 25!.

C H A P T E R R E V I E W

Things To Know

Sequence (p. 616)	A function whose domain is the set of positive integers.
Factorials (p. 619)	$0! = 1, 1! = 1, n! = n(n-1) \cdot \ldots \cdot 3 \cdot 2 \cdot 1$ if $n \ge 2$
Arithmetic sequence (pp. 626 and 627)	$a_1 = a, a_n = a_{n-1} + d$, where $a =$ first term, $d =$ common difference, $a_n = a + (n-1)d$
Sum of the first n terms of an arithmetic sequence (p. 629)	$S_n = \dfrac{n}{2}[2a + (n-1)d] = \dfrac{n}{2}(a + a_n)$
Geometric sequence (pp. 632 and 633)	$a_1 = a, \quad a_n = ra_{n-1}$, where $a =$ first term, $r =$ common ratio, $a_n = ar^{n-1}, \quad r \ne 0$
Sum of the first n terms of a geometric sequence (p. 634)	$S_n = a\dfrac{1 - r^n}{1 - r}, \quad r \ne 0, 1$
Infinite geometric series (p. 636)	$a + ar + \cdots + ar^{n-1} + \cdots = \displaystyle\sum_{k=1}^{\infty} ar^{k-1}$
Sum of an infinite geometric series (p. 636)	$\displaystyle\sum_{k=1}^{\infty} ar^{k-1} = \dfrac{a}{1-r}, \quad |r| < 1$
Amount of an annuity (p. 639)	$A = P\dfrac{(1+i)^n - 1}{i}$
Principle of Mathematical Induction (p. 644)	Suppose the following two conditions are satisfied. Condition I: The statement is true for the natural number 1. Condition II: If the statement is true for some natural number k, it is also true for $k + 1$. Then the statement is true for all natural numbers n.
Binomial coefficient (p. 648)	$\dbinom{n}{j} = \dfrac{n!}{j!(n-j)!}$
Pascal triangle (p. 649)	See Figure 13.
Binomial Theorem (p. 650)	$(x+a)^n = \dbinom{n}{0}x^n + \dbinom{n}{1}ax^{n-1} + \cdots + \dbinom{n}{j}a^j x^{n-j} + \cdots + \dbinom{n}{n}a^n$

Objectives

You should be able to:

Write the first several terms of a sequence (p. 617)

Write the terms of a sequence defined by a recursive formula (p. 620)

Use summation notation (p. 621)

Find the sum of a sequence (p. 622)

Determine if a sequence is arithmetic (p. 626)

Find a formula for an arithmetic sequence (p. 627)

Find the sum of an arithmetic sequence (p. 628)

Determine if a sequence is geometric (p. 632)

Find a formula for a geometric sequence (p. 633)

Find the sum of a geometric sequence (p. 634)

Find the sum of a geometric series (p. 636)

Solve annuity problems (p. 638)

Prove statements using mathematical induction (p. 643)

Evaluate a binomial coefficient (p. 648)

Expand a binomial (p. 650)

Fill-in-the-Blank Items

1. A(n) _____ is a function whose domain is the set of positive integers.

2. In a(n) _____ sequence, the difference between successive terms is always the same number.

3. In a(n) _____ sequence, the ratio of successive terms is always the same number.

4. The _____ _____ is a triangular display of the binomial coefficients.

5. $\binom{6}{2} =$ _____.

True/False Items

T F   **1.** A sequence is a function.

T F   **2.** For arithmetic sequences, the difference of successive terms is always the same number.

T F   **3.** For geometric sequences, the ratio of successive terms is always the same number.

T F   **4.** Mathematical induction can sometimes be used to prove theorems that involve natural numbers.

T F   **5.** $\binom{n}{j} = \dfrac{j!}{n!(n-j)!}$

T F   **6.** The expansion of $(x + a)^n$ contains n terms.

T F   **7.** $\sum\limits_{i=1}^{n+1} i = 1 + 2 + 3 + \cdots + n$

Review Exercises

Blue problem numbers indicate the author's suggestions for use in a Practice Test.

In Problems 1–8, write down the first five terms of each sequence.

1. $\left\{(-1)^n\left(\dfrac{n+3}{n+2}\right)\right\}$ **2.** $\{(-1)^{n+1}(2n+3)\}$ **3.** $\left\{\dfrac{2^n}{n^2}\right\}$ **4.** $\left\{\dfrac{e^n}{n}\right\}$

5. $a_1 = 3;\ a_n = \frac{2}{3}a_{n-1}$ **6.** $a_1 = 4;\ a_n = -\frac{1}{4}a_{n-1}$ **7.** $a_1 = 2;\ a_n = 2 - a_{n-1}$ **8.** $a_1 = -3;\ a_n = 4 + a_{n-1}$

In Problems 9–20, determine whether the given sequence is arithmetic, geometric, or neither. If the sequence is arithmetic, find the common difference and the sum of the first n terms. If the sequence is geometric, find the common ratio and the sum of the first n terms.

9. $\{n + 5\}$

10. $\{4n + 3\}$

11. $\{2n^3\}$

12. $\{2n^2 - 1\}$

13. $\{2^{3n}\}$

14. $\{3^{2n}\}$

15. $0, 4, 8, 12, \ldots$

16. $1, -3, -7, -11, \ldots$

17. $3, \frac{3}{2}, \frac{3}{4}, \frac{3}{8}, \frac{3}{16}, \ldots$

18. $5, -\frac{5}{3}, \frac{5}{9}, -\frac{5}{27}, \frac{5}{81}, \ldots$

19. $\frac{2}{3}, \frac{3}{4}, \frac{4}{5}, \frac{5}{6}, \ldots$

20. $\frac{3}{2}, \frac{5}{4}, \frac{7}{6}, \frac{9}{8}, \frac{11}{10}, \ldots$

In Problems 21–26, evaluate each sum.

21. $\sum_{k=1}^{5} (k^2 + 12)$

22. $\sum_{k=1}^{3} (k + 2)^2$

23. $\sum_{k=1}^{10} (3k - 9)$

24. $\sum_{k=1}^{9} (-2k + 8)$

25. $\sum_{k=1}^{7} \left(\frac{1}{3}\right)^k$

26. $\sum_{k=1}^{10} (-2)^k$

In Problems 27–32, find the indicated term in each sequence.

27. 9th term of $3, 7, 11, 15, \ldots$

28. 8th term of $1, -1, -3, -5, \ldots$

29. 11th term of $1, \frac{1}{10}, \frac{1}{100}, \ldots$

30. 11th term of $1, 2, 4, 8, \ldots$

31. 9th term of $\sqrt{2}, 2\sqrt{2}, 3\sqrt{2}, \ldots$

32. 9th term of $\sqrt{2}, 2, 2^{3/2}, \ldots$

In Problems 33–36, find a general formula for each arithmetic sequence.

33. 7th term is 31;   20th term is 96

34. 8th term is -20;   17th term is -47

35. 10th term is 0;   18th term is 8

36. 12th term is 30;   22nd term is 50

In Problems 37–42, find the sum of each infinite geometric series.

37. $3 + 1 + \frac{1}{3} + \frac{1}{9} + \cdots$

38. $2 + 1 + \frac{1}{2} + \frac{1}{4} + \cdots$

39. $2 - 1 + \frac{1}{2} - \frac{1}{4} + \cdots$

40. $6 - 4 + \frac{8}{3} - \frac{16}{9} + \cdots$

41. $\sum_{k=1}^{\infty} 4\left(\frac{1}{2}\right)^{k-1}$

42. $\sum_{k=1}^{\infty} 3\left(-\frac{3}{4}\right)^{k-1}$

In Problems 43–48, use the Principle of Mathematical Induction to show that the given statement is true for all natural numbers.

43. $3 + 6 + 9 + \cdots + 3n = \dfrac{3n}{2}(n + 1)$

44. $2 + 6 + 10 + \cdots + (4n - 2) = 2n^2$

45. $2 + 6 + 18 + \cdots + 2 \cdot 3^{n-1} = 3^n - 1$

46. $3 + 6 + 12 + \cdots + 3 \cdot 2^{n-1} = 3(2^n - 1)$

47. $1^2 + 4^2 + 7^2 + \cdots + (3n - 2)^2 = \frac{1}{2}n(6n^2 - 3n - 1)$

48. $1 \cdot 3 + 2 \cdot 4 + 3 \cdot 5 + \cdots + n(n + 2) = \dfrac{n}{6}(n + 1)(2n + 7)$

In Problems 49–52, expand each expression using the Binomial Theorem.

49. $(x + 2)^5$

50. $(x - 3)^4$

51. $(2x + 3)^5$

52. $(3x - 4)^4$

---

53. Find the coefficient of x^7 in the expansion of $(x + 2)^9$.

54. Find the coefficient of x^3 in the expansion of $(x - 3)^8$.

55. Find the coefficient of x^2 in the expansion of $(2x + 1)^7$.

56. Find the coefficient of x^6 in the expansion of $(2x + 1)^8$.

57. Constructing a Brick Staircase   A brick staircase has a total of 25 steps. The bottom step requires 80 bricks. Each successive step requires three less bricks than the prior step.
   (a) How many bricks are required for the top step?
   (b) How many bricks are required to build the staircase?

58. Creating a Floor Design   A mosaic tile floor is designed in the shape of a trapezoid 30 feet wide at the base and 15 feet wide at the top. The tiles, 12 inches by 12 inches, are to be placed so that each successive row contains one less tile than the row below. How many tiles will be required? [**Hint:** Refer to Figure 8 on page 630]

59. Retirement Planning   Chris gets paid once a month and contributes $200 each pay period into his 401(k). If Chris plans on retiring in 20 years, what will be the value of his 401(k) if the per annum rate of return of the 401(k) is 10% compounded monthly?

60. Retirement Planning Jacky contributes $500 every quarter to an IRA. If Jacky plans on retiring in 30 years, what will be the value of the IRA if the per annum rate of return of the IRA is 8% compounded quarterly?

61. Bouncing Balls A ball is dropped from a height of 20 feet. Each time it strikes the ground, it bounces up to three-quarters of the previous height.
   (a) What height will the ball bounce up to after it strikes the ground for the third time?
   (b) How high will it bounce after it strikes the ground for the *n*th time?
   (c) How many times does the ball need to strike the ground before its bounce is less than 6 inches?
   (d) What total distance does the ball travel before it stops bouncing?

62. Salary Increases Your friend has just been hired at an annual salary of $20,000. If she expects to receive annual increases of 4%, what will be her salary as she begins her fifth year?

Project at Motorola

Digital wireless communications

Motorola uses digital communication technology in the design of cellular phones. In digital communications, a speech or data signal is converted into bits by a mobile phone and transmitted over the air to a base station. A bit is a unit of information that can be either 0 or 1. Thus, a single bit can represent two levels of information. A speech signal is converted into a digital signal by taking samples of the speech signal at specified time intervals, and then quantizing these samples. A *sample* is simply the value of the speech signal at a single time instant. To *quantize* the sample, a number of acceptable levels is defined, and the level that is closest to the value of the speech signal at a sampling instant is assigned to that sample. A symbol consisting of multiple bits is assigned to each level, and these resulting symbols are transmitted over the air.

The number of levels is determined by the length of the symbols. For instance, a symbol of length one bit can only represent two levels (0 and 1). A symbol of length two bits represents 4 levels (00, 01, 10, 11).

1. A code has two quantization levels when the symbol length is one, and four levels when the symbol length is two. Thus, if the symbol lengths are $1, 2, 3, 4, \ldots$, then the number of quantization levels of the corresponding codes form a sequence where the first two terms are 2 and 4. Write the next four terms of this sequence.

2. Write an expression for this sequence, where the number of bits per symbol (i.e., the symbol length) is *n*. What type of sequence is this? Write a recursive expression for the sequence.

3. How many bits per symbol are needed to have 256 quantization levels?

Counting and Probability

Field Trip to Motorola

How would you like to design products that allow you to talk on a cellular phone from anywhere in the world or surf the web while riding in a car? Motorola designs and manufactures products that do these things and more. Probability of error is a performance measure that is important to the design of wireless communication systems and products. Many parts of a wireless system are designed to decrease the probability of error. This is what allows you to hear someone clearly on your cellular phone or surf the web more quickly. In the chapter project, you will learn how speech signals are coded for transmission and error control.

10.1 SETS AND COUNTING

OBJECTIVES
1. Find All the Subsets of a Set
2. Find the Intersection and Union of Sets
3. Find the Complement of a Set
4. Count the Number of Elements in a Set

SETS

A **set** is a well-defined collection of distinct objects. The objects of a set are called its **elements.** By **well-defined,** we mean that there is a rule that enables us to determine whether a given object is an element of the set. If a set has no elements, it is called the **empty set,** or **null set,** and is denoted by the symbol $\emptyset$.

Because the elements of a set are distinct, we never repeat elements. For example, we would never write $\{1, 2, 3, 2\}$; the correct listing is $\{1, 2, 3\}$. Because a set is a collection, the order in which the elements are listed is immaterial. $\{1, 2, 3\}$, $\{1, 3, 2\}$, $\{2, 1, 3\}$, and so on, all represent the same set.

EXAMPLE 1 | **Writing the Elements of a Set**

Write the set consisting of the possible results (outcomes) from tossing a coin twice. Use H for *heads* and T for *tails*.

Solution In tossing a coin twice, we can get heads each time, HH; or heads the first time and tails the second, HT; or tails the first time and heads the second, TH; or tails each time, TT. Because no other possibilities exist, the set of outcomes is

$$\{HH, HT, TH, TT\}$$ ∎

 If two sets A and B have precisely the same elements, we say that A and B are **equal** and write $A = B$.

If each element of a set A is also an element of a set B, we say that A is a **subset** of B and write $A \subseteq B$.

If $A \subseteq B$ and $A \neq B$, then we say that A is a **proper subset** of B and write $A \subset B$.

If $A \subseteq B$, every element in set A is also in set B, but B may or may not have additional elements. If $A \subset B$, every element in A is also in B, and B has at least one element not found in A.

Finally, we agree that the empty set is a subset of every set; that is,

$$\varnothing \subseteq A, \qquad \text{for any set } A$$

EXAMPLE 2

Finding All the Subsets of a Set

Write down all the subsets of the set $\{a, b, c\}$.

Solution To organize our work, we write down all the subsets with no elements, then those with one element, then those with two elements, and finally those with three elements. These will give us all the subsets. Do you see why?

0 Elements	**1 Element**	**2 Elements**	**3 Elements**
$\varnothing$	$\{a\}, \{b\}, \{c\}$	$\{a, b\}, \{b, c\}, \{a, c\}$	$\{a, b, c\}$

➤ NOW WORK PROBLEM **21**.

2 If A and B are sets, the **intersection** of A with B, denoted $A \cap B$, is the set consisting of elements that belong to both A and B. The **union** of A with B, denoted $A \cup B$, is the set consisting of elements that belong to either A or B, or both.

EXAMPLE 3

Finding the Intersection and Union of Sets

Let $A = \{1, 3, 5, 8\}$, $B = \{3, 5, 7\}$, and $C = \{2, 4, 6, 8\}$. Find:

(a) $A \cap B$ (b) $A \cup B$ (c) $B \cap (A \cup C)$

Solution (a) $A \cap B = \{1, 3, 5, 8\} \cap \{3, 5, 7\} = \{3, 5\}$
(b) $A \cup B = \{1, 3, 5, 8\} \cup \{3, 5, 7\} = \{1, 3, 5, 7, 8\}$
(c) $B \cap (A \cup C) = \{3, 5, 7\} \cap \big[\{1, 3, 5, 8\} \cup \{2, 4, 6, 8\} \big]$
$\qquad\qquad\qquad = \{3, 5, 7\} \cap \{1, 2, 3, 4, 5, 6, 8\} = \{3, 5\}$

➤ NOW WORK PROBLEM **5**.

3 Usually, in working with sets, we designate a **universal set** U, the set consisting of all the elements that we wish to consider. Once a universal set has been designated, we can consider elements of the universal set not found in a given set.

If A is a set, the **complement** of A, denoted $\overline{A}$, is the set consisting of all the elements in the universal set that are not in A.

NOTE: Some books use the notation A' for the complement of A.

EXAMPLE 4

Finding the Complement of a Set

If the universal set is $U = \{1, 2, 3, 4, 5, 6, 7, 8, 9\}$, and if $A = \{1, 3, 5, 7, 9\}$, then $\overline{A} = \{2, 4, 6, 8\}$.

It follows that $A \cup \overline{A} = U$ and $A \cap \overline{A} = \varnothing$. Do you see why?

NOW WORK PROBLEM 13.

It is often helpful to draw pictures of sets. Such pictures, called **Venn diagrams,** represent sets as circles enclosed in a rectangle, which represents the universal set. Such diagrams often help us to visualize various relationships among sets. See Figure 1.

If we know that $A \subset B$, we might use the Venn diagram in Figure 2(a). If we know that A and B have no elements in common, that is, if $A \cap B = \varnothing$, we might use the Venn diagram in Figure 2(b). The sets A and B in Figure 2(b) are said to be **disjoint.**

Figure 1

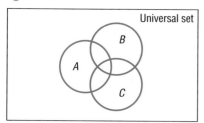

Figure 2

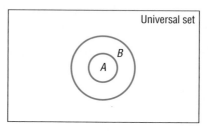

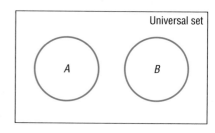

(a) $A \subset B$
proper subset

(b) $A \cap B = \varnothing$
disjoint sets

Figures 3(a), 3(b), and 3(c) use Venn diagrams to illustrate the definitions of intersection, union, and complement, respectively.

Figure 3

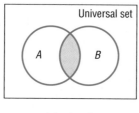

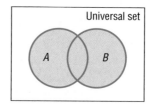

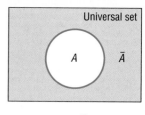

(a) $A \cap B$
intersection

(b) $A \cup B$
union

(c) $\overline{A}$
complement

COUNTING

4

As you count the number of students in a classroom or the number of pennies in your pocket, what you are really doing is matching, on a one-to-one basis, each object to be counted with the set of counting numbers $1, 2, 3, \ldots, n$, for some number n. If a set A matched up in this fashion with the set $\{1, 2, \ldots, 25\}$, you would conclude that there are 25 elements in the set A. We use the notation $n(A) = 25$ to indicate that there are 25 elements in the set A.

Because the empty set has no elements, we write

$$n(\varnothing) = 0$$

If the number of elements in a set is a nonnegative integer, we say that the set is **finite.** Otherwise, it is **infinite.** We shall concern ourselves only with finite sets.

Look again at Example 2. A set with 3 elements has $2^3 = 8$ subsets. This result can be generalized.

> If A is a set with n elements, then A has 2^n subsets.

For example, the set $\{a, b, c, d, e\}$ has $2^5 = 32$ subsets.

EXAMPLE 5   **Analyzing Survey Data**

In a survey of 100 college students, 35 were registered in College Algebra, 52 were registered in Computer Science I, and 18 were registered in both courses.

(a) How many students were registered in College Algebra or Computer Science I?

(b) How many were registered in neither course?

Solution   (a) First, let A = set of students in College Algebra

B = set of students in Computer Science I

Then the given information tells us that

$$n(A) = 35 \qquad n(B) = 52 \qquad n(A \cap B) = 18$$

Refer to Figure 4. Since $n(A \cap B) = 18$, we know that the common part of the circles representing set A and set B has 18 elements. In addition, we know that the remaining portion of the circle representing set A will have $35 - 18 = 17$ elements. Similarly, we know that the remaining portion of the circle representing set B has $52 - 18 = 34$ elements. We conclude that $17 + 18 + 34 = 69$ students were registered in College Algebra or Computer Science I.

(b) Since 100 students were surveyed, it follows that $100 - 69 = 31$ were registered in neither course. ∎

Figure 4

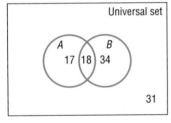

NOW WORK PROBLEM 35.

The solution to Example 5 contains the basis for a general counting formula. If we count the elements in each of two sets A and B, we necessarily count twice any elements that are in both A and B, that is, those elements in $A \cap B$. To count correctly the elements that are in A or B, that is, to find $n(A \cup B)$, we need to subtract those in $A \cap B$ from $n(A) + n(B)$.

Theorem   **Counting Formula**

If A and B are finite sets, then

$$n(A \cup B) = n(A) + n(B) - n(A \cap B) \qquad \textbf{(1)}$$

∎

Refer back to Example 5. Using (1), we have

$$n(A \cup B) = n(A) + n(B) - n(A \cap B)$$
$$= 35 + 52 - 18$$
$$= 69$$

There are 69 students registered in College Algebra or Computer Science I.

A special case of the counting formula (1) occurs if A and B have no elements in common. In this case, $A \cap B = \varnothing$, so $n(A \cap B) = 0$.

Theorem

Addition Principle of Counting

If two sets A and B have no elements in common, that is

$$\text{if } A \cap B = \varnothing, \text{ then } n(A \cup B) = n(A) + n(B) \qquad \textbf{(2)}$$

We can generalize formula (2).

Theorem

General Addition Principle of Counting

If, for n sets $A_1, A_2, \ldots, A_n$, no two have elements in common, then

$$n(A_1 \cup A_2 \cup \cdots \cup A_n) = n(A_1) + n(A_2) + \cdots + n(A_n) \qquad \textbf{(3)}$$

EXAMPLE 6

Counting

In 1996 there were 738,028 full-time sworn law-enforcement officers in the United States. Table 1 lists the type of law-enforcement agencies and the corresponding number of full-time sworn officers from each agency. No officer is classified in more than one type of agency.

TABLE 1	
Type of Agency	**Number of Full-Time Sworn Officers**
Local police	410,956
Sheriff	152,922
State police	54,587
Special police	43,082
Texas constable	1,988
Federal	74,493

Source: Bureau of Justice Statistics

(a) How many full-time sworn law-enforcement officers in the United States were local police or sheriffs?

(b) How many full-time sworn law-enforcement officers in the United States were local police, sheriffs, or state police?

Solution Let A represent the set of local police, B represent the set of sheriffs, and C represent the set of state police. No two of the sets A, B, and C have elements in common since a single officer cannot be classified in more than one type of agency.

(a) Using formula (2), we have

$$n(A \cup B) = n(A) + n(B) = 410{,}956 + 152{,}922 = 563{,}878$$

There were 563,878 officers that were local police or sheriffs.

(b) Using formula (3), we have

$$n(A \cup B \cup C) = n(A) + n(B) + n(C) = 410{,}956 + 152{,}922 + 54{,}587 = 618{,}465$$

There were 618,465 officers that were local police, sheriffs, or state police. ■

NOW WORK PROBLEM **39**.

10.1 EXERCISES

In Problems 1–10, use A = {1, 3, 5, 7, 9}, B = {1, 5, 6, 7}, and C = {1, 2, 4, 6, 8, 9} to find each set.

1. $A \cup B$
2. $A \cup C$
3. $A \cap B$
4. $A \cap C$
5. $(A \cup B) \cap C$

6. $(A \cap C) \cup (B \cap C)$
7. $(A \cap B) \cup C$
8. $(A \cup B) \cup C$
9. $(A \cup C) \cap (B \cup C)$
10. $(A \cap B) \cap C$

In Problems 11–20, use U = universal set = {0, 1, 2, 3, 4, 5, 6, 7, 8, 9}, A = {1, 3, 4, 5, 9}, B = {2, 4, 6, 7, 8}, and C = {1, 3, 4, 6} to find each set.

11. $\overline{A}$
12. $\overline{C}$
13. $\overline{A \cap B}$
14. $\overline{B \cup C}$
15. $\overline{A} \cup \overline{B}$

16. $\overline{B} \cap \overline{C}$
17. $\overline{A} \cap \overline{C}$
18. $\overline{B} \cup \overline{C}$
19. $\overline{A \cup B \cup C}$
20. $\overline{A \cap B \cap C}$

21. Write down all the subsets of $\{a, b, c, d\}$.

22. Write down all the subsets of $\{a, b, c, d, e\}$.

23. If $n(A) = 15, n(B) = 20$, and $n(A \cap B) = 10$, find $n(A \cup B)$.

24. If $n(A) = 20, n(B) = 40$, and $n(A \cup B) = 35$, find $n(A \cap B)$.

25. If $n(A \cup B) = 50, n(A \cap B) = 10$, and $n(B) = 20$, find $n(A)$.

26. If $n(A \cup B) = 60, n(A \cap B) = 40$, and $n(A) = n(B)$, find $n(A)$.

In Problems 27–34, use the information given in the figure.

27. How many are in set A?

28. How many are in set B?

29. How many are in A or B?

30. How many are in A and B?

31. How many are in A but not C?

32. How many are not in A?

33. How many are in A and B and C?

34. How many are in A or B or C?

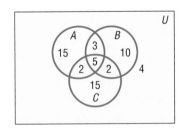

35. **Analyzing Survey Data**   In a consumer survey of 500 people, 200 indicated that they would be buying a major appliance within the next month; 150 indicated that they would buy a car, and 25 said that they would purchase both a major appliance and a car. How many will purchase neither? How many will purchase only a car?

36. **Analyzing Survey Data**   In a student survey, 200 indicated that they would attend Summer Session I and 150 indicated Summer Session II. If 75 students plan to attend both summer sessions and 275 indicated that they would attend neither session, how many students participated in the survey?

37. Analyzing Survey Data In a survey of 100 investors in the stock market,

> 50 owned shares in IBM
> 40 owned shares in AT&T
> 45 owned shares in GE
> 20 owned shares in both IBM and GE
> 15 owned shares in both AT&T and GE
> 20 owned shares in both IBM and AT&T
> 5 owned shares in all three

(a) How many of the investors surveyed did not have shares in any of the three companies?
(b) How many owned just IBM shares?
(c) How many owned just GE shares?
(d) How many owned neither IBM nor GE?
(e) How many owned either IBM or AT&T but no GE?

38. Classifying Blood Types Human blood is classified as either Rh+ or Rh−. Blood is also classified by type: A, if it contains an A antigen; B, if it contains a B antigen; AB, if it contains both A and B antigens; and O, if it contains neither antigen. Draw a Venn diagram illustrating the various blood types. Based on this classification, how many different kinds of blood are there?

39. The following data represent the marital status of males 18 years old and older in March 1997.

Marital Status	Number (in thousands)
Married, spouse present	54,654
Married, spouse absent	3,232
Widowed	2,686
Divorced	8,208
Never married	25,375

Source: Current Population Survey

(a) Determine the number of males 18 years old and older who are married.
(b) Determine the number of males 18 years old and older who are widowed or divorced.
(c) Determine the number of males 18 years old and older who are married, spouse absent, widowed, or divorced.

40. The following data represent the marital status of females 18 years old and older in March 1997.

Marital Status	Number (in thousands)
Married, spouse present	54,626
Married, spouse absent	4,122
Widowed	11,056
Divorced	11,107
Never married	20,503

Source: Current Population Survey

(a) Determine the number of females 18 years old and older who are married.
(b) Determine the number of females 18 years old and older who are widowed or divorced.
(c) Determine the number of females 18 years old and older who are married, spouse absent, widowed, or divorced.

41. Make up a problem different from any found in the text that requires the addition principle of counting to solve. Give it to a friend to solve and critique.

42. Investigate the notion of counting as it relates to infinite sets. Write an essay on your findings.

P R E P A R I N G   F O R   T H I S   S E C T I O N

Before getting started, review the following concept:

✓ Factorial (Section 9.1, p. 619)

10.2 PERMUTATIONS AND COMBINATIONS

OBJECTIVES
1. Solve Counting Problems Using the Multiplication Principle
2. Solve Counting Problems Using Permutations
3. Solve Counting Problems Using Combinations
4. Solve Counting Problems Using Permutations Involving *n* Non-Distinct Objects

1   Counting plays a major role in many diverse areas, such as probability, statistics, and computer science; counting techniques are a part of a branch of mathematics called **combinatorics.** In this section we shall look at special types of counting problems and develop general formulas for solving them.

We begin with an example that will demonstrate a general counting principle.

EXAMPLE 1   Counting the Number of Possible Meals

The fixed-price dinner at Mabenka Restaurant provides the following choices:

Appetizer:	soup or salad
Entree:	baked chicken, broiled beef patty, baby beef liver, or roast beef au jus
Dessert:	ice cream or cheese cake

How many different meals can be ordered?

Solution   Ordering such a meal requires three separate decisions:

Choose an Appetizer	**Choose an Entree**	**Choose a Dessert**
2 choices	4 choices	2 choices

Look at the **tree diagram** in Figure 5. We see that, for each choice of appetizer, there are 4 choices of entrees. And for each of these $2 \cdot 4 = 8$ choices, there are 2 choices for dessert. A total of

$$2 \cdot 4 \cdot 2 = 16$$

different meals can be ordered.

Figure 5

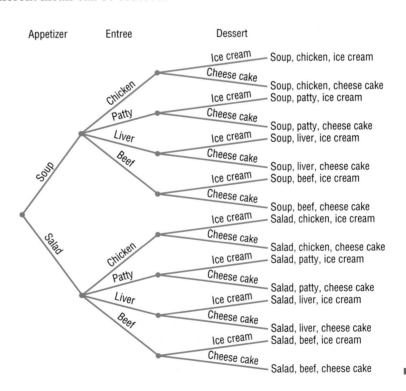

Theorem   **Multiplication Principle of Counting**

If a task consists of a sequence of choices in which there are p selections for the first choice, q selections for the second choice, r selections for the third choice, and so on, then the task of making these selections can be done in

$$p \cdot q \cdot r \cdot \ ...$$

different ways.

◼

EXAMPLE 2   **Forming Codes**

How many two-symbol codewords can be formed if the first symbol is a letter (uppercase) and the second symbol is a digit?

Solution   It sometimes helps to begin by listing some of the possibilities. The code consists of a letter (uppercase) followed by a digit, so some possibilities are A1, A2, B3, X0, and so on. The task consists of making two selections: the first selection requires choosing an uppercase letter (26 choices) and the second task requires choosing a digit (10 choices). By the Multiplication Principle, there are

$$26 \cdot 10 = 260$$

different codewords of the type described.   ◼

NOW WORK PROBLEM **25**.

Permutations

2   We begin with the definition

A **permutation** is an ordered arrangement of r objects chosen from n objects.

We discuss three types of permutations:

1. The n objects are distinct (different), and repetition is allowed in the selection of r of them. [Distinct, with repetition]
2. The n objects are distinct (different), and repetition is not allowed in the selection of r of them, where $r \leq n$. [Distinct, without repetition]
3. The n objects are not distinct, and we use all of them in the arrangement. [Not distinct]

We take up the first two types here and deal with the third type at the end of this section.

The first type of permutation is handled using the Multiplication Principle.

| EXAMPLE 3 | **Counting Airport Codes**
[Permutation: Distinct, with Repetition] |

The International Airline Transportation Association (IATA) assigns three-letter codes to represent airport locations. For example, the airport code for Ft. Lauderdale, Florida is FLL. Notice that repetition is allowed in forming this code. How many airport codes are possible?

Solution   We are choosing 3 letters from 26 letters and arranging them in order. In the ordered arrangement a letter may be repeated. This is an example of a permutation with repetition in which 3 objects are chosen from 26 distinct objects.

  The task of counting the number of such arrangements consists of making three selections. Each selection requires choosing a letter of the alphabet (26 choices). By the Multiplication Principle, there are

$$26 \cdot 26 \cdot 26 = 17{,}576$$

different airport codes.   ■

The solution given to Example 3 can be generalized.

| **Theorem** | **Permutations: Distinct Objects, With Repetition**

The number of ordered arrangements of r objects chosen from n objects, in which the n objects are distinct and repetition is allowed, is n^r.

                           ■ |

       ✏ NOW WORK PROBLEM **29**.

  We begin the discussion of permutations in which the objects are distinct and repetition is not allowed with an example.

| EXAMPLE 4 | **Forming Codes**
[Permutation: Distinct, without Repetition] |

Suppose that we wish to establish a three-letter code using any of the 26 uppercase letters of the alphabet, but we require that no letter be used more than once. How many different three-letter codes are there?

Solution   Some of the possibilities are: ABC, ABD, ABZ, ACB, CBA, and so on. The task consists of making three selections. The first selection requires choosing from 26 letters. Because no letter can be used more than once, the second selection requires choosing from 25 letters. The third selection requires choosing from 24 letters. (Do you see why?) By the Multiplication Principle, there are

$$26 \cdot 25 \cdot 24 = 15{,}600$$

different three-letter codes with no letter repeated.   ■

For the second type of permutation, we introduce the following symbol.

> The symbol $P(n, r)$ represents the number of ordered arrangements of r objects chosen from n distinct objects, where $r \leq n$ and repetition is not allowed.

For example, the question posed in Example 4 asks for the number of ways that the 26 letters of the alphabet can be arranged in order using three nonrepeated letters. The answer is

$$P(26, 3) = 26 \cdot 25 \cdot 24 = 15{,}600$$

EXAMPLE 5 **Lining Up People**

In how many ways can 5 people be lined up?

Solution The 5 people are distinct. Once a person is in line, that person will not be repeated elsewhere in the line; and, in lining up people, order is important. We have a permutation of 5 objects taken 5 at a time. We can line up 5 people in

$$P(5, 5) = \underbrace{5 \cdot 4 \cdot 3 \cdot 2 \cdot 1}_{5 \text{ factors}} = 120 \text{ ways}$$

∎

NOW WORK PROBLEM **31**.

To arrive at a formula for $P(n, r)$, we note that the task of obtaining an ordered arrangement of n objects in which only $r \leq n$ of them are used, without repeating any of them, requires making r selections. For the first selection, there are n choices; for the second selection, there are $n - 1$ choices; for the third selection, there are $n - 2$ choices; ... ; for the rth selection, there are $n - (r - 1)$ choices. By the Multiplication Principle, we have

$$
\begin{array}{cccc}
\text{1st} & \text{2nd} & \text{3rd} & r\text{th} \\
\end{array}
$$
$$P(n, r) = n \cdot (n - 1) \cdot (n - 2) \cdot \ldots \cdot \left[n - (r - 1)\right]$$
$$= n \cdot (n - 1) \cdot (n - 2) \cdot \ldots \cdot (n - r + 1)$$

This formula for $P(n, r)$ can be compactly written using factorial notation.*

$$P(n, r) = n \cdot (n - 1) \cdot (n - 2) \cdot \ldots \cdot (n - r + 1)$$

$$= n \cdot (n - 1) \cdot (n - 2) \cdot \ldots \cdot (n - r + 1) \cdot \frac{(n - r) \cdot \ldots \cdot 3 \cdot 2 \cdot 1}{(n - r) \cdot \ldots \cdot 3 \cdot 2 \cdot 1} = \frac{n!}{(n - r)!}$$

---

*Recall that $0! = 1, 1! = 1, 2! = 2 \cdot 1, \ldots, n! = n(n - 1) \cdot \ldots \cdot 3 \cdot 2 \cdot 1$.

Theorem   **Permutations of *r* Objects Chosen from *n* Distinct Objects without Repetition**

The number of arrangements of *n* objects using $r \leq n$ of them, in which

1. the *n* objects are distinct,
2. once an object is used it cannot be repeated, and
3. order is important,

is given by the formula

$$P(n, r) = \frac{n!}{(n - r)!} \qquad \textbf{(1)}$$

EXAMPLE 6   **Computing Permutations**

Evaluate: (a) $P(7, 3)$ (b) $P(6, 1)$ (c) $P(52, 5)$

Solution   We shall work parts (a) and (b) in two ways.

(a) $P(7, 3) = \underbrace{7 \cdot 6 \cdot 5}_{3 \text{ factors}} = 210$

or

$$P(7, 3) = \frac{7!}{(7 - 3)!} = \frac{7!}{4!} = \frac{7 \cdot 6 \cdot 5 \cdot 4!}{4!} = 210$$

(b) $P(6, 1) = \underbrace{6}_{1 \text{ factor}} = 6$

Figure 6

or

52 nPr 5
311875200

$$P(6, 1) = \frac{6!}{(6 - 1)!} = \frac{6!}{5!} = \frac{6 \cdot 5!}{5!} = 6$$

(c) Figure 6 shows the solution using a TI-83 graphing calculator: $P(52, 5) = 311,875,200$.

NOW WORK PROBLEM 1.

EXAMPLE 7   **The Birthday Problem**

All we know about Shannon, Patrick, and Ryan is that they have different birthdays. If we listed all the possible ways this could occur, how many would there be? Assume that there are 365 days in a year.

Solution   This is an example of a permutation in which 3 birthdays are selected from a possible 365 days, and no birthday may repeat itself. The number of ways that this can occur is

$$P(365, 3) = \frac{365!}{(365 - 3)!} = \frac{365 \cdot 364 \cdot 363 \cdot 362!}{362!} = 365 \cdot 364 \cdot 363 = 48{,}228{,}180$$

There are 48,228,180 ways in a group of three people that each has a different birthday.

NOW WORK PROBLEM 47.

COMBINATIONS

③ In a permutation, order is important; for example, the arrangements ABC, CAB, BAC,... are considered different arrangements of the letters A, B, and C. In many situations, though, order is unimportant. For example, in the card game of poker, the order in which the cards are received does not matter; it is the *combination* of the cards that matters.

A **combination** is an arrangement, without regard to order, of r objects selected from n distinct objects without repetition, where $r \leq n$. The symbol $C(n, r)$ represents the number of combinations of n distinct objects using r of them.

EXAMPLE 8 **Listing Combinations**

List all the combinations of the 4 objects a, b, c, d taken 2 at a time. What is $C(4, 2)$?

Solution One combination of a, b, c, d taken 2 at a time is

$$ab$$

We exclude ba from the list because order is not important in a combination. The list of all such combinations (convince yourself of this) is

$$ab, \quad ac, \quad ad, \quad bc, \quad bd, \quad cd$$

so,

$$C(4, 2) = 6 \qquad \blacksquare$$

We can find a formula for $C(n, r)$ by noting that the only difference between a permutation of type 2 and a combination is that we disregard order in combinations. To determine $C(n, r)$, we need only eliminate from the formula for $P(n, r)$ the number of permutations that were simply rearrangements of a given set of r objects. This can be determined from the formula for $P(n, r)$ by calculating $P(r, r) = r!$. So, if we divide $P(n, r)$ by $r!$, we will have the desired formula for $C(n, r)$:

$$C(n, r) = \frac{P(n, r)}{r!} = \underset{\substack{\uparrow \\ \text{Use formula (1).}}}{\frac{n!/(n - r)!}{r!}} = \frac{n!}{(n - r)!r!}$$

We have proved the following result:

Theorem **Number of Combinations of n Distinct Objects Taken r at a Time**

The number of arrangements of n objects using $r \leq n$ of them, in which

1. the n objects are distinct,
2. once an object is used, it cannot be repeated, and
3. order is not important,

is given by the formula

$$C(n, r) = \frac{n!}{(n - r)!r!} \qquad \qquad \textbf{(2)}$$

Based on formula (2), we discover that the symbol $C(n, r)$ and the symbol $\binom{n}{r}$ for the binomial coefficients are, in fact, the same. The Pascal triangle (see Section 9.5) can be used to find the value of $C(n, r)$. However, because it is more practical and convenient, we will use formula (2) instead.

EXAMPLE 9

Using Formula (2)

Use formula (2) to find the value of each expression.

(a) $C(3, 1)$ (b) $C(6, 3)$ (c) $C(n, n)$ (d) $C(n, 0)$  (e) $C(52, 5)$

Solution   (a) $C(3, 1) = \dfrac{3!}{(3 - 1)!1!} = \dfrac{3!}{2!1!} = \dfrac{3 \cdot 2 \cdot 1}{2 \cdot 1 \cdot 1} = 3$

Figure 7

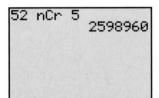

(b) $C(6, 3) = \dfrac{6!}{(6 - 3)!3!} = \dfrac{6 \cdot 5 \cdot 4 \cdot 3!}{3! \cdot 3!} = \dfrac{6 \cdot 5 \cdot 4}{6} = 20$

(c) $C(n, n) = \dfrac{n!}{(n - n)!n!} = \dfrac{n!}{0! n!} = \dfrac{1}{1} = 1$

(d) $C(n, 0) = \dfrac{n!}{(n - 0)!0!} = \dfrac{n!}{n! 0!} = \dfrac{1}{1} = 1$

(e) Figure 7 shows the solution using a TI-83 graphing calculator: $C(52, 5) = 2{,}598{,}960$. ∎

NOW WORK PROBLEM **9**.

EXAMPLE 10

Forming Committees

How many different committees of 3 people can be formed from a pool of 7 people?

Solution   The 7 people are distinct. More important, though, is the observation that the order of being selected for a committee is not significant. The problem asks for the number of combinations of 7 objects taken 3 at a time.

$$C(7, 3) = \frac{7!}{4!3!} = \frac{7 \cdot 6 \cdot 5 \cdot 4!}{4! 3!} = \frac{7 \cdot 6 \cdot 5}{6} = 35$$ ∎

EXAMPLE 11

Forming Committees

In how many ways can a committee consisting of 2 faculty members and 3 students be formed if 6 faculty members and 10 students are eligible to serve on the committee?

Solution   The problem can be separated into two parts: the number of ways that the faculty members can be chosen, $C(6, 2)$, and the number of ways that the student members can be chosen, $C(10, 3)$. By the Multiplication Principle, the committee can be formed in

$$C(6, 2) \cdot C(10, 3) = \frac{6!}{4!2!} \cdot \frac{10!}{7!3!} = \frac{6 \cdot 5 \cdot 4!}{4! 2!} \cdot \frac{10 \cdot 9 \cdot 8 \cdot 7!}{7! 3!}$$

$$= \frac{30}{2} \cdot \frac{720}{6} = 1800 \text{ ways}$$ ∎

NOW WORK PROBLEM **49**.

PERMUTATIONS INVOLVING n OBJECTS THAT ARE NOT DISTINCT

④ We begin with an example.

EXAMPLE 12 **Forming Different Words**

How many different words (real or imaginary) can be formed using all the letters in the word REARRANGE?

Solution Each word formed will have 9 letters: 3 R's, 2 A's, 2 E's, 1 N, and 1 G. To construct each word, we need to fill in 9 positions with the 9 letters:

$$\overline{1}\ \ \overline{2}\ \ \overline{3}\ \ \overline{4}\ \ \overline{5}\ \ \overline{6}\ \ \overline{7}\ \ \overline{8}\ \ \overline{9}$$

The process of forming a word consists of five tasks:

Task 1: Choose the positions for the 3 R's.
Task 2: Choose the positions for the 2 A's.
Task 3: Choose the positions for the 2 E's.
Task 4: Choose the position for the 1 N.
Task 5: Choose the position for the 1 G.

Task 1 can be done in $C(9, 3)$ ways. There then remain 6 positions to be filled, so Task 2 can be done in $C(6, 2)$ ways. There remain 4 positions to be filled, so Task 3 can be done in $C(4, 2)$ ways. There remain 2 positions to be filled, so Task 4 can be done in $C(2, 1)$ ways. The last position can be filled in $C(1, 1)$ way. Using the Multiplication Principle, the number of possible words that can be formed is

$$C(9, 3) \cdot C(6, 2) \cdot C(4, 2) \cdot C(2, 1) \cdot C(1, 1) = \frac{9!}{3! \cdot \not{6!}} \cdot \frac{\not{6!}}{2! \cdot \not{4!}} \cdot \frac{\not{4!}}{2! \cdot \not{2!}} \cdot \frac{\not{2!}}{1! \cdot \not{1!}} \cdot \frac{\not{1!}}{0! \cdot 1!}$$

$$= \frac{9!}{3! \cdot 2! \cdot 2! \cdot 1! \cdot 1!}$$

∎

The form of the answer to Example 12 is suggestive of a general result. Had the letters in REARRANGE each been different, there would have been $P(9, 9) = 9!$ possible words formed. This is the numerator of the answer. The presence of 3 R's, 2 A's, and 2 E's reduces the number of different words, as the entries in the denominator illustrate. We are led to the following result:

Theorem **Permutations Involving n Objects That Are Not Distinct**

The number of permutations of n objects of which n_1 are of one kind, n_2 are of a second kind, ..., and n_k are of a kth kind is given by

$$\frac{n!}{n_1! \cdot n_2! \cdot \ \ldots \ \cdot n_k!} \tag{3}$$

where $n = n_1 + n_2 + \cdots + n_k$.

∎

| EXAMPLE 13 | Arranging Flags |

How many different vertical arrangements are there of 8 flags if 4 are white, 3 are blue, and 1 is red?

Solution   We seek the number of permutations of 8 objects, of which 4 are of one kind, 3 of a second kind, and 1 of a third kind. Using formula (3), we find that there are

$$\frac{8!}{4! \cdot 3! \cdot 1!} = \frac{8 \cdot 7 \cdot 6 \cdot 5 \cdot \cancel{4!}}{\cancel{4!} \cdot 3! \cdot 1!} = 280 \text{ different arrangements} \quad \blacksquare$$

NOW WORK PROBLEM **51**.

10.2   EXERCISES

In Problems 1–8, find the value of each permutation.

1. $P(6, 2)$ **2.** $P(7, 2)$ **3.** $P(4, 4)$ **4.** $P(8, 8)$

5. $P(7, 0)$ **6.** $P(9, 0)$ **7.** $P(8, 4)$ **8.** $P(8, 3)$

In Problems 9–16, use formula (2) to find the value of each combination.

9. $C(8, 2)$ **10.** $C(8, 6)$ **11.** $C(7, 4)$ **12.** $C(6, 2)$

13. $C(15, 15)$ **14.** $C(18, 1)$ **15.** $C(26, 13)$ **16.** $C(18, 9)$

17. List all the ordered arrangements of 5 objects $a, b, c, d,$ and e choosing 3 at a time without repetition. What is $P(5, 3)$?

18. List all the ordered arrangements of 5 objects $a, b, c, d,$ and e choosing 2 at a time without repetition. What is $P(5, 2)$?

19. List all the ordered arrangements of 4 objects 1, 2, 3, and 4 choosing 3 at a time without repetition. What is $P(4, 3)$?

20. List all the ordered arrangements of 6 objects 1, 2, 3, 4, 5, and 6 choosing 3 at a time without repetition. What is $P(6, 3)$?

21. List all the combinations of 5 objects $a, b, c, d,$ and e taken 3 at a time. What is $C(5, 3)$?

22. List all the combinations of 5 objects $a, b, c, d,$ and e taken 2 at a time. What is $C(5, 2)$?

23. List all the combinations of 4 objects 1, 2, 3, and 4 taken 3 at a time. What is $C(4, 3)$?

24. List all the combinations of 6 objects 1, 2, 3, 4, 5, and 6 taken 3 at a time. What is $C(6, 3)$?

25. Shirts and Ties   A man has 5 shirts and 3 ties. How many different shirt and tie arrangements can he wear?

26. Blouses and Skirts   A woman has 3 blouses and 5 skirts. How many different outfits can she wear?

27. Forming Codes   How many two-letter codes can be formed using the letters $A, B, C,$ and D? Repeated letters are allowed.

28. Forming Codes   How many two-letter codes can be formed using the letters $A, B, C, D,$ and E? Repeated letters are allowed.

29. Forming Numbers   How many three-digit numbers can be formed using the digits 0 and 1? Repeated digits are allowed.

30. Forming Numbers   How many three-digit numbers can be formed using the digits 0, 1, 2, 3, 4, 5, 6, 7, 8, and 9? Repeated digits are allowed.

31. Lining People Up   In how many ways can 4 people be lined up?

32. Stacking Boxes   In how many ways can 5 different boxes be stacked?

33. Forming Codes   How many different three-letter codes are there if only the letters $A, B, C, D,$ and E can be used and no letter can be used more than once?

34. Forming Codes   How many different four-letter codes are there if only the letters $A, B, C, D, E,$ and F can be used and no letter can be used more than once?

35. Stocks on the NYSE   Companies whose stocks are listed on the New York Stock Exchange (NYSE) have their company name represented by either 1, 2, or 3 letters (repetition of letters is allowed). What is the maximum number of companies that can be listed on the NYSE?

36. Stocks on the NASDAQ Companies whose stocks are listed on the NASDAQ stock exchange have their company name represented by either 4 or 5 letters (repetition of letters is allowed). What is the maximum number of companies that can be listed on the NASDAQ?

37. Establishing Committees In how many ways can a committee of 4 students be formed from a pool of 7 students?

38. Establishing Committees In how many ways can a committee of 3 professors be formed from a department having 8 professors?

39. Possible Answers on a True/False Test How many arrangements of answers are possible for a true/false test with 10 questions?

40. Possible Answers on a Multiple-choice Test How many arrangements of answers are possible in a multiple-choice test with 5 questions, each of which has 4 possible answers?

41. Four-Digit Numbers How many four-digit numbers can be formed using the digits 0, 1, 2, 3, 4, 5, 6, 7, 8, and 9 if the first digit cannot be 0? Repeated digits are allowed.

42. Five-Digit Numbers How many five-digit numbers can be formed using the digits 0, 1, 2, 3, 4, 5, 6, 7, 8, and 9 if the first digit cannot be 0 or 1? Repeated digits are allowed.

43. Arranging Books Five different mathematics books are to be arranged on a student's desk. How many arrangements are possible?

44. Forming License Plate Numbers How many different license plate numbers can be made using 2 letters followed by 4 digits selected from the digits 0 through 9, if
(a) Letters and digits may be repeated?
(b) Letters may be repeated, but digits may not be repeated?
(c) Neither letters nor digits may be repeated?

45. Stock Portfolios As a financial planner, you are asked to select one stock each from the following groups: 8 DOW stocks, 15 NASDAQ stocks, and 4 global stocks. How many different portfolios are possible?

46. Combination Locks A combination lock displays 50 numbers. To open it, you turn to a number, then rotate

clockwise to a second number, and then counterclockwise to the third number. How many different lock combinations are there?

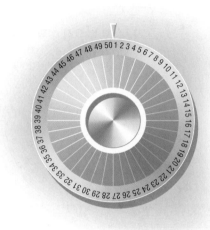

47. Birthday Problem In how many ways can 2 people each have different birthdays? Assume that there are 365 days in a year.

48. Birthday Problem In how many ways can 5 people each have different birthdays? Assume that there are 365 days in a year.

49. Forming a Committee A student dance committee is to be formed consisting of 2 boys and 3 girls. If the membership is to be chosen from 4 boys and 8 girls, how many different committees are possible?

50. Forming a Committee The student relations committee of a college consists of 2 administrators, 3 faculty members, and 5 students. Four administrators, 8 faculty members, and 20 students are eligible to serve. How many different committees are possible?

51. Forming Words How many different 9-letter words (real or imaginary) can be formed from the letters in the word ECONOMICS?

52. Forming Words How many different 11-letter words (real or imaginary) can be formed from the letters in the word MATHEMATICS?

53. Selecting Objects An urn contains 7 white balls and 3 red balls. Three balls are selected. In how many ways can the 3 balls be drawn from the total of 10 balls:
(a) If 2 balls are white and 1 is red?
(b) If all 3 balls are white?
(c) If all 3 balls are red?

54. Selecting Objects An urn contains 15 red balls and 10 white balls. Five balls are selected. In how many ways can the 5 balls be drawn from the total of 25 balls:
(a) If all 5 balls are red?
(b) If 3 balls are red and 2 are white?
(c) If at least 4 are red balls?

55. Senate Committees The U.S. Senate has 100 members. Suppose that it is desired to place each senator on exactly 1 of 7 possible committees. The first committee has 22 members, the second has 13, the third has 10, the fourth has 5, the fifth has 16, and the sixth and seventh have 17 apiece. In how many ways can these committees be formed?

56. Football Teams A defensive football squad consists of 25 players. Of these, 10 are linemen, 10 are linebackers, and 5 are safeties. How many different teams of 5 linemen, 3 linebackers, and 3 safeties can be formed?

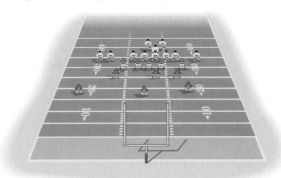

57. Baseball In the American Baseball League, a designated hitter may be used. How many batting orders is it possible for a manager to use? (There are 9 regular players on a team.)

58. Baseball In the National Baseball League, the pitcher usually bats ninth. If this is the case, how many batting orders is it possible for a manager to use?

59. Baseball Teams A baseball team has 15 members. Four of the players are pitchers, and the remaining 11 mem-

bers can play any position. How many different teams of 9 players can be formed?

60. World Series In the World Series the American League team (A) and the National League team (N) play until one team wins four games. If the sequence of winners is designated by letters (for example, *NAAAA* means that the National League team won the first game and the American League won the next four), how many different sequences are possible?

61. Basketball Teams A basketball team has 6 players who play guard (2 of 5 starting positions). How many different teams are possible, assuming that the remaining 3 positions are filled and it is not possible to distinguish a left guard from a right guard?

62. Basketball Teams On a basketball team of 12 players, 2 only play center, 3 only play guard, and the rest play forward (5 players on a team: 2 forwards, 2 guards, and 1 center). How many different teams are possible, assuming that it is not possible to distinguish left and right guards and left and right forwards?

63. Make up a problem different from any found in the text that requires the Multiplication Principle of counting to solve. Give it to a friend to solve and critique.

64. Make up a problem different from any found in the text that requires a permutation to solve. Give it to a friend to solve and critique.

65. Make up a problem different from any found in the text that requires a combination to solve. Give it to a friend to solve and critique.

66. Explain the difference between a permutation and a combination. Give an example to illustrate your explanation.

10.3 PROBABILITY

OBJECTIVES 1 Construct Probability Models
2 Compute Probabilities of Equally Likely Outcomes
3 Utilize the Addition Rule to Find Probabilities
4 Utilize the Complement Rule to Find Probabilities
5 Compute Probabilities Using Permutations and Combinations

Probability is an area of mathematics that deals with experiments that yield random results, yet admit a certain regularity. Such experiments do not always produce the same result or outcome, so the result of any one observation is not predictable. However, the results of the experiment over a long period do produce regular patterns that enable us to predict with remarkable accuracy.

| EXAMPLE 1 | Tossing a Fair Coin |

In tossing a fair coin, we know that the outcome is either a head or a tail. On any particular throw, we cannot predict what will happen, but, if we toss the coin many times, we observe that the number of times that a head comes up is approximately equal to the number of times that we get a tail. It seems reasonable, therefore, to assign a probability of $\frac{1}{2}$ that a head comes up and a probability of $\frac{1}{2}$ that a tail comes up. ▬

PROBABILITY MODELS

① The discussion in Example 1 constitutes the construction of a **probability model** for the experiment of tossing a fair coin once. A probability model has two components: a sample space and an assignment of probabilities. A **sample space** S is a set whose elements represent all the possibilities that can occur as a result of the experiment. Each element of S is called an **outcome.** To each outcome, we assign a number, called the **probability** of that outcome, which has two properties:

1. The probability assigned to each outcome is nonnegative.
2. The sum of all the probabilities equals 1.

If a probability model has the sample space

$$S = \{e_1, e_2, \ldots, e_n\}$$

where $e_1, e_2, \ldots, e_n$ are the possible outcomes, and if $P(e_1), P(e_2), \ldots, P(e_n)$ denote the respective probabilities of these outcomes, then

$$P(e_1) \geq 0, P(e_2) \geq 0, \ldots, P(e_n) \geq 0 \qquad \textbf{(1)}$$

$$\sum_{i=1}^{n} P(e_i) = P(e_1) + P(e_2) + \cdots + P(e_n) = 1 \qquad \textbf{(2)}$$

| EXAMPLE 2 | Determining Probability Models |

In a bag of M&Ms, the candies are colored red, green, blue, brown, yellow, and orange. Suppose that a candy is drawn from the bag and the color is recorded. The sample space of this experiment is {red, green, blue, brown, yellow, orange}. Determine which of the following are probability models.

(a)

Outcome	Probability
{red}	0.3
{green}	0.15
{blue}	0
{brown}	0.15
{yellow}	0.2
{orange}	0.2

(b)

Outcome	Probability
{red}	0.1
{green}	0.1
{blue}	0.1
{brown}	0.4
{yellow}	0.2
{orange}	0.3

(c) **Outcome**	**Probability**	(d) **Outcome**	**Probability**
{red}	0.3	{red}	0
{green}	−0.3	{green}	0
{blue}	0.2	{blue}	0
{brown}	0.4	{brown}	0
{yellow}	0.2	{yellow}	1
{orange}	0.2	{orange}	0

Solution   (a) This model is a probability model since all the outcomes have probabilities that are nonnegative and the sum of the probabilities is 1.

(b) This model is not a probability model because the sum of the probabilities is not 1.

(c) This model is not a probability model because $P(\text{green})$ is less than 0. Recall, all probabilities must be nonnegative.

(d) This model is a probability model because all the outcomes have probabilities that are nonnegative, and the sum of the probabilities is 1. Notice that $P(\text{yellow}) = 1$, meaning that this outcome will occur with 100% certainty each time that the experiment is repeated. This means that the entire bag of M&Ms has yellow candies.   ■

NOW WORK PROBLEM 3.

Let's look at an example of constructing a probability model.

EXAMPLE 3	Constructing a Probability Model

An experiment consists of rolling a fair die once.* Construct a probability model for this experiment.

Solution   A sample space S consists of all the possibilities that can occur. Because rolling the die will result in one of six faces showing, the sample space S consists of

Figure 8

$$S = \{1, 2, 3, 4, 5, 6\}$$

Because the die is fair, one face is no more likely to occur than another. As a result, our assignment of probabilities is

$$P(1) = \tfrac{1}{6} \qquad P(2) = \tfrac{1}{6}$$

$$P(3) = \tfrac{1}{6} \qquad P(4) = \tfrac{1}{6}$$

$$P(5) = \tfrac{1}{6} \qquad P(6) = \tfrac{1}{6} \qquad ■$$

Now suppose that a die is loaded (weighted) so that the probability assignments are

$$P(1) = 0, \quad P(2) = 0, \quad P(3) = \frac{1}{3}, \quad P(4) = \frac{2}{3}, \quad P(5) = 0, \quad P(6) = 0$$

This assignment would be made if the die were loaded so that only a 3 or 4 could occur and the 4 is twice as likely as the 3 to occur. This assignment is

---

*A die is a cube with each face having either 1, 2, 3, 4, 5, or 6 dots on it. See Figure 8.

consistent with the definition, since each assignment is nonnegative and the sum of all the probability assignments equals 1.

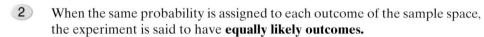 NOW WORK PROBLEM **19.**

| EXAMPLE 4 | **Constructing a Probability Model** |

An experiment consists of tossing a coin. The coin is weighted so that heads (H) is three times as likely to occur as tails (T). Construct a probability model for this experiment.

Solution   The sample space S is $S = \{H, T\}$. If x denotes the probability that a tail occurs, then

$$P(T) = x \quad \text{and} \quad P(H) = 3x$$

Since the sum of the probabilities of the possible outcomes must equal 1, we have

$$P(T) + P(H) = x + 3x = 1$$
$$4x = 1$$
$$x = \frac{1}{4}$$

We assign the probabilities

$$P(T) = \frac{1}{4} \qquad P(H) = \frac{3}{4}$$ ∎

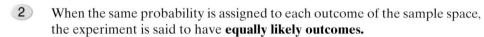 NOW WORK PROBLEM **23.**

In working with probability models, the term **event** is used to describe a set of possible outcomes of the experiment. An event E is some subset of the sample space S. The **probability of an event** $E, E \neq \varnothing$, denoted by $P(E)$, is defined as the sum of the probabilities of the outcomes in E. We can also think of the probability of an event E as the likelihood that the event E occurs. If $E = \varnothing$, then $P(E) = 0$; if $E = S$, then $P(E) = P(S) = 1$.

EQUALLY LIKELY OUTCOMES

⟨2⟩   When the same probability is assigned to each outcome of the sample space, the experiment is said to have **equally likely outcomes.**

Theorem   **Probability for Equally Likely Outcomes**

If an experiment has n equally likely outcomes and if the number of ways that an event E can occur is m, then the probability of E is

$$P(E) = \frac{\text{Number of ways that } E \text{ can occur}}{\text{Number of all logical possibilities}} = \frac{m}{n} \qquad \textbf{(3)}$$

If S is the sample space of this experiment, then

$$P(E) = \frac{n(E)}{n(S)} \qquad \textbf{(4)}$$

| EXAMPLE 5 | Calculating Probabilities of Events Involving Equally Likely Outcomes |

Calculate the probability that in a 3-child family there are 2 boys and 1 girl. Assume equally likely outcomes.

Solution

Figure 9

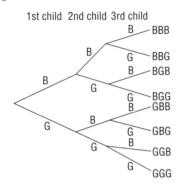

1st child 2nd child 3rd child

We begin by constructing a tree diagram to help in listing the possible outcomes of the experiment. See Figure 9, where B stands for boy and G for girl. The sample space S of this experiment is

$$S = \{BBB, BBG, BGB, BGG, GBB, GBG, GGB, GGG\}$$

so $n(S) = 8$.

We wish to know the probability of the event E: "having two boys and one girl." From Figure 9, we conclude that $E = \{BBG, BGB, GBB\}$, so $n(E) = 3$. Since the outcomes are equally likely, the probability of E is

$$P(E) = \frac{n(E)}{n(S)} = \frac{3}{8}$$

∎

NOW WORK PROBLEM **33**.

COMPOUND PROBABILITIES

So far, we have calculated probabilities of single events. We will now compute probabilities of multiple events, called **compound probabilities.**

| EXAMPLE 6 | Computing Compound Probabilities |

Consider the experiment of rolling a single fair die. Let E represent the event "roll an odd number" and let F represent the event "roll a 1 or 2."

(a) Write the event E and F.　　　(b) Write the event E or F.
(c) Compute $P(E)$ and $P(F)$.　　　(d) Compute $P(E \cap F)$.
(e) Compute $P(E \cup F)$.

Solution

The sample space S of the experiment is $\{1, 2, 3, 4, 5, 6\}$, so $n(S) = 6$. Since the die is fair, the outcomes are equally likely. The event E: "roll an odd number" is $\{1, 3, 5\}$, and the event F: "roll a 1 or 2" is $\{1, 2\}$, so $n(E) = 3$ and $n(F) = 2$.

(a) The word *and* in probability means the intersection of two events. The event E and F is

$$E \cap F = \{1, 3, 5\} \cap \{1, 2\} = \{1\} \qquad n(E \cap F) = 1$$

(b) The word *or* in probability means the union of the two events. The event E or F is

$$E \cup F = \{1, 3, 5\} \cup \{1, 2\} = \{1, 2, 3, 5\} \qquad n(E \cup F) = 4$$

(c) We use formula (4).

$$P(E) = \frac{n(E)}{n(S)} = \frac{3}{6} = \frac{1}{2} \qquad P(F) = \frac{n(F)}{n(S)} = \frac{2}{6} = \frac{1}{3}$$

(d) $P(E \cap F) = \dfrac{n(E \cap F)}{n(S)} = \dfrac{1}{6}$

(e) $P(E \cup F) = \dfrac{n(E \cup F)}{n(S)} = \dfrac{4}{6} = \dfrac{2}{3}$

∎

③   The **Addition Rule** can be used to find the probability of the union of two events.

Theorem   Addition Rule

For any two events E and F,

$$P(E \cup F) = P(E) + P(F) - P(E \cap F) \qquad \textbf{(5)}$$

For example, we can use the Addition Rule to find $P(E \cup F)$ in Example 6(e). Then

$$P(E \cup F) = P(E) + P(F) - P(E \cap F) = \frac{1}{2} + \frac{1}{3} - \frac{1}{6} = \frac{3}{6} + \frac{2}{6} - \frac{1}{6} = \frac{4}{6} = \frac{2}{3}$$

as before.

EXAMPLE 7   **Computing Probabilities of Compound Events Using the Addition Rule**

If $P(E) = 0.2$, $P(F) = 0.3$, and $P(E \cap F) = 0.1$, find the probability of E or F, that is find $P(E \cup F)$.

Solution   We use the Addition Rule, formula (5).

$$\text{probability of } E \text{ or } F = P(E \cup F) = P(E) + P(F) - P(E \cap F)$$
$$= 0.2 + 0.3 - 0.1 = 0.4$$

A Venn diagram can sometimes be used to obtain probabilities. To construct a Venn diagram representing the information in Example 7, we draw two sets E and F. We begin with the fact that $P(E \cap F) = 0.1$. See Figure 10(a). Then, since $P(E) = 0.2$ and $P(F) = 0.3$, we fill in E with $0.2 - 0.1 = 0.1$ and F with $0.3 - 0.1 = 0.2$. See Figure 10(b). Since $P(S) = 1$, we complete the diagram by inserting $1 - [0.1 + 0.1 + 0.2] = 0.6$. See Figure 10(c). Now it is easy to see, for example, that the probability of F, but not E, is 0.2. Also, the probability of neither E nor F is 0.6.

Figure 10

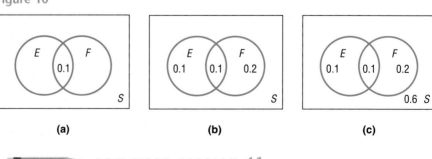

(a)   (b)   (c)

✏ NOW WORK PROBLEM **41.**

If events E and F are disjoint so that $E \cap F = \varnothing$, we say they are **mutually exclusive.** In this case, $P(E \cap F) = 0$, and the Addition Rule takes the following form:

Theorem   **Mutually Exclusive Events**

If E and F are **mutually exclusive events,** then

$$P(E \cup F) = P(E) + P(F) \qquad \textbf{(6)}$$

EXAMPLE 8   **Computing Compound Probabilities**
of Mutually Exclusive Events

If $P(E) = 0.4$ and $P(F) = 0.25$, and E and F are mutually exclusive, find $P(E \cup F)$.

Solution   Since E and F are mutually exclusive, we use formula (6).

$$P(E \cup F) = P(E) + P(F) = 0.4 + 0.25 = 0.65$$

NOW WORK PROBLEM 43.

COMPLEMENTS

④   Recall, if A is a set, the complement of A, denoted $\overline{A}$, is the set of all elements in the universal set U not in A. We similarly define the complement of an event.

Complement of an Event

Let S denote the sample space of an experiment and let E denote an event. The **complement of E,** denoted $\overline{E}$, is the set of all outcomes in the sample space S that are not outcomes in the event E.

The complement of an event E, that is, $\overline{E}$, in a sample space S has the following two properties:

$$E \cap \overline{E} = \varnothing \qquad E \cup \overline{E} = S$$

Since E and $\overline{E}$ are mutually exclusive, it follows from (6) that

$$P(E \cup \overline{E}) = P(S) = 1 \qquad P(E) + P(\overline{E}) = 1 \qquad P(\overline{E}) = 1 - P(E)$$

We have the following result:

Theorem   **Computing Probabilities of Complementary Events**

If E represents any event and $\overline{E}$ represents the complement of E, then

$$P(\overline{E}) = 1 - P(E) \qquad \textbf{(7)}$$

| EXAMPLE 9 | **Computing Probabilities Using Complements** |

On the local news the weather reporter stated that the probability of rain tomorrow is 40%. What is the probability that it will not rain?

Solution The complement of the event "rain" is "no rain."

$$P(\text{no rain}) = 1 - P(\text{rain}) = 1 - 0.4 = 0.6$$

There is a 60% chance of no rain tomorrow.

NOW WORK PROBLEM **47.**

| EXAMPLE 10 | **Birthday Problem** |

What is the probability that in a group of 10 people at least 2 people have the same birthday? Assume that there are 365 days in a year.

Solution We assume that a person is as likely to be born on one day as another, so we have equally likely outcomes.

We first determine the number of outcomes in the sample space S. There are 365 possibilities for each person's birthday. Since there are 10 people in the group, there are 365^{10} possibilities for the birthdays. [For one person in the group, there are 365 days on which his or her birthday can fall; for two people, there are $(365)(365) = 365^2$ pairs of days; and, in general, using the Multiplication Principle, for n people there are 365^n possibilities.] So

$$n(S) = 365^{10}$$

We wish to find the probability of the event E: "at least two people have the same birthday." It is difficult to count the elements in this set; it is much easier to count the elements of the complementary event $\bar{E}$: "no two people have the same birthday."

We find $n(\bar{E})$ as follows: Choose one person at random. There are 365 possibilities for his or her birthday. Choose a second person. There are 364 possibilities for this birthday, if no two people are to have the same birthday. Choose a third person. There are 363 possibilities left for this birthday. Finally, we arrive at the tenth person. There are 356 possibilities left for this birthday. By the Multiplication Principle, the total number of possibilities is

$$n(\bar{E}) = 365 \cdot 364 \cdot 363 \cdot \ldots \cdot 356$$

Hence, the probability of event $\bar{E}$ is

$$P(\bar{E}) = \frac{n(\bar{E})}{n(S)} = \frac{365 \cdot 364 \cdot 363 \cdot \ldots \cdot 356}{365^{10}} \approx 0.883$$

The probability of two or more people in a group of 10 people having the same birthday is then

$$P(E) = 1 - P(\bar{E}) = 1 - 0.883 = 0.117$$

The birthday problem can be solved for any group size. The following table gives the probabilities for two or more people having the same birthday for various group sizes. Notice that the probability is greater than $\frac{1}{2}$ for any group of 23 or more people.

	Number of People															
	5	10	15	20	21	22	23	24	25	30	40	50	60	70	80	90
Probability That Two or More Have the Same Birthday	0.027	0.117	0.253	0.411	0.444	0.476	0.507	0.538	0.569	0.706	0.891	0.970	0.994	0.99916	0.99991	0.99999

> NOW WORK PROBLEM **65.**

⑤ PROBABILITIES INVOLVING COMBINATIONS AND PERMUTATIONS

EXAMPLE 11   **Computing Probabilities**

Because of a mistake in packaging, 5 defective phones were packaged with 15 good ones. All phones look alike and have equal probability of being chosen. Three phones are selected.

(a) What is the probability that all 3 are defective?

(b) What is the probability that exactly 2 are defective?

(c) What is the probability that at least 2 are defective?

Solution   The number of outcomes in the sample space S consists of the number of ways that 3 objects can be selected from 20 objects, that is, the number of combinations of 20 things taken 3 at a time.

$$n(S) = C(20, 3) = \frac{20!}{17! \cdot 3!} = \frac{20 \cdot 19 \cdot 18}{6} = 1140$$

Each of these outcomes is equally likely to occur.

(a) If E is the event "3 are defective," the number of elements in E is the number of ways that the 3 defective phones can be chosen from the 5 defective phones: $C(5, 3) = 10$. The probability of E is

$$P(E) = \frac{n(E)}{n(S)} = \frac{C(5, 3)}{C(20, 3)} = \frac{10}{1140} \approx 0.0088$$

(b) If F is the event "exactly 2 are defective" and 3 phones are selected, the number of elements in F is the number of ways to select 2 defective phones from the 5 defective phones and 1 good phone from the 15 good ones. The first of these can be done in $C(5, 2)$ ways and the second in $C(15, 1)$ ways. By the Multiplication Principle, the event F can occur in

$$C(5, 2) \cdot C(15, 1) = \frac{5!}{3! \cdot 2!} \cdot \frac{15!}{14! \cdot 1!} = 10 \cdot 15 = 150 \text{ ways}$$

The probability of F is therefore

$$P(F) = \frac{n(F)}{n(S)} = \frac{C(5, 2) \cdot C(15, 1)}{C(20, 3)} = \frac{150}{1140} \approx 0.1316$$

(c) The event G, "at least 2 are defective," when 3 are chosen is equivalent to requiring that either exactly 2 defective are chosen or exactly 3 defective are chosen. That is, $G = E \cup F$. Since E and F are mutually exclusive (it is not possible to select 2 defective phones and, at the same time, select 3 defective phones), we find that

$$P(G) = P(E) + P(F) = \frac{C(5,2)}{C(20,3)} + \frac{C(5,2) \cdot C(15,1)}{C(20,3)} \approx 0.0088 + 0.1316 = 0.1404 \quad \blacksquare$$

NOW WORK PROBLEM **73**.

EXAMPLE 12   **Tossing a Coin**

A fair coin is tossed 6 times.

(a) What is the probability of obtaining exactly 5 heads and 1 tail?

(b) What is the probability of obtaining between 4 and 6 heads, inclusive?

Solution   The number of elements in the sample space S is found using the Multiplication Principle. Each toss results in a head (H) or a tail (T). Since the coin is tossed 6 times, we have

$$n(S) = \underbrace{2 \cdot 2 \cdot \ldots \cdot 2}_{6 \text{ tosses}} = 2^6 = 64$$

The outcomes are equally likely since the coin is fair.

(a) Any sequence that contains 5 heads and 1 tail is determined once the position of the 5 heads (or 1 tail) is known. The number of ways that we can position 5 heads in a sequence of 6 slots is $C(6, 5) = 6$. The probability of the event E, exactly 5 heads and 1 tail, is

$$P(E) = \frac{n(E)}{n(S)} = \frac{C(6, 5)}{2^6} = \frac{6}{64} \approx 0.0938$$

(b) Let F be the event "between 4 and 6 heads, inclusive." To obtain between 4 and 6 heads is equivalent to the event "either 4 heads or 5 heads or 6 heads." Since each of these is mutually exclusive (it is impossible to obtain both 4 heads and 5 heads when tossing a coin 6 times), we have

$$P(F) = P(4 \text{ heads or } 5 \text{ heads or } 6 \text{ heads})$$

$$= P(4 \text{ heads}) + P(5 \text{ heads}) + P(6 \text{ heads})$$

We proceed as in part (a).

$$P(F) = \frac{C(6, 4)}{2^6} + \frac{C(6, 5)}{2^6} + \frac{C(6, 6)}{2^6} = \frac{15}{64} + \frac{6}{64} + \frac{1}{64} = \frac{22}{64} \approx 0.3438 \quad \blacksquare$$

HISTORICAL FEATURE

Blaise Pascal (1623–1662)

Set theory, counting, and probability first took form as a systematic theory in an exchange of letters (1654) between Pierre de Fermat (1601–1665) and Blaise Pascal (1623–1662). They discussed the problem of how to divide the stakes in a game that is interrupted before completion, knowing how many points each player needs to win. Fermat solved the problem by listing all possibilities and counting the favorable ones, whereas Pascal made use of the triangle that now bears his name. As mentioned in the text, the entries in Pascal's triangle are equivalent to $C(n, r)$. This recognition of the role of $C(n, r)$ in counting is the foundation of all further developments.

The first book on probability, the work of Christian Huygens (1629–1695), appeared in 1657. In it, the notion of mathematical expectation is explored. This allows the calculation of the profit or loss that a gambler might expect, knowing the probabilities involved in the game (see the Historical Problems that follow).

Although Girolamo Cardano (1501–1576) wrote a treatise on probability, it was not published until 1663 in Cardano's collected works, and this was too late to have any effect on the development of the theory.

In 1713, the posthumously published *Ars Conjectandi* of Jakob Bernoulli (1654–1705) gave the theory the form it would have until 1900. Recently, both combinatorics (counting) and probability have undergone rapid development due to the use of computers.

A final comment about notation. The notations $C(n, r)$ and $P(n, r)$ are variants of a form of notation developed in England after 1830. The notation $\binom{n}{r}$ for $C(n, r)$ goes back to Leonhard Euler (1707–1783), but is now losing ground because it has no clearly related symbolism of the same type for permutations. The set symbols $\cup$ and $\cap$ were introduced by Giuseppe Peano (1858–1932) in 1888 in a slightly different context. The inclusion symbol $\subset$ was introduced by E. Schroeder (1841–1902) about 1890. The treatment of set theory in the text is due to George Boole (1815–1864), who wrote $A + B$ for $A \cup B$ and AB for $A \cap B$ (statisticians still use AB for $A \cap B$).

HISTORICAL PROBLEMS

1. *The Problem Discussed by Fermat and Pascal*   A game between two equally skilled players, A and B, is interrupted when A needs 2 points to win and B needs 3 points. In what proportion would the stakes be divided?

   [**NOTE:**   If each play results in 1 point for either player, at most four more plays will decide the game.]

   (a) *Fermat's solution*   List all possible outcomes that will end the game to form the sample space (for example, $ABA, ABBB$, etc.). The probabilities for A to win and B to win then determine how the stakes should be divided.

   (b) *Pascal's solution*   Use combinations to determine the number of ways that the 2 points needed for A to win could occur in four plays. Then use combinations to determine the number of ways that the 3 points needed for B to win could occur. This is trickier than it looks, since A can win with 2 points in either two plays, three plays, or four plays. Compute the probabilities and compare with the results in part (a).

2. *Huygen's Mathematical Expectation*   In a game with n possible outcomes with probabilities $p_1, p_2, \ldots, p_n$, suppose that the *net* winnings are $w_1, w_2, \ldots, w_n$, respectively. Then the mathematical expectation is

   $$E = p_1 w_1 + p_2 w_2 + \cdots + p_n w_n$$

   The number E represents the profit or loss per game in the long run. The following problems are a modification of those of Huygens.

   (a) A fair die is tossed. A gambler wins \$3 if he throws a 6 and \$6 if he throws a 5. What is his expectation?

   [**NOTE:** $w_1 = w_2 = w_3 = w_4 = 0$]

   (b) A gambler plays the same game as in part (a), but now the gambler must pay \$1 to play. This means that $w_5 = \$5$, $w_6 = \$2$, and $w_1 = w_2 = w_3 = w_4 = -\$1$. What is the expectation?

10.3 EXERCISES

1. In a probability model, which of the following numbers could be the probability of an outcome:
$$0, \quad 0.01, \quad 0.35, \quad -0.4, \quad 1, \quad 1.4?$$

2. In a probability model, which of the following numbers could be the probability of an outcome:
$$1.5, \quad \tfrac{1}{2}, \quad \tfrac{3}{4}, \quad \tfrac{2}{3}, \quad 0, \quad -\tfrac{1}{4}?$$

3. Determine whether the following is a probability model.

Outcome	Probability
{1}	0.2
{2}	0.3
{3}	0.1
{4}	0.4

4. Determine whether the following is a probability model.

Outcome	Probability
{Jim}	0.4
{Bob}	0.3
{Faye}	0.1
{Patricia}	0.2

5. Determine whether the following is a probability model.

Outcome	Probability
{Linda}	0.3
{Jean}	0.2
{Grant}	0.1
{Ron}	0.3

6. Determine whether the following is a probability model.

Outcome	Probability
{Lanny}	0.3
{Joanne}	0.2
{Nelson}	0.1
{Rich}	0.5
{Judy}	-0.1

In Problems 7–12, construct a probability model for each experiment.

7. Tossing a fair coin twice

8. Tossing two fair coins once

9. Tossing two fair coins, then a fair die

10. Tossing a fair coin, a fair die, and then a fair coin

11. Tossing three fair coins once

12. Tossing one fair coin three times

In Problems 13–18, use the following spinners to construct a probability model for each experiment.

Spinner I   Spinner II   Spinner III

13. Spin spinner I, then spinner II. What is the probability of getting a 2 or a 4, followed by Red?

14. Spin spinner III, then spinner II. What is the probability of getting Forward, followed by Yellow or Green?

15. Spin spinner I, then II, then III. What is the probability of getting a 1, followed by Red or Green, followed by Backward?

16. Spin spinner II, then I, then III. What is the probability of getting Yellow, followed by a 2 or a 4, followed by Forward?

17. Spin spinner I twice, then spinner II. What is the probability of getting a 2, followed by a 2 or a 4, followed by Red or Green?

18. Spin spinner III, then spinner I twice. What is the probability of getting Forward, followed by a 1 or a 3, followed by a 2 or a 4?

In Problems 19–22, consider the experiment of tossing a coin twice. The table at the top of page 689 lists six possible assignments of probabilities for this experiment. Using this table, answer the following questions.

19. Which of the assignments of probabilities are consistent with the definition of a probability model?

20. Which of the assignments of probabilities should be used if the coin is known to be fair?

21. Which of the assignments of probabilities should be used if the coin is known to always come up tails?

22. Which of the assignments of probabilities should be used if tails is twice as likely as heads to occur?

	Sample Space			
Assignments	**HH**	**HT**	**TH**	**TT**
A	$\frac{1}{4}$	$\frac{1}{4}$	$\frac{1}{4}$	$\frac{1}{4}$
B	0	0	0	1
C	$\frac{3}{16}$	$\frac{5}{16}$	$\frac{5}{16}$	$\frac{3}{16}$
D	$\frac{1}{2}$	$\frac{1}{2}$	$-\frac{1}{2}$	$\frac{1}{2}$
E	$\frac{1}{4}$	$\frac{1}{4}$	$\frac{1}{4}$	$\frac{1}{8}$
F	$\frac{1}{9}$	$\frac{2}{9}$	$\frac{2}{9}$	$\frac{4}{9}$

23. **Assigning Probabilities**   A coin is weighted so that heads is four times as likely as tails to occur. What probability should we assign to heads? to tails?

24. **Assigning Probabilities**   A coin is weighted so that tails is twice as likely as heads to occur. What probability should we assign to heads? to tails?

25. **Assigning Probabilities**   A die is weighted so that an odd-numbered face is twice as likely to occur as an even-numbered face. What probability should we assign to each face?

26. **Assigning Probabilities**   A die is weighted so that a six cannot appear. The other faces occur with the same probability. What probability should we assign to each face?

For Problems 27–30, let the sample space be $S = \{1, 2, 3, 4, 5, 6, 7, 8, 9, 10\}$. *Suppose that the outcomes are equally likely.*

27. Compute the probability of the event $E = \{1, 2, 3\}$.

28. Compute the probability of the event $F = \{3, 5, 9, 10\}$.

29. Compute the probability of the event E: "an even number."

30. Compute the probability of the event F: "an odd number."

For Problems 31 and 32, an urn contains 5 white marbles, 10 green marbles, 8 yellow marbles, and 7 black marbles.

31. If one marble is selected, determine the probability that it is white.

32. If one marble is selected, determine the probability that it is black.

In Problems 33–36, assume equally likely outcomes.

33. Determine the probability of having 3 boys in a 3-child family.

34. Determine the probability of having 3 girls in a 3-child family.

35. Determine the probability of having 1 girl and 3 boys in a 4-child family.

36. Determine the probability of having 2 girls and 2 boys in a 4-child family.

For Problems 37–40, two fair dice are rolled.

37. Determine the probability that the sum of the two dice is 7.

38. Determine the probability that the sum of the two dice is 11.

39. Determine the probability that the sum of the two dice is 3.

40. Determine the probability that the sum of the two dice is 12.

In Problems 41–44, find the probability of the indicated event if $P(A) = 0.25$ *and* $P(B) = 0.45$.

41. $P(A \cup B)$ if $P(A \cap B) = 0.15$

42. $P(A \cap B)$ if $P(A \cup B) = 0.6$

43. $P(A \cup B)$ if A, B are mutually exclusive

44. $P(A \cap B)$ if A, B are mutually exclusive

45. If $P(A) = 0.60$, $P(A \cup B) = 0.85$, and $P(A \cap B) = 0.05$, find $P(B)$.

46. If $P(B) = 0.30$, $P(A \cup B) = 0.65$, and $P(A \cap B) = 0.15$, find $P(A)$.

47. According to the Federal Bureau of Investigation, in 1997 there was a 25.3% probability of theft involving a motor vehicle. If a victim of theft is randomly selected, what is the probability that he or she was not a victim of motor vehicle theft?

48. According to the Federal Bureau of Investigation, in 1997 there was a 5.6% probability of theft involving a bicycle. If a victim of theft is randomly selected, what is the probability that he or she was not a victim of bicycle theft?

49. In Chicago, there is a 30% probability that Memorial Day will have a high temperature in the 70s. What is the probability that next Memorial Day will not have a high temperature in the 70s in Chicago?

50. In Chicago, there is a 4% probability that Memorial Day will have a low temperature in the 30s. What is the probability that next Memorial Day will not have a low temperature in the 30s in Chicago?

For Problems 51–54, a golf ball is selected at random from a container. If the container has 9 white balls, 8 green balls, and 3 orange balls, find the probability of each event.

51. The golf ball is white or green.

52. The golf ball is white or orange.

53. The golf ball is not white.

54. The golf ball is not green.

55. On the "Price is Right" there is a game in which a bag is filled with 3 strike chips and 5 numbers. Let's say that the numbers in the bag are 0, 1, 3, 6, and 9. What is the probability of selecting a strike chip or the number 1?

56. Another game on the "Price is Right" requires the contestant to spin a wheel with numbers 5, 10, 15, 20, ... , 100. What is the probability that the contestant spins 100 or 30?

Problems 57–60 are based on a consumer survey of annual incomes in 100 households. The following table gives the data.

Income	$0–9999	$10,000–19,999	$20,000–29,999	$30,000–39,999	$40,000 or more
Number of households	5	35	30	20	10

57. What is the probability that a household has an annual income of $30,000 or more?

58. What is the probability that a household has an annual income between $10,000 and $29,999, inclusive?

59. What is the probability that a household has an annual income of less than $20,000?

60. What is the probability that a household has an annual income of $20,000 or more?

61. Surveys In a survey about the number of TV sets in a house, the following probability table was constructed:

Number of TV sets	0	1	2	3	4 or more
Probability	0.05	0.24	0.33	0.21	0.17

Find the probability of a house having:
(a) 1 or 2 TV sets
(b) 1 or more TV sets
(c) 3 or fewer TV sets
(d) 3 or more TV sets
(e) Less than 2 TV sets
(f) Less than 1 TV set
(g) 1, 2, or 3 TV sets
(h) 2 or more TV sets

62. Checkout Lines Through observation it has been determined that the probability for a given number of people waiting in line at the "5 items or less" checkout register of a supermarket is as follows:

Number waiting in line	0	1	2	3	4 or more
Probability	0.10	0.15	0.20	0.24	0.31

Find the probability of:
(a) At most 2 people in line
(b) At least 2 people in line
(c) At least 1 person in line

63. In a certain College Algebra class, there are 18 freshmen and 15 sophomores. Of the 18 freshmen, 10 are male, and of the 15 sophomores, 8 are male. Find the probability that a randomly selected student is:
(a) A freshman or female
(b) A sophomore or male

64. The faculty of the mathematics department at Joliet Junior College is composed of 4 females and 9 males. Of the 4 females, 2 are under the age of 40, and 3 of the males are under age 40. Find the probability that a randomly selected faculty member is:
(a) Female or under age 40
(b) Male or over age 40

65. Birthday Problem What is the probability that at least 2 people have the same birthday in a group of 12 people? Assume that there are 365 days in a year.

66. Birthday Problem What is the probability that at least 2 people have the same birthday in a group of 35 people? Assume that there are 365 days in a year.

67. Winning a Lottery In a certain lottery, there are ten balls, numbered 1, 2, 3, 4, 5, 6, 7, 8, 9, 10. Of these, five are drawn in order. If you pick five numbers that match those drawn in the correct order, you win $1,000,000. What is the probability of winning such a lottery?

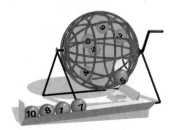

68. A committee of 6 people is to be chosen at random from a group of 14 people consisting of 2 supervisors, 5 skilled laborers, and 7 unskilled laborers. What is the probability that the committee chosen consists of 2 skilled and 4 unskilled laborers?

69. A fair coin is tossed 5 times.
  (a) Find the probability that exactly 3 heads appear.
  (b) Find the probability that no heads appear.

70. A fair coin is tossed 4 times.
  (a) Find the probability that exactly 1 tail appears.
  (b) Find the probability that no more than 1 tail appears.

71. A pair of fair dice is tossed 3 times.
  (a) Find the probability that the sum of 7 appears 3 times.
  (b) Find the probability that a sum of 7 or 11 appears at least twice.

72. A pair of fair dice is tossed 5 times.
  (a) Find the probability that the sum is never 2.
  (b) Find the probability that the sum is never 7.

73. Through a mix-up on the production line, 5 defective TVs were shipped out with 25 good ones. If 5 are selected at random, what is the probability that all 5 are defective? What is the probability that at least 2 of them are defective?

74. In a shipment of 50 transformers, 10 are known to be defective. If 30 transformers are picked at random, what is the probability that all 30 are nondefective? Assume that all transformers look alike and have an equal probability of being chosen.

75. In a promotion, 50 silver dollars are placed in a bag, one of which is valued at more than $10,000. The winner of the promotion is given the opportunity to reach into the bag, while blindfolded, and pull out 5 coins. What is the probability that 1 of the 5 coins is the one valued at more than $10,000?

CHAPTER REVIEW

Things To Know

Set (p. 660)		Well-defined collection of distinct objects, called elements
Null set (p. 660)	$\varnothing$	Set that has no elements
Equality (p. 660)	$A = B$	A and B have the same elements.
Subset (p. 660)	$A \subseteq B$	Each element of A is also an element of B.
Intersection (p. 661)	$A \cap B$	Set consisting of elements that belong to both A and B
Union (p. 661)	$A \cup B$	Set consisting of elements that belong to either A or B, or both
Universal set (p. 661)	U	Set consisting of all the elements that we wish to consider
Complement (p. 661)	$\overline{A}$	Set consisting of elements of the universal set that are not in A
Finite set (p. 662)		The number of elements in the set is a nonnegative integer.
Infinite set (p. 662)		A set that is not finite
Counting formula (p. 663)		$n(A \cup B) = n(A) + n(B) - n(A \cap B)$
Addition Principle (p. 664)		If $A \cap B = \varnothing$, then $n(A \cup B) = n(A) + n(B)$.
Multiplication Principle (p. 668)		If a task consists of a sequence of choices in which there are p selections for the first choice, q selections for the second choice, and so on, then the task of making these selections can be done in $p \cdot q \cdot \ldots$ different ways.
Permutation (p. 668)		An ordered arrangement of r objects chosen from n objects.
Type 1 (p. 669)	n^r	The n objects are distinct (different) and repetition is allowed in the selection of r of them.
Type 2 (p. 671)	$P(n, r) = n(n - 1) \cdot \ldots \cdot [n - (r - 1)]$ $= \dfrac{n!}{(n - r)!}$	The n objects are distinct (different) and repetition is not allowed in the selection of r of them, where $r \leq n$.
Type 3 (p. 674)	$\dfrac{n!}{n_1! n_2! \cdots n_k!}$	The number of permutations of n objects of which n_1 are of one kind, n_2 are of a second kind, ..., and n_k are of a kth kind, where $n = n_1 + n_2 + \cdots + n_k$

Combination (p. 672)	$C(n, r) = \dfrac{P(n, r)}{r!}$ $= \dfrac{n!}{(n - r)!r!}$	An arrangement, without regard to order, of r objects selected from n distinct objects without repetition, where $r \leq n$.
Sample space (p. 678)		Set whose elements represent all the logical possibilities that can occur as a result of an experiment
Probability (p. 678)		A nonnegative number assigned to each outcome of a sample space; the sum of all the probabilities of the outcomes equals 1
Equally likely outcomes (p. 680)	$P(E) = \dfrac{n(E)}{n(S)}$	The same probability is assigned to each outcome.
Addition Rule (p. 682)	$P(E \cup F) = P(E) + P(F) - P(E \cap F)$	
Complement of an event (p. 683)	$P(\bar{E}) = 1 - P(E)$	

Objectives

You should be able to:

Find all the subsets of a set (p. 660)

Find the intersection and union of sets (p. 661)

Find the complement of a set (p. 661)

Count the number of elements in a set (p. 662)

Solve counting problems using the Multiplication Principle (p. 666)

Solve counting problems using permutations (p. 668)

Solve counting problems using combinations (p. 672)

Solve counting problems using permutations involving non-distinct objects (p. 674)

Construct probability models (p. 678)

Compute probabilities of equally likely outcomes (p. 680)

Utilize the Addition Rule to find probabilities (p. 682)

Utilize the Complement Rule to find probabilities (p. 683)

Compute probabilities using permutations and combinations (p. 685)

Fill-in-the-Blank Items

1. The _____ of A with B consists of all elements in either A or B or both; the _____ of A with B consists of all elements in both A and B.

2. $P(5, 2) = $ _____ ; $C(5, 2) = $ _____ .

3. A(n) _____ is an ordered arrangement of r objects chosen from n objects.

4. A(n) _____ is an arrangement of n distinct objects without regard to order.

5. When the same probability is assigned to each outcome of a sample space, the experiment is said to have _____ _____ outcomes.

6. The _____ of an event E is the set of all outcomes in the sample space S that are not outcomes in the event E.

True/False Items

T F **1.** The intersection of two sets is always a subset of their union.

T F **2.** $P(n, r) = \dfrac{n!}{r!}$

T F **3.** In a combination problem, order is not important.

T F **4.** A permutation is an ordered arrangement of r objects chosen from n objects.

T F **5.** The probability of an event can never equal 0.

T F **6.** In a probability model, the sum of all probabilities is 1.

Review Exercises

Blue problem numbers indicate the author's suggestions for use in a Practice Test.

In Problems 1–8, use U = universal set = $\{1, 2, 3, 4, 5, 6, 7, 8, 9\}$, A = $\{1, 3, 5, 7\}$, B = $\{3, 5, 6, 7, 8\}$, and C = $\{2, 3, 7, 8, 9\}$ to find each set.

1. $A \cup B$

2. $B \cup C$

3. $A \cap C$

4. $A \cap B$

5. $\bar{A} \cup \bar{B}$

6. $\bar{B} \cap \bar{C}$

7. $\overline{B \cap C}$

8. $\overline{A \cup B}$

9. If $n(A) = 8$, $n(B) = 12$, and $n(A \cap B) = 3$, find $n(A \cup B)$.

10. If $n(A) = 12$, $n(A \cup B) = 30$, and $n(A \cap B) = 6$, find $n(B)$.

In Problems 11–16, use the information supplied in the figure:

11. How many are in A?

12. How many are in A or B?

13. How many are in A and C?

14. How many are not in B?

15. How many are in neither A nor C?

16. How many are in B but not in C?

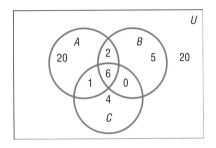

In Problems 17–22, compute the given expression.

17. $5!$

18. $6!$

19. $P(8, 3)$

20. $P(7, 3)$

21. $C(8, 3)$

22. $C(7, 3)$

---

23. Displaying Suits A clothing store sells pure wool and polyester–wool suits. Each suit comes in 3 colors and 10 sizes. How many suits are required for a complete assortment?

24. Connecting Wires In connecting a certain electrical device, 5 wires are to be connected to 5 different terminals. How many different wirings are possible if 1 wire is connected to each terminal?

25. Baseball On a given day, the American Baseball League schedules 7 games. How many different outcomes are possible, assuming that each game is played to completion?

26. Baseball On a given day, the National Baseball League schedules 6 games. How many different outcomes are possible, assuming that each game is played to completion?

27. Filling Seats If 4 people enter a bus having 9 vacant seats, in how many ways can they be seated?

28. Arranging Letters How many different arrangements are there of the letters in the word ROSE?

29. Relay Teams In how many ways can a squad of 4 relay runners be chosen from a track team of 8 runners?

30. Designing Tests A professor has 10 similar problems to put on a test with 3 problems. How many different tests can she design?

31. Baseball In how many different ways can 14 baseball teams in the American League be paired without regard to which team is at home?

32. Arranging Books on a Shelf How many ways are there to arrange 5 different French books and 5 different Spanish books on a shelf if:
(a) Books of the same language must be grouped together, French on the left, Spanish on the right?
(b) French and Spanish books must alternate in the grouping, beginning with a French book?

33. Telephone Numbers Using the digits $0, 1, 2, \ldots, 9$, how many 7-digit numbers can be formed if the first digit cannot be 0 or 9 and if the last digit is greater than or equal to 2 and less than or equal to 3? Repeated digits are allowed.

34. Home Choices A contractor constructs homes with 5 different choices of exterior finish, 3 different roof arrangements, and 4 different window designs. How many different types of homes can be built?

35. License Plate Possibilities A license plate consists of 1 letter, excluding O and I, followed by a 4-digit number that cannot have a 0 in the lead position. How many different plates are possible?

36. Using the digits 0 and 1, how many different numbers consisting of 8 digits can be formed?

37. Forming Different Words How many different 7-letter words (real or imaginary) can be formed using all the letters in the word MISSING?

38. Arranging Flags How many different vertical arrangements are there of 10 flags if 4 are white, 3 are blue, 2 are green, and 1 is red?

39. Forming Committees A group of 9 people is going to be formed into committees of 4, 3, and 2 people. How many committees can be formed if:
(a) A person can serve on any number of committees?
(b) No person can serve on more than one committee?

40. Forming Committees A group consists of 5 men and 8 women. A committee of 4 is to be formed from this group, and policy dictates that at least 1 woman be on this committee.
(a) How many committees can be formed that contain exactly 1 man?
(b) How many committees can be formed that contain exactly 2 women?
(c) How many committees can be formed that contain at least 1 man?

41. Birthday Problem For this problem, assume that the year has 365 days.
(a) How many ways can 18 people have different birthdays?
(b) What is the probability that nobody has the same birthday in a group of 18 people?
(c) What is the probability in a group of 18 people that at least 2 people have the same birthday?

42. Death Rates According to the U.S. National Center for Health Statistics, 32.1% of all deaths in 1994 were due to heart disease.
(a) What is the probability that a randomly selected death in 1994 was due to heart disease?
(b) What is the probability that a randomly selected death in 1994 was not due to heart disease?

43. Unemployment According to the U.S. Bureau of Labor Statistics, 5.4% of the U.S. labor force was unemployed in 1996.
(a) What is the probability that a randomly selected member of the U.S. labor force was unemployed in 1996?
(b) What is the probability that a randomly selected member of the U.S. labor force was not unemployed in 1996?

44. From a box containing three 40-watt bulbs, six 60-watt bulbs, and eleven 75-watt bulbs, a bulb is drawn at random. What is the probability that the bulb is 40 watts? What is the probability that it is not a 75-watt bulb?

45. You have four $1 bills, three $5 bills, and two $10 bills in your wallet. If you pick a bill at random, what is the probability that it will be a $1 bill?

46. Each of the letters in the word ROSE is written on an index card and the cards are then shuffled. What is the probability that, when the cards are dealt out, they spell the word ROSE?

47. Each of the numbers, 1, 2, ..., 100 is written on an index card and the cards are then shuffled. If a card is selected at random, what is the probability that the number on the card is divisible by 5? What is the probability that the card selected is either a 1 or names a prime number?

48. Computing Probabilities Because of a mistake in packaging, a case of 12 bottles of red wine contained 5 Merlot and 7 Cabernet, each without labels. All the bottles look alike and have equal probability of being chosen. Three bottles are selected.
(a) What is the probability that all 3 are Merlot?
(b) What is the probability that exactly 2 are Merlot?
(c) What is the probability that none is a Merlot?

49. Tossing a Coin A fair coin is tossed 10 times.
(a) What is the probability of obtaining exactly 5 heads?
(b) What is the probability of obtaining all heads?

50. At the Milex tune-up and brake repair shop, the manager has found that a car will require a tune-up with a probability of 0.6, a brake job with a probability of 0.1, and both with a probability of 0.02.
(a) What is the probability that a car requires either a tune-up or a brake job?
(b) What is the probability that a car requires a tune-up but not a brake job?
(c) What is the probability that a car requires neither type of repair?

Project at Motorola

Probability of error in digital wireless communications

Motorola uses digital communication technology in the design of cellular phones. In digital communications, a speech or data signal is converted into bits by a mobile phone and transmitted over the air to a base station. A bit is a unit of information that can be either 0 or 1. Thus, a single bit can represent two levels of information. A speech signal is converted into a digital signal by taking samples of the speech signal at specified time intervals and then quantizing these samples. A *sample* is simply the value of the speech signal at a single time instant. To *quantize* the sample, a number of acceptable levels is defined, and the level that is closest to the value of the speech signal at a sampling instant is assigned to that sample. A symbol consisting of multiple bits is assigned to each level, and these resulting symbols are transmitted over the air. The number of levels is determined by the length of the symbols. For instance, a symbol of length 1 bit can only represent two levels (0 or 1). A symbol of length 2 bits represents 4 levels (00, 01, 10, or 11).

Each symbol is sent by the cellular phone over the air and received by the base station. Because of many factors, including other cellular phone transmissions and obstacles in the path of the transmitted signal, the symbols may be received incorrectly. If one or more bits of the symbol are changed (that is, a 0 becomes a 1, or vice versa), the symbol is received in error. If a symbol is 2 bits in length and the probability that a single bit is received in error is equal to k, the probability that both bits are received in error is equal to the probability that the first bit is received in error multiplied by the probability that the second bit is received in error: $k \cdot k = k^2$.

1. Assume that the transmitted symbol is 1011. Construct the sample space of possible symbols received, including symbols with 0, 1, 2, 3, or 4 bit errors.

2. Suppose that the transmitted symbol is 1011 and the probability that a single bit is changed is equal to $1/3$. What is the probability that the symbol is received correctly?
   [**Hint:** What is the probability that the first bit is correct? The second? The third? The fourth?]

3. Assume that a symbol is 8 bits long. How many combinations of received symbols with 2 bit errors are there? If the probability that a single bit is changed is equal to $1/3$, what is the probability that the message is received correctly? Based on this result, what is the probability that it is received incorrectly?

In some cases, there is a probability that a transmitted symbol is received in error. Error-detection coding is used to determine whether errors have occurred in the received symbol. In some cases, detected errors can also be corrected so that the decoded message is exactly what was sent. A simple type of error-detection code adds a single bit to the end of each transmitted symbol, to make the number of 1's in the symbol an even number. This is called an **even parity** code. At the receiver, the number of 1's in the received symbol is counted. If this number is odd, the receiver knows that there was at least one error in the symbol transmission. This type of code can detect an odd number of bit errors.

4. Assume that a symbol is 7 bits long, and a parity code bit is added so that the transmitted symbol is 8 bits in length. The probability that a single bit is changed is equal to $1/3$. What is the probability that an error is detected in the received symbol? What is the probability that an error occurred, but is not detected?
   [**Hint:** Research the Binomial Probability Theorem in a text on finite mathematics or elementary statistics.]

Graphing Utilities

1   THE VIEWING RECTANGLE

Figure 1 $y = 2x$

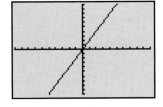

Figure 2

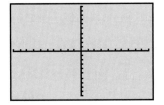

All graphing utilities, that is, all graphing calculators and all computer software graphing packages, graph equations by plotting points on a screen. The screen itself actually consists of small rectangles, called **pixels.** The more pixels the screen has, the better the resolution. Most graphing calculators have 48 pixels per square inch; most computer screens have 32 to 108 pixels per square inch. When a point to be plotted lies inside a pixel, the pixel is turned on (lights up). The graph of an equation is a collection of pixels. Figure 1 shows how the graph of $y = 2x$ looks on a TI-83 graphing calculator.

The screen of a graphing utility will display the coordinate axes of a rectangular coordinate system. However, you must set the scale on each axis. You must also include the smallest and largest values of x and y that you want included in the graph. This is called **setting the viewing rectangle** or **viewing window.** Figure 2 illustrates a typical viewing window.

To select the viewing window, we must give values to the following expressions:

Xmin: the smallest value of x

Xmax: the largest value of x

Xscl: the number of units per tick mark on the x-axis

Ymin: the smallest value of y

Ymax: the largest value of y

Yscl: the number of units per tick mark on the y-axis

Figure 3 illustrates these settings and their relation to the Cartesian coordinate system.

Figure 3

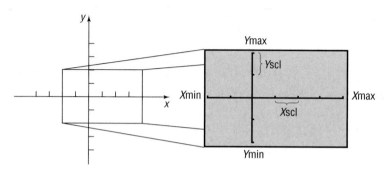

If the scale used on each axis is known, we can determine the minimum and maximum values of x and y shown on the screen by counting the tick marks. Look again at Figure 2. For a scale of 1 on each axis, the minimum and maximum values of x are -10 and 10, respectively; the minimum and maximum values of y are also -10 and 10. If the scale is 2 on each axis, then the minimum and maximum values of x are -20 and 20, respectively; and the minimum and maximum values of y are -20 and 20, respectively.

Conversely, if we know the minimum and maximum values of x and y, we can determine the scales being used by counting the tick marks displayed. We shall follow the practice of showing the minimum and maximum values of x and y in our illustrations so that you will know how the viewing window was set. See Figure 4.

Figure 4

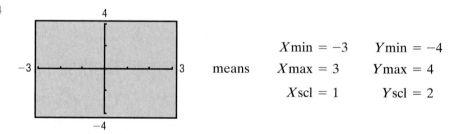

means

$$X\text{min} = -3 \qquad Y\text{min} = -4$$
$$X\text{max} = 3 \qquad Y\text{max} = 4$$
$$X\text{scl} = 1 \qquad Y\text{scl} = 2$$

EXAMPLE 1

Finding the Coordinates of a Point Shown on a Graphing Utility Screen

Find the coordinates of the point shown in Figure 5. Assume that the coordinates are integers.

Figure 5

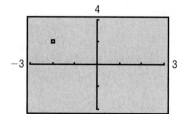

Solution First we note that the viewing window used in Figure 5 is

$$X\text{min} = -3 \qquad Y\text{min} = -4$$
$$X\text{max} = 3 \qquad Y\text{max} = 4$$
$$X\text{scl} = 1 \qquad Y\text{scl} = 2$$

The point shown is 2 tick units to the left on the horizontal axis (scale = 1) and 1 tick up on the vertical axis (scale = 2). The coordinates of the point shown are $(-2, 2)$. ∎

1   EXERCISES

In Problems 1–4, determine the coordinates of the points shown. Tell in which quadrant each point lies. Assume that the coordinates are integers.

1.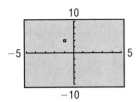

2.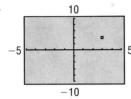

3.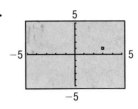

4.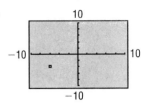

In Problems 5–10, determine the viewing window used.

5.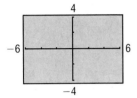

6.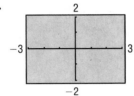

7.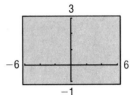

8.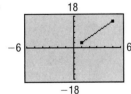

9.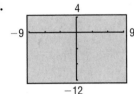

10.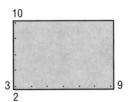

In Problems 11–16, select a setting so that each of the given points will lie within the viewing rectangle.

11. $(-10, 5), (3, -2), (4, -1)$

12. $(5, 0), (6, 8), (-2, -3)$

13. $(40, 20), (-20, -80), (10, 40)$

14. $(-80, 60), (20, -30), (-20, -40)$

15. $(0, 0), (100, 5), (5, 150)$

16. $(0, -1), (100, 50), (-10, 30)$

In Problems 17–20, find the length of the line segment. Assume that the endpoints of each line segment have integer coordinates.

17.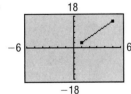

18.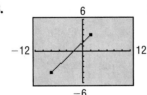

19.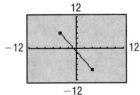

20.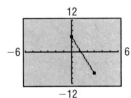

---

2   USING A GRAPHING UTILITY TO GRAPH EQUATIONS

From Examples 2 and 3 of Section 2.2, we see that a graph can be obtained by plotting points in a rectangular coordinate system and connecting them. Graphing utilities perform these same steps when graphing an equation. For example, the TI-83 determines 95 evenly spaced input values,* uses the equation to determine the output values, plots these points on the screen, and finally (if in the connected mode) draws a line between consecutive points.

*These input values depend on the values of Xmin and Xmax. For example, if Xmin $= -10$ and Xmax $= 10$, then the first input value will be -10 and the next input value will be $-10 + (10 - (-10))/94 = -9.7872$, and so on.

To graph an equation in two variables x and y using a graphing utility requires that the equation be written in the form $y = \{$expression in $x\}$. If the original equation is not in this form, replace it by equivalent equations until the form $y = \{$expression in $x\}$ is obtained. In general, there are four ways to obtain equivalent equations.

PROCEDURES THAT RESULT IN EQUIVALENT EQUATIONS

1. Interchange the two sides of the equation:

    Replace $3x + 5 = y$ by $y = 3x + 5$

2. Simplify the sides of the equation by combining like terms, eliminating parentheses, and so on:

    Replace $(2y + 2) + 6 = 2x + 5(x + 1)$

    by $2y + 8 = 7x + 5$

3. Add or subtract the same expression on both sides of the equation:

    Replace $y + 3x - 5 = 4$

    by $y + 3x - 5 + 5 = 4 + 5$

4. Multiply or divide both sides of the equation by the same nonzero expression:

    Replace $3y = 6 - 2x$

    by $\dfrac{1}{3} \cdot 3y = \dfrac{1}{3}(6 - 2x)$

EXAMPLE 1

Expressing an Equation in the Form $y = \{$expression in $x\}$

Solve for y: $2y + 3x - 5 = 4$

Solution   We replace the original equation by a succession of equivalent equations.

$$2y + 3x - 5 = 4$$

$$2y + 3x - 5 + 5 = 4 + 5 \qquad \text{Add 5 to both sides.}$$

$$2y + 3x = 9 \qquad \text{Simplify.}$$

$$2y + 3x - 3x = 9 - 3x \qquad \text{Subtract } 3x \text{ from both sides.}$$

$$2y = 9 - 3x \qquad \text{Simplify.}$$

$$\frac{2y}{2} = \frac{9 - 3x}{2} \qquad \text{Divide both sides by 2.}$$

$$y = \frac{9 - 3x}{2} \qquad \text{Simplify.} \qquad ■$$

Now we are ready to graph equations using a graphing utility. Most graphing utilities require the following steps:

STEPS FOR GRAPHING AN EQUATION
USING A GRAPHING UTILITY

STEP 1: Solve the equation for y in terms of x.

STEP 2: Get into the graphing mode of your graphing utility. The screen will usually display $y =$ ____ , prompting you to enter the expression involving x that you found in Step 1. (Consult your manual for the correct way to enter the expression; for example, $y = x^2$ might be entered as $x^\wedge 2$ or as $x*x$ or as $x \ x^Y \ 2$).

STEP 3: Select the viewing window. Without prior knowledge about the behavior of the graph of the equation, it is common to select the **standard viewing window*** initially. The viewing window is then adjusted based on the graph that appears. In this text the standard viewing window will be

$$X\text{min} = -10 \qquad Y\text{min} = -10$$
$$X\text{max} = 10 \qquad Y\text{max} = 10$$
$$X\text{scl} = 1 \qquad Y\text{scl} = 1$$

STEP 4: Execute.

STEP 5: Adjust the viewing window until a complete graph is obtained.

EXAMPLE 2

Graphing an Equation on a Graphing Utility

Graph the equation: $6x^2 + 3y = 36$

Solution   STEP 1: We solve for y in terms of x.

$$6x^2 + 3y = 36$$

$$3y = -6x^2 + 36 \qquad \text{Subtract } 6x^2 \text{ from both sides of the equation.}$$

$$y = -2x^2 + 12 \qquad \text{Divide both sides of the equation by 3 and simplify.}$$

STEP 2: From the graphing mode, enter the expression $-2x^2 + 12$ after the prompt $y =$ ____ .

STEP 3: Set the viewing window to the standard viewing window.

STEP 4: Execute. The screen should look like Figure 6.

STEP 5: The graph of $y = -2x^2 + 12$ is not complete. The value of Ymax must be increased so that the top portion of the graph is visible. After increasing the value of Ymax to 12, we obtain the graph in Figure 7. The graph is now complete.

Figure 6

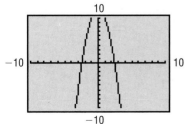

Figure 7

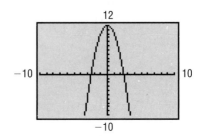

*Some graphing utilities have a ZOOM-STANDARD feature that automatically sets the viewing window to the standard viewing window and graphs the equation.

Look again at Figure 7. Although a complete graph is shown, the graph might be improved by adjusting the values of Xmin and Xmax. Figure 8 shows the graph of $y = -2x^2 + 12$ using Xmin $= -4$ and Xmax $= 4$. Do you think this is a better choice for the viewing window?

Figure 8

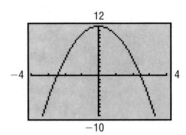

EXAMPLE 3   Creating a Table and Graphing an Equation

Create a table and graph the equation: $y = x^3$

Solution   Most graphing utilities have the capability of creating a table of values for an equation. (Check your manual to see if your graphing utility has this capability.) Table 1 illustrates a table of values for $y = x^3$ on a TI-83. See Figure 9 for the graph.

Figure 9

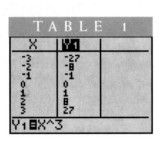

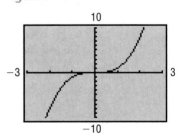

---

2 EXERCISES

In Problems 1–16, graph each equation using the following viewing windows:

(a) Xmin $= -5$	(b) Xmin $= -10$	(c) Xmin $= -10$	(d) Xmin $= -5$
Xmax $= 5$	Xmax $= 10$	Xmax $= 10$	Xmax $= 5$
Xscl $= 1$	Xscl $= 1$	Xscl $= 2$	Xscl $= 1$
Ymin $= -4$	Ymin $= -8$	Ymin $= -8$	Ymin $= -20$
Ymax $= 4$	Ymax $= 8$	Ymax $= 8$	Ymax $= 20$
Yscl $= 1$	Yscl $= 1$	Yscl $= 2$	Yscl $= 5$

1. $y = x + 2$ **2.** $y = x - 2$ **3.** $y = -x + 2$ **4.** $y = -x - 2$

5. $y = 2x + 2$ **6.** $y = 2x - 2$ **7.** $y = -2x + 2$ **8.** $y = -2x - 2$

9. $y = x^2 + 2$ **10.** $y = x^2 - 2$ **11.** $y = -x^2 + 2$ **12.** $y = -x^2 - 2$

13. $3x + 2y = 6$ **14.** $3x - 2y = 6$ **15.** $-3x + 2y = 6$ **16.** $-3x - 2y = 6$

17.–32. *For each of the above equations, create a table, $-3 \le x \le 3$, and list points on the graph.*

3 USING A GRAPHING UTILITY TO LOCATE INTERCEPTS AND CHECK FOR SYMMETRY

VALUE AND ZERO (OR ROOT)

Most graphing utilities have an eVALUEate feature that, given a value of x, determines the value of y for an equation. We can use this feature to evaluate an equation at $x = 0$ to determine the y-intercept. Most graphing utilities also have a ZERO (or ROOT) feature that can be used to determine the x-intercept(s) of an equation.

EXAMPLE 1   **Finding Intercepts Using a Graphing Utility**

Use a graphing utility to find the intercepts of the equation $y = x^3 - 8$.

Solution   Figure 10(a) shows the graph of $y = x^3 - 8$.

Figure 10

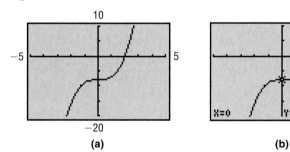
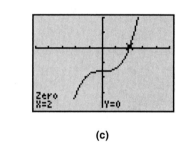

(a)   (b)   (c)

The eVALUEate feature of a TI-83 graphing calculator accepts as input a value of x and determines the value of y. If we let $x = 0$, we find that the y-intercept is -8. See Figure 10(b).

The ZERO feature of a TI-83 is used to find the x-intercept(s). See Figure 10(c). The x-intercept is 2.   ∎

TRACE

Most graphing utilities allow you to move from point to point along the graph, displaying on the screen the coordinates of each point. This feature is called TRACE.

EXAMPLE 2   **Using TRACE to Locate Intercepts**

Graph the equation $y = x^3 - 8$. Use TRACE to locate the intercepts.

Solution   Figure 11 shows the graph of $y = x^3 - 8$.

Figure 11

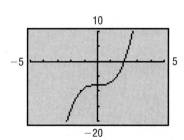

Activate the TRACE feature. As you move the cursor along the graph, you will see the coordinates of each point displayed. When the cursor is on the *y*-axis, we find that the *y*-intercepts is −8. See Figure 12.

Figure 12

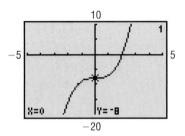

Continue moving the cursor along the graph. Just before you get to the *x*-axis, the display will look like the one in Figure 13(a). (Due to differences in graphing utilities, your display may be slightly different from the one shown here.)

Figure 13

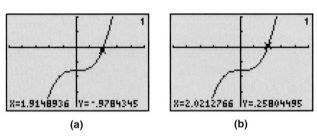

(a)   (b)

In Figure 13(a), the negative value of the *y*-coordinate indicates that we are still below the *x*-axis. The next position of the cursor is shown in Figure 13(b). The positive value of the *y*-coordinate indicates that we are now above the *x*-axis. This means that between these two points the *x*-axis was crossed. The *x*-intercept lies between 1.9148936 and 2.0212766. ∎

EXAMPLE 3

Graphing the Equation $y = \dfrac{1}{x}$

Figure 14

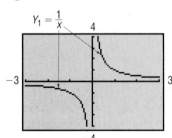

Graph the equation: $y = \dfrac{1}{x}$

With the viewing window set as

$$X\text{min} = -3 \qquad Y\text{min} = -4$$
$$X\text{max} = 3 \qquad Y\text{max} = 4$$
$$X\text{scl} = 1 \qquad Y\text{scl} = 1$$

Use TRACE to infer information about the intercepts and symmetry.

Solution Figure 14 illustrates the graph. We infer from the graph that there are no intercepts; we may also infer that symmetry with respect to the origin is a possibility. The TRACE feature on a graphing utility can provide further evidence of symmetry with respect to the origin. Using TRACE, we observe that for any ordered pair (x, y) the ordered pair $(-x, -y)$ is also a point on the graph. For example, the points $(0.95744681, 1.0444444)$ and $(-0.95744681, -1.0444444)$ both lie on the graph. ∎

3   EXERCISES

In Problems 1–6, use ZERO (or ROOT) to approximate the smaller of the two x-intercepts of each equation. Express the answer rounded to two decimal places.

1. $y = x^2 + 4x + 2$

2. $y = x^2 + 4x - 3$

3. $y = 2x^2 + 4x + 1$

4. $y = 3x^2 + 5x + 1$

5. $y = 2x^2 - 3x - 1$

6. $y = 2x^2 - 4x - 1$

*In Problems 7–14, use ZERO (or ROOT) to approximate the **positive** x-intercepts of each equation. Express each answer rounded to two decimal places.*

7. $y = x^3 + 3.2x^2 - 16.83x - 5.31$

8. $y = x^3 + 3.2x^2 - 7.25x - 6.3$

9. $y = x^4 - 1.4x^3 - 33.71x^2 + 23.94x + 292.41$

10. $y = x^4 + 1.2x^3 - 7.46x^2 - 4.692x + 15.2881$

11. $y = \pi x^3 - (8.88\pi + 1)x^2 - (42.066\pi - 8.88)x + 42.066$

12. $y = \pi x^3 - (5.63\pi + 2)x^2 - (108.392\pi - 11.26)x + 216.784$

13. $y = x^3 + 19.5x^2 - 1021x + 1000.5$

14. $y = x^3 + 14.2x^2 - 4.8x - 12.4$

In Problems 15–18, the graph of an equation is given.
  (a) *List the intercepts of the graph.*
  (b) *Based on the graph, tell whether the graph is symmetric with respect to the x-axis, y-axis, and/or origin.*

15.
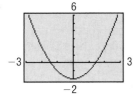

16.
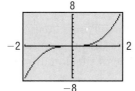

17.
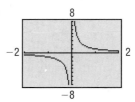

18.
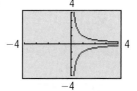

---

4   USING A GRAPHING UTILITY TO SOLVE EQUATIONS

For many equations, there are no algebraic techniques that lead to a solution. For such equations, a graphing utility can often be used to investigate possible solutions. When a graphing utility is used to solve an equation, usually *approximate* solutions are obtained. Unless otherwise stated, we shall follow the practice of giving approximate solutions *rounded to two decimal places*.

The ZERO (or ROOT) feature of a graphing utility can be used to find the solutions of an equation when one side of the equation is 0. In using this feature to solve equations, we make use of the fact that the *x*-intercepts (or zeros) of the graph of an equation are found by letting $y = 0$ and solving the equation for *x*. Solving an equation for *x* when one side of the equation is 0 is equivalent to finding where the graph of the corresponding equation crosses or touches the *x*-axis.

EXAMPLE 1   **Using ZERO (or ROOT) to Approximate Solutions of an Equation**

Find the solution(s) of the equation $x^2 - 6x + 7 = 0$. Round answers to two decimal places.

Solution   The solutions of the equation $x^2 - 6x + 7 = 0$ are the same as the x-intercepts of the graph of $Y_1 = x^2 - 6x + 7$. We begin by graphing the equation. See Figure 15(a).

Figure 15

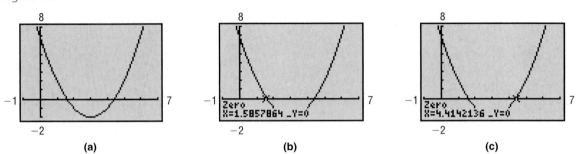

(a)   (b)   (c)

From the graph there appear to be two x-intercepts (solutions to the equation): one between 1 and 2, the other between 4 and 5.

Using the ZERO (or ROOT) feature of our graphing utility, we determine that the x-intercepts, and so the solutions to the equation, are $x = 1.59$ and $x = 4.41$, rounded to two decimal places. See Figures 15(b) and (c). ■

A second method for solving equations using a graphing utility involves the INTERSECT feature of the graphing utility. This feature is used most effectively when one side of the equation is not 0.

| EXAMPLE 2 | **Using INTERSECT to Approximate Solutions of an Equation** |

Find the solution(s) to the equation $3(x - 2) = 5(x - 1)$. Round answers to two decimal places.

Solution   We begin by graphing each side of the equation as follows: graph $Y_1 = 3(x - 2)$ and $Y_2 = 5(x - 1)$. See Figure 16(a).

Figure 16

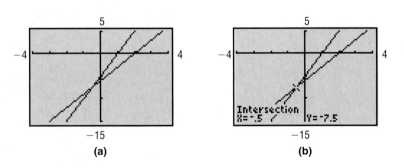

(a)   (b)

At the point of intersection of the graphs, the value of the y-coordinate is the same. We conclude that the x-coordinate of the point of intersection represents the solution to the equation. Do you see why? The INTERSECT feature on a graphing utility determines the point of intersection of the graphs. Using this feature, we find that the graphs intersect at $(-0.5, -7.5)$. See Figure 16(b). The solution of the equation is therefore $x = -0.5$. ■

Check: We can verify our solution by evaluating each side of the equation with -0.5 STOred in x. See Figure 17. Since the left side of the equation equals the right side of the equation, the solution checks.

Figure 17

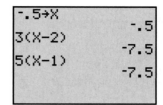

SUMMARY

The steps to follow for approximating solutions of equations are given next.

STEPS FOR APPROXIMATING SOLUTIONS OF EQUATIONS USING ZERO (OR ROOT)

STEP 1:   Write the equation in the form {expression in x} = 0.

STEP 2:   Graph Y_1 = {expression in x}.

STEP 3:   Use ZERO (or ROOT) to determine each x-intercept of the graph.

STEPS FOR APPROXIMATING SOLUTIONS OF EQUATIONS USING INTERSECT

STEP 1:   Graph Y_1 = {expression in x on left side of equation}.

Graph Y_2 = {expression in x on right side of equation}.

STEP 2:   Use INTERSECT to determine each x-coordinate of the point(s) of intersection, if any.

EXAMPLE 3	Solving a Radical Equation

Find the real solutions of the equation $\sqrt[3]{2x - 4} - 2 = 0$.

Solution   Figure 18 shows the graph of the equation $Y_1 = \sqrt[3]{2x - 4} - 2$. From the graph, we see one x-intercept near 6. Using ZERO (or ROOT), we find that the x-intercept is 6. The only solution is $x = 6$.

Figure 18

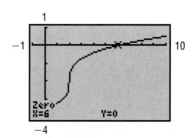

5 SQUARE SCREENS

Figure 19

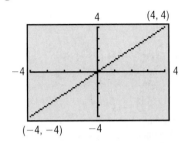

Most graphing utilities have a rectangular screen. Because of this, using the same settings for both x and y will result in a distorted view. For example, Figure 19 shows the graph of the line $y = x$ connecting the points $(-4, -4)$ and $(4, 4)$.

We expect the line to bisect the first and third quadrants, but it doesn't. We need to adjust the selections for Xmin, Xmax, Ymin, and Ymax so that a **square screen** results. On most graphing utilities, this is accomplished by setting the ratio of x to y at $3:2$.* In other words,

$$2(X\text{max} - X\text{min}) = 3(Y\text{max} - Y\text{min})$$

EXAMPLE 1	**Examples of Viewing Rectangles That Result in Square Screens**

(a) Xmin $= -3$ (b) Xmin $= -6$ (c) Xmin $= -6$

      Xmax $=$ 3 Xmax $=$ 6 Xmax $=$ 6

      Xscl $=$ 1 Xscl $=$ 1 Xscl $=$ 2

Figure 20 Ymin $= -2$ Ymin $= -4$ Ymin $= -4$

      Ymax $=$ 2 Ymax $=$ 4 Ymax $=$ 4

      Yscl $=$ 1 Yscl $=$ 1 Yscl $=$ 1 ■

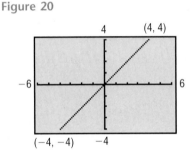

Figure 20 shows the graph of the line $y = x$ on a square screen using the viewing rectangle given in Example 1(b). Notice that the line now bisects the first and third quadrants. Compare this illustration to Figure 19.

*Some graphing utilities have a built-in function that automatically squares the screen. For example, the TI-85 has a ZSQR function that does this. Some graphing utilities require a ratio other than $3:2$ to square the screen. For example, the HP 48G requires the ratio of x to y to be $2:1$ for a square screen. Consult your manual.

5 EXERCISES

In Problems 1–8, determine which of the given viewing rectangles result in a square screen.

1. Xmin $= -3$

    Xmax $=$ 3

    Xscl $=$ 2

    Ymin $= -2$

    Ymax $=$ 2

    Yscl $=$ 2

2. Xmin $= -5$

    Xmax $=$ 5

    Xscl $=$ 1

    Ymin $= -4$

    Ymax $=$ 4

    Yscl $=$ 1

3. Xmin $=$ 0

    Xmax $=$ 9

    Xscl $=$ 3

    Ymin $= -2$

    Ymax $=$ 4

    Yscl $=$ 2

4. Xmin $= -6$

    Xmax $=$ 6

    Xscl $=$ 1

    Ymin $= -4$

    Ymax $=$ 4

    Yscl $=$ 2

5. Xmin $= -6$

    Xmax $=$ 6

    Xscl $=$ 1

    Ymin $= -2$

    Ymax $=$ 2

    Yscl $=$ 0.5

6. Xmin $= -6$

    Xmax $=$ 6

    Xscl $=$ 2

    Ymin $= -4$

    Ymax $=$ 4

    Yscl $=$ 1

7. Xmin $=$ 0

    Xmax $=$ 9

    Xscl $=$ 1

    Ymin $= -2$

    Ymax $=$ 4

    Yscl $=$ 1

8. Xmin $= -6$

    Xmax $=$ 6

    Xscl $=$ 2

    Ymin $= -4$

    Ymax $=$ 4

    Yscl $=$ 2

9. If $X\min = -4$, $X\max = 8$, and $X\mathrm{scl} = 1$, how should $Y\min$, $Y\max$, and $Y\mathrm{scl}$ be selected so that the viewing rectangle contains the point $(4, 8)$ and the screen is square?

10. If $X\min = -6$, $X\max = 12$, and $X\mathrm{scl} = 2$, how should $Y\min$, $Y\max$, and $Y\mathrm{scl}$ be selected so that the viewing rectangle contains the point $(4, 8)$ and the screen is square?

---

6 | USING A GRAPHING UTILITY TO GRAPH INEQUALITIES

It is easiest to begin with an example.

| EXAMPLE 1 | **Graphing an Inequality Using a Graphing Utility** |

Use a graphing utility to graph: $3x + y - 6 \leq 0$

Solution   We begin by graphing the equation $3x + y - 6 = 0$ $(Y_1 = -3x + 6)$. See Figure 21.

Figure 21

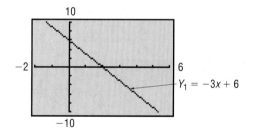

As with graphing by hand, we need to test points selected from each region and determine whether they satisfy the inequality. To test the point $(-1, 2)$, for example, enter $3(-1) + 2 - 6 \leq 0$. See Figure 22(a). The 1 that appears indicates that the statement entered (the inequality) is true. When the point $(5, 5)$ is tested, a 0 appears, indicating that the statement entered is false. Thus, $(-1, 2)$ is a part of the graph of the inequality and $(5, 5)$ is not. Figure 22(b) shows the graph of the inequality on a TI-83.*

Figure 22

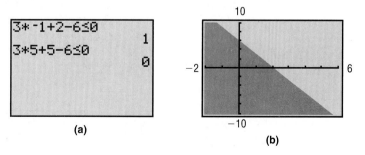

(a)

(b)

The steps to follow to graph an inequality using a graphing utility are given next.

*Consult your owner's manual for shading techniques.

STEPS FOR GRAPHING AN INEQUALITY
USING A GRAPHING UTILITY

STEP 1: Replace the inequality symbol by an equal sign and graph the resulting equation, $y =$.

STEP 2: In each of the regions, select a test point P.

    (a) Use a graphing utility to determine if the test point P satisfies the inequality. If the test point satisfies the inequality, then so do all the points in the region. Indicate this by using the graphing utility to shade the region.

    (b) If the coordinates of P do not satisfy the inequality, then none of the points in that region do.

7 USING A GRAPHING UTILITY TO SOLVE SYSTEMS OF LINEAR EQUATIONS

Most graphing utilities have the capability to put the augmented matrix of a system of linear equations in row echelon form. The next example, Example 6 from Section 8.3, demonstrates this feature using a TI-83 graphing calculator.

EXAMPLE 1

Solving a System of Linear Equations Using a Graphing Utility

Solve: $\begin{cases} x - y + z = 8 & (1) \\ 2x + 3y - z = -2 & (2) \\ 3x - 2y - 9z = 9 & (3) \end{cases}$

Solution The augmented matrix of the system is

$$\begin{bmatrix} 1 & -1 & 1 & | & 8 \\ 2 & 3 & -1 & | & -2 \\ 3 & -2 & -9 & | & 9 \end{bmatrix}$$

We enter this matrix into our graphing utility and name it A. See Figure 23(a). Using the REF (row echelon form) command on matrix A, we obtain the results shown in Figure 23(b). Since the entire matrix does not fit on the screen, we need to scroll right to see the rest of it. See Figure 23(c).

Figure 23

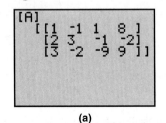

(a)

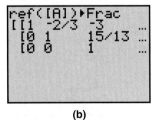

(b)   (c)

The system of equations represented by the matrix in row echelon form is

$$\begin{bmatrix} 1 & -\frac{2}{3} & -3 & \Big| & 3 \\ 0 & 1 & \frac{15}{13} & \Big| & -\frac{24}{13} \\ 0 & 0 & 1 & \Big| & 1 \end{bmatrix}$$

$$\begin{cases} x - \dfrac{2}{3}y - 3z = 3 & (1) \\[2mm] y + \dfrac{15}{13}z = -\dfrac{24}{13} & (2) \\[2mm] z = 1 & (3) \end{cases}$$

Using $z = 1$, we back-substitute to get

$$\begin{cases} x - \dfrac{2}{3}y - 3(1) = 3 & (1) \\[2mm] y + \dfrac{15}{13}(1) = -\dfrac{24}{13} & (2) \end{cases} \xrightarrow[\text{Simplify.}]{} \begin{cases} x - \dfrac{2}{3}y = 6 & (1) \\[2mm] y = \dfrac{-39}{13} = -3 & (2) \end{cases}$$

Solving the second equation for y, we find that $y = -3$. Back-substituting $y = -3$ into $x - \frac{2}{3}y = 6$, we find that $x = 4$. The solution of the system is $x = 4$, $y = -3$, $z = 1$. ∎

Notice that the row echelon form of the augmented matrix using the graphing utility differs from the row echelon form in our solution (p. 540), yet both matrices provide the same solution! This is because the two solutions used different row operations to obtain the row echelon form. In all likelihood, the two solutions parted ways in Step 4 of the algebraic solution, where we avoided introducing fractions by interchanging rows 2 and 3.

Most graphing utilities also have the ability to put a matrix in reduced row echelon form. Figure 24 shows the reduced row echelon form of the augmented matrix from Example 1 using the RREF command on a TI-83 graphing calculator. Using this command, we see that the solution of the system is $x = 4$, $y = -3$, $z = 1$.

Figure 24

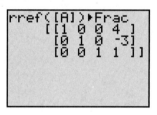

ANSWERS

CHAPTER R Review

Historical Problems (page 14)

1. (a) $1, 20$ (b) $2, 50$

1 Exercises (page 14)

1. (a) $\{2, 5\}$ (b) $\{-6, 2, 5\}$ (c) $\left\{-6, \dfrac{1}{2}, -1.333..., 2, 5\right\}$ (d) $\{\pi\}$ (e) $\left\{-6, \dfrac{1}{2}, -1.333..., \pi, 2, 5\right\}$ **3.** (a) $\{1\}$ (b) $\{0, 1\}$

(c) $\left\{0, 1, \dfrac{1}{2}, \dfrac{1}{3}, \dfrac{1}{4}\right\}$ (d) None (e) $\left\{0, 1, \dfrac{1}{2}, \dfrac{1}{3}, \dfrac{1}{4}\right\}$ **5.** (a) None (b) None (c) None (d) $\left\{\sqrt{2}, \pi, \sqrt{2} + 1, \pi + \dfrac{1}{2}\right\}$

(e) $\left\{\sqrt{2}, \pi, \sqrt{2} + 1, \pi + \dfrac{1}{2}\right\}$ **7.** (a) 18.953 (b) 18.952 **9.** (a) 28.653 (b) 28.653 **11.** (a) 0.063 (b) 0.062 **13.** (a) 9.999

(b) 9.998 **15.** (a) 0.429 (b) 0.428 **17.** (a) 34.733 (b) 34.733 **19.** $3 + 2 = 5$ **21.** $x + 2 = 3 \cdot 4$ **23.** $3 \cdot y = 1 + 2$

25. $x - 2 = 6$ **27.** $\dfrac{x}{2} = 6$ **29.** 7 **31.** 6 **33.** 1 **35.** $\dfrac{13}{3}$ **37.** -11 **39.** 11 **41.** -4 **43.** 1 **45.** 6 **47.** $\dfrac{2}{7}$ **49.** $\dfrac{4}{45}$ **51.** $\dfrac{23}{20}$ **53.** $\dfrac{79}{30}$

55. $\dfrac{13}{36}$ **57.** $-\dfrac{16}{45}$ **59.** $\dfrac{1}{60}$ **61.** $\dfrac{15}{22}$ **63.** $6x + 24$ **65.** $x^2 - 4x$ **67.** $x^2 + 6x + 8$ **69.** $x^2 - x - 2$ **71.** $x^2 - 10x + 16$ **73.** $x^2 - 4$

75. $2x + 3x = (2 + 3)x = 5x$ **77.** $2(3 \cdot 4) = 2 \cdot 12 = 24; (2 \cdot 3) \cdot (2 \cdot 4) = 6 \cdot 8 = 48$ **79.** No; $2 - 3 \neq 3 - 2$

81. No; $\dfrac{2}{3} \neq \dfrac{3}{2}$ **83.** Symmetric property **85.** No; no **87.** 1

2 Exercises (page 21)

1.

3. $>$ **5.** $>$ **7.** $>$ **9.** $=$ **11.** $<$ **13.** $x > 0$ **15.** $x < 2$ **17.** $x \leq 1$ **19.** ⟶ **21.** ⟶ **23.** 1 **25.** 2 **27.** 6 **29.** 4 **31.** -28

33. $\dfrac{4}{5}$ **35.** 0 **37.** 1 **39.** 5 **41.** 1 **43.** 22 **45.** 2 **47.** $x = 0$ **49.** $x = 3$ **51.** None **53.** $x = 1, x = 0, x = -1$ **55.** $\{x | x \neq 5\}$

57. $\{x | x \neq -4\}$ **59.** $0°C$ **61.** $25°C$ **63.** $A = lw$; all the variables are positive real numbers

65. $C = \pi d$; d and C are positive real numbers **67.** $A = \dfrac{\sqrt{3}}{4}x^2$; x and A are positive real numbers

69. $V = \dfrac{4}{3}\pi r^3$; r and V are positive real numbers **71.** $V = x^3$; x and V are positive real numbers **73.** (a) $\$6000$ (b) $\$8000$

75. (a) $2 \leq 5$ (b) $6 > 5$ **77.** (a) Acceptable (b) Not acceptable **79.** No; $\dfrac{1}{3}$ is larger; $0.000333...$ **81.** No

3 Exercises (page 26)

1. 13 **3.** 26 **5.** 25 **7.** Right triangles; 5 **9.** Not a right triangle **11.** Right triangle; 25 **13.** Not a right triangle **15.** $8\ \text{in}^2$ **17.** $4\ \text{in}^2$

19. $A = 25\pi\ \text{m}^2$; $C = 10\pi\ \text{m}$ **21.** $224\ \text{ft}^3$ **23.** $V = \dfrac{256}{3}\pi\ \text{cm}^3$; $S = 64\pi\ \text{cm}^2$ **25.** $648\pi\ \text{in}^3$ **27.** π square units **29.** 2π square units

31. About $16.8\ \text{ft}$ **33.** $64\ \text{ft}^2$ **35.** $24 + 2\pi \approx 30.28\ \text{ft}^2$; $16 + 2\pi \approx 22.28\ \text{ft}^2$ **37.** About $5.477\ \text{mi}$

39. From 100 ft: 12.247 mi; From 150 ft: 15 mi

4 Exercises (page 35)

1. 16 **3.** $\dfrac{1}{16}$ **5.** $-\dfrac{1}{16}$ **7.** $\dfrac{1}{8}$ **9.** $\dfrac{1}{4}$ **11.** $\dfrac{1}{9}$ **13.** $\dfrac{81}{64}$ **15.** $\dfrac{27}{8}$ **17.** $\dfrac{81}{64}$ **19.** $\dfrac{4}{81}$ **21.** $\dfrac{1}{12}$ **23.** $-\dfrac{2}{3}$ **25.** y^2 **27.** $\dfrac{x}{y^2}$ **29.** $\dfrac{1}{64x^6}$ **31.** $-\dfrac{4}{x}$

33. 3 **35.** $\dfrac{1}{x^3 y}$ **37.** $\dfrac{1}{xy}$ **39.** $\dfrac{y}{x}$ **41.** $\dfrac{25x^2}{16y^2}$ **43.** $\dfrac{1}{x^2 y^2}$ **45.** $\dfrac{1}{x^3 y^3}$ **47.** $-\dfrac{8x^3}{9yz^2}$ **49.** $\dfrac{y^3}{x^8}$ **51.** $\dfrac{16x^2}{9y^2}$ **53.** $\dfrac{1}{x^3 y}$ **55.** $\dfrac{y^2}{x^2}$ **57.** $10; 0$ **59.** 81

61. $304,006.671$ **63.** 0.004 **65.** 481.890 **67.** 0.000 **69.** 4.542×10^2 **71.** 1.3×10^{-2} **73.** 3.2155×10^4 **75.** 4.23×10^{-4} **77.** $61,500$

79. 0.001214 **81.** $110,000,000$ **83.** 0.081 **85.** $400,000,000\ \text{m}$ **87.** $0.0000005\ \text{m}$ **89.** $5 \times 10^{-4}\ \text{in}$ **91.** 1.92×10^9 barrels

93. $5.865696 \times 10^{12}\ \text{mi}$

5 Exercises *(page 46)*

1. Monomial; Variable: x; Coefficient: 2; Degree: 3   **3.** Not a monomial   **5.** Monomial; Variables: x, y; Coefficient: -2; Degree: 3
7. Not a monomial   **9.** Not a monomial   **11.** Yes; 2   **13.** Yes; 0   **15.** No   **17.** Yes; 3   **19.** No   **21.** $x^2 + 7x + 2$
23. $x^3 - 4x^2 + 9x + 7$ **25.** $6x^5 + 5x^4 + 3x^2 + x$ **27.** $7x^2 - x - 7$ **29.** $-2x^3 + 18x^2 - 18$ **31.** $2x^2 - 4x + 6$
33. $15y^2 - 27y + 30$ **35.** $x^3 + x^2 - 4x$ **37.** $-8x^5 - 10x^2$ **39.** $x^3 + 3x^2 - 2x - 4$ **41.** $x^2 + 6x + 8$ **43.** $2x^2 + 9x + 10$
45. $x^2 - 2x - 8$ **47.** $x^2 - 5x + 6$ **49.** $2x^2 - x - 6$ **51.** $-2x^2 + 11x - 12$ **53.** $2x^2 + 8x + 8$ **55.** $x^2 - xy - 2y^2$
57. $-6x^2 - 13xy - 6y^2$ **59.** $x^2 - 49$ **61.** $4x^2 - 9$ **63.** $x^2 + 8x + 16$ **65.** $x^2 - 8x + 16$ **67.** $9x^2 - 16$ **69.** $4x^2 - 12x + 9$
71. $x^2 - y^2$ **73.** $9x^2 - y^2$ **75.** $x^2 + 2xy + y^2$ **77.** $x^2 - 4xy + 4y^2$ **79.** $x^3 - 6x^2 + 12x - 8$ **81.** $8x^3 + 12x^2 + 6x + 1$
83. $4x^2 - 3x + 1$; remainder 1   **85.** $4x^2 - 11x + 23$; remainder -45 **87.** $4x - 3$; remainder $x + 1$ **89.** $4x - 3$; remainder $-7x + 7$
91. 2; remainder $-3x^2 + x + 3$ **93.** $2x - \dfrac{5}{2}$; remainder $\dfrac{3}{2}x + \dfrac{7}{2}$ **95.** $-4x^2 - 3x - 3$; remainder -7
97. $x^2 - x - 1$; remainder $2x + 2$ **99.** $x^2 + ax + a^2$; remainder 0   **101.** $a = 1, b = -4, c = 11, d = -17; a + b + c + d = -9$

6 Exercises *(page 54)*

1. $3(x + 2)$ **3.** $a(x^2 + 1)$ **5.** $x(x^2 + x + 1)$ **7.** $2x(x - 1)$ **9.** $3xy(x - 2y + 4)$ **11.** $(x - 1)(x + 1)$ **13.** $(2x + 1)(2x - 1)$
15. $(x + 4)(x - 4)$ **17.** $(5x + 2)(5x - 2)$ **19.** $(x + 1)^2$ **21.** $(x + 2)^2$ **23.** $(x - 5)^2$ **25.** $(2x + 1)^2$ **27.** $(4x + 1)^2$
29. $(x - 3)(x^2 + 3x + 9)$ **31.** $(x + 3)(x^2 - 3x + 9)$ **33.** $(2x + 3)(4x^2 - 6x + 9)$ **35.** $(x + 2)(x + 3)$ **37.** $(x + 6)(x + 1)$
39. $(x + 5)(x + 2)$ **41.** $(x - 8)(x - 2)$ **43.** $(x - 8)(x + 1)$ **45.** $(x + 8)(x - 1)$ **47.** $(x + 2)(2x + 3)$ **49.** $(x - 2)(2x + 1)$
51. $(2x + 3)(3x + 2)$ **53.** $(3x + 1)(x + 1)$ **55.** $(z + 1)(2z + 3)$ **57.** $(x + 2)(3x - 4)$ **59.** $(x - 2)(3x + 4)$ **61.** $(x + 4)(3x + 2)$
63. $(x + 4)(3x - 2)$ **65.** $(x + 6)(x - 6)$ **67.** $2(1 + 2x)(1 - 2x)$ **69.** $(x + 2)(x + 5)$ **71.** $(x - 7)(x - 3)$ **73.** $4(x^2 - 2x + 8)$
75. Prime   **77.** $-(x - 5)(x + 3)$ **79.** $3(x + 2)(x - 6)$ **81.** $y^2(y + 5)(y + 6)$ **83.** $(2x + 3)^2$ **85.** $2(3x + 1)(x + 1)$
87. $(x - 3)(x + 3)(x^2 + 9)$ **89.** $(x - 1)^2(x^2 + x + 1)^2$ **91.** $x^5(x - 1)(x + 1)$ **93.** $(4x + 3)^2$ **95.** $-(4x - 5)(4x + 1)$
97. $(2y - 5)(2y - 3)$ **99.** $-(3x - 1)(3x + 1)(x^2 + 1)$ **101.** $(x + 3)(x - 6)$ **103.** $(x + 2)(x - 3)$ **105.** $(3x - 5)(9x^2 - 3x + 7)$
107. $(x + 5)(3x + 11)$ **109.** $(x - 1)(x + 1)(x + 2)$ **111.** $(x - 1)(x + 1)(x^2 - x + 1)$
113. The possibilities are $(x \pm 1)(x \pm 4) = x^2 \pm 5x + 4$ or $(x \pm 2)(x \pm 2) = x^2 \pm 4x + 4$, none of which equals $x^2 + 4$.

7 Exercises *(page 65)*

1. $\dfrac{3}{x - 3}$ **3.** $\dfrac{x}{3}$ **5.** $\dfrac{4x}{2x - 1}$ **7.** $\dfrac{y + 5}{2(y + 1)}$ **9.** $\dfrac{x + 5}{x - 1}$ **11.** $-(x + 7)$ **13.** $\dfrac{3}{5x(x - 2)}$ **15.** $\dfrac{2x}{x + 4}$ **17.** $\dfrac{8}{3x}$ **19.** $\dfrac{x - 3}{x + 7}$
21. $\dfrac{4x}{(x - 2)(x - 3)}$ **23.** $\dfrac{4}{5(x - 1)}$ **25.** $\dfrac{(4 - x)(x - 4)}{4x}$ **27.** $\dfrac{(x + 3)^2}{(x - 3)^2}$ **29.** $\dfrac{(x - 4)(x + 3)}{(x - 1)(2x + 1)}$ **31.** $\dfrac{x + 5}{2}$ **33.** $\dfrac{(x - 2)(x + 2)}{2x - 3}$
35. $\dfrac{3x - 2}{x - 3}$ **37.** $\dfrac{x + 9}{2x - 1}$ **39.** $\dfrac{4 - x}{x - 2}$ **41.** $\dfrac{2(x + 5)}{(x - 1)(x + 2)}$ **43.** $\dfrac{3x^2 - 2x - 3}{(x + 1)(x - 1)}$ **45.** $\dfrac{-11x - 2}{(x + 2)(x - 2)}$ **47.** $\dfrac{2(x^2 - 2)}{x(x - 2)(x + 2)}$
49. $(x - 2)(x + 2)(x + 1)$ **51.** $x(x - 1)(x + 1)$ **53.** $x^3(2x - 1)^2$ **55.** $x(x - 1)^2(x + 1)(x^2 + x + 1)$
57. $\dfrac{5x}{(x - 6)(x - 1)(x + 4)}$ **59.** $\dfrac{2(2x^2 + 5x - 2)}{(x - 2)(x + 2)(x + 3)}$ **61.** $\dfrac{5x + 1}{(x - 1)^2(x + 1)^2}$ **63.** $\dfrac{-x^2 + 3x + 13}{(x - 2)(x + 1)(x + 4)}$
65. $\dfrac{x^3 - 2x^2 + 4x + 3}{x^2(x + 1)(x - 1)}$ **67.** $\dfrac{-1}{x(x + h)}$ **69.** $\dfrac{x + 1}{x - 1}$ **71.** $\dfrac{(x - 1)(x + 1)}{x^2 + 1}$ **73.** $\dfrac{2(5x - 1)}{(x - 2)(x + 1)^2}$ **75.** $\dfrac{-2x(x^2 - 2)}{(x + 2)(x^2 - x - 3)}$
77. $\dfrac{-1}{x - 1}$ **79.** $f = \dfrac{R_1 \cdot R_2}{(n - 1)(R_1 + R_2)}$; $\dfrac{2}{15}$ m

8 Exercises *(page 73)*

1. 5   **3.** 3   **5.** -4 **7.** $\dfrac{1}{3}$ **9.** $5x^2$ **11.** $2(1 + x)$ **13.** $2\sqrt{2}$ **15.** $5\sqrt{2}$ **17.** $2\sqrt[3]{2}$ **19.** $-2\sqrt[3]{2}$ **21.** $\dfrac{5}{3}|x|$ **23.** x^3y^2 **25.** $6\sqrt{x}$ **27.** $6x\sqrt{x}$
29. 1   **31.** $\dfrac{4y^2}{3x}$ **33.** $15\sqrt[3]{3}$ **35.** $\dfrac{1}{x(2x + 3)}$ **37.** 1   **39.** $7\sqrt{2}$ **41.** $\sqrt{2}$ **43.** $-\sqrt[3]{2}$ **45.** $(2x - 15)\sqrt{2x}$ **47.** $(-x - 5y)\sqrt[3]{2xy}$
49. $36\sqrt{2}$ **51.** $3 - 4\sqrt{3}$ **53.** $42 + 9\sqrt{7}$ **55.** $3 - 2\sqrt{2}$ **57.** $1 - 3\sqrt[3]{4} + 3\sqrt[3]{2}$ **59.** $4x + 4\sqrt{x} - 15$ **61.** $\dfrac{-x^2}{\sqrt{1 - x^2}}$ **63.** $\dfrac{2\sqrt{5}}{5}$
65. $\dfrac{4\sqrt{6}}{3}$ **67.** $\dfrac{\sqrt{x}}{x}$ **69.** $\dfrac{15 - 3\sqrt{2}}{23}$ **71.** $\dfrac{4 - \sqrt{7}}{3}$ **73.** $\dfrac{15 - 2\sqrt{5}}{41}$ **75.** $5 - 2\sqrt{6}$ **77.** $\dfrac{\sqrt{x} - 2}{x - 4}$ **79.** 1.41   **81.** 1.59   **83.** 4.89
85. 2.15   **87. (a)** 15,660.4 gal   **(b)** 390.7 gal   **89.** $2\sqrt{2}\pi \approx 8.89$ s   **91.** $\dfrac{\pi\sqrt{3}}{6} \approx 0.91$ s

9 Exercises (page 77)

1. 4 **3.** 9 **5.** $\frac{1}{8}$ **7.** $\frac{1}{27}$ **9.** $\frac{27}{8}$ **11.** $\frac{27}{8}$ **13.** 8 **15.** 8 **17.** 27 **19.** $\frac{1}{5}$ **21.** 9 **23.** $\frac{1}{7}$ **25.** 2 **27.** 3 **29.** $\frac{1}{2}$ **31.** $6^{2/3}$ **33.** $2^{5/6}$

35. $\sqrt{x}$ **37.** $x^{7/4}$ **39.** x **41.** $x^2 y^4$ **43.** $x^{4/3} y^{5/3}$ **45.** $\frac{8x^{3/2}}{y^{1/4}}$ **47.** $\frac{x^{11}}{y^3}$ **49.** $\frac{3x+2}{(1+x)^{1/2}}$ **51.** $\frac{x(3x^2+2)}{(x^2+1)^{1/2}}$ **53.** $\frac{22x+5}{10\sqrt{x}-5\sqrt{4x+3}}$

55. $\frac{2+x}{2(1+x)^{3/2}}$ **57.** $\frac{4-x}{(x+4)^{3/2}}$ **59.** $\frac{1}{x^2(x^2-1)^{1/2}}$ **61.** $\frac{1-3x^2}{2\sqrt{x}(1+x^2)^2}$ **63.** $\frac{1}{2}(5x+2)(x+1)^{1/2}$ **65.** $2x^{1/2}(3x-4)(x+1)$

67. $(x^2+4)^{1/3}(11x^2+12)$ **69.** $(3x+5)^{1/3}(2x+3)^{1/2}(17x+27)$ **71.** $\frac{3(x+2)}{2x^{1/2}}$

Fill-in-the-Blank Items (page 80)

1. symmetric **3.** distributive **5.** 3; leading **7.** hypotenuse

True/False Items (page 80)

1. T **3.** F **5.** F

Review Exercises (page 80)

1. -11 **3.** $\frac{1}{6}$ **5.** $\frac{93}{8}$ **7.** -29 **9.** $\frac{1}{128}$ **11.** 16 **13.** $\frac{9}{4}$ **15.** $\frac{1}{2}$ **17.** 4 **19.** 4 **21.** $\frac{y^2}{x^2}$ **23.** $\frac{1}{x^5 y}$ **25.** xy^4 **27.** $\frac{y^2}{x^2+y^2}$ **29.** $\frac{125}{x^2 y}$

31. $\frac{x^2}{16y^3}$ **33.** $-8x^2+16x-6$ **35.** $9x^3-20x^2+6x+13$ **37.** $6x^3-15x^2+4x-10$ **39.** x^3-7x-6

41. $3x^2+8x+25$; remainder 79 **43.** $-3x^2+4$; remainder -2 **45.** $8x^2+24x+62$; remainder $167x-61$

47. $x^4-x^3+x^2-x+1$; remainder 0 **49.** $3x^4-2x^2+1$; remainder 0 **51.** $(x+7)(x-2)$ **53.** $(3x+2)(2x-3)$

55. $3(x+2)(x-7)$ **57.** $(2x+1)(4x^2-2x+1)$ **59.** $(2x+3)(x-1)(x+1)$ **61.** $(5x-2)(5x+2)$ **63.** Prime

65. $\frac{2x+7}{x-2}$ **67.** $\frac{3(3x-1)}{(x+3)(3x+1)}$ **69.** $\frac{4x}{(x+1)(x-1)}$ **71.** $\frac{x^2+17x+2}{(x-2)(x+2)^2}$ **73.** $\frac{4\sqrt{5}}{5}$ **75.** $-2(1+\sqrt{2})$ **77.** $-\frac{3+\sqrt{5}}{2}$

79. $\frac{2(1+x^2)}{(2+x^2)^{1/2}}$ **81.** $\frac{x(3x+16)}{2(x+4)^{3/2}}$ **83. (a)** $8000 **(b)** $12,000 **85.** About 37.81 ft²; 25 ft **87.** $16\pi \approx 50.27$ ft²; $10\pi \approx 31.4$ ft

C H A P T E R 1 Equations and Inequalities

1.1 Exercises (page 93)

1. 3 **3.** -5 **5.** $\frac{3}{2}$ **7.** $\frac{5}{4}$ **9.** $\{-2\}$ **11.** $\{3\}$ **13.** $\{-1\}$ **15.** $\{-2\}$ **17.** $\{-18\}$ **19.** $\{-4\}$ **21.** $\left\{-\frac{3}{4}\right\}$ **23.** $\{-20\}$ **25.** $\{2\}$ **27.** $\{0.5\}$

29. $\left\{\frac{29}{10}\right\}$ **31.** $\{2\}$ **33.** $\{8\}$ **35.** $\{2\}$ **37.** $\{-1\}$ **39.** $\{3\}$ **41.** No solution **43.** $\{0,9\}$ **45.** $\{0,9\}$ **47.** No solution **49.** $\{-6\}$

51. $\{34\}$ **53.** $\left\{-\frac{20}{39}\right\}$ **55.** $\{-1\}$ **57.** $\left\{-\frac{11}{6}\right\}$ **59.** $\{-6\}$ **61.** $\{5.91\}$ **63.** $\{0.41\}$ **65.** $\{3,4\}$ **67.** $\left\{-3,\frac{1}{2}\right\}$ **69.** $\{-3,0,3\}$

71. $\{-5,0,4\}$ **73.** $\{-1,1\}$ **75.** $\{-2,2,3\}$ **77.** $x=\frac{b+c}{a}$ **79.** $x=\frac{abc}{a+b}$ **81.** $x=a^2$ **83.** $a=3$ **85.** $R=\frac{R_1 R_2}{R_1+R_2}$

87. $R=\frac{mv^2}{F}$ **89.** $r=\frac{S-a}{S}$ **91.** In obtaining step (7) we divided by $x-2$. Since $x=2$ from step (1), we actually divided by 0.

1.2 Exercises (page 102)

1. $A=\pi r^2$; $r=$ radius, $A=$ area **3.** $A=s^2$; $A=$ area, $s=$ length of a side **5.** $F=ma$; $F=$ force, $m=$ mass, $a=$ acceleration
7. $W=Fd$; $W=$ work, $F=$ force, $d=$ distance **9.** $C=150x$; $C=$ total variable cost, $x=$ number of dishwashers
11. $11,500 will be invested in bonds and $8500 in CDs. **13.** Scott will receive $400,000, Alice $300,000, and Tricia $200,000.
15. The regular hourly rate is $8.50. **17.** Brooke needs a score of 85. **19.** The original price was $147,058.82; purchasing the model
saves $22,058.82. **21.** The book store pays $41.48 for the book. **23.** There were 3260 adults. **25.** The length is 19 ft; the width is 11 ft.
27. Invest $31,250 in bonds and $18,750 in CDs. **29.** $11,600 was loaned out at 8%. **31.** Mix 75 lb of Earl Grey tea with 25 lb of Orange
Pekoe tea. **33.** Mix 40 lb of cashews with the peanuts. **35.** The speed of the current is 2.286 mi/hr.
37. The Metra commuter averages 30 mi/hr; the Amtrak averages 80 mi/hr. **39.** Working together, it takes 12 min.
41. (a) The dimensions are 10 ft by 5 ft. **(b)** The area is 50 sq ft. **(c)** The dimensions would be 7.5 ft by 7.5 ft.
(d) The area would be 56.25 sq ft. **43.** The defensive back catches up to the tight end at the tight end's 45 yd line.

45. Add $\frac{2}{3}$ gal of water. **47.** Evaporate 10.67 oz of water. **49.** 40 g of 12 karat gold should be mixed with 20 g of pure gold.

51. Mike passes Dan $\frac{1}{3}$ mi from the start, 2 min from the time Mike started to race.   **53.** Start the auxiliary pump at 9:45 AM.

55. The tub will fill in 1 hr.   **57.** Lewis would beat Burke by 16.75 m.

Historical Problems *(page 116)*

1. The area of each shaded square is 9, so the larger square will have area $85 + 4(9) = 121$. The area of the larger square is also given by the expression $(x + 6)^2$, so $(x + 6)^2 = 121$. Taking the positive square root of each side, $x + 6 = 11$ or $x = 5$.

3.
$$\left(x + \frac{b}{2a}\right)^2 = \left(\frac{\sqrt{b^2 - 4ac}}{2a}\right)^2$$
$$\left(x + \frac{b}{2a}\right)^2 - \left(\frac{\sqrt{b^2 - 4ac}}{2a}\right)^2 = 0$$
$$\left(x + \frac{b}{2a} - \frac{\sqrt{b^2 - 4ac}}{2a}\right)\left(x + \frac{b}{2a} + \frac{\sqrt{b^2 - 4ac}}{2a}\right) = 0$$
$$\left(x + \frac{b - \sqrt{b^2 - 4ac}}{2a}\right)\left(x + \frac{b + \sqrt{b^2 - 4ac}}{2a}\right) = 0$$
$$x = \frac{-b + \sqrt{b^2 - 4ac}}{2a} \text{ or } x = \frac{-b - \sqrt{b^2 - 4ac}}{2a}$$

1.3 Exercises *(page 116)*

1. $\{0, 9\}$ **3.** $\{-5, 5\}$ **5.** $\{-3, 2\}$ **7.** $\left\{-\frac{1}{2}, 3\right\}$ **9.** $\{-4, 4\}$ **11.** $\{2, 6\}$ **13.** $\left\{\frac{3}{2}\right\}$ **15.** $\left\{-\frac{2}{3}, \frac{3}{2}\right\}$ **17.** $\left\{-\frac{2}{3}, \frac{3}{2}\right\}$ **19.** $\left\{-\frac{3}{4}, 2\right\}$

21. $\{-5, 5\}$ **23.** $\{-1, 3\}$ **25.** $\{-3, 0\}$ **27.** 16   **29.** $\frac{1}{16}$ **31.** $\frac{1}{9}$ **33.** $\{-7, 3\}$ **35.** $\left\{-\frac{1}{4}, \frac{3}{4}\right\}$ **37.** $\left\{\frac{-1 - \sqrt{7}}{6}, \frac{-1 + \sqrt{7}}{6}\right\}$

39. $\{2 - \sqrt{2}, 2 + \sqrt{2}\}$ **41.** $\{2 - \sqrt{5}, 2 + \sqrt{5}\}$ **43.** $\left\{1, \frac{3}{2}\right\}$ **45.** No real solution   **47.** $\left\{\frac{-1 - \sqrt{5}}{4}, \frac{-1 + \sqrt{5}}{4}\right\}$ **49.** $\left\{0, \frac{9}{4}\right\}$

51. $\left\{\frac{1}{3}\right\}$ **53.** $\left\{-\frac{2}{3}, 1\right\}$ **55.** $\left\{\frac{1 - \sqrt{33}}{8}, \frac{1 + \sqrt{33}}{8}\right\}$ **57.** No real solution   **59.** $\{0.63, 3.47\}$ **61.** $\{-2.80, 1.07\}$ **63.** $\{-0.85, 1.17\}$

65. $\{-8.16, -0.22\}$ **67.** $\{-\sqrt{5}, \sqrt{5}\}$ **69.** $\left\{\frac{1}{4}\right\}$ **71.** $\left\{-\frac{3}{5}, \frac{5}{2}\right\}$ **73.** $\left\{-\frac{1}{2}, \frac{2}{3}\right\}$ **75.** $\left\{\frac{-\sqrt{2} + 2}{2}, \frac{-\sqrt{2} - 2}{2}\right\}$

77. $\left\{\frac{-1 - \sqrt{17}}{2}, \frac{-1 + \sqrt{17}}{2}\right\}$ **79.** No real solution   **81.** Repeated real solution   **83.** Two unequal real solutions

85. The dimensions are 11 ft by 13 ft.   **87.** The dimensions are 5 m by 8 m.   **89.** The dimensions should be 4 ft by 4 ft.

91. (a) The ball strikes the ground after 6 sec.   **(b)** The ball passes the top of the building on its way down after 5 sec.

93. 175 boxes were ordered.   **95.** The border will be 2.56 ft wide.   **97.** The dimensions should be 11.55 cm by 6.55 cm by 3 cm.

99. The border will be 2.71 ft wide.   **101.** The speed of the current is 5 mi/hr.

103. $\dfrac{-b + \sqrt{b^2 - 4ac}}{2a} + \dfrac{-b - \sqrt{b^2 - 4ac}}{2a} = \dfrac{-2b}{2a} = -\dfrac{b}{a}$ **105.** $k = \dfrac{1}{2}$ or $k = -\dfrac{1}{2}$

107. $ax^2 + bx + c = 0, x = \dfrac{-b \pm \sqrt{b^2 - 4ac}}{2a}; ax^2 - bx + c = 0, x = \dfrac{b \pm \sqrt{(-b)^2 - 4ac}}{2a} = \dfrac{b \pm \sqrt{b^2 - 4ac}}{2a} = -\dfrac{-b \pm \sqrt{b^2 - 4ac}}{2a}$

109. 36 consecutive integers must be added.   **111.** The average speed is 49.5 mi/hr.

1.4 Exercises *(page 123)*

1. $\{1\}$ **3.** No real solution   **5.** $\{-13\}$ **7.** $\{0, 64\}$ **9.** $\{3\}$ **11.** $\{2\}$ **13.** $\left\{-\frac{8}{5}\right\}$ **15.** $\{8\}$ **17.** $\{-1, 3\}$ **19.** $\{1, 5\}$ **21.** $\{1\}$

23. $\{5\}$ **25.** $\{2\}$ **27.** $\{-4, 4\}$ **29.** $\{0, 3\}$ **31.** $\{-2, -1, 1.2\}$ **33.** $\{-1, 1\}$ **35.** $\{-2, 1\}$ **37.** $\{-6, -5\}$ **39.** $\left\{-\frac{1}{3}\right\}$ **41.** $\left\{-\frac{3}{2}, 2\right\}$

43. $\left\{0, \frac{1}{16}\right\}$ **45.** $\{16\}$ **47.** $\{1\}$ **49.** $\left\{\left(\frac{9 - \sqrt{17}}{8}\right)^4, \left(\frac{9 + \sqrt{17}}{8}\right)^4\right\}$ **51.** $\{\sqrt{2}, \sqrt{3}\}$ **53.** $\{-4, 1\}$ **55.** $\left\{-2, -\frac{1}{2}\right\}$ **57.** $\left\{-\frac{3}{2}, \frac{1}{3}\right\}$

59. $\left\{-\frac{1}{8}, 27\right\}$ **61.** $\left\{-2, -\frac{4}{5}\right\}$ **63.** $\{0.34, 11.66\}$ **65.** $\{-1.03, 1.03\}$ **67.** $\{-1.85, 0.17\}$ **69.** $\left\{\frac{3}{2}, 5\right\}$ **71.** The depth of the well is 229.94 ft.

1.5 Exercises *(page 133)*

1. $[0, 2]; 0 \le x \le 2$ **3.** $(-1, 2); -1 < x < 2$ **5.** $[0, 3); 0 \le x < 3$ **7. (a)** $6 < 8$ **(b)** $-2 < 0$ **(c)** $9 < 15$ **(d)** $-6 > -10$

9. (a) $7 > 0$ **(b)** $-1 > -8$ **(c)** $12 > -9$ **(d)** $-8 < 6$ **11. (a)** $2x + 4 < 5$ **(b)** $2x - 4 < -3$ **(c)** $6x + 3 < 6$ **(d)** $-4x - 2 > -4$

13. $[0, 4]$ **15.** $[4, 6)$ **17.** $[4, \infty)$ **19.** $(-\infty, -4)$

21. $2 \le x \le 5$

23. $-3 < x < -2$

25. $x \ge 4$

27. $x < -3$

29. $<$ **31.** $>$ **33.** $\ge$ **35.** $<$ **37.** $\le$ **39.** $>$ **41.** $\ge$

43. $\{x | x < 4\}$ or $(-\infty, 4)$

45. $\{x | x \ge -1\}$ or $[-1, \infty)$

47. $\{x | x > 3\}$ or $(3, \infty)$

49. $\{x | x \ge 2\}$ or $[2, \infty)$

51. $\{x | x > -7\}$ or $(-7, \infty)$

53. $\left\{x \middle| x \le \dfrac{2}{3}\right\}$ or $\left(-\infty, \dfrac{2}{3}\right]$

55. $\{x | x < -20\}$ or $(-\infty, -20)$

57. $\left\{x \middle| x \ge \dfrac{4}{3}\right\}$ or $\left[\dfrac{4}{3}, \infty\right)$

59. $\{x | 3 \le x \le 5\}$ or $[3, 5]$

61. $\left\{x \middle| \dfrac{2}{3} \le x \le 3\right\}$ or $\left[\dfrac{2}{3}, 3\right]$

63. $\left\{x \middle| -\dfrac{11}{2} < x < \dfrac{1}{2}\right\}$ or $\left(-\dfrac{11}{2}, \dfrac{1}{2}\right)$

65. $\{x | -6 < x < 0\}$ or $(-6, 0)$

67. $\{x | x < -5\}$ or $(-\infty, -5)$

69. $\{x | x \ge -1\}$ or $[-1, \infty)$

71. $\left\{x \middle| \dfrac{1}{2} \le x < \dfrac{5}{4}\right\}$ or $\left[\dfrac{1}{2}, \dfrac{5}{4}\right)$

73. $\left\{x \middle| x < -\dfrac{1}{2}\right\}$ or $\left(-\infty, -\dfrac{1}{2}\right)$

75. $\left\{x \middle| x > \dfrac{10}{3}\right\}$ or $\left(\dfrac{10}{3}, \infty\right)$

77. $\{x | x > 3\}$ or $(3, \infty)$

79. $a = 3, b = 5$

81. $a = -12, b = -8$

83. $a = 3, b = 11$

85. $a = \dfrac{1}{4}, b = 1$ **87.** $a = 4, b = 16$ **89.** $\{x | x \ge -2\}$ **91.** $21 < \text{Age} < 30$ **93. (a)** Male ≥ 73.4 **(b)** Female ≥ 79.7

(c) A female can expect to live 6.3 yr longer. **95.** The agent's commission ranges from \$45,000 to \$95,000, inclusive. As a percent of selling price, the commission ranges from 5% to 8.6%, inclusive. **97.** The amount withheld varies from \$72.14 to \$93.14, inclusive.
99. The usage varies from 675.43 kW · hr to 2500.86 kW · hr, inclusive.
101. The dealer's cost varies from \$7457.63 to \$7857.14, inclusive. **103.** You need at least a 74 on the fifth test.
105. $\dfrac{a+b}{2} - a = \dfrac{a+b-2a}{2} = \dfrac{b-a}{2} > 0$; therefore, $a < \dfrac{a+b}{2}$

$b - \dfrac{a+b}{2} = \dfrac{2b-a-b}{2} = \dfrac{b-a}{2} > 0$; therefore, $b > \dfrac{a+b}{2}$

107. $(\sqrt{ab})^2 - a^2 = ab - a^2 = a(b - a) > 0$; thus, $(\sqrt{ab})^2 > a^2$ and $\sqrt{ab} > a$

$b^2 - (\sqrt{ab})^2 = b^2 - ab = b(b - a) > 0$; thus $b^2 > (\sqrt{ab})^2$ and $b > \sqrt{ab}$

109. $h - a = \dfrac{2ab}{a+b} - a = \dfrac{ab - a^2}{a+b} = \dfrac{a(b - a)}{a+b} > 0$; thus, $h > a$

$b - h = b - \dfrac{2ab}{a+b} = \dfrac{b^2 - ab}{a+b} = \dfrac{b(b - a)}{a+b} > 0$; thus $h < b$

1.6 Exercises *(page 138)*

1. $\{-3, 3\}$ **3.** $\{-4, 1\}$ **5.** $\left\{-1, \dfrac{3}{2}\right\}$ **7.** $\{-4, 4\}$ **9.** $\{2\}$ **11.** $\left\{-\dfrac{27}{2}, \dfrac{27}{2}\right\}$ **13.** $\left\{-\dfrac{36}{5}, \dfrac{24}{5}\right\}$ **15.** No solution **17.** $\left\{-\dfrac{1}{2}, \dfrac{1}{2}\right\}$
19. $\{-3, 3\}$ **21.** $\{-1, 3\}$ **23.** $\{-2, -1, 0, 1\}$ **25.** $\{x | -4 < x < 4\}; (-4, 4)$ **27.** $\{x | x < -4 \text{ or } x > 4\}; (-\infty, -4) \text{ or } (4, \infty)$
29. $\{x | 1 < x < 3\}; (1, 3)$ **31.** $\left\{x \middle| -\dfrac{2}{3} \le x \le 2\right\}; \left[-\dfrac{2}{3}, 2\right]$ **33.** $\{x | x \le 1 \text{ or } x \ge 5\}; (-\infty, 1] \text{ or } [5, \infty)$ **35.** $\left\{x \middle| -1 < x < \dfrac{3}{2}\right\}; \left(-1, \dfrac{3}{2}\right)$
37. $\{x | x < -1 \text{ or } x > 2\}; (-\infty, -1) \text{ or } (2, \infty)$ **39.** No solution **41.** $\left\{x \middle| x < -\dfrac{3}{2} \text{ or } x > \dfrac{3}{2}\right\}; \left(-\infty, -\dfrac{3}{2}\right) \text{ or } \left(\dfrac{3}{2}, \infty\right)$
43. $\{x | -1 \le x \le 2\}; [-1, 2]$ **45.** $a = 2, b = 8$ **47.** $a = -15, b = -7$ **49.** $a = -1, b = -\dfrac{1}{15}$ **51.** $\left|\dfrac{a}{b}\right| = \sqrt{\left(\dfrac{a}{b}\right)^2} = \sqrt{\dfrac{a^2}{b^2}} = \dfrac{\sqrt{a^2}}{\sqrt{b^2}} = \dfrac{|a|}{|b|}$
53. $(a + b)^2 = a^2 + 2ab + b^2 \le |a|^2 + 2|a||b| + |b|^2 = (|a| + |b|)^2$; thus, $a + b \le |a| + |b|$ **55.** $|x - 3| < \dfrac{1}{2}; \dfrac{5}{2} < x < \dfrac{7}{2}$
57. $|x + 3| > 2; x < -5 \text{ or } x > -1$ **59.** $|x - 98.6| \ge 1.5; x \le 97.1 \text{ or } x \ge 100.1$ **61.** $x^2 - a < 0; (x - \sqrt{a})(x + \sqrt{a}) < 0$; therefore, $-\sqrt{a} < x < \sqrt{a}$ **63.** $-1 < x < 1$ **65.** $x \le -3 \text{ or } x \ge 3$ **67.** $-4 \le x \le 4$ **69.** $x < -2 \text{ or } x > 2$ **71.** $\{-1, 5\}$

Fill-in-the-Blank Items *(page 140)*

1. equivalent   **3.** add; $\dfrac{25}{4}$ **5.** $-a$ **7.** extraneous   **9.** $-2; 2$

True/False Items *(page 141)*

1. T   **3.** F   **5.** T   **7. (a)** F   **(b)** T   **(c)** T

Review Exercises *(page 141)*

1. $\{-18\}$ **3.** $\{6\}$ **5.** $\left\{\dfrac{1}{5}\right\}$ **7.** $\{6\}$ **9.** No real solution   **11.** $\left\{\dfrac{11}{8}\right\}$ **13.** $\left\{-2, \dfrac{3}{2}\right\}$ **15.** $\left\{\dfrac{1 - \sqrt{13}}{4}, \dfrac{1 + \sqrt{13}}{4}\right\}$ **17.** $\{-3, 3\}$

19. No real solution   **21.** $\{-2, -1, 1, 2\}$ **23.** $\{2\}$ **25.** $\{0\}$ **27.** $\left\{\dfrac{\sqrt{5}}{2}\right\}$ **29.** $\left\{-\dfrac{1}{8}, 1\right\}$ **31.** $\left\{-1, \dfrac{1}{2}\right\}$ **33.** $\left\{\dfrac{m}{1-n}, \dfrac{m}{1+n}\right\}$

35. $\left\{-\dfrac{9b}{5a}, \dfrac{2b}{a}\right\}$ **37.** $\left\{-\dfrac{9}{5}\right\}$ **39.** $\{-5, 2\}$ **41.** $\left\{-\dfrac{5}{3}, 3\right\}$

43. $\{x \mid x \geq 14\}; [14, \infty)$

45. $\left\{x \mid -\dfrac{31}{2} \leq x \leq \dfrac{33}{2}\right\}; \left[-\dfrac{31}{2}, \dfrac{33}{2}\right]$

47. $\{x \mid -23 < x < -7\}; (-23, -7)$

49. $\left\{x \mid -\dfrac{3}{2} < x < -\dfrac{7}{6}\right\}; \left(-\dfrac{3}{2}, -\dfrac{7}{6}\right)$

51. $\{x \mid x \leq -2 \text{ or } x \geq 7\}; (-\infty, -2] \text{ or } [7, \infty)$

53. $\left\{x \mid 0 \leq x \leq \dfrac{4}{3}\right\}; \left[0, \dfrac{4}{3}\right]$

55. $\left\{x \mid x < -1 \text{ or } x > \dfrac{7}{3}\right\}; (-\infty, -1) \text{ or } \left(\dfrac{7}{3}, \infty\right)$

57. 9   **59.** $\dfrac{4}{9}$ **61.** The storm is 3300 ft away.

63. The search plane can go as far as 616 mi.

65. The helicopter will reach the life raft in a little less than 1 hr 35 min.

67. It takes Clarissa 10 days by herself.   **69.** Mix 90 cm^3 of 15% HCl to obtain 150 cm^3 of 25% HCl.   **71.** Add 256 oz of water.
73. 5 cm and 12 cm   **75.** The freight train is 190.67 ft long.   **77.** It will take the smaller pump 2 hr.
79. 36 seniors went on the trip; each one paid $13.40.

81. (a) No   **(b)** Todd wins again.   **(c)** Todd wins by $\dfrac{1}{4}$ m.   **(d)** Todd should line up 5.26 m behind the start line.   **(e)** Yes

CHAPTER 2   Graphs

2.1 Exercises *(page 151)*

1. (a) Quadrant II   **(b)** Positive x-axis   **(c)** Quadrant III
(d) Quadrant I   **(e)** Negative y-axis   **(f)** Quadrant IV

3. The points will be on a vertical line that is 2 units to the right of the y-axis

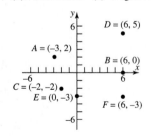

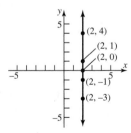

5. $\sqrt{5}$ **7.** $\sqrt{10}$ **9.** $2\sqrt{17}$ **11.** $\sqrt{85}$ **13.** $\sqrt{53}$ **15.** $\sqrt{6.89} \approx 2.625$ **17.** $\sqrt{a^2 + b^2}$

19. $d(A, B) = \sqrt{13}$
$d(B, C) = \sqrt{13}$
$d(A, C) = \sqrt{26}$
$(\sqrt{13})^2 + (\sqrt{13})^2 = (\sqrt{26})^2$
Area $= \dfrac{13}{2}$ square units

21. $d(A, B) = \sqrt{130}$
$d(B, C) = \sqrt{26}$
$d(A, C) = 2\sqrt{26}$
$(\sqrt{26})^2 + (2\sqrt{26})^2 = (\sqrt{130})^2$
Area $= 26$ square units

23. $d(A, B) = 4$
$d(B, C) = \sqrt{41}$
$d(A, C) = 5$
$4^2 + 5^2 = (\sqrt{41})^2$
Area $= 10$ square units

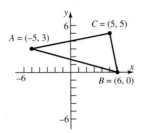

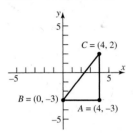

25. $(2, 2); (2, -4)$ **27.** $(0, 0); (8, 0)$ **29.** $(4, -1)$ **31.** $\left(\dfrac{3}{2}, 1\right)$ **33.** $(5, -1)$ **35.** $(1.05, 0.7)$ **37.** $\left(\dfrac{a}{2}, \dfrac{b}{2}\right)$ **39.** $\sqrt{17}; 2\sqrt{5}; \sqrt{29}$

41. $d(P_1, P_2) = 6; d(P_2, P_3) = 4; d(P_1, P_3) = 2\sqrt{13}$; right triangle

43. $d(P_1, P_2) = 2\sqrt{17}; d(P_2, P_3) = \sqrt{34}; d(P_1, P_3) = \sqrt{34}$; isosceles right triangle   **45.** $4\sqrt{10}$ **47.** $2\sqrt{65}$ **49.** $\left(\dfrac{s}{2}, \dfrac{s}{2}\right)$

51. $90\sqrt{2} \approx 127.28$ ft   **53.** **(a)** $(90, 0), (90, 90), (0, 90)$ **(b)** $5\sqrt{2161}$ ft or ≈ 232.4 ft   **(c)** $30\sqrt{149}$ ft or ≈ 366.2 ft   **55.** $d = 50t$

2.2 Exercises *(page 162)*

1.
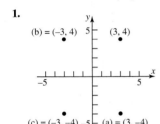

3.
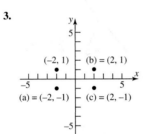

5.
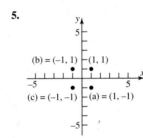

7.
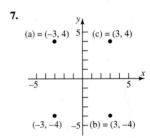

9.
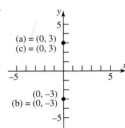

11. **(a)** $(-1, 0), (1, 0)$ **(b)** x-axis, y-axis, origin   **13.** **(a)** $\left(-\dfrac{\pi}{2}, 0\right), (0, 1), \left(\dfrac{\pi}{2}, 0\right)$ **(b)** y-axis

15. **(a)** $(0, 0)$ **(b)** x-axis   **17.** **(a)** $(1, 0)$ **(b)** none   **19.** **(a)** $(-1.5, 0), (0, -2), (1.5, 0)$ **(b)** y-axis

21. **(a)** none   **(b)** origin   **23.** $(0, 0)$ is on the graph.   **25.** $(0, 3)$ is on the graph.

27. $(0, 2)$ and $(\sqrt{2}, \sqrt{2})$ are on the graph.   **29.** $(0, 0)$; symmetric with respect to the y-axis

31. $(0, 0)$; symmetric with respect to the origin

33. $(0, 9), (3, 0), (-3, 0)$; symmetric with respect to the y-axis

35. $(-2, 0), (2, 0), (0, -3), (0, 3)$; symmetric with respect to the x-axis, y-axis, and origin

37. $(0, -27), (3, 0)$; no symmetry   **39.** $(0, -4), (4, 0), (-1, 0)$; no symmetry

41. $(0, 0)$; symmetric with respect to the origin   **43.** $(0, 0)$; symmetric with respect to the origin.

45.
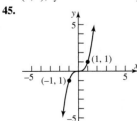

47.
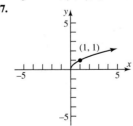

49. $a = -1$ **51.** $2a + 3b = 6$

53. (a)

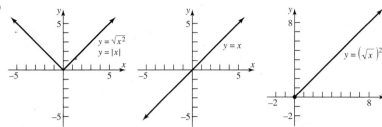

(b) Since $\sqrt{x^2} = |x|$, for all x, the graphs of $y = \sqrt{x^2}$ and $y = |x|$ are the same.

(c) For $y = (\sqrt{x})^2$, the domain of the variable x is $x \geq 0$; for $y = x$, the domain of the variable x is all real numbers. Thus, $(\sqrt{x})^2 = x$ only for $x \geq 0$.

(d) For $y = \sqrt{x^2}$, the range of the variable y is $y \geq 0$; for $y = x$, the range of the variable y is all real numbers. Also, $\sqrt{x^2} = |x|$, which equals x only if $x \geq 0$.

2.3 Exercises *(page 172)*

1. (a) $\dfrac{1}{2}$ **(b)** For every 2 unit increase in x, y will increase by 1 unit. **3. (a)** $-\dfrac{1}{3}$ **(b)** For every 3 unit increase in x, y will decrease by 1 unit.

5. Slope $= -\dfrac{3}{2}$

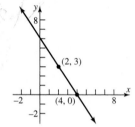

7. Slope $= -\dfrac{1}{2}$

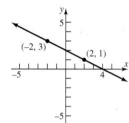

9. Slope $= 0$

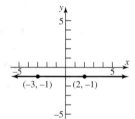

11. Slope undefined

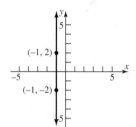

13.

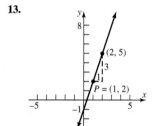

15.

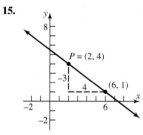

17.

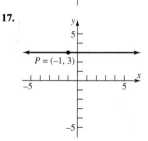

19.

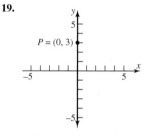

21. $(2, 6); (3, 10); (4, 14)$ **23.** $(4, -7); (6, -10); (8, -13)$ **25.** $(-1, -5); (0, -7); (1, -9)$ **27.** $x - 2y = 0$ or $y = \dfrac{1}{2}x$

29. $x + 3y = 4$ or $y = -\dfrac{1}{3}x + \dfrac{4}{3}$ **31.** $3x - y = -9$ or $y = 3x + 9$ **33.** $2x + 3y = -1$ or $y = -\dfrac{2}{3}x - \dfrac{1}{3}$

35. $x - 2y = -5$ or $y = \dfrac{1}{2}x + \dfrac{5}{2}$ **37.** $3x + y = 3$ or $y = -3x + 3$ **39.** $x - 2y = 2$ or $y = \dfrac{1}{2}x - 1$ **41.** $x = 2$; no slope–intercept form

43. Slope $= 2$; y-intercept $= 3$

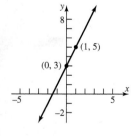

45. $y = 2x - 2$; Slope $= 2$; y-intercept $= -2$

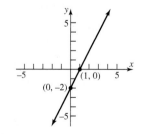

47. Slope $= \dfrac{1}{2}$; y-intercept $= 2$

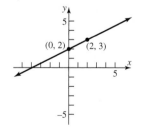

49. $y = -\dfrac{1}{2}x + 2$; Slope $= -\dfrac{1}{2}$; y-intercept $= 2$

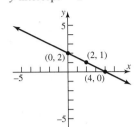

51. $y = \dfrac{2}{3}x - 2$; Slope $= \dfrac{2}{3}$; y-intercept $= -2$

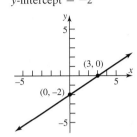

53. $y = -x + 1$; Slope $= -1$; y-intercept $= 1$

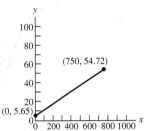

55. Slope undefined; no y-intercept

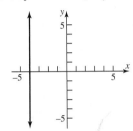

57. Slope $= 0$; y-intercept $= 5$

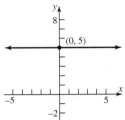

59. $y = x$; Slope $= 1$; y-intercept $= 0$

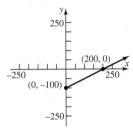

61. $y = \dfrac{3}{2}x$; Slope $= \dfrac{3}{2}$; y-intercept $= 0$

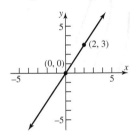

63. $y = 0$ **65.** $°C = \dfrac{5}{9}(°F - 32)$; approximately 21 °C

67. (a) $P = 0.5x - 100$
(b) $400   **(c)** $2400

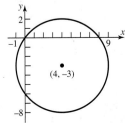

69. $C = 0.06543x + 5.65$; $C = 25.28; $C = 54.72

71. (b)   73. (d)   75. $x - y = -2$ or $y = x + 2$ **77.** $x + 3y = 3$ or $y = -\dfrac{1}{3}x + 1$ **79.** b, c, e, g

83. No, if the intercepts are $(0, 0)$. No, every line crosses at least one axis.   **85.** They are the same line.   **87.** Yes, if the y-intercept $= 0$.

2.4 Exercises (page 182)

1. (a) 6 **(b)** $-\dfrac{1}{6}$ **3. (a)** $-\dfrac{1}{2}$ **(b)** 2 **5. (a)** $\dfrac{1}{2}$ **(b)** -2 **7. (a)** $-\dfrac{3}{5}$ **(b)** $\dfrac{5}{3}$ **9. (a)** Undefined **(b)** 0 **11.** $2x - y = 3$ or $y = 2x - 3$

13. $x + 2y = 5$ or $y = -\dfrac{1}{2}x + \dfrac{5}{2}$ **15.** $2x - y = -4$ or $y = 2x + 4$ **17.** $2x - y = 0$ or $y = 2x$ **19.** $x = 4$; no slope–intercept form

21. $2x + y = 0$ or $y = -2x$ **23.** $x - 2y = -3$ or $y = \dfrac{1}{2}x + \dfrac{3}{2}$ **25.** $y = 4$ or $y = 4$

27. Center $(2, 1)$; Radius 2; $(x - 2)^2 + (y - 1)^2 = 4$ **29.** Center $\left(\dfrac{5}{2}, 2\right)$; Radius $\dfrac{3}{2}$; $\left(x - \dfrac{5}{2}\right)^2 + (y - 2)^2 = \dfrac{9}{4}$

31. $x^2 + y^2 = 4$; $x^2 + y^2 - 4 = 0$

33. $(x - 1)^2 + (y + 1)^2 = 1$; $x^2 + y^2 - 2x + 2y + 1 = 0$

35. $x^2 + (y - 2)^2 = 4$; $x^2 + y^2 - 4y = 0$

37. $(x - 4)^2 + (y + 3)^2 = 25$; $x^2 + y^2 - 8x + 6y = 0$

39. $(h, k) = (0, 0); r = 2$ **41.** $(h, k) = (3, 0); r = 2$ **43.** $(h, k) = (-2, 2); r = 3$ **45.** $(h, k) = \left(\dfrac{1}{2}, -1\right); r = \dfrac{1}{2}$

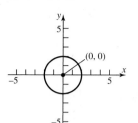

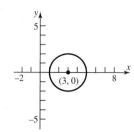

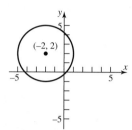

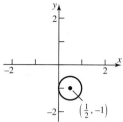

47. $(h, k) = (3, -2); r = 5$ **49.** $x^2 + y^2 - 13 = 0$ **51.** $x^2 + y^2 - 4x - 6y + 4 = 0$ **53.** $x^2 + y^2 + 2x - 6y + 5 = 0$

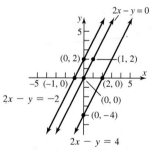

55. $P_1 = (-2, 5), P_2 = (1, 3), m_1 = -\dfrac{2}{3}; P_2 = (1, 3), P_3 = (-1, 0), m_2 = \dfrac{3}{2};$ because $m_1 m_2 = -1,$ the lines are perpendicular; thus, the points $P_1, P_2,$ and P_3 are the vertices of a right triangle.

57. $P_1 = (-1, 0), P_2 = (2, 3), m = 1; P_3 = (1, -2), P_4 = (4, 1), m = 1; P_1 = (-1, 0),$ $P_3 = (1, -2), m = -1; P_2 = (2, 3), P_4 = (4, 1), m = -1;$ opposite sides are parallel, and adjacent sides are perpendicular; the points are the vertices of a rectangle.

59. c   **61.** b   **63.** $(x + 3)^2 + (y - 1)^2 = 16$ **65.** $(x - 2)^2 + (y - 2)^2 = 9$ **67.** b, c, e, g   **69.** c

71. Refer to Figure 49; $m_1 m_2 = -1; d(A, B) = \sqrt{(m_2 - m_1)^2}; d(O, A) = \sqrt{1 + m_2^2};$ $d(O, B) = \sqrt{1 + m_1^2}.$ Now show that $[d(O, B)]^2 + [d(O, A)]^2 = [d(A, B)]^2.$

73. (a) $x^2 + (mx + b)^2 = r^2$
$(1 + m^2)x^2 + 2mbx + b^2 - r^2 = 0$
One solution if and only if discriminant $= 0$
$(2mb)^2 - 4(1 + m^2)(b^2 - r^2) = 0$
$-4b^2 + 4r^2 + 4m^2 r^2 = 0$
$r^2(1 + m^2) = b^2$

(b) $x = \dfrac{-2mb}{2(1 + m^2)} = \dfrac{-2mb}{2b^2/r^2} = -\dfrac{r^2 m}{b}$
$y = m\left(-\dfrac{r^2 m}{b}\right) + b = -\dfrac{r^2 m^2}{b} + b = \dfrac{-r^2 m^2 + b^2}{b} = \dfrac{r^2}{b}$

(c) Slope of tangent line $= m$
Slope of line joining center to point of tangency $= \dfrac{r^2/b}{-r^2 m/b} = -\dfrac{1}{m}$

75. $\sqrt{2}x + 4y = 11\sqrt{2} - 12$ **77.** $x + 5y = -13$

79. All have the same slope, 2; the lines are parallel.   **81.** $y = 2$

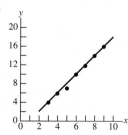

2.5 Exercises *(page 189)*

1. Linear   **3.** Linear   **5.** Nonlinear

7. (a), (c)

(b) Using $(3, 4)$ and $(9, 16), y = 2x - 2.$
(d) $y \approx 2.0357x - 2.3571$ **(e)**

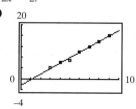

9. (a), (c)

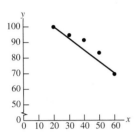

(b) Using $(-2, -4)$ and $(2, 5)$, $y = \dfrac{9}{4}x + \dfrac{1}{2}$.

(d) $y = 2.2x + 1.2$ **(e)**

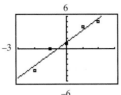

11. (a), (c)

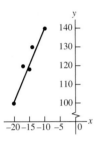

(b) Using $(20, 100)$ and $(60, 70)$, $y = -\dfrac{3}{4}x + 115$.

(d) $y = -0.72x + 116.6$ **(e)**

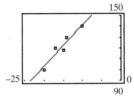

13. (a), (c)

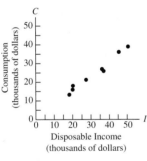

(b) Using $(-20, 100)$ and $(-10, 140)$, $y = 4x + 180$.

(d) $y \approx 3.8613x + 180.292$ **(e)**

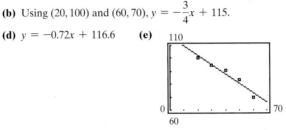

15. (a)

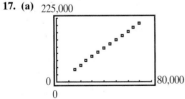

(b) Using $(20, 16)$ and $(50, 39)$, $C = \dfrac{23}{30}I + \dfrac{2}{3}$.

(c) As disposable income increases by \$1, consumption increases by about \$0.77.

(d) \$32,867

(e) $C \approx 0.755I + 0.6266$

17. (a)

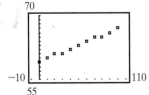

(b) $L \approx 2.98I - 76.11$

(c)

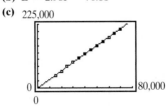

(d) For each additional dollar of income, the amount that the institution will loan you increases by \$2.98.

(e) \$125,143

19. (a)

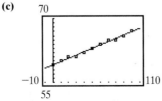

(b) $T \approx 0.0782h + 59.091$

(c)

(d) If relative humidity increases by 1%, the apparent temperature increases by 0.0782 degrees Fahrenheit.

(e) about 65°F

21. (a)

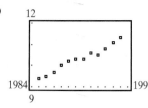

(b) $M \approx 0.1633x - 314.7139$

(c)

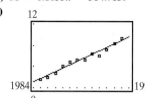

(d) For each year that passes, the average miles driven per car increases by 0.1633 thousand miles (163.3 miles).

(e) about 11.4 thousand miles

2.6 Exercises *(page 196)*

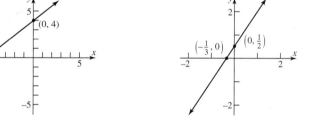

1. $y = \frac{1}{5}x$ **3.** $A = \pi x^2$ **5.** $F = \frac{250}{d^2}$ **7.** $z = \frac{1}{5}(x^2 + y^2)$ **9.** $M = \frac{9d^2}{2\sqrt{x}}$ **11.** $T^2 = \frac{8a^3}{d^2}$ **13.** $V = \frac{4\pi}{3}r^3$ **15.** $A = \frac{1}{2}bh$

17. $V = \pi r^2 h$ **19.** $F = 6.67 \times 10^{-11}\left(\frac{mM}{d^2}\right)$ **21.** 144 ft; 2 sec   **23.** 2.25   **25.** 54.86 lb   **27.** $\sqrt[3]{6} \approx 1.82$ in.   **29.** 900 ft-lb   **31.** 384 psi

33. $\frac{720}{49} \approx 14.69$ ohm   **35.** $v = \sqrt{gr}$ **37.** 18,001 mi/hr   **39.** 545 mi   **41.** $F = \frac{mv^2}{r}$ **43.** by 21%   **45.** 9 times

Fill-in-the-Blank Items *(page 199)*

1. x-coordinate or abscissa; y-coordinate or ordinate   **3.** y-axis   **5.** undefined; 0   **7.** $\frac{kx^2y^3}{\sqrt{t}}$

True/False Items *(page 200)*

1. F   **3.** T   **5.** F   **7.** T

Review Exercises *(page 200)*

1. Symmetric with respect to the x-axis   **3.** Symmetric with respect to the x-axis, y-axis, and origin   **5.** Symmetric with respect to the y-axis.
7. No symmetry   **9.** $2x + y = 5$ or $y = -2x + 5$ **11.** $x = -3$; no slope-intercept form

13. $x + 5y = -10$ or $y = -\frac{1}{5}x - 2$ **15.** $2x - 3y = -19$ or $y = \frac{2}{3}x + \frac{19}{3}$ **17.** $x - y = 7$ or $y = x - 7$

19. $4x - 5y = -20$

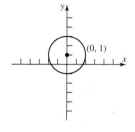

21. $\frac{1}{2}x - \frac{1}{3}y = -\frac{1}{6}$

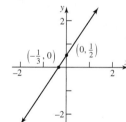

23. $\sqrt{2}x + \sqrt{3}y = \sqrt{6}$

25. Center $(0, 1)$; Radius $= 2$

27. Center $(1, -2)$; Radius $= 3$

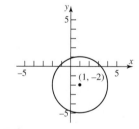

29. Center $(1, -2)$; Radius $= \sqrt{5}$

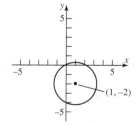

31. Slope $= \frac{1}{5}$; distance $= 2\sqrt{26}$; midpoint $= (2, 3)$

33. (a) $d(A, B) = 2\sqrt{5}; d(B, C) = \sqrt{145}; d(A, C) = 5\sqrt{5}; [d(B, C)]^2 = [d(A, B)]^2 + [d(A, C)]^2$

(b) Slope from A to B is -2; slope from A to C is $\frac{1}{2}$.

35. Slope from A to B is -1; slope from A to C is -1.   **37.** $A = \frac{\sqrt{3}}{4}s^2; s = \sqrt{\frac{64}{\sqrt{3}}} = \frac{8}{\sqrt[4]{3}} = \frac{8\sqrt[4]{27}}{3}$ cm   **39.** $a \approx 36$ million mi

41. (a)

(b) $-\dfrac{0.82}{3} \approx -0.27$

(c) Between the years 1987 and 1990, as x increases by 1 year, the level of carbon monoxide decreases by about 0.27 ppm.

(d) $-\dfrac{0.99}{3} = -0.33$

(e) Between the years 1990 and 1993, as x increases by 1 year, the level of carbon monoxide decreases by 0.33 ppm.

(f) ≈ -0.308

(g) Between the years 1987 and 1993, as x increases by 1 year, the level of carbon monoxide decreases by about 0.308 ppm.

(h) The slopes found in (b) and (d) each consider only two of the points; the slope found in (f) uses all seven points.

(i) As time passes, the average concentration of carbon monoxide is decreasing. Due to their commitment to protect the environment, companies are developing new technology to reduce carbon monoxide emissions.

43. Find the distance between the points $(-3, 4)$ and $(6, 1)$; find the midpoint of the line segment with endpoints $(-3, 4)$ and $(6, 1)$; find the slope of the line passing through the points $(-3, 4)$ and $(6, 1)$; find an equation of the line passing through the points $(-3, 4)$ and $(6, 1)$.

C H A P T E R   3   Functions and Their Graphs

3.1 Exercises (page 218)

1. Function; Domain: {Dad, Colleen, Kaleigh, Marissa}, Range: {January 8, March 15, September 17} **3.** Not a function **5.** Not a function
7. Function; Domain: {1, 2, 3, 4}; Range: {3} **9.** Not a function **11.** Function; Domain: {−2, −1, 0, 1}, Range: {0, 1, 4}
13. (a) −4 **(b)** 1 **(c)** −3 **(d)** $3x^2 - 2x - 4$ **(e)** $-3x^2 - 2x + 4$ **(f)** $3x^2 + 8x + 1$ **(g)** $12x^2 + 4x - 4$

(h) $3x^2 + 6xh + 3h^2 + 2x + 2h - 4$ **15. (a)** 0 **(b)** $\dfrac{1}{2}$ **(c)** $-\dfrac{1}{2}$ **(d)** $\dfrac{-x}{x^2 + 1}$ **(e)** $\dfrac{-x}{x^2 + 1}$ **(f)** $\dfrac{x + 1}{x^2 + 2x + 2}$ **(g)** $\dfrac{2x}{4x^2 + 1}$

(h) $\dfrac{x + h}{x^2 + 2xh + h^2 + 1}$ **17. (a)** 4 **(b)** 5 **(c)** 5 **(d)** $|x| + 4$ **(e)** $-|x| - 4$ **(f)** $|x + 1| + 4$ **(g)** $2|x| + 4$ **(h)** $|x + h| + 4$

19. (a) $-\dfrac{1}{5}$ **(b)** $-\dfrac{3}{2}$ **(c)** $\dfrac{1}{8}$ **(d)** $\dfrac{-2x + 1}{-3x - 5}$ **(e)** $\dfrac{-2x - 1}{3x - 5}$ **(f)** $\dfrac{2x + 3}{3x - 2}$ **(g)** $\dfrac{4x + 1}{6x - 5}$ **(h)** $\dfrac{2x + 2h + 1}{3x + 3h - 5}$ **21.** Function **23.** Function
25. Not a function **27.** Not a function **29.** Function **31.** Not a function **33.** All real numbers **35.** All real numbers
37. $\{x | x \neq -4, x \neq 4\}$ **39.** $\{x | x \neq 0\}$ **41.** $\{x | x \geq 4\}$ **43.** $\{x | x > 9\}$ **45.** $\{x | x > 1\}$ **47. (a)** $f(0) = 3; f(-6) = -3$
(b) $f(6) = 0; f(11) = 1$ **(c)** Positive **(d)** Negative **(e)** −3, 6, and 10 **(f)** $-3 < x < 6; 10 < x \leq 11$ **(g)** $\{x | -6 \leq x \leq 11\}$
(h) $\{y | -3 \leq y \leq 4\}$ **(i)** −3, 6, 10 **(j)** 3 **(k)** 3 times **(l)** once **(m)** 0, 4 **(n)** −5, 8 **49.** Not a function

51. Function **(a)** Domain: $\{x | -\pi \leq x \leq \pi\}$; Range: $\{y | -1 \leq y \leq 1\}$ **(b)** $\left(-\dfrac{\pi}{2}, 0\right), \left(\dfrac{\pi}{2}, 0\right), (0, 1)$ **(c)** y-axis **53.** Not a function

55. Function **(a)** Domain: $\{x | x > 0\}$; Range: all real numbers **(b)** $(1, 0)$ **(c)** None
57. Function **(a)** Domain: all real numbers; Range: $\{y | y \leq 2\}$ **(b)** $(-3, 0), (3, 0), (0, 2)$ **(c)** y-axis
59. Function **(a)** Domain: all real numbers; Range: $\{y | y \geq -3\}$ **(b)** $(1, 0), (3, 0), (0, 9)$ **(c)** None

61. (a) Yes **(b)** $f(-2) = 9; (-2, 9)$ **(c)** $0, \dfrac{1}{2}; (0, -1), \left(\dfrac{1}{2}, -1\right)$ **(d)** All real numbers **(e)** $-\dfrac{1}{2}, 1$ **(f)** −1

63. (a) No **(b)** $f(4) = -3; (4, -3)$ **(c)** $14; (14, 2)$ **(d)** $\{x | x \neq 6\}$ **(e)** −2 **(f)** $-\dfrac{1}{3}$ **65. (a)** Yes **(b)** $f(2) = \dfrac{8}{17}; \left(2, \dfrac{8}{17}\right)$

(c) $-1, 1; (-1, 1), (1, 1)$ **(d)** All real numbers **(e)** 0 **(f)** 0 **67.** $A = -\dfrac{7}{2}$ **69.** $A = -4$ **71.** $A = 8$; undefined at $x = 3$

73. (a) III **(b)** IV **(c)** I **(d)** V **(e)** II
75.

77. (a) 2 hr ellapsed during which Kevin was between 0 and 3 mi from home.
(b) 0.5 hr ellapsed during which Kevin was 3 mi from home.
(c) 0.3 hr ellapsed during which Kevin was between 0 and 3 mi from home.
(d) 0.2 hr ellapsed during which Kevin was 0 mi from home.
(e) 0.9 hr ellapsed during which Kevin was between 0 and 2.8 mi from home.
(f) 0.3 hr ellapsed during which Kevin was 2.8 mi from home.
(g) 1.1 hr ellapsed during which Kevin was between 0 and 2.8 mi from home.
(h) 3 mi
(i) 2 times

n, 14.07 m, 12.94 m, 11.72 m
ec, 1.43 sec, 1.75 sec

81. (a) 81.1 ft
(b) 129.6 ft
(c) 26.6 ft
(d) 528.125 ft

83. (a) $222 **(b)** $225 **(c)** $220 **(d)** $230
(e)

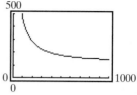

500

0   1000
0

(f) The cost varies from $220–$230.

85. $A(x) = \frac{1}{2}x^2$ **87.** $G(x) = 10x$

89. (a) No **(b)**

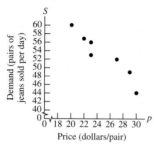

(c) For the points $(20, 60)$ and $(30, 44)$, $D = -1.6p + 92$
(d) If the price increases $1, the quantity demanded decreases by 1.6.
(e) $D(p) = -1.6p + 92$
(f) $\{p|p > 0\}$
(g) 47 pairs
(h) $D \approx -1.34p + 86.20$

91. (a) Yes **(b)**

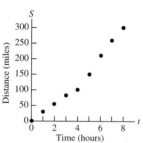

(c) For the points $(0, 0)$ and $(8, 300)$, $s = \frac{75}{2}t$
(d) If t increases by 1 hr, then distance increases by 37.5 mi.
(e) $s(t) = \frac{75}{2}t$
(f) $\{t|t \geq 0\}$
(g) 412.5 mi
(h) $s \approx 37.78t - 19.13$

93. Only $h(x) = 2x$ **95.** No; f has a domain of all real numbers, while g has a domain of $\{x|x \neq -1\}$.

3.2 Exercises *(page 231)*

1. Yes **3.** No **5.** $(-8, -2)$; $(0, 2)$; $(5, \infty)$ **7.** Yes; 10 **9.** $-2, 2$; 6, 10 **11. (a)** $(-2, 0)$, $(0, 3)$, $(2, 0)$ **(b)** Domain: $\{x|-4 \leq x \leq 4\}$ or $[-4, 4]$; Range: $\{y|0 \leq y \leq 3\}$ or $[0, 3]$ **(c)** Increasing on $(-2, 0)$ and $(2, 4)$; Decreasing on $(-4, -2)$ and $(0, 2)$ **(d)** Even
13. (a) $(0, 1)$ **(b)** Domain: all real numbers; Range: $\{y|y > 0\}$ or $(0, \infty)$. **(c)** Increasing on $(-\infty, \infty)$ **(d)** Neither
15. (a) $(-\pi, 0)$, $(0, 0)$, $(\pi, 0)$ **(b)** Domain: $\{x|-\pi \leq x \leq \pi\}$ or $[-\pi, \pi]$; Range: $\{y|-1 \leq y \leq 1\}$ or $[-1, 1]$
(c) Increasing on $\left(-\frac{\pi}{2}, \frac{\pi}{2}\right)$; Decreasing on $\left(-\pi, -\frac{\pi}{2}\right)$ and $\left(\frac{\pi}{2}, \pi\right)$ **(d)** Odd **17. (a)** $\left(0, \frac{1}{2}\right)$, $\left(\frac{1}{2}, 0\right)$, $\left(\frac{5}{2}, 0\right)$
(b) Domain: $\{x|-3 \leq x \leq 3\}$ or $[-3, 3]$; Range: $\{y|-1 \leq y \leq 2\}$ or $[-1, 2]$ **(c)** Increasing on $(2, 3)$; Decreasing on $(-1, 1)$;
Constant on $(-3, -1)$ and $(1, 2)$ **(d)** Neither **19. (a)** $(-2, 0)$, $(0, 2)$, $(2, 0)$ **(b)** Domain: $\{x|-4 \leq x \leq 4\}$ or $[-4, 4]$;
Range: $\{y|0 \leq y \leq 2\}$ or $[0, 2]$ **(c)** Increasing on $(-2, 0)$ and $(2, 4)$; Decreasing on $(-4, -2)$ and $(0, 2)$. **(d)** Even

21. (a) 0; 3 **(b)** $-2, 2$; 0, 0 **23. (a)** $\frac{\pi}{2}$; 1 **(b)** $-\frac{\pi}{2}$; -1 **25. (a)** 5 **(b)** 5 **(c)** $y = 5x$ **27. (a)** -3 **(b)** -3 **(c)** $y = -3x + 1$

29. (a) $x - 1$ **(b)** 1 **(c)** $y = x - 2$ **31. (a)** $x(x + 1)$ **(b)** 6 **(c)** $y = 6x - 6$ **33. (a)** $\frac{-1}{x + 1}$ **(b)** $-\frac{1}{3}$ **(c)** $y = -\frac{1}{3}x + \frac{4}{3}$

35. (a) $\frac{\sqrt{x} - 1}{x - 1}$ **(b)** $\sqrt{2} - 1$ **(c)** $(\sqrt{2} - 1)x - \sqrt{2} + 2$ **37.** Odd **39.** Even **41.** Odd **43.** Neither **45.** Even **47.** Odd **49.** 2

51. $2x + h + 2$ **53.** $4x + 2h - 3$ **55.** $\frac{-1}{x(x + h)}$

57. (a), (b), (e)

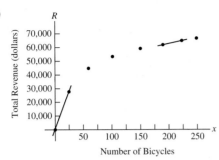

(c) 1120 dollars/bicycle
(d) For each additional bicycle sold between 0 and 25 bicycles, total revenue increases, on average, by $1120.
(f) 75 dollars/bicycle
(g) For each additional bicycle sold between 190 and 223 bicycles, total revenue increases, on average, by $75.

59. (a), (b), (e)

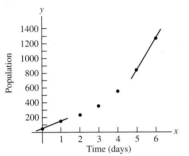

(c) 103 bacteria/day
(d) The population is increasing at an average rate of 103 bacteria per day between day 0 and day 1.
(f) 441 bacteria/day
(g) The population is increasing at an average rate of 441 bacteria per day between day 5 and day 6.
(h) The average rate of change is increasing.

61. At most one

3.3 Exercises *(page 240)*

1. C   **3.** E   **5.** B   **7.** F

9.

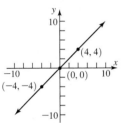

11.

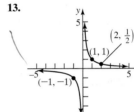

13.

15. (a) 4   **(b)** 2   **(c)** 5
17. (a) 2   **(b)** 3   **(c)** −4

19. (a) All real numbers
(b) $(0, 1)$
(c)

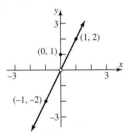

(d) $\{y | y \neq 0\}; (-\infty, 0)$ or $(0, \infty)$

21. (a) All real numbers
(b) $(0, 3)$
(c)

(d) $\{y | y \geq 1\}; [1, \infty)$

23. (a) $\{x | x \geq -2\}; [-2, \infty)$
(b) $(0, 3), (2, 0)$
(c)

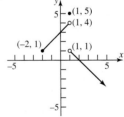

(d) $\{y | y < 4, y = 5\}; (-\infty, 4)$ or $\{5\}$

25. (a) All real numbers
(b) $(-1, 0), (0, 0)$
(c)

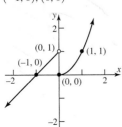

(d) All real numbers

27. (a) $\{x | x \geq -2\}; [-2, \infty)$
(b) $(0, 1)$
(c)

(d) $\{y | y > 0\}; (0, \infty)$

29. (a) All real numbers
(b) $(x, 0)$ for $0 \leq x < 1$
(c)

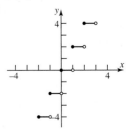

(d) Set of even integers

31. $f(x) = \begin{cases} -x & \text{if } -1 \le x \le 0 \\ \frac{1}{2}x & \text{if } 0 < x \le 2 \end{cases}$ (Other answers are possible.)   **33.** $f(x) = \begin{cases} -x & \text{if } x \le 0 \\ -x + 2 & \text{if } 0 < x \le 2 \end{cases}$ (Other answers are possible.)

35. (a) \$43.28 **(b)** \$235.54

(c) $C = \begin{cases} 0.67655x + 9.45 & \text{if } 0 \le x \le 50 \\ 0.42725x + 21.915 & \text{if } x > 50 \end{cases}$

(d)

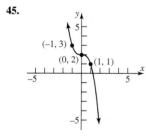

Monthly Charge (dollars) vs Therms graph showing points (0, 9.45), (50, 43.28), (500, 235.54).

37. (a) 10°C **(b)** 4°C **(c)** −3°C **(d)** −4°C
(e) The wind chill is equal to the air temperature.
(f) At wind speed greater than 20 m/sec, the wind chill factor depends only on the air temperature.
39. Each graph is that of $y = x^2$, but shifted vertically.
If $y = x^2 + k, k > 0$, the shift is up k units;
if $y = x^2 + k, k < 0$, the shift is down $|k|$ units.
41. Each graph is that of $y = |x|$, but either compressed or stretched.
If $y = k|x|$ and $k > 1$, the graph is stretched vertically;
if $y = k|x|, 0 < k < 1$, the graph is compressed vertically.
43. The graph of $y = f(-x)$ is the reflection about the y-axis of the graph of $y = f(x)$.

45. They are all ∪-shaped and open upward. All three go through the points $(-1, 1)$, $(0, 0)$ and $(1, 1)$. As the exponent increases, the steepness of the curve increases (except near $x = 0$).

3.4 Exercises *(page 252)*

1. B **3.** H **5.** I **7.** L **9.** F **11.** G **13.** $y = (x - 4)^3$ **15.** $y = x^3 + 4$ **17.** $y = -x^3$ **19.** $y = 4x^3$ **21.** (1) $y = \sqrt{x} + 2$;
(2) $y = -(\sqrt{x} + 2)$; (3) $y = -(\sqrt{-x} + 2)$ **23.** (1) $y = -\sqrt{x}$; (2) $y = -\sqrt{x} + 2$; (3) $y = -\sqrt{x + 3} + 2$ **25.** (c) **27.** (c)

29.

31.

33.

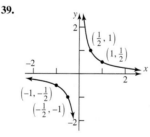

35.

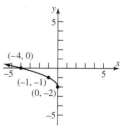

37.

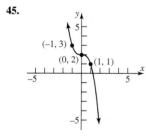

39.

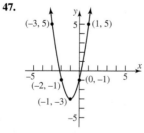

41.

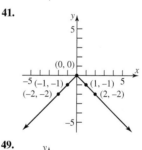

43.

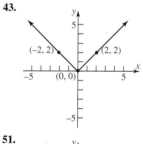

45.

47.

49.

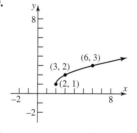

51.

53.

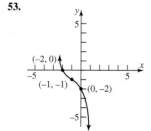

55.

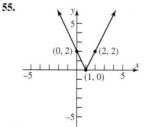

57.

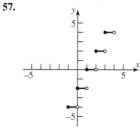

59. (a) $F(x) = f(x) + 3$ **(b)** $G(x) = f(x + 2)$ **(c)** $P(x) = -f(x)$ **(d)** $H(x) = f(x + 1) - 2$

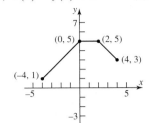   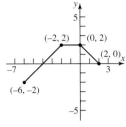   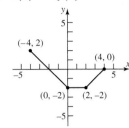   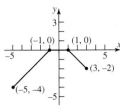

(e) $Q(x) = \frac{1}{2}f(x)$ **(f)** $g(x) = f(-x)$ **(g)** $h(x) = f(2x)$

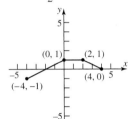   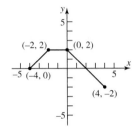   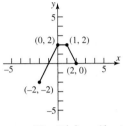

61. (a) $F(x) = f(x) + 3$ **(b)** $G(x) = f(x + 2)$ **(c)** $P(x) = -f(x)$ **(d)** $H(x) = f(x + 1) - 2$

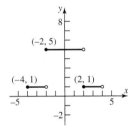   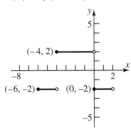   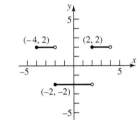   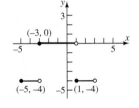

(e) $Q(x) = \frac{1}{2}f(x)$ **(f)** $g(x) = f(-x)$ **(g)** $h(x) = f(2x)$

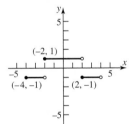   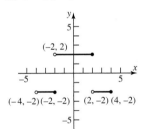   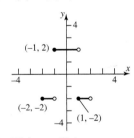

63. (a) $F(x) = f(x) + 3$ **(b)** $G(x) = f(x + 2)$ **(c)** $P(x) = -f(x)$

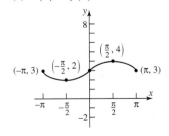   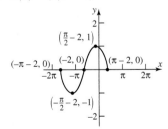   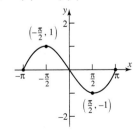

(d) $H(x) = f(x + 1) - 2$ **(e)** $Q(x) = \frac{1}{2}f(x)$ **(f)** $g(x) = f(-x)$ **(g)** $h(x) = f(2x)$

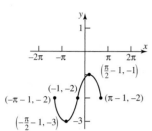 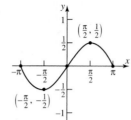 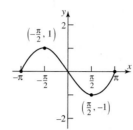 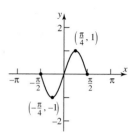

65. (a)

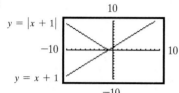

$y = |x + 1|$

$y = x + 1$

(b) $y = |4 - x^2|$

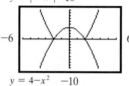

$y = 4 - x^2$

(c) $y = |x^3 + x|$

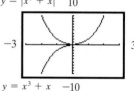

$y = x^3 + x$

(d) Any part of the graph of $y = f(x)$ that lies below the x-axis is reflected about the x-axis to obtain the graph of $y = |f(x)|$.

67. (a)

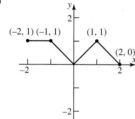

(b)

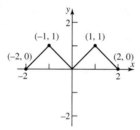

69. $f(x) = (x + 1)^2 - 1$

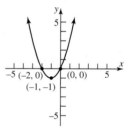

71. $f(x) = (x - 4)^2 - 15$

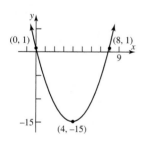

73. $f(x) = \left(x + \frac{1}{2}\right)^2 + \frac{3}{4}$

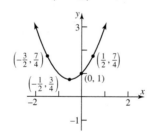

75.

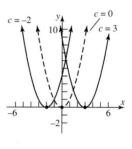

77.

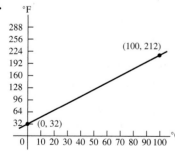

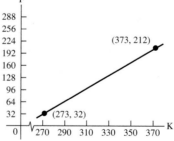

79. (a)

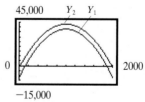

(b) 10% tax

(c) Y_1 is the graph of $p(x)$ shifted down vertically 10,000 units. Y_2 is the graph of $p(x)$ vertically compressed by a factor of 0.9.

(d) 10% tax

3.5 Exercises (page 262)

1. (a) $(f + g)(x) = 5x + 1$; All real numbers   **(b)** $(f - g)(x) = x + 7$; All real numbers

(c) $(f \cdot g)(x) = 6x^2 - x - 12$; All real numbers   **(d)** $\left(\dfrac{f}{g}\right)(x) = \dfrac{3x + 4}{2x - 3}$; $\left\{x \middle| x \neq \dfrac{3}{2}\right\}$

3. (a) $(f + g)(x) = 2x^2 + x - 1$; All real numbers   **(b)** $(f - g)(x) = -2x^2 + x - 1$; All real numbers

(c) $(f \cdot g)(x) = 2x^3 - 2x^2$; All real numbers   **(d)** $\left(\dfrac{f}{g}\right)(x) = \dfrac{x - 1}{2x^2}$; $\{x | x \neq 0\}$ **5. (a)** $(f + g)(x) = \sqrt{x} + 3x - 5$; $\{x | x \geq 0\}$

(b) $(f - g)(x) = \sqrt{x} - 3x + 5$; $\{x | x \geq 0\}$ **(c)** $(f \cdot g)(x) = 3x\sqrt{x} - 5\sqrt{x}$; $\{x | x \geq 0\}$ **(d)** $\left(\dfrac{f}{g}\right)(x) = \dfrac{\sqrt{x}}{3x - 5}$; $\left\{x | x \geq 0, x \neq \dfrac{5}{3}\right\}$

7. (a) $(f + g)(x) = 1 + \dfrac{2}{x}$; $\{x | x \neq 0\}$ **(b)** $(f - g)(x) = 1$; $\{x | x \neq 0\}$ **(c)** $(f \cdot g)(x) = \dfrac{1}{x} + \dfrac{1}{x^2}$; $\{x | x \neq 0\}$

(d) $\left(\dfrac{f}{g}\right)(x) = x + 1$; $\{x | x \neq 0\}$ **9. (a)** $(f + g)(x) = \dfrac{6x + 3}{3x - 2}$; $\left\{x \middle| x \neq \dfrac{2}{3}\right\}$ **(b)** $(f - g)(x) = \dfrac{-2x + 3}{3x - 2}$; $\left\{x \middle| x \neq \dfrac{2}{3}\right\}$

(c) $(f \cdot g)(x) = \dfrac{8x^2 + 12x}{(3x - 2)^2}$; $\left\{x \middle| x \neq \dfrac{2}{3}\right\}$ **(d)** $\left(\dfrac{f}{g}\right)(x) = \dfrac{2x + 3}{4x}$; $\left\{x \middle| x \neq 0, x \neq \dfrac{2}{3}\right\}$ **11.** $g(x) = 5 - \dfrac{7}{2}x$ **13. (a)** 98   **(b)** 49   **(c)** 4

(d) 4   **15. (a)** 97   **(b)** $-\dfrac{163}{2}$ **(c)** 1   **(d)** $-\dfrac{3}{2}$ **17. (a)** $2\sqrt{2}$ **(b)** $2\sqrt{2}$ **(c)** 1   **(d)** 0   **19. (a)** $\dfrac{1}{17}$ **(b)** $\dfrac{1}{5}$ **(c)** 1   **(d)** $\dfrac{1}{2}$

21. (a) $\dfrac{3}{\sqrt[3]{4} + 1}$ **(b)** 1   **(c)** $\dfrac{6}{5}$ **(d)** 0   **23.** $\{x | x \neq 0, x \neq 2\}$ **25.** $\{x | x \neq -4, x \neq 0\}$ **27.** $\left\{x \middle| x \geq -\dfrac{3}{2}\right\}$ **29.** $\{x | x \geq 1\}$

31. (a) $(f \circ g)(x) = 6x + 3$; All real numbers   **(b)** $(g \circ f)(x) = 6x + 9$; All real numbers   **(c)** $(f \circ f)(x) = 4x + 9$; All real numbers

(d) $(g \circ g)(x) = 9x$; All real numbers   **33. (a)** $(f \circ g)(x) = 3x^2 + 1$; All real numbers   **(b)** $(g \circ f)(x) = 9x^2 + 6x + 1$; All real numbers

(c) $(f \circ f)(x) = 9x + 4$; All real numbers   **(d)** $(g \circ g)(x) = x^4$; All real numbers   **35. (a)** $(f \circ g)(x) = x^4 + 8x^2 + 16$; All real numbers

(b) $(g \circ f)(x) = x^4 + 4$; All real numbers   **(c)** $(f \circ f)(x) = x^4$; All real numbers   **(d)** $(g \circ g)(x) = x^4 + 8x^2 + 20$; All real numbers

37. (a) $(f \circ g)(x) = \dfrac{3x}{2 - x}$; $\{x | x \neq 0, x \neq 2\}$ **(b)** $(g \circ f)(x) = \dfrac{2(x - 1)}{3}$; $\{x | x \neq 1\}$ **(c)** $(f \circ f)(x) = \dfrac{3(x - 1)}{4 - x}$; $\{x | x \neq 1, x \neq 4\}$

(d) $(g \circ g)(x) = x$; $\{x | x \neq 0\}$ **39. (a)** $(f \circ g)(x) = \dfrac{4}{4 + x}$; $\{x | x \neq -4, x \neq 0\}$ **(b)** $(g \circ f)(x) = \dfrac{-4(x - 1)}{x}$; $\{x | x \neq 0, x \neq 1\}$

(c) $(f \circ f)(x) = x$; $\{x | x \neq 1\}$ **(d)** $(g \circ g)(x) = x$; $\{x | x \neq 0\}$ **41. (a)** $(f \circ g)(x) = \sqrt{2x + 3}$; $\left\{x \middle| x \geq -\dfrac{3}{2}\right\}$

(b) $(g \circ f)(x) = 2\sqrt{x} + 3$; $\{x | x \geq 0\}$ **(c)** $(f \circ f)(x) = \sqrt[4]{x}$; $\{x | x \geq 0\}$ **(d)** $(g \circ g)(x) = 4x + 9$; All real numbers

43. (a) $(f \circ g)(x) = x$; $\{x | x \geq 1\}$ **(b)** $(g \circ f)(x) = |x|$; All real numbers   **(c)** $(f \circ f)(x) = x^4 + 2x^2 + 2$; All real numbers

(d) $(g \circ g)(x) = \sqrt{\sqrt{x - 1} - 1}$; $\{x | x \geq 2\}$ **45. (a)** $(f \circ g)(x) = acx + ad + b$; All real numbers

(b) $(g \circ f)(x) = acx + bc + d$; All real numbers   **(c)** $(f \circ f)(x) = a^2x + ab + b$; All real numbers

(d) $(g \circ g)(x) = c^2x + cd + d$; All real numbers

47. $(f \circ g)(x) = f(g(x)) = f\left(\dfrac{1}{2}x\right) = 2\left(\dfrac{1}{2}x\right) = x$; $(g \circ f)(x) = g(f(x)) = g(2x) = \dfrac{1}{2}(2x) = x$

49. $(f \circ g)(x) = f(g(x)) = f(\sqrt[3]{x}) = (\sqrt[3]{x})^3 = x$; $(g \circ f)(x) = g(f(x)) = g(x^3) = \sqrt[3]{x^3} = x$

51. $(f \circ g)(x) = f(g(x)) = f\left(\dfrac{1}{2}(x + 6)\right) = 2\left[\dfrac{1}{2}(x + 6)\right] - 6 = x + 6 - 6 = x$;

$(g \circ f)(x) = g(f(x)) = g(2x - 6) = \dfrac{1}{2}(2x - 6 + 6) = x$

53. $(f \circ g)(x) = f(g(x)) = f\left(\dfrac{1}{a}(x - b)\right) = a\left[\dfrac{1}{a}(x - b)\right] + b = x$; $(g \circ f)(x) = g(f(x)) = g(ax + b) = \dfrac{1}{a}(ax + b - b) = x$

55. $f(x) = x^4$; $g(x) = 2x + 3$ (Other answers are possible.)   **57.** $f(x) = \sqrt{x}$; $g(x) = x^2 + 1$ (Other answers are possible.)

59. $f(x) = |x|$; $g(x) = 2x + 1$ (Other answers are possible.)   **61.** $(f \circ g)(x) = 11$; $(g \circ f)(x) = 2$ **63.** $-3, 3$ **65.** $S(t) = \dfrac{16}{9}\pi t^6$

67. $C(t) = 15{,}000 + 800{,}000t - 40{,}000t^2$ **69.** $C(p) = \dfrac{2\sqrt{100 - p}}{25} + 600$, $0 \leq p \leq 100$ **71.** $V(r) = 2\pi r^3$

3.6 Exercises (page 269)

1. $V(r) = 2\pi r^3$

3. **(a)** $R(x) = -\dfrac{1}{6}x^2 + 100x$

   (b) \$13,333

   (c)
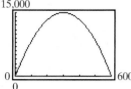
15,000 ... 600 ... 0 ... 0

   (d) 300; \$15,000   **(e)** \$50

5. **(a)** $R(x) = -\dfrac{1}{5}x^2 + 20x$

   (b) \$255

   (c)
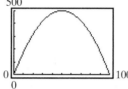
500 ... 100 ... 0 ... 0

   (d) 50; \$500   **(e)** \$10

7. **(a)** $A(x) = -x^2 + 200x$

   (b) $0 < x < 200$

   (c) A is largest when $x = 100$ yd.

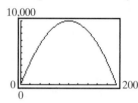
10,000 ... 200 ... 0 ... 0

9. **(a)** $d(x) = \sqrt{x^4 - 15x^2 + 64}$

   (b) $d(0) = 8$

   (c) $d(1) = \sqrt{50} \approx 7.07$

   (d)
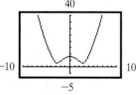
40 ... −10 ... 10 ... −5

   (e) d is smallest when $x \approx -2.74$
     or $x \approx 2.74$.

11. **(a)** $d(x) = \sqrt{x^2 - x + 1}$

   (b)
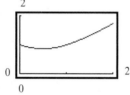
2 ... 0 ... 2 ... 0

   (c) $x = \dfrac{1}{2}$

13. $A(x) = \dfrac{1}{2}x^4$

15. **(a)** $A(x) = x(16 - x^2)$

   (b) Domain: $\{x \mid 0 < x < 4\}$

   (c) The area is largest when $x \approx 2.31$.

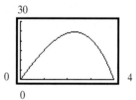
30 ... 0 ... 4 ... 0

17. **(a)** $A(x) = 4x\sqrt{4 - x^2}$

   (b) $p(x) = 4x + 4\sqrt{4 - x^2}$

   (c) A is largest when $x \approx 1.41$.

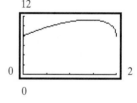
10 ... 0 ... 2 ... 0

   (d) p is largest when $x \approx 1.41$.

12 ... 0 ... 2 ... 0

19. **(a)** $A(x) = x^2 + \dfrac{25 - 20x + 4x^2}{\pi}$

   (b) Domain: $\{x \mid 0 < x < 2.5\}$

   (c) A is smallest when $x \approx 1.40$ m.

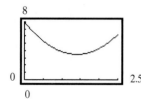
8 ... 0 ... 2.5 ... 0

21. **(a)** $C(x) = x$ **(b)** $A(x) = \dfrac{x^2}{4\pi}$

23. **(a)** $A(r) = 2r^2$ **(b)** $p(r) = 6r$

25. $A(x) = \left(\dfrac{\pi}{3} - \dfrac{\sqrt{3}}{4}\right)x^2$

27. $C = \begin{cases} 95 & \text{if } x = 7 \\ 119 & \text{if } 7 < x \leq 8 \\ 143 & \text{if } 8 < x \leq 9 \\ 167 & \text{if } 9 < x \leq 10 \\ 190 & \text{if } 10 < x \leq 14 \end{cases}$

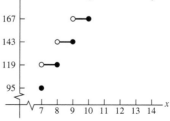
C ... 190 ... 167 ... 143 ... 119 ... 95 ... 7 8 9 10 11 12 13 14 ... x

29. $d(t) = 50t$

31. (a) $V(x) = x(24 - 2x)^2$

(b) 972 in^3 **(c)** 160 in^3

(d) V is largest when $x = 4$.

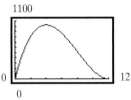

33. $V(h) = \pi h \left(R^2 - \dfrac{h^2}{4} \right)$

35. (a) $C(x) = 10x + 14\sqrt{x^2 - 10x + 29}, 0 \leq x \leq 5$

(b) $C(1) = \$72.61$ **(c)** $C(3) = \$69.60$

(d)

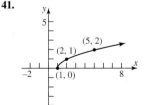

(e) The least cost occurs when $x \approx 2.96$ mi.

37. $V(h) = \dfrac{\pi}{48} h^3$

Fill-in-the-Blank Items *(page 275)*

1. independent; dependent   **3.** slope   **5.** horizontal; right

True/False Items *(page 275)*

1. F   **3.** T   **5.** F   **7.** F

Review Exercises *(page 275)*

1. $f(x) = -2x + 3$ **3.** $A = 11$ **5.** b, c, d   **7. (a)** $f(-x) = \dfrac{-3x}{x^2 - 4}$ **(b)** $-f(x) = \dfrac{-3x}{x^2 - 4}$ **(c)** $f(x + 2) = \dfrac{3x + 6}{x^2 + 4x}$

(d) $f(x - 2) = \dfrac{3x - 6}{x^2 - 4x}$ **(e)** $f(2x) = \dfrac{6x}{4x^2 - 4}$ **9. (a)** $f(-x) = \sqrt{x^2 - 4}$ **(b)** $-f(x) = -\sqrt{x^2 - 4}$ **(c)** $f(x + 2) = \sqrt{x^2 + 4x}$

(d) $f(x - 2) = \sqrt{x^2 - 4x}$ **(e)** $f(2x) = 2\sqrt{x^2 - 1}$ **11. (a)** $f(-x) = \dfrac{x^2 - 4}{x^2}$ **(b)** $-f(x) = -\dfrac{x^2 - 4}{x^2}$ **(c)** $f(x + 2) = \dfrac{x^2 + 4x}{x^2 + 4x + 4}$

(d) $f(x - 2) = \dfrac{x^2 - 4x}{x^2 - 4x + 4}$ **(e)** $f(2x) = \dfrac{x^2 - 1}{x^2}$ **13.** $\{x | x \neq -3, x \neq 3\}$ **15.** $\{x | x \leq 2\}$ **17.** $\{x | x > 0\}$ **19.** $\{x | x \neq -3, x \neq 1\}$

21. (a) $\{x | x > -2\}; (-2, \infty)$

(b) $(0, 0)$

(c)

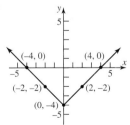

(d) $\{y | y > -6\}; (-6, \infty)$

23. (a) $\{x | x \geq -4\}; [-4, \infty)$

(b) $(0, 1)$

(c)

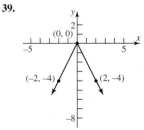

(d) $\{y | y \geq -4\}; [-4, \infty)$

25. -5 **27.** $-4x - 5$

29. Odd   **31.** Even

33. Neither   **35.** Odd

37.

Intercepts: $(-4, 0), (4, 0), (0, -4)$
Domain: all real numbers
Range: $\{y | y \geq -4\}$ or $[-4, \infty)$

39.

Intercept: $(0, 0)$
Domain: all real numbers
Range: $\{y | y \leq 0\}$ or $(-\infty, 0]$

41.

Intercept: $(1, 0)$
Domain: $\{x | x \geq 1\}$ or $[1, \infty)$
Range: $\{y | y \geq 0\}$ or $[0, \infty)$

43.

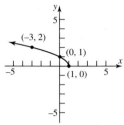

Intercepts: $(0, 1)$, $(1, 0)$

Domain: $\{x \mid x \le 1\}$ or $(-\infty, 1]$

Range: $\{y \mid y \ge 0\}$ or $[0, \infty)$

45.

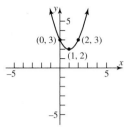

Intercept: $(0, 3)$

Domain: all real numbers

Range: $\{y \mid y \ge 2\}$ or $[2, \infty)$

47.

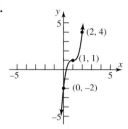

Intercepts: $(0, -2)$,

$1 + \dfrac{\sqrt[3]{-9}}{3}$ or about $(0.3, 0)$

Domain: all real numbers

Range: all real numbers

49. (a) -26 **(b)** -241 **(c)** 16 **(d)** -1 **51. (a)** $\sqrt{11}$ **(b)** 1 **(c)** $\sqrt{\sqrt{6} + 2}$ **(d)** 19 **53. (a)** $\dfrac{1}{20}$ **(b)** $-\dfrac{13}{8}$ **(c)** $\dfrac{400}{1601}$ **(d)** -17

55. $(f \circ g)(x) = 1 - 3x$; All real numbers; $(g \circ f)(x) = 7 - 3x$; All real numbers; $(f \circ f)(x) = x$; All real numbers; $(g \circ g)(x) = 9x + 4$; All real numbers

57. $(f \circ g)(x) = 27x^2 + 3|x| + 1$; All real numbers; $(g \circ f)(x) = 3|3x^2 + x + 1|$; All real numbers; $(f \circ f)(x) = 27x^4 + 18x^3 + 24x^2 + 7x + 5$; All real numbers; $(g \circ g)(x) = 9|x|$; All real numbers

59. $(f \circ g)(x) = \dfrac{1 + x}{1 - x}$; $\{x \mid x \ne 0, x \ne 1\}$; $(g \circ f)(x) = \dfrac{x - 1}{x + 1}$; $\{x \mid x \ne -1, x \ne 1\}$; $(f \circ f)(x) = x$; $\{x \mid x \ne 1\}$; $(g \circ g)(x) = x$; $\{x \mid x \ne 0\}$

61. (a)

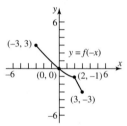

(b)

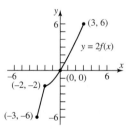

(c)

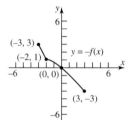

(d)

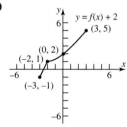

(e)

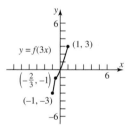

(f)

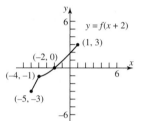

63. $T(h) = -0.0025h + 30, 0 \le x \le 10,000$ **65.** $S(x) = kx(36 - x^2)^{3/2}$; Domain: $\{x \mid 0 < x < 6\}$

67. (a), (b), (e)

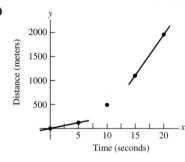

(c) 22.5 ft/sec

(d) Between 0 and 5 sec, the average speed of the parachutist is 22.5 ft/sec.

(f) 171.5 ft/sec

(g) Between 15 and 20 sec, the average speed of the parachutist is 171.5 ft/sec.

(h) It is increasing.

69. (a) $C(r) = 0.12\pi r^2 + \dfrac{40}{r}$ **(b)** \$16.03

(c) \$29.13

(d) The cost is least for $r \approx 3.76$ cm.

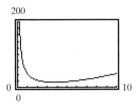

71. $V(S) = \dfrac{S}{6}\sqrt{\dfrac{S}{\pi}}$; If the surface area doubles, the volume increases by a factor of $2\sqrt{2}$.

C H A P T E R 4   Polynomial and Rational Functions

4.1 Exercises (page 295)

1. D **3.** A **5.** B **7.** E

9.

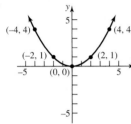

11.

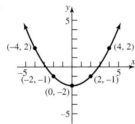

13.

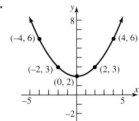

15.

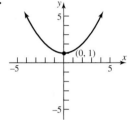

17. $f(x) = (x + 2)^2 - 2$

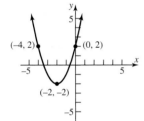

19. $f(x) = 2(x - 1)^2 - 1$

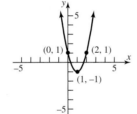

21. $f(x) = -(x + 1)^2 + 1$

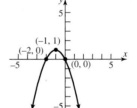

23. $f(x) = \dfrac{1}{2}(x + 1)^2 - \dfrac{3}{2}$

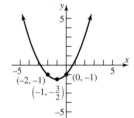

25.

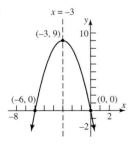

27.

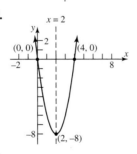

29.

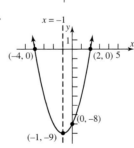

31.

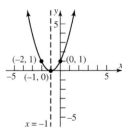

33.

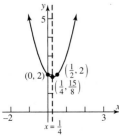

35.

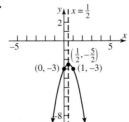

37.

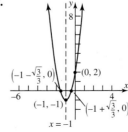

39.

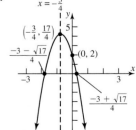

41. Minimum value; -18 **43.** Minimum value; -21 **45.** Maximum value; 21 **47.** Maximum value; 13

49. (a) $a = 1: f(x) = (x + 3)(x - 1) = x^2 + 2x - 3$ (c) The value of a does not affect the axis of symmetry. It is $x = -1$
      $a = 2: f(x) = 2(x + 3)(x - 1) = 2x^2 + 4x - 6$ for all values of a.
      $a = -2: f(x) = -2(x + 3)(x - 1) = -2x^2 - 4x + 6$ (d) The value of a does not affect the x-coordinate of the vertex.
      $a = 5: f(x) = 5(x + 3)(x - 1) = 5x^2 + 10x - 15$ However, the y-coordinate of the vertex is multiplied by a.
   (b) The value of a does not affect the intercepts   (e) The midpoint of the x-intercepts is the x-coordinate of the vertex.

51. 500; $1,000,000   **53.** (a) $R(x) = -\dfrac{1}{6}x^2 + 100x$ (b) $13,333   (c) 300; $15,000   (d) $50   **55.** (a) $R(x) = -\dfrac{1}{5}x^2 + 20x$

(b) $255   (c) 50; $500   (d) $10   **57.** (a) $A(x) = -x^2 + 200x$ (b) A is largest when $x = 100$ yd.   (c) $10,000$ sq yd   **59.** $2,000,000$ m^2

61. (a) $\dfrac{625}{16} \approx 39$ ft   (b) 219.5 ft   (c) 170 ft   **63.** 18.75 m   **65.** 3 in.   **67.** $\dfrac{750}{\pi}$ by 375 m

   (d)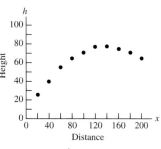

   (e) When the height is 100 ft, the
   projectile is 135.7 ft from the cliff.

69. (a) Quadratic, $a < 0$.   (b) 44.7 yr old
   (c) $46,484
   (e)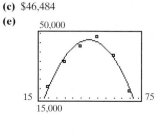

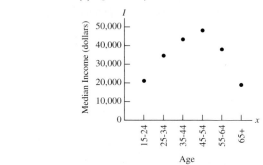

71. (a) Quadratic, $a > 0$ (b) Midway between 1981 and 1982
                                        (c) 8,867,000 barrels per day
                                        (e)

73. (a) Quadratic, $a < 0$.   (b) 139.2 ft
                                    (c) 77.4 ft
                                    (e)

75. $\left.\begin{array}{r} ah^2 - bh + c = y_0 \\ c = y_1 \\ ah^2 + bh + c = y_2 \end{array}\right\}$ $\left.\begin{array}{r} y_0 + y_2 = 2ah^2 + 2c \\ 4y_1 = 4c \end{array}\right\}$ Area $= \dfrac{h}{3}(2ah^2 + 6c) = \dfrac{h}{3}(y_0 + 4y_1 + y_2)$ **77.** $\dfrac{128}{3}$ **79.** $\dfrac{22}{3}$

81. If x is even, then ax^2 and bx are even and $ax^2 + bx$ is even, which means $ax^2 + bx + c$ is odd. If x is odd, then ax^2 and bx are odd and $ax^2 + bx$ is even, which means $ax^2 + bx + c$ is odd. In either case, $f(x)$ is odd.

4.2 Exercises *(page 314)*

1. Yes; degree 3   **3.** Yes; degree 2   **5.** No; x is raised to the -1 power.   **7.** No; x is raised to the $\dfrac{3}{2}$ power.   **9.** Yes; degree 4

11. Yes; degree 4

13.
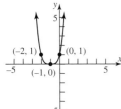

15.
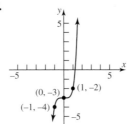

17.
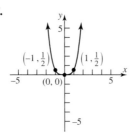

19.
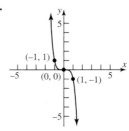

21.
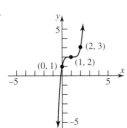

23.
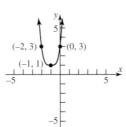

25.
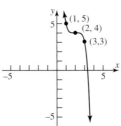

27. $f(x) = x^3 - 3x^2 - x + 3$
for $a = 1$

29. $f(x) = x^3 - x^2 - 12x$
for $a = 1$

31. $f(x) = x^4 - 15x^2 + 10x + 24$
for $a = 1$

33. (a) 7, multiplicity 1; −3, multiplicity 2   **(b)** graph touches the x-axis at −3 and crosses it at 7   **(c)** $y = 3x^3$

35. (a) 2, multiplicity 3   **(b)** graph crosses the x-axis at 2   **(c)** $y = 4x^5$ **37. (a)** $-\dfrac{1}{2}$, multiplicity 2   **(b)** graph touches the x-axis at $-\dfrac{1}{2}$

(c) $y = -2x^6$ **39. (a)** 5, multiplicity 3; −4, multiplicity 2   **(b)** graph touches the x-axis at −4 and crosses it at 5   **(c)** $y = x^5$

41. (a) No real zeros   **(b)** graph neither crosses nor touches the x-axis   **(c)** $y = 3x^6$

43. (a) 0, multiplicity 2; $-\sqrt{2}$, $\sqrt{2}$, multiplicity 1   **(b)** graph touches the x-axis at 0 and crosses at $-\sqrt{2}$ and $\sqrt{2}$ **(c)** $y = -2x^4$

45. (a) x-intercept: 1; y-intercept: 1
(b) Touches at 1
(c) $y = x^2$
(d) 1
(e)

(f)
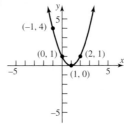

Interval	$(-\infty, 1)$	$(1, \infty)$
Number Chosen	-1	2
Value of f	$f(-1) = 4$	$f(2) = 1$
Location of Graph	Above x-axis	Above x-axis
Point on Graph	$(-1, 4)$	$(2, 1)$

47. (a) x-intercepts: 0, 3; y-intercept: 0
(b) Touches at 0; crosses at 3
(c) $y = x^3$
(d) 2
(e)

(f)
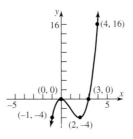

Interval	$(-\infty, 0)$	$(0, 3)$	$(3, \infty)$
Number Chosen	-1	2	4
Value of f	$f(-1) = -4$	$f(2) = -4$	$f(4) = 16$
Location of Graph	Below x-axis	Below x-axis	Above x-axis
Point on Graph	$(-1, -4)$	$(2, -4)$	$(4, 16)$

49. (a) x-intercepts: −4, 0; y-intercept: 0
(b) Crosses at −4, 0
(c) $y = 6x^4$
(d) 3
(e)

(f)
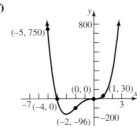

Interval	$(-\infty, -4)$	$(-4, 0)$	$(0, \infty)$
Number Chosen	-5	-2	1
Value of f	$f(-5) = 750$	$f(-2) = -96$	$f(1) = 30$
Location of Graph	Above x-axis	Below x-axis	Above x-axis
Point on Graph	$(-5, 750)$	$(-2, -96)$	$(1, 30)$

51. (a) x-intercepts: $-2, 0$; y-intercept: 0

(b) Crosses at -2; touches at 0

(c) $y = -4x^3$

(d) 2

(e)

Interval	$(-\infty, -2)$	$(-2, 0)$	$(0, \infty)$
Number Chosen	-3	-1	1
Value of f	$f(-3) = 36$	$f(-1) = -4$	$f(1) = -12$
Location of Graph	Above x-axis	Below x-axis	Below x-axis
Point on Graph	$(-3, 36)$	$(-1, -4)$	$(1, -12)$

(f)

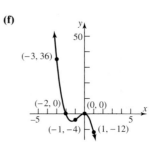

53. (a) x-intercepts: $-4, 0, 2$; y-intercept: 0

(b) Crosses at $-4, 0, 2$

(c) $y = x^3$

(d) 2

(e)

Interval	$(-\infty, -4)$	$(-4, 0)$	$(0, 2)$	$(2, \infty)$
Number Chosen	-5	-2	1	3
Value of f	$f(-5) = -35$	$f(-2) = 16$	$f(1) = -5$	$f(3) = 21$
Location of Graph	Below x-axis	Above x-axis	Below x-axis	Above x-axis
Point on Graph	$(-5, -35)$	$(-2, 16)$	$(1, -5)$	$(3, 21)$

(f)

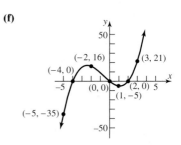

55. $f(x) = 4x - x^3 = -x(x^2 - 4)$
$= -x(x + 2)(x - 2)$

(a) x-intercepts: $-2, 0, 2$; y-intercept: 0

(b) Crosses at $-2, 0, 2$

(c) $y = -x^3$

(d) 2

(e)

Interval	$(-\infty, -2)$	$(-2, 0)$	$(0, 2)$	$(2, \infty)$
Number Chosen	-3	-1	1	3
Value of f	$f(-3) = 15$	$f(-1) = -3$	$f(1) = 3$	$f(3) = -15$
Location of Graph	Above x-axis	Below x-axis	Above x-axis	Below x-axis
Point on Graph	$(-3, 15)$	$(-1, -3)$	$(1, 3)$	$(3, -15)$

(f)

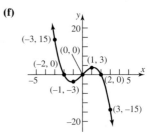

57. (a) x-intercepts: $-2, 0, 2$; y-intercept: 0

(b) Crosses at $-2, 2$; touches at 0

(c) $y = x^4$

(d) 3

(e)

Interval	$(-\infty, -2)$	$(-2, 0)$	$(0, 2)$	$(2, \infty)$
Number Chosen	-3	-1	1	3
Value of f	$f(-3) = 45$	$f(-1) = -3$	$f(1) = -3$	$f(3) = 45$
Location of Graph	Above x-axis	Below x-axis	Below x-axis	Above x-axis
Point on Graph	$(-3, 45)$	$(-1, -3)$	$(1, -3)$	$(3, 45)$

(f)

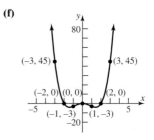

59. (a) x-intercepts: $0, 2$; y-intercept: 0

(b) Touches at $0, 2$

(c) $y = x^4$

(d) 3

(e)

Interval	$(-\infty, 0)$	$(0, 2)$	$(2, \infty)$
Number Chosen	-1	1	3
Value of f	$f(-1) = 9$	$f(1) = 1$	$f(3) = 9$
Location of Graph	Above x-axis	Above x-axis	Above x-axis
Point on Graph	$(-1, 9)$	$(1, 1)$	$(3, 9)$

(f)

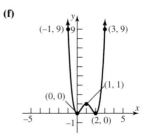

61. (a) x-intercepts: $-1, 0, 3$; y-intercept: 0

(b) Crosses at $-1, 3$; touches at 0

(c) $y = x^4$

(d) 3

(e)

Interval	$(-\infty, -1)$	$(-1, 0)$	$(0, 3)$	$(3, \infty)$
Number Chosen	-2	$-\frac{1}{2}$	2	4
Value of f	$f(-2) = 20$	$f(-\frac{1}{2}) = -\frac{7}{16}$	$f(2) = -12$	$f(4) = 80$
Location of Graph	Above x-axis	Below x-axis	Below x-axis	Above x-axis
Point on Graph	$(-2, 20)$	$(-\frac{1}{2}, -\frac{7}{16})$	$(2, -12)$	$(4, 80)$

(f)
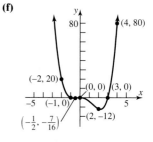

63. (a) x-intercepts: $-2, 4$; y-intercept: 64

(b) Touches at -2 and 4

(c) $y = x^4$

(d) 3

(e)

Interval	$(-\infty, -2)$	$(-2, 4)$	$(4, \infty)$
Number Chosen	-3	0	5
Value of f	$f(-3) = 49$	$f(0) = 64$	$f(5) = 49$
Location of Graph	Above x-axis	Above x-axis	Above x-axis
Point on Graph	$(-3, 49)$	$(0, 64)$	$(5, 49)$

(f)
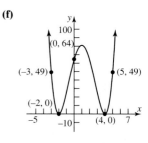

65. (a) x-intercepts: $0, 2$; y-intercept: 0

(b) Touches at 0; crosses at 2

(c) $y = x^5$

(d) 4

(e)

Interval	$(-\infty, 0)$	$(0, 2)$	$(2, \infty)$
Number Chosen	-1	1	3
Value of f	$f(-1) = -12$	$f(1) = -4$	$f(3) = 108$
Location of Graph	Below x-axis	Below x-axis	Above x-axis
Point on Graph	$(-1, -12)$	$(1, -4)$	$(3, 108)$

(f)
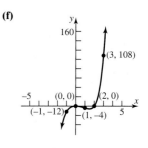

67. (a) x-intercepts: $-1, 0, 1$; y-intercept: 0

(b) Crosses at 1; touches at -1 and 0

(c) $y = -x^5$

(d) 4

(e)

Interval	$(-\infty, -1)$	$(-1, 0)$	$(0, 1)$	$(1, \infty)$
Number Chosen	-2	$-\frac{1}{2}$	$\frac{1}{2}$	2
Value of f	$f(-2) = 12$	$f(-\frac{1}{2}) = \frac{3}{32}$	$f(\frac{1}{2}) = \frac{9}{32}$	$f(2) = -36$
Location of Graph	Above x-axis	Above x-axis	Above x-axis	Below x-axis
Point on Graph	$(-2, 12)$	$(-\frac{1}{2}, \frac{3}{32})$	$(\frac{1}{2}, \frac{9}{32})$	$(2, -36)$

(f)
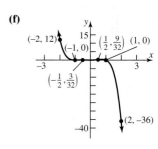

69. c, e, f   **71.** c, e

73. x-intercepts: $-1.26, -0.20, 1.26$
Turning points: $(0.66, -0.99)$;
$(-0.80, 0.57)$

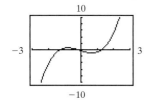

75. x-intercepts: $-3.56, 0.50$
Turning points: $(0.50, 0)$;
$(-2.21, 9.91)$

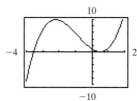

77. x-intercepts: $-1.50, -0.50, 0.50, 1.50$
Turning points: $(-1.12, -1), (1.12, -1)$,
$(0, 0.5625)$

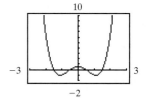

79. *x*-intercepts: −4.78, 0.45, 3.23
Turning points: (−3.32, −135.92),
(2.38, −22.67); (0.45, 0)

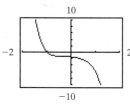

81. *x*-intercept: 0.84
Turning points: (−0.51, −1.54);
(0.21, −2.12)

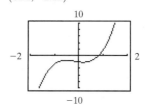

83. *x*-intercepts: −1.07, 1.62
Turning point: (−0.42, −4.64)

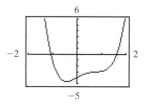

85. *x*-intercept: −0.98
No turning points

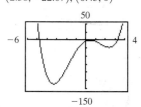

87. (a) Cubic, *a* > 0

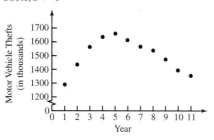

(b) ≈ 1,524,220 motor vehicle thefts

(d) 1700

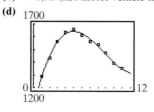

89. (a) Cubic, *a* > 0

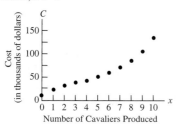

(b) 7 thousand dollars
(c) 20 thousand dollars
(d) ≈ $155,000
(f) 150

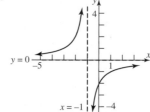

(g) Fixed costs of $10,200.

4.3 Exercises *(page 326)*

1. All real numbers except 3; $\{x | x \neq 3\}$ **3.** All real numbers except 2 and −4; $\{x | x \neq 2, x \neq -4\}$

5. All real numbers except $-\frac{1}{2}$ and 3; $\left\{x \middle| x \neq -\frac{1}{2}, x \neq 3\right\}$ **7.** All real numbers except 2; $\{x | x \neq 2\}$ **9.** All real numbers

11. All real numbers except −3 and 3; $\{x | x \neq -3, x \neq 3\}$ **13. (a)** Domain: $\{x | x \neq 2\}$; Range: $\{y | y \neq 1\}$ **(b)** (0, 0)   **(c)** $y = 1$

(d) $x = 2$ **(e)** None   **15. (a)** Domain: $\{x | x \neq 0\}$; Range: all real numbers   **(b)** (−1, 0), (1, 0)   **(c)** None   **(d)** $x = 0$ **(e)** $y = 2x$

17. (a) Domain: $\{x | x \neq -2, x \neq 2\}$; Range: $\{y | y \leq 0, y > 1\}$ **(b)** (0, 0)   **(c)** $y = 1$ **(d)** $x = -2, x = 2$ **(e)** None

19.

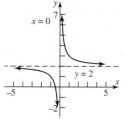

21.

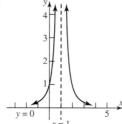

23.

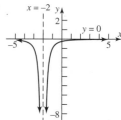

25.

27.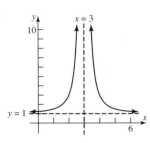

29.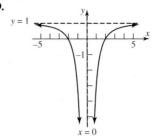

31. Horizontal asymptote: $y = 3$; vertical asymptote: $x = -4$
33. No asymptotes
35. Horizontal asymptote: $y = 0$;
vertical asymptotes: $x = 1$, $x = -1$
37. Horizontal asymptote: $y = 0$; vertical asymptote: $x = 0$
39. Oblique asymptote: $y = 3x$; vertical asymptote: $x = 0$
41. Oblique asymptote: $y = -(x + 1)$;
vertical asymptote: $x = 0$

43. (a) 9.82 m/sec^2 **(b)** 9.8195 m/sec^2 **(c)** 9.7936 m/sec^2 **(d)** h-axis   **(e)**

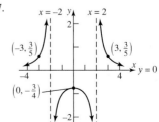

4.4 Exercises *(page 339)*

1. 1. Domain: $\{x \mid x \neq 0, x \neq -4\}$
   2. x-intercept: -1; no y-intercept
   3. No symmetry
   4. Vertical asymptotes: $x = 0$, $x = -4$
   5. Horizontal asymptote: $y = 0$, intersected at $(-1, 0)$
   6.

7.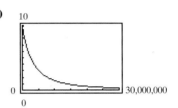

Interval	$(-\infty, -4)$	$(-4, -1)$	$(-1, 0)$	$(0, \infty)$
Number Chosen	-5	-2	$-\frac{1}{2}$	1
Value of R	$R(-5) = -\frac{4}{5}$	$R(-2) = \frac{1}{4}$	$R(-\frac{1}{2}) = -\frac{2}{7}$	$R(1) = \frac{2}{5}$
Location of Graph	Below x-axis	Above x-axis	Below x-axis	Above x-axis
Point on Graph	$(-5, -\frac{4}{5})$	$(-2, \frac{1}{4})$	$(-\frac{1}{2}, -\frac{2}{7})$	$(1, \frac{2}{5})$

3. 1. Domain: $\{x \mid x \neq -2\}$
   2. x-intercept: -1; y-intercept: $\dfrac{3}{4}$
   3. No symmetry
   4. Vertical asymptote: $x = -2$
   5. Horizontal asymptote: $y = \dfrac{3}{2}$, not intersected
   6.

7.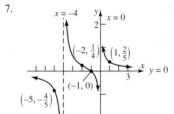

Interval	$(-\infty, -2)$	$(-2, -1)$	$(-1, \infty)$
Number Chosen	-3	$-\frac{3}{2}$	0
Value of R	$R(-3) = 3$	$R(-\frac{3}{2}) = -\frac{3}{2}$	$R(0) = \frac{3}{4}$
Location of Graph	Above x-axis	Below x-axis	Above x-axis
Point on Graph	$(-3, 3)$	$(-\frac{3}{2}, -\frac{3}{2})$	$(0, \frac{3}{4})$

5. 1. Domain: $\{x \mid x \neq -2, x \neq 2\}$
   2. No x-intercept; y-intercept: $-\dfrac{3}{4}$
   3. Symmetric with respect to y-axis
   4. Vertical asymptotes: $x = 2$, $x = -2$
   5. Horizontal asymptote: $y = 0$, not intersected
   6.

7.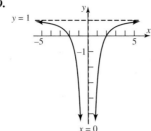

Interval	$(-\infty, -2)$	$(-2, 2)$	$(2, \infty)$
Number Chosen	-3	0	3
Value of R	$R(-3) = \frac{3}{5}$	$R(0) = -\frac{3}{4}$	$R(3) = \frac{3}{5}$
Location of Graph	Above x-axis	Below x-axis	Above x-axis
Point on Graph	$(-3, \frac{3}{5})$	$(0, -\frac{3}{4})$	$(3, \frac{3}{5})$

7. 1. Domain: $\{x | x \neq -1, x \neq 1\}$
 2. No x-intercept; y-intercept: -1
 3. Symmetric with respect to y-axis
 4. Vertical asymptotes: $x = -1, x = 1$
 5. No horizontal or oblique asymptotes
 6.

Interval	$(-\infty, -1)$	$(-1, 1)$	$(1, \infty)$
Number Chosen	-2	0	2
Value of P	$P(-2) = 7$	$P(0) = -1$	$P(2) = 7$
Location of Graph	Above x-axis	Below x-axis	Above x-axis
Point on Graph	$(-2, 7)$	$(0, -1)$	$(2, 7)$

7.

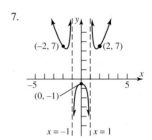

9. 1. Domain: $\{x | x \neq -3, x \neq 3\}$
 2. x-intercept: 1; y-intercept: $\dfrac{1}{9}$
 3. No symmetry
 4. Vertical asymptotes: $x = 3, x = -3$
 5. Oblique asymptote: $y = x$, intersected at $\left(\dfrac{1}{9}, \dfrac{1}{9}\right)$
 6.

Interval	$(-\infty, -3)$	$(-3, 1)$	$(1, 3)$	$(3, \infty)$
Number Chosen	-4	0	2	4
Value of H	$H(-4) \approx -9.3$	$H(0) = \frac{1}{9}$	$H(2) = -1.4$	$H(4) = 9$
Location of Graph	Below x-axis	Above x-axis	Below x-axis	Above x-axis
Point on Graph	$(-4, -9.3)$	$(0, \frac{1}{9})$	$(2, -1.4)$	$(4, 9)$

7.

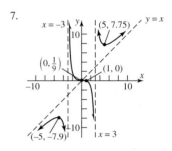

11. 1. Domain: $\{x \neq -3, x \neq 2\}$
 2. Intercept: $(0, 0)$
 3. No symmetry
 4. Vertical asymptotes: $x = 2, x = -3$
 5. Horizontal asymptote: $y = 1$, intersected at $(6, 1)$
 6.

Interval	$(-\infty, -3)$	$(-3, 0)$	$(0, 2)$	$(2, \infty)$
Number Chosen	-6	-1	1	3
Value of R	$R(-6) = 1.5$	$R(-1) = -\frac{1}{6}$	$R(1) = -0.25$	$R(3) = 1.5$
Location of Graph	Above x-axis	Below x-axis	Below x-axis	Above x-axis
Point on Graph	$(-6, 1.5)$	$(-1, -\frac{1}{6})$	$(1, -0.25)$	$(3, 1.5)$

7.

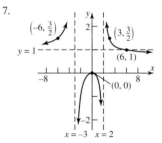

13. 1. Domain: $\{x | x \neq -2, x \neq 2\}$
 2. Intercept: $(0, 0)$
 3. Symmetry with respect to origin
 4. Vertical asymptotes: $x = -2, x = 2$
 5. Horizontal asymptote: $y = 0$, intersected at $(0, 0)$
 6.

Interval	$(-\infty, -2)$	$(-2, 0)$	$(0, 2)$	$(2, \infty)$
Number Chosen	-3	-1	1	3
Value of G	$G(-3) = -\frac{3}{5}$	$G(-1) = \frac{1}{3}$	$G(1) = -\frac{1}{3}$	$G(3) = \frac{3}{5}$
Location of Graph	Below x-axis	Above x-axis	Below x-axis	Above x-axis
Point on Graph	$(-3, -\frac{3}{5})$	$(-1, \frac{1}{3})$	$(1, -\frac{1}{3})$	$(3, \frac{3}{5})$

7.

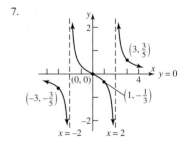

15. 1. Domain: $\{x|x \neq 1, x \neq -2, x \neq 2\}$

2. No x-intercept; y-intercept: $\dfrac{3}{4}$

3. No symmetry

4. Vertical asymptotes: $x = -2, x = 1, x = 2$

5. Horizontal asymptote: $y = 0$, not intersected

6.

Interval	$(-\infty, -2)$	$(-2, 1)$	$(1, 2)$	$(2, \infty)$
Number Chosen	-3	0	1.5	3
Value of R	$R(-3) = -\frac{3}{20}$	$R(0) = \frac{3}{4}$	$R(1.5) = -\frac{24}{7}$	$R(3) = \frac{3}{10}$
Location of Graph	Below x-axis	Above x-axis	Below x-axis	Above x-axis
Point on Graph	$(-3, -\frac{3}{20})$	$(0, \frac{3}{4})$	$(1.5, -\frac{24}{7})$	$(3, \frac{3}{10})$

7.
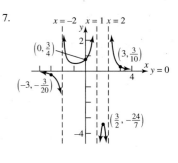

17. 1. Domain: $\{x|x \neq -2, x \neq 2\}$

2. x-intercepts: $-1, 1$; y-intercept: $\dfrac{1}{4}$

3. Symmetry with respect to y-axis

4. Vertical asymptotes: $x = -2, x = 2$

5. Horizontal asymptote: $y = 0$, intersected at $(-1, 0)$ and $(1, 0)$

6.

Interval	$(-\infty, -2)$	$(-2, -1)$	$(-1, 1)$	$(1, 2)$	$(2, \infty)$
Number Chosen	-3	-1.5	0	1.5	3
Value of H	$H(-3) \approx 0.49$	$H(-1.5) \approx -0.46$	$H(0) = \frac{1}{4}$	$H(1.5) \approx -0.46$	$H(3) \approx 0.49$
Location of Graph	Above x-axis	Below x-axis	Above x-axis	Below x-axis	Above x-axis
Point on Graph	$(-3, 0.49)$	$(-1.5, -0.46)$	$(0, \frac{1}{4})$	$(1.5, -0.46)$	$(3, 0.49)$

7.
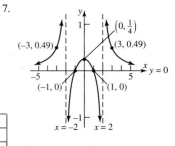

19. 1. Domain: $\{x|x \neq -2\}$

2. x-intercepts: $-1, 4$; y-intercept: -2

3. No symmetry

4. Vertical asymptote: $x = -2$

5. Oblique asymptote: $y = x - 5$, not intersected

6.

Interval	$(-\infty, -2)$	$(-2, -1)$	$(-1, 4)$	$(4, \infty)$
Number Chosen	-3	-1.5	0	5
Value of F	$F(-3) = -14$	$F(-1.5) = 5.5$	$F(0) = -2$	$F(5) \approx 0.86$
Location of Graph	Below x-axis	Above x-axis	Below x-axis	Above x-axis
Point on Graph	$(-3, -14)$	$(-1.5, 5.5)$	$(0, -2)$	$(5, 0.86)$

7.
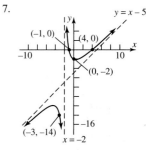

21. 1. Domain: $\{x|x \neq 4\}$

2. x-intercepts: $-4, 3$; y-intercept: 3

3. No symmetry

4. Vertical asymptote: $x = 4$

5. Oblique asymptote: $y = x + 5$, not intersected

6.

Interval	$(-\infty, -4)$	$(-4, 3)$	$(3, 4)$	$(4, \infty)$
Number Chosen	-5	0	3.5	5
Value of R	$R(-5) = -\frac{8}{9}$	$R(0) = 3$	$R(3.5) = -7.5$	$R(5) = 18$
Location of Graph	Below x-axis	Above x-axis	Below x-axis	Above x-axis
Point on Graph	$(-5, -\frac{8}{9})$	$(0, 3)$	$(3.5, -7.5)$	$(5, 18)$

7.
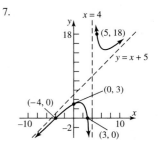

23. 1. Domain: $\{x \mid x \neq -2\}$
2. x-intercepts: $-4, 3$; y-intercept: -6
3. No symmetry
4. Vertical asymptote: $x = -2$
5. Oblique asymptote: $y = x - 1$, not intersected
6.

	-4	-2	3
Interval	$(-\infty, -4)$	$(-4, -2)$	$(-2, 3)$	$(3, \infty)$
Number Chosen	-5	-3	0	4
Value of F	$F(-5) = -\frac{8}{3}$	$F(-3) = 6$	$F(0) = -6$	$F(4) = \frac{4}{3}$
Location of Graph	Below x-axis	Above x-axis	Below x-axis	Above x-axis
Point on Graph	$(-5, -\frac{8}{3})$	$(-3, 6)$	$(0, -6)$	$(4, \frac{4}{3})$

7.
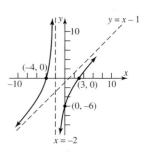

25. 1. Domain: $\{x \mid x \neq -3\}$
2. x-intercepts: $0, 1$; y-intercept: 0
3. No symmetry
4. Vertical asymptote: $x = -3$
5. Horizontal asymptote: $y = 1$, not intersected
6.

	-3	0	1
Interval	$(-\infty, -3)$	$(-3, 0)$	$(0, 1)$	$(1, \infty)$
Number Chosen	-4	-1	$\frac{1}{2}$	2
Value of R	$R(-4) = 100$	$R(-1) = -0.5$	$R(\frac{1}{2}) \approx 0.003$	$R(2) = 0.016$
Location of Graph	Above x-axis	Below x-axis	Above x-axis	Above x-axis
Point on Graph	$(-4, 100)$	$(-1, -0.5)$	$(\frac{1}{2}, 0.003)$	$(2, 0.016)$

7.
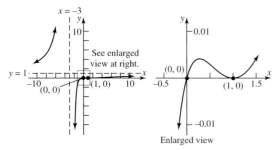

Enlarged view

27. 1. Domain: $\{x \mid x \neq -2, x \neq 3\}$
2. x-intercept: -4; y-intercept: 2
3. No symmetry
4. Vertical asymptote: $x = -2$; hole at $\left(3, \frac{7}{5}\right)$
5. Horizontal asymptote: $y = 1$, not intersected
6.

	-4	-2	3
Interval	$(-\infty, -4)$	$(-4, -2)$	$(-2, 3)$	$(3, \infty)$
Number Chosen	-5	-3	0	4
Value of R	$R(-5) = \frac{1}{3}$	$R(-3) = -1$	$R(0) = 2$	$R(4) = \frac{4}{3}$
Location of Graph	Above x-axis	Below x-axis	Above x-axis	Above x-axis
Point on Graph	$(-5, \frac{1}{3})$	$(-3, -1)$	$(0, 2)$	$(4, \frac{4}{3})$

7.
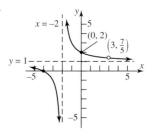

29. 1. Domain: $\left\{ x \middle| x \neq \dfrac{3}{2}, x \neq 2 \right\}$

2. x-intercept: $-\dfrac{1}{3}$; y-intercept: $-\dfrac{1}{2}$

3. No symmetry

4. Vertical asymptote: $x = 2$; hole at $\left(\dfrac{3}{2}, -11 \right)$

5. Horizontal asymptote: $y = 3$, not intersected

6.

Interval	$\left(-\infty, -\frac{1}{3}\right)$	$\left(-\frac{1}{3}, \frac{3}{2}\right)$	$\left(\frac{3}{2}, 2\right)$	$(2, \infty)$
Number Chosen	-1	0	1.7	6
Value of R	$R(-1) = \frac{2}{3}$	$R(0) = -\frac{1}{2}$	$R(1.7) \approx -20.3$	$R(6) = 4.75$
Location of Graph	Above x-axis	Below x-axis	Below x-axis	Above x-axis
Point on Graph	$\left(-1, \frac{2}{3}\right)$	$\left(0, -\frac{1}{2}\right)$	$(1.7, -20.3)$	$(6, 4.75)$

7.

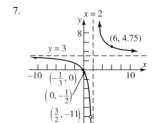

31. 1. Domain: $\{x | x \neq -3\}$

2. x-intercept: -2; y-intercept: 2

3. No symmetry

4. Vertical asymptote: None

5. Oblique asymptote: $y = x + 2$

6.

Interval	$(-\infty, -3)$	$(-3, -2)$	$(-2, \infty)$
Number Chosen	-4	-2.5	0
Value of R	$R(-4) = -2$	$R(-2.5) = -\frac{1}{2}$	$R(0) = 2$
Location of Graph	Below x-axis	Below x-axis	Above x-axis
Point on Graph	$(-4, -2)$	$(-2.5, -\frac{1}{2})$	$(0, 2)$

7.

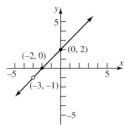

33. 1. Domain: $\{x | x \neq 0\}$

2. No x-intercepts; no y-intercepts

3. Symmetric about the origin

4. Vertical asymptote: $x = 0$

5. Oblique asymptotes: $y = x$, not intersected

6.

Interval	$(-\infty, 0)$	$(0, \infty)$
Number Chosen	-1	1
Value of f	$f(-1) = -2$	$f(1) = 2$
Location of Graph	Below x-axis	Above x-axis
Point on Graph	$(-1, -2)$	$(1, 2)$

7.

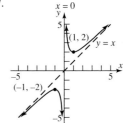

35. 1. Domain: $\{x | x \neq 0\}$

2. x-intercept: -1; no y-intercepts

3. No symmetry

4. Vertical asymptote: $x = 0$

5. No horizontal or oblique asymptotes

6.

Interval	$(-\infty, -1)$	$(-1, 0)$	$(0, \infty)$
Number Chosen	-2	$-\frac{1}{2}$	1
Value of f	$f(-2) = 3.5$	$f(-\frac{1}{2}) = -1.75$	$f(1) = 2$
Location of Graph	Above x-axis	Below x-axis	Above x-axis
Point on Graph	$(-2, 3.5)$	$(-\frac{1}{2}, -1.75)$	$(1, 2)$

7.

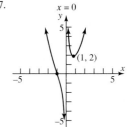

37. 1. Domain: $\{x \mid x \neq 0\}$
2. No x-intercepts; no y-intercepts
3. Symmetric about the origin
4. Vertical asymptote: $x = 0$
5. Oblique asymptote: $y = x$, not intersected
6.

Interval	$(-\infty, 0)$	$(0, \infty)$
Number Chosen	-1	1
Value of f	$f(-1) = -2$	$f(1) = 2$
Location of Graph	Below x-axis	Above x-axis
Point on Graph	$(-1, -2)$	$(1, 2)$

7.

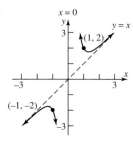

39. One possibility: $R(x) = \dfrac{x^2}{x^2 - 4}$

41. One possibility: $R(x) = \dfrac{(x-1)(x-3)(x^2 + \frac{4}{3})}{(x+1)^2(x-2)^2}$

43. (a) t-axis; $C(t) \to 0$

(b)

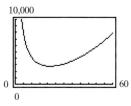

(c) 0.71 hr

45. (a) $\overline{C}(x) = \dfrac{0.2x^3 - 2.3x^2 + 14.3x + 10.2}{x}$

(b) $\overline{C}(6) = \$9400$

(c) $\overline{C}(9) \approx \$10{,}933$

(d)

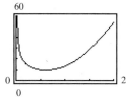

(e) 6
(f) $9400

47. (a) $S(x) = 2x^2 + \dfrac{40{,}000}{x}$

(b)

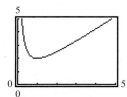

(c) 2784.95 sq in.
(d) 21.54 in. $\times$ 21.54 in. $\times$ 21.54 in.
(e) To minimize the cost of material needed for construction.

49. (a) $C(r) = 12\pi r^2 + \dfrac{4000}{r}$

(b) The cost is least for r about 3.76 cm.

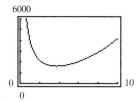

51. No. Each of the functions is a quotient of polynomials, but not written in lowest terms. Each function is undefined for $x = 1$; each graph has a hole at $x = 1$.
53. Minimum value: 2.00 at $x = 1.00$ **55.** Minimum value: 1.89 at $x = 0.79$ **57.** Minimum value: 1.75 at $x = 1.32$

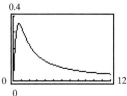 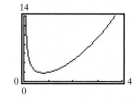

4.5 Exercises *(page 347)*

1. $\{x \mid -2 < x < 5\}; (-2, 5)$ **3.** $\{x \mid x \leq 0 \text{ or } x \geq 4\}; (-\infty, 0] \text{ or } [4, \infty)$ **5.** $\{x \mid -3 < x < 3\}; (-3, 3)$ **7.** $\{x \mid x \leq -2 \text{ or } x \geq 1\};$
$(-\infty, -2] \text{ or } [1, \infty)$ **9.** $\left\{ x \mid -\dfrac{1}{2} \leq x \leq 3 \right\}; \left[-\dfrac{1}{2}, 3 \right]$ **11.** $\{x \mid x < -1 \text{ or } x > 8\}; (-\infty, -1) \text{ or } (8, \infty)$ **13.** No real solution
15. $\left\{ x \mid x < -\dfrac{2}{3} \text{ or } x > \dfrac{3}{2} \right\}; \left(-\infty, -\dfrac{2}{3} \right) \text{ or } \left(\dfrac{3}{2}, \infty \right)$ **17.** $\{x \mid x \geq 1\}; [1, \infty)$ **19.** $\{x \mid x \leq 1 \text{ or } 2 \leq x \leq 3\}; (-\infty, 1] \text{ or } [2, 3]$

21. $\{x|-1 < x < 0 \text{ or } x > 3\}; (-1, 0) \text{ or } (3, \infty)$ **23.** $\{x|x < -1 \text{ or } x > 1\}; (-\infty, -1) \text{ or } (1, \infty)$ **25.** $\{x|x = 0 \text{ or } x \geq 4\}; 0 \text{ or } [4, \infty)$
27. $\{x|x < -1 \text{ or } x > 1\}; (-\infty, -1) \text{ or } (1, \infty)$ **29.** $\{x|x < -1 \text{ or } x > 1\}; (-\infty, -1) \text{ or } (1, \infty)$ **31.** $\{x|x \leq -1 \text{ or } 0 < x \leq 1\}; (-\infty, -1] \text{ or } (0, 1]$
33. $\{x|x < -1 \text{ or } x > 1\}; (-\infty, -1) \text{ or } (1, \infty)$ **35.** $\left\{x\left|x < -\dfrac{2}{3} \text{ or } 0 < x < \dfrac{3}{2}\right.\right\}; \left(-\infty, -\dfrac{2}{3}\right) \text{ or } \left(0, \dfrac{3}{2}\right)$ **37.** $\{x|x < 2\}; (-\infty, 2)$
39. $\{x|-2 < x \leq 9\}; (-2, 9]$ **41.** $\{x|x < 2 \text{ or } 3 < x < 5\}; (-\infty, 2) \text{ or } (3, 5)$ **43.** $\{x|x < -3 \text{ or } -1 < x < 1 \text{ or } x > 2\};$
$(-\infty, -3) \text{ or } (-1, 1) \text{ or } (2, \infty)$ **45.** $\{x|x < -5 \text{ or } -4 \leq x \leq -3 \text{ or } x = 0 \text{ or } x > 1\}; (-\infty, -5) \text{ or } [-4, -3] \text{ or } 0 \text{ or } (1, \infty)$
47. $\{x|x > 4\}; (4, \infty)$ **49.** $\{x|x \leq -4 \text{ or } x \geq 4\}; (-\infty, -4] \text{ or } [4, \infty)$ **51.** $\{x|x < -4 \text{ or } x \geq 2\}; (-\infty, -4) \text{ or } [2, \infty)$
53. The ball is more than 96 ft above the ground for time t between 2 and 3 sec, $2 < t < 3$
55. For a profit of at least \$50, between 8 and 32 watches must be sold, $8 \leq x \leq 32$ **57.** $-2 < k < 2$

Fill-in-the-Blank Items *(page 349)*

1. parabola **3.** zero or root **5.** $x = -1$

True/False Items *(page 349)*

1. T **3.** T **5.** F

Review Exercises *(page 349)*

1.
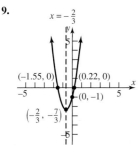

3.
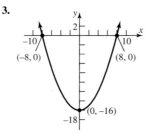

5.
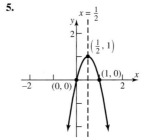

7.
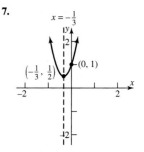

9.
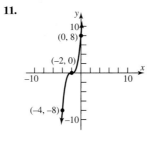

11.

13.
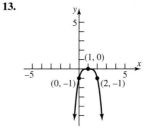

15.
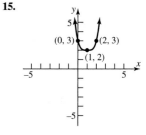

17. Minimum value; 1   **19.** Maximum value; 12   **21.** Maximum value; 16
23. **(a)** x-intercepts: $-4, -2, 0$; y-intercept: 0
  (b) Crosses at $-4, -2, 0$
  (c) $y = x^3$
  (d) 2
  (e)

(f)
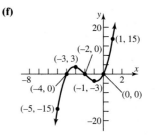

Interval	$(-\infty, -4)$	$(-4, -2)$	$(-2, 0)$	$(0, \infty)$
Number Chosen	-5	-3	-1	1
Value of f	$f(-5) = -15$	$f(-3) = 3$	$f(-1) = -3$	$f(1) = 15$
Location of Graph	Below x-axis	Above x-axis	Below x-axis	Above x-axis
Point on Graph	$(-5, -15)$	$(-3, 3)$	$(-1, -3)$	$(1, 15)$

25. (a) x-intercepts: $-4, 2$; y-intercept: 16

(b) Crosses at -4; touches at 2

(c) $y = x^3$

(d) 2

(e)

Interval	$(-\infty, -4)$	$(-4, 2)$	$(2, \infty)$
Number Chosen	-5	-2	3
Value of f	$f(-5) = -49$	$f(-2) = 32$	$f(3) = 7$
Location of Graph	Below x-axis	Above x-axis	Above x-axis
Point on Graph	$(-5, -49)$	$(-2, 32)$	$(3, 7)$

(f)

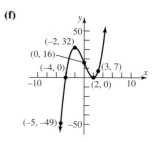

27. (a) x-intercepts: $0, 2$; y-intercept: 0

(b) Touches at 0; crosses at 2

(c) $y = -2x^3$

(d) 2

(e)

Interval	$(-\infty, 0)$	$(0, 2)$	$(2, \infty)$
Number Chosen	-1	1	3
Value of f	$f(-1) = 6$	$f(1) = 2$	$f(3) = -18$
Location of Graph	Below x-axis	Above x-axis	Below x-axis
Point on Graph	$(-1, 6)$	$(1, 2)$	$(3, -18)$

(f)

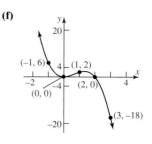

29. (a) x-intercepts: $-3, -1, 1$; y-intercept: 3

(b) Crosses at $-3, -1$; touches at 1

(c) $y = x^4$

(d) 3

(e)

Interval	$(-\infty, -3)$	$(-3, -1)$	$(-1, 1)$	$(1, \infty)$
Number Chosen	-4	-2	0	2
Value of f	$f(-4) = 75$	$f(-2) = -9$	$f(0) = 3$	$f(2) = 15$
Location of Graph	Above x-axis	Below x-axis	Above x-axis	Above x-axis
Point on Graph	$(-4, 75)$	$(-2, -9)$	$(0, 3)$	$(2, 15)$

(f)

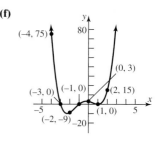

31. 1. Domain: $\{x \mid x \neq 0\}$

2. x-intercept: 3; no y-intercept

3. No symmetry

4. Vertical asymptote: $x = 0$

5. Horizontal asymptote: $y = 2$; not intersected

6.

Interval	$(-\infty, 0)$	$(0, 3)$	$(3, \infty)$
Number Chosen	-2	1	4
Value of R	$R(-2) = 5$	$R(1) = -4$	$R(4) = \frac{1}{2}$
Location of Graph	Above x-axis	Below x-axis	Above x-axis
Point on Graph	$(-2, 5)$	$(1, -4)$	$(4, \frac{1}{2})$

7.

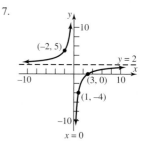

33. 1. Domain: $\{x \mid x \neq 0, x \neq 2\}$

2. x-intercept: -2; no y-intercept

3. No symmetry

4. Vertical asymptotes: $x = 0, x = 2$

5. Horizontal asymptote: $y = 0$; intersected at $(-2, 0)$

6.

Interval	$(-\infty, -2)$	$(-2, 0)$	$(0, 2)$	$(2, \infty)$
Number Chosen	-3	-1	1	3
Value of H	$H(-3) = -\frac{1}{15}$	$H(-1) = \frac{1}{3}$	$H(1) = -3$	$H(3) = \frac{5}{3}$
Location of Graph	Below x-axis	Above x-axis	Below x-axis	Above x-axis
Point on Graph	$(-3, -\frac{1}{15})$	$(-1, \frac{1}{3})$	$(1, -3)$	$(3, \frac{5}{3})$

7.

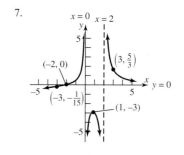

35. 1. Domain: $\{x | x \neq -2, x \neq 3\}$
2. x-intercepts: -3, 2; y-intercept: 1
3. No symmetry
4. Vertical asymptote: $x = -2$, $x = 3$
5. Horizontal asymptote: $y = 1$; intersected at $(0, 1)$
6.

7.

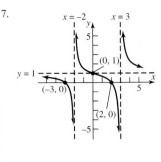

Interval	$(-\infty, -3)$	$(-3, -2)$	$(-2, 2)$	$(2, 3)$	$(3, \infty)$
Number Chosen	-4	-2.5	0	2.5	4
Value of R	$R(-4) \approx 0.43$	$R(-2.5) \approx -0.82$	$R(0) = 1$	$R(2.5) \approx -1.22$	$R(4) = \frac{7}{3}$
Location of Graph	Above x-axis	Below x-axis	Above x-axis	Below x-axis	Above x-axis
Point on Graph	$(-4, 0.43)$	$(-2.5, -0.82)$	$(0, 1)$	$(2.5, -1.22)$	$(4, \frac{7}{3})$

37. 1. Domain: $\{x | x \neq -2, x \neq 2\}$
2. x-intercept: 0; y-intercept: 0
3. Symmetric with respect to the origin
4. Vertical asymptotes: $x = -2$, $x = 2$
5. Oblique asymptote: $y = x$; intersected at $(0, 0)$
6.

7.

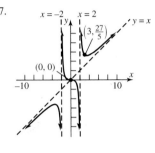

Interval	$(-\infty, -2)$	$(-2, 0)$	$(0, 2)$	$(2, \infty)$
Number Chosen	-3	-1	1	3
Value of F	$F(-3) = -\frac{27}{5}$	$F(-1) = \frac{1}{3}$	$F(1) = -\frac{1}{3}$	$F(3) = \frac{27}{5}$
Location of Graph	Below x-axis	Above x-axis	Below x-axis	Above x-axis
Point on Graph	$(-3, -\frac{27}{5})$	$(-1, \frac{1}{3})$	$(1, -\frac{1}{3})$	$(3, \frac{27}{5})$

39. 1. Domain: $\{x | x \neq 1\}$
2. x-intercept: 0; y-intercept: 0
3. No symmetry
4. Vertical asymptote: $x = 1$
5. No oblique or horizontal asymptote
6.

7.

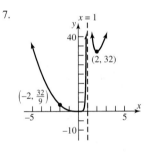

Interval	$(-\infty, 0)$	$(0, 1)$	$(1, \infty)$
Number Chosen	-2	$\frac{1}{2}$	2
Value of R	$R(-2) \approx \frac{32}{9}$	$R(\frac{1}{2}) = \frac{1}{2}$	$R(2) = 32$
Location of Graph	Above x-axis	Above x-axis	Above x-axis
Point on Graph	$(-2, \frac{32}{9})$	$(\frac{1}{2}, \frac{1}{2})$	$(2, 32)$

41. 1. Domain: $\{x | x \neq -1, x \neq 2\}$
2. x-intercept: -2; y-intercept: 2
3. No symmetry
4. Vertical asymptote: $x = -1$; hole at $\left(2, \frac{4}{3}\right)$
5. Horizontal asymptote: $y = 1$, not intersected
6.

7.

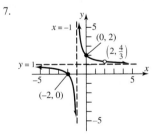

Interval	$(-\infty, -2)$	$(-2, -1)$	$(-1, 2)$	$(2, \infty)$
Number Chosen	-3	-1.5	0	3
Value of G	$G(-3) = \frac{1}{2}$	$G(-1.5) = -1$	$G(0) = 2$	$G(3) = 1.25$
Location of Graph	Above x-axis	Below x-axis	Above x-axis	Above x-axis
Point on Graph	$(-3, \frac{1}{2})$	$(-1.5, -1)$	$(0, 2)$	$(3, 1.25)$

43. $\left\{x \middle| -4 < x < \frac{3}{2}\right\}; \left(-4, \frac{3}{2}\right)$ **45.** $\{x | -3 < x \leq 3\}; (-3, 3]$ **47.** $\{x | x < 1 \text{ or } x > 2\}; (-\infty, 1) \text{ or } (2, \infty)$

49. $\{x | 1 \leq x \leq 2 \text{ or } x > 3\}; [1, 2] \text{ or } (3, \infty)$ **51.** $\{x | x < -4 \text{ or } 2 < x < 4 \text{ or } x > 6\}; (-\infty, -4) \text{ or } (2, 4) \text{ or } (6, \infty)$ **53.** $(2, 2)$

55. 50 ft by 50 ft   **57.** 4,166,666.7 m^2 **59.** The side with the semi-circles should be $\dfrac{50}{\pi}$ ft; the other side should be 25 ft.

61. (a)

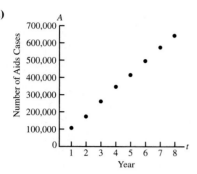

(b) 796,999   **(d)** 680,000

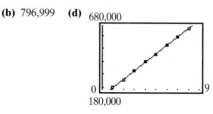

C H A P T E R   5   The Zeros of a Polynomial Function

5.1 Exercises *(page 357)*

1. $q(x) = x^2 + x + 4; R = 12$ **3.** $q(x) = 3x^2 + 11x + 32; R = 99$ **5.** $q(x) = x^4 - 3x^3 + 5x^2 - 15x + 46; R = -138$
7. $q(x) = 4x^5 + 4x^4 + x^3 + x^2 + 2x + 2; R = 7$ **9.** $q(x) = 0.1x^2 - 0.11x + 0.321; R = -0.3531$
11. $q(x) = x^4 + x^3 + x^2 + x + 1; R = 0$ **13.** No   **15.** Yes   **17.** Yes   **19.** No   **21.** Yes   **23.** $a + b + c + d = -9$

Historical Problems *(page 370)*

1.
$$\left(x - \frac{b}{3}\right)^3 + b\left(x - \frac{b}{3}\right)^2 + c\left(x - \frac{b}{3}\right) + d = 0$$
$$x^3 - bx^2 + \frac{b^2x}{3} - \frac{b^3}{27} + bx^2 - \frac{2b^2x}{3} + \frac{b^3}{9} + cx - \frac{bc}{3} + d = 0$$
$$x^3 + \left(c - \frac{b^2}{3}\right)x + \left(\frac{2b^3}{27} - \frac{bc}{3} + d\right) = 0$$
Let $p = c - \frac{b^2}{3}$ and $q = \frac{2b^3}{27} - \frac{bc}{3} + d$. Then $x^3 + px + q = 0$.

3.
$$3HK = -p$$
$$K = -\frac{p}{3H}$$
$$H^3 + \left(-\frac{p}{3H}\right)^3 = -q$$
$$H^3 - \frac{p^3}{27H^3} = -q$$
$$27H^6 - p^3 = -27qH^3$$
$$27H^6 + 27qH^3 - p^3 = 0$$
$$H^3 = \frac{-27q \pm \sqrt{(27q)^2 - 4(27)(-p^3)}}{2 \cdot 27}$$
$$H^3 = \frac{-q}{2} \pm \sqrt{\frac{27^2q^2}{2^2(27^2)} + \frac{4(27)p^3}{2^2(27^2)}}$$
$$H^3 = \frac{-q}{2} \pm \sqrt{\frac{q^2}{4} + \frac{p^3}{27}} \qquad \text{Choose the positive root for now.}$$
$$H = \sqrt[3]{\frac{-q}{2} + \sqrt{\frac{q^2}{4} + \frac{p^3}{27}}}$$

5. $x = H + K$
$$x = \sqrt[3]{\frac{-q}{2} + \sqrt{\frac{q^2}{4} + \frac{p^3}{27}}} + \sqrt[3]{\frac{-q}{2} - \sqrt{\frac{q^2}{4} + \frac{p^3}{27}}} \quad \text{(Note that if we had used the negative root in 3, the result would be the same.)}$$
7. $x = 2$

5.2 Exercises *(page 370)*

1. No; $f(2) = 8$ **3.** Yes; $f(2) = 0$ **5.** Yes; $f(-3) = 0$ **7.** No; $f(-4) = 1$ **9.** Yes; $f\left(\dfrac{1}{2}\right) = 0$ **11.** 7; 3 or 1 positive; 2 or 0 negative

13. 6; 2 or 0 positive; 2 or 0 negative   **15.** 3; 2 or 0 positive; 1 negative   **17.** 4; 2 or 0 positive; 2 or 0 negative   **19.** 5; 0 positive; 3 or 1 negative

21. 6; 1 positive; 1 negative   **23.** $\pm 1, \pm \dfrac{1}{3}$ **25.** $\pm 1, \pm 3$ **27.** $\pm 1, \pm 2, \pm \dfrac{1}{4}, \pm \dfrac{1}{2}$ **29.** $\pm 1, \pm 3, \pm 9, \pm \dfrac{1}{2}, \pm \dfrac{1}{3}, \pm \dfrac{1}{6}, \pm \dfrac{3}{2}, \pm \dfrac{9}{2}$

31. $\pm 1, \pm 2, \pm 3, \pm 4, \pm 6, \pm 12, \pm \dfrac{1}{2}, \pm \dfrac{3}{2}$ **33.** $\pm 1, \pm 2, \pm 4, \pm 5, \pm 10, \pm 20, \pm \dfrac{1}{2}, \pm \dfrac{5}{2}, \pm \dfrac{1}{3}, \pm \dfrac{2}{3}, \pm \dfrac{4}{3}, \pm \dfrac{5}{3}, \pm \dfrac{10}{3}, \pm \dfrac{20}{3}, \pm \dfrac{1}{6}, \pm \dfrac{5}{6}$

35. $-3, -1, 2; f(x) = (x + 3)(x + 1)(x - 2)$ **37.** $\dfrac{1}{2}; f(x) = 2\left(x - \dfrac{1}{2}\right)(x^2 + 1)$ **39.** $-1, 1; f(x) = (x + 1)(x - 1)(x^2 + 2)$

41. $-\dfrac{1}{2}, \dfrac{1}{2}; f(x) = 4\left(x + \dfrac{1}{2}\right)\left(x - \dfrac{1}{2}\right)(x^2 + 2)$ **43.** 1, multiplicity 2; $-2, -1; f(x) = (x + 2)(x + 1)(x - 1)^2$

45. $-\dfrac{\sqrt{2}}{2}, \dfrac{\sqrt{2}}{2}, 2; f(x) = 4\left(x + \dfrac{\sqrt{2}}{2}\right)\left(x - \dfrac{\sqrt{2}}{2}\right)(x - 2)\left(x^2 + \dfrac{1}{2}\right)$ **47.** $\{-1, 2\}$ **49.** $\left\{\dfrac{2}{3}, -1 + \sqrt{2}, -1 - \sqrt{2}\right\}$ **51.** $\left\{\dfrac{1}{3}, \sqrt{5}, -\sqrt{5}\right\}$

53. $\{-3, -2\}$ **55.** $\left\{-\dfrac{1}{3}\right\}$ **57.** $\left\{\dfrac{1}{2}, 2, 5\right\}$

59. y-intercept: -6; x-intercepts: $-3, -1, 2$

$(-\infty, -3), f(-4) = -18$, below x-axis

$(-3, -1), f(-2) = 4$, above x-axis

$(-1, 2), f(0) = -6$, below x-axis
$(2, \infty), f(3) = 24$, above x-axis

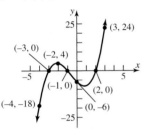

61. y-intercept: -1; x-intercept: $\dfrac{1}{2}$

$\left(-\infty, \dfrac{1}{2}\right), f(0) = -1$, below x-axis

$\left(\dfrac{1}{2}, \infty\right), f(1) = 2$, above x-axis

63. y-intercept: -2; x-intercepts: $-1, 1$

$(-\infty, -1), f(-2) = 18$, above x-axis

$(-1, 1), f(0) = -2$, below x-axis

$(1, \infty), f(2) = 18$, above x-axis

(symmetric with respect to the y-axis)

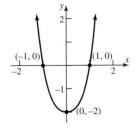

65. y-intercept: -2; x-intercepts: $-\dfrac{1}{2}, \dfrac{1}{2}$

$\left(-\infty, -\dfrac{1}{2}\right), f(-1) = 9$, above x-axis

$\left(-\dfrac{1}{2}, \dfrac{1}{2}\right), f(0) = -2$, below x-axis

$\left(\dfrac{1}{2}, \infty\right), f(1) = 9$, above x-axis

(symmetric with respect to the y-axis)

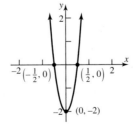

67. y-intercept: 2; x-intercepts: $-2, -1, 1$

$(-\infty, -2), f(-3) = 32$, above x-axis

$(-2, -1), f(-1.5) \approx -1.6$, below x-axis

$(-1, 1), f(0) = 2$, above x-axis

$(1, \infty), f(2) = 12$, above x-axis

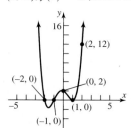

69. y-intercept: 2; x-intercepts: $-\dfrac{\sqrt{2}}{2}, \dfrac{\sqrt{2}}{2}, 2$

$\left(-\infty, -\dfrac{\sqrt{2}}{2}\right), f(-1) = -9$, below x-axis

$\left(-\dfrac{\sqrt{2}}{2}, \dfrac{\sqrt{2}}{2}\right), f(0) = 2$, above x-axis

$\left(\dfrac{\sqrt{2}}{2}, 2\right), f(1) = -3$, below x-axis

$(2, \infty), f(3) = 323$, above x-axis

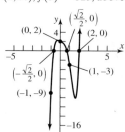

71. 5   **73.** 2   **75.** 5   **77.** $\dfrac{3}{2}$ **79.** $f(0) = -1; f(1) = 10$ **81.** $f(-5) = -58; f(-4) = 2$ **83.** $f(1.4) = -0.17536; f(1.5) = 1.40625$

85. 0.21   **87.** -4.04 **89.** 1.15   **91.** 2.53   **93.** $k = 5$ **95.** -7 **97.** If $f(x) = x^n - c^n$, then $f(c) = c^n - c^n = 0$ so $x - c$ is a factor of f.

99. 5   **101.** No, by the Rational Zeros Theorem, $\dfrac{1}{3}$ is not a potential rational zero.   **103.** No, by the Rational Zeros Theorem, $\dfrac{3}{5}$ is not a

potential rational zero.   **105.** 7 in.   **107.** All the potential rational zeros are integers. Hence, r is either an integer or is not a rational zero (and is therefore irrational).

5.3 Exercises *(page 380)*

1. $8 + 5i$ **3.** $-7 + 6i$ **5.** $-6 - 11i$ **7.** $6 - 18i$ **9.** $6 + 4i$ **11.** $10 - 5i$ **13.** 37   **15.** $\dfrac{6}{5} + \dfrac{8}{5}i$ **17.** $1 - 2i$ **19.** $\dfrac{5}{2} - \dfrac{7}{2}i$

21. $-\dfrac{1}{2} + \dfrac{\sqrt{3}}{2}i$ **23.** $2i$ **25.** $-i$ **27.** i **29.** -6 **31.** $-10i$ **33.** $-2 + 2i$ **35.** 0   **37.** 0   **39.** $2i$ **41.** $5i$ **43.** $5i$ **45.** $\{-2i, 2i\}$

47. $\{-4, 4\}$ **49.** $\{3 - 2i, 3 + 2i\}$ **51.** $\{3 - i, 3 + i\}$ **53.** $\left\{\dfrac{1}{4} - \dfrac{1}{4}i, \dfrac{1}{4} + \dfrac{1}{4}i\right\}$ **55.** $\left\{\dfrac{1}{5} - \dfrac{2}{5}i, \dfrac{1}{5} + \dfrac{2}{5}i\right\}$ **57.** $\left\{-\dfrac{1}{2} - \dfrac{\sqrt{3}}{2}i, -\dfrac{1}{2} + \dfrac{\sqrt{3}}{2}i\right\}$

59. $\{2, -1 - \sqrt{3}i, -1 + \sqrt{3}i\}$ **61.** $\{-2, 2, -2i, 2i\}$ **63.** $\{-3i, -2i, 2i, 3i\}$ **65.** Two complex solutions that are conjugates of each other.

67. Two unequal real solutions.   **69.** A repeated real solution.   **71.** $2 - 3i$ **73.** 6   **75.** 25

77. $z + \bar{z} = (a + bi) + (a - bi) = 2a; z - \bar{z} = (a + bi) - (a - bi) = 2bi$

79. $\overline{z + w} = \overline{(a + bi) + (c + di)} = \overline{(a + c) + (b + d)i} = (a + c) - (b + d)i = (a - bi) + (c - di) = \bar{z} + \bar{w}$

5.4 Exercises *(page 386)*

1. $4 + i$ **3.** $-i, 1 - i$ **5.** $-i, -2i$ **7.** $-i$ **9.** $2 - i, -3 + i$ **11.** $f(x) = x^4 - 14x^3 + 77x^2 - 200x + 208; a = 1$

13. $f(x) = x^5 - 4x^4 + 7x^3 - 8x^2 + 6x - 4; a = 1$ **15.** $f(x) = x^4 - 6x^3 + 10x^2 - 6x + 9; a = 1$ **17.** $-2i, 4$ **19.** $2i, -3, \dfrac{1}{2}$

21. $3 + 2i, -2, 5$ **23.** $4i, -\sqrt{11}, \sqrt{11}, -\dfrac{2}{3}$ **25.** $1, -\dfrac{1}{2} - \dfrac{\sqrt{3}}{2}i, -\dfrac{1}{2} + \dfrac{\sqrt{3}}{2}i; f(x) = (x - 1)\left(x + \dfrac{1}{2} + \dfrac{\sqrt{3}}{2}i\right)\left(x + \dfrac{1}{2} - \dfrac{\sqrt{3}}{2}i\right)$

27. $2, 3 - 2i, 3 + 2i; f(x) = (x - 2)(x - 3 + 2i)(x - 3 - 2i)$ **29.** $-i, i, -2i, 2i; f(x) = (x + i)(x - i)(x + 2i)(x - 2i)$

31. $-5i, 5i, -3, 1; f(x) = (x + 5i)(x - 5i)(x + 3)(x - 1)$ **33.** $-4, \dfrac{1}{3}, 2 - 3i, 2 + 3i;$

$f(x) = 3(x + 4)\left(x - \dfrac{1}{3}\right)(x - 2 + 3i)(x - 2 - 3i)$ **35.** Zeros that are complex numbers must occur in conjugate pairs; or a

polynomial with real coefficients of odd degree must have at least one real zero.   **37.** If the remaining zero were a complex number, then its conjugate would also be a zero, creating a polynomial of degree 5.

Fill-in-the-Blank Items *(page 387)*

1. remainder; dividend   **3.** $f(c) = 0$ **5.** $\pm 1, \pm \dfrac{1}{2}$ **7.** $3 - 4i$ **9.** -4

True/False Items *(page 388)*

1. F   **3.** T   **5.** T

Review Exercises *(page 388)*

1. $q(x) = 8x^2 + 5x + 6; R = 10$ **3.** $q(x) = x^3 - 4x^2 + 8x - 15; R = 29$ **5.** $f(4) = 47{,}105$ **7.** 4, 2, or 0 positive; 2 or 0 negative

9. $\pm\dfrac{1}{12}, \pm\dfrac{1}{6}, \pm\dfrac{1}{4}, \pm\dfrac{1}{3}, \pm\dfrac{1}{2}, \pm\dfrac{3}{4}, \pm1, \pm\dfrac{3}{2}, \pm3$ **11.** $-2, 1, 4; f(x) = (x+2)(x-1)(x-4)$ **13.** $\dfrac{1}{2}$, multiplicity 2; -2; $f(x) = 4\left(x - \dfrac{1}{2}\right)^2(x+2)$

15. 2, multiplicity 2; $f(x) = (x-2)^2(x^2+5)$ **17.** $\{-3, 2\}$ **19.** $\left\{-3, -1, -\dfrac{1}{2}, 1\right\}$ **21.** $-2, 1, 4; f(x) = (x+2)(x-1)(x-4)$

23. $-2; \dfrac{1}{2}$ (multiplicity 2); $f(x) = 4(x+2)\left(x - \dfrac{1}{2}\right)^2$ **25.** 2 (multiplicity 2), $-\sqrt{5}i, \sqrt{5}i; f(x) = (x + \sqrt{5}i)(x - \sqrt{5}i)(x-2)^2$

27. $-3, 2, -\dfrac{\sqrt{2}}{2}i, \dfrac{\sqrt{2}}{2}i; f(x) = 2(x+3)(x-2)\left(x + \dfrac{\sqrt{2}}{2}i\right)\left(x - \dfrac{\sqrt{2}}{2}i\right)$ **29.** $-3, -1, -\dfrac{1}{2}, 1; f(x) = 2(x+3)(x+1)\left(x + \dfrac{1}{2}\right)(x-1)$

31. 5   **33.** $\dfrac{37}{2}$ **35.** $f(0) = -1; f(1) = 1$ **37.** $f(0) = -1; f(1) = 1$ **39.** 1.52   **41.** 0.94   **43.** $4 + 7i$ **45.** $-3 + 2i$ **47.** $\dfrac{9}{10} - \dfrac{3}{10}i$

49. -1 **51.** $-46 + 9i$ **53.** $4 - i$ **55.** $-i, 1 - i$ **57.** $\left\{-\dfrac{1}{2} - \dfrac{\sqrt{3}}{2}i, -\dfrac{1}{2} + \dfrac{\sqrt{3}}{2}i\right\}$ **59.** $\left\{\dfrac{-1 - \sqrt{17}}{4}, \dfrac{-1 + \sqrt{17}}{4}\right\}$

61. $\left\{\dfrac{1}{2} - \dfrac{\sqrt{11}}{2}i, \dfrac{1}{2} + \dfrac{\sqrt{11}}{2}i\right\}$ **63.** $\left\{\dfrac{1}{2} - \dfrac{\sqrt{23}}{2}i, \dfrac{1}{2} + \dfrac{\sqrt{23}}{2}i\right\}$ **65.** $\{-\sqrt{2}, \sqrt{2}, -2i, 2i\}$ **67.** $\{-3, 2\}$ **69.** $\left\{\dfrac{1}{3}, 1, -i, i\right\}$

C H A P T E R   6   Exponential and Logarithmic Functions

6.1 Exercises *(page 401)*

1. (a)

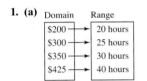

(b) Inverse is a function

3. (a)

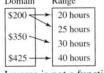

(b) Inverse is not a function

5. (a) $\{(6, 2), (6, -3), (9, 4), (10, 1)\}$
 (b) Inverse is not a function
7. (a) $\{(0, 0), (1, 1), (16, 2), (81, 3)\}$
 (b) Inverse is a function
9. One-to-one   **11.** Not one-to-one   **13.** One-to-one

15.

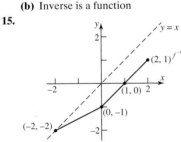

17.

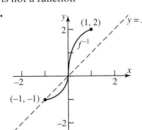

19.

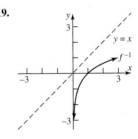

21. $f(g(x)) = f\left(\dfrac{1}{3}(x - 4)\right) = 3\left[\dfrac{1}{3}(x - 4)\right] + 4$
$= (x - 4) + 4 = x;$
$g(f(x)) = g(3x + 4) = \dfrac{1}{3}[(3x + 4) - 4] = \dfrac{1}{3}3x = x$

23. $f(g(x)) = 4\left[\dfrac{x}{4} + 2\right] - 8 = (x + 8) - 8 = x;$
$g(f(x)) = \dfrac{4x - 8}{4} + 2 = (x - 2) + 2 = x$

25. $f(g(x)) = (\sqrt[3]{x + 8})^3 - 8 = (x + 8) - 8 = x;$
$g(f(x)) = \sqrt[3]{(x^3 - 8) + 8} = \sqrt[3]{x^3} = x$

27. $f(g(x)) = \dfrac{1}{\left(\dfrac{1}{x}\right)} = x; g(f(x)) = \dfrac{1}{\left(\dfrac{1}{x}\right)} = x$

29. $f(g(x)) = \dfrac{2\left(\dfrac{4x-3}{2-x}\right) + 3}{\dfrac{4x-3}{2-x} + 4} = \dfrac{2(4x-3) + 3(2-x)}{4x - 3 + 4(2-x)}$

$= \dfrac{5x}{5} = x;$

$g(f(x)) = \dfrac{4\left(\dfrac{2x+3}{x+4}\right) - 3}{2 - \dfrac{2x+3}{x+4}} = \dfrac{4(2x+3) - 3(x+4)}{2(x+4) - (2x+3)}$

$= \dfrac{5x}{5} = x$

31. $f^{-1}(x) = \dfrac{1}{3}x$

$f(f^{-1}(x)) = 3\left(\dfrac{1}{3}x\right) = x$

$f^{-1}(f(x)) = \dfrac{1}{3}(3x) = x$

Domain f = Range f^{-1} = All real numbers
Range f = Domain f^{-1} = All real numbers

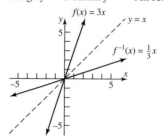

33. $f^{-1}(x) = \dfrac{x}{4} - \dfrac{1}{2}$

$f(f^{-1}(x)) = 4\left(\dfrac{x}{4} - \dfrac{1}{2}\right) + 2 = (x - 2) + 2 = x$

$f^{-1}(f(x)) = \dfrac{4x+2}{4} - \dfrac{1}{2} = \left(x + \dfrac{1}{2}\right) - \dfrac{1}{2} = x$

Domain f = Range f^{-1} = All real numbers
Range f = Domain f^{-1} = All real numbers

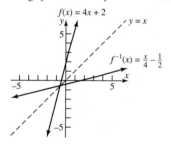

35. $f^{-1}(x) = \sqrt[3]{x+1}$

$f(f^{-1}(x)) = (\sqrt[3]{x+1})^3 - 1 = x$

$f^{-1}(f(x)) = \sqrt[3]{(x^3 - 1) + 1} = x$

Domain f = Range f^{-1} = All real numbers
Range f = Domain f^{-1} = All real numbers

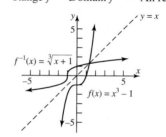

37. $f^{-1}(x) = \sqrt{x-4}$

$f(f^{-1}(x)) = (\sqrt{x-4})^2 + 4 = x$

$f^{-1}(f(x)) = \sqrt{(x^2+4) - 4} = \sqrt{x^2} = x, x \geq 0$

Domain f = Range f^{-1} = $\{x | x \geq 0\}$ or $[0, \infty)$
Range f = Domain f^{-1} = $\{x | x \geq 4\}$ or $[4, \infty)$

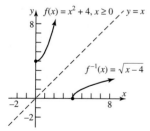

39. $f^{-1}(x) = \dfrac{4}{x}$

$f(f^{-1}(x)) = \dfrac{4}{\left(\dfrac{4}{x}\right)} = x$

$f^{-1}(f(x)) = \dfrac{4}{\left(\dfrac{4}{x}\right)} = x$

Domain f = Range f^{-1} = All real numbers except 0
Range f = Domain f^{-1} = All real numbers except 0

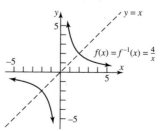

41. $f^{-1}(x) = \dfrac{2x + 1}{x}$

$$f(f^{-1}(x)) = \dfrac{1}{\dfrac{2x + 1}{x} - 2} = \dfrac{x}{(2x + 1) - 2x} = x$$

$$f^{-1}(f(x)) = \dfrac{2\left(\dfrac{1}{x - 2}\right) + 1}{\dfrac{1}{x - 2}} = \dfrac{2 + (x - 2)}{1} = x$$

Domain f = Range f^{-1} = All real numbers except 2
Range f = Domain f^{-1} = All real numbers except 0

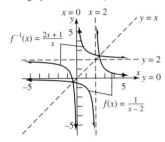

43. $f^{-1}(x) = \dfrac{2 - 3x}{x}$

$$f(f^{-1}(x)) = \dfrac{2}{3 + \dfrac{2 - 3x}{x}} = \dfrac{2x}{3x + 2 - 3x} = \dfrac{2x}{2} = x$$

$$f^{-1}(f(x)) = \dfrac{2 - 3\left(\dfrac{2}{3 + x}\right)}{\dfrac{2}{3 + x}} = \dfrac{2(3 + x) - 3 \cdot 2}{2} = \dfrac{2x}{2} = x$$

Domain f = All real numbers except -3
Range f = Domain f^{-1} = All real numbers except 0

45. $f^{-1}(x) = \dfrac{-2x}{x - 3}$

$$f(f^{-1}(x)) = \dfrac{3\left(\dfrac{-2x}{x - 3}\right)}{\dfrac{-2x}{x - 3} + 2} = \dfrac{3(-2x)}{-2x + 2(x - 3)} = \dfrac{-6x}{-6} = x$$

$$f^{-1}(f(x)) = \dfrac{-2\left(\dfrac{3x}{x + 2}\right)}{\dfrac{3x}{x + 2} - 3} = \dfrac{-2(3x)}{3x - 3(x + 2)} = \dfrac{-6x}{-6} = x$$

Domain f = All real numbers except -2
Range f = Domain f^{-1} = All real numbers except 3

47. $f^{-1}(x) = \dfrac{x}{3x - 2}$

$$f(f^{-1}(x)) = \dfrac{2\left(\dfrac{x}{3x - 2}\right)}{3\left(\dfrac{x}{3x - 2}\right) - 1} = \dfrac{2x}{3x - (3x - 2)} = \dfrac{2x}{2} = x$$

$$f^{-1}(f(x)) = \dfrac{\dfrac{2x}{3x - 1}}{3\left(\dfrac{2x}{3x - 1}\right) - 2} = \dfrac{2x}{6x - 2(3x - 1)} = \dfrac{2x}{2} = x$$

Domain f = All real numbers except $\dfrac{1}{3}$

Range f = Domain f^{-1} = All real numbers except $\dfrac{2}{3}$

49. $f^{-1}(x) = \dfrac{3x + 4}{2x - 3}$

$$f(f^{-1}(x)) = \dfrac{3\left(\dfrac{3x + 4}{2x - 3}\right) + 4}{2\left(\dfrac{3x + 4}{2x - 3}\right) - 3} = \dfrac{3(3x + 4) + 4(2x - 3)}{2(3x + 4) - 3(2x - 3)} = \dfrac{17x}{17} = x$$

$$f^{-1}(f(x)) = \dfrac{3\left(\dfrac{3x + 4}{2x - 3}\right) + 4}{2\left(\dfrac{3x + 4}{2x - 3}\right) - 3} = \dfrac{3(3x + 4) + 4(2x - 3)}{2(3x + 4) - 3(2x - 3)} = \dfrac{17x}{17} = x$$

Domain f = All real numbers except $\dfrac{3}{2}$

Range f = Domain f^{-1} = All real numbers except $\dfrac{3}{2}$

51. $f^{-1}(x) = \dfrac{-2x + 3}{x - 2}$

$$f(f^{-1}(x)) = \frac{2\left(\dfrac{-2x + 3}{x - 2}\right) + 3}{\dfrac{-2x + 3}{x - 2} + 2} = \frac{2(-2x + 3) + 3(x - 2)}{-2x + 3 + 2(x - 2)} = \frac{-x}{-1} = x$$

$$f^{-1}(f(x)) = \frac{-2\left(\dfrac{2x + 3}{x + 2}\right) + 3}{\dfrac{2x + 3}{x + 2} - 2} = \frac{-2(2x + 3) + 3(x + 2)}{2x + 3 - 2(x + 2)} = \frac{-x}{-1} = x$$

Domain f = All real numbers except -2
Range f = Domain f^{-1} = All real numbers except 2

53. $f^{-1}(x) = \dfrac{2}{\sqrt{1 - 2x}}$

$$f(f^{-1}(x)) = \frac{\dfrac{4}{1 - 2x} - 4}{2 \cdot \dfrac{4}{1 - 2x}} = \frac{4 - 4(1 - 2x)}{2 \cdot 4} = \frac{8x}{8} = x$$

$$f^{-1}(f(x)) = \frac{2}{\sqrt{1 - 2\left(\dfrac{x^2 - 4}{2x^2}\right)}} = \frac{2}{\sqrt{\dfrac{4}{x^2}}} = \sqrt{x^2} = x, \text{ since } x > 0.$$

Domain f = $\{x | x > 0\}$ or $(0, \infty)$

Range f = Domain f^{-1} = $\left\{x \middle| x < \dfrac{1}{2}\right\}$ or $\left(-\infty, \dfrac{1}{2}\right)$

55. $f^{-1}(x) = \dfrac{1}{m}(x - b), m \neq 0$ **57.** Quadrant I   **59.** $f(x) = |x|, x \geq 0$, is one-to-one; $f^{-1}(x) = x, x \geq 0$

61. $f(g(x)) = \dfrac{9}{5}\left[\dfrac{5}{9}(x - 32)\right] + 32 = x; g(f(x)) = \dfrac{5}{9}\left[\left(\dfrac{9}{5}x + 32\right) - 32\right] = x$ **63.** $l(T) = \dfrac{gT^2}{4\pi^2}, T > 0$

6.2 Exercises *(page 414)*

1. (a) 11.212   **(b)** 11.587   **(c)** 11.664   **(d)** 11.665   **3. (a)** 8.815   **(b)** 8.821   **(c)** 8.824   **(d)** 8.825
5. (a) 21.217   **(b)** 22.217   **(c)** 22.440   **(d)** 22.459   **7.** 3.320   **9.** 0.427   **11.** B   **13.** D   **15.** A   **17.** E

19.

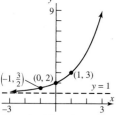

Domain: All real numbers
Range: $\{y | y > 1\}$ or $(1, \infty)$
Horizontal asymptote: $y = 1$

21.

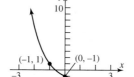

Domain: All real numbers
Range: $\{y | y > -2\}$ or $(-2, \infty)$
Horizontal asymptote: $y = -2$

23.

Domain: All real numbers
Range: $\{y | y > 2\}$ or $(2, \infty)$
Horizontal asymptote: $y = 2$

25.

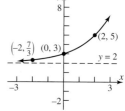

Domain: All real numbers
Range: $\{y | y > 2\}$ or $(2, \infty)$
Horizontal asymptote: $y = 2$

27.

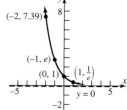

Domain: All real numbers
Range: $\{y | y > 0\}$ or $(0, \infty)$
Horizontal asymptote: $y = 0$

29.

Domain: All real numbers
Range: $\{y | y > 0\}$ or $(0, \infty)$
Horizontal asymptote: $y = 0$

31.

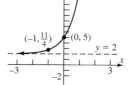

Domain: All real numbers
Range: $\{y | y < 5\}$ or $(-\infty, 5)$
Horizontal asymptote: $y = 5$

33.

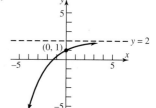

Domain: All real numbers
Range: $\{y | y < 2\}$ or $(-\infty, 2)$
Horizontal asymptote: $y = 2$

35. $\dfrac{1}{2}$ **37.** $\{-\sqrt{2}, 0, \sqrt{2}\}$ **39.** $\left\{1 - \dfrac{\sqrt{6}}{3}, 1 + \dfrac{\sqrt{6}}{3}\right\}$ **41.** 0   **43.** 4   **45.** $\dfrac{3}{2}$ **47.** $\{1, 2\}$ **49.** $\dfrac{1}{49}$ **51.** $\dfrac{1}{4}$

53.
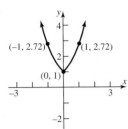
Domain: All real numbers
Range: $\{y | y \geq 1\}$ or $[1, \infty)$
Intercept: $(0, 1)$

55.
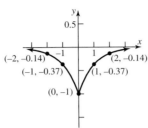
Domain: All real numbers
Range: $\{y | -1 \leq y < 0\}$ or $[-1, 0)$
Intercept: $(0, -1)$

57. (a) 74%   **(b)** 47%
59. (a) 44.3 watts   **(b)** 11.6 watts
61. 3.35 mg; 0.45 mg
63. (a) 0.63   **(b)** 0.98   **(c)** 1
(d)   **(e)** About 7 min

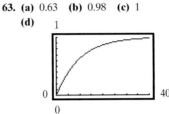

65. (a) 5.16%   **(b)** 8.88%   **67. (a)** 71%   **(b)** 73%   **(c)** 100%

69. (a) 5.414 amp, 7.585 amp, 10.376 amp   **(b)** 12 amp   **(d)** 3.343 amp, 5.309 amp, 9.443 amp   **(e)** 24 amp

(c), (f)
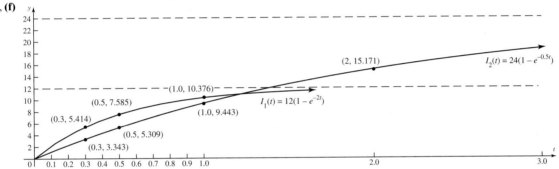

71. $n = 4: 2.7083; n = 6: 2.7181; n = 8: 2.7182788; n = 10: 2.7182818$ **73.** $\dfrac{f(x + h) - f(x)}{h} = \dfrac{a^{x+h} - a^x}{h} = \dfrac{a^x a^h - a^x}{h} = \dfrac{a^x(a^h - 1)}{h}$

75. $f(-x) = a^{-x} = \dfrac{1}{a^x} = \dfrac{1}{f(x)}$ **77. (a)** 9.23×10^{-3}, or about 0   **79.** 59 min
(b) 0.81, or about 1
(c) 5.01, or about 5
(d) $57.91°, 43.99°, 30.07°$

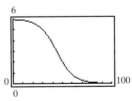

6.3 Exercises   (page 425)

1. $2 = \log_3 9$ **3.** $2 = \log_a 1.6$ **5.** $2 = \log_{1.1} M$ **7.** $x = \log_2 7.2$ **9.** $\sqrt{2} = \log_x \pi$ **11.** $x = \ln 8$ **13.** $2^3 = 8$ **15.** $a^6 = 3$ **17.** $3^x = 2$

19. $2^{1.3} = M$ **21.** $(\sqrt{2})^x = \pi$ **23.** $e^x = 4$ **25.** 0   **27.** 2   **29.** -4 **31.** $\dfrac{1}{2}$ **33.** 4   **35.** $\dfrac{1}{2}$ **37.** $\{x | x > 3\}; (3, \infty)$

39. All real numbers except 0; $\{x | x \neq 0\}$ **41.** All real numbers except 1; $\{x | x \neq 1\}$ **43.** $\{x | x > -1\}; (-1, \infty)$
45. $\{x | x < -1 \text{ or } x > 0\}; (-\infty, -1) \text{ or } (0, \infty)$ **47.** 0.511   **49.** 30.099   **51.** $\sqrt{2}$ **53.** B   **55.** D   **57.** A   **59.** E

61.

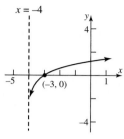

Domain: $\{x \mid x > -4\}$ or $(-4, \infty)$
Range: All real numbers
Vertical asymptote: $x = -4$

63.

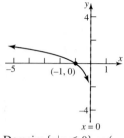

Domain: $\{x \mid x < 0\}$ or $(-\infty, 0)$
Range: All real numbers
Vertical asymptote: $x = 0$

65.

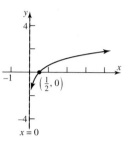

Domain: $\{x \mid x > 0\}$ or $(0, \infty)$
Range: All real numbers
Vertical asymptote: $x = 0$

67.

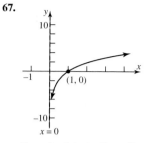

Domain: $\{x \mid x > 0\}$ or $(0, \infty)$
Range: All real numbers
Vertical asymptote: $x = 0$

69.

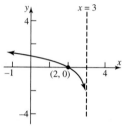

Domain: $\{x \mid x < 3\}$ or $(-\infty, 3)$
Range: All real numbers
Vertical asymptote: $x = 3$

71.

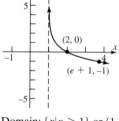

Domain: $\{x \mid x > 1\}$ or $(1, \infty)$
Range: All real numbers
Vertical asymptote: $x = 1$

73. 9   **75.** $\dfrac{7}{2}$ **77.** 2   **79.** 5   **81.** 3

83. 2   **85.** $\dfrac{\ln 10}{3}$ **87.** $\dfrac{\ln 8 - 5}{2}$

89. $\{-2\sqrt{2}, 2\sqrt{2}\}$ **91.** -1

93.

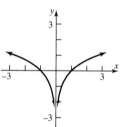

Domain: All real numbers except
0 or $\{x \mid x \neq 0\}$
Range: All real numbers
Intercepts: $(-1, 0), (1, 0)$

95.

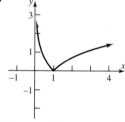

Domain: $\{x \mid x > 0\}$ or $(0, \infty)$
Range: $\{y \mid y \geq 0\}$ or $[0, \infty)$
Intercept: $(1, 0)$

97. (a) $n \approx 6.93$ so 7 panes are necessary
   (b) $n \approx 13.86$ so 14 panes are necessary
99. (a) $d \approx 127.7$ so it takes about 128 days
   (b) $d \approx 575.6$ so it takes about 576 days
101. (a) 6.93 min
   (b) 16.09 min
   (c) No, since $F(t)$ can never equal one.
103. $h \approx 2.29$ so the time between injections is about 2 hr, 17 min

105. 0.2695 sec
0.8959 sec

(graph: Amperes vs. Seconds, points (0.8959, 1), (0.2695, 0.5))

107. (a) $k = 20.07$ **(b)** 91%   **(c)** 0.175   **(d)** 0.08

6.4 Exercises   *(page 438)*

1. 71   **3.** -4 **5.** 7   **7.** 1   **9.** 1   **11.** 2   **13.** $\dfrac{5}{4}$ **15.** 4   **17.** $a + b$ **19.** $b - a$ **21.** $3a$ **23.** $\dfrac{1}{5}(a + b)$ **25.** $2 \log_a u + 3 \log_a v$

27. $-3 \log M$ **29.** $\dfrac{1}{2}[3 \log_5 a - \log_5 b]$ **31.** $2 \ln x + \dfrac{1}{2}\ln(1 - x)$ **33.** $3 \log_2 x - \log_2(x - 3)$ **35.** $\log x + \log(x + 2) - 2 \log(x + 3)$

37. $\dfrac{1}{3}\ln(x - 2) + \dfrac{1}{3}\ln(x + 1) - \dfrac{2}{3}\ln(x + 4)$ **39.** $\ln 5 + \ln x + \dfrac{1}{2}\ln(1 - 3x) - 3 \ln(x - 4)$ **41.** $\log_5 u^3 v^4$ **43.** $-\dfrac{5}{2}\log_{1/2} x$

45. $-2\ln(x-1)$ **47.** $\log_2[x(3x-2)^4]$ **49.** $\log_a\left(\dfrac{25x^6}{\sqrt{2x+3}}\right)$ **51.** 2.771   **53.** -3.880 **55.** 5.615   **57.** 0.874   **59.** 3   **61.** 1

63. $y=\left(\dfrac{\log x}{\log 4}\right)=\left(\dfrac{\ln x}{\ln 4}\right)$

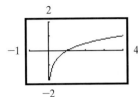

65. $y=\left(\dfrac{\log(x+2)}{\log 2}\right)=\left(\dfrac{\ln(x+2)}{\ln 2}\right)$

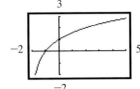

67. $y=\dfrac{\log(x+1)}{\log(x-1)}=\dfrac{\ln(x+1)}{\ln(x-1)}$

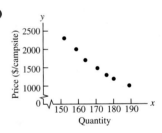

69. $y=Cx$ **71.** $y=Cx(x+1)$ **73.** $y=Ce^{3x}$ **75.** $y=Ce^{-4x}+3$ **77.** $y=\dfrac{\sqrt[3]{C}(2x+1)^{1/6}}{(x+4)^{1/9}}$

79. $\log_a(x+\sqrt{x^2-1})+\log_a(x-\sqrt{x^2-1})=\log_a[(x+\sqrt{x^2-1})(x-\sqrt{x^2-1})]=\log_a[x^2-(x^2-1)]=\log_a 1=0$

81. $\ln(1+e^{2x})=\ln(e^{2x}(e^{-2x}+1))=\ln e^{2x}+\ln(e^{-2x}+1)=2x+\ln(1+e^{-2x})$

83. $y=f(x)=\log_a x; a^y=x$ implies $a^y=\left(\dfrac{1}{a}\right)^{-y}=x$, so $-y=\log_{1/a}x=-f(x)$.

85. $f(x)=\log_a x; f\left(\dfrac{1}{x}\right)=\log_a\dfrac{1}{x}=\log_a 1-\log_a x=-f(x)$

87. If $A=\log_a M$ and $B=\log_a N$, then $a^A=M$ and $a^B=N$.

Then $\log_a\left(\dfrac{M}{N}\right)=\log_a\left(\dfrac{a^A}{a^B}\right)=\log_a a^{A-B}=A-B=\log_a M-\log_a N$.

89. (a)

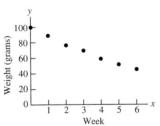

91. (a)

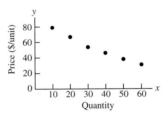

93. (a)

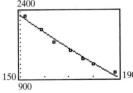

(b) $A=100e^{-0.128t}$
(c) 5.4 weeks
(d) 0.17 g
(f)

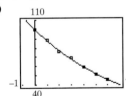

(b) $y=96e^{-0.02x}$
(c) 23.5 units
(e)

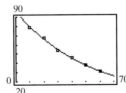

(b) 168 computers
(d)

6.5 Exercises *(page 445)*

1. 6   **3.** 16   **5.** 8   **7.** 3   **9.** 5   **11.** $\{-1+\sqrt{1+e^4},-1-\sqrt{1+e^4}\}\approx\{6.456,-8.456\}$ **13.** $\dfrac{\ln 3}{\ln 2}\approx 1.585$ **15.** 0   **17.** $\dfrac{\ln 10}{\ln 2}\approx 3.322$

19. $-\dfrac{\ln 1.2}{\ln 8}\approx -0.088$ **21.** $\dfrac{\ln 3}{2\ln 3+\ln 4}\approx 0.307$ **23.** $\dfrac{\ln 7}{\ln 0.6+\ln 7}\approx 1.356$ **25.** 0   **27.** $\dfrac{\ln \pi}{1+\ln \pi}\approx 0.534$ **29.** $\dfrac{\ln 1.6}{3\ln 2}\approx 0.226$ **31.** $\dfrac{9}{2}$

33. 2   **35.** 1   **37.** 16   **39.** $\left\{-1,\dfrac{2}{3}\right\}$ **41.** 0   **43.** $\ln(2+\sqrt5)\approx 1.444$ **45.** 1.92   **47.** 2.79   **49.** -0.57 **51.** -0.70 **53.** 0.57

55. $\{0.39,1.00\}$ **57.** 1.32   **59.** 1.31

6.6 Exercises *(page 453)*

1. $108.29   **3.** $609.50   **5.** $697.09   **7.** $12.46   **9.** $125.23   **11.** $88.72   **13.** $860.72   **15.** $554.09   **17.** $59.71   **19.** $361.93   **21.** 5.35%

23. 26%   **25.** $6\dfrac{1}{4}$% compounded annually   **27.** 9% compounded monthly   **29.** 104.32 months (about 8.7 yr); 103.97 mo (about 8.66 yr)

31. 61.02 mo; 60.82 mo   **33.** 15.27 yr or 15 yr, 4 mo   **35.** $104,335   **37.** $12,910.62   **39.** About $30.17 per share or $3017   **41.** 9.35%

43. Not quite. Jim will have $1057.60. The second bank gives a better deal, since Jim will have $1060.62 after 1 yr.   **45.** Will has $11,632.73; Henry has $10,947.89.   **47. (a)** Interest is $30,000   **(b)** Interest is $38,613.59   **(c)** Interest is $37,752.73. Simple interest at 12% is best.

49. (a) $1364.62   **(b)** $1353.35   **51.** $4631.93   **53. (a)** 6.1 yr   **(b)** 18.45 yr   **(c)** $mP = P\left(1 + \dfrac{r}{n}\right)^{nt}$

$$m = \left(1 + \frac{r}{n}\right)^{nt}$$

$$\ln m = \ln\left(1 + \frac{r}{n}\right)^{nt} = nt\ln\left(1 + \frac{r}{n}\right)$$

$$t = \frac{\ln m}{n\ln\left(1 + \dfrac{r}{n}\right)}$$

6.7 Exercises *(page 462)*

1. 34.7 days; 69.3 days   **3. (a)** 28.4 yr   **(b)** 94.4 yr   **5.** 5832; 3.9 days   **7.** 25,198   **9.** 9.797 g   **11.** 9727 yr ago
13. (a) 5:18 P.M.   **(b)** After 14.3 min, the pizza will be 160°F.   **(c)** As time passes, the temperature of the pizza gets closer to 70°F.
15. 18.63°C; 25.1°C   **17.** 7.34 kg; 76.6 hr   **19.** 26.6 days   **21. (a)** 0.1286   **(b)** 0.9   **(c)** 1996   **23. (a)** 1000   **(b)** 30   **(c)** 11.076 hr

6.8 Exercises *(page 467)*

1. 70 decibels   **3.** 111.76 decibels   **5.** 10 watt/m^2 **7.** 4.0 on the Richter scale   **9.** 125,892.54 mm; the Mexico City earthquake was 15.85 times as intense as the one in San Francisco.   **11.** Crowd noise was 31.6 times as intense as guidelines allow.

Fill-in-the Blank Items *(page 469)*

1. One-to-one   **3.** $(0, 1), (1, a),$ and $\left(-1, \dfrac{1}{a}\right)$ **5.** 4   **7.** 1   **9.** All real numbers greater than 0   **11.** 1

True/False Items *(page 470)*

1. F   **3.** T   **5.** F   **7.** T   **9.** F

Review Exercises *(page 470)*

1. $f^{-1}(x) = \dfrac{2x + 3}{5x - 2}$; $f(f^{-1}(x)) = \dfrac{2\left(\dfrac{2x+3}{5x-2}\right) + 3}{5\left(\dfrac{2x+3}{5x-2}\right) - 2}$
$= \dfrac{2(2x + 3) + 3(5x - 2)}{5(2x + 3) - 2(5x - 2)} = \dfrac{19x}{19} = x;$

$f^{-1}(f(x)) = \dfrac{2\left(\dfrac{2x+3}{5x-2}\right) + 3}{5\left(\dfrac{2x+3}{5x-2}\right) - 2} = \dfrac{2(2x + 3) + 3(5x - 2)}{5(2x + 3) - 2(5x - 2)} = \dfrac{19x}{19} = x;$

Domain f = Range f^{-1} = All real numbers except $\dfrac{2}{5}$;

Range f = Domain f^{-1} = All real numbers except $\dfrac{2}{5}$

3. $f^{-1}(x) = \dfrac{x + 1}{x}$; $f(f^{-1}(x)) = \dfrac{1}{\dfrac{x+1}{x} - 1}$
$= \dfrac{x}{x + 1 - x} = x;$

$f^{-1}(f(x)) = \dfrac{\dfrac{1}{x-1} + 1}{\dfrac{1}{x-1}} = \dfrac{1 + x - 1}{1} = x;$

Domain f = Range f^{-1} = All real numbers except 1;
Range f = Domain f^{-1} = All real numbers except 0

5. $f^{-1}(x) = \dfrac{27}{x^3}$; $f(f^{-1}(x)) = \dfrac{3}{\left(\dfrac{27}{x^3}\right)^{1/3}} = \dfrac{3}{\left(\dfrac{3}{x}\right)} = x;$

$f^{-1}(f(x)) = \dfrac{27}{\left(\dfrac{3}{x^{1/3}}\right)^3} = \dfrac{27}{\left(\dfrac{27}{x}\right)} = x;$

Domain f = Range f^{-1} = All real numbers except 0;
Range f = Domain f^{-1} = All real numbers except 0

7. -3 **9.** $\sqrt{2}$ **11.** 0.4   **13.** $\log_3 u + 2\log_3 v - \log_3 w$ **15.** $2\log x + \dfrac{1}{2}\log(x^3 + 1)$ **17.** $\ln x + \dfrac{1}{3}\ln(x^2 + 1) - \ln(x - 3)$

19. $\dfrac{25}{4}\log_4 x$ **21.** $-2\ln(x + 1)$ **23.** $\log\left(\dfrac{4x^3}{[(x + 3)(x - 2)]^{1/2}}\right)$ **25.** 2.124   **27.** $y = Ce^{2x^2}$ **29.** $y = \sqrt{e^{x+C} + 9}$
31. $y = \ln(x^2 + 4) - C$

33.

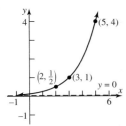

Domain: All real numbers
Range: $\{y|y > 0\}$ or $(0, \infty)$
Horizontal asymptote: $y = 0$

35.

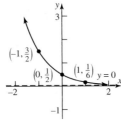

Domain: All real numbers
Range: $\{y|y > 0\}$ or $(0, \infty)$
Horizontal asymptote: $y = 0$

37.

Domain: All real numbers
Range: $\{y|y < 1\}$ or $(-\infty, 1)$
Horizontal asymptote: $y = 1$

39.

Domain: $\{x|x > 0\}$ or $(0, \infty)$
Range: All real numbers
Vertical asymptote: $x = 0$

41.

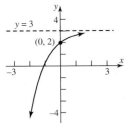

Domain: All real numbers
Range: $\{y|y < 3\}$ or $(-\infty, 3)$
Horizontal asymptote: $y = 3$

43. $\dfrac{1}{4}$ **45.** $\left\{\dfrac{-1-\sqrt{3}}{2}, \dfrac{-1+\sqrt{3}}{2}\right\}$ **47.** $\dfrac{1}{4}$ **49.** $\dfrac{2\ln 3}{\ln 5 - \ln 3} \approx 4.301$ **51.** $\dfrac{12}{5}$

53. 83 **55.** $\left\{\dfrac{1}{2}, -3\right\}$ **57.** -1 **59.** $1 - \ln 5 \approx -0.609$ **61.** $\dfrac{\ln 3}{3\ln 2 - 2\ln 3} \approx -9.327$

63. 3229.5 m **65.** 7.6 mm of mercury **67. (a)** 37.3 watt **(b)** 6.9 decibels

69. (a) 71% **(b)** 85.5% **(c)** 90% **(d)** About 1.6 mo **(e)** About 4.8 mo

71. (a) 9.85 yr **(b)** 4.27 yr **73.** $41,669 **75.** 80 decibels **77.** 24,203 yr ago

79. (a)

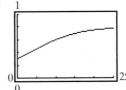

(b) $N = 1000e^{0.346574t}$

(c) 11,314

(e)

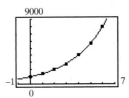

81. 6,078,190,457 **83. (a)** 0.3 **(b)** 0.8 **(c)**

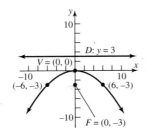

(d) When $t \approx 20.1$ or about 2008.

C H A P T E R 7   The Conics

7.2 Exercises (page 484)

1. B **3.** E **5.** H **7.** C

9. $y^2 = 16x$

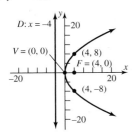

11. $x^2 = -12y$

13. $y^2 = -8x$

15. $x^2 = 2y$

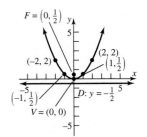
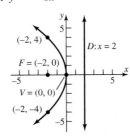

17. $(x - 2)^2 = -8(y + 3)$

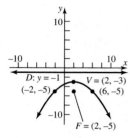

19. $x^2 = \dfrac{4}{3}y$

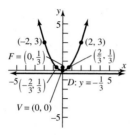

21. $(x + 3)^2 = 4(y - 3)$

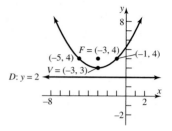

23. $(y + 2)^2 = -8(x + 1)$

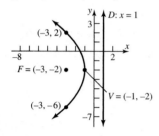

25. Vertex: $(0, 0)$; Focus: $(0, 1)$; Directrix: $y = -1$

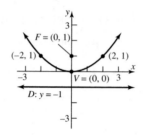

27. Vertex: $(0, 0)$; Focus: $(-4, 0)$; Directrix: $x = 4$

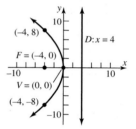

29. Vertex: $(-1, 2)$; Focus: $(1, 2)$;

Directrix: $x = -3$

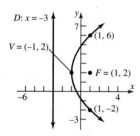

31. Vertex: $(3, -1)$; Focus: $\left(3, -\dfrac{5}{4}\right)$;

Directrix: $y = -\dfrac{3}{4}$

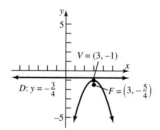

33. Vertex: $(2, -3)$; Focus: $(4, -3)$;

Directrix: $x = 0$

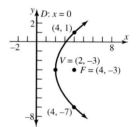

35. Vertex: $(0, 2)$; Focus: $(-1, 2)$;

Directrix: $x = 1$

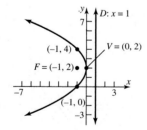

37. Vertex: $(-4, -2)$; Focus: $(-4, -1)$;

Directrix: $y = -3$

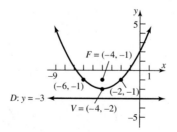

39. Vertex: $(-1, -1)$; Focus: $\left(-\dfrac{3}{4}, -1\right)$;

Directrix: $x = -\dfrac{5}{4}$

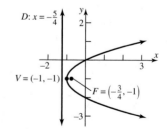

41. Vertex: $(2, -8)$; Focus: $\left(2, -\dfrac{31}{4}\right)$;

Directrix: $y = -\dfrac{33}{4}$

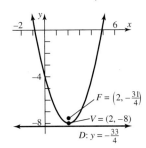

$F = \left(2, -\dfrac{31}{4}\right)$
$V = (2, -8)$
$D: y = -\dfrac{33}{4}$

43. $(y - 1)^2 = x$ **45.** $(y - 1)^2 = -(x - 2)$

47. $x^2 = 4(y - 1)$ **49.** $y^2 = \dfrac{1}{2}(x + 2)$

51. 1.5625 ft from the base of the dish, along the axis of symmetry
53. 1 in. from the vertex **55.** 20 ft **57.** 0.78125 ft
59. 4.17 ft from the base along the axis of symmetry
61. 24.31 ft, 18.75 ft, 7.64 ft

63. $Ax^2 + Ey = 0, A \neq 0, E \neq 0$

$$Ax^2 = -Ey$$

$$x^2 = -\dfrac{E}{A}y$$

This is the equation of a parabola with vertex at $(0, 0)$ and axis of symmetry the y-axis.
The focus is $\left(0, -\dfrac{E}{4A}\right)$; the directrix is the line $y = \dfrac{E}{4A}$. The parabola opens up if $-\dfrac{E}{A} > 0$

and down if $-\dfrac{E}{A} < 0$.

65. $Ax^2 + Dx + Ey + F = 0, A \neq 0$

$$Ax^2 + Dx = -Ey - F$$

$$x^2 + \dfrac{D}{A}x = -\dfrac{E}{A}y - \dfrac{F}{A}$$

$$\left(x + \dfrac{D}{2A}\right)^2 = -\dfrac{E}{A}y - \dfrac{F}{A} + \dfrac{D^2}{4A^2}$$

$$\left(x + \dfrac{D}{2A}\right)^2 = -\dfrac{E}{A}y + \dfrac{D^2 - 4AF}{4A^2}$$

(a) If $E \neq 0$, then the equation may be written as

$$\left(x + \dfrac{D}{2A}\right)^2 = -\dfrac{E}{A}\left(y - \dfrac{D^2 - 4AF}{4AE}\right)$$

This is the equation of a parabola with vertex at

$\left(-\dfrac{D}{2A}, \dfrac{D^2 - 4AF}{4AE}\right)$ and axis of symmetry parallel to the y-axis.

(b)–(d) If $E = 0$, the graph of the equation contains no points if
$D^2 - 4AF < 0$, is a single vertical line if $D^2 - 4AF = 0$, and is
two vertical lines if $D^2 - 4AF > 0$.

7.3 Exercises *(page 495)*

1. C **3.** B

5. Vertices: $(-5, 0), (5, 0)$

Foci: $(-\sqrt{21}, 0), (\sqrt{21}, 0)$

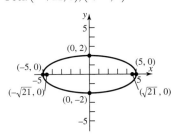

$(0, 2)$
$(-5, 0)$ $(5, 0)$
$(-\sqrt{21}, 0)$ $(\sqrt{21}, 0)$
$(0, -2)$

7. Vertices: $(0, -5), (0, 5)$

Foci: $(0, -4), (0, 4)$

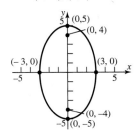

$(0, 5)$
$(0, 4)$
$(-3, 0)$ $(3, 0)$
$(0, -4)$
$(0, -5)$

9. $\dfrac{x^2}{4} + \dfrac{y^2}{16} = 1$

Vertices: $(0, -4), (0, 4)$
Foci: $(0, -2\sqrt{3}), (0, 2\sqrt{3})$

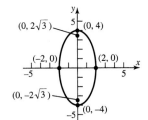

$(0, 2\sqrt{3})$ $(0, 4)$
$(-2, 0)$ $(2, 0)$
$(0, -2\sqrt{3})$
$(0, -4)$

11. $\dfrac{x^2}{8} + \dfrac{y^2}{2} = 1$

Vertices: $(-2\sqrt{2}, 0), (2\sqrt{2}, 0)$
Foci: $(-\sqrt{6}, 0), (\sqrt{6}, 0)$

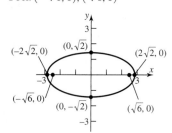

13. $\dfrac{x^2}{16} + \dfrac{y^2}{16} = 1$

Vertices: $(-4, 0), (4, 0), (0, -4), (0, 4)$
Focus: $(0, 0)$

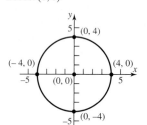

15. $\dfrac{x^2}{25} + \dfrac{y^2}{16} = 1$

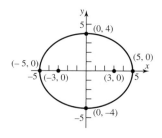

17. $\dfrac{x^2}{9} + \dfrac{y^2}{25} = 1$

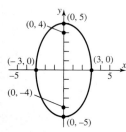

19. $\dfrac{x^2}{9} + \dfrac{y^2}{5} = 1$

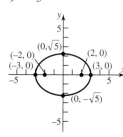

21. $\dfrac{x^2}{4} + \dfrac{y^2}{13} = 1$

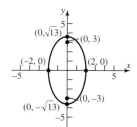

23. $x^2 + \dfrac{y^2}{16} = 1$

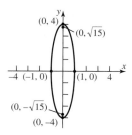

25. $\dfrac{(x+1)^2}{4} + (y-1)^2 = 1$ **27.** $(x-1)^2 + \dfrac{y^2}{4} = 1$

29. Center: $(3, -1)$; Vertices: $(3, -4), (3, 2)$
Foci: $(3, -1 - \sqrt{5}), (3, -1 + \sqrt{5})$

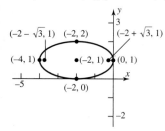

31. $\dfrac{(x+5)^2}{16} + \dfrac{(y-4)^2}{4} = 1$

Center: $(-5, 4)$; Vertices: $(-9, 4), (-1, 4)$
Foci: $(-5 - 2\sqrt{3}, 4), (-5 + 2\sqrt{3}, 4)$

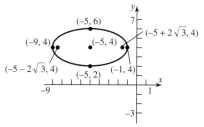

33. $\dfrac{(x+2)^2}{4} + (y-1)^2 = 1$

Center: $(-2, 1)$; Vertices: $(-4, 1), (0, 1)$
Foci: $(-2 - \sqrt{3}, 1), (-2 + \sqrt{3}, 1)$

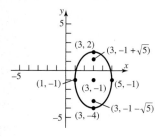

35. $\dfrac{(x-2)^2}{3} + \dfrac{(y+1)^2}{2} = 1$

Center: $(2, -1)$; Vertices: $(2 - \sqrt{3}, -1)$,
$(2 + \sqrt{3}, -1)$; Foci: $(1, -1), (3, -1)$

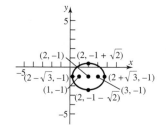

37. $\dfrac{(x-1)^2}{4} + \dfrac{(y+2)^2}{9} = 1$

Center: $(1, -2)$; Vertices: $(1, -5), (1, 1)$
Foci: $(1, -2 - \sqrt{5}), (1, -2 + \sqrt{5})$

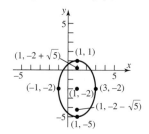

39. $x^2 + \dfrac{(y+2)^2}{4} = 1$

Center: $(0, -2)$; Vertices: $(0, -4)$, $(0, 0)$
Foci: $(0, -2 - \sqrt{3})$, $(0, -2 + \sqrt{3})$

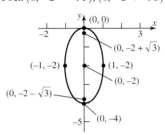

41. $\dfrac{(x-2)^2}{25} + \dfrac{(y+2)^2}{21} = 1$

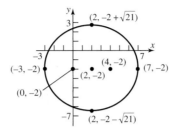

43. $\dfrac{(x-4)^2}{5} + \dfrac{(y-6)^2}{9} = 1$

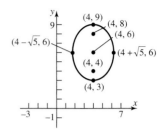

45. $\dfrac{(x-2)^2}{16} + \dfrac{(y-1)^2}{7} = 1$

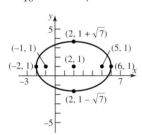

47. $\dfrac{(x-1)^2}{10} + (y-2)^2 = 1$

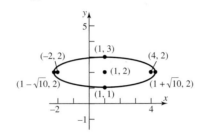

49. $\dfrac{(x-1)^2}{9} + (y-2)^2 = 1$

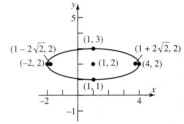

51.

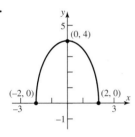

53.

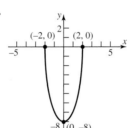

55. $\dfrac{x^2}{100} + \dfrac{y^2}{36} = 1$

57. 43.3 ft

59. 24.65 ft, 21.65 ft, 13.82 ft

61. 0 ft, 12.99 ft, 15 ft, 12.99 ft, 0 ft

63. 91.5 million mi; $\dfrac{x^2}{(93)^2} + \dfrac{y^2}{8646.75} = 1$

65. perihelion: 460.6 million mi; mean distance: 483.8 million mi; $\dfrac{x^2}{(483.8)^2} + \dfrac{y^2}{233{,}524.2} = 1$ **67.** 30 ft

69. (a) $Ax^2 + Cy^2 + F = 0$

$Ax^2 + Cy^2 = -F$

If A and C are of the same and F is of opposite sign, then the equation takes the form

$\dfrac{x^2}{\left(-\dfrac{F}{A}\right)} + \dfrac{y^2}{\left(-\dfrac{F}{C}\right)} = 1$, where $-\dfrac{F}{A}$ and $-\dfrac{F}{C}$ are positive. This is the equation of an ellipse with center at $(0, 0)$.

(b) If $A = C$, the equation may be written as $x^2 + y^2 = -\dfrac{F}{A}$.

This is the equation of a circle with center at $(0, 0)$ and radius equal to $\sqrt{-\dfrac{F}{A}}$.

7.4 Exercises *(page 509)*

1. B **3.** A

5. $x^2 - \dfrac{y^2}{8} = 1$

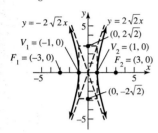

7. $\dfrac{y^2}{16} - \dfrac{x^2}{20} = 1$

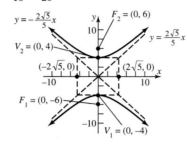

9. $\dfrac{x^2}{9} - \dfrac{y^2}{16} = 1$

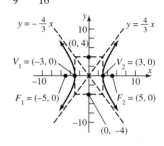

11. $\dfrac{y^2}{36} - \dfrac{x^2}{9} = 1$

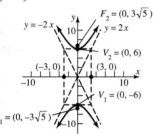

13. $\dfrac{x^2}{8} - \dfrac{y^2}{8} = 1$

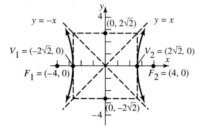

15. $\dfrac{x^2}{25} - \dfrac{y^2}{9} = 1$

Center: $(0,0)$
Transverse axis: x-axis
Vertices: $(-5, 0), (5, 0)$
Foci: $(-\sqrt{34}, 0), (\sqrt{34}, 0)$
Asymptotes: $y = \pm\dfrac{3}{5}x$

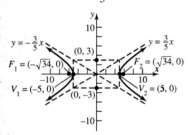

17. $\dfrac{x^2}{4} - \dfrac{y^2}{16} = 1$

Center: $(0,0)$
Transverse axis: x-axis
Vertices: $(-2, 0), (2, 0)$
Foci: $(-2\sqrt{5}, 0), (2\sqrt{5}, 0)$
Asymptotes: $y = \pm 2x$

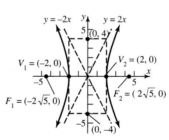

19. $\dfrac{y^2}{9} - x^2 = 1$

Center: $(0,0)$
Transverse axis: y-axis
Vertices: $(0, -3), (0, 3)$
Foci: $(0, -\sqrt{10}), (0, \sqrt{10})$
Asymptotes: $y = \pm 3x$

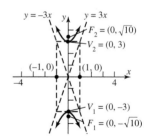

21. $\dfrac{y^2}{25} - \dfrac{x^2}{25} = 1$

Center: $(0,0)$
Transverse axis: y-axis
Vertices: $(0, -5), (0, 5)$
Foci: $(0, -5\sqrt{2}), (0, 5\sqrt{2})$
Asymptotes: $y = \pm x$

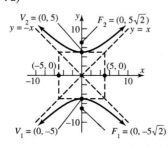

23. $x^2 - y^2 = 1$ **25.** $\dfrac{y^2}{36} - \dfrac{x^2}{9} = 1$

27. $\dfrac{(x-4)^2}{4} - \dfrac{(y+1)^2}{5} = 1$

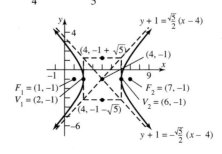

29. $\dfrac{(y+4)^2}{4} - \dfrac{(x+3)^2}{12} = 1$

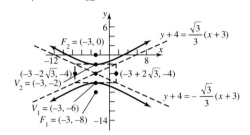

31. $(x-5)^2 - \dfrac{(y-7)^2}{3} = 1$

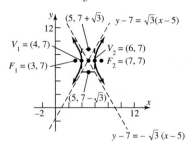

33. $\dfrac{(x-1)^2}{4} - \dfrac{(y+1)^2}{9} = 1$

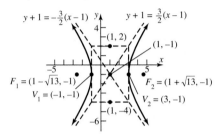

35. $\dfrac{(x-2)^2}{4} - \dfrac{(y+3)^2}{9} = 1$

Center: $(2, -3)$

Transverse axis: Parallel to x-axis

Vertices: $(0, -3), (4, -3)$

Foci: $(2 - \sqrt{13}, -3), (2 + \sqrt{13}, -3)$

Asymptotes: $y + 3 = \pm\dfrac{3}{2}(x - 2)$

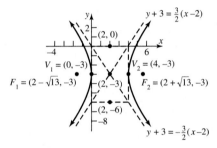

37. $\dfrac{(y-2)^2}{4} - (x+2)^2 = 1$

Center: $(-2, 2)$

Transverse axis: Parallel to y-axis

Vertices: $(-2, 0), (-2, 4)$

Foci: $(-2, 2 - \sqrt{5}), (-2, 2 + \sqrt{5})$

Asymptotes: $y - 2 = \pm 2(x + 2)$

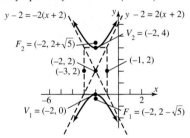

39. $\dfrac{(x+1)^2}{4} - \dfrac{(y+2)^2}{4} = 1$

Center: $(-1, -2)$

Transverse axis: Parallel to x-axis

Vertices: $(-3, -2), (1, -2)$

Foci: $(-1 - 2\sqrt{2}, -2), (-1 + 2\sqrt{2}, -2)$

Asymptotes: $y + 2 = \pm(x + 1)$

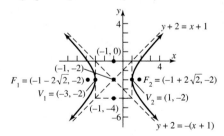

41. $(x - 1)^2 - (y + 1)^2 = 1$

Center: $(1, -1)$
Transverse axis: Parallel to x-axis
Vertices: $(0, -1), (2, -1)$
Foci: $(1 - \sqrt{2}, -1), (1 + \sqrt{2}, -1)$
Asymptotes: $y + 1 = \pm(x - 1)$

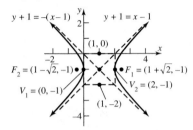

43. $\dfrac{(y - 2)^2}{4} - (x + 1)^2 = 1$

Center: $(-1, 2)$
Transverse axis: Parallel to y-axis
Vertices: $(-1, 0), (-1, 4)$
Foci: $(-1, 2 - \sqrt{5}), (-1, 2 + \sqrt{5})$
Asymptotes: $y - 2 = \pm 2(x + 1)$

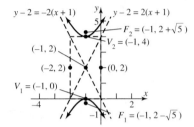

45. $\dfrac{(x - 3)^2}{4} - \dfrac{(y + 2)^2}{16} = 1$

Center: $(3, -2)$
Transverse axis: Parallel to x-axis
Vertices: $(1, -2), (5, -2)$
Foci: $(3 - 2\sqrt{5}, -2), (3 + 2\sqrt{5}, -2)$
Asymptotes: $y + 2 = \pm 2(x - 3)$

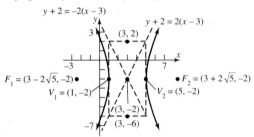

47. $\dfrac{(y - 1)^2}{4} - (x + 2)^2 = 1$

Center: $(-2, 1)$
Transverse axis: Parallel to y-axis
Vertices: $(-2, -1), (-2, 3)$
Foci: $(-2, 1 - \sqrt{5}), (-2, 1 + \sqrt{5})$
Asymptotes: $y - 1 = \pm 2(x + 2)$

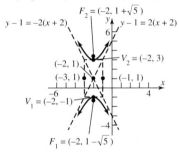

49.

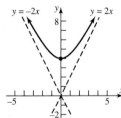

51.

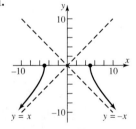

53. (a) The ship will reach shore at a point 64.66 miles from the master station.
(b) 0.00086 sec
(c) $(104, 50)$

55. (a) 450 ft

57. If e is close to 1, narrow hyperbola; if e is very large, wide hyperbola

59. $\dfrac{x^2}{4} - y^2 = 1$; asymptotes $y = \pm\dfrac{1}{2}x$, $y^2 - \dfrac{x^2}{4} = 1$; asymptotes $y = \pm\dfrac{1}{2}x$

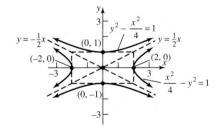

61. $Ax^2 + Cy^2 + F = 0$ If A and C are of opposite sign and $F \neq 0$, this equation may be written as $\dfrac{x^2}{\left(-\dfrac{F}{A}\right)} + \dfrac{y^2}{\left(-\dfrac{F}{C}\right)} = 1$,

$Ax^2 + Cy^2 = -F$ where $-\dfrac{F}{A}$ and $-\dfrac{F}{C}$ are opposite in sign. This is the equation of a hyperbola with center $(0, 0)$.

The transverse axis is the x-axis if $-\dfrac{F}{A} > 0$; the transverse axis is the y-axis if $-\dfrac{F}{A} < 0$.

Fill-in-the-Blank Items *(page 511)*

1. parabola   **3.** hyperbola   **5.** y-axis

True/False Items *(page 511)*

1. T   **3.** T   **5.** T

Review Exercises *(page 512)*

1. Parabola; vertex $(0, 0)$, focus $(-4, 0)$, directrix $x = 4$ **3.** Hyperbola; center $(0, 0)$, vertices $(5, 0)$ and $(-5, 0)$, foci $(\sqrt{26}, 0)$ and $(-\sqrt{26}, 0)$, asymptotes $y = \dfrac{1}{5}x$ and $y = -\dfrac{1}{5}x$ **5.** Ellipse; center $(0, 0)$, vertices $(0, 5)$ and $(0, -5)$, foci $(0, 3)$ and $(0, -3)$

7. $x^2 = -4(y - 1)$: Parabola; vertex $(0, 1)$, focus $(0, 0)$, directrix $y = 2$ **9.** $\dfrac{x^2}{2} - \dfrac{y^2}{8} = 1$: Hyperbola; center $(0, 0)$, vertices $(\sqrt{2}, 0)$ and $(-\sqrt{2}, 0)$, foci $(\sqrt{10}, 0)$ and $(-\sqrt{10}, 0)$, asymptotes $y = 2x$ and $y = -2x$ **11.** $(x - 2)^2 = 2(y + 2)$: Parabola; vertex $(2, -2)$, focus $\left(2, -\dfrac{3}{2}\right)$, directrix $y = -\dfrac{5}{2}$ **13.** $\dfrac{(y - 2)^2}{4} - (x - 1)^2 = 1$: Hyperbola; center $(1, 2)$, vertices $(1, 4)$ and $(1, 0)$, foci $(1, 2 + \sqrt{5})$ and $(1, 2 - \sqrt{5})$, asymptotes $y - 2 = \pm 2(x - 1)$ **15.** $\dfrac{(x - 2)^2}{9} + \dfrac{(y - 1)^2}{4} = 1$: Ellipse; center $(2, 1)$, vertices $(5, 1)$ and $(-1, 1)$, foci $(2 + \sqrt{5}, 1)$ and $(2 - \sqrt{5}, 1)$ **17.** $(x - 2)^2 = -4(y + 1)$: Parabola; vertex $(2, -1)$, focus $(2, -2)$, directrix $y = 0$

19. $\dfrac{(x - 1)^2}{4} + \dfrac{(y + 1)^2}{9} = 1$: Ellipse; center $(1, -1)$, vertices $(1, 2)$ and $(1, -4)$, foci $(1, -1 + \sqrt{5})$ and $(1, -1 - \sqrt{5})$

21. $y^2 = -8x$

23. $\dfrac{y^2}{4} - \dfrac{x^2}{12} = 1$

25. $\dfrac{x^2}{16} + \dfrac{y^2}{7} = 1$

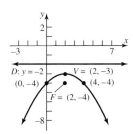

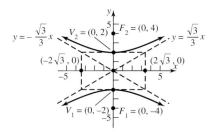

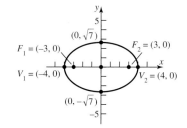

27. $(x - 2)^2 = -4(y + 3)$

29. $(x + 2)^2 - \dfrac{(y + 3)^2}{3} = 1$

31. $\dfrac{(x + 4)^2}{16} + \dfrac{(y - 5)^2}{25} = 1$

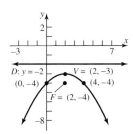

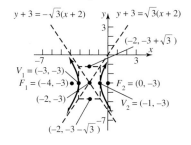

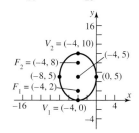

33. $\dfrac{(x+1)^2}{9} - \dfrac{(y-2)^2}{7} = 1$

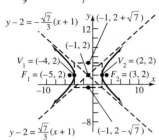

35. $\dfrac{(x-3)^2}{9} - \dfrac{(y-1)^2}{4} = 1$

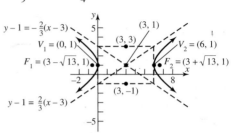

37. $\dfrac{x^2}{5} - \dfrac{y^2}{4} = 1$ **39.** The ellipse $\dfrac{x^2}{16} + \dfrac{y^2}{7} = 1$ **41.** $\dfrac{1}{4}$ ft or 3 in. **43.** 19.72 ft, 18.86 ft, 14.91 ft

45. (a) 45.24 mi from the master station **(b)** 0.000645 sec **(c)** $(66, 20)$

C H A P T E R   8   Systems of Equations and Inequalities

8.1 Exercises (page 525)

1. $2(2) - (-1) = 5$ and $5(2) + 2(-1) = 8$ **3.** $3(2) - 4\left(\dfrac{1}{2}\right) = 4$ and $\dfrac{1}{2}(2) - 3\left(\dfrac{1}{2}\right) = -\dfrac{1}{2}$ **5.** $2^2 - 1^2 = 3$ and $(2)(1) = 2$

7. $\dfrac{0}{1+0} + 3(2) = 6$ and $0 + 9(2)^2 = 36$ **9.** $x = 6, y = 2$ **11.** $x = 3, y = 2$ **13.** $x = 8, y = -4$ **15.** $x = \dfrac{1}{3}, y = -\dfrac{1}{6}$

17. Inconsistent **19.** $x = 1, y = 2$ **21.** $x = 4 - 2y$, y is any real number **23.** $x = 1, y = 1$ **25.** $x = \dfrac{3}{2}, y = 1$ **27.** $x = 4, y = 3$

29. $x = \dfrac{4}{3}, y = \dfrac{1}{5}$ **31.** $x = \dfrac{1}{5}, y = \dfrac{1}{3}$ **33.** $x = 48.15, y = 15.18$ **35.** $x = -21.48, y = 16.12$ **37.** $x = 0.26, y = 0.07$

39. $p = \$16, Q_d = Q_s = 600$ T-shirts **41.** Length 30 ft; width 15 ft **43.** Cheeseburger \$1.55; shake \$0.85 **45.** 22.5 lb
47. Average wind speed 25 mi/hr; average airspeed 175 mi/hr **49.** 80 \$25 sets and 120 \$45 sets **51.** \$5.56

53. Mix 50 mg of the first liquid with 75 mg of the second. **55.** 50,000 **57.** $x = \dfrac{b_1 - b_2}{m_2 - m_1}, y = \dfrac{m_2 b_1 - m_1 b_2}{m_2 - m_1}$

59. $y = mx + b$, x is any real number

8.2 Exercises (page 532)

1. $3(1) + 3(-1) + 2(2) = 4, 1 - (-1) - 2 = 0$, and $2(-1) - 3(2) = -8$ **3.** $x = 8, y = 2, z = 0$ **5.** $x = 2, y = -1, z = 1$
7. Inconsistent **9.** $x = 5z - 2, y = 4z - 3$, where z is any real number **11.** Inconsistent **13.** $x = 1, y = 3, z = -2$

15. $x = -3, y = \dfrac{1}{2}, z = 1$ **17.** $a = \dfrac{4}{3}, b = -\dfrac{5}{3}, c = 1$ **19.** $I_1 = \dfrac{10}{71}, I_2 = \dfrac{65}{71}, I_3 = \dfrac{55}{71}$

21. 100 orchestra, 210 main, and 190 balcony seats **23.** 1.5 chicken, 1 corn, 2 milk

25. If x = price of hamburgers, y = price of fries, z = price of colas, then $x = 2.75 - z, y = 0.68 + \dfrac{1}{3}z, z$ is any real number.

There is not sufficient information:

x	\$2.15	\$2.00	\$1.85
y	\$0.88	\$0.93	\$0.98
z	\$0.60	\$0.75	\$0.90

27. It will take Beth 30 hr, Bill 24 hr, and Edie 40 hr.

8.3 Exercises (page 545)

1. $\begin{bmatrix} 1 & -5 & | & 5 \\ 4 & 3 & | & 6 \end{bmatrix}$ **3.** $\begin{bmatrix} 2 & 3 & | & 6 \\ 4 & -6 & | & -2 \end{bmatrix}$ **5.** $\begin{bmatrix} 0.01 & -0.03 & | & 0.06 \\ 0.13 & 0.10 & | & 0.20 \end{bmatrix}$ **7.** $\begin{bmatrix} 1 & -1 & 1 & | & 10 \\ 3 & 3 & 0 & | & 5 \\ 1 & 1 & 2 & | & 2 \end{bmatrix}$ **9.** $\begin{bmatrix} 1 & 1 & -1 & | & 2 \\ 3 & -2 & 0 & | & 2 \\ 5 & 3 & -1 & | & 1 \end{bmatrix}$

11. $\begin{bmatrix} 1 & -1 & -1 & | & 10 \\ 2 & 1 & 2 & | & -1 \\ -3 & 4 & 0 & | & 5 \\ 4 & -5 & 1 & | & 0 \end{bmatrix}$ **13.** $\begin{bmatrix} 1 & -3 & | & -2 \\ 0 & 1 & | & 9 \end{bmatrix}$ **15. (a)** $\begin{bmatrix} 1 & -3 & 4 & | & 3 \\ 0 & 1 & -2 & | & 0 \\ -3 & 3 & 4 & | & 6 \end{bmatrix}$ **(b)** $\begin{bmatrix} 1 & -3 & 4 & | & 3 \\ 2 & -5 & 6 & | & 6 \\ 0 & -6 & 16 & | & 15 \end{bmatrix}$

17. (a) $\begin{bmatrix} 1 & -3 & 2 & | & -6 \\ 0 & 1 & -1 & | & 8 \\ -3 & -6 & 4 & | & 6 \end{bmatrix}$ **(b)** $\begin{bmatrix} 1 & -3 & 2 & | & -6 \\ 2 & -5 & 3 & | & -4 \\ 0 & -15 & 10 & | & -12 \end{bmatrix}$ **19. (a)** $\begin{bmatrix} 1 & -3 & 1 & | & -2 \\ 0 & 1 & 4 & | & 2 \\ -3 & 1 & 4 & | & 6 \end{bmatrix}$ **(b)** $\begin{bmatrix} 1 & -3 & 1 & | & -2 \\ 2 & -5 & 6 & | & -2 \\ 0 & -8 & 7 & | & 0 \end{bmatrix}$

21. $\begin{cases} x = 5 \\ y = -1 \end{cases}$
consistent; $x = 5, y = -1$

23. $\begin{cases} x = 1 \\ y = 2 \\ 0 = 3 \end{cases}$
inconsistent

25. $\begin{cases} x + 2z = -1 \\ y - 4z = -2 \\ 0 = 0 \end{cases}$
consistent;
$x = -1 - 2z,$
$y = -2 + 4z,$
z is any real number

27. $\begin{cases} x_1 = 1 \\ x_2 + x_4 = 2 \\ x_3 + 2x_4 = 3 \end{cases}$
consistent;
$x_1 = 1, x_2 = 2 - x_4,$
$x_3 = 3 - 2x_4,$
x_4 is any real number

29. $\begin{cases} x_1 + 4x_4 = 2 \\ x_2 + x_3 + 3x_4 = 3 \\ 0 = 0 \end{cases}$
consistent;
$x_1 = 2 - 4x_4,$
$x_2 = 3 - x_3 - 3x_4,$
x_3, x_4 are any real numbers

31. $\begin{cases} x_1 + x_4 = -2 \\ x_2 + 2x_4 = 2 \\ x_3 - x_4 = 0 \end{cases}$
consistent;
$x_1 = -2 - x_4,$
$x_2 = 2 - 2x_4,$
$x_3 = x_4,$
x_4 is any real number

33. $x = 6, y = 2$ **35.** $x = \dfrac{1}{2}, y = \dfrac{3}{4}$

37. $x = 4 - 2y, y$ is any real number

39. $x = \dfrac{3}{2}, y = 1$ **41.** $x = \dfrac{4}{3}, y = \dfrac{1}{5}$

43. $x = 8, y = 2, z = 0$ **45.** $x = 2, y = -1, z = 1$

47. Inconsistent

49. $x = 5z - 2, y = 4z - 3$, where z is any real number **51.** Inconsistent **53.** $x = 1, y = 3, z = -2$ **55.** $x = -3, y = \dfrac{1}{2}, z = 1$

57. $x = \dfrac{1}{3}, y = \dfrac{2}{3}, z = 1$ **59.** $x = 1, y = 2, z = 0, w = 1$ **61.** $y = 0, z = 1 - x, x$ is any real number **63.** $x = 2, y = z - 3,$

z is any real number **65.** $x = \dfrac{13}{9}, y = \dfrac{7}{18}, z = \dfrac{19}{18}$ **67.** $x = \dfrac{7}{5} - \dfrac{3}{5}z - \dfrac{2}{5}w, y = -\dfrac{8}{5} + \dfrac{7}{5}z + \dfrac{13}{5}w$, where z and w are any real numbers

69. $y = -2x^2 + x + 3$ **71.** $f(x) = 3x^3 - 4x^2 + 5$ **73.** 1.5 salmon steak, 2 baked eggs, 1 acorn squash **75.** \$4000 in Treasury bills,

\$4000 in Treasury bonds, \$2000 in corporate bonds **77.** 8 Deltas, 5 Betas, 10 Sigmas **79.** $I_1 = \dfrac{44}{23}, I_2 = 2, I_3 = \dfrac{16}{23}, I_4 = \dfrac{28}{23}$

81. (a)

Amount Invested At		
7%	**9%**	**11%**
0	10,000	10,000
1000	8000	11,000
2000	6000	12,000
3000	4000	13,000
4000	2000	14,000
5000	0	15,000

(b)

Amount Invested At		
7%	**9%**	**11%**
12,500	12,500	0
14,500	8500	2000
16,500	4500	4000
18,750	0	6250

(c) All the money invested at 7% provides \$2100, more than what is required.

83.

First Liquid	**Second Liquid**	**Third Liquid**
50 mg	75 mg	0 mg
36 mg	76 mg	8 mg
22 mg	77 mg	16 mg
8 mg	78 mg	24 mg

8.4 Exercises *(page 557)*

1. 2 **3.** 22 **5.** -2 **7.** 10 **9.** -26 **11.** $x = 6, y = 2$ **13.** $x = 3, y = 2$ **15.** $x = 8, y = -4$ **17.** $x = 4, y = -2$ **19.** Not applicable

21. $x = \dfrac{1}{2}, y = \dfrac{3}{4}$ **23.** $x = \dfrac{1}{10}, y = \dfrac{2}{5}$ **25.** $x = \dfrac{3}{2}, y = 1$ **27.** $x = \dfrac{4}{3}, y = \dfrac{1}{5}$ **29.** $x = 1, y = 3, z = -2$ **31.** $x = -3, y = \dfrac{1}{2}, z = 1$

33. Not applicable **35.** $x = 0, y = 0, z = 0$ **37.** Not applicable **39.** $x = \dfrac{1}{5}, y = \dfrac{1}{3}$ **41.** -5 **43.** $\dfrac{13}{11}$ **45.** 0 or -9 **47.** -4 **49.** 12

51. 8 **53.** 8 **55.** $(y_1 - y_2)x - (x_1 - x_2)y + (x_1y_2 - x_2y_1) = 0$
$$(y_1 - y_2)x + (x_2 - x_1)y = x_2y_1 - x_1y_2$$
$$(x_2 - x_1)y - (x_2 - x_1)y_1 = (y_2 - y_1)x + x_2y_1 - x_1y_2 - (x_2 - x_1)y_1$$
$$(x_2 - x_1)(y - y_1) = (y_2 - y_1)x - (y_2 - y_1)x_1$$
$$y - y_1 = \frac{y_2 - y_1}{x_2 - x_1}(x - x_1)$$

57. $\begin{vmatrix} x^2 & x & 1 \\ y^2 & y & 1 \\ z^2 & z & 1 \end{vmatrix} = x^2 \begin{vmatrix} y & 1 \\ z & 1 \end{vmatrix} - x \begin{vmatrix} y^2 & 1 \\ z^2 & 1 \end{vmatrix} + \begin{vmatrix} y^2 & y \\ z^2 & z \end{vmatrix} = x^2(y - z) - x(y^2 - z^2) + yz(y - z)$

$= (y - z)[x^2 - x(y + z) + yz] = (y - z)[(x^2 - xy) - (xz - yz)] = (y - z)[x(x - y) - z(x - y)]$
$= (y - z)(x - y)(x - z)$

59. $\begin{vmatrix} a_{13} & a_{12} & a_{11} \\ a_{23} & a_{22} & a_{21} \\ a_{33} & a_{32} & a_{31} \end{vmatrix} = a_{13}(a_{22}a_{31} - a_{32}a_{21}) - a_{12}(a_{23}a_{31} - a_{33}a_{21}) + a_{11}(a_{23}a_{32} - a_{33}a_{22})$

$$= -[a_{11}(a_{22}a_{33} - a_{32}a_{23}) - a_{12}(a_{21}a_{33} - a_{31}a_{23}) + a_{13}(a_{21}a_{32} - a_{31}a_{22})] = -\begin{vmatrix} a_{11} & a_{12} & a_{13} \\ a_{21} & a_{22} & a_{23} \\ a_{31} & a_{32} & a_{33} \end{vmatrix}$$

61. $\begin{vmatrix} a_{11} & a_{12} & a_{11} \\ a_{21} & a_{22} & a_{21} \\ a_{31} & a_{32} & a_{31} \end{vmatrix} = a_{11}(a_{22}a_{31} - a_{32}a_{21}) - a_{12}(a_{21}a_{31} - a_{31}a_{21}) + a_{11}(a_{21}a_{32} - a_{31}a_{22})$

$$= a_{11}a_{22}a_{31} - a_{11}a_{32}a_{21} - a_{12}(0) + a_{11}a_{21}a_{32} - a_{11}a_{31}a_{22} = 0$$

Historical Problems (page 573)

1. (a) $2 - 5i \longleftrightarrow \begin{bmatrix} 2 & -5 \\ 5 & 2 \end{bmatrix}, 1 + 3i \longleftrightarrow \begin{bmatrix} 1 & 3 \\ -3 & 1 \end{bmatrix}$ **(b)** $\begin{bmatrix} 2 & -5 \\ 5 & 2 \end{bmatrix}\begin{bmatrix} 1 & 3 \\ -3 & 1 \end{bmatrix} = \begin{bmatrix} 17 & 1 \\ -1 & 17 \end{bmatrix}$ **(c)** $17 + i$ **(d)** $17 + i$

8.5 Exercises (page 573)

1. $\begin{bmatrix} 4 & 4 & -5 \\ -1 & 5 & 4 \end{bmatrix}$ **3.** $\begin{bmatrix} 0 & 12 & -20 \\ 4 & 8 & 24 \end{bmatrix}$ **5.** $\begin{bmatrix} -8 & 7 & -15 \\ 7 & 0 & 22 \end{bmatrix}$ **7.** $\begin{bmatrix} 28 & -9 \\ 4 & 23 \end{bmatrix}$ **9.** $\begin{bmatrix} 1 & 14 & -14 \\ 2 & 22 & -18 \\ 3 & 0 & 28 \end{bmatrix}$ **11.** $\begin{bmatrix} 15 & 21 & -16 \\ 22 & 34 & -22 \\ -11 & 7 & 22 \end{bmatrix}$

13. $\begin{bmatrix} 25 & -9 \\ 4 & 20 \end{bmatrix}$ **15.** $\begin{bmatrix} -13 & 7 & -12 \\ -18 & 10 & -14 \\ 17 & -7 & 34 \end{bmatrix}$ **17.** $\begin{bmatrix} -2 & 4 & 2 & 8 \\ 2 & 1 & 4 & 6 \end{bmatrix}$ **19.** $\begin{bmatrix} 9 & 2 \\ 34 & 13 \\ 47 & 20 \end{bmatrix}$ **21.** $\begin{bmatrix} 1 & -1 \\ -1 & 2 \end{bmatrix}$ **23.** $\begin{bmatrix} 1 & -\frac{5}{2} \\ -1 & 3 \end{bmatrix}$

25. $\begin{bmatrix} 1 & \frac{-1}{a} \\ -1 & \frac{2}{a} \end{bmatrix}$ **27.** $\begin{bmatrix} 3 & -3 & 1 \\ -2 & 2 & -1 \\ -4 & 5 & -2 \end{bmatrix}$ **29.** $\begin{bmatrix} -\frac{5}{7} & \frac{1}{7} & \frac{3}{7} \\ \frac{9}{7} & \frac{1}{7} & -\frac{4}{7} \\ \frac{3}{7} & -\frac{2}{7} & \frac{1}{7} \end{bmatrix}$ **31.** $x = 3, y = 2$ **33.** $x = -5, y = 10$ **35.** $x = 2, y = -1$

37. $x = \frac{1}{2}, y = 2$ **39.** $x = -2, y = 1$ **41.** $x = \frac{2}{a}, y = \frac{3}{a}$ **43.** $x = -2, y = 3, z = 5$ **45.** $x = \frac{1}{2}, y = -\frac{1}{2}, z = 1$

47. $x = -\frac{34}{7}, y = \frac{85}{7}, z = \frac{12}{7}$ **49.** $x = \frac{1}{3}, y = 1, z = \frac{2}{3}$ **51.** $\begin{bmatrix} 4 & 2 & 1 & 0 \\ 2 & 1 & 0 & 1 \end{bmatrix} \rightarrow \begin{bmatrix} 1 & \frac{1}{2} & \frac{1}{4} & 0 \\ 2 & 1 & 0 & 1 \end{bmatrix} \rightarrow \begin{bmatrix} 1 & \frac{1}{2} & \frac{1}{4} & 0 \\ 0 & 0 & -\frac{1}{2} & 1 \end{bmatrix}$

53. $\begin{bmatrix} 15 & 3 & 1 & 0 \\ 10 & 2 & 0 & 1 \end{bmatrix} \rightarrow \begin{bmatrix} 1 & \frac{1}{5} & \frac{1}{15} & 0 \\ 10 & 2 & 0 & 1 \end{bmatrix} \rightarrow \begin{bmatrix} 1 & \frac{1}{5} & \frac{1}{15} & 0 \\ 0 & 0 & -\frac{2}{3} & 1 \end{bmatrix}$

55. $\begin{bmatrix} -3 & 1 & -1 & 1 & 0 & 0 \\ 1 & -4 & -7 & 0 & 1 & 0 \\ 1 & 2 & 5 & 0 & 0 & 1 \end{bmatrix} \rightarrow \begin{bmatrix} 1 & 2 & 5 & 0 & 0 & 1 \\ 1 & -4 & -7 & 0 & 1 & 0 \\ -3 & 1 & -1 & 1 & 0 & 0 \end{bmatrix} \rightarrow \begin{bmatrix} 1 & 2 & 5 & 0 & 0 & 1 \\ 0 & -6 & -12 & 0 & 1 & -1 \\ 0 & 7 & 14 & 1 & 0 & 3 \end{bmatrix} \rightarrow \begin{bmatrix} 1 & 2 & 5 & 0 & 0 & 1 \\ 0 & 1 & 2 & 0 & -\frac{1}{6} & \frac{1}{6} \\ 0 & 1 & 2 & \frac{1}{7} & 0 & \frac{3}{7} \end{bmatrix}$

$\rightarrow \begin{bmatrix} 1 & 2 & 5 & 0 & 0 & 1 \\ 0 & 1 & 2 & 0 & -\frac{1}{6} & \frac{1}{6} \\ 0 & 0 & 0 & \frac{1}{7} & \frac{1}{6} & \frac{11}{42} \end{bmatrix}$ **57.** $\begin{bmatrix} 0.01 & 0.05 & -0.01 \\ 0.01 & -0.02 & 0.01 \\ -0.02 & 0.01 & 0.03 \end{bmatrix}$ **59.** $\begin{bmatrix} 0.02 & -0.04 & -0.01 & 0.01 \\ -0.02 & 0.05 & 0.03 & -0.03 \\ 0.02 & 0.01 & -0.04 & 0.00 \\ -0.02 & 0.06 & 0.07 & 0.06 \end{bmatrix}$

61. $x = 4.57, y = -6.44, z = -24.07$ **63.** $x = -1.19, y = 2.46, z = 8.27$

65. (a) $\begin{bmatrix} 500 & 350 & 400 \\ 700 & 500 & 850 \end{bmatrix}; \begin{bmatrix} 500 & 700 \\ 350 & 500 \\ 400 & 850 \end{bmatrix}$ **(b)** $\begin{bmatrix} 15 \\ 8 \\ 3 \end{bmatrix}$ **(c)** $\begin{bmatrix} 11{,}500 \\ 17{,}050 \end{bmatrix}$ **(d)** $[0.10 \quad 0.05]$ **(e)** $\$2002.50$

8.6 Exercises (page 582)

1. Proper **3.** Improper; $1 + \dfrac{9}{x^2 - 4}$ **5.** Improper; $5x + \dfrac{22x - 1}{x^2 - 4}$ **7.** Improper; $1 + \dfrac{-2(x - 6)}{(x + 4)(x - 3)}$ **9.** $\dfrac{-4}{x} + \dfrac{4}{x - 1}$

11. $\dfrac{1}{x} + \dfrac{-x}{x^2 + 1}$ **13.** $\dfrac{-1}{x - 1} + \dfrac{2}{x - 2}$ **15.** $\dfrac{\frac{1}{4}}{x + 1} + \dfrac{\frac{3}{4}}{x - 1} + \dfrac{\frac{1}{2}}{(x - 1)^2}$ **17.** $\dfrac{\frac{1}{12}}{x - 2} + \dfrac{-\frac{1}{12}(x + 4)}{x^2 + 2x + 4}$

19. $\dfrac{\frac{1}{4}}{x - 1} + \dfrac{\frac{1}{4}}{(x - 1)^2} + \dfrac{-\frac{1}{4}}{x + 1} + \dfrac{\frac{1}{4}}{(x + 1)^2}$ **21.** $\dfrac{-5}{x + 2} + \dfrac{5}{x + 1} + \dfrac{-4}{(x + 1)^2}$ **23.** $\dfrac{\frac{1}{4}}{x} + \dfrac{1}{x^2} + \dfrac{-\frac{1}{4}(x + 4)}{x^2 + 4}$

25. $\dfrac{\frac{2}{3}}{x + 1} + \dfrac{\frac{1}{3}(x + 1)}{x^2 + 2x + 4}$ **27.** $\dfrac{\frac{2}{7}}{3x - 2} + \dfrac{\frac{1}{7}}{2x + 1}$ **29.** $\dfrac{\frac{3}{4}}{x + 3} + \dfrac{\frac{1}{4}}{x - 1}$ **31.** $\dfrac{1}{x^2 + 4} + \dfrac{2x - 1}{(x^2 + 4)^2}$ **33.** $\dfrac{-1}{x} + \dfrac{2}{x - 3} + \dfrac{-1}{x + 1}$

35. $\dfrac{4}{x-2} + \dfrac{-3}{x-1} + \dfrac{-1}{(x-1)^2}$ **37.** $\dfrac{x}{(x^2+16)^2} + \dfrac{-16x}{(x^2+16)^3}$ **39.** $\dfrac{-\frac{8}{7}}{2x+1} + \dfrac{\frac{4}{7}}{x-3}$ **41.** $\dfrac{-\frac{2}{9}}{x} + \dfrac{-\frac{1}{3}}{x^2} + \dfrac{\frac{1}{6}}{x-3} + \dfrac{\frac{1}{18}}{x+3}$

Historical Problems *(page 588)*

1. $x = 6$ units, $y = 8$ units

8.7 Exercises *(page 588)*

1.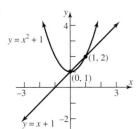

3.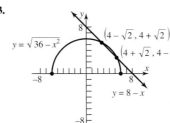

5.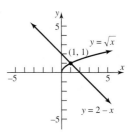

7.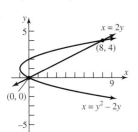

9.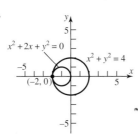

11.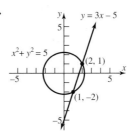

13.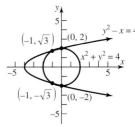

15.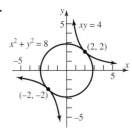

17. No points of intersection.

19.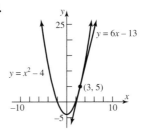

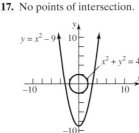

21. $x = 1, y = 4; x = -1, y = -4; x = 2\sqrt{2}, y = \sqrt{2}; x = -2\sqrt{2}, y = -\sqrt{2}$ **23.** $x = 0, y = 1; x = -\dfrac{2}{3}, y = -\dfrac{1}{3}$

25. $x = 0, y = -1; x = \dfrac{5}{2}, y = -\dfrac{7}{2}$ **27.** $x = 2, y = \dfrac{1}{3}; x = \dfrac{1}{2}, y = \dfrac{4}{3}$ **29.** $x = 3, y = 2; x = 3, y = -2; x = -3, y = 2; x = -3, y = -2$

31. $x = \dfrac{1}{2}, y = \dfrac{3}{2}; x = \dfrac{1}{2}, y = -\dfrac{3}{2}; x = -\dfrac{1}{2}, y = \dfrac{3}{2}; x = -\dfrac{1}{2}, y = -\dfrac{3}{2}$ **33.** $x = \sqrt{2}, y = 2\sqrt{2}; x = -\sqrt{2}, y = -2\sqrt{2}$

35. No real solution exists **37.** $x = \dfrac{8}{3}, y = \dfrac{2\sqrt{10}}{3}; x = -\dfrac{8}{3}, y = \dfrac{2\sqrt{10}}{3}; x = \dfrac{8}{3}, y = -\dfrac{2\sqrt{10}}{3}; x = -\dfrac{8}{3}, y = -\dfrac{2\sqrt{10}}{3}$

39. $x = 1, y = \dfrac{1}{2}; x = -1, y = \dfrac{1}{2}; x = 1, y = -\dfrac{1}{2}; x = -1, y = -\dfrac{1}{2}$ **41.** No real solution exists **43.** $x = \sqrt{3}, y = \sqrt{3}; x = -\sqrt{3},$

$y = -\sqrt{3}; x = 2, y = 1; x = -2, y = -1$ **45.** $x = 0, y = -2; x = 0, y = 1; x = 2, y = -1$ **47.** $x = 2, y = 8$ **49.** $x = 81, y = 3$

51.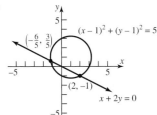

53.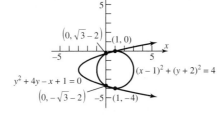

55.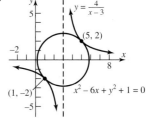

57. $x = 0.48, y = 0.62$ **59.** $x = -1.65, y = -0.89$ **61.** $x = 0.58, y = 1.86; x = 1.81, y = 1.05; x = 0.58, y = -1.86; x = 1.81, y = -1.05$

63. $x = 2.35, y = 0.85$ **65.** 3 and 1; -3 and -1 **67.** 2 and 2; -2 and -2 **69.** $\dfrac{1}{2}$ and $\dfrac{1}{3}$ **71.** 5   **73.** 5 in. by 3 in.   **75.** 2 cm and 4 cm

77. tortoise: 7 mi/hr, hare: $7\dfrac{1}{2}$ mi/hr   **79.** 12 cm by 18 cm   **81.** $x = 60$ ft; $y = 30$ ft   **83.** $l = \dfrac{P + \sqrt{P^2 - 16A}}{4}$; $w = \dfrac{P - \sqrt{P^2 - 16A}}{4}$

85. $y = 4x - 4$ **87.** $y = 2x + 1$ **89.** $y = -\dfrac{1}{3}x + \dfrac{7}{3}$ **91.** $y = 2x - 3$ **93.** $r_1 = \dfrac{-b + \sqrt{b^2 - 4ac}}{2a}$; $r_2 = \dfrac{-b - \sqrt{b^2 - 4ac}}{2a}$

95. (a) 4.274 ft by 4.274 ft or 0.093 ft by 0.093 ft

8.8 Exercises *(page 598)*

1.
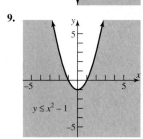

3.
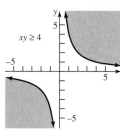

5.
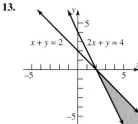

7.
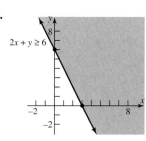

9.
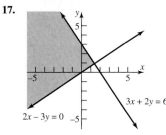

11.
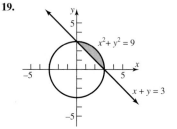

13.
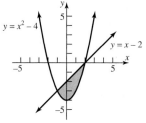

15.
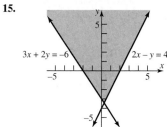

17.
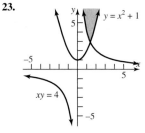

19.
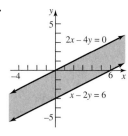

21.
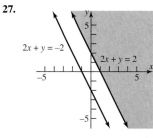

23.
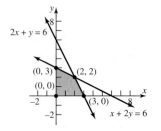

25.
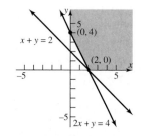

27.
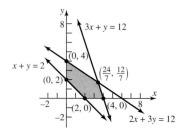

29. No solution
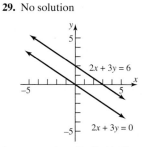

31. Bounded; corner points

$(0, 0), (3, 0), (2, 2), (0, 3)$

33. Unbounded; corner points

$(2, 0), (0, 4)$

35. Bounded; corner points $(2, 0), (4, 0),$

$\left(\dfrac{24}{7}, \dfrac{12}{7}\right), (0, 4), (0, 2)$

37. Bounded; corner points $(2, 0)$, $(5, 0)$,
$(2, 6)$, $(0, 8)$, $(0, 2)$

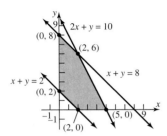

39. Bounded; corner points
$(1, 0)$, $(10, 0)$, $(0, 5)$, $\left(0, \dfrac{1}{2}\right)$

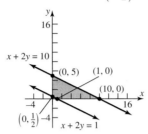

41. $\begin{cases} x \leq 4 \\ x + y \leq 6 \\ x \geq 0 \\ y \geq 0 \end{cases}$

43. $\begin{cases} x \leq 20 \\ y \geq 15 \\ x + y \leq 50 \\ x - y \leq 0 \\ x \geq 0, y \geq 15 \end{cases}$

45. (a) $\begin{cases} x + y \leq 50{,}000 \\ x \geq 35{,}000 \\ y \leq 10{,}000 \\ x \geq 0 \\ y \geq 0 \end{cases}$

(b)

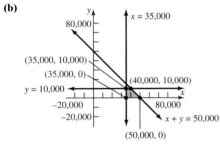

47. (a) $\begin{cases} x \geq 0 \\ y \geq 0 \\ x + 2y \leq 300 \\ 3x + 2y \leq 480 \end{cases}$

(b)

49. (a) $\begin{cases} 3x + 2y \leq 160 \\ 2x + 3y \leq 150 \\ x \geq 0 \\ y \geq 0 \end{cases}$

(b)

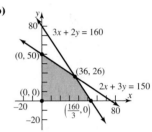

8.9 Exercises *(page 605)*

1. Maximum value is 11; minimum value is 3.   **3.** Maximum value is 65; minimum value is 4.   **5.** Maximum value is 67; minimum value is 20.
7. The maximum value of z is 12, and it occurs at the point $(6, 0)$.   **9.** The minimum value of z is 4, and it occurs at the point $(2, 0)$.
11. The maximum value of z is 20, and it occurs at the point $(0, 4)$.   **13.** The minimum value of z is 8, and it occurs at the point $(0, 2)$.
15. The maximum value of z is 50, and it occurs at the point $(10, 0)$.   **17.** 8 downhill, 24 cross-country; $1760; $1920

19. 30 acres of soybeans and 10 acres of corn.   **21.** $\dfrac{1}{2}$ hr on machine 1; $5\dfrac{1}{4}$ hr on machine 2   **23.** 100 lb of ground beef and 50 lb of pork

25. 10 racing skates, 15 figure skates   **27.** 2 metal samples, 4 plastic samples; $34   **29. (a)** 10 first class, 120 coach   **(b)** 15 first class, 120 coach

Fill-in-the-Blank Items *(page 608)*

1. inconsistent   **3.** determinants   **5.** inverse   **7.** identity   **9.** half-plane   **11.** feasible point

True/False Items *(page 609)*

1. F   **3.** F   **5.** F   **7.** F   **9.** T   **11.** T

Review Exercises *(page 609)*

1. $x = 2, y = -1$ **3.** $x = 2, y = \dfrac{1}{2}$ **5.** $x = 2, y = -1$ **7.** $x = \dfrac{11}{5}, y = -\dfrac{3}{5}$ **9.** $x = -\dfrac{8}{5}, y = \dfrac{12}{5}$ **11.** $x = 6, y = -1$

13. $x = -4, y = 3$ **15.** $x = 2, y = 3$ **17.** Inconsistent   **19.** $x = -1, y = 2, z = -3$

21. $\begin{bmatrix} 4 & -4 \\ 3 & 9 \\ 4 & 0 \end{bmatrix}$ **23.** $\begin{bmatrix} 6 & 0 \\ 12 & 24 \\ -6 & 12 \end{bmatrix}$ **25.** $\begin{bmatrix} 4 & -3 & 0 \\ 12 & -2 & -8 \\ -2 & 5 & -4 \end{bmatrix}$ **27.** $\begin{bmatrix} 8 & -13 & 8 \\ 9 & 2 & -10 \\ 18 & -17 & 4 \end{bmatrix}$ **29.** $\begin{bmatrix} \frac{1}{2} & -1 \\ -\frac{1}{6} & \frac{2}{3} \end{bmatrix}$ **31.** $\begin{bmatrix} -\frac{5}{7} & \frac{9}{7} & \frac{3}{7} \\ \frac{1}{7} & \frac{1}{7} & -\frac{2}{7} \\ \frac{3}{7} & -\frac{4}{7} & \frac{1}{7} \end{bmatrix}$ **33.** Singular

35. $x = \dfrac{2}{5}, y = \dfrac{1}{10}$ **37.** $x = \dfrac{1}{2}, y = \dfrac{2}{3}, z = \dfrac{1}{6}$ **39.** $x = -\dfrac{1}{2}, y = -\dfrac{2}{3}, z = -\dfrac{3}{4}$ **41.** $z = -1, x = y + 1$, where y is any real number

43. $x = 1, y = 2, z = -3, t = 1$ **45.** 5   **47.** 108   **49.** -100 **51.** $x = 2, y = -1$ **53.** $x = 2, y = 3$ **55.** $x = -1, y = 2, z = -3$

57. $\dfrac{-\frac{3}{2}}{x} + \dfrac{\frac{3}{2}}{x - 4}$ **59.** $\dfrac{-3}{x - 1} + \dfrac{3}{x} + \dfrac{4}{x^2}$ **61.** $\dfrac{-\frac{1}{10}}{x + 1} + \dfrac{\frac{1}{10}x + \frac{9}{10}}{x^2 + 9}$ **63.** $\dfrac{x}{x^2 + 4} + \dfrac{-4x}{(x^2 + 4)^2}$ **65.** $\dfrac{\frac{1}{2}}{x^2 + 1} + \dfrac{\frac{1}{4}}{x - 1} + \dfrac{-\frac{1}{4}}{x + 1}$

67. $x = -\dfrac{2}{5}, y = -\dfrac{11}{5}; x = -2, y = 1$ **69.** $x = 2\sqrt{2}, y = \sqrt{2}; x = -2\sqrt{2}, y = -\sqrt{2}$ **71.** $x = 0, y = 0; x = -3, y = 3; x = 3, y = 3$

73. $x = \sqrt{2}, y = -\sqrt{2}; x = -\sqrt{2}, y = \sqrt{2}; x = \dfrac{4}{3}\sqrt{2}, y = -\dfrac{2}{3}\sqrt{2}; x = -\dfrac{4}{3}\sqrt{2}, y = \dfrac{2}{3}\sqrt{2}$ **75.** $x = 1, y = -1$

77. Unbounded;
corner point $(0, 2)$

79. Bounded;
corner points $(0, 0), (0, 2), (3, 0)$

81. Bounded;
corner points $(0, 1), (0, 8), (4, 0), (2, 0)$

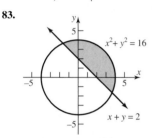

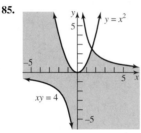

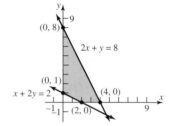

83.

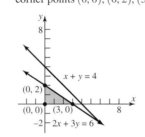

85.

87. The maximum value is 32 when $x = 0$ and $y = 8$.
89. The minimum value is 3 when $x = 1$ and $y = 0$.
91. The maximum value is $\dfrac{108}{7}$ when $x = \dfrac{12}{7}$ and $y = \dfrac{12}{7}$.

93. 10   **95.** $y = x^2 - \dfrac{1}{2}x + \dfrac{3}{2}$ **97.** $y = -\dfrac{1}{3}x^2 - \dfrac{2}{3}x + 1$

99. 70 lb of \$3 coffee and 30 lb of \$6 coffee
101. 1 small, 5 medium, 2 large

103. 24 ft by 10 ft   **105.** $4 + \sqrt{2}$ in. and $4 - \sqrt{2}$ in.   **107.** $100\sqrt{10}$ ft   **109.** speedboat, 36.7 km/hr; Aguarico, 3.3 km/hr
111. Bruce: 4 hr; Bryce: 2 hr; Marty: 8 hr   **113.** 35 gasoline engines, 15 diesel engines; 15 gasoline engines, 0 diesel engines

C H A P T E R   9   Sequences; Induction; The Binomial Theorem

9.1 Exercises (page 624)

1. $1, 2, 3, 4, 5$ **3.** $\dfrac{1}{3}, \dfrac{1}{2}, \dfrac{3}{5}, \dfrac{2}{3}, \dfrac{5}{7}$ **5.** $1, -4, 9, -16, 25$ **7.** $\dfrac{1}{2}, \dfrac{2}{5}, \dfrac{2}{7}, \dfrac{8}{41}, \dfrac{8}{61}$ **9.** $-\dfrac{1}{6}, \dfrac{1}{12}, -\dfrac{1}{20}, \dfrac{1}{30}, -\dfrac{1}{42}$ **11.** $\dfrac{1}{e}, \dfrac{2}{e^2}, \dfrac{3}{e^3}, \dfrac{4}{e^4}, \dfrac{5}{e^5}$ **13.** $a_n = \dfrac{n}{n+1}$

15. $a_n = \dfrac{1}{2^{n-1}}$ **17.** $a_n = (-1)^{n+1}$ **19.** $a_n = (-1)^{n+1}n$ **21.** $a_1 = 2, a_2 = 5, a_3 = 8, a_4 = 11, a_5 = 14$

23. $a_1 = -2, a_2 = 0, a_3 = 3, a_4 = 7, a_5 = 12$ **25.** $a_1 = 5, a_2 = 10, a_3 = 20, a_4 = 40, a_5 = 80$ **27.** $a_1 = 3, a_2 = \dfrac{3}{2}, a_3 = \dfrac{1}{2}, a_4 = \dfrac{1}{8}, a_5 = \dfrac{1}{40}$

29. $a_1 = 1, a_2 = 2, a_3 = 2, a_4 = 4, a_5 = 8$ **31.** $a_1 = A, a_2 = A + d, a_3 = A + 2d, a_4 = A + 3d, a_5 = A + 4d$

33. $a_1 = \sqrt{2}, a_2 = \sqrt{2 + \sqrt{2}}, a_3 = \sqrt{2 + \sqrt{2 + \sqrt{2}}}, a_4 = \sqrt{2 + \sqrt{2 + \sqrt{2 + \sqrt{2}}}}, a_5 = \sqrt{2 + \sqrt{2 + \sqrt{2 + \sqrt{2 + \sqrt{2}}}}}$

35. 50   **37.** 21   **39.** 90   **41.** 26   **43.** 42   **45.** 96   **47.** $3 + 4 + \cdots + (n + 2)$ **49.** $\dfrac{1}{2} + 2 + \dfrac{9}{2} + \cdots + \dfrac{n^2}{2}$ **51.** $1 + \dfrac{1}{3} + \dfrac{1}{9} + \cdots + \dfrac{1}{3^n}$

53. $\dfrac{1}{3} + \dfrac{1}{9} + \cdots + \dfrac{1}{3^n}$ **55.** $\ln 2 - \ln 3 + \ln 4 - \cdots + (-1)^n \ln n$ **57.** $\displaystyle\sum_{k=1}^{20} k$ **59.** $\displaystyle\sum_{k=1}^{13} \dfrac{k}{k+1}$ **61.** $\displaystyle\sum_{k=0}^{6} (-1)^k \left(\dfrac{1}{3^k}\right)$ **63.** $\displaystyle\sum_{k=1}^{n} \dfrac{3^k}{k}$

65. $\displaystyle\sum_{k=0}^{n} (a + kd)$ or $\displaystyle\sum_{k=1}^{n+1} [a + (k-1)d]$ **67.** \$2930   **69.** 2162   **71.** 21   **73.** A Fibonacci sequence

75. $2S = \underbrace{(1 + n) + (1 + n) + \cdots + (n + 1)}_{n \text{ terms}} = n(n + 1)\,; S = \dfrac{1}{2}n(n + 1)$

9.2 Exercises (page 630)

1. $d = 1; 5, 6, 7, 8$ **3.** $d = 2; -3, -1, 1, 3$ **5.** $d = -2; 4, 2, 0, -2$ **7.** $d = -\dfrac{1}{3}; \dfrac{1}{6}, -\dfrac{1}{6}, -\dfrac{1}{2}, -\dfrac{5}{6}$ **9.** $d = \ln 3; \ln 3, 2\ln 3, 3\ln 3, 4\ln 3$

11. $a_n = 3n - 1; a_5 = 14$ **13.** $a_n = 8 - 3n; a_5 = -7$ **15.** $a_n = \dfrac{1}{2}(n - 1); a_5 = 2$ **17.** $a_n = \sqrt{2}n; a_5 = 5\sqrt{2}$ **19.** $a_{12} = 24$

21. $a_{10} = -26$ **23.** $a_8 = a + 7b$ **25.** $a_1 = -13; d = 3; a_n = a_{n-1} + 3$ **27.** $a_1 = -53; d = 6; a_n = a_{n-1} + 6$

29. $a_1 = 28; d = -2; a_n = a_{n-1} - 2$ **31.** $a_1 = 25; d = -2; a_n = a_{n-1} - 2$ **33.** n^2 **35.** $\dfrac{n}{2}(9 + 5n)$ **37.** 1260   **39.** 324

41.

43.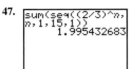

45.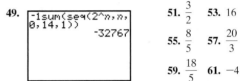

47. $-\dfrac{3}{2}$

49. 1185 seats

51. 210 of beige and 190 blue

53. 30 rows

Historical Problems *(page 640)*

1. $1\dfrac{2}{3}$ loaves, $10\dfrac{5}{6}$ loaves, 20 loaves, $29\dfrac{1}{6}$ loaves, $38\dfrac{1}{3}$ loaves

9.3 Exercises *(page 640)*

1. $r = 3; 3, 9, 27, 81$ **3.** $r = \dfrac{1}{2}; -\dfrac{3}{2}, -\dfrac{3}{4}, -\dfrac{3}{8}, -\dfrac{3}{16}$ **5.** $r = 2; \dfrac{1}{4}, \dfrac{1}{2}, 1, 2$ **7.** $r = 2^{1/3}; 2^{1/3}, 2^{2/3}, 2, 2^{4/3}$ **9.** $r = \dfrac{3}{2}; \dfrac{1}{2}, \dfrac{3}{4}, \dfrac{9}{8}, \dfrac{27}{16}$

11. Arithmetic; $d = 1$ **13.** Neither **15.** Arithmetic; $d = -\dfrac{2}{3}$ **17.** Neither **19.** Geometric; $r = \dfrac{2}{3}$ **21.** Geometric; $r = 2$

23. Geometric; $r = 3^{1/2}$ **25.** $a_5 = 162; a_n = 2 \cdot 3^{n-1}$ **27.** $a_5 = 5; a_n = 5 \cdot (-1)^{n-1}$ **29.** $a_5 = 0; a_n = 0$ **31.** $a_5 = 4\sqrt{2}; a_n = (\sqrt{2})^n$

33. $a_7 = \dfrac{1}{64}$ **35.** $a_9 = 1$ **37.** $a_8 = 0.00000004$ **39.** $-\dfrac{1}{4}(1 - 2^n)$ **41.** $2\left[1 - \left(\dfrac{2}{3}\right)^n\right]$ **43.** $1 - 2^n$

45.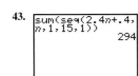

47. *(calculator screen)* sum(seq((2/3)^n, n,1,15,1)) 1.995432683

49. *(calculator screen)* -1sum(seq(2^n,n, 0,14,1)) -32767

51. $\dfrac{3}{2}$ **53.** 16

55. $\dfrac{8}{5}$ **57.** $\dfrac{20}{3}$

59. $\dfrac{18}{5}$ **61.** -4

63. \$21,879.11 **65. (a)** 0.775 ft **(b)** 8th **(c)** 15.88 ft **(d)** 20 ft **67.** \$349,496.41 **69.** \$96,885.98 **71.** \$305.10

73. *A*: \$25,250 per year in 5th year, \$112,742 total; *B*: \$24,761 per year in 5th year, \$116,801 total

75. Option 2 results in the most: \$16,038, 304; Option 1 results in the least: \$14,700,000. **77.** 1.845×10^{19} **79.** 10 **81.** \$72.67 per share

83. Yes. A constant sequence is both arithmetic and geometric. For example, $3, 3, 3, \ldots$ is an arithmetic sequence with $a_1 = 3$ and $d = 0$ and is a geometric sequence with $a_1 = 3$ and $r = 1$.

9.4 Exercises *(page 646)*

1. (I) $n = 1: 2(1) = 2$ and $1(1 + 1) = 2$

(II) If $2 + 4 + 6 + \cdots + 2k = k(k + 1)$, then $2 + 4 + 6 + \cdots + 2k + 2(k + 1) = (2 + 4 + 6 + \cdots + 2k) + 2(k + 1)$
$= k(k + 1) + 2(k + 1) = k^2 + 3k + 2 = (k + 1)(k + 2) = (k + 1)[(k + 1) + 1]$.

3. (I) $n = 1: 1 + 2 = 3$ and $\dfrac{1}{2}(1)(1 + 5) = \dfrac{1}{2}(6) = 3$

(II) If $3 + 4 + 5 + \cdots + (k + 2) = \dfrac{1}{2}k(k + 5)$, then $3 + 4 + 5 + \cdots + (k + 2) + [(k + 1) + 2]$

$= [3 + 4 + 5 + \cdots + (k + 2)] + (k + 3) = \dfrac{1}{2}k(k + 5) + k + 3 = \dfrac{1}{2}(k^2 + 7k + 6) = \dfrac{1}{2}(k + 1)(k + 6)$

$= \dfrac{1}{2}(k + 1)[(k + 1) + 5]$.

5. (I) $n = 1: 3(1) - 1 = 2$ and $\dfrac{1}{2}(1)[3(1) + 1] = \dfrac{1}{2}(4) = 2$

(II) If $2 + 5 + 8 + \cdots + (3k - 1) = \dfrac{1}{2}k(3k + 1)$, then $2 + 5 + 8 + \cdots + (3k - 1) + [3(k + 1) - 1]$

$= [2 + 5 + 8 + \cdots + (3k - 1)] + (3k + 2) = \dfrac{1}{2}k(3k + 1) + (3k + 2) = \dfrac{1}{2}(3k^2 + 7k + 4) = \dfrac{1}{2}(k + 1)(3k + 4)$

$= \dfrac{1}{2}(k + 1)[3(k + 1) + 1]$.

7. (I) $n = 1: 2^{1-1} = 1$ and $2^1 - 1 = 1$

(II) If $1 + 2 + 2^2 + \cdots + 2^{k-1} = 2^k - 1$, then $1 + 2 + 2^2 + \cdots + 2^{k-1} + 2^{((k+1)-1)} = (1 + 2 + 2^2 + \cdots + 2^{k-1}) + 2^k$
$= 2^k - 1 + 2^k = 2(2^k) - 1 = 2^{k+1} - 1$.

9. (I) $n = 1: 4^{1-1} = 1$ and $\frac{1}{3}(4^1 - 1) = \frac{1}{3}(3) = 1$

(II) If $1 + 4 + 4^2 + \cdots + 4^{k-1} = \frac{1}{3}(4^k - 1)$, then $1 + 4 + 4^2 + \cdots + 4^{k-1} + 4^{(k+1)-1} = (1 + 4 + 4^2 + \cdots + 4^{k-1}) + 4^k$

$= \frac{1}{3}(4^k - 1) + 4^k = \frac{1}{3}[4^k - 1 + 3(4^k)] = \frac{1}{3}[4(4^k) - 1] = \frac{1}{3}(4^{k+1} - 1)$.

11. (I) $n = 1: \dfrac{1}{1 \cdot 2} = \dfrac{1}{2}$ and $\dfrac{1}{1 + 1} = \dfrac{1}{2}$

(II) If $\dfrac{1}{1 \cdot 2} + \dfrac{1}{2 \cdot 3} + \dfrac{1}{3 \cdot 4} + \cdots + \dfrac{1}{k(k + 1)} = \dfrac{k}{k + 1}$, then $\dfrac{1}{1 \cdot 2} + \dfrac{1}{2 \cdot 3} + \dfrac{1}{3 \cdot 4} + \cdots + \dfrac{1}{k(k + 1)} + \dfrac{1}{(k + 1)[(k + 1) + 1]}$

$= \left[\dfrac{1}{1 \cdot 2} + \dfrac{1}{2 \cdot 3} + \dfrac{1}{3 \cdot 4} + \cdots + \dfrac{1}{k(k + 1)}\right] + \dfrac{1}{(k + 1)(k + 2)} = \dfrac{k}{k + 1} + \dfrac{1}{(k + 1)(k + 2)} = \dfrac{k(k + 2) + 1}{(k + 1)(k + 2)}$

$= \dfrac{k^2 + 2k + 1}{(k + 1)(k + 2)} = \dfrac{(k + 1)^2}{(k + 1)(k + 2)} = \dfrac{k + 1}{k + 2} = \dfrac{k + 1}{(k + 1) + 1}$.

13. (I) $n = 1: 1^2 = 1$ and $\frac{1}{6} \cdot 1 \cdot 2 \cdot 3 = 1$

(II) If $1^2 + 2^2 + 3^2 + \cdots + k^2 = \frac{1}{6}k(k + 1)(2k + 1)$, then $1^2 + 2^2 + 3^2 + \cdots + k^2 + (k + 1)^2$

$= (1^2 + 2^2 + 3^2 + \cdots + k^2) + (k + 1)^2 = \frac{1}{6}k(k + 1)(2k + 1) + (k + 1)^2 = \frac{1}{6}(2k^3 + 9k^2 + 13k + 6)$

$= \frac{1}{6}(k + 1)(k + 2)(2k + 3) = \frac{1}{6}(k + 1)[(k + 1) + 1][2(k + 1) + 1]$.

15. (I) $n = 1: 5 - 1 = 4$ and $\frac{1}{2}(1)(9 - 1) = \frac{1}{2} \cdot 8 = 4$

(II) If $4 + 3 + 2 + \cdots + (5 - k) = \frac{1}{2}k(9 - k)$, then $4 + 3 + 2 + \cdots + (5 - k) + [5 - (k + 1)]$

$= [4 + 3 + 2 + \cdots + (5 - k)] + 4 - k = \frac{1}{2}k(9 - k) + 4 - k = \frac{1}{2}(9k - k^2 + 8 - 2k) = \frac{1}{2}(-k^2 + 7k + 8)$

$= \frac{1}{2}(k + 1)(8 - k) = \frac{1}{2}(k + 1)[9 - (k + 1)]$.

17. (I) $n = 1: 1 \cdot (1 + 1) = 2$ and $\frac{1}{3} \cdot 1 \cdot 2 \cdot 3 = 2$

(II) If $1 \cdot 2 + 2 \cdot 3 + 3 \cdot 4 + \cdots + k(k + 1) = \frac{1}{3}k(k + 1)(k + 2)$, then $1 \cdot 2 + 2 \cdot 3 + 3 \cdot 4 + \cdots + k(k + 1)$

$+ (k + 1)[(k + 1) + 1] = [1 \cdot 2 + 2 \cdot 3 + 3 \cdot 4 + \cdots + k(k + 1)] + (k + 1)(k + 2)$

$= \frac{1}{3}k(k + 1)(k + 2) + (k + 1)(k + 2) = \frac{1}{3}(k + 1)(k + 2)(k + 3) = \frac{1}{3}(k + 1)[(k + 1) + 1][(k + 1) + 2]$.

19. (I) $n = 1: 1^2 + 1 = 2$ which is divisible by 2.
(II) If $k^2 + k$ is divisible by 2, then $(k + 1)^2 + (k + 1) = k^2 + 2k + 1 + k + 1 = (k^2 + k) + 2k + 2$. Since $k^2 + k$ is divisible by 2 and $2k + 2$ is divisible by 2, $(k + 1)^2 + (k + 1)$ is divisible by 2.

21. (I) $n = 1: 1^2 - 1 + 2 = 2$ which is divisible by 2.
(II) If $k^2 - k + 2$ is divisible by 2, then $(k + 1)^2 - (k + 1) + 2 = k^2 + 2k + 1 - k - 1 + 2 = (k^2 - k + 2) + 2k$. Since $k^2 - k + 2$ is divisible by 2 and $2k$ is divisible by 2, $(k + 1)^2 - (k + 1) + 2$ is divisible by 2.

23. (I) $n = 1$: If $x > 1$, then $x^1 = x > 1$.
(II) Assume, for any natural number k, that if $x > 1$, then $x^k > 1$. Multiply both sides of the inequality $x^k > 1$ by x. Then $x^{k+1} > x > 1$.

25. (I) $n = 1: a - b$ is a factor of $a^1 - b^1 = a - b$.
(II) If $a - b$ is a factor of $a^k - b^k$, show that $a - b$ is a factor of $a^{k+1} - b^{k+1}: a^{k+1} - b^{k+1} = a(a^k - b^k) + b^k(a - b)$. Since $a - b$ is a factor of $a^k - b^k$ and $a - b$ is a factor of $a - b$, then $a - b$ is a factor of $a^{k+1} - b^{k+1}$.

27. $n = 1: 1^2 - 1 + 41 = 41$ which is a prime number.
$n = 41: 41^2 - 41 + 41 = 1681 = 41^2$ which is not prime.

29. (I) $n = 1: ar^{1-1} = a \cdot 1 = a$ and $a \cdot \dfrac{1 - r^1}{1 - r} = a$, because $r \neq 1$.

(II) If $a + ar + ar^2 + \cdots + ar^{k-1} = a\left(\dfrac{1 - r^k}{1 - r}\right)$, then $a + ar + ar^2 + \cdots + ar^{k-1} + ar^{(k+1)-1} = (a + ar + ar^2 + \cdots + ar^{k-1}) + ar^k$

$= a\left(\dfrac{1 - r^k}{1 - r}\right) + ar^k = \dfrac{a(1 - r^k) + ar^k(1 - r)}{1 - r} = \dfrac{a - ar^k + ar^k - ar^{k+1}}{1 - r} = a\left(\dfrac{1 - r^{k+1}}{1 - r}\right)$.

31. (I) $n = 4$: The number of diagonals in a convex polygon of 4 sides is 2 and $\frac{1}{2} \cdot 4 \cdot (4 - 3) = 2$

(II) If the number of diagonals in a convex polygon of k sides is $\frac{1}{2}k(k - 3)$ then that of $(k + 1)$ sides is increased by

$(k + 1) - 2 = k - 1$. Thus, the number of diagonals in a convex polygon of $(k + 1)$ sides is $\frac{1}{2}k(k - 3) + (k - 1)$

$= \frac{1}{2}[k^2 - 3k + 2k - 2] = \frac{1}{2}[k^2 - k - 2] = \frac{1}{2}(k + 1)(k - 2) = \frac{1}{2}(k + 1)[(k + 1) - 3]$.

9.5 Exercises (page 654)

1. 10 **3.** 21 **5.** 50 **7.** 1 **9.** 1.866×10^{15} **11.** 1.483×10^{13} **13.** $x^5 + 5x^4 + 10x^3 + 10x^2 + 5x + 1$
15. $x^6 - 12x^5 + 60x^4 - 160x^3 + 240x^2 - 192x + 64$ **17.** $81x^4 + 108x^3 + 54x^2 + 12x + 1$
19. $x^{10} + 5x^8y^2 + 10x^6y^4 + 10x^4y^6 + 5x^2y^8 + y^{10}$ **21.** $x^3 + 6\sqrt{2}x^{5/2} + 30x^2 + 40\sqrt{2}x^{3/2} + 60x + 24\sqrt{2}x^{1/2} + 8$
23. $a^5x^5 + 5a^4bx^4y + 10a^3b^2x^3y^2 + 10a^2b^3x^2y^3 + 5ab^4xy^4 + b^5y^5$ **25.** 17,010 **27.** $-101,376$ **29.** 41,472 **31.** $2835x^3$ **33.** $314,928x^7$
35. 495 **37.** 3360 **39.** 1.00501

41. $\binom{n}{n - 1} = \frac{n!}{(n - 1)![n - (n - 1)]!} = \frac{n!}{(n - 1)!1!} = \frac{n \cdot (n - 1)!}{(n - 1)!} = n; \binom{n}{n} = \frac{n!}{n!(n - n)!} = \frac{n!}{n!0!} = \frac{n!}{n!} = 1$

43. $2^n = (1 + 1)^n = \binom{n}{0}1^n + \binom{n}{1}(1)^{n-1}(1) + \cdots + \binom{n}{n}1^n = \binom{n}{0} + \binom{n}{1} + \cdots + \binom{n}{n}$ **45.** 1

Fill-in-the-Blank Items (page 656)

1. sequence **3.** geometric **5.** 15

True/False Items (page 656)

1. T **3.** T **5.** F **7.** F

Review Exercises (page 656)

1. $-\frac{4}{3}, \frac{5}{4}, -\frac{6}{5}, \frac{7}{6}, -\frac{8}{7}$ **3.** $2, 1, \frac{8}{9}, 1, \frac{32}{25}$ **5.** $3, 2, \frac{4}{3}, \frac{8}{9}, \frac{16}{27}$ **7.** $2, 0, 2, 0, 2$ **9.** Arithmetic; $d = 1$; $S_n = \frac{n}{2}(n + 11)$ **11.** Neither

13. Geometric; $r = 8$; $S_n = \frac{8}{7}(8^n - 1)$ **15.** Arithmetic; $d = 4$; $S_n = 2n(n - 1)$ **17.** Geometric; $r = \frac{1}{2}$; $S_n = 6\left[1 - \left(\frac{1}{2}\right)^n\right]$

19. Neither **21.** 115 **23.** 75 **25.** 0.49977 **27.** 35 **29.** $\frac{1}{10^{10}}$ **31.** $9\sqrt{2}$ **33.** $a_n = 5n - 4$ **35.** $a_n = n - 10$ **37.** $\frac{9}{2}$ **39.** $\frac{4}{3}$ **41.** 8

43. (I) $n = 1$: $3 \cdot 1 = 3$ and $\frac{3}{2}(2) = 3$

(II) If $3 + 6 + 9 + \cdots + 3k = \frac{3k}{2}(k + 1)$, then $3 + 6 + 9 + \cdots + 3k + 3(k + 1) = (3 + 6 + 9 + \cdots + 3k) + (3k + 3)$

$= \frac{3k}{2}(k + 1) + (3k + 3) = \frac{3k^2}{2} + \frac{9k}{2} + \frac{6}{2} = \frac{3}{2}(k^2 + 3k + 2) = \frac{3}{2}(k + 1)(k + 2) = \frac{3(k + 1)}{2}[(k + 1) + 1]$.

45. (I) $n = 1$: $2 \cdot 3^{1-1} = 2$ and $3^1 - 1 = 2$
(II) If $2 + 6 + 18 + \cdots + 2 \cdot 3^{k-1} = 3^k - 1$, then $2 + 6 + 18 + \cdots + 2 \cdot 3^{k-1} + 2 \cdot 3^{(k+1)-1} = (2 + 6 + 18 + \cdots + 2 \cdot 3^{k-1}) + 2 \cdot 3^k$
$= 3^k - 1 + 2 \cdot 3^k = 3 \cdot 3^k - 1 = 3^{k+1} - 1$.

47. (I) $n = 1$: $1^2 = 1$ and $\frac{1}{2}(6 - 3 - 1) = \frac{1}{2}(2) = 1$

(II) If $1^2 + 4^2 + 7^2 + \cdots + (3k - 2)^2 = \frac{1}{2}k(6k^2 - 3k - 1)$, then $1^2 + 4^2 + 7^2 + \cdots + (3k - 2)^2 + [3(k + 1) - 2]^2$

$= [1^2 + 4^2 + 7^2 + \cdots + (3k - 2)^2] + (3k + 1)^2 = \frac{1}{2}k(6k^2 - 3k - 1) + (3k + 1)^2 = \frac{1}{2}(6k^3 - 3k^2 - k) + (9k^2 + 6k + 1)$

$= \frac{1}{2}(6k^3 + 15k^2 + 11k + 2) = \frac{1}{2}(k + 1)(6k^2 + 9k + 2) = \frac{1}{2}(k + 1)[6(k + 1)^2 - 3(k + 1) - 1]$.

49. $x^5 + 10x^4 + 40x^3 + 80x^2 + 80x + 32$ **51.** $32x^5 + 240x^4 + 720x^3 + 1080x^2 + 810x + 243$ **53.** 144 **55.** 84

57. (a) 8 **(b)** 1100 **59.** \$151,873.77 **61. (a)** $20\left(\frac{3}{4}\right)^3 = \frac{135}{16}$ ft **(b)** $20\left(\frac{3}{4}\right)^n$ ft **(c)** 13 times **(d)** 140 ft

C H A P T E R   1 0   Counting and Probability

10.1 Exercises (page 665)

1. $\{1, 3, 5, 6, 7, 9\}$ **3.** $\{1, 5, 7\}$ **5.** $\{1, 6, 9\}$ **7.** $\{1, 2, 4, 5, 6, 7, 8, 9\}$ **9.** $\{1, 2, 4, 5, 6, 7, 8, 9\}$ **11.** $\{0, 2, 6, 7, 8\}$ **13.** $\{0, 1, 2, 3, 5, 6, 7, 8, 9\}$
15. $\{0, 1, 2, 3, 5, 6, 7, 8, 9\}$ **17.** $\{0, 1, 2, 3, 4, 6, 7, 8\}$ **19.** $\{0\}$ **21.** $\varnothing, \{a\}, \{b\}, \{c\}, \{d\}, \{a, b\}, \{a, c\}, \{a, d\}, \{b, c\}, \{b, d\}, \{c, d\},$
$\{a, b, c\}, \{b, c, d\}, \{a, c, d\}, \{a, b, d\}, \{a, b, c, d\}$ **23.** 25   **25.** 40   **27.** 25   **29.** 37   **31.** 18   **33.** 5   **35.** 175; 125
37. (a) 15   **(b)** 15   **(c)** 15   **(d)** 25   **(e)** 40   **39. (a)** 57,886 thousand   **(b)** 10,894 thousand   **(c)** 14,126 thousand

10.2 Exercises (page 675)

1. 30   **3.** 24   **5.** 1   **7.** 1680   **9.** 28   **11.** 35   **13.** 1   **15.** 10,400,600   **17.** $\{abc, abd, abe, acb, acd, ace, adb, adc, ade, aeb, aec, aed$
$bac, bad, bae, bca, bcd, bce, bda, bdc, bde, bea, bec, bed$
$cab, cad, cae, cba, cbd, cbe, cda, cdb, cde, cea, ceb, ced$
$dab, dac, dae, dba, dbc, dbe, dca, dcb, dce, dea, deb, dec$
$eab, eac, ead, eba, ebc, ebd, eca, ecb, ecd, eda, edb, edc\}; 60$
19. $\{123, 124, 132, 134, 142, 143, 213, 214, 231, 234, 241, 243, 312, 314, 321, 324, 341, 342, 412, 413, 421, 423, 431, 432\}; 24$
21. $\{abc, abd, abe, acd, ace, ade, bcd, bce, bde, cde\}; 10$ **23.** $\{123, 124, 134, 234\}; 4$ **25.** 15   **27.** 16   **29.** 8   **31.** 24   **33.** 60   **35.** 18,278
37. 35   **39.** 1024   **41.** 9000   **43.** 120   **45.** 480   **47.** 132,860   **49.** 336   **51.** 90,720   **53. (a)** 63   **(b)** 35   **(c)** 1   **55.** 1.157×10^{76}
57. 362,880   **59.** 660   **61.** 15

Historical Problems (page 687)

1. (a) $\{AA, ABA, BAA, ABBA, BBAA, BABA, BBB, ABBB, BABB, BBAB\}$

(b) $P(A \text{ wins}) = \dfrac{C(4, 2) + C(4, 3) + C(4, 4)}{2^4} = \dfrac{6 + 4 + 1}{16} = \dfrac{11}{16}$

$P(B \text{ wins}) = \dfrac{C(4, 3) + C(4, 4)}{2^4} = \dfrac{4 + 1}{16} = \dfrac{5}{16}$

The outcomes listed in part (a) are not equally likely.

10.3 Exercises (page 688)

1. $0, 0.01, 0.35, 1$ **3.** Probability model   **5.** Not a probability model

7. $S = \{\text{HH, HT, TH, TT}\}; P(\text{HH}) = \dfrac{1}{4}, P(\text{HT}) = \dfrac{1}{4}, P(\text{TH}) = \dfrac{1}{4}, P(\text{TT}) = \dfrac{1}{4}$

9. $S = \{\text{HH1, HH2, HH3, HH4, HH5, HH6, HT1, HT2, HT3, HT4, HT5, HT6, TH1, TH2, TH3, TH4, TH5, TH6, TT1, TT2, TT3, TT4, TT5, TT6}\};$
each outcome has the probability of $\dfrac{1}{24}$.

11. $S = \{\text{HHH, HHT, HTH, HTT, THH, THT, TTH, TTT}\};$ each outcome has the probability of $\dfrac{1}{8}$.

13. $S = \{1 \text{ Yellow}, 1 \text{ Red}, 1 \text{ Green}, 2 \text{ Yellow}, 2 \text{ Red}, 2 \text{ Green}, 3 \text{ Yellow}, 3 \text{ Red}, 3 \text{ Green}, 4 \text{ Yellow}, 4 \text{ Red}, 4 \text{ Green}\};$ each outcome has the
probability of $\dfrac{1}{12}$; thus, $P(2 \text{ Red}) + P(4 \text{ Red}) = \dfrac{1}{12} + \dfrac{1}{12} = \dfrac{1}{6}$.

15. $S = \{1 \text{ Yellow Forward}, 1 \text{ Yellow Backward}, 1 \text{ Red Forward}, 1 \text{ Red Backward}, 1 \text{ Green Forward}, 1 \text{ Green Backward}, 2 \text{ Yellow Forward},$
$2 \text{ Yellow Backward}, 2 \text{ Red Forward}, 2 \text{ Red Backward}, 2 \text{ Green Forward}, 2 \text{ Green Backward}, 3 \text{ Yellow Forward}, 3 \text{ Yellow Backward},$
$3 \text{ Red Forward}, 3 \text{ Red Backward}, 3 \text{ Green Forward}, 3 \text{ Green Backward}, 4 \text{ Yellow Forward}, 4 \text{ Yellow Backward}, 4 \text{ Red Forward},$
$4 \text{ Red Backward}, 4 \text{ Green Forward}, 4 \text{ Green Backward}\};$ each outcome has the probability of $\dfrac{1}{24}$; thus,

$P(1 \text{ Red Backward}) + P(1 \text{ Green Backward}) = \dfrac{1}{24} + \dfrac{1}{24} = \dfrac{1}{12}$.

17. $S = \{11 \text{ Red}, 11 \text{ Yellow}, 11 \text{ Green}, 12 \text{ Red}, 12 \text{ Yellow}, 12 \text{ Green}, 13 \text{ Red}, 13 \text{ Yellow}, 13 \text{ Green}, 14 \text{ Red}, 14 \text{ Yellow}, 14 \text{ Green}, 21 \text{ Red},$
$21 \text{ Yellow}, 21 \text{ Green}, 22 \text{ Red}, 22 \text{ Yellow}, 22 \text{ Green}, 23 \text{ Red}, 23 \text{ Yellow}, 23 \text{ Green}, 24 \text{ Red}, 24 \text{ Yellow}, 24 \text{ Green}, 31 \text{ Red}, 31 \text{ Yellow},$
$31 \text{ Green}, 32 \text{ Red}, 32 \text{ Yellow}, 32 \text{ Green}, 33 \text{ Red}, 33 \text{ Yellow}, 33 \text{ Green}, 34 \text{ Red}, 34 \text{ Yellow}, 34 \text{ Green}, 41 \text{ Red}, 41 \text{ Yellow}, 41 \text{ Green}, 42 \text{ Red},$
$42 \text{ Yellow}, 42 \text{ Green}, 43 \text{ Red}, 43 \text{ Yellow}, 43 \text{ Green}, 44 \text{ Red}, 44 \text{ Yellow}, 44 \text{ Green}\};$ each outcome has the probability of $\dfrac{1}{48}$;

thus, $E = \{22 \text{ Red}, 22 \text{ Green}, 24 \text{ Red}, 24 \text{ Green}\}; P(E) = \dfrac{n(E)}{n(S)} = \dfrac{4}{48} = \dfrac{1}{12}$.

19. A, B, C, F **21.** B **23.** $P(H) = \dfrac{4}{5}; P(T) = \dfrac{1}{5}$ **25.** $P(1) = P(3) = P(5) = \dfrac{2}{9}; P(2) = P(4) = P(6) = \dfrac{1}{9}$ **27.** $\dfrac{3}{10}$ **29.** $\dfrac{1}{2}$ **31.** $\dfrac{1}{6}$

33. $\frac{1}{8}$ **35.** $\frac{1}{4}$ **37.** $\frac{1}{6}$ **39.** $\frac{1}{18}$ **41.** 0.55 **43.** 0.70 **45.** 0.30 **47.** 0.747 **49.** 0.7 **51.** $\frac{17}{20}$ **53.** $\frac{11}{20}$ **55.** $\frac{1}{2}$ **57.** $\frac{3}{10}$ **59.** $\frac{2}{5}$

61. (a) 0.57 **(b)** 0.95 **(c)** 0.83 **(d)** 0.38 **(e)** 0.29 **(f)** 0.05 **(g)** 0.78 **(h)** 0.71 **63. (a)** $\frac{25}{33}$ **(b)** $\frac{25}{33}$ **65.** 0.167 **67.** 0.000033069

69. (a) $\frac{5}{16}$ **(b)** $\frac{1}{32}$ **71. (a)** 0.00463 **(b)** 0.126 **73.** $7.02 \times 10^{-6}; 0.183$ **75.** 0.1

Fill-in-the-Blank Items *(page 692)*

1. union; intersection **3.** permutation **5.** equally likely

True/False Items *(page 692)*

1. T **3.** T **5.** F

Review Exercises *(page 693)*

1. $\{1, 3, 5, 6, 7, 8\}$ **3.** $\{3, 7\}$ **5.** $\{1, 2, 4, 6, 8, 9\}$ **7.** $\{1, 2, 4, 5, 6, 9\}$ **9.** 17 **11.** 29 **13.** 7 **15.** 25 **17.** 120 **19.** 336 **21.** 56 **23.** 60
25. 128 **27.** 3024 **29.** 70 **31.** 91 **33.** 1,600,000 **35.** 216,000 **37.** 1260 **39. (a)** 381,024 **(b)** 1260

41. (a) $8.634628387 \times 10^{45}$ **(b)** 65.31% **(c)** 34.69% **43. (a)** 5.4% **(b)** 94.6% **45.** $\frac{4}{9}$ **47.** 0.2; 0.26 **49. (a)** 0.246 **(b)** 9.8×10^{-4}

A P P E N D I X   Graphing Utilities

1 Exercises *(page 699)*

1. $(-1, 4)$; II **3.** $(3, 1)$; I **5.** $X\text{min} = -6, X\text{max} = 6, X\text{scl} = 2, Y\text{min} = -4, Y\text{max} = 4, Y\text{scl} = 2$ **7.** $X\text{min} = -6, X\text{max} = 6,$
$X\text{scl} = 2, Y\text{min} = -1, Y\text{max} = 3, Y\text{scl} = 1$ **9.** $X\text{min} = 3, X\text{max} = 9, X\text{scl} = 1, Y\text{min} = 2, Y\text{max} = 10, Y\text{scl} = 2$
11. $X\text{min} = -11, X\text{max} = 5, X\text{scl} = 1, Y\text{min} = -3, Y\text{max} = 6, Y\text{scl} = 1$ **13.** $X\text{min} = -30, X\text{max} = 50, X\text{scl} = 10, Y\text{min} = -90,$
$Y\text{max} = 50, Y\text{scl} = 10$ **15.** $X\text{min} = -10, X\text{max} = 110, X\text{scl} = 10, Y\text{min} = -10, Y\text{max} = 160, Y\text{scl} = 10$ **17.** $4\sqrt{10}$ **19.** $2\sqrt{65}$

2 Exercises *(page 702)*

1. (a) 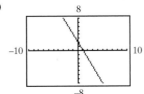 **(b)** 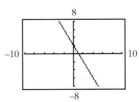 **(c)** 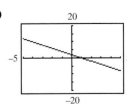 **(d)**

3. (a) 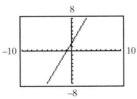 **(b)** 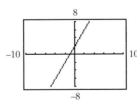 **(c)** 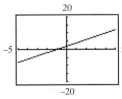 **(d)**

5. (a) 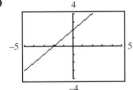 **(b)** 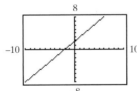 **(c)** 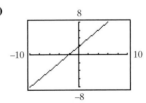 **(d)**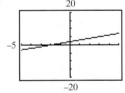

7. (a) 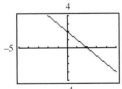 **(b)** 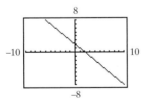 **(c)** 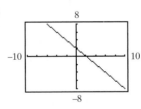 **(d)**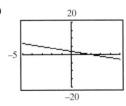

9. (a) 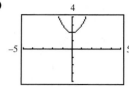 **(b)** 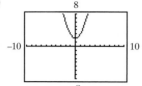 **(c)** 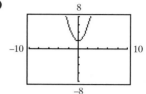 **(d)**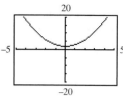

11. (a) 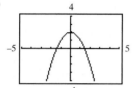 **(b)** 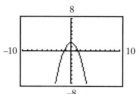 **(c)** 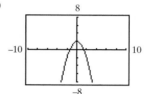 **(d)**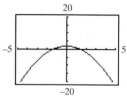

13. (a) 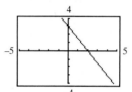 **(b)** 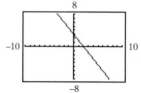 **(c)**  **(d)**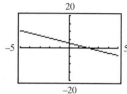

15. (a) 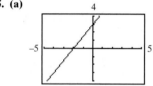 **(b)** 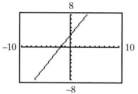 **(c)** 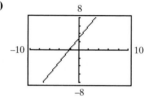 **(d)**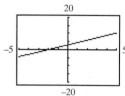

17.

X	Y1
-3	-1
-2	0
-1	1
0	2
1	3
2	4
3	5

Y1▊X+2

19.

X	Y1
-3	5
-2	4
-1	3
0	2
1	1
2	0
3	-1

Y1▊-X+2

21.

X	Y1
-3	-4
-2	-2
-1	0
0	2
1	4
2	6
3	8

Y1▊2X+2

23.

X	Y1
-3	8
-2	6
-1	4
0	2
1	0
2	-2
3	-4

Y1▊-2X+2

25.

X	Y1
-3	11
-2	6
-1	3
0	2
1	3
2	6
3	11

Y1▊X²+2

27.

X	Y1
-3	-7
-2	-2
-1	1
0	2
1	1
2	-2
3	-7

Y1▊-X²+2

29.

X	Y1
-3	7.5
-2	6
-1	4.5
0	3
1	1.5
2	0
3	-1.5

Y1▊-(3/2)X+3

31.

X	Y1
-3	-1.5
-2	0
-1	1.5
0	3
1	4.5
2	6
3	7.5

Y1▊(3/2)X+3

3 Exercises *(page 705)*

1. -3.41 **3.** -1.71 **5.** -0.28 **7.** 3.00 **9.** 4.50 **11.** $0.32, 12.30$ **13.** $1.00, 23.00$ **15. (a)** $(-1.5, 0), (0, -2), (1.5, 0)$ **(b)** y-axis
17. (a) none   **(b)** origin

5 Exercises *(page 708)*

1. Yes   **3.** Yes   **5.** No   **7.** Yes   **9.** $Y\text{min} = 4$ Other answers are possible.
$Y\text{max} = 12$
$Y\text{scl} = 1$

Index

CONICS

Parabola

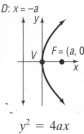

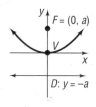

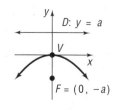

$$y^2 = 4ax \qquad\qquad y^2 = -4ax \qquad\qquad x^2 = 4ay \qquad\qquad x^2 = -4ay$$

Ellipse

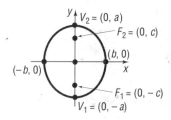

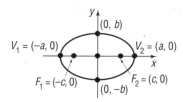

$$\frac{x^2}{a^2} + \frac{y^2}{b^2} = 1, \quad c^2 = a^2 - b^2 \qquad\qquad \frac{x^2}{b^2} + \frac{y^2}{a^2} = 1, \quad c^2 = a^2 - b^2$$

Hyperbola

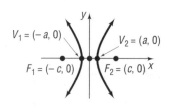

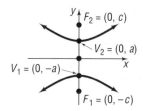

$$\frac{x^2}{a^2} - \frac{y^2}{b^2} = 1, \quad c^2 = a^2 + b^2 \qquad\qquad \frac{y^2}{a^2} - \frac{x^2}{b^2} = 1, \quad c^2 = a^2 + b^2$$

Asymptotes: $y = \dfrac{b}{a}x, \quad y = -\dfrac{b}{a}x$ $\qquad$ Asymptotes: $y = \dfrac{a}{b}x, \quad y = -\dfrac{a}{b}x$